Students, eGrade Plus Allows You to:

Study More Effectively

Get Immediate Feedback When You Practice on Your Own

eGrade Plus problems link directly to relevant sections of the **electronic book content,** so that you can review the text while you study and complete homework online. Additional resources include **interactive simulations, animated figures, extensive hyperlinks to appropriate examples and equations, answers to selected exercises,** and other problem-solving resources.

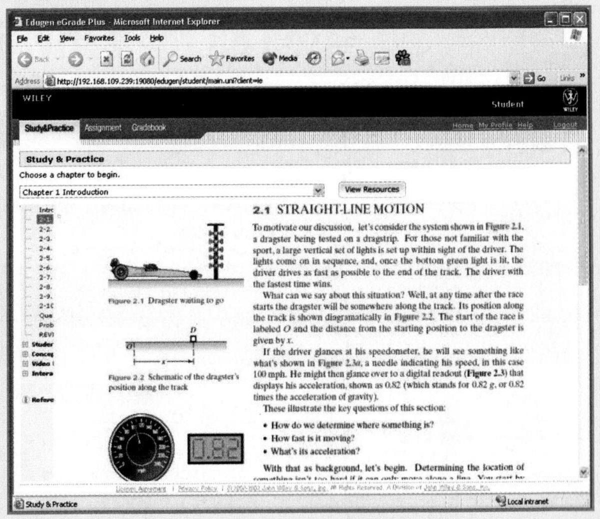

Complete Assignments and Get Help with Problem Solving

An **Assignment** area keeps all your assigned work in one location, making it easy for you to stay "on task."

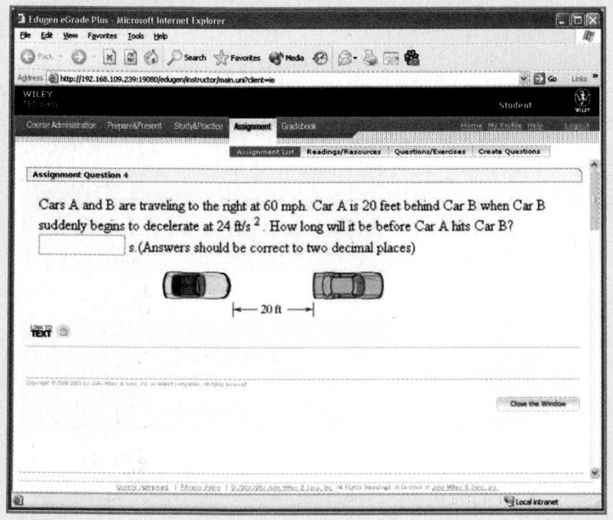

In addition, many homework problems contain a link to the relevant section of the **multimedia book,** providing you with a text explanation to help you conquer problem-solving obstacles as they arise. You will have access to a variety of **interactive problem-solving tools,** as well as other resources for building your confidence and understanding.

Keep Track of How You're Doing

A **Personal Gradebook** allows you to view your results from past assignments at any time.

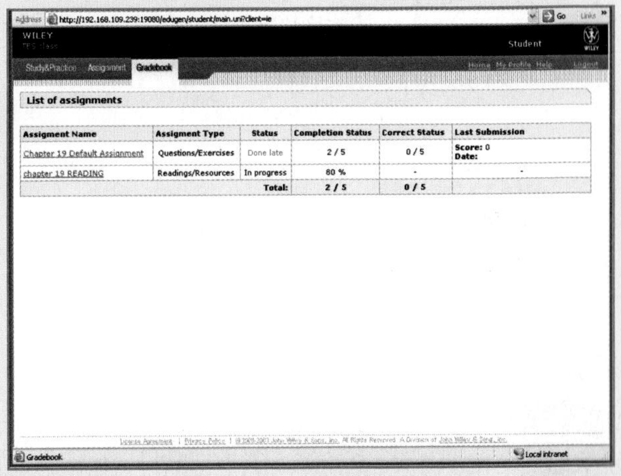

www.wiley.com/college/sheppard

Box 1.1: Overview of Engineering Analysis Procedure

Goal: Formulate the question to be answered by the analysis; what is the analysis to find? In many textbook problems, the question is provided in the problem statement, so this step involves restating the problem in your own words to be sure you understand what is being asked. Make sure that your restatement mentions every final result you should have once you finish working the problem. In engineering practice, the question is often whether a design meets a specific requirement.

Given: Summarize and record what is known. For textbook problems, this may mean restating what is given in the problem, including creating a sketch of the situation. In engineering practice, the source of information might be a design drawing or specification, previous analysis, or a standard reference source.

Assume: Make assumptions about the behavior of the system under consideration to create a simplified representation or model that can be analyzed. This is sometimes referred to as system modeling.

Draw: Draw any diagrams necessary to clarify the model. In statics, a free-body diagram is used to clarify the assumptions made in modeling the system under consideration.

Formulate Equations: Apply engineering principles, generally in mathematical form, to set up equations that represent the model's behavior. In statics, these principles are Newton's laws expressed as equilibrium conditions.

Solve: Solve the resulting equations. In some cases, this can be done by hand. In other cases, the solution requires the use of appropriate software. Clearly state how numerical answers address goal in undertaking the analysis.

Check: Check the results using technical knowledge, engineering judgment, and common sense.

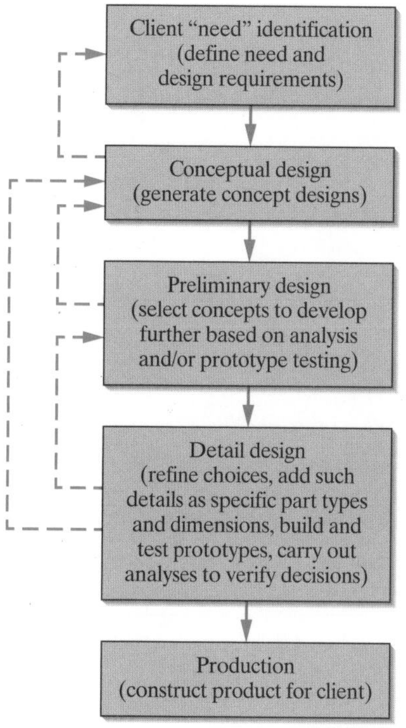

Figure 1.2 Product realization process flow chart

A2.3 Conversion Factors from U.S. Customary Units

PHYSICAL QUANTITY	U.S. CUSTOMARY UNIT	= SI EQUIVALENT
	BASIC UNITS	
Length	1 foot (ft)	= $3.048(10^{-1})$ meter (m)*
	1 inch (in.)	= $2.54(10^{-2})$ meter (m)*
	1 mile (U.S. statute)	= $1.6093(10^{3})$ meter (m)
Mass	1 slug ($lb \cdot s^2/ft$)	= $1.4594(10)$ kilogram (kg)
	1 pound mass (lbm)	= $4.5359(10^{-1})$ kilogram (kg)
	DERIVED UNITS	
Acceleration	1 foot/second^2 (ft/s^2)	= $3.048(10^{-1})$ meter/second^2 (m/s^2)*
	1 inch/second^2 ($in./s^2$)	= $2.54(10^{-2})$ meter/second^2 (m/s^2)*
Area	1 foot^2 (ft^2)	= $9.2903(10^{-2})$ meter^2 (m^2)
	1 inch^2 ($in.^2$)	= $6.4516(10^{-2})$ meter^2 (m^2)*
Density	1 slug/foot^3 ($lb \cdot s^2/ft^4$)	= $5.1537(10^{2})$ kilogram/meter^3 (kg/m^3)
	1 pound mass/foot^3 (lbm/ft^3)	= $1.6018(10)$ kilogram/meter^3 (kg/m^3)
Energy and Work	(1 joule ≡ 1 meter-newton)	
	1 foot-pound ($ft \cdot lb$)	= 1.3558 joules (J)
	1 kilowatt-hour ($kW \cdot hr$)	= $3.60(10^{6})$ joules (J)*
	1 British thermal unit (Btu)	= $1.0551(10^{3})$ joules (J)
Force	(1 newton ≡ 1 kilogram-meter/second^2)	
	1 pound (lb)	= 4.4482 newtons (N)
	1 kip (1000 lb)	= $4.4482(10^{3})$ newtons (N)
Power	(1 watt ≡ 1 joule/second)	
	1 foot-pound/second ($ft \cdot lb/s$)	= 1.3558 watt (W)
	1 horsepower (hp)	= $7.4570(10^{2})$ watt (W)
Pressure and Stress	(1 pascal ≡ 1 newton/meter^2)	
	1 pound/foot^2 (lb/ft^2)	= $4.7880(10)$ pascal (Pa)
	1 pound/inch^2 ($lb/in.^2$)	= $6.8948(10^{3})$ pascal (Pa)
	1 atmosphere (standard, 14.7 $lb/in.^2$)	= $1.0133(10^{5})$ pascal (Pa)
Speed	1 foot/second (ft/s)	= $3.048(10^{-1})$ meter/second (m/s)*
	1 mile/hr	= $4.4704(10^{-1})$ meter/second (m/s)
	1 mile/hr	= 1.6093 kilometer/hour (km/hr)
Volume	1 foot^3 (ft^3)	= $2.8317(10^{-2})$ meter^3 (m^3)
	1 inch^3 ($in.^3$)	= $1.6387(10^{-5})$ meter^3 (m^3)
	1 gallon (U.S. liquid)	= $3.7854(10^{-3})$ meter^3 (m^3)

*Denotes an exact factor.

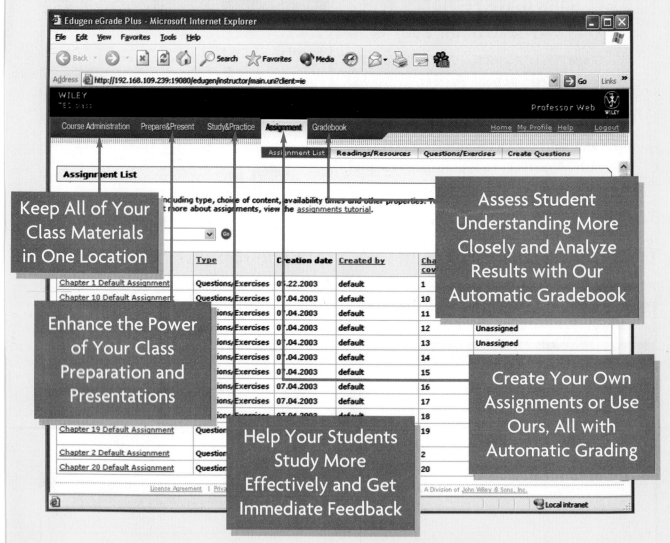

A letter to professors:

It is a rare pleasure when a project like this comes across an editor's desk. While there are many historically sound books in mechanics, we recognized that more could be done to motivate and involve students. In 1998 Sheri Sheppard proposed writing a book that taught mechanics principles through the study of a bicycle. Through various concept iterations (including an idea for a "hands-on" approach to statics), and with the addition of Benson Tongue in 2000, the project evolved into an introduction to engineering mechanics from a very practical and applied standpoint.

From the beginning we received enthusiastic and positive response from reviewers and students. We were delighted to find that many faculty were interested in contributing exercises and cases that fit the approach of the project—read about them in the Contributor section of the Preface. Over 250 reviewers provided both supportive and critical comments, and the authors carefully incorporated their input. Sheri and Benson also bravely put their ideas to the test by class-testing various versions of the manuscript with faculty and students at Stanford, Berkeley, Colorado School of Mines, University of Iowa, Cal Poly—SLO, and University of South Florida (thanks to all those reviewers and students!).

Because many professors are concerned with possible errors in first editions, we have taken extraordinary steps to ensure accuracy. More than fifteen professors (one of whom worked on the project while on assignment in the oil fields of Iraq), a developmental editor, several graduate students, and several hundred undergraduate students all had a hand in checking for errors during development, including a group of Benson's students who were paid per error found (this turned out to be a very small sum)! Read more about these efforts in the Commitment to Accuracy section of the Preface.

We invite you to browse through the Preface to learn about all the innovative features of this new project. It is with great pleasure that we offer to you this new text by Sheri D. Sheppard of Stanford University and Benson H. Tongue of University of California at Berkeley.

Sincerely,

Joseph P. Hayton
Editor

Advance Praise for Sheppard/Tongue, STATICS

"I like the down-to-earth format! The illustrations coupled with the dialog makes this a great text!"

Duane Jardin, University of New Orleans

"[The greatest strength of the text is] the sample sketch in the example. They are more like handwritten and are great guide for student doing homework."

Anonymous

"This is the only [statics] book I have come across that aims to be almost a self-teaching book. It does a great job of explaining what's going on—something other books are traditionally weak at."

Colin Ratcliffe, U.S. Naval Academy

"I also like the six-step process and how it is highlighted throughout."

Lisa Hailey, Brookdale Community College

"This author speaks to the students in a manner that engages their minds in today's world. Students will grasp how statics enables us to analyze practical, everyday problems and as well as advanced designs. It is much more practical than similar texts."

Roy Henk, LeTourneau University

"Based on the two chapters I have reviewed, this text is well-written and organized in an appropriate manner for effective instruction. The inclusion of a separate chapter on free-body diagrams and loads/supports is very good. The quality of examples and exercises is another strong point. They convey the body of knowledge very effectively."

Manoj Chopra, University of Central Florida

"I would rate all the chapters as excellent. The examples in this book and the approach are very good."

M. Zikry, North Carolina State University

"I like the idea that students start with a concrete experience (bicycle). That will help them understand why we are presenting what we are presenting . . ."

Paul Barr, New Mexico State University

"As the free-body diagram is the most important tool for the equilibrium analysis, it is an excellent idea to give it a full chapter to strengthen the student's concept and methodology."

George Weng, Rutgers University

"I really like the simple experiments that illustrate concepts from the material and the system analysis exercises at the end of the chapters."

John Krohn, Arkansas Tech

STATICS

ANALYSIS AND DESIGN OF SYSTEMS IN EQUILIBRIUM

Sheri D. Sheppard
Stanford University

Benson H. Tongue
University of California at Berkeley

With special contributions by:
Thalia Anagnos
San Jose State University

WILEY John Wiley & Sons, Inc.

This book is dedicated to Ed Carryer, Portia Carryer, and Rolf Faste—
my teachers in looking, seeing, and drawing,

to my dear friends and family who have encouraged me through its long gestation,

and

to students everywhere.

sds

ACQUISITIONS EDITOR	Joseph Hayton
MARKETING MANAGER	Jennifer Powers
SENIOR PRODUCTION EDITOR	Sujin Hong
TEXT DESIGNER	Madelyn Lesure
COVER DESIGNER	Norm Christiansen
SENIOR ILLUSTRATION EDITOR	Sigmund Malinowski
ILLUSTRATION ASSISTANT	Eugene Aiello
SENIOR PHOTO EDITOR	Jennifer MacMillan
PHOTO RESEARCHER	Elyse Rieder
FRONT COVER PHOTO	ImageState–Pictor/PictureQuest
ELECTRONIC ILLUSTRATIONS	Precision Graphics

This book was set in Times Ten by GGS Book Services, Atlantic Highlands and printed and bound by Von Hoffmann. The cover was printed by Von Hoffmann.

This book is printed on acid-free paper. ∞

Library of Congress Cataloging in Publication Data:

Sheppard, S. (Sheri)
Statics : analysis and design of systems in equilibrium / Sheri D. Sheppard, Benson H. Tongue, with special contributions by Thalia Anagnos.
p. cm.
Includes index.
ISBN 0-471-37299-4 (acid-free paper)
1. Mechanics. 2. Statics. 3. Equilibrium. I. Tongue, Benson H. II. Anagnos, Thalia. III. Title.

QA821.S44 2005
531′.12—dc22

2004042292

Printed in the United States of America

10 9 8 7 6 5 4 3 2 1

Mechanics courses have historically confronted engineering students with a precise, mathematical, and, dare we say it, less than engaging treatment of the material. This approach has appeal in that it presents mechanics as a relatively uncluttered "science," but the material often comes across as a somewhat mysterious body of facts and "tricks" that allow idealized cases to be solved. When confronted with more realistic systems, students are often at a loss as to how to proceed. What is lacking is an appreciation for and understanding of the material that will empower the students to tackle meaningful problems at an early stage in their undergraduate education.

In statics, we have tried to present the best of both worlds. Chapters 1–3 present a readable overview of the concepts of mechanics. While we introduce important equations, the emphasis is on developing a "feel" for forces and moments, and for how loads are transferred through structures and machines. This introduction of the material helps lay a motivational framework for the more mathematically complete presentation of statics found in Chapters 4–10.

Throughout this volume, our emphasis is to present and illustrate:

a. *The physical principles* and concepts that describe non-accelerating objects. These principles and concepts are grounded in the reader's own experiences—this approach serves to motivate and provide a context for formal mathematical representations.

b. *An analytical problem-solving methodology* for describing and assessing physical systems, so that the reader is able to apply the principles in a systematic manner in evaluating engineered systems. Furthermore, throughout the book, the methodology and its application are framed within the context of broader engineering practice.

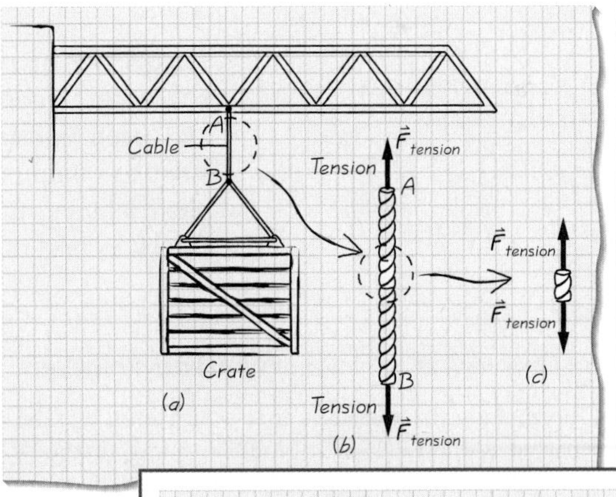

Figure 4.13 (b) Looking
transmitted a

Features

The goals outlined above are supported by a number of unique features in this text:

Emphasis on sketching: The importance of communicating solutions through graphics is continuously emphasized. Most engineering students are visual learners.[1] This, coupled with the importance of graphical information and communication in engineering practice (e.g., the use of sketching during conceptual design), makes graphical representation of information an inviting and key element of the book. Chapter 1 discusses the importance of visualization and sketching skills in successful implementation of structured evaluation procedures, and provides some guidelines for sketching objects. Other elements reinforcing the importance of drawing include:

a. A full chapter (Chapter 6) devoted to the key skill of drawing correct free-body diagrams.

b. An innovative illustration program that uses engineering graph paper background and a ***hand-sketched*** look that shows

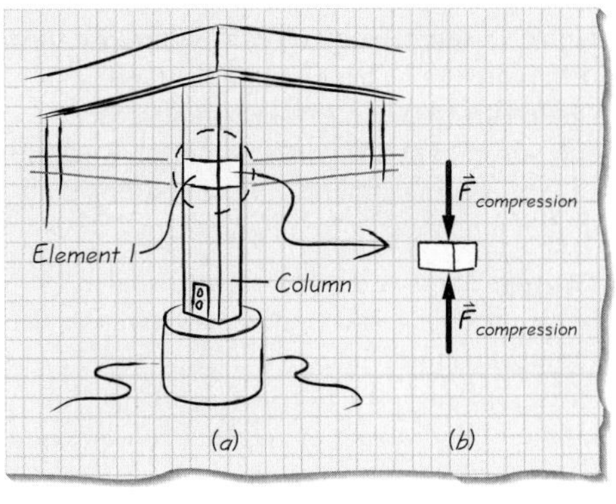

Figure 4.14 (a) A column holds up a deck; (b) Compression is transmitted along the length of the column

[1] Felder, Richard, "Reaching the Second Tier: Learning and Teaching Styles in College Science Education." *J. College Science Teaching*, 23(5), 286–290 (1993).

students how they should be documenting their solutions. An ideal response from a reader regarding a graphical element of the book would be, "the sketch in Figures 4.13 and 4.14 made the concept more understandable AND I think that I could create a similar drawing to illustrate the concept to someone else."

c. A **Draw** step included in every worked example.

To reinforce the drawing concept, vectors in the "hand-drawn" figures appear with an "arrow over top" notation, mirroring how they would be drawn in a hand-written solution.

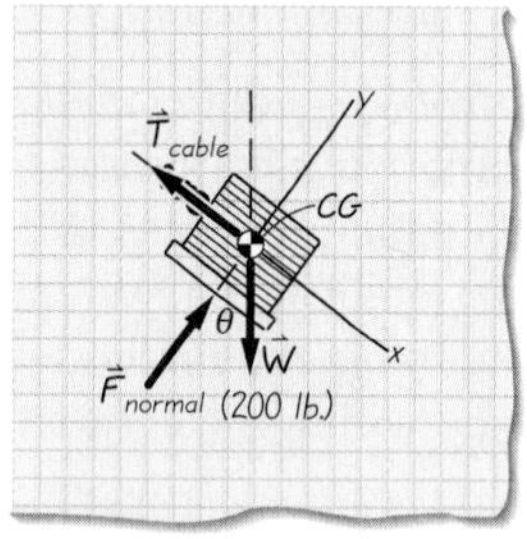

Figure 4.25

Draw The external loads acting on the structure are the cable tension ($\boldsymbol{T}_{\text{cable}}$), the gravitational force acting on the pallet and tiles ($\boldsymbol{W}$), and the normal contact force exerted by the roof on the pallet ($\boldsymbol{F}_{\text{normal}}$).

See Figure 4.25 (which is the answer to (a)). In creating this drawing, we assumed that the normal contact force acts at the center of the bottom of the pallet. We have drawn each force in the direction we think it acts on the structure. Finally, we have placed a set of coordinate axes with the origin at the center of gravity. We could have oriented these axes horizontally and vertically, but as we will see in the next step, orienting them along the roof pitch will make force addition easier. **Figure 4.25** is a free-body diagram of the pallet–tile unit.

Development of structured problem solving procedures: A consistent analysis procedure is introduced early in the text and used consistently throughout all worked examples. Several key steps are emphasized here more than in most other texts, including explicitly listing the **Assumptions** made and the importance of **Draw** and **Check** as part of the solution.

Box 1.1: Overview of Engineering Analysis Procedure

Goal: Formulate the question to be answered by the analysis; what is the analysis to find? In many textbook problems, the question is provided in the problem statement, so this step involves restating the problem in your own words to be sure you understand what is being asked. Make sure that your restatement mentions every final result you should have once you finish working the problem. In engineering practice, the question is often whether a design meets a specific requirement.

Given: Summarize and record what is known. For textbook problems, this may mean restating what is given in the problem, including creating a sketch of the situation. In engineering practice, the source of information might be a design drawing or specification, previous analysis, or a standard reference source.

Assume: Make assumptions about the behavior of the system under consideration to create a simplified representation or model that can be analyzed. This is sometimes referred to as system modeling.

Draw: Draw any diagrams necessary to clarify the model. In statics, a free-body diagram is used to clarify the assumptions made in modeling the system under consideration.

Formulate Equations: Apply engineering principles, generally in mathematical form, to set up equations that represent the model's behavior. In statics, these principles are Newton's laws expressed as equilibrium conditions.

Solve: Solve the resulting equations. In some cases, this can be done by hand. In other cases, the solution requires the use of appropriate software. Clearly state how numerical answers address goal in undertaking the analysis.

Check: Check the results using technical knowledge, engineering judgment, and common sense.

Application of principles to engineering systems: End-of-chapter **System Analysis (SA) Exercises** offer students the opportunity to apply mechanics principles to broader systems. These exercises are more open-ended than those in other parts of the text, and sometimes have more than one "correct" answer. We hope that these exercises will provide opportunities for group work, exploration of similar systems near the students' own campus, and in general show how the principles in the text apply to analysis of real artifacts.

SYSTEM ANALYSIS (SA) EXERCISES

SA 4.3 Problem: Forces to Hold the Scoreboard in Place

The basketball facility inside the Reynolds Coliseum at North Carolina State University contains a heavy scoreboard that is suspended from the ceiling with two cables, shown in Figure SA4.3.1. Two electric winches make it possible to lower the entire structure to the floor where signs can be changed or maintained.

The 4500-N scoreboard can be lowered using two winches, A and B, which are attached to the bottom flange of a main roof beam girder. Figure SA4.3.2 presents a view of the key suspension elements when standing directly underneath.

Assume that you are attending a basketball game with your buddy George, who has not studied statics. Before the game, you are explaining how the heavy scoreboard is being secured in the air, which stimulates him to ask whether the maximum tension in the cables that run from E, F, G, and H to tie cable locations C and D will be 1/4 of the weight of the scoreboard. He also wonders what the purpose is of the tie cable CD. Having just studied this chapter, you should have no problem addressing his questions. Consider George's curiosity as a wonderful opportunity to understand the material, since we all know that the best way to learn something well is by teaching it. Figure SA4.3.3 provides you with the dimensions and labels that you need to "teach."

Figure SA4.3.1 View of the 49-meter-wide field with suspended scoreboard

Here is how you work with George to address his questions:

1. Figure SA4.3.4 shows the forces acting at point D of the rigging.
   (a) Find forces F_{DC}, F_{DB}, F_{DG}, and F_{DH}. Write in vector notation.
   (b) If the magnitude of F_{DB} is 4500/2 N (is this reasonable?) and the sum of the forces acting at point D is zero, what are the magnitudes of F_{DC}, F_{DG}, and F_{DH}?

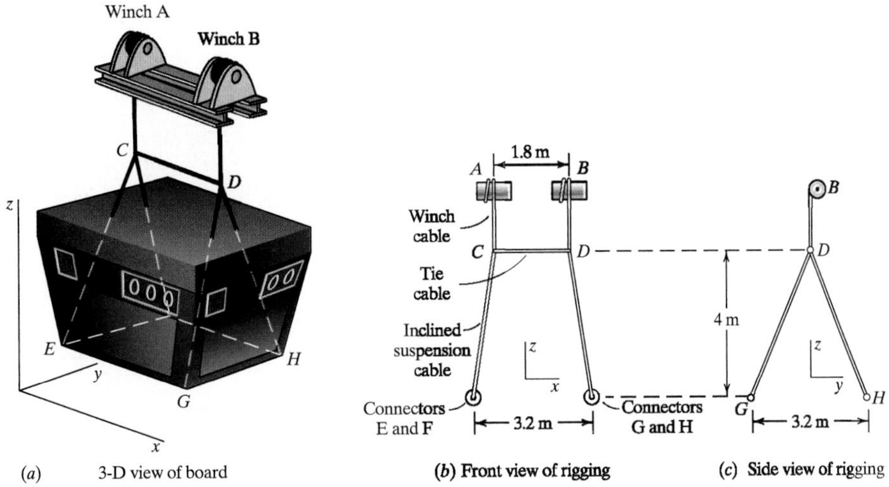

Figure SA4.3.3 Models of the rigging system suspending the scoreboard from the roof

Use of case studies to motivate discussion of principles: Chapters 2 and 3 of this volume contain two case studies that illustrate the application of statics principles in understanding how and why artifacts behave the way they do—one is the bicycle and the other is San Francisco's Golden Gate Bridge. These two cases preview many of the key concepts in statics, concepts that are developed in detail in subsequent chapters. These cases give a real-world feel and motivation for the study of mechanics, and are revisited throughout the book in examples and homework problems.

One or both of these cases can be used in conjunction with example and exercises on the bicycle and the Golden Gate Bridge contained in Chapters 4–10. In this manner the student will gain an increasingly sophisticated understanding of the role of analysis in understanding particular systems and an appreciation of the appropriateness (or inappropriateness) of various simplifications. The instructor's manual provides an example syllabus—listing topics, examples, and exercises in each chapter that use the bridge or bicycle as a theme. However, if time does not allow, both chapters can be skipped (or assigned as background reading) without loss of continuity.

Inclusion of useful study tools: Most students will read and review the text to find key information as quickly as possible. To facilitate speedy access to key content, we have included review and study tools, such as **Chapter Objectives** at the start of each chapter, and a **Just the Facts** section at the end of each chapter that summarizes key terms, key equations, and key concepts from the chapter. To the greatest extent possible, all in-text figures include ***descriptive figure captions*** that show at a glance what is being illustrated. ***Key equations*** are highlighted in yellow, and ***key terms*** are in bold blue type when they first appear.

This text has been written very explicitly with the student in mind. We are not trying to talk to the professors teaching the course but rather to those in the class who are trying to get their minds around the material. Mechanics can sometimes be counterintuitive, and it can be a major frustration to those students who don't immediately relate to the logic behind the material (and this includes many of them!). Thus the presentation is a very personalized one—one in which the students feel that they are having a one-on-one discussion with the authors. We do not skimp on rigor but also try and make the material as accessible as possible and, as far as we can, make it fun to learn.

Instructor Resources

The following resources are available to faculty using this text in their courses:

Grade Plus: A complete online learning system to help prepare and present lectures, assign and manage homework, keep track of student progress, and customize your course content and course delivery. See the two-page description in front of the book for more information regarding eGrade Plus, and talk to your local Wiley representative for details on setting up your eGrade Plus course.

Solutions Manual: Fully worked solutions to all exercises in the text, using the same solution procedure as the worked examples.

Electronic figures: All figures from the text are available electronically, for use in creating your own lectures.

Animations and simulations: A collection of animations and simulations that enhance visualization skills and allow "what if" analysis are available from the text website, www.wiley.com/college/sheppard, and are also available as part of the eGrade Plus package.

Student Resources

The following resources are available to students:

Answers to selected exercises: The text website, www.wiley.com/college/sheppard, includes answers to selected exercises from the text, to help students check that they have solved the exercise correctly.

Animations and simulations: A collection of animations and simulations that enhance visualization skills and allow "what if" analysis are available from the text website, and are also available as part of the eGrade Plus package.

Commitment to Accuracy

From the beginning we have committed to providing accurate and error-free coverage of the material. In this mission we have benefited from the help of many, many people.

While writing the book, eighty-seven faculty provided detailed content reviews on individual chapters—some of it quite pointed, and all of it helpful. More than three hundred students class-tested full manuscripts at Stanford and UC—Berkeley in the last two years, reporting and helping to correct any typographical errors found.

While writing solutions (approximately 1,800 exercises and solutions were written!), each solution was solved and checked at least twice, and often three or more times, by a combination of authors, accuracy checkers, and graduate students.

During production, more than 15 faculty served as accuracy checkers, and were specifically tasked with reviewing pages at two stages to check for accuracy of text, equations, examples, figures, and exercises—we specifically recognize them below. All text and art were reviewed line by line by a developmental editor. A proofreader compared all corrections to final pages to confirm that any and all corrections were made. Finally, and certainly not least, the authors themselves spent countless hours checking all elements of the project at every step of the way to guarantee accuracy.

Acknowledgments

Contributors: The following faculty contributed content, System Analysis Exercises, and/or additional assignments, and we would like to recognize their important contributions to this project.

Thalia Anagnos is Professor of Civil and Environmental Engineering and Director of Assessment at San Jose State University. She has taught classes in statics, strength of materials, structural analysis, concrete design, probability, finite elements, and catastrophic events. Thalia received her Ph.D. from Stanford University, and has been involved in NSF-funded research on educational applications of the George E. Brown Network for Earthquake Engineering Simulation (NEES), and improving science education at the middle school level. Thalia is co-author or contributor on three texts, including this one.

Leonhard E. Bernold is Associate Professor of Civil Engineering at North Carolina State University, and founder and director of the Construction Automation & Robotics Laboratory there. Leonhard received his Ph.D. from Georgia Institute of Technology. He teaches courses in civil engineering and conducts NSF-sponsored research related to engineering education. He founded the international student competition "Lunar Construction," which is sponsored by the American Society of Civil Engineers.

George T. Flowers is Professor of Mechanical Engineering at Auburn University and is a licensed practicing engineer in the state of Alabama. He received his Ph.D. from Georgia Institute of Technology. His area of expertise is the dynamics, vibration, and control of rotating machinery.

Major Joseph (Joe) P. Hanus received an M.S. from University of Minnesota and is a member of the U.S. Army Corps of Engineers. He served as an Assistant Professor in the Civil Engineering Department at the U.S. Military Academy, West Point, New York, and is currently studying for his Ph.D. at the University of Wisconsin.

Nels H. Madsen is Associate Professor of Mechanical Engineering and Associate Dean for Assessment and Special Programs in the Samuel Ginn College of Engineering at Auburn University. He received his Ph.D. from University of Iowa in 1978. Nels's research area is biomechanics and motion capture. As Vice President for Research and Development at Motion Reality, Incorporated, he is among the principal developers of the motion capture system used in the *Lord of the Rings* movie trilogy. Nels has contributed throughout the project, and helped guide thinking on the development of the bridge case study material and System Analysis Exercises in this text.

Eric Nauman is an Assistant Professor of Mechanical Engineering at Purdue University. He received his Ph.D. from University of California—Berkeley. His current research interests include tissue engineering, dynamics of biological systems, and mechanics of materials with microstructure.

Brian Self is an Associate Professor of Engineering Mechanics at the U.S. Air Force Academy. He received his B.S. and M.S. in Engineering Mechanics from Virginia Tech and his Ph.D. in Bioengineering from University of Utah. He has four years of experience with the Air Force Research Laboratory and is in his fifth year of teaching in the Department of Engineering Mechanics at the U.S. Air Force Academy. His areas of research include impact injury mechanisms, sports biomechanics, and aerospace physiology.

Accuracy checkers: The following faculty were closely involved in reading and accuracy-checking the project throughout the production process. We could not have completed it without them.

Makola Abdullah, Florida State University
Mark Evans, U.S. Military Academy
George Flowers, Auburn University
Mark Hanson, University of Kentucky
Joe Hanus, U.S. Military Academy
Nels Madsen, Auburn University
M. Mahinfaleh, North Dakota State University
Masami Nakagawa, Colorado School of Mines
Eric Nauman, Purdue University
Wilfrid Nixon, University of Iowa
Nipon Raatanawangcharoen, University of Manitoba, Canada
Joe Slater, Wright State University
Ted Stathopoulos, Concordia University, Canada
Warren White, Kansas State University

We would also like to thank the students who accuracy-checked and provided us with the student perspective to this process.

Shannon Grady, FAMU–FSU
Xin Guo, Kansas State University
Rhett Larson, Kansas State University
Anil Valevate, Wright State University
Claudia Mara Dias Wilson, FAMU–FSU

Reviewers: Although almost too numerous to mention, we would like to recognize all the individuals who provided constructive feedback on the project and helped improve it through their insights and suggestions.

Bechara Abboud, Temple University
Tarek Abdel-Salam, Old Dominion University
Makola Abdullah, FAMU–FSU
Mohammad Alimi, North Dakota State University
Rafael Alpizar, Miami Dade Community College
Jeff E. Arrington, Abilene Christian University
Eric Austin, Clemson University
Pranab Banerjee, Community College of Rhode Island
Paul Barr, New Mexico State University
Christina Barsotti, Clark College
Olivier Bauchau, Georgia Institute of Technology
Steve Bechtel, Ohio State University
Kenneth Belanus, Oklahoma State University
Richard Bennett, University of Tennessee
Leonard Berkowitz, California Polytechnic and State University—Pomona
Edward Bernstein, Alabama A&M University
Dadbeh Bigonahy, Quinsigamond Community College
Kris Bishop, Colby Community College
Ashland O. Brown, University of the Pacific
Richard Budynas, Rochester Institute of Technology
Liang-Wu Cai, Kansas State University
Louis Caplan, Fort Hays State University
Claudius Carnegie, Florida International University
Cetin Cetinkaya, Clarkson University
Paul Chan, New Jersey Institute of Technology
Tao Chang, Iowa State University
KangPing Chen, Arizona State University
Tony Chen, James Madison University
Manoj Chopra, University of Central Florida
Karen Chou, Minnesota State University—Mankato
Brian Collar, University of Illinois—Chicago
Agamemnon Crassidis, Rochester Institute of Technology
Joe Cuscheri, Florida Atlantic University
Paul Dalessandris, Monroe Community College
Kurt DeGoede, Elizabethtown College
Fereidoon Delfanian, South Dakota State University
Laura Demsetz, College of San Mateo
James Devine, University of Southern Florida
Keith DeVries, University of Utah
Anna Dollar, Miami University—Ohio
Deanna Durnford, Colorado State University
Thomas Eason, University of South Florida
Nader Ebrahimi, University of New Mexico
Mark Evans, U.S. Military Academy
Ron Fannin, University of Missouri—Rolla
Joseph Farmer, College of the Desert
Al Ferri, Georgia Institute of Technology
David Fikes, University of Alabama—Huntsville
George Flowers, Auburn University
Betsy Fochs, University of Minnesota
Steven Folkman, Utah State University
John Gardner, Boise State University
Slade Gellin, Buffalo State College
Guy Genin, Washington University in St. Louis
Manouchehr Gorji, Portland State University
Kurt Gramoll, University of Oklahoma
Donald Grant, University of Maine
Ivan Griffin, Tulsa Community College
Karen Groppi, Cabrillo College
Abhinav Gupta, North Carolina State University
Lisa Hailey, Brookdale Community College
Stephen Hall, Pacific University
Dominic Halsmer, Oral Roberts University
Joe Hanus, U.S. Military Academy
Kent Harries, University of South Carolina
Roy Henk, LeTourneau University
Debra Hill, Sierra College
Jerre Hill, University of North Carolina at Charlotte
Ed Howard, Milwaukee School of Engineering
Joshua Hsu, University of California—Riverside
Hugh Huntley, University of Michigan—Dearborn
Syed I. Hussain, Oakland Community College—Highland Lakes
Ron Huston, University of Cincinnati
Irina Ionescu, University of Central Florida
Duane Jardine, University of New Orleans

Jim Jones, Purdue University
Thomas Jordan, Oklahoma State University
Jennifer Kadlowec, Rowan University
Yohannes Ketema, University of Minnesota
Lidvin Kjerengtroen, South Dakota School of Mines & Technology
Steve Klein, Yuba College
Peter Knipp, Christopher Newport University
John Koepke, Joliet Junior College
Sefa Koseoglu, Texas A&M University
George Krestas, De Anza College
Sundar Krishnamurty, University of Massachusetts—Amherst
John Krohn, Arkansas Technological University
David Kukulka, Buffalo State College
Jeff Kuo, California State University—Fullerton
Kent Ladd, Portland State University
John Ligon, Michigan Technological University
Ti Lin Liu, Rochester Institute of Technology
Roger Ludin, California Polytechnic and State University
Nels Madsen, Auburn University
Mohammad Mahinfalah, North Dakota State University
Frank Mahuta, Milwaukee School of Engineering
Masroor Malik, Cleveland State University
Ken Manning, Hudson Valley Community College
Francisco Manzo-Robledo, Washington State University
Christine Masters, Pennsylvania State University
Steve Mayes, Alfred University
Keith Mazachek, Washburn University
Gary McDonald, University of Tennessee at Chattanooga
Wayne McIntire, California Polytechnic and State University—Pomona
Robert McLauchlan, Texas A&M University—Kingsville
Howard Medoff, Pennsylvania State University
Sudhir Mehta, North Dakota State University
Robert Melendy, Heald College
Daniel Mendelsohn, Ohio State University
Robert Merrill, Rochester Institute of Technology
Najm Meshkati, University of Southern California
Geraldine Milano, New Jersey Institute of Technology
Paul Mitiguy, Stanford University
Soheil Mohajerjasbi, Drexel University
Karla Mossi, Virginia Commonwealth University
Masoud Naghedolfeizi, Fort Valley State University
Masami Nakagawa, Colorado School of Mines
R. Nathan, Villanova University
Eric Nauman, Tulane University
Saeed Niku, California Polytechnic and State University
Wilfrid Nixon, University of Iowa
Karim Nohra, University of South Florida
Robert Oakberg, Montana State University
David Olson, College of Dupage
Joseph Palladino, Trinity College
William Palm, University of Rhode Island
David Parish, North Carolina State University
Assimina Pelegri, Rutgers University
James Pitarresi, SUNY Binghamton
Kirstie Plantenberg, University of Detroit Mercy
Suzana Popovic, Trinity College
Hamid Rad, Portland Community College
Robert Rankin, Arizona State University
Jeff Raquet, University of North Carolina at Charlotte
Colin Ratcliffe, U.S. Naval Academy
David Reichard, College of Southern Maryland
Andy Rose, University of Pittsburgh—Johnstown
James Rybak, Mesa State College
Jose Saez, Loyola Marymount University
Bruce Savage, Bucknell University
Joseph Schaefer, Iowa State University
Scott Schiff, Clemson University
Kevin Schmaltz, Western Kentucky University
Will Schrank, Angelina College
Kenneth Schroeder, Pierce College
James Scudder, Rochester Institute of Technology
Patricia Shamamy, Lawrence Technical University
Mala Sharma-Judd, Pennsylvania State University
Geoffrey Shiflett, University of Southern California
Angela Shih, California Polytechnic and State University—Pomona
Joe Slater, Wright State University
Lisa Spainhour, Florida State University
Anne Spence, University of Maryland Baltimore County
Randy Stein, Ferris State University
Todd Swift, Loras College
Chandra Thamire, Frostburg State University
John Uicker, University of Wisconsin—Madison
Eduardo Velasco, Truman State University
Tim Veteto, Lower Columbia College
Arkady Voloshin, Lehigh University
Richard Wabrek, Idaho State University
Jonathan Weaver, University of Detroit Mercy
George Weng, Rutgers University
Matthew Werner, Webb Institute
Jeffrey Wharton, Shasta College
Warren White, Kansas State University
Bonnie Wilson, Auburn University
B. Wolf Yeigh, St. Louis University
Fred Young, Lamar University
Gary Young, Oklahoma State University
Jianping Yue, Essex County College
Mark Yuly, Houghton College
Hong Zhang, Rowan University
Weidong Zhu, University of Maryland Baltimore County
Mohammed Zikry, North Carolina State University

Additional Author Acknowledgments

Sheri wishes to acknowledge the contributions of:

- Laura Demsetz and Joe Hayton, who supported the idea of a "new and improved" statics book.
- Benson Tongue, who was willing to take the lead in writing a companion dynamics book based on the principle of "student accessibility."

• Wiley staff and consultants, who have worked on this project with good humor, creative ideas, patience, and professionalism (especially Irene Nunes, from whom I have learned a tremendous amount about clarity in writing).

• Lillian Chang, Denise Curti, Jonathan Dirrenberger, Robyn Dunbar, Jonathan Fiene, Eric Gutierrez, Natalie Jeremijenko, Antonio Layon, Diane Palme, Kim Shasby, Amy Sheng, and Joshua Webb for the many hours they devoted to creating homework and example problems, working solutions, and keeping me organized, on track, and enthused about the project.

• Chloe, Alexei, and Jeff Koseff, for their contributions to the photography in Chapter 3.

• My students in the Fall 2001 and 2002 offerings of E14 (Statics) at Stanford, who gave me valuable feedback on draft chapters of the book, and to my students (past and future) who are the reason and motivation for my work.

• My parents, who taught me to dream big, believe deeply, work hard, and not forget to have fun.

• My daughter Portia, who continues to be an inspiration to me—her excitement for learning and the written word are contagious.

• My best friend and husband Ed Carryer, who is always there with encouragement, thoughtful feedback, and incredible patience (through the many hours I devoted to writing and rewriting the ten chapters that make up this book, when we could have been out playing!).

Sheri D. Sheppard

Ph.D., P.E., is the Carnegie Foundation for the Advancement of Teaching Senior Scholar principally responsible for the Preparations for the Professions Program (PPP) engineering study. She is an Associate Professor of Mechanical Engineering at Stanford University. She received her Ph.D. from the University of Michigan in 1985. Besides teaching both undergraduate and graduate design-related classes at Stanford University, she conducts research on weld fatigue and impact failures, fracture mechanics, and applied finite element analysis.

Dr. Sheppard was recently named co-principal investigator on a NSF grant to form the Center for the Advancement of Engineering Education (CAEE), along with faculty at the University of Washington, Colorado School of Mines, and Howard University. She was co-principal investigator with Professor Larry Leifer on a multi-university NSF grant that was critically looking at engineering undergraduate curriculum (Synthesis). In 1999, Sheri was named a fellow of the American Society of Mechanical Engineering (ASME) and the American Association for the Advancement of Science (AAAS). Recently Sheri was awarded the 2004 ASEE Chester F. Carlson Award in recognition of distinguished accomplishments in engineering education. Before coming to Stanford University, she held several positions in the automotive industry, including senior research engineer at Ford Motor Company's Scientific Research Lab. She also worked as a design consultant, providing companies with structural analysis expertise.

In her spare time Sheri likes to build houses, hike, and travel.

Benson H. Tongue

Ph.D., is a Professor of Mechanical Engineering at University of California—Berkeley. He received his Ph.D. from Princeton University in 1988, and currently teaches graduate and undergraduate courses in dynamics, vibrations, and control theory. His research concentrates on the modeling and analysis of nonlinear dynamical systems and the control of both structural and acoustic systems. This work involves experimental, theoretical, and numerical analysis and has been directed toward helicopters, computer disk drives, robotic manipulators, and general structural systems. Most recently, he has been involved in a multidisciplinary study of automated highways and has directed research aimed at understanding the nonlinear behavior of vehicles traveling in platoons and in devising controllers that optimize the platoon's behavior in the face of non-nominal operating conditions. His most recent research has involved the active control of loudspeakers and biomechanical analysis of human fall dynamics.

Dr. Tongue is the author of *Principles of Vibration*, a senior/first-year graduate-level textbook. He has served as Associate Technical Editor of the *ASME Journal of Vibration and Acoustics* and is currently a member of the ASME Committee on Dynamics of Structures and Systems. He is the recipient of the NSF Presidential Young Investigator Award, the Sigma Xi Junior Faculty award, and the Pi Tau Sigma Excellence in Teaching award. He serves as a reviewer for numerous journals and funding agencies and is the author of more than sixty publications.

In his spare time Benson races his bikes up and down mountains, draws and paints, birdwatches, and creates latte art.

CONTENTS

INTRODUCTION

This book is about how to describe the forces that act on structures in equilibrium. Newton's Laws of Motion are used to establish mathematical relationships between the various quantities involved. These relationships enable us to predict how the quantities affect one another. After studying the material in this book, you should be able to use **static analysis**, which involves

1. looking at a structure and seeing how it resists loads,
2. creating a model of the structure,
3. evaluating the loads on the structure that keep it in equilibrium, and
4. postulating and answering "what if" questions about the structure.

This sequence of events is illustrated in Figure 1.1.

Static analysis is one example of **engineering analysis**. More generally, engineering analysis involves performing the calculations needed to assess the behavior of a system. The basis for these calculations is often physical principles from chemistry and physics.

This chapter presents background material for static analysis. We begin with a discussion of engineering practice and the role of analysis in this practice. Then we examine Newton's Laws of Motion, the basis of the analysis presented in this book. Next, we review units used in calculations, as well as coordinate systems and vectors. Then we look at techniques for drawing diagrams and solving problems. The chapter ends with a look at how the rest of the book is organized.

OBJECTIVES

On completion of this chapter, you will be able to:

- **Appreciate how analysis fits into engineering practice**
- **Follow the steps used in static analysis**
- **State Newton's three laws of motion**
- **Convert units and represent vectors**
- **Use good problem-solving habits, including drawing a structure**

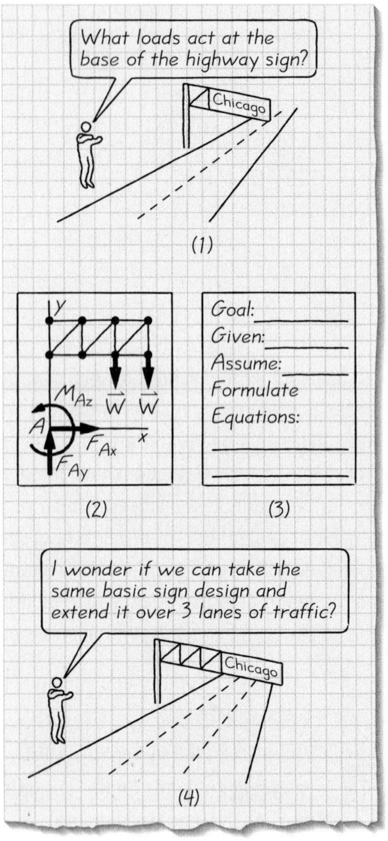

Figure 1.1 Engineer using analysis to answer a question

1.1 HOW DOES ENGINEERING ANALYSIS FIT INTO ENGINEERING PRACTICE?

There are 1.5 million practicing engineers in the United States; this is less than 1% of the U.S. population. Engineers create the products and systems that we interact with daily (bridges, roadways, buildings, VCRs, computers, coffee pots, airplanes). They create products that improve our quality of life (surgical devices, air-scrubbers in smoke stacks), entertain us (roller blades, roller coasters, electric trains, bikes), and educate us (LCD projection systems, computers). Engineers also create the systems that extend our reach from our planet's surface to the bottom of the ocean and to distant planets.

These products and systems don't just happen. The process by which engineers design and manufacture them is referred to as the **product realization process** and may extend over months (less than six months for disk drives), years (for automobiles or bridges), or even decades (as in the case of the space shuttle).

Any product or system begins with someone identifying an initial *client need* (the design problem). This need may arise from the market, the development of new technology, the demand for more sophisticated engineered systems or simply the President of the United States stating, "We will go to the moon before the end of the decade." Another way of thinking about this is that a product or system is made to solve some problem. Identification of a problem includes development of a list of design requirements. These design requirements are benchmarks used to evaluate progress toward a design solution, as well as the performance of the final design solution. They may have to do with, for example, the final design's performance, appearance, time-to-market, cost, ease of manufacture, safety, impact on the environment, or ability to meet national or international standards.

Enumeration of design requirements is followed by generation of ideas on how to address the need or problem. These early ideas are referred to as "design concepts," and this phase of the product realization process is known as *conceptual design*.

Conceptual design is followed by *preliminary design*, where some of the concepts are developed further and some are discarded. Often the decision to continue with or discard a concept is based on an evaluation of how well the concept meets the design requirements. Evaluation may involve calculations and/or building prototypes (physical or virtual) of the concept. Typically, preliminary design ends with the selection of a single concept that will be detailed and refined in the next phase of design (called detail design).

Decisions made during *detail design* about specific configurations of components, types of materials, size of connections, methods of manufacturing, and so on, are often based on analysis to confirm that design decisions and choices continue to meet the design requirements. The analysis may involve numerical modeling and simulation. Building and testing of prototypes may also be involved.

Detail design results in a *comprehensive description* of the product or system. This description consists of drawings, complete fabrication specifications, and supporting documentation that describes the design decisions. It should also include analysis details and test results that support these decisions. ☑

Detail design is followed by production, in which the product or system is constructed or manufactured. Here engineers oversee the process to verify that the final product meets the design requirements. Analysis may be used in this verification.

The product realization process that we have described is shown in **Figure 1.2**. We have presented it as a linear, sequential process, with one phase connecting to the start of another phase. In reality the process is a continuous loop. For example, new design requirements may be generated later in the process as additional details of the design are being worked out. Also the real problem being solved may not be identified until well into the conceptual phase of design, or two competing concepts may be carried into detail design before a decision is made as to which one will be produced.]note

Regardless of where in the product realization process flowchart an engineer is working, he or she is likely to be involved in verifying and justifying decisions about the product. Engineering analysis is one of the main tools the engineer will use. The major steps in engineering analysis are summarized as an **engineering analysis procedure** (see **Box 1.1** below).

A dual thought process is required in carrying out engineering analysis. It is necessary to think in terms of the physical situation and in terms

[GGADFSC]

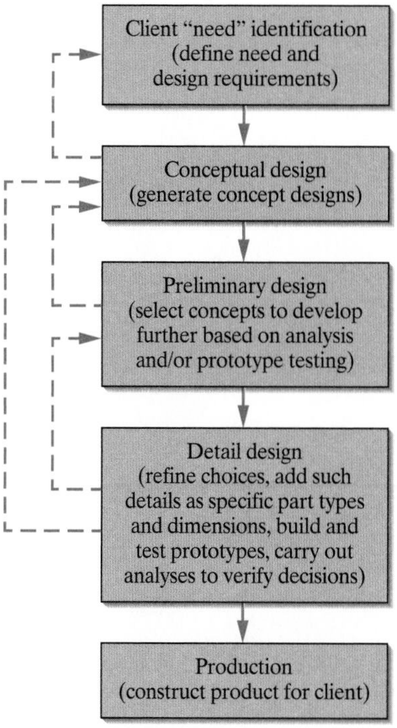

Figure 1.2 Product realization process flow chart ☑

☞ Memorize & use ☑

Box 1.1: Overview of Engineering Analysis Procedure

Goal: Formulate the question to be answered by the analysis; what is the analysis to find? In many textbook problems, the question is provided in the problem statement, so this step involves restating the problem in your own words to be sure you understand what is being asked. Make sure that your restatement mentions every final result you should have once you finish working the problem. In engineering practice, the question is often whether a design meets a specific requirement.

Given: Summarize and record what is known. For textbook problems, this may mean restating what is given in the problem, including creating a sketch of the situation. In engineering practice, the source of information might be a design drawing or specification, previous analysis, or a standard reference source.

Assume: Make assumptions about the behavior of the system under consideration to create a simplified representation or model that can be analyzed. This is sometimes referred to as system modeling.

Draw: Draw any diagrams necessary to clarify the model. In statics, a free-body diagram is used to clarify the assumptions made in modeling the system under consideration.

Formulate Equations: Apply engineering principles, generally in mathematical form, to set up equations that represent the model's behavior. In statics, these principles are Newton's laws expressed as equilibrium conditions.

Solve: Solve the resulting equations. In some cases, this can be done by hand. In other cases, the solution requires the use of appropriate software. Clearly state how numerical answers address goal in undertaking the analysis.

Check: Check the results using technical knowledge, engineering judgment, and common sense.

of the corresponding mathematical description. Analysis of every problem will require the repeated transition of thought between the physical and the mathematical. Without question, *one of the most important goals for your study of equilibrium is to develop the ability to make this transition of thought freely*. You should recognize that the mathematical formulation of a physical problem represents an ideal limiting description, or model, which approximates but never quite matches the actual physical situation.

1.2 PHYSICS PRINCIPLES: NEWTON'S LAWS REVIEWED

The physical principles that underlie engineering analysis in this book are Newton's three laws of motion[1]:

First Law: An object will remain at rest (if originally at rest) or will move with constant speed in a straight line (if originally in motion) if the resultant force acting on an object is zero. Another way of stating the same law is that an object originally at rest, or moving in a straight line with constant velocity, will remain in this state provided the object is acted on by balanced force.

Second Law: If the resultant force acting on an object is not zero, the object will have an acceleration proportional to the magnitude of the resultant force and in the direction of this resultant force.

Third Law: The forces exerted by two objects on each other are equal in magnitude and opposite in direction.

In this book we use the first and third laws extensively to describe situations where objects are at rest or are moving at constant velocity as a result of being acted on by balanced forces. We call these situations "static." This book is about static analysis, which is often referred to simply as **statics**.

Closely related to statics is **dynamics**, the area of engineering that also embodies analysis based on Newton's laws except that the object is moving at a nonconstant velocity, an acceleration, as described by Newton's second law. In mathematical terms, the second law says that if an object is acted upon by an unbalanced force $\boldsymbol{F}$, the object experiences acceleration $\boldsymbol{a}$ in the same direction as the force. The acceleration is

[1]The man most immediately responsible for what you'll be learning in this book is Sir Isaac Newton. Even among geniuses, Newton stands out. He needed a new mathematical approach to handle his investigations and so he invented calculus. That same year, he revolutionized optics by realizing that white light is made up of a spectrum of colors. And, to top it all off, he laid down his three laws of motion. Even more amazing, he did all of this when he was in his early twenties while taking a short break from London in order to avoid the plague.

He was one of the supreme scientists the world has seen but he was also a card carrying alchemist and spent quite a lot of his time doing things like searching for the philosopher's stone, a useful item (if you could make it) which supposedly turns base metals into gold. Quite the fascinating guy.

Interestingly, his laws of motion weren't overly complicated and you've undoubtedly run across them in a physics class.

proportional to the force (and the proportionality factor is the mass m of the object):

$$\boldsymbol{F} = m\boldsymbol{a} \tag{1.1}$$

The bold italic notations $\boldsymbol{F}$ and $\boldsymbol{a}$ denote that these are vector quantities, as discussed below. Dynamics is covered in Volume II.

Together statics and dynamics make up the study of "rigid body mechanics." A **rigid body** is a combination of a large number of particles in which all the particles remain at a fixed distance from one another before, during, and after a force is applied to the object. As a result, the material properties of any object that is assumed to be rigid will not be considered when analyzing the forces acting on the object. In most cases, the actual deformations occurring in structures, machines, mechanisms, and the like are relatively small, and the rigid-body assumption is suitable for analysis or preliminary design. Detail design requires full investigation of the deformations.

1.3 PROPERTIES AND UNITS IN ENGINEERING ANALYSIS

Static analysis involves quantifying, manipulating, and measuring properties of objects. The properties we are concerned with are length, time, mass, and force:

Length is a description of distance.

Time is conceived as a succession of events. Although the principles of statics are time-independent, this quantity does play an important role in the study of dynamics.

Mass is a property of matter by which the action of one object can be compared with the action of another. This property manifests itself as a gravitational attraction between two bodies and provides a quantitative measure of the resistance of matter to a change in velocity.

Force is considered as a push or pull exerted by one object on another.

In working with these quantities we need consistent and standard measures—these are provided by the **International System of Units** (abbreviated SI after the French Le Systeme International d'Unites) and the **U.S. Customary system** of units, as summarized in **Table 1.1**. The SI system is the accepted national standard of measurement in all countries except Myanmar (the former Burma), Liberia, and the United States. The U.S. Customary system is also known as the British system or the foot-pound-second (FPS) system.

SI Units

As shown in **Table 1.1**, the standard measure of length in the SI system is the **meter**. A meter is roughly the length from an adult's nose to his or her extended finger tips. Often engineers deal with lengths that are much larger (e.g., earth's radius) or smaller (e.g., the thickness of this

Table 1.1 Standard Measures

Name	Standard Unit of Length	Standard Unit of Time	Standard Unit of Mass	Standard Unit of Force
International System of Units (SI)	meter (m)	second (s)	kilogram (kg)	newton (N)*
US Customary (British System) (FPS)	foot (ft)	second (s)	slug**	pound (lb)

*derived quantity, based on meter, second, and kilogram, as discussed below ($N = \frac{kg \cdot m}{s^2}$)

**derived quantity, based on foot, second, and pound, as discussed below ($slug = \frac{lb \cdot s^2}{ft}$)

page) than a meter; therefore, it may be more appropriate to deal with multiples or submultiples of the meter. We denote these multiples or submultiples with the prefixes listed in Table 1.2. For example, the mean radius of the earth is 6.37e6 m or 6370 km, and this page is 1×10^{-4} m or 0.1 mm thick.

The standard measure of mass in the SI system is the **kilogram** (kg), defined as the mass of a particular platinum-iridium cylinder kept at the International Bureau of Weights and Measures near Paris. From **Table 1.2**, we see that the prefix of "kilo" means that this standard has a mass of 1000 grams. Engineers work with a range of mass sizes, from the very large (mass of a Boeing 777) to the very small (mass of a white blood cell).

Table 1.2 SI Prefixes*

Factor	Prefix	Symbol
10^{18}	exa-	E
10^{15}	peta-	P
$1\ 000\ 000\ 000\ 000 = 10^{12}$	tera-	T
$1\ 000\ 000\ 000 = 10^{9}$	giga-	G
$1\ 000\ 000 = 10^{6}$	mega-	M
$1\ 000 = 10^{3}$	**kilo-**	**k**
$100 = 10^{2}$	hecto-	h
$10 = 10^{1}$	deka-	da
$0.1 = 10^{-1}$	deci	d
$0.1 = 10^{-2}$	centi-	c
$0.001 = 10^{-3}$	**milli-**	**m**
$0.000\ 001 = 10^{-6}$	micro-	μ
$0.000\ 000\ 001 = 10^{-9}$	nano-	n
$0.000\ 000\ 000\ 001 = 10^{-12}$	pico-	p
$0.000\ 000\ 000\ 000\ 001 = 10^{-15}$	femto-	f
$0.000\ 000\ 000\ 000\ 000\ 001 = 10^{-18}$	atto-	a

*Prefixes commonly used in this book are shown in boldface type.

The standard measure of time is the **second** (s).

The standard unit of force in the SI system is the **newton** (N). One newton is equal to the force required to give 1 kilogram of mass an acceleration of 1 m/s^2. We will have a lot more to say about forces in Chapter 4.

In the SI system, length, mass, and time are the fundamental properties, and force is a derived quantity from Newton's second law. By Newton's second law (1.1), one newton (1 N) of force equals $[1\ kg][1 \frac{m}{s^2}] = [\frac{kg \cdot m}{s^2}]$. Guidelines for working with SI prefixes and units are given in **Box 1.2** below.

U.S. Customary Units

The standard measure of length in this system is the **foot**, as shown in **Table 1.1**. The standard measures for time and force are the **second** and **pound**, respectively.

In the U.S. Customary system, the fundamental properties are length, force, and time. The standard unit of mass in the U.S. Customary system is called the **slug** and is derived from the foot, second, and pound using Newton's second law. One slug is equal to the amount of matter that is accelerated at 1 ft/s^2 when acted upon by a force of 1 pound (1 slug = 1 $lb \cdot s^2/ft$).

No matter which system of units you are working with, it is imperative that you *use consistent units*. For example, if you are using kilometers as the measure of length, make sure that you use kilometers consistently for all measures of length in the problem. Do not mix with feet or miles. Sometimes you may need to convert quantities from one measurement system to another; **Table 1.3** lists some conversion factors for going between U.S. Customary units and SI units.

Box 1.2: Guidelines for Working with SI Prefixes and Units

1. Unit symbols are always written in lowercase letters, with the following exceptions: symbols for some prefixes and symbols named after an individual are capitalized (e.g., N for newton).
2. Unit symbols are never written with a plural "s" because this may be confused with the unit for second (s).
3. Compound prefixes should not be used. For example, k μm (kilo-micro-meter) should be expressed as mm (millimeter) since $1(10^3)(10^{-6})$ m = $1(10^{-3})$ m = 1 mm.
4. The exponential power given for a unit having a prefix refers to both the unit and its prefix (e.g., $mm^2 = (mm)^2 = mm \cdot mm$).
5. In engineering notation, exponents are generally displayed in multiples of three. This convention facilitates conversion to the appropriate prefix. For example, $4.0(10^3)$ N can be rewritten as 4.0 kN.
6. Quantities defined by several units that are multiples of one another are separated by a dot to avoid confusion with prefix notion (e.g., $N = kg \cdot m/s^2 = kg \cdot m \cdot s^{-2}$). The dot notation differentiates m·s (meter-second) from ms (millisecond).
7. Avoid prefixes in the denominator of composite units. For example, write kN/m rather than N/mm. The exception to this rule is the kilogram (kg); since it is the base unit of mass, it is fine to use it in the denominator (e.g., write Mm/kg rather than km/g).
8. When calculating, convert all prefixes to powers of 10. For example, (100 kN)(200 μm) = $[100(10^3)\ N][200(10^{-6})\ m] = 20{,}000(10^{-3})\ N \cdot m$. Then express the final result using a single prefix combined with a numerical value between 0.1 and 1000: $20{,}000(10^{-3})\ N \cdot m$ becomes 20 N·m.
9. Minutes, hours, days, and so forth are used for multiples of the second. Plane angular measurement is made using radians (rad) or degrees (°).

Table 1.3 Conversion Factors

Converting from U.S. Customary to SI				
Quantity	To convert from	U.S. Customary	to SI	Multiply by
Force		lb		4.4482 N
Mass		slug		14.5938 kg
Length		ft		0.3048 m
Converting from SI to U.S. Customary				
Quantity	To convert from	SI	U.S. Customary	Multiply by
Force		N		0.2248 lb
Mass		kg		0.06852 slug
Length		m		3.2808 ft

EXERCISES 1.3

1.3.1 Derive conversion factors for changing the following U.S. Customary units to their SI equivalents:

a. Pressure, $lb/in.^2$
b. Force, kip
c. Volume, ft^3
d. Area, $in.^2$

1.3.2 Derive conversion factors for changing the following SI units to their U.S. Customary equivalents:

a. Pressure, N/m^2 (pascal)
b. Pressure, MPa (MegaPascal)
c. Volume, m^3
d. Area, mm^2

1.3.3 Maurice Green set the world record for the 100-meter dash on June 16, 1999. His time was 9.79 seconds. Calculate his average speed in m/s, ft/s, and mph.

1.3.4 Calculate the percent difference between the mile and the metric mile (1500 meters).

1.3.5 The world best performance in the women's marathon is 2:18:47 by Catherine Ndereba from Kenya. (The race was run in Chicago on October 7, 2001.) On average, how long did it take her to run each mile? What was her average speed in m/s? A previous world best performance was turned in by Kenyan Tegla Laroupe. She finished her Berlin race on September 26, 1999, in 2:20:43. How much faster did Catherine Ndereba run each mile of the race?

1.3.6 Complete the following two tables:

Table 1.3.6a Men's World Records for Selected Field Events

Event	Meters	Centimeters	Inches	Feet	Miles
High jump	2.45		96.5		
Pole vault	6.14			20.14	3.82E-03
Long jump			352.4		
Triple jump			720.1	60.01	1.14E-02
Shot put	23.12			75.85	
Discus throw		7408	2916.5		
Hammer throw		8674		284.58	
Javelin throw	98.48		3877.2	323.10	

Table 1.3.6b Women's World Records for Selected Field Events

Event	Meters	Centimeters	Inches	Feet	Miles
High jump	2.09			6.86	
Pole vault		462	181.9		
Long jump				24.67	4.67E-03
Triple jump	15.5		610.2		9.63E-03
Shot put		2263		74.25	
Discus throw			3023.6	251.97	
Hammer throw			2994.9	249.57	
Javelin throw				220.11	

1.3.7 In the heavyweight division, East German Ronny Weller holds the world record for the clean and jerk. He lifted a mass of 262.5 kg. Calculate the mass in slugs. What is the corresponding weight in newtons and pounds? How many people would it take to clean and jerk a Porsche 911 if they were all as strong as Ronny Weller?

1.3.8 When a certain linear spring has a length of 180 mm, the tension in it is 170 N. For a length of 160 mm, the compressive force in the spring is 120 N.

a. What is the stiffness of the spring in SI units? In U.S. Customary units?

b. What is its unstretched length in SI units? In U.S. customary units?

1.4 COORDINATE SYSTEMS AND VECTORS

Coordinate Systems

In working with physical objects it is useful to specify information about them relative to a **Cartesian coordinate system**, which uses three axes that are orthogonal to one another, as shown in **Figure 1.3*a***. In addition, the system is **right-handed**. In a right-handed system, if you point the fingers of your right hand in the direction of the positive x axis and bend them (as in preparing to make a fist) toward the positive y axis, your thumb will point in the direction of the positive z axis, as shown in **Figure 1.3*b***.

The assignment of coordinate axes is often a matter of convenience, and the choice is frequently up to the engineer. The logical choice is usually indicated by the manner in which the geometry of the situation is specified. For example, when the principal dimensions of a system or structure are given in the horizontal and vertical directions, the assignment of coordinate axes in these directions is generally convenient (**Figure 1.4*a***). If, on the other hand, the structure and/or the forces are not aligned with the horizontal and vertical directions, alternative orientations of the coordinate axes may be appropriate, as shown in **Figure 1.4*b***.

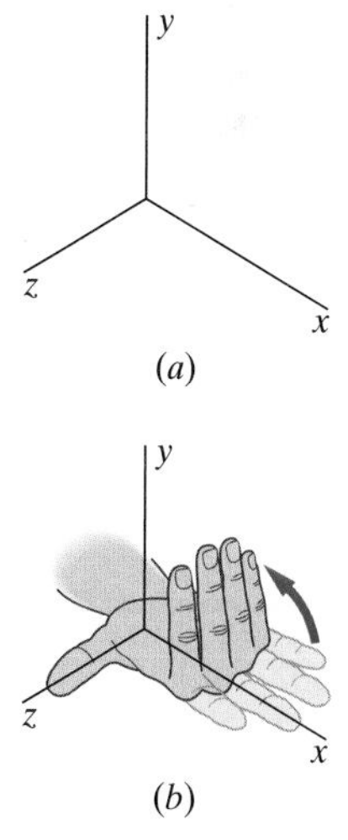

Figure 1.3 xyz coordinates arranged in right-handed manner

Scalar and Vector Quantities

Static analysis deals with two kinds of quantities—scalars and vectors. **Scalar quantities** can be completely described with a magnitude (number only) and associated units. Examples of scalar quantities are mass, density, length, area, volume, speed, energy, time, and temperature. In mathematical operations, scalars follow the rules of elementary algebra.

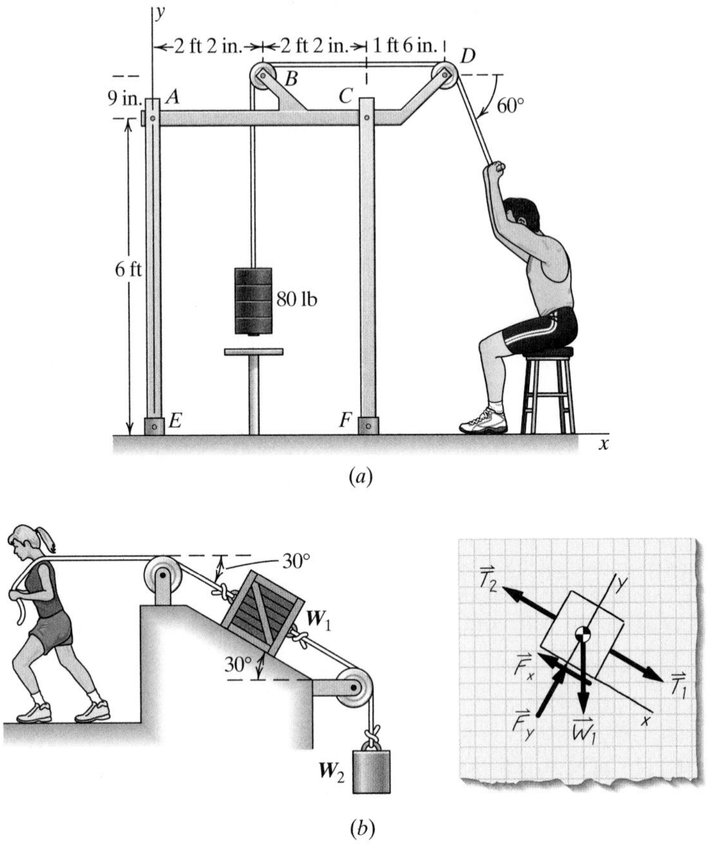

Figure 1.4 Various orientations of coordinate axes

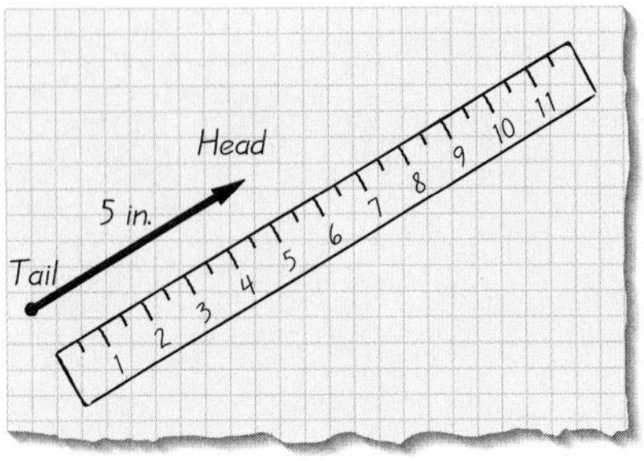

Figure 1.5 A position vector

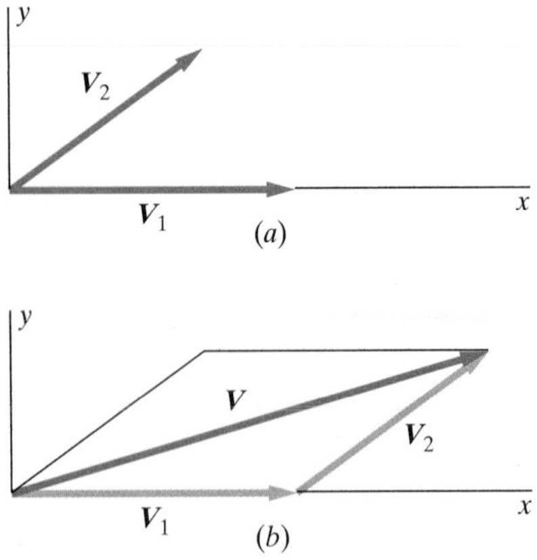

Figure 1.6 (*a*) Two vectors to be added; (*b*) vector addition using the parallelogram law

In contrast to scalars, **vector quantities** have both magnitude (with units) and direction, and obey the parallelogram law of addition, as described below. Examples of vector quantities are velocity, acceleration, momentum, force, moment, and position vector.

A vector is typically represented in drawings by an arrow with a head and a tail (**Figure 1.5**). The direction from the tail to the head of the arrow represents the direction of the vector, and the length of the arrow is often drawn proportional to the magnitude of the vector. The magnitude of the vector is generally written next to the arrow.

In this book, vector quantities are distinguished from scalar quantities through the use of boldface italic type ($\boldsymbol{V}$). In longhand writing, a vector may be denoted by drawing a "half arrow" above the letter, $\overset{\rightharpoonup}{\mathrm{V}}$. Euclidean norm bars surrounding the vector symbol are used to denote the *magnitude* of a vector. Thus, the magnitude of the vector $\boldsymbol{V}$ is denoted by $\|\boldsymbol{V}\|$, or $\|\overset{\rightharpoonup}{\mathrm{V}}\|$ (in longhand).

As mentioned above, vectors obey the **parallelogram law of addition**. This means that the two vectors $\boldsymbol{V}_1$ and $\boldsymbol{V}_2$ in **Figure 1.6*a*** can be replaced by an equivalent vector $\boldsymbol{V}$ that is the diagonal of a parallelogram having $\boldsymbol{V}_1$ and $\boldsymbol{V}_2$ as two of its sides (**Figure 1.6*b***). This combination, or *vector sum*, is represented by the vector equation

$$\boldsymbol{V} = \boldsymbol{V}_1 + \boldsymbol{V}_2$$

where the plus sign used in conjunction with the vector quantities (boldface italic type) means vector and not scalar addition. It is important to note that the vector sum of $\boldsymbol{V}_1$ and $\boldsymbol{V}_2$ is not equal to the scalar sum (which is $V_1 + V_2$).

Vectors are generally specified relative to a Cartesian coordinate system. The specification can be in terms of a magnitude plus angle representation or in terms of the vector components. The **magnitude–angle representation** involves specifying the vector's magnitude accompanied by angles indicating the vector's direction relative to right-handed coordinate axes. For example, the **space angles** θ_x, θ_y, and θ_z, as depicted in Figure 1.7, can be used to specify the direction of $\boldsymbol{V}$. In Chapter 4, we discuss in detail how to use angles to specify the direction of a vector.

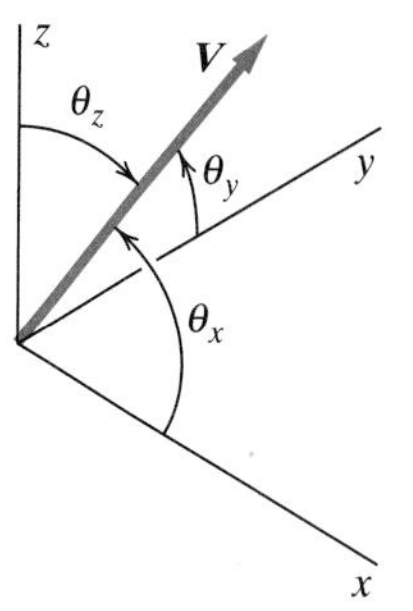

Figure 1.7 Space angles defined

A **vector component representation** of a vector $\boldsymbol{V}$ involves specifying the "hike" you would take in the x direction, the y direction, and the z direction to get from the tail of $\boldsymbol{V}$ to its head (Figure 1.8). The "hike" in each direction is a **component vector**. We indicate that $\boldsymbol{V}$ is the sum of the three component vectors $\boldsymbol{V}_x$, $\boldsymbol{V}_y$, and $\boldsymbol{V}_z$ by the expression

$$\boldsymbol{V} = \boldsymbol{V}_x + \boldsymbol{V}_y + \boldsymbol{V}_z \tag{1.2A}$$

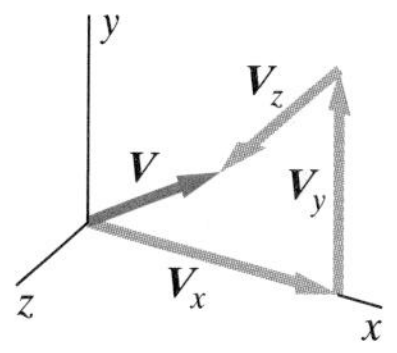

Figure 1.8 Vector represented in terms of its x, y, and z components

To define the component vectors further, we introduce the concept of a **unit vector**. By definition, a unit vector has magnitude 1. Align a unit vector with each of the axes, as shown in Figure 1.9*a*; let $\boldsymbol{i}$ be a unit vector aligned with the x axis, $\boldsymbol{j}$ a unit vector aligned with the y axis, and $\boldsymbol{k}$ a unit vector aligned with the z axis.

The component vector $\boldsymbol{V}_x$ can then be written as $V_x\boldsymbol{i}$, where V_x is the distance we hike in the direction of $\boldsymbol{i}$. We will refer to V_x as the x component. If V_x is positive, we hike in the positive x direction; if V_x is negative, we hike in the negative x direction. Similarly, $\boldsymbol{V}_y$ can be written as $V_y\boldsymbol{j}$, and $\boldsymbol{V}_z$ as $V_z\boldsymbol{k}$. Therefore, $\boldsymbol{V}$ can be written in terms of unit vectors:

$$\boldsymbol{V} = V_x\boldsymbol{i} + V_y\boldsymbol{j} + V_z\boldsymbol{k} \tag{1.2B}$$

We can also interpret the vector components V_x, V_y, and V_z as scale factors of the unit vectors $\boldsymbol{i}$, $\boldsymbol{j}$, and $\boldsymbol{k}$.

Finally, if we align a unit vector with $\boldsymbol{V}$ (call this unit vector $\boldsymbol{u}$, where $\boldsymbol{u} = \cos\theta_x\boldsymbol{i} + \cos\theta_y\boldsymbol{j} + \cos\theta_z\boldsymbol{k}$; see **Figure 1.9*b***), we are able to rewrite $\boldsymbol{V}$ as $\|\boldsymbol{V}\|\,\mathbf{u}$, where $\|\boldsymbol{V}\|$ is the magnitude (size) of the vector. Therefore, (1.2B) becomes

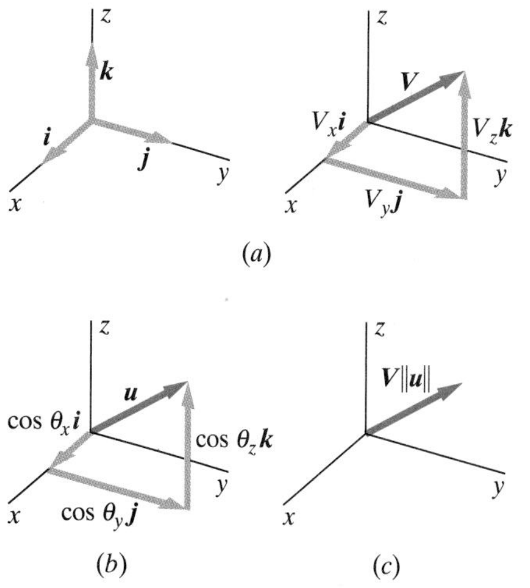

Figure 1.9 Unit vectors

$$\boldsymbol{V} = \|\boldsymbol{V}\|\,(\cos\theta_x\boldsymbol{i} + \cos\theta_y\boldsymbol{j} + \cos\theta_z\boldsymbol{k}) \tag{1.2C}$$

This expression is depicted graphically in **Figure 1.9*c***.

EXERCISES 1.4

1.4.1 Determine whether the missing axis in each case in E1.4.1 is oriented into or out of the page for a right-handed coordinate system.

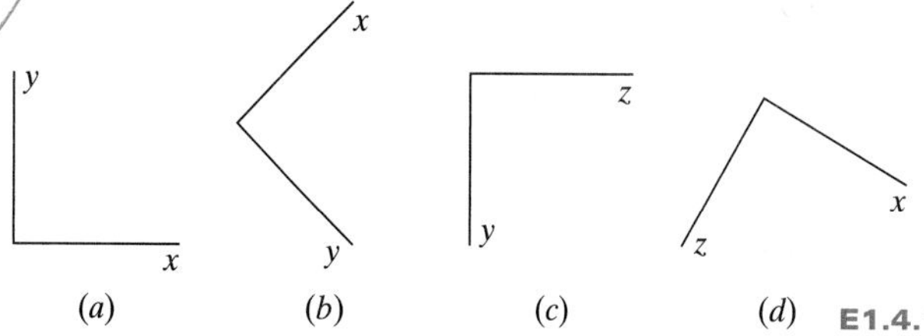

E1.4.1

1.4.2 Which of the coordinate systems in **E1.4.2** are right-handed? Clearly state your assumptions about which axes go into or out of the page, or are in the plane of the page.

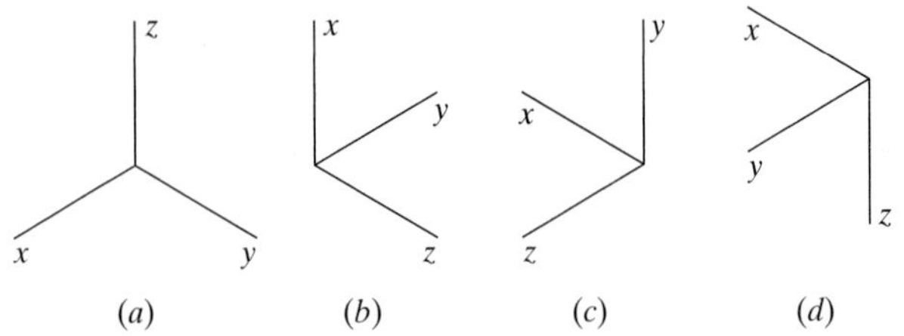

E1.4.2

1.5 DRAWING

Graphical representation is used to communicate information that would be difficult (if not impossible) to communicate between engineers and users about a product or system with words alone. For example, graphics in the form of drawings can be used to document the size and configuration of a product or how it is assembled. We will refer to these types of drawings as **formal engineering drawings**. In addition, graphics in the form of charts, schematics, and diagrams are used to convey information about the sequence of steps in a product's manufacture or how it was modeled for the purpose of analysis. We will refer to these types of representations as **engineering diagrams**. The intent of both formal engineering drawings and engineering diagrams is to transmit information about the product to someone else. Examples are shown in **Figure 1.10**.

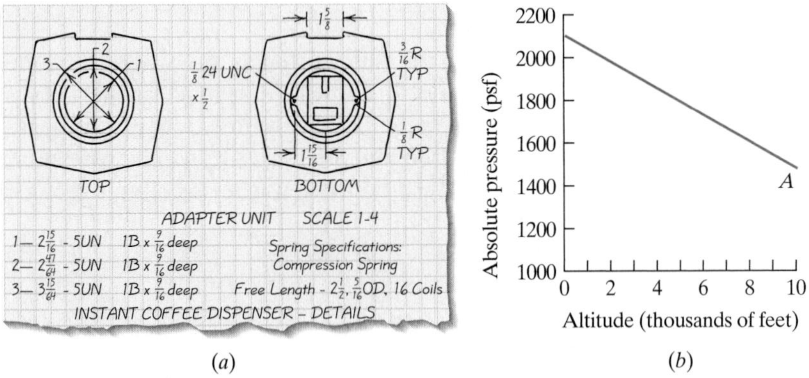

Figure 1.10 Examples of formal engineering drawings

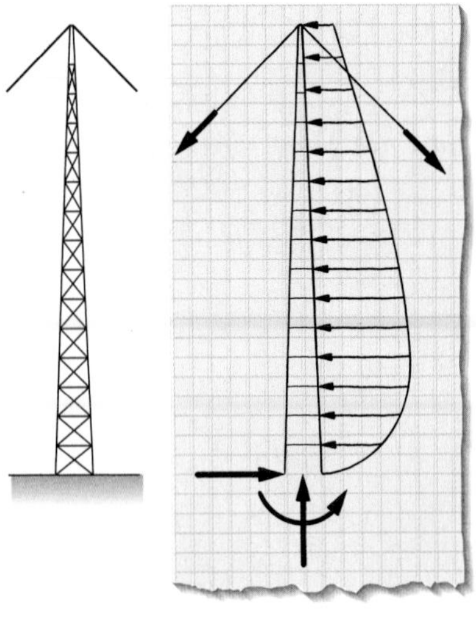

Figure 1.11 A system represented as a free-body diagram

Engineering diagrams play a role in engineering analysis. The engineering diagram created in static analysis is the **free-body diagram** (**Figure 1.11**). This diagram describes a structure in terms of its size, the relationships between its important components, and how the rest of the world interacts with it. Many details of the real structure are simplified in this diagram and are represented in terms of their function. For example, a joint between two components that allows their relative rotation may be represented by a circle, and a distributed load may be represented as a point force. Implicit in these representations is the assumption that for the purpose of this static analysis, it is reasonable to assume that these simplified representations adequately model the real structure's behavior. A free-body diagram is a model, or representation, of the real structure. More details on creating free-body diagrams are given in Chapter 6. It will be very important in our studies!

Beyond formal engineering drawings and engineering diagrams, drawings are also created by engineers as "think pieces." For example, in the conceptual stage of design, an engineer draws a thumbnail sketch of an idea as part of explaining the idea to a colleague, and the very process of representing the idea on paper spawns other ideas in both of their imaginations. Or the process of committing a design solution to paper may help an engineer identify and work out problems toward a solution. The value of these types of drawings is often in the act of their creation. We will refer to this type as **informal engineering drawing** (**Figure 1.12**).

Informal drawing plays a role in helping you to visualize, understand, and define the overall structure and relationships between components. It is drawing done as part of a visual study of the structure and in preparation for creating a free-body diagram. Informal drawing is typically self-intended or directed to a small group and is often concerned more with overall features than with details. This form of drawing generally assumes that additional information will be added verbally—hence they do not stand alone in communicating.

The process of informal drawing starts with a "picture" of the structure that may be a photograph, the actual structure, another drawing, or even an image from your imagination. The process cycles through seeing–deciding–drawing, beginning at a large scale and then moving through progressively smaller scales:

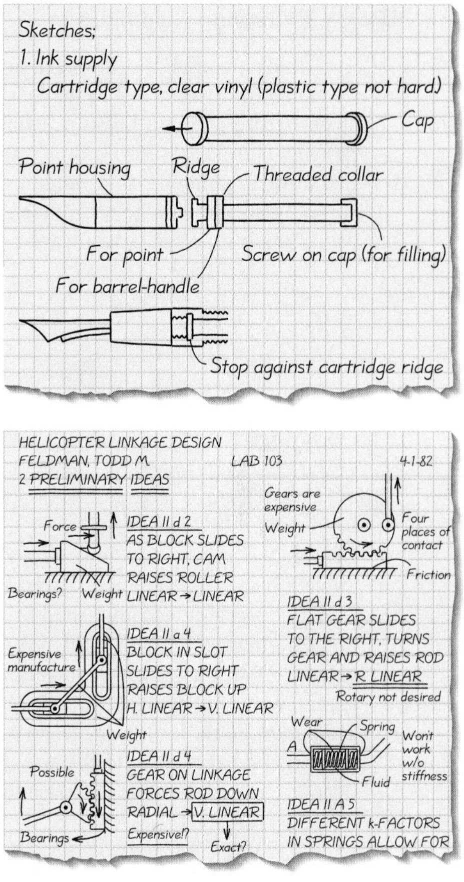

Figure 1.12 Examples of informal engineering drawings

Seeing means scanning the overall structure, noticing details, shapes and spaces, relationships, scales, sizes, patterns, proportions, and connections. It is a focused visual study (of course, if you can actually touch, move, and/or manipulate the structure, you are urged to do this too). Avoid applying labels as you go about a visual study. Contrary to common belief, seeing is an active art to be developed, not a passive experience to be taken for granted.

Deciding means making choices about which features are important, which can be simplified and which can be ignored, what is inside the structure and what is outside. These choices should be based on what you saw in your visual study. Deciding filters your seeing. Deciding also means coming up with labels for features or components to make reasoning about them easier and planning the overall size of the drawing. Use labels and words carefully; "don't chip away at things to make them fit words, but instead conscientiously use words to try to make them fit things" [*Experiences in Visual Thinking*, Robert H. McKim (Monterey; Calif.: Brooks/Cole, 1972), page 62].

Drawing is putting down on paper what you decided was important. Continue to look at the structure as you draw (in fact, your eyes should move often between your paper and the structure), but now you should be able to focus on those features that you have decided should be included. Some specific guidelines for putting line on paper are included in **Box 1.3** on the following page.

We could end this section on drawing by simply telling you to make sure to do both informal and formal drawings as part of carrying out static analysis. Some of you, however, may need a pep talk regarding drawing. Most adults have little confidence in their ability to draw, with their

Box 1.3: Drawing Guidelines

1. The **tools** needed for drawing are commonplace, so don't feel that you need to make a trip to an art supply store before you can begin drawing. The tools include pencils, pens, an eraser, a straightedge or ruler with a metal edge, unlined paper, and engineering grid paper. Keep your drawing tools materials in good working order in a box so that you always know where they are.

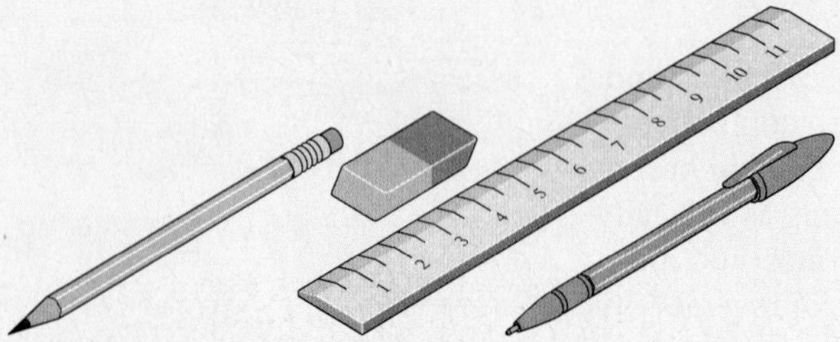

2. Create drawings with a sense of **proportion**. This means depicting the right relative size of components and features. This can be accomplished by taking a few relative measurements of features of the device either with a ruler or other aid (pencil or fingers). These relative measurements then become guidemarks or landmarks on the paper. Here, for instance, the person is drawn much too large in (*a*) relative to the size of the equipment; the properly scaled drawing of (*b*) is much more useful.

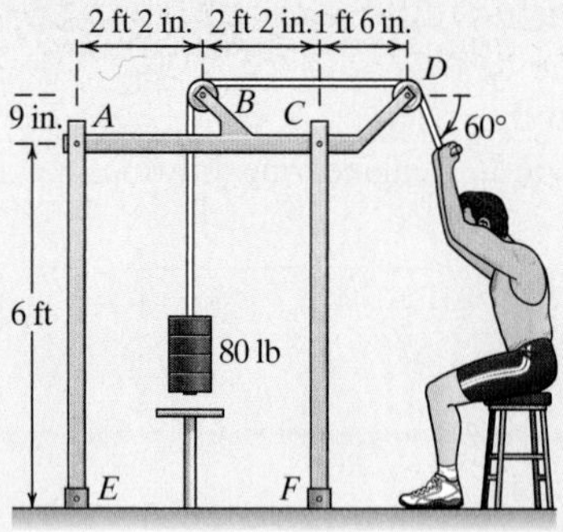

(*a*)

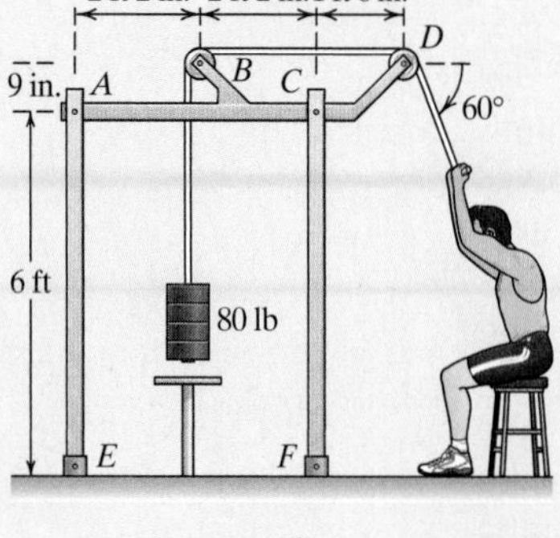

(*b*)

3. Create drawings with a sense of **scale**. This doesn't mean including all dimensions, but it does mean including some way of knowing how big the object is. This can be accomplished by including:
   - Something of known size in the drawing (e.g., a person)
   - A textual note that states the scale (e.g., the drawing is 3 times actual size)
   - A background grid with scale noted
   - A few dimensions

4. **Symbols** are useful for conveying standard information compactly. Alphanumeric characters and words are an example of this. Along with word and numeric labels in drawings, you will probably use arrows, people, circles, ellipses, and boxes. It is worth practicing these symbols so that they become second nature when drawing.

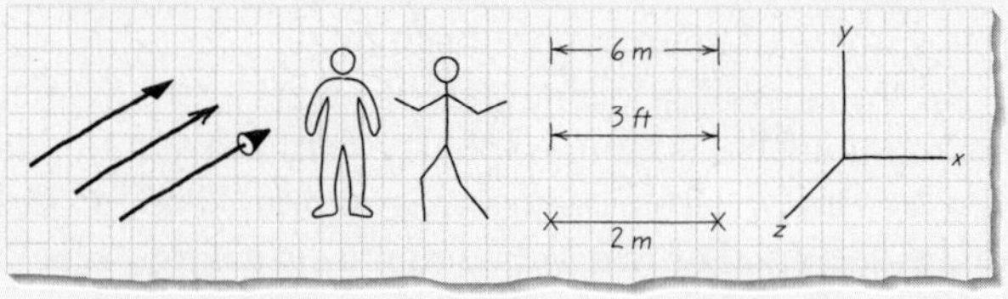

5. **Plan** your drawing by considering overall size relative to your paper. If you like to create large drawings that do not fit well into your homework solution, consider drawing large, then using a copy machine to **shrink** your final drawing for the purpose of your homework solution. Sometimes you may need **multiple views** of an object to convey it. As you go through the see–decide–draw cycle, use tracing paper over prior drawings or photographs to **trace** key lines. Outlining in pen will help you to see those lines through the tracing paper.

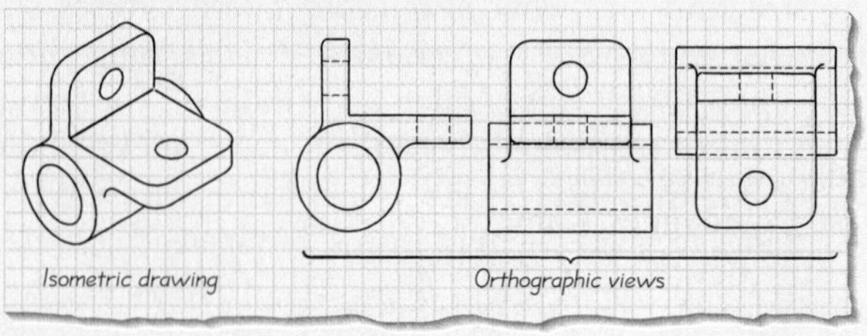

From *Drawing on the Right Side of the Brain* by Betty Edwards, Copyright © 1979, 1989, 1999 by Betty Edwards. Used by permission of Jeremy P. Tarcher, an imprint of Penguin Groups (USA) Inc.

Figure 1.13 Examples of how a little practice (and instruction) in drawing can go a long way. (Left image is before practice and instruction, right image is after.)

confidence starting to wane as early as elementary school. Many people think that the ability to create a drawing is a gift and that only talented individuals can draw. But like reading, writing, and riding a bicycle, with a few guidelines, some practice, and a little encouragement, everyone (yes, *everyone*) can create drawings that embody what things look like. As evidence, consider the examples in **Figure 1.13** from Betty Edwards' book *Drawing on the Right Side of the Brain*.

EXERCISES 1.5

1.5.1 Identify three devices that involve a conversion of human input (i.e., forces and movements) into some other force or movement. Create sketches of each device that show how you think it works and what forces are involved. Examples to get your thinking jump-started are piano, typewriter, house window crank, foot-actuated garbage can, and bicycle pump. Do not draw those.

1.5.2 Find an interesting artifact in your kitchen, garage, or dorm room. Create a storyboard (cartoon-like description) of how the artifact works or how it is operated. Examples of artifacts include hand mixer, hole punch, and nail clipper. Do not use these examples.

1.5.3 Given the front and side views in the three multi-view drawings in **E1.5.3**, sketch the missing top view in each case.

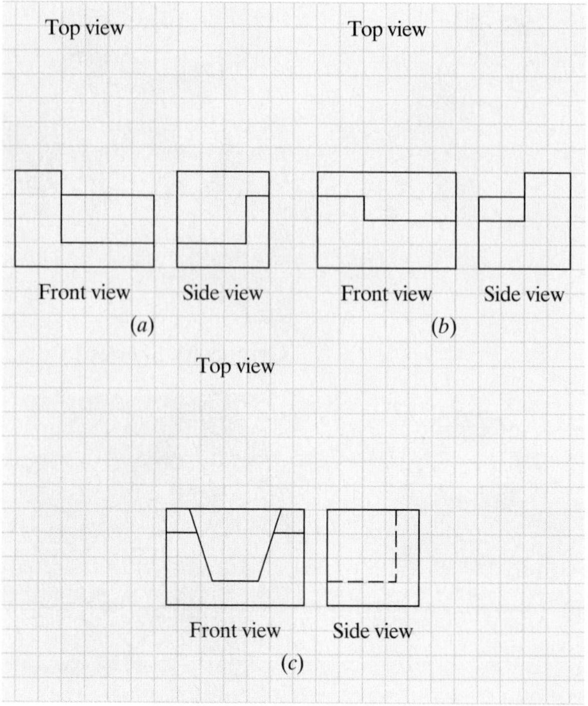

E1.5.3

1.5.4 For each of the five objectsin **E1.5.4**, create a multi-view drawing showing separate front, side, and top views.

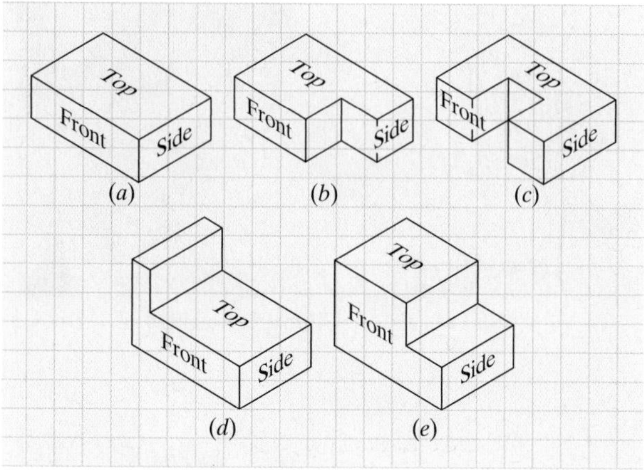

E1.5.4

1.6 PROBLEM SOLVING

These comments are about good analysis habits. These habits do not guarantee that your solutions will be consistent and correct, but they do make it more likely! Neat and well-compiled analyses are also a means of engineers communicating with one another about design decisions. It is not unusual for an analysis to be reexamined when there is a design change or a problem in the field.

Static analysis is generally carried out on paper using the steps outlined in **Box 1.1**. Make sure that your written work is neat. It should also be complete—a reader should be able to follow your work from the drawings, equations, and words that are on the paper (and should not need to go to another reference). There should be a clear layout of steps so another engineer can easily follow your thinking. This means that there should be a neatly drawn free-body diagram (with multiple views if needed to convey the structure and loads) and a list of assumptions. The algebra that is part of the *formulate and solve equations* steps should be included as part of the solution. The values of the unknowns that result from the algebra should be boxed so that they are easy to find. It is also useful to draw the correct direction of a load next to its boxed value. Finally, revisit the free-body diagram and write next to the variable labels their now-known values.

Good analysis habits also consist of ensuring that the analysis (1) maintains dimensional consistency, (2) solves any algebraic equation for the desired variable before plugging in known numeric values, (3) maintains consistent numerical accuracy and the correct number of significant figures, and (4) is based on reasonable assumptions. Each of these elements is discussed below:

1. In setting up and solving equations, make sure that all terms in an equation are in the same units. This is called **dimensional consistency**. For example, the value of the magnitude of the total force $\boldsymbol{F}_{\text{total}}$ acting vertically on the overhang in **Figure 1.14** is

$$\| \boldsymbol{F}_{\text{total}} \| = \underbrace{pA}_{\substack{\textit{force from} \\ \textit{snow pressure}}} + \underbrace{\frac{Gm_1m_2}{r^2}}_{\substack{\textit{gravitational} \\ \textit{force}}} \tag{1.3}$$

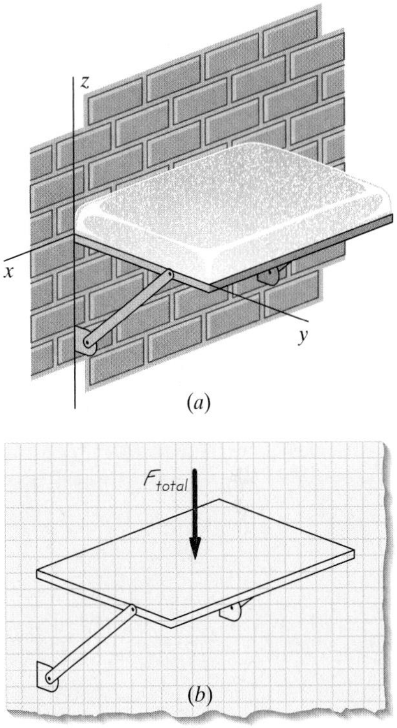

Figure 1.14 (*a*) Snow on top of an overhang; (*b*) forces acting on the overhang

where

p is the pressure of snow pressing down on the structure
A is the area of the upper surface of the structure
G is the universal gravitational constant
m_1 is the mass of the structure
m_2 is the mass of the earth
r is the distance between the centers of mass of the structure and earth

The SI units in the various quantities are $\| \boldsymbol{F}_{\text{total}} \|$ in newtons, p in newtons per meter squared, A in meters squared, m_1 and m_2 in kilograms, r in meters, and G in $[\text{m}^3/(\text{kg} \cdot \text{s}^2)]$. Thus in (1.3), each term is expressed in newtons:

$$\| \boldsymbol{F}_{\text{total}} \| : \quad \mathbf{N}$$

$$\underbrace{pA}_{\substack{\textit{force from} \\ \textit{snow pressure}}} : \quad \frac{\mathbf{N}}{\mathbf{m}^2}\mathbf{m}^2 = \mathbf{N}$$

$$\underbrace{\frac{Gm_1m_2}{r^2}}_{\substack{\textit{gravitational} \\ \textit{force}}} : \quad \frac{\dfrac{\mathbf{m}^3}{\mathbf{kg \cdot s}^2}(\mathbf{kg})(\mathbf{kg})}{\mathbf{m}^2} = \frac{\mathbf{kg \cdot m}}{\mathbf{s}} = \mathbf{N}$$

Each term maintains consistency. *Because static analysis involves the solution of equations that must remain dimensionally consistent, the fact that all terms of an equation you might formulate are represented by a consistent set of units can be used as a partial check.*

2. In setting up and solving equations, you can either insert known numerical values right from the beginning or solve the equation for the desired variable before inserting any numbers. For example, if the position s of an object is at 10 m, and its current speed v is 15 m/s at $t = 20$ s, you could substitute these numerical values directly into the expression that relates distance traveled to velocity, time, and acceleration—namely,

$$s = vt + \tfrac{1}{2}at^2$$

$$10 \text{ m} = (15 \text{ m/s})20 \text{ s} + \tfrac{1}{2}a(20 \text{ s})^2$$

$[s = v \cdot t + \frac{at^2}{2}]$

which you can solve for a,

$$a = \frac{2(10 \text{ m} - (15 \text{ m/s})20 \text{ s})}{(20 \text{ s})^2} = -1.45 \text{ m/s}^2$$

With this approach, the magnitude of each quantity expressed in its particular units is evident at each stage of the calculation. The main advantage of the approach is that the practical significance of the magnitude of each term can be assessed.

Alternatively, you could first solve the equation for the acceleration to obtain a **symbolic solution** and then make numerical substitutions:

$$\left[s = vt + \frac{1}{2}at^2 \right]$$

$s = $ distance
$v = $ velocity
$t = $ time
$a = $ acceleration

$$a = 2\frac{s - vt}{t^2}$$

$$a = \frac{2(10 \text{ m} - (15 \text{ m/s})20 \text{ s})}{(20 \text{ s})^2} = -1.45 \text{ m/s}^2$$

This symbolic solution has several advantages. First, the abbreviation achieved by the use of symbols aids in focusing your attention on the connection between the physical situation and its mathematical description. Second, a symbolic solution allows you to make a dimensional check at every step, whereas dimensional consistency may be lost when only numerical values are used. Third, you may use a symbolic solution repeatedly for obtaining answers to the same problem when different data sets are given.

3. In calculations, you should consider the *accuracy* of the numbers that you are working with. The accuracy of the solution depends on the accuracy of the given data and of the computations performed. For example, if the load on a bridge is known to be 120,000 N with a possible error of 240 N either way, the **relative error** that measures the degree of accuracy of the data is

$$\frac{240 \text{ N}}{120,000 \text{ N}} = 0.002 = 0.2\%$$

In engineering problems, the data are seldom known with an accuracy greater than 0.2%, so it is seldom justified to write the answers with any greater accuracy. What this means for the bridge example is that if we compute the force acting on the bridge at one of its supports to be 43,625 N, the force is actually somewhere between 43,537 N (= 0.998 × 43,625 N) and 43,712 N (= 1.002 × 43,625 N). What answer do we give if we want it to reflect an accuracy of 0.2%?

To answer this question consider that the accuracy of a number is reflected in its number of **significant figures**. A significant figure is any digit in a number that we know with certainty (including zero, provided it is not used to specify the location of the decimal point for the number) plus one uncertain digit. The rule for counting the number of significant figures is to count digits from the left and ignore leading zeros, and keep all digits up to and including the first doubtful one. For example, $x = 3$ m has only one significant figure, and expressing this value as $x = 0.003$ km does not change the number of significant figures. If we instead wrote $x = 3.0$ m (or, equivalently, $x = 0.0030$ km), we would imply that we know the value of x to two significant figures.

Be careful of ambiguity; $x = 300$ m does not indicate whether there are one, two, or three significant figures; we don't know whether the zeros are carrying information or merely serving as place holders. Instead, we should write $x = 3(10^2)$, $x = 3.0(10^2)$, or $x = 3.00(10^2)$, to indicate one, two, or three significant figures, respectively.

Let's return to the question of recording the bridge reaction force of 43,625 N. How many significant figures should be retained to reflect the 0.2% accuracy? Following the rule above, the first doubtful digit from the left is the 6 (in the hundreds slot), since it might be a 5 or a 7. Therefore, recording any digits to the right of the 6 is meaningless. We should record the answer as $43.6(10^3)$ N, which is three significant figures. Even though your calculator display may show 9 or 10 digits, you are not justified by the accuracy of the input data to record more than three significant figures.

A practical rule in engineering calculations is to use four figures to record numbers with a leading "1" and three figures in all other cases in presenting your final answer. Intermediate calculation steps should retain more significant figures. With this rule, a force of 40 is 40.0 N, and a force of 15 is 15.00 N. Numbers are generally **rounded** (as opposed to **truncated**) in reporting values to the correct number of significant figures. For example, 29.694 N would be written with three significant figures as 29.7 N with rounding (and not as 29.6 N, which is what we would get if we truncated the answer).

4. When we construct an idealized mathematical model for a given engineering problem, certain approximations will always be involved. Some of these approximations may be mathematical. For instance, we may ignore one term in an expression if that term is small relative to others in the expression, or we may approximate a trigonometric function for small angles, such as $\tan\theta \approx \theta$, for θ very small. Some approximations may be physical; for instance, it is often necessary to neglect distances, angles, weights, or forces that are much smaller than other distances, angles, weights, or forces. We must be constantly alert to the various assumptions called for in the formulation of real problems. The ability to understand and make use of the appropriate assumptions in the formulation and solution of engineering problems is certainly one of the most important characteristics of a successful engineer. One of the major aims of this book is to provide the maximum opportunity to develop this ability.

EXERCISES 1.6

1.6.1 Round off the numbers listed below to three significant figures.

a. 0.015362 d. 26.39473

b. 0.837482 e. 374.9371

c. 1.839462 f. 6471.907

1.6.2 When an object moves through a fluid, the magnitude of the drag force $\boldsymbol{F}_{\text{drag}}$ acting on the object is given by $\frac{1}{2}C_D\rho V^2 A$, where ρ is the density of the fluid, V is the velocity of the object relative to the fluid, and A is the cross-sectional area of the object. What are the dimensions of the drag coefficient C_D?

1.6.3 The pressure within objects subjected to forces is called stress and is given the symbol σ. The equation for stress in an eccentrically loaded short column is

$$\sigma = -\frac{P}{A} - \frac{Pey}{I}$$

where P is force, A is area, and e and y are lengths. What are the dimensions of the stress σ and the second moment of area I?

1.6.4 In the expressions that follow, c_1 and c_2 are constants, and θ is an angle. Determine the dimensions of these constants if the formula is to be dimensionally correct.

a. $a = c_1 \frac{v^2}{x}$
b. $\frac{1}{2}mv^2 = c_1 x^2$
c. $x = c_1 v + c_2 a^2$
d. θ (degrees) $= c_1 \theta$ (radians)

1.6.5 Being able to make good educated guesses (often called engineering estimation or intuition) is an important aspect of engineering. It is a skill that can be practiced. The goal of this problem is to determine how far an average individual would have to run or jog in order to burn off the calories found in a typical candy bar. **Table 1.6.5** contains information about some of the more popular brands.

The "Nutritional Facts" that the FDA requires on food packages always provide the number of Calories in each serving. It is worth noting that Calories with a capital C is an abbreviation for kilocalories (kcal), where the calorie is the standard unit of heat that you study in chemistry and biology. Typically we do not make value judgments on the unit that is used to measure a given quantity, but it seems inherently better to think of a Snickers bar as having 280 Calories rather than 280,000 calories.

Choose one of the candy bars in the table and estimate how far an average individual would have to run in order to burn off its calories. Next, solve the problem assuming that a typical runner or jogger burns 100 kcal for every mile he or she travels. In fact, this is a pretty good approximation and is not strongly affected by how fast the person moves. Calculate the distance in meters and miles and compare with your initial estimate. What if the person decides not to run that day and all of the calories in the candy bar are converted to fat (9.4 kcal yields 1 g fat)? Calculate the weight gain in newtons and pounds. Is it significant? Did this analysis affect your appetite in any way?

Table 1.6.5 Nutritional Content of Assorted Candy Bars

Maker	Candy Bar	Calories	Fat Grams	Protein (g)
Nestle	Crunch	230	12	2
Nestle	100 Grand	190	8	1
Nestle	Butterfinger	270	11	3
Nestle	Kit-Kat	220	11	3
M&M/Mars	3 Musketeers	260	8	2
M&M/Mars	Twix	280	14	3
M&M/Mars	Snickers	280	14	4
M&M/Mars	Milky Way	270	10	2
M&M/Mars	Milky Way—Lite	170	5	2
M&M/Mars	Milky Way—Midnight	220	8	1
	Average	**239**	**10.1**	**2.3**
	Standard Deviation	**39**	**2.9**	**0.9**

1.7 A MAP OF THIS BOOK

Figure 1.15 shows an example of a well-laid-out static analysis. The solution follows the analysis steps outlined in **Box 1.1**. This example also illustrates the application of the physical principles that form the basis for all static analysis—namely, Newton's laws of motion. Also, this is the sort of problem that you probably handled in physics class.

At this point you may be wondering why this book has nine more chapters on performing static analysis if we have been able to review the physical principles and key skills in the first chapter. This is a good question; the answer is basically that you probably need more experience in

Goal Find the forces at A and B that act on the hopper shown on the right.

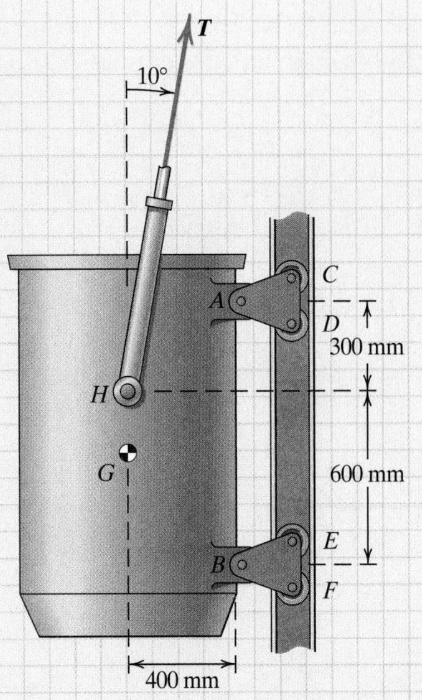

Given Combined mass of hopper and its contents is 4000 kg and its center of mass, location where cable is applied and its orientation angle relative to the vertical (10°).

Assume This is a planar system. A and B are rollers in a vertical track. Hopper is in equilibrium.

Draw Free body diagram of hopper.

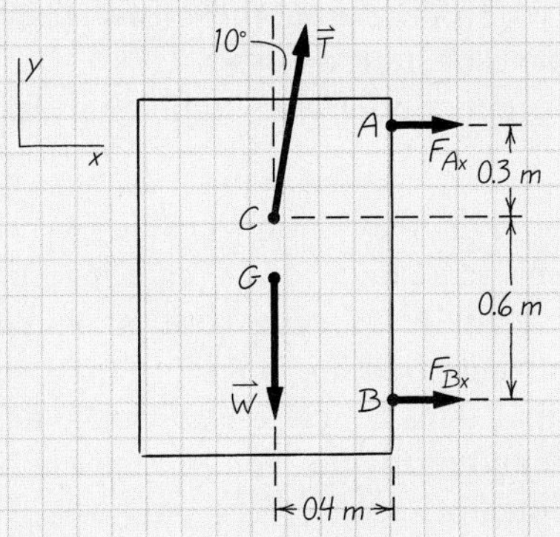

Formulate Equations Based on equilibrium conditions (Newton's First Law) and the free-body diagram, we write:

$$\Sigma F_x = T \sin 10° + F_{Ax} + F_{Bx} + 0 \qquad \text{(A)}$$

$$\Sigma F_y = T \cos 10° - W = 0, \qquad \text{(B)}$$

where $W = (4000\text{ kg})(9.8\text{ m/s}^2) = 39.2\text{ kN}$

With moment center at C, we write:

$$\Sigma M_{z@C} = (-F_{Ax})(0.3\text{ m}) + (F_{Bx})(0.6\text{ m}) = 0 \qquad \text{(C)}$$

Solve

From (B), $T \cos 10° = 39.2\text{ kN} \Rightarrow T = 39.8\text{ kN}$ (D)

From (C), $F_{Ax} = 2F_{Bx}$ (E)

Substitute (D) and (E) into (A):

$$39.8\text{ kN}(\sin 10°) + 2F_{Bx} + F_{Bx} = 0 \Rightarrow F_{Bx} = -2.30\text{ kN}$$

Substitute this into (E) to find:

$$F_{Ax} = -4.61\text{ kN};\ F_{Bx} = -2.30\text{ kN};\ F_{Ax} = -4.61\text{ kN}$$

or in vector notation,

$$\vec{F}_{Bx} = -2.30\text{ kN}\,\vec{i};\ \vec{F}_{Ax} = -4.61\text{ kN}\,\vec{i}$$

Answer $F_{Bx} = 2.30\text{ kN}$ $\quad \vec{F}_B = -2.30\text{ kN}\,\vec{i}$

$F_{Ax} = -4.61\text{ kN}$ $\quad \vec{F}_A = -4.61\text{ kN}\,\vec{i}$

Check It makes sense that A bears more of the load than B since it is closer to the point of application of the force.

Figure 1.15 A well-laid-out example

applying Newton's laws to real engineering situations. With experience you will be able to look at a structure and identify its significant features and forces, study the loads "within" a structure, be comfortable with the vocabulary engineers use to describe engineering structures, and be able to reason how changes to a structure will affect its performance. That's what the remaining chapters are about.

This book has two parts. Part I consists of an introduction and a review of prerequisite material (this chapter) and overviews of two structures familiar to most readers (the bicycle in Chapter 2 and the Golden Gate Bridge in Chapter 3). In discussing these two structures we introduce many of the basic ideas of static analysis.

In Part II the key concepts underlying static analysis are presented through

1. *Concrete experience.* This may entail recalling some prior experience (e.g., see-sawing as a child, removing a lug nut from a wheel assembly), and/or actually carrying out a simple physical or thought experiment (e.g., building a simple truss structure out of straws and paperclips). This concrete experience helps give a context for the concept and motivates you to read a more formal representation of the concept.
2. *Mathematics and physics.* The experience exercise is immediately reinforced by application to solving straightforward examples. Structured problem-solving approaches are introduced at this point.
3. *The application* to a real engineered component and/or system. This includes showing you the assumptions and idealizations that engineers must make in applying a structured problem-solving approach. Two of the applications visited repeatedly in Part II are the bicycle and the Golden Gate Bridge.

In addition, Part II develops in much greater detail the analysis steps in **Box 1.1**.

1.8 JUST THE FACTS

In this chapter we discussed what **engineering analysis** is and how it fits into engineering practice. The particular type of engineering analysis that this book is about is **static analysis**, which is based on Newton's first and third laws. The steps of static analysis consist of defining the goals of the analysis, along with the information given and any necessary assumptions. This is followed by creating a drawing of the physical system (typically a **free-body diagram**), then formulating and solving equations that describe the system (making sure to maintain dimensional consistency and appropriate significant figures). The final step is to check your numerical answers for accuracy and reasonableness.

CHAPTER 2

THE BICYCLE ("STATIC" DOESN'T MEAN THAT YOU AREN'T MOVING)

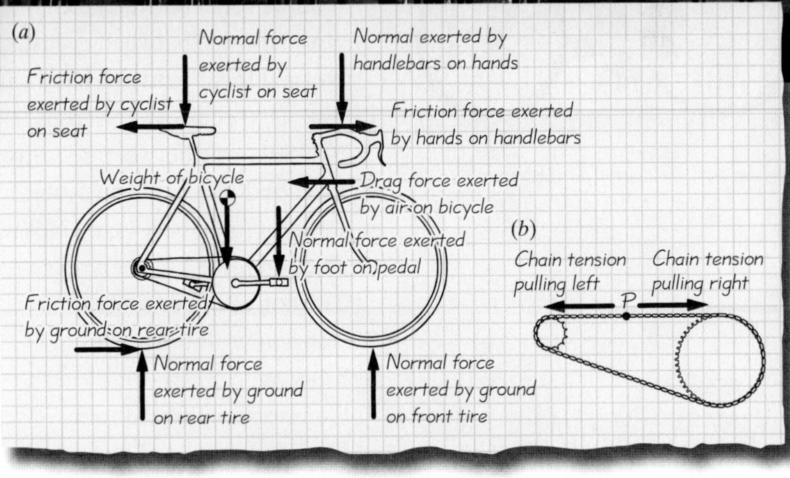

In this chapter we illustrate how the basic concepts of statics help us understand how a bicycle moves. We've chosen the bicycle because most people have had firsthand experience with it and because its "machinery" is so visible. Engineers concerned with modifying parts of the bicycle base their work on the ideas presented in this chapter—an overview of how a bicycle works and a description of how to quantify the forces acting on one. These engineers might be motivated by a need to eliminate a part failure, to reduce manufacturing costs, or to improve performance.

All of the major statics concepts introduced in this chapter are covered in greater detail in subsequent chapters. The two main purposes of this chapter are to give you an overview of these concepts in the context of an engineered artifact you are likely to be familiar with and to illustrate the utility of these concepts in describing the artifact in a quantitative manner. In other words, we want to whet your appetite for the rest of the book!

Before reading further, we urge you to put down this book and

1. draw a bicycle from memory, labeling all of the components you know, then
2. study a bicycle, taking particular note of the relationships between pedals, chain, and wheels.

Completing these two tasks is likely to make our discussion of how a bicycle works all the more real to you. You may even want to see the 1979 Academy Award-winning movie *Breaking Away* (directed by Peter Yates) to get into a bicycling frame of mind.

OBJECTIVES

On completion of this chapter, you will be able to:

- **Describe the types of forces that act on a moving bicycle**
- **Isolate a cyclist and bicycle from the rest of the world and identify the external forces acting on them**
- **Use Newton's first law to answer questions about the performance of a cyclist and bicycle**
- **Appreciate the usefulness of a procedure known as static analysis**

Figure 2.1 A cyclist and bicycle

2.1 THE FORCES OF BICYCLING

All of us experience forces—all the pushes, pulls, tugs, and shoves of everyday life. Some of these forces are welcome—for example, the up-down-sideways jostles of a roller coaster ride—and some are unwelcome—like the thud and bump when two cars roll into each other. Forces operate can openers, automobiles, ski lifts, and airplanes and are what buildings, bridges, and ships must stand up to.

Forces are also what operate a bicycle. So what forces are involved? To answer this question, take a minute to list as many as you can of the forces acting in **Figure 2.1** for a bicycle moving along level ground at a constant velocity.

Your list probably includes most (or all, if you're very good) of the following:

Weight of bicycle
Weight of cyclist
Normal force between foot and pedal
Normal force between cyclist and seat
Normal force between hands and handlebars
Normal force between front tire and ground
Normal force between rear tire and ground
Friction force between cyclist and seat
Friction force between hands and handlebars
Friction force between rear tire and ground[1]
Tension in brake cables
Tension in shifter cables
Tension in chain
Compression in seat tube
Drag force exerted by air on cyclist
Drag force exerted by air on bicycle

Note that this list is organized by categories of forces: **weight**, **normal force**, **friction force**, **tension**, **compression**, and **drag force**. We shall have a lot more to say about all these categories in Chapter 4.

Which forces from the list are relevant to our analysis depends on which portion of the system we are interested in. We draw a boundary around the portion of interest and consider only those forces that act ei-

[1]If the bicycle was traveling up or down a slope or was accelerating, there would also be a friction force between the front tire and the ground.

Table 2.1 Forces Involved in Cycling

All forces	A. External forces acting on cyclist	B. External forces acting on bicycle	C. External forces acting on cyclist–bicycle
Weight of bicycle	~~Weight of bicycle~~	Weight of bicycle	Weight of bicycle
Weight of cyclist	Weight of cyclist	~~Weight of cyclist~~	Weight of cyclist
Normal force between foot and pedal	Normal force between foot and pedal	Normal force between foot and pedal	~~Normal force between foot and pedal~~
Normal force between cyclist and seat	Normal force between cyclist and seat	Normal force between cyclist and seat	~~Normal force between cyclist and seat~~
Normal force between hands and handlebars	Normal force between hands and handlebars	Normal force between hands and handlebars	~~Normal force between hands and handlebars~~
Normal force between front tire and ground	~~Normal force between fronttire and ground~~	Normal force between front tire and ground	Normal force between front tire and ground
Normal force between rear tire and ground	~~Normal force between rear tire and ground~~	Normal force between rear tire and ground	Normal force between rear tire and ground
Friction force between cyclist and seat	Friction force between cyclist and seat	Friction force between cyclist and seat	~~Friction force between cyclist and seat~~
Friction force between hands and handlebars	Friction force between hands and handlebars	Friction force between hands and handlebars	~~Friction force between hands and handlebars~~
Friction force between rear tire and ground	~~Friction force between rear tire and ground~~	Friction force between rear tire and ground	Friction force between rear tire and ground
Tension in brake cables	~~Tension in brake cables~~	~~Tension in brake cables~~	~~Tension in brake cables~~
Tension in shifter cables	~~Tension in shifter cables~~	~~Tension in shifter cables~~	~~Tension in shifter cables~~
Tension in chain	~~Tension in chain~~	~~Tension in chain~~	~~Tension in chain~~
Compression in seat tube	~~Compression in seat tube~~	~~Compression in seat tube~~	~~Compression in seat tube~~
Drag force exerted by air on cyclist	Drag force exerted by air on cyclist	~~Drag force exerted by air on cyclist~~	Drag force exerted by air on cyclist
Drag force exerted by air on bicycle	~~Drag force exerted by air on bicycle~~	Drag force exerted by air on bicycle	Drag force exerted by air on bicycle

ther on the boundary or across the boundary. We call these the **external forces**. In analyzing a static system, only external forces are relevant.

To illustrate the process of identifying external forces, let's suppose we are interested only in the cyclist. The external forces acting on her are listed in column A of **Table 2.1** and shown in the free-body diagram of **Figure 2.2**. All the forces crossed off in column A are not acting on the boundary surrounding the cyclist and therefore do not act on her. Because they are not acting on her, they are not shown in the free-body diagram.

Alternatively, say we are interested in the bicycle. The external forces are listed in column B of **Table 2.1** and presented in a free-body diagram in **Figure 2.3*a***. Two of the forces crossed off in column B are *beyond* the boundary surrounding the bicycle (weight of cyclist and drag force exerted by air on cyclist). The others crossed off are completely *within* the boundary and are therefore called **internal forces**. Internal forces come in equal and opposite force pairs (as described by Newton's third law), and therefore cancel each another. For example, at point P in the chain in **Figure 2.3*b***, the tension of the chain pulling to the right at P is canceled by the tension of the chain pulling to the left, as illustrated in the figure.

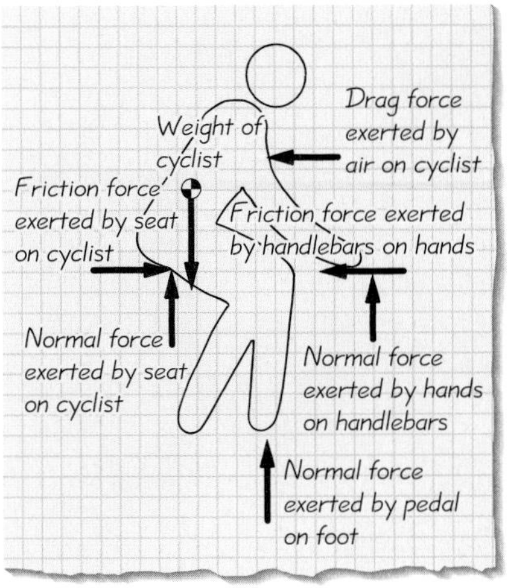

Figure 2.2 Free-body diagram of cyclist

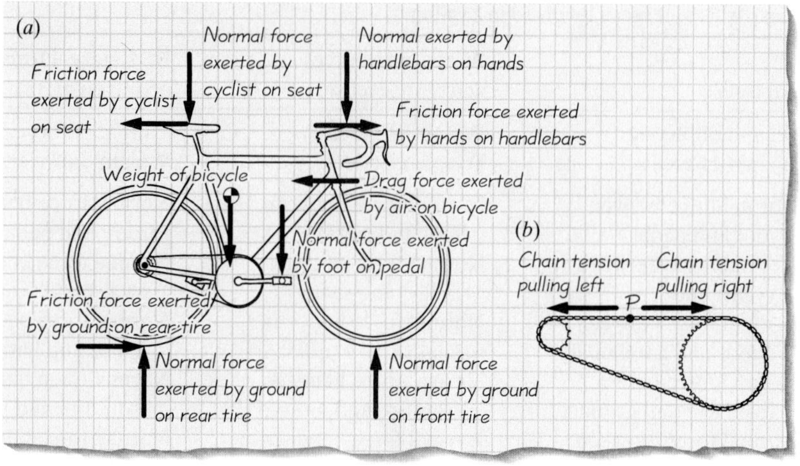

Figure 2.3 (*a*) Free-body diagram for bicycle; (*b*) internal chain forces at *P*

Finally, suppose we are interested in the combination of cyclist and bicycle. The external forces for this system are listed in column C of **Table 2.1** and presented in a free-body diagram in **Figure 2.4*a***. Any forces shown in **Figure 2.2** or **Figure 2.3** but not here are completely *within* the boundary and so are internal forces. For example, the force between the cyclist and the bicycle seat consists of a downward normal force exerted by the cyclist on the bicycle seat canceled by an equal and opposite upward normal force exerted by the seat on the cyclist (**Figure 2.4*b***).

Summary

In this section the key ideas are:

1. The forces involved in bicycling are weight, normal, friction, tension, compression, and drag forces.

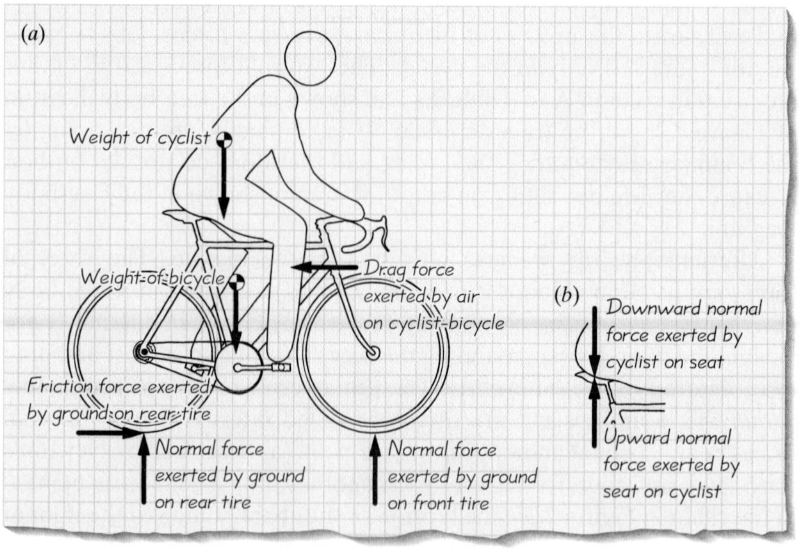

Figure 2.4 (*a*) Free-body diagram for cyclist–bicycle combination; (*b*) internal seat–cyclist forces

2. We represent an object and the forces acting on it in a free-body diagram. The forces in the diagram are those that act on or across a boundary surrounding the object; these are the external forces.

3. Forces that are within the boundary come in pairs, and we call them internal forces. Each force pair consists of equal and opposite forces that cancel each other. Internal forces are not shown in a free-body diagram.

4. Forces that are internal to one object may be external to parts of the object—for example, the normal force between foot and pedal is an external force when we consider either the cyclist alone or the bicycle alone but is an internal force when we consider the cyclist–bicycle combination.

Note:

2.2 WHAT IS THE MAXIMUM SPEED?

Now that we have introduced the major forces involved in bicycling, we consider how forces affect bicycle performance. More specifically, suppose that an average engineering student named Merrill is riding the bicycle shown in **Figure 2.5** in a race. *How fast could he sprint toward the finish line?*

One way to answer this question is to have Merrill perform the task and for us to measure his speed. Several trials would be necessary, and we would need to factor in how representative his performance was on that particular day. Another approach is for us to come up with an estimate based on the application of Newton's first law; in other words, we could perform static analysis. We now step through this latter approach, making the assumptions listed in **Figure 2.5**.

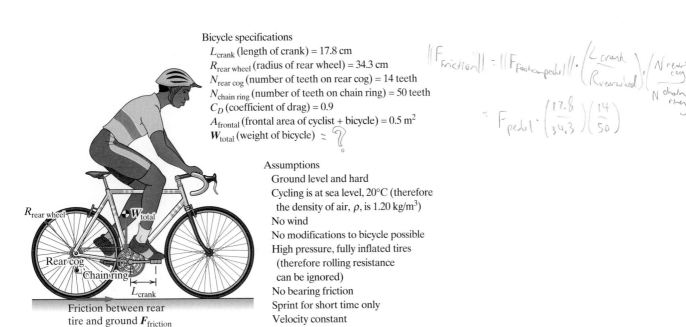

Bicycle specifications

L_{crank} (length of crank) = 17.8 cm
$R_{rear\ wheel}$ (radius of rear wheel) = 34.3 cm
$N_{rear\ cog}$ (number of teeth on rear cog) = 14 teeth
$N_{chain\ ring}$ (number of teeth on chain ring) = 50 teeth
C_D (coefficient of drag) = 0.9
$A_{frontal}$ (frontal area of cyclist + bicycle) = 0.5 m²
W_{total} (weight of bicycle) = ?

$$\|F_{friction}\| = \|F_{foot\ on\ pedal}\| \cdot \left(\frac{L_{crank}}{R_{rear\ wheel}}\right) \cdot \left(\frac{N_{rear\ cog}}{N_{chain\ ring}}\right)$$

$$= F_{pedal} \cdot \left(\frac{17.8}{34.3}\right)\left(\frac{14}{50}\right)$$

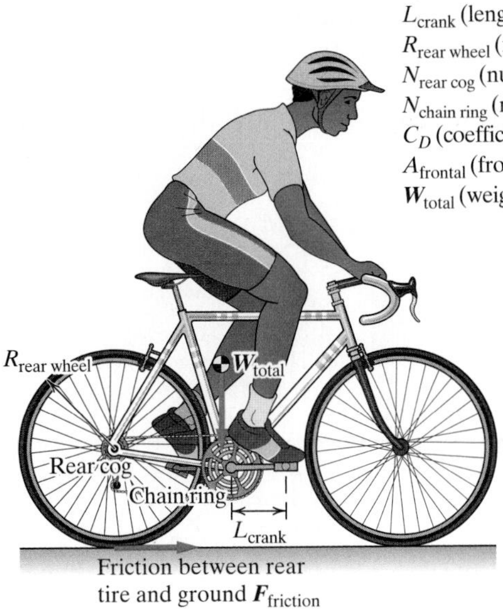

Assumptions

Ground level and hard
Cycling is at sea level, 20°C (therefore the density of air, ρ, is 1.20 kg/m³)
No wind
No modifications to bicycle possible
High pressure, fully inflated tires (therefore rolling resistance can be ignored)
No bearing friction
Sprint for short time only
Velocity constant
Cyclist in semi-standing position

$R_{rear\ wheel}$ W_{total} Rear cog Chain ring L_{crank}

Friction between rear tire and ground $F_{friction}$

Figure 2.5 Cyclist–bicycle combination with assumptions

Our analysis to find out how fast Merrill can sprint toward the finish line consists of addressing three interrelated subquestions:

1. What is the maximum force Merrill can apply to each pedal?
2. How is this force related to the friction force between the rear tire and the ground?
3. How does the friction force relate to the drag force on the bicycle?

In answering each subquestion, we will identify external forces, draw a free-body diagram, and apply Newton's first law.

1. What Is the Maximum Force Merrill Can Apply to Each Pedal? An initial estimate might be Merrill's weight based on him not sitting on the bicycle but rather standing on the pedals with all of his weight on one foot. In order to come up with a better estimate, consider the setup in **Figure 2.6*a***. A student stands on a bathroom scale; if she is on earth, the scale should read her weight on earth. Now she pushes up on the lip of a cabinet next to her (**Figure 2.6*b***)—does the reading on the scale go up, go down, or remain the same? To answer this question, consider the free-body diagram of the student, shown in **Figure 2.6*c***. Since she pushes up on the lip, we show how the lip pushes down on her (therefore $\boldsymbol{F}_{\substack{\text{cabinet pushing}\\ \text{on student}}}$ is in the downward direction). Newton's first law tells us that because the student is not moving vertically, the sum of the forces acting in the vertical direction is zero. This means that

$$\boldsymbol{F}_{\substack{\text{scale pushing}\\ \text{on stude}}} - \boldsymbol{W}_{\text{student}} - \boldsymbol{F}_{\substack{\text{cabinet pushing}\\ \text{on studen}}} = 0$$

$$\boldsymbol{F}_{\substack{\text{scale pushing}\\ \text{on student}}} = \boldsymbol{W}_{\text{student}} + \boldsymbol{F}_{\substack{\text{cabinet pushing}\\ \text{on student}}} \tag{2.1}$$

Therefore, the reading on the scale, which is equal to $\boldsymbol{F}_{\substack{\text{scale pushing}\\ \text{on student}}}$, increases as the student pushes up on the cabinet.

The connection between this bathroom scale example and Merrill's pedal force is that, if he is pulling up on the bicycle handlebars and not sitting on the bicycle seat, the magnitude of the force he applies to the pedal is greater than his weight. To see how this is so, look at

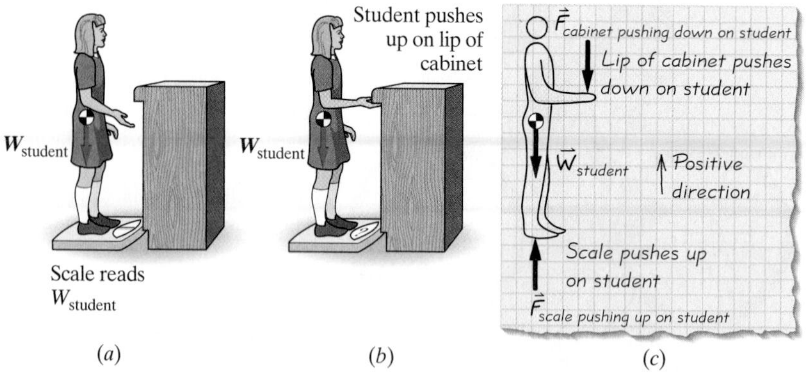

Figure 2.6 (*a*) Student standing on bathroom scale; (*b*) student pushes up a cabinet; (*c*) free-body diagram of student

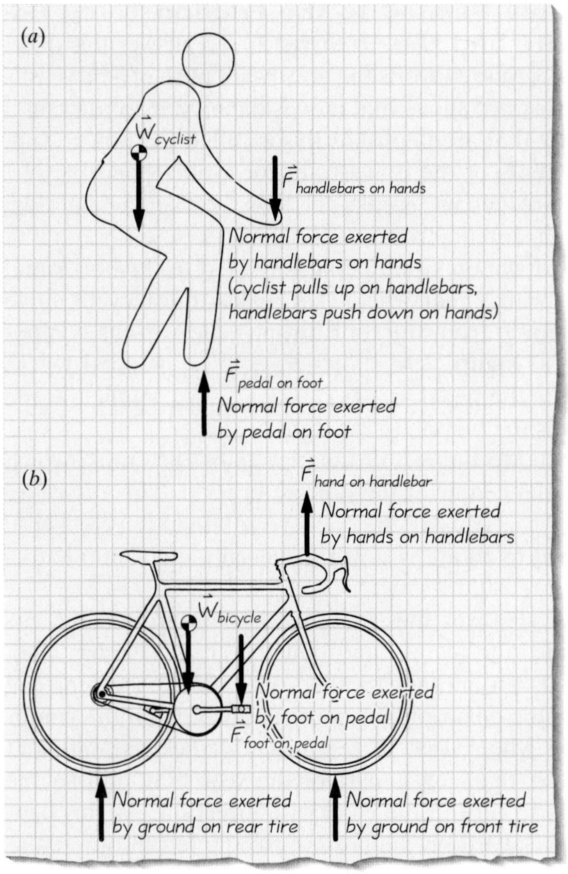

Figure 2.7 (*a*) Free-body diagram of Merrill; (*b*) free-body diagram of Merrill's bicycle

Figure 2.7*a*. As with the student standing on the scale, the sum of the forces acting on Merrill in the vertical direction must be zero because he's stationary in that direction. Thus

$$\boldsymbol{F}_{\text{pedal on foot}} = \boldsymbol{W}_{\text{cyclist}} + \boldsymbol{F}_{\text{handlebars on hands}}$$

Because $\boldsymbol{F}_{\text{foot on pedal}}$, the force Merrill exerts on the pedal, is the normal force opposing the force $\boldsymbol{F}_{\text{pedal on foot}}$, their magnitudes are the same, meaning that Merrill's force $\boldsymbol{F}_{\text{foot on pedal}}$ on the pedal is greater than his weight!

Answer to Question 1

We estimate that Merrill is able to apply to the pedal a force that is greater than his weight. What makes this additional force possible is the force he can exert on the handlebars, and a ballpark estimate of the magnitude of that force is how much weight he can lift. Since an average engineering student weighs 750 N and can repetitively lift 400 N,* we estimate that Merrill is able to apply at most 1150 N (= 750 N + 400 N) to the pedal with his foot. If Merrill does not realize that by pulling up on the handlebars he can increase the force he can apply to the pedal, our estimate of how much force he can apply to the pedal is simply his weight of 750 N. Therefore our complete estimate of the force Merrill is able to apply to the pedal is between 750 N and 1150 N (assuming that Merrill is in a semi-erect position, not sitting on the seat). In other words, $\|\boldsymbol{F}_{\text{foot on pedal}}\|$ in **Figure 2.7*b*** is estimated to be between 750 and 1150 N. Its exact size will depend on how skilled Merrill is in combining the action of legs and arms in cycling.

*Based on data from 70 engineering students, fall 2001.

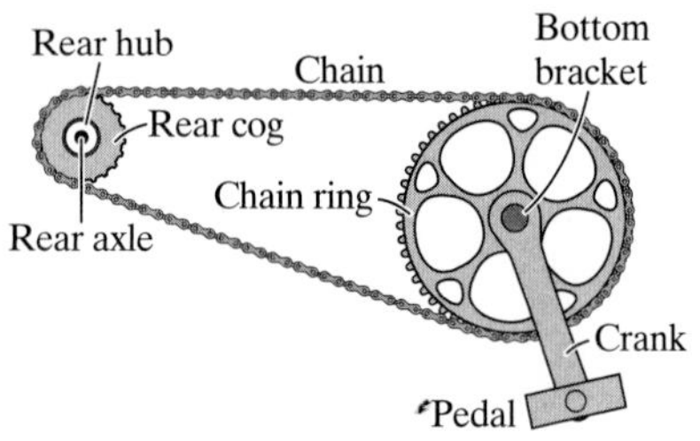

Figure 2.8 Bicycle powertrain and its components

2. How Is This Force Related to the Friction Force Between the Rear Tire and the Ground? To answer this question, we consider how a bicycle converts the downward push of a bicyclist's foot on the pedal to a forward shove that moves the bicycle (and cyclist) forward. The machinery that accomplishes this is called the powertrain. The important components of the powertrain for a one-speed bicycle, shown in Figure 2.8, are pedals, cranks, bottom bracket, chain ring, chain, rear cog, rear hub, and rear axle.

To see how pushing down on either pedal is related to the horizontal forward push exerted on the rear wheel that makes the bicycle move forward, we trace the force from the foot through the various components of the powertrain. Figure 2.9*a* shows $\boldsymbol{F}_{\text{foot on pedal}}$ applied where the cyclist's foot engages the pedal and $\boldsymbol{F}_{\text{friction, ground on tire}}$ applied at the rear wheel, which moves the bicycle to the right.

The pedal is connected to the rest of the bicycle by the crank, which offsets the pedal from the center of the bottom bracket. This offset results in the force $\boldsymbol{F}_{\text{foot on pedal}}$ causing the crank to rotate in the clockwise direction. This tendency to rotate is given a special name—**moment**—and is illustrated in **Figure 2.9*b***.

As the crank rotates, it turns the chain ring. Teeth around the circumference of the chain ring fit into and pull the chain (**Figure 2.9*c***). The chain, in turn, pulls on another toothed disk, called the rear cog (**Figure 2.9*d***). Because the chain pulling force is offset from the center of rotation of the rear wheel, that force causes a clockwise moment about the rear hub. The rear cog is part of the rear hub, and so the whole rear wheel tends to turn in the clockwise direction. As this happens, the tire pushes backward on the ground (to the left in our example). In response, the ground pushes forward on the tire, as indicated by the force $\boldsymbol{F}_{\text{friction, ground on tire}}$ in **Figure 2.9*d***. This is how the vertical input force $\boldsymbol{F}_{\text{foot on pedal}}$ gets converted into a horizontal push force $F_{\text{friction, ground on tire}}$.

By drawing free-body diagrams for the powertrain components and then evaluating the forces in these diagrams, we can convert the description of the powertrain given above into a mathematical relationship between $\|\boldsymbol{F}_{\text{foot on pedal}}\|$ and $\|\boldsymbol{F}_{\text{friction, ground on tire}}\|$ (we present these details in Chapter 9):

$$\|\boldsymbol{F}_{\text{friction, ground on tire}}\| = \|\boldsymbol{F}_{\text{foot on pedal}}\| \left(\frac{L_{\text{crank}}}{R_{\text{wheel}}}\right)\left(\frac{N_{\text{rear cog}}}{N_{\text{chain ring}}}\right) \tag{2.2}$$

where

$\|\boldsymbol{F}_{\text{friction, ground on tire}}\|$ is the magnitude of the friction force where the ground pushes on the rear tire

$\|\boldsymbol{F}_{\text{foot on pedal}}\|$ is the magnitude of the cyclist's foot pressing on the pedal

L_{crank} is the length of the crank

R_{wheel} is the radius of the rear wheel

$N_{\text{rear cog}}$ is the number of teeth around the circumference of the rear cog

$N_{\text{chain ring}}$ is the number of teeth around the circumference of the chain ring

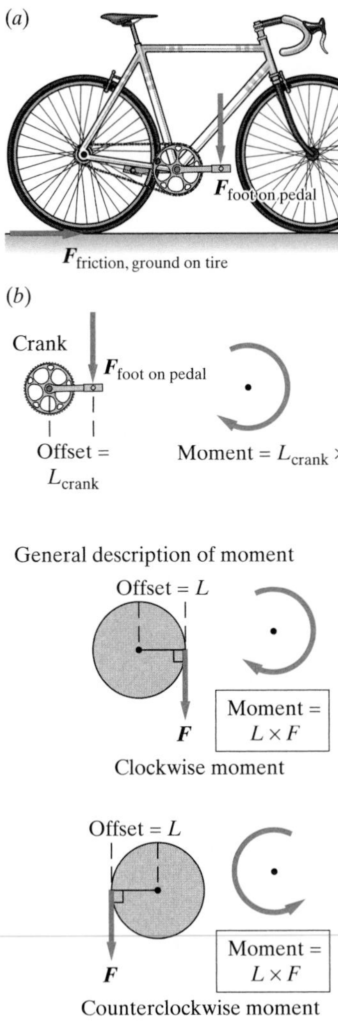

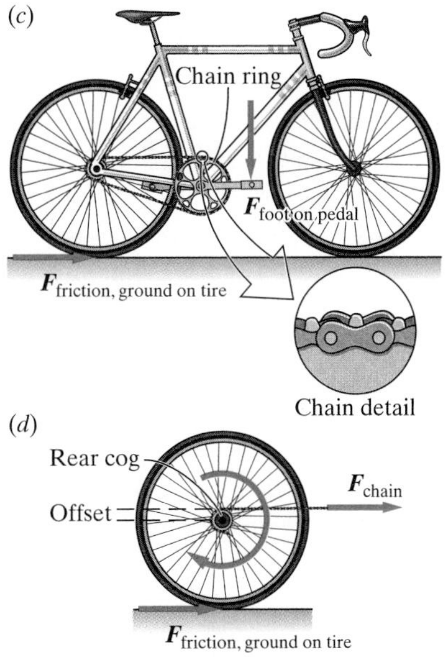

Figure 2.9 (*a*) The force of the foot acting on a pedal and the friction force at the rear wheel; (*b*) the foot force creates a clockwise moment; (*c*) chain tension is produced; (*d*) the chain pulls on the rear cog

Notice in (2.2) that $\| \boldsymbol{F}_{\text{foot on pedal}} \|$ is multiplied by a factor that is a product of sizes of bicycle components and therefore is a function only of the geometry of the bicycle. Based on the information in **Figure 2.5**, this factor is

$$\left(\frac{17.8\ \text{cm}}{34.3\ \text{cm}}\right)\left(\frac{14\ \text{teeth}}{50\ \text{teeth}}\right) = 0.145$$

for Merrill's bicycle; therefore, the magnitude of $\boldsymbol{F}_{\text{friction, ground on tire}}$ (the force that pushes the bicycle forward) is $0.145 \| \boldsymbol{F}_{\text{foot on pedal}} \|$.

Answer to Question 2

A bicycle's powertrain, comprising pedals, cranks, toothed cogs, chain, and tire-wheel assembly, converts the force of a foot pushing down on a pedal. The conversion includes changing the direction from a downward force ($\boldsymbol{F}_{\text{foot on pedal}}$) to a forward force ($\boldsymbol{F}_{\text{friction, ground on tire}}$) and changing the magnitude of the force. In the case of Merrill and his bicycle, the magnitude of the force is reduced by a factor of 0.145: $\| \boldsymbol{F}_{\text{friction, ground on tire}} \| = 0.145 \| \boldsymbol{F}_{\text{foot on pedal}} \|$. We estimated in question 1 that $750\ \text{N} < \| \boldsymbol{F}_{\text{foot on pedal}} \| < 1150\ \text{N}$. This means that $109\ \text{N} < \| \boldsymbol{F}_{\text{friction, ground on tire}} \| < 167\ \text{N}$.

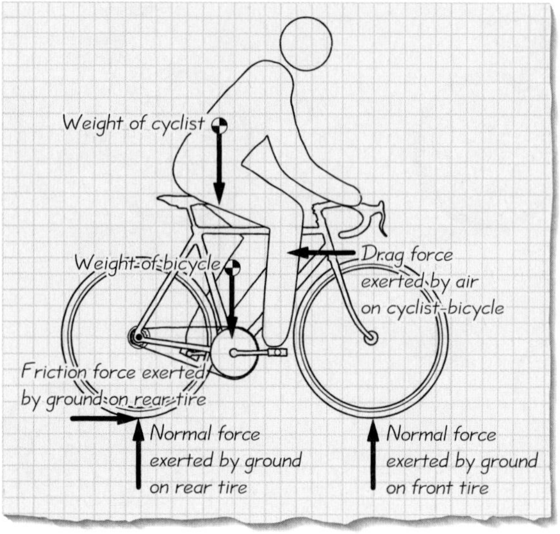

Figure 2.10 Free-body diagram of Merrill and his bicycle

3. How Does the Friction Force Relate to the Drag Force on the Bicycle? Figure 2.10 is a free-body diagram of the cyclist–bicycle combination. Newton's first law applied to this diagram tells us that because Merrill and his bicycle are moving at a constant velocity (an assumption we've made; see **Figure 2.5**), there is no net force in the horizontal direction. This means that $\boldsymbol{F}_{\text{friction, ground on tire}}$, which is pushing the bicycle forward, is canceled by the drag force $\boldsymbol{F}_{\text{drag}}$ on the bicycle:

$$\begin{aligned} \boldsymbol{F}_{\text{friction}} - \boldsymbol{F}_{\text{drag}} &= 0 \\ \boldsymbol{F}_{\text{friction}} &= \boldsymbol{F}_{\text{drag}} \end{aligned} \tag{2.3}$$

Based on our answer to question 2, we estimate that this drag force is $109\ \text{N} < \|\boldsymbol{F}_{\text{drag}}\| < 167\ \text{N}$.

Our final step is to relate the drag force to the velocity of the air moving around the bicycle. For blunt objects moving a relatively low velocity near the ground (e.g., bicycles and cyclists), this relationship takes the form[2]

$$\|\boldsymbol{F}_{\text{drag}}\| = \left(\frac{C_d \rho A}{2}\right) V^2 \tag{2.4}$$

where

C_d is the drag coefficient
ρ is the density of air (kg/m^3)
A is the frontal area of the cyclist–bicycle (m^2)
V is the velocity of air moving around the bicycle (m/s)

[2]See *Bicycling Science*, 2nd edition, R. Rowland Whitt and D. G. Wilson (Cambridge, Mass.: The MIT Press, 1990).

Answer to Question 3

Newton's first law was used to equate the horizontal forces acting on the bicycle–cyclist combination. The horizontal forces are friction force where the rear tire contacts the road and a drag force. In calculating the answer to question 2 we found that 109 N $< \| \boldsymbol{F}_{\text{friction, ground on tire}} \| <$ 167 N. Therefore, the drag force has this same range in magnitude. Finally, we modeled the drag force as a function of V^2, the air velocity squared in (2.4), to find the velocity of Merrill and his bicycle.

Our answer to the question of how fast Merrill could sprint toward the finish line is 20.1 m/s $< V <$ 24.9 m/s (or 43 mph $< V <$ 54 mph). The answer is a range because Merrill's speed depends on how skilled he is in pulling up on the handlebars. In addition, the answer is dependent on the particular values of C_d, ρ, and A, and the assumptions listed in **Figure 2.5**.

We should not just accept this answer, but should compare it to our expectations and experience. Do you think that 43 mph is a reasonable answer? What about 54 mph? Also, are there any additional factors that we should consider? We will come back to this point in the next section.

Figure 2.11 shows a plot of (2.4) for Merrill's bicycle. With 109 N $< \| \boldsymbol{F}_{\text{drag}} \| <$ 167 N, we read off the plot that 20.1 m/s $< V <$ 24.9 m/s (or 43 mph $< V <$ 54 mph). Since Merrill is bicycling on a windless day (one of our assumptions), the magnitude of the wind velocity is equal to the magnitude of Merrill's velocity.

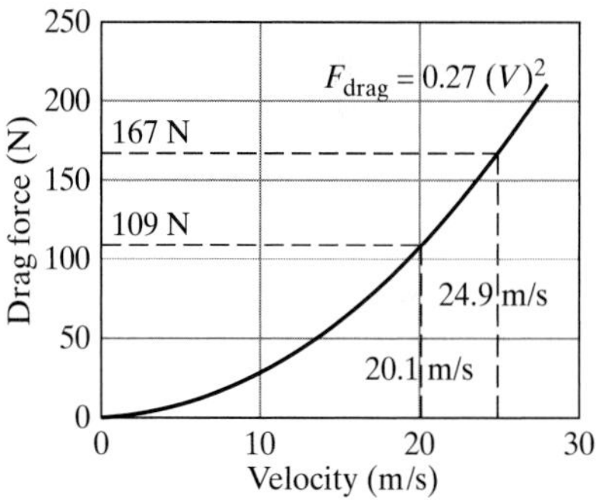

Figure 2.11 Drag force versus velocity of air moving past bicycle

$F_{drag} = V^2 \cdot \left[\frac{(0.9)(1.20)(0.5)}{2}\right] = 0.27(V)^2$

2.3 ADDING MORE REALITY

Our discussion up to now has assumed a constant foot force always oriented at a right angle to the crank, as **Figures 2.7*b*** and **2.9*a*** show. In real bicycling, the push on the pedal is not constant as the cyclist's foot moves 360°. **Figure 2.12*a*** shows, for a cyclist's right foot, the force applied to the pedal as it travels around the 360° pedaling cycle, and **Figure 2.12*b*** shows a graph of this force during the rotation. There are three things to note from these figures:

1. The force vector hardly ever has the vertical orientation we have assumed up to now in our discussion.
2. The applied force varies greatly around the 360° of travel, and we can examine this force by looking at the crank angle. This angle is defined as being 0° when the crank points straight up (with the pedal end of the crank being the "point" end) and increases as the pedal moves clockwise. At a crank angle of 90°, the applied force is maximum, while near top dead center (the top position), there is no applied force.
3. On the upstroke between 180° and 360°, the cyclist's input does not contribute at all to the input moment that propels the bicycle forward.

The pedaling force exerted by the cyclist's left foot is shown in **Figure 2.12*c***. Superimposing the action of the right foot and the left foot, and factoring in that only the component of force perpendicular to the crank creates a moment about the bottom bracket (we will have a lot more to say about this in Chapter 5), we get the effective input force profile shown in **Figure 2.12*d***. Also shown in the figure is the average value of the input force, which is about $\frac{2}{3}$ of the peak force. This force level represents the average force applied over a 360° leg cycle.

$F_{foot\ pedal} = \frac{2}{3} \cdot F_{foot\ pedal\ (max)}$

We could now redo our calculations and include this average force applied over a 360° leg cycle to find Merrill's velocity; this is one of the exercises included at the end of the chapter.

In estimating Merrill's velocity from the data given in **Figure 2.5**, we did not consider the rate at which his legs would need to be spinning in order to achieve a given velocity. The next step of our analysis should include a check to answer the questions:

- When traveling at 20.1 m/s, at what rate are his legs spinning?
- Could Merrill actually accomplish this rate for the given leg force?

This check is included as an exercise at the end of the chapter.

2.4 JUST THE FACTS

In this chapter we have examined the question: How fast could Merrill, racing on a bicycle, sprint toward the finish line? By looking at how forces are transferred from the pedal to the chain ring through the chain to the rear cog to the wheel to the ground, we were able to relate the downward push exerted by the cyclist on the pedal ($\boldsymbol{F}_{\text{foot}}$) to the forward push force exerted on the rear wheel ($\boldsymbol{F}_{\text{friction, ground on tire}}$). Then we related this push force to the drag force and velocity. This examination involved making assumptions, creating free-body diagrams, and applying Newton's first law. This process of assuming, drawing, and applying Newton's first law is the heart of static analysis.

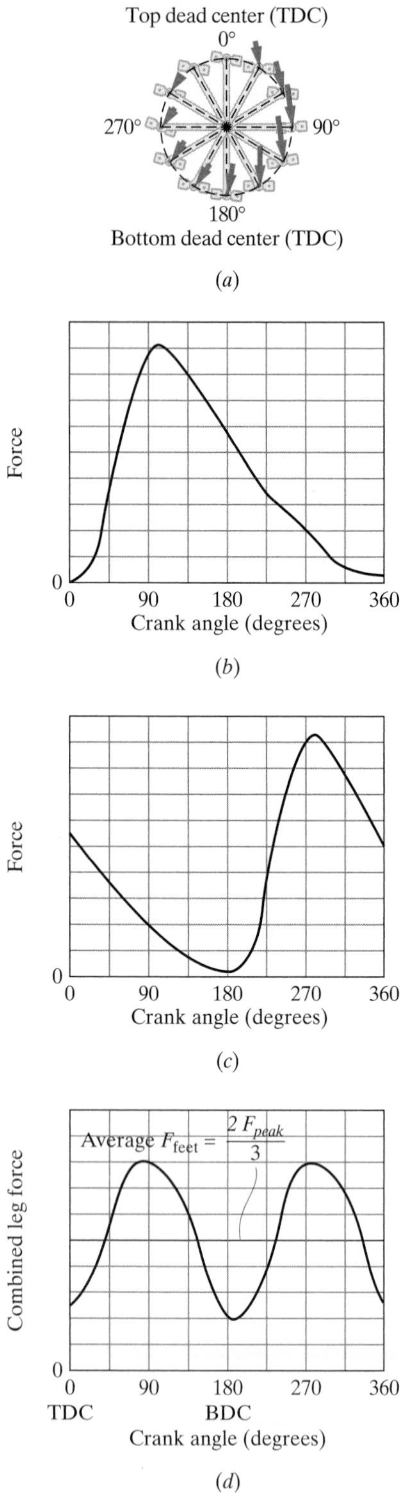

Figure 2.12 (*a*) Force direction as right foot travels 360°; (*b*) right foot force; (*c*) left foot force; (*d*) combined foot force

SYSTEM ANALYSIS (SA) EXERCISES

SA2.1 Exploring a Bicycle

1. Through drawings engineers convey ideas about objects. The following drawing exercises are about gaining comfort in conveying information about a bicycle wheel.
   (a) Draw as many circles as you can on an $8.5 \times 11''$ sheet of paper in 15 seconds, then convert one of the circles into a bicycle wheel.
   (b) Draw as many rectangles as you can on an $8.5 \times 11''$ sheet of paper in 15 seconds, then convert one of the rectangles into a bicycle wheel.
   (c) Inspect a bicycle wheel. How representative of the structure of a bicycle wheel are your drawings in (a) and (b)? (If you had trouble with (b), imagine what a bicycle wheel must look like to a bird flying overhead.)
2. Examine the following on a bicycle:
   (a) *Chain and chain ring:* How many teeth are on the largest chain ring? What is its diameter? Create a sketch that shows the interaction of the chain and chain ring. State any assumptions that you make in taking the measurements.
   (b) *Chain and the smallest rear cog:* How many teeth are on the smallest rear cog? What is its diameter? Create a sketch that shows the interaction of the chain and rear cog. State any assumptions that you make in taking the measurements.
   (c) *Pedal and crank:* What is the distance from the center of the pedal pivot point to the center of the bottom bracket? Would you describe the connection between the pedal and crank as (I) fixed (the pedal and crank always keep the same orientation relative to one another), or (II) connected but not fixed relative to one another?
   (d) *Rear wheel assembly:* What is the diameter of the rear wheel? State any assumptions that you make in taking the measurements.
3. Consider the force profile presented in **Figure 2.12*d***. When the right pedal is in the configurations listed below, how much moment is created about the bottom bracket and what is the direction (clockwise or counterclockwise) of the moment? F_{peak} in **Figure 2.12*d*** is at 800 N. Show all work.
   (a) 90° configuration
   (b) 180° configuration
   (c) 270° configuration
4. Following the same reasoning that was used to generate the free-body diagrams associated with **Table 2.1**, generate a list of the external forces acting on each of the following systems when a cyclist and bicycle are traveling at constant velocity. Also draw a free-body diagram of each system:
   (a) Front wheel assembly
   (b) Rear wheel assembly
   (c) Front fork
5. When Merrill is traveling at 20.1 m/s, at what rate are his legs spinning (in revolutions/minute)? Is this a sustainable pedaling rate?
6. (a) In Section 2.3, we discuss an average pedaling force of $\frac{2}{3}$ the peak. If this is indeed the average pedaling force, calculate the maximum velocity you estimate that Merrill will be able to travel. Include all supporting calculations.
   (b) When traveling at the velocity you calculated in (a), at what rpm (revolutions per minute) will Merrill's legs be rotating? How does the rpm you just calculated compare with the preferred rpm of 90–110 rev/min? Include all supporting calculations.

SA2.2 Analysis of Bicycle Performance

This exercise is about describing the powertrain of a bicycle. Flip a bicycle upside down on a table, as shown in **Figure SA2.2.1*a***. While turning one of the pedals, use your other hand to shift through the front and rear gears (as defined in **Figure SA2.2.1*b***) in order to answer the following two questions:

1. Qualitative analysis
   (a) In low gear (smallest chain ring, largest rear cog), is the rear wheel rotating faster or slower than in high gear (largest chain ring, smallest rear cog)?
   (b) In high gear (largest chain ring, smallest rear cog), is more or less friction force applied to the ground than in low gear (smallest chain ring, largest rear cog)?[3]

[3]While pedaling, put your bicycle in high gear. Place a 2×4 on top of the rear wheel in order to create resistance—note how difficult it is to increase pedaling to reach a rate of one revolution per second. Repeat in low gear.

(*a*)

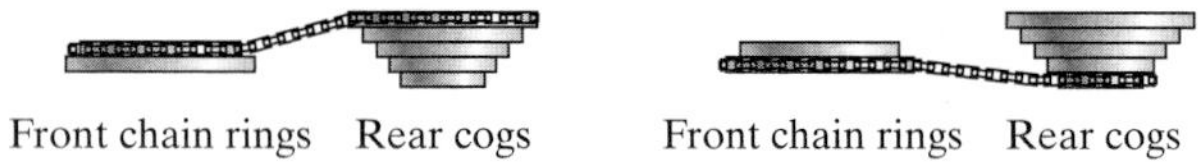

Rear cogs and front chain rings when viewed from above in (*a*).

(*b*)

Figure SA2.2.1 Experimenting with gear ratios

2. Quantitative analysis

(a) For your bicycle, record the following information:

Bicycle make: ______________

Number of front chain rings: ______________

Number of teeth on each front chain ring ($N_{\text{chain ring}}$):

____

Approximate diameter of each front chain ring:

____

Number of rear sprockets: ______________

Number of teeth on each rear cog ($N_{\text{rear cog}}$):

____

Length of crank arm: ______________

Diameter of rear tire: ______________

Given here is some useful background material on bicycle gearing, gear ratios, and gear selection. Let:

L_{crank} be the length of the crank

$R_{\text{chainwheel}}$ be the radius of the chain ring (front sprocket)

$R_{\text{rear cog}}$ be the radius of the rear cog (rear sprocket)

R_{wheel} be the radius of the rear wheel (= $D/2$)

(A) We are able to determine that the rotational speed of the rear wheel assembly ($\omega_{\text{rear wheel}}$) is related to the rotational speed of the front sprocket ($\omega_{\text{chain ring}}$) by:

$$\omega_{\text{rear wheel}} = \omega_{\text{chain ring}}(R_{\text{chain ring}}/R_{\text{rear cog}}) \quad (1a)$$

Furthermore, we can also write that

$$(R_{\text{chain ring}}/R_{\text{rear cog}}) = (N_{\text{chain ring}}/N_{\text{rear cog}}) \quad (1b)$$

which, when substituted into (1a), gives us:

$$\omega_{\text{rear wheel}}/\omega_{\text{chain ring}} = N_{\text{chain ring}}/N_{\text{rear cog}} \quad (1c)$$

(B) The product ($N_{\text{chain ring}}D/N_{\text{rear cog}}$) is commonly called the **gear-inch number *G***. Therefore we can write (2.2) as

$$\|\boldsymbol{F}_{\text{friction, ground on tire}}\| = \|\boldsymbol{F}_{\text{foot on pedal}}\| \frac{2L_{\text{crank}}}{G} \quad (2a)$$

or

$$\frac{\|\boldsymbol{F}_{\text{friction, ground on tire}}\|}{\|\boldsymbol{F}_{\text{foot on pedal}}\|} = \frac{2L_{\text{crank}}}{G} \quad (2b)$$

Let's apply (1c) and (2b) for the author's 10-speed bicycle:

The chain ring has sprockets with 40 and 52 teeth.

The rear cogs have 14, 17, 20, 24, and 28 teeth.

The crank is 6 inches long, and the diameter of the rear tire is 27 inches.

In low gear ($G = 39$), $\|\boldsymbol{F}_{\text{friction, ground on tire}}\|$ is 31% of $\|\boldsymbol{F}_{\text{foot on pedal}}\|$ and the rear wheel spins 1.43 times for every revolution of her legs. In high gear ($G = 100$), $\|\boldsymbol{F}_{\text{friction, ground on tire}}\|$ is 12% of $\|\boldsymbol{F}_{\text{foot on pedal}}\|$ and the rear wheel spins 3.71 times for every revolution of her legs.

(b) Use the data recorded in (a) to complete a table similar to **Table SA2.2.1** for your bicycle. You may find it useful to create a spreadsheet. Include a printout of your table.

(c) In high gear, what is the value of G (the gear-inch number)? What is the ratio of $\|\boldsymbol{F}_{\text{friction, ground on tire}}\|$ to $\|\boldsymbol{F}_{\text{foot on pedal}}\|$? What is the ratio of $\omega_{\text{rear wheel}}$ to $\omega_{\text{chain ring}}$?

(d) In low gear, what is the value of G (the gear-inch number)? What is the ratio of $\|\boldsymbol{F}_{\text{friction, ground on tire}}\|$ to $\|\boldsymbol{F}_{\text{foot on pedal}}\|$? What is the ratio of $\omega_{\text{rear wheel}}$ to $\omega_{\text{chain ring}}$?

(e) Based on the data you presented in (c) and (d), complete the following statement for your bicycle: In low gear (G = __________), $\|\boldsymbol{F}_{\text{friction, ground on tire}}\|$ is __________% of $\|\boldsymbol{F}_{\text{foot on pedal}}\|$ and the rear wheel spins __________ times for every revolution of my legs. In high gear (G = __________), $\|\boldsymbol{F}_{\text{friction, ground on tire}}\|$ is __________% of $\|\boldsymbol{F}_{\text{foot on pedal}}\|$ and the rear wheel spins __________ times for every revolution of my legs.

(f) Write up a brief comparison of the author's 10-speed bicycle and the bicycle you just explored.

Table SA2.2.1 Performance Data on Author's Bicycle

	$N_{\text{rear cog}}$	14	17	20	24	28
$N_{\text{chain ring}} = 40$	$G = (N_{\text{chain ring}} D / N_{\text{rear cog}})$	77	64	54	45	39
	$\omega_{\text{rear wheel}}/\omega_{\text{chain ring}}$ (1c)	2.86	2.35	2.00	1.67	1.43
	$\dfrac{\lVert \boldsymbol{F}_{\text{friction}} \rVert}{\lVert \boldsymbol{F}_{\text{foot on pedal}} \rVert}$ (2b)	0.16	0.19	0.22	0.27	0.31
$N_{\text{chain ring}} = 52$	$G = (N_{\text{chain ring}} D / N_{\text{rear cog}})$	100	83	70	59	50
	$\omega_{\text{rear wheel}}/\omega_{\text{chain ring}}$ (1c)	3.71	3.06	2.60	2.17	1.86
	$\dfrac{\lVert \boldsymbol{F}_{\text{friction}} \rVert}{\lVert \boldsymbol{F}_{\text{foot on pedal}} \rVert}$ (2b)	0.12	0.15	0.17	0.21	0.24

CHAPTER 3

THE GOLDEN GATE BRIDGE

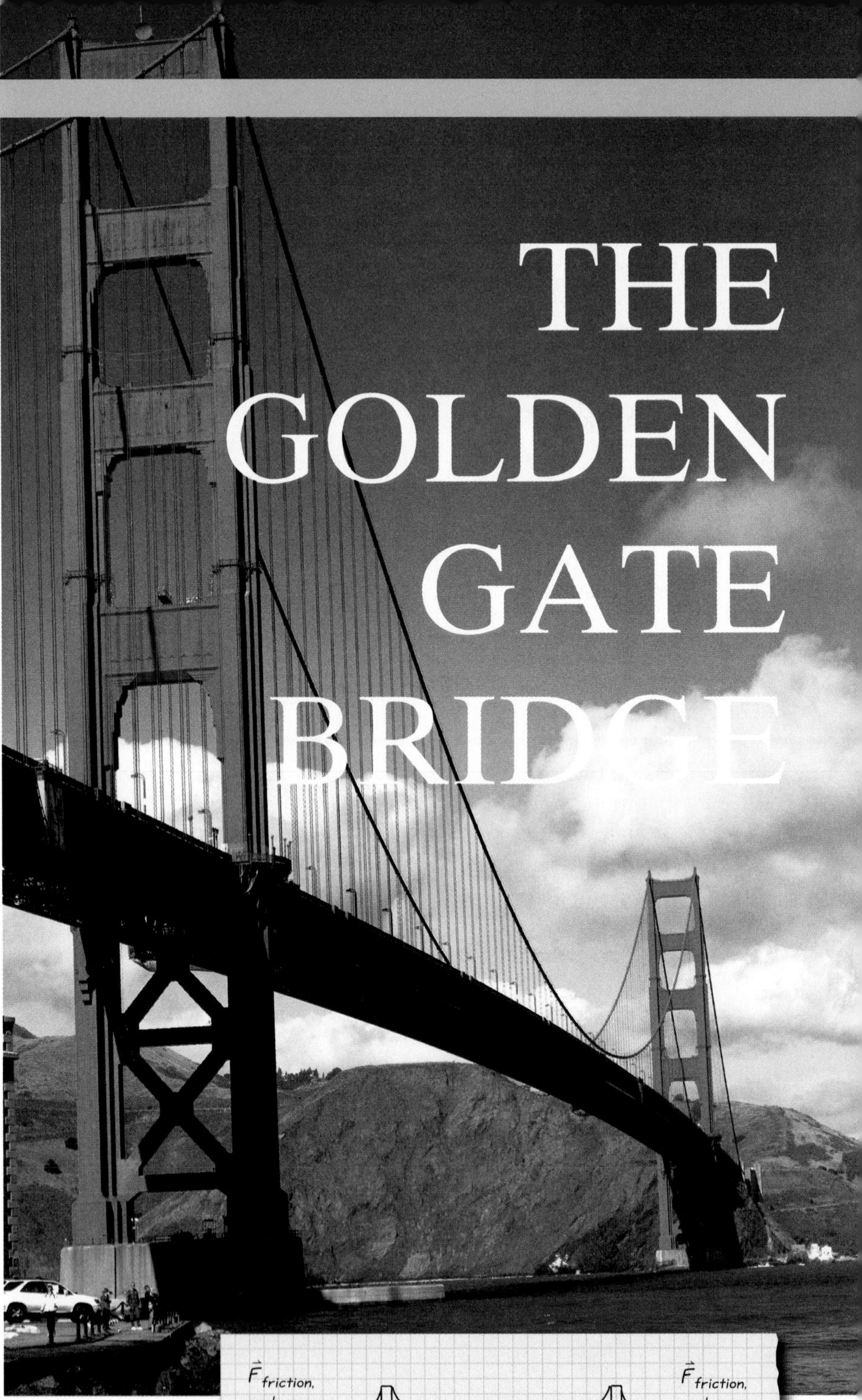

In this chapter, we use statics concepts to understand how a bridge functions and why the Golden Gate Bridge is shaped the way it is.[1] We have chosen the Golden Gate Bridge because it is one of the most recognized and beautiful structures in the world. Its graceful lines, Art Deco details, and spectacular views make it a popular stop for tourists from around the world. When it was completed in 1937, the 1280-meter suspension span was the longest in the world (now the seventh longest). Today more than 100,000 vehicles cross the bridge every day. It has been subjected to earthquakes, strong winds, and swift tides, and yet it continues to perform its function of linking the headlands on the two sides of the entrance to San Francisco Bay.[2] Engineers designing the bridge used statics concepts first to evaluate all the loads the bridge could potentially experience and then to design the members to resist these loads.

Before reading further, think about different types of bridges you have seen.

1. Sketch at least two of them and label the parts you know.
2. Identify the locations where the forces exerted on the bridge are transferred to the ground.

You may want to go to a bridge near you and study how it is built and, particularly, how it is attached to the ground. There are also many excellent pictures of bridges on the Internet. This task will provide good background for the discussion in this chapter.

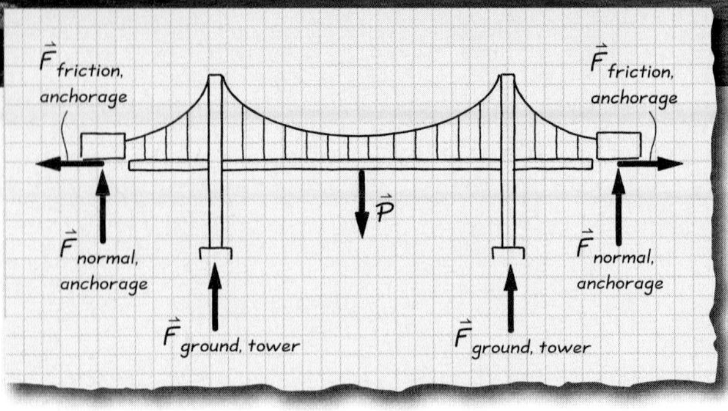

[1]The bridge is named after the entrance to San Francisco Bay, which was named the Golden Gate by John Charles Fremont in the mid-1800s.

[2]The Golden Gate Bridge District maintains a Web site (http://www.goldengate.org/) if you are interested in more information about this bridge.

OBJECTIVES

On completion of this chapter, you will be able to:

- **Describe the types of forces that act on a bridge**
- **Isolate components of a suspension bridge and identify the external forces**
- **Use Newton's first law to answer questions about the transfer of loads from one bridge component to another**
- **Compare the analysis of an approximate model with a more "exact" model**

3.1 A WALK ACROSS THE BRIDGE

The main components of a suspension bridge that carry the loads from the bridge deck down to the ground are shown in **Figure 3.1**. Cars, trucks, trains, and people travel on the bridge deck, which hangs from suspenders hung from main cables that are draped over towers and attached to anchorages at each end of the bridge. The towers are embedded deep in the ground and supported by massive concrete foundations.

Imagine that you are standing in the middle of the Golden Gate Bridge. How would your weight be transferred to the ground? To get a feel for how the bridge works, let's build a model. You'll need two large paper clips, a piece of string about 3 meters long, a pencil, and six heavy books. Set aside two of the books that are about the same size (you will use them in a moment for the towers). Tie one end of the string around two of the remaining four books, tie the other end around the other two books, and place the two piles about 2 meters apart with the string lying slack between them. These piles are the two anchorages. Stand the two books you set aside on end, one about 30 cm to the right of the left anchorage and the other about 30 cm to the left of the right anchorage, as in **Figure 3.2*a***. Drape the string (the main cable) over the top of the towers, hook the two paper clips (the suspenders) onto the string, and slide the pencil (the bridge deck) into the paper clips. To represent yourself standing on the bridge, push down on the pencil.

Now modify the model by removing the anchorages and tying the string directly around the two towers (**Figure 3.2*b***). What happens when you push down on the pencil this time? Does the system collapse? Yes, the towers fall over and the bridge collapses because the towers are being pulled toward each other (inward) by the force in the main cable. In the first bridge model, the cable tied to the anchorage provides an outward force on the towers to balance the inward force. This outward

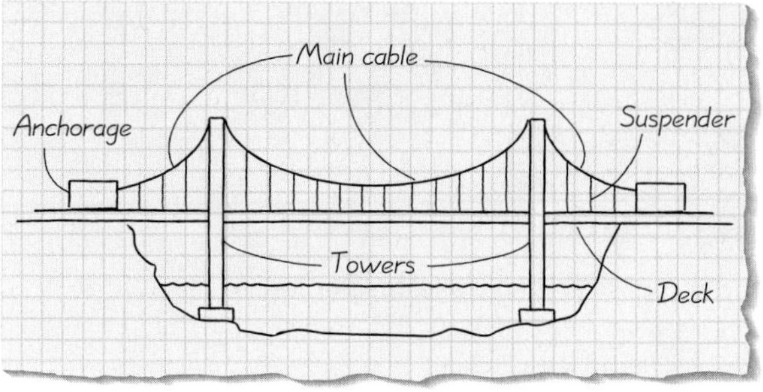

Figure 3.1 The basic components of a suspension bridge

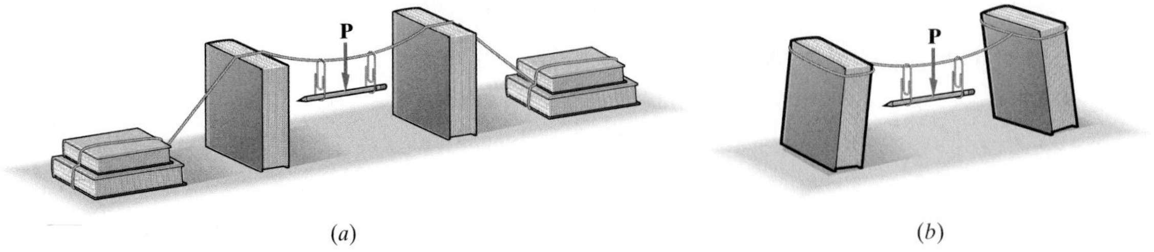

Figure 3.2 A model of a suspension bridge made from string, pencil, paper clips, and books (*a*) with anchorages and (*b*) without anchorages

force is transferred to the table at the anchorages through friction. When the anchorages are removed and the main cable is tied directly to the towers, there is no outward force pulling on the towers to balance the inward force. Furthermore, unless we glue the upright books to the table, there is no way for the table to pull on the bottom of the towers to keep them from tipping. These two models demonstrate how all the bridge components work together to make a complete "load path" to transfer loads from bridge to ground.[3]

For a more systematic view of how the loads exerted on the bridge deck are related to the forces exerted by the ground on each tower base ($\boldsymbol{F}_{\text{ground, tower}}$) and on each anchorage ($\boldsymbol{F}_{\text{ground, anchorage}}$ and $\boldsymbol{F}_{\text{friction, anchorage}}$), shown in **Figure 3.3**, we will trace the forces through all the bridge components. We will think in terms of the **free-body diagram** for each component of our model when we put a load on it.

1. Start with the deck. You push down on the deck (pencil), and it pulls down on the suspenders (paper clips). As a result, tension is developed in the suspenders as they pull up on the deck ($\boldsymbol{T}_{\text{suspender, deck}}$ in **Figure 3.4*a***) and down on the main cable ($\boldsymbol{T}_{\text{suspender, cable}}$ in **Figure 3.4*b***), just as in a game of tug-of-war the rope pulls on the two teams at its ends. As the suspenders pull down on the main cable (string), a tension force is created in the cable throughout its length. (The tension force is what took the slack out of the string in your model.)

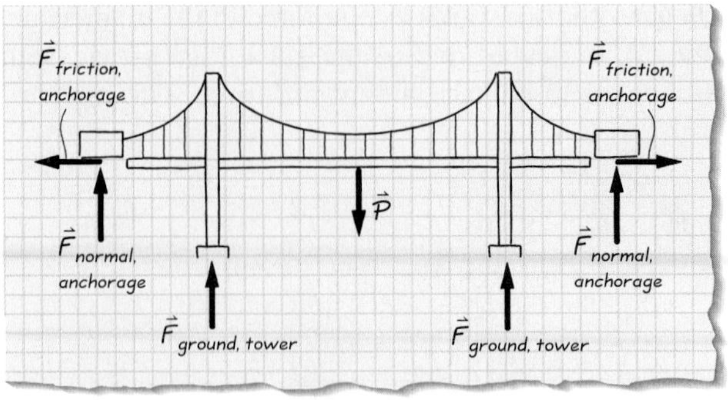

Figure 3.3 Forces on the suspension bridge as a result of a load ***P*** exerted on the deck

[3]Exercise modified from http://www.pbs.org/wgbh/nova/bridge/meetsusp.html.

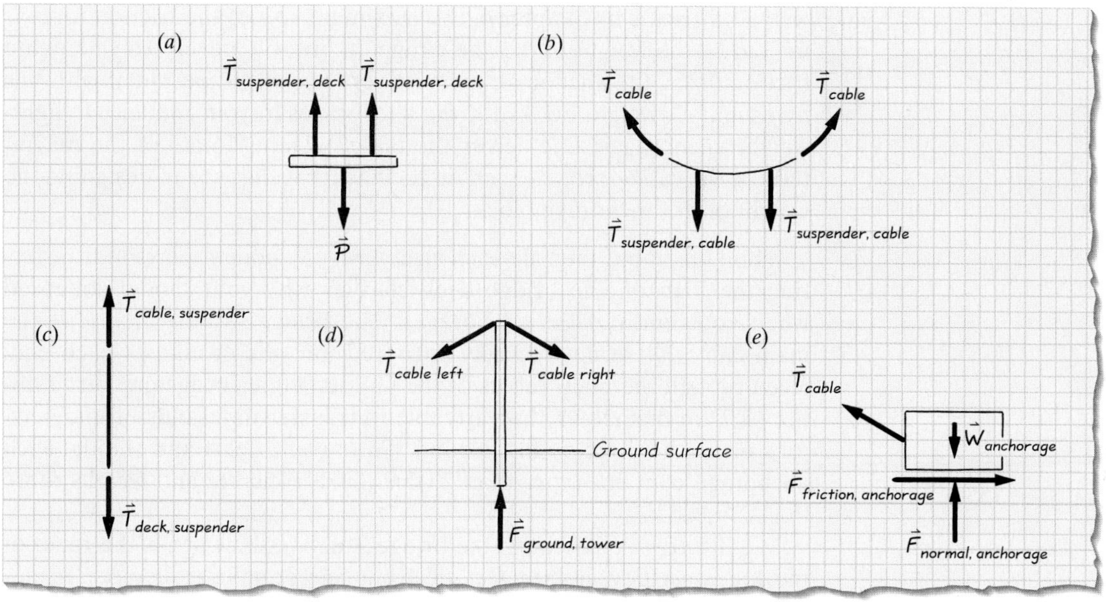

Figure 3.4 Free-body diagrams for the bridge components: (*a*) The suspenders pull up on the bridge deck with a force $\boldsymbol{T}_{suspender, deck}$; (*b*) the suspenders pull down on the main cable, causing tension $\boldsymbol{T}_{cable}$ to develop in the cable; (*c*) the suspender is in tension with the main cable pulling up and the deck pulling down; (*d*) the main cable forces are counteracted by the force $\boldsymbol{F}_{ground, tower}$ exerted by the ground on the tower; (*e*) the main cable force $\boldsymbol{T}_{cable}$ is counteracted by the weight of the anchorage $\boldsymbol{W}_{anchorage}$ and the forces that the ground exerts on the anchorage $\boldsymbol{F}_{normal, anchorage}$ and $\boldsymbol{F}_{friction, anchorage}$.

Note that as the suspender pulls down on the main cable, the cable pulls up on the suspender with an equal and opposite force. **Figure 3.4*c*** shows a free-body diagram of a suspender. The main cable pulls up on the suspender with a force $\boldsymbol{T}_{cable, suspender}$ and the deck pulls down with a force $\boldsymbol{T}_{deck, suspender}$. Since these are the only two forces acting on the suspender, Newton's first law requires that they must be equal. For simplicity, from now on we will call these tensile forces $\boldsymbol{T}_{suspender}$.

2. Now follow the main cable to the towers. Where the main cable passes over the top of a tower (upright book), the cable is sloping away from the tower on both sides. This orientation causes the cable to pull down on the tower (**Figure 3.4*d***). To counteract this, the tower pushes up on the cable. The force of the cable pulling down on the towers is transferred to the ground through the towers and creates the force $\boldsymbol{F}_{ground, tower}$ shown in **Figures 3.3** and **3.4*d***. Because the tower is being pushed on at the top and bottom, it is in compression.

3. Continue tracing the main cable to the right anchorage. Because the cable is embedded in the anchorage (stacked books), it pulls on the anchorage with a large force ($\boldsymbol{T}_{cable}$ in **Figure 3.4*e***). Though $\boldsymbol{T}_{cable}$ is pulling upward, the anchorage is kept from lifting off the ground by its enormous weight ($\boldsymbol{W}_{anchorage}$). The ground exerts both a normal contact force ($\boldsymbol{F}_{normal, anchorage}$) and a friction force ($\boldsymbol{F}_{friction, anchorage}$) on the anchorage, and the anchorage is kept from sliding by this friction force. As

we will discover in Chapter 6, the maximum size of the friction force is limited by the weight of the anchorage. In order to develop an adequate friction force, the anchorage must be very heavy. (Experiment with this by using lightweight books for the anchorages in your model. Do lightweight anchorages slide across the table?) The estimate of the magnitude of the cable force exerted on the anchorage is used in designing the weight of the anchorage block.

Thinking about the role each structural component plays in transferring the load on the bridge deck clarifies why your model collapsed when you removed the anchorages. The bridge would collapse if the main cables were not securely anchored into the ground at each end. Even though the main cables of the Golden Gate Bridge are very slender (0.92 m diameter), they are able to transfer thousands of kiloNewtons of tensile force to the ground. To prevent uplift and sliding, each anchorage contains more than 20,000 m^3 of concrete and weighs more than 530,000 kN.

We just traveled through the bridge's **load path**, which is the route of the loads as they are transferred from one structural member to another. In studying the load path of any structure, think of the structure as a series of interconnected pipes and imagine pouring water into one end and watching the water exit at the other end. In the case of a suspension bridge, you pour water into the deck, and it flows from deck to suspenders to cables to towers and anchorages and then to the ground, where it exits the "pipe."

Summary

In this section the key ideas are:

1. A simple physical model can be used to gain an understanding of a complex structure.
2. Forces acting on the bridge are transferred from one component to another and then to the ground. The load path is the route of the loads as they are transferred from one component to another.
3. The cables and suspenders on the bridge are in tension. The bridge towers are in compression. The anchorages are kept from sliding through friction forces.

3.2 HOW HEAVY SHOULD THE ANCHORAGES BE?

Now that we have laid out in a general manner the forces acting on the components of the Golden Gate Bridge and how those forces are transferred to the ground, we will answer the same question Joseph B. Strauss and his team of engineers had to answer when they designed the bridge—how heavy should the anchorages be?

As is common in engineering analysis, we will make several assumptions to create a simplified analytical model. This will allow us to develop some equations that provide reasonable estimates of the forces acting on the components. Later in the book, we will use more complex

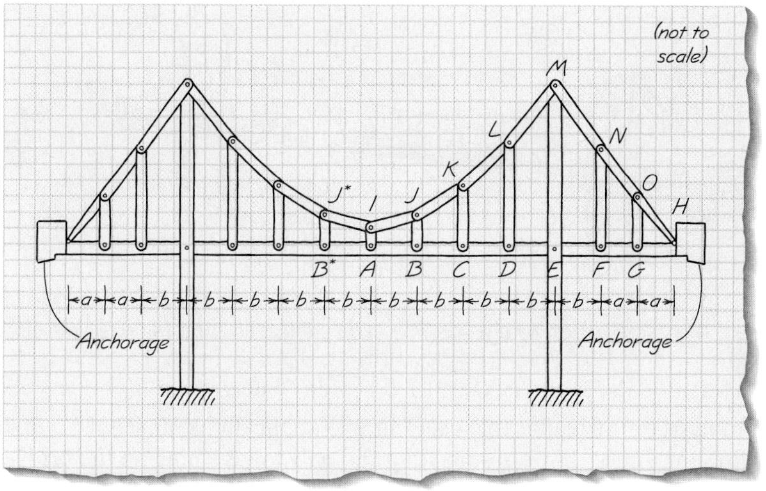

Figure 3.5 Approximate representation of the Golden Gate Bridge using pinned links to model the main cables and suspenders

assumptions and equations to perform a more exact analysis. We can then compare our approximate analysis with the more exact analysis to investigate how our simplifying assumptions have affected our results.

In order to answer the question about the weight of the anchorages, we must work our way through the load path and answer three interrelated subquestions:

1. How large is the force pulling on each of the suspenders?
2. What is the tension force in each main cable?
3. What forces on the anchorage would cause uplift or sliding?

What Assumptions Are We Making?

First, we replace each main cable by a series of links that mimic the geometry of the Golden Gate Bridge and are connected to one another with pins (**Figure 3.5** and **Table 3.1**). This will allow us to complete an approxi-

Table 3.1 Geometry of "Main Cable" in Our Approximate Model

Pin	Height above Bridge Deck (meters)	Distance from Center of Bridge (meters)
I	10.0	0
J and J*	18.9	160
K	45.8	320
L	90.4	480
M	153.0	640
N	71.4	800
O	32.8	891.5
H	0	983

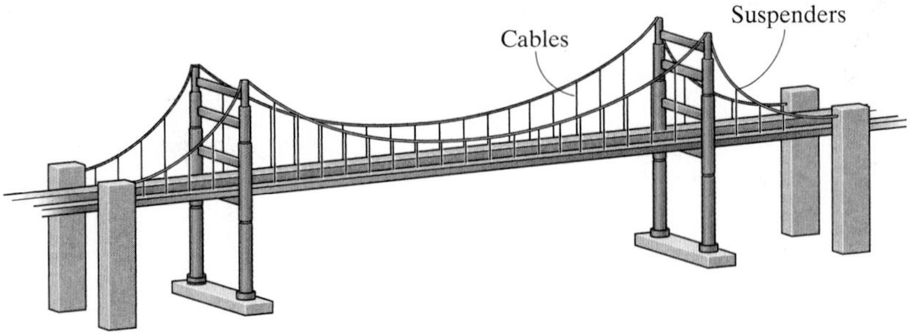

Figure 3.6 A suspension bridge has two main cables, each taking one-half of the load

mate analysis of the main cable using simple applications of Newton's first and third laws. Second, we assume that the only loads acting on the bridge are **gravity loads** (vertical loads exerted by people, vehicles, and the weight of the bridge). This allows us to ignore such horizontal loads as winds or earthquakes. Third, we assume that the weight of the main cable is much less than the weight of the bridge deck and vehicles and therefore its weight can be ignored in our analysis. This assumption allows us to assume that the main cable is in the shape of a parabola. (For the Golden Gate Bridge, the combined weight of the deck and vehicles is more than seven times the weight of the main cables, so ignoring the weight of the cables is a reasonable assumption for a preliminary analysis.)

1. How Large Is the Force Pulling on Each of the Suspenders? When we look at the bridge from the orientation shown in Figure 3.6, we are reminded that there are two main cables, each attached to the deck with suspenders. When the total load is distributed uniformly across the width of the bridge, each main cable supports one half the load.

To calculate the force exerted by the deck on one suspender, we start by calling the weight per unit length of deck $\boldsymbol{w}$. We assume that all the suspenders on each side are evenly spaced along the length of the deck and that the distance between any two suspenders on the same side of the bridge is b. We then slice a length of deck out of our model (**Figure 3.7*a***), making our first cut halfway between two suspenders and our second cut a distance b away from the first cut. Thus the length of the deck slice is b, and there is one suspender attached on each side at the midpoint of length b (**Figure 3.7*b***). Applying Newton's first law, we can say that because the bridge is not moving, the sum of the forces in the vertical direction will be zero. This means that the force exerted by the suspenders pulling up ($2\boldsymbol{T}_{\text{suspender}}$) must equal the weight of the deck slice pulling down ($\boldsymbol{W}_{\text{deck slice}}$):

$$\| 2\boldsymbol{T}_{\text{suspender}} \| - \| \boldsymbol{W}_{\text{deck slice}} \| = 0$$

$$\| 2\boldsymbol{T}_{\text{suspender}} \| - \| \boldsymbol{w} \| \, b = 0$$

$$\| \boldsymbol{T}_{\text{suspender}} \| = \frac{\| \boldsymbol{W} \| \, b}{2} \qquad (3.1)$$

We can now use this relationship to calculate the force on each suspender in our model.

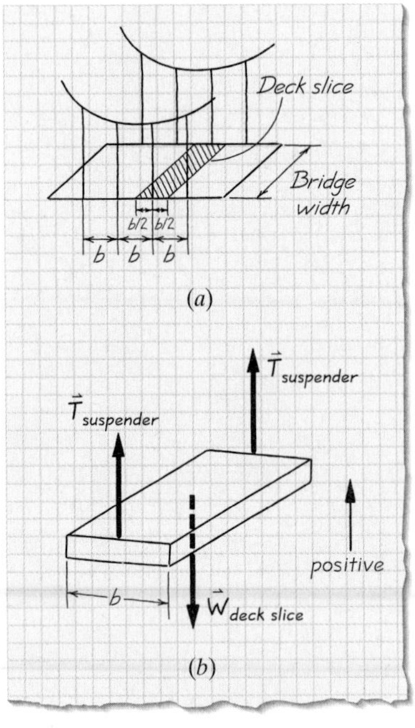

Figure 3.7 (*a*) A slice of bridge deck of length b is used for analysis. (*b*) The weight of the slice of bridge deck, pulling down, is balanced by the force exerted by the suspenders, pulling up.

Answer to Question 1

For our simplified model, the suspenders are 160 m apart and the deck weighs 330 kN/m (**Table 3.2**). The total weight of the slice we are analyzing is 52,800 kN = (330 kN/m)(160 m). Then the force of each suspender pulling up on the bridge deck is 26,400 kN = (330 kN/m)(160 m/2). We write our final answer in meganewtons (see **Table 1.2**): $\| \boldsymbol{T}_{\text{suspender}} \|$ = 26.4 MN.

Table 3.2   Properties of the Golden Gate Bridge[*]

Structural Property	Quantity
Length of main span (distance between towers)	1280 m
Length of one side span	343 m
Width of bridge	27 m
Height of each tower above road deck	152 m
Height of each tower	227 m
Maximum sag in main cable	144 m
Weight of cable per one horizontal meter	48.7 kN/m
Diameter of one main cable with wrapping	0.92 m
Number of wires in each cable	27,572
Hanger spacing	15.2 m
Weight per unit length of bridge deck	330 kN/m
Weight of one tower	196,000 kN
Weight of one anchorage	530,000 kN

[*]Data from www.goldengatebridge.org/research/factsGGBDesign.html and Abdel-Ghaffar and Scanlan (1985).

2. What Is the Tension Force in Each Main Cable?

In our model, we are representing each of the main cables by a series of links attached by pins. **Figure 3.8** shows the forces acting on the pin at the bridge center, the pin labeled *I* in **Figure 3.5**. We use two-letter subscripts to identify the link forces acting on any pin: the first letter indicates the pin we are currently evaluating, and the second letter indicates the other pin the link is attached to. For example, $\boldsymbol{F}_{IJ}$ symbolizes the force that link *IJ* exerts on pin *I* and $\boldsymbol{F}_{JI}$ symbolizes the force that link *IJ* exerts on pin *J*. A free-body diagram of link *IJ* would show that $\| \boldsymbol{F}_{IJ} \| = \| \boldsymbol{F}_{JI} \|$. In our analysis here of pin *I*, therefore, the two link forces are $\boldsymbol{F}_{IJ*}$ and $\boldsymbol{F}_{IJ}$.

The suspender force $\boldsymbol{T}_{\text{suspender}}$ pulls down on the pin, and the forces $\boldsymbol{F}_{IJ}$ and $\boldsymbol{F}_{IJ*}$ in the links *IJ* and *IJ** each pull away from the pin along the long axes of the links. Because the pin's state of motion is not changing, Newton's first law requires that the sum of the forces in the horizontal direction as well as the sum of the forces in the vertical direction be zero (**Figure 3.9**). Looking first at the horizontal forces:

$$\| \boldsymbol{F}_{IJ \text{ horizontal}} \| - \| \boldsymbol{F}_{IJ* \text{ horizontal}} \| = 0$$
$$\| \boldsymbol{F}_{IJ} \| \ \cos \alpha - \| \boldsymbol{F}_{IJ*} \| \ \cos \alpha = 0$$
$$\| \boldsymbol{F}_{IJ*} \| = \| \boldsymbol{F}_{IJ} \| \qquad (3.2A)$$

Pin *I*

Note

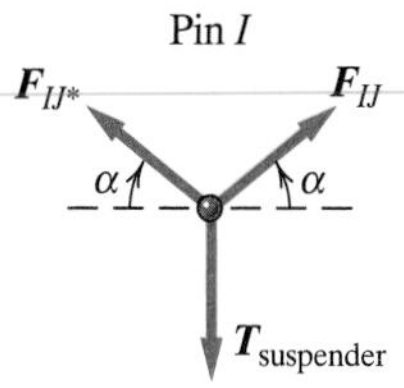

Figure 3.8 Free-body diagram of pin *I*

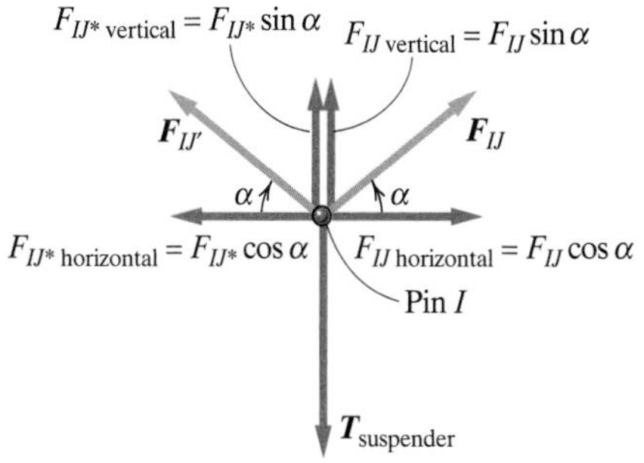

Figure 3.9 Force balance: Equilibrium means that the forces pulling to the left equal the forces pulling to the right on pin *I*. Similarly, the forces pulling down equal the forces pulling up.

Now looking at the vertical forces:

$$\| \boldsymbol{F}_{IJ^* \text{ vertical}} \| + \| \boldsymbol{F}_{IJ \text{ vertical}} \| - \| \boldsymbol{T}_{\text{suspender}} \| = 0$$
$$\| \boldsymbol{F}_{IJ^*} \| \sin\alpha + \| \boldsymbol{F}_{IJ} \| \sin\alpha - \| \boldsymbol{T}_{\text{suspender}} \| = 0 \tag{3.2B}$$

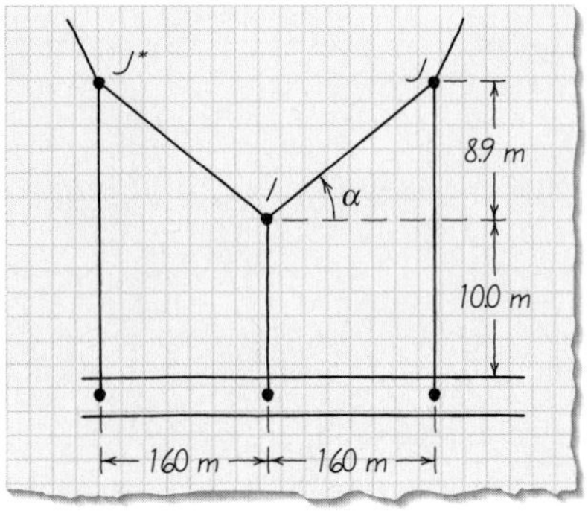

Figure 3.10 Blow-up of a segment at the center of the bridge

Substituting from (3.2A) into (3.2B) and rearranging gives

$$\| \boldsymbol{F}_{IJ} \| = \frac{\| \boldsymbol{T}_{\text{suspender}} \|}{2 \sin\alpha} \tag{3.3}$$

, T_sus = 26.4 MN
∴ F_IJ = 237 MN

The angle α can be determined from the geometry of the bridge shown in **Figure 3.5** and in **Table 3.1** data.[4] **Figure 3.10** shows a blow-up of a segment cut out of the center of the bridge. From the dimensions shown in the figure, $\cos\alpha = 160/160.25 = 0.998$ and $\sin\alpha = 8.9/160.25 = 0.0556$. From (3.3) and our known value $\| \boldsymbol{T}_{\text{suspender}} \| = 26.4$ MN, we see that $\| \boldsymbol{F}_{IJ} \| = \| \boldsymbol{F}_{IJ^*} \| = 237$ MN. The result of this calculation indicates that the force in the main cable is quite large—a force of 237 MN is equivalent to the weight of about 18,000 automobiles.

The next step is to draw a free-body diagram of the pin at J, as shown in **Figure 3.11**. Once again, requiring the sum of the forces in the horizontal direction to be zero gives

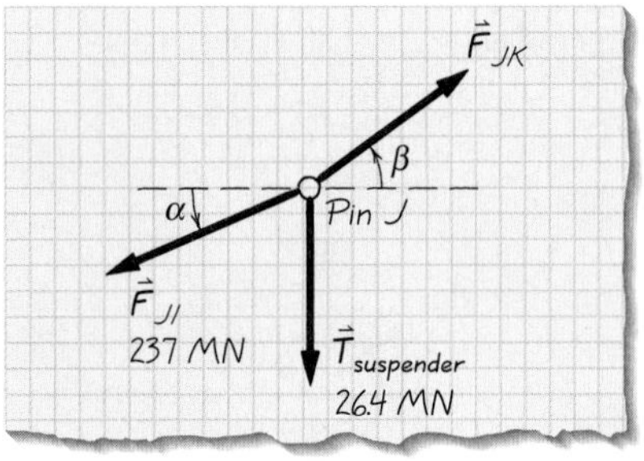

Figure 3.11 Free-body diagram of pin J

$$\| \boldsymbol{F}_{JK \text{ horizontal}} \| - \| \boldsymbol{F}_{JI \text{ horizontal}} \| = 0$$
$$\| \boldsymbol{F}_{JK} \| \cos\beta - \| \boldsymbol{F}_{JI} \| \cos\alpha = 0$$

We determine β from **Table 3.1**, which shows that the vertical distance from J to K is 45.8 m − 18.9 m = 26.9 m. This distance is the length of the side opposite β in the right triangle suggested in **Figure 3.11**. **Table 3.1** also shows that the side adjacent to β is 320 m − 160 m = 160 m. Using these values in the hypotenuse formula from footnote 5 gives 162.25 m for the hypotenuse length in the right triangle suggested in **Figure 3.11**. Therefore $\cos\beta$ = adjacent side/hypotenuse = 160/162.25 = 0.986. This gives

$$\| \boldsymbol{F}_{JK} \| = \| \boldsymbol{F}_{JI} \| \frac{\cos\alpha}{\cos\beta} = (237 \text{ MN}) \frac{0.998}{0.986} = 240 \text{ MN}$$

[4]For a right triangle, the lengths of the two sides and the hypotenuse are sufficient to determine angles and sines and cosines of those angles.

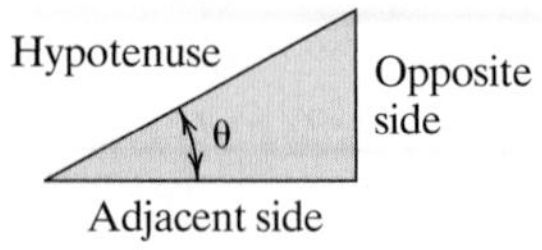

For the triangle shown here, "adjacent side" is the length of the side adjacent to the angle θ, "opposite side" is the length of the side opposite the angle θ, and "hypotenuse" is the length of the hypotenuse = (adjacent side^2 + opposite side^2)$^{1/2}$. Then $\sin\theta$ = opposite side/hypotenuse and $\cos\theta$ = adjacent side/hypotenuse. , tan θ = opposite/adjacent

To complete the analysis, we repeat the same type of calculation for each pin as we move across the bridge. The next pin is pin K. Note that as we move along the center span of the bridge from the center toward the towers, the force on each successive link increases, with the largest force being on link LM. On the side span, we find the force decreasing as we move from the tower toward the anchorage.

Answer to Question 2

Figure 3.12 compares our simplified analytic model (red lines) with the results of a more exact analysis (green curves) as presented in Chapter 10. This figure shows that our simplified analysis provides a good estimate of the force exerted on the main cable. Near the center of the bridge, the results vary by less than 0.1%, and near either tower the variation is no more than 3%, Variations occur because the links, being a series of straight lines, cannot exactly duplicate the geometry of the parabolic cable. If we were to modify our approximate model by making the links shorter and the suspenders closer together, we could more closely approximate the actual bridge geometry and would converge to the parabolic cable solution in **Figure 3.12**.

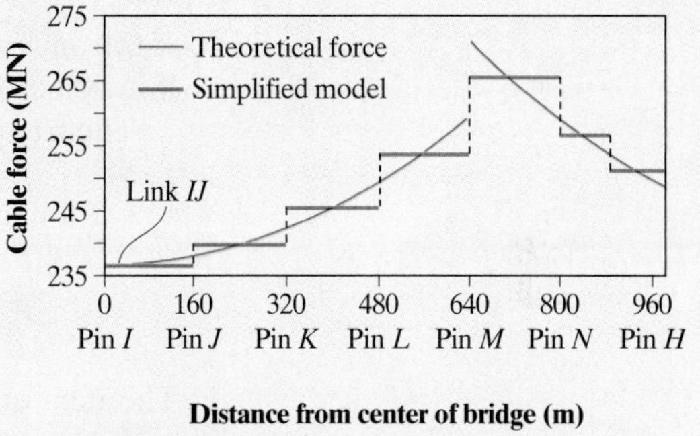

Figure 3.12 Forces calculated from the simplified analytical model of the main cable are a good approximation of the actual forces exerted on the cable.

Why do you think there is a discontinuity in the theoretical curve in **Figure 3.12** at the tower (pin M)? This discontinuity occurs because of the change in the orientation of the cable as it is draped over the top of the tower. The horizontal component (T_h) of the cable force remains constant throughout the length of the cable. The magnitude of the cable force at any location is $T = T_h/\cos\theta$, where θ is the angle between the cable and the horizontal. At the tower, the angle changes abruptly and consequently so does the cable force.

3. What Forces on the Anchorage Would Cause Uplift or Sliding? The main cable is pulling on the anchorage at the right end of the bridge with a very large force directed upward and to the left (**Figure 3.4e** and **Figure 3.13**). In our model this force is represented by $\boldsymbol{F}_{HO}$, the force of link HO pulling on the anchorage at H. The anchorage is kept from sliding to the left by the large friction force $\boldsymbol{F}_{\text{friction, anchorage}}$ developed between it and the ground. It is kept from lifting off the ground by its heavy weight. In designing the Golden Gate Bridge anchorages, engineers had to make each anchorage heavy enough to prevent either sliding or uplift.

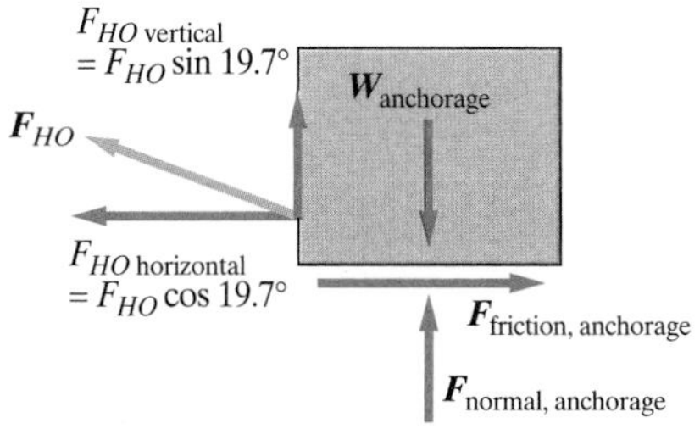

Figure 3.13 Friction forces and the weight of the anchorage counteract the pull of the main cable.

What minimum weight of the anchorage will prevent it from lifting? Newton's first law tells us that if the anchorage is stationary, the sum of the forces in the vertical and horizontal directions must be zero. Looking at **Figure 3.13**, this means that

$$\| \boldsymbol{F}_{HO \text{ vertical}} \| + \| \boldsymbol{F}_{\text{normal, anchorage}} \| - \boldsymbol{W}_{\text{anchorage}} = 0 \qquad (3.4)$$

$$- \| \boldsymbol{F}_{HO \text{ horizontal}} \| + \| \boldsymbol{F}_{\text{friction, anchorage}} \| = 0 \qquad (3.5)$$

If the anchorage were to lift off the ground, the normal force shown in **Figure 3.13** would be zero. We can calculate the weight of the anchorage when the normal force is zero by equating the forces in the vertical direction to zero and eliminating the normal force from (3.4):

$$\begin{aligned} \| \boldsymbol{F}_{HO \text{ vertical}} \| - \| \boldsymbol{W}_{\text{anchorage at uplift}} \| &= 0 \qquad (3.6) \\ \| \boldsymbol{W}_{\text{anchorage at uplift}} \| &= \| \boldsymbol{F}_{HO \text{ vertical}} \| \\ \| \boldsymbol{W}_{\text{anchorage at uplift}} \| &= (251.2 \text{ MN})(\sin 19.7°) = 84.7 \text{ MN} \end{aligned}$$

If the anchorage weighs less than 84.7 MN, it will lift off the ground, and if it weighs more it will not.

The friction force required to prevent sliding can be determined from (3.5):

$$\begin{aligned} \| \boldsymbol{F}_{\text{friction, anchorage}} \| &= \| \boldsymbol{F}_{HO \text{ horizontal}} \| \\ \| \boldsymbol{F}_{\text{friction, anchorage}} \| &= (251.2 \text{ MN})(\cos 19.7°) = 236.5 \text{ MN} \qquad (3.7) \end{aligned}$$

The next question we want to ask is how heavy the anchorage must be to develop this large friction force. As presented in Chapter 6, the maximum friction force ($\boldsymbol{F}_{\text{friction max}}$) that can be produced between the ground and the anchorage depends on normal force between the anchorage and the ground and the roughness of contact between the two surfaces, reflected in the **coefficient of friction**, μ_{static}. For rough materials such as rock and concrete, μ_{static} could be in the range from 0.5 to 0.7. For our example we shall use $\mu_{\text{static}} = 0.6$. The relationship is expressed mathematically as

$$\boldsymbol{F}_{\text{friction max}} = \mu_{\text{static}} \boldsymbol{F}_{\text{normal, anchorage}} \qquad (3.8)$$

In (3.7) we determined that the friction force needed to prevent sliding is 236.5 MN, which must be developed by the roughness between the anchorage and the ground as expressed by (3.8). Therefore

$$\begin{aligned} \| \boldsymbol{F}_{\text{friction, anchorage}} \| = \| \boldsymbol{F}_{\text{friction max}} \| = \mu_{\text{static}} \| \boldsymbol{F}_{\text{normal, anchorage}} \| &= 236.5 \text{ MN} \\ \| \boldsymbol{F}_{\text{normal, anchorage}} \| = \frac{236.5 \text{ MN}}{0.6} &= 394 \text{ MN} \end{aligned}$$

Finally, the weight of anchorage required to produce a normal force of 394 MN is found from (3.4):

$$\begin{aligned} \| \boldsymbol{W}_{\text{anchorage required}} \| &= \| \boldsymbol{F}_{\text{normal, anchorage}} \| + \| \boldsymbol{F}_{HO \text{ vertical}} \| \qquad (3.9) \\ &= 394 \text{ MN} + 84.7 \text{ MN} = 479 \text{ MN} \end{aligned}$$

Answer to Question 3

This tells us that the anchorage must weigh more than 84.7 MN to prevent uplift and more than 479 MN to prevent sliding. In fact, on the Golden Gate Bridge each anchorage weighs about 530 MN, which satisfies both conditions.

3.3 ADDING MORE REALITY

Not all bridges are suspension bridges. Engineers use different design solutions after considering many issues, such as distance to be spanned, types of loads to be carried, strength of the rock available for the foundation, type of material to be used, aesthetics, and cost. Suspension bridges are typically used for spanning large distances, on the order of 600 to 2000 meters. For shorter distances, designers might use beam, arch, or truss bridges. Beam bridges, typically seen as freeway overpasses, are inexpensive to build and efficient for spanning distances of 75 meters or less. Arch bridges, developed by the ancient Romans, are useful for spanning distances from 100 to 400 meters. Truss bridges have the advantage of being lightweight and can be built up from a series of short members.

If you study the Golden Gate Bridge, you will see that it is made up of several types of bridges. For example, the south approach consists of a steel arch, five truss spans, and a series of steel beam bridges. Each bridge type has a different mechanism for transferring loads to the ground.

Up to this point we have assumed that the Golden Gate Bridge is not moving and that only gravity forces act on it. In fact, the bridge is moving all the time and is subjected to a number of dynamic loads, including earthquakes, wind loads, vehicle loads, and the action of strong tidal currents. The currents and the wind impart sideways loads on the towers, causing them to sway approximately 0.3 m from side to side. Vehicular traffic is another source of bridge movement, and as you stand on the bridge sidewalk, you can feel the vibrations of the deck as the cars and trucks drive by.

In extreme cases, wind loads can cause a bridge deck to oscillate and twist wildly, possibly leading to a collapse, as was the case in 1940 on the Tacoma Narrows Bridge. The deck acts like an airfoil as the wind passes by and causes the deck to lift and fall. As the deck goes up and down, changes in the geometry of the main cables cause the towers to sway shoreward and channelward as much as 0.5 m. Thermal expansion and contraction of the main cables also causes the deck to move. Design calculations indicate that at its center, the deck of the Golden Gate Bridge can deflect downward 3.3 m and upward 1.8 m as a result of temperature and other loading.

Because the Golden Gate Bridge is not far from the San Andreas and Hayward faults, it is periodically subjected to earthquakes. During an earthquake, the ground accelerates vertically and horizontally, causing **inertial forces** to act on the bridge. The inertia of a structure causes it to resist any sudden movement of its base, so that the upper parts of the structure deform relative to the base (Figure 3.14). A unique feature of inertial forces caused by earthquakes is that they are proportional to the weight of the structure—the heavier the structure, the larger the forces. The motion of the bridge during an earthquake is very complex, consisting of horizontal and vertical vibrations as well a twisting of the deck and towers. Calculation of bridge deflections and the resulting forces requires a **dynamic analysis**. Although a static analysis can provide preliminary

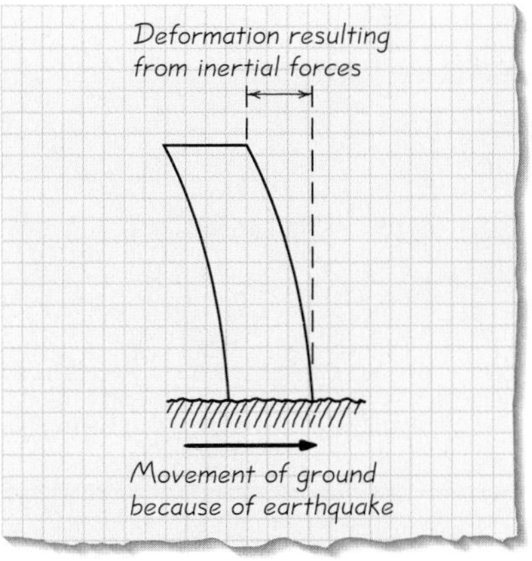

Figure 3.14 Deformation of a structure subjected to earthquake inertial forces

estimates of the earthquake and wind forces acting on each bridge component, a dynamic analysis will provide more accurate results.

3.4 JUST THE FACTS

In this chapter we examined the question of how heavy the anchorages should be for the Golden Gate Bridge. By analyzing the loads acting on a suspension bridge as they are transferred from the bridge deck to the suspenders and then through the main cables to the anchorages, we were able to find the forces of the cables pulling on the anchorages. We used a simplified analytical model to calculate an approximate solution to the forces in the bridge's main cable. We then compared our approximate analysis with a more exact solution. We examined how heavy the anchorages must be to prevent both uplift and sliding. The analysis involved making assumptions, creating free-body diagrams, and then applying Newton's first law.

3.5 REFERENCES

A. M. Abdel-Ghaffar and R. H. Scanlan, "Ambient Vibration Studies of Golden Gate Bridge: I. Suspended Structure," *Journal of Engineering Mechanics*, vol. 111, no. 4, pp. 463–482 (1985).

Joseph Gies, *Bridges and Men* (New York: Grosset & Dunlap, 1963).

Golden Gate Bridge, Highway and Transportation District: www.goldengatebridge.org/research/

SYSTEM ANALYSIS (SA) EXERCISES

SA3.1 Exploring a Suspension Bridge

1. Reconstruct the model of a suspension bridge in Section 3.1 (**Figure 3.2*a***). Push down on the pencil and feel how much force the system can resist before the anchorages start to slide.

   Now remove one of the books from each of the anchorages. Push on the pencil again. Can the system resist more or less force than before? How does the system fail?

   Try making other alterations to the model to examine the effect on the system capacity (i.e., the force it can support) and the failure mechanism. Examples of alterations you can implement include:

   (a) Adding a book to each of the anchorages so there are three books for each

   (b) Inserting a shiny (slippery) piece of paper between the table surface and the books that serve as anchorages

   (c) Inserting a rough piece of cloth or carpet between the table surface and the books that serve as anchorages

   (d) Shortening the string so that it is tight across the books that serve as the towers

   (e) Moving the towers very close to the anchorages or very close together

2. Assuming that the geometry of the Golden Gate Bridge remains unchanged, double the weight per unit length of the bridge deck to 660 kN/m and calculate the force $\boldsymbol{F}_{IJ}$ acting on member *IJ* (**Figures 3.5** and **3.8**).

   (a) How much does $\boldsymbol{F}_{IJ}$ change?

   (b) How much will doubling the weight per unit length of the bridge deck change the force $\boldsymbol{F}_{HO}$ pulling on the anchorage? Explain your answer.

3. If $\boldsymbol{F}_{HO}$ is doubled (**Figure 3.12**), by how much will the required weight of the anchorage increase? Explain your answer.

SA3.2 Exploring a Beam Bridge

Whereas suspension bridges are efficient for spanning distances of about 600 m to 2000 m, beam bridges are often used to span short distances. A common example of a beam bridge is a freeway overpass. To model a beam bridge, you need three books. Place two books on end about 20 cm apart to serve as the piers and lay the third book across the two to create the bridge deck as shown in **Figure SA3.2.1**.

Load the bridge in two ways:

1. Push straight down on top of the deck with your hand.
2. Push horizontally on the deck with your hand.

For each of these loading cases:

(a) Draw a free-body diagram for each component of the bridge to trace the load path as $\boldsymbol{F}_{\text{hand}}$ is transferred to the ground. State any assumptions you are making in drawing the diagrams.

(b) Explain why the bridge in (*b*) of the figure fell over and how you might alter the design so that it wouldn't.

(c) Now replace the book that is modeling the bridge deck with a piece of cardboard or thick paper. Push straight down on the deck with your hand. Describe the behavior of the deck.

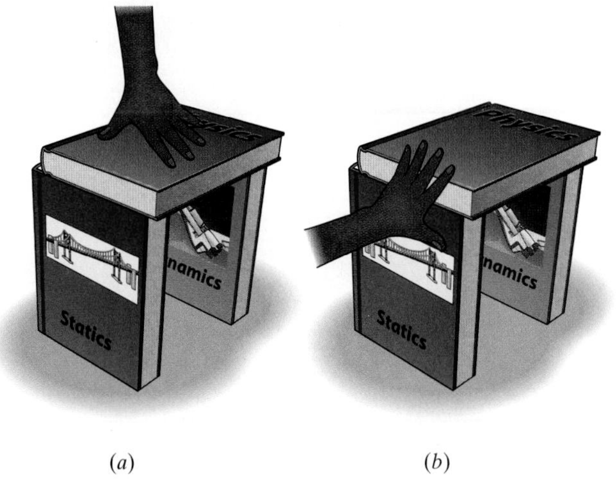

Figure SA3.2.1 Beam bridge modeled with books (*a*) loaded with a vertical force and (*b*) loaded with a horizontal force.

SA3.3 Exploring an Arch Bridge

An arch bridge is made up of a bridge deck supported by an arch that is connected at both ends to supports called abutments. The load on the bridge is transferred along the curve of the arch to the abutments. The arch bridge can span larger distances than a beam bridge (100 to 400 meters). To understand the load transfer in an arch bridge and the function of the abutments, you can build a model.

- You need a one-pint (or larger) container like the type used to package cottage cheese, sour cream, or delicatessen food.
- Cut the container in half along its diameter so that it makes two semicircular pieces.
- To complete the arch, cut off the bottom of the container and the stiffening ring (or lip) at the top (**Figure SA3.3.1*a***).

Load the arch by pressing down on the center as shown in **Figure SA3.3.1*b***.

(a) Is the arch in tension or compression? What happens to the ends of the arch?

To prevent the arch from collapsing, you must add abutments. To model your abutments, have a friend place her or his hands at the intersections of the arch and the ground as shown in **Figure SA3.3.1*c***. Again load the arch by pressing down on the center as shown in **Figure SA3.3.1*d***.

(b) How do the abutments affect how the ends of the arch move?

(c) Is the arch pushing or pulling on the abutments?

(d) What prevents the abutments from sliding?

(e) With the abutments in place, does the bridge provide more or less resistance to the push of your finger?

(f) Draw a free-body diagram showing all of the forces acting on one abutment.

(g) Review the analysis of the Golden Gate Bridge anchorage (**Figure 3.12** and (3.4) to (3.9)) and explain how the weight of the arch bridge abutment is important in preventing it from sliding. To test your reasoning, model the abutments using something relatively lightweight such as CD-ROM cases. When you load the bridge, do the lightweight abutments move? Now press down on the lightweight abutments and load the bridge. Do the abutments move when the extra weight is on them?

(h) Are there any forces pulling up on the abutment? To determine the required weight of an arch bridge abutment, how would the analysis differ from the analysis of the suspension bridge anchorage?

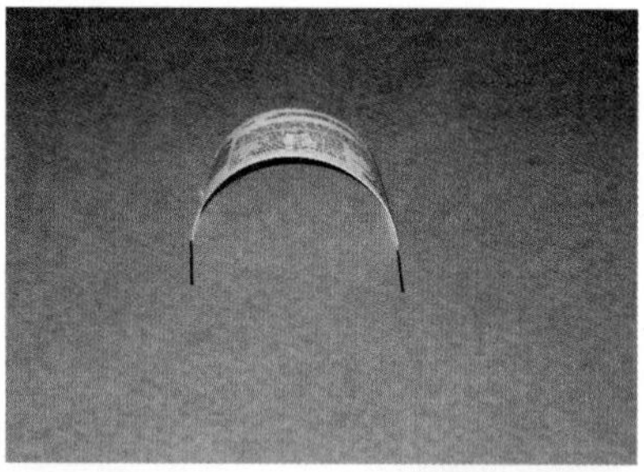

(*a*) Arch bridge made from cottage cheese container

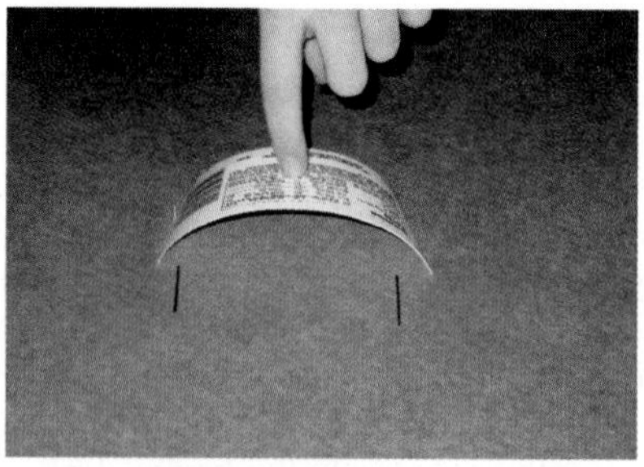

(*b*) Loaded arch without abutments

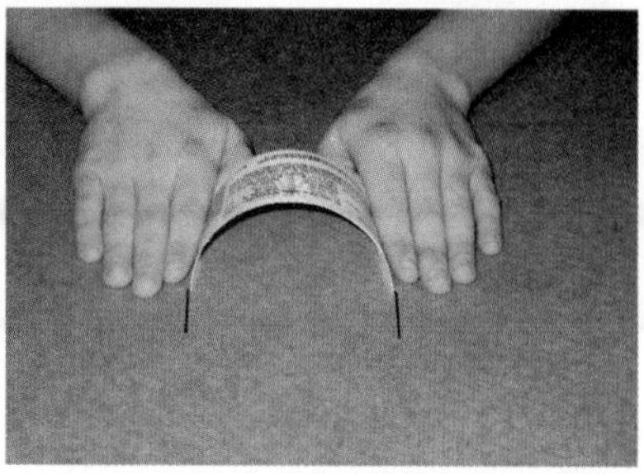

(*c*) Abutments added to arch bridge

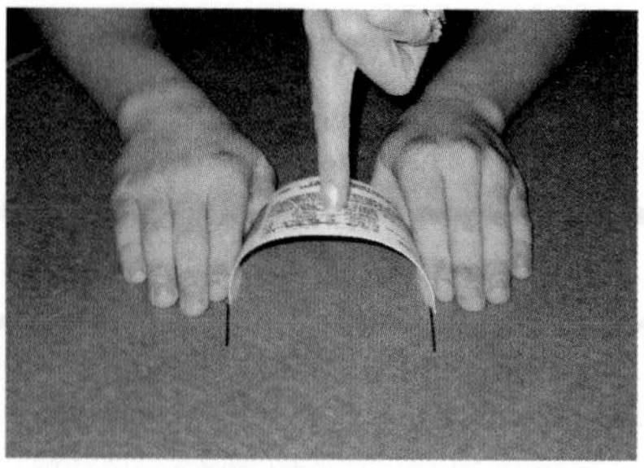

(*d*) Loaded arch with abutments

Figure SA3.3.1

SA3.4 Exploring a Cable-Stayed Bridge

It is possible to confuse cable-stayed bridges with suspension bridges because both types use cables to hold up the bridge deck. However, the two bridge types have different mechanisms for transferring loads to the ground. In a suspension bridge, the main cables are draped over the tops of the towers and pull on the anchorages. Both the towers and the anchorages transfer the loads to the ground. In a cable-stayed bridge, the cables are attached directly to the towers and only the towers transfer the loads to the ground. As shown in **Figure SA3.4.1**, the cables, which are in tension, pull up on the deck and down on the tower. The tower, which is in compression, transfers a downward force to the ground.

To understand the load transfer in a cable-stayed bridge, we can build a model.[5] The cables can be attached to the towers in a number of patterns, but for our model we will use a fan pattern. In this pattern all of the cables are attached to the top of the tower, and then each cable is attached to a different point along the length of the bridge deck.

You need two pieces of string, one about 1.5 meters long and the other 2 meters long, and a partner to help you. Use your arms to model the bridge deck by holding both arms out horizontally to the side. You should be able to feel your muscles holding up your arms. Model the bridge tower with the trunk of your body and your head, and use the tower to support the cables that support the bridge deck. Have your partner tie the 1.5-m piece of string to each of your elbows, with the middle of the string lying on top of your head. The string acts as a stay-cable and holds your elbows up, but your hands and lower arms are hanging downward with little support. You should feel less stress on your muscles.

Have your partner tie the 2-m piece of string to each wrist, with the middle of the string lying on top of your head, making sure both strings are still taut. The bridge is now supported by two stay-cables, and your lower arms are also supported as shown in **Figure SA3.4.2**.

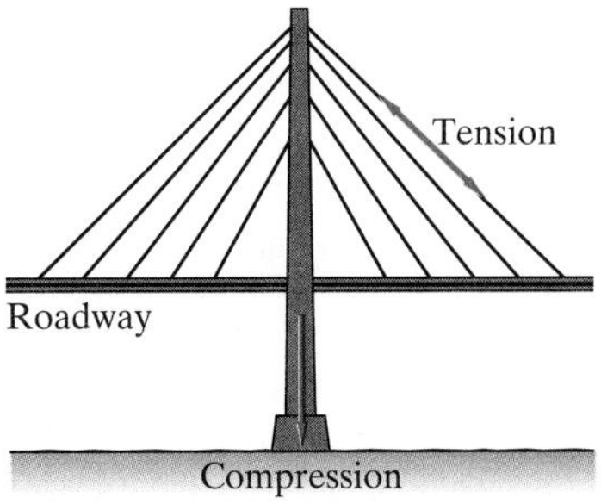

Figure SA3.4.1 Load path for cable-stayed bridge

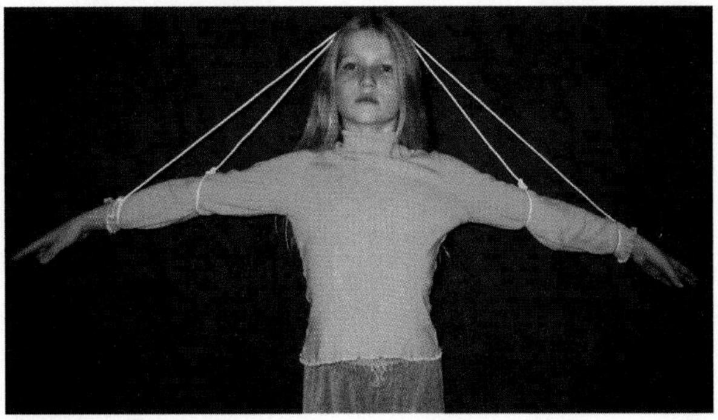

Figure SA3.4.2 Model of cable-stayed bridge

Describe the load path of the cable-stayed bridge by answering these questions:

(a) What forces acting on the bridge deck are being transferred to the ground?

(b) Which components of the bridge are in tension?

(c) In which parts of your body do you feel a compression force? (**Figure SA3.4.1** does not show all of the compressive forces acting on the bridge.)

(d) How is the load on the top of your head transferred to the ground?

Assuming that the cables are attached to the deck at equal intervals and each carries an equal portion of the weight of bridge deck, the cables attached farther from the tower are subjected to larger tension forces than those attached closer to the tower. **Figure SA3.4.3** is an incomplete free-body diagram of a deck slice showing the force pulling on cable stay 1 and the weight of the slice.

(e) Using the variables shown in **Figure SA3.4.3**, convince yourself that the steeper cables, which are attached closer to the tower, are subjected to smaller tension forces. For this exercise, the cable number increases as you approach the tower.

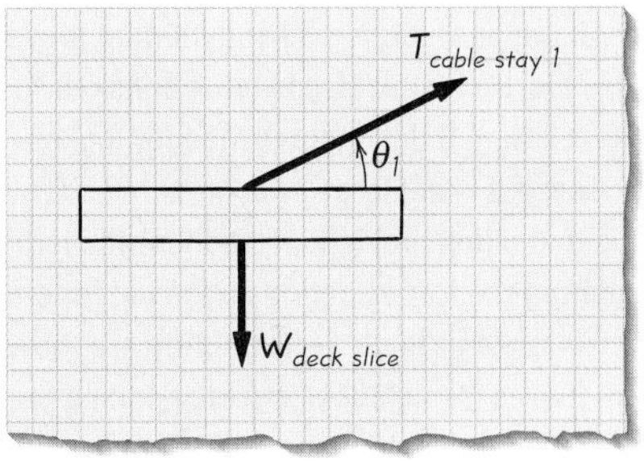

Figure SA3.4.3 Tension force on cable stay 1 and the weight of the bridge deck slice

[5]Adapted from http://www.pbs.org/wgbh/nova/bridge/meetcable.html.

CHAPTER 4

FORCES

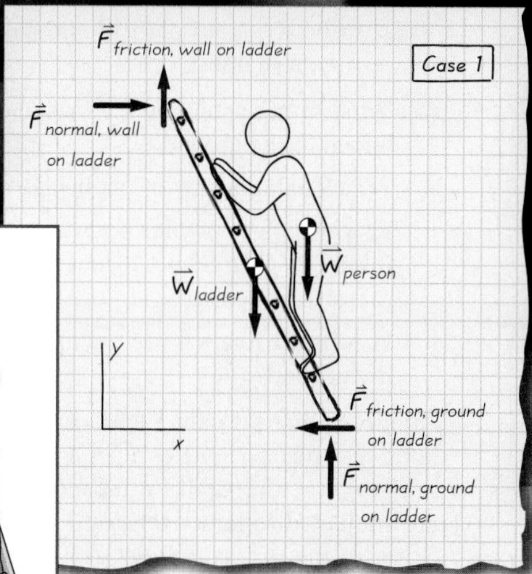

In the previous two chapters we discussed the bicycle and Golden Gate Bridge and how forces act on them. In the case of the bicycle we considered the question of how large a pedal force you apply while cycling, and how this relates to maximum velocity. In the case of the Golden Gate Bridge we considered the question of how the weight of a vehicle or the weight of the bridge deck gets transferred into the ground. In this chapter we discuss forces in a more formal manner.

Engineers must consider how forces affect the structures and machines they design. For example, a civil engineer[1] designing a dam would think about the water pushing against the dam and ask "Will the steel tie-downs connected to the bedrock be strong enough?" A mechanical engineer[2] designing landing gear for an airplane would think about the forces applied during landing and ask "Will the size of the gear forging be sufficient to prevent failure after repeated landings?"

In everyday life, for example, you must consider forces whenever you prop a ladder against a house to wash a window or against a tree to rescue a kitten. If the angle the ladder makes relative to the ground is too large, the top of the ladder will tend to tip away from the house or tree; if it is not large enough, the ladder will tend to slide away from the house or tree. The question that you might ask is, "Will I be able to get just the right position for the ladder so that I can safely accomplish my task? " In asking this question, you are implicitly considering the forces acting on the ladder and their relationship to one another.

This chapter looks at forces—what they are, how we categorize them, and how we represent them. Learning how to categorize and represent forces is the first step in developing the analytic skills you must have in order to ask the right questions about any structure or machine you are working on and come up with the right answers to those questions.

[1]*Civil engineers* are responsible for planning, designing, and constructing the infrastructure of our civilization—buildings, bridges, power plants, transportation systems, water systems, and much more. The civil engineer is called on to apply physical (and, in some cases, chemical and biological) principles, assess social and environmental impact, and evaluate the costs and benefits of infrastructure projects.

[2]*Mechanical engineers* work in a variety of industries, including transportation, product manufacturing, energy generation, consumer products, and applied research. Their work involves the design, manufacture, and maintenance of products or systems to meet human needs. The mechanical engineer is called on to use knowledge of physical principles, understanding of existing products, and imagination of what products might be.

OBJECTIVES

On completion of this chapter, you will be able to:

- **Define the types of forces that act on systems**
- **Isolate an object from the rest of the world and identify the types of external forces acting on it**
- **Represent a force mathematically and be able to convert between various representations**
- **Manipulate forces using vector addition and the dot product**
- **Appreciate how forces factor into static analysis**

4.1 WHAT ARE FORCES?

A **force** is any interaction between an object and the rest of the world that tends to affect the state of motion of the object. The strength of a force is related to the extent of its effect on the object. You cannot see forces, but if you've ever seen a car crash or felt a rush of air escaping from a balloon, you have seen the effects forces have.

We specify force in **units** of newtons (N) in the SI system and of pounds (lb) in the U.S. Customary system. The conversion between the two systems is 4.4482 N = 1 lb or, equivalently, 1 N = 0.2248 lb. Forces range from very small (e.g., 0.000 000 5 N for the gravitational pull exerted by Mars on an earth-bound engineering student) to very large (e.g., 10 tons = 88 960 N for the weight of a large farm tractor). Forces are vector quantities; this means they have both **magnitude** (size) and **direction** associated with them. Graphically, we represent a force by an arrow (**Figure 4.1**). The direction from tail to head represents the direction of the force, and we commonly draw the length of the arrow proportional to the magnitude of the force. If we know the magnitude of the force, we write it (including units) next to the arrow. The line along which the force acts is called the **line of action** of the force.

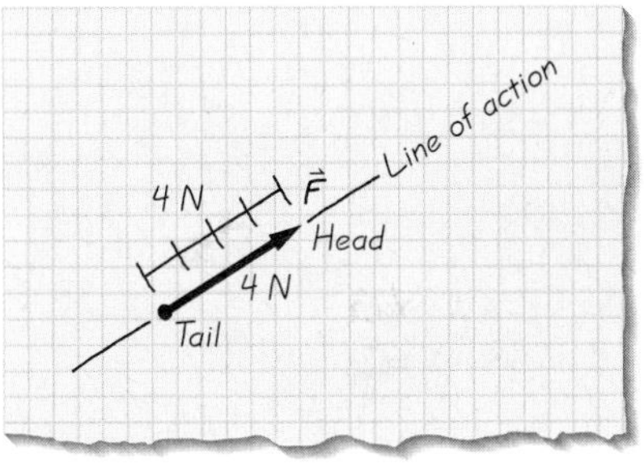

Figure 4.1 Drawing a force

In printed material, we show the symbol for a vector in boldface italic—$\boldsymbol{F}$—and with the Euclidean norm — $\| \boldsymbol{F} \|$ —when only the magnitude is of concern. For instance, you might read that a force $\boldsymbol{F}_1$ is exerted in a northerly direction, a force $\boldsymbol{F}_2$ is exerted in an easterly direction, the magnitude $\| \boldsymbol{F}_1 \|$ of the first force is known, and you are to determine the magnitude $\| \boldsymbol{F}_2 \|$ of the second force. In handwritten work, a vector is generally indicated with an overbar or underbar—$\overline{\text{F}}$ or $\underline{\text{F}}$.

Physicists have traditionally identified four fundamental forces: gravitational, electromagnetic, weak, and strong. The relative strengths of these forces are strong 1, electromagnetic 10^{-2}, weak 10^{-7}, and gravitational 10^{-38}. Generally, engineers considering equilibrium of systems are concerned with **gravitational forces**. They are also concerned with the electromagnetic forces that result from the interaction of electrical and magnetic fields at the atomic and subatomic levels—we refer to these as **contact forces**. The strong force (which keeps every atomic nucleus intact) and the weak force (a factor in radioactive decay) are significant only at the subatomic level, so we will not consider them further in this book.

In the remainder of this chapter, we look at these two fundamental forces that engineers must deal with every day. The gravitational force exerted by the earth must be considered every time a civil engineer designs a skyscraper, say, or a crane hoists a 2-ton beam 200 feet over a city street. And of course it's a matter of life and death that the mechanical engineer has an intimate knowledge of the contact forces keeping the crane's cable attached both to the crane and to the beam.

4.2 GRAVITATIONAL FORCES

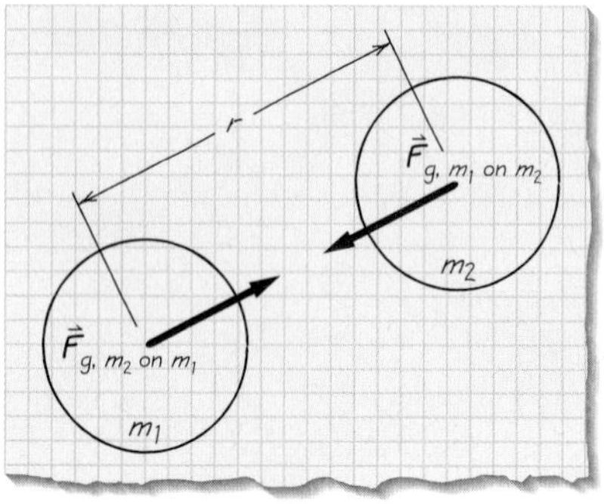

Figure 4.2 Gravitational force between two objects

Every object in the universe exerts a force on every other object in the universe, and we call this force a gravitational force. Because every object exerts such a force on every other object, gravitational forces always come in pairs, as Figure 4.2 shows. There you see two objects 1 and 2 and the gravitational force exerted by each on the other. The gravitational force is always an attractive force. As a result, the direction of the force exerted by object 2 on object 1 is from 1 to 2 and the direction of the force exerted by object 1 on object 2 is from 2 to 1. The magnitude of the gravitational force between two objects is directly proportional to the product of their masses and inversely proportional to the square of the distance between them. The **universal gravitational constant** G changes the proportionality to an equality:

$$\| \boldsymbol{F}_g \| = \frac{G m_1 m_2}{r^2} \tag{4.1}$$

where

- $\| \boldsymbol{F}_g \|$ is the magnitude of the gravitational force (in newtons)
- G is the universal gravitational constant ($G = 6.673 \times 10^{-11}$ m^3/kg · s^2 in SI units)[3]
- m_1 and m_2 are the masses of the two bodies (in kilograms)
- r is the distance between their **centers of mass** (in meters), as shown in **Figure 4.2**)

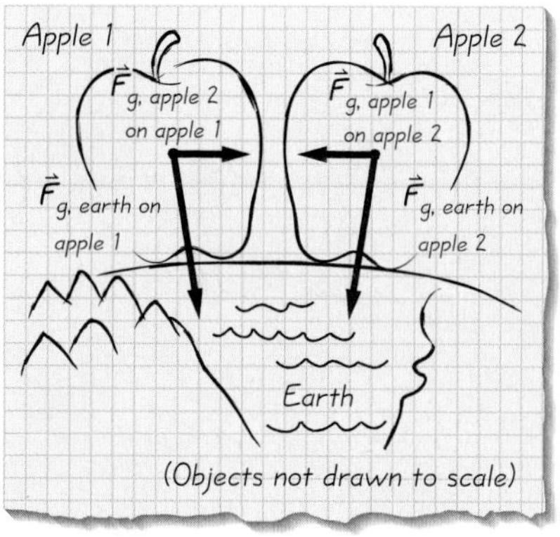

Figure 4.3 Gravitational forces between apples and earth

Because the mass of the earth is at least 20 orders of magnitude greater than the mass of most objects on the planet, the gravitational attraction between any two objects at or near the earth's surface is negligible relative to the gravitational attraction between either object and the earth. For example, the magnitude of the gravitational force between two average-size apples is a mere 0.000 000 000 066 7 N[4], compared with the 0.98 N gravitational force between one apple and the earth[5] (Figure 4.3). Therefore, the gravitational force exerted by the earth is an important force acting on all objects at or near the earth's surface.

When analyzing a gravitational force where one of the two interacting objects is the earth, we can simplify (4.1) by realizing that (a) the mass of the earth is constant and (b) for any object either on or not far from the earth's surface, the distance from the earth's center of mass to

[3] $G = 3.439 \times 10^{-8}$ ft^3/slug · s^2 in U.S. Customary units.

[4] This force was calculated using (4.1), with $m_1 = m_2 = 0.1$ kg, r = radius of apple 1 + radius of apple 2 = 0.050 m + 0.050 m = 0.100 m.

[5] This force was found using (4.1), with m_1 = mass of apple = 0.1 kg, m_2 = mass of earth = 5.976×10^{24} kg, r = radius of earth + radius of apple = 6.371×10^6 m + 0.05 m = 6.371×10^6 m.

the object's center of mass can be taken to be the earth's average radius. In other words, once we arbitrarily say that m_1 in (4.1) is the mass of the earth, we can group our three constants—G, m_1, and r^2—into a new constant g. Therefore, (4.1) can be rewritten as

$$\| \boldsymbol{F}_g \| = \frac{Gm_1m_2}{r^2} = m_2\left(\frac{Gm_1}{r^2}\right) = mg \tag{4.2}$$

where

$g = \dfrac{Gm_1}{r^2}$ and is called the gravitational constant

m $(= m_2)$ is the mass in kilograms of the object feeling the gravitational force exerted by the earth

By substituting the values of G, m_1 for the earth $(= 5.976 \times 10^{24}$ kg), and the mean radius of the earth $(= 6.371 \times 10^6$ m)[6] into (4.2), we find that at or near the earth's surface the magnitude of the gravitational force (in newtons) exerted by the earth on the object of mass m (in kilograms) is

$$\| \boldsymbol{F}_g \| = (9.807m) \text{ m/s}^2 \tag{4.3A}$$

This gravitational force exerted by the earth on an object is given a special name—the object's **weight on earth** ($\boldsymbol{W}_{\text{earth}}$). The magnitude of this weight force is the product of the mass of the object and the constant g $(= 9.807 \text{ m/s}^2 = 32.2 \text{ ft/s}^2)$:

$$\| \boldsymbol{W}_{\text{earth}} \| = \| \boldsymbol{F}_g \| = (9.807m) \text{ m/s}^2 \tag{4.3B}$$

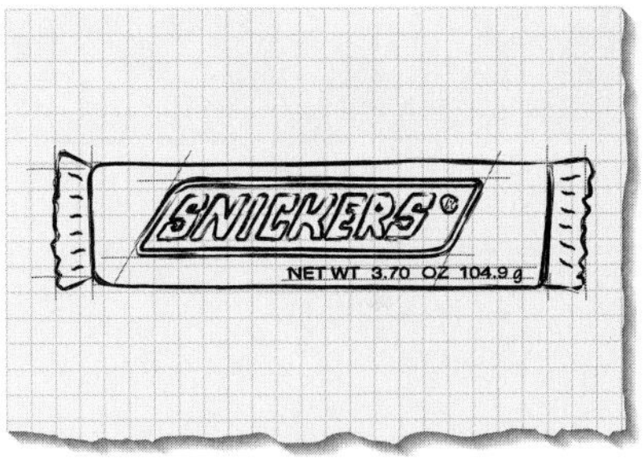

Figure 4.4 Product label contains weight and mass information

Occasionally you may see mass and force units seemingly used to mean the same thing. An example is the label on the candy bar shown in Figure 4.4, where the net weight is given as 104.9 grams and 3.70 ounces. Gram is a mass unit, and ounce is a force unit. The manufacturer assumes that its candy will be consumed on the earth and is saying that the gravitational force exerted by the earth on this 104.9 grams of candy is 3.70 ounces when the candy is near the earth's surface.

Weight is a **body force**. This means that the gravitational force exerted by the earth on an object exists between every atom in the object and every atom in the earth (Figure 4.5*a*). There will be times when the distributed nature of the gravitational force needs to be considered in engineering practice, and Chapter 8 deals with formal procedures for doing this. Far more frequently, however, it is sufficient to lump all the distributed gravitational forces acting on an object into a single force. This single force, which represents the weight of the object, is directed from the center of mass of the object to the center of mass of the earth (**Figure 4.5*b***).

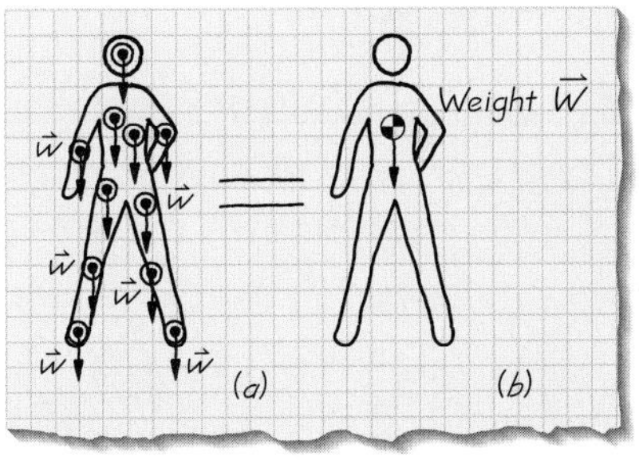

Figure 4.5 (*a*) The weights associated with Adam's atoms; (*b*) the total weight of Adam

[6]The earth is not a perfect sphere. Rather, it is an ellipsoid, flattened at the poles and bulging at the equator. Its equatorial radius is greater than its polar radius by 21 km (see Halliday, Resnick, and Krane, *Physics*, 4th Edition, John Wiley & Sons). What this means is that the gravitational force exerted by the earth is slightly greater at the poles ($g = 9.835$ m/s^2) than at the equator (9.78 m/s^2). For most engineering work, this difference is insignificant, and we use a value of g based on the mean radius of the earth.

EXAMPLE 4.1 GRAVITY, WEIGHT, AND MASS (A)

Figure 4.6

Figure 4.7

Consider the NASA Mars rover Spirit (**Figure 4.6**). Spirit was launched toward Mars on June 10, 2003, in search of answers about the history of water on Mars. It landed on Mars on January 3, 2004, and in its wanderings on the Red Planet sent back pictures such as the one shown in **Figure 4.7**. Spirit is about the size of a golf cart and has a mass of 174 kg.

Determine

(a) the weight of Spirit on the earth in newtons and in pounds (Name two objects that weigh approximately the same amount on the earth as Spirit.)

(b) the weight of Spirit on Mars in newtons

(c) the mass of Spirit in slugs

Goal We are asked to find the weight of the rover Spirit on the earth and on Mars. We are also asked to specify the mass of Spirit in slugs.

Given We are given the mass of Spirit in kilograms.

Assume Assumptions related to distances, as stated below.

Draw No drawings are required to address this problem.

Formulate Equations and Solve **(a)** *The weight of Spirit on the earth* can be found using (4.3B):

$$\| \boldsymbol{W}_{\text{earth}} \| = \| \boldsymbol{F}_g \| = (9.807m) \text{ m/s}^2 \tag{4.3B}$$

where the mass m is in kilograms and the weight W is in newtons. Since the mass of Spirit is 174 kg, its weight is

$$\| \boldsymbol{W}_{\text{earth}} \| = \| \boldsymbol{F}_g \| = (9.807 \text{ m/s}^2)(174. \text{ kg}) = 1706 \text{ N}$$

To convert from newtons to pounds we multiply by 0.2248 lb/N (**Table 1.3**). Therefore

$$\| \boldsymbol{W}_{\text{earth}} \| = 1796 \text{ N } (0.2248 \text{ lb/N}) = 384 \text{ lb}$$

Answer to (a) The weight on the earth of the rover Spirit is 1706 N. This is 384 lb, which is approximately the weight of a motorcycle or baby grand piano.

(b) *The weight of Spirit on Mars* can be found using (4.1):

$$\| \boldsymbol{F}_g \| = \frac{G m_1 m_2}{r^2} \tag{4.1}$$

where

G is the universal gravitational constant (6.673×10^{-11} m^3/kg · s^2)
m_1 is the mass of Spirit (174 kg)
m_2 is the mass of Mars (6.39×10^{23} kg)[7]
r is the mean radius of Mars (3.394×10^6 m).[7] We ignore the distance from the surface of Mars to the center of Spirit. Because the radius of Mars is so much larger than the distance from the surface of Mars to the center of mass of Spirit (which is approximately 0.5 m), we can take r as approximately (3.394×10^6 m).

Substituting these values, we find that Spirit weighs 644 N on Mars. This is the force exerted by Mars on Spirit as it sits on Mars' surface.

We have ignored the pull of the other plants on Spirit as it sits on Mars. Is this a reasonable assumption? How could you verify this assumption?

Answer to (b) The weight on Mars of Spirit is 644 N.

(c) *The mass of Spirit in slugs.* Using the units conversion in **Table 1.3**, we write that

$$(174 \text{ kg})(1 \text{ slug}/14.5938 \text{ kg}) = 11.92 \text{ slugs}$$

Answer to (c) The mass of Spirit is 11.92 slugs.

Check Our results say that an object that weighs 1706 N on the earth weighs 644 N on Mars. These weights are in the ratio of 1.00:0.377. Now we arrange the original data on the mass and radius of the earth and Mars in a table to confirm this ratio. The last column of the table gives us these same ratios and provides a check to the calculations presented above.

	Mean radius (r in km)	**Mass** (m in kg)	**Value of** m/r^2 (in kg/m^2, as seen in (4.1))	Ratio m/r^2 to earth value
Mars	3394	6.394×10^{23}	5.55×10^{16}	0.377
Earth	6371	5.976×10^{24}	14.7×10^{17}	1.00

We could also check another source of data to confirm that we are using correct values for the mass and radius of the earth and the value of the universal gravitational constant.

[7]Solar system constants are from Appendix A2.2.

EXAMPLE 4.2 GRAVITY, WEIGHT, AND MASS (B)

Consider a mission specialist on the space shuttle who has a mass of 65 kg (**Figure 4.8**).

(a) Find the weight of the specialist on the earth (express in newtons and pounds).

(b) When the space shuttle is orbiting 200 km above the earth's surface, what is the force of gravity acting on the specialist (express in newtons and pounds)?

Figure 4.8

Goal We are to find the weight of a mission specialist on the earth, and the force of gravity acting on the specialist when orbiting 200 km above the earth's surface.

Given We are told the mass of the specialist (65 kg).

Assume We will assume that it is appropriate to work with the mean radius of the earth (which according to Appendix A2.2 is 6371 km).

Draw A drawing is not required to find the requested quantities. You are urged to think about where you would be if you drove 200 km in any direction from where you are currently sitting; does it seem like a long distance?

Formulate Equations and Solve The weight of the mission specialist on the earth can be found using (4.3B):

$$\| \boldsymbol{W}_{\text{earth}} \| = \| \boldsymbol{F}_g \| = (9.807m) \text{ m/s}^2 \tag{4.3B}$$

where m is in kilograms and the weight W is in newtons. Therefore

$$\| \boldsymbol{W}_{\text{earth}} \| = \| \boldsymbol{F}_g \| = (65 \text{ kg})(9.807 \text{ m/s}^2) = 637 \text{ N}$$

To convert from N to lb, we look up the conversation factor in **Table 1.3**; it is 0.2248. Therefore, the mission specialist's weight on the earth in pounds is 637 N (0.2248 lb/N) = 143 lb.

Answer to (a) The weight on the earth of the mission specialist is 637 N or 143 lb.

The force of gravity acting on the mission specialist when in orbit 200 km above the earth's surface can be found using (4.1):

$$F_g = \frac{Gm_1m_2}{r^2}$$

where

G is the universal gravitational constant (6.673×10^{-11} m^3/(kg · s^2))
m_1 is the mass of the mission specialist (65 kg)
m_2 is the mass of the earth (5.976×10^{24} kg)
r is the mean radius of the earth plus the space shuttle's orbit ($6371 + 200 = 6571$ km $= 6.571 \times 10^6$ m)

Substituting in these values, we find that the force of gravity acting on the mission specialist when in orbit is 600 N.

Answer to (b) The force of gravity acting on the mission specialist is 600 N (135 lb). This is 5.8% less than the force found in (a).

Check A check would include repeating the calculations to confirm that all of the numbers were properly entered into a calculator. We could also check another source of data to confirm that we are using correct values for the mass and radius of the earth, and the value of the universal gravitational constant.

Comment: A mission specialist orbiting the earth is acted on by the force of gravity. This force is less than it would be if the specialist were on the surface of the earth; in this particular case it is 5.8% less. So why might we think of the mission specialist as being "weightless" when orbiting the earth?

EXAMPLE 4.3 GRAVITY, WEIGHT, AND MASS (C)

In traveling from the earth to Mars, determine at what distance (as measured from the earth) the gravitational force of the earth acting on Spirit is equal to the gravitational force of Mars on Spirit.

Goal We are to determine the distance from the earth at which the gravitational force of the earth acting on the rover Spirit is equal to the gravitational force of Mars acting on Spirit.

Given Since we are not given any mass or distance information in the statement of this exercise, we will use data contained in Appendix A2.2. The mass of Spirit was given in Example 4.1 (174 kg).

Assume We will assume that it is appropriate to work with the mean distances of Mars and the earth from the sun. Therefore, our calculation is, strictly speaking, correct only when the sun, earth, and Mars are aligned (in that order).

Draw We make a sketch of the various bodies and note their masses and distances (as found in Appendix A2.2) in **Figure 4.9**.

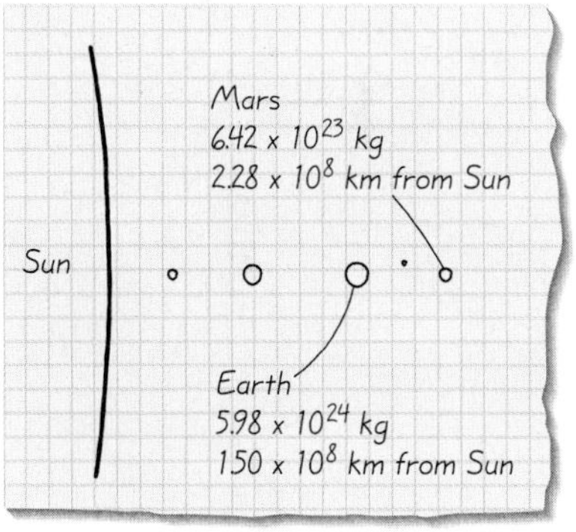

Figure 4.9

Formulate Equations and Solve Equation (4.1) describes the gravitational force between any two objects:

$$\| \boldsymbol{F}_g \| = \frac{G m_1 m_2}{r^2} \tag{4.1}$$

We first write (4.1) for the gravitational force of the earth acting on Spirit, where

- G is the universal gravitational constant
- m_1 is the mass of the earth (5.976×10^{24} kg)
- m_2 is the mass of Spirit (174 kg)
- r is the distance from the center of mass of the earth to the center of mass of Spirit

Substituting these values into (4.1), we find that the gravitational attraction (force) between the earth and Spirit is

$$\| \boldsymbol{F}_{g,\text{ earth on Spirit}} \| = \frac{G(5.976 \times 10^{24}\text{ kg})(174\text{ kg})}{r^2} \quad (1)$$

Now we write (4.1) for the gravitational force of Mars acting on Spirit,

where

G is the universal gravitational constant
m_1 is the mass of Mars (6.394×10^{23} kg)
m_2 is the mass of Spirit (174 kg)

We take the distance from the center of mass of Mars to the center of mass of Spirit (the denominator in (4.1)) as the distance between Mars and the earth (78.3×10^6 kg) minus the distance between the earth and Spirit.[8] Therefore we write (4.1) as

$$\| F_{g,\text{ Mars on Spirit}} \| = \frac{G(6.394 \times 10^{23}\text{ kg})(174\text{ kg})}{(78.3 \times 10^6\text{ km} - r)^2} \quad (2)$$

Since we are interested in finding where the gravitational force of the earth on Spirit is equal to the gravitational force of Mars on Spirit, we equate (1) and (2):

$$\frac{G(5.976 \times 10^{24}\text{ kg})(174\text{ kg})}{r^2} = \frac{G(6.394 \times 10^{23}\text{ kg})(174\text{ kg})}{(78.3 \times 10^6\text{ km} - r)^2} \quad (3)$$

We solve (3) for r to find that at $r = 59.0 \times 10^6$ km, the forces are equal.

Answer At a distance of 59.0×10^6 km, from the center of mass of the earth in a direction toward Mars the gravitational forces of the earth and Mars on Spirit are equal. To get a feel for how far this is, it is about 1474 trips around the circumference of the earth!

Check A check would include repeating the calculations to confirm that all of the numbers were properly entered into a calculator. We could also use another source of data to confirm that we are using correct values for the masses and distances in the exercise.

Note: The distance we calculated is independent of the mass of the rover.

[8] Appendix A2.2 says that the mean distance between Mars and the sun is 227.9×10^6 km and that between the earth and the sun is 149.6×10^6 km.

EXERCISES 4.2

4.2.1. Investigate the gravity force at the earth's surface acting on different wheeled vehicles by answering the following questions. Make sure to identify the source of your information and the model of the vehicle. Express your answer both in pounds and in newtons.

a. How much does a typical racing bicycle weigh?
b. How much does a typical mountain bicycle weigh?
c. How much does a child's tricycle weigh?
d. How much does a typical automobile weigh?

4.2.2. The planet Venus has a diameter of 7700 miles and a mass of $3.34(10^{23})$ slugs.

a. Determine the gravitational acceleration at the surface of the planet. Express your answer in units of (ft/s^2).

b. Your answer in **a** is what fraction of the gravitational acceleration at the surface of the earth?

4.2.3. Determine the gravitational force exerted by:

a. The moon on the earth, using the following data. Make sure that you show your work.

Mass of moon: m_{moon}	$7.350(10^{22})$ kg
Mass of earth: m_{earth}	$5.976(10^{24})$ kg
Radius of moon: r_{moon}	$1.738(10^{6})$ m
Radius of earth: r_{earth}	$6.371(10^{6})$ m
Distance between moon and earth	$3.844(10^{8})$ m

b. The sun on the earth, using the following data. Make sure that you show your work.

Mass of sun: m_{sun}	$1.990(10^{30})$ kg
Mass of earth: m_{earth}	$5.976(10^{24})$ kg
Radius of sun: r_{sun}	$6.960(10^{8})$ m
Radius of earth: r_{earth}	$6.371(10^{6})$ m
Distance between sun and earth	$1.496(10^{11})$ m

4.2.4. Determine the force of gravity acting on a satellite when it is in orbit 20.2×10^6 m above the surface of the earth. Its weight when on the surface of the earth is 8450 N. Use the tables in **Exercise 4.2.3** as needed.

4.2.5. At what distance, in kilometers, from the surface of the earth on a line from center to center would the gravitational force of the earth on a body be exactly balanced by the gravitational force of the moon on the body? Use the tables in **Exercise 4.2.3** as needed.

4.3 CONTACT FORCES

Contact forces result from the electrical and magnetic interactions that are responsible for the bonding of atoms. Under this general heading is the force that prevents one solid object from moving through another (normal contact force), the force that results when one solid object slides or tends to slide across another (friction force), the force that results from the interaction between a solid object and a fluid (fluid contact force), and the force that results when molecules in a solid object are pulled relative to one another (tension force), pushed relative to one another (compression force), or shifted relative to one another (shear force).

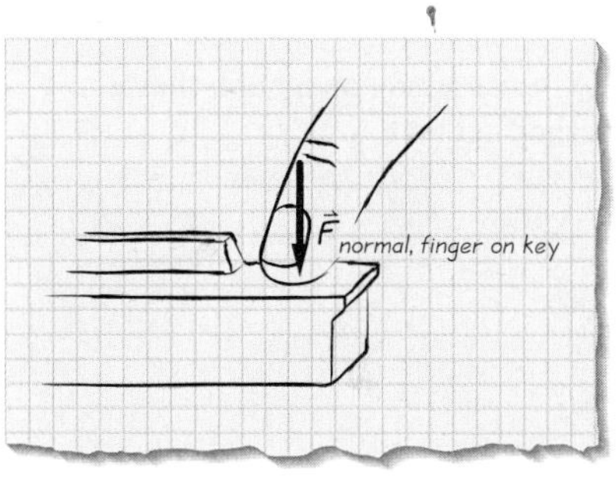

Figure 4.10 A fingertip pushing on a piano key

Normal Contact Force

Whenever two solid objects are in contact with each other, each exerts on the other a force that is *perpendicular* to the two contacting surfaces and is called a **normal contact force**. For example, when a pianist hits a piano key, his fingertip exerts a normal contact force on the key and the key exerts an equal and opposite normal contact force on his fingertip (**Figure 4.10**). Similarly, this book lying on your desk exerts a normal contact force on your desk, and the desk exerts an equal and opposite normal contact force on the book. Every normal contact force is directed so as to bring the two solids together. What this means in practical terms is that a clean fingertip contacting the top surface of a piano key can push on the key but can't pull it, as illustrated in **Figure 4.11**.

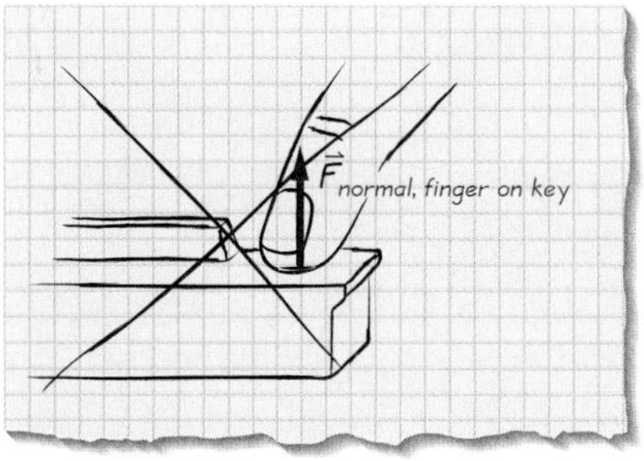

Figure 4.11 A finger cannot pull on a piano key

Friction Force

If you attempt to slide one solid object over another, the motion is resisted by interactions between the surfaces of the two objects. This resistance is a **friction force** and is oriented *parallel* to the two contacting surfaces in a direction opposite the direction of (pending) motion. For example, if you push on an edge of this book as it rests on a table, as in

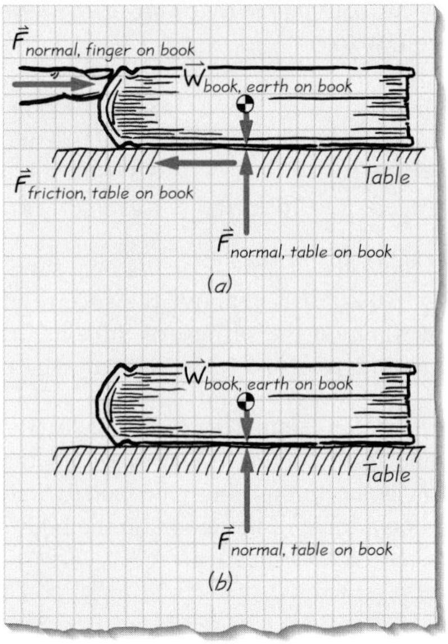

Figure 4.12 (*a*) Forces acting on a book that is pushed across a table; (*b*) forces acting on a book that is not pushed across a table

Figure 4.12*a*, the friction force exerted by the table on the book is in the direction opposite the sliding direction. An equal and opposite friction force is exerted by the book on the table (per Newton's third law). Friction forces are related to and limited by the normal contact forces acting and the characteristics (e.g., smoothness) of the objects in contact. A normal contact force must be present in order for a friction force to be present (but not vice versa; compare **Figure 4.12*a*** with **Figure 4.12*b***).

Fluid Contact Force

As a fluid presses on or moves past a solid object, the fluid exerts a force on the surface of the object; we call this force a **fluid contact force**. (Fluid is the general term for gases and liquids—substances that change shape to fill a volume.) You have experienced a fluid contact force if you have ever put your hand out the window of a moving car—there is a definite force pushing on your hand. When we refer to the interaction between a fluid and a solid, we will typically be talking in terms of the fluid contact pressure, which describes the fluid contact force acting over a surface area. The dimensions of pressure are force/area, and so pressure units are N/m^2 in the SI system and $lb/in.^2$ (sometimes written as psi) in the U.S. Customary system.

The fluid contact pressures engineers work with may be very small (1000 N/m^2 for the water pressure exerted on the bottom of a full tea kettle), medium size (500,000 N/m^2 for the air pressure in a high-performance bicycle tire), or very large (4,000,000 N/m^2 for the air pressure in a scuba tank).

Tension Force

A cable attached to a solid object and pulled taut is said to be under **tension**, which is another contact force we will be studying. For example, consider a cable holding up a crate, as in Figure 4.13*a*. The tension force in the cable is transmitted along the cable (**Figure 4.13*b***). Microscopically, each atom of the cable "pulls" on the atom next to it and is in turn

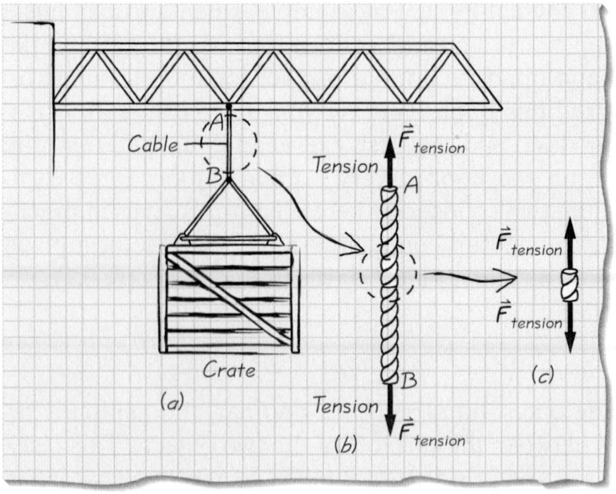

Figure 4.13 (*a*) A crane holds a crate with a cable; (*b*) looking more closely at the cable; (*c*) tension is transmitted along the length of the cable

pulled by that atom, according to Newton's third law (**Figure 4.13*c***). As a result of this atom-to-atom contact within the solid, the force pulling on one end of the cable is transmitted to the object on the other end. If we were to cut the cable at any point and insert a spring scale at the cut ends, the spring scale would read the tension force $\boldsymbol{F}_{\text{tension}}$ directly. Tension forces may be very small (0.001 N for a spider swinging on its web) or very large (on the order of 300,000,000 N tension in the main cables of the Golden Gate Bridge). Tension forces are also present in ropes, chains, bicycle spokes, rubber bands, and bungee cords.

Compressive Force

When the atoms in a solid object are pushed closer together, they experience a contact force called **compressive force** (or simply **compression**). For example, consider a vertical column holding up a wooden deck, as in Figure 4.14*a*. Compression is transmitted along the column as the deck pushes down from the top and the support pillar pushes up from the bottom. As with a cable in tension, adjacent atoms in the column exert forces on each other. In the case of the column in compression, the atoms "push" on each other with compression force $\boldsymbol{F}_{\text{compression}}$ (**Figure 4.14*b***). Compression forces in objects may be very small (0.5 N compression applied by household tweezers) to very large (1,000,000 N compression applied during sheet metal stamping). Do not confuse compressive forces with normal contact forces. A compressive force is *within* an object and is due to the atoms that make up the object pushing on one another. A normal contact force acts on an object's surface and comes about when that object is pushed on by another object.

Shear Force

When the atoms that make up a solid object are shifted relative to one another, they experience a contact force called a **shear force** (or simply **shear**). For example, consider a rock climber standing on a small rock toehold (Figure 4.15*a*). More specifically, consider the interface between the toehold and the larger rock mass the toehold is part of (**Figure 4.15*b***). At the interface, a shear force is transmitted. The atoms on the right of the

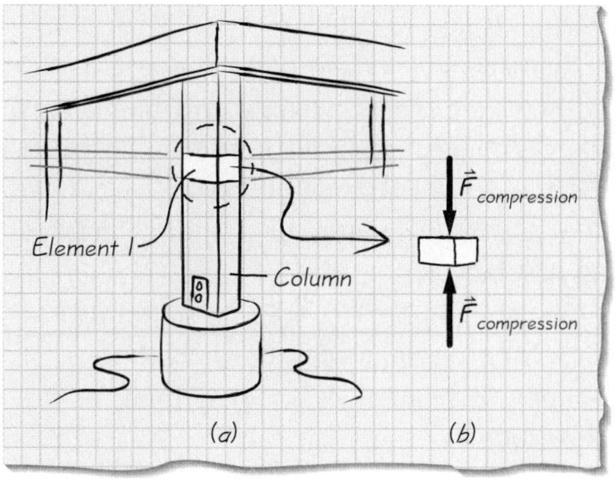

Figure 4.14 (*a*) A column holds up a deck; (*b*) compression is transmitted along the length of the column

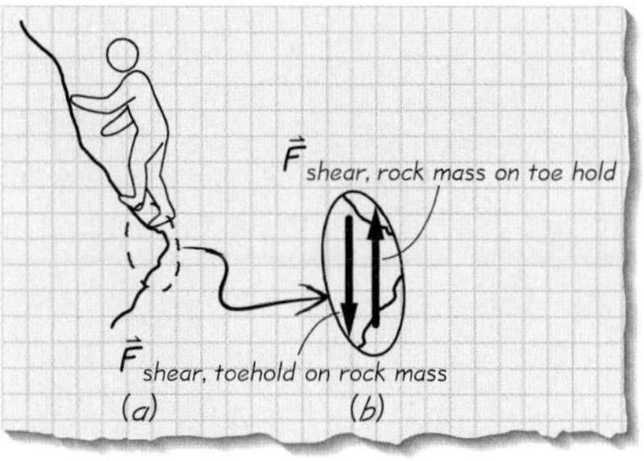

Figure 4.15 (*a*) A rock outcropping holds up a climber; (*b*) shear forces are transmitted across the interface

interface are pushed downward (ever so slightly) relative to the atoms on the left. This shift results in an upward shear force acting on the toehold and an equal and opposite downward shear force acting on the larger rock mass. Notice that the shear force is parallel to the interface. Do not confuse shear forces with friction forces. A shear force is *within* an object and is due to the atoms that make up the object tending to shift relative to one another. A friction force acts on the object's surface and comes about when that object is positioned to slide relative to another object.

EXAMPLE 4.4 TYPES OF FORCES

Identify some of the gravity and contact forces associated with the bicycle shown in **Figure 4.16**.

Figure 4.16

Goal We are to identify some of the forces associated with a bicycle and categorize them as gravity or contact forces.

Given We are given a picture of a bicycle.

Assume We assume that the bicycle is moving along the road and that the cyclist is seated and is holding onto the handlebars. We also assume that it is okay for us to examine a real bicycle as we identify the various forces.

Draw and Solve To complete this exercise, we do not need to set up any equations. We do need to consider various components of a bicycle–cyclist system and the forces present. We use **Figure 4.16** (or even better, examine a real bicycle) to prompt our thinking, then put labels on the drawing. Although we are not explicitly told how many forces we should identify, we choose to find at least two examples of each type of force. Based on **Figure 4.16**, we then compile a list:

Answers *Gravity:* Weight of cyclist; weight of bicycle. The weight of the bicycle could be broken out into the weights of the various components that make up the bicycle.

Contact Forces:

Normal Contact Force—between road and tire, between cyclist and seat, between cyclist's feet and pedals, between cyclist's hands and handlebars

Friction Force—between cyclist and seat (prevents sliding on the seat), between cyclist's hands and handlebars (prevents sliding on the handlebars), between rear wheel and ground tangent to wheel circumference

Fluid Contact Force—in air-inflated tires, in drag created by air moving past bicycle

Tension Force—in chain, in brake cables, in shifter cables, in muscles in cyclist's legs and arms, in wheel spokes

Compressive Force—in front fork, on seat tube, in tire where it meets ground

Shear Force—in crank arm adjacent to bottom bracket, in handlebars adjacent to front fork tube

EXERCISES 4.3

4.3.1. Make a sketch of each of the objects shown in E4.3.1, then note on the sketch which components are in tension and which are in compression.

a. Bicycle
b. Table
c. Pendulum
d. Portable camp stool
e. Pulley system
f. Hiking boot
g Clothes line tree
h. Inverted open umbrella

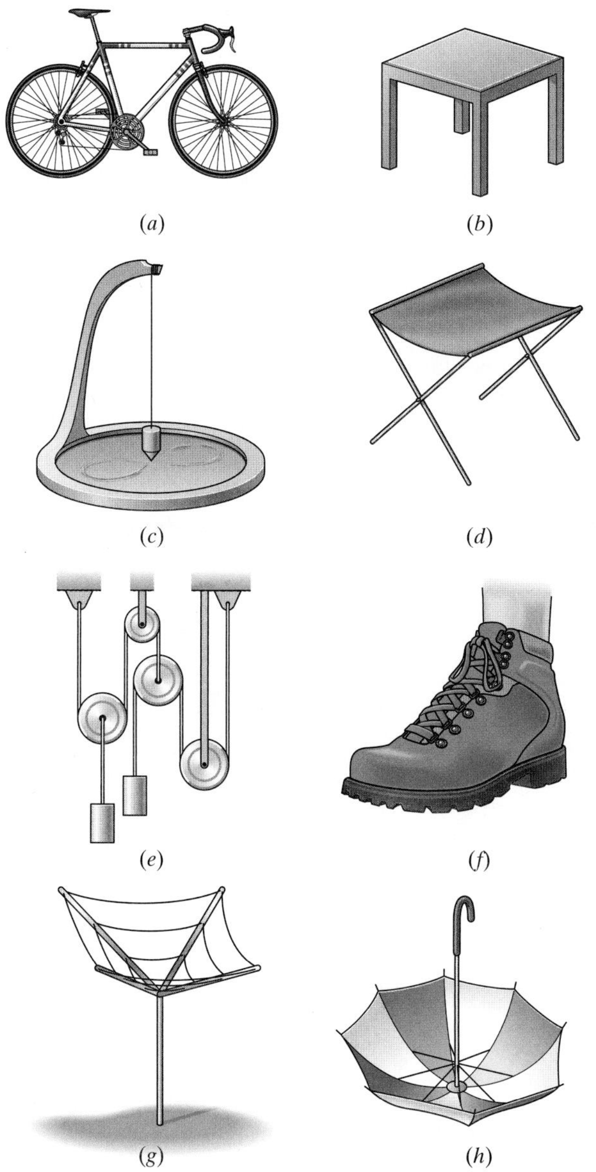

E4.3.1

4.3.2. Consider the situations depicted in E4.3.2*a*, *b*, and *c*. Identify at least five of the forces involved in each situation as a gravity force, normal contact force, friction force, fluid contact force, tension force, compression force, or shear force. (An example is given in each situation.)

a. Between the rider's bottom and the bicycle seat there are normal contact and friction forces. List at least five other forces involved.

b. There is normal force between cable AB and the metal eyelet at A. List at least five other forces involved.

c. There is fluid contact force (hot air) inside the balloon. List at least five other forces involved.

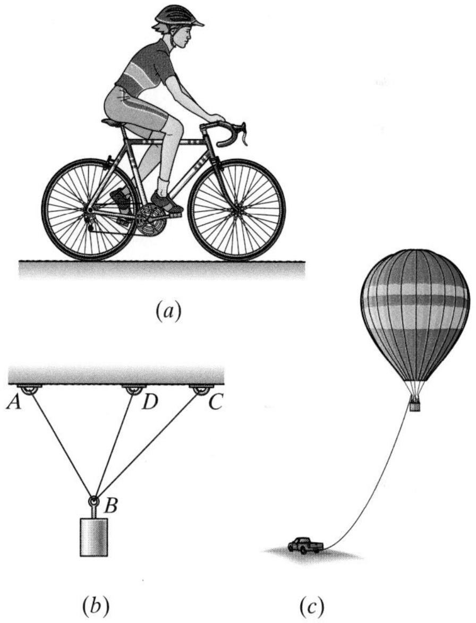

E4.3.2

4.3.3. Listed below are some real-world examples of forces. List three additional examples of each type of force:

a. Tension—in string being pulled by a kite, in a bungee cord when someone is bungee jumping

b. Normal contact—between a parked car and the road, between a gymnast's hand and the vault

c. Friction—between the roller ball in a mouse and the mouse pad, between bicycle brakes and the wheels

d. Fluid contact—the air in a balloon, the wind on a sail, the soda in a can

e. Compression—in the cookie dough under a rolling pin, in the road under a parked car, in the columns on the front of the White House

4.3.4. Friction forces and shear forces are parallel to interfaces. Describe how these forces are different from each other. Use examples to illustrate your description.

4.3.5. Normal contact forces, tension forces, and compression forces are perpendicular to interfaces. Describe how these forces are different from one another. Use examples to illustrate your description.

4.4 ANALYZING FORCES

Which Forces Are Important—Zooming In

The ability to identify one small part of the world that is relevant to some particular engineering problem and then to identify the forces acting on that part are key points when addressing questions concerning structural integrity and performance. We refer to the part of the world being studied as the **system** and to the forces acting ON the system as **external forces**. External forces may be any of the types we have discussed above—gravity, normal contact, friction, fluid contact, tension, compression, or shear. **Internal forces** are those that exist WITHIN the system in equal and opposite pairs (this is Newton's third law).

Whether a particular force we are looking at in a problem is external or internal depends on how the system is defined. For example, if we define our system as being two stacked books resting on a table (**Figure 4.17*a***), the external forces acting on the system are the weights of the two books and the normal contact force exerted by the table on the lower book. The normal contact forces between the two books (the push exerted by the lower book on the upper book and the equal and opposite push exerted by the upper book on the lower book) are internal forces (**Figure 4.17*b***). Because they are equal in magnitude and opposite in direction, the two members of any pair of internal forces cancel each other.

If, on the other hand, we define our system to be only the upper book, the external forces acting on the system now are the weight of the upper book and the normal contact force exerted by the lower book on the upper book (**Figure 4.17*c***). Finally, if we define our system to be only the lower book, the external forces acting on this system are the weight of the lower book and the normal contact forces exerted by upper book and by the table on the lower book (**Figure 4.17*d***). Notice that if we overlay **Figures 4.17*c*** and **4.17*d*** we get the two-book system in

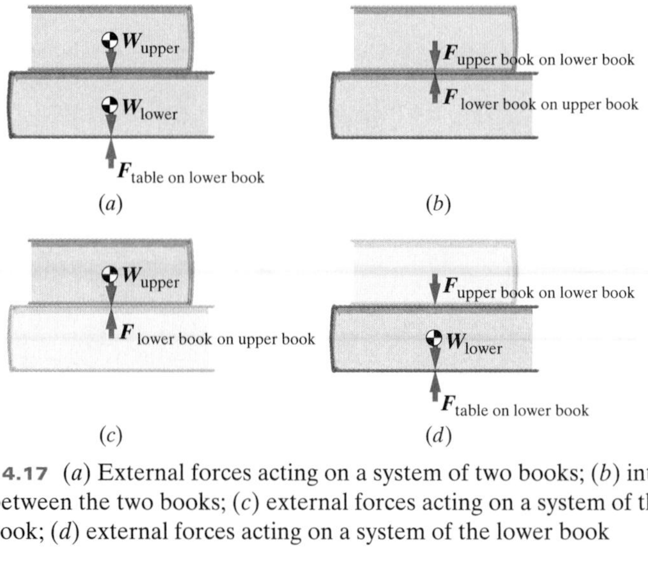

Figure 4.17 (*a*) External forces acting on a system of two books; (*b*) internal forces between the two books; (*c*) external forces acting on a system of the upper book; (*d*) external forces acting on a system of the lower book

Figure 4.17*a* (remember that the normal forces between the two books cancel each other).

Which forces are external and which are internal in a given situation is further illustrated in the following example.

Consider a person standing on a ladder that is leaning against a building. The ladder consists of two long stringers connected by eight rungs (**Figure 4.18**). Normal contact forces and friction forces exist between ladder and wall and between ladder and ground. Normal contact forces and friction forces are also present between the person's hands and one of the rungs and between the person's feet and another rung. Which of these forces we need to consider depends on what we want to know about the situation. For example:

Case 1: If we want to know whether the ladder feet will begin sliding away from the building (not a desirable state of affairs!), we could define our system to be person + ladder. The external forces acting on this system are the weights of the person and ladder and the normal contact forces and friction forces that the wall and ground exert on the ladder. The normal contact forces and friction forces between the person and the ladder are internal to our system. **Figure 4.19** depicts the external forces in Case 1.

Case 2: Alternatively, we could take the ladder alone as our system in determining whether the ladder will slide. The external forces acting on this system are the weight of the ladder, the normal contact forces and friction forces exerted by hands and feet on the rungs, and the normal contact forces and friction forces exerted by the wall and ground on the ladder. **Figure 4.20** depicts the external forces in Case 2.

Case 3: If we want to know whether the connections between a rung and the stringers are strong enough, we would take the system to be the rung on which the person is standing. The external forces

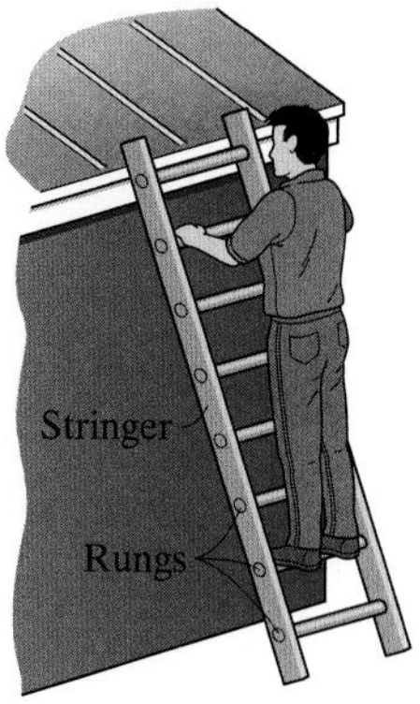

Figure 4.18 A person climbing a ladder

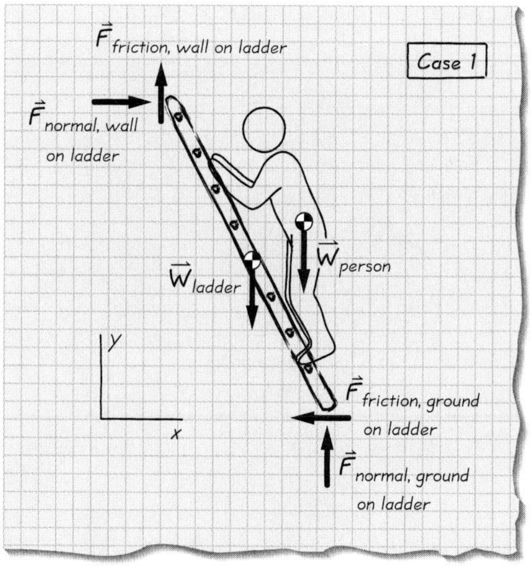

Figure 4.19 Case 1: External forces acting on the person-ladder system

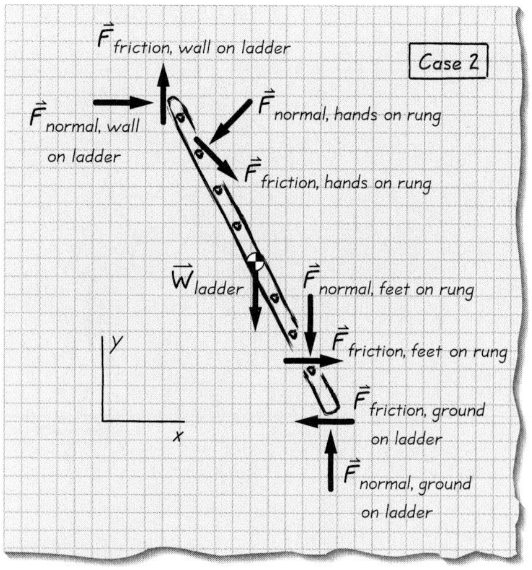

Figure 4.20 Case 2: External forces acting on the ladder system

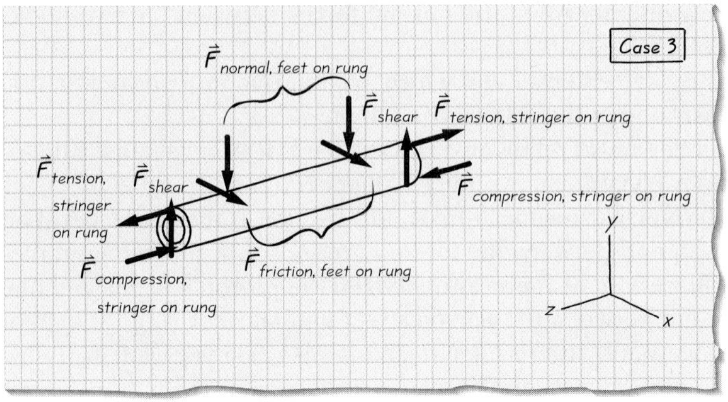

Figure 4.21 Case 3: External forces acting on the rung system

exerted on this rung are the normal contact forces and friction forces exerted by the feet plus the tension, compression, and shear forces that the stringers apply to the rung. **Figure 4.21** depicts the external forces in Case 3.

Case 4: If we want to know about the forces exerted on the person's lower back, we could begin by defining the system as the person (Case 4A). The external forces exerted on this system are the person's weight and the normal contact forces and friction forces exerted by the ladder on the hands and feet (**Figure 4.22*a***). After analyzing this system, we would analyze another system—the upper body of the person (Case 4B). The external forces exerted on this system (shown in **Figure 4.22*b***) are the weight of the upper body, the normal contact forces and friction forces exerted by the ladder on the hands, and the normal contact forces and tension forces exerted by the lower body on the upper body. It is these latter external forces that are carried by the lower back. This case illustrates that sometimes we may need to take an iterative approach to analysis, first looking at one system and then looking at a second system that is some portion of the first system.

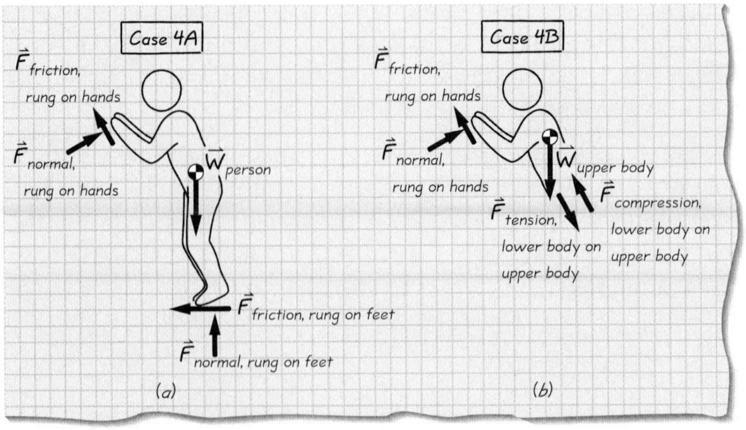

Figure 4.22 (*a*) Case 4A: External forces acting on the person system; (*b*) Case 4B: External forces acting on the upper body of the person

In each of these four cases, we zoomed in and isolated a part of the world that was relevant to the question we were asking. We called this part the system. We then identified the forces acting on that system and called these the external forces. It is usually most convenient to define your system so that the forces you are trying to find are external ones. As noted earlier, internal forces, because they come in pairs, cancel each other, and so we do not consider them when analyzing a system.

Notice that as we went from Case 1 to Case 4, forces that were internal to some systems became external to others. Once again, whether a force is external or internal depends on the system of interest.

General Steps for Analysis

The process of defining a system and then identifying external forces acting on that system is critical in evaluating how the system performs. We presented this process more formally in the Goal, Given, Assume . . . steps laid out in the Engineering Analysis Procedure overview of Section 1.1 (**Box 1.1**).

Zooming in and identifying external forces leads to the Draw step of the procedure. The drawing we commonly create is called a **free-body diagram**—*free* because an imaginary boundary around our system cuts off the system from the world around it, *body* because we have defined a specific system to focus on, and *diagram* to emphasize the importance of a visual representation. A free-body diagram uses vector arrows to represent the external forces acting on the system. Each force is shown on the diagram at its **point of application**; this is the point on the system where the force acts. **Figures 4.19** through **4.22** are examples of free-body diagrams. We will have a lot more to say about creating free-body diagrams in Chapter 6.

EXAMPLE 4.5 PRACTICE IN APPLYING THE ENGINEERING ANALYSIS PROCEDURE

A pallet of tiles sits on a roof with a 3:4 pitch (rise over run) and is held in place with a cable attached to the upper roof as shown in **Figure 4.23**. The pallet is sitting on loose tar paper, so it is reasonable to assume there is no friction force between the pallet and the roof.

(a) Draw a free-body diagram of the pallet–tile unit by following the analysis steps presented in Section 1.1 (**Box 1.1**).

(b) If the sum of the forces acting on the pallet is zero and the pallet and tiles have a combined weight of 200 lb, what is the magnitude of the force exerted by the cable on the pallet?

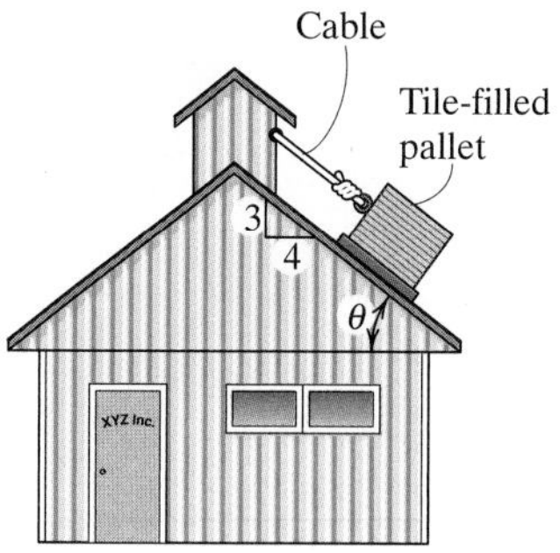

Figure 4.23

Goal We are asked first to (a) draw a free-body diagram of the pallet–tile unit and then (b) find the force exerted by the cable on this unit. Once we complete our task, we'll have a drawing and a numeric answer in pounds.

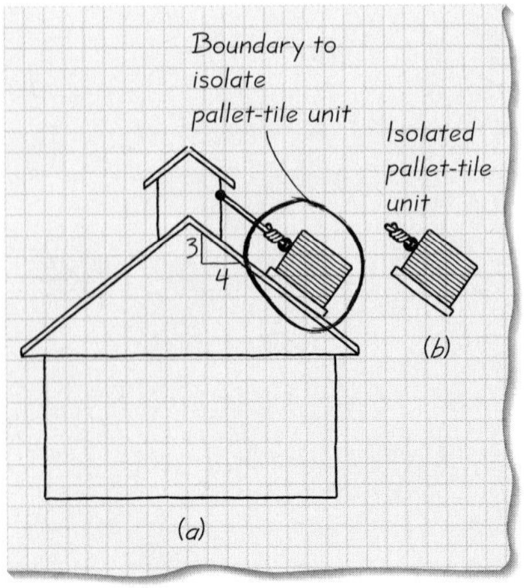

Figure 4.24

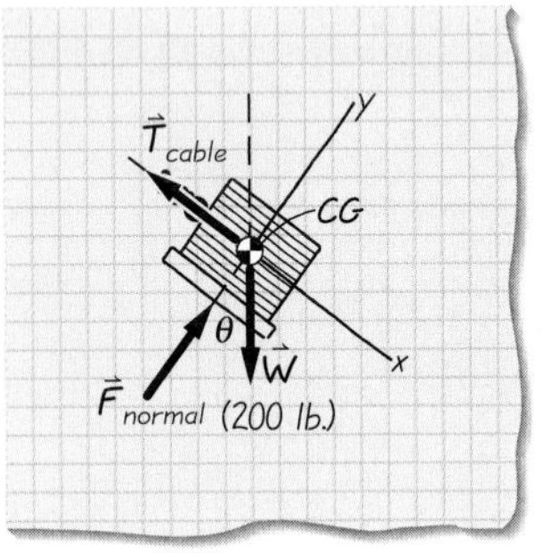

Figure 4.25

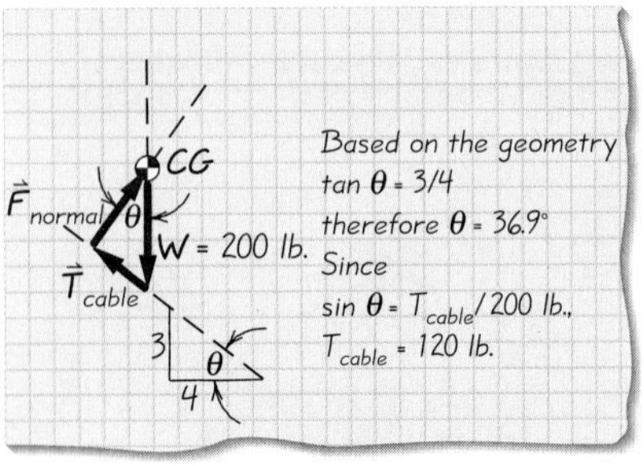

Figure 4.26

Given

- The ratio of the rise to the run of the roof is 3 : 4.
- Combined weight of the pallet and tiles is 200 lb.
- Sum of the forces acting on the pallet–tile unit is zero.

Assume We assume that:

1. The forces acting on the pallet–tile unit lie in a single plane.
2. The portion of the system of interest is the pallet–tile unit (isolated as shown in Figure 4.24*a*). Notice that the boundary cuts through the cable and runs between the pallet and the roof. **Figure 4.24*b*** is the isolated structure.
3. Since the tar paper is loose, there is no friction between the roof and the pallet. (We are actually told to assume this in the example statement.)
4. The center of gravity of the pallet is at its geometric center. (We would need more information to make any other assumption.)

Draw The external loads acting on the structure are the cable tension ($\boldsymbol{T}_{\text{cable}}$), the gravitational force acting on the pallet and tiles ($\boldsymbol{W}$), and the normal contact force exerted by the roof on the pallet ($\boldsymbol{F}_{\text{normal}}$).

See Figure 4.25 (which is the answer to (a)). In creating this drawing, we assumed that the normal contact force acts at the center of the bottom of the pallet. We have drawn each force in the direction we think it acts on the structure. Finally, we have placed a set of coordinate axes with the origin at the center of gravity. We could have oriented these axes horizontally and vertically, but as we will see in the next step, orienting them along the roof pitch will make force addition easier. **Figure 4.25** is a free-body diagram of the pallet–tile unit.

Answer to (a) See **Figure 4.25.**

Formulate Equations and Solve Apply the requirement that the sum of the forces is zero.

To find the magnitude $\| \boldsymbol{T}_{\text{cable}} \|$, we could use graphical force addition, trigonometry, or component addition; all three of these approaches are covered in detail in Section 4.6. Using trigonometry on the force triangle formed by $\boldsymbol{F}_{\text{normal}}$, $\boldsymbol{T}_{\text{cable}}$, and $\boldsymbol{W}$ (see Figure 4.26), we find that $\| \boldsymbol{T}_{\text{cable}} \| = 120$ lb.

Answer to (b) The tension in the cable is $\| \boldsymbol{T}_{\text{cable}} \| = 120$ lb.

Check One way to check your answer is to use two different approaches to adding forces, then compare the answers. Answers should also be checked for reasonableness; does it seem reasonable that the cable tension would be 120 lb for a pallet weighing 200 lb?

Comment: Would the presence of friction between the pallet–tile unit and the roof increase, decrease, or leave $\| \boldsymbol{T}_{\text{cable}} \|$ unchanged?

EXERCISES 4.4

4.4.1. A person grips a pair of locking jaw pliers (more commonly known as vise-grips), as shown in E4.4.1, in order to tighten a nut onto a bolt. Three cases (systems) are described in the top row of Table 4.4.1. For each system, identify the external (E) and internal (I) forces by completing **Table 4.4.1**. Leave blank those squares associated with a force not applied to the system under consideration.

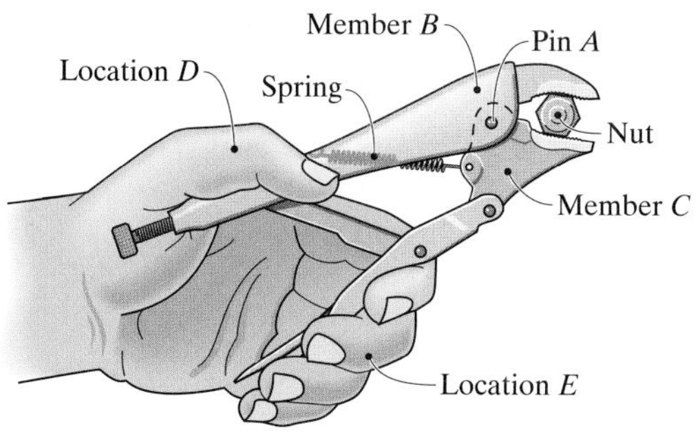

E4.4.1

4.4.2. A cyclist is bicycling down the road. Five cases (systems) are described in the top row of Table 4.4.2 and in E4.4.2. For each system, identify the external (E) and internal (I) forces by completing **Table 4.4.2**. Leave blank the squares associated with a force not applied to the system under consideration.

E4.4.2

Table 4.4.1

Forces:	**Case 1** Vise-grips and nut defined as system.	**Case 2** Vise-grips defined as system.	**Case 3** Nut defined as system.
Tension in spring			
Normal contact force between pin *A* and member *B*			
Normal contact force between nut and member *B*			
Normal contact force between nut and member *C*			
Normal contact and friction forces between thumb and vise-grips at *D*			
Normal and friction forces between fingers and vise-grips at *E*			
Weight of vise-grips			
Weight of nut			
Weight of bolt			
Normal contact and friction forces between nut and bolt			

Table 4.4.2

Forces	**Case 1** (*a*) Cyclist (including back-pack) and bicycle defined as system.	**Case 2** (*b*) Cyclist (not including back-pack) and bicycle defined as system.	**Case 3** (*c*) Bicycle defined as system.	**Case 4** (*d*) Cyclist defined as system.	**Case 5** (*e*) Backpack defined as system.
Chain tension					
Weight of bicycle					
Weight of cyclist					
Weight of backpack					
Normal contact and friction forces between rider's hands and handlebars					
Normal contact and friction forces between rider's seat and bicycle seat					
Normal contact force where rider's foot presses on pedal					
Normal contact force between front fork and front wheel hub					
Normal contact force between cyclist's back and backpack					
Tension in backpack shoulder strap					
Normal contact forces between tires and road					
Wind force acting on bicycle					
Wind force acting on cyclist					
Tension in cyclist's back muscles					

4.5 MAGNITUDE AND DIRECTION DEFINE A FORCE

Working with free-body diagrams involves representing and manipulating forces. Therefore we now consider how to formally work with the vector quantity of force. Although this section and the next are framed in terms of force, our comments apply to any vector quantity.

Suppose you are asked to remove a tent stake from the ground, as shown in **Figure 4.27*a***. How would you pull on the rope? It is likely you would pull with both arms so that the rope was aligned with the long axis of the stake and the magnitude of your arm pull would be about 50% of your weight.[9] We can represent this pulling force graphically (**Figure 4.27*b***), but we can also represent it mathematically. We now present several math-based approaches to representing the magnitude and direction of a force.

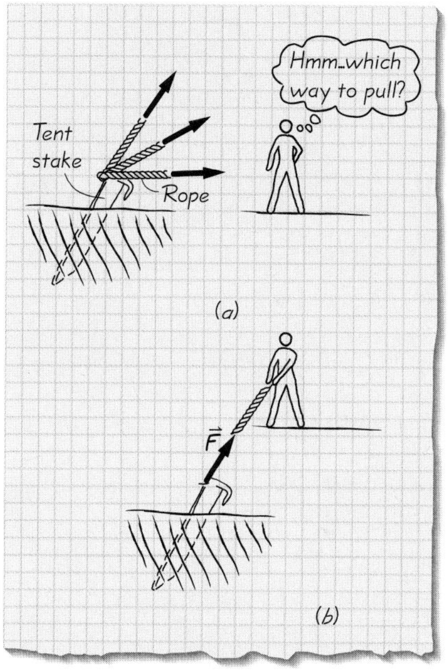

Figure 4.27 (*a*) Person considering how to pull on a stake; (*b*) person pulls along the stake's long axis

Magnitude-Angle Representation with Space Angles

Consider a force $\boldsymbol{F}$ of known magnitude $\|\boldsymbol{F}\|$. The direction of this force can be specified by its angular orientation relative to a set of right-handed coordinate axes. We specify its angular orientation with the angles θ_x, θ_y, and θ_z from the x, y, and z axes, respectively, as illustrated in **Figure 4.28**, and refer to them as the **space angles**. They can be specified either in degrees or in radians, and their numeric values can generally be found from the geometry of the situation. For example, in **Figure 4.29*a*** a cable runs from a hook at A, around a pulley at B, and to another hook at C. If we are given the coordinates of A and B, we can find the space angles of the cable force acting on the hook at A (**Figure 4.29*b***) from these relationships:

$$\theta_x = \cos^{-1}\left(\frac{x_2 - x_1}{L}\right)$$
$$\theta_y = \cos^{-1}\left(\frac{y_2 - y_1}{L}\right) \tag{4.4}$$
$$\theta_z = \cos^{-1}\left(\frac{z_2 - z_1}{L}\right)$$

where $L = \sqrt{(x_2 - x_1)^2 + (y_2 - y_1)^2 + (z_2 - z_1)^2}$

$L = \overline{AB}$

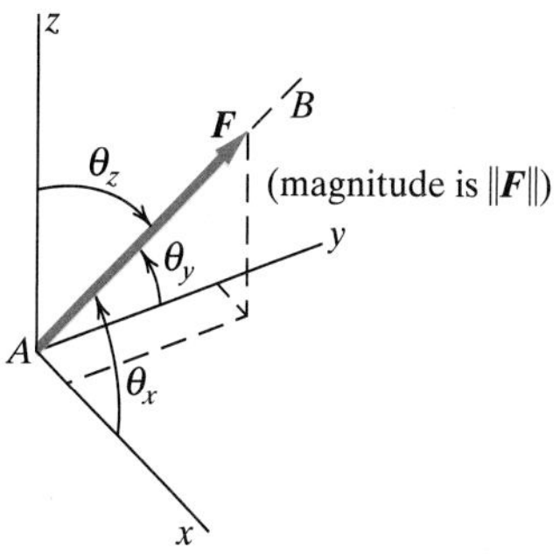

Figure 4.28 Space angles defined

When working with space angles, consider that

1. The angles θ_x, θ_y, and θ_z are not independent of one another. They are related by the expression

$$\sqrt{(\cos\theta_x)^2 + (\cos\theta_y)^2 + (\cos\theta_z)^2} = 1 \tag{4.5A}$$

where $\cos\theta_x$, $\cos\theta_y$, and $\cos\theta_z$ are the **direction cosines** and are defined (based on 4.4)) as

$$\cos\theta_x = \left(\frac{x_2 - x_1}{L}\right)$$
$$\cos\theta_y = \left(\frac{y_2 - y_1}{L}\right) \tag{4.5B}$$
$$\cos\theta_z = \left(\frac{z_2 - z_1}{L}\right)$$

Therefore, if we know two of the space angles, the third angle is automatically defined by (4.5A).

[9]Based on data gathered on nine college sophomores in October 2002.

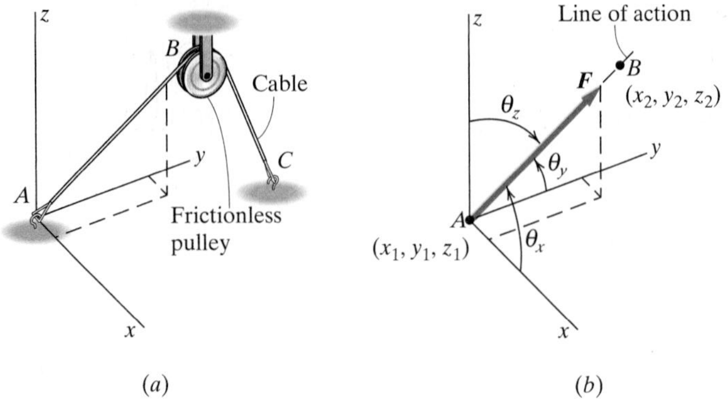

Figure 4.29 (*a*) A cable passes around a frictionless pulley; (*b*) the cable force acting at *A*

2. The space angles are always defined as positive angles between zero and 180°.

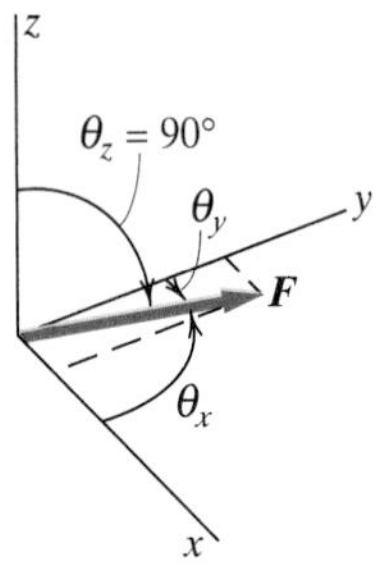

Figure 4.30 Space angles for a force in the *xy* plane

A Note on Planar and Nonplanar Forces. If a force is in the plane of two of the coordinate axes, one of the space angles is 90°. For example, if a force is in the *xy* plane, the angle θ_z between the *z* axis and the force is 90°, as depicted in **Figure 4.30**. We call a force in the plane of two of the reference axes either a **planar force** or a **two-dimensional force**; otherwise it is called either a **nonplanar force** or a **three-dimensional force**. For a planar force in the *xy* plane, (4.5A) simplifies to

$$\sqrt{(\cos\theta_x)^2 + (\cos\theta_y)^2} = 1 \tag{4.6}$$

since $\theta_z = 90°$ and therefore $\cos\theta_z = 0$.

Examples of planar and nonplanar forces specified with space angles are shown in **Figures 4.31*a*** and **4.31*b***, respectively.

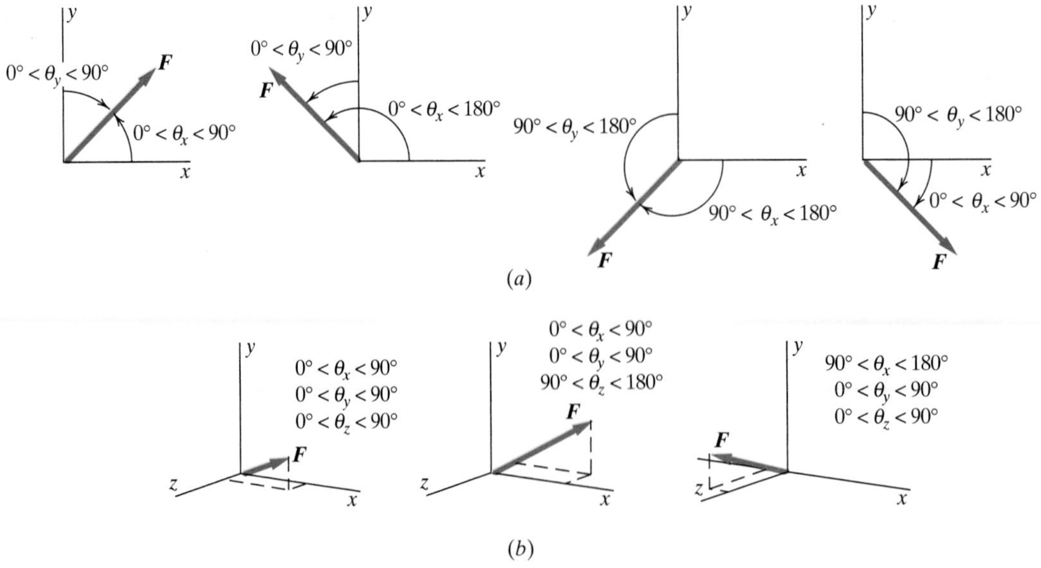

Figure 4.31 (*a*) Examples of planar forces with $\theta_z = 90°$; (*b*) examples of three-dimensional forces

Magnitude-Angle Representation with Spherical Coordinate Angles

Another way to describe the direction of a force $\boldsymbol{F}$ is to use the two angles ϕ and θ associated with spherical coordinates, as illustrated in Figure 4.32*a*. The angle θ defines the sweep from the x axis to the projection of $\boldsymbol{F}$ onto the xy plane. (You can think of the projection as the "shadow" that $\boldsymbol{F}$ would cast on the xy plane if a light source were sitting far out on the z axis. This shadow is shown as a slender line in the figure.) The angle ϕ defines the sweep from the z axis to $\boldsymbol{F}$, and you may recognize that it is the space angle θ_z. The angles ϕ and θ can be specified either in degrees or in radians.

When working with angles ϕ and θ, consider that:

1. The angle θ is a positive angle between zero and 360° ($0 \leq \theta \leq 2\pi$), measured counterclockwise from the x axis. The angle ϕ is between zero and 180° ($0 \leq \phi \leq \pi$), and is measured from the z axis. An example of a force directed with θ between zero and 90° and ϕ between 90° and 180° is shown in **Figure 4.32*b***.
2. The angles ϕ and θ and the space angles θ_x, θ_y, and θ_z are related to one another by the expressions

$$\begin{aligned} \cos\theta_x &= (\sin\phi)(\cos\theta) \\ \cos\theta_y &= (\sin\phi)(\sin\theta) \\ \cos\theta_z &= \cos\phi \end{aligned} \tag{4.7A}$$

$$\begin{aligned} \cos\phi &= \cos\theta_z \\ \tan\theta &= \frac{\cos\theta_y}{\cos\theta_x} \end{aligned} \tag{4.7B}$$

If the force is in the xy plane (and therefore is a planar force), ϕ is 90°. Examples of planar forces are illustrated in Figure 4.33.

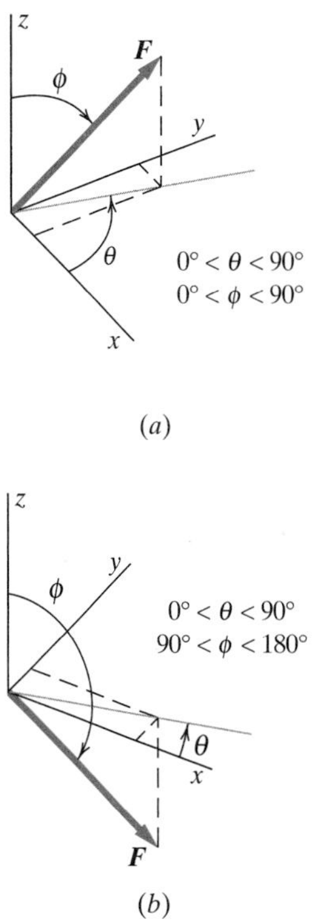

Figure 4.32 (*a*) Spherical angles (θ, ϕ) defined; (*b*) another example of spherical angles defining the direction of a force

Rectangular Component Representation

Another way of representing the magnitude and direction of a force $\boldsymbol{F}$ is to specify its three **rectangular component vectors $\boldsymbol{F}_x$, $\boldsymbol{F}_y$,** and $\boldsymbol{F}_z$ relative to a right-handed coordinate system. This is the same as specifying the "hike" you would take in the x, y, and z directions to get from the

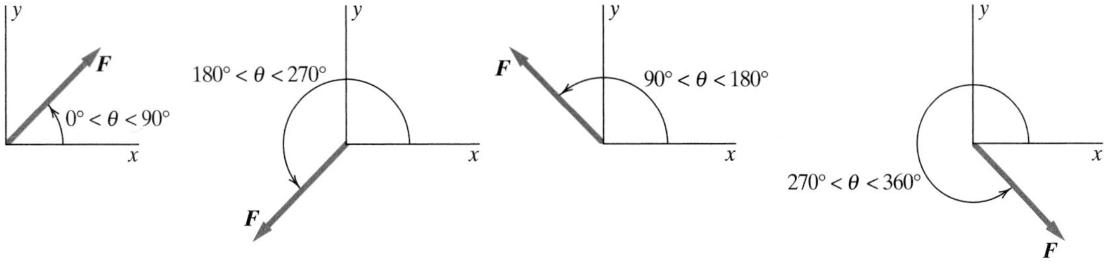

Figure 4.33 Examples of planar forces defined in terms of spherical angles. All of these forces are in the xy plane; therefore ϕ = 90°.

tail of the force vector to its head (**Figure 4.34*a***). The force is the vector sum of its component vectors:

$$\boldsymbol{F} = \boldsymbol{F}_x + \boldsymbol{F}_y + \boldsymbol{F}_z \tag{4.8A}$$

This equation can be rewritten in terms of the unit vectors $\boldsymbol{i}, \boldsymbol{j}$, and $\boldsymbol{k}$ directions (unit vectors aligned with the x, y, and z axes) as

$$\boldsymbol{F} = F_x\boldsymbol{i} + F_y\boldsymbol{j} + F_z\boldsymbol{k} \tag{4.8B}$$

where F_x, F_y, and F_z are the **scalar components** of $\boldsymbol{F}$. Each scalar component tells us, for a particular direction, the length of a leg of the "hike" described in **Figure 4.34*a***. Unlike magnitude, a scalar component has a sign associated with it.

The scalar components F_x, F_y, and F_z are the projections of $\boldsymbol{F}$ onto the x, y, and z axes, respectively. If we know the space angles $(\theta_x, \theta_y, \theta_z)$ and magnitude $\|\boldsymbol{F}\|$ of $\boldsymbol{F}$, we can write these projections as:

$$F_x = \|\boldsymbol{F}\| \cos\theta_x \tag{4.9A}$$
$$F_y = \|\boldsymbol{F}\| \cos\theta_y \tag{4.9B}$$
$$F_z = \|\boldsymbol{F}\| \cos\theta_z \tag{4.9C}$$

(see **Figure 4.34*b***). We now write $\boldsymbol{F}$ in terms of its component vectors by substituting from (4.9A–C) into (4.8B):

$$\boldsymbol{F} = \|\boldsymbol{F}\| \cos\theta_x\boldsymbol{i} + \|\boldsymbol{F}\| \cos\theta_y\boldsymbol{j} + \|\boldsymbol{F}\| \cos\theta_z\boldsymbol{k} \tag{4.10A}$$

which can be rearranged to

$$\boldsymbol{F} = \|\boldsymbol{F}\| (\cos\theta_x\boldsymbol{i} + \cos\theta_y\boldsymbol{j} + \cos\theta_z\boldsymbol{k}) \tag{4.10B}$$

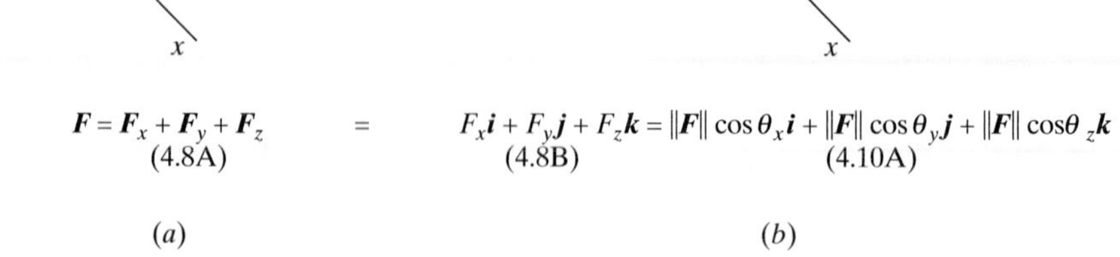

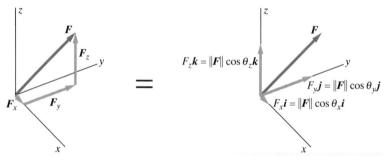

Figure 4.34 (*a*) $\boldsymbol{F}$ defined in terms of rectangular components $\boldsymbol{F}_x$, $\boldsymbol{F}_y$, and $\boldsymbol{F}_z$; (*b*) the rectangular components are defined in terms of unit vectors $\boldsymbol{i}, \boldsymbol{j}$, and $\boldsymbol{k}$, and in terms of the space angles

The term in parentheses defines a unit vector that is aligned with the line of action of $\boldsymbol{F}$. We call this unit vector $\boldsymbol{u}$, where

$$\boldsymbol{u} = \cos\theta_x\boldsymbol{i} + \cos\theta_y\boldsymbol{j} + \cos\theta_z\boldsymbol{k} \tag{4.10C}$$

as illustrated in **Figure 4.35**. Therefore (4.10B) can be rewritten in terms of $\boldsymbol{u}$:

$$\boldsymbol{F} = \|\boldsymbol{F}\|(\cos\theta_x\boldsymbol{i} + \cos\theta_y\boldsymbol{j} + \cos\theta_z\boldsymbol{k}) = \|\boldsymbol{F}\|\boldsymbol{u} \tag{4.11}$$

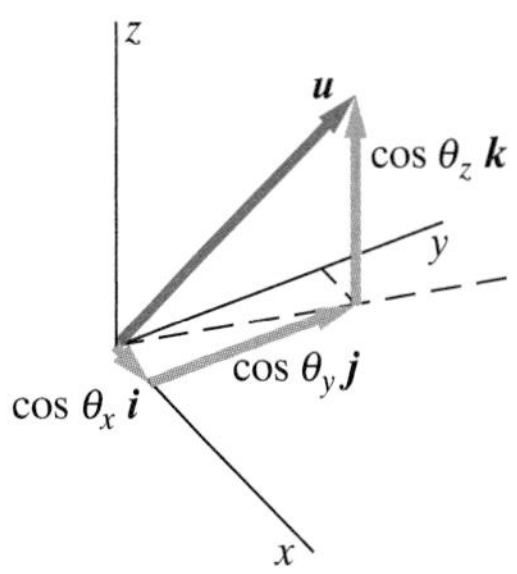

Figure 4.35 Rectangular components of the unit vector $\boldsymbol{u}$ aligned with $\boldsymbol{F}$

Equations (4.8A) to (4.11) are concerned with writing $\boldsymbol{F}$ in terms of **vector notation**. This means writing $\boldsymbol{F}$ in terms of its components in a right-handed coordinate system. Each rectangular component vector ($\boldsymbol{F_x}$, $\boldsymbol{F_y}$, $\boldsymbol{F_z}$) that makes up $\boldsymbol{F}$ consists of a scalar component (F_x, F_y, F_z) and a unit vector ($\boldsymbol{i}, \boldsymbol{j}, \boldsymbol{k}$). Each scalar component can be written in terms of the space angles (θ_x, θ_y, θ_z). Furthermore, the space angles can be combined into a unit vector that is aligned with $\boldsymbol{F}$ (we called this unit vector $\boldsymbol{u}$). Multiplying this unit vector $\boldsymbol{u}$ by $\|\boldsymbol{F}\|$ (the magnitude of $\boldsymbol{F}$) gives us another form of $\boldsymbol{F}$. All of the expressions for $\boldsymbol{F}$ presented above are equivalent. Which form you use in calculations will largely be dictated by the information available.

In engineering analysis, you are just as likely to be required to combine component vectors $\boldsymbol{F_x}$, $\boldsymbol{F_y}$, $\boldsymbol{F_z}$ into their resultant force $\boldsymbol{F}$ as to decompose $\boldsymbol{F}$ into its component vectors $\boldsymbol{F_x}$, $\boldsymbol{F_y}$, $\boldsymbol{F_z}$. The scalar components can be combined to determine the magnitude of the resultant $\|\boldsymbol{F}\|$:

$$\|\boldsymbol{F}\| = \sqrt{F_x^2 + F_y^2 + F_z^2} \tag{4.12}$$

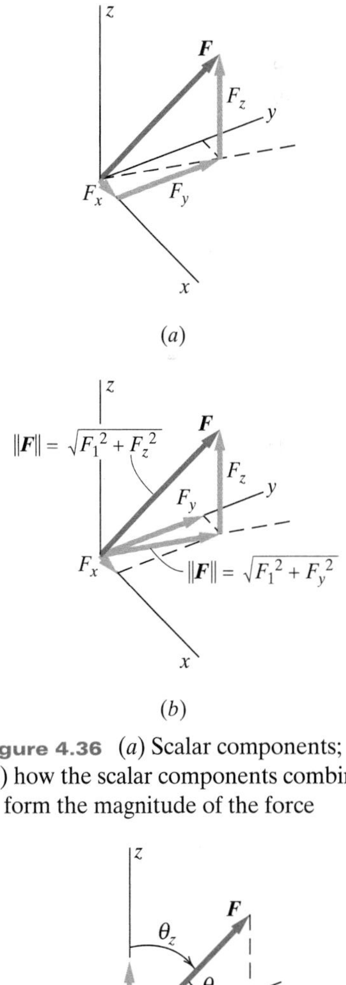

Figure 4.36 (*a*) Scalar components; (*b*) how the scalar components combine to form the magnitude of the force

as illustrated in **Figure 4.36**. Equation (4.12) says that the magnitude of $\boldsymbol{F}$ is equal to the positive square root of the sum of the squares of its scalar components.

It is also common in engineering analysis to use the scalar components F_x, F_y, F_z of a force to find the direction cosines of the force. Rearranging the expression in (4.9), we find

$$\begin{aligned} \cos\theta_x &= \frac{F_x}{\|\boldsymbol{F}\|} \\ \cos\theta_y &= \frac{F_y}{\|\boldsymbol{F}\|} \\ \cos\theta_z &= \frac{F_z}{\|\boldsymbol{F}\|} \end{aligned} \tag{4.13}$$

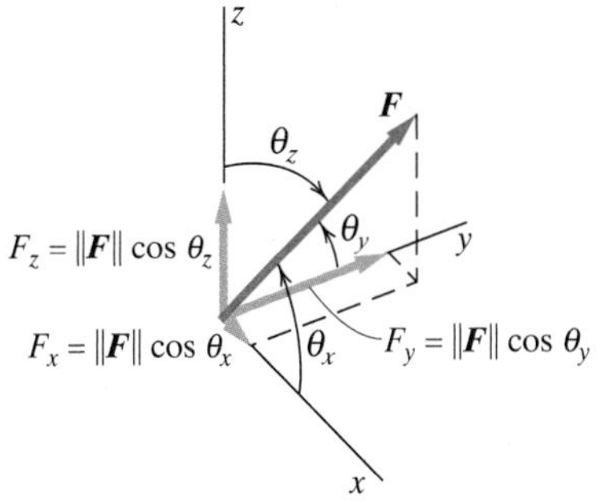

Figure 4.37 Illustration of (4.13)

where $\|\boldsymbol{F}\|$ is given in (4.12). The relationships given in (4.13) are shown graphically in **Figure 4.37**.

You will use (4.4) to (4.13) in various ways as you work with forces and free-body diagrams. Sometimes the scalar force components will be known, and you will be interested in finding the magnitude of the resultant

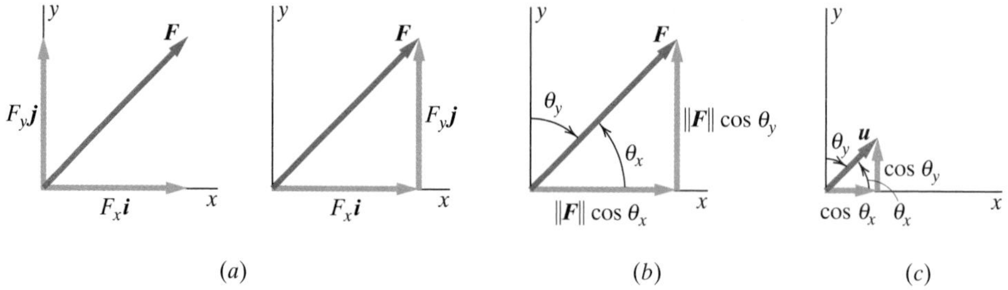

Figure 4.38 Illustration of rectangular components for a planar force

force (where (4.12) will come in handy). At other times, the force direction and magnitude will be known, and you will need to find the scalar components (using (4.9)). Either way, you need to feel comfortable manipulating forces and their components. If you understand the principles behind the various ways of representing a force in terms of its components, finding magnitudes and directions will become straightforward with a little practice.

A Note on Planar and Nonplanar Forces. The discussion of rectangular representation of forces up to this point has been in terms of nonplanar forces. If, on the other hand, a force is planar (meaning that it lies in the plane of two of the coordinate axes), then (4.8B), (4.10A), and (4.10C) can be simplified. For example, if the force is contained in the xy plane, $F_z = 0$, $\theta_z = 90°$, and $\cos\theta_z = 0$. Therefore, (4.8B), (4.10A), and (4.10C) simplify to

$$\boldsymbol{F} = F_x\boldsymbol{i} + F_y\boldsymbol{j} \quad \text{(planar version of 4.8B)} \quad (4.14A)$$

$$\boldsymbol{F} = \|\boldsymbol{F}\| \cos\theta_x\boldsymbol{i} + \|\boldsymbol{F}\| \cos\theta_y\boldsymbol{j} \quad \text{(planar version of 4.10A)} \quad (4.14B)$$

$$\boldsymbol{u} = \cos\theta_x\boldsymbol{i} + \cos\theta_y\boldsymbol{j} \quad \text{(planar version of 4.10C)} \quad (4.14C)$$

as illustrated in Figure 4.38. Furthermore, for a planar force in the xy plane, $\cos\theta_y = \pm\sin\theta_x$ (the plus or minus sign depending on which quadrant the force lies in).

EXAMPLE 4.6 REPRESENTING PLANAR FORCES (A)

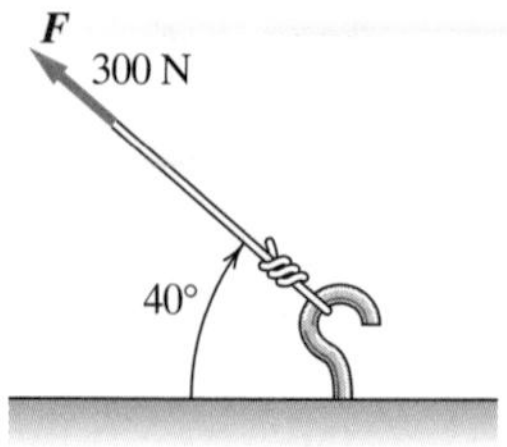

Figure 4.39

A cable force $\boldsymbol{F}$ acts on a hook, as shown in Figure 4.39. Describe the force relative to horizontal and vertical axes in terms of

(a) its direction cosines
(b) a unit vector along its line of action
(c) vector notation
(d) its scalar components
(e) the angles ϕ and θ

Goal We are to find the direction cosines (a), unit vector (b), force in vector notation (c), scalar components (d), and spherical angles (e) associated with the force $\boldsymbol{F}$ shown in **Figure 4.39**.

Given We are given the magnitude of the force (300 N) and that it is oriented at 40° relative to the horizontal axis (as shown in **Figure 4.39**).

Assume We will assume that $\boldsymbol{F}$ is a planar force (in the plane of the paper) because we are not given any information about the out-of-plane direction.

Draw We create a drawing of the force and hook, defining x along the horizontal axis and y along the vertical axis (**Figure 4.40**). We arbitrarily place the origin at the center of the hook along the line of action of the force. We draw $\boldsymbol{F}$, label its magnitude, and show the 40° angle. The angles θ_x and θ_y are also identified in the figure. Since the force is assumed to be planar and in the xy plane, $\theta_z = 90°$.

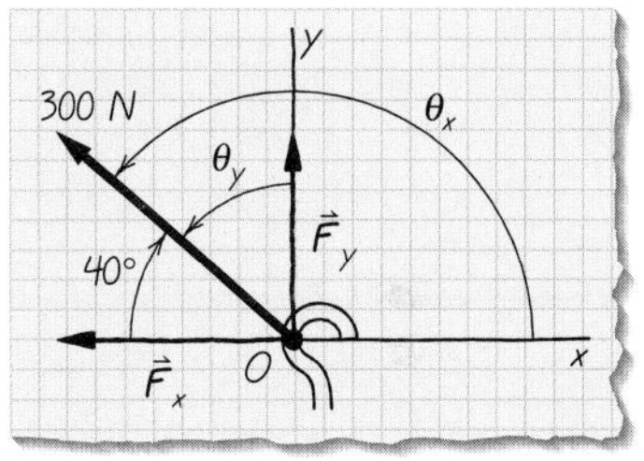

Figure 4.40

Formulate Equations, Solve, and Check **(a)** We can read off the angles θ_x and θ_y from **Figure 4.40** and take the cosines:

θ_x, the angle between the positive x axis and $\boldsymbol{F}$, is 140°; therefore $\cos\theta_x = \cos 140° = -0.766$.

θ_y, the angle between the positive y axis and the force, is $140° - 90° = 50°$; therefore $\cos\theta_y = 50° = 0.643$.

θ_z, the angle between the positive z axis and the force, is 90° (because the force is in the xy plane); therefore $\cos\theta_z = \cos 90° = 0$.

Answer to (a) $\cos\theta_x = -0.766$,
$\cos\theta_y = 0.643$,
$\cos\theta_z = 0$

We can check that the direction cosines obey (4.5A): $\sqrt{(-0.766)^2 + (0.643)^2 + (0)^2} = 1$

(b) By (4.10C), a unit vector $\boldsymbol{u}$ along the line of action of a force is

$$\boldsymbol{u} = \cos\theta_x\boldsymbol{i} + \cos\theta_y\boldsymbol{j} + \cos\theta_z\boldsymbol{k}$$

We then use the answer for (a) to write $\boldsymbol{u} = -0.766\boldsymbol{i} + 0.643\boldsymbol{j}$.

Answer to (b) $\boldsymbol{u} = -0.766\boldsymbol{i} + 0.643\boldsymbol{j}$

We can check that the magnitude of $\boldsymbol{u}$ is indeed unity: $\sqrt{(-0.766)^2 + (0.643)^2 + (0)^2} = 1$. A qualitative check involves observing that the x component of $\boldsymbol{u}$ ($= -0.766\boldsymbol{i} + 0.643\boldsymbol{j}$) is in the negative x direction and the y component is in the positive y direction, and that the x component (in terms of magnitude) is slightly larger than the y component. Inspection of **Figure 4.39** or **4.40** shows that these observations are consistent with the physical problem.

(c) Because $\boldsymbol{F}$ lies in the xy plane, its z component vector is zero ($\boldsymbol{F}_z = 0$). Its two nonzero component vectors are $\boldsymbol{F}_x$ and $\boldsymbol{F}_y$. We have added these two component vectors to our drawing (**Figure 4.40**).

It is not uncommon for there to be more than one approach to setting up and solving equations. The selection of an approach is often based on personal preference. As you will see, some approaches involve less work than others. Regardless of which approach you use, you should get the same answer.

Approach 1: Based on answer (a), we can write that $\theta_x = 140°$, $\theta_y = 50°$, $\theta_z = 90°$, and therefore by (4.9)

$$F_x = \| \boldsymbol{F} \| \cos 140° = (300 \text{ N})(-0.766) = -230 \text{ N}$$
$$F_y = \| \boldsymbol{F} \| \cos 50° = (300 \text{ N})(0.643) = 193 \text{ N}$$
$$F_z = \| \boldsymbol{F} \| \cos 90° = (300 \text{ N})(0) = 0 \text{ N}$$

Therefore, based on (4.8A), $\boldsymbol{F} = -230 \text{ N}\boldsymbol{i} + 193 \text{ N}\boldsymbol{j}$.

Approach 2: Based on (4.11), the product of the magnitude of $\boldsymbol{F}$ and the unit vector $\boldsymbol{u}$ (found in (b)) expresses the force in terms of its rectangular components.

$$\boldsymbol{F} = \| \boldsymbol{F} \| \boldsymbol{u}$$
$$\boldsymbol{F} = (300 \text{ N})(-0.766\boldsymbol{i} + 0.643\boldsymbol{j})$$
$$\boldsymbol{F} = -230 \text{ N}\boldsymbol{i} + 193 \text{ N}\boldsymbol{j}$$

Approach 3: Drawing $\boldsymbol{F}$ to scale and in its proper orientation, we can then read off the values of F_x and F_y (**Figure 4.41**) as −230 N and 193 N, respectively. The accuracy of this graphical approach depends on the precision of the drawing.

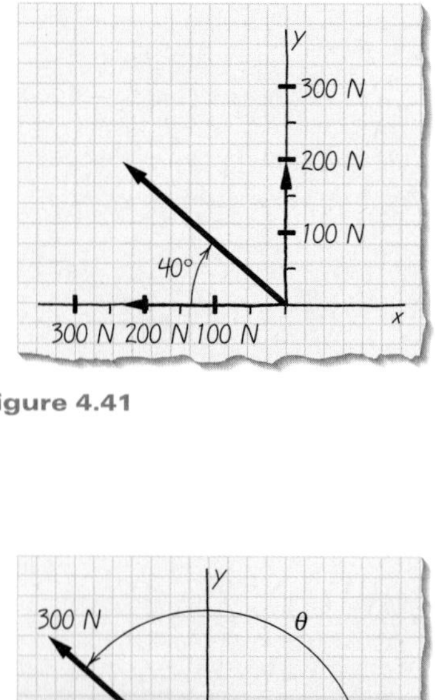

Figure 4.41

Answer to (c) (from Approaches 1, 2, and 3) $\boldsymbol{F} = -230 \text{ N}\boldsymbol{i} + 193 \text{ N}\boldsymbol{j}$

This answer can be checked by calculating the magnitude of $\boldsymbol{F}$ from its components. Using (4.12), we find $\sqrt{(-230 \text{ N})^2 + (193 \text{ N})^2 + (0)^2} = 300$ N, which is the magnitude of $\boldsymbol{F}$ given in the problem statement. A less quantitative check would be to look at the relative sizes of $\| \boldsymbol{F}_x \|$ and $\| \boldsymbol{F}_y \|$; both are smaller than $\| \boldsymbol{F} \|$ (which is to be expected), and $\| \boldsymbol{F}_x \| > \| \boldsymbol{F}_y \|$ (which is also to be expected because we are dealing with an angle that, relative to the horizontal, is less than 45°).

(d) We do not need to formulate and solve any addition equations to find the scalar components of $\boldsymbol{F}$. We can simply take them from the answer for (c).

Answer to (d) $F_x = -230 \text{ N}$, $F_y = 193 \text{ N}$, $F_z = 0 \text{ N}$

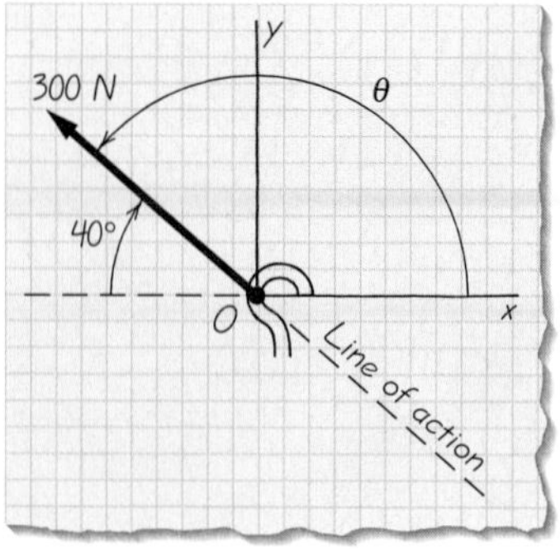

Figure 4.42

(e) We do not need to formulate and solve any additional equations to find the spherical coordinate angles ϕ and θ. A drawing that shows these answers is useful (**Figure 4.42**). Because $\boldsymbol{F}$ is a planar force, $\phi = 90°$. The angle θ is measured from the positive x axis. From the geometry shown in the figure, we see that $\theta = 180° - 40° = 140°$.

The magnitude of the force is given in the figure: $\|\boldsymbol{F}\| = 300$ N

Answer to (e) $\theta = 140°$, $\phi = 90°$, $\|\boldsymbol{F}\| = 300$ N

Alternately, and as a check, (4.7B) could be used to convert from the space angles θ_x, θ_y and θ_z found in part (a) to the angles ϕ and θ.

Comment: The force $\boldsymbol{F}$ we are dealing with in this example is a force in a coordinate plane (in this case, the xy plane) and so is a planar force. Therefore, we could have used the vector notation equations for a planar force represented in (4.14A–D). We chose not to do this and instead worked with the equations for nonplanar force representation, noting where some terms are zero because the force is in the xy plane The advantage of this approach is that with one general set of equations (4.4)–(4.13) we can write any force in vector notation, regardless of whether it is nonplanar or planar in any of the three coordinate planes.

EXAMPLE 4.7 REPRESENTING PLANAR FORCES (B)

A force $\boldsymbol{F}$, applied via a frictionless pulley, acts on a bracket as shown in **Figure 4.43**.

(a) Resolve the 500-N force into its scalar components in the x and y directions. Also represent $\boldsymbol{F}$ in vector notation.

(b) Repeat (a) for the x^* and y^* directions shown in **Figure 4.43**.

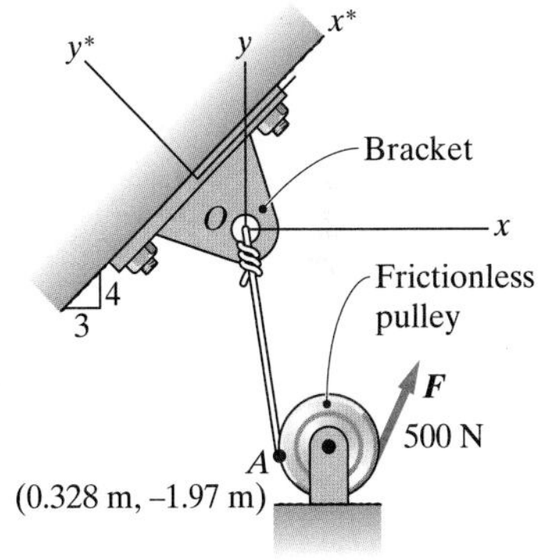

Figure 4.43

Goal We are to represent the force $\boldsymbol{F}$ in terms of scalar components relative to the x and y axes (a) and the x^* and y^* axes (b).

Given: We are given coordinates along the line of action of the force where it acts on the bracket (the origin and point A). In addition, we are given the 4/3 orientation of the bracket and the magnitude of the force (500 N), and are told that the pulley is frictionless.

Assume Since the problem statement gives no indication that $\boldsymbol{F}$ is out of the plane of the paper, we will assume that $\boldsymbol{F}$ lies in the xy plane; therefore, it is a planar force and $F_z = 0$. Since the pulley is frictionless, there is a 500-N force along the length of the cable (we will prove this in Chapter 7). Therefore the cable applies a 500-N force to the bracket.

Draw We draw the force applied to the bracket in **Figure 4.44**. Also shown are its scalar components F_x and F_y.

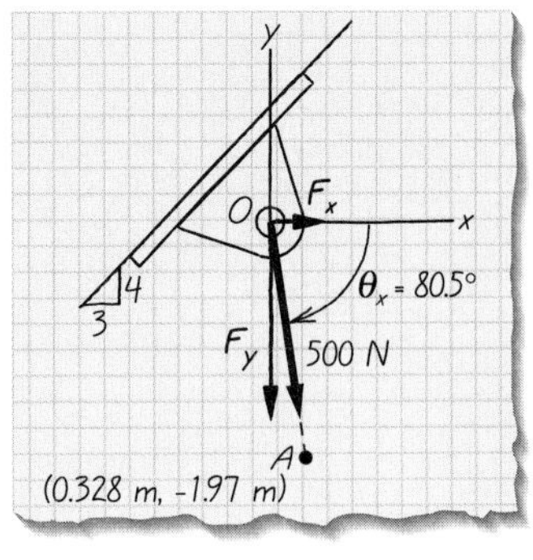

Figure 4.44

Formulate Equations, Solve, and Check **(a)** Based on the coordinates on the line of action of $\boldsymbol{F}$, we use (4.10C) to define a unit vector $\boldsymbol{u}$ along the line of action. We will need to know the distance between the origin (O) and A; it is

$$L_{OA} = \sqrt{(0.328 - 0)^2 + (-1.97 - 0)^2}\text{ m} = 2.00\text{ m}$$

The direction cosines of the unit vector $\boldsymbol{u}$, as defined in (4.5B), are then

$$\cos\theta_x = \frac{(0.328 - 0)\text{ m}}{2.00\text{ m}} = 0.164$$

$$\cos\theta_y = \frac{(-1.97 - 0)\text{ m}}{2.00\text{ m}} = -0.985$$

$$\cos\theta_z = \frac{(0 - 0)\text{ m}}{2.00\text{ m}} = 0$$

and the unit vector is therefore

$$\boldsymbol{u} = 0.164\boldsymbol{i} - 0.985\boldsymbol{j}$$

The product of the magnitude of $\boldsymbol{F}$ and the unit vector $\boldsymbol{u}$ expresses the force in terms of its rectangular component vectors (4.11):

$$\begin{aligned}\boldsymbol{F} &= \|\boldsymbol{F}\|\,\boldsymbol{u}\\ &= (500\text{ N})(0.164\boldsymbol{i} - 0.985\boldsymbol{j})\\ &= 82.0\text{ N}\boldsymbol{i} - 493\text{ N}\boldsymbol{j}\end{aligned}$$

Alternately, since $\theta_x = \cos^{-1}(0.164) = 80.5°$, we can use (4.9A) to write

$$F_x = \|\boldsymbol{F}\|\cos 80.5° = (500\text{ N})(0.164) = 82.0\text{ N}$$

For F_y, we note that for this particular problem, $\cos\theta_y = -\sin\theta_x$ and from (4.9B) we have

$$\begin{aligned}F_y &= \|\boldsymbol{F}\|\cos\theta_y = -\|\boldsymbol{F}\|\sin\theta_x\\ &= -\|\boldsymbol{F}\|\sin 80.5° = -(500\text{ N})(0.986) = -493\text{ N}\end{aligned}$$

Answer (a) $F_x = 82.0$ N; $F_y = -493$ N; $F_z = 0$ N.
In vector notation: $\boldsymbol{F} = 82.0\text{ N}\boldsymbol{i} - 493\text{ N}\boldsymbol{j}$

This answer can be checked using (4.12) to calculate the magnitude of $\boldsymbol{F}$ from its components: $\sqrt{(82.0\text{ N})^2 + (-493\text{ N})^2 + (0\text{ N})^2} = 500$ N, which is the magnitude given in the problem statement. Also note that $|F_y| >> |F_x|$, F_x is positive and F_y negative; based on the orientation of $\boldsymbol{F}$ shown in **Figure 4.44**, this seems reasonable.

(b) Now we go about finding the components of $\boldsymbol{F}$ in an alternate coordinate system. First we establish some angles. Based on (a), θ_x, the angle between $\boldsymbol{F}$ and the x axis, is 80.5°. The angle between the horizontal and the x^* axis (call this angle θ) is found based on the 4/3 orientation of the bracket (see **Figure 4.45**):

$$\theta = \tan^{-1}\left(\tfrac{4}{3}\right) = 53.1°$$

Knowing θ_x and θ, we can find the angles ω ($= 36.9° = 180° - 53.1° - 90°$) and λ ($= 46.6° = 36.9° + 9.5°$) (**Figure 4.45**). Here again, as in part (c) of Example 4.6, we have a choice in how we solve this problem.

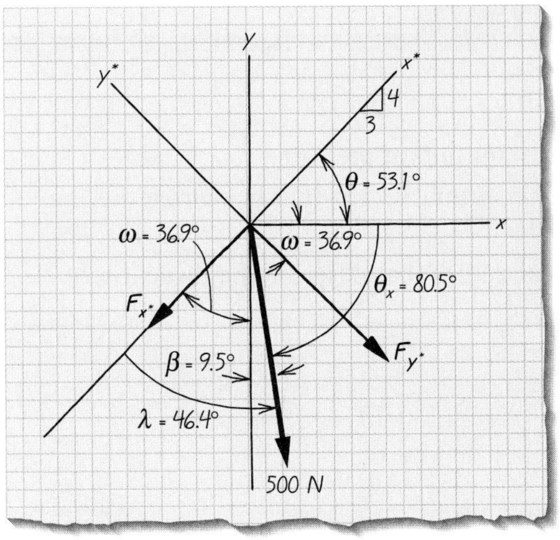

Figure 4.45

Approach 1: From the geometry of **Figure 4.45**, we can write the starred components in terms of $\| \boldsymbol{F} \|$:

$$F_{x*} = -\| \boldsymbol{F} \| \cos\lambda = -\| \boldsymbol{F} \| \cos 46.9° = (-500 \text{ N})(0.683) = -342 \text{ N}$$
$$F_{y*} = -\| \boldsymbol{F} \| \sin\lambda = -\| \boldsymbol{F} \| \sin 46.9° = (-500 \text{ N})(0.730) = -365 \text{ N}$$

Approach 2: We can use the scalar components found in part (a) ($F_x = 82.0$ N and $F_y = -493$ N) to find F_{x*} and F_{y*}, as in **Figure 4.45**:

$$F_{x*} = +F_x \cos\theta + F_y \cos\omega = (82.0 \text{ N}) \cos 53.1° + (-493 \text{ N}) \cos 36.9° = -345 \text{ N}$$
$$F_{y*} = -F_x \sin\theta + F_y \sin\omega = (-82.0 \text{ N}) \sin 53.1° + (-493 \text{ N}) \sin 36.9° = -362 \text{ N}$$

Answer to (b) $F_{x*} = -345$ N; $F_{y*} = -362$ N; $F_z = 0$ N.

We define unit vectors $\boldsymbol{i}^*$ and $\boldsymbol{j}^*$ aligned with the x^* and y^* axes, respectively. Therefore, we write $\boldsymbol{F}$ in vector notation as

$\boldsymbol{F} = -345 \text{ N}\boldsymbol{i}^* - 362 \text{ N}\boldsymbol{j}^*$.

Check If one approach is used to solve the example, an alternate approach can be used as a check. For example, Approach 2 could be used as a check for the answer from Approach 1. The answer could also be checked using (4.12) to calculate the magnitude of $\boldsymbol{F}$ from its components: $\sqrt{(-345 \text{ N})^2 + (-362 \text{ N})^2 + (0 \text{ N})^2} = 500$ N, which is the magnitude given in the problem statement. We can also qualitatively check the relative magnitudes of F_{x*} and F_{y*}. Because we are resolving $\boldsymbol{F}$ into two components that are roughly 45° to $\boldsymbol{F}$, we could expect the two components to be of similar size, which they are!

EXAMPLE 4.8 REPRESENTING NONPLANAR FORCES (A)

A force is specified as $\boldsymbol{F} = -10\ \text{N}\boldsymbol{i} + 15\ \text{N}\boldsymbol{j} + 5\ \text{N}\boldsymbol{k}$.

(a) Draw a graphical representation of the force.

(b) Write, in vector notation, a unit vector that lies along the line of action of $\boldsymbol{F}$.

(c) Determine the spherical angles that describe the direction of $\boldsymbol{F}$.

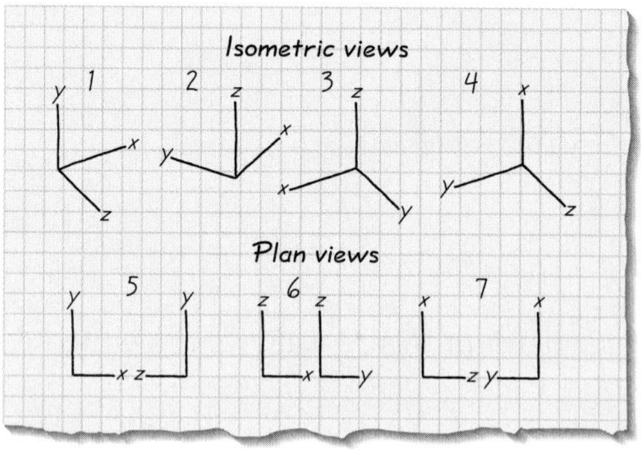

Figure 4.46

Goal We are to draw the force (a), find a unit vector along its line of action (b), and determine angles that define the direction of $\boldsymbol{F}$ (c).

Assume No assumptions are needed.

Draw A vector representation of $\boldsymbol{F}$ can be drawn either as a single vector arrow representing $\boldsymbol{F}$ or as three vector arrows representing the component vectors $\boldsymbol{F}_x$, $\boldsymbol{F}_y$, and $\boldsymbol{F}_z$. We will show both.

First we need to decide which view to show. In other words, how should the x, y, and z axes be oriented on the paper to best illustrate the force? A number of possible orientations of right-handed coordinate systems are shown in **Figure 4.46**. Orientations 1–4 are called isometric views, and orientations 5–7 are plan views.

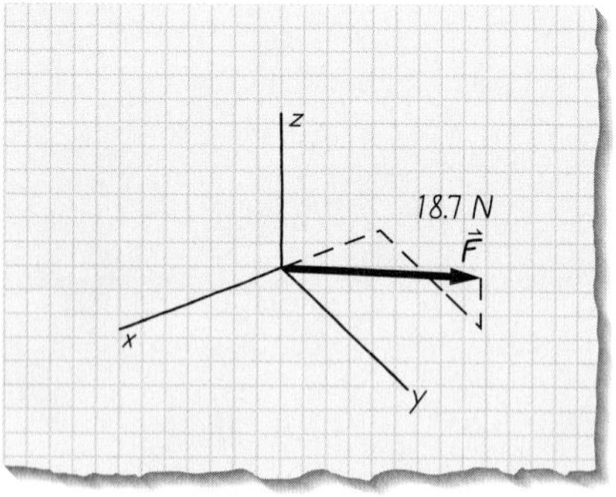

Figure 4.47

Approach 1 (Showing Force *F*): **(a)** Several axis orientations would work for clearly showing $\boldsymbol{F}$, and we have decided on orientation 3. A few grid marks aid in correctly orienting $\boldsymbol{F}$. The magnitude of $\boldsymbol{F}$ is given by (4.12) as

$$\|\boldsymbol{F}\| = \sqrt{(-10.0\ \text{N})^2 + (15.0\ \text{N})^2 + (5.00\ \text{N})^2} = 18.7\ \text{N}$$

Answer to (a) See **Figure 4.47**.

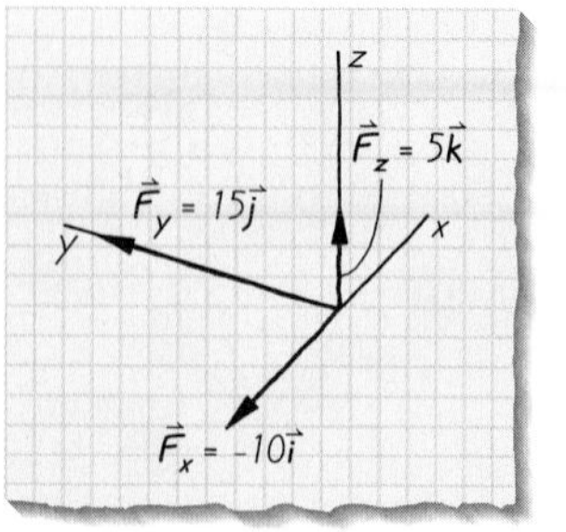

Figure 4.48

Approach 2 (Showing Components in Isometric View): **(a)** Several of the axis orientations shown in **Figure 4.46** would work for showing components; we have decided on orientation 2.

Answer to (a, alternative) See **Figure 4.48**.

Approach 3 (Showing Components in Plan View): **(a)** Because this is a nonplanar force, two plan views are needed, with the y component vector (15 N$\boldsymbol{j}$) shown in both. We choose orientation 5.

Answer to (a, another alternative) See **Figure 4.49**.

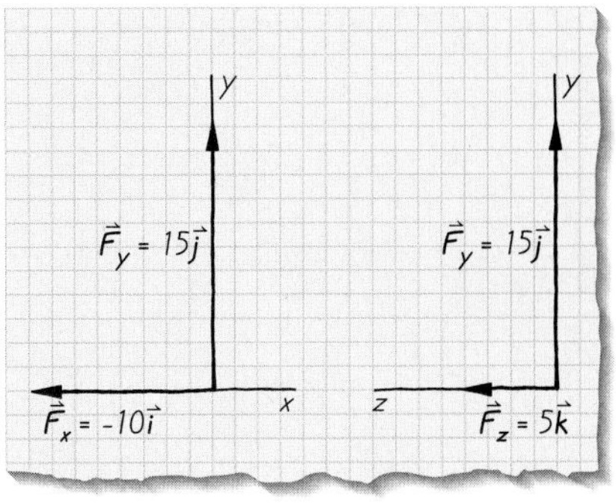

Figure 4.49

All three answers convey the same information and are all equally valid vector representations of the force.

Formulate Equations, Solve, and Check **(b)** Write, in vector notation, a unit vector that lies along the line of action of $\boldsymbol{F}$. We call the requested unit vector $\boldsymbol{u}$, shown in Figure 4.50. The key to finding $\boldsymbol{u}$ is to recognize that the direction cosines are its scalar components. First, we find the magnitude of $\boldsymbol{F}$:

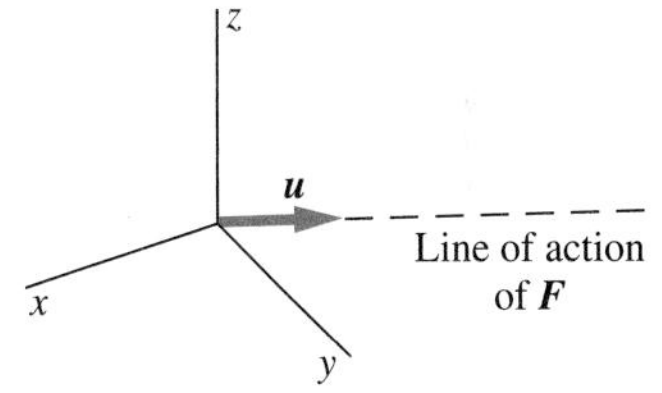

Figure 4.50

$$\| \boldsymbol{F} \| = \sqrt{(-10.0\text{ N})^2 + (15.0\text{ N})^2 + (5.00\text{ N})^2} = 18.7\text{ N}$$

Based on (4.13) we then write:

$$\cos\theta_x = \frac{F_x}{\|\boldsymbol{F}\|} = \frac{-10\text{ N}}{18.7\text{ N}} = -0.534 \quad \Rightarrow \theta_x = 122°$$

$$\cos\theta_y = \frac{F_y}{\|\boldsymbol{F}\|} = \frac{15\text{ N}}{18.7\text{ N}} = 0.802 \quad \Rightarrow \theta_y = 36.7°$$

$$\cos\theta_z = \frac{F_z}{\|\boldsymbol{F}\|} = \frac{5\text{ N}}{18.7\text{ N}} = 0.267 \quad \Rightarrow \theta_z = 74.5°$$

Therefore, the unit vector $\boldsymbol{u}$ is

$$\boldsymbol{u} = -0.534\boldsymbol{i} + 0.802\boldsymbol{j} + 0.267\boldsymbol{k}$$

Answer to (b) $\boldsymbol{u} = -0.534\boldsymbol{i} + 0.802\boldsymbol{j} + 0.267\boldsymbol{k}$

If $\boldsymbol{u}$ is a unit vector, (4.5A) tells us we should be able to take the square root of the sum of its square and get unity: $\sqrt{(-0.534)^2 + (0.802)^2 + (0.267)^2} = 1$! We should also check that the values of the space angles seem reasonable given the orientation of $\boldsymbol{F}$ shown in **Figure 4.47**; they do.

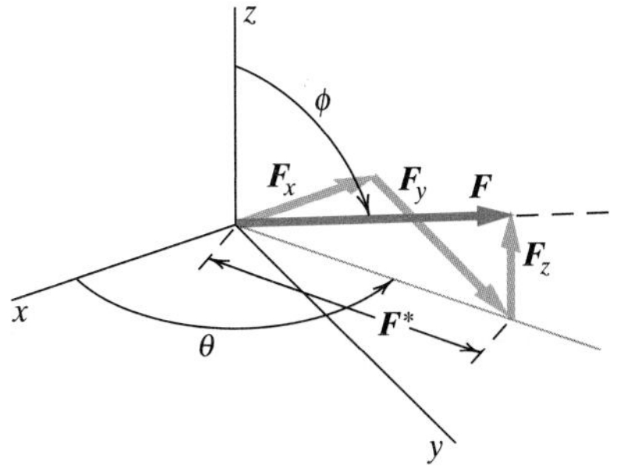

Figure 4.51

(c) We are to determine the angles ϕ and θ. **Figure 4.47** tells us to expect $90° < \theta < 180°$ and $0° < \phi < 90°$, where θ and ϕ are as defined in **Figure 4.32*a*** and shown here in Figure 4.51. Both approaches shown below require the magnitude of $\boldsymbol{F}$, which we found above to be 18.7 N.

Approach 1 (Find θ and ϕ Directly): Based on the geometry shown in **Figure 4.51**, we can write

$$\theta = \cos^{-1}\left(\frac{F_x}{F^*}\right) = \cos^{-1}\left(\frac{F_x}{\sqrt{F_x^2 + F_y^2}}\right)$$

$$= \cos^{-1}\left(\frac{-10\text{ N}}{\sqrt{(-10\text{ N})^2 + (15\text{ N})^2}}\right) = 124°$$

$$\phi = \cos^{-1}\left(\frac{F_z}{\|\boldsymbol{F}\|}\right) = \cos^{-1}\left(\frac{5\text{ N}}{18.7\text{ N}}\right) = 74.5°$$

Approach 2 (Find θ and ϕ Based on the Direction Cosines): In part (b) we found the space angles to be

$$\theta_x = 122°, \theta_y = 36.7°, \theta_z = 74.5°$$

Now we use (4.7B) to convert θ_x, θ_y, θ_z to the angles θ and ϕ:

$$\theta = \tan^{-1}\left(\frac{\cos 36.7°}{\cos 122°}\right) = \tan^{-1}\left(\frac{0.802}{-0.534}\right)$$
$$= 124° \text{ (Remember that } 0 \le \theta \le 2\pi.)$$
$$\phi = \theta_z = 74.5°$$

Answer to (c) $\| \boldsymbol{F} \| = 18.7 \text{ N}, \theta = 124°, \phi = 74.5°$

Check The values of θ and ϕ are consistent with our expectations that $90° < \theta < 180°$ and $0° < \phi < 90°$. The two approaches can be used as cross-checks for each other.

EXAMPLE 4.9 REPRESENTING NONPLANAR FORCES (B)

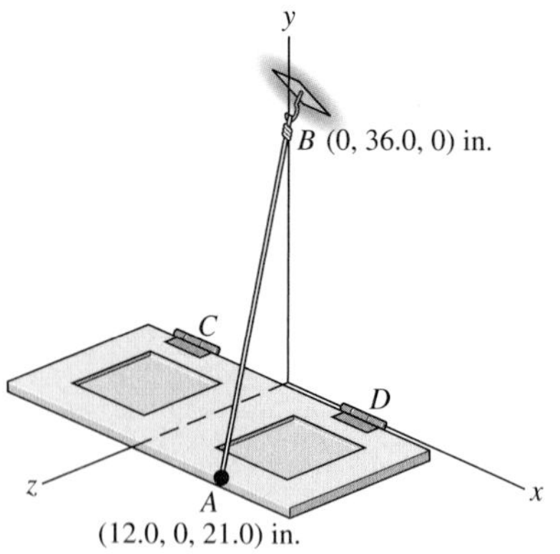

Figure 4.52

The cable in Figure 4.52 holds up a hinged door. The tension force $\boldsymbol{F}$ in the cable has magnitude of 500 lb and acts along line AB. Consider the cable force acting on the hook at B.

(a) Show the force vector components graphically.
(b) Write the force in vector notation.

Goal We are to show the force $\boldsymbol{F}$ graphically (a) and write it in vector notation (b).

Given We are given the coordinates of two points along the line of action of the force and the magnitude of the force (500 lb). A right-handed coordinate system has been established.

Assume No assumptions are needed.

Draw We draw $\boldsymbol{F}$, as applied to the hook at B. Also shown are the line of action of the force and point A (Figure 4.53). Inspection of this figure shows us that F_x will be positive, F_y negative, and F_z positive.

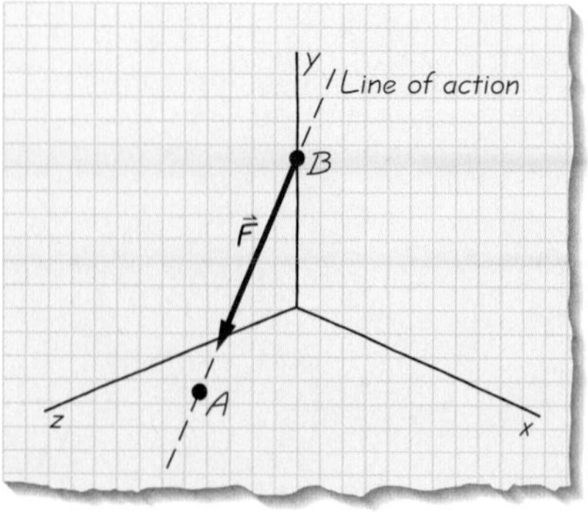

Figure 4.53

Formulate Equations, Solve, and Check **(a)** For both of the approaches presented below for finding the scalar components, we need to find L_{AB} (the distance between points A and B):

$$L_{AB} = \sqrt{(x_A - x_B)^2 + (y_A - y_B)^2 + (z_A - z_B)^2}$$
$$L_{AB} = \sqrt{(12.0 - 0)^2 + (0 - 36.0)^2 + (21.0 - 0)^2} \text{ in.} = 43.4 \text{ in.}$$

Approach 1 (Use Direction Cosines): Before we can create a drawing that shows vector components, we need to calculate the scalar components. We use (4.9A–C) to find the scalar components:

$$F_x = \|\boldsymbol{F}\| \cos\theta_x \qquad (4.9A)$$
$$F_y = \|\boldsymbol{F}\| \cos\theta_y \qquad (4.9B)$$
$$F_z = \|\boldsymbol{F}\| \cos\theta_z \qquad (4.9C)$$

In order to apply (4.9), we must know the direction cosines of $\boldsymbol{F}$. To find them, we use (4.5B):

$$\cos\theta_x = \frac{(x_A - x_B)}{L_{AB}} = \frac{12.0 \text{ in.}}{43.4 \text{ in.}} = 0.277$$
$$\cos\theta_y = \frac{(y_A - y_B)}{L_{AB}} = \frac{-36.0 \text{ in.}}{43.4 \text{ in.}} = -0.830$$
$$\cos\theta_z = \frac{(z_A - z_B)}{L_{AB}} = \frac{21.0 \text{ in.}}{43.4 \text{ in.}} = 0.485$$

Therefore, application of (4.9) with $\|\boldsymbol{F}\| = 500$ lb results in these scalar components:

$$F_x = (500 \text{ lb})(0.277) = +139 \text{ lb}$$
$$F_y = (500 \text{ lb})(-0.830) = -415 \text{ lb}$$
$$F_z = (500 \text{ lb})(0.485) = +243 \text{ lb}$$

Approach 2 (Use a Unit Vector): We will find the unit vector $\boldsymbol{u}$ along the line of action of $\boldsymbol{F}$, then multiply this unit vector by $\|\boldsymbol{F}\|$. The unit vector is

$$\boldsymbol{u} = u_x\boldsymbol{i} + u_y\boldsymbol{j} + u_z\boldsymbol{k}$$

where

$$u_x = \frac{(x_A - x_B)}{L_{AB}} = \frac{12.0 \text{ in.}}{43.4 \text{ in.}} = 0.277$$
$$u_y = \frac{(y_A - y_B)}{L_{AB}} = \frac{-36.0 \text{ in.}}{43.4 \text{ in.}} = -0.830$$
$$u_z = \frac{(z_A - z_B)}{L_{AB}} = \frac{21.0 \text{ in.}}{43.4 \text{ in.}} = 0.485$$

Therefore $\boldsymbol{u} = 0.277\boldsymbol{i} - 0.830\boldsymbol{j} + 0.485\boldsymbol{k}$, which means we can say, from (4.10C),

$$\cos\theta_x = 0.277 \qquad \cos\theta_y = -0.830 \qquad \cos\theta_z = 0.485$$

Inserting these values in (4.11) gives us

$$\boldsymbol{F} = 500 \text{ lb}(0.277\boldsymbol{i} - 0.830\boldsymbol{j} + 0.485\boldsymbol{k})$$
$$\boldsymbol{F} = 139 \text{ lb}\boldsymbol{i} - 415 \text{ lb}\boldsymbol{j} + 243 \text{ lb}\boldsymbol{k}$$

From this we read: $F_x = 139$ lb, $F_y = -415$ lb, and $F_z = 243$ lb.

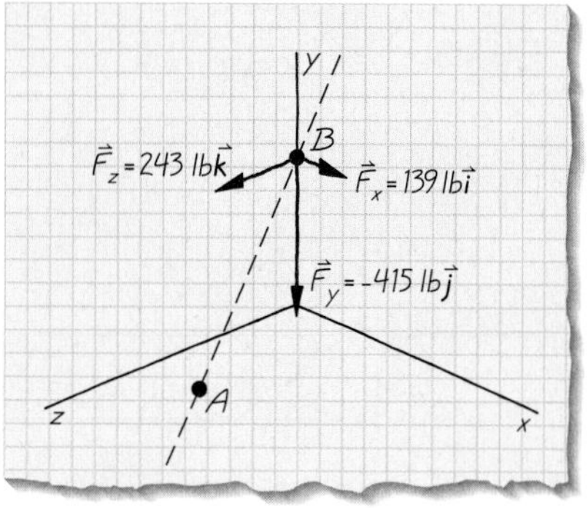

Figure 4.54

Now that we know the scalar components of $\boldsymbol{F}$ (using either Approach 1 or Approach 2), we can show these three scalar components in terms of vector components in the $\boldsymbol{i}, \boldsymbol{j}$, and $\boldsymbol{k}$ directions graphically.

Answer to (a) See **Figure 4.54.**

We could use Approaches 1 and 2 to cross-check the answer. Another check is that the square root of the sum of the squares of the components should be equal to the magnitude of the force:

$$\sqrt{(139 \text{ lb})^2 + (-415 \text{ lb})^2 + (243 \text{ lb})^2} = 500 \text{ lb}$$

(b) Based on the data in **Figure 4.54**, we can write $\boldsymbol{F}$ in vector notation.

Answer to (b) $\boldsymbol{F} = 139 \text{ lb}\boldsymbol{i} - 415 \text{ lb}\boldsymbol{j} + 243 \text{ lb}\boldsymbol{k}$

EXERCISES 4.5

Use E4.5.1 to answer **Exercises 4.5.1–4.5.4**. The line of action of the 4.0-kN force $\boldsymbol{F}$ runs through the points A and B as shown.

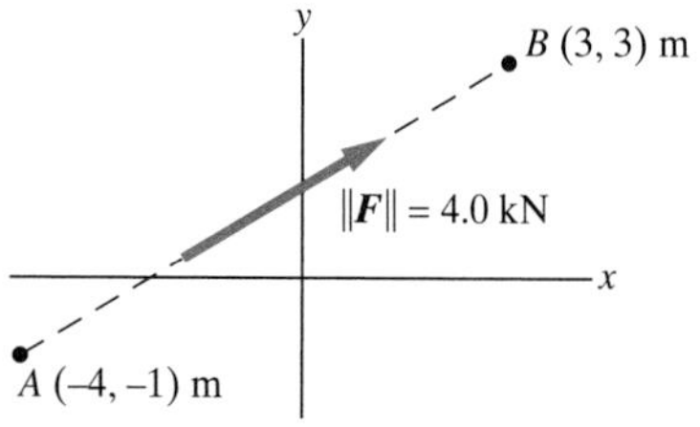

E4.5.1

4.5.1. Find the unit vector along the line of action of the force.

4.5.2. Express $\boldsymbol{F}$ in vector notation.

4.5.3. Find the space angles that describe the direction of the force vector.

4.5.4. Express $\boldsymbol{F}$ in terms of its magnitude and the angles θ and ϕ.

Use E4.5.5 to answer **Exercises 4.5.5–4.5.8**. The line of action of the 350 lb force $\boldsymbol{F}$ is oriented at 30° to the horizontal as shown.

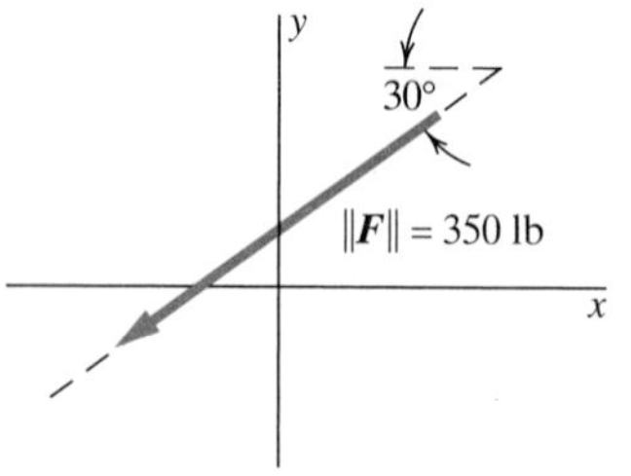

E4.5.5

4.5.5. Find the unit vector along the line of action of the force.

4.5.6. Express $\boldsymbol{F}$ in terms of the unit vectors $\boldsymbol{i}$ and $\boldsymbol{j}$.

4.5.7. Find the space angles that describe the direction of the force vector.

4.5.8. Express $\boldsymbol{F}$ in terms of its magnitude and the angles θ and ϕ.

Use E4.5.9 to answer **Exercises 4.5.9–4.5.12**. The slope of the 1500-N force is specified as shown in **E.4.5.9**.

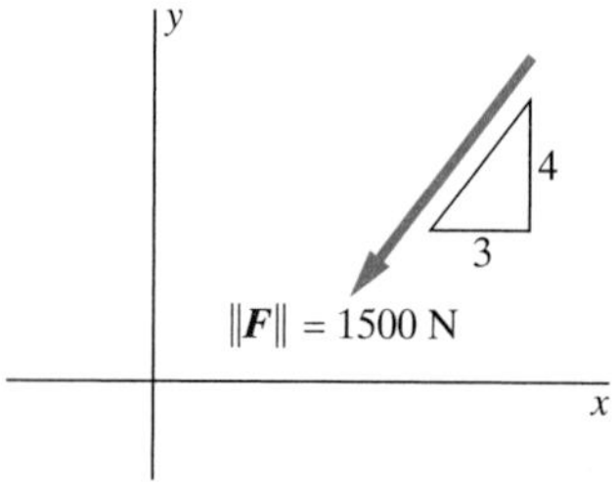

E4.5.9

4.5.9. Find the unit vector along the line of action of the force.

4.5.10. Express $\boldsymbol{F}$ in terms of the unit vectors $\boldsymbol{i}$ and $\boldsymbol{j}$.

4.5.11. Find the space angles that describe the direction of the force vector.

4.5.12. Express $\boldsymbol{F}$ in terms of its magnitude and the angles θ and ϕ.

Use **E4.5.13** to answer **Exercises 4.5.13–4.5.16**. The 40-kip force is oriented at 40° relative to the horizontal, as shown in **E.4.5.13** (1 kip = 1000 lb).

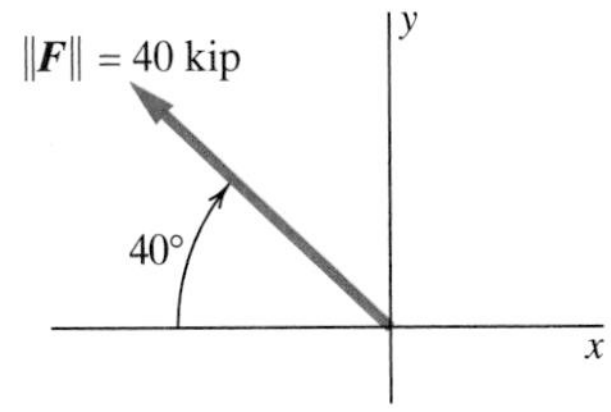

E4.5.13

4.5.13. Find the unit vector along the line of action of the force.

4.5.14. Express $\boldsymbol{F}$ in terms of the unit vectors $\boldsymbol{i}$ and $\boldsymbol{j}$.

4.5.15. Find the space angles that describe the direction of the force vector.

4.5.16. Express $\boldsymbol{F}$ in terms of its magnitude and the angles θ and ϕ.

Use **E4.5.17** to answer **Exercises 4.5.17–4.5.20**. The cable pulls on the eyebolt with a tensile force of 120 N, and $\theta = 30°$.

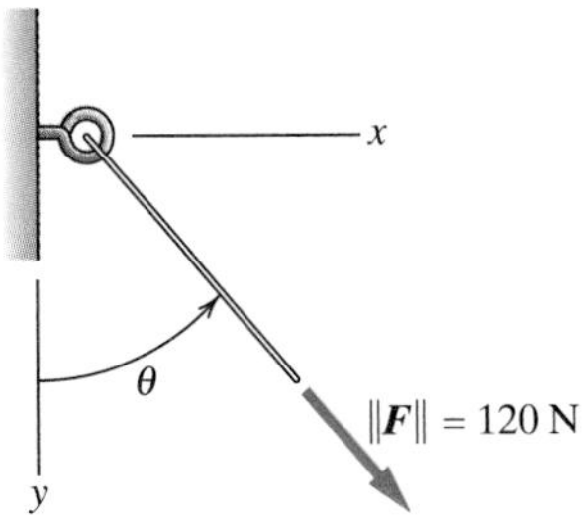

E4.5.17

4.5.17. Determine the scalar components of the force with respect to the xyz coordinate system.

4.5.18. Redraw the figure, showing the tension force in terms of its rectangular component vectors.

4.5.19. Find the unit vector along the line of action of the force.

4.5.20. Determine the space angles that describe the direction of the force vector.

Use **E4.5.21** to answer **Exercises 4.5.21–4.5.24**.

4.5.21. Determine the unit vector along the line of action of the force.

4.5.22. Determine the scalar components of the force with respect to the xyz coordinate system.

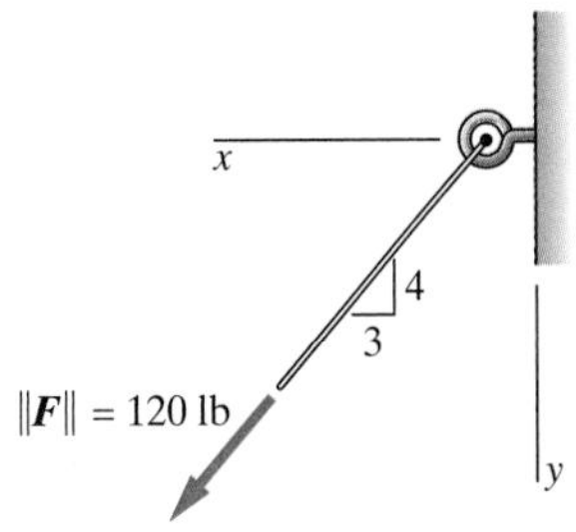

E4.5.21

4.5.23. Redraw the figure, showing the tension force in terms of its rectangular component vectors.

4.5.24. Find the space angles that describe the direction of the force vector.

Use **E4.5.25** to answer **Exercises 4.5.25–4.5.27**. The cable pulls on the eyebolt with a tensile force of 120 N. Coordinates along the line of action of the force are as given in **E4.5.25**.

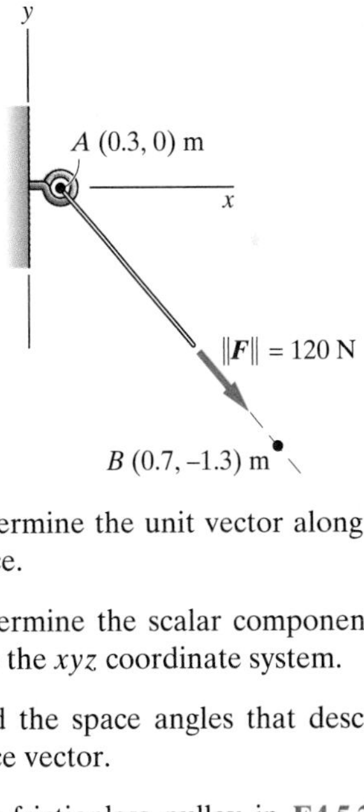

E4.5.25

4.5.25. Determine the unit vector along the line of action of the force.

4.5.26. Determine the scalar components of the force with respect to the xyz coordinate system.

4.5.27. Find the space angles that describe the direction of the force vector.

4.5.28. The frictionless pulley in **E4.5.28** is such that the magnitude of $\boldsymbol{T}_1$ is equal to the magnitude of $\boldsymbol{T}_2$. For $\boldsymbol{T}_1 = 25.9\text{ kN}\boldsymbol{i} + 96.6\text{ kN}\boldsymbol{j}$ and $\boldsymbol{T}_2 = -90.6\text{ kN}\boldsymbol{i} + 42.3\text{ kN}\boldsymbol{j}$, prove that the magnitudes of these two forces are equal.

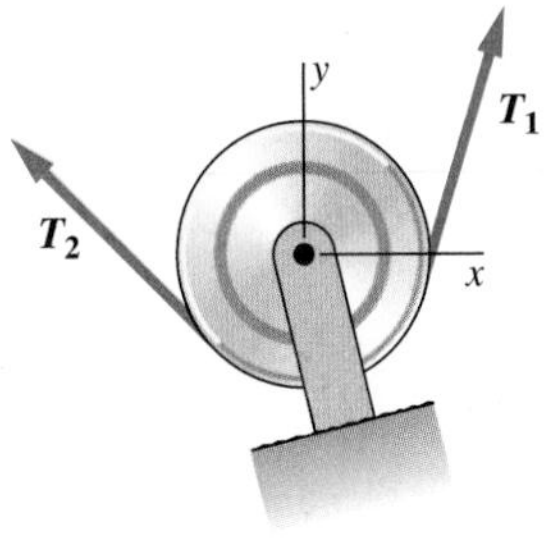

E4.5.28

4.5.29. Table 4.5.29 contains a list of forces.

a. Which forces in the table have the same magnitude?

b. Which items in the table represent the same force?

Table 4.5.29

Item	Force
(1)	$\boldsymbol{F} = 10 \text{ kN}\boldsymbol{i} - 25 \text{ kN}\boldsymbol{j}$
(2)	$\boldsymbol{F} = 10 \text{ kN}\boldsymbol{i} + 25 \text{ kN}\boldsymbol{k}$
(3)	$\boldsymbol{F} = -30 \text{ N}\boldsymbol{i} - 40 \text{ N}\boldsymbol{j}$
(4)	$\lVert \boldsymbol{F} \rVert = 26.9 \text{ kN}; \theta_x = 68.0°, \theta_y = 158°$
(5)	$\lVert \boldsymbol{F} \rVert = 26.9 \text{ kN}; \theta = 0°, \phi = 158°$
(6)	$\lVert \boldsymbol{F} \rVert = 26.9 \text{ kN}; \theta = 291.8°, \phi = 90°$
(7)	$\lVert \boldsymbol{F} \rVert = 50 \text{ N}; \theta_x = 127°, \theta_y = 143°, \theta_z = 90°$
(8)	$\lVert \boldsymbol{F} \rVert = 26.9 \text{ kN}; \theta_x = 68.0°, \theta_y = 90°, \theta_z = 158°$

4.5.30. Create a drawing of each unique force in **Table 4.5.29** that clearly shows the direction and magnitude of the force.

4.5.31. The x component of a 120-lb force $\boldsymbol{F}$ is twice as large as the y component.

a. Find the unit vector along the line of action of the force.

b. Express $\boldsymbol{F}$ in terms of the unit vectors $\boldsymbol{i}$ and $\boldsymbol{j}$.

c. Find the space angles that describe the direction of the force vector.

d. Express $\boldsymbol{F}$ in terms of its magnitude and the angles θ and ϕ.

4.5.32. A 500-N force $\boldsymbol{F}$ is applied to the post in **E4.5.32**.

a. Determine the magnitude of the x and y components of $\boldsymbol{F}$.

b. Determine the magnitude of the u and v components of $\boldsymbol{F}$.

c. Comment on whether the following are true or false mathematical statements:

$$\lVert \boldsymbol{F} \rVert = \sqrt{(F_x)^2 + (F_y)^2}$$

$$\lVert \boldsymbol{F} \rVert = \sqrt{(F_u)^2 + (F_v)^2}$$

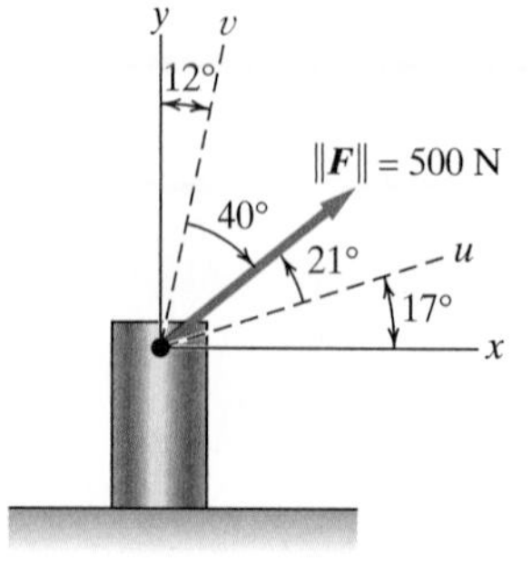

E4.5.32

4.5.33. If the magnitude of the x component of a 120-N force is three times as large as the magnitude of the y component and the force lies in the xy plane, determine all possible values of space angle θ_x.

4.5.34. Cables A and B act on the hook at C_1, as shown in **E4.5.34**. The tension in Cable A is 500 lb and the tension in Cable B is 400 lb.

a. Express in vector notation the force that Cable A applies on the hook at C_1.

b. Express in vector notation the force that Cable B applies on the bracket at C_2.

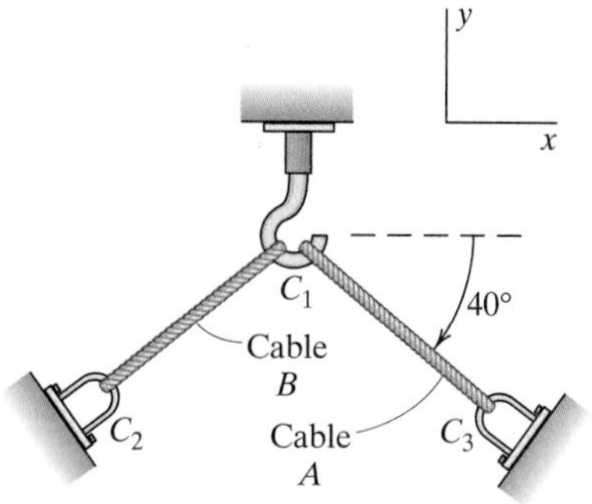

E4.5.34

4.5.35. For each of the partial coordinate systems shown in **E4.5.35**, draw the missing axis so as to form a right-handed coordinate system. An example is given in the top row of the figure.

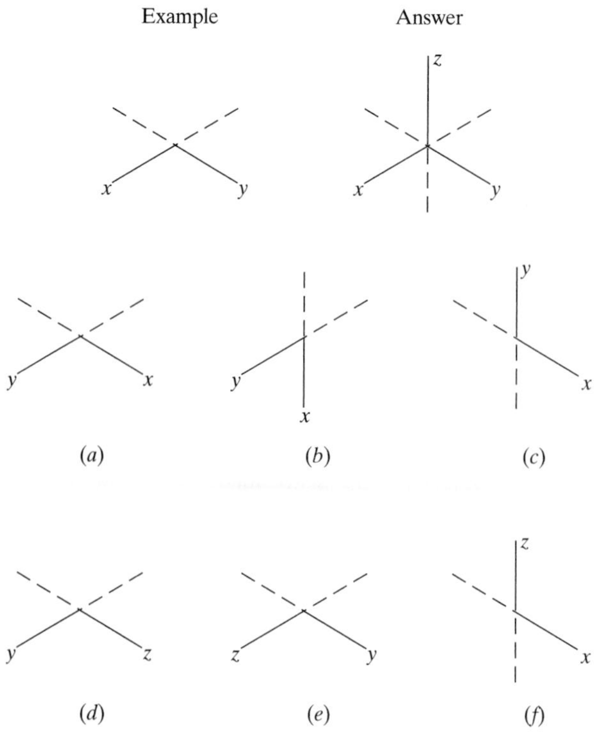

E4.5.35

4.5.36. Consider (4.4), (4.5), (4.7), (4.9), and (4.10C). Simplify these equations when

a. there is a force $\boldsymbol{F}$ in the xy plane

b. there is a force $\boldsymbol{F}$ in the yz plane

c. there is a force $\boldsymbol{F}$ in the zx plane

4.5.37. Prove that (4.7A) and (4.7B) are true.

4.5.38. Find θ_z for each combination of space angles shown in Table 4.5.38.

Table 4.5.38

	θ_x	θ_y	θ_z
Vector 1	15°	75°	90°
Vector 2	120°	150°	90°
Vector 3	30°	100°	
Vector 4	90°	30°	

4.5.39. Match vectors 1–4 from **Exercise 4.5.38** with their corresponding representations in E4.5.39 by completing Table 4.5.39.

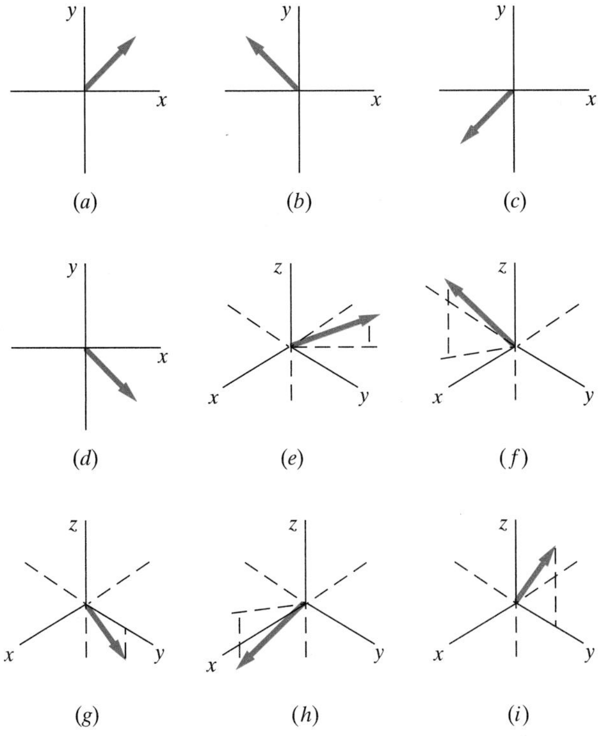

E4.5.39

Table 4.5.39

Represented by this part letter in **E4.5.39**:
Vector 1
Vector 2
Vector 3
Vector 4

4.5.40. Match each force in Table 4.5.40 with the appropriate graphical representations in E4.5.40.

Table 4.5.40

	Force	Represented numerically as:	Represented graphically by this part letter in **E4.5.40**:
(a)	$\boldsymbol{F}_1$	$\boldsymbol{F}_1 = -20\ \text{kN}\boldsymbol{i} + 30\ \text{kN}\boldsymbol{j} + 10\ \text{kN}\boldsymbol{k}$	
(b)	$\boldsymbol{F}_2$	$\lVert \boldsymbol{F}_2 \rVert = 40\ \text{N}, \theta_x = 125°, \theta_y = 145°$	
(c)	$\boldsymbol{F}_3$	$\boldsymbol{F}_3 = 20\ \text{kN}\boldsymbol{j} + 40\ \text{kN}\boldsymbol{k}$	

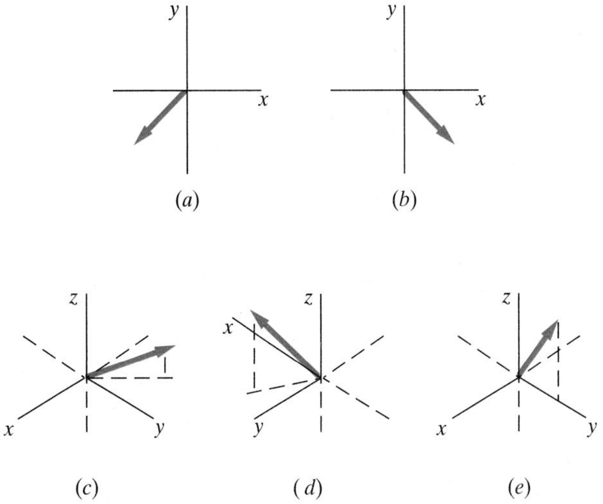

E4.5.40

4.5.41. Classify each force shown in E4.5.41 as planar or nonplanar. If the force is planar, specify the coordinate plane that contains the force.

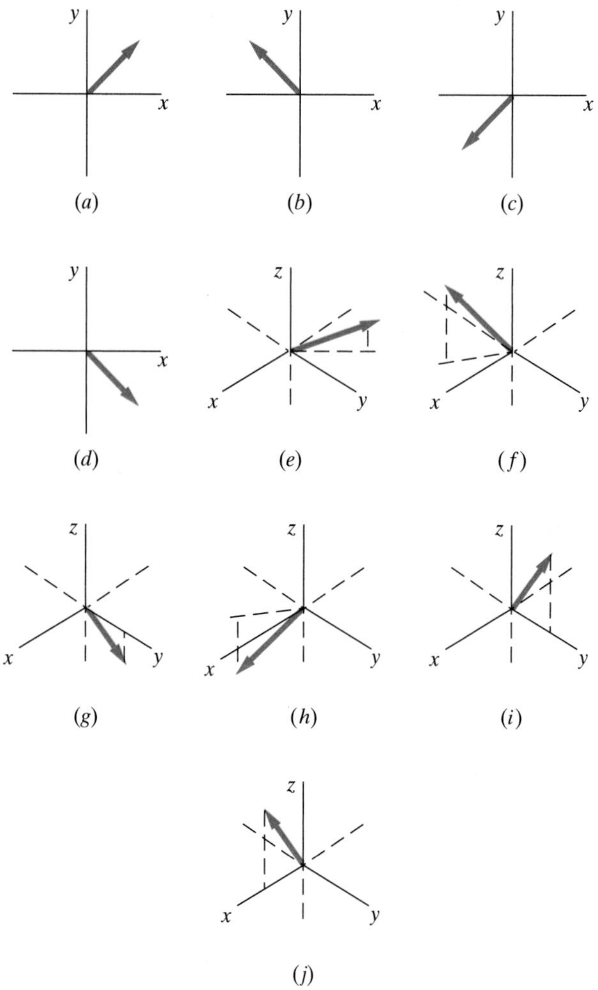

E4.5.41

4.5.42. Based on E4.5.42, write in vector notation a unit vector that is along a line between

a. points A and B
b. points A and C
c. points A and D

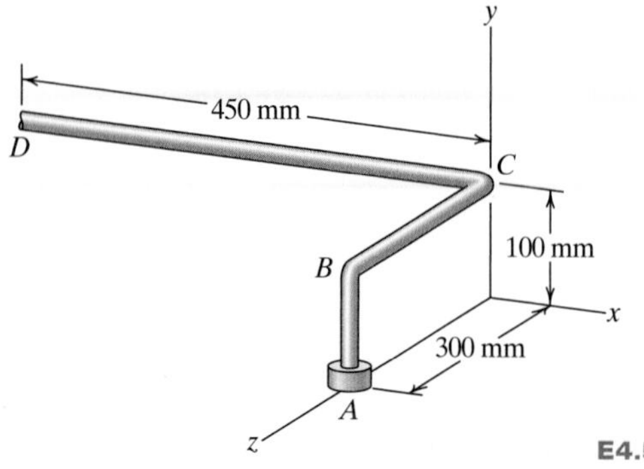

E4.5.42

4.5.43. A person applies a force $\boldsymbol{F} = (-16\text{ lb}\boldsymbol{i} + 4\text{ lb}\boldsymbol{j} + 6\text{ lb}\boldsymbol{k})$ to the top of a fence post in E4.5.43.

a. Find the unit vector along the line of action of the force.

b. Find the space angles that describe the direction of the force vector.

c. Determine the magnitude of the force. Does this value of force seem comparable to the force that you could exert with your arms?

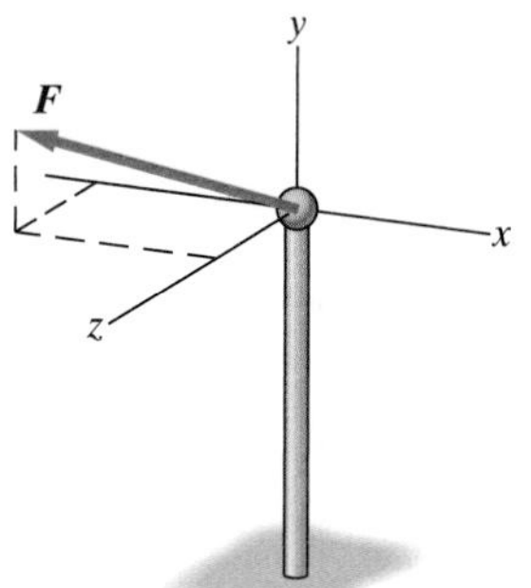

E4.5.43

4.5.44. Two forces $\boldsymbol{F}_1$ and $\boldsymbol{F}_2$ are applied to the hook, as shown in E4.5.44. Express each force in vector notation.

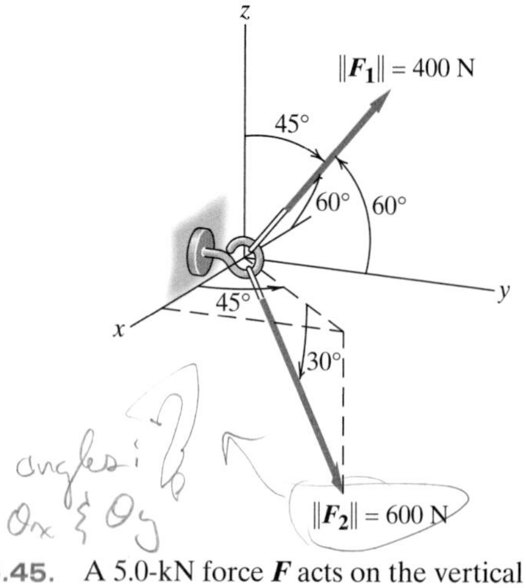

E4.5.44

4.5.45. A 5.0-kN force $\boldsymbol{F}$ acts on the vertical pole shown in E4.5.45.

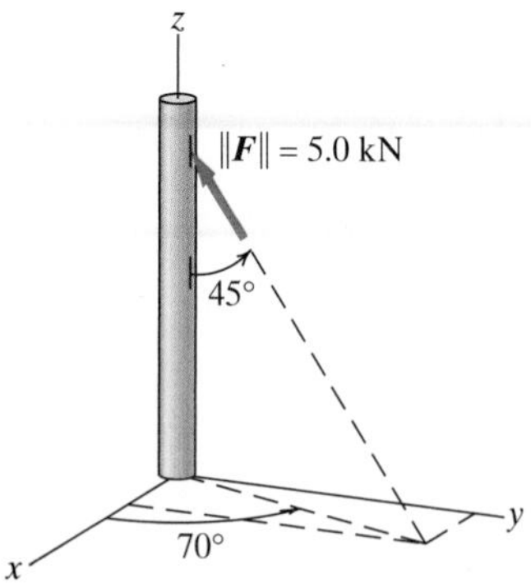

E4.5.45

a. Determine the unit vector along the line of action of $\boldsymbol{F}$.

b. Express $\boldsymbol{F}$ in vector notation (use your answer from **a** as the starting point).

c. Find the space angles that describe the direction of $\boldsymbol{F}$.

4.5.46. Consider the 1-kip force $\boldsymbol{F}$ in E4.5.46. Write this force

a. in terms of a unit vector along the line of action of $\boldsymbol{F}$, and a magnitude

b. in vector notation

c. as magnitude and space angles

d. in terms of its magnitude and the angles θ and ϕ

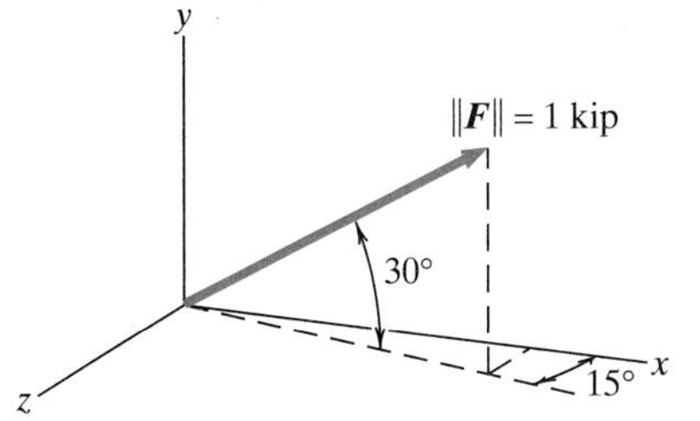

E4.5.46

4.5.47. The x, y, and z scalar components of a 140-lb force are in the proportion of $3{:}-2{:}6$. Determine these components and the space angles between the force and the reference axes.

4.5.48. Table 4.5.48 contains a list of forces.

a. Which of the forces in the table have the same magnitude?

b. Which items in the table represent the same force?

Table 4.5.48

Item	Force
(1)	$\boldsymbol{F} = 28\text{ kN}\boldsymbol{i} + 18\text{ kN}\boldsymbol{j} - 38\text{ kN}\boldsymbol{k}$
(2)	$\lVert \boldsymbol{F} \rVert = 50.5\text{ kN}$, $\cos\theta_x = -0.554$, $\cos\theta_y = 0.356$, $\cos\theta_z = 0.752$
(3)	$\lVert \boldsymbol{F} \rVert = 50.5\text{ kN}$, $\theta_x = 56.4°$, $\theta_y = 69.1°$, $\theta_z = 138.8°$
(4)	$\boldsymbol{F} = -28\text{ kN}\boldsymbol{i} + 18\text{ kN}\boldsymbol{j} + 38\text{ kN}\boldsymbol{k}$
(5)	$\lVert \boldsymbol{F} \rVert = 50.5\text{ kN}$, $\cos\theta_x = 0.554$, $\cos\theta_y = 0.356$, $\cos\theta_z = -0.752$
(6)	$\lVert \boldsymbol{F} \rVert = 50.5\text{ kN}$, $\theta_x = 124°$, $\theta_y = 69.1°$, $\theta_z = 41.2°$

4.5.49. **Table 4.5.48** contains a list of forces. Create a drawing of each unique force that clearly shows the direction and magnitude of the force.

4.5.50. The tension in the supporting cable AB in E4.5.50 is 20 kip. Write the cable tension force in vector notation.

4.5.51. The tension in the supporting cable AD in **E4.5.50** is 10 kN. Write the cable tension force in vector notation.

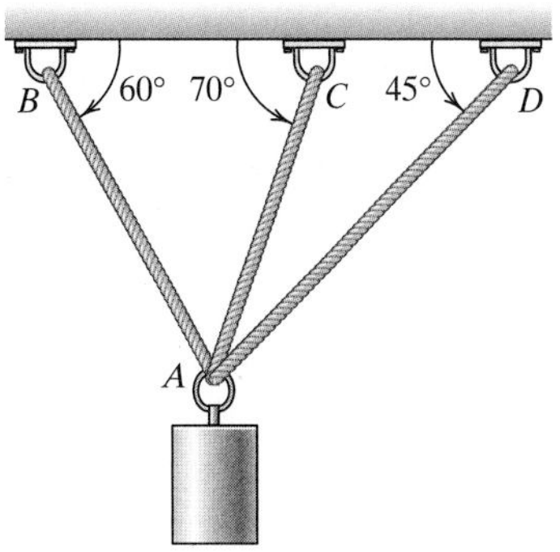

E4.5.50

4.5.52. Three forces are applied with cables to the anchor block as shown in E4.5.52. Write each force in vector notation.

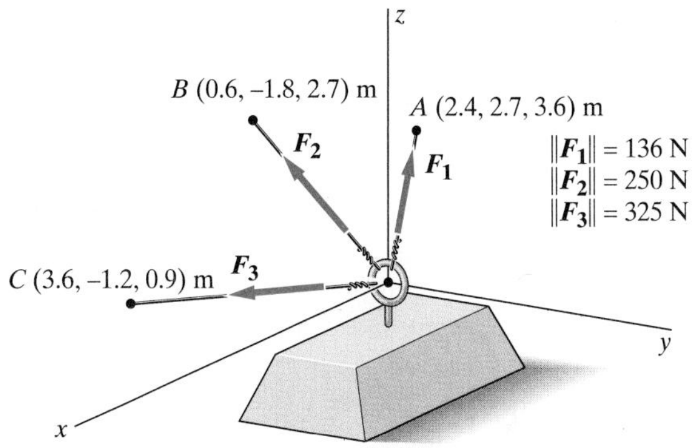

E4.5.52

For **Exercises 4.5.53 and 4.5.54**, a cart weighing 400 N is pushed up a 15% grade by the force $\boldsymbol{F}$ shown in E4.5.53. "Percent grade" defines the rise (upward movement) relative to the run (horizontal movement). For example, for a 15% grade, for every 100 m of horizontal movement (run) there is 15 m of upward movement (rise). The cart travels at a constant speed.

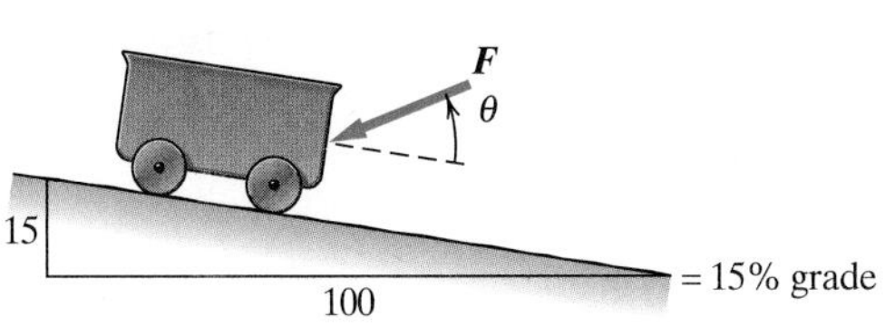

E4.5.53

4.5.53. Find the scalar components of the 400-N cartweight that are parallel and perpendicular to the slope.

4.5.54. If $\boldsymbol{F}$ has magnitude of 200 N, determine the angle θ such that the scalar component of $\boldsymbol{F}$ parallel to the slope is equal in magnitude to the parallel component found in **Exercise 4.5.53** but points in the opposite direction.

4.5.55. The bicycle and cyclist in E4.5.55 have a combined weight of 200 lb. As the bicycle moves up the 25% grade, how much of the 200-lb gravity force is directed down the incline? How much is directed perpendicular to the incline? "Percent grade" defines the rise (upward movement) relative to the run (horizontal movement). For example, for a 25% grade, for every 100 m of horizontal movement (run) there is 25 m of upward movement (rise).

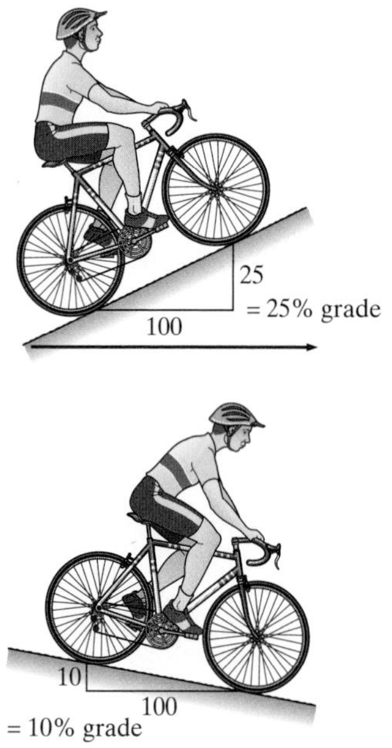

E4.5.55

4.5.56. As the bicycle moves down the 10% grade (**E4.5.55**), how much of the 200-lb gravity force is directed down the incline? How much is directed perpendicular to the incline?

4.5.57. While pushing a copy machine up an incline at constant speed, a person exerts a 180-N force $\boldsymbol{F}$, as shown in E4.5.57. Determine the components of $\boldsymbol{F}$ that are parallel and perpendicular to the incline in **E4.5.57**.

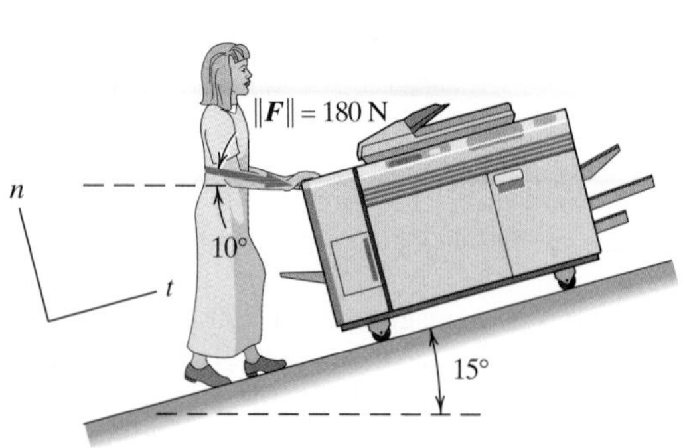

E4.5.57

4.5.58. For the person and machine (**E4.5.57**) to move at constant speed, the magnitude of the component of $\boldsymbol{F}$ parallel to the incline is equal to the magnitude of the component of the weight of the machine parallel to the incline. What is the weight of the machine?

4.5.59. A spring whose stiffness is 250 N/m and whose unstretched length is 400 mm (0.400 m) is stretched between ends A and C of a bent rod ABC, as shown in E4.5.59. If the magnitude of the spring force is the product of the spring stiffness and the displacement, determine the force exerted by the spring on end C of the rod. Write your answer in vector notation.

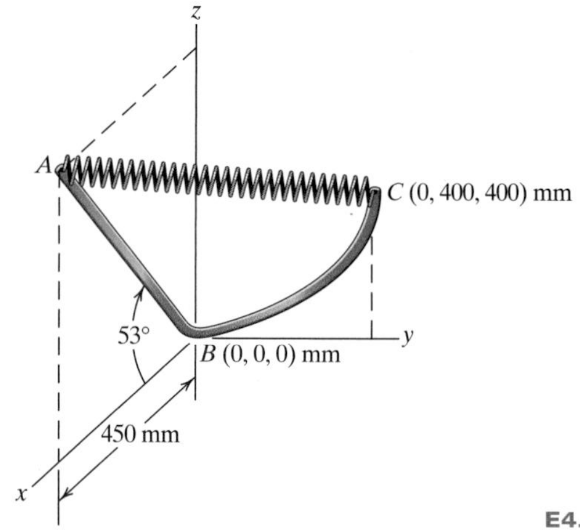

E4.5.59

4.5.60. The turnbuckle in E4.5.60 is tightened until the tension in cable AB equals 3.0 kN.

a. Write an expression for the cable force pulling on member AD. Call this force $\boldsymbol{F}_1$.

b. Write an expression for the cable force pulling on the ground. Call this force $\boldsymbol{F}_2$.

c. Show that $\boldsymbol{F}_1 = -\boldsymbol{F}_2$.

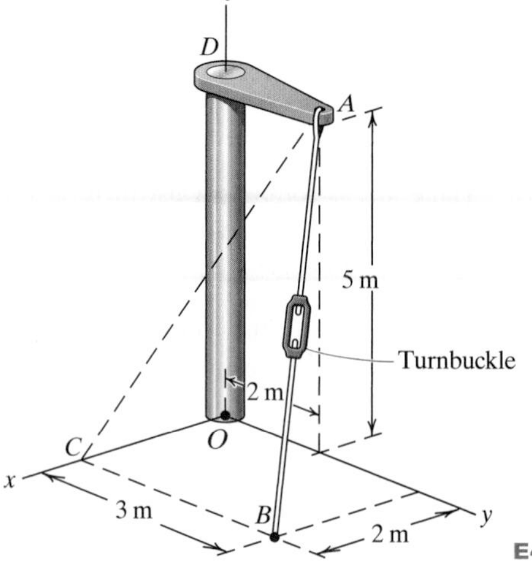

E4.5.60

4.5.61. The free-body diagram in E4.5.61 shows the forces acting on links AB and BC. Express each force in terms of its rectangular component vectors.

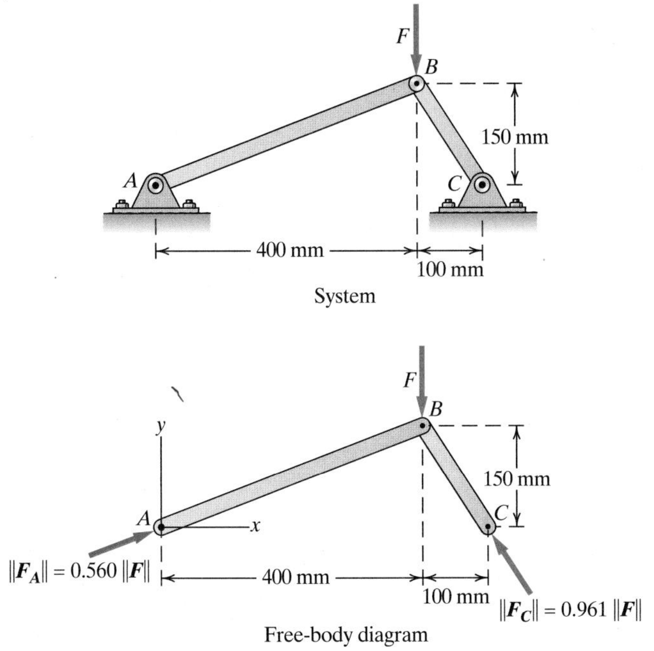

E4.5.61

4.5.62. The free-body diagram in E4.5.62 shows the forces acting on the bed of a dump truck. Express each force in terms of its rectangular component vectors.

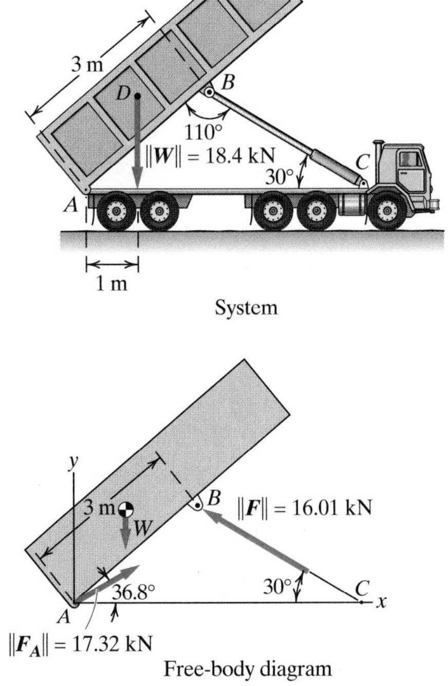

E4.5.62

4.5.63. The free-body diagram in E4.5.63 shows the forces acting on a frictionless pulley. Express each force in terms of its rectangular component vectors.

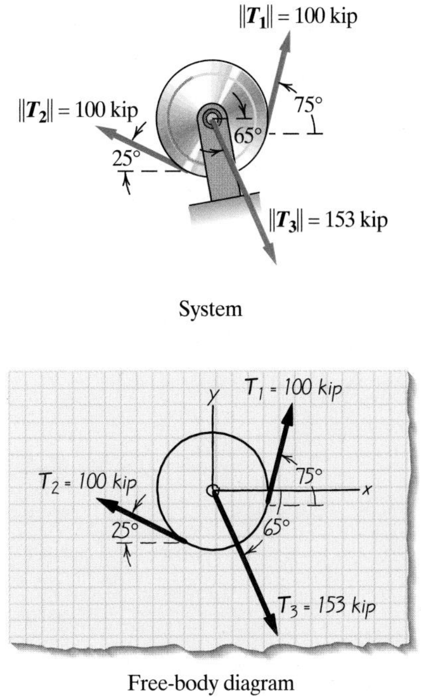

E4.5.63

4.5.64. Express each force shown in E4.5.64 in terms of rectangular component vectors.

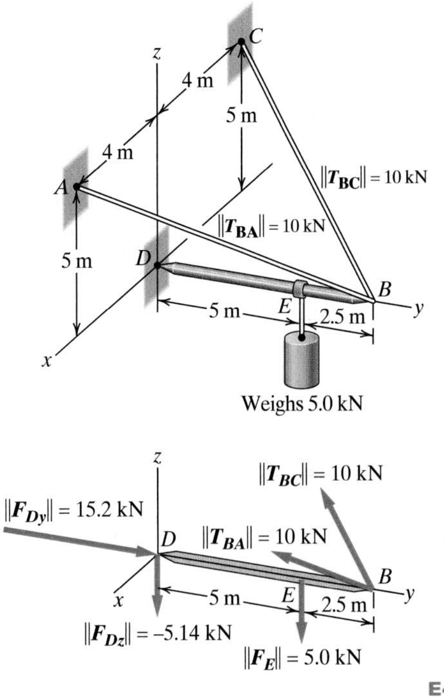

E4.5.64

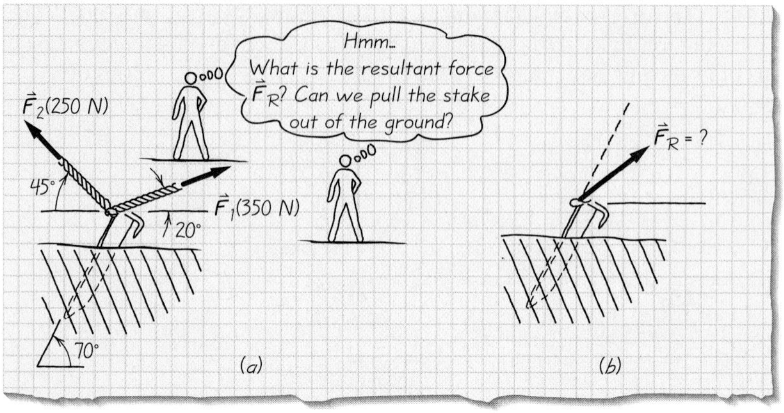

Figure 4.55 (*a*) Two people considering how their pulls will affect the stake; (*b*) the resultant force from the two pulls

4.6 RESULTANT FORCE (VECTOR ADDITION)

Consider a situation in which you and a friend are pulling on a tent stake, as shown in Figure 4.55*a*. The total force exerted on the stake is the vector sum of these two forces; we call this total force the resultant force $\boldsymbol{F_R}$, as indicated in **Figure 4.55*b***. What are the magnitude and direction of $\boldsymbol{F_R}$? In this section we present the vector techniques for answering this question.

Two forces with the same point of application are called **concurrent forces**—$\boldsymbol{F_1}$ and $\boldsymbol{F_2}$ in **Figure 4.55a** are examples of concurrent forces. More generally, concurrent forces are any number of forces such that their lines of action intersect at a single point, as illustrated in Figure 4.56. Any number of concurrent forces can be added together and expressed as a single resultant force, and we now present several methods for doing this.

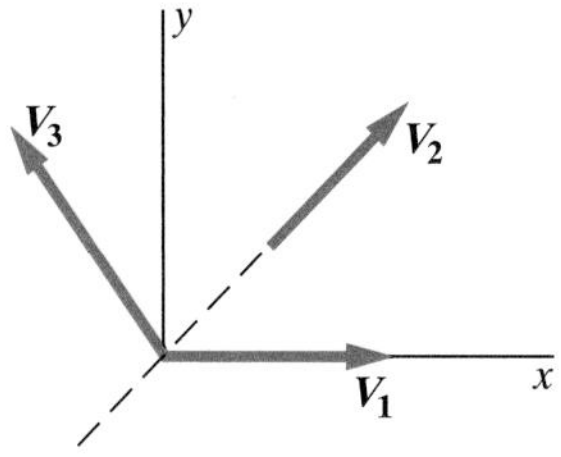

Figure 4.56 Three forces to be added together

Graphical Vector Addition

Adding vectors graphically involves laying out the vectors to be added to scale on graph paper. For example, suppose we want to find the resultant vector $\boldsymbol{V_R}$ of adding the vectors $\boldsymbol{V_1}$, $\boldsymbol{V_2}$, and $\boldsymbol{V_3}$ shown in **Figure 4.56.**

One way to add vectors graphically is to "hike" along the vectors by measuring the combined length of the vectors after laying them out head to tail. We call this the **head-to-tail approach**. Start by laying out $\boldsymbol{V_1}$ in its correct orientation and to scale on graph paper with its tail at the origin, as in Figure 4.57. This vector $\boldsymbol{V_1}$ becomes the first leg of the hike (1). Then slide $\boldsymbol{V_2}$ parallel to itself until its tail touches the head of $\boldsymbol{V_1}$, and slide $\boldsymbol{V_3}$ parallel to itself until its tail touches the head of $\boldsymbol{V_2}$. If you are working with more than three vectors, continue this process until all the vectors you are adding have been drawn. Assign the label *A* to the head of the final vector you've laid out. The resultant $\boldsymbol{V_R}$ is a vector drawn from the origin *O* to *A*. Its length is the magnitude $\|\boldsymbol{V_R}\|$ of $\boldsymbol{V_R}$, and its angular orientation (measured with a protractor from a set of reference axes) defines its direction. You may recall that we also used

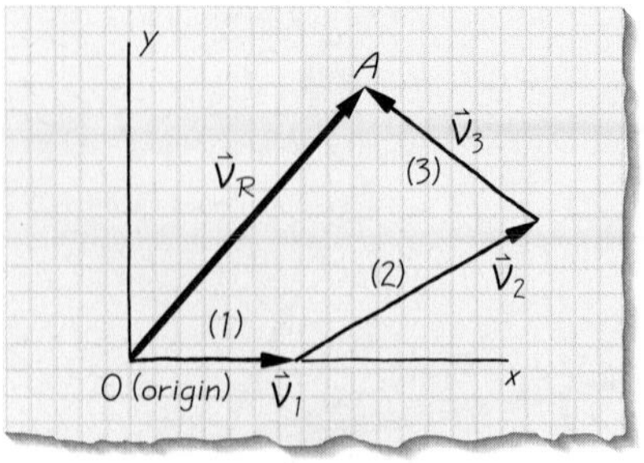

Figure 4.57 The head-to-tail approach to vector addition

this hiking idea when we discussed representing a force in terms of components. In that case we were adding the rectangular component vectors in order to find the resultant force.

An alternative to the head-to-tail approach is the **parallelogram law of addition**. First select any two of the vectors to be added. Let's begin with $\boldsymbol{V_1}$ and $\boldsymbol{V_2}$ in Figure 4.58*a*. Keeping one of the vectors in its original position, slide the other vector parallel to itself until its tail touches the tail of the first vector, assigning the label A to the point where the two tails touch (**Figure 4.58*b***). Then, beginning at the head of $\boldsymbol{V_2}$, draw a line that extends in the direction of $\boldsymbol{V_1}$ and is parallel to $\boldsymbol{V_1}$. Draw a second line that begins at the head of $\boldsymbol{V_1}$, extends in the direction of $\boldsymbol{V_2}$, and is parallel to $\boldsymbol{V_2}$. You have now formed a parallelogram with sides $\boldsymbol{V_1}$ and $\boldsymbol{V_2}$ plus the two lines parallel to these vectors. Call the point where these two lines intersect A^*. A vector extending from A to A^* is the interim resultant $\boldsymbol{V}_{\text{inter}}$ of $\boldsymbol{V_1} + \boldsymbol{V_2}$, and we now use this interim resultant to find the resultant of $\boldsymbol{V_1} + \boldsymbol{V_2} + \boldsymbol{V_3}$. Leaving $\boldsymbol{V}_{\text{inter}}$ where it is, slide $\boldsymbol{V_3}$ parallel to itself until its tail is touching the tail of $\boldsymbol{V}_{\text{inter}}$, and construct another parallelogram just as you did when adding $\boldsymbol{V_1}$ and $\boldsymbol{V_2}$ (**Figure 4.58*c***). Call the intersection of these two lines point A^{**}. The final resultant of adding $\boldsymbol{V_1}$, $\boldsymbol{V_2}$, and $\boldsymbol{V_3}$ (actually the resultant of $\boldsymbol{V}_{\text{inter}}$ and $\boldsymbol{V_3}$, which amounts to the same thing) is a vector from A to A^{**}, as shown in **Figure 4.58c**.

Now back to the tent stake example. The forces $\boldsymbol{F}_1$ and $\boldsymbol{F}_2$ exerted by you and your friend are shown in Figure 4.59. Graphical addition of these forces is illustrated using the head-to-tail approach in Figure 4.60*a* and using the parallelogram law in **Figure 4.60*b***. The length of $\boldsymbol{F_R}$ is the magnitude of the resultant force (333 N), and the angles θ_x and θ_y define its direction ($\theta_x = 62.8°$, $\theta_y = 27.2°$).

One advantage of adding force vectors graphically is that doing so gives you a feel for the physical situation described in the problem and for how the individual forces contribute to the resultant force. Furthermore, with proper drawing/drafting tools and skills, the approach is straightforward when adding planar forces. The main disadvantages of graphical force addition are that the resultant is only as accurate as the drawing, the process becomes complex when more than three or four forces must be added or when the forces are nonplanar, and it is difficult to generalize into an algorithmic/procedural form appropriate for computers.

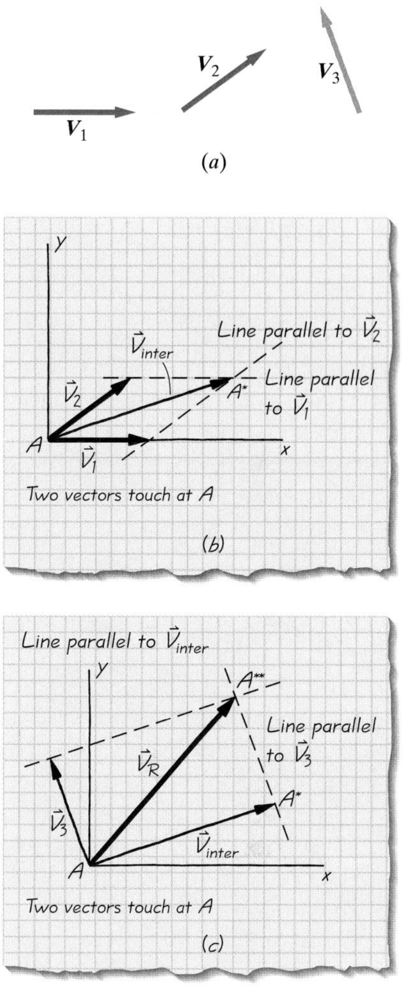

Figure 4.58 The parallelogram law of addition

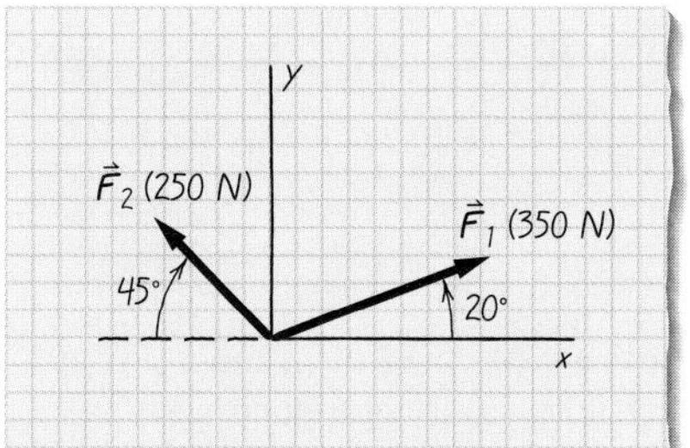

Figure 4.59 Forces $\boldsymbol{F}_1$ and $\boldsymbol{F}_2$ applied to the tent stake

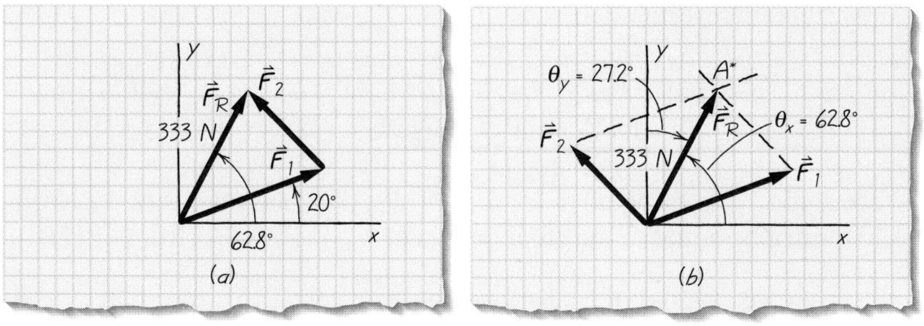

Figure 4.60 Forces $\boldsymbol{F}_1$ and $\boldsymbol{F}_2$ to be added together

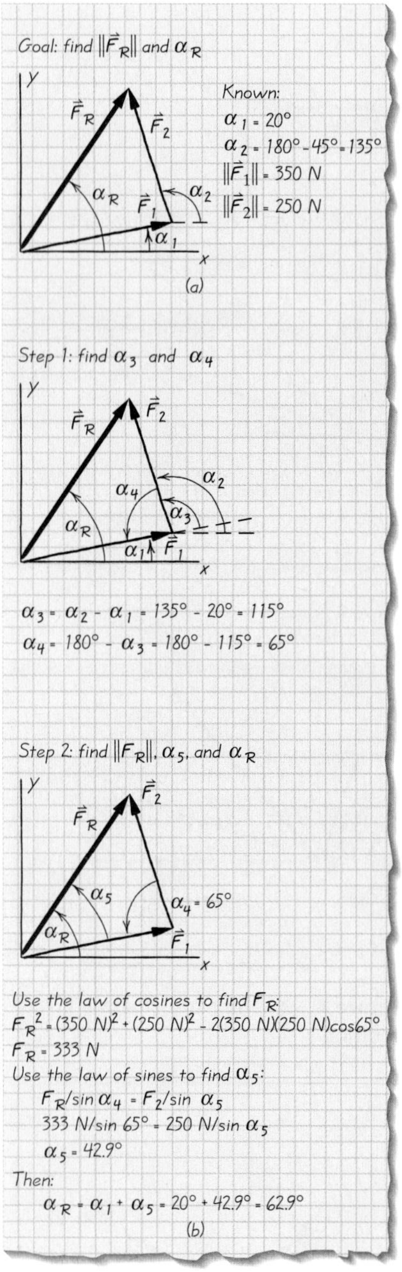

Figure 4.61 (*a*) $\boldsymbol{F_1}$, $\boldsymbol{F_2}$, and $\boldsymbol{F_R}$ form a triangle; (*b*) using trigonometry to find $\boldsymbol{F_R}$

Geometric/Trigonometric Addition

We can also add vectors by taking advantage of trigonometric identities. With this approach to addition of vectors, it is useful to begin by sketching the vectors head to tail, as with the graphical method. Now, however, it is far less critical that the drawing be precise because the sketch serves only to show the spatial relationships between the vectors.

Figure 4.61*a* shows that $\boldsymbol{F_1}$, $\boldsymbol{F_2}$, and $\boldsymbol{F_R}$ from the tent stake example form a triangle. Using the law of sines and the law of cosines, we can "operate" on the triangle to determine the magnitude and direction of $\boldsymbol{F_R}$, as shown in **Figure 4.61*b***. The laws of sines and cosines, along with other useful trigonometric identities, are presented in Appendix A1.3.

The main advantage of geometric/trigonometric addition is that it is straightforward when we are adding two planar forces, particularly if the directions of the two forces are defined in terms of angular orientation. The main disadvantages are that the procedure becomes complex with more than two forces and with nonplanar forces, and it is difficult to generalize into an algorithmic/procedural form appropriate for computers.

Component Force Addition

Component addition is based on the idea that if two vectors, such as $\boldsymbol{V_R}$ and $\boldsymbol{V_1} + \boldsymbol{V_2} + \boldsymbol{V_3}$ are equal, they must have the same magnitude and must point in the same direction. This can happen only if their components are equal. We now illustrate the idea of component equality.

A resultant general vector $\boldsymbol{V_R}$ is composed of component vectors

$$\boldsymbol{V_R} = V_{Rx}\boldsymbol{i} + V_{Ry}\boldsymbol{j} + V_{Rz}\boldsymbol{k} \tag{4.15}$$

and $\boldsymbol{V_1}$, $\boldsymbol{V_2}$, and $\boldsymbol{V_3}$ are composed of the component vectors

$$\boldsymbol{V_1} = V_{1x}\boldsymbol{i} + V_{1y}\boldsymbol{j} + V_{1z}\boldsymbol{k} \tag{4.16A}$$

$$\boldsymbol{V_2} = V_{2x}\boldsymbol{i} + V_{2y}\boldsymbol{j} + V_{2z}\boldsymbol{k} \tag{4.16B}$$

$$\boldsymbol{V_3} = V_{3x}\boldsymbol{i} + V_{3y}\boldsymbol{j} + V_{3z}\boldsymbol{k} \tag{4.16C}$$

To add $\boldsymbol{V_1}$, $\boldsymbol{V_2}$, and $\boldsymbol{V_3}$, we write

$$\boldsymbol{V_1} + \boldsymbol{V_2} + \boldsymbol{V_3} = (V_{1x}\boldsymbol{i} + V_{1y}\boldsymbol{j} + V_{1z}\boldsymbol{k}) + (V_{2x}\boldsymbol{i} + V_{2y}\boldsymbol{j} + V_{2z}\boldsymbol{k}) + (V_{3x}\boldsymbol{i} + V_{3y}\boldsymbol{j} + V_{3z}\boldsymbol{k}) \tag{4.17A}$$

which can be rearranged to

$$\boldsymbol{V_1} + \boldsymbol{V_2} + \boldsymbol{V_3} = (V_{1x} + V_{2x} + V_{3x})\boldsymbol{i} + (V_{1y} + V_{2y} + V_{3y})\boldsymbol{j} + (V_{1z} + V_{2z} + V_{3z})\boldsymbol{k} \tag{4.17B}$$

In order for $\boldsymbol{V_R}$ to be equal to $\boldsymbol{V_1} + \boldsymbol{V_2} + \boldsymbol{V_3}$, the components of $\boldsymbol{V_R}$ (from (4.15)) must be equal to the components of $\boldsymbol{V_1} + \boldsymbol{V_2} + \boldsymbol{V_3}$ (from (4.17B)):

$$V_{Rx}\boldsymbol{i} + V_{Ry}\boldsymbol{j} + V_{Rz}\boldsymbol{k} = (V_{1x} + V_{2x} + V_{3x})\boldsymbol{i} + (V_{1y} + V_{2y} + V_{3y})\boldsymbol{j} + (V_{1z} + V_{2z} + V_{3z})\boldsymbol{k}$$

Therefore, equating the scalar components, we get

$$\begin{aligned} V_{Rx} &= (V_{1x} + V_{2x} + V_{3x}) \\ V_{Ry} &= (V_{1y} + V_{2y} + V_{3y}) \\ V_{Rz} &= (V_{1z} + V_{2z} + V_{3z}) \end{aligned} \tag{4.18}$$

This result implies that if we know the scalar components of the vectors we want to add, we can add these components to find the scalar components of the resultant. This method works for any number of vectors being added. In Figure 4.62 it shows the addition of the two forces you and your friend apply to the stake.

The main advantage of component addition is that it is straightforward for adding any number of planar and nonplanar forces. It is also easy to generalize into an algorithmic/procedural form appropriate for computers.

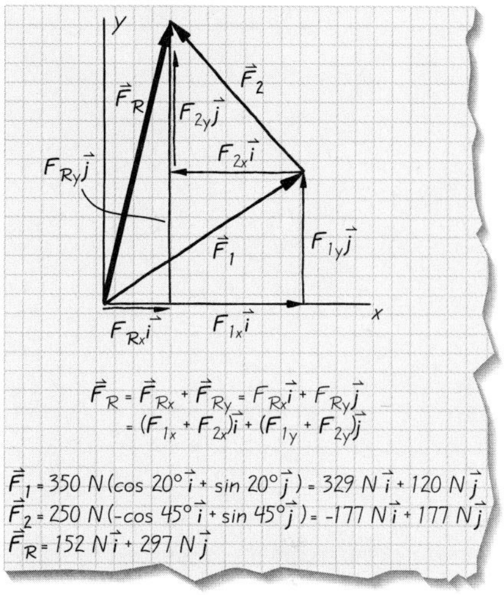

Figure 4.62 Using trigonometric identities to add forces

EXAMPLE 4.10 VECTOR ADDITION PRACTICE (A)

Two forces act on a gusset plate as shown in Figure 4.63. The forces lie in the plane of the paper. Find the resultant $\boldsymbol{F}_R$ of these two forces, using

(a) graphical addition
(b) geometric/trigonometric addition
(c) component addition

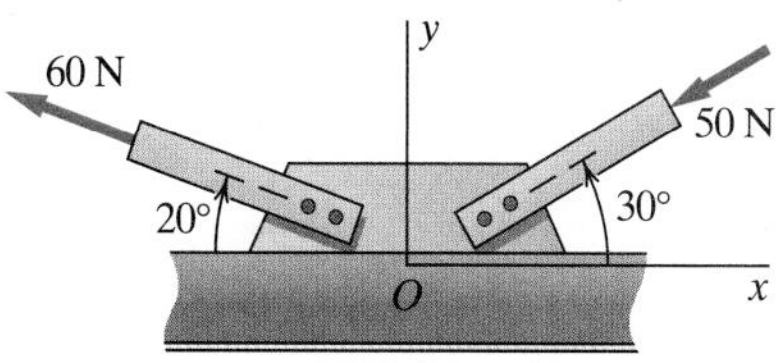

Figure 4.63 Component addition

Goal We are to find the sum of two forces using three specified approaches to vector addition.

Given We are presented with a physical situation of two forces applied to a gusset plate. We know the magnitudes of the forces and their directions. Axes have been specified. Because the forces are in the plane of the paper, they are planar.

Assume We will assume that it is satisfactory to present the result either in terms of magnitude and spherical angles or in vector notation.

Draw In order to add the forces graphically, we lay them out in a scaled drawing. We show graphical addition using the head-to-tail approach and the parallelogram law approach.

Approach 1 (Head-to-Tail Approach, as in Figure 4.64): **(a)** First, lay out a horizontal x axis on a sheet of graph paper, assign a unit length along this axis, and establish an origin O along it. To the right of this origin is the $+x$ axis, and to the left is the $-x$ axis. Also establish a scale (e.g., 1 cm = 25 N). Then draw a vector arrow representing either the 60-N or 50-N force. It doesn't matter which you begin with, so let's arbitrarily consider the 60-N force first. Place the tail of the arrow at the origin and use a protractor to measure 20° up from the $-x$ axis. Now draw the 60-N force from the origin at the 20° orientation. Next slide the 50 N force at 30° to the horizontal so that its tail coincides with the head

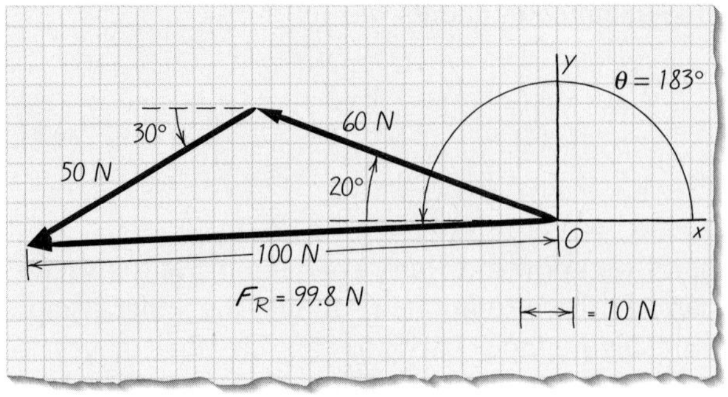

Figure 4.64 Head-to-Tail approach

of the 60-N force. Make sure that this 50-N arrow remains in its same orientation with respect to a horizontal line. Draw the 50-N force.

The resultant force is represented by an arrow drawn from the tail of the first vector to the head of the second. The length of this arrow is the magnitude of the resultant force acting on the plate (100 N), and the angle θ of the arrow (as measured with a protractor) is the direction of that force (183° counterclockwise from the $+x$ axis). Alternately, we could measure the x and y scalar components of $\boldsymbol{F_R}$ from the figure and would find that $F_{Rx} = -99.7$ N and $F_{Ry} = -4.5$ N.

Approach 2 (Parallelogram Law Approach): **(a)** Lay out arrows representing the forces as shown in Figure 4.65—tail to tail and each making the proper angle with the x axis. Now create a parallelogram having these two arrows as two of its sides. The resultant force is the diagonal of this parallelogram that originates at O. It is 100 N and is oriented at $\theta = 183°$ (as measured with a protractor). Alternately, we could measure the x and y scalar components of $\boldsymbol{F_R}$ from the figure and would find that $F_{Rx} = -99.7$ N and $F_{Ry} = -4.5$ N.

Answer to (a) $\|\boldsymbol{F_R}\| = 100.0\text{ N}, \theta = 183°$ or $\boldsymbol{F_R} = -99.7\text{ N}\boldsymbol{i} - 4.5\text{ N}\boldsymbol{j}$

Comment: It would take very careful layout work to get this level of accuracy in the answer using these graphical approaches.

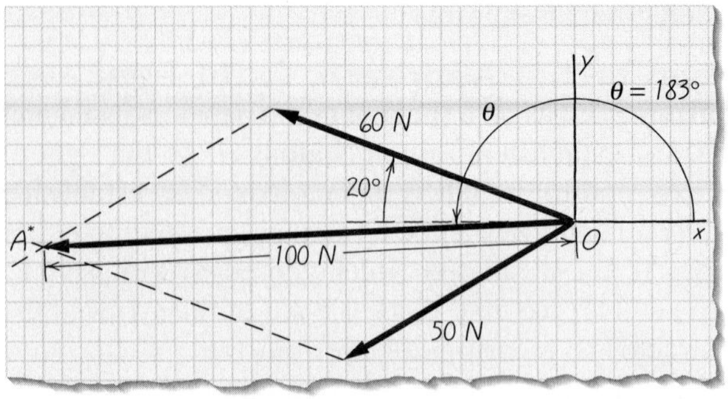

Figure 4.65 Parallelogram Law approach

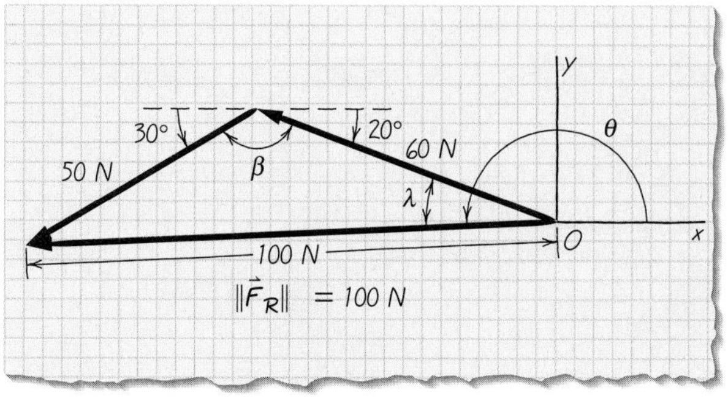

Figure 4.66 Sketch supporting geometric/trigonometric force addition

Formulate Equations and Solve **(b)** We will use geometric/trigonometric addition to find the resultant $\boldsymbol{F_R}$. We start by drawing the vector arrows head to tail, as in **Figure 4.66**. The 50-N, 60-N, and resultant forces form a triangle, as shown. Based on the geometry, the angle β between the two force arrows when they are laid out head to tail is 130° (= 180° − 20° − 30°). Now we use the law of cosines to find the magnitude of the resultant force,

$$\| \boldsymbol{F_R} \| = \sqrt{(60\text{ N})^2 + (50\text{ N})^2 - 2(60\text{ N})(50\text{ N})\cos 130^\circ} = 99.8\text{ N}$$

and the law of sines to find λ (the angle between the 60-N force and the resultant),

$$\frac{\sin 130^\circ}{99.8\text{ N}} = \frac{\sin\lambda}{50\text{ N}}$$

from which we find $\lambda = 22.5^\circ$. Therefore $\theta = 160^\circ + 22.5^\circ = 182.5^\circ$.

Alternately, we could find the x and y components of $\boldsymbol{F_R}$:

$$F_{Rx} = \| \boldsymbol{F_R} \| \cos\theta = (99.8\text{ N})\cos 182.5^\circ = -99.7\text{ N}$$
$$F_{Ry} = \| \boldsymbol{F_R} \| \sin\theta = (-99.7\text{ N})\sin 182.5^\circ = -4.5\text{ N}$$

Answer to (b) $\| \boldsymbol{F_R} \| = 99.8\text{ N}, \theta = 182.5^\circ$ or $\boldsymbol{F_R} = -99.7\text{ N}\boldsymbol{i} - 4.5\text{ N}\boldsymbol{j}$

Comment: Notice that we are able to get more accuracy in our answer using geometric/trigonometric addition than with graphical methods.

Formulate Equations and Solve **(c)** In order to use component addition, we first need to find the scalar components of each force. For the 60-N force (which we will call $\boldsymbol{F_{60}}$), we find its x component $F_{60,x}$ and its y component $F_{60,y}$ (**Figure 4.67*a***):

$$F_{60,x} = (-60\text{ N})\cos 20^\circ = -56.4\text{ N}$$
$$F_{60,y} = (60\text{ N})\sin 20^\circ = 20.5\text{ N}$$

For the 50-N force (which we will call $\boldsymbol{F_{50}}$), the components are (**Figure 4.67*b***):

$$F_{50,x} = (-50\text{ N})\cos 30^\circ = -43.3\text{ N}$$
$$F_{50,y} = (-50\text{ N})\sin 30^\circ = -25.0\text{ N}$$

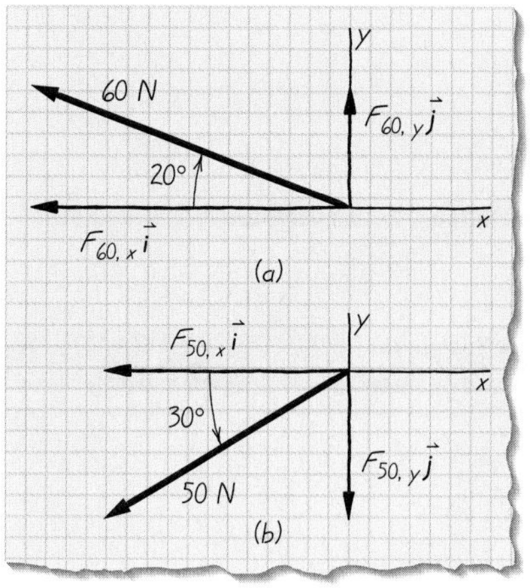

Figure 4.67 Sketch supporting component addition

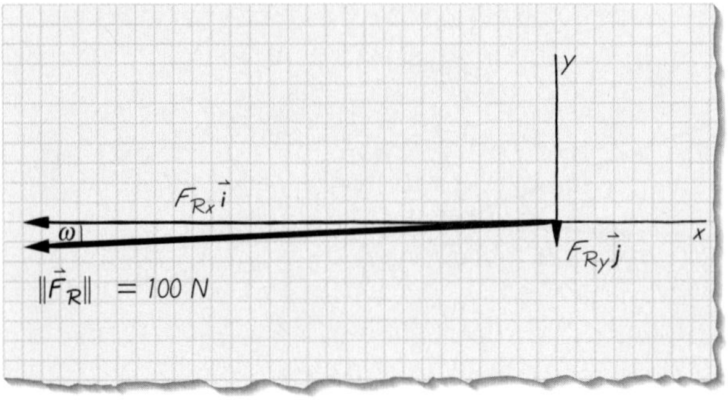

Figure 4.68

Next we combine these components to find the components of the resultant force $\boldsymbol{F_R}$:

$$F_{R,x} = F_{60,x} + F_{50,x} = -56.4\text{ N} - 43.3\text{ N} = -99.7\text{ N}$$
$$F_{R,y} = F_{60,y} + F_{50,y} = 20.5\text{ N} - 25.0\text{ N} = -4.5\text{ N}$$

The magnitude of $\boldsymbol{F_R}$ is found using (4.12):

$$\| \boldsymbol{F_R} \| = \sqrt{(-99.7\text{ N})^2 + (-4.5\text{ N})^2} = 99.8\text{ N}$$

From Figure 4.68, we find $\omega = \tan^{-1}(F_{R,y}/F_{R,x}) = 2.58°$. Therefore $\theta = 180° + 2.58° = 182.6°$.

Answer to (c) $\| \boldsymbol{F_R} \| = 99.8\text{ N}, \theta = 182.5°$ or $\boldsymbol{F_R} = -99.7\text{ N}\boldsymbol{i} - 4.5\text{ N}\boldsymbol{j}$

Check We could check the answer by using two different methods of addition. A more qualitative check involves noting that, given the orientations and magnitudes of the 50-N and 60-N forces, the x and y components of the resultant should both be negative (they are: −99.7 N and −4.5 N), and the magnitude of the x component should be significantly bigger than the magnitude of the y component (it is: 99.7 N vs. 4.5 N).

Comment: Using geometric/trigonometric addition or component addition gives a more accurate solution than graphical addition because the accuracy of the graphical solution is limited by the accuracy of the drawings. However, drawings are useful and recommended no matter which approach you use.

EXAMPLE 4.11 VECTOR ADDITION PRACTICE (B)

Two forces act on a frame as shown in Figure 4.69. Find the resultant force due to these two forces.

Goal We are to find the resultant force due to the forces $\boldsymbol{F_1}$ and $\boldsymbol{F_2}$. In other words, we are to find $\boldsymbol{F_R}$, where $\boldsymbol{F_R} = \boldsymbol{F_1} + \boldsymbol{F_2}$.

Given We are given the magnitudes of the two forces. (Note that 1 kip = 1000 lb; a "kip" is a "*kilo*pound.") Based on the geometry in **Figure 4.69**, we are able to find the directions of the two forces.

Assume We assume that the frame and forces all lie in the plane of the paper.

Draw At first glance, $\boldsymbol{F_1}$ and $\boldsymbol{F_2}$ appear not to be concurrent forces, but by extending the line of action of $\boldsymbol{F_1}$ toward $\boldsymbol{F_2}$, you can see that they are concurrent. We draw the diagram of **Figure 4.70**, establishing a right-handed coordinate system with its origin arbitrarily placed at A.

Formulate Equations and Solve Now we go about adding $\boldsymbol{F_1}$ and $\boldsymbol{F_2}$ to find their resultant. We could add them graphically, or by using geometry and trigonometry, or by adding components. We will illustrate these latter two approaches. Both require that we define the forces relative to the coordinate system. Therefore, based on the geometry in **Figure 4.70**, the space angles for $\boldsymbol{F_1}$ are

$$\theta_{x,F_1} = 30°, \theta_{y,F_1} = 60°, \theta_{z,F_1} = 90° \qquad \text{(because } \boldsymbol{F_1} \text{ is in the } xy \text{ plane)}$$

and the space angles for $\boldsymbol{F_2}$ are

$$\theta_{x,F_2} = 30°, \theta_{y,F_2} = 120°, \theta_{z,F_2} = 90° \qquad \text{(because } \boldsymbol{F_2} \text{ is in the } xy \text{ plane)}$$

Approach 1 (Adding the Forces Using Geometry and Trigonometry): We lay out the forces head to tail in a sketch (Figure 4.71). Because the addition of vectors is commutative, either of the sketches in **Figure 4.71*a*** or ***b*** can be used; we arbitrarily choose the sketch in **Figure 4.71*a***. Based on the geometry in **Figure 4.71*a***, the included angle β_1 between $\boldsymbol{F_1}$ and $\boldsymbol{F_2}$ is 120° (**Figure 4.71*c***). Now we use the law of cosines to find the magnitude of the resultant $\boldsymbol{F_R}$:

$$\| \boldsymbol{F_R} \|^2 = \| \boldsymbol{F_1} \|^2 + \| \boldsymbol{F_2} \|^2 - 2 \| \boldsymbol{F_1} \| \, \| \boldsymbol{F_2} \| \cos \beta_1$$
$$\| \boldsymbol{F_R} \|^2 = (5 \text{ kip})^2 + (3 \text{ kip})^2 - 2(5 \text{ kip})(3 \text{ kip}) \cos 120°$$

Therefore

$$\| \boldsymbol{F_R} \| = 7 \text{ kip}$$

Using the law of sines, we find the direction (in terms of the space angles) of $\boldsymbol{F_R}$ (**Figure 4.71a**):

$$\frac{\sin \beta_1}{\| \boldsymbol{F_R} \|} = \frac{\sin \beta_2}{\| \boldsymbol{F_2} \|}$$
$$\frac{\sin 120°}{7 \text{ kip}} = \frac{\sin \beta_2}{3 \text{ kip}}$$

from which we calculate $\beta_2 = 21.8°$. Therefore, the space angles associated with $\boldsymbol{F_R}$ are

$$\theta_{x,F_R} = 30° - 21.8° = 8.2°$$
$$\theta_{y,F_R} = 90° - 8.21° = 81.8°$$

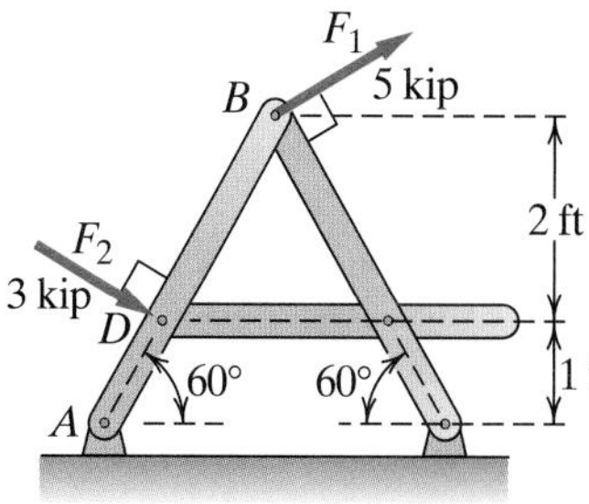

Figure 4.69

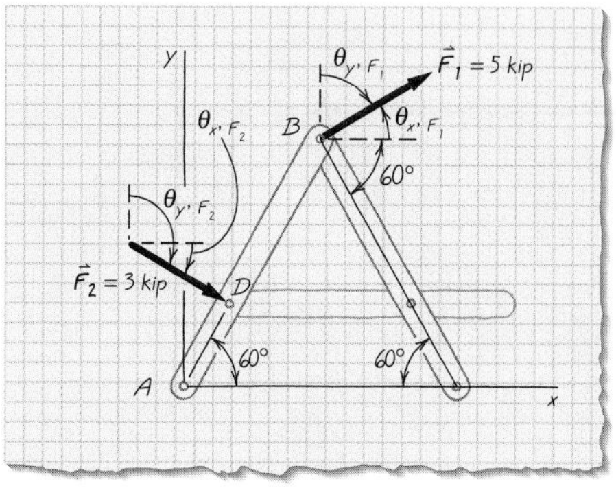

Figure 4.70

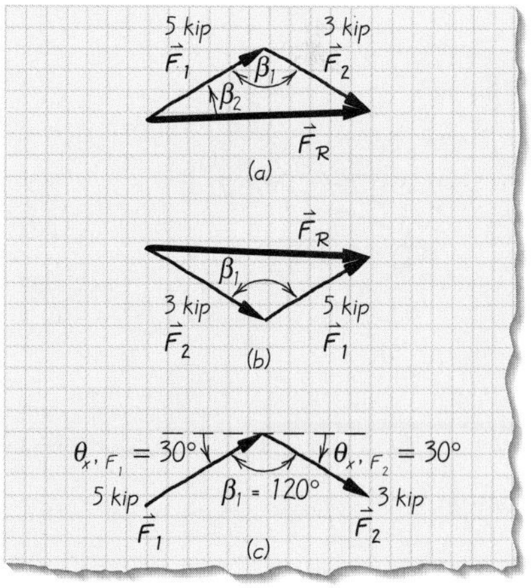

Figure 4.71

We can specify $\boldsymbol{F_R}$ in terms of magnitude and space angles:

$$\| \boldsymbol{F_R} \| = 7 \text{ kip}, \qquad \theta_{x,F_R} = 8.2°, \theta_{y,F_R} = 81.8°, \theta_{z,F_R} = 90°$$

Alternatively, we can specify $\boldsymbol{F_R}$ in vector notation:

$$\boldsymbol{F_R} = 7 \text{ kip}(\cos 8.2° \, \boldsymbol{i} + \cos 81.8° \, \boldsymbol{j})$$
$$\boldsymbol{F_R} = 6.93 \text{ kip } \boldsymbol{i} + 0.998 \text{ kip } \boldsymbol{j}$$

Approach 2 (Component Addition): With component addition, we first recall from **Figure 4.70** and the text accompanying this illustration that

$$F_{1x} = (5 \text{ kip}) \cos 30° = +4.33 \text{ kip}$$
$$F_{1y} = (5 \text{ kip}) \sin 30° = +2.50 \text{ kip}$$
$$F_{2x} = (3 \text{ kip}) \cos 30° = +2.60 \text{ kip}$$
$$F_{2y} = (-3 \text{ kip}) \sin 30° = -1.50 \text{ kip}$$

Combining these components to find the resultant force $\boldsymbol{F_R} = F_{Rx}\boldsymbol{i} + F_{Ry}\boldsymbol{j}$, we get

$$F_{Rx} = F_{1x} + F_{2x} = 4.33 \text{ kip} + 2.60 \text{ kip} = 6.93 \text{ kip}$$
$$F_{Ry} = F_{1y} + F_{2y} = +2.50 \text{ kip} - 1.50 \text{ kip} = 1.00 \text{ kip}$$
$$\boldsymbol{F_R} = 6.93 \text{ kip } \boldsymbol{i} + 1.00 \text{ kip } \boldsymbol{j}$$

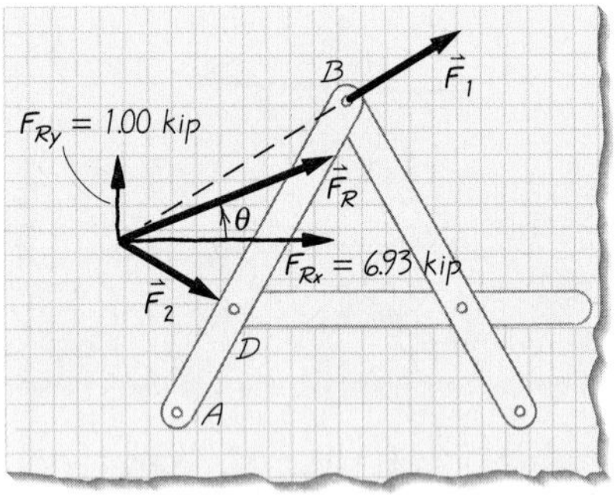

Figure 4.72

as shown in Figure 4.72 in vector notation. We now use (4.12) to get the magnitude of $\boldsymbol{F_R}$:

$$\| \boldsymbol{F_R} \| = \sqrt{6.93^2 + 1.00^2} \text{ kip} = 7.00 \text{ kip}$$

From **Figure 4.72**, we find that $\theta_{x,F_R} = \tan^{-1}(F_{Ry}/F_{Rx}) = 8.2°$ and $\theta_{y,F_R} = 90° - 8.2° = 81.8°$.

If you are wondering why the tail of $\boldsymbol{F_R}$ is located at what seems like an arbitrary point in **Figure 4.72**, recall that forces must be concurrent if we are to add them. As **Figure 4.72** shows, the tail of $\boldsymbol{F_R}$ lies at the point where the lines of action of concurrent forces $\boldsymbol{F_1}$ and $\boldsymbol{F_2}$ of **Figure 4.69** intersect.

Answer In terms of magnitude and space angles:
$\| \boldsymbol{F_R} \| = 7$ kip, $\theta_{x,F_R} = 8.2°$, $\theta_{y,F_R} = 81.8°$, $\theta_{z,F_R} = 90.0°$.

In terms of vector notation: $\boldsymbol{F_R} = 6.93 \text{ kip}\boldsymbol{i} + 1.00 \text{ kip}\boldsymbol{j}$.

Check We can use the answer from Approach 1 as a check for the answer from Approach 2. In addition, we can check the answer for reasonableness. Given the orientations of $\boldsymbol{F_1}$ and $\boldsymbol{F_2}$ in **Figure 4.69**, we would expect their resultant to have positive x and y components; it does. Also based on the sketch, we would expect θ_{x,F_R} to be a small angle; it is.

EXAMPLE 4.12 VECTOR ADDITION PRACTICE (C)

The boom BD in Figure 4.73 is held up by a continuous cable that runs from A to B to C. The cable is attached to the wall via hooks at A and C. At B the cable passes through an eyelet (see detail). The tension in the cable is 5 kN. Find

(a) the resultant force the cable exerts on the boom (in vector notation)
(b) the direction cosines of this resultant force

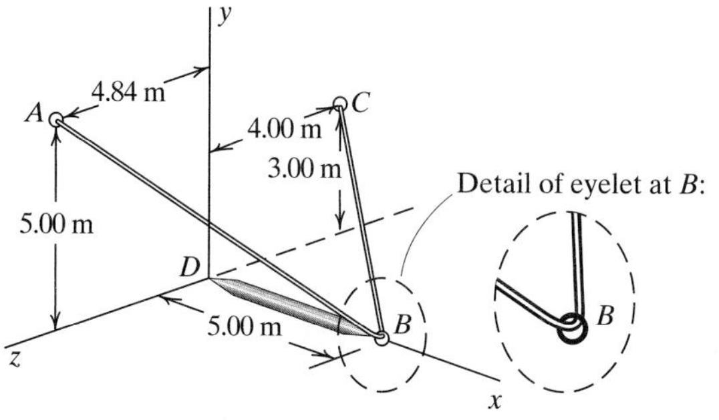

Figure 4.73

Goal We are to find the resultant force from the cable attached to the boom at B (a). In addition, we are to find the direction cosines of this resultant force (b).

Given We are given the locations of points A, B, and C. We are also told that the tension in the continuous cable that runs from A to B to C is 5 kN.

Assume We assume that the tension in the portion of the cable that runs from B to A is 5 kN. Similarly, since the cable is continuous, we assume the tension in the cable between B and C is also 5 kN.

Draw The cable exerts two forces at B—one with a line of action that runs from B to A, and the other with a line of action that runs from B to C (as shown in Figure 4.74). Each force is of magnitude 5 kN. Notice that $\boldsymbol{T_{BA}}$ and $\boldsymbol{T_{BC}}$ are drawn to indicate tension in the cable.

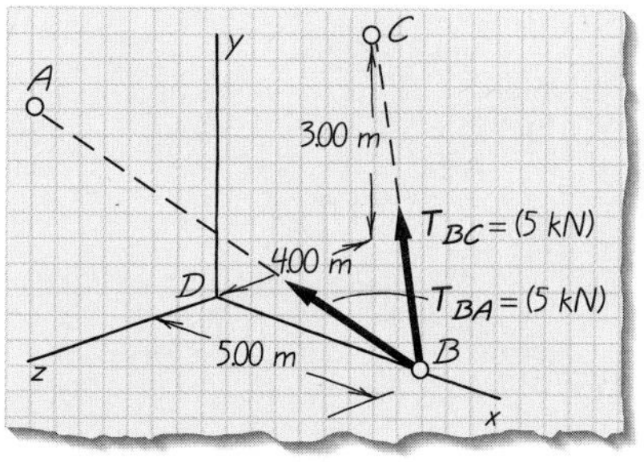

Figure 4.74

Formulate Equations and Solve **(a)** We will find the scalar components of the two forces acting at B ($\boldsymbol{T_{BA}}$ and $\boldsymbol{T_{BC}}$), then add the components to find the resultant force at B.

First consider $\boldsymbol{T_{BC}}$. Its line of action is along BC. First we find the direction cosines for this line of action, then use them to find the scalar components of $\boldsymbol{T_{BC}}$. The length L_{BC} from B to C is (Figure 4.75*a*):

$$L_{BC} = \sqrt{(x_C - x_B)^2 + (y_C - y_B)^2 + (z_C - z_B)^2}$$
$$L_{BC} = \sqrt{(0 - 5.00)^2 + (3.00 - 0)^2 + (-4.00 - 0)^2}\ \text{m} = 7.07\ \text{m}$$

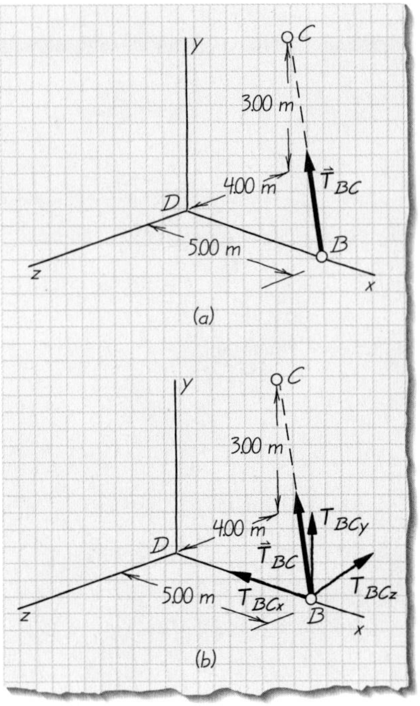

Figure 4.75

We can write $\boldsymbol{T_{BC}}$ as

$$\boldsymbol{T_{BC}} = \| \boldsymbol{T_{BC}} \| \boldsymbol{u_{BC}}$$

where $\boldsymbol{u_{BC}}$ is a unit vector in the direction from B to C. The scalar components of $\boldsymbol{u_{BC}}$ are

$$u_{BCx} = \frac{x_C - x_B}{L_{BC}} = \frac{-5.00 \text{ m}}{7.07 \text{ m}} = -0.707$$

$$u_{BCy} = \frac{y_C - y_B}{L_{BC}} = \frac{3.00 \text{ m}}{7.07 \text{ m}} = 0.424$$

$$u_{BCz} = \frac{z_C - z_B}{L_{BC}} = \frac{-4.00 \text{ m}}{7.07 \text{ m}} = -0.566$$

Therefore we can write $\boldsymbol{T_{BC}}$ in vector notation as

$$\boldsymbol{T_{BC}} = 5 \text{ kN} \,(-0.707\boldsymbol{i} + 0.424\,\boldsymbol{j} - 0.566\boldsymbol{k})$$

$$\boldsymbol{T_{BC}} = -3.54 \text{ kN}\boldsymbol{i} + 2.12 \text{ kN}\,\boldsymbol{j} - 2.83 \text{ kN}\boldsymbol{k}$$

See **Figure 4.75*b***, and check **Figure 4.74** to confirm that the size and direction of these components make physical sense.

The line of action of $\boldsymbol{T_{BA}}$ is along BA. First we find the direction cosines for this line of action, then use them to find the scalar components of $\boldsymbol{T_{BA}}$. The length L_{BA} from B to A is (**Figure 4.76*a***):

$$L_{BA} = \sqrt{(x_A - x_B)^2 + (y_A - y_B)^2 + (z_A - z_B)^2}$$

$$L_{BA} = \sqrt{(0 - 5.00)^2 + (5.00 - 0)^2 + (4.84 - 0)^2} \text{ m} = 8.57 \text{ m}$$

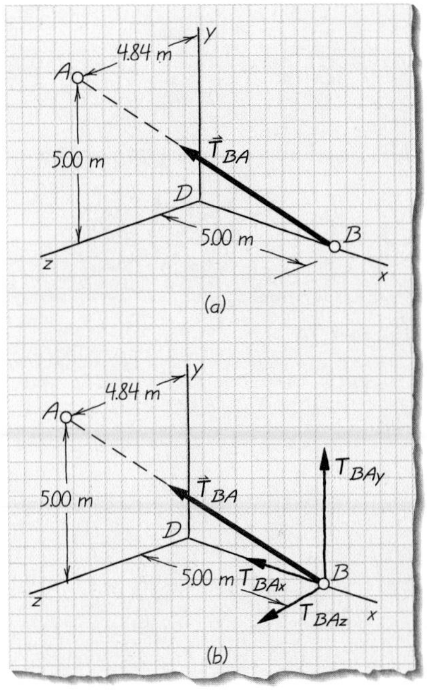

Figure 4.76

We can write $\boldsymbol{T_{BA}}$ as

$$\boldsymbol{T_{BA}} = \| \boldsymbol{T_{BA}} \| \boldsymbol{u_{BA}}$$

where $\boldsymbol{u_{BA}}$ is a unit vector in the direction from B to A. The scalar components of $\boldsymbol{u_{BA}}$ are

$$u_{BAx} = \frac{x_A - x_B}{L_{BA}} = \frac{-5.00 \text{ m}}{8.57 \text{ m}} = -0.584$$

$$u_{BAy} = \frac{y_A - y_B}{L_{BA}} = \frac{5.00 \text{ m}}{8.57 \text{ m}} = 0.584$$

$$u_{BAz} = \frac{z_A - z_B}{L_{BA}} = \frac{4.84 \text{ m}}{8.57 \text{ m}} = 0.565$$

Therefore we can write $\boldsymbol{T_{BC}}$ in vector notation as

$$\boldsymbol{T_{BA}} = 5 \text{ kN} \,(-0.584\boldsymbol{i} + 0.584\boldsymbol{j} + 0.565\,\boldsymbol{k})$$

$$\boldsymbol{T_{BA}} = -2.92 \text{ kN}\boldsymbol{i} + 2.92 \text{ kN}\boldsymbol{j} + 2.83 \text{ kN}\boldsymbol{k}$$

See **Figure 4.76*b***.

Now we add the scalar components of $\boldsymbol{T}_{BC}$ and $\boldsymbol{T}_{BA}$ to find the resultant force $\boldsymbol{T}_R$ at B:

$$T_{Rx} = T_{BCx} + T_{BAx} = -3.54 \text{ kN} - 2.92 \text{ kN} = -6.46 \text{ kN}$$
$$T_{Ry} = T_{BCy} + T_{BAy} = 2.12 \text{ kN} + 2.92 \text{ kN} = 5.04 \text{ kN}$$
$$T_{Rz} = T_{BCz} + T_{BAz} = -2.83 \text{ kN} + 2.83 \text{ kN} = 0 \text{ kN}$$

The resultant vector (**Figure 4.77**) is

$$\boldsymbol{T}_R = -6.46 \text{ kN}\boldsymbol{i} + 5.04 \text{ kN}\boldsymbol{j}$$

Answer to (a) $\boldsymbol{T}_R = -6.46 \text{ kN}\boldsymbol{i} + 5.04 \text{ kN}\boldsymbol{j}$

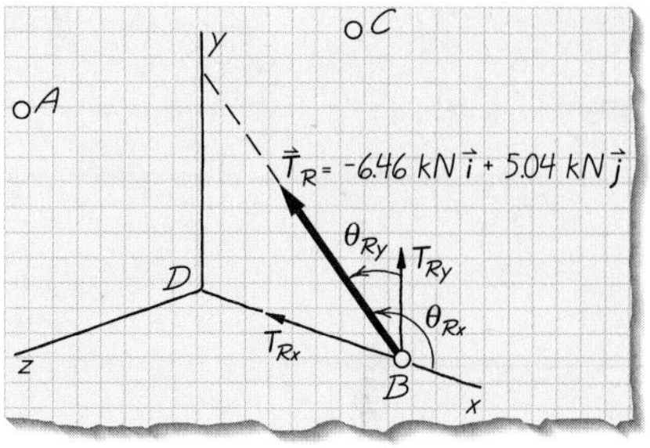

Figure 4.77

(b) Now we find the direction cosines of this resultant force. Since we know the scalar components of $\boldsymbol{T}_R$ from (a), we can use them in (4.13) to find the direction cosines of $\boldsymbol{T}_R$. To use (4.13) we need to know the magnitude of $\boldsymbol{T}_R$.

$$\| \boldsymbol{T}_R \| = \sqrt{T_{Rx}^2 + T_{Ry}^2 + T_{Rz}^2} = \sqrt{(-6.46)^2 + (5.04)^2 + (0)^2} \text{ kN} = 8.19 \text{ kN}$$

$$\cos\theta_{Rx} = \frac{T_{Rx}}{\| \boldsymbol{T}_R \|} = \frac{-6.46 \text{ kN}}{8.19 \text{ kN}} = -0.789 \text{ (therefore } \theta_{Rx} = 142°)$$
$$\cos\theta_{Ry} = \frac{T_{Ry}}{\| \boldsymbol{T}_R \|} = \frac{5.04 \text{ kN}}{8.19 \text{ kN}} = 0.615 \text{ (therefore } \theta_{Ry} = 52.0°)$$
$$\cos\theta_{Rz} = \frac{T_{Rz}}{\| \boldsymbol{T}_R \|} = \frac{0 \text{ kN}}{8.19 \text{ kN}} = 0 \text{ (therefore } \theta_{Rz} = 90.0°)$$

Answer to (b) $\cos\theta_{Rx} = -0.789$, $\cos\theta_{Ry} = 0.615$, $\cos\theta_{Rz} = 0$

Check Revisit **Figure 4.74** to confirm that the sizes of the space angles make physical sense.

Summary

We have presented three methods of vector addition for finding either the magnitude and direction of a resultant force or the scalar components of a resultant force. These methods can also be used when two forces are being added, but all you know are the magnitude and direction of one force, the direction of the other force, and the direction of the resultant. With only these four pieces of information, you can find the two unknown magnitudes. In fact, all these addition methods work any time there are at most two unknowns when adding planar forces and at most three unknowns when adding nonplanar forces.

EXAMPLE 4.13 VECTOR ADDITION PRACTICE (D)

A boat is to be pulled onto the shore by two ropes. The force $\boldsymbol{T}$ exerted by one of the ropes acts at 30° to the keel line *a-a*, as shown in **Figure 4.78**. Call the force exerted by the other rope $\boldsymbol{P}$. The resultant of forces $\boldsymbol{T}$ and $\boldsymbol{P}$ has a magnitude of 300 lb and is directed along the keel line. Find the magnitudes of $\boldsymbol{T}$ and $\boldsymbol{P}$ and the angle θ at which $\| \boldsymbol{P} \|$ is a minimum.

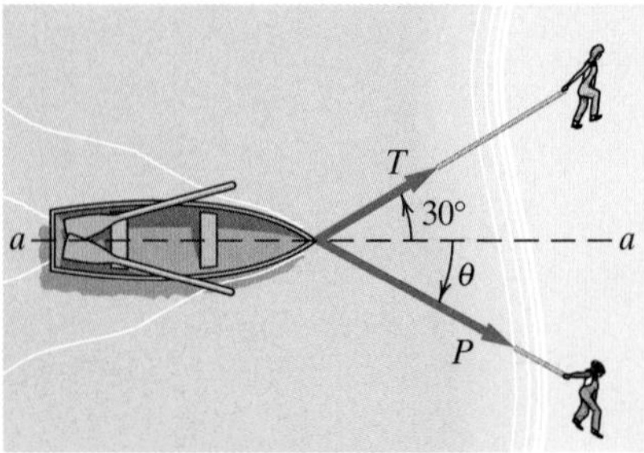

Figure 4.78

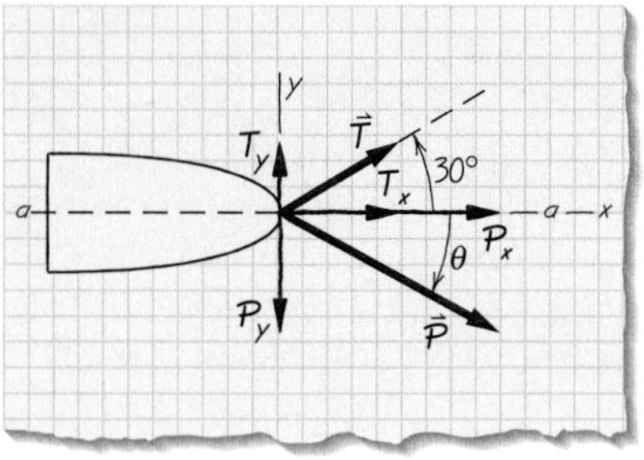

Figure 4.79

Goal We are to find the magnitudes of the tensions in the two ropes such that the resultant of the two tensions has a magnitude of 300 lb and is directed along the keel line. Furthermore, we are to find θ (the angle between $\boldsymbol{P}$ and the keel line) such that $\|\boldsymbol{P}\|$ is to be as small as possible.

Given We are given the direction of $\boldsymbol{T}$ relative to the keel line and that the resultant of $\boldsymbol{T}$ and $\boldsymbol{P}$ has a magnitude of 300 lb.

Assume Assume that $\boldsymbol{P}$, $\boldsymbol{T}$, and the keel line all lie in the same plane.

Draw Define a set of reference axes, as shown in Figure 4.79. Aligning the x axis with the resultant force and placing the origin at the point where the ropes are attached make setting up the equations straightforward.

Formulate Equations and Solve We are looking for three quantities: $\|\boldsymbol{T}\|$, $\|\boldsymbol{P}\|$, and θ. The three conditions of the problem are that:

A. The resultant has a magnitude of 300 lb
B. The resultant is directed along the x axis (i.e., it has no y component because it is along the keel line, which is aligned with the x axis)
C. $\|\boldsymbol{P}\|$ is as small as possible while still meeting conditions A and B

By enforcing these three conditions, we will be able to find the three unknown quantities.

We start by considering the x and y components of $\boldsymbol{P}$ and $\boldsymbol{T}$. We want the sum of the x components to be 300 lb (condition A) and the sum of the y components to be zero (condition B). These two conditions give us two equations:

$$T_x + P_x = \|\boldsymbol{T}\| \cos 30° + \|\boldsymbol{P}\| \cos\theta = 300 \text{ lb} \quad (1)$$
$$T_y + P_y = \|\boldsymbol{T}\| \sin 30° - \|\boldsymbol{P}\| \sin\theta = 0 \text{ lb} \quad (2)$$

We can meet condition C by first solving these two equations for $\|\boldsymbol{P}\|$ as a function of θ. We can either solve (1) for $\|\boldsymbol{T}\|$ as a function of (θ, $\|\boldsymbol{P}\|$), then substitute this value for $\|\boldsymbol{T}\|$ into (2) and find an expression for $\|\boldsymbol{P}\|$ as a function of θ, or we can solve (2) for $\|\boldsymbol{T}\|$ as a function of (θ, $\|\boldsymbol{P}\|$), then substitute this value for $\|\boldsymbol{T}\|$ into (1) and find an expression for $\|\boldsymbol{P}\|$ as a function of θ. We arbitrarily take the latter approach:

Solve (2) for $\|\boldsymbol{T}\|$:
$$\|\boldsymbol{T}\| = \frac{\|\boldsymbol{P}\| \sin\theta}{\sin 30°} \quad (2')$$

Substitute (2′) into (1) and rearrange:

$$\|\boldsymbol{P}\| = \frac{300 \text{ lb}}{\left(\dfrac{\sin\theta}{\tan 30°} + \cos\theta\right)} \quad (3)$$

We now show two approaches to finding the combination of θ and $\|\boldsymbol{P}\|$ that minimizes $\|\boldsymbol{P}\|$ (condition C).

Approach 1 (Graphing): Substitute various values of θ into equation (3) and solve for $\| \boldsymbol{P} \|$. We expect $0° < \theta \le 90°$. (Why don't we look at $\theta = 0$?) Plot these combinations of θ and $\| \boldsymbol{P} \|$ to see where $\| \boldsymbol{P} \|$ is a minimum:

θ (degrees)	$\| \boldsymbol{P} \|$ from (3) (lb)
10	233
20	196
30	173
40	160
50	152
60	150
70	152
80	160
90	173

From the graph in **Figure 4.80** we see that $\| \boldsymbol{P} \|$ is smallest when $\theta = 60°$.

We find $\| \boldsymbol{T} \|$ by substituting the minimum value of $\| \boldsymbol{P} \| = 150$ lb and $\theta = 60°$ into (2′):

$$\| \boldsymbol{T} \| = \frac{(150 \text{ lb}) \sin 60°}{\sin 30°} = 260 \text{ lb}$$

Answer $\theta = 60°$, $\| \boldsymbol{P} \| = 150$ lb, $\| \boldsymbol{T} \| = 260$ lb.

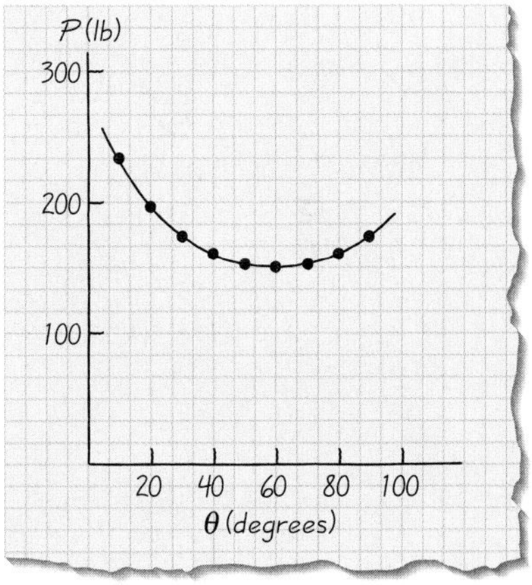

Figure 4.80

Approach 2 (Using Calculus): If the function described by (3) has a local minimum (or maximum) for θ in the range $0° < \theta \le 90°$, the derivative of the function will be zero at this minimum (or maximum). Take the derivative of (3) with respect to θ and set it equal to zero (because the derivative of a function is zero at a minimum or maximum):

$$\frac{dP}{d\theta} = \frac{-300 \text{ lb}\left(\dfrac{\cos\theta}{\tan 30°} - \sin\theta\right)}{\left(\dfrac{\sin\theta}{\tan 30°} + \cos\theta\right)^2} = 0 \qquad (4)$$

Multiplying both sides of (4) by

$$\left(\frac{\sin\theta}{\tan 30°} + \cos\theta\right)^2$$

and dividing by 300 lb, we have

$$\frac{\cos\theta}{\tan 30°} - \sin\theta = 0$$

Next we can divide both sides by $\cos\theta$ to get

$$\frac{1}{\tan 30^\circ} - \tan\theta = 0$$

which means

$$\tan\theta = \frac{1}{\tan 30^\circ}$$

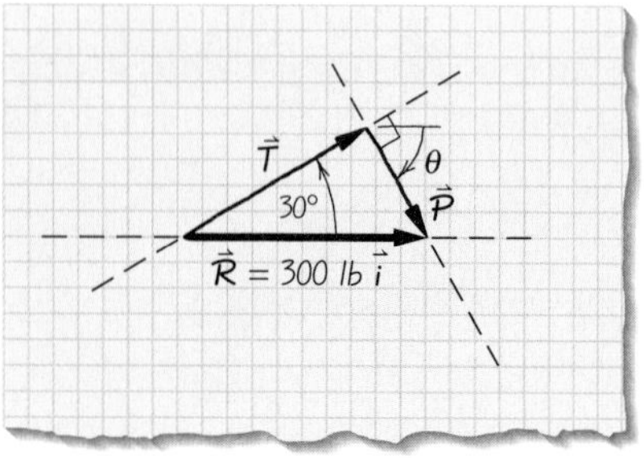

Figure 4.81

Solving for θ, we find that $\theta = 60^\circ$. When $\theta = 60^\circ$ is substituted back into (3), we find that $\|\boldsymbol{P}\| = 150$ lb.

We find $\|\boldsymbol{T}\|$ by substituting $\|\boldsymbol{P}\| = 150$ lb and $\theta = 60^\circ$ into (2):

$$\|\boldsymbol{T}\| = (150 \text{ lb})\frac{\sin 60^\circ}{\sin 30^\circ} = 260 \text{ lb}$$

Answer $\theta = 60^\circ$, $\|\boldsymbol{P}\| = 150$ lb, $\|\boldsymbol{T}\| = 260$ lb.

Figure 4.81 shows $\boldsymbol{P}$, $\boldsymbol{T}$, and their resultant $\boldsymbol{R}$.

Check The answers can be checked by substituting the values of θ, $\|\boldsymbol{P}\|$, and $\|\boldsymbol{T}\|$ back into (1) and (2). Also notice from **Figure 4.81** that the angle between $\boldsymbol{T}$ and $\boldsymbol{P}$ is 90°; does this make physical sense? If $\theta < 60^\circ$, $\|\boldsymbol{P}\|$ will need to be larger than 150 lb so that its y component cancels the y component of $\boldsymbol{T}$ (because of condition B). At $\theta > 60^\circ$, $\|\boldsymbol{P}\|$ will need to be larger than 150 lb so that its x component when added to the x component of $\boldsymbol{T}$ sums to 300 lb (because of condition A). Always check your answers to make sure they make physical sense.

Comments on Approaches and Answers: The two approaches use different skills (e.g., drawing, calculus, algebra). Both approaches are based on writing the problem statement requirements in numerical form. The check presented requires insight into the nature of the problem and how forces act.

EXERCISES 4.6

Use E4.6.1 to answer **Exercises 4.6.1–4.6.3**. Two forces $\boldsymbol{F}_1$ and $\boldsymbol{F}_2$ are applied to the bracket as shown. Write the resultant $\boldsymbol{F}_R$ (where $\boldsymbol{F}_R = \boldsymbol{F}_1 + \boldsymbol{F}_2$) in terms of $\boldsymbol{i}$ and $\boldsymbol{j}$.

4.6.1. Use graphical force addition.

4.6.2. Use geometric/trigonometric force addition.

4.6.3. Use component force addition.

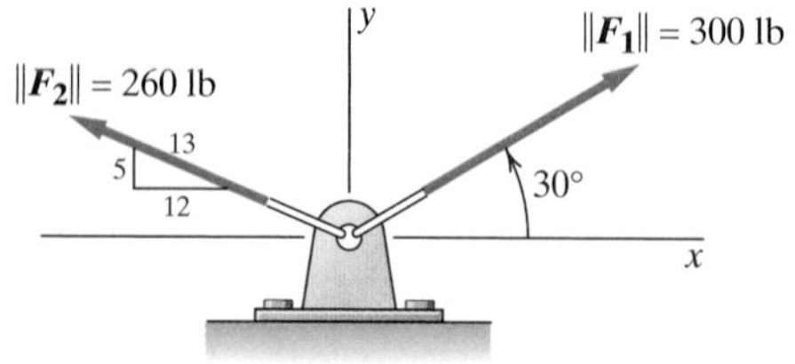

E4.6.1

Use E4.6.4 to answer **Exercises 4.6.4–4.6.6**. Two forces $\boldsymbol{F}_1$ and $\boldsymbol{F}_2$ are applied to the bracket as shown. Write the resultant $\boldsymbol{F}_R$ (where $\boldsymbol{F}_R = \boldsymbol{F}_1 + \boldsymbol{F}_2$) in terms of $\boldsymbol{i}$ and $\boldsymbol{j}$.

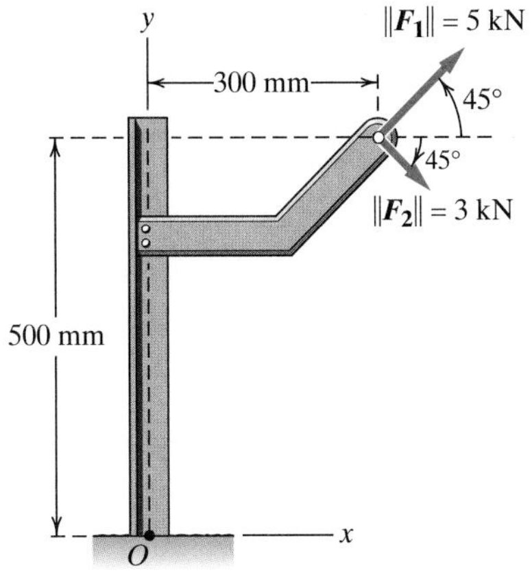

E4.6.4

4.6.4. Use graphical force addition.

4.6.5. Use geometric/trigonometric force addition.

4.6.6. Use component force addition.

4.6.7. Two forces $\boldsymbol{F}_1$ and $\boldsymbol{F}_2$ are applied to the bracket as shown in E4.6.7. Write the resultant $\boldsymbol{F}_R$ (where $\boldsymbol{F}_R = \boldsymbol{F}_1 + \boldsymbol{F}_2$) in terms of $\boldsymbol{i}$ and $\boldsymbol{j}$.

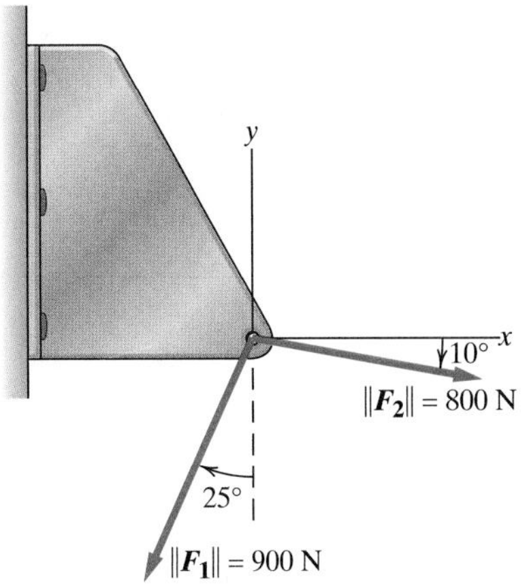

E4.6.7

4.6.8. Two structural members are pinned to the support at O. Tension forces $\boldsymbol{T}_1$ and $\boldsymbol{T}_2$ act on the members, as shown in E4.6.8.

a. Determine the magnitude of the resultant $\boldsymbol{T}_R$ (where $\boldsymbol{T}_R = \boldsymbol{T}_1 + \boldsymbol{T}_2$) and its space angles.

b. Write $\boldsymbol{T}_R$ in vector notation.

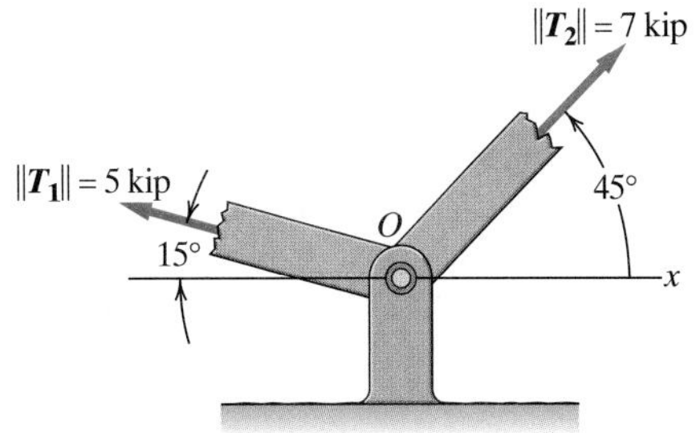

E4.6.8

4.6.9. Two structural members are pinned to the support at O. Tension forces $\boldsymbol{T}_1$ and $\boldsymbol{T}_2$ act on the members as shown in E4.6.9.

a. Write $\boldsymbol{T}_R = \boldsymbol{T}_1 + \boldsymbol{T}_2$ in vector notation based on the x and y axes shown.

b. Write $\boldsymbol{T}_{R^*} = \boldsymbol{T}_1 + \boldsymbol{T}_2$ in vector notation based on the x^* and y^* axes shown. (Call the unit vector aligned with x^*, $\boldsymbol{i}^*$, and the unit vector aligned with y^*, $\boldsymbol{j}^*$.)

c. Show that the magnitude of $\boldsymbol{T}_R$ is equal to the magnitude of $\boldsymbol{T}_{R^*}$.

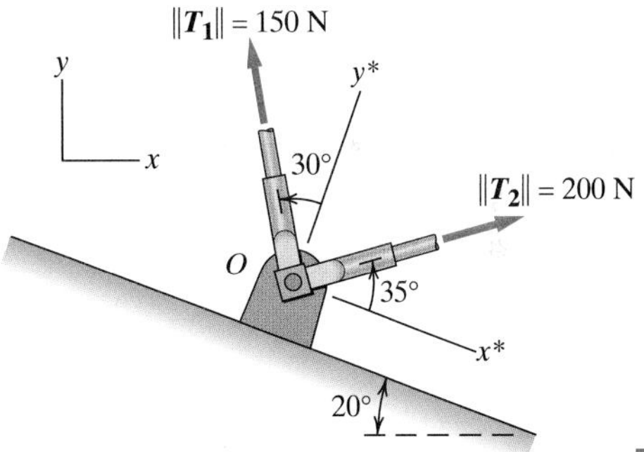

E4.6.9

4.6.10. A 2.5-kip horizontal force $\boldsymbol{P}$ and a 1.8-kip cable tension $\boldsymbol{T}$ act on the top of the wall at point B as shown in E4.6.10. What is their resultant $\boldsymbol{R}$?

a. Express $\boldsymbol{R}$ in terms of its components along the x and y axes shown. Include a sketch of the components.

b. Express $\boldsymbol{R}$ in terms of its magnitude and space angles. Include a sketch that shows the space angles.

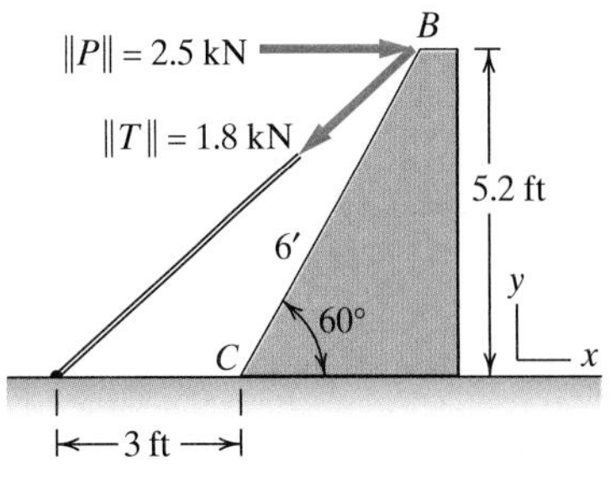

E4.6.10

4.6.11. Consider the frictionless pulleys shown in E4.6.11. Because the pulleys are frictionless, $\| \boldsymbol{T}_1 \| = \| \boldsymbol{T}_2 \|$ and $\| \boldsymbol{T}_3 \| = \| \boldsymbol{T}_4 \|$.

a. Express the resultant $\boldsymbol{R}_a$ of $\boldsymbol{T}_1 + \boldsymbol{T}_2$ in terms of rectangular components. Also show that the line of action of $\boldsymbol{R}_a$ bisects the angle between $\boldsymbol{T}_1$ and $\boldsymbol{T}_2$.

b. Express the resultant $\boldsymbol{R}_b$ of $\boldsymbol{T}_3 + \boldsymbol{T}_4$ in terms of rectangular components. Also show that the line of action of $\boldsymbol{R}_b$ bisects the angle between $\boldsymbol{T}_3$ and $\boldsymbol{T}_4$.

c. If $\| \boldsymbol{T}_1 \| = \| \boldsymbol{T}_2 \| = \| \boldsymbol{T}_3 \| = \| \boldsymbol{T}_4 \|$, show that $\| \boldsymbol{R}_a \| = \| \boldsymbol{R}_b \|$.

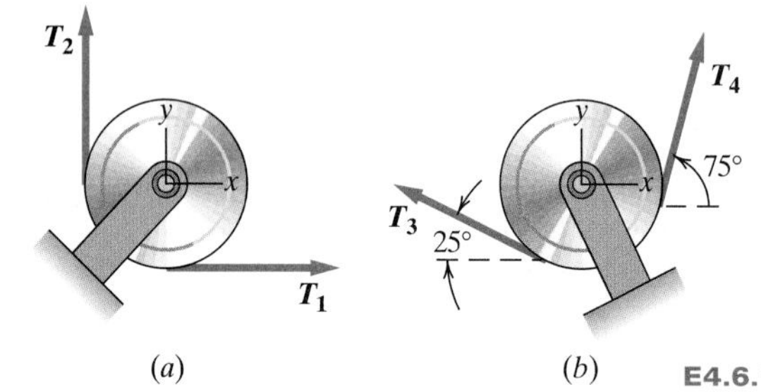

E4.6.11

4.6.12. If $\boldsymbol{F}_A = 1.00 \text{ kN}\boldsymbol{i} - 4.5 \text{ kN}\boldsymbol{j}$ and $\boldsymbol{F}_B = -2 \text{ kN}\boldsymbol{i} - 2 \text{ kN}\boldsymbol{j}$, what is the magnitude of $\boldsymbol{F} = 6\boldsymbol{F}_A + 4\boldsymbol{F}_B$?

For **Exercises 4.6.13–4.6.15**, two forces $\boldsymbol{F}_1$ and $\boldsymbol{F}_2$ are applied to a post as shown in E4.6.13. Write the resultant $\boldsymbol{F}_R$ $(= \boldsymbol{F}_1 + \boldsymbol{F}_2)$ in the following ways:

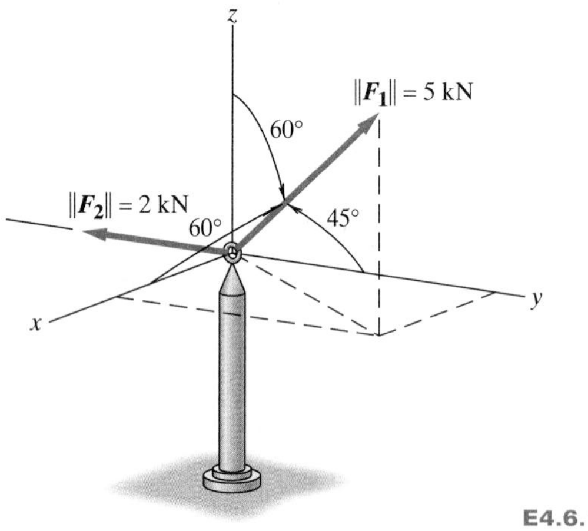

E4.6.13

4.6.13. Write in terms of vector notation.

4.6.14. Write in terms of magnitude and space angles.

4.6.15. Write the expression for the unit vector along the line of action of $\boldsymbol{F}_R$.

4.6.16. Two forces $\boldsymbol{F}_1$ and $\boldsymbol{F}_2$ are applied to an eyelet as shown in E4.6.16. Determine the resultant $\boldsymbol{F}_R$ $(= \boldsymbol{F}_1 + \boldsymbol{F}_2)$, and write it in vector notation.

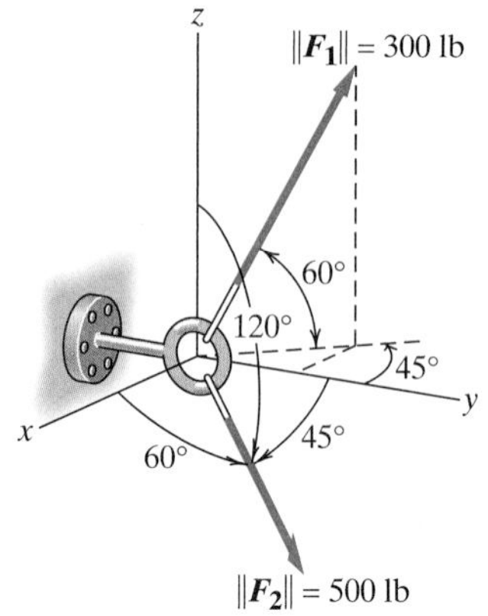

E4.6.16

4.6.17. A radio antenna is supported by three guy wires (E4.6.17). The tensile force in wires AB, AC, and AD are 20 kN, 25 kN, and 30 kN, respectively. Determine the resultant $\boldsymbol{T}_R$ $(= \boldsymbol{T}_{AB} + \boldsymbol{T}_{AC} + \boldsymbol{T}_{AD})$ acting at A, and write it in vector notation.

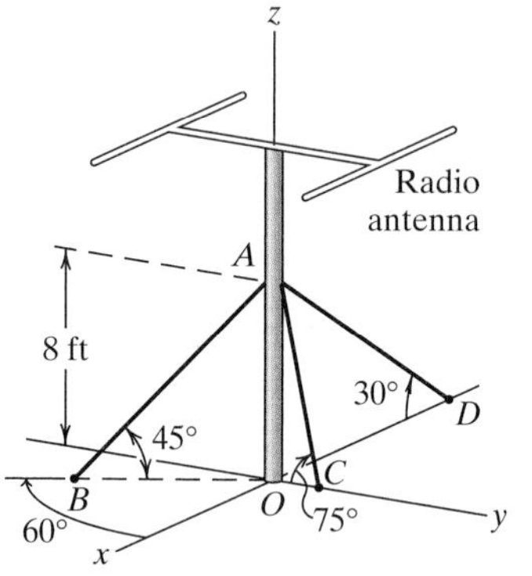

E4.6.17

4.6.18. Three cables brace a newly planted tree in E4.6.18. Each cable has a tension of 250 N. Determine the resultant force $\boldsymbol{F}_R$ acting on the tree as a result of the three cables. Express $\boldsymbol{F}_R$ in vector notation.

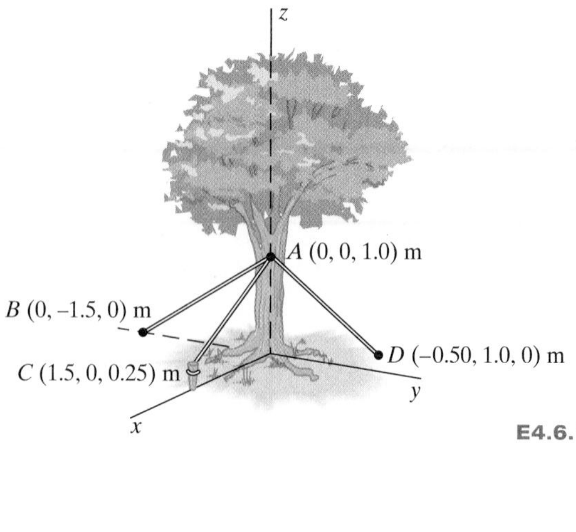

E4.6.18

For **Exercises 4.6.19 and 4.6.20** three forces act as shown in E4.6.19. Use rectangular component addition.

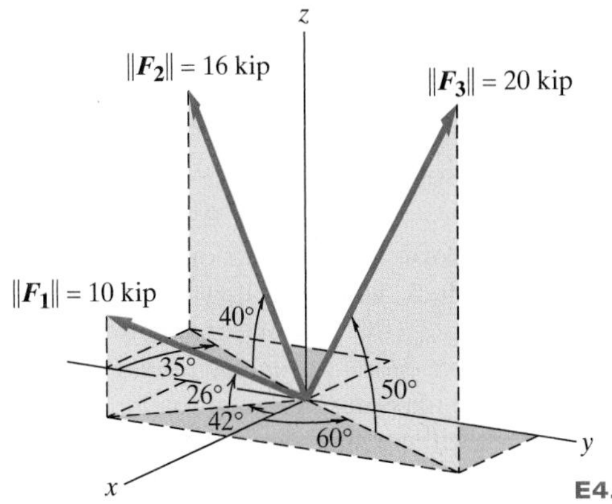

E4.6.19

4.6.19. Determine the magnitude of the resultant F_R of the three forces.

4.6.20. Determine the space angles θ_x, θ_y, and θ_z between the line of action of the resultant and the positive x, y, and z coordinate axes.

For **Exercises 4.6.21 and 4.6.22** four forces act as shown in E4.6.21. Using rectangular component addition,

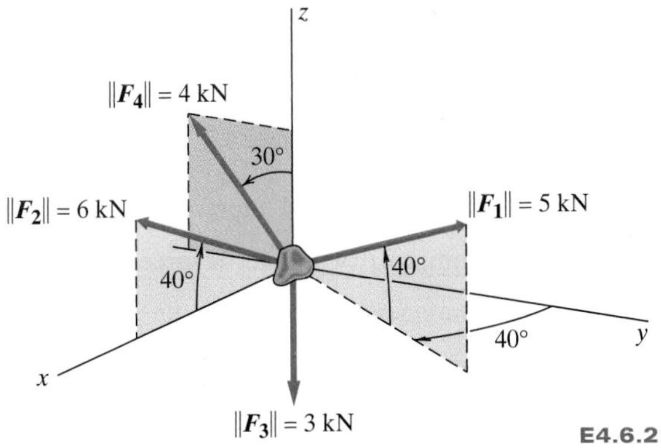

E4.6.21

4.6.21. Determine the magnitude of the resultant F_R of the four forces.

4.6.22. Determine the space angles θ_x, θ_y, and θ_z between the line of action of the resultant and the positive x, y, and z coordinate axes.

4.6.23. The tension in each of the two supporting cables AB and AC in E4.6.23 is 10 lb. Determine the resultant force acting at A due to the two cables.

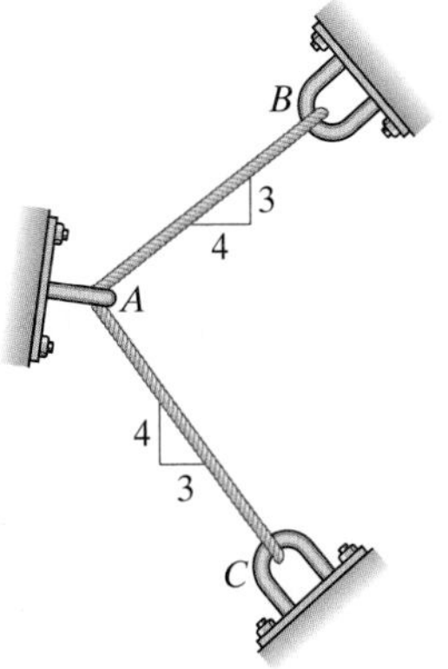

E4.6.23

4.6.24. Two forces act on an eyebolt as shown in E4.6.24.

a. Find the resultant of the two forces and express it in vector notation. Also, find the space angles associated with the resultant.

b. Find the magnitudes of two other forces F_u and F_v that would have the same resultant as found in **a**.

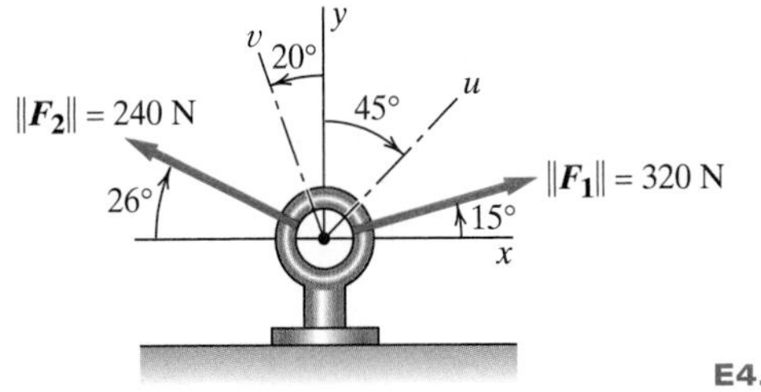

E4.6.24

4.6.25. The cables A and B in E4.6.25 exert forces F_A and F_B on the hook. The magnitude of F_A is 100 N. The tension in cable B has been adjusted so that the resultant force $(F_A + F_B)$ is perpendicular to the wall to which the hook is attached.

a. Find the magnitude of F_B.

b. Specify the resultant force F_R in vector notation.

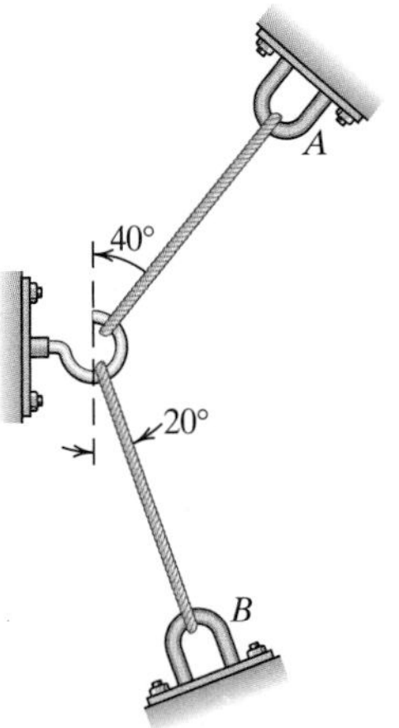

E4.6.25

4.6.26. A bracket is bolted to a wall as shown in E4.6.26. The cables A and B exert forces $\boldsymbol{F}_A$ and $\boldsymbol{F}_B$ on the bracket. The magnitude of $\boldsymbol{F}_A$ is 200 lb. The tension in cable B has been adjusted so that the resultant force $(\boldsymbol{F}_A + \boldsymbol{F}_B)$ is parallel to the wall to which the bracket is attached.

a. Find the magnitude of $\boldsymbol{F}_B$.

b. Specify the resultant force $\boldsymbol{F}_R$ in vector notation.

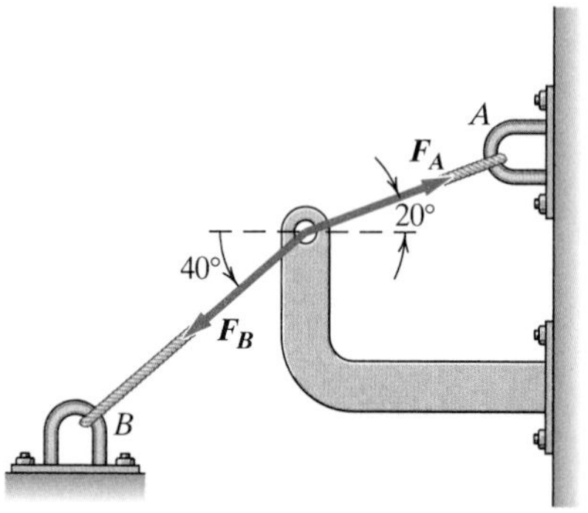

E4.6.26

4.6.27. Cables A and B act on a bracket as shown in E4.6.27. The tension in cable A is 500 N, oriented as shown. The orientation of $\boldsymbol{F}_B$ is at $\theta = 20°$, as shown.

a. Write an expression (in vector notation) for the resultant force $\boldsymbol{F}_R$ acting on the bracket.

b. If the bracket will pull away from the wall when the component of $\boldsymbol{F}_R$ parallel to the wall is greater than or equal to 550 N and the component of $\boldsymbol{F}_R$ perpendicular to the wall is greater than or equal to 600 N, how large can the magnitude of $\boldsymbol{F}_B$ be?

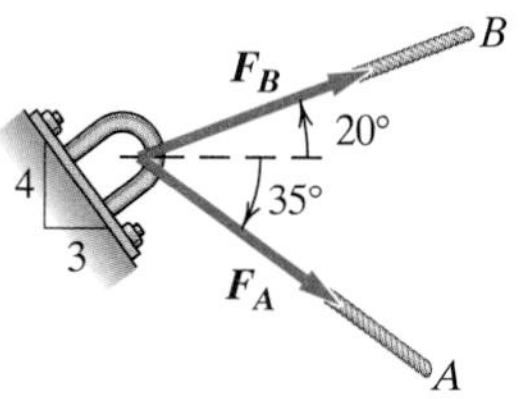

E4.6.27

4.6.28. The frictionless pulley in E4.6.28 is such that the magnitude of $\boldsymbol{T}_1$ is equal to the magnitude of $\boldsymbol{T}_2$. Based on the information given in the figure, what are the scalar components of $\boldsymbol{T}_2$?

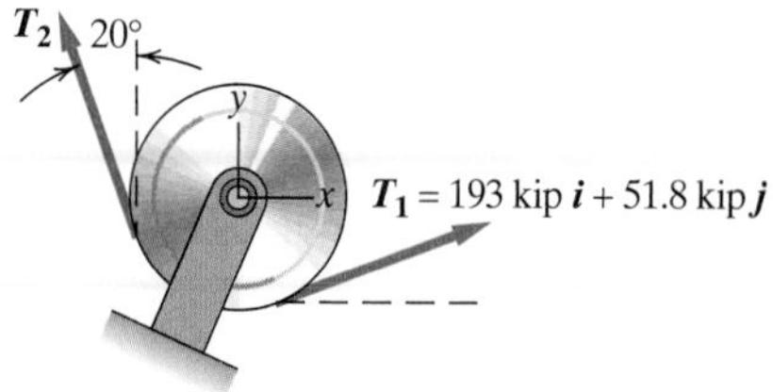

E4.6.28

4.6.29. A bracket is glued to the floor, and a cable is attached as shown in E4.6.29. If $\boldsymbol{F}_{\text{shear}}$ (the shear force between the bracket and the floor) is not to exceed 50 N and $\boldsymbol{F}_{\text{tension}}$ (the tension force between the bracket and the floor) is not to exceed 100 N, what is the maximum allowable tension in the cable? (*Hint:* Graphical addition of $\boldsymbol{F}_{\text{shear}}$, $\boldsymbol{F}_{\text{tension}}$, and $\boldsymbol{F}_{\text{cable}}$ should result in a triangle.)

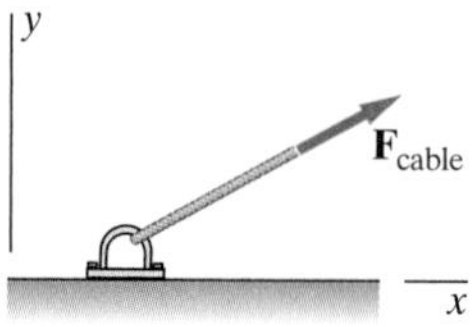

E4.6.29

4.6.30. A block of weight W sits on an inclined surface (E4.6.30). The block will slide down the surface if the magnitude of $\boldsymbol{F}_{\text{friction}}$ (the friction force between the block and the surface) exceeds $\frac{1}{3}$ of the magnitude of the normal force. What is the maximum value of the incline angle θ at which the block will not slide down the surface?

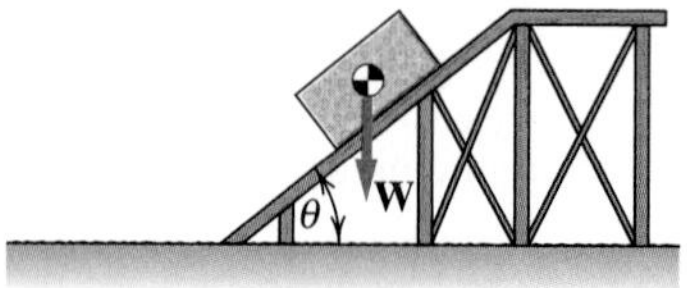

E4.6.30

Use E4.6.31 to answer **Exercises 4.6.31–4.6.33**. The forces acting on a block are gravity ($\boldsymbol{W}$, magnitude 10 lb), a normal contact force ($\boldsymbol{F}_N$), and a push force ($\boldsymbol{F}_{\text{push}}$), as shown. The surface on which the block rests is smooth, so there is no friction force. If the sum of the forces acting on the block is zero, what are the magnitudes of $\boldsymbol{F}_N$ and $\boldsymbol{F}_{\text{push}}$?

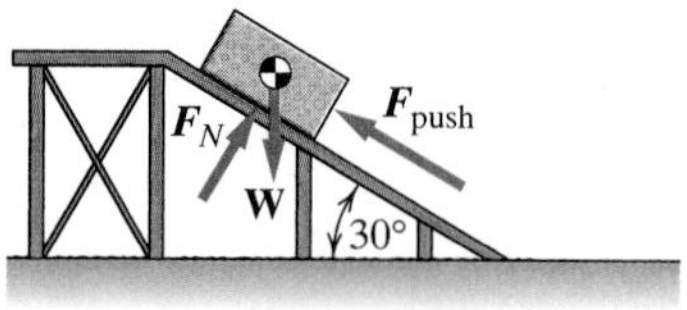

E4.6.31

4.6.31. Use graphical force addition.

4.6.32. Use geometric/trigonometric force addition.

4.6.33. Use component force addition.

4.6.34. You have designed a bracket to support a force $\boldsymbol{F} = F_x\boldsymbol{i} + F_y\boldsymbol{j}$ (E4.6.34). The design is such that the magnitude of $\boldsymbol{F}$ should not exceed 1000 N. If F_x ranges from $-750 \text{ N} \leq F_x \leq 500 \text{ N}$, what is the range of F_y that can be safely supported by the bracket? (Assume that F_x and F_y are independent of each other.)

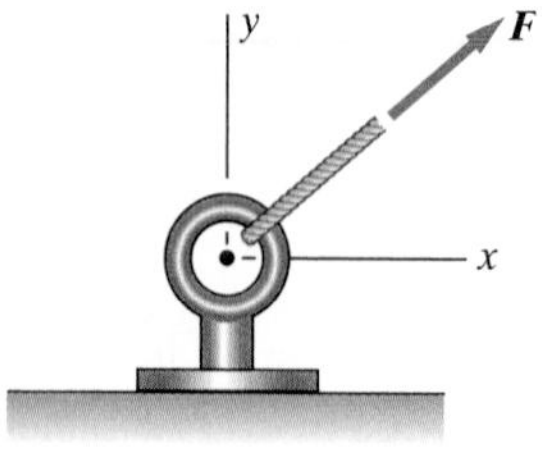

E4.6.34

4.6.35. The top of the tower in E4.6.35 is subjected to a horizontal force of 60 kN and a tension $\boldsymbol{T}$ in the heavy, flexible cable. The cable is tightened by the winch B. If the net effect of the two forces is to produce a downward compression of 40 kN on the tower at A, determine the magnitude of the tension in the cable at A and the angle θ the cable makes with the horizontal.

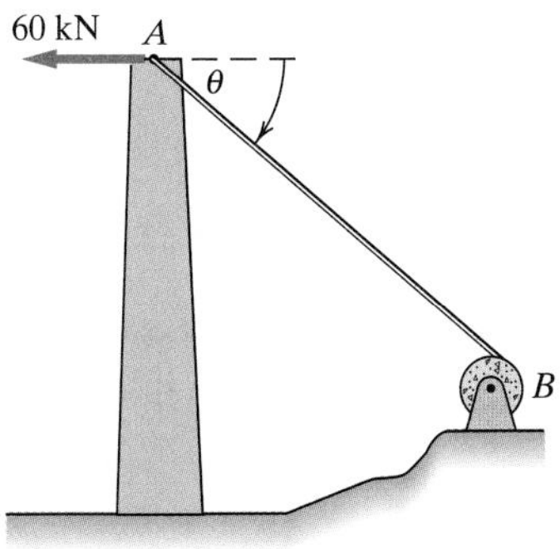

E4.6.35

4.6.36. Three ropes attached to a stalled car apply the forces shown in E4.6.36. Determine the magnitude of $\boldsymbol{F}_3$ and the magnitude of the resultant $\boldsymbol{F}_R$ if the line of action of the resultant is along the x axis.

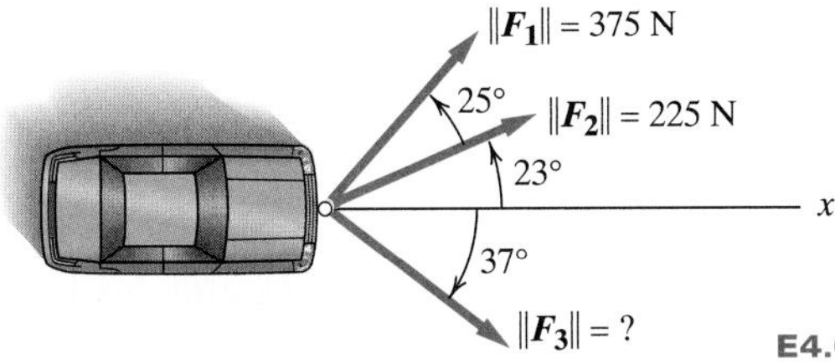

E4.6.36

4.6.37. A 2000-lb force is resisted by two pipe struts as shown in E4.6.37. Determine

a. the component F_u of the force along the axis of strut AB

b. the component F_v of the force along the axis of strut BC

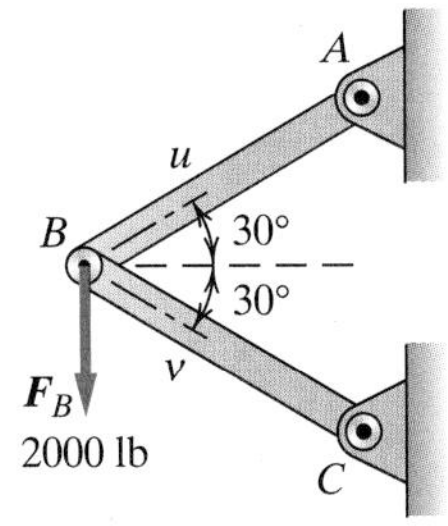

E4.6.37

4.6.38. A man is pushing a copy machine that weighs 600 N up an incline. A free-body diagram of the man is shown in E4.6.38.

a. If the net force in direction t is zero, what is the value of $\boldsymbol{F}_{\text{friction}}$?

b. If the net force in direction n is zero, what is the value of $\boldsymbol{F}_{\text{normal}}$?

c. What is the value of the ratio $F_{\text{friction}}/F_{\text{normal}}$? (Note: The friction coefficient between the man's shoes and the incline must be at least as great as this ratio if his feet do not slip as he walks up the incline.)

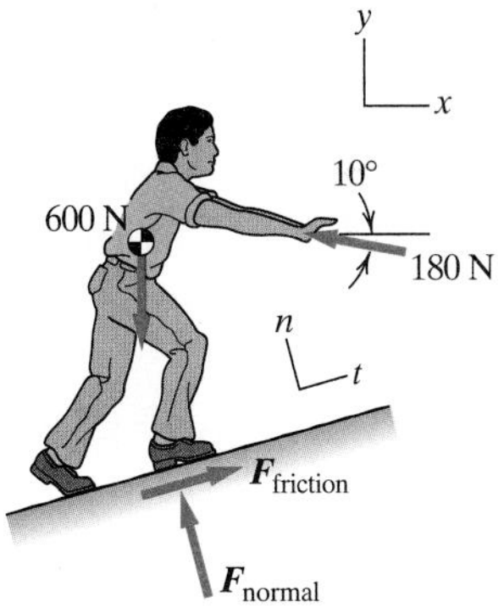

E4.6.38

4.6.39. Redraw the free-body diagram in E4.6.39 to show $\boldsymbol{T}_{BC}$ and $\boldsymbol{F}_B$ as a single force acting at B and $\boldsymbol{F}_{Ax}$ and $\boldsymbol{F}_{Ay}$ as a single force acting at A.

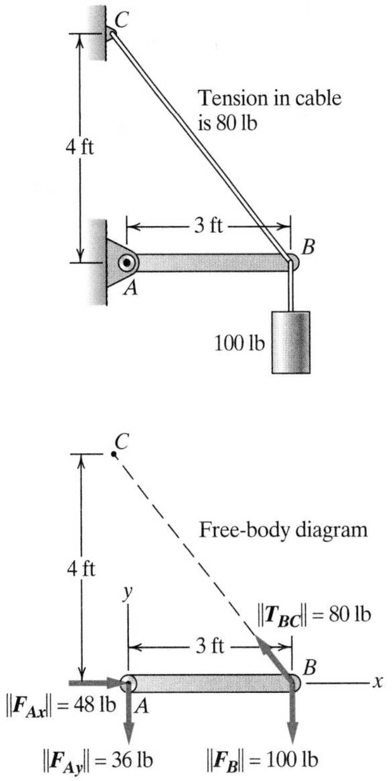

E4.6.39

4.6.40. Redraw the free-body diagram in E4.6.40 to show $\boldsymbol{F_1}$ and $\boldsymbol{F_2}$ as a single force acting at C and $\boldsymbol{F_{Ax}}$ and $\boldsymbol{F_{Ay}}$ as a single force acting at A.

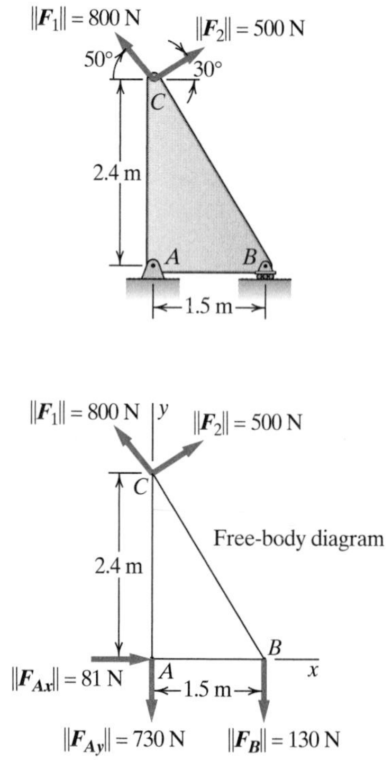

E4.6.40

4.6.41. Three forces are applied with cables to the anchor block of E4.6.41. Determine

a. the resultant $\boldsymbol{F_R}$ in terms of its rectangular components

b. the magnitude of the force that tends to pull the block upward (in the z direction)

c. the magnitude of the force that tends to slide the block along the ground (in the xy plane)

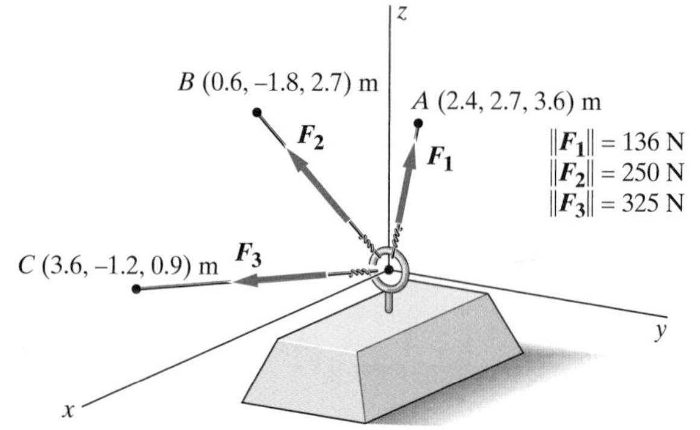

E4.6.41

4.7 ANGLE BETWEEN TWO FORCES (THE DOT PRODUCT)

Let's return to the situation of you and your friend pulling on the tent stake. We found the resultant force $\boldsymbol{F_R}$ by adding $\boldsymbol{F_1}$ and $\boldsymbol{F_2}$, by using graphical addition, geometric/trigonometric addition, and component addition. In this section we ask the question, "Will the force you and your friend exert be greater than the gripping force holding the stake in the ground?"

We can answer this question if we assume that it is the scalar component of the resultant force $\boldsymbol{F_R}$ along the axis of the stake that pulls the stake out of the ground. (Is this a reasonable assumption?) As noted in Figure 4.82, let us call this scalar component F_p, where the subscript "p" stands for projection. If we define a unit vector $\boldsymbol{m}$ along the stake axis,

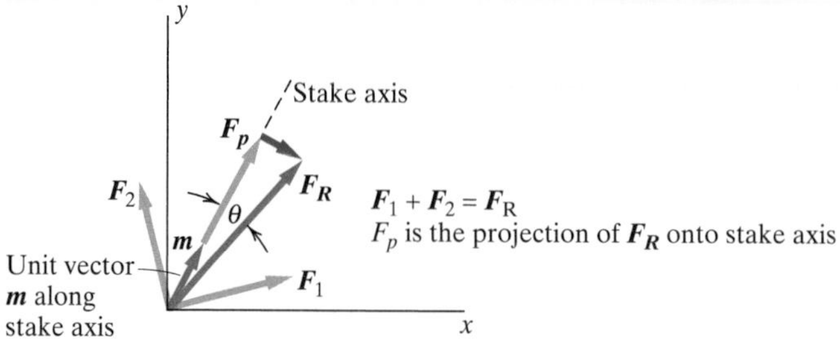

Figure 4.82 Finding the projection of $\boldsymbol{F_R}$ onto $\boldsymbol{m}$

the projection of $\boldsymbol{F_R}$ onto $\boldsymbol{m}$ is F_p. The geometry in **Figure 4.82** shows that the size of this projection is

$$F_p = \| \boldsymbol{F_R} \| \cos\theta \tag{4.19}$$

where θ is the angle between $\boldsymbol{m}$ and $\boldsymbol{F_R}$.

In terms of formal mathematical constructs, we have just taken the **dot product**, or **scalar product**, of vectors $\boldsymbol{F_R}$ and $\boldsymbol{m}$. For any two vectors $\boldsymbol{V_1}$ and $\boldsymbol{V_2}$, the dot product can be interpreted as a means of finding the component of $\boldsymbol{V_1}$ in the direction of $\boldsymbol{V_2}$ multiplied by the magnitude of $\boldsymbol{V_2}$ (or of finding the component of $\boldsymbol{V_2}$ in the direction of $\boldsymbol{V_1}$ multiplied by the magnitude of $\boldsymbol{V_1}$). The dot product of two vectors $\boldsymbol{V_1}$ and $\boldsymbol{V_2}$ is a scalar quantity and is formally written

$$\text{dot product} = \boldsymbol{V_1} \cdot \boldsymbol{V_2} = \boldsymbol{V_2} \cdot \boldsymbol{V_1} = \| \boldsymbol{V_1} \| \; \| \boldsymbol{V_2} \| \cos\theta \tag{4.20}$$

where

$\| \boldsymbol{V_1} \|$ and $\| \boldsymbol{V_2} \|$ are the magnitudes of $\boldsymbol{V_1}$ and $\boldsymbol{V_2}$, respectively, and θ is the angle between them and is between 0° and 180°.

In the case of you and your friend pulling on the tent stake, we are interested in

$$\text{dot product} = F_p = \boldsymbol{F_R} \cdot \boldsymbol{m} = \| \boldsymbol{F_R} \| \; \| \boldsymbol{m} \| \cos\theta$$

This expression is the same as (4.19) because $\| \boldsymbol{m} \| = 1$.

The dot product of two vectors $\boldsymbol{V_1}$ and $\boldsymbol{V_2}$ can also be found by considering the vectors in terms of their scalar components. For example, if

$$\boldsymbol{V_1} = V_{1x}\boldsymbol{i} + V_{1y}\boldsymbol{j} + V_{1z}\boldsymbol{k} \quad \text{and} \quad \boldsymbol{V_2} = V_{2x}\boldsymbol{i} + V_{2y}\boldsymbol{j} + V_{2z}\boldsymbol{k} \tag{4.21}$$

then by the distributive law,[10]

$$\begin{aligned}\boldsymbol{V_1} \cdot \boldsymbol{V_2} &= (V_{1x}\boldsymbol{i} + V_{1y}\boldsymbol{j} + V_{1z}\boldsymbol{k}) \cdot (V_{2x}\boldsymbol{i} + V_{2y}\boldsymbol{j} + V_{2z}\boldsymbol{k}) \\ &= V_{1x}V_{2x}(\boldsymbol{i} \cdot \boldsymbol{i}) + V_{1x}V_{2y}(\boldsymbol{i} \cdot \boldsymbol{j}) + V_{1x}V_{2z}(\boldsymbol{i} \cdot \boldsymbol{k}) + \\ &\quad V_{1y}V_{2x}(\boldsymbol{j} \cdot \boldsymbol{i}) + V_{1y}V_{2y}(\boldsymbol{j} \cdot \boldsymbol{j}) + V_{1y}V_{2z}(\boldsymbol{j} \cdot \boldsymbol{k}) + \\ &\quad V_{1z}V_{2x}(\boldsymbol{k} \cdot \boldsymbol{i}) + V_{1z}V_{2y}(\boldsymbol{k} \cdot \boldsymbol{j}) + V_{1z}V_{2z}(\boldsymbol{k} \cdot \boldsymbol{k})\end{aligned} \tag{4.22}$$

Once we carry out the dot-product operations,[11] (4.22) becomes

$$\boldsymbol{V_1} \cdot \boldsymbol{V_2} = V_{1x}V_{2x} + V_{1y}V_{2y} + V_{1z}V_{2z} \tag{4.23}$$

Equations (4.20) and (4.23) are equivalent expressions for the dot product of two vectors that we can set equal to one another:

$$\boldsymbol{V_1} \cdot \boldsymbol{V_2} = \| \boldsymbol{V_1} \| \; \| \boldsymbol{V_2} \| \cos\theta = V_{1x}V_{2x} + V_{1y}V_{2y} + V_{1z}V_{2z} \tag{4.24}$$

[10]The distributive law is $\boldsymbol{A} \cdot (\boldsymbol{B} + \boldsymbol{C}) = (\boldsymbol{A} \cdot \boldsymbol{B}) + (\boldsymbol{A} \cdot \boldsymbol{C})$.

[11]$\boldsymbol{i} \cdot \boldsymbol{i} = \boldsymbol{j} \cdot \boldsymbol{j} = \boldsymbol{k} \cdot \boldsymbol{k} = (1)(1)\cos 0° = 1$
$\boldsymbol{i} \cdot \boldsymbol{j} = \boldsymbol{i} \cdot \boldsymbol{k} = \boldsymbol{j} \cdot \boldsymbol{i} = \boldsymbol{j} \cdot \boldsymbol{k} = \boldsymbol{k} \cdot \boldsymbol{i} = \boldsymbol{k} \cdot \boldsymbol{j} = (1)(1)\cos 90° = 0$

Rearranging this expression and solving for the angle θ between the tails of the two vectors, we have

$$\cos\theta = \frac{V_{1x}V_{2x} + V_{1y}V_{2y} + V_{1z}V_{2z}}{\| \mathbf{V_1} \| \; \| \mathbf{V_2} \|} \qquad \text{with } 0^\circ \leq \theta \leq 180^\circ \quad (4.25)$$

If we know $\mathbf{V_1}$ and $\mathbf{V_2}$ in terms of their scalar components, this expression can be used to find the angle θ between the two vectors.

Now back to the tent stake and our desire to find F_p, the scalar component that is the projection of $\mathbf{F_R}$ onto the tent stake axis. If we know θ, the angle between $\mathbf{F_R}$ and the stake axis, it makes most sense to use (4.20) to find F_p. If, instead, we know the scalar components of $\mathbf{F_R}$ and the components of $\mathbf{m}$, it makes more sense to use (4.23) to find F_p. How to use the dot product to find F_p is illustrated in Figure 4.83, based on our previously found value $\| \mathbf{F_R} \| = 333$ N (see, for example, **Figure 4.61**). Once we have found F_p, we can compare its absolute value $|F_p|$ with the absolute value $|F_{grip}|$ of the grip force $\mathbf{F}_{grip}$ exerted on the stake by the ground (**Figure 4.83**). If $|F_p|$ is greater than or equal to $|F_{grip}|$, you and your friend should be able to remove the stake from the ground simply by pulling. If $|F_p|$ is less than $|F_{grip}|$ and you still want to remove the stake from the ground, you could increase $|F_p|$ and/or reduce $|F_{grip}|$. How might you do this? Or what else might you do?

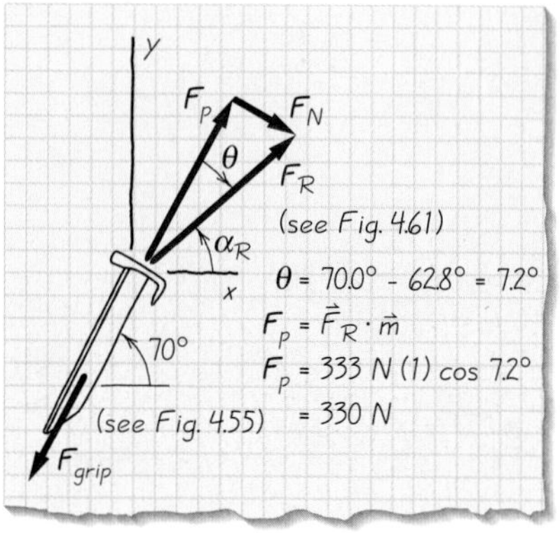

Figure 4.83

EXAMPLE 4.14 VECTOR ADDITION PRACTICE (E)

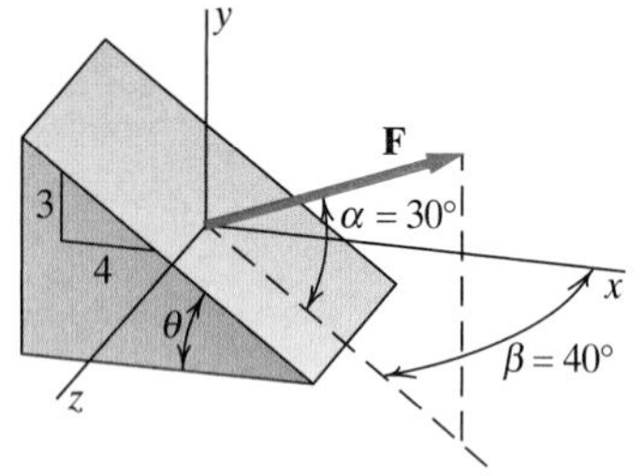

Figure 4.84

It is common practice in describing the resultant force acting on a surface to refer to the component vector normal to the surface as a pull force ($\mathbf{F}_{pull}$) if it points away from the surface and as a push force ($\mathbf{F}_{push}$) if it points toward the surface. The component vector of the resultant force parallel to the surface is commonly called the tangential force ($\mathbf{F}_{tang}$). Consider the force $\mathbf{F}$ acting on the surface shown in Figure 4.84. Its magnitude is 6.00 kN.

Determine

(a) the magnitudes of the pull (or push) and tangential forces acting on the sloped rectangular surface

(b) the angle between the line of action of $\mathbf{F}$ and the x axis

Goal We are to find the components of $\mathbf{F}$ that are perpendicular to and in the plane of the sloped surface. In addition, we are to find the angle between the line of action of $\mathbf{F}$ and the x axis.

Given We are given the magnitude of $\mathbf{F}$ (6.00 kN) as well as its direction relative to a right-handed coordinate system. In addition, we are given information on the orientation of the surface relative to this same coordinate system.

Assume No assumptions are required.

Draw We create a drawing of $\boldsymbol{F}$ in terms of its scalar component F_x, F_y, F_z (**Figure 4.85**) and a drawing that shows the direction perpendicular to the sloped surface (**Figure 4.86**).

Formulate Equations and Solve **(a)** We are asked to find the component of $\boldsymbol{F}$ perpendicular to the surface and the component parallel to the surface. Because we are given orientation information on $\boldsymbol{F}$ based on a horizontal x axis and a vertical y axis, we will start by finding the components of $\boldsymbol{F}$ relative to the x, y, and z axes shown in **Figure 4.85**. Then we will use these components to find the pull or push force and the tangential force relative to the surface.

We begin by expressing $\boldsymbol{F}$ in terms of its scalar components. As shown in **Figure 4.84**, the force is at $\alpha = 30°$, $\beta = 40°$. Based on the geometry in **Figure 4.85**, we start by finding the projection of $\boldsymbol{F}$ onto the xz plane. Call this projection F^*:

$$F^* = F \cos \alpha = (6.00 \text{ kN})(\cos 30°) = 5.20 \text{ kN}$$

(What we have just done is take the dot product of $\boldsymbol{F}$ and a unit vector along $\boldsymbol{F^*}$.) Next we find the scalar components of $\boldsymbol{F^*}$ in the x and z directions:

$$F_x = F^* \cos \beta = (5.20 \text{ kN}) \cos 40° = 3.98 \text{ kN}$$
$$F_z = F^* \sin \beta = (5.20 \text{ kN}) \sin 40° = 3.34 \text{ kN}$$

In addition, the component of $\boldsymbol{F}$ in the y direction is

$$F_y = F \sin \alpha = (6.00 \text{ kN})(\sin 30°) = 3.00 \text{ kN}$$

Therefore, we can write $\boldsymbol{F}$ in vector notation as

$$\begin{aligned}\boldsymbol{F} &= F_x\boldsymbol{i} + F_y\boldsymbol{j} + F_z\boldsymbol{k}\\ &= 3.98 \text{ kN}\boldsymbol{i} + 3.00 \text{ kN}\boldsymbol{j} + 3.34 \text{ kN}\boldsymbol{k}\end{aligned}$$

Now we find the projection of $\boldsymbol{F}$ onto an axis perpendicular to the sloped surface. As shown in **Figure 4.86**, the surface is at an angle θ of 36.9° to the horizontal. The direction perpendicular to the surface is defined by the unit vector $\boldsymbol{m}$. The scalar components of $\boldsymbol{m}$ are

$$m_x = \sin \theta = \sin 36.9° = 0.600$$
$$m_y = \cos \theta = \cos 36.9° = 0.800$$
$$m_z = 0$$

or in vector notation

$$\boldsymbol{m} = m_x\boldsymbol{i} + m_y\boldsymbol{j} + m_z\boldsymbol{k} = 0.600\boldsymbol{i} + 0.800\boldsymbol{j}$$

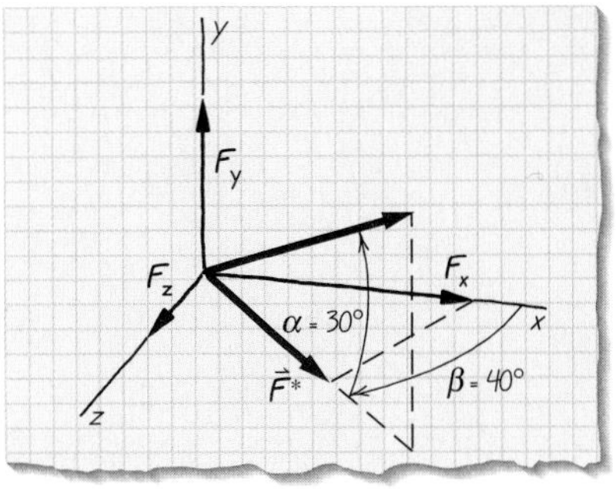

Figure 4.85 The component of the resultant force along the long axis of the stake

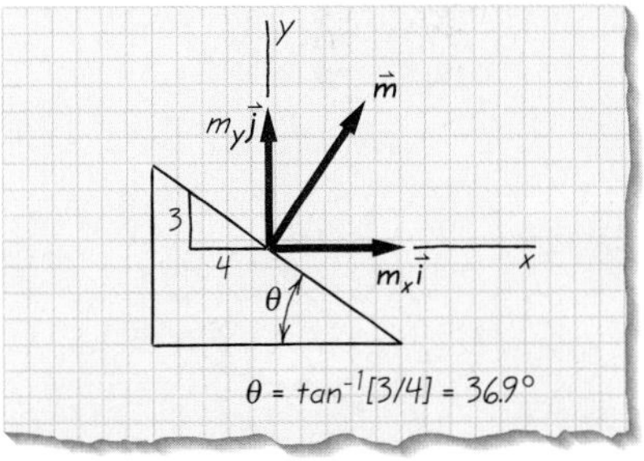

Figure 4.86 (*a*) Finding the component along the long axis of the stake; (*b*) Is this component large enough to break the stake away from the ground?

The projection of $\boldsymbol{F}$ onto $\boldsymbol{m}$ is the dot product $\boldsymbol{F} \cdot \boldsymbol{m}$. Based on (4.23) the dot product is given as

$$\begin{aligned} \boldsymbol{F} \cdot \boldsymbol{m} &= F_x m_x + F_y m_y + F_z m_z \\ &= 3.98 \text{ kN}(0.600) + 3.00 \text{ kN}(0.800) + 3.34 \text{ kN}(0) \\ &= 4.79 \text{ kN} \end{aligned}$$

This dot product has a magnitude of 4.79 kN and is positive (meaning it is along the $+\boldsymbol{m}$ direction and, as such, indicates a pull force). Therefore the force that is perpendicular to the surface pulls on the surface, and we denote it as $\boldsymbol{F}_{\text{pull}}$, with $\| \boldsymbol{F}_{\text{pull}} \| = 4.79$ kN.

The tangential force, $\boldsymbol{F}_{\text{tang}}$, is the component vector of $\boldsymbol{F}$ that is parallel to the sloped surface and therefore perpendicular to the direction of $\boldsymbol{F}_{\text{pull}}$. Using the Pythagorean theorem, we can write

$$\| \boldsymbol{F}_{\text{tang}} \| = \sqrt{\| \boldsymbol{F} \|^2 - \| \boldsymbol{F}_{\text{pull}} \|^2} = \sqrt{(6.00 \text{ kN})^2 - (4.79 \text{ kN})^2} = 3.62 \text{ kN}$$

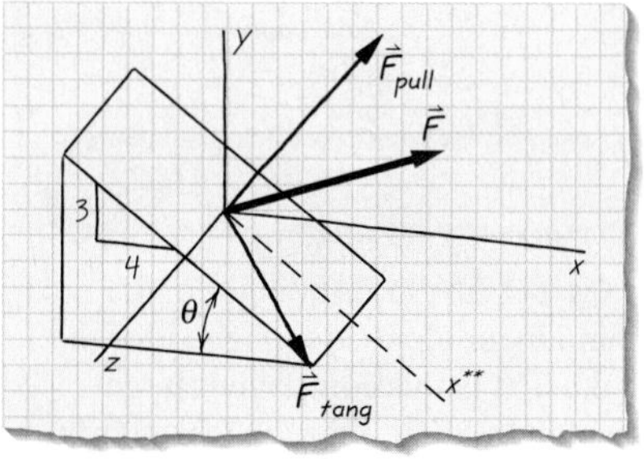

Figure 4.87

Alternately, we could have found $\| \boldsymbol{F}_{\text{tang}} \|$ by finding the projections of $\boldsymbol{F}$ onto the z and x^{**} axes (as defined in **Figure 4.87**) using the dot product two times, then summing these projections using the Pythagorean theorem—but this would have involved a lot more work.

Answer to (a) $\| \boldsymbol{F}_{\text{pull}} \| = 4.79$ kN (points away from surface), $\| \boldsymbol{F}_{\text{tang}} \| = 3.62$ kN

(b) We are asked to find the angle between the x axis and $\boldsymbol{F}$. Another way of stating this is that we are to find the angle between the vector $\boldsymbol{i}$ (a unit vector along the x axis) and $\boldsymbol{F}$. Equation (4.24) enables us to find the angle between two vectors based on their components. Therefore, the angle θ between $\boldsymbol{F}$ $(= 3.98 \text{ kN}\boldsymbol{i} + 3.00 \text{ kN}\boldsymbol{j} + 3.34 \text{ kN}\boldsymbol{k})$ from (a)) and $\boldsymbol{i} = 1\boldsymbol{i} + 0\boldsymbol{j} + 0\boldsymbol{k}$ is

$$\begin{aligned} \theta &= \cos^{-1}\left(\frac{F_x(1) + F_y(0) + F_z(0)}{\| \boldsymbol{F} \| (1)}\right) \\ &= \cos^{-1}\left(\frac{3.98 \text{ kN}(1) + 3.00 \text{ kN}(0) + 3.34(0)}{6.00 \text{ kN}(1)}\right) \\ &= 48.4^\circ \end{aligned}$$

Answer to (b) $\theta = 48.4^\circ$

Check Inspection of **Figure 4.87** shows that θ is the space angle θ_x, and so we can calculate this space angle using the scalar component F_x to confirm that it is 48.4°:

$$\cos \theta_x = \frac{F_x}{\| \boldsymbol{F} \|} = \frac{3.98 \text{ N}}{6.00 \text{ N}} \Rightarrow \theta_x = 48.4^\circ$$

This is the same answer we found in (b).

EXERCISES 4.7

4.7.1 Determine the dot product of the two forces shown in E4.7.1 by

a. using the definition given in (4.20)

b. using the definition given in (4.23)

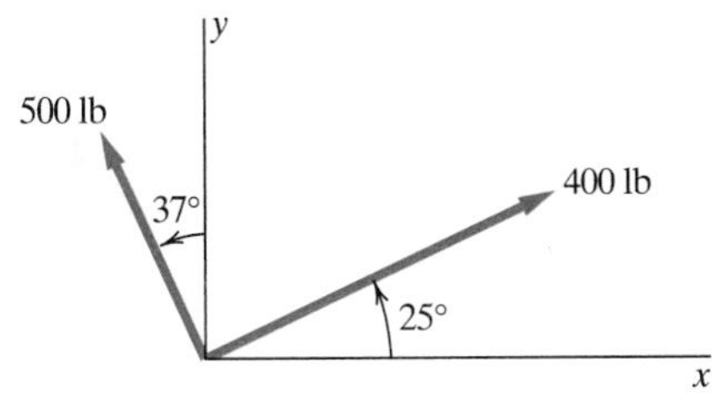

E4.7.1

4.7.2. Three cables are attached to a tree as shown in E4.7.2. Use (4.25) to determine the angles between

a. cables AB and AC

b. cables AB and AD

c. cables AC and AD

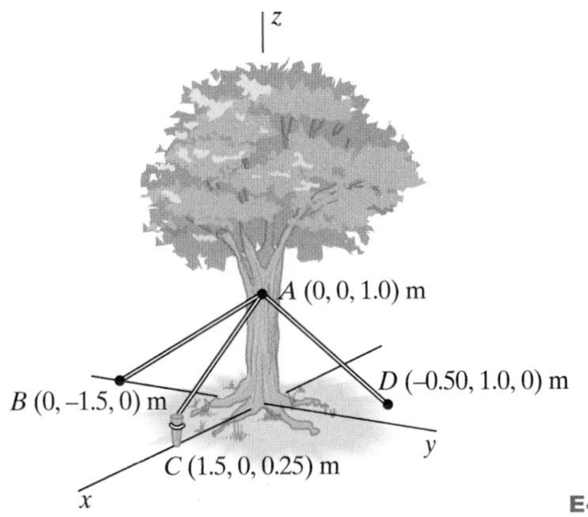

E4.7.2

4.7.3. Three cables are attached to a block as shown in E4.7.3. Use (4.25) to determine the angles between

a. cables OA and OB

b. cables OA and OC

c. cables OB and OC

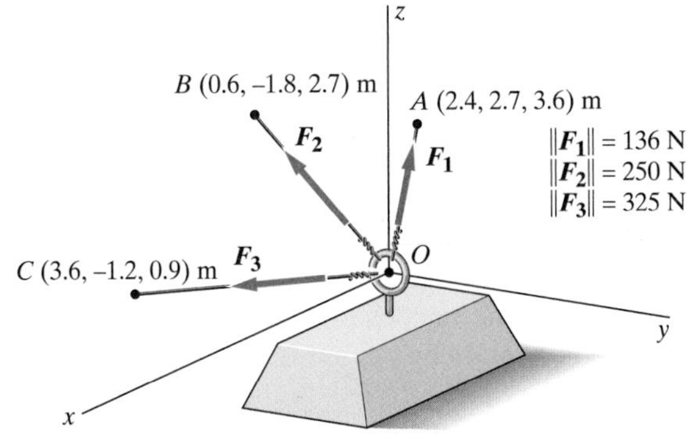

E4.7.3

4.7.4. Consider a force $\boldsymbol{F}$ in E4.7.4.

a. Write an expression for the projection of $\boldsymbol{F}$ in the direction OA. If the magnitude of $\boldsymbol{F}$ is 800 lb, determine the value of the projection.

b. Write an expression for the projection of $\boldsymbol{F}$ in the direction OB. If the magnitude of $\boldsymbol{F}$ is 800 lb, determine the value of the projection.

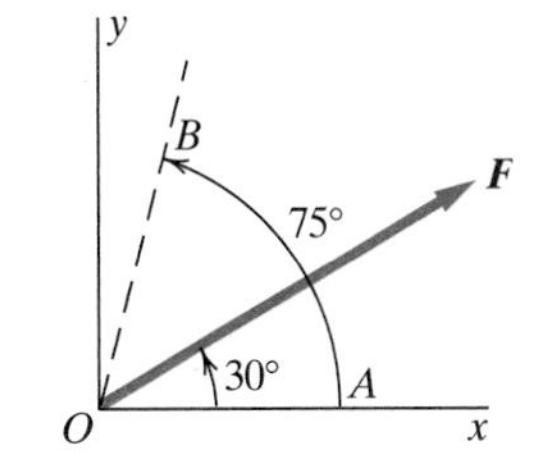

E4.7.4

4.7.5. Consider the 2.0-kN force $\boldsymbol{F}$ acting on the end of the I-beam in E4.7.5.

a. Determine the value of the projection of $\boldsymbol{F}$ in the direction AB. Call this value F_{AB}.

b. Determine the value of the projection of $\boldsymbol{F}$ in the direction perpendicular to AB. Call this value F_{per}.

c. For the design to be acceptable, F_{AB} should not exceed 1.3 kN. Is the design acceptable?

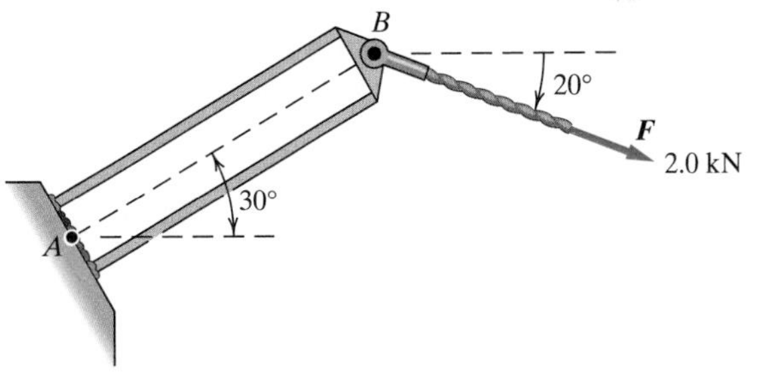

E4.7.5

4.7.6. A force $\boldsymbol{F}$ lies in the plane defined by the lines L_A and L_B in E4.7.6. Its magnitude is 400 N. Resolve $\boldsymbol{F}$ into components parallel to L_A and L_B.

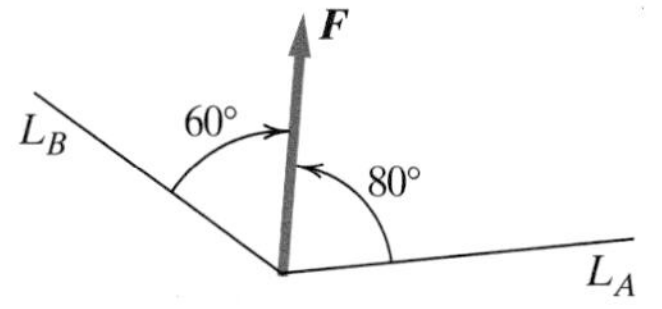

E4.7.6

4.7.7. Reconsider the structure in **E4.5.60**, and again denote as $\boldsymbol{F}_1$ the force applied by cable AB on member AD. Determine the magnitude of the projection of $\boldsymbol{F}_1$ along line AC. The tension in the cable is 3.0 kN.

4.7.8. Consider the structure in E4.7.8.

a. Determine the magnitude of the projection of $\boldsymbol{F}_1$ along A-a.

b. Determine the magnitude of the projection of $\boldsymbol{F}_2$ along A-a.

c. Use the results from **a** and **b** to find the net force at A in the A-a direction due to $\boldsymbol{F}_1$ and $\boldsymbol{F}_2$.

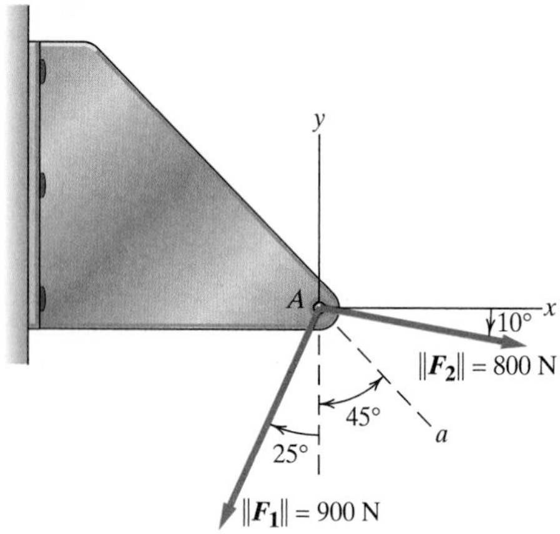

E4.7.8

4.7.9. A cable is attached at B to the right-angle pipe OAB in E4.7.9. The tension in the cable is 750 lb.

a. Determine the angle between the cable and member BA.

b. Determine the components of $\boldsymbol{F}_{BC}$ (the cable force acting at B) parallel and perpendicular to member BA.

4.7.10. Let $\boldsymbol{A} = t\boldsymbol{i} - 3\boldsymbol{j}$ and $\boldsymbol{B} = 5\boldsymbol{i} + 7\boldsymbol{j}$, where t is a scalar. Find t so that $\boldsymbol{A}$ and $\boldsymbol{B}$ are perpendicular to each other.

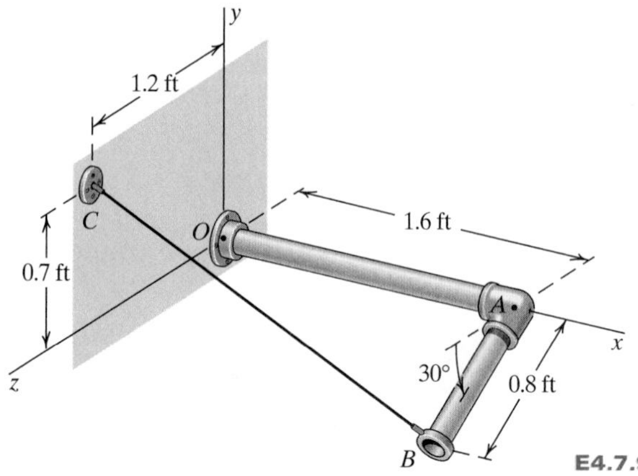

E4.7.9

4.7.11. Let $\boldsymbol{A} = 3\boldsymbol{i} + 6\boldsymbol{j} - 2\boldsymbol{k}$, $\boldsymbol{B} = -3\boldsymbol{j} + 4\boldsymbol{k}$, and $\boldsymbol{C} = 1\boldsymbol{i} - 7\boldsymbol{j} + 4\boldsymbol{k}$. Determine

a. $(\boldsymbol{A} \cdot \boldsymbol{B})\boldsymbol{C}$

b. $\boldsymbol{A} \cdot (\boldsymbol{B} + \boldsymbol{C})(-\boldsymbol{A}) \cdot (3\boldsymbol{B} - 5\boldsymbol{C})$

4.7.12. Let $\boldsymbol{A} = \boldsymbol{i} + 5\boldsymbol{j} - 3\boldsymbol{k}$ and $\boldsymbol{B} = 4\boldsymbol{i} - \boldsymbol{j} + 7\boldsymbol{k}$. Determine the scalar component of $\boldsymbol{A}$ in the direction of $\boldsymbol{B}$.

4.7.13. Use the properties of the dot product to prove the identity

$$(\boldsymbol{A} + \boldsymbol{B}) \cdot (\boldsymbol{A} - \boldsymbol{B}) = \|\boldsymbol{A}\|^2 - \|\boldsymbol{B}\|^2$$

4.7.14. Use the dot product to show that the four points $P = (1, 2)$, $Q = (2, 3)$, $R = (1, 4)$, and $S = (0, 3)$, are the vertices of a square.

4.8 JUST THE FACTS

This chapter looks at forces—what they are, how to categorize them, and how to represent them. A **force** is any interaction between an object and the rest of the world that tends to affect the state of motion of the object. It is a vector quantity and is therefore specified in terms of a **magnitude** (size and associated **unit**) and a **direction**. In this text we denote a force in boldface italic (e.g., $\boldsymbol{F}$) and its magnitude as $\|\boldsymbol{F}\|$. In drawings we use an arrow to represent a force. The line along which the force acts is called its **line of action**. The point on which it acts on an object is the force's **point of application**.

Force Categories

We discussed two categories of forces: **gravitational forces** and **contact forces**. Gravitational force is the attraction force between any two objects. Its magnitude is related to the masses of the two objects and the distance between their **centers of mass** by

$$\|\boldsymbol{F}_g\| = \frac{Gm_1m_2}{r^2} \tag{4.1}$$

and its line of action is along a line connecting the centers of mass. This expression for gravitational force is used to determine an expression for the gravitational force acting on an object on or near the earth. We refer to this gravitational force as an object's **weight** on the earth, and it is given in newtons as

$$\| \boldsymbol{W}_{\text{earth}} \| = \| \boldsymbol{F}_g \| = (9.807m) \text{ m/s}^2 \tag{4.3B}$$

The second category of forces is contact forces. They include **normal contact forces** (those that prevent one object from moving through another), **friction forces** (those that tend to prevent one object from sliding relative to another object), **fluid contact forces** (those exerted by fluids acting on some surface), **tension forces** (forces from atoms within an object pulling on one another), **compressive forces** (forces from atoms within an object pushing on one another), and **shear forces** (forces from atoms within an object tending to shift relative to one another).

The normal contact force, friction force, and fluid contact force act on the surface of an object and result from contact between the object and the rest of the world. The tension force, compressive force, and shear force occur within an object and result from the interactions of the atoms that make up the object.

In this chapter we also looked at zooming in on a part of the world (we call that small part our **system**), drawing an imaginary boundary around the system to isolate it from the rest of the world, and identifying the **external forces** acting on it, which may be gravitational forces and/or contact forces. We reviewed how the process of zooming in and isolating the system, identifying the external forces, and then drawing a **free-body diagram** of the system fits into the Engineering Analysis Procedure outlined in Chapter 1.

Force Representation

A major portion of this chapter was devoted to representing and manipulating forces mathematically. The direction of a force can be specified in terms of space angles relative to a right-handed coordinate system. **Space angles** are given in terms of two points along the line of action of the force as

$$\theta_x = \cos^{-1}\left(\frac{x_2 - x_1}{L}\right)$$
$$\theta_y = \cos^{-1}\left(\frac{y_2 - y_1}{L}\right) \tag{4.4}$$
$$\theta_z = \cos^{-1}\left(\frac{z_2 - z_1}{L}\right)$$

where $L = \sqrt{(x_2 - x_1)^2 + (y_2 - y_1)^2 + (z_2 - z_1)^2}$.

The angles θ_x, θ_y, and θ_z are not independent of one another. They are related by their **direction cosines** by the expression

$$\sqrt{(\cos\theta_x)^2 + (\cos\theta_y)^2 + (\cos\theta_z)^2} = 1 \tag{4.5A}$$

where $\cos\theta_x$, $\cos\theta_y$, and $\cos\theta_z$ are the direction cosines and are defined as

$$\cos\theta_x = \left(\frac{x_2 - x_1}{L}\right)$$
$$\cos\theta_y = \left(\frac{y_2 - y_1}{L}\right) \qquad (4.5B)$$
$$\cos\theta_z = \left(\frac{z_2 - z_1}{L}\right)$$

Furthermore, the direction cosines define a unit vector $\boldsymbol{u}$ in the same direction as $\boldsymbol{F}$:

$$\boldsymbol{u} = \cos\theta_x\boldsymbol{i} + \cos\theta_y\boldsymbol{j} + \cos\theta_z\boldsymbol{k} \qquad (4.10C)$$

Another way of representing the magnitude and direction of a force $\boldsymbol{F}$ is to specify its three **rectangular component vectors** $\boldsymbol{F_x}$, $\boldsymbol{F_y}$, and $\boldsymbol{F_z}$ relative to a right-handed coordinate system. These component vectors consist of **scalar components** and unit vectors in the $\boldsymbol{i}$, $\boldsymbol{j}$, and $\boldsymbol{k}$ directions. We write $\boldsymbol{F}$ as

$$\boldsymbol{F} = F_x\boldsymbol{i} + F_y\boldsymbol{j} + F_z\boldsymbol{k} \qquad (4.8B)$$

The scalar components F_x, F_y, and F_z are the projections of $\boldsymbol{F}$ onto the x, y, and z axes, respectively, and the sum of their squares is related to the magnitude of the force:

$$\|\boldsymbol{F}\| = \sqrt{F_x^2 + F_y^2 + F_z^2} \qquad (4.12)$$

The scalar components are also related to the space angles by

$$F_x = \|\boldsymbol{F}\|\cos\theta_x \qquad (4.9A)$$
$$F_y = \|\boldsymbol{F}\|\cos\theta_y \qquad (4.9B)$$
$$F_z = \|\boldsymbol{F}\|\cos\theta_z \qquad (4.9C)$$

Substituting (4.9A–C) into (4.8B) and rearranging, we write

$$\boldsymbol{F} = \|\boldsymbol{F}\|(\cos\theta_x\boldsymbol{i} + \cos\theta_y\boldsymbol{j} + \cos\theta_z\boldsymbol{k}) = \|\boldsymbol{F}\|\boldsymbol{u} \qquad (4.11)$$

We call a force that lies in the plane of two of the reference axes either a **planar force** or a **two-dimensional force**; otherwise it is called either a **nonplanar force** or a **three-dimensional force**. For a planar force, one of the space angles is 90°, and therefore its cosine is zero and one of its scalar components is zero.

Force Addition and Dot Product

Engineering analysis commonly involves adding forces (or vectors more generally) to find their resultant. We presented three approaches to vector addition: **graphical vector addition**, **geometric/trigonometric addition**, and **component force addition**. Graphical vector addition involves laying out the vectors to be added to scale on graph paper. Geometric/trigonometric addition is based on the idea that two forces added together and their resultant form a triangle. Component force addition involves adding the appropriate scalar components of the forces together in order to find the resultant's scalar components.

We also presented the idea of the **dot product** in the chapter. The dot project is a convenient mathematical construct for finding the projection of a vector in a particular direction and for finding the angle between two vectors. The dot product of vectors $\boldsymbol{V_1}$ and $\boldsymbol{V_2}$ is

$$\boldsymbol{V_1} \cdot \boldsymbol{V_2} = \| \boldsymbol{V_1} \| \, \| \boldsymbol{V_2} \| \cos\theta = V_{1x}V_{2x} + V_{1y}V_{2y} + V_{1z}V_{2z} \quad (4.24)$$

Rearranging this expression and solving for the angle θ between the tails of the two vectors, we have

$$\cos\theta = \left(\frac{V_{1x}V_{2x} + V_{1y}V_{2y} + V_{1z}V_{2z}}{\| \boldsymbol{V_1} \| \, \| \boldsymbol{V_2} \|} \right) \quad \text{with } 0° \le \theta \le 180° \quad (4.25)$$

SYSTEM ANALYSIS (SA) EXERCISES

SA 4.1 Calibrating Your Capacity

It is important for engineers to have a sense of how large forces are. One way of developing this sense is to develop reference frames for force comparison; one such reference frame is your physical capacity. To this end, find a weight room on campus, and report the following information. If you are personally not able to complete the exercises, observe someone else engaged in them and record their data. Before attempting any of these exercises, make sure that you have received instruction on how properly to operate the equipment.

(a) Location of weight room

(b) Maximum weight (in newtons) you are able to bench press (without hurting yourself)

(c) Weight (in newtons) you are able to bench press repetitively (without hurting yourself) 20 times

(d) Maximum weight (in newtons) you are able to leg press (without hurting yourself)

(e) Weight (in newtons) you are able to leg press repetitively (without hurting yourself) 20 times

(f) Maximum weight (in newtons) you are able to arm curl (without hurting yourself)

(g) Weight (in newtons) you are able to arm curl repetitively (without hurting yourself) 20 times

Complete the table below with your measurements from (b)–(g):

Exercise	Average College Sophomore Male (N)	Average College Sophomore Female (N)	Your measurements (N)	Your measurements (lb)	Your data expressed as a percentage of average college sophomore (select male or female, as appropriate)
Bench press (max. weight)	753*	328*			
Bench press (repetitive)	609*	243*			
Leg press (max. weight)	1459*	633*			
Leg press (repetitive)	1168*	504*			
Arm curl (max. weight)	104**	57**			
Arm curl (repetitive)	—	—			

*Based on data gathered from 177 college students in the fall of 2001 and 2002. The complete data set is shown in **Figure SA4.1.1**.

**Based on data gathered from 12 college students in the fall of 2002.

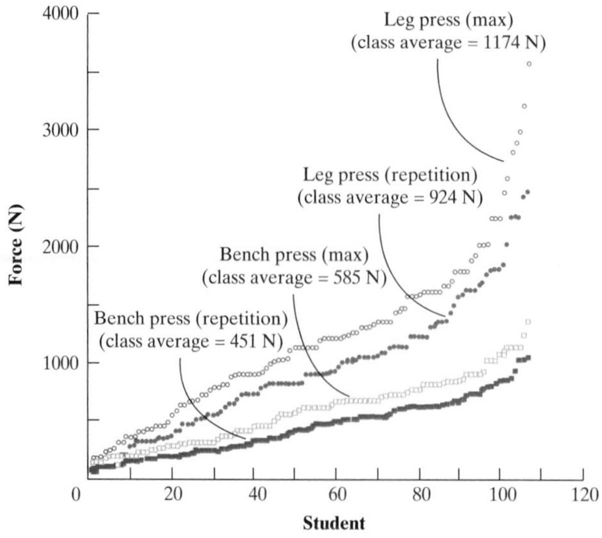

Figure SA4.1.1 Force capacity for college students (ordered from least to greatest capacity)

SA 4.2 Estimating Force Values

It is important for engineers to be able to estimate the forces involved in various situations. In this exercise you are asked to estimate forces. If you are not familiar with a particular situation, you may want to consult with experts, visit the library, and/or research the Web for information. Make sure to record your sources of information.

(a) Order the following list of forces from smallest to largest magnitude.

(b) If possible, represent the size of each force by an order of magnitude.

	System	Situation
A.	Adult 1	Gravitational attraction force of adult 2 on adult 1 (in close proximity to each other)
B.	Bicycle pedal	Force exerted by trained athlete on a bicycle pedal
C.	Bus	Drag force on a bus when bus is traveling at 50 mph
D.	Fully loaded commercial aircraft	Weight of fully loaded commercial aircraft
E.	People held by Golden Gate Bridge	Weight of people on Golden Gate Bridge for 50th anniversary celebration
F.	Leg press	Force exerted on leg press by trained athlete
G.	Leg press	Force exerted on leg press by average engineering student
H.	Leg press	Total force applied by all the students in your statics class acting simultaneously on the leg press
I.	Ninety-fifth percentile adult female in the United States	Weight
J.	Ninety-fifth percentile adult male in the United States	Weight
K.	Sports car	Drag force when sports car is traveling at 20 mph
L.	Sports car	Wind drag force on a sports car when sports car is traveling at 60 mph
M.	Touring bicycle	Drag force on a bicycle when bicycle is traveling at 20 mph

SA 4.3 Problem: Forces to Hold the Scoreboard in Place

The basketball facility inside the Reynolds Coliseum at North Carolina State University contains a heavy scoreboard that is suspended from the ceiling with two cables, shown in **Figure SA4.3.1**. Two electric winches make it possible to lower the entire structure to the floor where signs can be changed or maintained.

The 4500-N scoreboard can be lowered using two winches, A and B, which are attached to the bottom flange of a main roof beam girder. **Figure SA4.3.2** presents a view of the key suspension elements when standing directly underneath.

Assume that you are attending a basketball game with your buddy George, who has not studied statics. Before the game, you are explaining how the heavy scoreboard is being secured in the air, which stimulates him to ask

Figure SA4.3.1 View of the 49-meter-wide field with suspended scoreboard

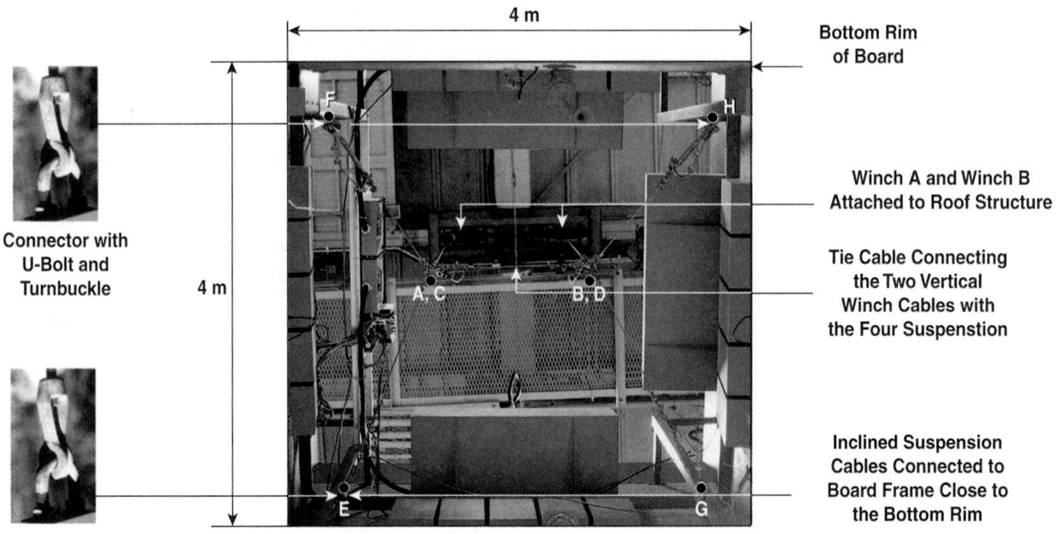

Figure SA4.3.2 View into the scoreboard from the floor

whether the maximum tension in the cables that run from E, F, G, and H to tie cable locations C and D will be 1/4 of the weight of the scoreboard. He also wonders what the purpose is of the tie cable CD. Having just studied this chapter, you should have no problem addressing his questions. Consider George's curiosity as a wonderful opportunity to understand the material, since we all know that the best way to learn something well is by teaching it. **Figure SA4.3.3** provides you with the dimensions and labels that you need to "teach."

Here is how you work with George to address his questions:

1. **Figure SA4.3.4** shows the forces acting at point D of the rigging.
   (a) Find forces $\boldsymbol{F}_{DC}$, $\boldsymbol{F}_{DB}$, $\boldsymbol{F}_{DG}$, and $\boldsymbol{F}_{DH}$. Write in vector notation.
   (b) If the magnitude of $\boldsymbol{F}_{DB}$ is 4500/2 N (is this reasonable?) and the sum of the forces acting at point

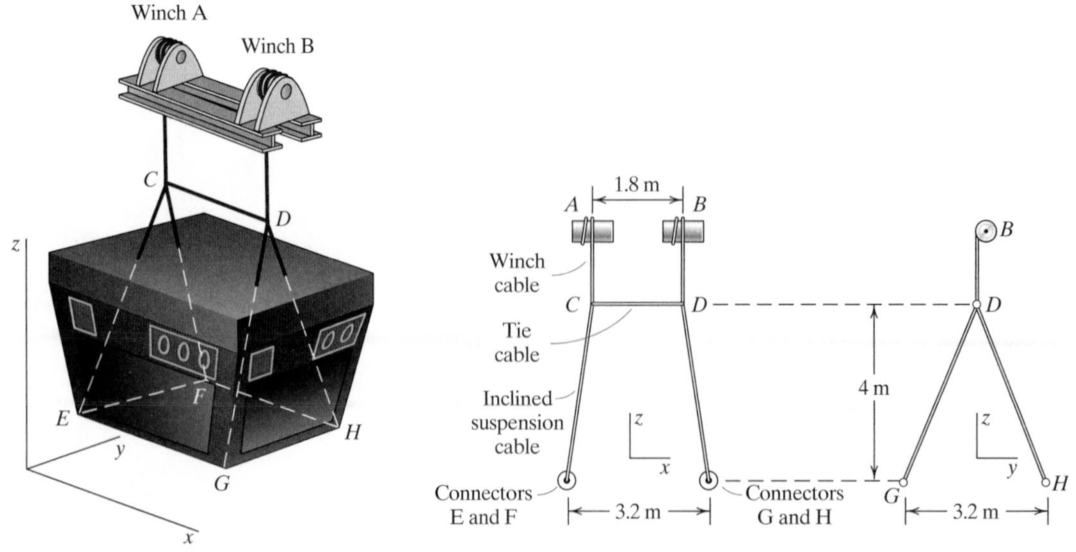

(*a*) 3-D view of board (*b*) Front view of rigging (*c*) Side view of rigging

Figure SA4.3.3 Models of the rigging system suspending the scoreboard from the roof

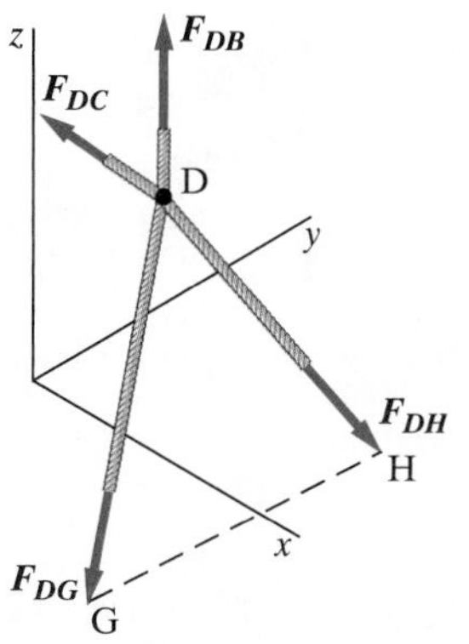

Figure SA4.3.4 Free-body diagram for Point D

D is zero, what are the magnitudes of $\boldsymbol{F}_{DC}$, $\boldsymbol{F}_{DG}$, and $\boldsymbol{F}_{DH}$?

(c) Based on your findings in (b), help George answer his question about whether the maximum tension in the cables that run from E, F, G, and H to the winches connections at C and D will simply be 1/4 of the weight of the scoreboard.

2. What do you think the function of the tie cable is? If the tie cable were not present, would the maximum tension in the cables that run from E, F, G, and H to tie cable locations C and D increase, decrease, or remain the same? Why?

CHAPTER 5

MOMENTS

In Chapter 4 we considered the forces that push and pull on a system. In this chapter we consider how forces not only push and pull but also tend to twist, tip, turn, and rock the systems on which they act. This tendency is illustrated in **Figure 5.1**. If you place your hand near the bottom of the can and push, the can slides across the table. If you place your hand near the top of the can and push, the can turns over. This chapter is about this turning tendency of applied forces, which is called a moment.

Some designs function because of moments—see-saws, balance scales, can openers, and torsion-bar suspensions are a few examples. Other designs, such as skyscrapers, diving boards, and airplane wings, must be designed to withstand moments.

We begin this chapter by presenting the properties and characteristics of moments. Then we outline two formal mathematical methods for calculating moments and address situations involving multiple forces that cause moments.

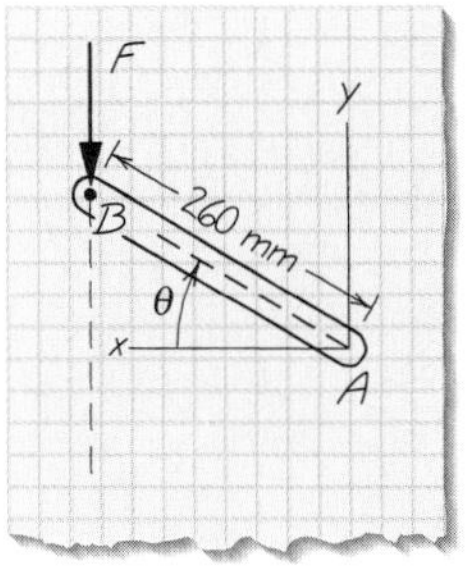

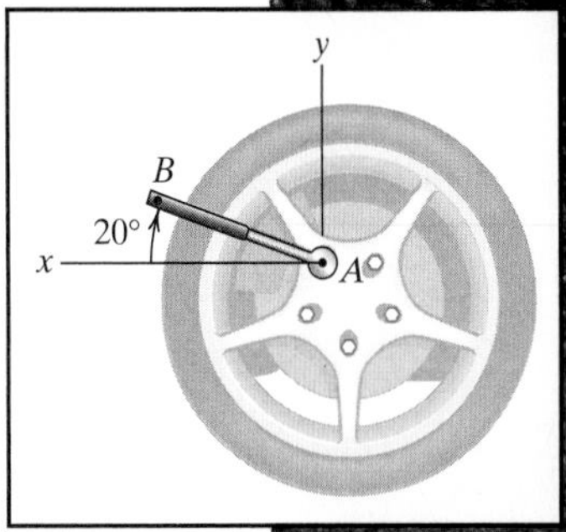

OBJECTIVES

On completion of this chapter, you will be able to:

- **Define a moment both qualitatively and mathematically**
- **Calculate the moment created by a force, and interpret the physical significance of the result**
- **Find the equivalent moment and equivalent force due to a number of loads acting on a system**

5.1 WHAT ARE MOMENTS?

Imagine you have just changed a flat tire on your car and are replacing the lug nuts that bolt the wheel to the hub. You use a wrench to tighten each nut onto a bolt, as in Figure 5.2. By pushing downward on the wrench, you are applying a **moment**[1] to the nut. This moment is created by the force exerted by your hand on the wrench and is about the axis of the bolt (which in the figure is perpendicular and "into" the paper). The magnitude of the moment is the product of the magnitude of the applied force and the distance from the center of the bolt to your hand.

A moment is created by a force that is offset relative to a point in space. It is a vector quantity and therefore has both magnitude and direction. In working with moments, we must concern ourselves with the force's position relative to this point in space, the magnitude, direction, and sense of the moment, and its graphical depiction. Let's look at these elements one at a time.

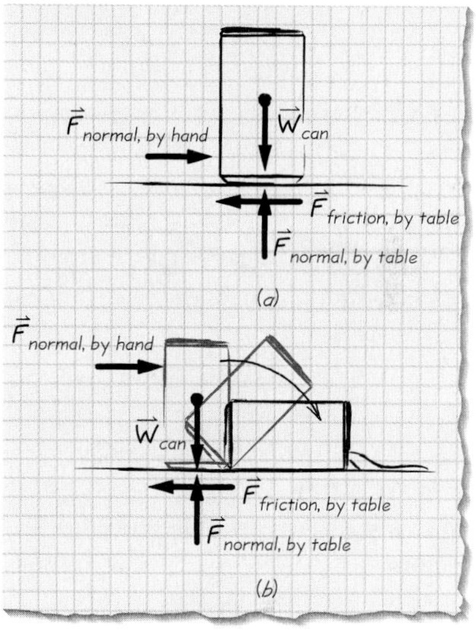

Figure 5.1 (*a*) When push force is applied near the bottom of the can, can slides. (*b*) When push force is applied near the top, can turns over.

Moment Center

A moment is created by a force that is offset relative to a point in space—we call this point the **moment center** (or MC for short). The moment center can be specified by x, y, and z coordinates in a Cartesian coordinate system.

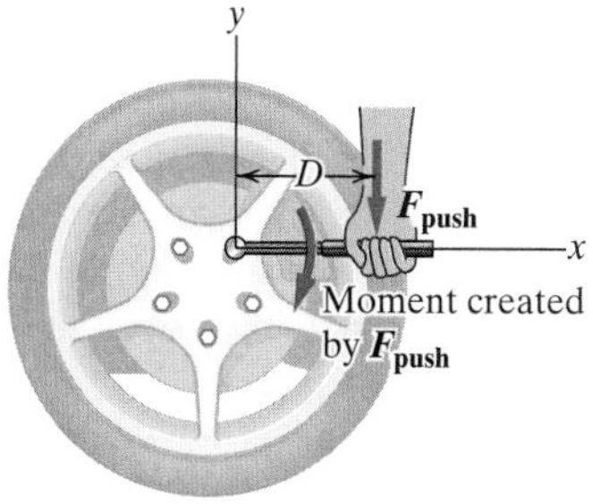

Figure 5.2 Pushing down on the wrench creates a moment

Position Vector

A **position vector** is any vector that runs from the moment center to a point on the line of action of the force (Figure 5.3). Position vectors are commonly specified in vector notation (e.g., position vector $\boldsymbol{r}$ is specified

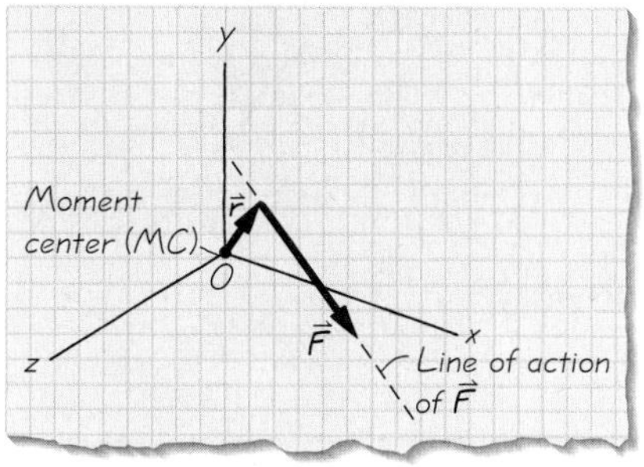

Figure 5.3 Position vector runs from moment center to line of action of the force

[1]Some physics textbooks use the word **torque** for what we are calling moment—namely, the tendency of a force to cause tipping or turning. As we will see in Chapter 9, engineers generally use the word torque to describe a moment created in conjunction with a machine.

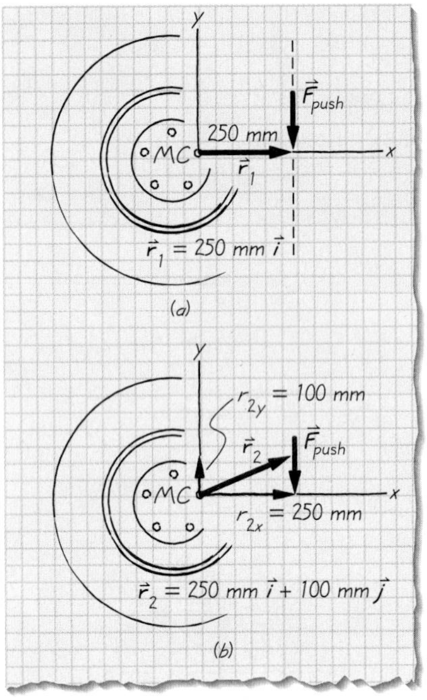

Figure 5.4 Two possible position vectors

as $r_x\boldsymbol{i} + r_y\boldsymbol{j} + r_z\boldsymbol{k}$, where r_x, r_y, and r_z are scalar components in the $\boldsymbol{i}$, $\boldsymbol{j}$, and $\boldsymbol{k}$ directions, respectively, from the MC to the point on the line of action).

A position vector for the wrench force $\boldsymbol{F}_{push}$ relative to a moment center at the center of the bolt is illustrated in **Figure 5.4*a*** as a vector that runs from the moment center (MC) to the point of application of $\boldsymbol{F}_{push}$. With the coordinate system shown in **Figure 5.4**, this vector is defined as $\boldsymbol{r}_1 = 250 \text{ mm } \boldsymbol{i}$. It is just one of a family of position vectors because *any* vector from the moment center to *any* point on the line of action of the force is a position vector. For example, another position vector for $\boldsymbol{F}_{push}$ is $\boldsymbol{r}_2 = 250 \text{ mm } \boldsymbol{i} + 100 \text{ mm } \boldsymbol{j}$ (**Figure 5.4*b***).

We now consider defining a position vector for the situation depicted in **Figure 5.5*a***, where an individual pulls on a rope tied to a branch. (He is attempting to pull down the nearly-sawn-through branch. Note this is sort of dangerous, maybe he should wear a helmet!). We can represent this force as $\boldsymbol{F} = \| \boldsymbol{F} \| (0.116\boldsymbol{i} + 0.349\boldsymbol{j} - 0.930\boldsymbol{k})$, as shown in **Figure 5.5*b***. If we are interested in the moment created by this force at the saw cut, we establish the saw cut as the moment center (MC). Possible position vectors from the MC to the line of action of the force include $\boldsymbol{r}_1 = 2 \text{ m } \boldsymbol{i} + 6 \text{ m } \boldsymbol{j} - 11 \text{ m } \boldsymbol{k}$ (which runs from MC to B) and $\boldsymbol{r}_2 = 0.5 \text{ m } \boldsymbol{i} + 1.5 \text{ m } \boldsymbol{j} + 1.0 \text{ m } \boldsymbol{k}$ (which runs from MC to A), as shown in **Figure 5.5*c***.

When choosing a position vector, keep in mind that it need not lie along a physical connection between the MC and the line of action of the force. Generally, the position vector you should choose is the one that is easiest to define with the information you know about the physical system. **Figure 5.6** shows position vectors in two systems; in each case, the position vector that is easiest to define with available dimensions is labeled.

Synonyms commonly used in engineering practice for the position vector that is perpendicular to the line of action of the force are moment arm vector, moment arm, lever arm, and offset.

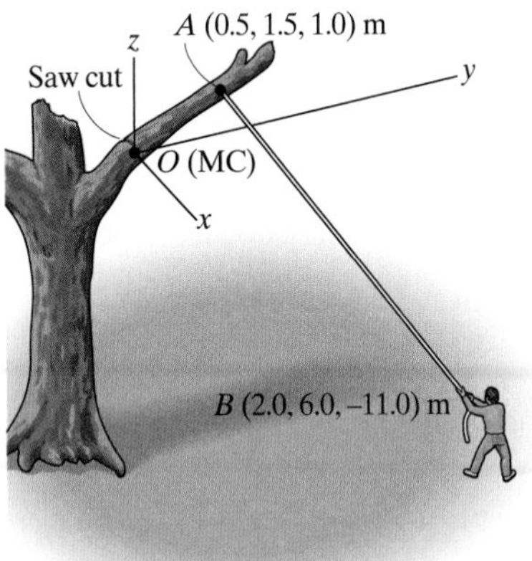

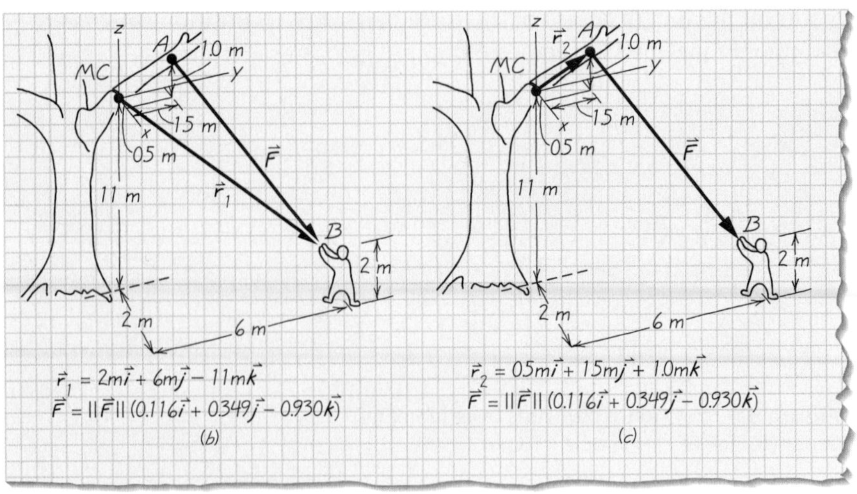

Figure 5.5 A person pulling on a rope attached to a tree branch and two possible position vectors

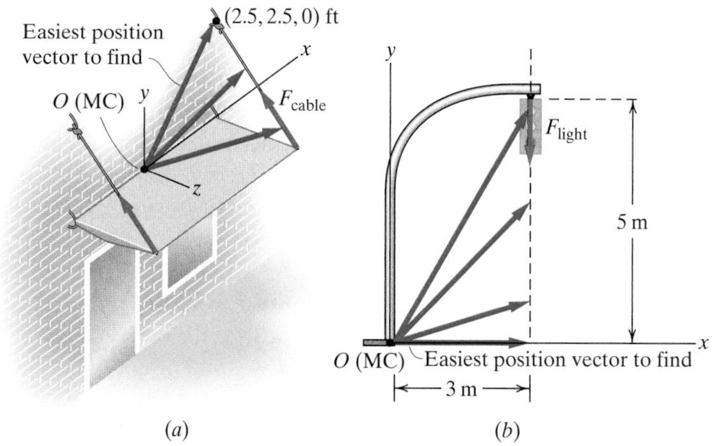

Figure 5.6 There are an infinite number of position vectors, use the one that is easiest to find.

Magnitude of Moment

The **magnitude of a moment** $\|\boldsymbol{M}\|$ is the product of the magnitudes of the position vector $\|\boldsymbol{r}\|$ and the *force component perpendicular to the position vector*. This force component is $\|\boldsymbol{F}\| \sin\theta$, where θ is the angle between the position vector and the line of action of the force, as illustrated in **Figure 5.7**. Therefore, we write the magnitude of the moment as

$$\|\boldsymbol{M}\| = \|\boldsymbol{r}\|\,(\,\|\boldsymbol{F}\| \sin\theta) \tag{5.1}$$

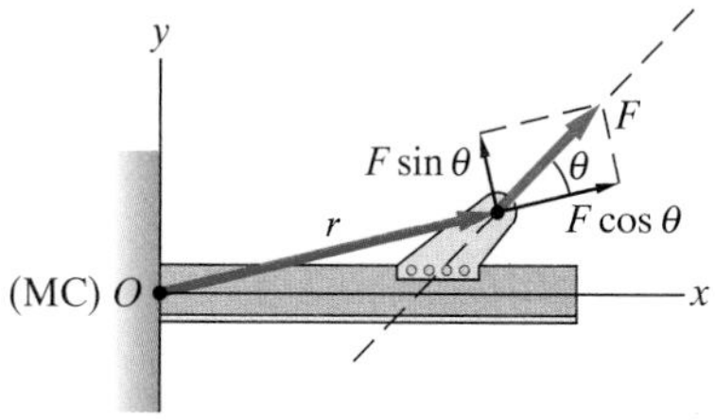

Figure 5.7 The component of the force that is perpendicular to the position vector creates a moment.

where $\|\boldsymbol{r}\|$ is the magnitude of the position vector,[2] $\|\boldsymbol{F}\|$ is the magnitude of the force vector,[3] and θ is the angle between these two vectors when they are placed tail to tail such that $0° < \theta < 180°$.

Figure 5.8 illustrates how to use (5.1) to determine the magnitude of the moment for our lug nut example from **Figure 5.4**, with two position vectors $\boldsymbol{r}_1$ and $\boldsymbol{r}_2$. With $\boldsymbol{r}_1$, the angle θ is 90°, which means the component

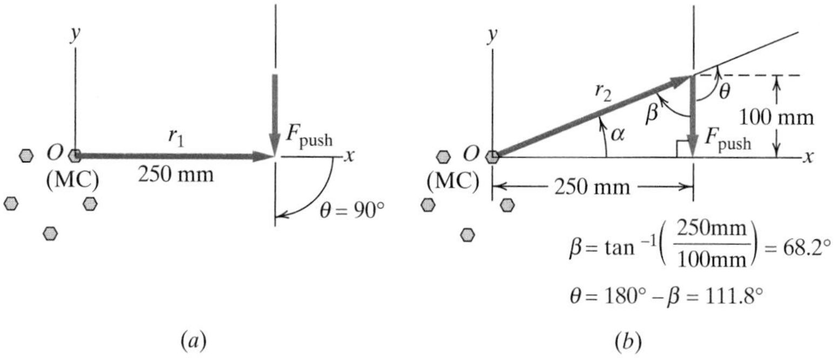

Figure 5.8 Calculating the moment based on (a) r_1 and (b) r_2

[2]For a position vector expressed in scalar components, $\|\boldsymbol{r}\| = \sqrt{r_x^2 + r_y^2 + r_z^2}$.

[3]For a force vector expressed in scalar components, $\|\boldsymbol{F}\| = \sqrt{F_x^2 + F_y^2 + F_z^2}$.

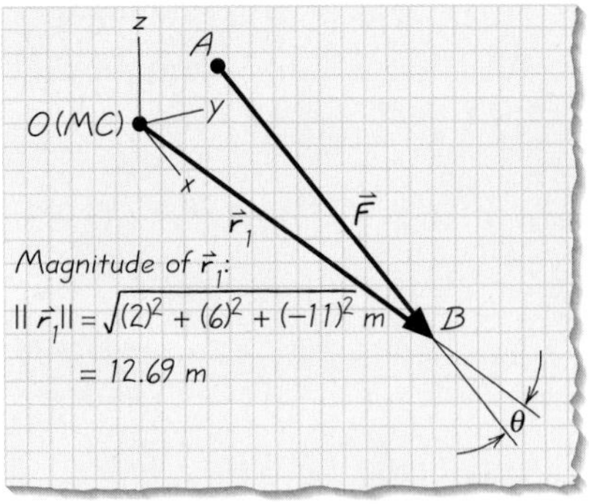

Figure 5.9 Calculating the moment based on $\boldsymbol{r}_1$

of $\boldsymbol{F}_{\text{push}}$ perpendicular to $\boldsymbol{r}_1$ is simply $\| \boldsymbol{F}_{\text{push}} \|$. With $\boldsymbol{r}_2$, the angle θ is 111.8°, which means the component of $\boldsymbol{F}_{\text{push}}$ perpendicular to $\boldsymbol{r}_2$ is $\| \boldsymbol{F} \|$ sin 111.8°. *Remember that only the component of* ***F*** *perpendicular to* ***r*** *creates the moment. The component of* ***F*** *along* ***r*** *does* not *create a moment.*

We end up with the same value $\| \boldsymbol{M} \|$ no matter which position vector we used. This leads us to conclude that *the magnitude of the moment created by a force is independent of the choice of position vector used to calculate the magnitude.*

Figure 5.9 shows the angle θ in (5.1) for the moment created by the rope tension pulling on the branch in **Figure 5.5**. With $\boldsymbol{F}$ and $\boldsymbol{r}_1$ known in terms of their components, the dot product would be a straightforward way to find θ. Recall that the dot product was introduced in Chapter 4, and it allows us to find the projection of one vector onto another.

Sense and Direction of Moment

Now back to our tire iron. To tighten the lug nut onto the bolt, a force is applied at the end of the wrench so that the resulting moment twists the nut clockwise about a z axis, as viewed from in front of the wheel. To loosen the nut, we would apply the force so that the resulting moment twists the nut counterclockwise about a z axis. The terms *clockwise* and *counterclockwise* refer to the rotational **sense** of the moment about an axis. If we establish a local coordinate system at the moment center, we define the sense of the moment by standing on the positive axis (in this case, the positive z axis) and looking back at the moment center, as illustrated in **Figure 5.10**. The axis that we are standing on defines the **direction** of the moment, as detailed below.

A right-hand rule enables us to formalize a procedure for determining the direction of a moment. Consider a force applied to the wrench that results in a counterclockwise (*loosening*) rotation (**Figure 5.11*a***). Align the fingers of your right hand with the position vector (**Figure 5.11*b***), with your palm in the position that allows you to curl your four fingers so that they point in the direction of the force (**Figure 5.11c**). Now move your thumb to make a 90° angle with your palm. The direction in which your thumb points is the direction of $\boldsymbol{M}$, which in this ex-

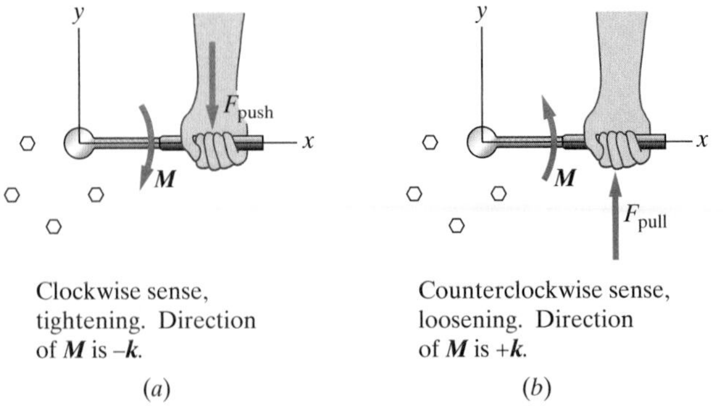

Figure 5.10 (*a*) Clockwise sense, tightening, direction is $-\boldsymbol{k}$; (*b*) counterclockwise sense, loosening, direction is $+\boldsymbol{k}$

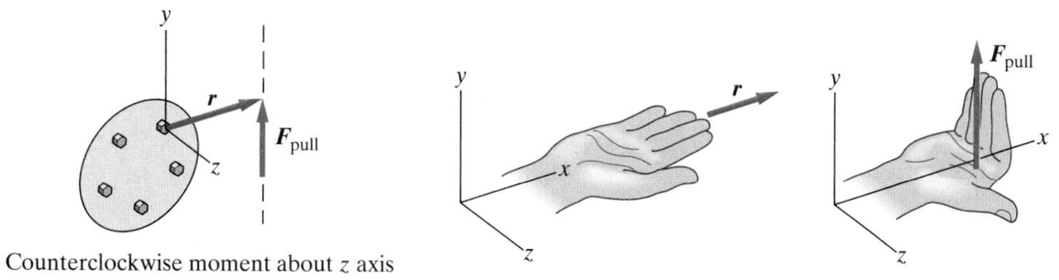

(a) (b) (c)

Figure 5.11 Orienting right-hand figures to show moment sense is counterclockwise

ample is the positive ***k*** (+***k***) direction. Consistent with this, we say that the sense of the moment is counterclockwise about a *z* axis.

When you follow the right-hand rule for a force applied to the wrench that results in a clockwise (*tightening*) rotation, you find you have to rotate your hand such that your thumb aligns with the negative *z* axis (**Figure 5.12**). Therefore, the moment is in the negative ***k*** (−***k***) direction. Consistent with this, we say that the sense of the moment is clockwise about a *z* axis.

The right-hand rule also works when determining the direction of moments about the *x* and *y* axes. For example, a moment direction of positive ***i*** (+***i***) or negative ***i*** (−***i***) refers to a counterclockwise or clockwise sense, respectively, about the *x* axis, as illustrated in **Figure 5.13*a***. Similarly, positive ***j*** (+***j***) or negative ***j*** (−***j***) refers to a counterclockwise or clockwise sense, respectively, about the *y* axis, as illustrated in **Figure 5.13*b***.

The position vector and force vector define a plane, and the moment is in the direction perpendicular to this plane (the direction of your thumb). Therefore, the direction of the moment is perpendicular to both the direction of the position vector and the direction of the force vector.

Graphical Representation of a Moment

We have been representing moments in drawings (e.g., **Figure 5.13**) with an arrow-headed arc. The arrow shows the sense of the moment. The magnitude of the moment (if known) is written next to the arc, as illustrated in **Figure 5.14*a***.

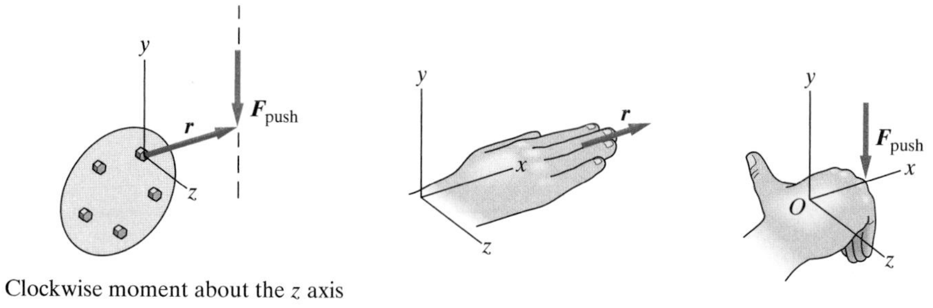

(a) (b) (c)

Figure 5.12 Orienting right-hand figures to show moment sense is clockwise

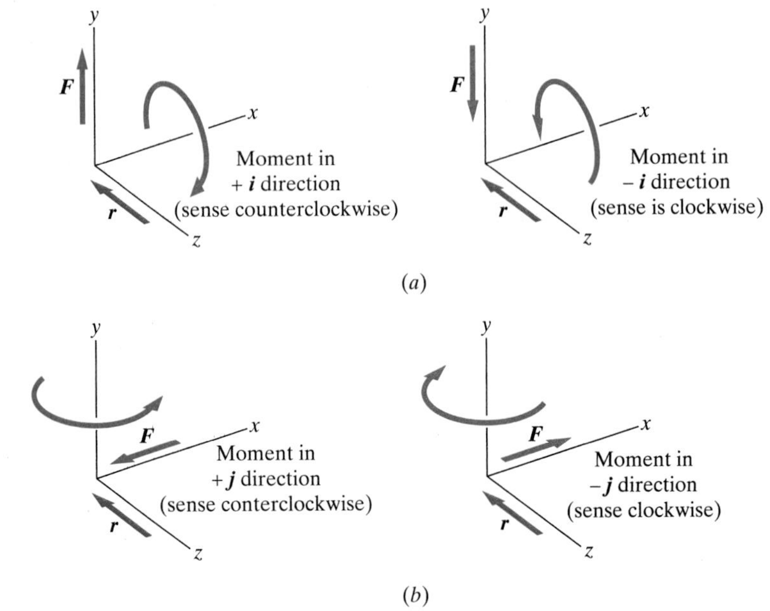

Figure 5.13 (*a*) Moments about the *x* axis; (*b*) moments about the *y* axis

A moment can also be represented with a double-headed arrow (**Figure 5.14*b***). The direction of the arrow represents the direction of the moment; the moment is positive if the arrow points along the positive rotational axis and negative if the arrow points along the negative rotational axis. If the magnitude of the moment is known, it is written next to the arrow and/or the length of the arrow is drawn in proportion to the magnitude.

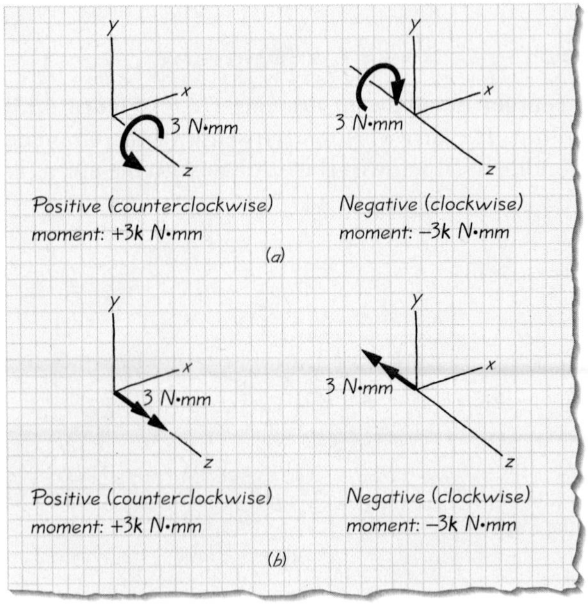

Figure 5.14 Representing moments graphically

These various representations of moments in **Figure 5.14** enable moments to be readily distinguished from forces (which are depicted as single-headed, straight arrows).

EXAMPLE 5.1 SPECIFYING THE POSITION VECTOR (A)

Consider two gears that contact at point P (**Figure 5.15*a***). Gear B pushes on gear A with a force of 100 N at P (**Figure 5.15*b***). The radius of gear A is 100 mm, and the radius of gear B is 80 mm. Coordinate axes are located as shown. Specify in vector notation the position vectors shown in **Figure 5.16**:

(a) Position vector $\boldsymbol{r}_1$

(b) Position vector $\boldsymbol{r}_2$

(c) Position vector $\boldsymbol{r}_3$, which has a magnitude of 150 mm

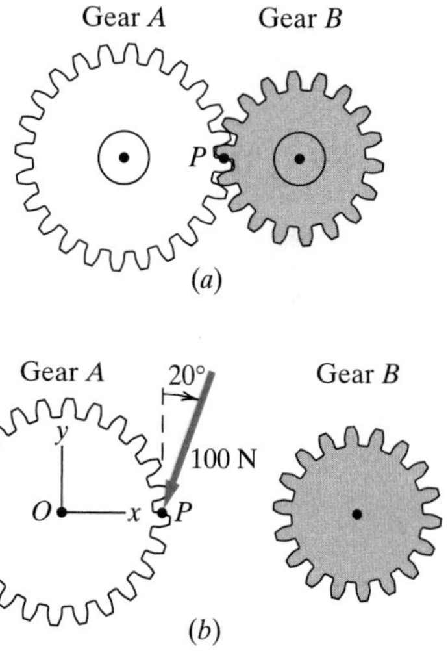

Figure 5.15

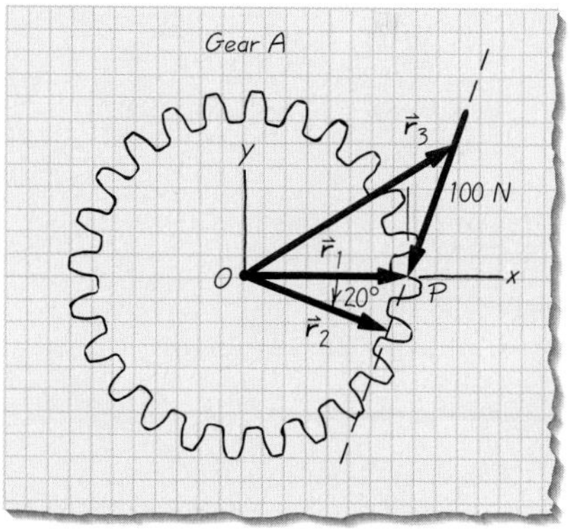

Figure 5.16

Goal We are to define three position vectors ($\boldsymbol{r}_1$, $\boldsymbol{r}_2$, $\boldsymbol{r}_3$) in vector notation for a gear force relative to a moment center at the center of one of the gears.

Given We are told that gear B pushes on gear A with a force of magnitude 100 N at an angle of 20° relative to the vertical. This force acts at point P. The x and y axes have been specified. In addition, the radius of gear A is 100 mm, and the radius of gear B is 80 mm. Position vectors are defined in **Figure 5.16**. Position vector $\boldsymbol{r}_3$ has a magnitude of 150 mm.

Assume We assume that the gear force lies in the xy plane.

Draw, Formulate Equations, and Solve **(a)** Position vector $\boldsymbol{r}_1$ is aligned with the x axis and goes from the moment center at O to point P. The distance from the moment center to P is 100 mm. Therefore, by inspection of **Figure 5.16**, we can write

Answer $\boldsymbol{r}_1 = 100.0 \text{ mm } \boldsymbol{i}$

(b) Position vector $\boldsymbol{r}_2$ goes from the moment center at O to the line of action of the force and is perpendicular to the line of action. Based on the drawing of $\boldsymbol{r}_2$ in **Figure 5.17*a***, we can write

$$\|\boldsymbol{r}_2\| = 100 \text{ mm } \cos 20° = 94.0 \text{ mm}$$

Based on **Figure 5.17*b***, the scalar components can be written as

$$r_{2x} = \|\boldsymbol{r}_2\| \cos 20° = 94.0 \text{ mm } \cos 20° = 88.3 \text{ mm}$$
$$r_{2y} = -\|\boldsymbol{r}_2\| \sin 20° = -94.0\,0 \text{ mm } \sin 20° = -32.1 \text{ mm}$$

We use these components to write $\boldsymbol{r}_2$ in vector notation as

Answer $\boldsymbol{r}_2 = 88.3 \text{ mm } \boldsymbol{i} - 32.1 \text{ mm } \boldsymbol{j}$

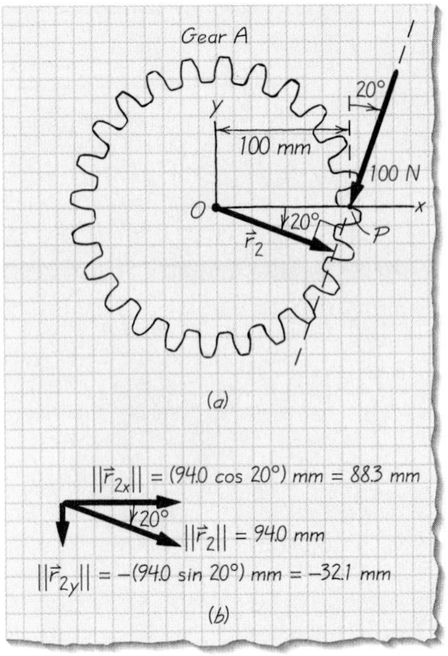

Figure 5.17

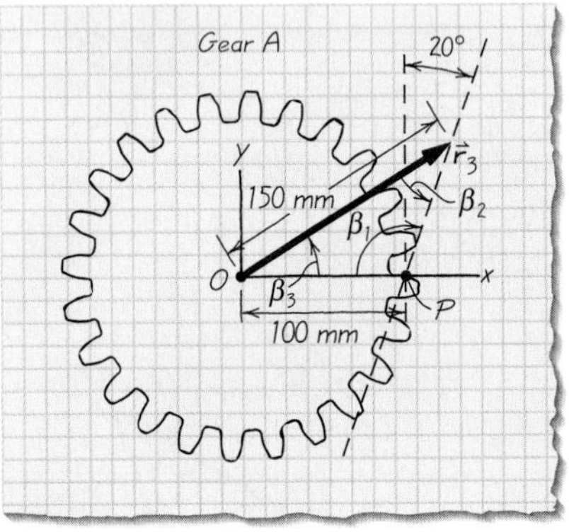

Figure 5.18

(c) Position vector $\boldsymbol{r}_3$ goes from the moment center at O to the line of action of the force. We are told it has a magnitude of 150 mm. Based on **Figure 5.18**, we determine that the angle β_1 between the x axis and the line of action of the force is 110° (= 90° + 20°). Therefore, using the law of sines we find β_2:

$$\frac{\sin 110°}{150\text{ mm}} = \frac{\sin \beta_2}{100\text{ mm}}$$

or

$$\beta_2 = 38.8°$$

Furthermore

$$\beta_3 = 180° - \beta_1 - \beta_2 = 31.2°$$

Finally, we can write

$$r_{3x} = 150\text{ mm}\ \cos\ 31.2° = 128.3\text{ mm}$$
$$r_{3y} = 150\text{ mm}\ \sin\ 31.2° = 77.7\text{ mm}$$

We use these components to write $\boldsymbol{r}_3$ in vector notation as

Answer $\boldsymbol{r}_3 = 128.3\text{ mm}\,\boldsymbol{i} + 77.7\text{ mm}\,\boldsymbol{j}$

Check This answer for (a) can be checked by inspection; when we look at **Figure 5.16** we see that $\boldsymbol{r}_1$ is aligned with the positive x axis. Furthermore, it is along a radius of gear A. Therefore, the answer of $\boldsymbol{r}_1 = 100.0\text{ mm}\,\boldsymbol{i}$ makes physical sense.

The answer for (b) can be checked by calculating the magnitude of $\boldsymbol{r}_2$ from its components:

$$\sqrt{(88.3\text{ mm})^2 + (-32.1\text{ mm})^2} = 94.0\text{ mm}$$

The answer for (c) can be checked by calculating the magnitude of $\boldsymbol{r}_3$ from its components:

$$\sqrt{(128.3\text{ mm})^2 + (77.7\text{ mm})^2} = 150\text{ mm}$$

Comment: The position vectors $\boldsymbol{r}_1$, $\boldsymbol{r}_2$, and $\boldsymbol{r}_3$ are all valid for locating $\boldsymbol{F}$ relative to a moment center at the center of gear A since each of them goes from the moment center to the line of action of $\boldsymbol{F}$. They are only three examples of the infinite number of position vectors for locating $\boldsymbol{F}$ relative to O.

EXAMPLE 5.2 SPECIFYING THE POSITION VECTOR (B)

A cable exerts a tension force $\boldsymbol{F}$ of 500 lb on the right-angle pipe shown in **Figure 5.19**. Specify position vectors in vector notation for the force relative to the moment center at O:

(a) Position vector $\boldsymbol{r}_1$

(b) Position vector $\boldsymbol{r}_2$

Goal We are to define two position vectors ($\boldsymbol{r}_1$, $\boldsymbol{r}_2$) in vector notation for a force applied at the end of a right-angle pipe.

Given We are told the magnitude of the force (500 lb) and that it is applied at the end of the right-angle pipe at point A. Furthermore, based on the information in **Figure 5.19**, we can determine the coordinates of points A and C.

Assume No assumptions are required.

Draw The sketch of pipe element BA in **Figure 5.20** will facilitate our finding the coordinates of point A.

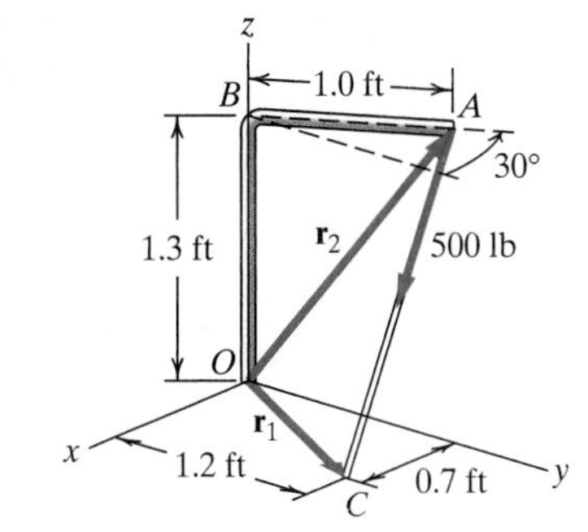

Figure 5.19

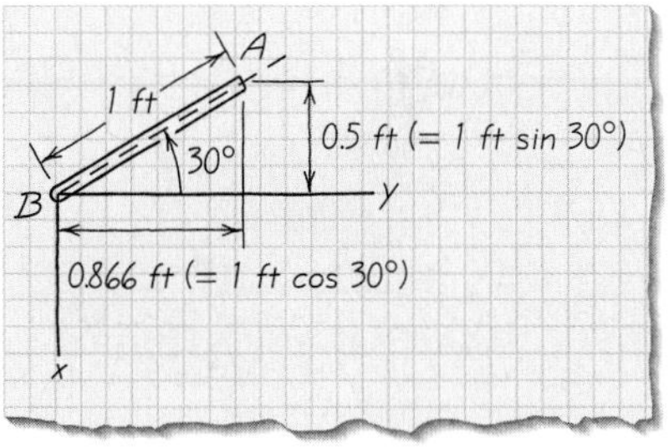

Figure 5.20

Formulate Equations and Solve **(a)** Position vector $\boldsymbol{r}_1$ in **Figure 5.19** goes from the moment center to point C on the line of action of the force. By inspection of **Figure 5.19**, we can determine that

- the coordinates of the origin are $x = 0$ ft, $y = 0$ ft, and $z = 0$ ft, and
- the coordinates of point C are $x = 0.70$ ft, $y = 1.20$ ft, and $z = 0.0$ ft.

Based on this information, we can write $\boldsymbol{r}_1 = (0.70 \text{ ft} - 0 \text{ ft})\boldsymbol{i} + (1.20 \text{ ft} - 0 \text{ ft})\boldsymbol{j}$, which simplifies to

Answer to (a) $\boldsymbol{r}_1 = 0.70 \text{ ft } \boldsymbol{i} + 1.20 \text{ ft } \boldsymbol{j}$

(b) Position vector $\boldsymbol{r}_2$ in **Figure 5.19** goes from the moment center O to the point of application of $\boldsymbol{F}$ at A. By inspection of **Figures 5.19** and **5.20**, we can determine that

- the coordinates of the origin are $x = 0$ ft, $y = 0$ ft, and $z = 0$ ft, and
- the coordinates of point A are $x = -1.0 \text{ ft} \sin 30° = -0.5$ ft, $y = 1.0 \text{ ft} \cos 30° = 0.866$ ft, and $z = 1.3$ ft.

Based on this information, we can write

$$\boldsymbol{r}_2 = (-0.5 \text{ ft} - 0 \text{ ft})\boldsymbol{i} + (0.866 \text{ ft} - 0 \text{ ft})\boldsymbol{j} + (1.3 \text{ ft} - 0 \text{ ft})\boldsymbol{k}$$

which simplifies to

Answer to (b) $\boldsymbol{r}_2 = -0.50 \text{ ft } \boldsymbol{i} + 0.87 \text{ ft } \boldsymbol{j} + 1.30 \text{ ft } \boldsymbol{k}$

Check We will check our answers by inspection. Looking at **Figure 5.19**, we see that to go from O (the moment center) to C (a point on the line of action of the force) we "walk" 0.70 ft in the $\boldsymbol{i}$ direction and 1.20 ft in the $\boldsymbol{j}$ direction; therefore, our answer of $\boldsymbol{r}_1 = 0.70 \text{ ft } \boldsymbol{i} + 1.20 \text{ ft } \boldsymbol{j}$ makes sense.

Looking at **Figures 5.19** and **5.20**, we see that to go from O to A (the point of application of the 500-lb force) we "walk" in the negative $\boldsymbol{i}$ direction a distance of 1.0 ft sin 30°, in the positive $\boldsymbol{j}$ direction a distance of 1.0 ft cos 30°, and in the positive $\boldsymbol{k}$ direction a distance of 1.3 ft; therefore, our answer of $\boldsymbol{r}_2 = -0.5 \text{ ft } \boldsymbol{i} + 0.866 \text{ ft } \boldsymbol{j} + 1.3 \text{ ft } \boldsymbol{k}$ makes sense.

EXAMPLE 5.3 THE MAGNITUDE OF THE MOMENT (A)

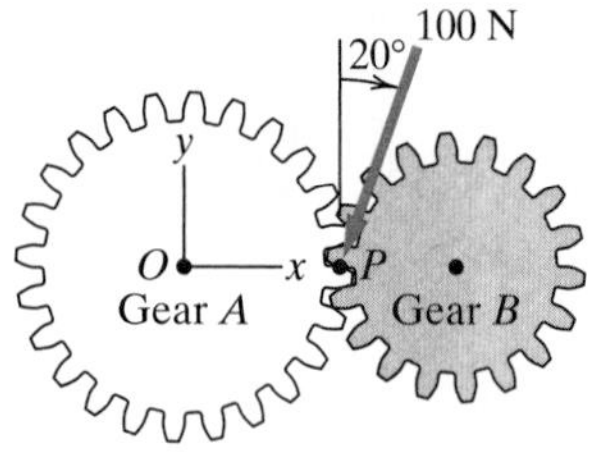

Figure 5.21

Gear B pushes on gear A with a force of 100 N at point P (Figure 5.21). The radius of gear A is 100 mm, and the radius of gear B is 80 mm. Use (5.1) to find the magnitude of the moment that the force acting on gear A creates about an axis perpendicular to the page that passes through the center of gear A.

Goal We are to use (5.1) to calculate the magnitude of the moment created at a moment center at O by the force of gear B acting on gear A.

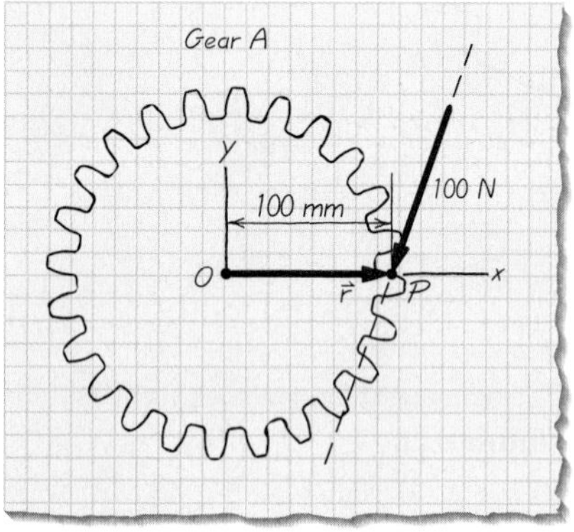

Figure 5.22

Given We are told that gear B pushes on gear A with a force of magnitude 100 N at an angle of 20° relative to the vertical. This force acts at point P. The x and y axes have been specified. In addition, the radius of gear A is 100 mm and the radius of gear B is 80 mm.

Assume No assumptions are required.

Draw We sketch gear A, force $\boldsymbol{F}$ (100 N), and its line of action (Figure 5.22). Inspection of this figure shows that the most straightforward position vector to consider will be from the moment center (O, the center of gear A) to point P; call this position vector $\boldsymbol{r}$.

Formulate Equations and Solve Equation (5.1) states

$$\| \boldsymbol{M} \| = \| \boldsymbol{r} \| \left(\| \boldsymbol{F} \| \sin \theta \right) \tag{5.1}$$

To apply (5.1), we must find $\| \boldsymbol{r} \|$ (the magnitude of the position vector) and the angle θ.

The magnitude of *the position vector* $\boldsymbol{r}$ is simply the radius of gear A, which is 100 mm. Therefore,

$$\| \boldsymbol{r} \| = 100 \text{ mm}$$

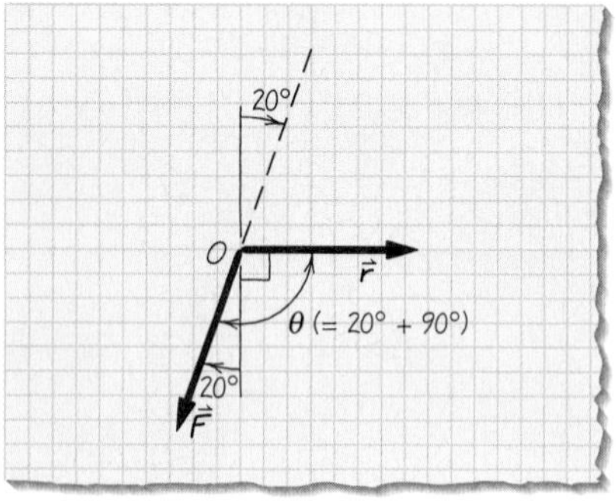

Figure 5.23

We could have used any of the position vectors we defined in Example 5.1. The decision to work with the one from O to P was a pragmatic one; we wanted to make the moment calculation as simple as possible!

The angle θ is the angle between the position vector and the force when they are placed end to end. Based on the geometry depicted in Figure 5.23,

$$\theta = 110°$$

The magnitude of the moment (5.1) yields

$$\| \boldsymbol{M} \| = \| \boldsymbol{r} \| \left(\| \boldsymbol{F} \| \sin \theta \right) = (100 \text{ mm})(100 \text{ N}) \sin 110° = 9400 \text{ N} \cdot \text{mm}$$

Answer The magnitude of the moment is 9400 N · mm (which can also be written as 9.40 N · m).

Check We present a check to confirm that the calculations were set up and carried out correctly; we refer to this as an "accuracy check." This can be done by finding the magnitude of the moment using a different position vector to confirm the 9400 N · mm answer. For example, if r_2 found in Example 5.1 is used,

$$\| \boldsymbol{M} \| = \| \boldsymbol{r}_2 \| (\| \boldsymbol{F} \| \sin 90°) = \sqrt{88.3 \text{ mm}^2 + 32.1 \text{ mm}^2}\,(100 \text{ N}) \sin 90° = 9400 \text{ N} \cdot \text{mm}$$

Therefore our answer of 9400 N · mm is confirmed, and our calculations are correct.

EXAMPLE 5.4 THE MAGNITUDE OF THE MOMENT (B)

A cable exerts a tension force $\boldsymbol{F}$ of 500 lb on the right-angle pipe shown in Figure 5.24. Use (5.1) to find the moment the force creates at a moment center:

(a) At O based on position vector $\boldsymbol{r}_1$
(b) At O based on position vector $\boldsymbol{r}_2$
(c) At A
(d) At C

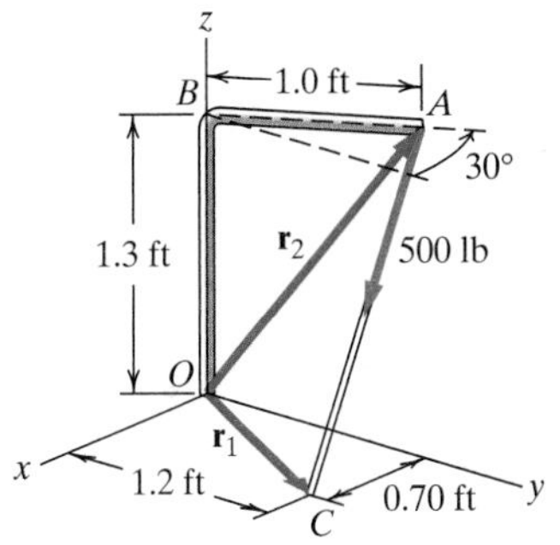

Figure 5.24

Goal We are to use (5.1) to calculate the magnitude of the moment created by the 500-lb force acting at A on a right-angle pipe at a moment center at O using two different position vectors (a, b), at a moment center at A (c), and at a moment center at C (d).

Given We are given the dimensions of the right-angle pipe, an xyz coordinate system, and the magnitude of a force (as well as information to find its direction) acting at the end of the pipe.

Assume No assumptions are required.

Formulate Equations and Solve

Comment: Before we actually start any calculations, we note that we expect the answers for parts (a) and (b) to be the same—remember that the moment a force creates at a particular moment center is independent of the position vector. We can use this as a check of our calculations in (a) and (b).

(a) To apply (5.1) to find the magnitude of the moment created by $\boldsymbol{F}$ at a moment center at O, we must find $\| \boldsymbol{r}_1 \|$ (the magnitude of the position vector) and the angle θ between $\boldsymbol{r}_1$ and $\boldsymbol{F}$.

The position vector $\boldsymbol{r}_1$ was defined in Example 5.2a as $\boldsymbol{r}_1 = 0.70 \text{ ft}\boldsymbol{i} + 1.20 \text{ ft}\boldsymbol{j}$. Its magnitude is therefore

$$\| \boldsymbol{r}_1 \| = \sqrt{(0.70 \text{ ft})^2 + (1.20 \text{ ft})^2} = 1.39 \text{ ft}$$

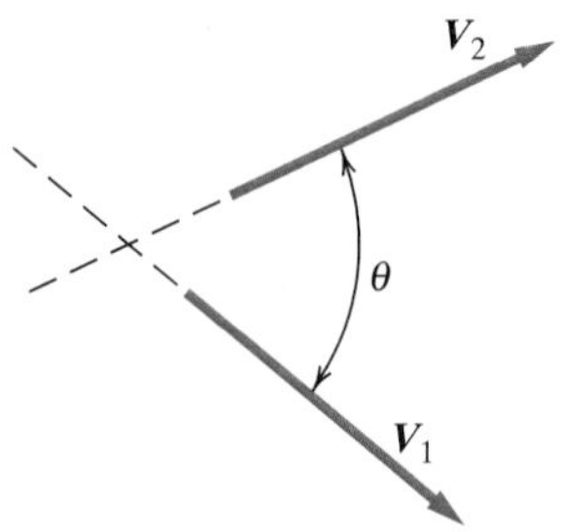

Figure 5.25

The angle θ is the angle between $\boldsymbol{r}_1$ and $\boldsymbol{F}$ when they are placed end to end. We cannot simply read this angle from the figure as we did in Example 5.3. We can, however, use (4.25), which says that the angle between the lines of action of two vectors $\boldsymbol{V}_1 = V_{1x}\boldsymbol{i} + V_{1y}\boldsymbol{j} + V_{1z}\boldsymbol{k}$ and $\boldsymbol{V}_2 = V_{2x}\boldsymbol{i} + V_{2y}\boldsymbol{j} + V_{2z}\boldsymbol{k}$ is

$$\cos\theta = \frac{V_{1x}V_{2x} + V_{1y}V_{2y} + V_{1z}V_{2z}}{\| \boldsymbol{V}_1 \| \, \| \boldsymbol{V}_2 \|} \tag{4.25}$$

where $0° \leq \theta \leq 180°$, as illustrated in Figure 5.25.

In the current problem we use (4.25) to find the angle between the two vectors $\boldsymbol{r}_1$ and $\boldsymbol{F}$. Based on Example 5.2a, we know $\boldsymbol{r}_1 = 0.70 \text{ ft } \boldsymbol{i} + 1.2 \text{ ft } \boldsymbol{j}$. We need to write $\boldsymbol{F}$ in vector notation. Begin by finding the vector that goes from A to C. Based on the geometry in **Figure 5.24**,

Point A is at	Point C is at
$x_A = -(1.0 \text{ ft}) \sin 30° = -0.50 \text{ ft}$	$x_c = 0.70 \text{ ft}$
$y_A = (1.0 \text{ ft}) \cos 30° = 0.866 \text{ ft}$	$y_C = 1.20 \text{ ft}$
$z_A = 1.3 \text{ ft}$	$z_C = 0.0 \text{ ft}$

The vector that runs from A to C is therefore

$$\begin{aligned} \boldsymbol{V}_{AC} &= (x_C - x_A)\boldsymbol{i} + (y_C - y_A)\boldsymbol{j} + (z_C - z_A)\boldsymbol{k} \\ &= 1.20 \text{ ft } \boldsymbol{i} + 0.33 \text{ ft } \boldsymbol{j} - 1.30 \text{ ft } \boldsymbol{k} \end{aligned}$$

with a magnitude of

$$\| \boldsymbol{V}_{AC} \| = \sqrt{(1.20 \text{ ft})^2 + (0.33 \text{ ft})^2 + (1.30 \text{ ft})^2} = 1.80 \text{ ft}$$

The vector $\boldsymbol{V}_{AC}$ can be converted into a unit vector $\boldsymbol{n}_{AC}$ by dividing each scalar component by the magnitude V_{AC}:

$$\begin{aligned} \boldsymbol{n}_{AC} &= \frac{\boldsymbol{V}_{AC}}{\| \boldsymbol{V}_{AC} \|} \\ &= \frac{1.20 \text{ ft}}{1.80 \text{ ft}}\boldsymbol{i} + \frac{0.33 \text{ ft}}{1.80 \text{ ft}}\boldsymbol{j} - \frac{1.30 \text{ ft}}{1.80 \text{ ft}}\boldsymbol{k} \\ &= 0.667\boldsymbol{i} + 0.183\boldsymbol{j} - 0.722\boldsymbol{k} \end{aligned}$$

The force $\boldsymbol{F}$ can now be written as the product of the magnitude of the force (which we are told is 500 lb) and $\boldsymbol{n}_{AC}$:

$$\boldsymbol{F} = (500 \text{ lb})\boldsymbol{n}_{AC} = 334 \text{ lb } \boldsymbol{i} + 91.5 \text{ lb } \boldsymbol{j} - 361 \text{ lb } \boldsymbol{k}$$

We are now (finally!) in a position to apply (4.25) to find the angle θ between $\boldsymbol{r}_1 = 0.70 \text{ ft } \boldsymbol{i} + 1.20 \text{ ft } \boldsymbol{j}$ and $\boldsymbol{F} = 334 \text{ lb } \boldsymbol{i} + 91.5 \text{ lb } \boldsymbol{j} - 361 \text{ lb } \boldsymbol{k}$:

$$\theta = \cos^{-1}\left[\frac{(0.70 \text{ ft})(334 \text{ lb}) + (1.20 \text{ ft})(91.5 \text{ lb}) + (0 \text{ ft})(-361 \text{ lb})}{(1.39 \text{ ft})(500 \text{ lb})}\right]$$

$$\theta = 60.4°$$

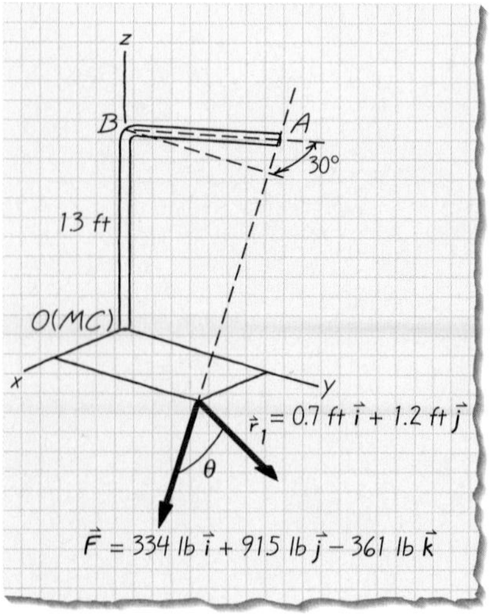

Figure 5.26

This angle is illustrated in Figure 5.26.

Intermediate check[4]: Based on the geometry depicted in **Figure 5.26**, we would expect $0° < \theta < 90°$. This is indeed what we found with $\theta = 60.4°$.

The magnitude of the moment: Substituting for $\| \boldsymbol{r}_1 \|$, $\| \boldsymbol{F} \|$, and θ in (5.1) yields

$$\| \boldsymbol{M} \| = \| \boldsymbol{r}_1 \| (\| \boldsymbol{F} \| \sin \theta) = (1.39 \text{ ft})(500 \text{ lb}) \sin 60.4° = 604 \text{ ft} \cdot \text{lb}$$

Answer to (a) The magnitude of the moment created at a moment center at O by the force is 604 ft · lb.

(b) We use (5.1) (again) to determine the magnitude of the moment created at a moment center at O by a force acting at A. We are to use position vector $\boldsymbol{r}_2$ in this calculation.

The position vector $\boldsymbol{r}_2$ was defined in Example 5.2. Its magnitude is

$$\| \boldsymbol{r}_2 \| = \sqrt{(-0.50 \text{ ft})^2 + (0.87 \text{ ft})^2 + (1.30 \text{ ft})^2} = 1.64 \text{ ft}$$

The angle θ*:* We can follow the same procedure that we used in (a) to find the angle between position vector $\boldsymbol{r}_2 = -0.50 \text{ ft } \boldsymbol{i} + 0.87 \text{ ft } \boldsymbol{j} + 1.30 \text{ ft } \boldsymbol{k}$ and $\boldsymbol{F} = 334 \text{ lb } \boldsymbol{i} + 91.5 \text{ lb } \boldsymbol{j} - 361 \text{ lb } \boldsymbol{k}$:

$$\theta = \cos^{-1}\left[\frac{(-0.50 \text{ ft})(334 \text{ lb}) + (0.87 \text{ ft})(91.5 \text{ lb}) + (1.30 \text{ ft})(-361 \text{ lb})}{(1.64 \text{ ft})(500 \text{ lb})}\right]$$

$$\theta = 132.7°$$

Intermediate check: Based on the geometry depicted in **Figure 5.27**, we would expect $90° < \theta < 180°$. This is indeed what we found with $\theta = 132.7°$.

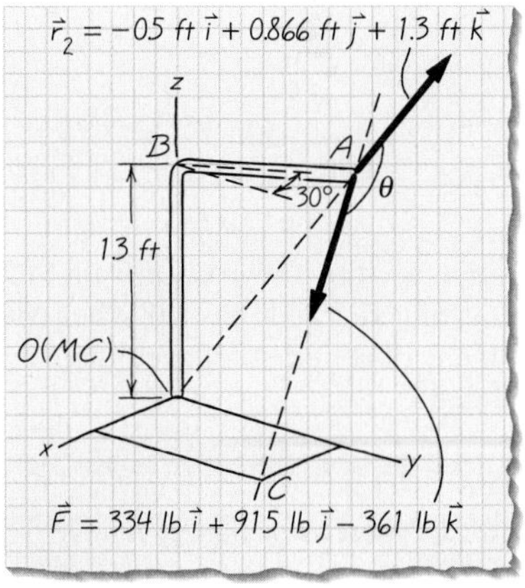

Figure 5.27

The magnitude of the moment: Substituting for $\| \boldsymbol{r}_2 \|$, $\| \boldsymbol{F} \|$, and θ in (5.1) yields

$$\| \boldsymbol{M} \| = \| \boldsymbol{r}_2 \| (\| \boldsymbol{F} \| \sin \theta) = (1.64 \text{ ft})(500 \text{ lb}) \sin 132.7° = 603 \text{ ft} \cdot \text{lb}$$

Answer to (b) The magnitude of the moment created by the force about the moment center at O is 603 ft · lb.

Note: The slight differences in the answers from (a) and (b) are due to differences in round-off in intermediate calculation steps.

Check for (a) and (b): We calculated the same answer in parts (a) and (b). As we commented when we started the solution, this is as expected. Parts (a) and (b) can be used as checks for one another.

(c) We are to find the magnitude of the moment the force creates at a moment center at A. Considering **Figure 5.24**, if we attempt to draw a position vector from the moment center (at point A) to the point of application of the force (also point A), we find that the position vector has zero length—in other words, $\| \boldsymbol{r} \| = 0$. Substituting this into (5.1), we

[4] As you go along, check your intermediate calculations whenever possible. This will both increase the likelihood that you will get the right answer and make it easier to find where you went wrong if you get an incorrect answer.

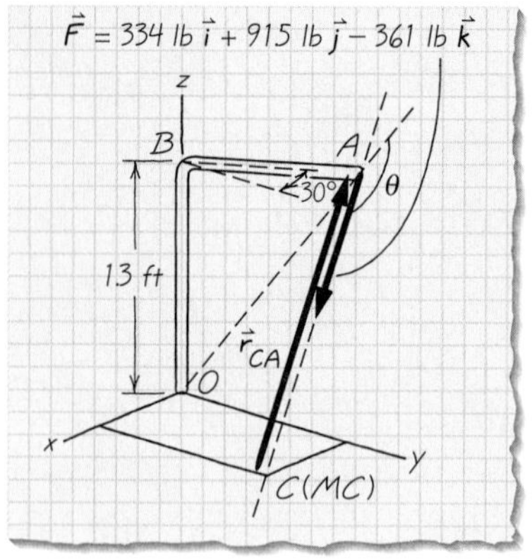

Figure 5.28

find that the magnitude of the moment created at a moment center at A by the force is zero.

Answer to (c) The magnitude of the moment created at A by the force is 0 ft · lb.

(d) We are to find the magnitude of the moment the force creates at a moment center at C. Considering **Figure 5.28**, we draw a position vector from the moment center (at point C) to the point of application of the force (point A). Based on the work in (a), this position vector is

$$\begin{aligned} \boldsymbol{r}_{CA} &= (x_A - x_C)\boldsymbol{i} + (y_A - y_C)\boldsymbol{j} + (z_A - z_C)\boldsymbol{k} \\ &= -1.20 \text{ ft } \boldsymbol{i} - 0.33 \text{ ft } \boldsymbol{j} + 1.30 \text{ ft } \boldsymbol{k} \end{aligned}$$

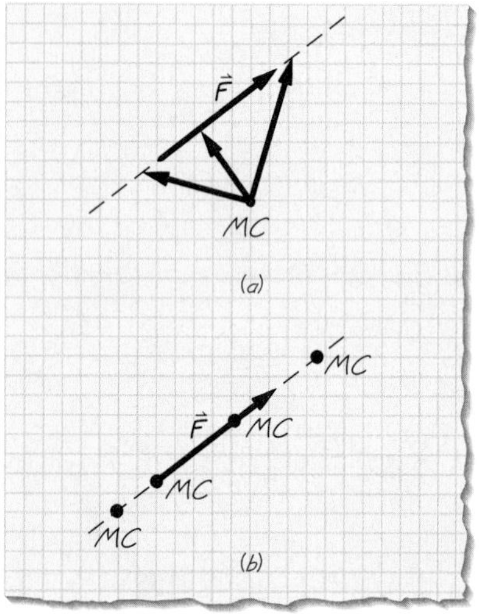

Figure 5.29 (*a*) Examples of valid position vectors—any vector from MC to line of action of force; (*b*) When MC is on line of action of force no moment is created.

Now, using expressions for the position vector ($\boldsymbol{r}_{CA} = -1.20$ ft $\boldsymbol{i}$ – 0.33 ft $\boldsymbol{j}$ + 1.30 ft $\boldsymbol{k}$) and force ($\boldsymbol{F}$ = 334 lb $\boldsymbol{i}$ + 91.5 lb $\boldsymbol{j}$ – 3.61 lb $\boldsymbol{k}$) in (4.25) or by inspection of **Figure 5.28**, we find that $\theta = 180°$. Since $\sin 180° = 0$, substituting into (5.1) leads us to the result that the moment created at a moment center at C by the force is zero.

Alternately, we could avoid all calculations by noting that a position vector drawn from point C (the moment center) to the point C (a point on the line of action of the force) has zero length. Then, using the same reasoning as in (c) we can conclude that the moment created at a moment center at C by the force is zero.

Answer to (d) The magnitude of the moment created at C by the force is 0 ft · lb.

Comment: This example illustrates that

- for a particular moment center, the magnitude of the moment is independent of the position vector, and
- a force creates no moment at a moment center located anywhere on the line of action of the force.

These two key ideas are illustrated in **Figure 5.29**.

EXAMPLE 5.5 INCREASING THE MAGNITUDE OF THE MOMENT (A)

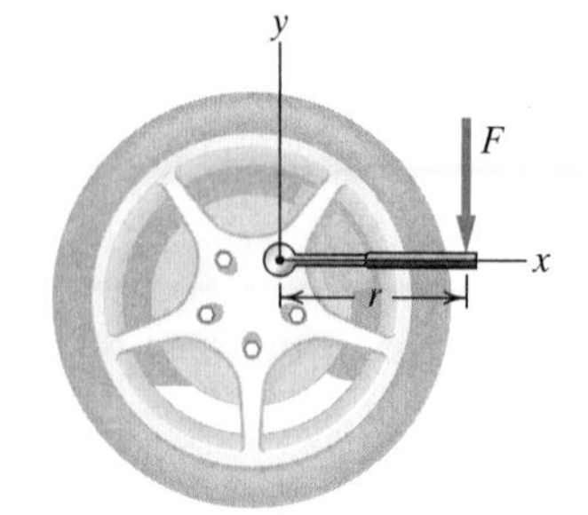

Figure 5.30

A wrench is attached to one of the lug nuts holding a tire in place, as shown in **Figure 5.30**. A downward force of 20 N is applied to the wrench, and the wrench is 250 mm in length.

(a) Find the moment applied by the wrench on the lug nut for the given situation.

(b) Determine the increase in the moment if the length of the wrench is increased by a factor of two.

(c) Determine the increase in the moment if the magnitude of the force is increased by a factor of two.

(d) Comment on the advantages and disadvantages of the two approaches of increasing the moment.

Goal We are asked to calculate the moment for the given physical situation (a). In addition, we are asked to calculate the moment if the length of the wrench is increased by two (b), the force is increased by two (c), and to comment on the advantages and disadvantages of these two changes (d).

Given We are given a picture of the physical situation—a force acting on the end of a wrench. Furthermore, the force is perpendicular to the length of the wrench and has a magnitude of 20 N.

Assume We assume that wrench and force lie in the plane of the paper (call this the xy plane) and that the axis of the lug nut bolt is perpendicular to this plane (the z axis). The origin of the axes is placed at the center of the lug nut. Furthermore, we assume that in (a) "find the moment applied by the wrench *on the lug nut*" really means "find the moment created at a moment center at the origin by the 20-N force."

Draw **Figure 5.30** is sufficient.

Formulate Equations and Solve **(a)** To find the magnitude of the moment created at a moment center at the origin by the 20-N force, we use (5.1). Substituting $\|\boldsymbol{r}\| = 250$ mm, $\|\boldsymbol{F}\| = 20$ N, and $\theta = 90°$ into (5.1) results in

$$\|\boldsymbol{M}\| = \|\boldsymbol{r}_1\|(\|\boldsymbol{F}\|\sin\theta)$$
$$\|\boldsymbol{M}\| = 250\text{ mm}(20\text{ N})(\sin 90°) = 5000\text{ N}\cdot\text{mm}$$

If you use the right-hand rule, you will find that your thumb points into the paper and your fingers curl in the clockwise direction; this is the negative z direction.

Answer to (a) $\|\boldsymbol{M}\| = 5000\text{ N}\cdot\text{mm} = 5\text{ kN}\cdot\text{m}$. The direction is $-z$, and the sense is clockwise about the z axis.

(b) If we increase the length of the wrench by a factor of two, $\|\boldsymbol{r}\| = 500$ mm. Again using (5.1), we can find the new moment:

$$\|\boldsymbol{M}\| = \|\boldsymbol{r}\|(\|\boldsymbol{F}\|\sin\theta)$$
$$\|\boldsymbol{M}\| = 500\text{ mm}(20\text{ N})(\sin 90°) = 10{,}000\text{ N}\cdot\text{mm}$$

Answer to (b) $\|\boldsymbol{M}\| = 10{,}000\text{ N}\cdot\text{mm} = 10\text{ kN}\cdot\text{m}$, with the same direction as in (a).

This is why using a wrench to break the torque on a nut works better than using your fingers directly on the nut!

(c) Now, if we increase the magnitude of the force by a factor of two, $\|\boldsymbol{F}\| = 40$ N. Again using (5.1), we can find the new moment:

$$\|\boldsymbol{M}\| = \|\boldsymbol{r}\|(\|\boldsymbol{F}\|\sin\theta)$$
$$\|\boldsymbol{M}\| = 250\text{ mm}(40\text{ N})(\sin 90°) = 10{,}000\text{ N}\cdot\text{mm}$$

Answer to (c) $\|\boldsymbol{M}\| = 10{,}000\text{ N}\cdot\text{mm} = 10\text{ kN}\cdot\text{m}$, with the same direction as in (a).

(d) The advantage of using a longer wrench is that no additional force is needed. On the other hand, depending on the situation, there may not be room to fit the longer wrench into the workspace, and the longer wrench may be difficult to store. The main disadvantage of the larger force is that the person needing to perform the task may not be capable of applying the additional force. If the situation would allow for it, the person might use his or her foot to push down on the end of the wrench, use a star wrench (which allows the user to simultaneously push and pull with both hands), or get someone else to help.

Check Review of (5.1) shows that we can double the magnitude of the moment by doubling the magnitude of the component of the position vector perpendicular to the force (which is what was done in part (b)) or by doubling the magnitude of the force (which is what was done in part (c)). Therefore, it makes sense that the answers are the same in parts (b) and (c).

EXAMPLE 5.6 INCREASING THE MAGNITUDE OF THE MOMENT (B)

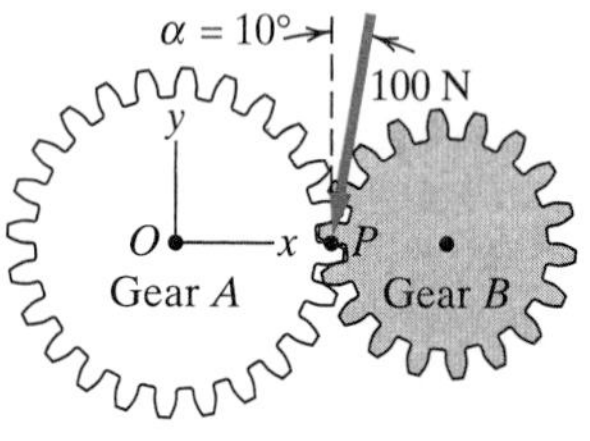

Figure 5.31

Consider two gears that contact at point P, as shown in **Figure 5.31**. Gear B pushes on gear A with a force of 100 N at point P. The radius of gear A is 100 mm, and the radius of gear B is 80 mm. Coordinate axes are located as shown.

(a) Find the moment applied to gear A by gear B given the force orientation (pressure angle) shown in **Figure 5.31**.

(b) Determine the new moment if the pressure angle is changed from 10° to 30°.

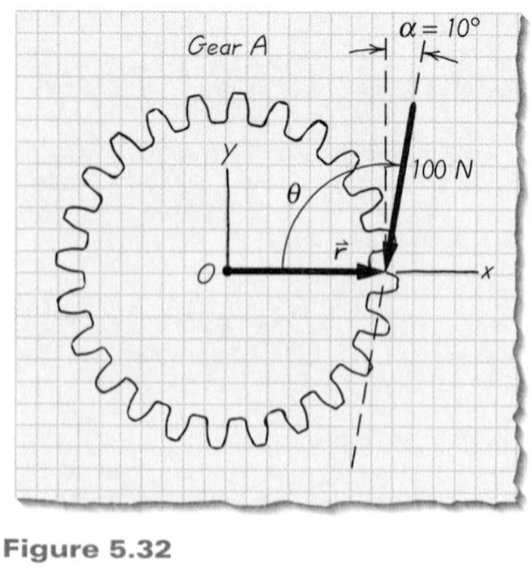

Figure 5.32

Goal We are to find the moment created at a moment center at O by the force that gear B applies to gear A when the pressure angle is 10°. Then we are to repeat the calculation for a pressure angle of 30°.

Given We are told that gear B pushes on gear A with a force of magnitude 100 N at an angle relative to the vertical. This force acts at point P. The angle relative to the vertical is called the pressure angle α. The x and y axes have been specified. In addition, the radius of gear A is 100 mm, and the radius of gear B is 80 mm.

Assume No assumptions are required.

Draw We make a sketch of the force and the position vector to facilitate our finding the angle θ between these two vectors; see **Figure 5.32**.

Formulate Equations and Solve **(a)** Moment with $\alpha = 10°$: In Example 5.3, the pressure angle[5] α was 20°, and we found the magnitude of the moment to be $\|\boldsymbol{M}\| = 9400\ \text{N}\cdot\text{mm}$. We now repeat this calculation for $\alpha = 10°$.

[5]The **pressure angle** is the angle between the tangent to the gear circle and the line of action of the force. See *Mechanical Engineering Design* by Shigley and Mischke, 7th edition, 2003, for more information.

First, we need to find the new angle between the position vector (from the moment center O to the point P) and the applied force. From the geometry in **Figure 5.33a**, $\theta = \alpha + 90° = 10° + 90° = 100°$. We can now solve for the magnitude of the moment using (5.1):

$$\| \boldsymbol{M} \| = \| \boldsymbol{r} \| (\| \boldsymbol{F} \| \sin \theta)$$
$$\| \boldsymbol{M} \| = 100 \text{ mm}(100 \text{ N})(\sin 100°) = 9848 \text{ N} \cdot \text{mm}$$

Answer to (a) The magnitude of the moment if the pressure angle of applied force is 10° is 9848 N · mm.

(b) Moment with $\alpha = 30°$: We can use the same equation as before, but now the angle between the applied force and position vector is $\theta = \alpha + 90° = 30° + 90° = 120°$. So the magnitude of the moment is

$$\| \boldsymbol{M} \| = \| \boldsymbol{r} \| (\| \boldsymbol{F} \| \sin \theta)$$
$$\| \boldsymbol{M} \| = 100 \text{ mm}(100 \text{ N})(\sin 120°) = 8660 \text{ N} \cdot \text{mm}$$

Answer to (b) The magnitude of the moment if the angle of applied force is changed to 30° is 8660 N · mm.

Comment: If we consider pressure angles in the range $0° < \alpha < 30°$, we find that the moment decreases as the pressure angle increases, as shown in **Figure 5.33b**. From this, we can see that an angle θ between a position vector and an applied force of 90° will result in the largest value of moment (since $\sin 90° = 1$). A 90° angle between a position vector and applied force yields a "maximum moment."[6] Because of how gear teeth are manufactured, it is not possible to cut them so that $\theta = 90°$ (and therefore have a pressure angle $\alpha = 0°$).

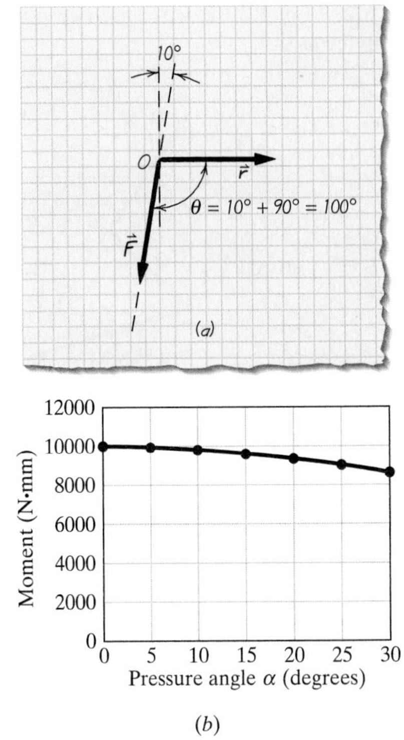

Figure 5.33

[6]Another way to convince yourself that there is a maximum moment at $\sin \theta = 90°$ is to take a calculus approach. To find a maximum of any equation, set the first derivative equal to zero and solve for the variable. At that point(s), there is either a local maximum or a local minimum. The derivative of (5.1) with respect to θ is simply $dM/d\theta = rF \cos \theta$. Letting $dM = 0$, we find that $\theta = 90°, 270°$. For $\theta = 90°$, $\sin \theta = 1$, and for $\theta = 270°$, $\sin \theta = -1$. So, the maximum moment is at $\theta = 90°$.

EXERCISES 5.1

5.1.1 For **E5.1.1**, write the two position vectors $\boldsymbol{r}_1$ and $\boldsymbol{r}_2$ in terms of $\boldsymbol{i}, \boldsymbol{j}, \boldsymbol{k}$.

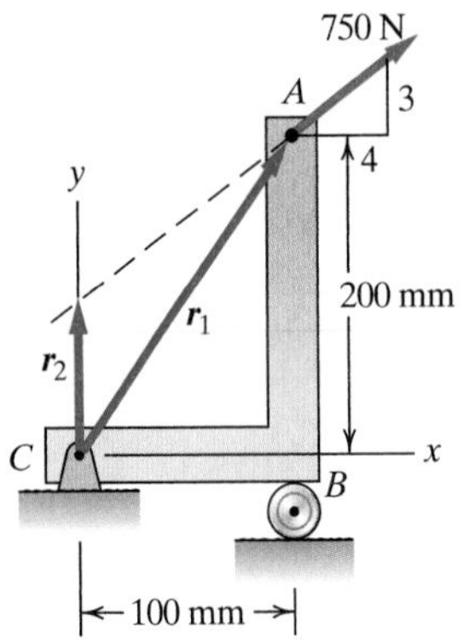

E5.1.1

5.1.2. For E5.1.2, write the two position vectors $\boldsymbol{r}_1$ and $\boldsymbol{r}_2$ in terms of $\boldsymbol{i}, \boldsymbol{j}, \boldsymbol{k}$.

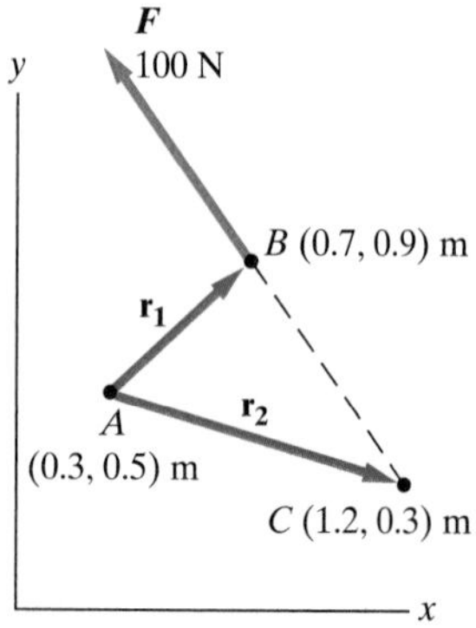

E5.1.2

5.1.3. For E5.1.3, write the two position vectors $\boldsymbol{r}_1$ and $\boldsymbol{r}_2$ in terms of $\boldsymbol{i}, \boldsymbol{j}, \boldsymbol{k}$.

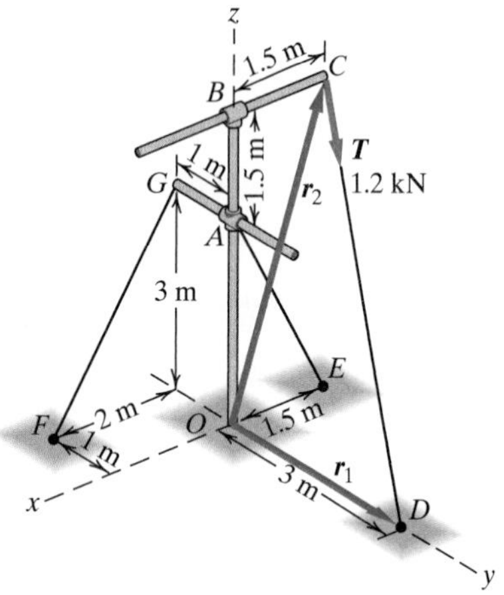

E5.1.3

5.1.4. For **E5.1.1**, find
  a. the magnitude of $\boldsymbol{r}_1$
  b. θ (the angle between $\boldsymbol{r}_1$ and the 750-N force)
  c. the magnitude of the moment $\boldsymbol{M}$ for a moment center at C

5.1.5. For **E5.1.2**, find
  a. the magnitude of $\boldsymbol{r}_1$
  b. θ (the angle between $\boldsymbol{r}_1$ and the 100-N force)
  c. the magnitude of the moment $\boldsymbol{M}$ for a moment center at A

5.1.6. For **E5.1.3**, find
  a. the magnitude of $\boldsymbol{r}_1$
  b. θ (the angle between $\boldsymbol{r}_1$ and the 1.2-kN tension force)
  c. the magnitude of the moment $\boldsymbol{M}$ for a moment center at the origin of the coordinate system shown

5.1.7. For **E5.1.1**, find
  a. the magnitude of $\boldsymbol{r}_2$
  b. θ (the angle between $\boldsymbol{r}_2$ and the 750-N force)
  c. the magnitude of the moment $\boldsymbol{M}$ for a moment center at C

5.1.8. For **E5.1.2**, find
  a. the magnitude of $\boldsymbol{r}_2$
  b. θ (the angle between $\boldsymbol{r}_2$ and the 100-N force)
  c. the magnitude of the moment $\boldsymbol{M}$ for a moment center at A

5.1.9. For **E5.1.3**, find
  a. the magnitude of $\boldsymbol{r}_2$
  b. θ (the angle between $\boldsymbol{r}_2$ and the 1.2-kN tension force)
  c. the magnitude of the moment $\boldsymbol{M}$ for a moment center at the origin of the coordinate system shown

5.1.10. For the moment representation in E5.1.10, specify
  a. the direction of the moment ($+\boldsymbol{i}$, $-\boldsymbol{i}$, $+\boldsymbol{j}$, $-\boldsymbol{j}$, $+\boldsymbol{k}$ or $-\boldsymbol{k}$)
  b. the sense of the moment (clockwise or counterclockwise)
  c. the axis (x, y, or z) about which the moment acts

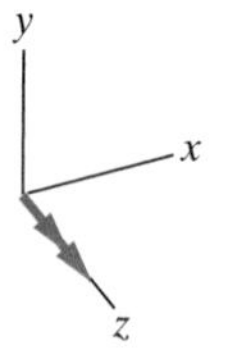

E5.1.10

5.1.11. For the moment representation in E5.1.11, specify
  a. the direction of the moment ($+\boldsymbol{i}$, $-\boldsymbol{i}$, $+\boldsymbol{j}$, $-\boldsymbol{j}$, $+\boldsymbol{k}$ or $-\boldsymbol{k}$)
  b. the sense of the moment (clockwise or counterclockwise)
  c. the axis (x, y, or z) about which the moment acts

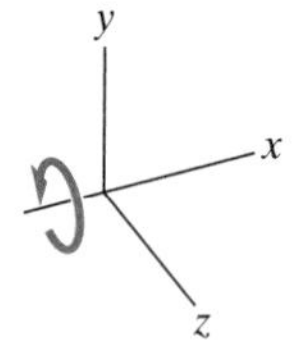

E5.1.11

5.1.12. For the moment representation in E5.1.12, specify
  a. the direction of the moment ($+\boldsymbol{i}$, $-\boldsymbol{i}$, $+\boldsymbol{j}$, $-\boldsymbol{j}$, $+\boldsymbol{k}$ or $-\boldsymbol{k}$)
  b. the sense of the moment (clockwise or counterclockwise)
  c. the axis (x, y, or z) about which the moment acts

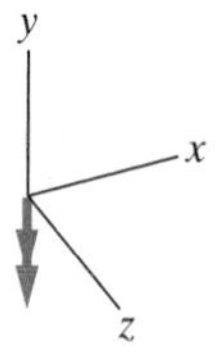

E5.1.12

5.1.13. For the moment representation in E5.1.13, specify
  a. the direction of the moment ($+\boldsymbol{i}$, $-\boldsymbol{i}$, $+\boldsymbol{j}$, $-\boldsymbol{j}$, $+\boldsymbol{k}$ or $-\boldsymbol{k}$)
  b. the sense of the moment (clockwise or counterclockwise)
  c. the axis (x, y, or z) about which the moment acts

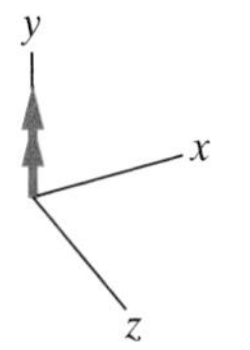

E5.1.13

5.1.14. For the moment representation in E5.1.14, specify

a. the direction of the moment ($+\boldsymbol{i}$, $-\boldsymbol{i}$, $+\boldsymbol{j}$, $-\boldsymbol{j}$, $+\boldsymbol{k}$ or $-\boldsymbol{k}$)

b. the sense of the moment (clockwise or counterclockwise)

c. the axis (x, y, or z) about which the moment acts

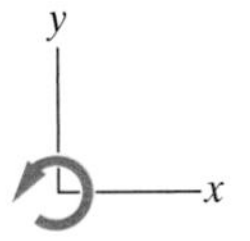

E5.1.14

5.1.15. For the moment representation in E5.1.15, specify

a. the direction of the moment ($+\boldsymbol{i}$, $-\boldsymbol{i}$, $+\boldsymbol{j}$, $-\boldsymbol{j}$, $+\boldsymbol{k}$ or $-\boldsymbol{k}$)

b. the sense of the moment (clockwise or counterclockwise)

c. the axis (x, y, or z) about which the moment acts

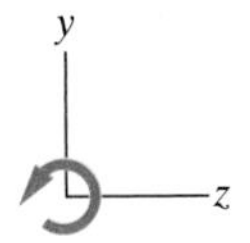

E5.1.15

5.1.16. For the moment representation in E5.1.16, specify

a. the direction of the moment ($+\boldsymbol{i}$, $-\boldsymbol{i}$, $+\boldsymbol{j}$, $-\boldsymbol{j}$, $+\boldsymbol{k}$ or $-\boldsymbol{k}$)

b. the sense of the moment (clockwise or counterclockwise)

c. the axis (x, y, or z) about which the moment acts

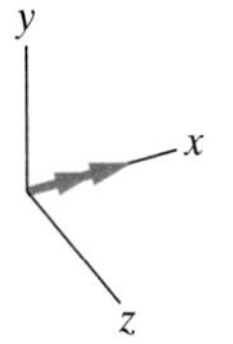

E5.1.16

5.1.17. Control bar AB in **E5.1.17** is inclined at 50° from the vertical. A 100-N force is applied to the bar at end B. Determine the magnitude and direction of the moment this force creates at a moment center at A if

a. $\alpha = 140°$

b. $\alpha = 210°$

c. $\alpha = 180°$

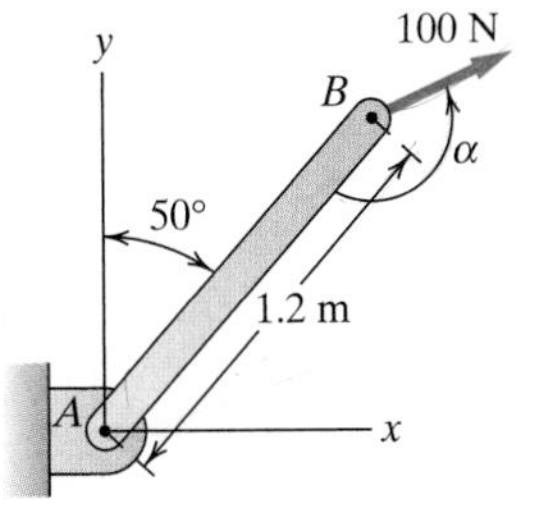

E5.1.17

5.1.18. For **E5.1.17**, determine the value of the angle α for which the 100-N force exerts

a. the largest counterclockwise moment at a moment center at A (Also determine the corresponding magnitude of the moment.)

b. the largest clockwise moment at a moment center at A (Also determine the corresponding magnitude of the moment.)

c. zero moment at a moment center at A

5.1.19. An 80-N force is applied to the special-purpose wrench shown in **E5.1.19.**

a. With a moment center at the center of the bolt, determine the moment this force creates when $\alpha = 75°$. (Remember that a moment is a vector quantity, requiring information on magnitude, direction, and units.)

b. Comment on whether 80 N is a large or small force relative to the arm strength of an average engineering student.

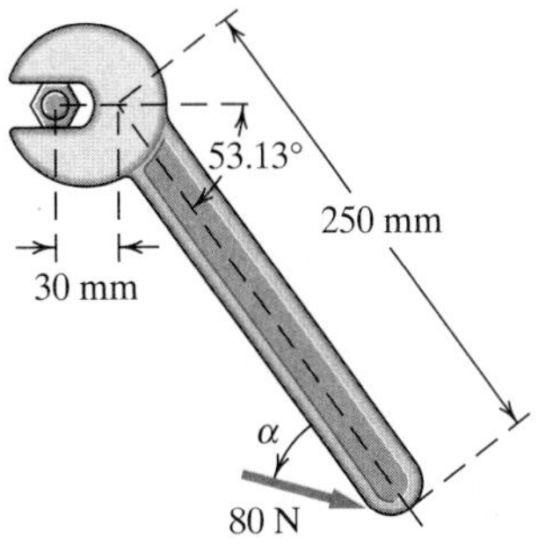

E5.1.19

5.1.20. For **E5.1.19**, determine the angle α for which the magnitude of the moment of the 80-N force at the center of the bolt is

a. a minimum

b. a maximum

c. Explain why it is or is not possible for the 80-N force to create zero moment at the center of the bolt.

5.1.21. A person applies a 200-N force to a wrench, as shown in **E5.1.21.**

a. What is the magnitude of the moment that this force creates at a moment center at A?

b. Suggest at least two ways that the wrench might be redesigned to allow the 200-N force to create a larger moment. For each design change, determine how much the moment will be increased, expressed as a percentage of your answer in **a**.

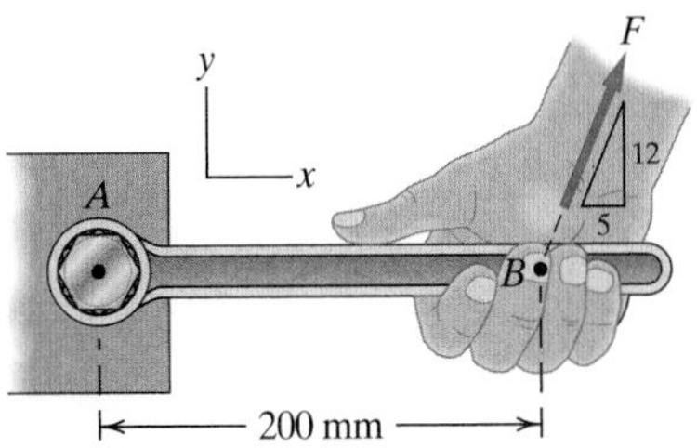

E5.1.21

5.1.22. A friend tells you that she had trouble pulling a tent stake out of the ground. The ground consists of firm mud. She tried pulling straight up on the stake (E5.1.22*a*), and she could not pull it out of the ground. Then she applied a back-and-forth force (**E5.1.22*b***). After this, she was able to remove the stake by pulling straight up (**E5.1.22*c***).

a. What did she accomplish by applying the back-and-forth force?

b. Speculate how effective this procedure would be for removing the stake if it had been in loose sand instead of firm mud.

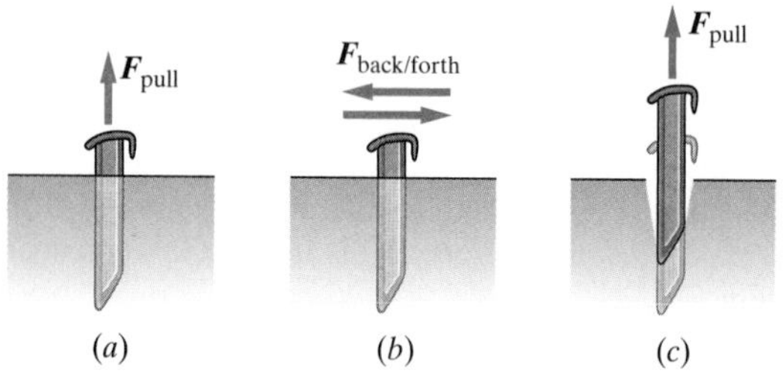

E5.1.22

5.1.23. Starting a lawn mower engine requires you to pull on a cord wrapped around a pulley, as shown in E5.1.23. The diameter of the pulley is 25 cm, and the moment required to start the mower is 5 N · m.

a. What force must be applied on the cord in order to start the mower?

b. Is more, less, or the sample magnitude of force required if you don't pull tangent to the circumference of the pulley to start the mower's engine? Why?

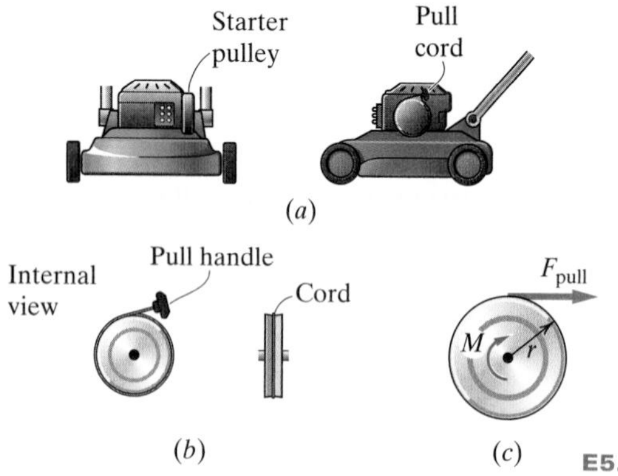

E5.1.23

5.1.24. Racing yachts have both large sails and deep keels (E5.1.24). Knowing that the rudder, rather than the keel, is used to steer the boat:

a. What is the purpose of the keel?

b. What do you think the relationship is between the depth of the keel, the size of the sail, and the speed of the boat?

E5.1.24

5.1.25. You are cycling down the road, and there's a big hill up ahead. Knowing that you have a long way to go to get home, you would like to have an easy ride up the hill.

a. Downshifting on your bike's rear derailleur makes pedaling easier. Explain in two or three sentences why this is so. Use sketches as necessary.

b. What is the side effect of downshifting? Explain in two or three sentences why this is so. Use sketches as necessary.

5.2 MATHEMATICAL REPRESENTATION OF A MOMENT

As discussed in Section 5.1, moment $\boldsymbol{M}$ is a vector quantity, which means it has both magnitude and direction. It is frequently convenient to represent a moment in terms of its scalar components:

$$\boldsymbol{M} = M_x\boldsymbol{i} + M_y\boldsymbol{j} + M_z\boldsymbol{k} \tag{5.2}$$

The scalar components define the magnitude of the moment $\| \boldsymbol{M} \|$, as shown in Figure 5.34*a*:

$$\| \boldsymbol{M} \| = \sqrt{M_x^2 + M_y^2 + M_z^2} \tag{5.3}$$

The direction of the moment can be defined in terms of the direction cosines (**Figure 5.34*b***):

$$\cos \theta_x = \frac{M_x}{\| \boldsymbol{M} \|} \qquad \cos \theta_y = \frac{M_y}{\| \boldsymbol{M} \|} \qquad \cos \theta_z = \frac{M_z}{\| \boldsymbol{M} \|} \tag{5.4}$$

The moment's direction can also be described by a unit vector $\boldsymbol{u}$:

$$\boldsymbol{u} = \cos \theta_x \boldsymbol{i} + \cos \theta_y \boldsymbol{j} + \cos \theta_z \boldsymbol{k} \tag{5.5}$$

The idea of representing the direction of a vector (in this case, $\boldsymbol{M}$) in terms of its direction cosines or in terms of a unit vector directed along its line of action was first presented in Chapter 4 (e.g., Section 4.5, (4.11)). This unit vector in (5.5) can also be interpreted as the rotational axis of the moment.

Our goal now is to rewrite the expressions for moment in (5.2), (5.3), and (5.4) in terms of the force vector and position vector. In other words, we will develop a procedure for finding the scalar components M_x, M_y, and M_z of a moment created by a force $\boldsymbol{F}$ offset from a moment center by position vector $\boldsymbol{r}$. We start by making the following observations about position vectors, forces, and moments:

- All are vector quantities.
- A force offset from a moment center creates a moment about an axis that is perpendicular to the plane defined by the position vector and the force vector (Figure 5.35*a*) and has a direction (+ or −) given by a right-hand rule.

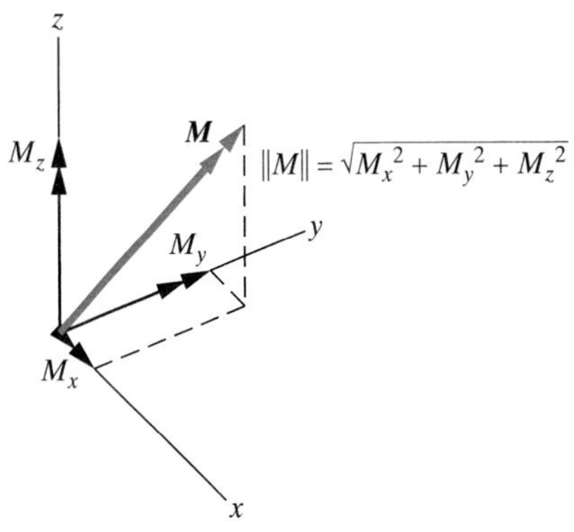

(*a*)

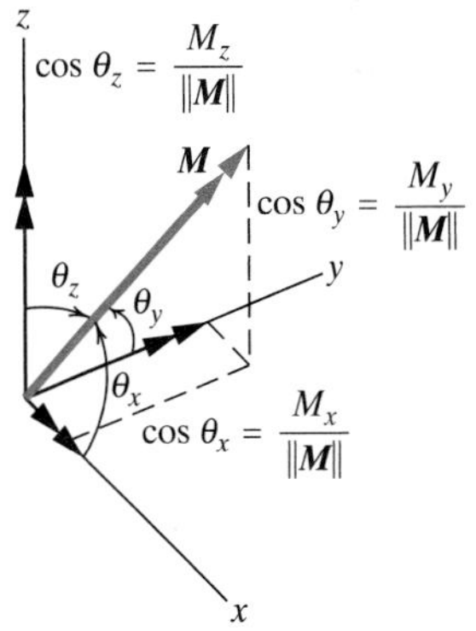

(*b*)

Figure 5.34 (*a*) Finding the magnitude of the moment; (*b*) moment direction defined by space angles

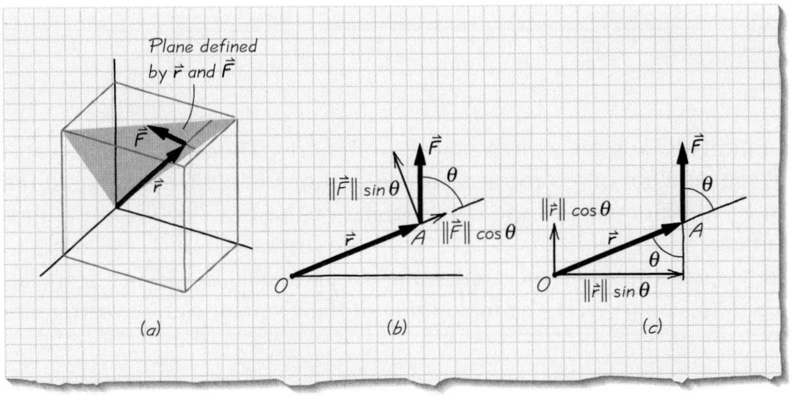

Figure 5.35 (*a*) Vectors $\boldsymbol{r}$ and $\boldsymbol{F}$ form a plane. Moment is perpendicular to this plane. (*b*) Component of force perpendicular to position vector creates moment. (*c*) Component of position vector perpendicular to force creates moment.

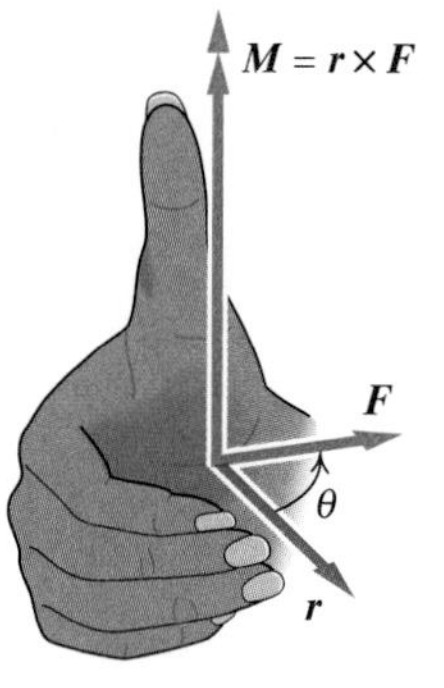

(a)

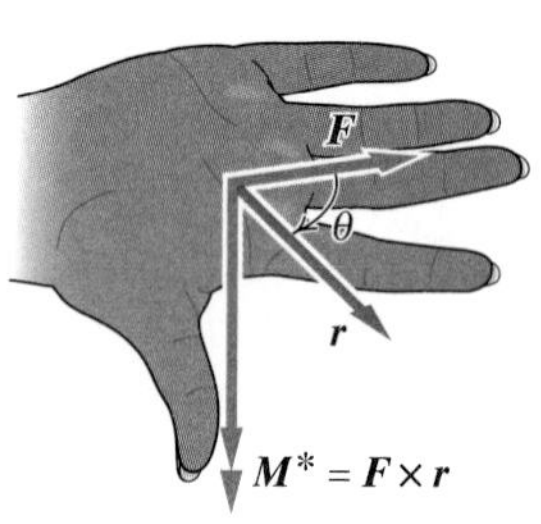

(b)

Figure 5.36 Right-hand rule used to find direction of moment

- Only the force component $\| \boldsymbol{F} \| \sin\theta$ perpendicular to the position vector creates a moment (**Figure 5.35*b***), and the magnitude of the moment is $\| \boldsymbol{M} \| = \| \boldsymbol{r} \| (\| \boldsymbol{F} \| \sin\theta)$. Another way of looking at this is that only the position vector component $\| \boldsymbol{r} \| \sin\theta$ perpendicular to the force creates a moment (**Figure 5.35*c***).

There is a mathematical construct that captures these observations—it is called the **vector product** or the **cross product**.

The cross product of two vectors (in the case at hand, $\boldsymbol{r}$ and $\boldsymbol{F}$, in that order) is, by definition, a third vector that is perpendicular to the plane defined by the two vectors. The third vector that we are dealing with is the moment $\boldsymbol{M}$. The magnitude of this vector is $(\| \boldsymbol{r} \| \; \| \boldsymbol{F} \| \sin\theta)$, where θ is the angle between $\boldsymbol{r}$ and the line of action of $\boldsymbol{F}$ when $\boldsymbol{r}$ and $\boldsymbol{F}$ are placed tail to tail:

$$\boldsymbol{M} = \boldsymbol{r} \times \boldsymbol{F} = (\| \boldsymbol{r} \| \; \| \boldsymbol{F} \| \sin\theta)\boldsymbol{u} \qquad (5.6\text{A})$$

where $\| \boldsymbol{r} \|$ and $\| \boldsymbol{F} \|$ are the magnitudes of the position vector and force vector, respectively, and $\boldsymbol{u}$ is a unit vector perpendicular to the plane defined by $\boldsymbol{r}$ and $\boldsymbol{F}$. This expression is read, "$\boldsymbol{r}$ crossed with $\boldsymbol{F}$" and is illustrated in **Figure 5.36*a***. The cross product is not communitive, meaning that $\boldsymbol{r} \times \boldsymbol{F}$ is not equal to $\boldsymbol{F} \times \boldsymbol{r}$. (Compare **Figure 5.36*a*** with **Figure 5.36*b***.)

It is generally more convenient to work with the position vector and force vectors in terms of their components. For the position and force vectors in **Figure 5.37*a***, (5.6A) can be rewritten as

$$\boldsymbol{M} = \boldsymbol{r} \times \boldsymbol{F} = (r_x\boldsymbol{i} + r_y\boldsymbol{j} + r_z\boldsymbol{k}) \times (F_x\boldsymbol{i} + F_y\boldsymbol{j} + F_z\boldsymbol{k}) \qquad (5.6\text{B})$$

By applying the distributive and associative laws of vector multiplication to (5.6B) and noting that $\boldsymbol{i} \times \boldsymbol{i} = 0, \boldsymbol{j} \times \boldsymbol{j} = 0, \boldsymbol{k} \times \boldsymbol{k} = 0, \boldsymbol{i} \times \boldsymbol{j} = \boldsymbol{k}$, $\boldsymbol{j} \times \boldsymbol{k} = \boldsymbol{i}$, and so on, we can write

$$\boldsymbol{M} = \boldsymbol{r} \times \boldsymbol{F} = \underbrace{(+r_yF_z - r_zF_y)}_{M_x}\boldsymbol{i} + \underbrace{(+r_zF_x - r_xF_z)}_{M_y}\boldsymbol{j} + \underbrace{(+r_xF_y - r_yF_x)}_{M_z}\boldsymbol{k} \qquad (5.7)$$

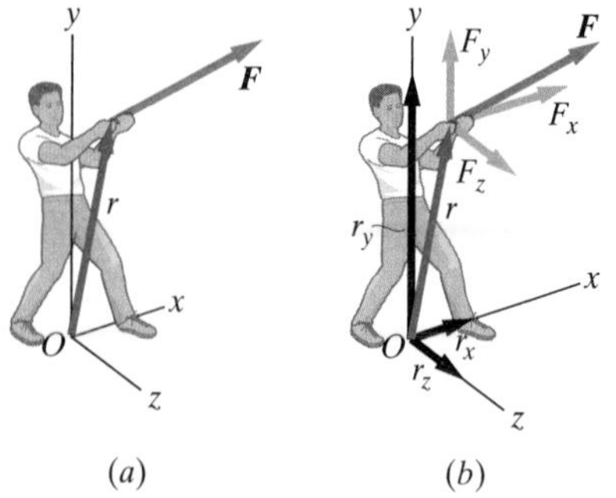

(a) (b)

Figure 5.37 (*a*) Position vector and force; (*b*) moment at moment center *O*

Equation (5.7) is an expression of **Varignon's Theorem**, which states that the moment of a force at a particular moment center is equal to the sum of the moments of the components of the force at the same moment center.

If we substitute the moment scalar components M_x, M_y, and M_z defined in (5.7) into (5.3), the magnitude of the moment can be written as

$$\| \boldsymbol{M} \| = \sqrt{\underbrace{(+r_yF_z - r_zF_y)^2}_{M_x} + \underbrace{(+r_zF_x - r_xF_z)^2}_{M_y} + \underbrace{(+r_xF_y - r_yF_x)^2}_{M_z}} \qquad (5.8)$$

We can also use the moment scalar components in (5.7) to rewrite the direction cosines in (5.4) that define the direction of the moment:

$$\cos\theta_x = \frac{M_x}{\|\boldsymbol{M}\|} = \frac{+r_yF_z - r_zF_y}{\|\boldsymbol{M}\|}$$
$$\cos\theta_y = \frac{M_y}{\|\boldsymbol{M}\|} = \frac{+r_zF_x - r_xF_z}{\|\boldsymbol{M}\|} \qquad (5.9)$$
$$\cos\theta_z = \frac{M_z}{\|\boldsymbol{M}\|} = \frac{+r_xF_y - r_yF_x}{\|\boldsymbol{M}\|}$$

The direction of the moment is defined by the unit vector $\boldsymbol{u}$ (see (5.5)) based on the direction cosines in (5.9) as

$$\boldsymbol{u} = \frac{+r_yF_z - r_zF_y}{\|\boldsymbol{M}\|}\boldsymbol{i} + \frac{+r_zF_x - r_xF_z}{\|\boldsymbol{M}\|}\boldsymbol{j} + \frac{+r_xF_y - r_yF_x}{\|\boldsymbol{M}\|}\boldsymbol{k} \qquad (5.10)$$

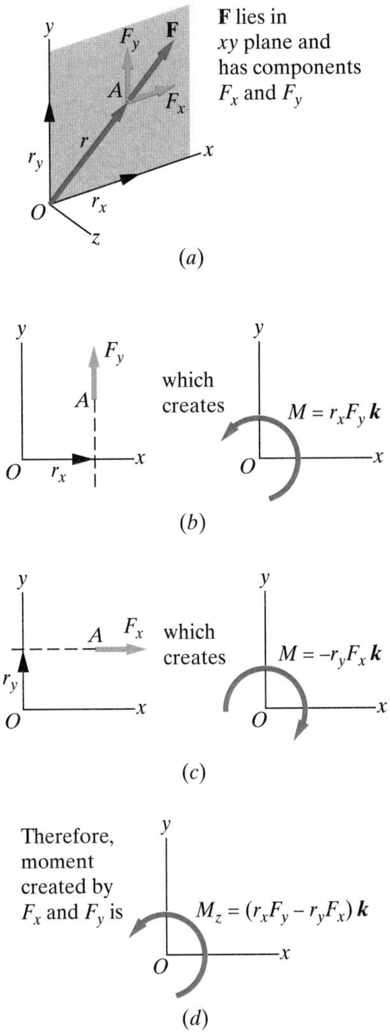

Figure 5.38 (*a*) Finding the moment created by a force with *x* and *y* components; (*b*) consider component F_y; (*c*) consider component F_x; (*d*) moment created by F_y and F_x

Notice that if the force and position vectors lie in the *xy*, *yz*, or *zx* plane, the moment found by applying (5.7) will be about the *z*, *x*, or *y* axis, respectively. More specifically:

For a force vector and position vector lying in the same xy *plane*, we can write

$$\boldsymbol{F} = F_x\boldsymbol{i} + F_y\boldsymbol{j}, \qquad \boldsymbol{r} = r_x\boldsymbol{i} + r_y\boldsymbol{j}$$

The moment at a moment center at O is about the z axis and is then (based on Figure 5.38):

$$\boldsymbol{M}_z = M_z\boldsymbol{k} = (+r_xF_y - r_yF_x)\boldsymbol{k} \qquad (5.11\text{A})$$

This is the third term on the right-hand side of (5.7). Because no moment is created about the x or y axis, $\boldsymbol{M}_x = \boldsymbol{M}_y = 0$. The magnitude of the moment (based on (5.3)) is $+r_xF_y - r_yF_x$. Also, because $\theta_z = 0°$, $\cos\theta_z = 1$.

For a force vector and position vector lying in the same yz *plane*, we can write

$$\boldsymbol{F} = F_y\boldsymbol{j} + F_z\boldsymbol{k}, \qquad \boldsymbol{r} = r_y\boldsymbol{j} + r_z\boldsymbol{k}$$

The moment at a moment center at O is about the x axis and is then (based on Figure 5.39):

$$\boldsymbol{M}_x = M_x\boldsymbol{i} = (+r_yF_z - r_zF_y)\boldsymbol{i} \qquad (5.11\text{B})$$

This is the first term on the right-hand side of (5.7). Because no moment is created about the y or z axis, $\boldsymbol{M}_y = \boldsymbol{M}_z = 0$. The magnitude of the moment (based on (5.3)) is $+r_yF_z - r_zF_y$. Also, because $\theta_x = 0°$, $\cos\theta_x = 1$.

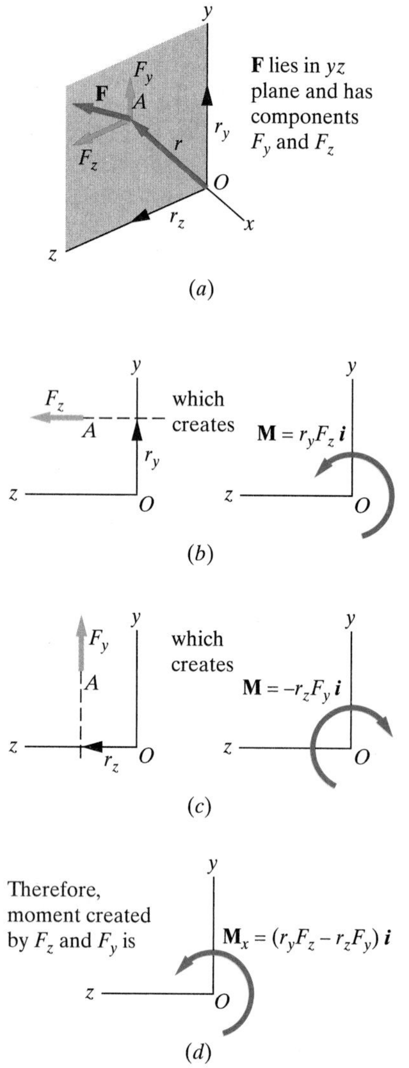

Figure 5.39 (*a*) Finding the moment created by a force with *y* and *z* components; (*b*) consider component F_z; (*c*) consider component F_y; (*d*) moment created by F_z and F_y

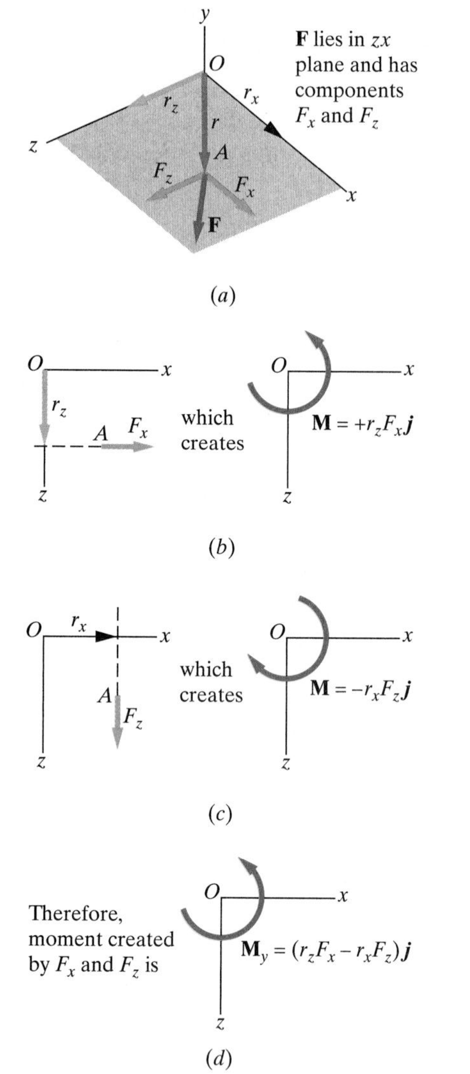

Figure 5.40 (*a*) Finding the moment created by a force with *z* and *x* components; (*b*) consider component F_x; (*c*) consider component F_z; (*d*) moment created by F_x and F_z

For a force vector and position vector lying in the same zx *plane,* we can write

$$\boldsymbol{F} = F_x\boldsymbol{i} + F_z\boldsymbol{k}, \qquad \boldsymbol{r} = r_x\boldsymbol{i} + r_z\boldsymbol{k}$$

The moment at a moment center O is about the y axis and is then (based on **Figure 5.40**):

$$\boldsymbol{M}_y = M_y\boldsymbol{j} = (+r_zF_x - r_xF_z)\boldsymbol{j} \tag{5.11C}$$

This is the second term on the right-hand side of (5.7). Because no moment is created about the z or x axis, $\boldsymbol{M}_x = \boldsymbol{M}_z = 0$. The magnitude of the moment (based on (5.3)) is $+r_z F_x - r_x F_z$. Also, because $\theta_y = 0°$, $\cos \theta_y = 1$.

Now we return to the cross-product form of moment calculation, as presented in (5.6B). More generally, the cross product of the position vector and the force can be written in matrix form as

$$\boldsymbol{M} = \boldsymbol{r} \times \boldsymbol{F} = \begin{vmatrix} \boldsymbol{i} & \boldsymbol{j} & \boldsymbol{k} \\ r_x & r_y & r_z \\ F_x & F_y & F_z \end{vmatrix} \tag{5.12}$$

The form in (5.12) specifies that the **determinant** of the matrix be taken. The reader unfamiliar with this form is referred to Appendix A1.4.

The expression for moment $\boldsymbol{M}$ in (5.12) is equivalent to the expression in (5.7). The beauty of (5.12) is that the various combinations of position vector and force components ($r_x, r_y, r_z, F_x, F_y, F_z$) needed to correctly calculate the moment come out naturally in taking the determinant, so there is no need to memorize (5.7). Simply by arranging the position vector and force vector in the matrix form in (5.12) and taking the determinant, we calculate the moment. Alternatively, for situations where the position vector and force are in the same plane, you might find application of one of the expressions in (5.11) more straightforward.

EXAMPLE 5.7 INCREASING THE MAGNITUDE OF THE MOMENT (C)

A cable is attached at B on the L-bracket shown in **Figure 5.41**. The tension $\boldsymbol{F}$ in the cable is 200 lb. Find the moment the cable force creates at a moment center at A.

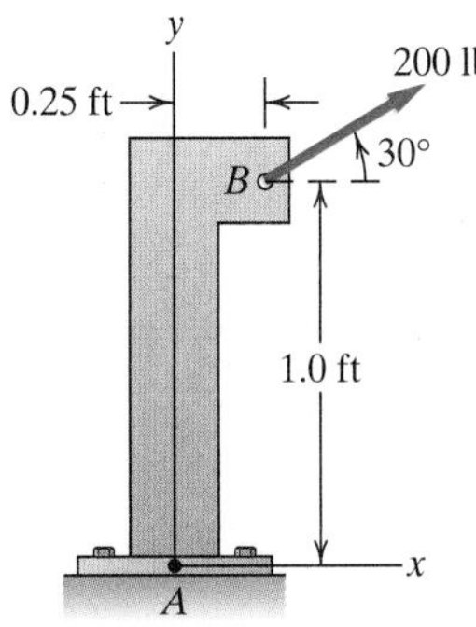

Figure 5.41

Goal We are to find the moment at a moment center at A created by a 200-lb force applied at B.

Given We are given the dimensions of the L-bracket and the locations of points A and B, and the xy axes are specified. In addition, we are given the magnitude of the force (200 lb) and angular information (30° from the horizontal) with which to find its direction.

Assume Based on the information given, we assume that the 200-lb force and the L-bracket lie in the xy plane.

Formulate Equations and Solve We are asked to find the moment at A created by the cable force. We start by inspecting the situation to see whether the force lies in a reference axes plane; if it does, it creates a moment about only one of the reference axes. Based upon the information provided in **Figure 5.41**, it is reasonable for us to assume that the line of action of the cable force lies in the xy plane.

We now show three approaches to finding the moment.

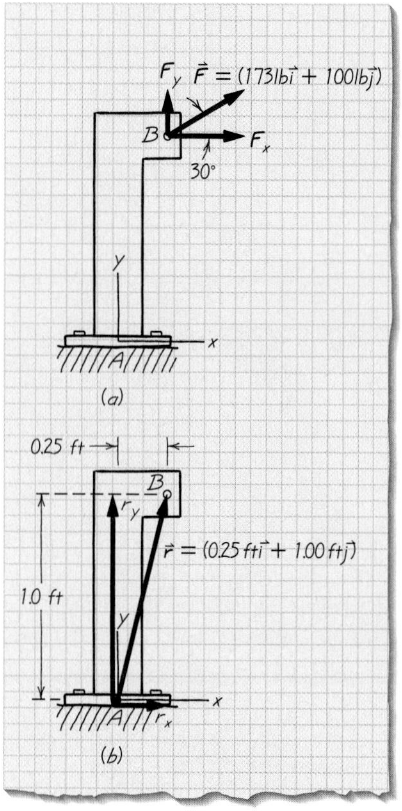

Figure 5.42

Approach 1 (Based on Vector Components)[7]**:** Since the line of action of $\boldsymbol{F}$ lies in the xy reference plane, we can use (5.11A) to find the moment created by $\boldsymbol{F}$:

$$\boldsymbol{M}_z = M_z\boldsymbol{k} = (+r_xF_y - r_yF_x)\boldsymbol{k} \tag{5.11A}$$

Recall that a "+" moment has a counterclockwise sense, and a "−" moment has clockwise sense about the z axis. In order to apply (5.11A), we need to write $\boldsymbol{F}$ and $\boldsymbol{r}$ in terms of their components.

The force $\boldsymbol{F}$ (**Figure 5.42*a***):

$$\begin{aligned}\boldsymbol{F} &= F_x\boldsymbol{i} + F_y\boldsymbol{j} = (200 \text{ lb})(\cos 30°)\boldsymbol{i} + (200 \text{ lb})(\sin 30°)\boldsymbol{j} \\ &= 173 \text{ lb } \boldsymbol{i} + 100 \text{ lb } \boldsymbol{j}\end{aligned}$$

Therefore $F_x = 173$ lb and $F_y = 100$ lb.

The position vector $\boldsymbol{r}$ (**Figure 5.42*b***)*:* We arbitrarily define the position vector as a vector from A (the moment center) to B, where B is the point of force application. This choice for the position vector makes it easy to write $\boldsymbol{r}$ in terms of its components based on the dimensions given in **Figure 5.42*b***:

$$\boldsymbol{r} = r_x\boldsymbol{i} + r_y\boldsymbol{j} = 0.25 \text{ ft } \boldsymbol{i} + 1.00 \text{ ft } \boldsymbol{j}$$

Therefore $r_x = 0.25$ ft and $r_y = 1.00$ ft.

The moment $\boldsymbol{M}_z$*:* Substituting these values for F_x, F_y, r_x, and r_y in (5.11A), we determine that

$$\begin{aligned}\boldsymbol{M}_z = M_z\boldsymbol{k} &= (+r_xF_y - r_yF_x)\boldsymbol{k} = [(0.25 \text{ ft})(100.0 \text{ lb}) - (1.00 \text{ ft})(173.2 \text{ lb})]\boldsymbol{k} \\ &= -148 \text{ ft}\cdot\text{lb } \boldsymbol{k}\end{aligned}$$

Answer The moment created by the cable force about A is -148 ft·lb $\boldsymbol{k}$. The moment is negative because the sense of the rotation caused by $\boldsymbol{M}_z$ is clockwise about the z axis. The negative sign also indicates that the moment is in the $-\boldsymbol{k}$ direction.

Check The negative sign for the moment can be confirmed by looking at the relative sizes of F_x, F_y and r_x, r_y. Looking at **Figure 5.42*a***, we see that F_x causes a negative (clockwise) moment about the z axis while F_y causes a positive (counterclockwise) moment. Because $F_x > F_y$ and $r_y > r_x$ (meaning that the position vector associated with F_x is larger than the one associated with F_y), the equilibrium moment resulting from F_x and F_y will be negative.[8]

[7]As we saw in Chapter 4, it is not uncommon for there to be more than one approach to solving a problem. We show three approaches for finding the moment vector. All three result in the same answer!

The selection of an approach is often up to personal preference. As you will see, some approaches involve less work. The key is that whatever approach you use, you should get the same answer.

[8]The check we have just presented did not involve any additional calculations, but involved reasoning about relative sizes of values. This type of check complements checks that require additional calculations. Remember that checks are done to convince yourself that you made reasonable assumptions, set the problem up properly, and carried out the calculations correctly.

Approach 2 (Based on (5.1)): The cable tension creates a moment about the z axis. Equation (5.1) says that the magnitude of this moment is

$$\| \boldsymbol{M} \| = \| \boldsymbol{r} \| (\| \boldsymbol{F} \| \sin \theta) \qquad (5.1)$$

Magnitude of the position vector: We arbitrarily choose the position vector to be from A (moment center) to B; therefore, its length is

$$\| \boldsymbol{r} \| = \sqrt{(0.25 \text{ ft})^2 + (1.00 \text{ ft})^2} = 1.031 \text{ ft}$$

The angle θ between the position vector and the force: Based on the geometry shown in Figure 5.43, we find that $\theta = 90° - \alpha - 30°$. Because $\alpha = \tan^{-1}(0.25 \text{ ft}/1.00 \text{ ft}) = 14°$, we have $\theta = 46°$.

Substituting these values of $\| \boldsymbol{r} \|$, $\| \boldsymbol{F} \|$, and θ into (5.1), we find that

$$\| \boldsymbol{M} \| = \| \boldsymbol{r} \| (\| \boldsymbol{F} \| \sin \theta) = (1.031 \text{ ft})(200 \text{ lb}) \sin 46° = 148 \text{ ft} \cdot \text{lb}$$

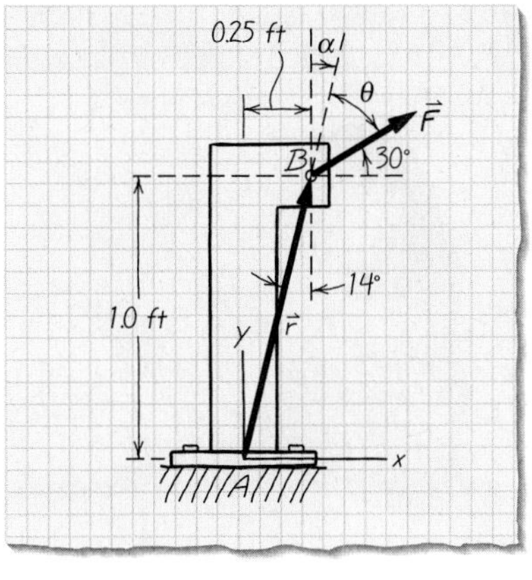

Figure 5.43

We still must determine whether the moment is in the counterclockwise (positive) or clockwise (negative) direction about the z axis. Use the right-hand rule to determine this (Figure 5.44); align your right-hand fingers with $\boldsymbol{r}$ in a way that will allow you to curl your fingers to align with $\boldsymbol{F}$, and then move your thumb to make a 90° angle with your palm. The "curl" sense is clockwise when viewed from far out on the z axis and defines the direction of the moment as being along an axis aligned with your thumb (which is in the negative z direction). Therefore, the direction of the moment is $-\boldsymbol{k}$, along the negative z axis.

The moment created by the cable force is described as $-148 \text{ ft} \cdot \text{lb}\, \boldsymbol{k}$.

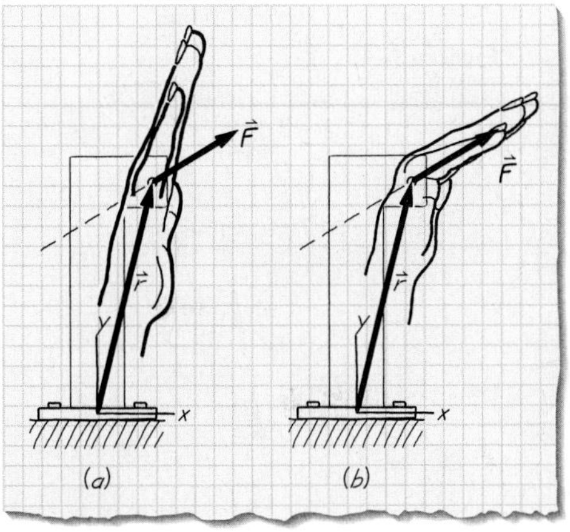

Figure 5.44

Approach 3 (Using the Cross Product): As shown in Approach 1,

$$\boldsymbol{F} = 173 \text{ lb}\, \boldsymbol{i} + 100 \text{ lb}\, \boldsymbol{j}$$

$$\boldsymbol{r} = r_x \boldsymbol{i} + r_y \boldsymbol{j} = 0.25 \text{ ft}\, \boldsymbol{i} + 1.00 \text{ ft}\, \boldsymbol{j}$$

Substituting these into the matrix form of the cross product in (5.12),

$$\boldsymbol{M} = \boldsymbol{r} \times \boldsymbol{F} = \begin{vmatrix} \boldsymbol{i} & \boldsymbol{j} & \boldsymbol{k} \\ r_x & r_y & r_z \\ F_x & F_y & F_z \end{vmatrix} = \begin{vmatrix} \boldsymbol{i} & \boldsymbol{j} & \boldsymbol{k} \\ 0.25 \text{ ft} & 1.00 \text{ ft} & 0 \text{ ft} \\ 173 \text{ lb} & 100 \text{ lb} & 0 \text{ lb} \end{vmatrix}$$

and finding the determinant,

$$\boldsymbol{M} = M_z \boldsymbol{k} = (+r_x F_y - r_y F_x)\boldsymbol{k} = [(0.25 \text{ ft})(100 \text{ lb}) - (1.00 \text{ ft})(173 \text{ lb})]\boldsymbol{k}$$
$$= 148 \text{ ft} \cdot \text{lb}\, \boldsymbol{k}$$

Answer The moment created by the cable force about A is $-148 \text{ ft} \cdot \text{lb}\, \boldsymbol{k}$. The moment is negative because the sense of the rotation caused by $\boldsymbol{M}$ is clockwise about the z axis. The negative sign also indicates that the force is in the $-\boldsymbol{k}$ direction.

Check The three approaches can be used as checks for one another. They yield the same answer.

EXAMPLE 5.8 CALCULATING THE MOMENT

Consider the apparatus for exercising the arm in Figure 5.45. When the user's elbow is in the flexed position shown, the cord pulls with a force $\boldsymbol{F}$ that has a magnitude of 400 N. Find the moment at E (the exerciser's elbow) created by the cord force.

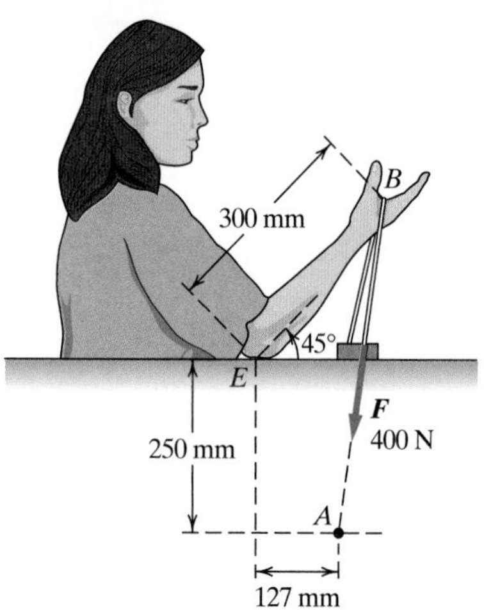

Figure 5.45

Goal We are to find the moment at a moment center at E created by the force applied at B.

Given We are given the dimensions of the physical situation. In addition, we are given the magnitude of the force (400 N) and information to find its direction.

Assume Based on the information provided, we assume that the cord and arm lie in a single plane (we arbitrarily call this the xy plane). We establish an origin for this plane at E, as shown in Figure 5.46.

Draw We create the drawing in **Figure 5.46** to facilitate writing a position vector and force in vector notation.

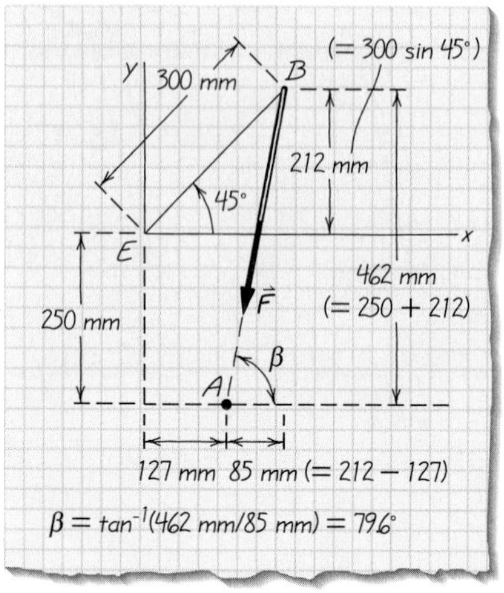

Figure 5.46

Formulate Equations and Solve Based on our assumption that the cord and arm lie in a single plane and that we have defined this as the xy plane, the force creates a moment about the z axis.

Approach 1 (Based on Vector Components): We have a force $\boldsymbol{F} = F_x\boldsymbol{i} + F_y\boldsymbol{j}$ applied at point B. A position vector $\boldsymbol{r} = r_x\boldsymbol{i} + r_y\boldsymbol{j}$ goes from E to anywhere along the line of action of $\boldsymbol{F}$. The force creates a moment of

$$\boldsymbol{M}_z = M_z\boldsymbol{k} = (+r_xF_y - r_yF_x)\boldsymbol{k} \tag{5.11A}$$

To apply (5.11A), we proceed to find $\boldsymbol{F}$ and $\boldsymbol{r}$.

The force $\boldsymbol{F}$*:* In order to write $\boldsymbol{F}$ in terms of its x and y components, we need to find the angle β, which is 79.6° (as shown **Figure 5.46**). Thus

$$\begin{aligned}\boldsymbol{F} = F_x\boldsymbol{i} + F_y\boldsymbol{j} &= (-400\text{ N})(\cos 79.6°)\boldsymbol{i} - (400\text{ N})(\sin 79.6°)\boldsymbol{j}\\ &= -72.2\text{ N}\,\boldsymbol{i} - 393\text{ N}\,\boldsymbol{j}\end{aligned}$$

Therefore $F_x = -72.2$ N and $F_y = -393$ N (Figure 5.47).

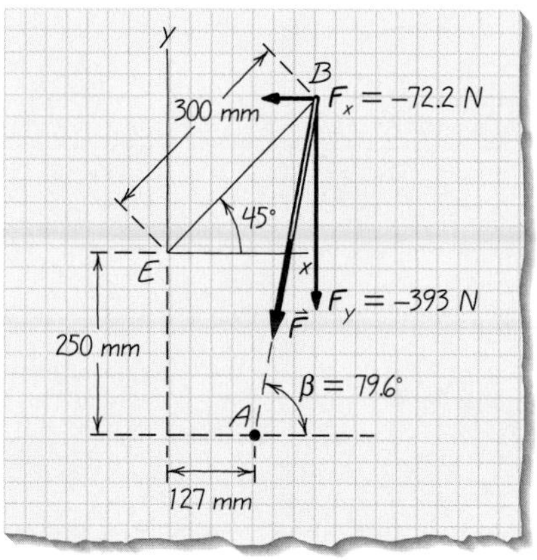

Figure 5.47

The position vector $\boldsymbol{r}$*:* We define the position vector as a vector from E (the moment center) to B (the point of force application). We could use *any* point along the line of action of $\boldsymbol{F}$, but we chose B because of the given information about B. The position vector from E to B, as shown in Figure 5.48, is

$$\begin{aligned}\boldsymbol{r} = r_x\boldsymbol{i} + r_y\boldsymbol{j} &= (300\text{ mm})(\cos 45°)\boldsymbol{i} + (300\text{ mm})(\sin 45°)\boldsymbol{j}\\ &= 212\text{ mm}\,\boldsymbol{i} + 212\text{ mm}\,\boldsymbol{j}\end{aligned}$$

Therefore $r_x = +212$ mm and $r_y = +212$ mm.

The moment M_z:

$$M_z = M_z k = (+r_x F_y - r_y F_x)k = [(212 \text{ mm})(-393 \text{ N}) - (212 \text{ mm})(-72.2 \text{ N})]k$$
$$= 68{,}000 \text{ N}\cdot\text{mm } k = 68.0 \text{ N}\cdot\text{m } k$$

Answer The moment at E (at the elbow) created by the cord force is -68.0 N · m $\boldsymbol{k}$. The sense of rotation caused by $\boldsymbol{M}$ is clockwise; the direction of $\boldsymbol{M}$ is $-\boldsymbol{k}$, about the negative z axis.

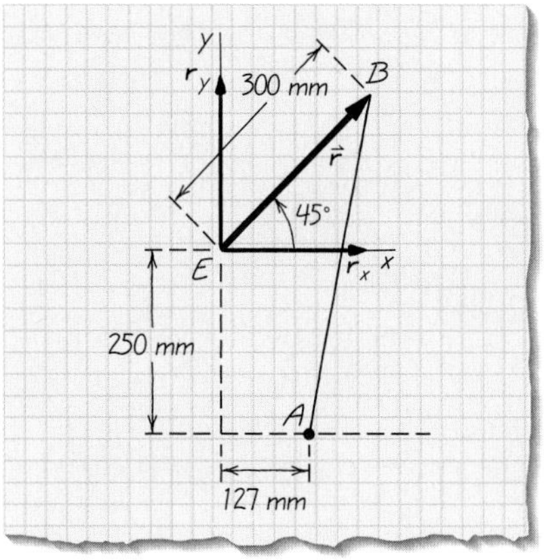

Figure 5.48

Approach 2 (Based on (5.1)): We will use (5.1) to find the magnitude of the moment:

$$\| \boldsymbol{M} \| = \| \boldsymbol{r} \| \,(\| \boldsymbol{F} \| \sin\theta) \tag{5.1}$$

In the particular situation under consideration, $\| \boldsymbol{F} \|$ is given as 400 N.

Magnitude of the position vector: We choose the position vector to be from E (the moment center) to B; therefore $\| \boldsymbol{r} \| = 300$ mm.

The angle θ between the position vector and the force: Based upon the geometry shown in Figure 5.49, we find that $\theta = 45° + 90° + 10.4° = 145°$. Thus

$$\| \boldsymbol{M} \| = \| \boldsymbol{r} \| \,(\| \boldsymbol{F} \| \sin\theta) = (300 \text{ mm})(400 \text{ N}) \sin 145°$$
$$= 68{,}000 \text{ N}\cdot\text{mm} = 68.0 \text{ N}\cdot\text{m}$$

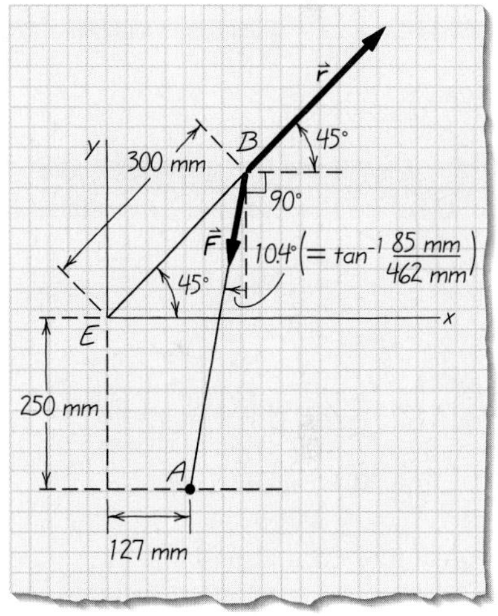

Figure 5.49

Since both $\boldsymbol{r}$ and $\boldsymbol{F}$ are in the xy plane, we can conclude that this moment is about the z axis. Furthermore, the right-hand rule is used to determine that this moment is in the $-\boldsymbol{k}$ direction. Therefore the moment created by the cord force is completely described as -68.0 N · m $\boldsymbol{k}$.

Answer The moment at E (at the elbow) created by the cord force is -68.0 N · m $\boldsymbol{k}$.

Approach 3 (Using the Cross Product): As shown in Approach 1,

$$\boldsymbol{F} = -72.2 \text{ N } \boldsymbol{i} - 393 \text{ N } \boldsymbol{j}$$
$$\boldsymbol{r} = 212 \text{ mm } \boldsymbol{i} + 212 \text{ mm } \boldsymbol{j}$$

Substituting these into the matrix form of the cross product (5.12), we get

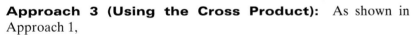

$$\boldsymbol{M} = \boldsymbol{r} \times \boldsymbol{F} = \begin{vmatrix} \boldsymbol{i} & \boldsymbol{j} & \boldsymbol{k} \\ r_x & r_y & r_z \\ F_x & F_y & F_z \end{vmatrix} = \begin{vmatrix} \boldsymbol{i} & \boldsymbol{j} & \boldsymbol{k} \\ 212 \text{ mm} & 212 \text{ mm} & 0 \text{ mm} \\ -72.2 \text{ N} & -393 \text{ N} & 0 \text{ N} \end{vmatrix}$$

and finding the determinant,

$$\boldsymbol{M}_z = [(212 \text{ mm})(-393 \text{ N}) - (212 \text{ mm})(-72.2 \text{ N})]\boldsymbol{k}$$
$$= -68{,}000 \text{ N}\cdot\text{mm } \boldsymbol{k} = -68.0 \text{ N}\cdot\text{m } \boldsymbol{k}$$

Answer The moment at E (at the elbow) created by the cord force is -68.0 N · m $\boldsymbol{k}$.

Check As expected, the three approaches result in the same answer. Sometimes solving the problem with two approaches is a way to check your answer (but since it can be laborious, it is recommended that it be done only in select cases).

If possible you should also confirm the reasonableness of the answer. From the discussion in Example 5.5, reasonableness involves confirming that the answer is consistent with the physical situation; in other words, does the problem describe a realistic physical situation? One check of reasonableness is to compare the size of the $-68.0\ \text{N}\cdot\text{m}\ \boldsymbol{k}$ moment with the one found with the person pushing on the wrench, since both involve human arm muscles. If we do this, we find that the moments are the same order of magnitude (68 N · m versus 10 N · m for Example 5.5 (b)); this adds some confidence that the problem is reasonable and is describing a realistic physical situation. One is still left wondering whether the 400-N arm force in this problem is reasonable; what do you think?

EXAMPLE 5.9 MOMENT CALCULATIONS FOR VARIOUS PHYSICAL SITUATIONS

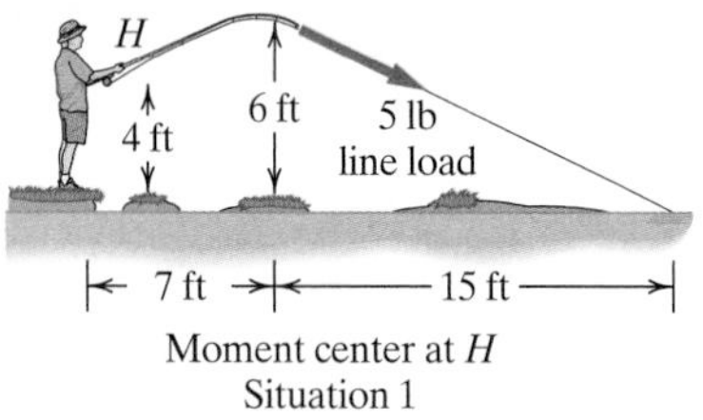

Figure 5.50 Situation 1

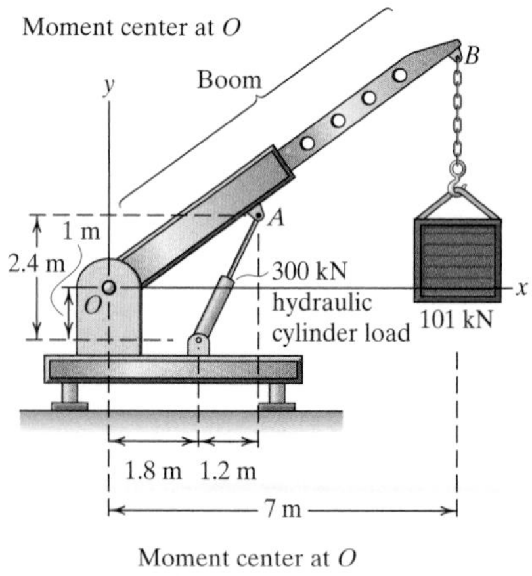

Figure 5.51 Situation 2

Figures 5.50 through 5.54 show physical situations consisting of a structure, force(s), and coordinate axes. For each situation:

(a) Write symbolic expressions for the force and position vector.

(b) Write an expression for the moment created by the force in (a) at a specified moment center. Equations (5.7), (5.11A), (5.11B), and (5.11C) may be useful.

Goal For each situation, **(a)**

- Inspect the force to determine which components are needed to describe it; write the force symbolically.
- Determine a position vector and write it symbolically.

(b) Write an expression for the moment created by each force, by

1. Using the *general expression* for moment in (5.7) and simplifying if possible:

$$\boldsymbol{M} = \underbrace{(+r_yF_z - r_zF_y)}_{M_x}\boldsymbol{i} + \underbrace{(+r_zF_x - r_xF_z)}_{M_y}\boldsymbol{j} + \underbrace{(+r_xF_y - r_yF_x)}_{M_z}\boldsymbol{k} \quad (5.7)$$

or

2. Determining whether the force and position vector lie in a plane defined by the *xy*, *yz*, or *zx* axes and then selecting the appropriate version of (5.11) to determine the moment.

For a force vector and position vector lying in the same *xy* plane, we can write

$$\boldsymbol{M}_z = M_z\boldsymbol{k} = (+r_xF_y - r_yF_x)\boldsymbol{k} \quad (5.11\text{A})$$

For a force vector and position vector lying in the same yz plane, we can write

$$\boldsymbol{M}_x = M_x\boldsymbol{i} = (+r_yF_z - r_zF_y)\boldsymbol{i} \qquad (5.11\text{B})$$

For a force vector and position vector lying in the same zx plane, we can write

$$\boldsymbol{M}_y = M_y\boldsymbol{j} = (+r_zF_x - r_xF_z)\boldsymbol{j} \qquad (5.11\text{C})$$

Draw, Formulate Equations, and Solve No calculations are involved with this problem; the emphasis is on reasoning about the situation in order to identify the key relationships among force, structure, coordinate axes, and moment. We set up the appropriate equations symbolically to represent the moment in each situation.

Situation 1 (Figure 5.50): The situation consists of the fishing rod (including the line that goes through the guides), a moment center located at the handle of the rod (H), and a 5-lb force ($\boldsymbol{F}$) acting at the tip of the rod. No information is given in **Figure 5.50** about whether the rod and the force are oriented out of the plane of the paper, and so based on the information provided, we assume that both lie in the same plane; we arbitrarily define this as the xy plane and the origin of the coordinate axes at the moment center H. We write the force as $\boldsymbol{F} = F_x\boldsymbol{i} + F_y\boldsymbol{j}$. The position vector is any vector that goes from the moment center to the line of action of the force; we select the position vector shown in **Figure 5.55** because of the dimensional information given in **Figure 5.50**; therefore $\boldsymbol{r} = r_x\boldsymbol{i} + r_y\boldsymbol{j}$.

Answer to Situation 1, (a) $\boldsymbol{F} = F_x\boldsymbol{i} + F_y\boldsymbol{j};\ \boldsymbol{r} = r_x\boldsymbol{i} + r_y\boldsymbol{j}.$

Based on these symbolic representations of $\boldsymbol{F}$ and $\boldsymbol{r}$, we can write the moment by using any one of the following three approaches.

- *Use the general expression in (5.7):*

$$\boldsymbol{M} = \underbrace{(+r_yF_z - r_zF_y)}_{M_x}\boldsymbol{i} + \underbrace{(+r_zF_x - r_xF_z)}_{M_y}\boldsymbol{j} + \underbrace{(+r_xF_y - r_yF_x)}_{M_z}\boldsymbol{k} \qquad (5.7)$$

Because $r_z = F_z = 0$, (5.7) simplifies to $(+r_xF_y - r_yF_x)\boldsymbol{k}$.

- *Note that the force lies in the* xy *reference plane*. Therefore (5.11A) applies:

$$\boldsymbol{M} = (+r_xF_y - r_yF_x)\boldsymbol{k} \qquad (5.11\text{A})$$

Answer to Situation 1, (b) $\boldsymbol{M} = (+r_xF_y - r_yF_x)\boldsymbol{k}$

These two approaches result in the same answer. In practice, you need only use a single approach but should use a second approach as a check.

If we had been asked to give a numerical answer for $\boldsymbol{M}$, based on the information in **Figure 5.50**, we would use $r_x = 7$ ft, $r_y = 2$ ft, $F_x = (5\text{ lb})(\cos 21.8°)$, and $F_y = (-5\text{ lb})(\sin 21.8°)$.

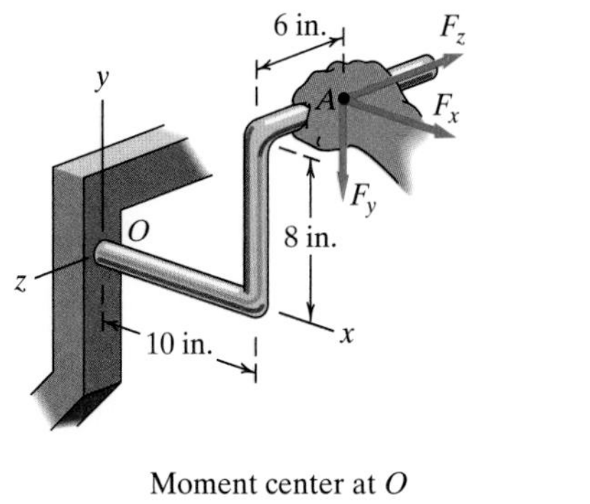

Figure 5.52 Situation 3

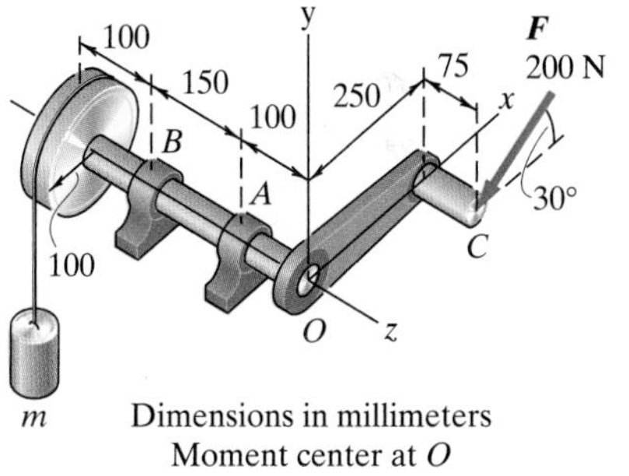

Figure 5.53 Situation 4

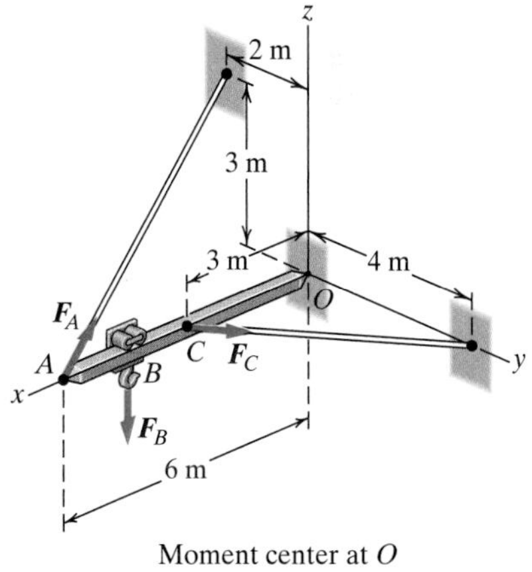

Figure 5.54 Situation 5

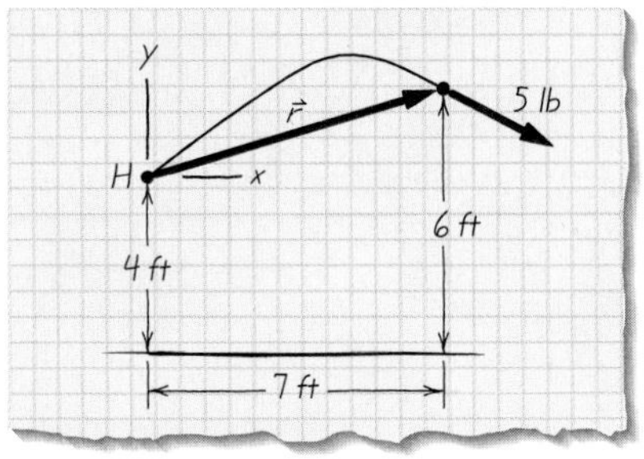

Figure 5.55 Situation 1

Situation 2 (Figure 5.51): The situation consists of the boom, moment center located at the base of the boom (O), and two forces. One of these forces is the 300-kN force the hydraulic cylinder applies to the boom ($\boldsymbol{F}_A$), and the other is the weight (10 kN) of the crate acting at the end of the boom ($\boldsymbol{F}_B$). No information is given in **Figure 5.51** about whether the boom and the forces are oriented out of the plane of the paper, and so based on the information provided, we assume that they are in an xy plane.

We write $\boldsymbol{F}_A$ as $\boldsymbol{F}_A = F_{Ax}\boldsymbol{i} + F_{Ay}\boldsymbol{j}$. The position vector is any vector that goes from the moment center to the line of action of $\boldsymbol{F}_A$; we arbitrarily select the one shown in **Figure 5.56**, which can be written as $\boldsymbol{r}_A = r_{Ax}\boldsymbol{i} + r_{Ay}\boldsymbol{j}$. Since $\boldsymbol{F}_A$ lies in the xy reference plane, we can use (5.11A) to find the moment it creates about axes at O:

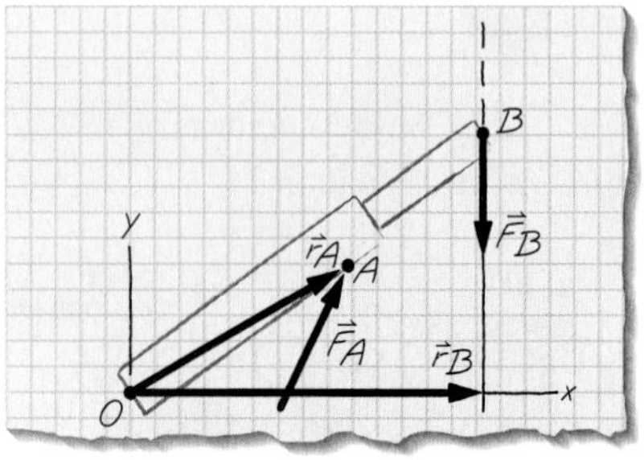

Figure 5.56 Situation 2

$$\boldsymbol{M}_A = (+r_{Ax}F_{Ay} - r_{Ay}F_{Ax})\boldsymbol{k}$$

We write $\boldsymbol{F}_B$ as $\boldsymbol{F}_B = F_{By}\boldsymbol{j}$. The position vector is any vector that goes from the moment center to the line of action of $\boldsymbol{F}_B$; we arbitrarily select the one shown in **Figure 5.56**, which can be written as $\boldsymbol{r}_B = r_{Bx}\boldsymbol{i}$. Since $\boldsymbol{F}_B$ lies in the xy reference plane, we can use (5.11A) to find the moment it creates at O:

$$\boldsymbol{M}_B = (+r_{Bx}F_{By} - r_{By}F_{Bx})\boldsymbol{k}$$

Because the only nonzero components are F_{By} and r_{Bx}, this expression can be simplified to

$$\boldsymbol{M}_B = (+r_{Bx}F_{By})\boldsymbol{k}$$

Answer to Situation 2, (a) $\boldsymbol{F}_A = F_{Ax}\boldsymbol{i} + F_{Ay}\boldsymbol{j}$; $\boldsymbol{r}_A = r_{Ax}\boldsymbol{i} + r_{Ay}\boldsymbol{j}$; and $\boldsymbol{F}_B = F_{By}\boldsymbol{j}$; $\boldsymbol{r}_B = r_{Bx}\boldsymbol{i}$

(b) $\boldsymbol{M}_A = (+r_{Ax}F_{Ay} - r_{Ay}F_{Ax})\boldsymbol{k}$ and $\boldsymbol{M}_B = (+r_{Bx}F_{By})\boldsymbol{k}$

If we had been asked to give numerical answers for $\boldsymbol{M}_A$ and $\boldsymbol{M}_B$, based on the information in **Figure 5.51**, we would use $r_{Ax} = 3.0$ m, $r_{Ay} = 2.4$ m, $F_{Ax} = +300$ (cos 26.6°) kN, $F_{Ay} = +300$ (sin 26.6°) kN, $r_{Bx} = 7.0$ m, and $F_{By} = -101$ kN.

Situation 3 (Figure 5.52): The system consists of the handlebar, a moment center at O located at the base of the handlebar, hand force ($\boldsymbol{F}$) acting at A, and a coordinate system. Based on the information supplied in **Figure 5.52**, $\boldsymbol{F}$ has components in the x, y, and z directions and is written as $\boldsymbol{F} = F_x\boldsymbol{i} + F_y\boldsymbol{j} + F_z\boldsymbol{k}$. A position vector from O to A is written as $\boldsymbol{r} = r_x\boldsymbol{i} + r_y\boldsymbol{j} + r_z\boldsymbol{k}$.

Since the force and position vector are not in an xy, yz, or zx plane, we use the general expression in (5.7) for finding the moment that $\boldsymbol{F}$ creates about the reference axes.

Answer to Situation 3, (a) $\boldsymbol{F} = F_x\boldsymbol{i} + F_y\boldsymbol{j} + F_z\boldsymbol{k}$ and $\boldsymbol{r} = r_x\boldsymbol{i} + r_y\boldsymbol{j} + r_z\boldsymbol{k}$

(b) $$\boldsymbol{M} = \underbrace{(+r_yF_z - r_zF_y)}_{M_x}\boldsymbol{i} + \underbrace{(+r_zF_x - r_xF_z)}_{M_y}\boldsymbol{j} + \underbrace{(+r_xF_y - r_yF_x)}_{M_z}\boldsymbol{k}$$

If we had been asked to give a numerical answer for $\boldsymbol{M}$ based on the information in **Figure 5.52**, we would use $r_x = 10$ in., $r_y = 8$ in., $r_z = -6$ in. Furthermore, the value of F_x would be positive, and the values of F_y and F_z would be negative.

Situation 4 (Figure 5.53): The system consists of the crank, a moment center located at the base of the crane arm, and a force ($\boldsymbol{F}$) acting at C. Based on the information supplied in **Figure 5.53**, $\boldsymbol{F}$ has x and y components and is written as $\boldsymbol{F} = F_x\boldsymbol{i} + F_y\boldsymbol{j}$. The position vector shown in **Figure 5.57**, from O to C, is written as $\boldsymbol{r} = r_x\boldsymbol{i} + r_z\boldsymbol{k}$.

Since the force and position vectors are not in an xy, yz, or zx plane, we use the general expression in (5.7) for finding the moment that $\boldsymbol{F}$ creates about the moment center. Because $F_z = 0$, we are able to simplify the expression.

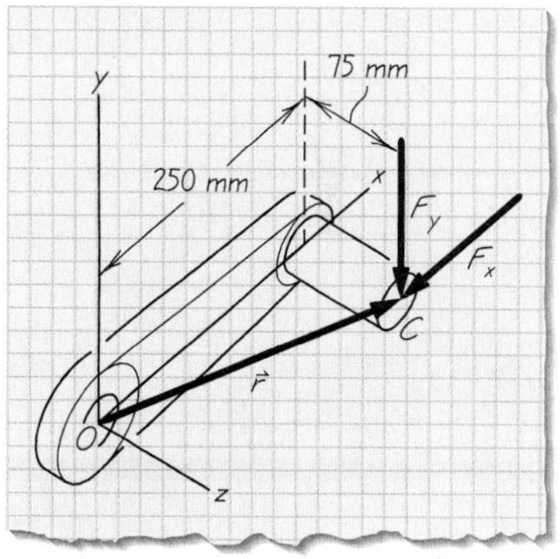

Figure 5.57 Situation 4

Answer to Situation 4, (a) $\boldsymbol{F} = F_x\boldsymbol{i} + F_y\boldsymbol{j}$; $\boldsymbol{r} = r_x\boldsymbol{i} + r_z\boldsymbol{k}$

(b) $$\boldsymbol{M} = \underbrace{(-r_zF_y)}_{M_x}\boldsymbol{i} + \underbrace{(+r_zF_x)}_{M_y}\boldsymbol{j} + \underbrace{(+r_xF_y - r_yF_x)}_{M_z}\boldsymbol{k}$$

If we had been asked to give a numerical answer for $\boldsymbol{M}$ based on the information in **Figure 5.53**, we would find that the value of F_x is negative and the value of F_y is negative. Furthermore, the values of r_x and r_z are 250 mm and 75 mm, respectively.

Situation 5 (Figure 5.54): The system consists of a boom, coordinate axes with an origin located at the base of the arm, and forces $\boldsymbol{F_A}$, $\boldsymbol{F_B}$, and $\boldsymbol{F_C}$ acting at A, B, and C, respectively. Forces $\boldsymbol{F_A}$ and $\boldsymbol{F_C}$ result from cables attached to the boom, and $\boldsymbol{F_B}$ results from a weight that hangs from the boom.

Based on the information supplied in **Figure 5.54**, $\boldsymbol{F_A}$ has components in the x, y, and z directions and is written as $\boldsymbol{F_A} = F_{Ax}\boldsymbol{i} + F_{Ay}\boldsymbol{j} + F_{Az}\boldsymbol{k}$. A position vector can be drawn from O to A and is written as $\boldsymbol{r_A} = r_{Ax}\boldsymbol{i}$ (**Figure 5.58**).

Since $\boldsymbol{F_A}$ and the position vector do not lie in an xy, yz, or xz plane, we use the general expression in (5.7) to find the moment $\boldsymbol{F_A}$ creates at the moment center. Because $r_{Ay} = r_{Az} = 0$, we are able to simplify the expression.

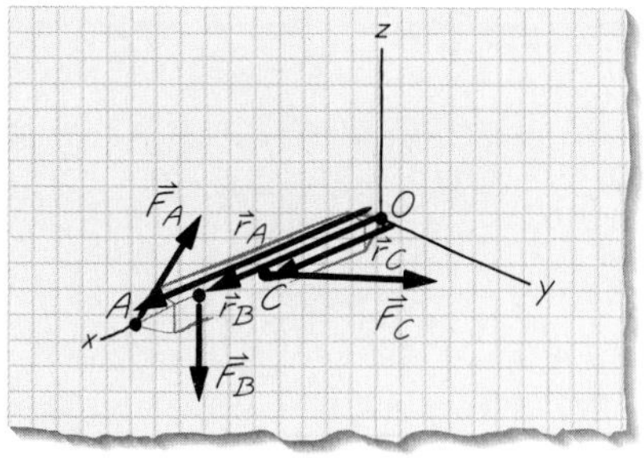

Figure 5.58 Situation 5

Answer to Situation 5 for F_A, (a) $\boldsymbol{F_A} = F_{Ax}\boldsymbol{i} + F_{Ay}\boldsymbol{j} + F_{Az}\boldsymbol{k}$; $\boldsymbol{r_A} = r_{Ax}\boldsymbol{i}$

(b) $$\boldsymbol{M} = \underbrace{(-r_{Ax}F_{Az})}_{M_y}\boldsymbol{j} + \underbrace{(+r_{Ax}F_{Ay})}_{M_z}\boldsymbol{k}$$

Based on the information supplied in **Figure 5.54**, $\boldsymbol{F}_B$ has a z component and is written as $\boldsymbol{F}_B = F_{Bz}\boldsymbol{j}$. A position vector can be drawn from O to B and is written as $\boldsymbol{r}_B = r_{Bx}\boldsymbol{i}$ (**Figure 5.58**).

Because $\boldsymbol{F}_B$ and the position vector are in the xz plane, we can use (5.11C) to find the moment the force creates at the moment center. In addition, because $F_{Bx} = r_{By} = 0$, we are able to simplify the expression.

Answer to Situation 5 for F_B, (a) $\boldsymbol{F}_B = F_{Bz}\boldsymbol{k}$; $\boldsymbol{r}_B = r_{Bx}\boldsymbol{i}$

(b) $\boldsymbol{M}_B = -r_{Bx}F_{Bz}\boldsymbol{j}$

Based on the information supplied in **Figure 5.54**, $\boldsymbol{F}_C$ has components in the x and y directions and is written as $\boldsymbol{F}_C = F_{cx}\boldsymbol{i} + F_{cy}\boldsymbol{j}$. A position vector can be drawn from O to C and is written as $\boldsymbol{r}_C = r_{Cx}\boldsymbol{i}$ (**Figure 5.58**).

Because $\boldsymbol{F}_C$ and the position vector are in the xy reference plane, we can use (5.11A) to find the moment $\boldsymbol{F}_C$ creates at the moment center. In addition, because $r_{Cy} = 0$, we are able to simplify the expression.

Answer to Situation 5 for F_C, (a) $\boldsymbol{F}_C = F_{cx}\boldsymbol{i} + F_{cy}\boldsymbol{j}$; $\boldsymbol{r}_C = r_{Cx}\boldsymbol{i}$

(b) $\boldsymbol{M}_C = +r_{Cx}F_{Cy}\boldsymbol{k}$

EXAMPLE 5.10 ANOTHER EXAMPLE OF FINDING THE MOMENT

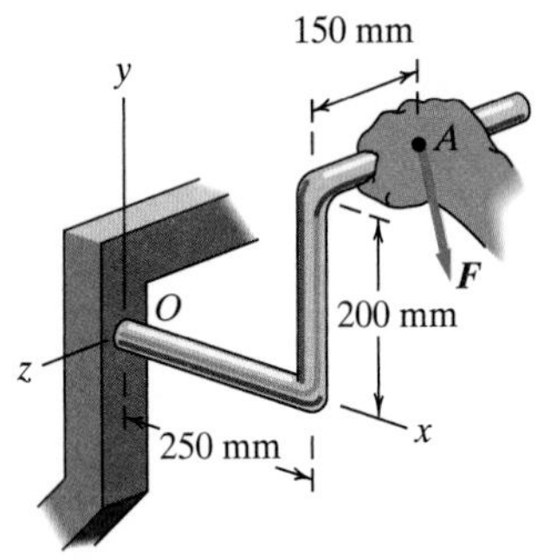

Figure 5.59

The hand in Figure 5.59 pulls on the handlebar with force $\boldsymbol{F}$. The point of application of the force is at A. The magnitude of the force is 110 N, and its direction is described by the unit vector $\boldsymbol{u} = (2/3)\boldsymbol{i} - (2/3)\boldsymbol{j} - (1/3)\boldsymbol{k}$.

(a) Determine the moment $\boldsymbol{M}$ created by $\boldsymbol{F}$ at a moment center at O. Present the answer in vector notation and graphically.

(b) Find the magnitude of the moment.

(c) Find the direction cosines associated with the moment, and the unit vector that describes the direction of the moment.

Goal We are to find (a) the moment at a specified moment center (O) created by a specified force and present our answer both in vector notation and graphically. In addition, we are asked to determine (b) the magnitude of the moment and (c) its direction cosines as well as the unit vector aligned with it.

Given We are given dimensions associated with the handlebar on which the hand is pulling. In addition, we are given the magnitude of the force (110 N), that it acts at point A, and that its direction is along a unit vector specified by $\boldsymbol{u} = (2/3)\boldsymbol{i} - (2/3)\boldsymbol{j} - (1/3)\boldsymbol{k}$. Finally, a coordinate system has been established.

Assume No assumptions are necessary.

Formulate Equations and Solve **(a)** We show two approaches to finding the moment; one uses the generalized expression in (5.7) for moment and the other uses the cross-product in (5.12).

Approach 1 (Using (5.7)): This is the same situation as in **Figure 5.52** of Example 5.9. There we determined that the moment created by the force has components M_x, M_y, and M_z and that (5.7) can be used to find $\boldsymbol{M}$:

$$\boldsymbol{M} = \underbrace{(+r_yF_z - r_zF_y)}_{M_x}\boldsymbol{i} + \underbrace{(+r_zF_x - r_xF_z)}_{M_y}\boldsymbol{j} + \underbrace{(+r_xF_y - r_yF_x)}_{M_z}\boldsymbol{k} \quad (5.7)$$

To apply (5.7), we find the scalar components of $\boldsymbol{F}$ and those of our position vector $\boldsymbol{r}$.

Force:

$$\boldsymbol{F} = 110 \text{ N } \boldsymbol{u} = 110 \text{ N}(\tfrac{2}{3}\boldsymbol{i} - \tfrac{2}{3}\boldsymbol{j} - \tfrac{1}{3}\boldsymbol{k}) \text{ N} = 73.3 \text{ N } \boldsymbol{i} - 73.3 \text{ N } \boldsymbol{j} - 36.7 \text{ N } \boldsymbol{k}$$

$$F_x = 73.3 \text{ N}$$
$$F_y = -73.3 \text{ N}$$
$$F_z = -36.7 \text{ N}$$

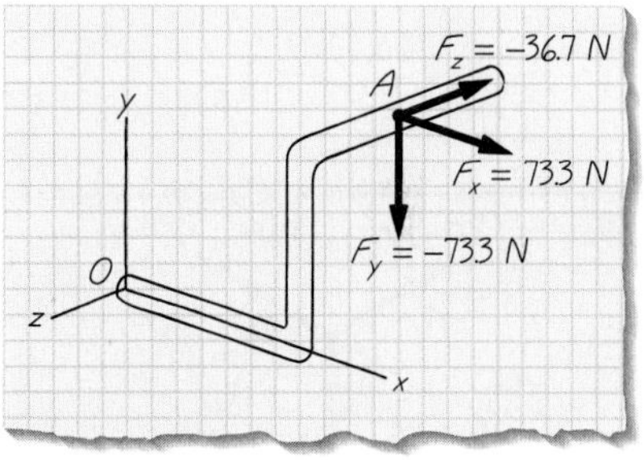

Figure 5.60

These scalar components are illustrated in **Figure 5.60**.

Position vector: The most straightforward position vector to define is one from O to A (**Figure 5.61**), where A is the point of application of the force. Reading off the dimensions given in the figure, we find

$$\boldsymbol{r} = 250 \text{ mm } \boldsymbol{i} + 200 \text{ mm } \boldsymbol{j} - 150 \text{ mm } \boldsymbol{k}$$

$$r_x = 250 \text{ mm}$$
$$r_y = 200 \text{ mm}$$
$$r_z = -150 \text{ mm}$$

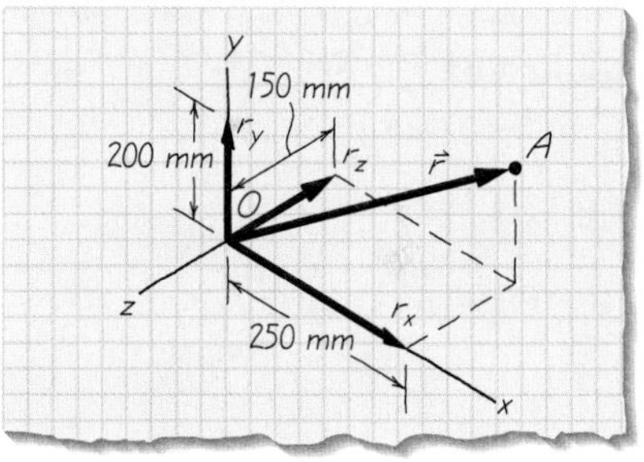

Figure 5.61

These scalar components are illustrated in **Figure 5.61**.

Moment components: M_x, M_y, and M_z can now be found based on (5.7).

$$\begin{aligned} M_x &= (+r_yF_z - r_zF_y) \\ &= (200 \text{ mm})(-36.7 \text{ N}) - (-150 \text{ mm})(-73.3 \text{ N}) \\ &= -18{,}330 \text{ N}\cdot\text{mm} \\ M_y &= (+r_zF_x - r_xF_z) \\ &= (-150 \text{ mm})(73.3 \text{ N}) - (250 \text{ mm})(-36.7 \text{ N}) \\ &= -1820 \text{ N}\cdot\text{mm} \\ M_z &= (+r_xF_y - r_yF_x) \\ &= (250 \text{ mm})(-73.3 \text{ N}) - (200 \text{ mm})(73.3 \text{ N}) \\ &= -33{,}000 \text{ N}\cdot\text{mm} \end{aligned}$$

Moment (in vector notation): The moment $\boldsymbol{M}$ created by $\boldsymbol{F}$ at a moment center at O is

$$\boldsymbol{M} = -18{,}330 \text{ N}\cdot\text{mm } \boldsymbol{i} - 1820 \text{ N}\cdot\text{mm } \boldsymbol{j} - 33{,}000 \text{ N}\cdot\text{mm } \boldsymbol{k}.$$

Approach 2 (Using (5.12)): **(a)** We can write $\boldsymbol{F}$ and $\boldsymbol{r}$ in vector notation as

$$\boldsymbol{F} = 73.3 \text{ N } \boldsymbol{i} - 73.3 \text{ N} \boldsymbol{j} - 36.7 \text{ N } \boldsymbol{k}$$
$$\boldsymbol{r} = 250 \text{ mm } \boldsymbol{i} + 200 \text{ mm} \boldsymbol{j} - 150 \text{ mm } \boldsymbol{k}$$

Now we write (5.12) for $\boldsymbol{M}$ in terms of $\boldsymbol{r}$ and $\boldsymbol{F}$:

$$\boldsymbol{M} = \boldsymbol{r} \times \boldsymbol{F} = \begin{vmatrix} \boldsymbol{i} & \boldsymbol{j} & \boldsymbol{k} \\ r_x & r_y & r_z \\ F_x & F_y & F_z \end{vmatrix} = \begin{vmatrix} \boldsymbol{i} & \boldsymbol{j} & \boldsymbol{k} \\ 250 \text{ mm} & 200 \text{ mm} & -150 \text{ mm} \\ 73.3 \text{ N} & -73.3 \text{ N} & -36.7 \text{ N} \end{vmatrix}$$

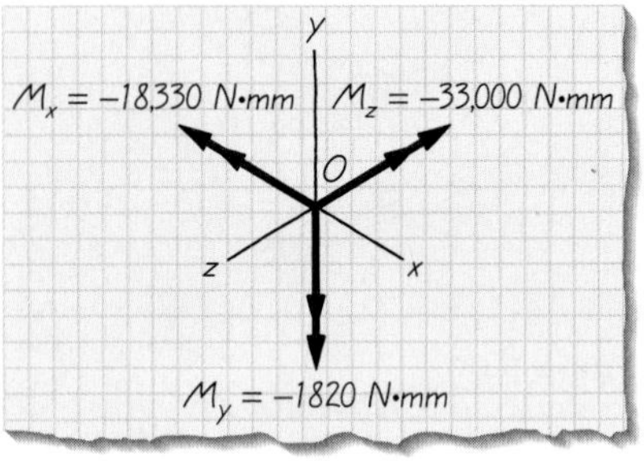

Figure 5.62

and use the rules of determinants and matrix algebra to solve this problem. Expanding the above matrix yields

$$\boldsymbol{M} = \boldsymbol{r} \times \boldsymbol{F} = \{[(200)(-36.7) - (-150)(-73.3)]\boldsymbol{i} - [(250)(-36.7) - (-150)(73.3)]\boldsymbol{j} + [(250)(-73.3) - (200)(73.3)]\boldsymbol{k}\} \text{ N}\cdot\text{mm}$$

Answer to (a) The moment $\boldsymbol{M}$ created by $\boldsymbol{F}$ about the coordinate axes is $\boldsymbol{M} = -18{,}330 \text{ N}\cdot\text{mm } \boldsymbol{i} - 1820 \text{ N}\cdot\text{mm} \boldsymbol{j} - 33{,}000 \text{ N}\cdot\text{mm } \boldsymbol{k}$ or $\boldsymbol{M} = -18.33 \text{ N}\cdot\text{m } \boldsymbol{i} - 1.82 \text{ N}\cdot\text{m} \boldsymbol{j} - 33.0 \text{ N}\cdot\text{m } \boldsymbol{k}$. It is shown graphically in **Figure 5.62**.

(b) Because we know the moment components from part (a), the most straightforward approach to finding the magnitude of the moment is to apply (5.3):

$$\begin{aligned} \|\boldsymbol{M}\| &= \sqrt{M_x^2 + M_y^2 + M_z^2} \\ &= \sqrt{(-18{,}300 \text{ N}\cdot\text{mm})^2 + (-1820 \text{ N}\cdot\text{mm})^2 + (-33{,}000 \text{ N}\cdot\text{mm})^2} \\ &= 37{,}800 \text{ N}\cdot\text{mm} \end{aligned}$$

Answer to (b) The magnitude of the moment is $37{,}800 \text{ N}\cdot\text{mm} = 37.8 \text{ N}\cdot\text{m}$

(c) The direction cosines of the moment describe the axis about which the moment acts are found based on (5.9):

$$\begin{aligned} \cos\theta_x &= \frac{M_x}{\|\boldsymbol{M}\|} = \frac{-18{,}330 \text{ N}\cdot\text{mm}}{37{,}800 \text{ N}\cdot\text{mm}} = -0.485 \\ \cos\theta_y &= \frac{M_y}{\|\boldsymbol{M}\|} = \frac{-1820 \text{ N}\cdot\text{mm}}{37{,}800 \text{ N}\cdot\text{mm}} = -0.0481 \\ \cos\theta_z &= \frac{M_z}{\|\boldsymbol{M}\|} = \frac{-33{,}000 \text{ N}\cdot\text{mm}}{37{,}800 \text{ N}\cdot\text{mm}} = -0.873 \end{aligned} \tag{5.9}$$

Now we use the direction cosines to describe a unit vector along the axis, per (5.5):

$$\begin{aligned} \boldsymbol{u}_M &= \cos\theta_x\boldsymbol{i} + \cos\theta_y\boldsymbol{j} + \cos\theta_z\boldsymbol{k} \\ &= -0.485\boldsymbol{i} - 0.0481\boldsymbol{j} - 0.873\boldsymbol{k} \end{aligned} \tag{5.5}$$

This unit vector is perpendicular to the plane defined by the position vector and the force.

Answer to (c) The direction cosines are $\cos\theta_x = -0.485$, $\cos\theta_y = -0.0481$, and $\cos\theta_z = -0.873$. The moment is about an axis defined by the unit vector $\boldsymbol{u}_M = -0.485\boldsymbol{i} - 0.0482\boldsymbol{j} - 0.873\boldsymbol{k}$. This same unit vector defines the direction of $\boldsymbol{M}$.

Check We can check our answer by reviewing the signs on each of the position vector and force components relative to **Figure 5.59** to make sure that we got all of the signs right in our calculations. We can also repeat the calculations to check that we correctly entered the numbers into the calculator. Finally, we review the sins and magnitudes of the scalar components of the moment. Does it make sense that the magnitude of M_z is almost twice that of M_x? Does it make sense that the magnitude of M_y is so much smaller than that of M_z and M_x?

Comment: Earlier in the chapter, it was mentioned that the cross product was not commutative. This means that $\boldsymbol{F} \times \boldsymbol{r} \neq \boldsymbol{r} \times \boldsymbol{F}$. This can be proved by finding $\boldsymbol{F} \times \boldsymbol{r}$ and comparing the answer to $\boldsymbol{r} \times \boldsymbol{F}$. Find $\boldsymbol{F} \times \boldsymbol{r}$:

$$\boldsymbol{F} \times \boldsymbol{r} = \begin{vmatrix} \boldsymbol{i} & \boldsymbol{j} & \boldsymbol{k} \\ F_x & F_y & F_z \\ r_x & r_y & r_z \end{vmatrix} = \begin{vmatrix} \boldsymbol{i} & \boldsymbol{j} & \boldsymbol{k} \\ 73.3\text{ N} & -73.3\text{ N} & -36.7\text{ N} \\ 250\text{ mm} & 200\text{ mm} & -150\text{ mm} \end{vmatrix}$$

Again using the rules of determinants and matrix algebra, we can expand this to yield

$$\begin{aligned} \boldsymbol{F} \times \boldsymbol{r} = {} & [(-73.3\text{ N})(-150\text{ mm}) - (-36.7\text{ N})(200\text{ mm})]\boldsymbol{i} \\ & - [(73.3\text{ N})(-150\text{ mm}) - (-36.7\text{ N})(250\text{ mm})]\boldsymbol{j} \\ & + [(73.3\text{ N})(200\text{ mm}) - (-73.3\text{ N})(250\text{ mm})]\boldsymbol{k} \end{aligned}$$

Carrying out the arithmetic gives us

$$\boldsymbol{F} \times \boldsymbol{r} = 18.3\text{ N}\cdot\text{m}\,\boldsymbol{i} + 1.82\text{ N}\cdot\text{m}\,\boldsymbol{j} + 33.0\text{ N}\cdot\text{m}\,\boldsymbol{k}$$

This is not the same as $\boldsymbol{r} \times \boldsymbol{F}$, but it does show that

$$-(\boldsymbol{F} \times \boldsymbol{r}) = \boldsymbol{r} \times \boldsymbol{F}$$

EXAMPLE 5.11 FINDING THE FORCE TO CREATE A MOMENT (A)

Suppose you need to change a tire. The owner's manual for your car states that the lug nuts were tightened to a moment of 12 N · m at the factory. You seat the tire iron on the tire, as shown in **Figure 5.63*a***. The tire iron is 260 mm long.

(a) Would you pull up or push down on the tire iron to break the factory moment?

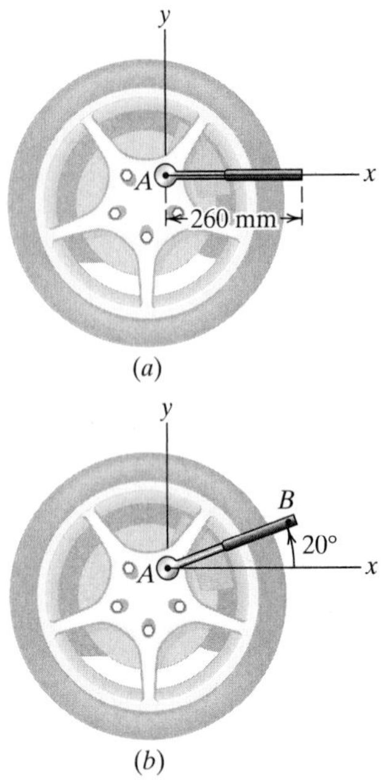

Figure 5.63

(b) How much force is required to break the factory moment on the nut if the tire iron is aligned with the x axis and the force is perpendicular to the tire iron?

(c) How much force is required to break the factory moment on the nut if the tire iron is oriented as shown in **Figure 5.63*b*** and the force is oriented vertically?

(d) Create a plot of the force versus the angle θ, where θ is the angle between the horizontal and the tire iron axis.

Goal We are asked whether we must pull up or push down on the tire iron depicted in **Figure 5.63*a*** to loosen the lug nut. In addition, we are asked to define the magnitude of the force necessary to "break the factory moment" of the nut. What this means is that we need to find the magnitude of a force acting a distance from the center of the lug nut that will create a moment of 12 N · m. We are to consider two orientations of the tire iron, as shown in **Figures 5.63**.

Given We are given that the tire iron is 260 mm long. In addition, we are told that a 12-N · m moment was used to install the lug nuts at the factory. The force $\boldsymbol{F}$ acts vertically at point B in **Figure 5.63*b***. A coordinate system has been established in **Figure 5.63*a*** that places the origin at A, the center of the lug nut. In **Figure 5.63*b*** we note that the tire iron is oriented at an angle of 20° relative to the horizontal.

Assume We assume that the force and tire iron are in the xy plane shown in **Figures 5.63**. We also assume that we are dealing with right-handed threads[9] on the lug nut and bolt. This means that rotating the nut clockwise will result in tightening the nut on the bolt, and rotating the nut counterclockwise will result in loosening the nut on the bolt.[10] This is true for most bolts in the U.S.

Draw In Figure 5.64*a* we sketch the basics for the situation of **Figure 5.63*b***, labeling the angle θ between the horizontal and the tire iron.

Formulate Equations and Solve **(a)** To loosen the nut we need to apply a counterclockwise moment. Therefore based on the physical situation in **Figure 5.63**, the force must act so as to pull up on the end of the tire iron. We add this information as a force arrow in **Figure 5.64*b***.

Answer to (a) In order to loosen the lug nut, the force must be oriented so as to pull up on the end of the tire iron, so that a counterclockwise (loosening) moment is created. The lug nut will therefore be backed off the lug bolt.

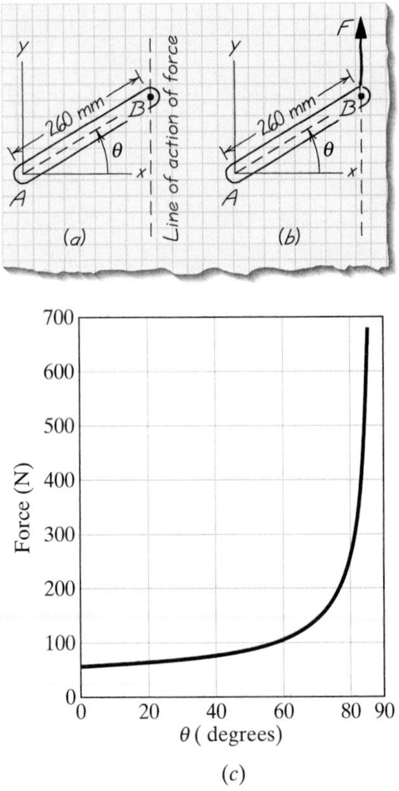

Figure 5.64

We could work parts (b) and (c) separately, *or* we could write all relevant equations in terms of θ and note that in the situation depicted in **Figure 5.63*a***, $\theta = 0°$ and in **Figure 5.63*b***, $\theta = 20°$. We take this latter approach.

[9]Right-handed threaded systems are very common. Left-handed threaded systems are rare and are used only under special conditions.

[10]A memory device for recalling this is "rightie tightie, leftie loosie."

We define a position vector from A to B.

$$\boldsymbol{r} = r_x\boldsymbol{i} + r_y\boldsymbol{j} = (260 \text{ mm}) \cos \theta\boldsymbol{i} + (260 \text{ mm}) \sin \theta\boldsymbol{j}$$

where θ is 0° for the configuration in **Figure 5.63*a*** and 20° for the configuration in **Figure 5.63*b***. The force is given as $\boldsymbol{F} = F_y\boldsymbol{j}$.

Since the position vector and the force are both in the xy plane, we can use (5.11A) to find the moment.

$$\boldsymbol{M}_z = (+r_xF_y - r_yF_x)\boldsymbol{k} \tag{5.11A}$$

Since we want to loosen the nut, the moment $\boldsymbol{M}_z$ must be acting in the positive (out of the paper) direction; therefore $M_z = +12 \text{ N}\cdot\text{m} = +12{,}000 \text{ N} \cdot \text{mm}$. Substituting this value and what we know about $\boldsymbol{r}$ into (5.11A), we get

$$+12{,}000 \text{ N} \cdot \text{mm}\,\boldsymbol{k} = [(260 \text{ mm})(\cos \theta)F_y - (260 \text{ mm})(\sin \theta)0]\boldsymbol{k}$$

Solving this for F_y gives us

$$F_y = \frac{12{,}000 \text{ N} \cdot \text{mm}}{(260 \text{ mm}) \cos \theta} \tag{1}$$

(b) With the configuration in **Figure 5.63*a***, θ is 0°. Substituting this into (1), we get

$$F_y = 46.2 \text{ N}$$

Answer to (b) $\boldsymbol{F} = F_y\boldsymbol{j} = 46.2 \text{ N}\,\boldsymbol{j}$

(c) With the configuration in **Figure 5.63*b***, θ is 20°. Substituting this into (1), we get

$$F_y = 49.1 \text{ N}$$

Answer to (c) $\boldsymbol{F} = F_y\boldsymbol{j} = 49.1 \text{ N}\,\boldsymbol{j}$

(d) We are to create a plot of required force versus angle θ.

Answer to (d) A plot of (1) is presented in **Figure 5.64c**.

Check We would expect the required force for breaking the factory moment to be the least when the force is perpendicular to the axis of the tire wrench ($\theta = 0°$); **Figure 5.64c** shows this to be the case. This figure also shows that as the axis of the tire iron approaches the vertical, the required force becomes very large—does this make sense? Why didn't we consider angles greater than 90° in creating **Figure 5.64**? As a final check, consider whether it is reasonable to expect someone to be able to apply somewhere between 40 and 50 N of force to loosen a lug nut.

EXAMPLE 5.12 FINDING THE FORCE TO CREATE A MOMENT (B)

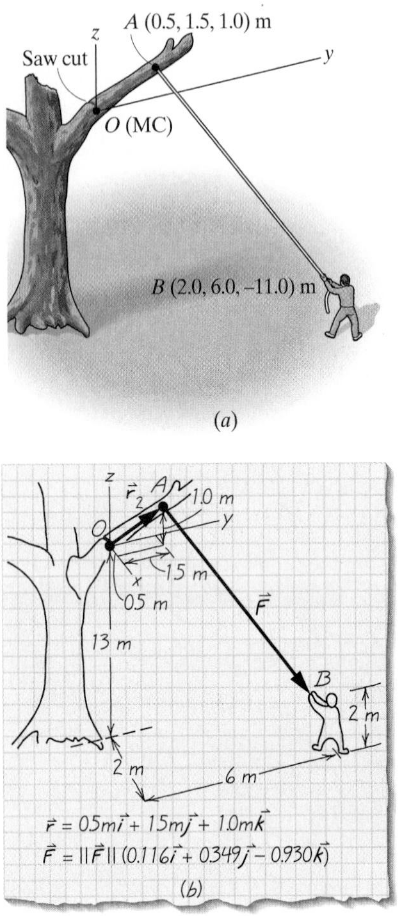

Figure 5.65

Consider someone trying to break off the tree branch from **Figure 5.65*a***.

Assuming that the magnitude of the moment necessary to break the branch is 300 N · m, what force is required to break the branch?

Goal We are to find the magnitude of force necessary to break the tree branch.

Given We are told that the magnitude of moment necessary to break the tree branch is 30 N · m. We are given the direction of the force, its point of application, and relevant dimensions.

Assume No assumptions are necessary.

Draw We draw **Figure 5.65*b***.

Formulate Equations and Solve First we write the force in terms of its rectangular components, and find the angle between the position vector and the force using (4.27). Then, using (5.1) and the required magnitude of moment, we can solve for the required force.

From **Figure 5.5** earlier in the chapter, we know that the applied force is along a unit vector of

$$\boldsymbol{u}_F = 0.116\boldsymbol{i} + 0.349\boldsymbol{j} - 0.930\boldsymbol{k}$$

Therefore, the force can be written as

$$\boldsymbol{F} = \| \boldsymbol{F} \| ((0.116\boldsymbol{i}) + (0.349\boldsymbol{j}) - (0.930\boldsymbol{k}))$$

We also need to define a position vector. Based on the information in **Figure 5.65**, we will use a position vector that runs from O to A. From **Figure 5.65*b***, we know that

$$\boldsymbol{r} = 0.5\text{ m }\boldsymbol{i} + 1.5\text{ m }\boldsymbol{j} + 1.0\text{ m }\boldsymbol{k}$$

The magnitude of the position vector is 1.87 m.

Now we can use (4.27) to determine the angle between $\boldsymbol{F}$ and $\boldsymbol{r}$:

$$\theta = \cos^{-1}\left[\frac{\| \boldsymbol{F} \| ((0.116)(0.5\text{ m}) + (0.349)(1.5\text{ m}) + (-0.930)(+1.0\text{ m}))}{\| \boldsymbol{F} \| (1.87\text{ m})}\right]$$
$$= 100.7^\circ$$

Given that the magnitude of the moment is 30 N · m, we can use (5.1) to determine the magnitude of the force:

$$\| \boldsymbol{M} \| = 300\text{ N}\cdot\text{m} = \| \boldsymbol{r} \| \; \| \boldsymbol{F} \| \sin\theta = (1.87\text{ m}) \| \boldsymbol{F} \| \sin 100.7^\circ$$

Solving for $\| \boldsymbol{F} \|$ we have:

Answer $\| \boldsymbol{F} \| = 163\text{ N}$

Check To check the answer we could use an alternative approach. For example, we could use (5.7) to find the moment based on the force and position vectors given above. We would then equate this expression to 300 N · m to find that $\| \boldsymbol{F} \| = 163$ N.

Comment: We would have found the same answer if we had used the other position vector, for example $\boldsymbol{r}_1$ in **Figure 5.5*b*** Is 163 N a small or large force? Would you be able to apply it to break off the branch? How should $\boldsymbol{F}$ be oriented so as to minimize its required magnitude?

EXERCISES 5.2

5.2.1. For each of two situations shown in **E5.2.1**, calculate the moment the force creates at the moment center (MC) indicated in the figure.

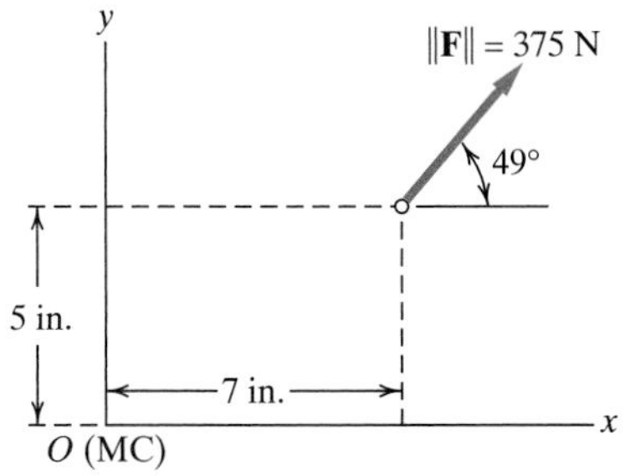

(*a*)

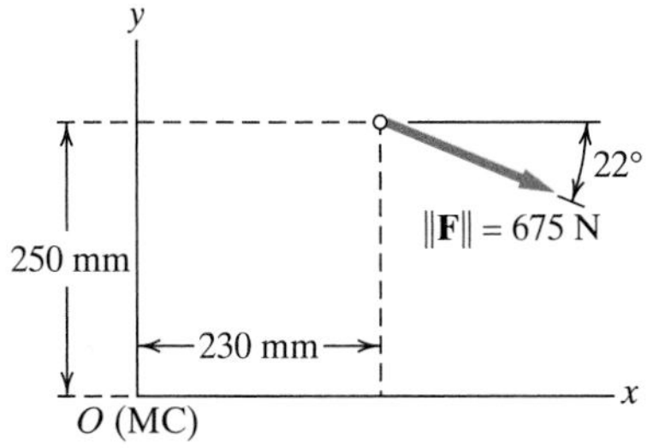

(*b*) **E5.2.1**

5.2.2. For each of two situations shown in **E5.2.2**, calculate the moment the force creates at the moment center (MC) indicated in the figure.

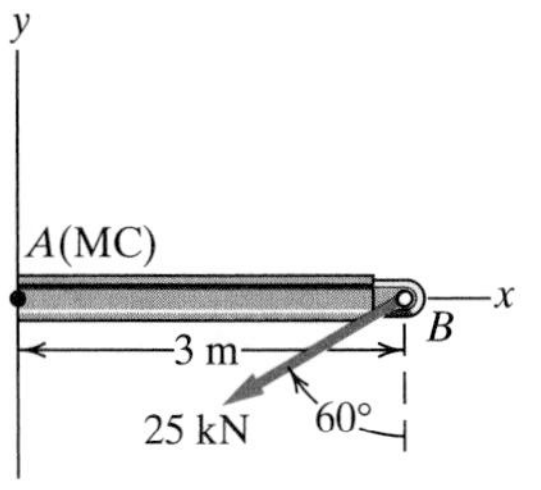

(*a*)

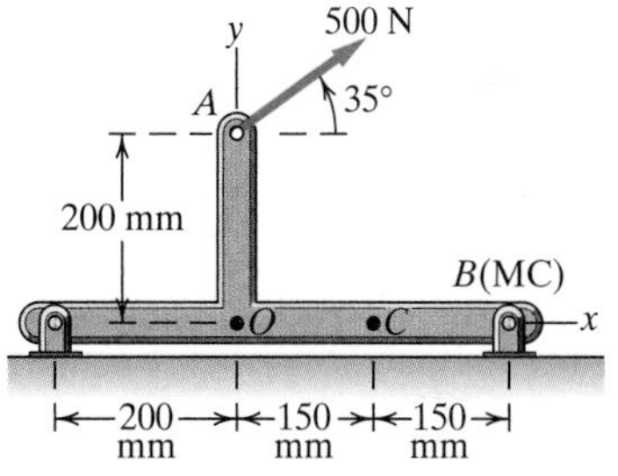

(*b*) **E5.2.2**

Exercises 5.2.3–5.2.6 may require use of (5.11A), (5.11B), or (5.11C).

5.2.3. For **E5.2.3**, consider $\boldsymbol{F}_1$ and moment center O. Write

a. a symbolic expression for the force and position vector

b. an expression for the moment created by the force at the moment center

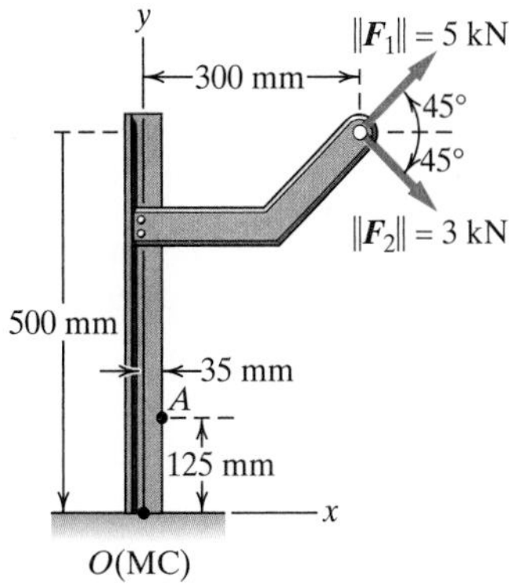

E5.2.3

5.2.4. For **E5.2.3**, consider $\boldsymbol{F}_2$ and moment center O. Write

a. a symbolic expression for the force and position vector

b. an expression for the moment created by the force at the moment center

5.2.5. For E5.2.5, consider $\boldsymbol{F}_1$ and moment center A. Write

a. a symbolic expression for the force and position vector

b. an expression for the moment created by the force at the moment center

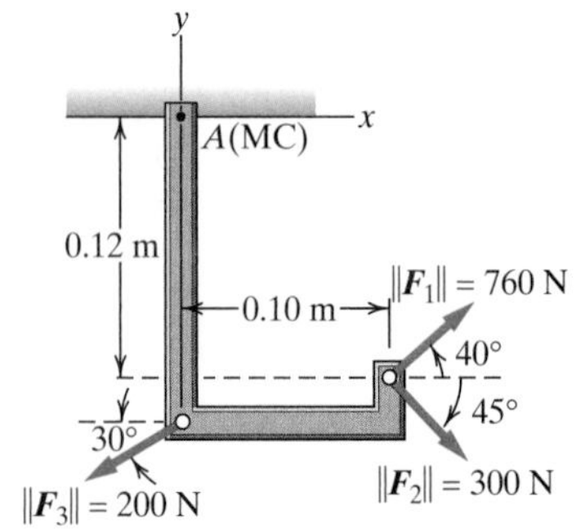

E5.2.5

5.2.6. For E5.2.6, consider $\boldsymbol{F}_A$ and moment center B. Write

a. a symbolic expression for the force and position vector

b. an expression for the moment created by the force at the moment center

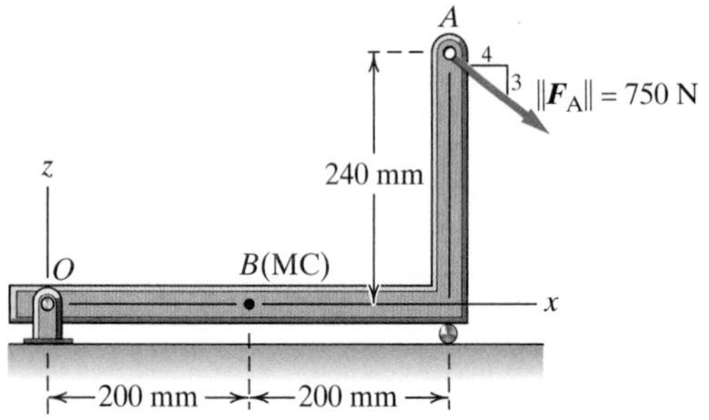

E5.2.6

5.2.7. A 5-kN force is applied to the end of an I-beam, as shown in E5.2.7.

a. Use the definition of $\| \boldsymbol{M} \|$ in (5.1) to determine the magnitude of the moment at A created by the force. About what axis is the moment?

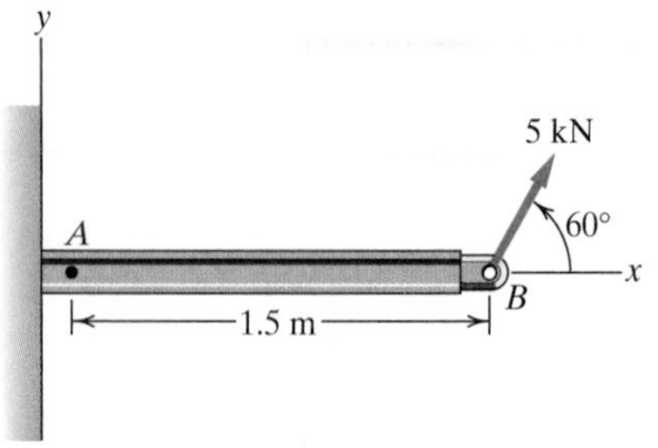

E5.2.7

b. Determine the moment $\boldsymbol{M}$ at A created by the force by using the appropriate expression from (5.11A), (5.11B), or (5.11C). Confirm that the magnitude of this moment is the same as found in **a**.

5.2.8. A 1000-N force acts on a bracket (E5.2.8).

a. Speculate on whether the force creates a greater moment about a moment center at B or a moment center at C.

b. Calculate the moment the force creates at a moment center at B.

c. Calculate the moment the force creates at a moment center at C.

d. Compare your answers in **b** and **c** with your speculation in **a**.

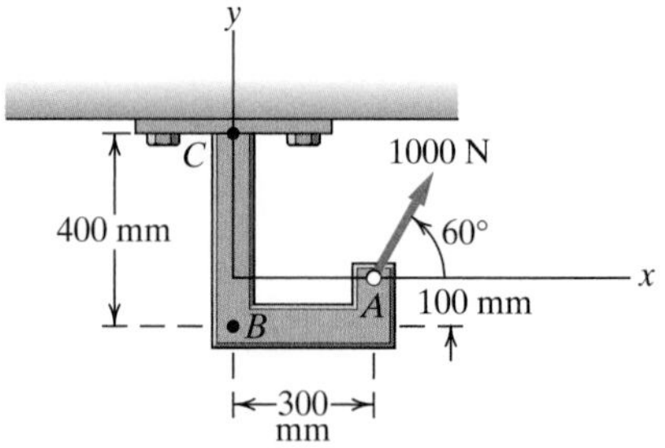

E5.2.8

5.2.9. A force with a magnitude of 750 N acts as shown in E5.2.9.

a. Speculate on where the force creates the largest moment (in terms of magnitude): at a moment center at A, at B, at C, or at D.

b. Calculate the moment the force creates at a moment center at A.

c. Calculate the moment the force creates at a moment center at B.

d. Calculate the moment the force creates at a moment center at C.

e. Calculate the moment the force creates at a moment center at D.

f. Compare your answers in **b–e** with your speculation in **a**.

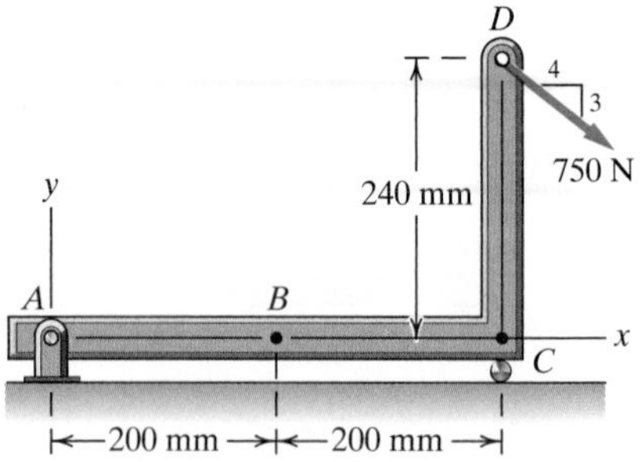

E5.2.9

5.2.10. A force with a magnitude of 580 N acts on a bracket as shown in E5.2.10. Determine the moment the force creates at a moment center at

a. point *B*
b. point *C*

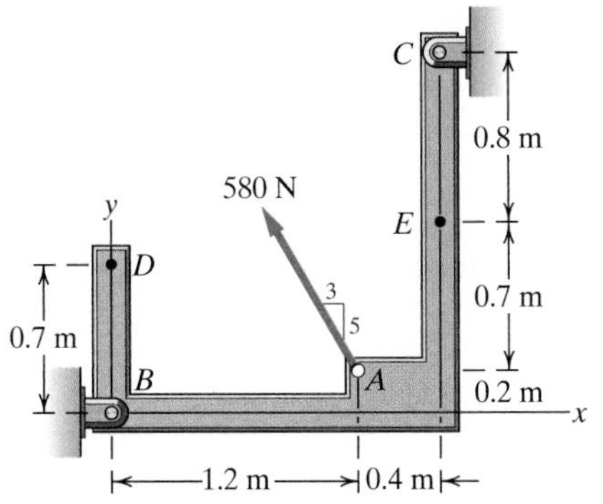

E5.2.10

The figures for **Exercises 5.2.11–5.2.14** show forces applied to systems. As with Example 5.9, write, for each situation, symbolic expressions for the force and position vector, then write an expression for the moment created by the force at the indicated moment center. Equation (5.7) may be useful.

5.2.11. For E5.2.11, consider $\boldsymbol{F}$ and a moment center at *A*.

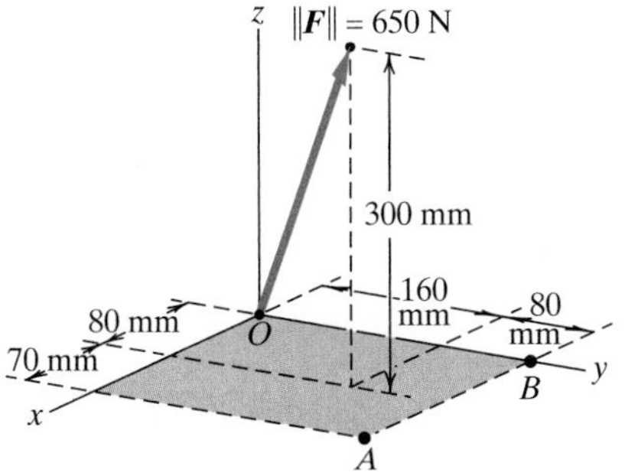

E5.2.11

5.2.12. For E5.2.12, consider $\boldsymbol{F}_1$ and a moment center at *A*.

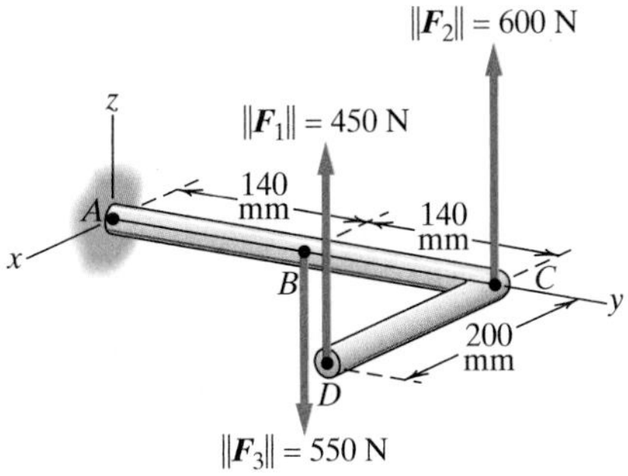

E5.2.12

5.2.13. For **E5.2.12**, consider $\boldsymbol{F}_2$ and a moment center at *A*.

5.2.14. For **E5.2.12**, consider $\boldsymbol{F}_3$ and a moment center at *A*.

5.2.15. A traffic light is steadied by cable *AB* (E5.2.15). If the tension in this cable is 4 kN, determine

a. the moment $\boldsymbol{M}$ the cable tension creates at *C*, expressed in vector notation

b. the moment $\boldsymbol{M}$ the cable tension creates at *C*, expressed in terms of a magnitude and a unit vector $\boldsymbol{u}$

c. the scalar component of the moment about the long axis of the pole, *CD*

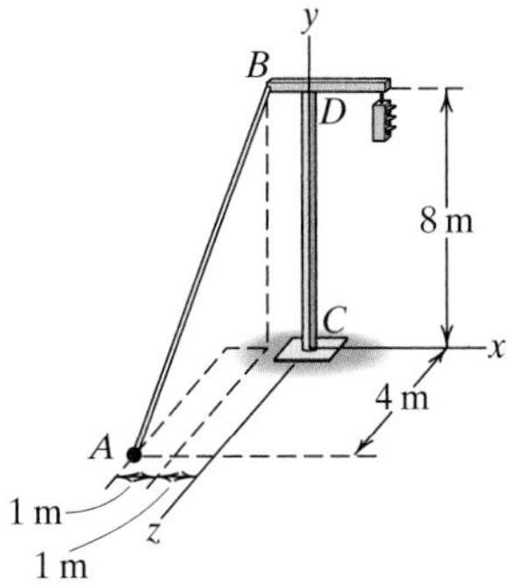

E5.2.15

5.2.16. The point of application of a force of magnitude 860 N is at *A* in E5.2.16. Determine the moment $\boldsymbol{M}_D$ the force creates at *D*. Express your answer in vector notation. In addition, write a unit vector that defines the axis of $\boldsymbol{M}_D$.

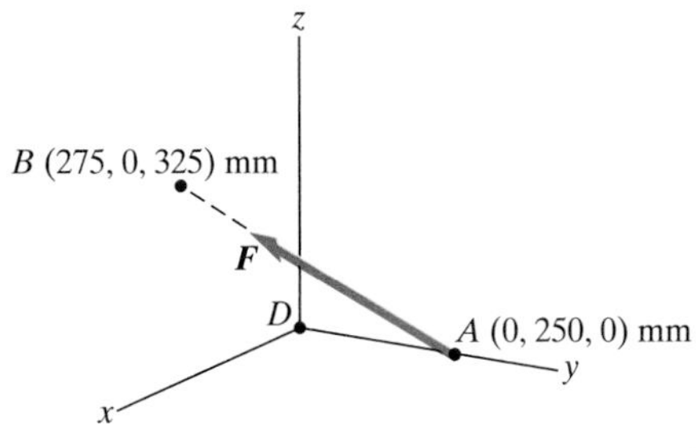

E5.2.16

5.2.17. For the situation shown in **E5.2.11**, determine

a. the moment $\boldsymbol{M}$ the force creates at *B*, expressed in vector notation

b. the moment $\boldsymbol{M}$ the force creates at *B*, expressed in terms of magnitude and a unit vector $\boldsymbol{u}$

c. the space angles associated with the moment vector

5.2.18. A 610-N force acts on a lever attached to a post as shown in E5.2.18. Determine

a. the moment $\boldsymbol{M}_O$ the force creates at a moment center at O

b. the space angles associated with the moment vector

c. the unit vector that defines the axis of $\boldsymbol{M}_O$

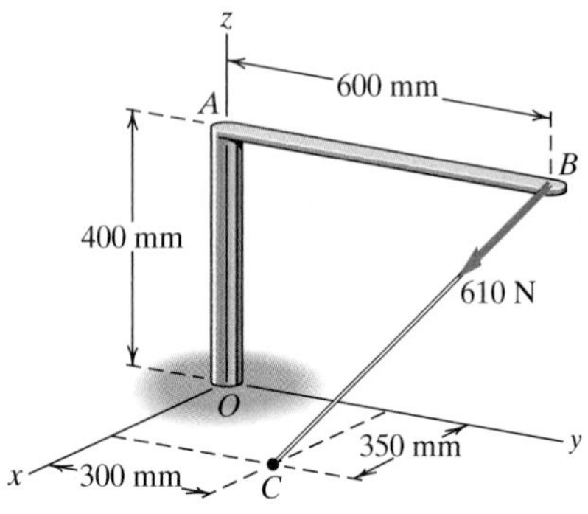

E5.2.18

5.2.19. Reconsider the force in **E5.2.18**.

a. Determine the moment $\boldsymbol{M}_A$ the force creates at a moment center at A.

b. Which moment ($\boldsymbol{M}_O$ from **Exercise 5.2.18** or $\boldsymbol{M}_A$) has the greater magnitude?

5.2.20. A portion of a mechanical coin sorter is shown in E5.2.20. Dimes and pennies roll down the 20° incline, the last portion of which pivots freely about a horizontal axis through O. Dimes are light enough (2.28 g) so that the triangular portion remains stationary, and the dimes roll into the right collection column. Pennies, on the other hand, are heavy enough (3.06 g) so that the triangular portion pivots clockwise, and the pennies roll into the left collection column. Write an expression for the moment created by a penny as it rolls down the incline. The expression should be written in terms of the slant distance s.

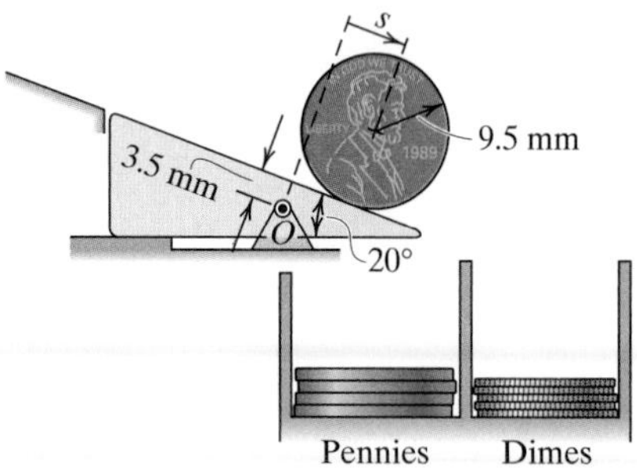

E5.2.20

5.2.21. The throttle-control sector in E5.2.21 pivots freely at O. An internal torsional spring exerts a return moment of magnitude 2 N · m when the sector is in the position shown. What is the necessary throttle-cable tension T so that the moment about O due to the spring moment and the moment caused by the cable tension is zero?

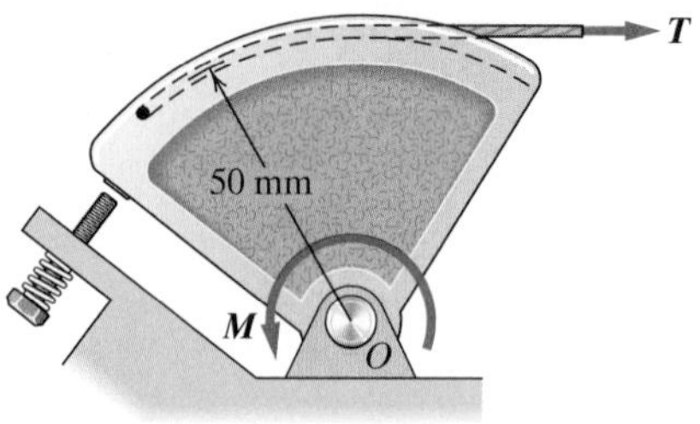

E5.2.21

Note: In working the following problems, remember that moments are vectors. Therefore, they can be added to one another using the vector addition rules used with forces in Chapter 4.

5.2.22. Consider the frame in **E5.2.3**. Determine

a. the moment $\boldsymbol{M}_1$ created by $\boldsymbol{F}_1$ at a moment center at O

b. the moment $\boldsymbol{M}_2$ created by $\boldsymbol{F}_2$ at a moment center at O

c. the sum of $\boldsymbol{M}_1$ and $\boldsymbol{M}_2$

5.2.23. Consider the frame in **E5.2.5**. Determine

a. the moment $\boldsymbol{M}_1$ created by $\boldsymbol{F}_1$ at a moment center at A

b. the moment $\boldsymbol{M}_2$ created by $\boldsymbol{F}_2$ at a moment center at A

c. the moment $\boldsymbol{M}_3$ created by $\boldsymbol{F}_3$ at a moment center at A

d. the sum $\boldsymbol{M}_1 + \boldsymbol{M}_2 + \boldsymbol{M}_2$

5.2.24. Consider the bracket shown in E5.2.24. Determine

a. the moment $\boldsymbol{M}_D$ created by $\boldsymbol{F}_D$ at a moment center at O

b. the moment $\boldsymbol{M}_A$ created by $\boldsymbol{F}_A$ at a moment center at O

c. the sum of $\boldsymbol{M}_D$ and $\boldsymbol{M}_A$

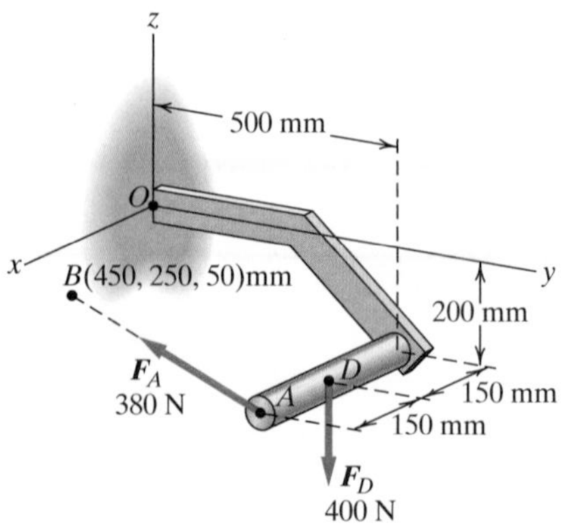

E5.2.24

5.2.25. In **E5.2.25**, $\boldsymbol{F}_2$ is parallel to and opposite $\boldsymbol{F}_1$. Determine

a. the moment $\boldsymbol{M}_1$ created by $\boldsymbol{F}_1$ at a moment center at A

b. the moment $\boldsymbol{M}_2$ created by $\boldsymbol{F}_2$ at a moment center at A

c. the sum of $\boldsymbol{M}_1$ and $\boldsymbol{M}_2$

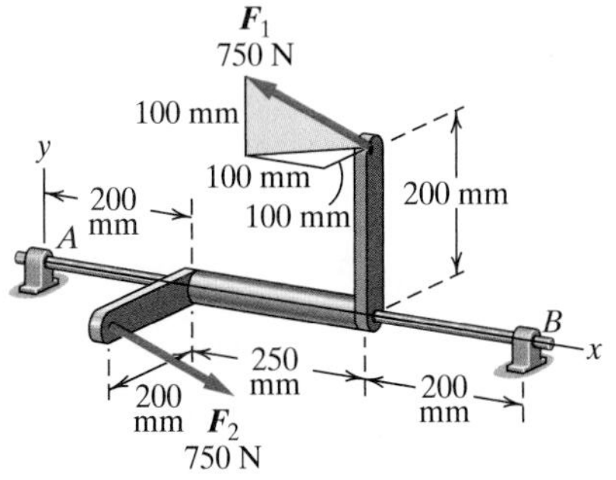

E5.2.25

5.2.26. The 25-N forces shown in **E5.2.26*a*** are applied to a bus's steering wheel. A side view of the wheel is shown in **E5.2.26*b***. Find

a. the moment created by force $\boldsymbol{F}_A$ at a moment center at O

b. the moment created by force $\boldsymbol{F}_B$ at a moment center at O

c. the sum of the moments found in **a** and **b**

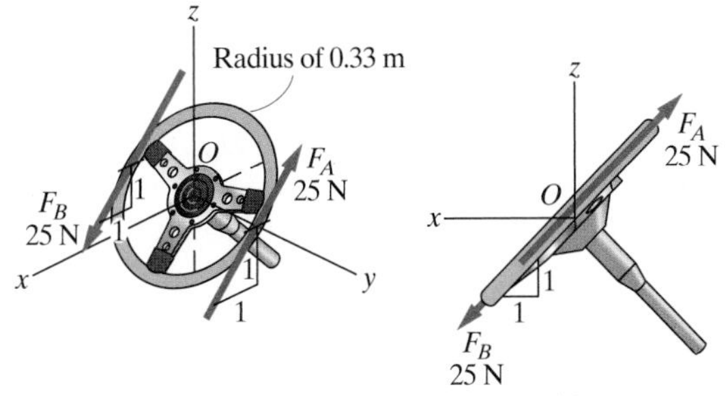

(*a*) (*b*)

E5.2.26

5.2.27. Repeat **Exercise 5.2.8** using the cross product.

5.2.28. Repeat **Exercise 5.2.9** using the cross product.

5.2.29. Repeat **Exercise 5.2.10** using the cross product.

5.2.30. A 5-kN force acts on the bent arm at point A in **E5.2.30**.

a. Use the cross product to write an expression for the moment $\boldsymbol{M}$ that the force creates at a moment center at C. This expression will contain the angle θ.

b. Determine the angle θ such that the force at A exerts the maximum moment at C.

c. What is the magnitude of the corresponding moment? Also write the corresponding moment in vector notation.

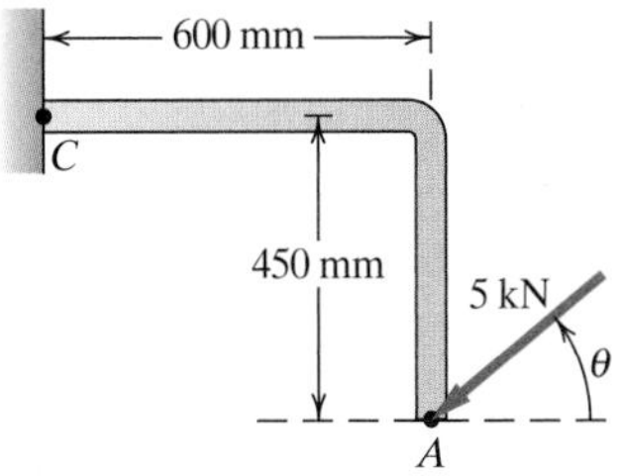

E5.2.30

5.2.31. Repeat **Exercise 5.2.18** using the cross product.

5.2.32. Use the cross product to determine the moment at O created by the 800-N force acting at C in **E5.2.32**.

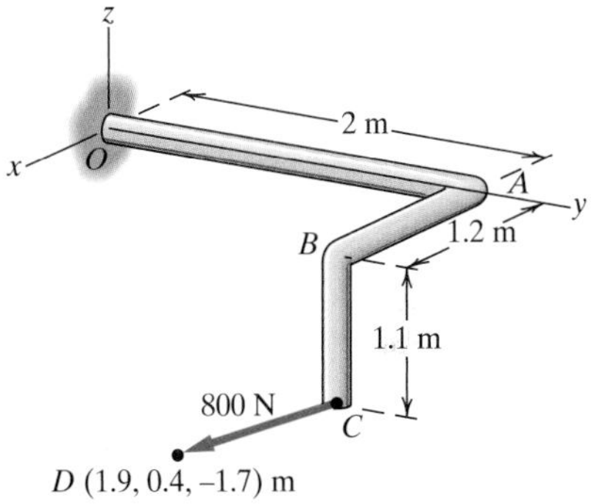

E5.2.32

5.2.33. Repeat **Exercise 5.2.24** using the cross product.

5.2.34. Repeat **Exercise 5.2.25** using the cross product.

5.2.35. A 10-kN force acts at point A in **E5.2.35**.

a. Use the cross product to determine the moment the 10-kN force creates at a moment center at C.

b. Find the space angles associated with the moment vector in **a**.

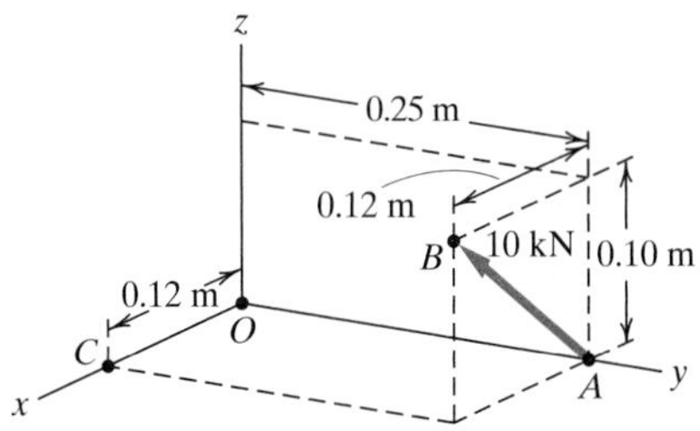

E5.2.35

5.2.36. The magnitude of the force ***F*** in E5.2.36 is 20 N.

a. Use the cross product to determine the moment the 20-N force creates at a moment center at *C*.

b. What are the space angles associated with the moment?

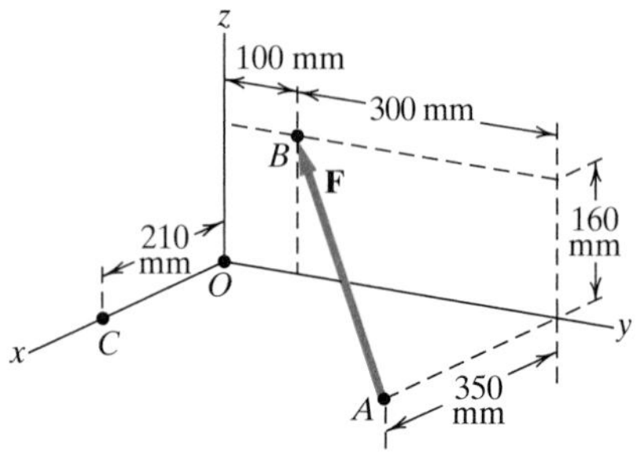

E5.2.36

5.2.37. A socket wrench is used to loosen a screw. A 120-N horizontal force is applied to the handle as shown in E5.2.37. Determine the moment this force creates at a moment center at *A* (the center of the head of the screw).

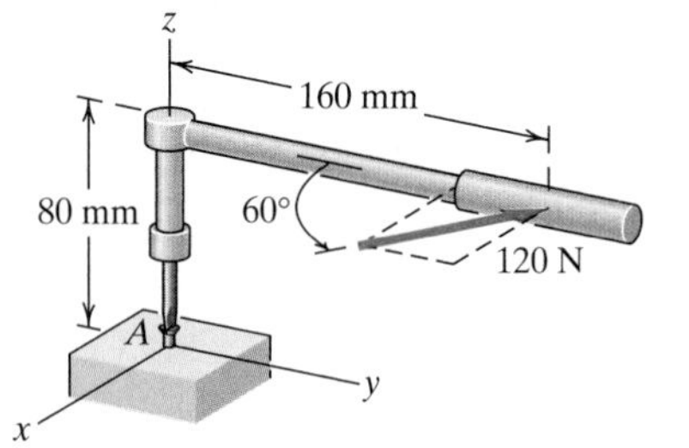

E5.2.37

5.2.38. A 500-N force acts at the end of a cable-supported bar as shown in E5.2.38.

a. Use the cross product to determine the moment the 500-N force creates at a moment center at *A*. Write the moment in vector notation based on the *xyz* coordinate axes.

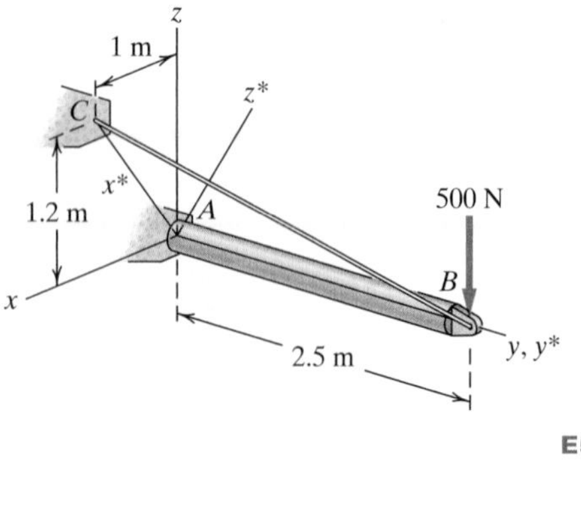

E5.2.38

b. Use the cross product to determine the moment the 500-N force creates at a moment center at *A*. Write the moment in vector notation based on the $x^*y^*z^*$ coordinate axes.

c. Confirm that the moments found in **a** and **b** have the same magnitude.

Use E5.2.39 for **Exercises 5.2.39–5.2.41**.

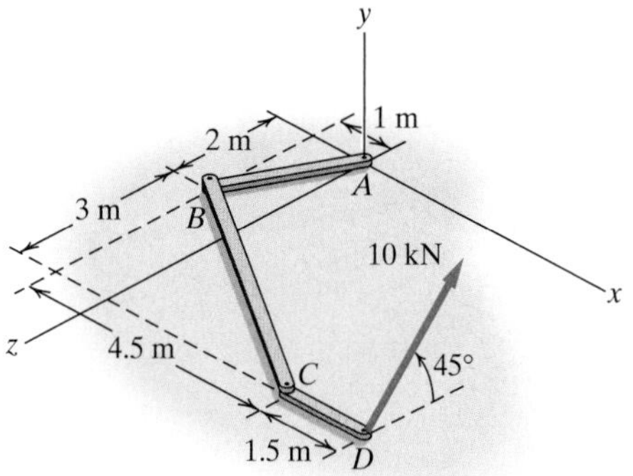

E5.2.39

5.2.39. A 10-kN force acts at point *D* in the *yz* plane on the linkage. Use the cross product to compute the moment this force creates at a moment center at *A*.

5.2.40. A 10-kN force acts at point *D* in the *yz* plane on the linkage. Use the cross product to compute the moment this force creates at a moment center at *B*.

5.2.41. A 10-kN force acts at point *D* in the *yz* plane on the linkage. Use the cross product to compute the moment this force creates at a moment center at *C*.

5.2.42. A 2-kN force acts on one end of the curved rod in E5.2.42. Section *AB* of the rod lies in the *xy* plane, and section *BC* lies in the *zy* plane. Use the cross product to determine the moment created by the force

a. at a moment center at *A*

b. at a moment center at the midpoint *B* of the rod

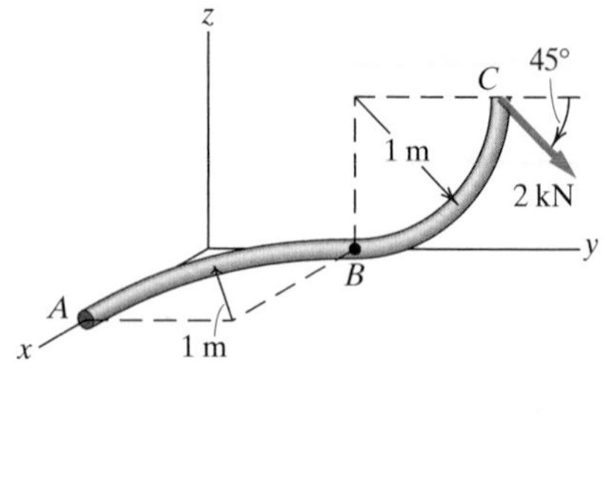

E5.2.42

5.2.43. A cable and weight act on a beam as shown in E5.2.43*a*. They can be replaced by the forces $\boldsymbol{F}$ and $\boldsymbol{W}$ (see **E5.2.43*b***).

a. Determine the moment $\boldsymbol{M}_1$ that $\boldsymbol{W}$ creates at A.
b. Determine the moment $\boldsymbol{M}_2$ that $\boldsymbol{F}$ creates at A.
c. Determine the sum of $\boldsymbol{M}_1$ and $\boldsymbol{M}_2$.
d. If the magnitude of $\boldsymbol{M}_1 + \boldsymbol{M}_2$ is limited to be less than or equal to 1 kN · m, what is the maximum allowable value of $\boldsymbol{W}$?

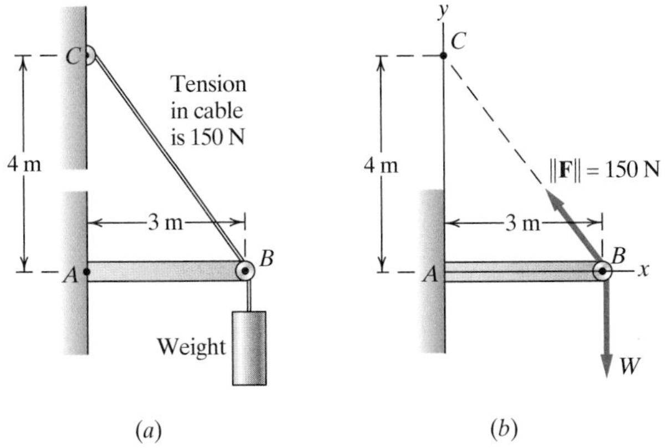

E5.2.43

5.2.44. The forces $\boldsymbol{F}_1$ and $\boldsymbol{F}_2$ act on a cantilever beam as shown in E5.2.44.

a. Determine the moment $\boldsymbol{M}_1$ that $\boldsymbol{F}_1$ creates at O.
b. Determine the moment $\boldsymbol{M}_2$ that $\boldsymbol{F}_2$ creates at O.
c. Determine the sum of $\boldsymbol{M}_1$ and $\boldsymbol{M}_2$.
d. If the magnitude of $\boldsymbol{M}_1 + \boldsymbol{M}_2$ is limited to be less than or equal to 100 kN · m, what is the maximum allowable magnitude of $\boldsymbol{F}_2$ given that the magnitude of $\boldsymbol{F}_1$ is fixed at 20 kN?

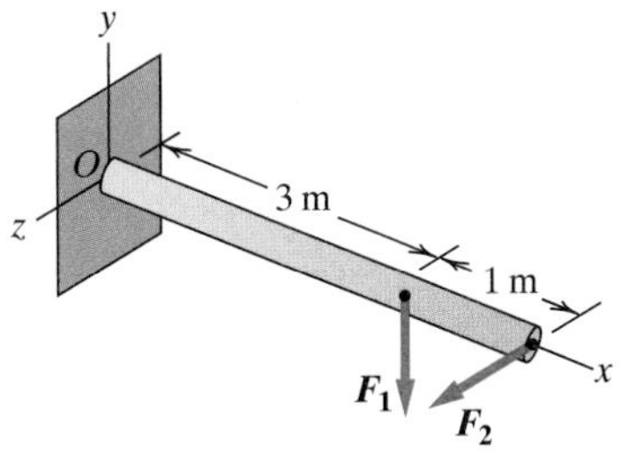

E5.2.44

5.2.45. Repeat **Exercise 5.2.26** using the cross product.

5.2.46. Two forces act on the handles of the pipe wrenches in E5.2.46. Use the cross product to determine

a. the moment created by $\boldsymbol{F}_A$ at O
b. the moment created by $\boldsymbol{F}_B$ at O
c. the sum of the moments found in **a** and **b**

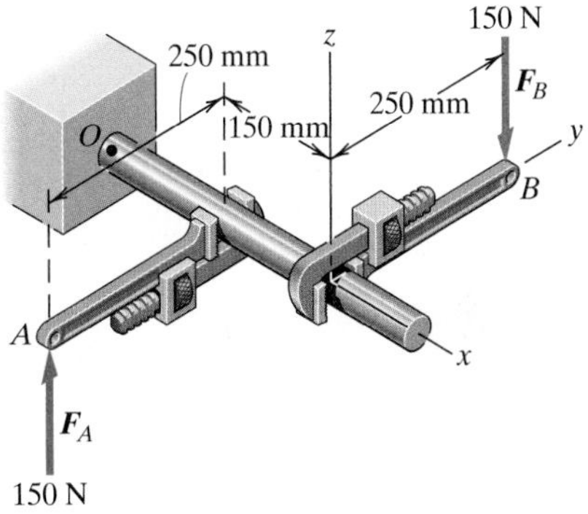

E5.2.46

5.3 FINDING MOMENT COMPONENT IN A PARTICULAR DIRECTION

In the previous section we presented relationships for calculating the moment created by a force offset from a moment center (MC). For example, we could use either (5.7) or (5.12) to find the moment $\boldsymbol{M}$ about a MC at A created by $\boldsymbol{F}_{\text{cable}}$ in Figure 5.66*a*.

Now let's say that we are interested in finding the component of $\boldsymbol{M}$ in the direction of the axis AB of the pole in **Figure 5.66*a*** because if the component becomes too large, the pole may be twisted out of its concrete mount. The component of $\boldsymbol{M}$ in the direction of AB is simply the dot product of $\boldsymbol{M}$ and $\boldsymbol{m}$, where $\boldsymbol{m}$ is a unit vector along AB (**Figure 5.66*b***). Calling the component of interest M_{AB}, we can write it as

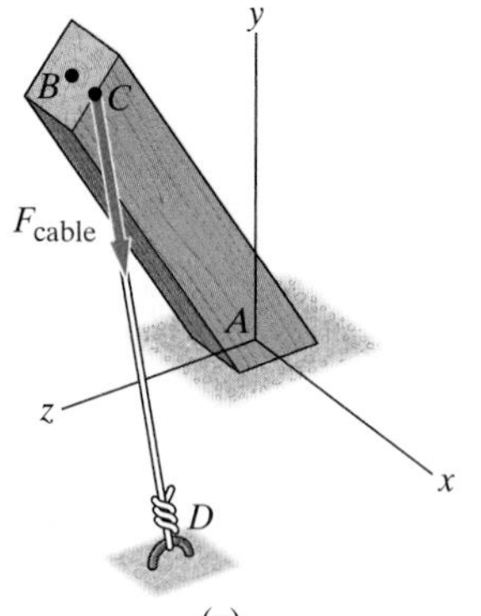

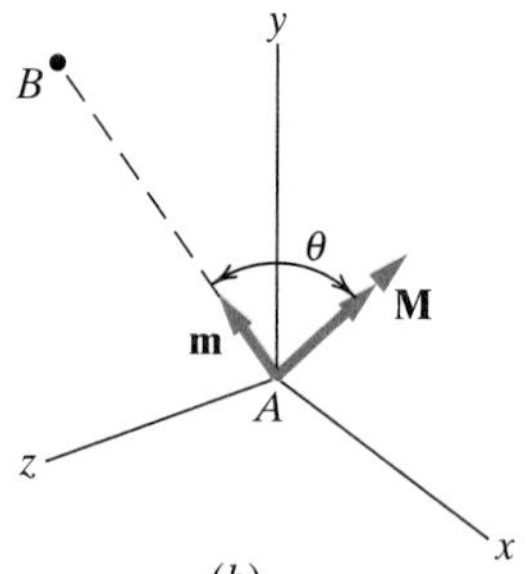

Figure 5.66 (*a*) Cable force creates a moment at A; (*b*) finding the component of that moment aligned with the pole

$$M_{AB} = \underbrace{\boldsymbol{M} \cdot \boldsymbol{m}}_{\substack{\text{projection of} \\ \boldsymbol{M} \text{ in the direction} \\ \text{of unit vector } \boldsymbol{m}}} = \| \boldsymbol{M} \| \; \| \boldsymbol{m} \| \cos\theta \tag{5.13}$$

where θ is the angle between $\boldsymbol{M}$ and $\boldsymbol{m}$, as indicated in **Figure 5.66*b***. The dot product, which enables us to find the component or projection of a vector in a particular direction, was introduced in Chapter 4.

If we can write $\boldsymbol{M}$ and $\boldsymbol{m}$ in terms of their scalar components, the dot product in (5.13) can be restated as

$$M_{AB} = \underbrace{\boldsymbol{M} \cdot \boldsymbol{m}}_{\substack{\text{projection of} \\ \boldsymbol{M} \text{ in the direction} \\ \text{of unit vector } \boldsymbol{m}}} = M_x m_x + M_y m_y + M_z m_z \tag{5.14}$$

and the angle θ, based on the development of the dot product in Chapter 4, is given as

$$\cos\theta = \frac{M_x m_x + M_y m_y + M_z m_z}{\| \boldsymbol{M} \| \; \| \boldsymbol{m} \|}, \qquad \text{with } 0° \le \theta \le 180° \tag{5.15}$$

The choice of using (5.13) or (5.14) to find the component of $\boldsymbol{M}$ in the direction of $\boldsymbol{m}$ will generally be a matter of convenience and/or personal preference.

EXAMPLE 5.13 USING THE DOT PRODUCT TO FIND THE MOMENT IN A PARTICULAR DIRECTION

Consider the moment $\boldsymbol{M} = 400 \text{ in.} \cdot \text{lb } \boldsymbol{i} + 250 \text{ in.} \cdot \text{lb } \boldsymbol{j} - 500 \text{ in.} \cdot \text{lb } \boldsymbol{k}$, as shown in **Figure 5.67*a***,

(a) Find the component of $\boldsymbol{M}$ in the direction defined by the unit vector $\boldsymbol{m} = 0.727\boldsymbol{i} + 0.364\boldsymbol{j} + 0.582\boldsymbol{k}$.

(b) Write the component vector of $\boldsymbol{M}$ in the direction of $\boldsymbol{m}$.

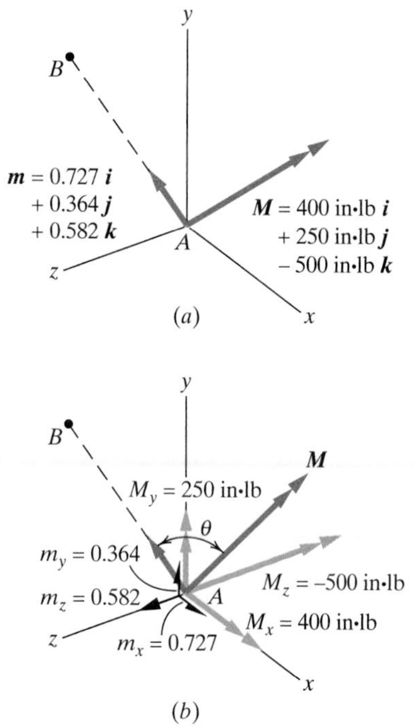

Figure 5.67

Goal We are to find the component of $\boldsymbol{M}$ in the direction specified by $\boldsymbol{m}$. Another way of saying this is that we are to find the projection of $\boldsymbol{M}$ in the direction of $\boldsymbol{m}$.

Given We are given $\boldsymbol{M}$ and $\boldsymbol{m}$ in vector notation.

Assume No assumptions are necessary.

Draw We redraw **Figure 5.67*a*** to show the angle between $\boldsymbol{M}$ and $\boldsymbol{m}$, and the components of these two vectors (**Figure 5.67*b***).

Formulate Equations and Solve **(a)** The dot product ($\boldsymbol{M} \cdot \boldsymbol{m}$) will give us the component of $\boldsymbol{M}$ in the direction of $\boldsymbol{m}$, multiplied by the magnitude of $\boldsymbol{m}$ (which is simply 1, since $\boldsymbol{m}$ is a unit vector).

Approach 1: Using (5.3)

(1) Find the magnitude of ***M*** using (5.3):

$$\| \boldsymbol{M} \| = \sqrt{M_x^2 + M_y^2 + M_z^2}$$
$$= \sqrt{(400 \text{ in.} \cdot \text{lb})^2 + (250 \text{ in.} \cdot \text{lb})^2 + (-500 \text{ in.} \cdot \text{lb})^2} = 687 \text{ in.} \cdot \text{lb} \quad (5.3)$$

The magnitude of ***m*** is unity (1).

(2) Find the angle θ between ***m*** and ***M*** when the two vectors are placed tail to tail, using (4.25):

$$\cos\theta = \frac{V_{1x}V_{2x} + V_{1y}V_{2y} + V_{1z}V_{2z}}{\| \boldsymbol{V} \| \; \| \boldsymbol{V}_2 \|}$$
$$= \frac{(400 \text{ in.} \cdot \text{lb}(0.727) + 250 \text{ in.} \cdot \text{lb}(0.364) - 500 \text{ in.} \cdot \text{lb}(0.582)}{687 \text{ in.} \cdot \text{lb}(1)} \quad (4.25)$$

from which we find $\theta = 82.4°$.

(3) Use the definition of the dot product given in (4.20) to find the component:

$$\boldsymbol{V}_2 \cdot \boldsymbol{V}_1 = \| \boldsymbol{V}_1 \| \; \| \boldsymbol{V}_2 \| \cos\theta$$
$$\boldsymbol{M} \cdot \boldsymbol{m} = (687 \text{ in.} \cdot \text{lb})(1) \cos 82.4° = 90.9 \text{ in.} \cdot \text{lb} \quad (4.20)$$

Approach 2: Since we are given ***M*** and ***m*** in terms of their components, a method more straightforward than Approach 1 is to find the dot product using the definition given in (4.23):

$$\boldsymbol{V}_1 \cdot \boldsymbol{V}_2 = V_{1x}V_{2x} + V_{1y}V_{2y} + V_{1z}V_{2z} \quad (4.23)$$

Substituting the components of $\boldsymbol{M}(M_x = 400$ in. · lb, $M_y = 250$ in. · lb, $M_z = -500$ in. · lb) and the components of ***m*** ($m_x = 0.727$, $m_y = 0.364$, $m_z = 0.582$) into (4.23) we find that $\boldsymbol{m} \cdot \boldsymbol{M} = 90.9$ in. · lb.

Answer (a) The component of ***M*** in the direction of the unit vector ***m*** is $\boldsymbol{m} \cdot \boldsymbol{M} = 90.9$ in. · lb.

(b) The component vector of ***M*** in the direction of ***m*** is

$$\boldsymbol{M}_{\text{in direction of } m} = 90.9 \text{ in.} \cdot \text{lb}(0.727\boldsymbol{i} + 0.364\boldsymbol{j} + 0.582\boldsymbol{k})$$
$$= 66.1 \text{ in.} \cdot \text{lb } \boldsymbol{i} + 33.1 \text{ in.} \cdot \text{lb } \boldsymbol{j} + 52.9 \text{ in.} \cdot \text{lb } \boldsymbol{k}$$

Answer (b) The component vector of ***M*** in the direction of the unit vector ***m*** is $\boldsymbol{M}_{\text{in direction of } m} = 66.1$ in. · lb $\boldsymbol{i}$ + 33.1 in. · lb $\boldsymbol{j}$ + 52.9 in. · lb $\boldsymbol{k}$

Check The two approaches in part (a) can be used as checks for each other. Also, inspection of **Figure 5.67*b*** (which was drawn to scale) shows that one would expect the projection of ***M*** onto ***m*** to be approximately 1/6 of the size of ***M***; indeed, 90.9 is 0.132 of 687.

Comment: Would it have made a difference in part (a) if we had taken $\boldsymbol{m} \cdot \boldsymbol{M}$ instead of $\boldsymbol{M} \cdot \boldsymbol{m}$?

EXERCISES 5.3

5.3.1. Consider E5.3.1.

a. Find the moment created at a moment center at C by the 20-lb force. Call this moment $\boldsymbol{M}_C$.

b. Determine the scalar component of $\boldsymbol{M}_C$ that is in the direction of CE.

c. Determine the scalar component of $\boldsymbol{M}_C$ that is in the direction of CD.

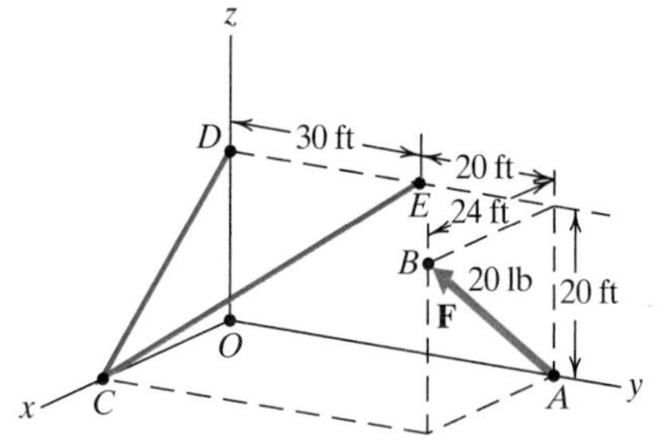

E5.3.1

5.3.2. Consider E5.3.2.

a. Find the moment created at C by the force $\boldsymbol{F}$. Call this moment $\boldsymbol{M}_C$.

b. Determine the scalar component of $\boldsymbol{M}_C$ that is in the direction of CE.

c. Determine the scalar component of $\boldsymbol{M}_C$ that is in the direction of CD.

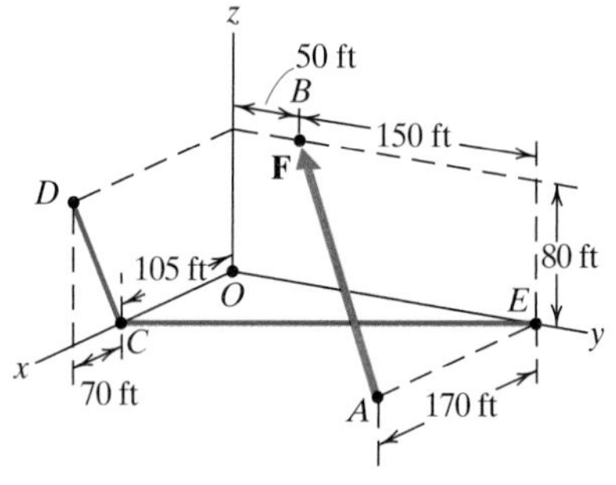

E5.3.2

5.3.3. **Exercise 5.2.18** asks for the moment $\boldsymbol{M}_O$ created by a 610-N force. Determine the scalar component of this moment about the z axis.

5.3.4. For the moment $\boldsymbol{M}_A$ created at O by the 380-N force $\boldsymbol{F}_A$ in **Exercise 5.2.24*b***, determine the scalar component of this moment about the

a. x axis

b. y axis

c. z axis

5.3.5. **Exercise 5.2.25** asks for the sum $\boldsymbol{M}_1 + \boldsymbol{M}_2$ for the moments created at A by the two forces $\boldsymbol{F}_1$ and $\boldsymbol{F}_2$ in **E5.2.25**. Call this sum $\boldsymbol{M}_3$.

a. What is the scalar component of $\boldsymbol{M}_3$ about the axis of the shaft AB?

b. Would your answer in **a** differ if $\boldsymbol{M}_1$ and $\boldsymbol{M}_2$ had been calculated for a moment center at B? (Present your answer in terms of the concepts presented in this chapter, no need to do additional calculations.)

5.3.6. **Exercise 5.2.46** asks for the sum of the moments created at O by $\boldsymbol{F}_A$ and $\boldsymbol{F}_B$ of **E5.2.46**. If the scalar component of that sum along the axis of the pipe must be at least 60 N·m in order to loosen the pipe from the threaded hole and if $\|\boldsymbol{F}_A\| = \|\boldsymbol{F}_B\|$, what is the minimum required magnitude of $\boldsymbol{F}_A$ and $\boldsymbol{F}_B$? Use the 60 N·m as a torque specification for tightening the fitting.

5.3.7. For the situation shown in **E5.2.42**, determine the magnitude of the moment created by the 2-kN force

a. about the line AB (One possible strategy is to find the moment at A using the cross product as was asked for in **Exercise 5.2.42*a***, and then find the dot product of the moment components in the direction defined by AB.)

b. about line BC (Why is this moment zero?)

5.3.8. Return to the 20-lb force acting at point A in **E5.3.1**.

a. Use the dot product to find the scalar component of the moment found in **Exercise 5.3.1*a*** that is parallel to line CD.

b. Find the component of the moment that is perpendicular to line CD.

5.3.9. The magnitude of the force $\boldsymbol{F}$ in **E5.3.2** is 10 lbs.

a. Use the dot product to find the component of the moment found in **Exercise 5.3.2*a*** that is parallel to line CE.

b. Find the component of the moment that is perpendicular to line CE.

5.3.10. A 120-N horizontal force is applied to the handle shown in **E5.2.37**. Determine the scalar component of the moment found in **Exercise 5.2.37** that has the effect of loosening the screw.

5.3.11. A 10-kN force acts at point D on the linkage shown in **E5.2.39**. Find the scalar component of the moment found in **Exercise 5.2.40** that is aligned with member BC.

5.4 EQUIVALENT LOADS

We have discussed the moment created by a force offset from a moment center. For example, $\boldsymbol{F}_{\text{push}}$ applied at A in **Figure 5.68*a*** creates a negative moment at a moment center at B of $-250 \text{ mm} \| \boldsymbol{F}_{\text{push}} \| \boldsymbol{k}$. Now we consider that we can replace $\boldsymbol{F}_{\text{push}}$ applied at A with a loading applied at B consisting of a force $\boldsymbol{F}_{\text{push}}$ and a moment $-250 \text{ mm} \| \boldsymbol{F}_{\text{push}} \| \boldsymbol{k}$ (**Figure 5.68*b***). The loading in **Figure 5.68*a*** is equivalent to the loading in **Figure 5.68*b***; we call them **equivalent loadings**. This means that these loadings have the same influence with regard to pushing, pulling, twisting, tipping, turning, and/or rocking the system on which they act. Once again, the loading in **Figure 5.68*b*** (consisting of a force and a moment applied at B) is equivalent to the loading in **Figure 5.68*a*** (consisting of a force and zero moment applied at A).

We can generalize the process that we went through with the lug nut example in **Figure 5.68** to say that

> *For any loading (consisting of forces and/or moments) applied to a system, we can replace that loading with an equivalent loading (consisting of a force and a moment) applied at a single point.*

We call this the **equivalent loading principle**. In the lug nut example, we applied this principle and replaced a loading consisting of a force applied at A with its equivalent loading (consisting of a force and a moment) applied at B.

Now let's apply the principle to consider a system with a number of forces acting on it. For example, the forces acting on the supporting frame for a traffic light (**Figure 5.69*a***) are the weight of the light, a drag force resulting from wind, and a tension force exerted by the cable attached to the sign (**Figure 5.69*b***). If we wish to find the loading applied at O that is equivalent to these three forces, we add the forces to come up with an **equivalent force**, and we add the moments created by the individual forces about a moment center at O to come up with an **equivalent moment**.

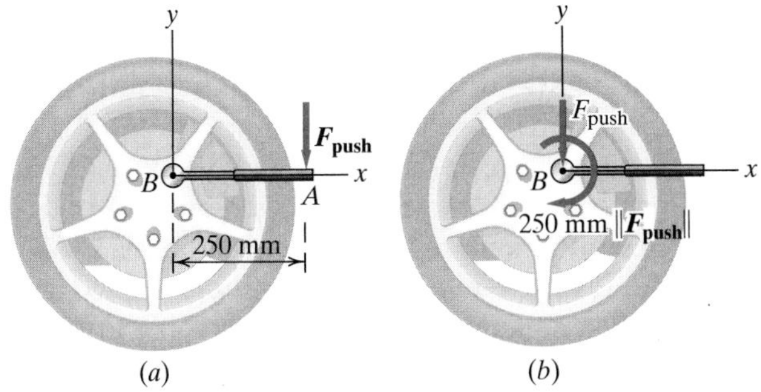

Figure 5.68 A force applied at the end of the wrench (*a*) is equivalent to a force and a moment applied at the center of the lug nut (*b*).

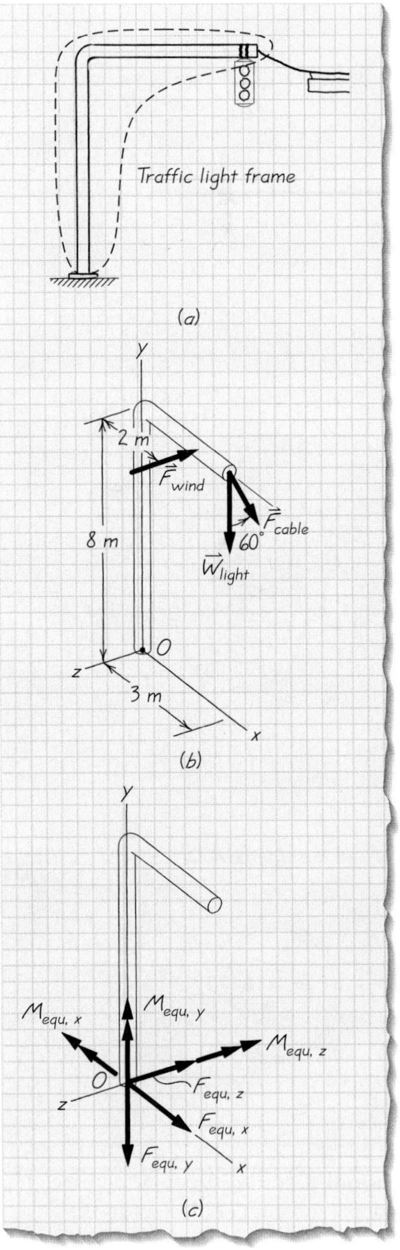

Figure 5.69 Example of finding equivalent loads

The equivalent force is simply the vector sum of the forces acting on the frame:

$$\boldsymbol{F}_{\text{equ@}O} = \underbrace{\|\boldsymbol{F}_{\text{cable}}\|\ \cos 30^\circ\, \boldsymbol{i}}_{F_{equ,\,x}} - \underbrace{(\|\boldsymbol{F}_{\text{cable}}\|\ \sin 30^\circ + \|\boldsymbol{W}_{\text{light}}\|)\boldsymbol{j}}_{F_{equ,\,y}} - \underbrace{\|\boldsymbol{F}_{\text{wind}}\|\,\boldsymbol{k}}_{F_{equ,\,z}}$$

The most straightforward approach to finding the equivalent moment at a moment center at O is to calculate the moment created by each force relative to this moment center, then add the moments together. We find that the equivalent moment due to the forces $\boldsymbol{F}_{\text{wind}}$, $\boldsymbol{F}_{\text{cable}}$, and $\boldsymbol{W}_{\text{light}}$ acting on the supporting frame is[11]

$$\boldsymbol{M}_{\text{equ@}O} = \underbrace{-8\text{ m}\ \|\boldsymbol{F}_{\text{wind}}\|\,\boldsymbol{i}}_{M_{equ,\,x}} \underbrace{+2\text{ m}\ \|\boldsymbol{F}_{\text{wind}}\|\,\boldsymbol{j}}_{M_{equ,\,y}} \underbrace{-\ (3\text{ m}\ \|\boldsymbol{W}_{\text{light}}\| + 8.48\text{ m}\ \|\boldsymbol{F}_{\text{cable}}\|)\boldsymbol{k}}_{M_{equ,\,z}}$$

We show the equivalent moment and equivalent force on a drawing of the system in **Figure 5.69*c***. Notice that the equivalent moment and equivalent force are placed at the moment center O.

The loading presented in **Figure 5.69*b*** is equivalent to the loading presented in **Figure 5.69*c***. More formally, we say that the loading in **Figure 5.69*b*** and those in **Figure 5.69*c*** are equivalent loadings.

As a final (and very important) example of finding equivalent loading, consider a lug wrench being used to tighten a nut onto a bolt (**Figure 5.70*a***); this type of wrench is called a star wrench and allows the user to simultaneously exert a downward force with one hand and an upward force with the other. Assume that $\boldsymbol{F}_{\text{LH}}$ and $\boldsymbol{F}_{\text{RH}}$ are equal in magnitude and opposite in direction. The perpendicular distance between them is $2D$. If we want to find the equivalent loading applied at O, we determine that the equivalent force (which is the vector sum of the forces acting on the system) is zero since the two forces are equal in magnitude and opposite in direction. The equivalent moment is found by calculating the moment created by each force relative to a moment center at O, then adding these moments; this results in an equivalent moment of $\boldsymbol{M}_{\text{equ}} = -(2D\ \|\boldsymbol{F}\|\ )\boldsymbol{k}$. We show the equivalent moment and equivalent force on a drawing of the system in **Figure 5.70*b***. Notice that the equivalent moment and equivalent force are placed at the moment center O.

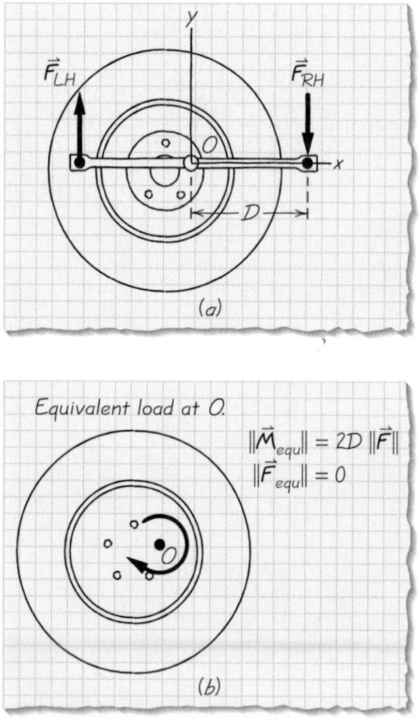

Figure 5.70 (*a*) A couple; (*b*) the couple moment equivalent to the couple in (*a*)

The loadings in **Figures 5.70*a*** and ***b*** are equivalent loadings and are given special labels—two parallel forces that are equal in magnitude and opposite in sign are referred to as a **couple** (**Figures 5.70*a***), and the equivalent moment and zero equivalent force created by a couple are referred to as a **couple moment** (or simply a moment) (**Figure 5.70*b***). The direction of the couple moment is perpendicular to the plane containing the forces that make up the couple. A notable property of a couple is that its equivalent couple moment is independent of the moment center used.

[11]Note that in the following equation, 8.48 m is the perpendicular distance between point O and the line of action of $\boldsymbol{F}_{\text{cable}}$.

EXAMPLE 5.14 EQUIVALENT MOMENT AND EQUIVALENT FORCE (A)

A beam is held in place by a pin at A and a rope attached at B and C. A block weighing 1000 lb hangs at end D, as shown in **Figure 5.71**.

(a) If the equivalent moment at A due to the rope tension and block weight is zero, what is the tension in the rope?

(b) What is the equivalent loading at A acting on the beam due to the rope tension and block weight?

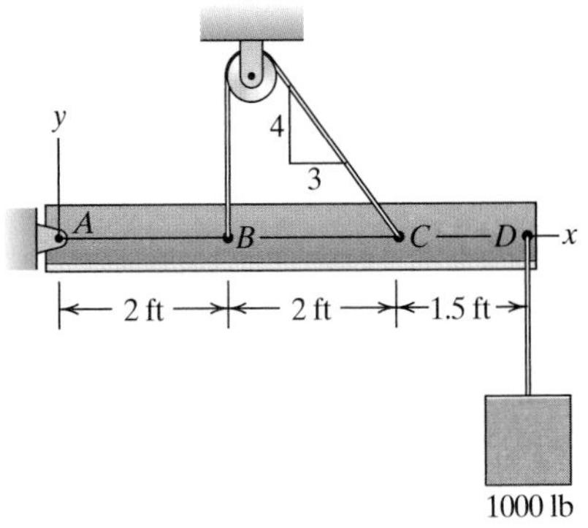

Figure 5.71

Goal We are to find the tension in the rope such that the equivalent moment at A due to the rope tension and the weight of the block acting on the beam has zero magnitude (a). Then using this rope force, we are to find the equivalent force (b).

Given We are given the dimensions of the beam, the weight of the block, a coordinate system, and the angular orientation of the rope that acts on the beam via a pulley. We are also told that the equivalent moment at A due to the rope tension and block weight is zero.

Assume We assume that the beam, block, rope, and pulley all lie in the xy plane. In addition, we assume that the pulley is frictionless.

Draw We begin by drawing a diagram of the beam, showing external loads $\boldsymbol{T}_1$, $\boldsymbol{T}_2$, and $\boldsymbol{W}$ (= 1000 lb $\boldsymbol{j}$); see **Figure 5.72*a***.

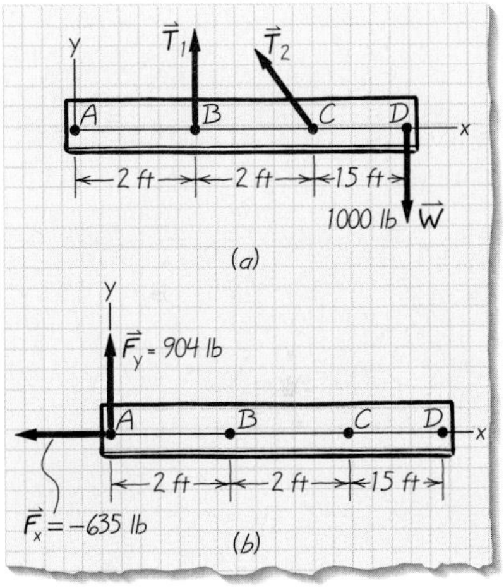

Figure 5.72

Formulate Equations and Solve (a) Forces acting on the beam are $\boldsymbol{T}_1$, $\boldsymbol{T}_2$, and $\boldsymbol{W}$. Because a single rope is attached at B and C and passes over a frictionless pulley, $\boldsymbol{T}_1$ and $\boldsymbol{T}_2$ have the same magnitude; call it $\|\boldsymbol{T}\|$.

We proceed by finding the moment at a moment center at A created by each force ($\boldsymbol{T}_1$, $\boldsymbol{T}_2$, and $\boldsymbol{W}$), then sum these moments; their sum should be equal to zero. Because $\boldsymbol{T}_1$, $\boldsymbol{T}_2$, and $\boldsymbol{W}$ lie in the xy plane, (5.11A) applies for finding the moment.

Moment $\boldsymbol{M}_1$ created by $\boldsymbol{T}_1$ at A: With $\boldsymbol{T}_1 = \|\boldsymbol{T}\|\boldsymbol{j}$ and $\boldsymbol{r}_1 = 2\text{ ft}\,\boldsymbol{i}$, application of (5.11A) leads to

$$\boldsymbol{M}_1 = 2\,\|\boldsymbol{T}\|\text{ ft}\,\boldsymbol{k}$$

Moment $\boldsymbol{M}_2$ created by $\boldsymbol{T}_2$ at A: With $\boldsymbol{T}_2 = -\frac{3}{5}\|\boldsymbol{T}\|\boldsymbol{i} + \frac{4}{5}\|\boldsymbol{T}\|\boldsymbol{j}$, and $\boldsymbol{r}_2 = 4\text{ ft}\,\boldsymbol{i}$, application of (5.11A) leads to

$$\boldsymbol{M}_2 = (4\text{ ft})(\tfrac{4}{5}\|\boldsymbol{T}\|)\boldsymbol{k} = \tfrac{16}{5}\text{ ft}\,\|\boldsymbol{T}\|\,\boldsymbol{k}$$

(Notice that the x component of $\boldsymbol{T}_2$, $-\frac{3}{5}\|\boldsymbol{T}\|\boldsymbol{i}$, does not create a moment at A because its line of action intersects A.)

Moment $\boldsymbol{M}_3$ created by $\boldsymbol{W}$ about A: With $\boldsymbol{W} = -1000\text{ lb}\,\boldsymbol{j}$ and $\boldsymbol{r}_3 = 5.5\text{ ft}\,\boldsymbol{i}$, application of (5.11A) leads to

$$\boldsymbol{M}_3 = -5500\text{ lb}\cdot\text{ft}\,\boldsymbol{k}$$

Equivalent moment at A: The equivalent moment $\boldsymbol{M}_{\text{equ}}$ is equal to $\boldsymbol{M}_1 + \boldsymbol{M}_2 + \boldsymbol{M}_3$:

$$\begin{aligned} \boldsymbol{M}_{\text{equ}@A} &= \boldsymbol{M}_1 + \boldsymbol{M}_2 + \boldsymbol{M}_3 \\ &= (2 \, \| \boldsymbol{T} \| \, + \tfrac{16}{5} \| \boldsymbol{T} \| \,)\boldsymbol{k} \text{ ft} - 5500 \text{ lb} \cdot \text{ft } \boldsymbol{k} \\ &= \| \boldsymbol{T} \| \, (2 + \tfrac{16}{5}) \text{ ft } \boldsymbol{k} - 5500 \text{ lb} \cdot \text{ft } \boldsymbol{k} \\ \boldsymbol{M}_{\text{equ}@A} &= (\| \boldsymbol{T} \| \, (2 + \tfrac{16}{5}) \text{ ft} - 5500 \text{ lb} \cdot \text{ft}) \, \boldsymbol{k} \end{aligned} \tag{1}$$

Because we are told that the equivalent moment is equal to zero based on the problem statement, we set (1) equal to zero and solve for $\| \boldsymbol{T} \|$:

$$\begin{aligned} (\| \boldsymbol{T} \| \, (2 + \tfrac{16}{5}) \text{ ft } \boldsymbol{k} - 5500 \text{ lb} \cdot \text{ft}) \, \boldsymbol{k} &= 0 \\ \| \boldsymbol{T} \| &= 1058 \text{ lb} \end{aligned}$$

Answer to (a) The tension in the rope is 1058 lb.

(b) The equivalent load at A due to the rope tension and block consists of an equivalent moment and an equivalent force. Based on our work in (a), the equivalent moment is zero for a rope tension of 1058 lb. The equivalent force is equal to $\boldsymbol{T}_1 + \boldsymbol{T}_2 + \boldsymbol{W}$:

$$\begin{aligned} \boldsymbol{F}_{\text{equ}} &= \| \boldsymbol{T} \| \boldsymbol{j} - \tfrac{3}{5} \| \boldsymbol{T} \| \, \boldsymbol{i} + \tfrac{4}{5} \| \boldsymbol{T} \| \boldsymbol{j} - 1000 \text{ lb } \boldsymbol{j} \\ &= 1058 \text{ lb } \boldsymbol{j} - 635 \text{ lb } \boldsymbol{i} + 846 \text{ lb } \boldsymbol{j} - 1000 \text{ lb } \boldsymbol{j} \\ &= -635 \text{ lb } \boldsymbol{i} + 904 \text{ lb } \boldsymbol{j} \end{aligned}$$

Answer The equivalent load at A consists of zero moment and a force due to the rope tension and block weight of $\boldsymbol{F}_{\text{equ}} = -635 \text{ lb } \boldsymbol{i} + 904 \text{ lb } \boldsymbol{j}$. We draw this equivalent load in **Figure 5.72*b***.

Check A good check is to substitute the value of the rope tension back into (1) to confirm that the equivalent moment due to the rope tension and block weight is indeed zero.

EXAMPLE 5.15 EQUIVALENT MOMENT AND EQUIVALENT FORCE (B)

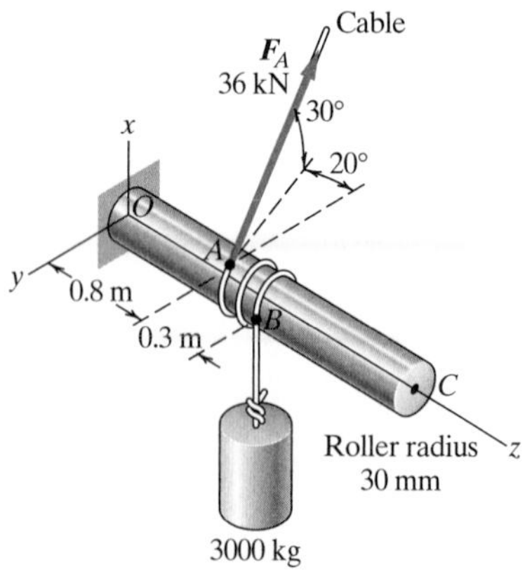

Figure 5.73

A 3000-kg cylinder hangs from a roller of radius 30 mm. In addition, a cable wound around the roller pulls as shown in **Figure 5.73**. The tension in the cable is 36 kN.

(a) Find the equivalent moment and equivalent force at O due to these two loads.

(b) Graphically show the equivalent loading.

Goal For a physical situation consisting of a cable wound around a roller and a hanging cylinder, we are to find the equivalent loading (moment and force) at a specified moment center (O).

Given We are given information on the dimensions of the roller, a coordinate system, and information on how the cable is wound around the roller. In addition, we are told that the tension in the cable is 36 kN and the mass of the cylinder is 3000 kg.

Assume We assume that earth's gravity acts in the $-x$ direction, and that we can ignore the weight of the roller.

Draw In Figure 5.74 we have drawn the roller and the forces acting on it due to the cable and the weight ($\boldsymbol{F}_A$ and $\boldsymbol{F}_B$, respectively). We determine the force due to the weight of the cylinder to be 29.4 kN (= 3000 kg × 9.81 m/s²).

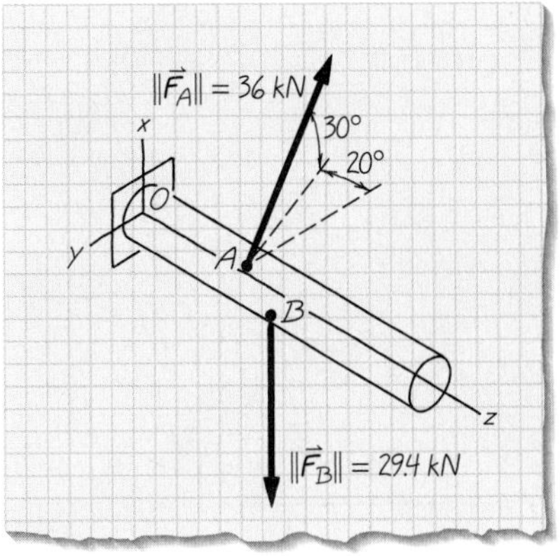

Figure 5.74

Formulate Equations and Solve (a) Since neither of these forces lies in a reference axes plane, we use (5.7) or the cross-product in (5.12) to find the moments.

Moment $\boldsymbol{M}_A$ created by $\boldsymbol{F}_A$ (36-kN force acting at A): We write this force in vector form as

$$\boldsymbol{F}_A = 18.0 \text{ kN } \boldsymbol{i} - 29.4 \text{ kN } \boldsymbol{j} - 10.7 \text{ kN } \boldsymbol{k}$$

(See calculation in Figure 5.75.) Now define a position vector from O to A: $\boldsymbol{r}_A = 0.03 \text{ m } \boldsymbol{i} + 0.8 \text{ m } \boldsymbol{k}$. We find the moment that $\boldsymbol{F}_A$ creates at a moment center at O to be

$$\boldsymbol{M}_A = \boldsymbol{r}_A \times \boldsymbol{F}_A = \begin{vmatrix} \boldsymbol{i} & \boldsymbol{j} & \boldsymbol{k} \\ 0.03 \text{ m} & 0.0 \text{ m} & 0.8 \text{ m} \\ 18.0 \text{ kN} & -29.4 \text{ kN} & -10.7 \text{ kN} \end{vmatrix}$$

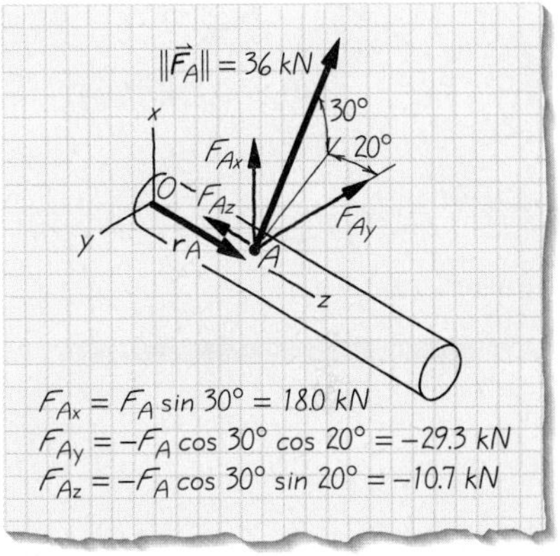

Figure 5.75

Expanded, this becomes

$$\boldsymbol{M}_A = 23.3 \text{ kN} \cdot \text{m } \boldsymbol{i} + 14.7 \text{ kN} \cdot \text{m } \boldsymbol{j} - 0.882 \text{ kN} \cdot \text{m } \boldsymbol{k}$$

Intermediate check: Even as you are proceeding through a calculation, check to make sure that the steps make sense. For example, we can check that the sign of each scalar component of $\boldsymbol{M}_A$ is reasonable, given the position of the force relative to the coordinate axes.

Moment $\boldsymbol{M}_B$ created by $\boldsymbol{F}_B$ (29.4-kN force due to gravity acting at B): We write this force in vector form as

$$\boldsymbol{F}_B = -29.4 \text{ kN } \boldsymbol{i}$$

(See Figure 5.76.) Define a position vector from O to B: $\boldsymbol{r}_B = 0.03 \text{ m } \boldsymbol{j} + 1.1 \text{ m } \boldsymbol{k}$. We find the moment that $\boldsymbol{F}_B$ creates at a moment center at O to be

$$\boldsymbol{M}_B = \boldsymbol{r}_B \times \boldsymbol{F}_B = \begin{vmatrix} \boldsymbol{i} & \boldsymbol{j} & \boldsymbol{k} \\ 0.0 \text{ m} & 0.03 \text{ m} & 1.1 \text{ m} \\ -29.3 \text{ kN} & 0.0 \text{ kN} & 0.0 \text{ kN} \end{vmatrix}$$

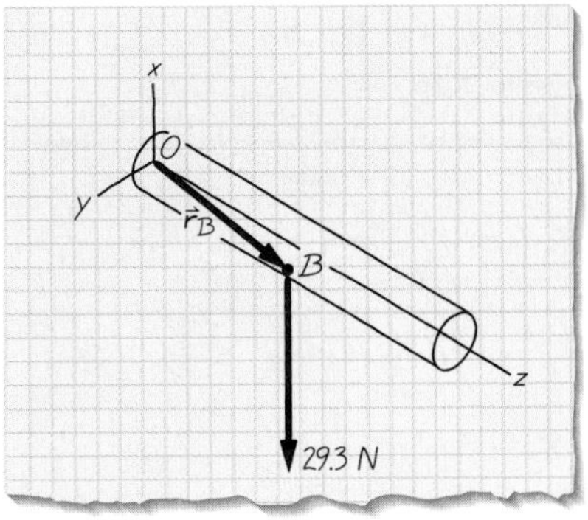

Figure 5.76

Expanding this becomes

$$\boldsymbol{M}_B = -32.3 \text{ kN} \cdot \text{m } \boldsymbol{j} + 0.882 \text{ kN} \cdot \text{m } \boldsymbol{k}$$

The equivalent moment:

$$\boldsymbol{M}_{\text{equ@}O} = \boldsymbol{M}_A + \boldsymbol{M}_B = \overbrace{(23.4\,\boldsymbol{i} + 14.7\boldsymbol{j} - 0.882\,\boldsymbol{k})\text{ kN}\cdot\text{m}}^{M_A}$$

$$+ \overbrace{(-32.3\,\boldsymbol{j} + 0.882\,\boldsymbol{k})\text{ kN}\cdot\text{m}}^{M_B}$$

$$= 23.4\text{ kN}\cdot\text{m}\,\boldsymbol{i} - 17.5\text{ kN}\cdot\text{m}\boldsymbol{j}$$

Notice that there is no scalar moment about the z axis because the z scalar components from $\boldsymbol{M}_A$ and $\boldsymbol{M}_B$ cancel each other.

The equivalent force:

$$\boldsymbol{F}_{\text{equ}} = \boldsymbol{F}_A + \boldsymbol{F}_B = \overbrace{(18.0\,\boldsymbol{i} - 29.3\boldsymbol{j} - 10.7\,\boldsymbol{k})\text{ kN}}^{F_A} + \overbrace{(-29.4\,\boldsymbol{i})\text{ kN}}^{F_B}$$

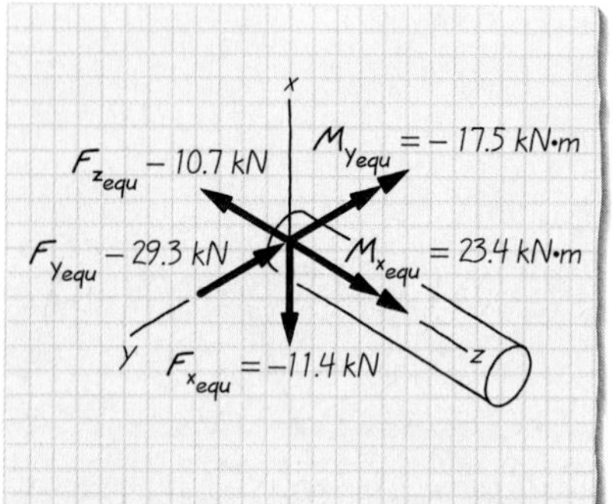

Figure 5.77

Answer to (a) The equivalent loading at O consists of the equivalent moment:

$\boldsymbol{M}_{\text{equ@}O} = 23.4\text{ kN}\cdot\text{m}\,\boldsymbol{i} - 17.5\text{ kN}\cdot\text{m}\,\boldsymbol{j}$

and the equivalent force:

$\boldsymbol{F}_{\text{equ}} = -11.4\text{ kN}\,\boldsymbol{i} - 29.3\text{ kN}\,\boldsymbol{j} - 10.7\text{ kN}\,\boldsymbol{k}$

(b) The equivalent loading consisting of the equivalent moment and equivalent force is shown in **Figure 5.77**.

EXAMPLE 5.16 WORKING WITH COUPLES (A)

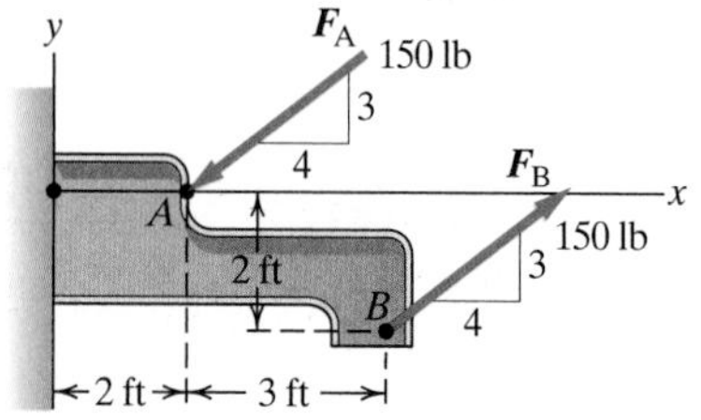

Figure 5.78

Two forces act on the member shown in **Figure 5.78**.
Find the equivalent moment created by the two forces. Do the two forces represent a couple?

Goal We are to find the equivalent moment created by two forces acting on a member.

Given We are given the dimensions of the member, coordinate axes, and information about the two forces (point of application, direction, magnitude).

Assume We assume that the two forces lie in the *xy* plane, as defined in **Figure 5.78**.

Formulate Equations and Solve Because the forces $\boldsymbol{F}_A$ and $\boldsymbol{F}_B$ are parallel, are of the same magnitude, and act in opposite directions, they form a couple. Recall that the moment created by this couple is independent of the location of the moment center. This means that we calculate the same equivalent moment regardless of where we choose to locate the moment center.

Approach 1: The magnitude of the couple is given by $\| \boldsymbol{M} \| = \| \boldsymbol{F} \| d$, where $\| \boldsymbol{F} \|$ is the magnitude of each force in the couple and d is the perpendicular distance between the two forces. In this case, the perpendicular distance is somewhat difficult to find. It is easier to break each force up into its horizontal and vertical components, find the moments associated with each, and then sum them together.

First, resolve the force $\boldsymbol{F}_A$ acting at point A into its scalar components:

$$F_{Ax} = -\tfrac{4}{5}(150 \text{ lb}) = -120 \text{ lb}$$
$$F_{Ay} = -\tfrac{3}{5}(150 \text{ lb}) = -90 \text{ lb}$$

Then, resolve the force $\boldsymbol{F}_B$ acting at point B into its scalar components:

$$F_{Bx} = \tfrac{4}{5}(150 \text{ lb}) = 120 \text{ lb}$$
$$F_{By} = \tfrac{3}{5}(150 \text{ lb}) = 90 \text{ lb}$$

We now redraw the figure as shown in Figure 5.79. (Note that the loads in **Figure 5.79** are equivalent to those in **Figure 5.78**.)

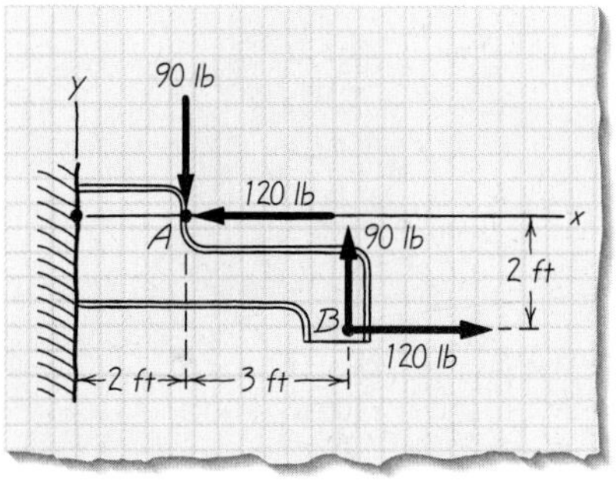

Figure 5.79

Forces F_{Ax} and F_{Bx} form a couple, as do F_{Ay} and F_{By}. We will find the moment associated with each of these couples, then sum them together using vector addition.

The magnitude of the moment associated with the couple that consists of F_{Ax} and F_{Bx} is the product of the perpendicular distance between F_{Ay} and F_{By} and the magnitude of F_{Ax} or F_{Bx}. It is in the $\boldsymbol{k}$ direction. We write this as

$$\boldsymbol{M}_1 = +(2 \text{ ft})(120 \text{ lb})\boldsymbol{k} = +240 \text{ ft} \cdot \text{lb } \boldsymbol{k}$$

The magnitude of the moment associated with the couple that consists of F_{Ay} and F_{By} is the product of the perpendicular distance between F_{Ay} and F_{By} and the magnitude of F_{Ay} or F_{By}. It is in the $\boldsymbol{k}$ direction. We write this as

$$\boldsymbol{M}_2 = +(3 \text{ ft})(90 \text{ lb})\boldsymbol{k} = +270 \text{ ft} \cdot \text{lb } \boldsymbol{k}$$

Finally, to find the overall moment, we simply sum $\boldsymbol{M}_1$ and $\boldsymbol{M}_2$.

Answer The equivalent moment on the system is $\boldsymbol{M}_{\text{equ}} = \boldsymbol{M}_1 + \boldsymbol{M}_2 = +510 \text{ ft} \cdot \text{lb } \boldsymbol{k}$ in the positive (counterclockwise) direction. The two forces $\boldsymbol{F}_A$ and $\boldsymbol{F}_B$ do form a couple; they are equal in magnitude and opposite in direction.

Approach 2 (MC at *A*): Since the moment equivalent to a couple is independent of the moment center, we arbitrarily select a moment center at A and proceed to find the moment due to $\boldsymbol{F}_A$ and $\boldsymbol{F}_B$. Notice that $\boldsymbol{F}_A$ creates zero moment at A. The equivalent moment $\boldsymbol{M}$ created by $\boldsymbol{F}_A$ and $\boldsymbol{F}_B$ is therefore

$$\begin{aligned} \boldsymbol{M}_{\text{equ}} &= [(2 \text{ ft})(F_{Bx}) + (3 \text{ ft})(F_{By})]\boldsymbol{k} \\ &= [(2 \text{ ft})(120 \text{ lb}) + (3 \text{ ft})(90 \text{ lb})]\boldsymbol{k} \\ &= +510 \text{ ft} \cdot \text{lb } \boldsymbol{k} \end{aligned}$$

This is the same answer we got with Approach 1.

Check Because the forces in this example form a couple, we could sum the moments created by $\boldsymbol{F}_A$ and $\boldsymbol{F}_B$ about any moment center and we should get the same answer as above—this can be used as a check.

Question: If forces $\boldsymbol{F}_A$ and $\boldsymbol{F}_B$ did not form a couple, what would be wrong with the statement "find the moment created by the two forces"?

EXAMPLE 5.17 WORKING WITH COUPLES (B)

A rigid structural member is subjected to two 100-N forces applied 100 mm apart, as shown in Figure 5.80. Replace the two 100-N forces with an equivalent load consisting of the two forces $\boldsymbol{P}_1$ and $\boldsymbol{P}_2$, each of which has a magnitude of 400 N. Find the orientation angle θ of these forces.

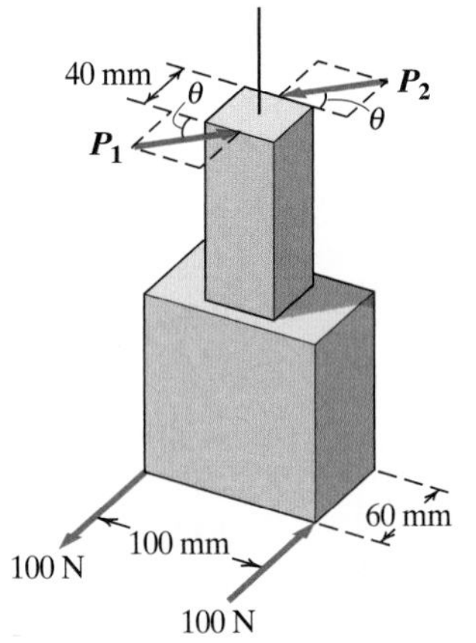

Figure 5.80

Goal We are asked find the orientation angle of two 400-N forces so that they are equivalent to a couple consisting of two 100-N forces spaced 100 mm apart.

Given We are told that two 100-N forces act on a rigid structure, and inspection shows that these two forces form a couple. We are also told that the equivalent loading we are to find will consist of two 400-N forces oriented as shown in **Figure 5.80**. Because $\boldsymbol{P}_1$ and $\boldsymbol{P}_2$ are equal in magnitude and opposite in direction, they constitute a couple.

Assume We assume the coordinate orientation shown in Figure 5.81.

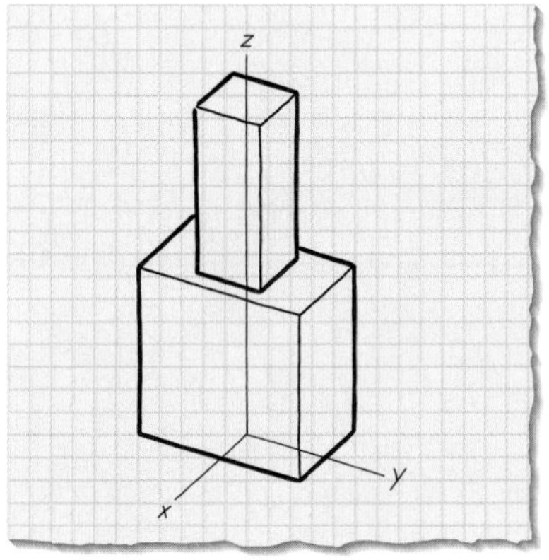

Figure 5.81

Formulate Equations and Solve We begin by looking at the couple consisting of the two 100-N forces from above and noting that the equivalent moment is acting counterclockwise about the z axis (Figure 5.82*a*). Given that, and knowing that the perpendicular distance between the two forces, d_1, is 100 mm or 0.1 m, we solve for the magnitude of the moment:

$$\|\boldsymbol{M}\| = \|\boldsymbol{F}\|\ d_1 = 100\ \text{N}\ (0.1\ \text{m}) = 10\ \text{N}\cdot\text{m} \tag{1}$$

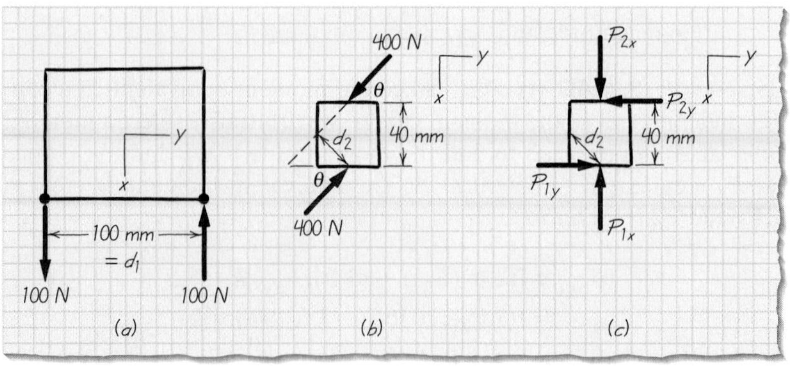

Figure 5.82

Looking at **Figure 5.80**, we can see that the forces $\boldsymbol{P}_1$ and $\boldsymbol{P}_2$ also produce a counterclockwise moment about the z axis (see **Figure 5.82*b***). We use trigonometry to determine the perpendicular distance d_2 between the two and thus write

$$\|\boldsymbol{M}\| = \|\boldsymbol{P}_1\|\, d_2 = 400 \text{ N } [(0.040 \text{ m}) \cos\theta] \qquad (2)$$

Since we are told that we want the loading in **Figure 5.82*a*** to be equivalent to the loading in **Figure 5.82*b***, we equate (1) and (2) and solve for the orientation angle θ.

Answer $\theta = \cos^{-1}\frac{10}{16} = 51.3°$

Check In **Figure 5.82*c*** the components of $\boldsymbol{P}_1$ and $\boldsymbol{P}_2$ are shown. The components in the x direction have the same line of action and therefore do not create a moment. The components in the y direction (which have magnitude of $\|\boldsymbol{P}\| \cos\theta = 400 \text{ N} \cos\theta$) form a couple, and the distance between these two components is 0.040 m. Since we want the magnitude of the moment equivalent to this couple to be 10 N · m, we write $0.04 \text{ m } \|\boldsymbol{P}\| \cos\theta = 10 \text{ N}\cdot\text{m}$, and find that $\theta = \cos^{-1}\frac{10}{16} = 51.3°$.

EXAMPLE 5.18 WORKING WITH COUPLES (C)

A T-joint is acted on by the loads shown in Figure 5.83. Determine the magnitude and direction of the equivalent moment and equivalent force.

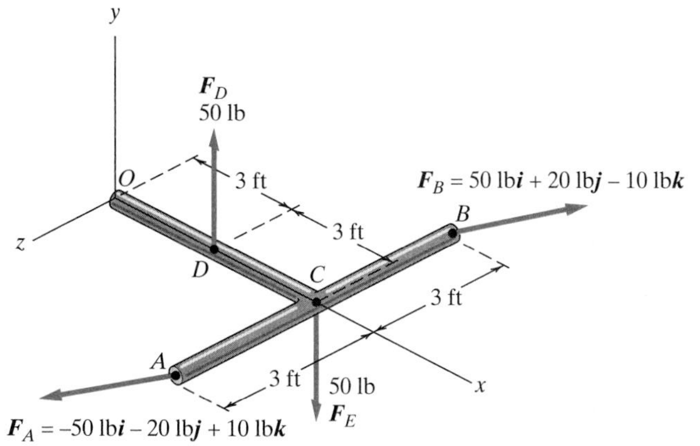

Figure 5.83

Goal We are asked to find the loading (consisting of a moment and a force) that is equivalent to the forces acting on the T-joint shown in **Figure 5.83**.

Given We are given dimensions of the T-joint, information of four forces acting on it (point of application, direction, and magnitude), and a coordinate system.

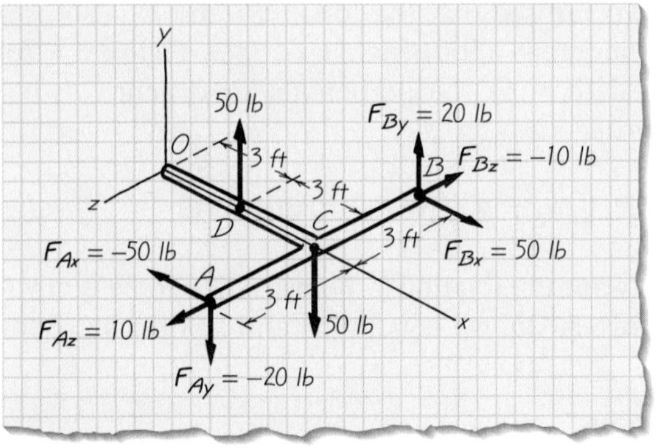

Figure 5.84

Assume No assumptions are necessary.

Draw We redraw the T-joint to show the forces at A and B in terms of scalar components; see **Figure 5.84**.

Formulate Equations and Solve Inspection of **Figure 5.83** or **5.84** shows that $\boldsymbol{F}_A$ and $\boldsymbol{F}_B$ constitute a couple, as do $\boldsymbol{F}_C$ and $\boldsymbol{F}_D$.

First we will consider the couple consisting of $\boldsymbol{F}_A$ and $\boldsymbol{F}_B$ and determine its equivalent moment and force (which will simply be a moment since we are dealing with a couple). As in Example 5.16, it is difficult to determine the perpendicular distance between these two forces. Therefore, in order to simplify the calculations, we will resolve each force into its x, y, and z components. We can do this easily since the forces are already written in vector notation.

$$\boldsymbol{F}_A = -50 \text{ lb } \boldsymbol{i} - 20 \text{ lb } \boldsymbol{j} + 10 \text{ lb } \boldsymbol{k} \qquad \boldsymbol{F}_B = 50 \text{ lb } \boldsymbol{i} + 20 \text{ lb } \boldsymbol{j} - 10 \text{ lb } \boldsymbol{k}$$

from which we can write

$$\begin{aligned} F_{Ax} &= -50 \text{ lb} \qquad & F_{Bx} &= 50 \text{ lb} \\ F_{Ay} &= -20 \text{ lb} \qquad & F_{By} &= 20 \text{ lb} \\ F_{Az} &= 10 \text{ lb} \qquad & F_{Bz} &= -10 \text{ lb} \end{aligned}$$

We can now consider the couple created by $\boldsymbol{F}_A$ and $\boldsymbol{F}_B$ to be three couples acting on the system, consisting of F_{Ax} and F_{Bx}, F_{Ay} and F_{By}, and F_{Az} and F_{Bz}. We will find the moment created by each of these couples and sum them together to find the total moment on the system due to $\boldsymbol{F}_A$ and $\boldsymbol{F}_B$. Arbitrarily choosing point C as where we locate our moment center, we calculate the following[12]:

Moment created by F_{Ax} and F_{Bx}:

$$\boldsymbol{M}_1 = [3 \text{ ft } (-50 \text{ lb}) - 3 \text{ ft } (50 \text{ lb})]\boldsymbol{j} = -300 \text{ ft} \cdot \text{lb } \boldsymbol{j}$$

[12]Since we are dealing with couples, we could have chosen any point as the moment center. The important thing to remember is that the position vector comes first and is defined from the moment center; the moment center is not required to be the origin of the coordinate axes.

Moment created by F_{Ay} *and* F_{By}:

$$\mathbf{M}_2 = [3\text{ ft }(20\text{ lb}) - 3\text{ ft }(-20\text{ lb})]\,\mathbf{i} = +120\text{ ft}\cdot\text{lb}\,\mathbf{i}$$

Moment created by F_{Az} *and* F_{Bz}:

$$\mathbf{M}_3 = [-0\text{ ft }(10\text{ lb}) - 0\text{ ft }(-10\text{ lb})]\,\mathbf{j} = 0\text{ ft}\cdot\text{lb}\,\mathbf{j}$$

($\mathbf{M}_3$ having zero magnitude makes sense because F_{Az} and F_{Bz} have the same line of action.)

The equivalent moment created by $\mathbf{F}_A$ *and* $\mathbf{F}_B$ is therefore

$$\begin{aligned}\mathbf{M}_4 &= \mathbf{M}_1 + \mathbf{M}_2 + \mathbf{M}_3\\ &= -300\text{ ft}\cdot\text{lb}\,\mathbf{j} + 120\text{ ft}\cdot\text{lb}\,\mathbf{i} + 0\text{ ft}\cdot\text{lb}\,\mathbf{j}\\ &= +120\text{ ft}\cdot\text{lb}\,\mathbf{i} - 300\text{ ft}\cdot\text{lb}\,\mathbf{j}\end{aligned}$$

The equivalent force created by $\mathbf{F}_A$ *and* $\mathbf{F}_B$ is, by definition, zero, as $\mathbf{F}_A$ and $\mathbf{F}_B$ are a couple.

Next we will consider the couple consisting of $\mathbf{F}_C$ and $\mathbf{F}_D$. Taking the moment about a moment center O (an arbitrary choice for the moment center), we find that:

The equivalent moment created by $\mathbf{F}_C$ *and* $\mathbf{F}_D$ is

$$\mathbf{M}_5 = -3\text{ ft }(50\text{ lb})\,\mathbf{k} = -150\text{ ft}\cdot\text{lb}\,\mathbf{k}$$

The equivalent force created by $\mathbf{F}_C$ *and* $\mathbf{F}_D$ is, by definition, zero, as $\mathbf{F}_C$ and $\mathbf{F}_D$ are a couple.

We are now (finally!) in a position to determine the magnitude and direction of the equivalent moment and equivalent force for the loading in **Figure 5.83**.

$$\begin{aligned}\mathbf{M}_{\text{equ}} &= \mathbf{M}_4 + \mathbf{M}_5\\ &= (+120\,\mathbf{i} - 300\,\mathbf{j})\text{ ft}\cdot\text{lb} - 150\text{ ft}\cdot\text{lb}\,\mathbf{k}\\ &= +120\text{ ft}\cdot\text{lb}\,\mathbf{i} - 300\text{ ft}\cdot\text{lb}\,\mathbf{j} - 150\text{ ft}\cdot\text{lb}\,\mathbf{k}\\ \mathbf{F}_{\text{equ}} &= 0\end{aligned}$$

Answer $\mathbf{M}_{\text{equ}} = +120\text{ ft}\cdot\text{lb}\,\mathbf{i} - 300\text{ ft}\cdot\text{lb}\,\mathbf{j} - 150\text{ ft}\cdot\text{lb}\,\mathbf{k}$, and $\mathbf{F}_{\text{equ}} = 0$

Check In solving this problem, we recognized that two couples were acting on the system and used that to our advantage. If we had not identified the couples, we would have proceeded by finding the moment created by each of the four forces at a common moment center (perhaps using the cross-product approach), then summing these moments together to find the equivalent moment. The same answer as above would have resulted.

Additional questions to consider: Which of the figures in **Figures 5.85 through 5.87** correctly depicts the equivalent moment and force found above? What conclusions can you draw about the equivalent loading associated with a couple? Would it have made sense for this example to have asked you to find the equivalent loading (without specifying a moment center) if the loadings acting at the T-joint did not consist of couples?

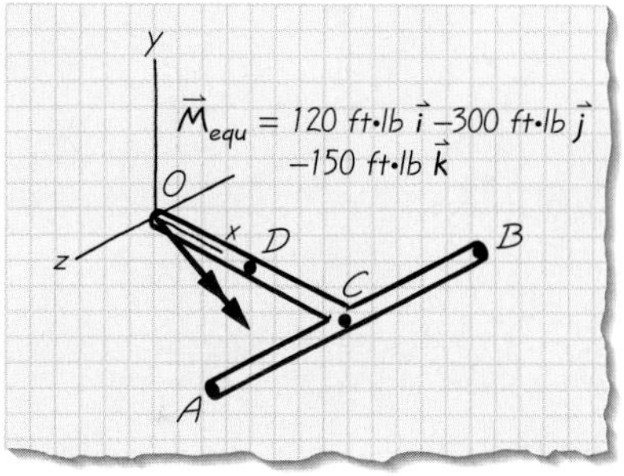

Figure 5.85

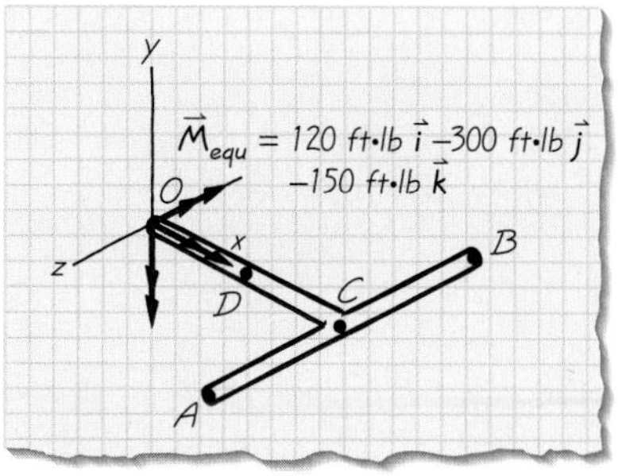

Figure 5.86

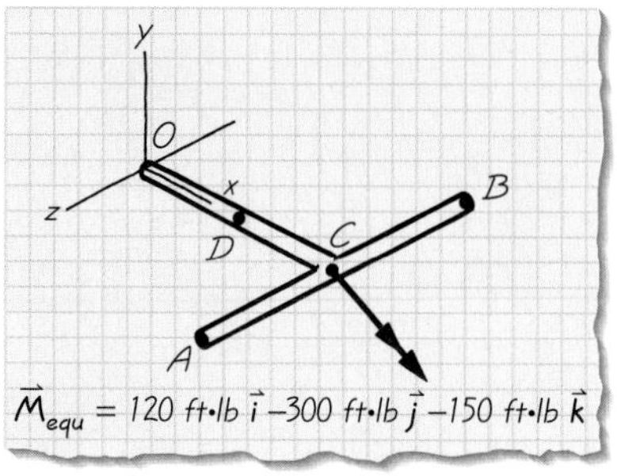

Figure 5.87

EXAMPLE 5.19 EQUIVALENT LOADINGS INCLUDING COUPLES

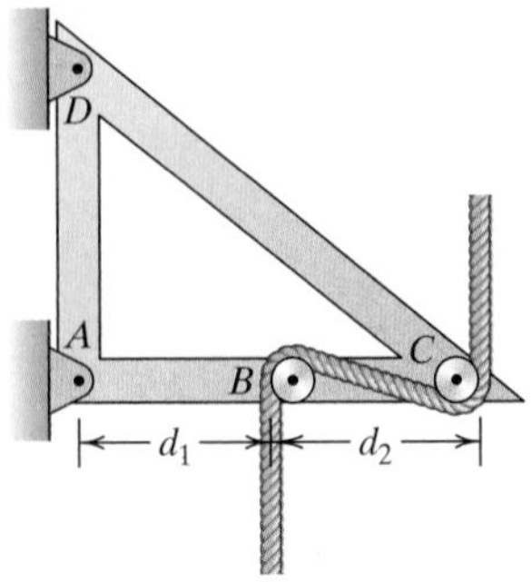

Figure 5.88

In **Figure 5.88**, a rope is wrapped around the frictionless pulleys at B and C attached to the bracket. The tension in the rope is $\boldsymbol{T}$.

(a) Show on the drawing in **Figure 5.89** the forces the rope applies to the bracket (including the two pulleys). Do these forces constitute a couple (why or why not?)

(b) What are the equivalent moment and equivalent force at a moment center at A resulting from the forces you drew in **Figure 5.89**? Show the equivalent loading graphically.

(c) What are the equivalent moment and equivalent force at a moment center midway between the two forces resulting from the forces you drew in **Figure 5.89**? Show the equivalent loading graphically.

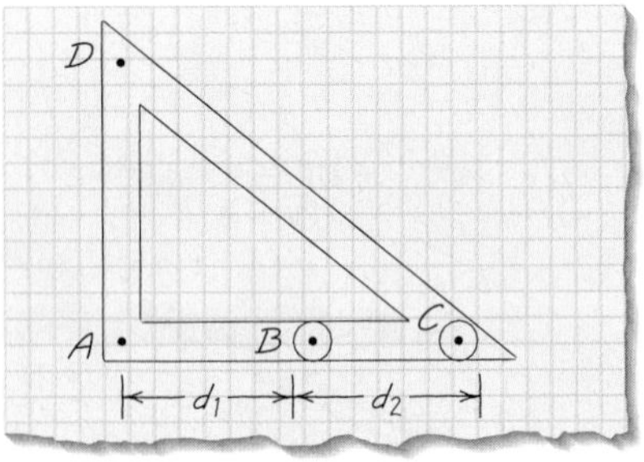

Figure 5.89

Goal We are asked to comment on whether the loading in **Figure 5.88** constitutes a couple (a), then determine the equivalent loading at a moment center at A (b) and at a moment center at the midpoint between the forces (c).

Given We are given information on some of the dimensions of the triangular bracket in terms of variables (d_1, d_2) We are also told that the pulleys are frictionless.

Assume We assume that the forces of the rope acting on the bracket act in the vertical direction and are in the plane of the bracket.

Draw Based on the assumptions, we redraw the bracket in **Figure 5.90** (which is part of the answer to (a)). Because the pulleys are frictionless, $\| \boldsymbol{T}_1 \| = \| \boldsymbol{T}_2 \| = \| \boldsymbol{T} \|$; therefore, these two forces are equal in magnitude and opposite in direction. As such, they constitute a couple.

Answer to (a) **Figure 5.90**; the two rope forces are equal in magnitude and opposite in direction—therefore they are, by definition, a couple.

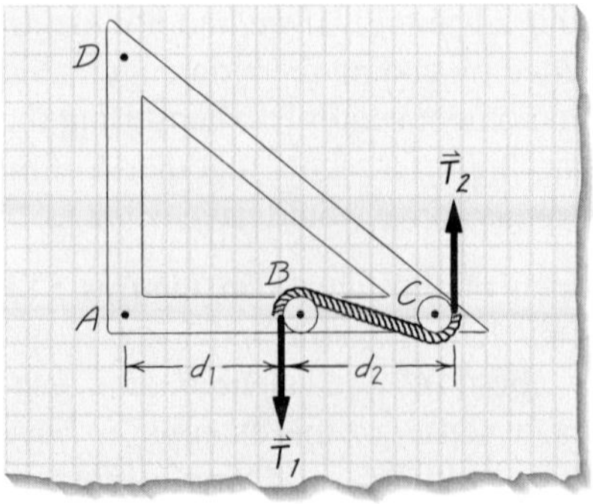

Figure 5.90

Formulate Equations and Solve **(b)** In Figure 5.91*a* we have drawn the bracket with $\boldsymbol{T}_1$ and $\boldsymbol{T}_2$ and coordinate axes having their origin at A. Force $\boldsymbol{T}_1$ creates a moment $\boldsymbol{M}_1$ at A of

$$\boldsymbol{M}_1 = -d_1 \ \|\boldsymbol{T}\| \ \boldsymbol{k}$$

and $\boldsymbol{T}_2$ creates a moment $\boldsymbol{M}_2$ at A of

$$\boldsymbol{M}_2 = +(d_1 + d_2) \ \|\boldsymbol{T}\| \ \boldsymbol{k}$$

The equivalent moment $\boldsymbol{M}_{\text{equ}}$ is

$$\boldsymbol{M}_{\text{equ}} = \boldsymbol{M}_1 + \boldsymbol{M}_2 = +d_2 \ \|\boldsymbol{T}\| \ \boldsymbol{k}$$

The equivalent force $\boldsymbol{F}_{\text{equ}} = \boldsymbol{T}_1 + \boldsymbol{T}_2$ is equal to zero; the two forces that make up a couple cancel each other and result in an equivalent force of zero.

Answer to (b) $\boldsymbol{M}_{\text{equ}} = +d_2 \ \|\boldsymbol{T}\| \ \boldsymbol{k}$, $\boldsymbol{F}_{\text{equ}} = 0$. See **Figure 5.91*b*** for the equivalent loading.

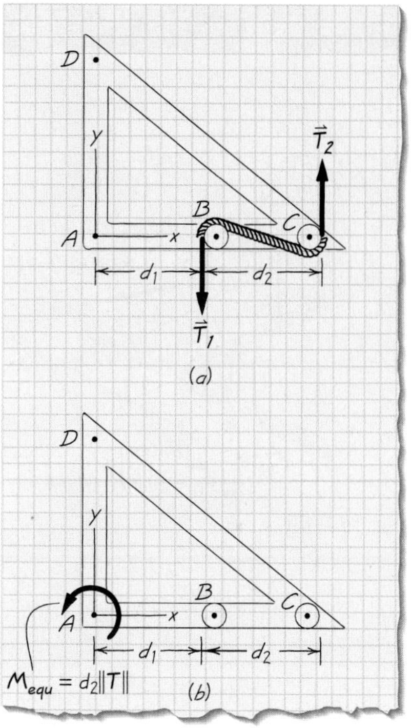

Figure 5.91

(c) In Figure 5.92*a* we have drawn the bracket with $\boldsymbol{T}_1$ and $\boldsymbol{T}_2$, and coordinate axes having their origin midway between the lines of action of the two forces (call this point O). We proceed by finding the moment created by $\boldsymbol{T}_1$ and $\boldsymbol{T}_2$ at a moment center at O. Force $\boldsymbol{T}_1$ creates a moment $\boldsymbol{M}_1$ at O of

$$\boldsymbol{M}_1 = +\frac{d_2}{2} \ \|\boldsymbol{T}\| \ \boldsymbol{k}$$

and $\boldsymbol{T}_2$ creates a moment $\boldsymbol{M}_2$ at O of

$$\boldsymbol{M}_2 = +\frac{d_2}{2} \ \|\boldsymbol{T}\| \ \boldsymbol{k}$$

The equivalent moment $\boldsymbol{M}_{\text{equ}}$ is equal to

$$\boldsymbol{M}_{\text{equ}} = \boldsymbol{M}_1 + \boldsymbol{M}_2 = +d_2 \ \|\boldsymbol{T}\| \ \boldsymbol{k}$$

The equivalent force $\boldsymbol{F}_{\text{equ}} = \boldsymbol{T}_1 + \boldsymbol{T}_2$ is equal to zero—not surprising since these two forces make up a couple. This is the same answer as in (b).

Answer to (c) $\boldsymbol{M}_{\text{equ}} = +d_2 \ \|\boldsymbol{T}\| \ \boldsymbol{k}$, $\boldsymbol{F}_{\text{equ}} = 0$. See **Figure 5.92*b*** for the equivalent loading. The equivalent loading (equivalent moment and equivalent force) represented by a couple consists of a moment and no force. This is true no matter where the moment center is located (and therefore we found the same answer in (b) and (c)).

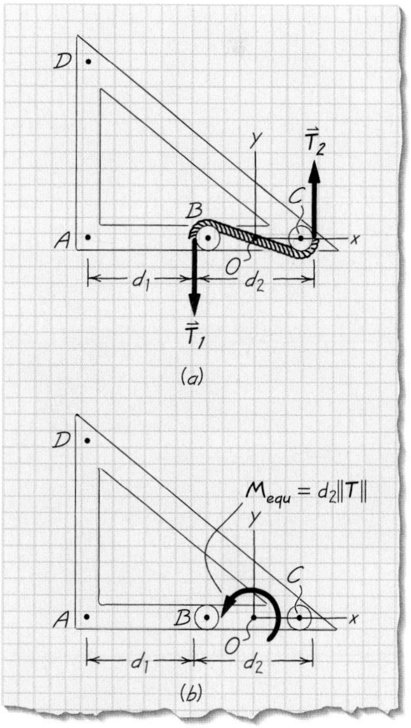

Figure 5.92

EXAMPLE 5.20 EQUIVALENT LOADS INCLUDING COUPLES

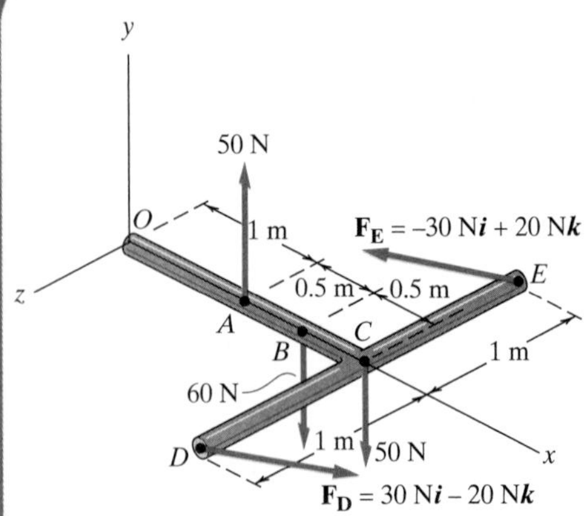

Figure 5.93

The loadings acting on the T-joint in **Figure 5.93** consist of two couples and a force.

(a) Find the equivalent force and equivalent moment about a set of axes at O for these loadings. Present your answer in vector and graphical forms.

(b) Find the equivalent force and equivalent moment about a set of axes at B for these loadings. Present your answer in vector and graphical forms.

Goal We are to find the equivalent loading for the forces acting on the T-joint for a moment center at O (a) and a moment center at B (b).

Given We are given dimensions of the T-joint, a coordinate system, and information on five forces acting on the T-joint.

Assume No assumptions are necessary.

Formulate Equations and Solve **(a)** The loadings acting on the T-joint are as follows:

- A couple consisting of 50-N forces spaced 1 m from each other
- A couple consisting of forces at D and E
- A 60-N force at B

To find the equivalent moment at O, we will find the moment at O created by each of these loadings, then sum the moments together using vector addition.

$\mathbf{M}_1$ resulting from two 50-N forces: Because these forces are equal in magnitude and opposite in sign, they create a moment of magnitude (50 N)(1 m) = 50 N · m. This moment is perpendicular to the plane of the two forces; this is the $-\mathbf{k}$ direction. Therefore,

$$\mathbf{M}_1 = -50\ \text{N} \cdot \text{m}\ \mathbf{k}$$

As illustrated in Example 5.18, this moment couple is independent of the moment center. Therefore, $\mathbf{M}_1$ is the moment the couple produces about O (and every other point in space since the equivalent moment created by a couple is independent of the location of the moment center).

$\mathbf{M}_2$ resulting from $\mathbf{F}_D$ and $\mathbf{F}_E$: Inspection of $\mathbf{F}_D$ and $\mathbf{F}_E$ (as given in **Figure 5.93**) shows that they are equal and opposite; therefore, they are a couple and create a moment. To find the magnitude of this moment, we could find the perpendicular distance between $\mathbf{F}_D$ and $\mathbf{F}_E$, which is shown in **Figure 5.94*a***; then $\|\mathbf{M}_2\|$ would be the product of this distance and the magnitude of the force.

Alternatively, we can break the two forces into their x and z components (**Figure 5.94*b***). By inspection, the z components do not contribute to the moment because they have the same line of action. The x components are a couple equivalent to a moment of (4 m) (30 N) = 120 N · m about an axis perpendicular to the plane of the two forces; this is in the

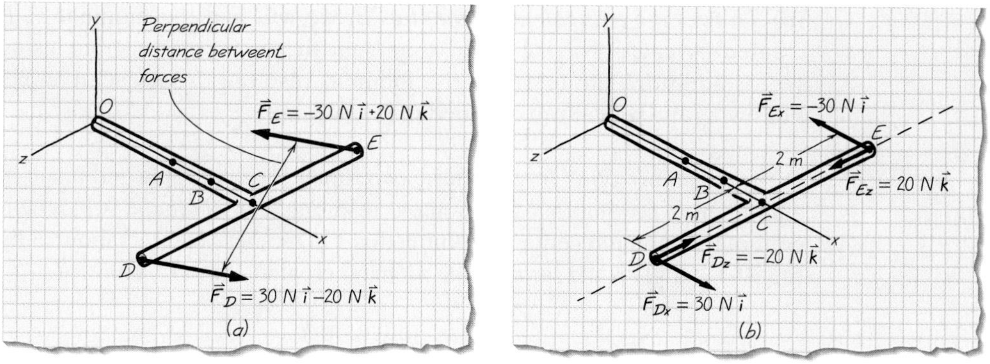

Figure 5.94

$\boldsymbol{j}$ direction (use the right-hand rule to confirm this). The rotation caused by this 120-N · m moment is counterclockwise about the y axis; therefore the moment is positive. We can now write the couple created by $\boldsymbol{F}_D$ and $\boldsymbol{F}_E$ as

$$\boldsymbol{M}_2 = +120\ \text{N}\cdot\text{m}\,\boldsymbol{j}$$

Using the same argument we used when considering the moment created by the couple consisting of the two 50-N forces, we know $\boldsymbol{M}_2$ is the moment the couple consisting of $\boldsymbol{F}_D$ and $\boldsymbol{F}_E$ produces about O (and every other point in space).

$\boldsymbol{M}_3$ created by $\boldsymbol{F}_B$: Force $\boldsymbol{F}_B$ is represented by the vector $\boldsymbol{F}_B = -60$ N $\boldsymbol{j}$. Its position vector relative to O is $\boldsymbol{r}_3 = 1.5$ m $\boldsymbol{i}$. Because this force and position vector lie in the xy plane, the moment the force creates can be calculated using (5.11A):

$$\begin{aligned}\boldsymbol{M}_3 &= (r_xF_y - r_yF_x)\,\boldsymbol{k}\\ &= [(1.5\ \text{m})(-60\ \text{N}) - (0\ \text{m})\,(0\ \text{N})]\,\boldsymbol{k} = -90\ \text{N}\cdot\text{m}\,\boldsymbol{k}\end{aligned}$$

Equivalent moment:

$$\boldsymbol{M}_{\text{equ@}O} = \boldsymbol{M}_1 + \boldsymbol{M}_2 + \boldsymbol{M}_3 = (-50\boldsymbol{k} + 120\boldsymbol{j} - 90\,\boldsymbol{k})\ \text{N}\cdot\text{m} = 120\ \text{N}\cdot\text{m}\,\boldsymbol{j} - 140\ \text{N}\cdot\text{m}\,\boldsymbol{k}$$

Equivalent force:

$$\begin{aligned}\boldsymbol{F}_{\text{equ}} &= \text{sum of forces} = [50\boldsymbol{j} - 50\boldsymbol{j}) + (30\boldsymbol{i} - 30\boldsymbol{i})\\ &\quad + (20\boldsymbol{k} - 20\boldsymbol{k}) - 60\boldsymbol{j}]\ \text{N} = -60\ \text{N}\,\boldsymbol{j}\end{aligned}$$

(We really did not need to include the couple forces in this calculation because, by definition of a couple, they cancel each other! We included them to show that they really do cancel.)

Answer to (a) The equivalent moment about O is $\boldsymbol{M}_{\text{equ@}O} = 120$ N · m $\boldsymbol{j}$ − 140 N · m $\boldsymbol{k}$. The equivalent force is $\boldsymbol{F}_{\text{equ}} = -60$ N $\boldsymbol{j}$. Graphically the equivalent moment and equivalent force are represented in **Figure 5.95**.

A final note on (a): The loadings depicted in **Figures 5.93** and **5.95** are equivalent to each other.

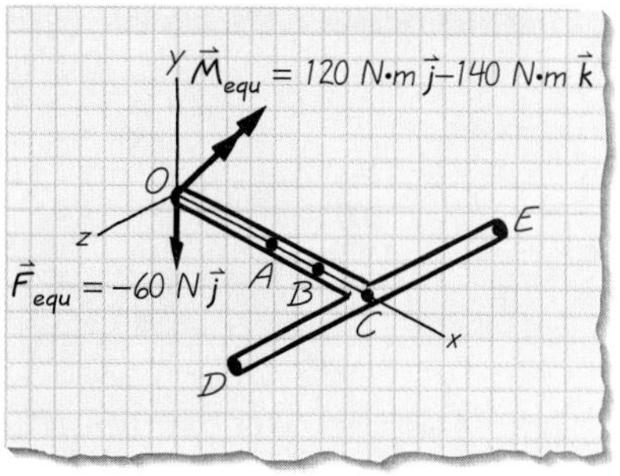

Figure 5.95

(b) *To find the equivalent moment at B*, we will find the moment at B created by each of these loads, then sum the moments together using vector addition. We first find the moment resulting from each couple. Following the procedure outlined in (a), we find:

$\boldsymbol{M}_1$ *due to two 50-N forces:* These two forces form a couple and result in a moment. Knowing that the moment resulting from a couple is independent of the moment center, we know that the moment created about B by the 50-N couple is the same as the moment created about O; therefore:

$$\boldsymbol{M}_1 = -50 \text{ N}\cdot\text{m } \boldsymbol{k}$$

$\boldsymbol{M}_2$ *resulting from* $\boldsymbol{F}_D$ *and* $\boldsymbol{F}_E$: Just as in (a),

$$\boldsymbol{M}_2 = +120 \text{ N}\cdot\text{m } \boldsymbol{j}$$

$\boldsymbol{M}_3$ *created by* $\boldsymbol{F}_B$: Because the position of this force relative to B is represented by the vector $\boldsymbol{r}_3 = 0\ \boldsymbol{i}$ m, the force creates zero moment about B.

Equivalent moment:

$$\boldsymbol{M}_{\text{equ@}B} = \boldsymbol{M}_1 + \boldsymbol{M}_2 + \boldsymbol{M}_3 = (-50\boldsymbol{k} + 120\boldsymbol{j} + 0\ \boldsymbol{k}) \text{ N}\cdot\text{m}$$
$$= -50 \text{ N}\cdot\text{m } \boldsymbol{k} + 120 \text{ N}\cdot\text{m } \boldsymbol{j}$$

Equivalent force:

$$\boldsymbol{F}_{\text{equ}} = \text{sum of forces} = [(50\boldsymbol{j} - 50\boldsymbol{j}) + (30\boldsymbol{i} - 30\boldsymbol{i}) + (20\boldsymbol{k} - 20\boldsymbol{k}) - 60\boldsymbol{j}] \text{ N} = -60 \text{ N } \boldsymbol{j}$$

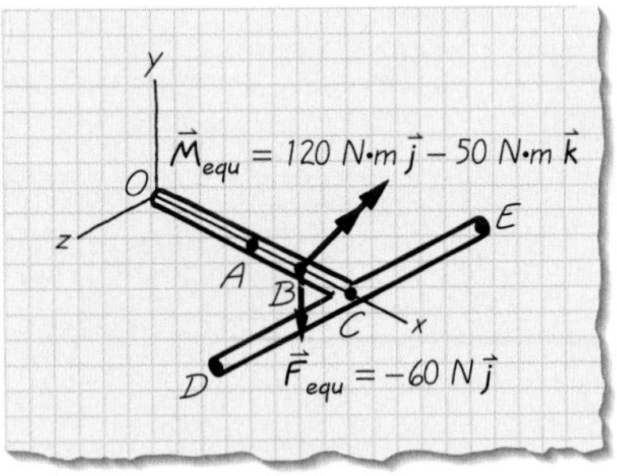

Figure 5.96

Answer to (b) $\boldsymbol{M}_{\text{equ@}B} = -50 \text{ N}\cdot\text{m } \boldsymbol{k} + 120 \text{ N}\cdot\text{m } \boldsymbol{j}$. The equivalent force is $\boldsymbol{F}_{\text{equ}} = -60 \text{ N } \boldsymbol{j}$. Graphically the equivalent moment and equivalent force are represented in **Figure 5.96.**

A final note on part (b): The loadings depicted in **Figures 5.93** and **5.96** are equivalent to each other. Furthermore, if the loading in **Figure 5.93** is equivalent to the loading in **Figure 5.95** (from part (a)) and the loading in **Figure 5.96**, we can conclude that the loadings in **Figures 5.95** and **5.96** are also equivalent to each other.

EXAMPLE 5.21 THE ANALYSIS PROCEDURE REVISITED

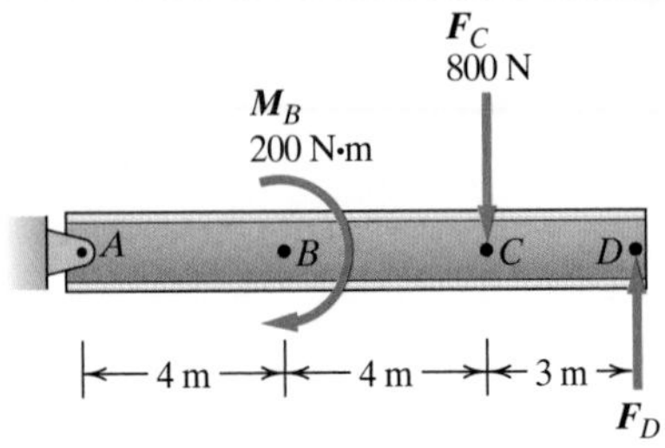

Figure 5.97

Consider the loadings (two forces and a moment) acting on the beam in **Figure 5.97.**

(a) If the equivalent moment at a moment center at A due to these loadings is zero, what is the magnitude of $\boldsymbol{F}_D$?

(b) What is the equivalent force due to the loads shown in **Figure 5.97**?

(c) Show the equivalent force and equivalent moment associated with (a) and (b) on a drawing of the beam.

Goal

(a) Calculate the magnitude of $\boldsymbol{F}_D$ (where $\boldsymbol{F}_D$ is the force applied at D).

(b) Calculate the equivalent force acting on the beam due to $\boldsymbol{F}_C$, $\boldsymbol{F}_D$, and moment $\boldsymbol{M}_B$.

(c) Draw a graphical representation of the equivalent force and equivalent moment due to the loadings shown in **Figure 5.97** at a moment center at A.

Given

1. The points of application of $\boldsymbol{F}_C$ and $\boldsymbol{F}_D$.
2. The magnitude and direction of $\boldsymbol{F}_C$, the magnitude and sense of $\boldsymbol{M}_B$, and the direction of $\boldsymbol{F}_D$.
3. The equivalent moment created by $\boldsymbol{F}_C$, $\boldsymbol{F}_D$, and $\boldsymbol{M}_B$ at a moment center at A is zero.

Assume

1. The system of interest is the beam that runs from A to D.
2. The coordinate axes we shall use have their origin at A.
3. The weight of the beam is small relative to loads $\boldsymbol{F}_C$ and $\boldsymbol{F}_D$ and therefore can be ignored.
4. Forces $\boldsymbol{F}_C$ and $\boldsymbol{F}_D$ and the beam all lie in the xy plane. Moment $\boldsymbol{M}_B$ is about an axis perpendicular to the xy plane.

Draw Isolate the beam and draw it, its dimensions, the coordinate axes, and the loads acting on the beam. See **Figure 5.98**.

Often in solving equilibrium problems the drawing created in this step of the analysis procedure will be a free-body diagram that shows the structure and the loads acting on it. Note, however, that **Figure 5.98*b*** is *not* a free-body diagram because it does not include the loads imposed on the beam at connection A by the rest of the world. As we will see in the next step, we do not consider these loads in answering the questions posed in this problem.

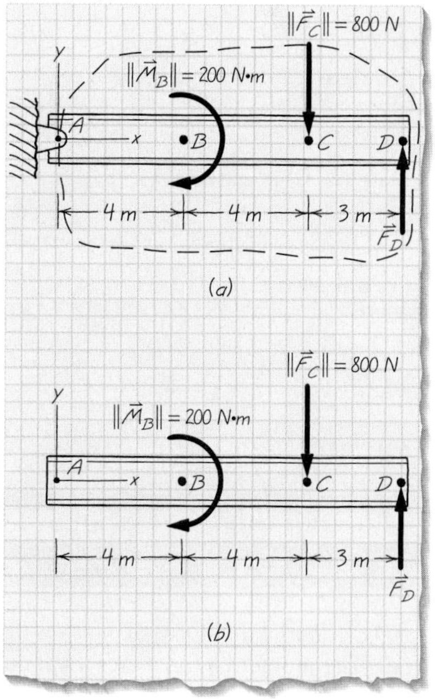

Figure 5.98

Formulate Equations and Solve **(a)** We proceed by finding the moments created by $\boldsymbol{F}_C$ and $\boldsymbol{F}_D$ about a moment center at A, then sum these to obtain the equivalent moment.

Moment $\boldsymbol{M}_C$ *created by* $\boldsymbol{F}_C$: For $\boldsymbol{F}_C = -800 \text{ N } \boldsymbol{j}$, the position vector from A to C is $\boldsymbol{r}_C = 8 \text{ m } \boldsymbol{i}$. Because $\boldsymbol{F}_C$ lies in the xy plane, the moment can be calculated using (5.11A):

$$\boldsymbol{M}_C = (r_C F_C)\,\boldsymbol{k} = (8 \text{ m})(-800 \text{ N})\,\boldsymbol{k} = -6400 \text{ N}\cdot\text{m } \boldsymbol{k}$$

Moment $\boldsymbol{M}_D$ *created by* $\boldsymbol{F}_D$: For $\boldsymbol{F}_D = F_D \boldsymbol{j}$, the position vector from A to D is $\boldsymbol{r}_D = 11 \text{ m } \boldsymbol{i}$. Because $\boldsymbol{F}_D$ lies in the xy plane, the moment can be calculated using (5.11A):

$$\boldsymbol{M}_D = (11 \text{ m})\ \|\boldsymbol{F}_D\|\ \boldsymbol{k}$$

Moment $\boldsymbol{M}_B$: At B there is a moment $\boldsymbol{M}_B = -200\ \text{N}\cdot\text{m}\ \boldsymbol{k}$. This moment is negative because the rotation it tends to cause is clockwise about the z axis. By definition of a couple (which is equivalent to a moment), this is equivalent to the same moment applied at A.

Equivalent moment: The equivalent moment at a moment center at A is

$$\boldsymbol{M}_{\text{equ@}A} = \boldsymbol{M}_C + \boldsymbol{M}_D + \boldsymbol{M}_B = [-6400\ \text{N}\cdot\text{m} + (11\ \text{m})\ \|\boldsymbol{F}_D\| - 200\ \text{N}\cdot\text{m}]\,\boldsymbol{k} \quad (1)$$

Find $\|\boldsymbol{F}_D\|$: We can solve (1) for $\|\boldsymbol{F}_D\|$:

$$\|\boldsymbol{F}_D\| = \frac{\|\boldsymbol{M}_{\text{equ@}A}\| + 6600\ \text{N}\cdot\text{m}}{11\ \text{m}}$$

Knowing from item 3 in our Given list that $\|\boldsymbol{M}_{\text{equ}}\| = 0$, we can say

$$\|\boldsymbol{F}_D\| = \frac{6600\ \text{N}\cdot\text{m}}{11\ \text{m}} = 600\ \text{N}$$

Answer to (a) The magnitude of the force at D is 600 N.

(b) To find the equivalent force, we will add the force components from the loadings acting on the beam.

$F_{\text{equ},x}$ = sum of forces in x direction = 0 (no forces acting in the x direction)

$F_{\text{equ},y}$ = sum of forces in y direction = $F_C + F_D = -800\ \text{N} + 600\ \text{N}$ = -200 N

$F_{\text{equ},y}$ = sum of forces in z direction = 0 (no forces acting in the z direction)

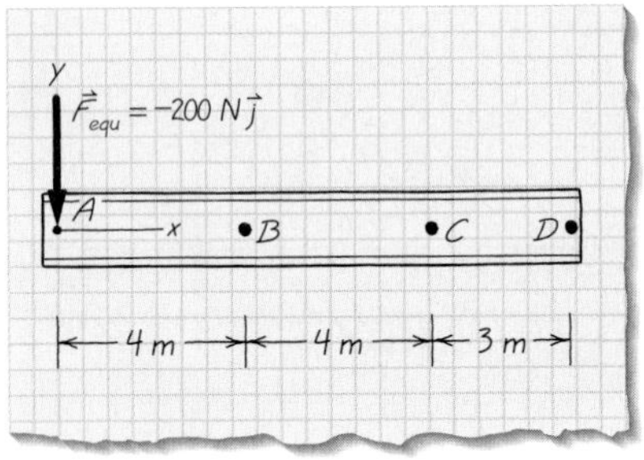

Figure 5.99

Answer to (b) $F_{\text{equ},x} = 0$, $F_{\text{equ},y} = -200$ N, $F_{\text{equ},z} = 0$ or, in vector notation, $\boldsymbol{F}_{\textbf{equ}} = -200\ \text{N}\,\boldsymbol{j}$.

(c) Recall that $\|\boldsymbol{F}_D\|$ was calculated in (a) for the case where the equivalent moment at A is zero. Therefore the loading equivalent to that shown in **Figure 5.97** is the loading shown in **Figure 5.99**.

Check The answer to (a) can be checked by reviewing each assumption and equation to confirm that correct data were used. In addition, by substituting the 600 N value of $\boldsymbol{F}_D$ found in (a) back into (1), we can confirm that the equivalent moment at A is zero.

EXERCISES 5.4

5.4.1. Choose the words that makes the statement correct.

a. The moment a force creates about a moment center is *independent of/dependent on* the position vector.

b. The moment a force creates about a moment center is *independent of/dependent on* the component of the position vector that is perpendicular to the line of action of the force.

c. The moment a force creates about a moment center is *independent of/dependent on* the component of the force that is parallel to the position vector.

d. A moment whose sense is *clockwise/counterclockwise* about the x axis is in the $+\boldsymbol{i}$ direction.

e. The loading equivalent to a couple *is/is not* a moment.

f. The moment equivalent to a couple is *independent of/dependent on* the moment center.

g. In calculating the equivalent moment from several forces acting on a system, *various moment centers are used/only one moment center is used.*

For **Exercises 5.4.2–5.4.4**, a cantilever beam acted on by a force $\boldsymbol{F}$ at point C is shown in E5.4.2*a*. The equivalent loading at A is shown in **E5.4.2*b***. Present your answer in vector notation and as a drawing of the cantilever beam.

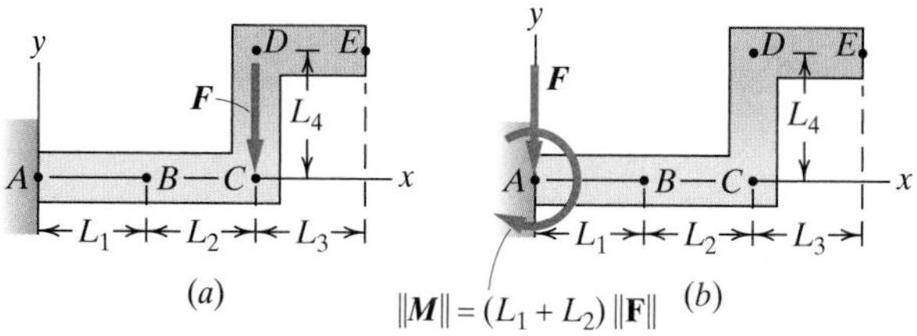

E5.4.2

5.4.2. Determine the equivalent loading at B.

5.4.3. Determine the equivalent loading at D.

5.4.4. Determine the equivalent loading at E.

5.4.5. A lug wrench is used to tighten a lug nut on an automobile wheel, as shown in E5.4.5. Replace the force $\boldsymbol{F}$ applied by the mechanic's hand to the lug wrench by an equivalent loading at the center of the lug nut. Present your answer in vector notation and as a diagram that shows the equivalent force and moment acting at the center of the lug nut.

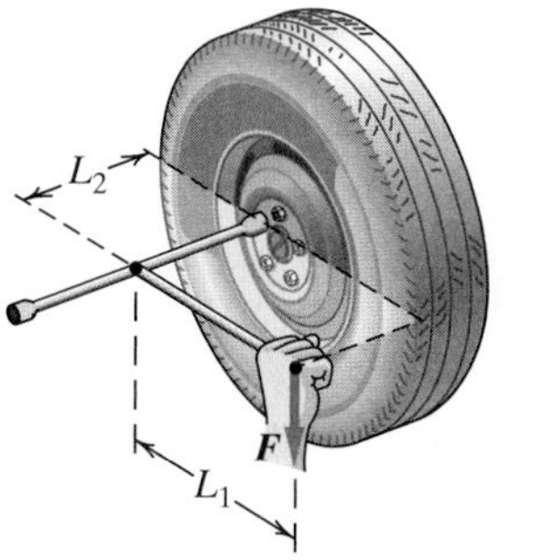

E5.4.5

5.4.6. Replace the three forces shown in E5.4.6 by the equivalent loading acting at O. Present your answer in vector notation and as a diagram that shows the equivalent force and moment acting at O.

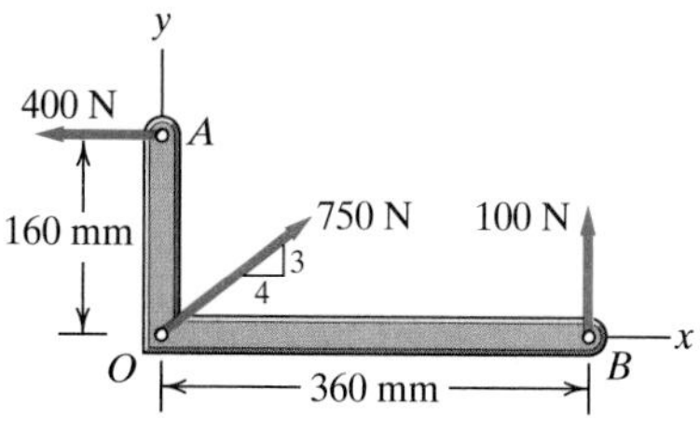

E5.4.6

5.4.7. Replace the three forces shown in E5.4.7 by the equivalent loading acting at A. Present your answer in vector notation and as a diagram that shows the equivalent force and moment acting at A.

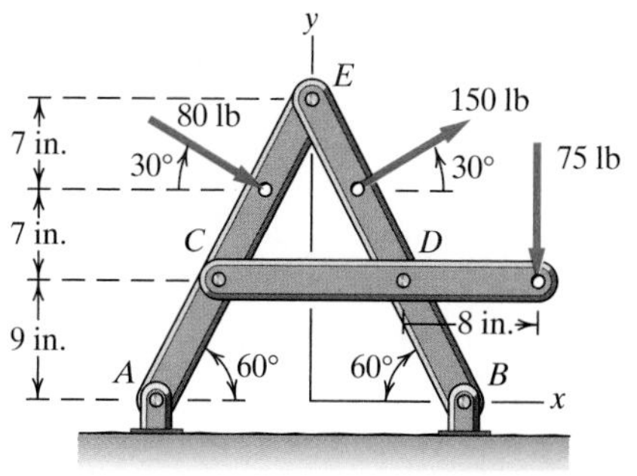

E5.4.7

5.4.8. For the cantilever beam acted on by $\boldsymbol{T}_{BC}$ and $\boldsymbol{F}_B$ in E.5.4.8 replace the forces by the equivalent loading acting at A. Present your answer in vector notation and as a diagram that shows the equivalent force and moment acting at A.

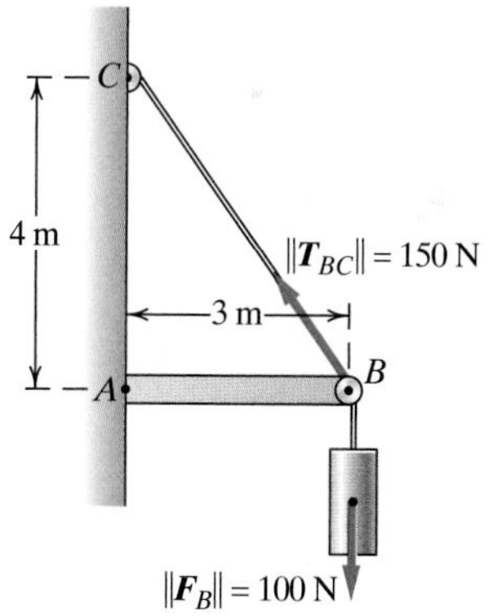

E5.4.8

5.4.9. For the beam pinned at A and acted on by $\boldsymbol{T}_{BC}$, $\boldsymbol{F}_B$, and $\boldsymbol{M}_C$ in E5.4.9, replace the loads by the equivalent loading acting at A. Present your answer in vector notation and as a diagram that shows the equivalent force and moment acting at A.

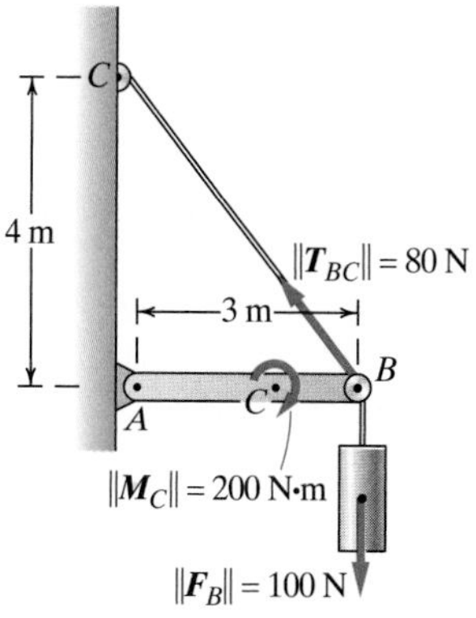

E5.4.9

5.4.10. For the cantilever beam acted on by $\boldsymbol{F}_B$ and $\boldsymbol{M}_C$ in E5.4.10, replace the loads by the equivalent loading acting at A. Present your answer in vector notation and as a diagram that shows the equivalent force and moment acting at A.

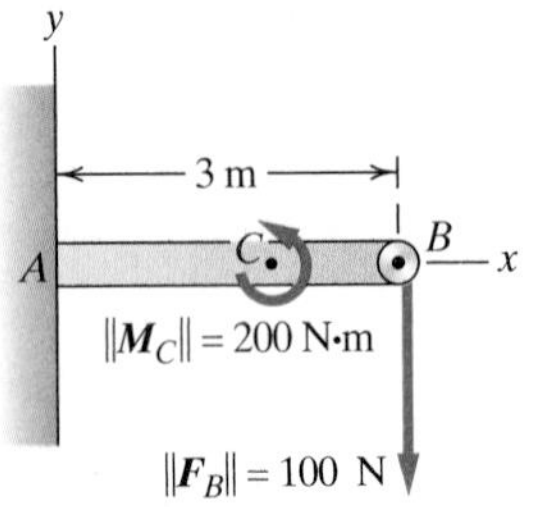

E5.4.10

5.4.11. For the cantilever beam acted on by moments $\boldsymbol{M}_B$ and $\boldsymbol{M}_C$ in E5.4.11, replace the loads by the equivalent loading acting at A. Present your answer in vector notation and as a diagram that shows the equivalent force and moment acting at A.

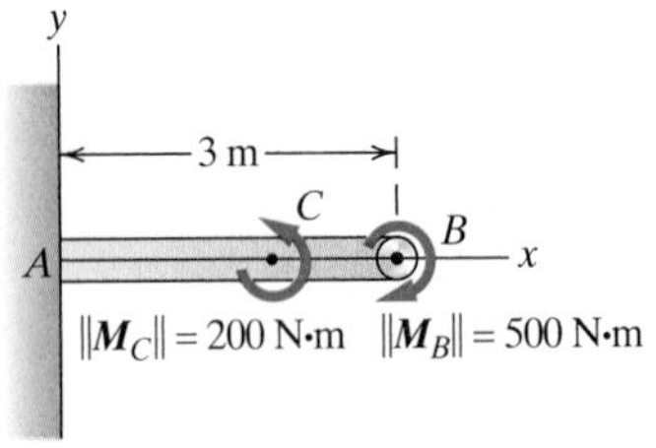

E5.4.11

5.4.12. A stepped gear is subjected to two forces as shown in E5.4.12. Determine the equivalent loading acting at the center C of the gear. Present your answer in vector notation and as a diagram that shows the equivalent force and moment acting at C.

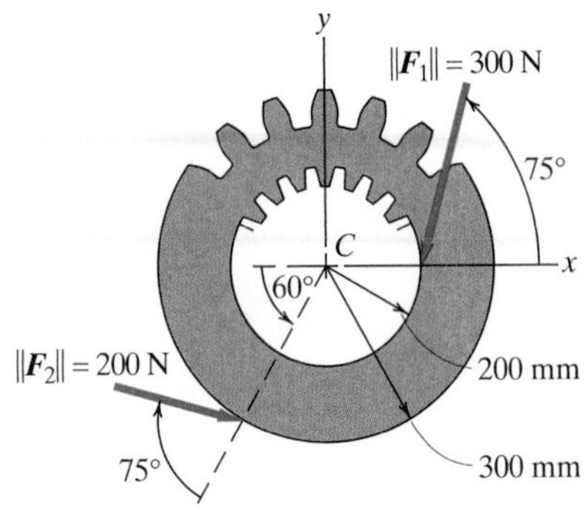

E5.4.12

5.4.13. The streetlight in E5.4.13 is supported by cable AC. The total weight of the structure is 1600 N acting at point G, and the tension in the cable is 3600 N. The 200-N horizontal force represents the effect of the wind. Determine the equivalent loading at the base B. Present your answer in vector notation and as a diagram that shows the equivalent force and moment acting at B.

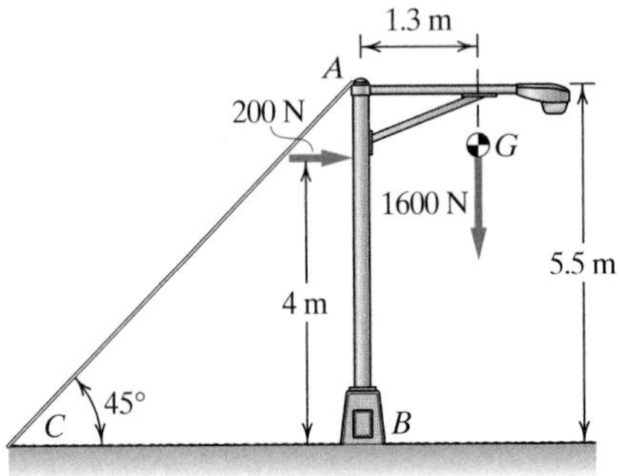

E5.4.13

5.4.14. For each of the right-pedal configurations shown in E5.4.14, determine the equivalent loading at the center of the bottom bracket axle. The length of the crank-arm is 170 mm. Present your answers in Table 5.4.14.

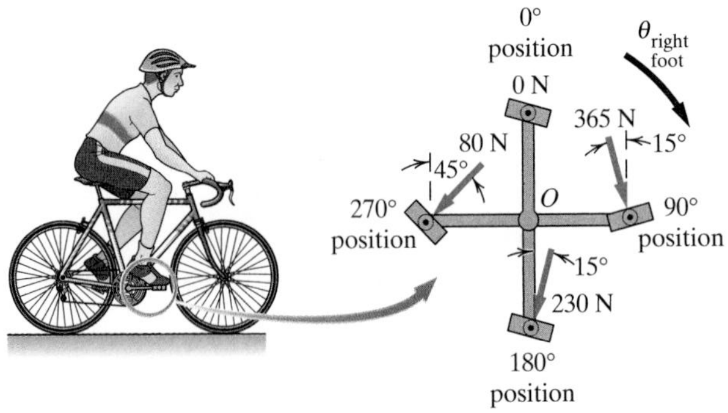

E5.4.14

Table 5.4.14

$\theta_{\text{right foot}}$	Moment at O (center of bottom bracket axle) created by right foot
90°	
180°	
270°	
0°	

5.4.15. Replace the two forces shown in **E5.2.44** by the equivalent loading acting at O. Present your answer in vector notation and as a diagram that shows the equivalent force and moment acting at O.

5.4.16. For the two pipe-wrench forces of **E5.2.46**, determine the equivalent loading at O. Present your answer in vector notation and as a diagram that shows the equivalent force and moment acting at O.

5.4.17. Three parallel forces act on the plate shown in **E5.4.17.**

a. Determine the equivalent loading at O. Present your answer in vector notation and as a diagram that shows the equivalent force and moment acting at O.

b. Find the location in terms of x and y coordinates of a loading equivalent to the one found in **a** that consists of an equivalent force and a *zero* equivalent moment. Present a drawing that shows your answer.

c. Explain why your answer in **b** makes intuitive sense.

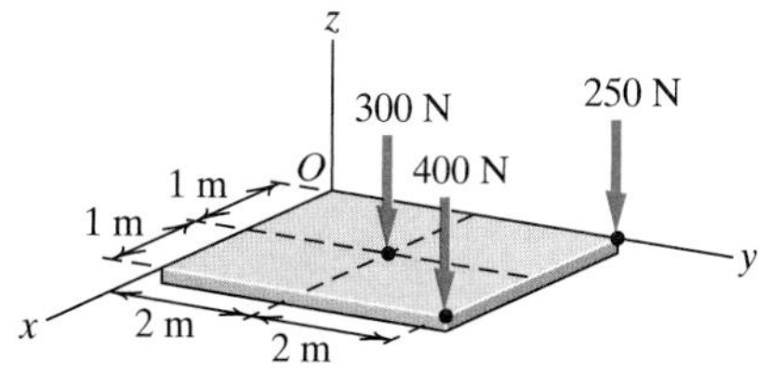

E5.4.17

5.4.18. Forces are applied at points A, B, and C on the bar shown in **E5.4.18**. Replace the forces with their equivalent loading acting at O. Present your answer in vector notation and as a diagram that shows the equivalent force and moment acting at O.

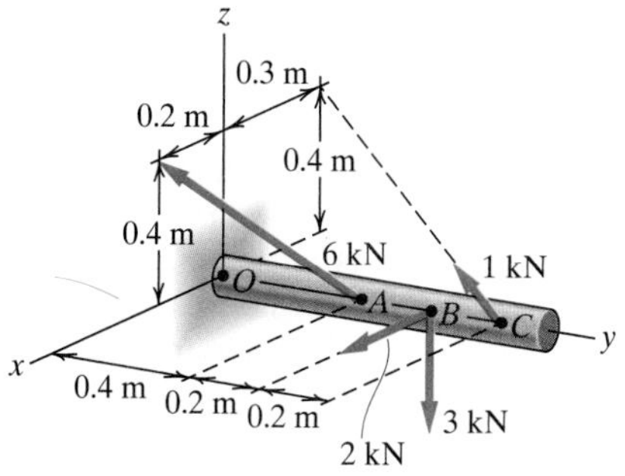

E5.4.18

5.4.19. A socket wrench is subjected to the forces shown in **E5.4.19**. Determine the equivalent loading acting at A. Present your answer in vector notation and as a diagram that shows the equivalent force and equivalent moment acting at A.

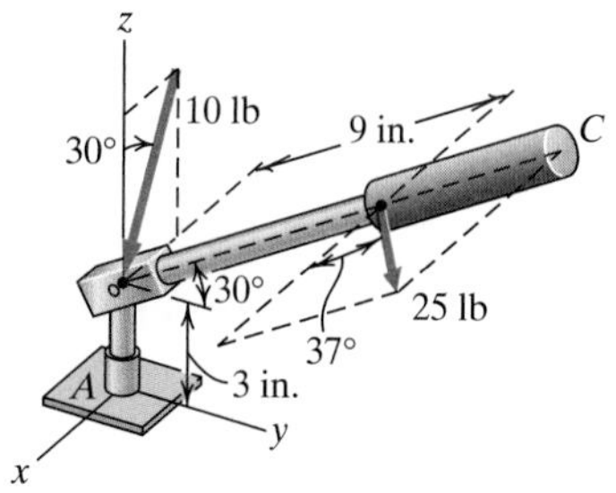

E5.4.19

5.4.20. A valve is loosened with the T-bar assembly of **E5.4.20** by applying a 75-N force at points B and C as shown. Determine

a. the equivalent loading acting at A

b. the portion of this equivalent moment that has the effect of turning the valve about axis AD

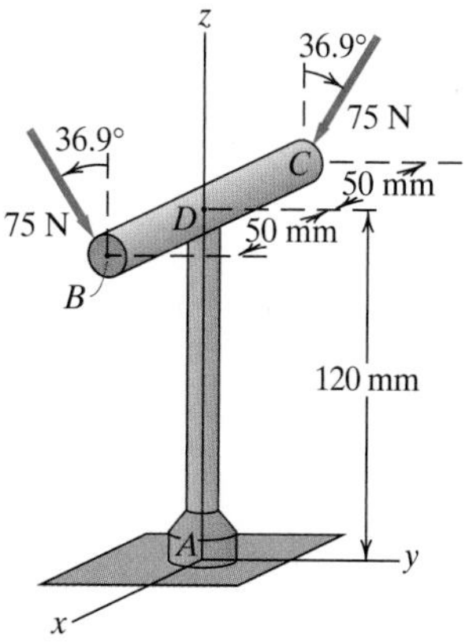

E5.4.20

For **Exercises 5.4.21–5.4.23**, consider the cantilever beam shown in **E5.4.21**. Acting on it are forces $\boldsymbol{F}_1$ and $\boldsymbol{F}_2$. Also $\| \boldsymbol{F}_1 \| = \| \boldsymbol{F}_2 \|$.

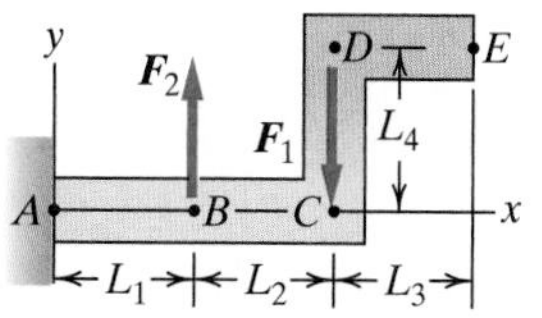

E5.4.21

5.4.21. Determine the equivalent loading at C. Present your answer in vector notation and as a diagram that shows the equivalent force and moment acting at C.

5.4.22. Determine the equivalent loading at B. Present your answer in vector notation and as a diagram that shows the equivalent force and moment acting at B.

5.4.23. Determine the equivalent loading that consists of horizontal forces applied at C and D.

5.4.24. A lug wrench is being used to tighten a lug nut on an automobile wheel, as shown in E5.4.24. Two equal-magnitude parallel forces of opposite sign are applied to the wrench.

a. Describe in a sentence the effect of the forces on the wheel.

b. Determine the equivalent loading at A due to $\boldsymbol{F}_1$ and $\boldsymbol{F}_2$.

c. Repeat **b** for moment centers at B and C. What do you notice about the equivalent loading at A, B, and C? What does this say about a couple?

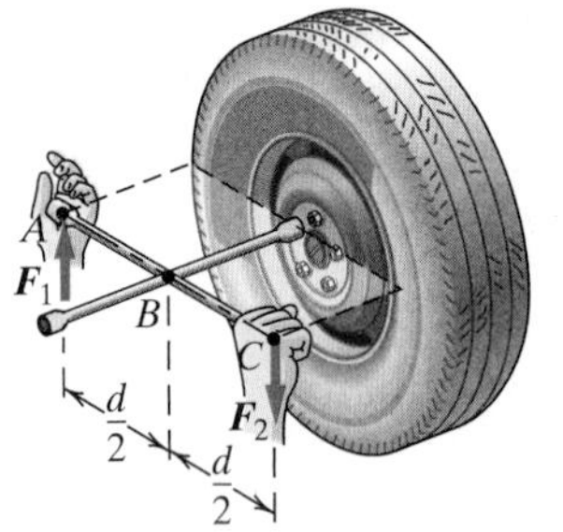

E5.4.24

5.4.25. Six forces act on the bracket shown in E5.4.25.

a. Determine the equivalent loading at D due to the six forces.

b. Identify and describe, in words, the three couples acting on the bracket.

c. Determine the equivalent loading at D due to the three couples acting on the bracket. Compare your answer with the one for **a**.

d. What is the equivalent loading at B? (You should be able to answer this without additional calculations.)

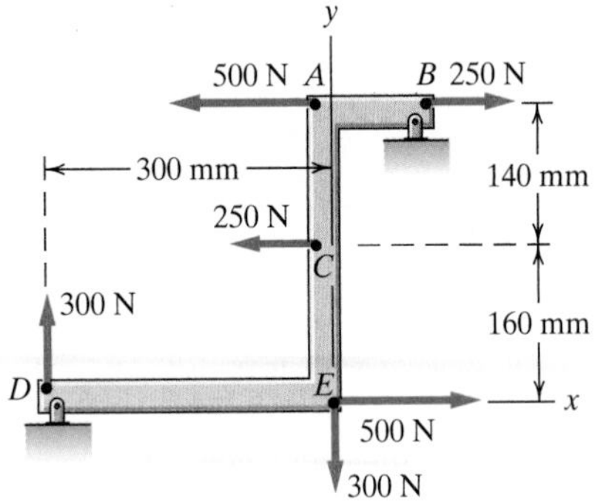

E5.4.25

5.4.26. A force acts at C on the cantilever beam in E5.4.26. Determine the equivalent loading at A that consists of:

a. An equivalent force and an equivalent moment. Present your answer in vector notation and as a drawing.

b. An equivalent force and a couple. (*Hint:* Convert the equivalent moment found in **a** into an equivalent couple.) Present your answer as a drawing.

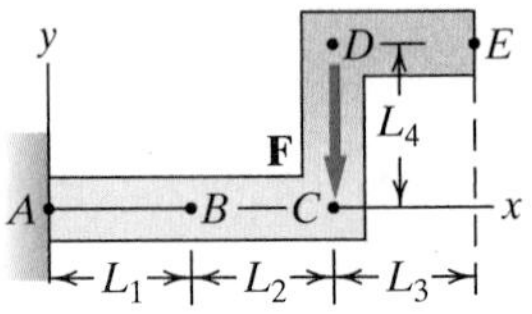

E5.4.26

5.4.27. A bus driver applies two 60-N forces to the steering wheel shown in E5.4.27. These two forces are a couple and are in the plane of the steering wheel.

a. What is the magnitude of the equivalent couple moment?

b. What is the magnitude of the component of this couple moment that has the effect of turning the steering wheel?

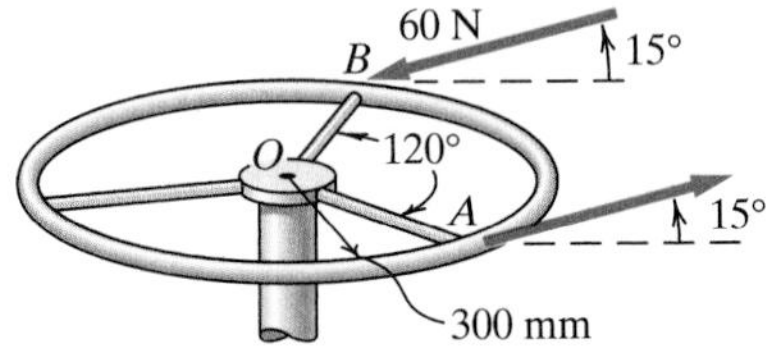

E5.4.27

5.4.28. The drive shaft from the engine of a small airplane applies to the propeller a couple moment that is counterclockwise as viewed from the front of the plane (E5.4.28). The magnitude of the moment is 1.8 kN · m.

a. Determine the couple consisting of vertical forces between the wheels and the ground points A and B that is equivalent to this couple moment.

b. Both of the vertical forces in **a** happen at the interface between the ground and the tires. One of the vertical forces—either the one at A or the one at B—seems to indicate that there is a tensile normal force between the ground and the tire. According to our discussion in Chapter 4, however, normal forces are compressive. How can this be?

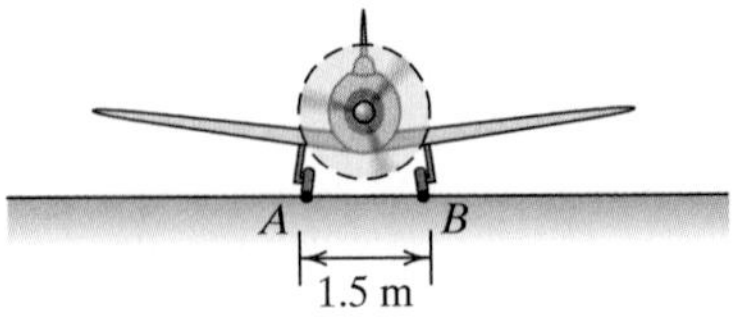

E5.4.28

5.4.29. Three couples are applied to a bent bar as shown in E5.4.29. Determine

a. the equivalent loading (Does it matter where the moment center is located?)

b. the scalar component of the equivalent moment about the line OA

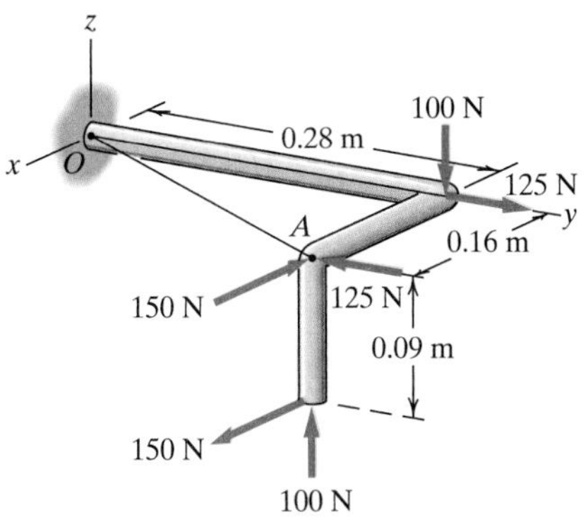

E5.4.29

5.4.30. **Exercise 5.2.26** asks for the moments created by $\boldsymbol{F}_A$ and $\boldsymbol{F}_B$ about O in **E5.2.26**. Now find the equivalent loading at O for $\boldsymbol{F}_A$ and $\boldsymbol{F}_B$. Specify both $\boldsymbol{F}_{\text{equ}}$ and $\boldsymbol{M}_{\text{equ}}$, and show the equivalent loading graphically.

5.4.31. The wind blowing on the freeway exit sign in E5.4.31 and the weight of the sign can be represented by point loadings acting at the center of the sign, as shown in the drawing. If the force of the wind is 12.6 kN and the sign weighs 10.5 kN, find the equivalent moment at O.

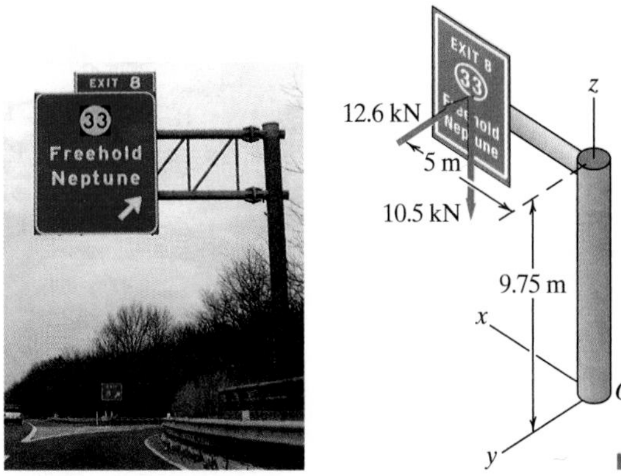

E5.4.31

5.4.32. The base of the freeway sign in **Exercise E5.4.31** is attached to its foundation with four bolts (E5.4.32). The four bolts must be able to resist the twisting of the pole that is caused by the wind. Assume each bolt carries the same force. If the diameter of the bolt circle is 0.9 m, calculate the horizontal (shear) force on each of the bolts due to the twisting moment about the y axis.

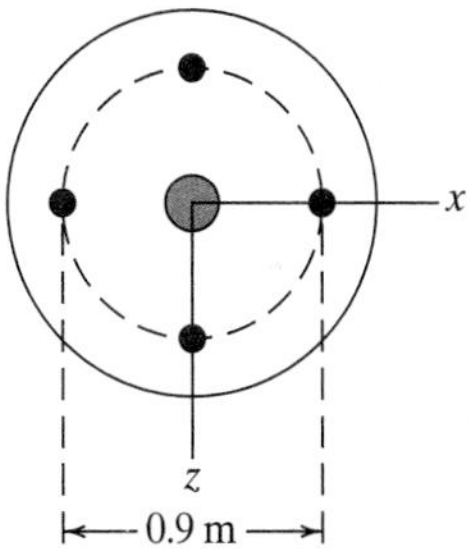

E5.4.32

5.5 JUST THE FACTS

This chapter looks at moments—what they are and how to represent them. A **moment** is created by a force offset from a point in space called a **moment center** and is the tendency of the force to cause rotation. Moment is a vector quantity and is specified in terms of magnitude and direction. In drawings we use an arc or double-headed arrow to denote a moment.

The magnitude of a moment is given by

$$\|\boldsymbol{M}\| = \|\boldsymbol{r}\| \left(\|\boldsymbol{F}\| \sin\theta \right) \tag{5.1}$$

where $\|\boldsymbol{r}\|$ is the magnitude of the **position vector** (any vector from the moment center to the line of action of $\boldsymbol{F}$), $\|\boldsymbol{F}\|$ is the magnitude of the force vector, and θ is the angle between the position vector and force vector when these two vectors are placed tail to tail such that $0° < \theta < 180°$.

The **direction** of the moment is perpendicular to the plane defined by the position vector $\boldsymbol{r}$ and the force $\boldsymbol{F}$, and is defined using the right-hand rule. This direction can also be interpreted as the axis about which the moment acts. If we establish a local coordinate system at the moment center, we define the **sense** of the moment (clockwise or counterclockwise) by standing on the positive axis and looking back at the moment center.

Mathematical Representation of a Moment, Including Cross Product and Dot Product

It is frequently convenient to represent a moment in terms of its scalar components:

$$\boldsymbol{M} = M_x\boldsymbol{i} + M_y\boldsymbol{j} + M_z\boldsymbol{k} \tag{5.2}$$

Furthermore, the magnitude of the moment $\| \boldsymbol{M} \|$ is

$$\| \boldsymbol{M} \| = \sqrt{M_x^2 + M_y^2 + M_z^2} \tag{5.3}$$

The moment's direction can be described by a unit vector $\boldsymbol{u}$:

$$\boldsymbol{u} = \cos\theta_x\boldsymbol{i} + \cos\theta_y\boldsymbol{j} + \cos\theta_z\boldsymbol{k} \tag{5.5}$$

with direction cosines defined as

$$\cos\theta_x = \frac{M_x}{\| \boldsymbol{M} \|} \qquad \cos\theta_y = \frac{M_y}{\| \boldsymbol{M} \|} \qquad \cos\theta_z = \frac{M_z}{\| \boldsymbol{M} \|} \tag{5.4}$$

We can rewrite the expressions for moment in (5.3), (5.4), and (5.5) in terms of the force vector ($\boldsymbol{F}$) and position vector ($\boldsymbol{r}$) as the vector product, or cross product, of the position vector and the force:

$$\boldsymbol{M} = \boldsymbol{r} \times \boldsymbol{F} = (\| \boldsymbol{r} \| \;\; \| \boldsymbol{F} \| \sin\theta)\boldsymbol{u} \tag{5.6A}$$

where $\boldsymbol{u}$ is a unit vector perpendicular to the plane defined by $\boldsymbol{r}$ and $\boldsymbol{F}$. This expression is read, "$\boldsymbol{r}$ crossed with $\boldsymbol{F}$." The cross product is not commutative, meaning that $\boldsymbol{r} \times \boldsymbol{F}$ is not equal to $\boldsymbol{F} \times \boldsymbol{r}$.

It is generally more convenient to work with the position vector and force in terms of their components.

$$\boldsymbol{M} = \boldsymbol{r} \times \boldsymbol{F} = \underbrace{(+r_yF_z - r_zF_y)}_{M_x}\boldsymbol{i} + \underbrace{(+r_zF_x - r_xF_z)}_{M_y}\boldsymbol{j} + \underbrace{(+r_xF_y - r_yF_x)}_{M_z}\boldsymbol{k} \tag{5.7}$$

Equation (5.7) is an expression of **Varignon's Theorem**, which states that the moment of a force at a particular moment center is equal to the sum of the moment of the components of the force at the same moment center.

More generally, the cross product of the position vector and the force can be written in matrix form as

$$\boldsymbol{M} = \boldsymbol{r} \times \boldsymbol{F} = \begin{vmatrix} \boldsymbol{i} & \boldsymbol{j} & \boldsymbol{k} \\ r_x & r_y & r_z \\ F_x & F_y & F_z \end{vmatrix} \tag{5.12}$$

The form in (5.12) specified that the **determinant** of the matrix be taken. The reader unfamiliar with this form is referred to Appendix A1.4.

We can use the moment scalar components in (5.7) to rewrite the direction cosines in (5.4) that define the direction of the moment:

$$\cos\theta_x = \frac{M_x}{\|\boldsymbol{M}\|} = \frac{+r_yF_z - r_zF_y}{\|\boldsymbol{M}\|}$$
$$\cos\theta_y = \frac{M_y}{\|\boldsymbol{M}\|} = \frac{+r_zF_x - r_xF_z}{\|\boldsymbol{M}\|} \qquad (5.9)$$
$$\cos\theta_z = \frac{M_z}{\|\boldsymbol{M}\|} = \frac{+r_xF_y - r_yF_x}{\|\boldsymbol{M}\|}$$

where the magnitude of the moment is

$$\|\boldsymbol{M}\| = \sqrt{\underbrace{(+r_yF_z - r_zF_y)^2}_{M_x} + \underbrace{(+r_zF_x - r_xF_z)^2}_{M_y} + \underbrace{(+r_xF_y - r_yF_x)^2}_{M_z}} \qquad (5.8)$$

The direction of the moment is defined by the unit vector $\boldsymbol{u}$, based on the direction cosines in (5.9), as

$$\boldsymbol{u} = \frac{+r_yF_z - r_zF_y}{\|\boldsymbol{M}\|}\boldsymbol{i} + \frac{+r_zF_x - r_xF_z}{\|\boldsymbol{M}\|}\boldsymbol{j} + \frac{+r_xF_y - r_yF_x}{\|\boldsymbol{M}\|}\boldsymbol{k} \qquad (5.10)$$

The expression for moment in (5.7) can be simplified if the position and force vectors lie in the same *xy*, *yz*, or *zx* plane:

For a force vector and position vector lying in the same *xy* plane, we can write

$$\boldsymbol{M}_z = M_z\boldsymbol{k} = (+r_xF_y - r_yF_x)\boldsymbol{k} \qquad (5.11\text{A})$$

For a force vector and position vector lying in the same *yz* plane, we can write

$$\boldsymbol{M}_x = M_x\boldsymbol{i} = (+r_yF_z - r_zF_y)\boldsymbol{i} \qquad (5.11\text{B})$$

For a force vector and position vector lying in the same *zx* plane, we can write

$$\boldsymbol{M}_y = M_y\boldsymbol{j} = (+r_zF_x - r_xF_z)\boldsymbol{j} \qquad (5.11\text{C})$$

The dot product is useful in finding the component of a moment in a particular direction (or, alternately worded, about a particular axis).

Equivalent Loadings

We introduced the ideas of equivalent loading and the **equivalent loading principle**. This principle states:

> *For any loading (consisting of forces and/or moments) applied to a system, we can replace that loading with an equivalent loading (consisting of a force and a moment) applied at a single point.*

We introduced an example of equivalent loading and the application of the principle that illustrated the equivalency of a loading consisting of forces of equal magnitude and opposite sign (which we called a **couple**) to a loading consisting of a **couple moment**.

SYSTEM ANALYSIS (SA) EXERCISES

SA5.1 Consideration of Left- and Right-Foot Pedaling

Multiple right pedal positions for a cyclist are shown in Figure SA5.1.1. The bicycle would be moving to the right.

(a) For $\theta_{\text{right foot}}$ equal to 90°, 180°, 270°, and 0°, determine the moment at O (which is the center of the bottom bracket) created by the foot force. Present your answers in column A of Table SA5.1.1.

(b) The action of the left pedal and foot is offset from that of the right pedal and foot by 180°. This means that, for example, when the right foot is applying 365 N at $\theta_{\text{right foot}} = 90°$, the left foot is at 270° applying 80 N. Based on this information, complete column B of **Table SA5.1.1**.

(c) Add columns A and B to determine the total moment acting at O due to simultaneously pedaling with the right foot and the left. Present your answers in column C.

(d) Based on your results in **Table SA5.1.1**, what is the maximum moment created in cycling? At what $\theta_{\text{right foot}}$ angle does it occur?

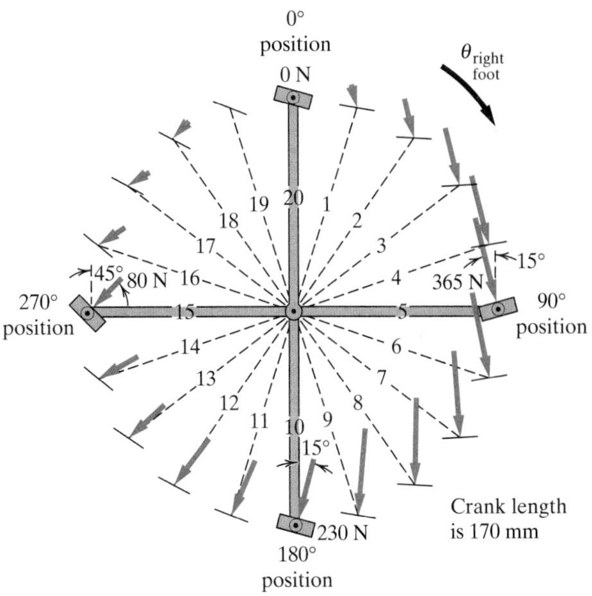

Figure SA5.1.1 Right foot force acting on bicycle pedal in various crank arm positions. Bicycle is moving to the right.

Table SA5.1.1 Summary of Moments in Various Angular Positions

Angular position $\theta_{\text{right foot}}$	A Moment at O created by right foot	B Moment at O created by left foot	C Total moment at O created by pedaling with both feet
90°			
180°			
270°			
0°			

SA5.2 Vehicle Recovery: Attempt 1

In the spring of 1996, an M1 U.S. Army tank (Figure SA5.2.1) slid off a mountain road while on a training exercise in the Republic of Korea. The road was an unimproved dirt trail along a steep slope. The tank became precariously balanced on the edge and had to be pulled back onto the road to be recovered. Two M88 recovery vehicles (Figure SA5.2.2) responded and began recovery operations.

Recovery operations such as this are dangerous and require careful planning. The M1 tank weighs approximately 70 tons, and the forces in the recovery cables are significant.

Figure SA5.2.1 M1 tank

Figure SA5.2.2 M88 recovery vehicle

(a) In **Figure SA5.2.3**, M88 1 is shown pulling with 10 kip of force at 70° from the horizontal; and M88 2 is pulling with 15 kip at 60° from the horizontal. Both cables are attached to the front left tow hook of the M1 tank (Point A). Calculate the resultant force vector due to the two M88 recovery vehicle forces.

(b) Assume that M88 1 cannot pull with a greater force because of its position on the road and M88 2 can pull with an increasingly greater force as needed. Graph the relationship between the magnitude of the resultant force and the force applied by M88 2. Assume that M88 2 can pull with 0 to 50 kip. Discuss the resulting graph.

(c) Assume that the applied forces by the recovery vehicles are limited to a constant magnitude. Recommend a physical change to the recovery operation that will increase the magnitude of the resultant force on the M1 tank.

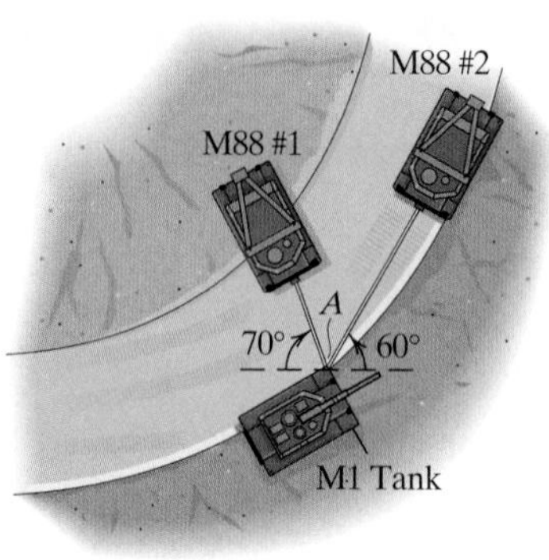

Figure SA5.2.3 Recovery operation

(d) Describe what you think will physically happen when the recovery vehicles actually apply their forces to the front left tow hook of the M1 tank. Assume that the applied forces are sufficient to cause the M1 tank to slide.

SA5.3 Vehicle Recovery: Attempt 2

Read **SA5.2** for background information.

The recovery operation from SA5.2 was unsuccessful. The M1 tank would rotate counterclockwise when the forces from the M88 recovery vehicles were applied to the front left tow hook of the tank, as shown in **Figure SA5.2.3**. This tendency for rotation caused the M1 tank to continue to slide off the road and potentially down the slope. A third M88 recovery vehicle was ordered to join the recovery operation. The third M88 was positioned behind the M1 tank as shown in **Figure SA5.3.1**.

(a) Discuss attaching the third tow cable to each of the four tow hooks A through D on the corners of the M1 tank. Which tow hook position do you recommend and why? Consider the mechanical advantage and practicality of each tow hook position.

(b) Assume that M88 1 applies a 10-kip force at 70° from the horizontal, M88 2 applies a 15-kip force at 60° from the horizontal, and M88 3 applies a 20-kip force along the horizontal through the M1 tank's center of gravity, G, as shown in **Figure SA5.3.2**. Calculate the tendency of each of the recovery forces to cause rotation about the M1 tank's center of gravity. Assume that the center of gravity is located at the geometric center of the tank's area as shown in **Figure SA5.3.2**.

(c) Describe what you think will physically happen when the forces shown in **Figure SA5.3.2** are applied to the M1 tank. Assume they are sufficient to cause the tank to slide.

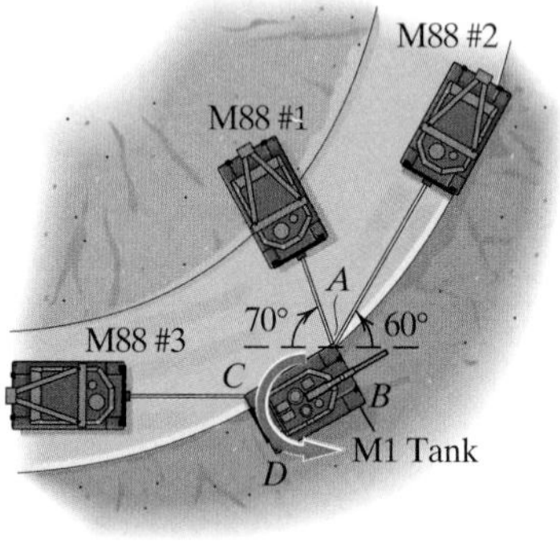

Figure SA5.3.1 Modified recovery operation

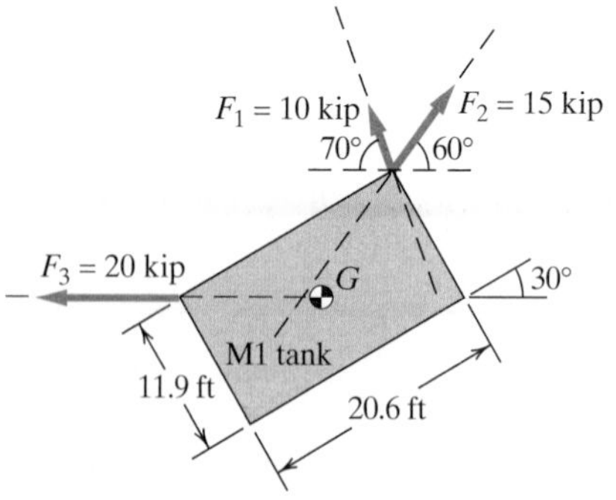

Figure SA5.3.2 Rotation about center of gravity

SA5.4 Too Much Moment Can Topple a Crane

History of Cranes

Cranes have long been used to assist humans in building structures. For example, the Ancient Egyptians, Greeks, and Romans created cranes with which to build bridges and statues. The first mobile cranes, made of wood and iron, were invented so that military personnel could get over large walls without having to climb them, as shown in Figure SA5.4.1*a*.

The Italian genius Leonardo Da Vinci (1452–1519) was extremely interested in cranes and helped advance the technology from simple tower cranes to large semi-mobile slewing cranes, all still built out of lumber and iron, as shown in **Figure SA5.4.1*b***. The rollers on which the slewing plate moved increased the productivity of those types of cranes dramatically—the cycle time for one lift was shortened due to the easy rotation of the jib around the main mast (= slewing).

The invention of steel, wire rope, and steam engines during the middle of the 19th century did much to advance transportation systems (e.g., canals, railways, harbors) and the construction technologies used to create these systems. In particular, the invention of wire rope in 1834 (also known as cable) and steam engines enabled the design of cranes with larger load capacities and increased mobility.

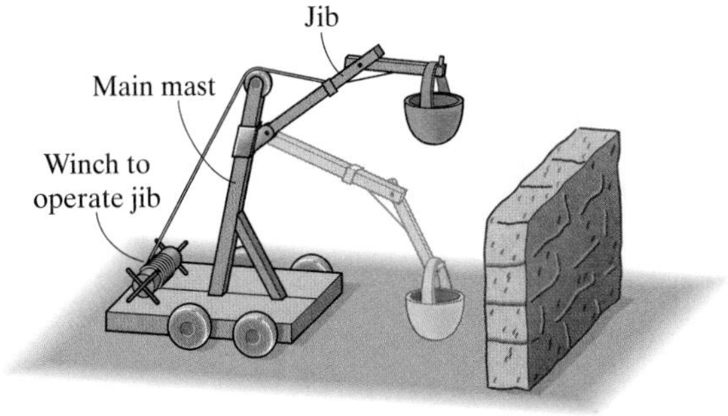

(*a*)

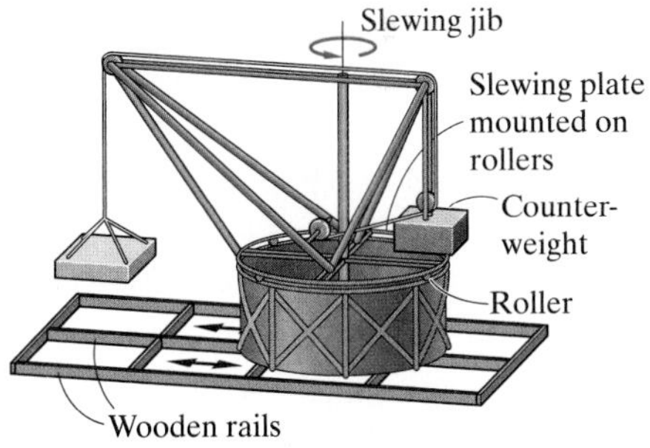

(*b*)

Figure SA5.4.1 Early engineered cranes for military and construction, made of wood and iron: (*a*) first mobile cranes in 15th century; (*b*) Da Vinci's first real slowing crane with counterweight

At the time the Reynolds Coliseum was constructed in 1949, derrick cranes were in use. The roof of the building includes heavy precast concrete panels placed on top of the steel beams that span the large open interior space, as shown in Figure SA5.4.2. The panels were hoisted from the ground and then stored on the center section of the roof before being placed one by one onto the beams by one of the two derrick cranes. Figure SA5.4.3 illustrates the basic elements and operation of a derrick crane. The placing sequence of each of the 4.0-kN precast concrete panels consisted of the following four steps: (1) hooking the panel to the spreader bar of the crane, (2) hoisting to the proper height and the start of slewing, (3) slewing, and (4) luffing of the jib (changing the boom angle) and placing the panel at the proper location.

Figure SA5.4.2 Placement on concrete panels with help of two derricks on March 17, 1949

Modern Technology

Construction technology has further evolved since 1949. One example is the emergence of the tower crane presented in Figure SA5.4.4. The horizontal jib, also called the saddle jib, is supported by fixed cables that tie into the counterjib with ballast. The entire saddle jib, which is able to cover a large circle, rests on a slewing ring, similar to Leonardo's design. In turn, the slewing ring, located at *C* in **Figure SA5.4.4**, sits atop a fixed tower that has the capability to climb higher as the building grows. This economical design became very popular in the 1960s in Europe for the construction of apartment buildings in tight spaces. This tower crane design has since spread all over the world.

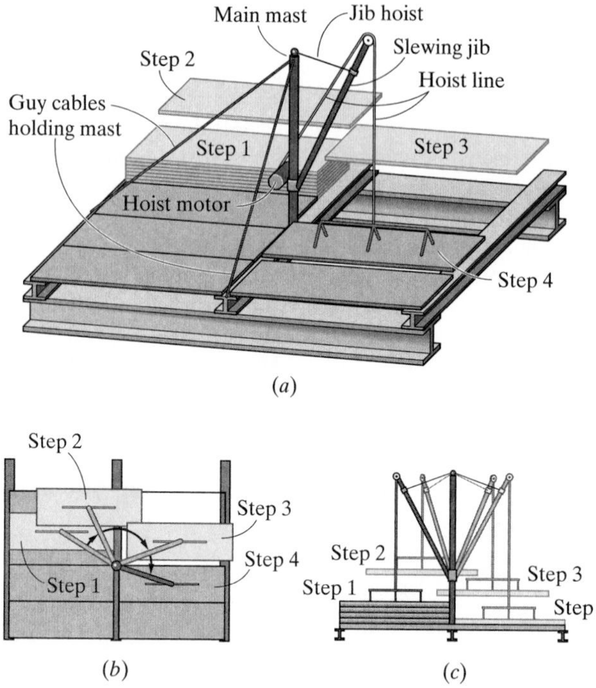

Figure SA5.4.3 Placing roof panels with a derrick crane: (*a*) layout of placing panels with derrick onto coliseum roof structure; (*b*) sequence of picking and placing one panel (top view); (*c*) sequence of picking and placing one panel (side view)

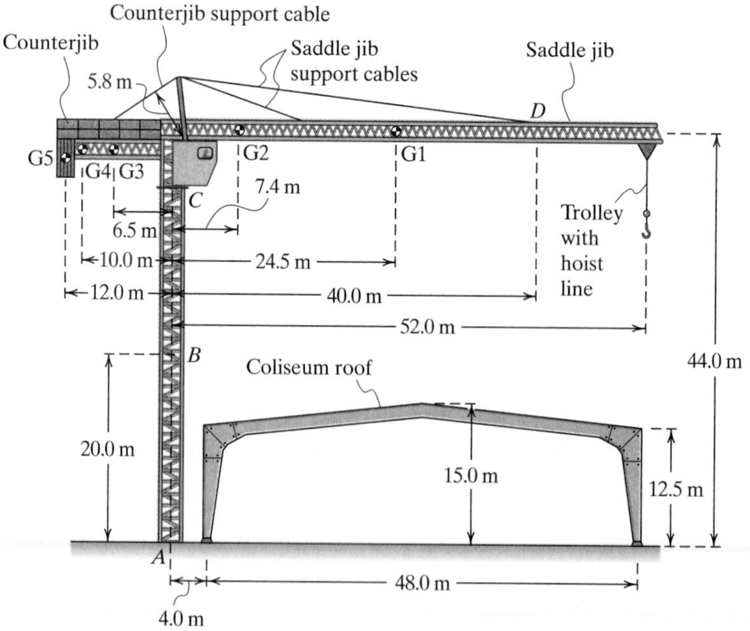

Figure SA5.4.4 Layout for placing concrete panels with fixed tower crane (Option A)

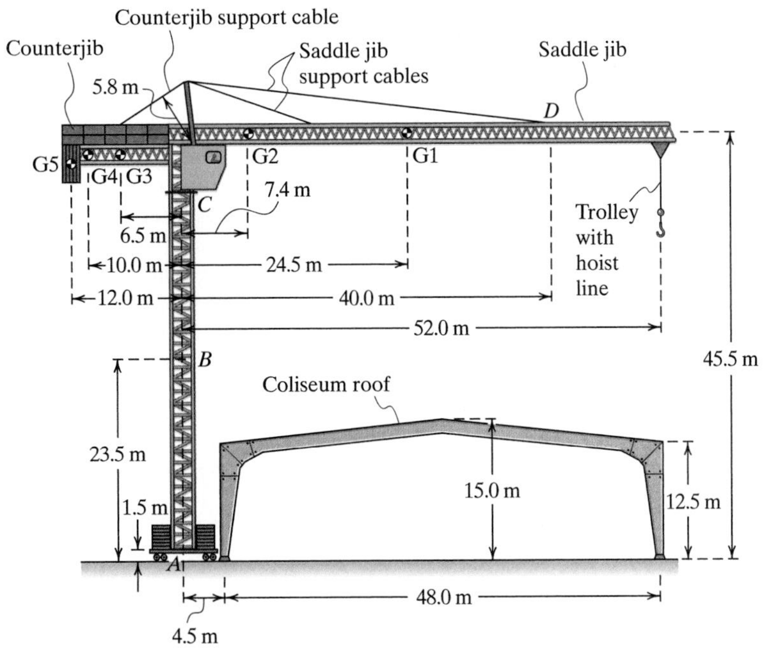

Figure SA5.4.5 Tower crane Option B on rails along the side of the building

Your Assignment

Assume that you are hired as a summer intern by a company that won a contract to replace all the concrete panels on the Reynolds Coliseum. Your knowledge of moments is critical for your next assignment. Your boss needs your help in deciding which crane arrangement he should select. He wants you to compare two crane setups, Option A (**Figure SA5.4.4**) and Option B (**Figure SA5.4.5**).

Each panel measures 2.0 m long by 0.8 m wide and weighs 4.0 kN. **Figures SA5.4.4** and **SA5.4.5** provide the dimensions for each crane option in meters. The allowable, or rated, loading is smaller than the theoretical lifting capacity by a safety factor to allow for unforeseen conditions and imperfections in real-world components and materials. The rated loadings at 52.0 m (end of the saddle jib) are 6.0 kN for Option A and 4.0 kN for Option B. **Table SA5.4.1** provides additional data for both cranes.

Table SA5.4.1 Weight of Various Crane Elements

		Option A Rated 6.0 kN at 52.0 m	Option B Rated 4.0 kN at 52.0 m
Label	Crane Element	Weight of Element (kN)	Weight of Element (kN)
G1	Saddle jib center of gravity	18.0	12.0
G2	Jib support cables center of gravity	9.4	6.3
G3	Counterjib center of gravity	17.1	11.3
G4	Counterjib winches and cables center of gravity	6.5	4.3
G5	Ballast center of gravity	45.0	30

Your boss would like to know which crane option he should use, considering that the building is approximately 98 m long. Since you are new to this job, he provides you with a step-by-step guideline on how to tackle this kind of problem. He expects you to do this on your own the next time around. Here is his "cookbook":

I. Assessment of Option A

1. Draw the area this crane option can cover, to scale, onto the "footprint" of the building. Determine what area you will be able to cover with option A. (*Hint:* You could place the base of crane A inside the building and remove its tower after completion of the coliseum with a separate truck crane.)
2. Where would you stage/store the precast panels before hooking them to the hoist line to be placed on the roof? Show this location on your footprint drawing.
3. Redraw the crane showing all the forces at G1–G5 acting on it (**Table SA5.4.1**). The forces at G1–G5 are referred to as the "dead weights" because they are due to the weight of the crane itself.
4. Based on your drawing in 3, calculate the equivalent loading (consisting of an equivalent force and an equivalent moment) at
   (a) a moment center at *A*
   (b) a moment center at *B*
   (c) a moment center at *C*
   (d) a moment center at *D*
5. Repeat the calculation in 4 to include the rated loading listed in **Table SA5.4.1** (in addition to the dead loads). Summarize your answers from 4 and 5 in Table SA5.4.2.
6. Based on your answers in **Table SA5.4.2**, calculate the ratio of M_{equ}/F_{equ} and add this to **Table SA5.4.2**. Would you expect this ratio to ever be greater than or equal to 52.0 m? Why or why not?

II. Assessment of Option B

Same questions as above.

III. Comparing the Two Options

1. Based on your footprint diagrams, would you recommend one option over the other? Why? What questions do you still have about the two options?
2. The crane represents an assembly of space frames held together by bolts and nuts. Locations *A* and *B* are most critical when dimensioning and maintaining the bolts. If the cross-sectional area of both towers is a square measuring 1.6 m by 1.6 m (Figure SA5.4.6), what is the maximum possible tension force that a bolt at each corner will experience beyond its preload (neglect the weight of the tower itself)? Which crane might possibly need two sets of bolts at each corner? Support your answer with your numerical calculations.

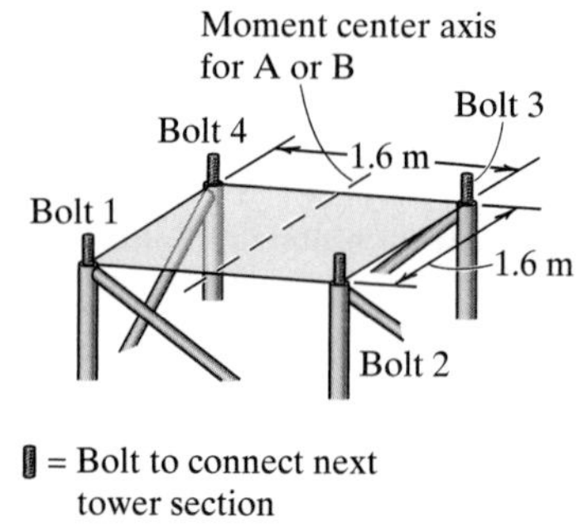

Figure SA5.4.6 Simple sketch of cross section A-A and B-B

Table SA5.4.2 Summary of Crane Option Calculations

	Option A			Option B		
Moment Center at	M_{equ} and F_{equ} due to Dead Loads	M_{equ} and F_{equ} due to Dead Loads and Rated Load	M_{equ}/F_{equ} due to Dead Loads and Rated Load	M_{equ} and F_{equ} due to Dead Loads	M_{equ} and F_{equ} due to Dead Loads and Rated Load	M_{equ}/F_{equ} due to Dead Loads and Rated Load
A						
B						
C						
D						

CHAPTER 6

DRAWING A FREE-BODY DIAGRAM

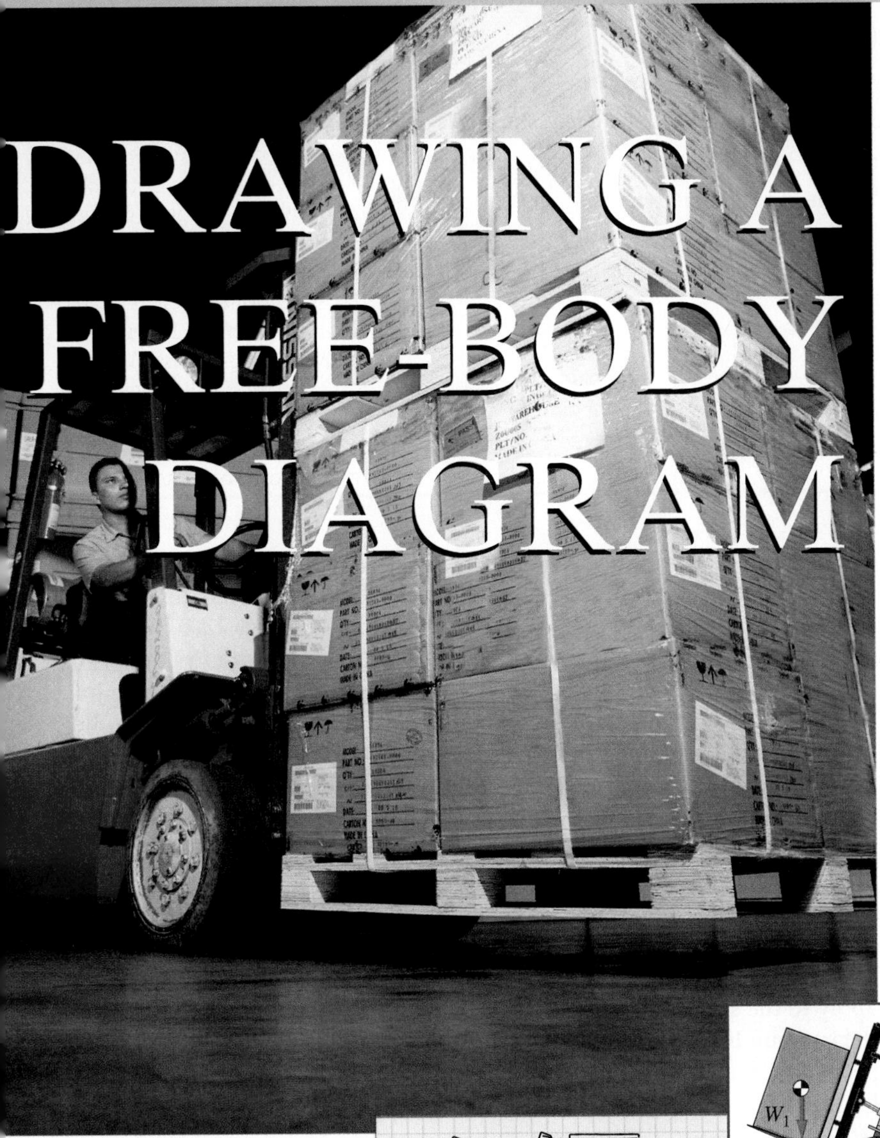

The **free-body diagram** is the most important tool in this book. It is a drawing of a system and the loads acting on it. Creating a free-body diagram involves mentally separating the system (the portion of the world you're interested in) from its surroundings (the rest of the world), and then drawing a simplified representation of the system. Next you identify all the loads (forces and moments) acting on the system and add them to the drawing.

This chapter is devoted exclusively to creating free-body diagrams. We build on the work in prior chapters on forces and moments and on the engineering analysis procedure presented in Chapter 1. Creating a free-body diagram is part of the DRAW step in the analysis procedure.

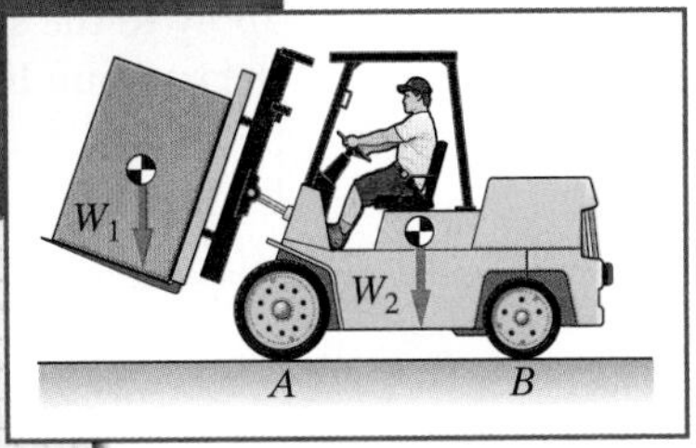

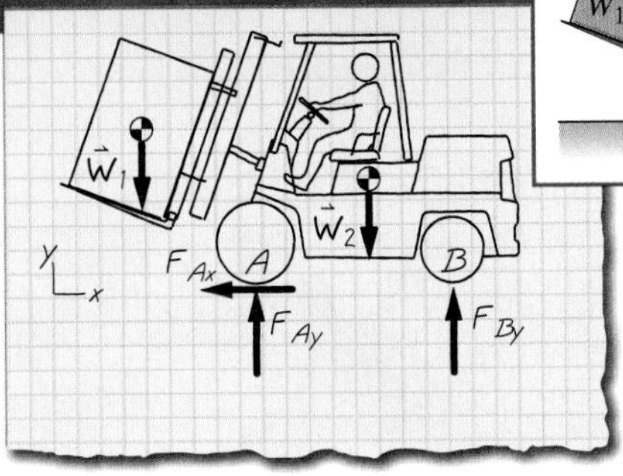

OBJECTIVES

On completion of this chapter, you will be able to:

- **Isolate a system from its surroundings and identify the supports**
- **Define the loads associated with supports and represent these loads in terms of vectors**
- **Inspect a system and determine whether it can be modeled as a planar system**
- **Represent the system and the external loads acting on it in a diagram called a free-body diagram**

As an example of creating a free-body diagram, consider the two friends leaning against each other in **Figure 6.1*a***. The free-body diagram of Friend 1 is shown in **Figure 6.1*c***. In going from **Figure 6.1*a*** to **6.1*c***, we zoomed in on and drew a boundary around the system (Friend 1) to isolate him from his surroundings, as shown in **Figure 6.1*b***. This boundary is an imaginary surface and the system is (by definition) the "stuff" inside this imaginary surface. The system's surroundings are everything else. You can think of the boundary as a shrink wrap around the system.

After drawing the boundary, we identified the external loads acting on the system either at or across this boundary and drew them at their points of application. These loads represent how the surroundings push, pull, and twist the system. In a free-body diagram we draw the system somewhat realistically and replace the surroundings with the loads they apply to the system. It is important to recognize that we are not ignoring the surroundings—we simply replace them with the loads the system experiences because of them. For example, in **Figure 6.1*c*** we replace the back of Friend 2 with the normal force he applies to the back of Friend 1 ($\boldsymbol{F}_{\text{normal, back 2 on back 1}}$).

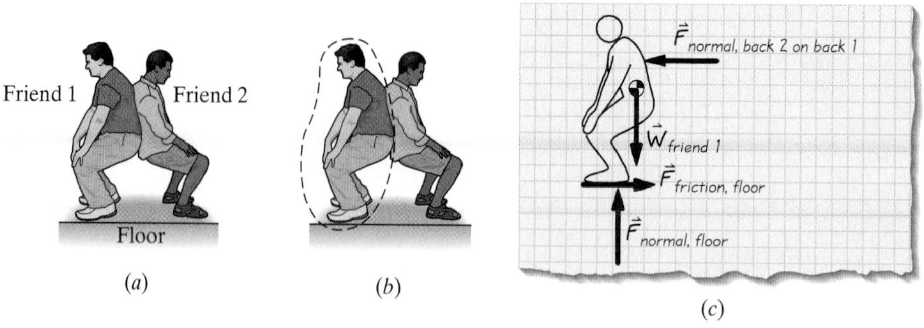

Figure 6.1 (*a*) Two friends leaning against one another; (*b*) Isolate Friend 1 by drawing a boundary. Friend 1 is the system; (*c*) A free-body diagram of Friend 1

6.1 TYPES OF EXTERNAL LOADS ACTING ON SYSTEMS

Some of the external loads acting on a system act *across* the system boundary; the principal example of this type of load is **gravity** (which manifests itself as weight). Another example is the magnetic force, which results from electromagnetic field interaction. The magnetic force is what turns a motor.

Other external loads act *directly on* the boundary. Consider where:

- The boundary passes through a connection between the system and its surroundings, commonly referred to as a **boundary support** (or **support** for short). A support may be, for example, a bolt, cable or a weld, or simply where the system rests against its surroundings. We replace each support with the loads it applies to the system. These loads consist of the contact forces discussed in Chapter 4 (normal contact, friction, tension, compression, and shear). Synonyms for the term *supports* are *reactions* and *boundary connections*.
- The boundary separates the system from fluid surroundings. We refer to this as a **fluid boundary**. We replace the fluid with the load the fluid applies to the system. This load consists of the pressure (force per unit area) of the fluid pressing on and/or moving along the boundary.

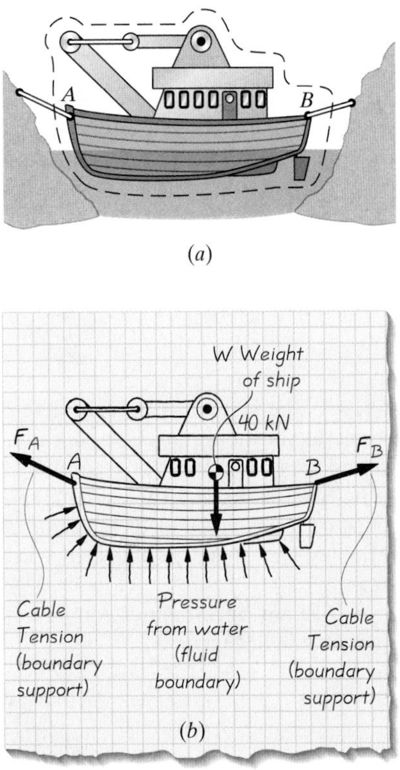

Figure 6.2 (*a*) Isolate the ship by drawing a boundary. The ship is the system; (*b*) A free-body diagram of the ship

In practice, a system may be acted on by a combination of cross-boundary loads (usually gravity), loads at supports, and loads at fluid boundaries, as illustrated in Figure 6.2.

Notice that at some boundary locations no loads act. At other locations there are what are called **known loads**—for example, in **Figure 6.2**, the 40-kN gravity force is a known load.

Depending on the nature of an external load acting on a system, when that load is drawn on a free-body diagram it is represented either by a vector acting at a point of application or as a distributed load acting on an area. The load is given a unique variable label, and its magnitude (if it is a known load) is written next to the vector.

EXERCISES 6.1

6.1.1. The system to be considered is a coat rack with some items hanging off of it as shown in E6.1.1. In your mind draw a boundary around the system to isolate it from its surroundings.

a. Make a sketch of the coat rack and the external loads acting on it. Show the loads as vectors and label them with variables, and where possible give word descriptions of the loads.

b. List any uncertainties you have about the free-body diagram you have created.

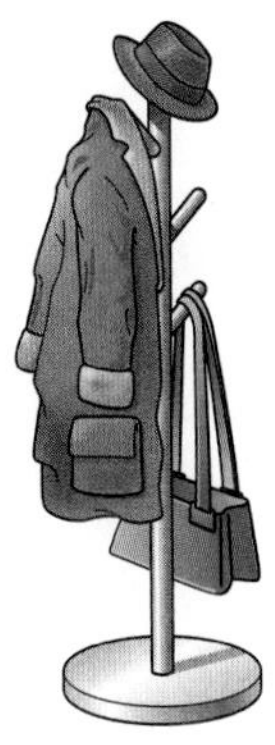

E6.1.1

6.1.2. The system to be considered is defined as a mobile as shown in E6.1.2. In your mind draw a boundary around the system to isolate it from its surroundings at point E.

a. Make a sketch of the mobile and the external loads acting on it. Show the loads as vectors and label them with variables, and where possible give word descriptions of the loads.

b. List any uncertainties you have about the free-body diagram you have created.

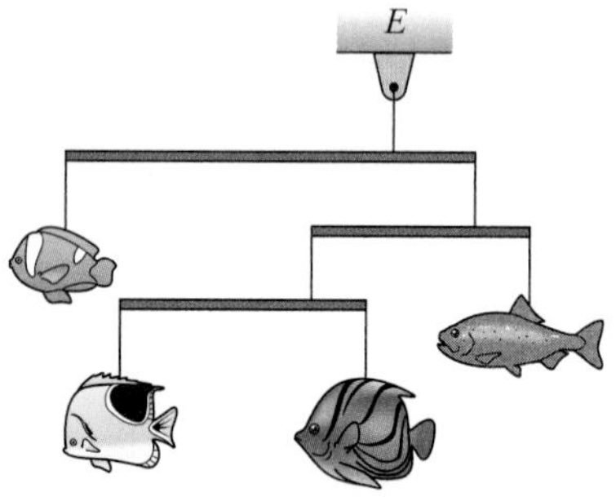

E6.1.2

6.1.3. The system to be considered is a person and a ladder, as shown in E6.1.3. In your mind draw a boundary around the system to isolate it from its surroundings.

a. Make a sketch of the system and the external loads acting on it. Show the loads as vectors and label them with variables, and where possible give word descriptions of the loads.

b. List any uncertainties you have about the free-body diagram you have created.

E6.1.3

6.1.4. Visit a weight room, and take a look at one of the exercise stations—preferably one that is in use!

a. Consider where the person is standing, hanging, laying, and/or pushing on it. In your mind, draw a boundary around *the person* to define him or her as the system. Make a sketch of the system and the external loads acting on it, showing the loads as vectors with variable labels. Where possible give word descriptions of the loads. List any uncertainties you have about the free-body diagram you have created.

b. Consider where the person is standing, hanging, laying, and/or pushing on it. In your mind, draw a boundary around the *exercise machine* to define it as the system. Make a sketch of the system and the external loads acting on it, showing the loads as vectors with variable labels. Where possible give word descriptions of the loads. List any uncertainties you have about the free-body diagram you have created.

6.1.5. Visit a local playground near campus, and take a look at a jungle gym—preferably one that is in use! Consider where the children (or adults!) are standing/hanging. In your mind, draw a boundary around the *jungle gym* to define it as your system. Make a sketch of the system and the external loads acting on it, showing the loads as vectors with variable labels. List any uncertainties you have about the free-body diagram you have created.

6.1.6. Visit a local pet store or zoo and look at the fish tanks.

a. In your mind, draw a boundary around the *fish tank including the water* to define it as your system. Make a sketch of the system and the external loads acting on it, showing the loads as vectors with variable labels. List any uncertainties you have about the free-body diagram you have created.

b. In your mind, draw a boundary around the *fish tank excluding the water* to define it as your system. Make a sketch of the system and the external loads acting on it, showing the loads as vectors with variable labels. List any uncertainties you have about the free-body diagram you have created.

6.2 PLANAR SYSTEM SUPPORTS

We now consider how to identify supports and represent the loads associated with them when working with **planar systems**. A system is planar if all the forces acting on it can be represented in a single plane and all moments are about an axis perpendicular to that plane. If a system is not planar it is a **nonplanar system**. In Section 6.4 we will lay out guidelines for identifying planar and nonplanar systems. For now, we assume that all of the systems we are dealing with in this section are planar.

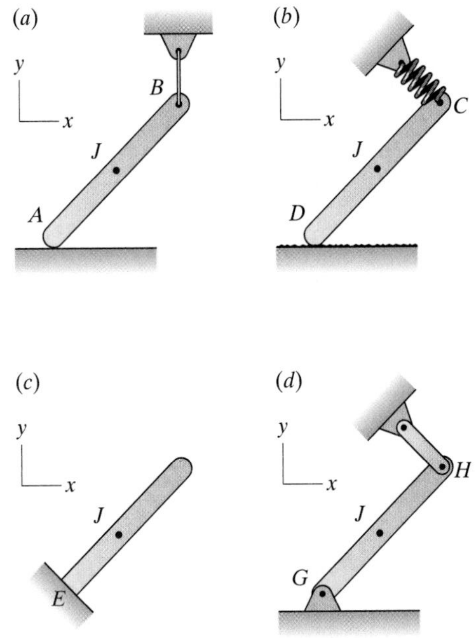

Figure 6.3 An object connected to its surroundings by various supports. The object can be modeled as a planar system.

Consider the systems in Figure 6.3 for which we want to draw fee-body diagrams. Each system consists of a uniform bar of weight W, oriented so that gravity acts in the negative y direction. However, each has different supports that connect it to its surroundings. For example:

In **Figure 6.3*a***, the supports consist of
- normal contact without friction at A, and
- cable attached to the system at B.

In **Figure 6.3*b***, the supports consist of
- a spring attached to the system at C, and
- normal contact with friction at D.

In **Figure 6.3*c***, the support consists of
- a system fixed to its surroundings at E.

In **Figure 6.3*d***, the supports consist of
- a system pinned to its surroundings at G, and
- a link attached to the system at H.

At each support we consider whether the surroundings act on the system with a force and/or a moment. As a general rule, *if a support prevents the translation of the system in a given direction, then a force acts on the system at the location of the support in the opposing direction. Likewise, if rotation is prevented, a moment opposing the rotation acts on the system at the location of the support.*

At Point *A* (Normal Contact Without Friction). At this support the system rests against a smooth, frictionless surface. A normal force prevents the system from moving into the surface and is oriented so

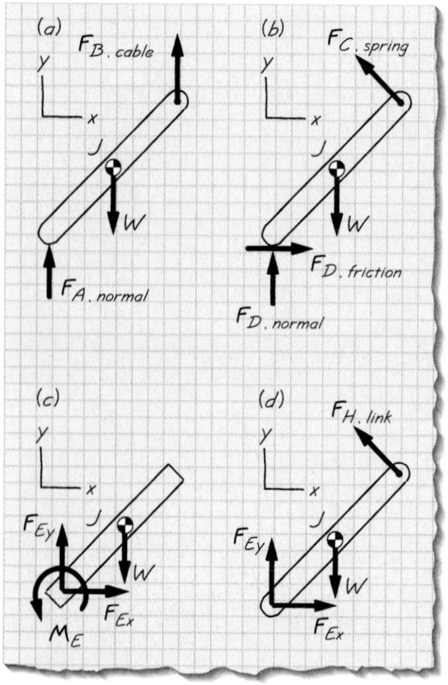

Figure 6.4 Free-body diagrams of the planar systems shown in Figure 6.3

as to push on the system. Because the supporting surface at A is smooth, friction between the system and its surroundings is nonexistent. Therefore there is no force component parallel to the surface. In **Figure 6.4*a*** (the free-body diagram of **Figure 6.3*a***) the force resulting from normal contact at A is represented by $\boldsymbol{F}_{A,\text{normal}}$; we know its direction is normal to the surface so as to *push* on the system.

At Point *B* (Cable). At this support a force acts on the system; the line of action of the force is along the cable. The force represents the cable pulling on the system because the cable can only act in tension. In **Figure 6.4*a*** the force from the cable at B is represented by $\boldsymbol{F}_{B,\text{cable}}$; we know its direction is along the cable axis in the direction that allows the cable to *pull* on the system.

At Point *C* (Spring). At this support there is a force that either pushes or pulls on the system; the line of action of the force is along the axis of the spring. If the spring is extended by an amount Δ, the spring is in tension and the force is oriented so as to pull on the system. If the spring is compressed by an amount Δ, the force is oriented so as to push on the system. The magnitude of the force is proportional to the amount of spring extension or compression, and the proportionality constant is the spring constant, k. In other words, the size of the force is equal to the product of k and the spring extension or compression:

$$F_C = k(\Delta) \tag{6.1}$$

where the dimensions of k are force/length (e.g., N/mm). The value of F_C in (6.1) will be positive when the spring is in tension (since Δ will be positive) and negative when the spring is compressed (since Δ will be negative).

In **Figure 6.4*b*** the spring force at C is represented by $\boldsymbol{F}_{C,\text{spring}}$; we know its direction is along the spring axis. If the spring is in tension, the force acts so as to *pull* on the system. If the spring is in compression, the force acts so as to push on the system. We have drawn the direction of $\boldsymbol{F}_{C,\text{spring}}$ to indicate that the spring is in tension. We could equally well have chosen the direction of $\boldsymbol{F}_{C,\text{spring}}$ to indicate that the spring is in compression, but as we will see in the next chapter, drawing the spring in tension will make interpreting numerical answers easier.

At Point *D* (Normal Contact with Friction). At this support there are two forces acting on the system. One is a normal force (just like normal contact without friction). The second is due to friction and is parallel to the surface against which the system rests—therefore it is perpendicular to the normal force.

The force due to friction ($\boldsymbol{F}_{\text{friction}}$) is related to and limited by normal contact force ($\boldsymbol{F}_{\text{normal}}$) and the characteristics of the contact. Often the relationship between $\boldsymbol{F}_{\text{friction}}$ and $\boldsymbol{F}_{\text{normal}}$ is represented in terms of the Coulomb Friction Model. This model states that if $\| \boldsymbol{F}_{\text{friction}} \| < \mu_{\text{static}} \| \boldsymbol{F}_{\text{normal}} \|$ the system will not slide relative to its surroundings, where

μ_{static} is the coefficient of static friction and typically ranges from 0.01 to 0.70, depending on the characteristics of the contact. If $\| \boldsymbol{F}_{\text{friction}} \| = \mu_{\text{static}} \| \boldsymbol{F}_{\text{normal}} \|$, the condition is that of impending motion.

In **Figure 6.4*b*** the normal force at D is represented by $\boldsymbol{F}_{D,\text{normal}}$; we know its direction is normal to the surface so as to push on the system. The friction force is represented by $\boldsymbol{F}_{D,\text{friction}}$ and is parallel to the surface and perpendicular to the normal force. We could have drawn $\boldsymbol{F}_{D,\text{friction}}$ to point to the right or to the left; we arbitrarily chose to draw it to the right.

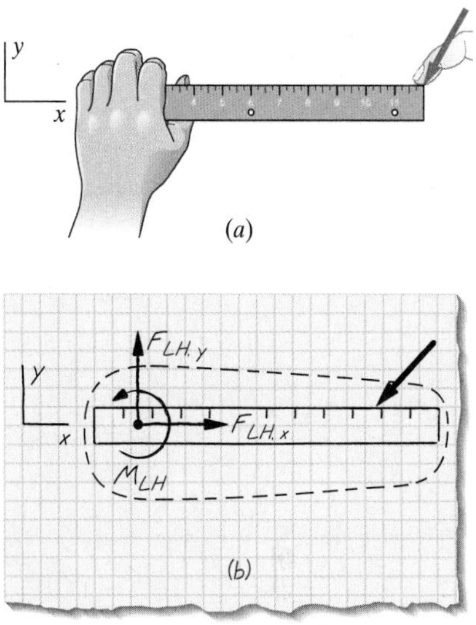

Figure 6.5 Illustrating a planar fixed boundary connection: (*a*) applying loads to a ruler; (*b*) the resulting free-body diagram if the ruler is defined as the system. Note how the loads the left-hand applies to the ruler are depicted.

At Point *E* (System Fixed to Surroundings, Referred to as a Fixed Support). At this support a force and a moment act on the system. To get a feeling for a fixed boundary, consider the setup shown in Figure 6.5 (better yet, reproduce it yourself). The ruler is the system, and your hands are the surroundings. Grip one end of the ruler firmly with your left hand and apply a force with a finger of your right hand, as shown in **Figure 6.5*a***. Notice that your left hand automatically applies a load (consisting of a force and a moment) to the ruler in order to keep the gripped end from translating and rotating; this load "fixes" the gripped end relative to your left hand. The load of your left hand acting on the ruler (a fixed support) can be represented as a force of $\boldsymbol{F}_{\text{LH}} = F_{\text{LH},x}\boldsymbol{i} + F_{\text{LH},y}\boldsymbol{j}$ and a moment of $\boldsymbol{M}_{\text{LH}} = M_{\text{LH},z}\boldsymbol{k}$ that are the net effect of your left hand gripping the ruler (see **Figure 6.5*b***).

Returning now to the system depicted in **Figure 6.3*c***, we can describe the loads acting at the fixed support at E as $\boldsymbol{F}_E = F_{Ex}\boldsymbol{i} + F_{Ey}\boldsymbol{j}$ and $\boldsymbol{M}_E = M_{Ez}\boldsymbol{k}$. These loads are shown in **Figure 6.4*c***.

At Point *G* (System Pinned to Its Surroundings, Referred to as a Pin Connection). A pin connection consists of a pin that is loosely fitted in a hole. At this support a force acts on the system. To get a feeling for the force at a pin connection consider the physical setup in Figure 6.6*a*. The ruler (which is the system) is lying on a flat surface in position 1. A pencil, which is acting like a pin, is placed in the hole in the ruler and is gripped firmly with your left hand. The pencil and your hands constitute the surroundings. Now load the system with your right hand as shown in **Figure 6.6*a***; notice how your left hand reacts with a force to counter the right-hand force. If you next orient the ruler and right hand load as shown in **Figure 6.6*b***, again your left hand counters with a force. Finally, load the ruler as shown in **Figure 6.6*c***, and notice that the ruler rotates because your left hand is unable to counter with an opposing moment. This exercise tells you that there is a force ($\boldsymbol{F}_{\text{LH}}$) acting on the system at the pin connection but no moment. The force $\boldsymbol{F}_{\text{LH}}$ lies in the plane perpendicular to the pencil's length. For the situation in **Figure 6.6**, this means that $\boldsymbol{F}_{\text{LH}}$ can be written $\boldsymbol{F}_{\text{LH}} = F_{\text{LH},x}\boldsymbol{i} + F_{\text{LH},y}\boldsymbol{j}$ (**Figure 6.6*d***).

For the system in **Figure 6.3*d***, we can describe the load acting at the pin connection at G as $\boldsymbol{F}_G = F_{Gx}\boldsymbol{i} + F_{Gy}\boldsymbol{j}$, as shown in **Figure 6.4*d***. We have arbitrarily chosen to draw both components in their respective positive direction.

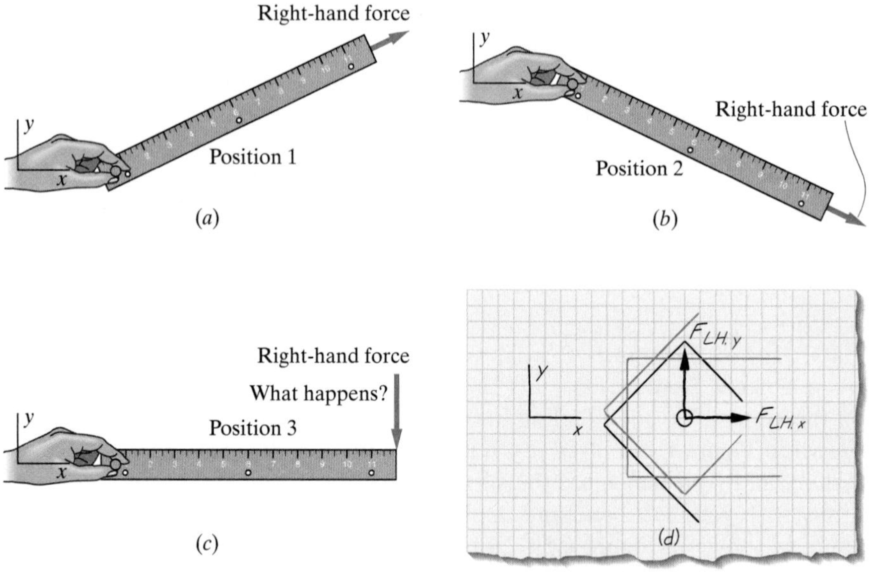

Figure 6.6 Illustrating a planar pin connection: (*a*) applying loads to a ruler (Position 1); (*b*) applying loads to a ruler (Position 2); (*c*) applying loads to a ruler (Position 3); (*d*) loads acting on the ruler at the pin connection

At Point *H* (Link). At this support there is a force that either pushes or pulls on the system; the line of action of the force is along the axis of the link. We shall have a lot more to say about links in the next chapter—for now we simply say that a link is a member with a pin connection at each end and no other loads acting on it.

In **Figure 6.4*d*** the force at *H* is represented by $\boldsymbol{F}_{H,\text{link}}$ acting along the long axis of the link. A link may either *push or pull* on the system, and here we have chosen to assume pulling. We could equally well have chosen the direction of $\boldsymbol{F}_{H,\text{link}}$ to indicate that the link is pushing, but as we will see in the next chapter, drawing the link as pulling will make interpreting numerical answers easier.

The free-body diagrams of the planar systems in **Figure 6.3** are presented in **Figure 6.4**. These diagrams include loads due to supports, as well as the load due to gravity acting at *J*. Each load is represented as a vector and is given a variable label. If the magnitude of a load is known, this value is included on the diagram.

Summary

Table 6.1 summarizes the loads associated with the planar supports discussed, along with some other commonly found planar supports. Don't feel that you need to memorize all the supports in this table—it is presented merely as a ready reference. On the other hand, you should be familiar with the loads associated with these standard planar supports.

Table 6.1 Standard Supports for Planar Systems

(A) Supports	Description of Loads	(B) Loads to Be Shown on Free-Body Diagram
1. **Normal contact without friction** 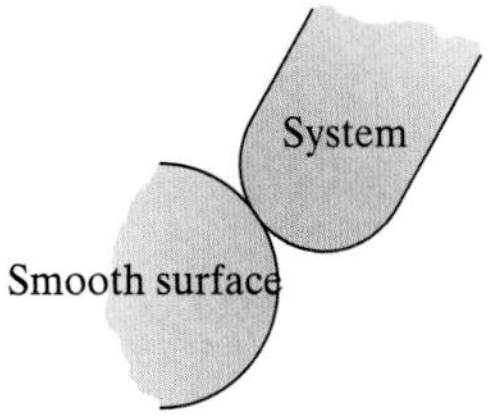	**Force** (***F***) oriented normal to surface on which system rests. Direction is such that force pushes on system.	***F*** 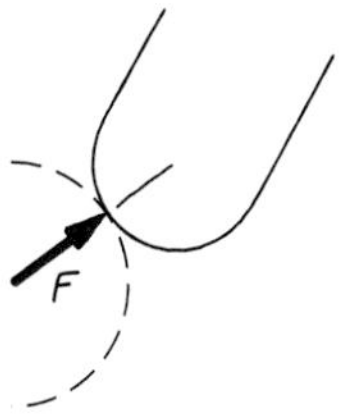
2. **Cable, rope, wire** 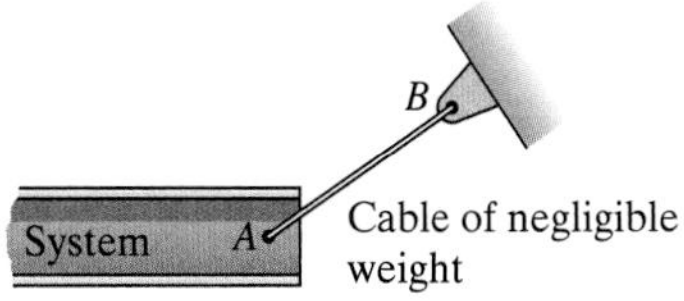	**Force** (***F***) oriented along cable. Direction is such that cable pulls on the system.	***F*** 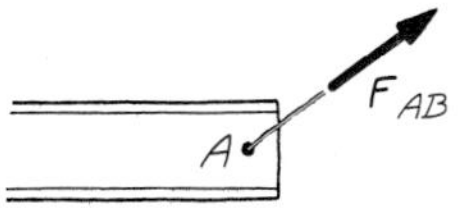
3. **Spring** 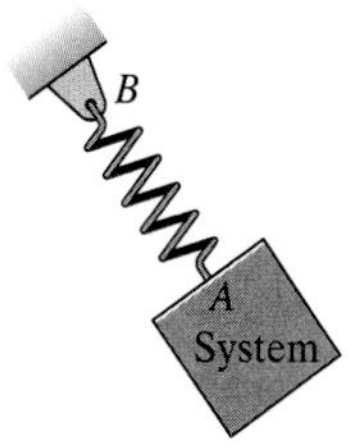	**Force** (***F***) oriented along long axis of spring. Direction is such that spring pulls on system if spring is in tension, and pushes if spring is in compression.	***F*** 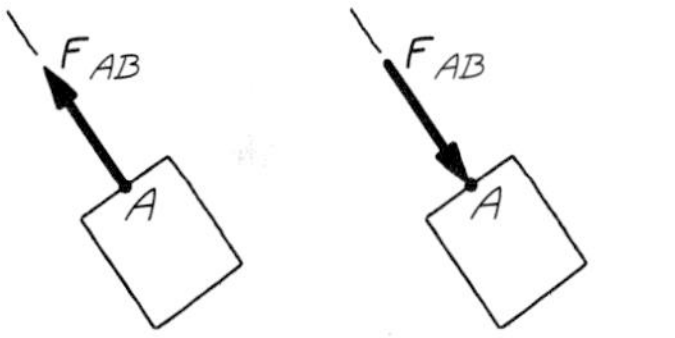 Extended spring Compressed spring
4. **Normal contact with friction** 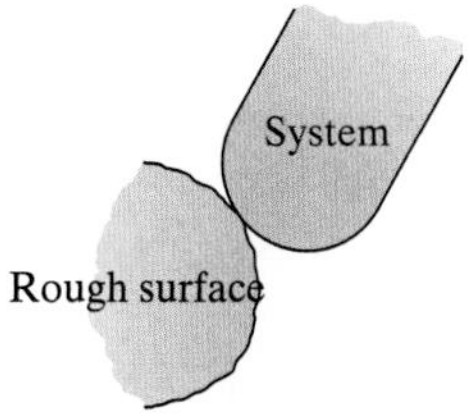	**Two forces**, one ($\boldsymbol{F}_y$) oriented normal to surface on which the system rests so as to push on system, other force ($\boldsymbol{F}_x$) is tangent to surface.	$\boldsymbol{F}_y$ $\boldsymbol{F}_x$ 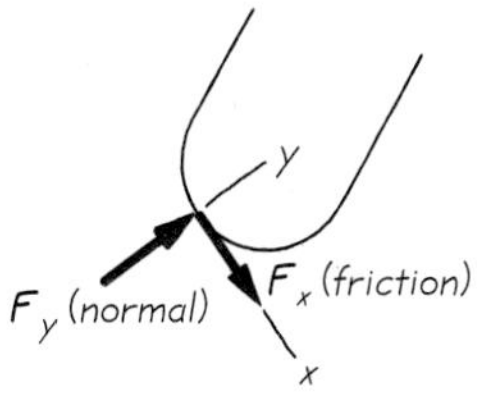
5. **Fixed support** 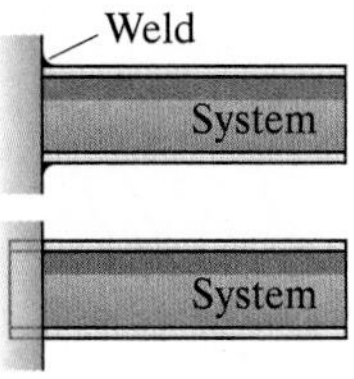	**Force** in *xy* plane represented in terms of components $\boldsymbol{F}_x$ and $\boldsymbol{F}_y$. **Moment** about *z* axis ($\boldsymbol{M}_z$).	$\boldsymbol{F}_x + \boldsymbol{F}_y$ $\boldsymbol{M}_z$ 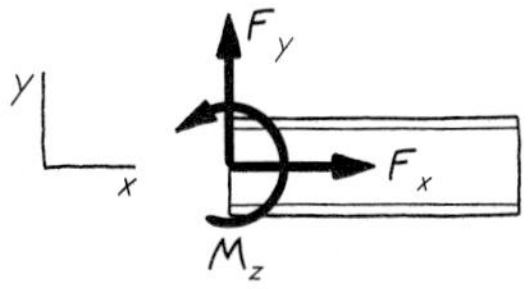

(Continued)

Table 6.1 (Cont.)

(A) Supports	Description of Loads	(B) Loads to Be Shown on Free-Body Diagram
6. **Pin connection** (pin or hole is part of system) 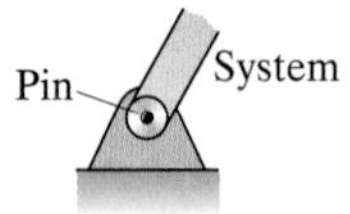	**Force** perpendicular to pin represented in terms of components F_x and F_y. Point of application is at center of pin.	$F_x + F_y$ 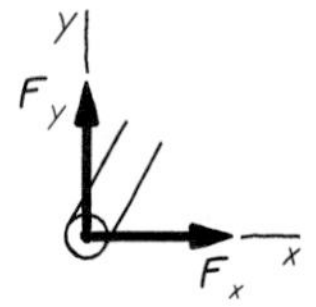
7. **Link** 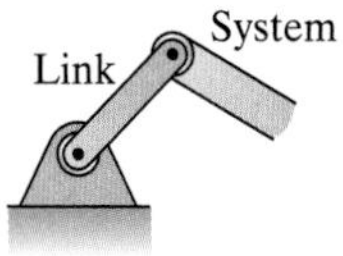	**Force** (F) oriented along link length; force can push or pull on the system.	F 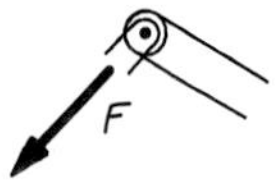
8. **Slot-on-pin** (slotted member is part of system) 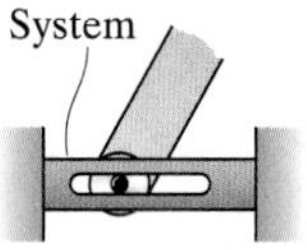	**Force** (F) oriented normal to long axis of slot. Direction is such that force can pull or push on system.	F 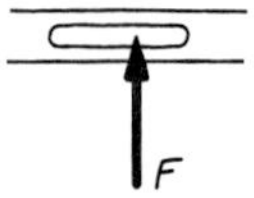
9. **Pin-in-slot** (pin is part of system) 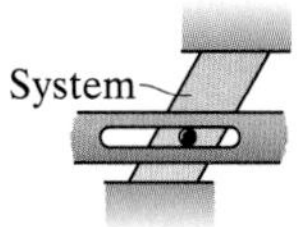	**Force** (F) oriented normal to long axis of slot. Direction is such that force can pull or push on system.	F 
10. **Smooth collar on smooth shaft** 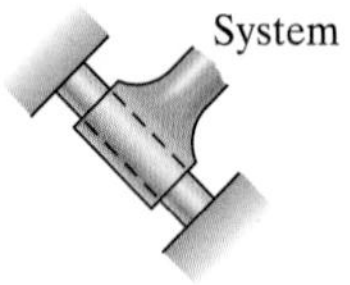	**Force** (F) oriented perpendicular to long axis of shaft. Direction is such that force can pull or push on system. **Moment** (M_z) about z axis.	F M_z 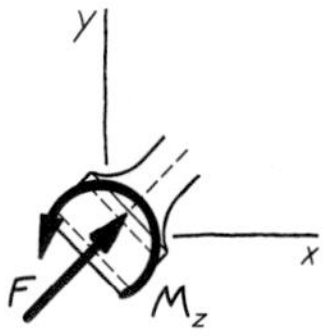
11. **Roller or rocker** 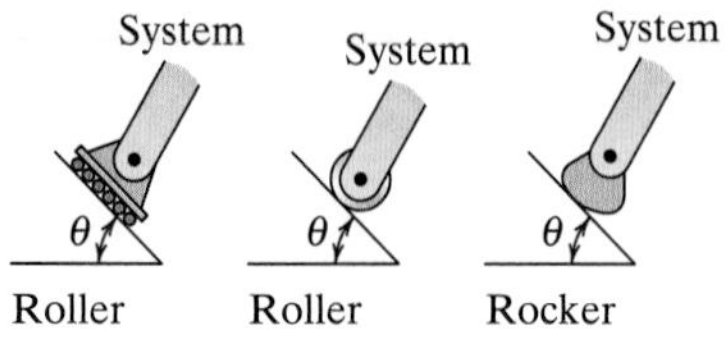	**Force** (F) oriented normal to surface on which system rests. Direction is such that force pushes on system.	F 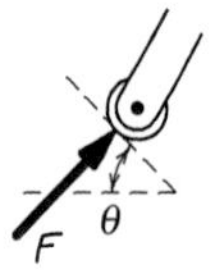

EXAMPLE 6.1 COMPLETE FREE-BODY DIAGRAMS

In Figure 6.7 and Figure 6.8, a block is supported at several points and the system is defined as the block. Gravity acts downward in the $-y$ direction at the indicated center of gravity (CG). In **Figure 6.7**, the surface at B is rough. As shown in **Figure 6.8**, the magnitudes of $\boldsymbol{F}_C$ and $\boldsymbol{M}_C$ are 10 lb and 40 in.-lb, respectively.

(a) Explain why these figures are not free-body diagrams.

(b) Create a free-body diagram of each system.

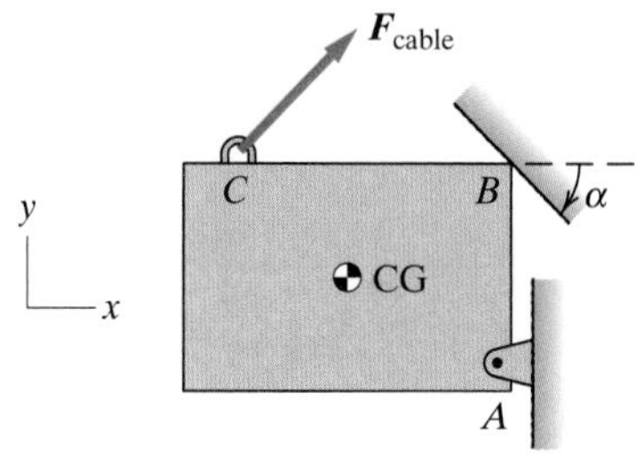

Figure 6.7

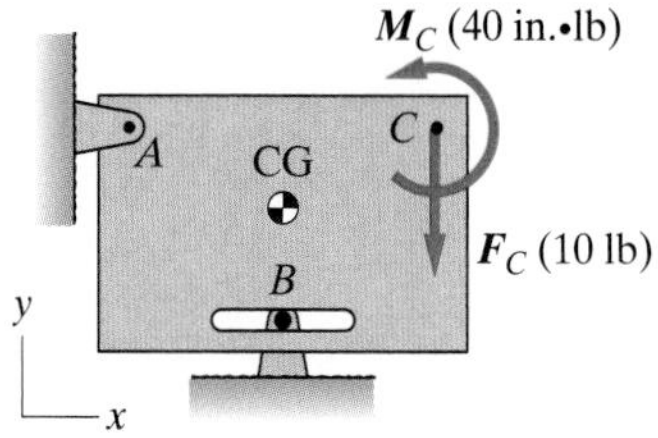

Figure 6.8

Goal Explain why the two figures are not free-body diagrams and create complete correct free-body diagrams.

Given We are given two systems with specified loads and supports.

Assume We assume that the system in each figure is planar because the known loads and the loads applied by supports all lie in a single plane. We also assume the slot-pin connection at B in **Figure 6.8** is smooth (frictionless).

Draw For each system, isolate it by drawing a boundary around the block as is done in Figure 6.9 and Figure 6.10. Then replace each support by its associated loads.

Solution (a) Neither figure is a free-body diagram because the system (the block) has not been isolated from its surroundings—at A and B each block is still shown connected to its surroundings.

(b) Figure 6.7: We isolate the block using the boundary shown in **Figure 6.9**, establish the xy coordinate system for the entire system and the x^*y^* coordinate system to simplify the representation of the support at B. Then we note the following:

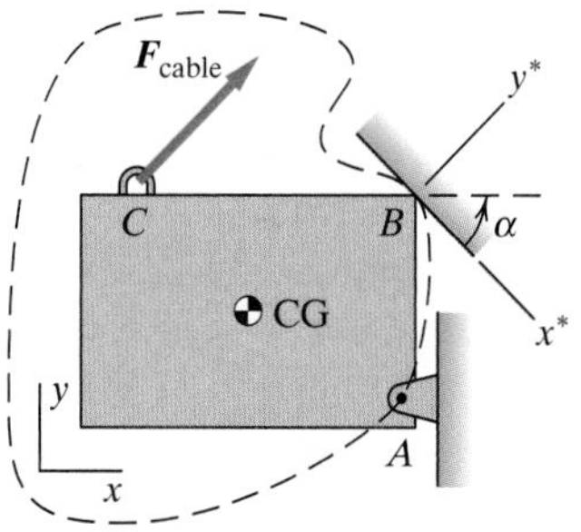

Figure 6.9

At A a pin connection attaches the system to its surroundings. According to **Table 6.1**, a pin connection applies a force to the system. As we do not know the direction or magnitude of this force, we represent it as two components, $\boldsymbol{F}_{Ax}$ and $\boldsymbol{F}_{Ay}$, which we arbitrarily draw in the positive x and y directions (Figure 6.11).

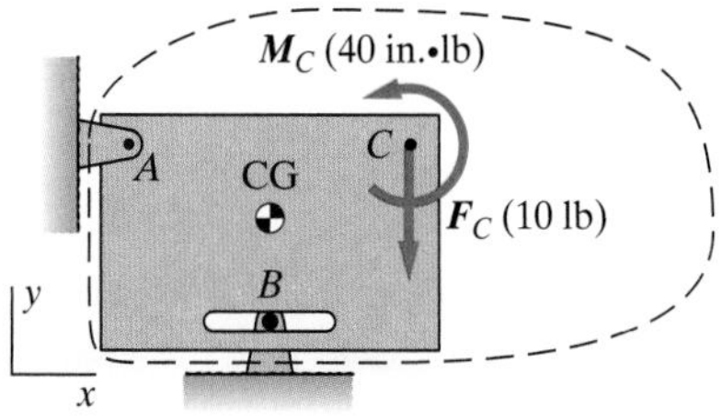

Figure 6.10

At B the system rests against a surface inclined at angle α relative to the horizontal. Since we know that the surface is rough, we must consider the friction between surface and system. According to **Table 6.1**, there will be a normal force $\boldsymbol{F}_{B,\text{normal}}$ acting on the system, oriented perpendicular to the surface so as to push on the system, as shown in **Figure 6.11**. We do not know its magnitude. There is also the friction force $\boldsymbol{F}_{B,\text{friction}}$, perpendicular to $\boldsymbol{F}_{B,\text{normal}}$. We do not know the magnitude of $\boldsymbol{F}_{B,\text{friction}}$ or whether it acts in the $+x^*$ or $-x^*$ direction, and so we arbitrarily draw it in the $+x^*$ direction.

At C a cable pulls on the system, which we represent as a force of unknown magnitude but known direction ($\boldsymbol{F}_C$).

At CG (the center of gravity) a force $\boldsymbol{W}$ acts in the $-y$ direction.

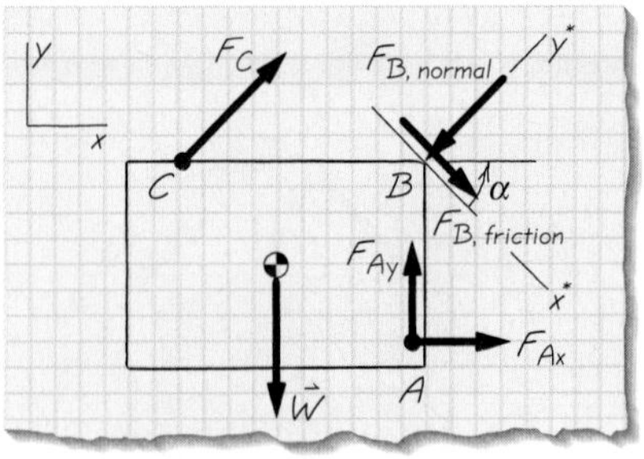

Figure 6.11 Free-body diagram of system in Figure 6.7

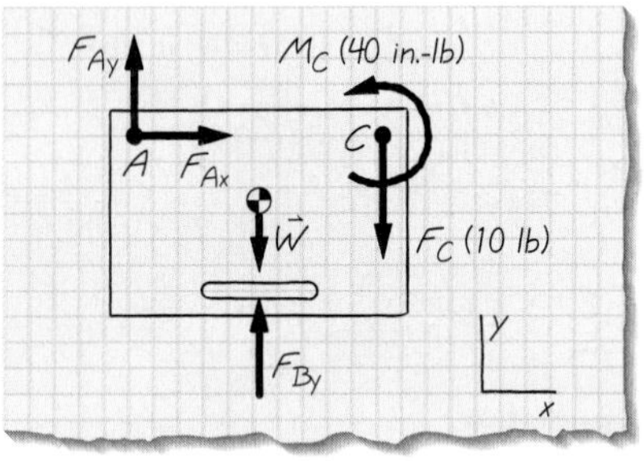

Figure 6.12 Free-body diagram of system in Figure 6.8

The block presented in **Figure 6.11**, with forces $\boldsymbol{F}_{Ax}$, $\boldsymbol{F}_{Ay}$, $\boldsymbol{F}_{B,\text{normal}}$, $\boldsymbol{F}_{B,\text{friction}}$, $\boldsymbol{F}_C$, and $\boldsymbol{W}$ each drawn at its point of application is a free-body diagram of the system in **Figure 6.7**.

Figure 6.8: We isolate the block using the boundary shown in **Figure 6.10**, establish an *xy* coordinate system as shown, and then we note the following:

At A a pin connection attaches the system to its surroundings. According to **Table 6.1**, a pin connection applies a force to the system. As we do not know the direction or magnitude of this force, we represent it as two components, $\boldsymbol{F}_{Ax}$ and $\boldsymbol{F}_{Ay}$, which we arbitrarily draw in the positive *x* and *y* directions (**Figure 6.12**).

At B a slot in the block is attached to a slider that allows the block to move in the *x* direction but prevents movement in the *y* direction. Therefore, we include a force $\boldsymbol{F}_{By}$, which we have arbitrarily drawn in the positive direction.

At C there is a known force ($\boldsymbol{F}_C$) and a moment ($\boldsymbol{M}_C$). Known values are written next to the vectors.

At CG (the center of gravity) a force $\boldsymbol{W}$ acts in the $-y$ direction.

The block presented in **Figure 6.12**, with forces $\boldsymbol{F}_{Ax}$, $\boldsymbol{F}_{Ay}$, $\boldsymbol{F}_{By}$, and $\boldsymbol{F}_{C,\text{known}}$, $\boldsymbol{W}$, and moment $\boldsymbol{M}_{C,\text{known}}$ each drawn at its point of application constitutes a free-body diagram of the system. Notice that this diagram includes the known magnitudes of $\boldsymbol{F}_C$ and $\boldsymbol{M}_C$.

Answer (a) **Figures 6.7** and **6.8** are not free-body diagrams of systems because each system (the block) has not been fully separated from all supports.

(b) The free-body diagrams of the systems in **Figures 6.7** and **6.8** are shown in **Figures 6.11** and **6.12**, respectively.

EXAMPLE 6.2 EVALUATING THE CORRECTNESS OF FREE-BODY DIAGRAMS

Consider the description of each planar system in **Figures 6.13–6.21** and determine whether the associated free-body diagram is correct. Unless otherwise stated, assume that the weight of the system is negligible and therefore can be ignored.

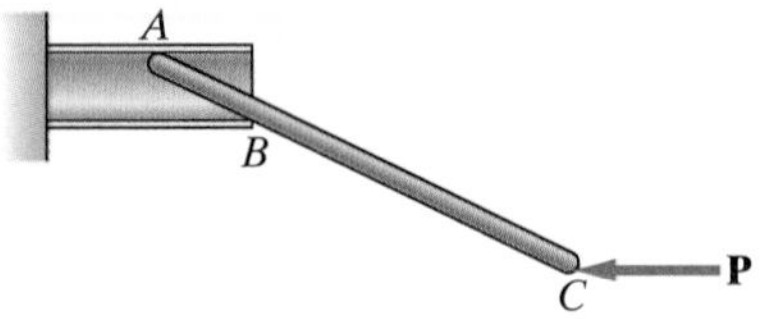

(*a*)

Figure 6.13

(a) A thin rod is supported by a smooth tube that is fixed to the wall (**Figure 6.13*a***). The rod is touching the tube interior at *A* and leaning on the tube end at *B*. A known force *P* is pushing on the rod at *C*. Assume no friction on the surface of the tube.

Answer *Tube:* The proposed free-body diagram of the tube in **Figure 6.13*b*** is *correct*. It accounts for the fixed end at the left and the normal contact between the tube and the rod.

Answer *Rod:* The proposed free-body diagram of the tube in **Figure 6.13*b*** is *not correct*. The force $\boldsymbol{F}_{B,\text{tube on rod}}$ in this diagram, representing the normal force of the tube pushing on the rod, should be in the other direction (Newton's third law). As shown, the tube is pulling on the rod which is not physically possible.

(b) A beam, pinned at A and resting against a roller at B, is loaded by a 2-kN force and a 2.4-kN · m moment (**Figure 6.14*a***).

Answer The proposed free-body diagram of the beam in **Figure 6.14*b*** is *not correct*. The normal force, $\boldsymbol{F}_{B,\text{normal}}$, acting on the beam at B should be oriented perpendicular to the inclined surface.

(c) A uniform beam weighing 200 lb is fixed at A. A 400-lb and 1400-lb load are applied at B and C as shown (**Figure 6.15*a***).

Answer The proposed free-body diagram in **Figure 6.15*b*** is *correct*. The weight of the beam is included at the $x = 5$ ft position since the beam is uniform.

(d) A door that weighs W hangs from hinges at A and B. Hinge A acts like a pin connection, and hinge B acts like a vertical slot support (**Figure 6.16*a***).

Answer The proposed free-body diagram in **Figure 6.16*b*** is *not correct*. The hinge at B does not apply a vertical force to the door, so this force should not be on the free-body diagram. The vertical force component at B ($\boldsymbol{F}_{By}$) is a thrust force.

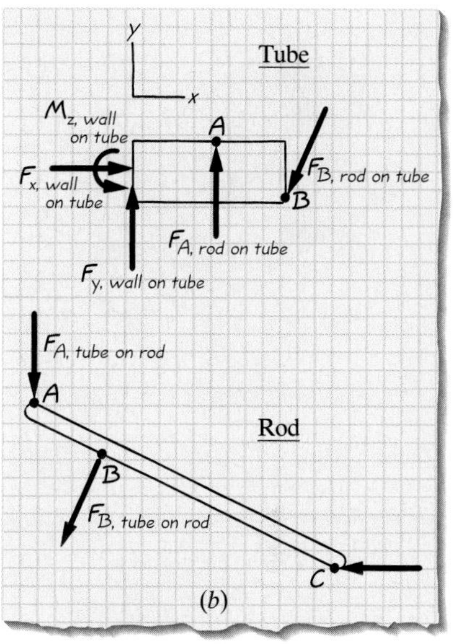

Proposed free-body diagram

Figure 6.13 (Cont.)

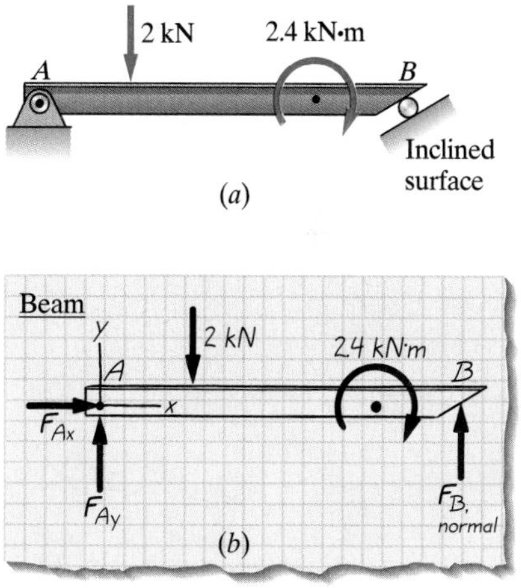

Figure 6.14

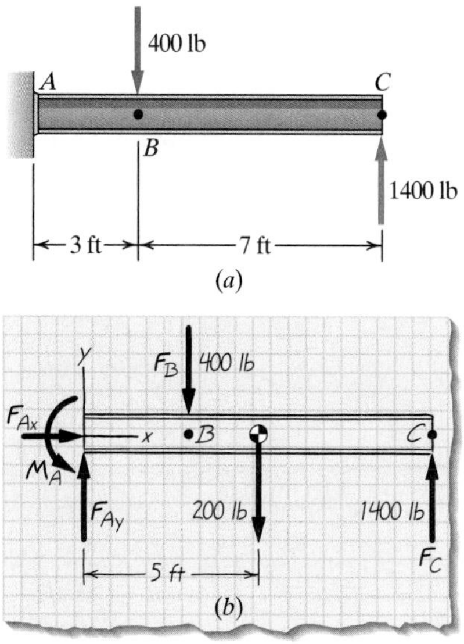

Figure 6.15

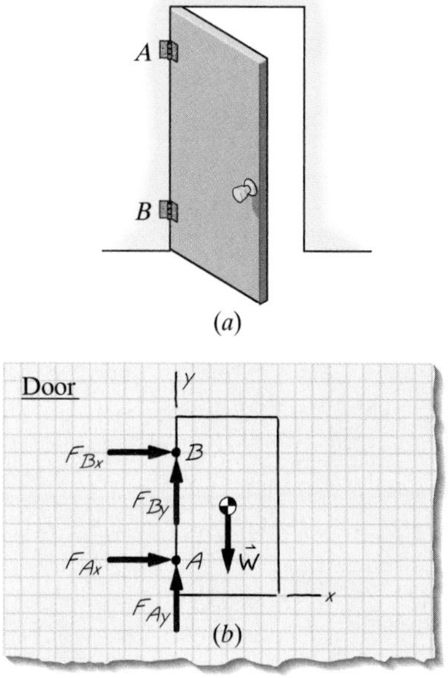

Figure 6.16

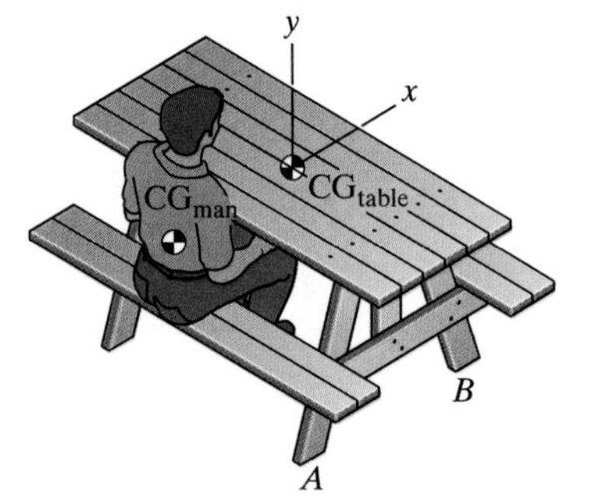

(a)

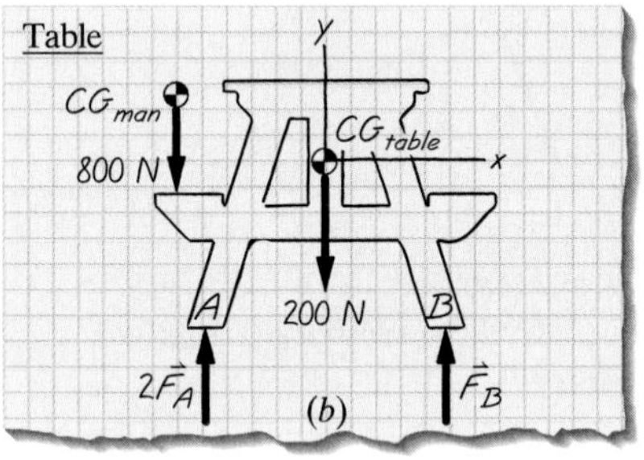

(b)

Figure 6.17

(e) A man weighing 800 N sits at the picnic table halfway between its two ends. His center of gravity (CG_{man}) is noted. The table weighs 200 N, with a center of gravity at CG_{table} (**Figure 6.17*a***). Assume that any friction between the legs and the ground can be neglected.

Answer The proposed free-body diagram in **Figure 6.17*b*** is *not correct.* The label for the normal force acting on the table at B should be $2\boldsymbol{F}_B$, (not $\boldsymbol{F}_B$), because the force vector at B represents the normal force for two legs.

(f) A frame used to lift a hatch is pinned to the hatch at A and B. A force $\boldsymbol{F}$ is applied to the frame at C (**Figure 6.18*a***).

Answer This free-body diagram in **Figure 6.18*b*** is *not correct.* At each pin connection force components in the x and y directions should be shown.

(g) A frame consisting of members AB and CD supports the pulleys, cable, and block L (**Figure 6.19*a***).

Answer *Whole frame:* The proposed free-body diagram in **Figure 6.19*b*** is *correct.* It includes the forces at pins A and C, the cable tension $\boldsymbol{F}_{cable}$ pulling on the frame, and the gravity force from block L ($\boldsymbol{W}_L$).

Member CD: The proposed free-body diagram in **Figure 6.19*b*** is *incorrect* because the forces at the pin connection at D have not been included.

Member AB: The proposed free-body diagram in **Figure 6.19*b*** is *correct.*

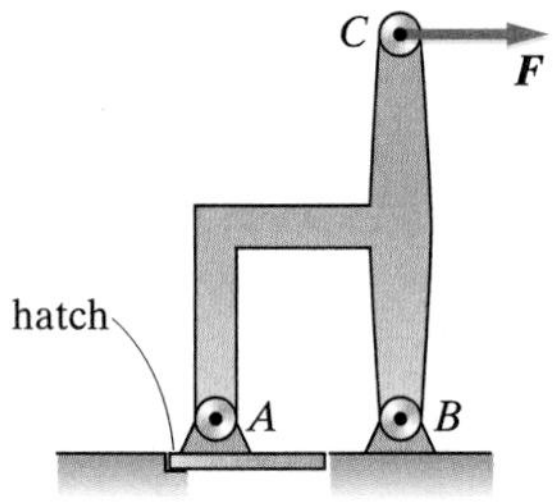

(a)

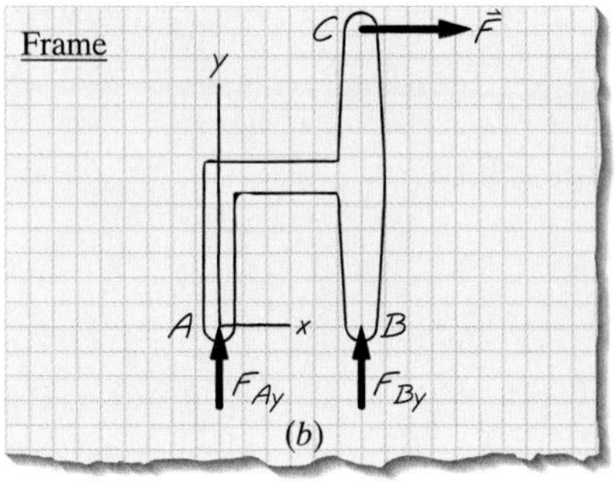

(b)

Figure 6.18

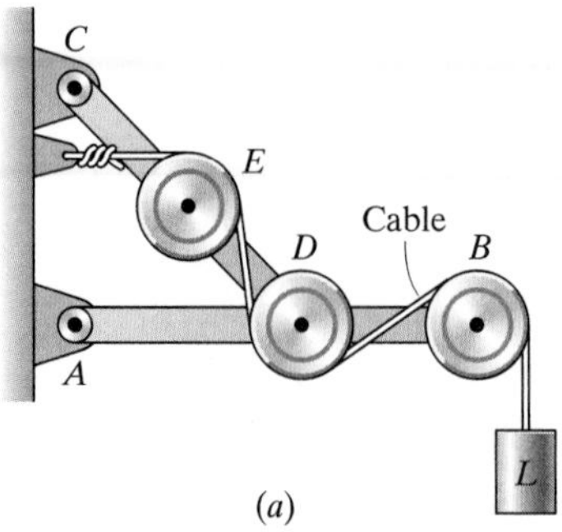

(a)

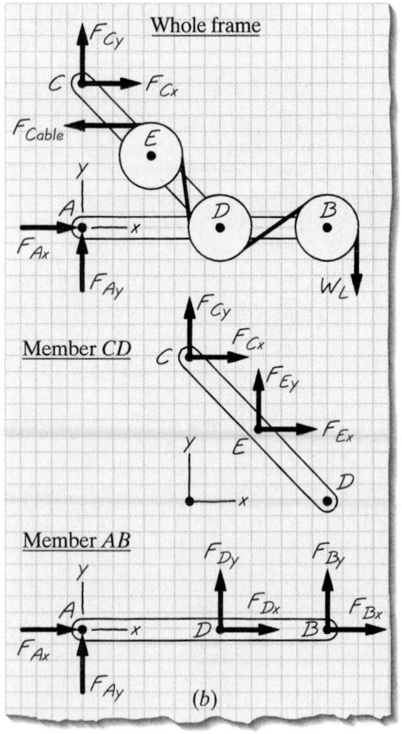

(b)

Figure 6.19

(h) A 50-kg roller is pulled up a 20° incline with a force P. As it is pulled over a smooth step, all of its weight rests against the step (**Figure 6.20*a***).

Answer The proposed free-body diagram in **Figure 6.20*b*** is *not correct* because the gravity force should be drawn in the negative y^* direction. The mass of the roller has been properly converted to weight on earth. The force $\boldsymbol{F}_B$ is drawn correctly.

(i) A 1200-lb object is held up by a force T on a rope threaded through a system of frictionless pulleys (**Figure 6.21*a***).

Answer *Pulley A:* The proposed free-body diagram in **Figure 6.21*b*** is *correct.*

Since the pulleys are frictionless, the force throughout the rope is constant (we will prove this in Chapter 7). The tension on the rope is T wherever one cuts the rope.

Pulley B: The proposed free-body diagram in **Figure 6.21*b*** is *correct.*

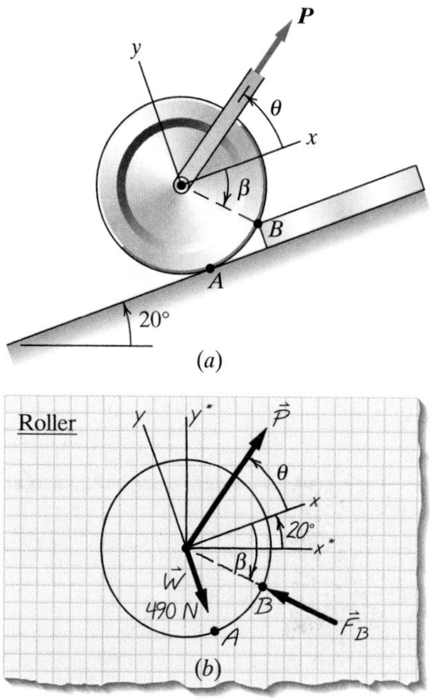

Figure 6.20

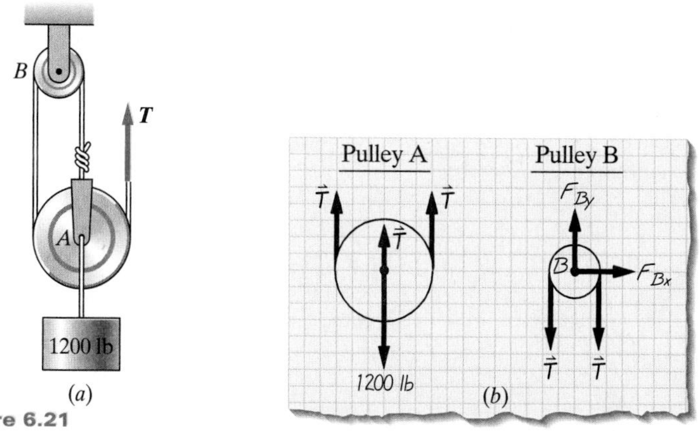

Figure 6.21

EXERCISES 6.2

6.2.1. Consider the description of each planar system and determine whether the proposed free-body diagram is correct.

a. A curved beam of weight W is supported at A by a pin connection and at B by a rocker, as shown in **E6.2.1*a***.

b. A beam is pinned at B and rests against a smooth incline at A as shown in **E6.2.1*b***. The total weight of the beam is W.

c. A forklift is lifting a crate of weight W_1 as shown in **E6.2.1*c***. The weight of the forklift is W_2. The front wheels are free to turn and the rear wheels are locked.

d. A mobile hangs from the ceiling from a cord as shown in **E6.2.1*d***.

e. A force acts on a brake pedal, as shown in **E6.2.1*e***.

f. A child balances on the beam as shown in **E6.2.1*f***. Planes A and B are smooth. The weight of the beam is negligible.

g. A beam is bolted to a wall at B as shown in **E6.2.1*g***. The weight of the beam is negligible.

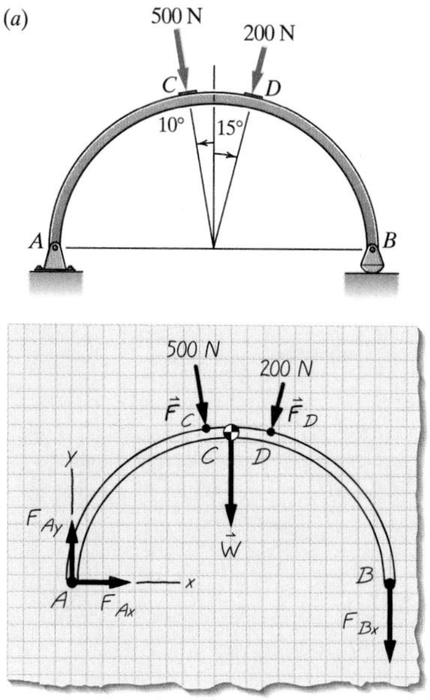

Proposed free-body diagram

E6.2.1

(b)

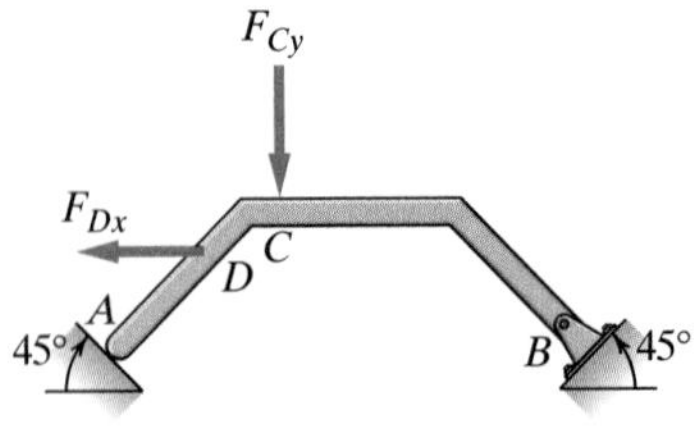

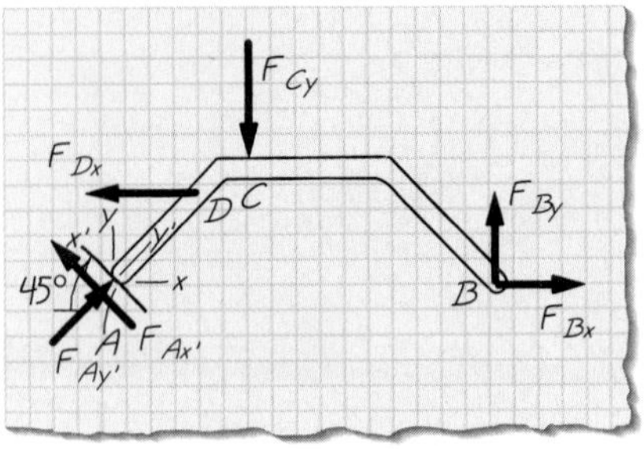

Proposed free-body diagram

(c)

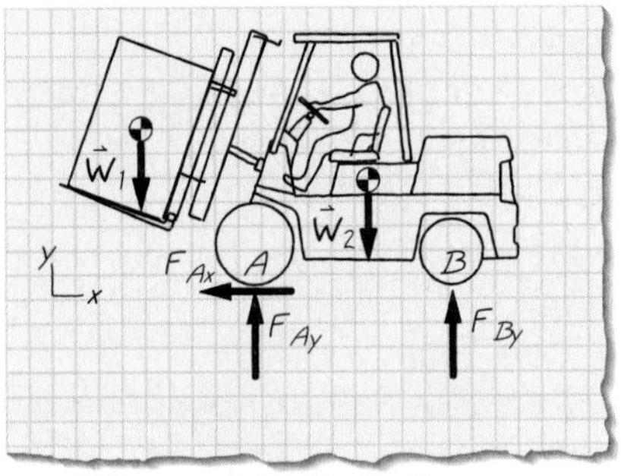

Proposed free-body diagram

(d)

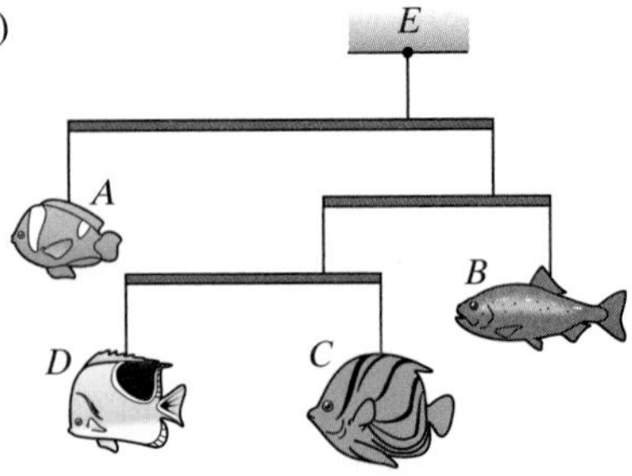

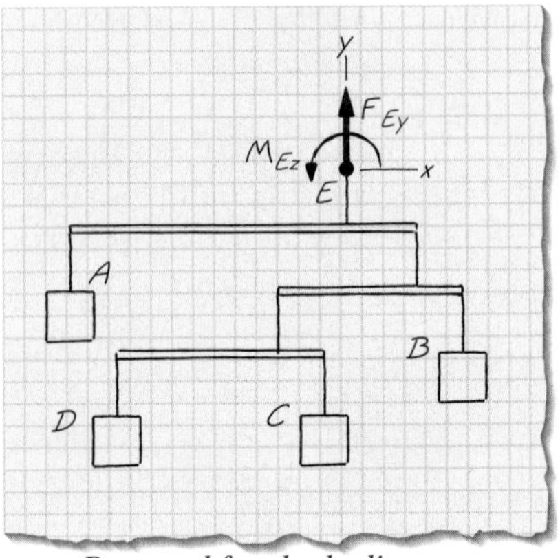

Proposed free-body diagram

(e)

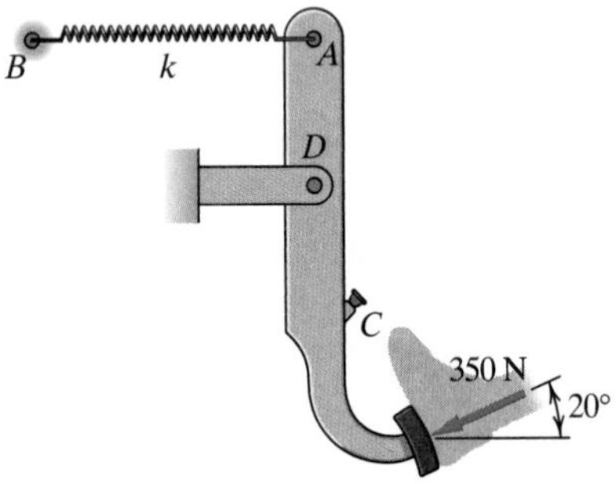

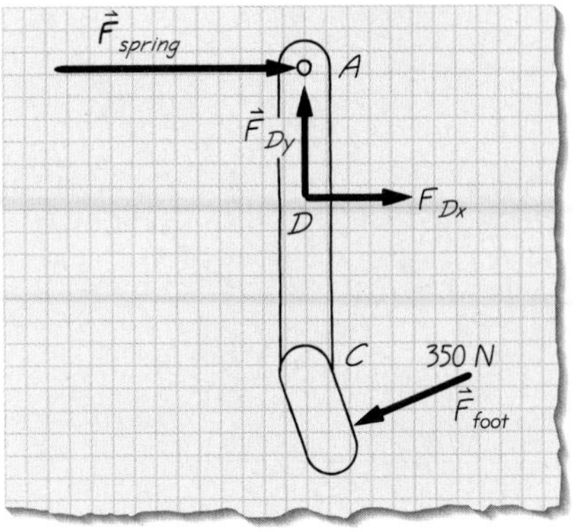

Proposed free-body diagram

(f)

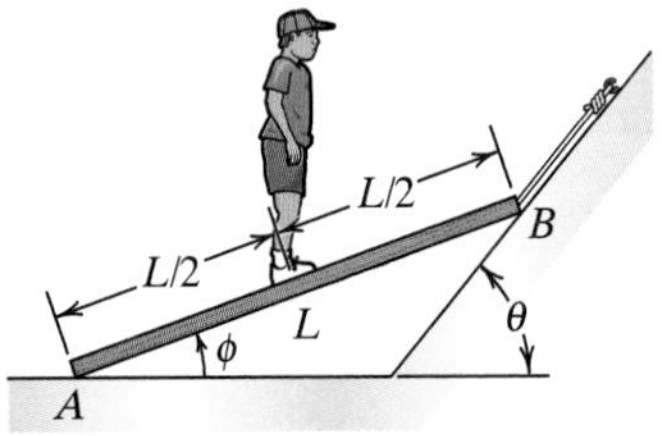

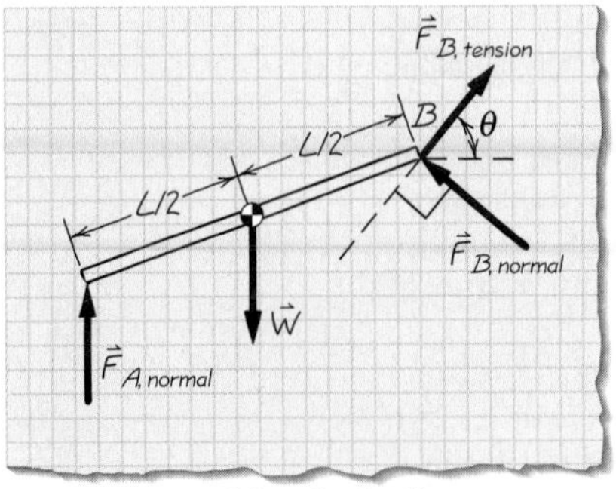

Proposed free-body diagram

(g)

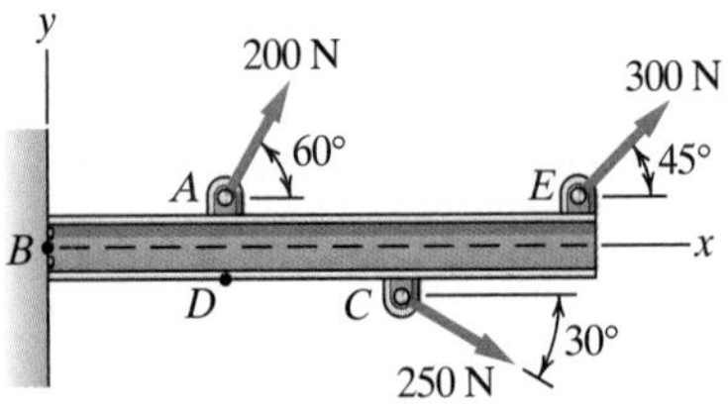

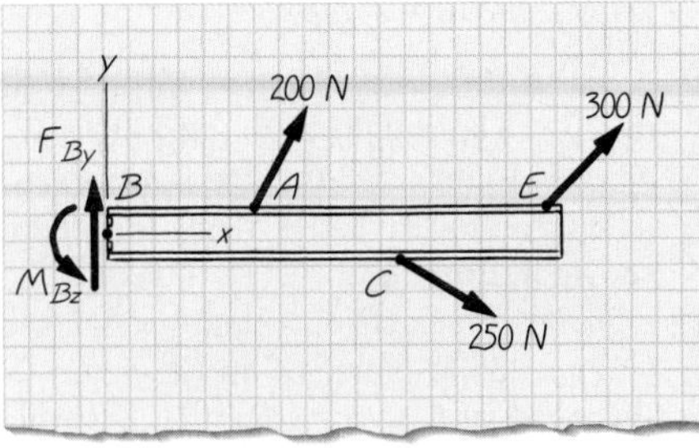

Proposed free-body diagram

E6.2.1 (Cont.)

6.2.2. The beam of uniform weight is fixed at C and rests against a smooth block at A (E6.2.2). In addition, a 100-N weight hangs from point B. Based on information in **Table 6.1**, what loads do you expect to act on the beam at C due to the fixed condition? What loads do you expect to act on the beam at A where it rests on the smooth block? Present your answer in terms of a sketch of the beam that shows the loads acting on it at A, B, and C. Also comment on whether the sketch you created is or is not a free-body diagram.

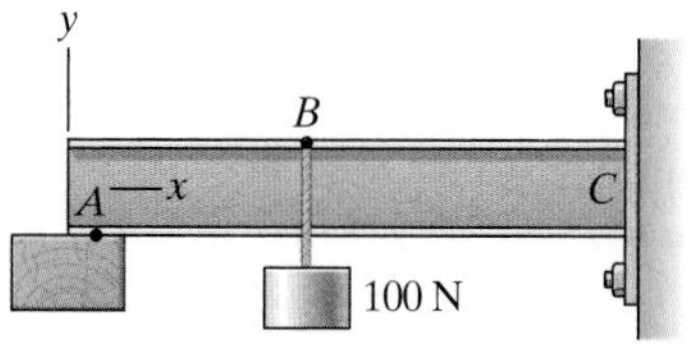

E6.2.2

6.2.3. The truss is attached to the ground at A with a rocker and at B with a pin connection (E6.2.3). Additional loads acting on the beam are as shown. Based on information in **Table 6.1**, what loads do you expect to act on the truss at A due to the rocker connection? What loads do you expect to act on the truss at B due to the pin connection? Present your answer in terms of a sketch of the truss that shows the loads acting on it at A and B. Also comment on whether the sketch you created is or is not a free-body diagram.

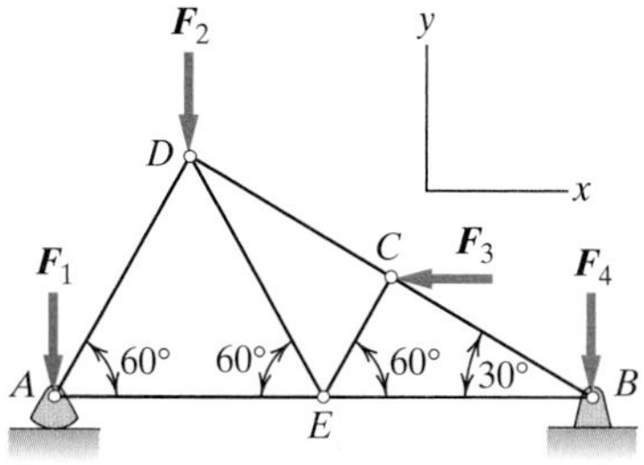

E6.2.3

6.2.4. The belt-tensioning device is as shown in E6.2.4. Pulley A is pinned to the L-arm at B. Based on information in **Table 6.1**, what loads do you expect to act on the pulley at B? What loads do you expect to act on the pulley due to the belt tension? Present your answer in terms of a sketch of the pulley that shows loads acting on it at B and due to the belt tension. Also comment on whether the sketch you created is or is not a free-body diagram.

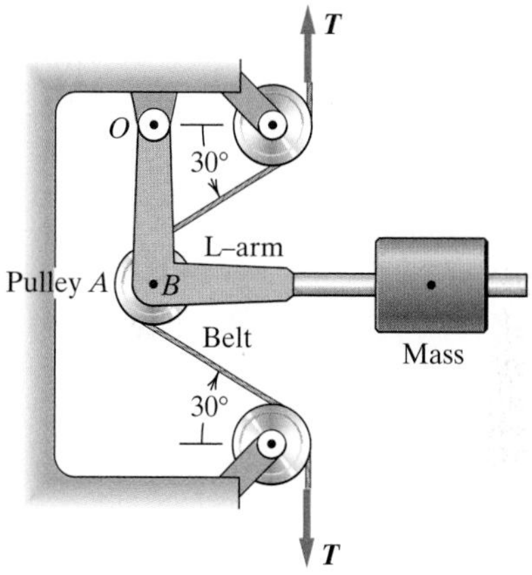

E6.2.4

6.2.5. Cable AB passes over the small frictionless pulley C without a change in tension and holds up the metal cylinder (E6.2.5). Based on information in **Table 6.1**, what loads do you expect to act on the cylinder? Present your answer in terms of a sketch of the cylinder that shows all the loads acting on it. Also comment on whether the sketch you created is or is not a free-body diagram.

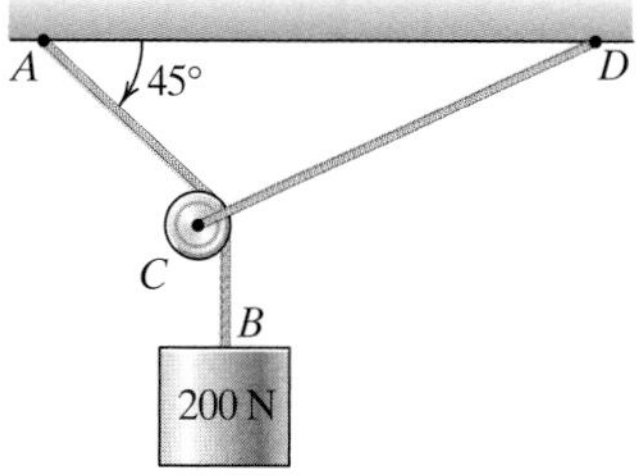

E6.2.5

6.3 NONPLANAR SYSTEM SUPPORTS

Now we consider how to identify nonplanar system supports and represent the loads associated with them. A system is nonplanar if all the forces acting on it *cannot* be represented in a single plane or all moments acting on it *are not* about an axis perpendicular to that plane. You will see similarities to our discussion in the prior section on planar systems *and* some important differences. As with planar systems, *if a boundary location prevents the translation of a nonplanar system in a given direction, then a force acts on the system at the location of the support in the opposite direction. Likewise, if rotation is prevented, a moment opposite the rotation acts on the system at the location of the support.*

Consider the systems in Figure 6.22, for which we want to draw free-body diagrams. The three nonplanar supports in **Figure 6.22*a*** (normal contact without friction, cable, spring) are identical to their planar counterparts. Associated with each support is a force acting along a known line of action, as depicted in Figure 6.23*a*.

The two nonplanar supports in **Figure 6.22*b*** are similar to their planar counterparts. **Normal contact with friction** involves normal and friction forces, where the friction force is in the plane perpendicular to the normal force; these are represented in **Figure 6.22*b*** as $\boldsymbol{F}_{4z}$, $\boldsymbol{F}_{4x}$, and $\boldsymbol{F}_{4y}$, respectively. The **fixed support** of a nonplanar system is able to prevent the system from translating along and rotating about any axis—therefore it involves a force with three components ($\boldsymbol{F}_5 = F_{5x}\boldsymbol{i} + F_{5y}\boldsymbol{j} + F_{5y}\boldsymbol{k}$) and a moment with three components ($\boldsymbol{M}_5 = M_{5x}\boldsymbol{i} + M_{5y}\boldsymbol{j} + M_{5z}\boldsymbol{k}$). The free-body diagram of the system in **Figure 6.22*b*** is depicted in **Figure 6.23*b***.

Another commonly found nonplanar support is a **single hinge** (Figure 6.24*a*). It does not restrict rotation of the system about the hinge

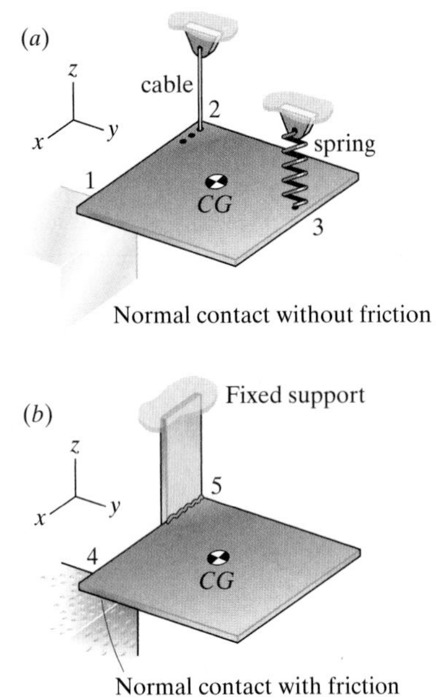

Figure 6.22 A plate connected to its surroundings by various supports. The plate must be modeled as a nonplanar system.

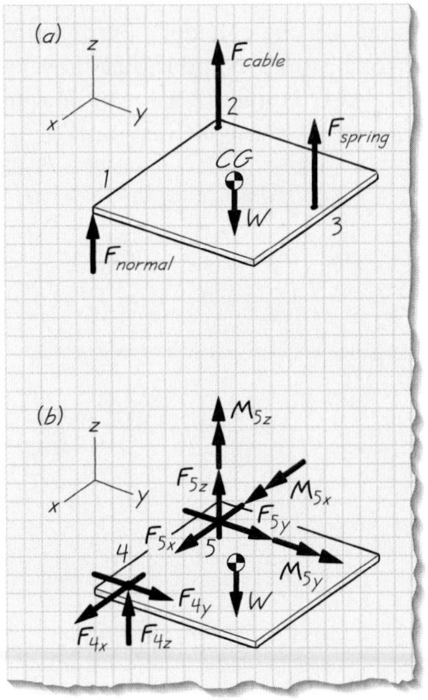

Figure 6.23 Free-body diagram of the nonplanar systems shown in Figure 6.22

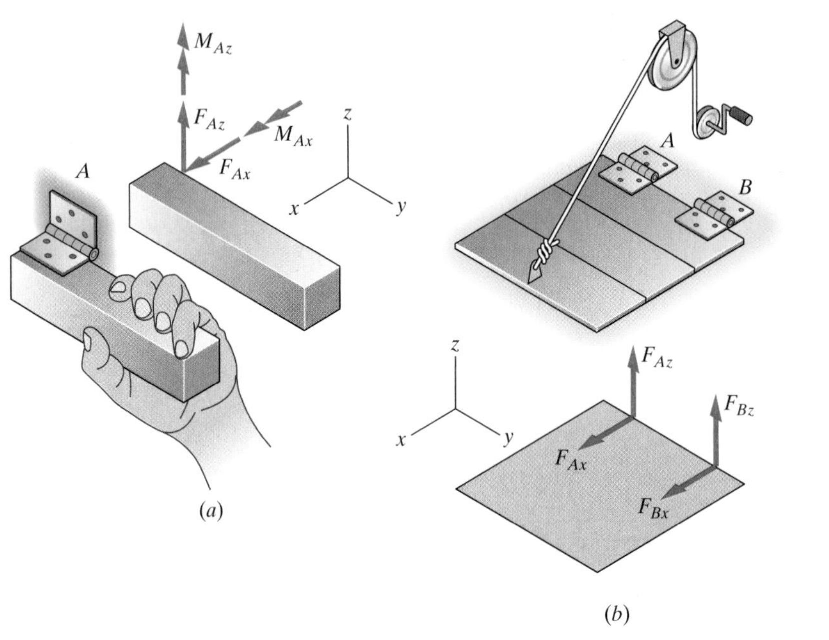

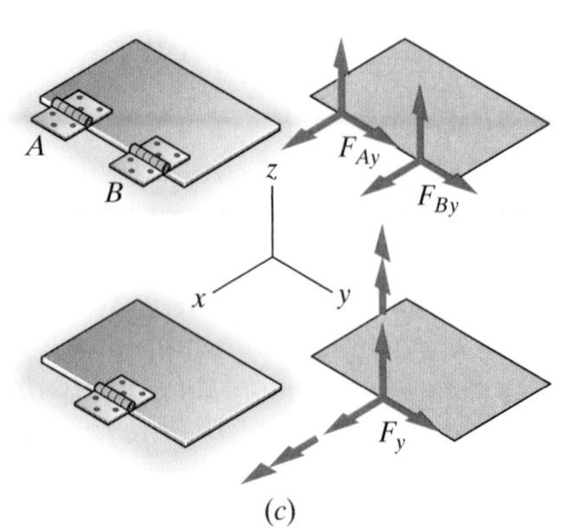

Figure 6.24 (*a*) The forces acting on the system (*A* block) at a single-hinge connection; (*b*) the forces acting on the system (a trap door) at multiple hinges; (*c*) hinges designed to prevent motion along the hinge axis. The force component F_z is commonly referred to as a thrust force.

pin. A single hinge applies a force (with two components) and a moment (with two components) perpendicular to the axis of the hinge. For the example in **Figure 6.24*a***, we represent the loads acting at the single hinge support as $\boldsymbol{F}_A = F_{Ax}\boldsymbol{i} + F_{Ay}\boldsymbol{j}$ and moment $\boldsymbol{M}_A = M_A\boldsymbol{i} + M_{Ay}\boldsymbol{j}$.

If a hinge is one of several properly aligned hinges attached to a system, each hinge applies a force perpendicular to the hinge axis (see **Figure 6.24*b***) and no moment. Depending on the design of a hinge (and regardless of whether it is a single hinge or one of several), it may also apply a force along the axis of the pin ($\boldsymbol{F}_z$ in **Figure 6.24*c***). The experiments outlined in Example 6.3 are intended to illustrate the difference in the loads involved with single versus multiple hinges.

Summary

Table 6.2 summarizes the loads associated with supports for nonplanar systems. Other supports commonly found with nonplanar systems are also included in the table. For example, the **ball-and-socket support** restricts all translations of the system by applying a force to the system, but it does not restrict rotation of the system about any axis. An example of a ball-and-socket support familiar to everyone is the human hip joint (Figure 6.25). As another example, a **journal bearing** does not restrict system rotation about one axis, while restricting translation in a plane perpendicular to the axis (**Figure 6.25*d***). Take a few minutes to study **Table 6.2** and notice the similarities and differences between hinges, journal bearings, and thrust bearings.

Table 6.2 is not an exhaustive list of nonplanar supports. It contains commonly found and representative examples. If you find yourself considering a support that is not neatly classified as one of these, remember that you can always return to the basic characteristics associated with any support: *If a support prevents the translation of the system in a given direction, then a force acts on the system in the opposing direction. If rotation is prevented, a moment opposing the rotation is exerted on the system.*

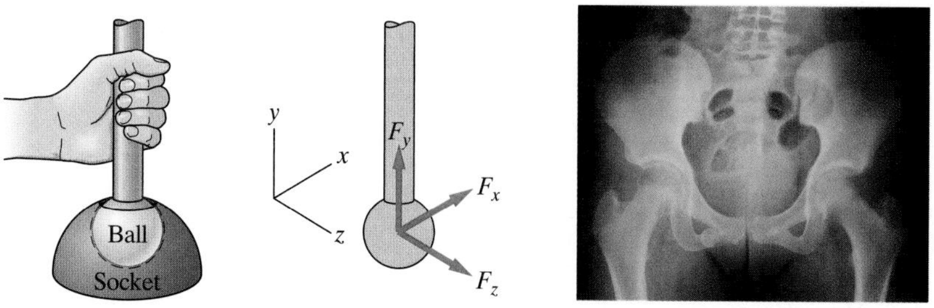

(*a*) (*b*) (*c*)

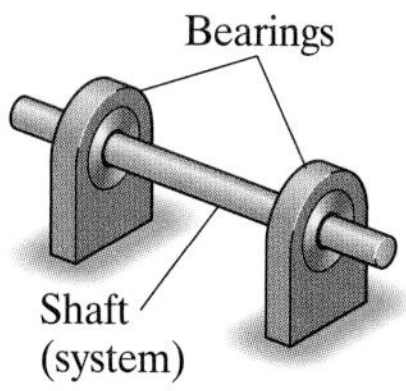

(*d*)

Figure 6.25 (*a*) A ball-and-socket connection; (*b*) the forces that the socket applies to the ball; (*c*) the hip joint is a ball-and-socket connection; (*d*) isometric view of a journal bearing with a shaft running through it

Table 6.2 Standard Supports for Nonplanar Systems

(A) Support	Description of Boundary Loads	(B) Loads to Be Shown in Free-Body Diagram
1. **Normal contact without friction** 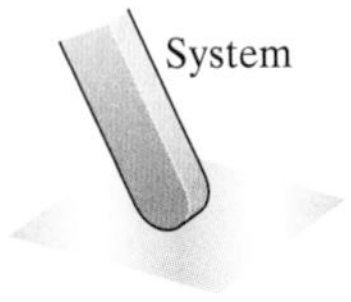	**Force** (***F***) oriented normal to surface on which system rests. Direction is such that force pushes on system.	***F*** 
2. **Cable, rope, wire** 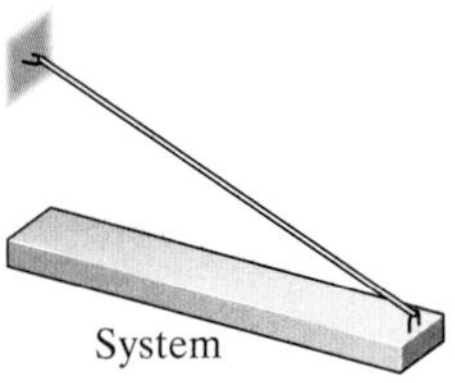	**Force** (***F***) oriented along cable. Direction is such that force pulls on system.	***F*** 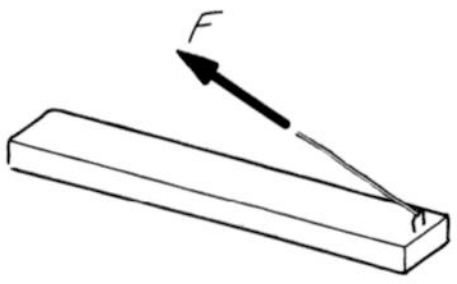
3. **Spring** 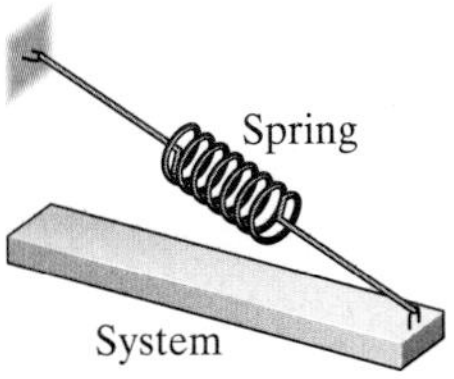	**Force** (***F***) oriented along long axis of spring. Direction is such that force pulls on system if spring is in tension and pushes if spring is in compression.	***F*** 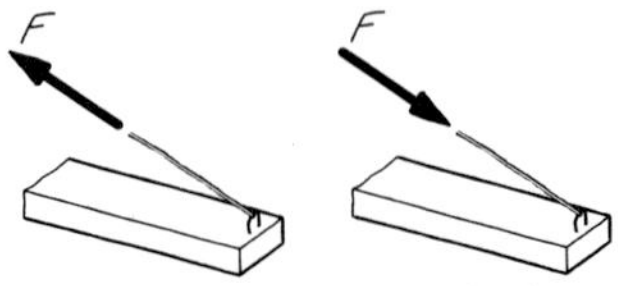 Spring in tension $+\Delta$; Spring in compression $-\Delta$
4. **Normal contact with friction** 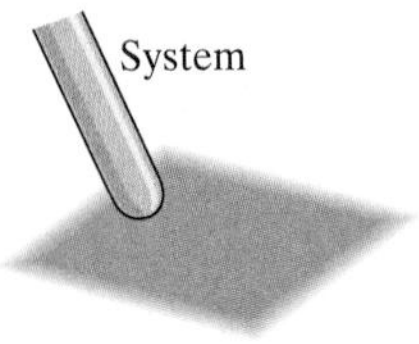	**Two forces**, one ($\boldsymbol{F}_z$) oriented normal to surface so as to push on system, other force is tangent to surface on which the system rests and is represented in terms of its components ($\boldsymbol{F}_x + \boldsymbol{F}_y$).	$\boldsymbol{F}_z$ $\boldsymbol{F}_x + \boldsymbol{F}_y$ 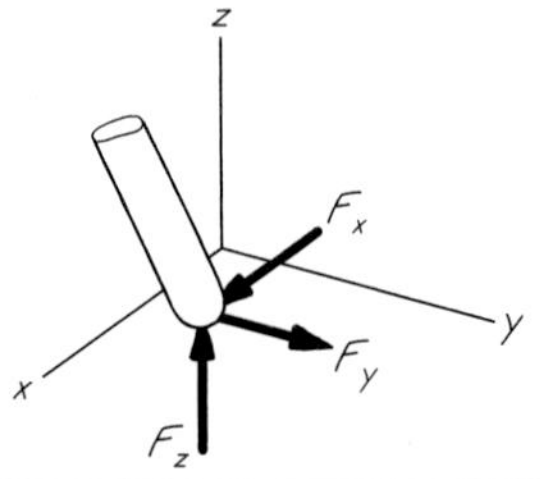
5. **Fixed support** 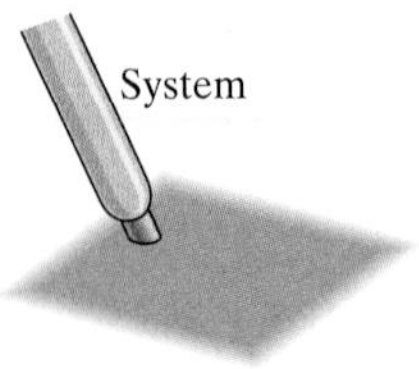	**Force** represented in terms of components ($\boldsymbol{F}_x + \boldsymbol{F}_y + \boldsymbol{F}_z$). **Moment** represented in terms of components ($\boldsymbol{M}_x + \boldsymbol{M}_y + \boldsymbol{M}_z$).	$\boldsymbol{F}_x + \boldsymbol{F}_y + \boldsymbol{F}_z$ $\boldsymbol{M}_x + \boldsymbol{M}_y + \boldsymbol{M}_z$ 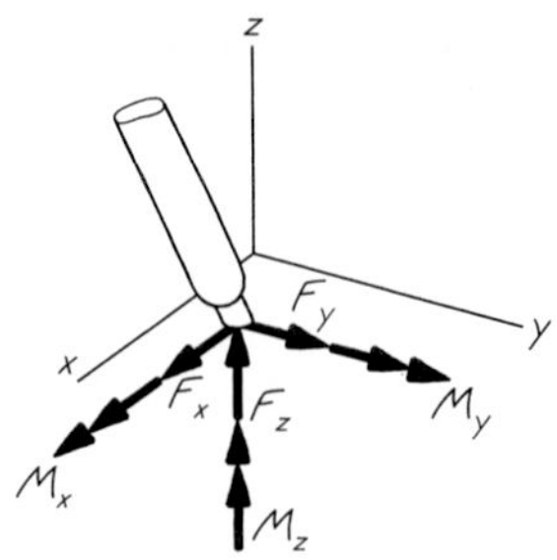

(A) Support	Description of Boundary Loads	(B) Loads to Be Shown in Free-Body Diagram
6A. **Single hinge** (shaft and articulated collar) 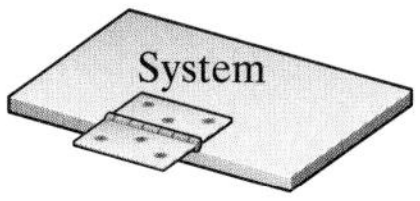	**Force** in plane perpendicular to shaft axis; represented as x and y components ($\boldsymbol{F}_x + \boldsymbol{F}_z$). **Moment** with components about axes perpendicular to shaft axis ($\boldsymbol{M}_x + \boldsymbol{M}_z$). Depending on the hinge design, may also apply **force** along axis of shaft, ($\boldsymbol{F}_y$).	$\boldsymbol{F}_x + \boldsymbol{F}_z$ $\boldsymbol{M}_x + \boldsymbol{M}_z$ OR $\boldsymbol{F}_x + \boldsymbol{F}_y + \boldsymbol{F}_z$ $\boldsymbol{M}_x + \boldsymbol{M}_z$ 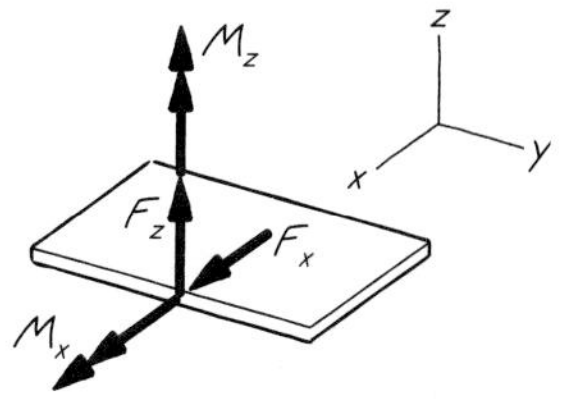
6B. **Multiple Hinges** (one of two or more properly aligned hinges) 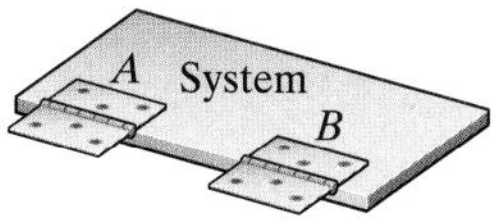	**Force** in plane normal to shaft axis represented in terms of components ($\boldsymbol{F}_x + \boldsymbol{F}_z$). Point of application at center of shaft. Depending on design, may also apply **force** along axis of shaft ($\boldsymbol{F}_y$).	At hinge A: $\boldsymbol{F}_{Ax} + \boldsymbol{F}_{Az}$ At hinge B: $\boldsymbol{F}_{Bx} + \boldsymbol{F}_{Bz}$ OR $\boldsymbol{F}_{Ax} + \boldsymbol{F}_{Ay} + \boldsymbol{F}_{Az}$ $\boldsymbol{F}_{Bx} + \boldsymbol{F}_{By} + \boldsymbol{F}_{Bz}$ 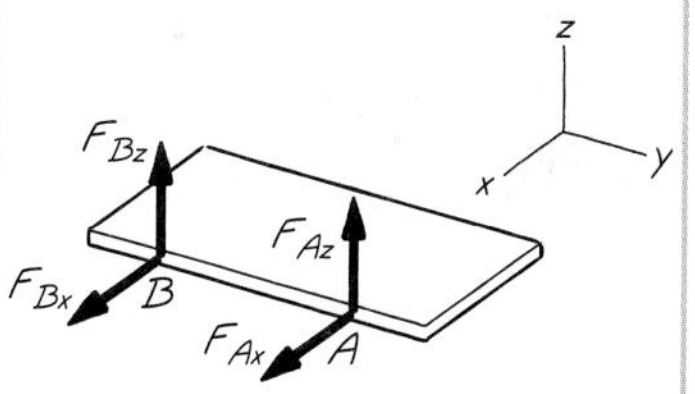
7. **Ball and socket support** (ball or socket as part of system) 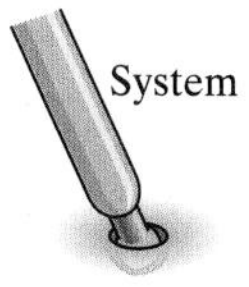	**Force** represented as three components.	$\boldsymbol{F}_x + \boldsymbol{F}_y + \boldsymbol{F}_z$ 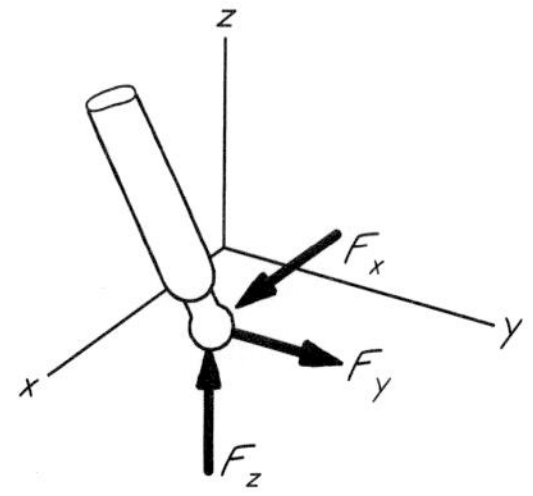
8A. **Single journal bearing** (frictionless collar that holds a shaft) 	**Force** in plane perpendicular to shaft axis; represented as x and z components ($\boldsymbol{F}_x + \boldsymbol{F}_z$). **Moment** with components about axes perpendicular to shaft axis ($\boldsymbol{M}_x + \boldsymbol{M}_z$).	$\boldsymbol{F}_x + \boldsymbol{F}_z$ $\boldsymbol{M}_x + \boldsymbol{M}_z$ 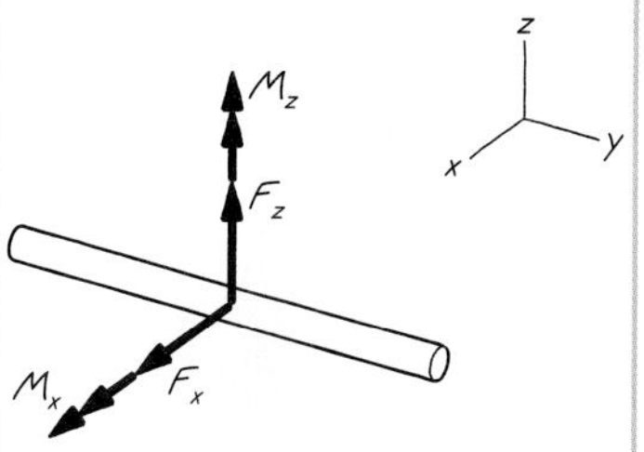

(*Continued*)

Table 6.2 (Cont.)

(A) Support	Description of Boundary Loads	(B) Loads to Be Shown in Free-Body Diagram
8B. **Multiple journal bearings** (two or more properly aligned journal bearings holding a shaft) 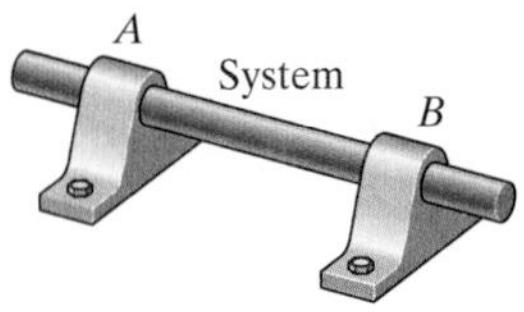	**Force** in plane perpendicular to shaft axis represented in terms of components ($\boldsymbol{F}_{Ax} + \boldsymbol{F}_{Az}$). Point of application at center of shaft.	At journal bearing A: $\boldsymbol{F}_{Ax} + \boldsymbol{F}_{Az}$ At journal bearing B: $\boldsymbol{F}_{Bx} + \boldsymbol{F}_{Bz}$ 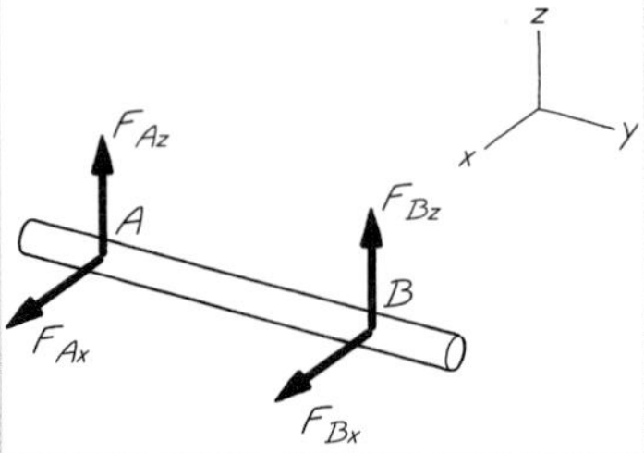
9A. **Single thrust bearing** (journal bearing that also restricts motion along axis of shaft) 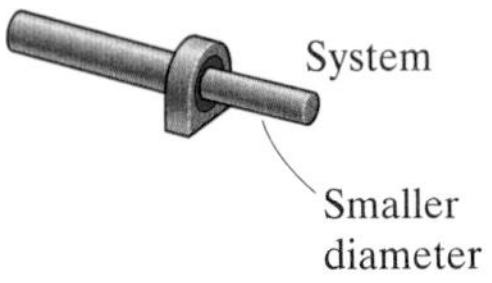	**Force** represented in terms of three components ($\boldsymbol{F}_x + \boldsymbol{F}_z + \boldsymbol{F}_z$). Component in direction of shaft axis ($\boldsymbol{F}_y$) is sometimes referred to as the "thrust force." Point of application is at center of shaft. **Moment** with components perpendicular to shaft axis ($\boldsymbol{M}_x + \boldsymbol{M}_z$).	$\boldsymbol{F}_x + \boldsymbol{F}_z + \boldsymbol{F}_z$ $\boldsymbol{M}_x + \boldsymbol{M}_z$ 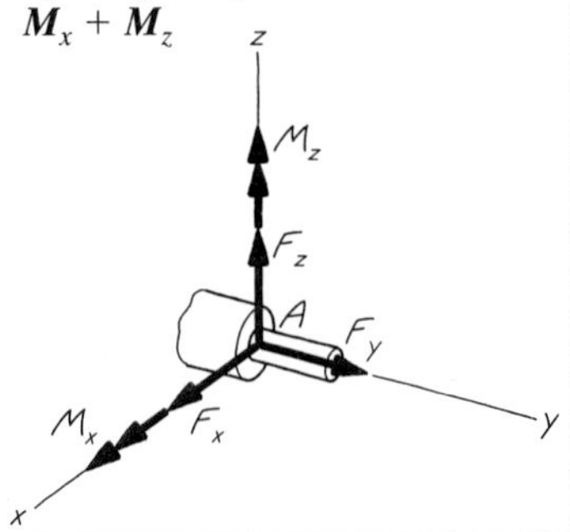
9B. **Multiple thrust bearings** (one of two or more properly aligned thrust bearings) 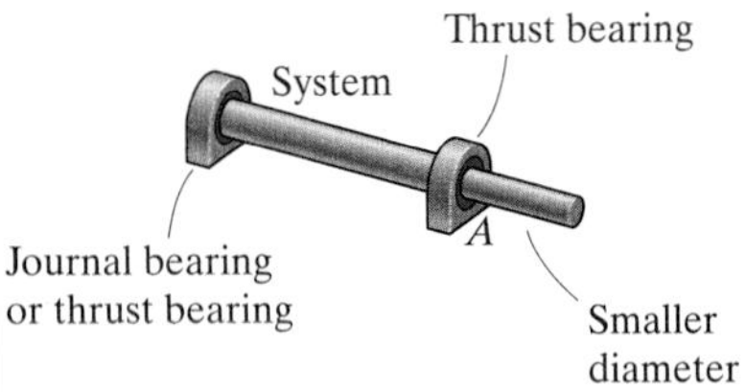	**Force** represented in terms of three components ($\boldsymbol{F}_x + \boldsymbol{F}_z + \boldsymbol{F}_z$). Component in direction of shaft axis ($\boldsymbol{F}_y$) is sometimes referred to as the "thrust force." Point of application is at center of shaft.	At thrust bearing A: $\boldsymbol{F}_{Ax} + \boldsymbol{F}_{Ay} + \boldsymbol{F}_{Az}$ 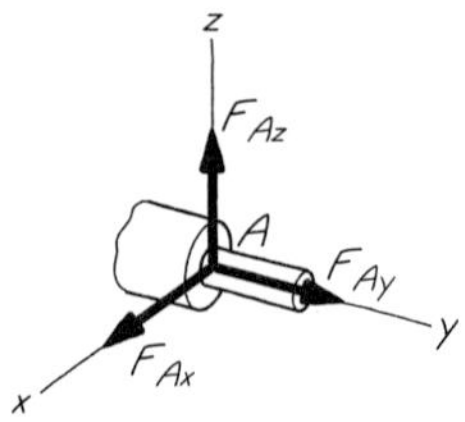
10. **Clevis: Collar on shaft with pin** (collar and shaft are part of system) 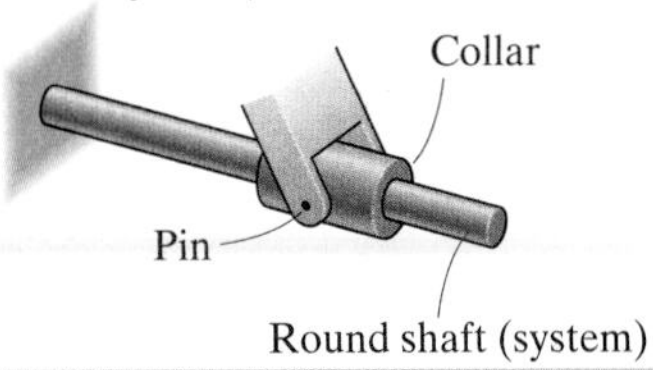	**Force** with components perpendicular to shaft axis ($\boldsymbol{F}_x + \boldsymbol{F}_z$). **Moment** with components perpendicular to shaft axis ($\boldsymbol{M}_z$).	$\boldsymbol{F}_x + \boldsymbol{F}_z$ $\boldsymbol{M}_z$ 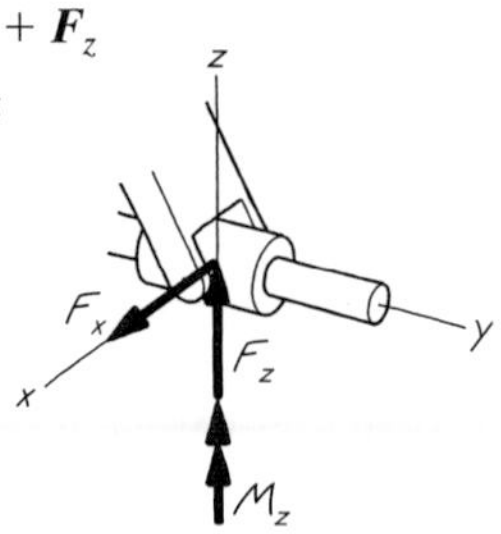
11. **Smooth roller in guide** 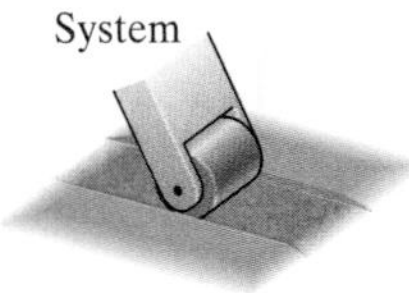	**Force** represented as two components. One component ($\boldsymbol{F}_z$) normal to surface on which system rests; the other is perpendicular to rolling direction ($\boldsymbol{F}_x$).	$\boldsymbol{F}_x + \boldsymbol{F}_z$ 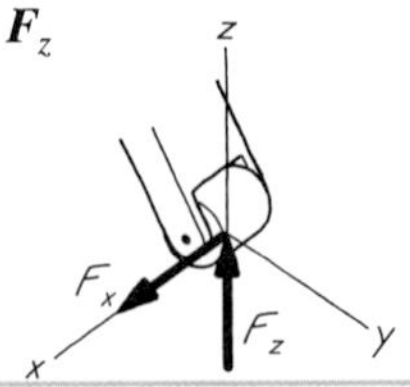

EXAMPLE 6.3 EXPLORING SINGLE AND DOUBLE BEARINGS AND HINGES

For each situation described below, draw a free-body diagram and describe the loads involved. To create the situations yourself, you will need a yardstick, a rubber band, and a candy bar (to serve as a weight). Your hands will serve as models of bearings and hinges. When considering the system, ignore the weight of the yardstick.

Situation 1: Hold the yardstick level as shown in **Figure 6.26*a***. The left hand is at $x = 0$ in. and the right hand is at $x = 18$ in. The candy bar is hanging at the far right.

Description The left-hand fingers push down on the top of the yardstick. Notice that you can move your left thumb away from the stick because there is no load on it.

The right thumb pushes up on the bottom of the yardstick. Notice that you can move your right-hand fingers away from the stick because there is no load on them.

Free-Body Diagram Consider how the hands apply loads to the yardstick. Defining the yardstick as the system, draw these loads on the yardstick to create the free-body diagram (**Figure 6.26*b***).

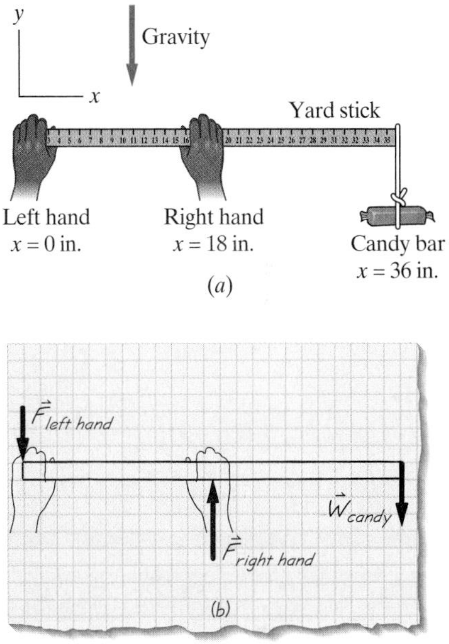

Figure 6.26 (*a*) Situation 1; (*b*) free-body diagram of Situation 1

Situation 2: Hold the yardstick level as shown in **Figure 6.27*a***. The left hand is at $x = 9$ in. and the right hand is at $x = 18$ in. The candy bar is hanging at the far right. Each hand acts like a bearing or hinge.

Description The description for (1) still holds. The difference is, that in order to keep the yardstick level, the forces involved in pushing down with the left fingers and up with the right thumb are larger in magnitude.

Free-Body Diagram Consider how the hands apply loads to the yardstick. Defining the yardstick as the system, draw these loads on the yardstick to create the free-body diagram (**Figure 6.27*b***).

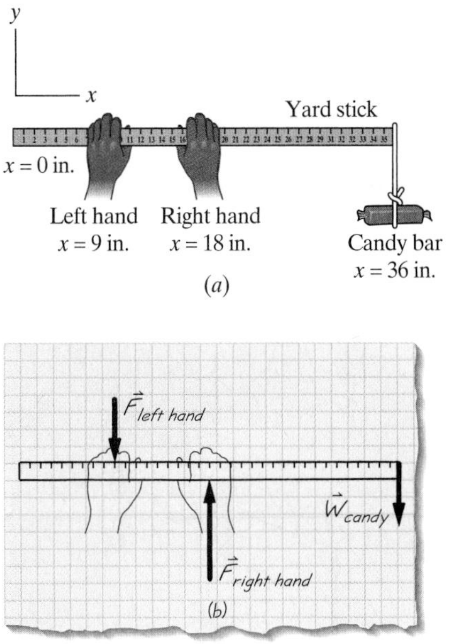

Figure 6.27 (*a*) Situation 2; (*b*) free-body diagram of Situation 2

Situation 3: Hold the yardstick level as shown in **Figure 6.28*a***. The right hand is at $x = 18$ in., and the candy bar is hanging at the far right. The right hand acts like a single bearing or hinge.

Description The thumb pushes up on the bottom of the yardstick and in conjunction with the right-hand fingers works to prevent the stick from rotating; in doing this, the right hand applies a moment and pushes upward with a force.

Free-Body Diagram Consider how the hand applies loads to the yardstick. Defining the yardstick as the system, draw these loads on the yardstick to create the free-body diagram (**Figure 6.28*b***).

Summary Situations 1 and 2 are analogous to systems with two properly aligned bearings or hinges, with the hands playing the role of bearings/hinges. Although each hand applies only a force, each does create an equivalent moment at a specified moment center (as introduced in Chapter 5). For example, if we call $x = 18$ in. the moment center (this is the point of

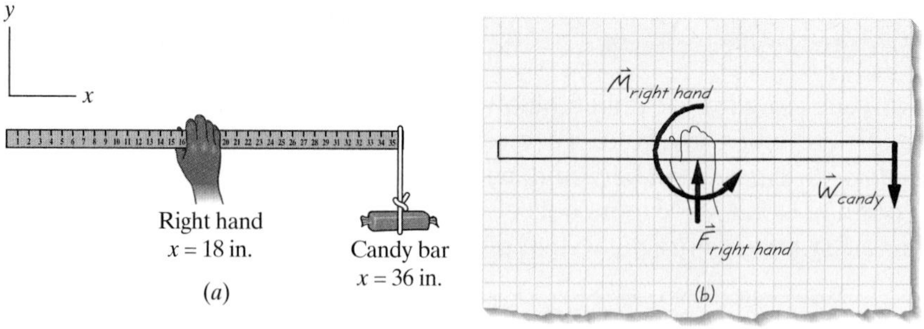

Figure 6.28 (*a*) Situation 3; (*b*) free-body diagram of Situation 3

application of $\boldsymbol{F}_{\text{right hand}}$) in Situation 1, $\boldsymbol{F}_{\text{left hand}}$ creates a counterclockwise equivalent moment of (18 in. $\|\boldsymbol{F}_{\text{right hand}}\|$) that counters the clockwise equivalent moment created by the dangling candy of (18 in. $\|\boldsymbol{W}_{\text{candy}}\|$).

In situation 2 the candy stays at the same position, creating the same clockwise equivalent moment of (18 in. $\|\boldsymbol{W}_{\text{candy}}\|$) about $x = 18$ in. Since the left hand is placed at $x = 9$ in., it must exert a larger force to maintain the same counterclockwise moment.

In situation 3 the right hand acts like a single bearing or hinge, and must apply a force and a moment to counter the clockwise moment created by the dangling candy that is 18 inches from the hand. Notice that in **Figure 6.28*b*** the right hand applied both a force and a moment to the yard stick.

EXAMPLE 6.4 EVALUATING THE CORRECTNESS OF FREE-BODY DIAGRAMS

Consider the description of each nonplanar system in **Figures 6.29–6.34** and determine whether the associated free-body diagram is correct or not correct. Unless stated otherwise, assume that gravity forces can be ignored.

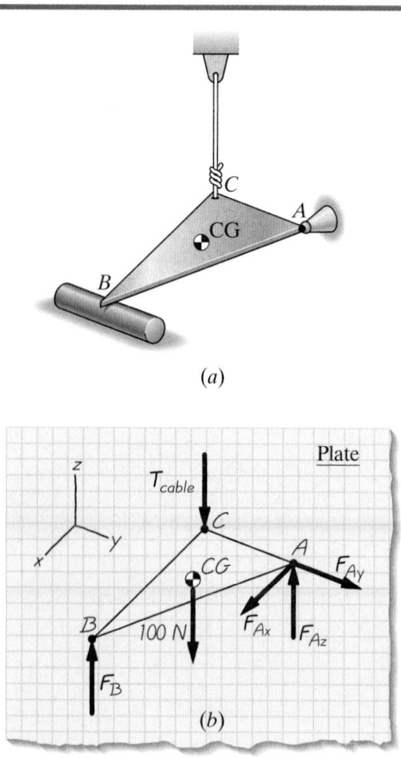

Figure 6.29

Situation A: The triangular plate *ABC* in **Figure 6.29*a*** is supported by a ball and socket at *A*, a roller at *B*, and a cable at *C*. The plate weighs 100 N.

Answer The proposed free-body diagram in **Figure 6.29*b*** is *not correct*. Because the cable is in tension, it will pull on the plate in the $+y$ direction (not push on it in the $-y$ direction, as shown).

Situation B: A bar is supported by three well-aligned journal bearings at *A*, *B*, and *C* and supports a 200-N load (**Figure 6.30*a***).

Answer The proposed free-body diagram in **Figure 6.30*b*** is *correct*. As summarized in **Table 6.2**, because there is more than one journal bearing supporting the system, each bearing applies a force (and no moment) to the system.

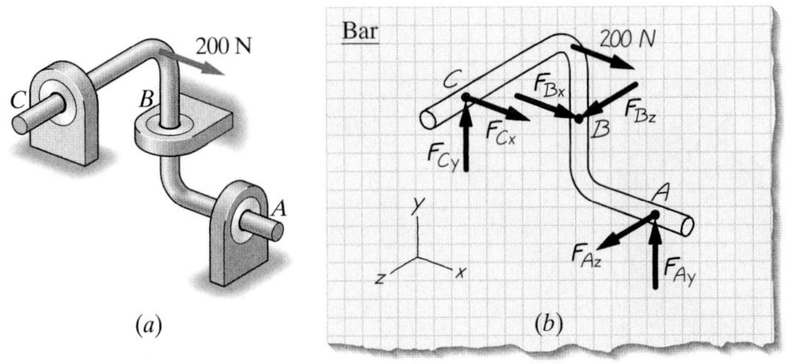

Figure 6.30

Situation C: A rod is supported by a thrust bearing at *A* and a cable that extends from *B* to *C*. A known force ***F*** is applied as shown in **Figure 6.31*a***.

Answer The proposed free-body diagram in **Figure 6.31*b*** is *not correct*. Because there is a single thrust bearing supporting the system, the bearing also applies moments about the *y* and *z* axes at *A*.

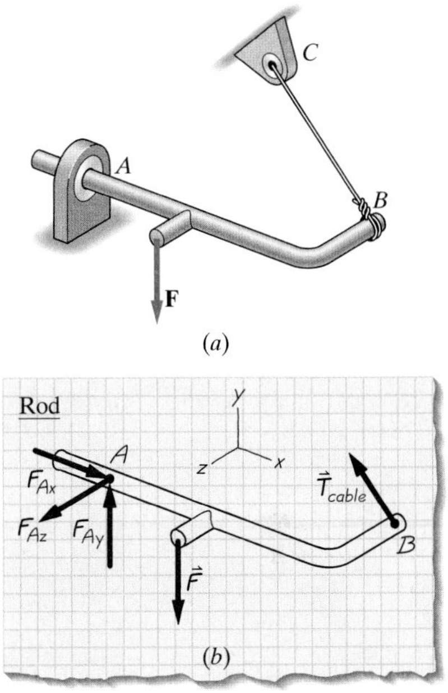

Figure 6.31

Situation D: A bar *ABC* has built-in support at *A* and loads applied at *B* and *C* as shown in **Figure 6.32*a***.

Answer The proposed free-body diagram in **Figure 6.32*b*** is *correct*.

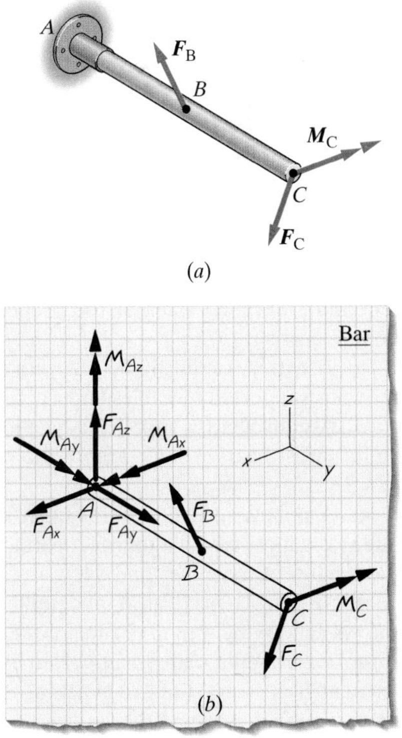

Figure 6.32

Situation E: The L-bar is supported at *B* by a cable and at *A* by a smooth square rod that just fits through the square hole of the collar. A known vertical load *F* is applied as shown in **Figure 6.33*a***.

Answer The proposed free-body diagram in **Figure 6.33*b*** is *not correct*. Since the rod is square and therefore prevents rotation of the collar, the connection at *A* also applies a moment about the *y* axis. Since the square rod is smooth, it cannot apply a force in the *y* direction at *A*.

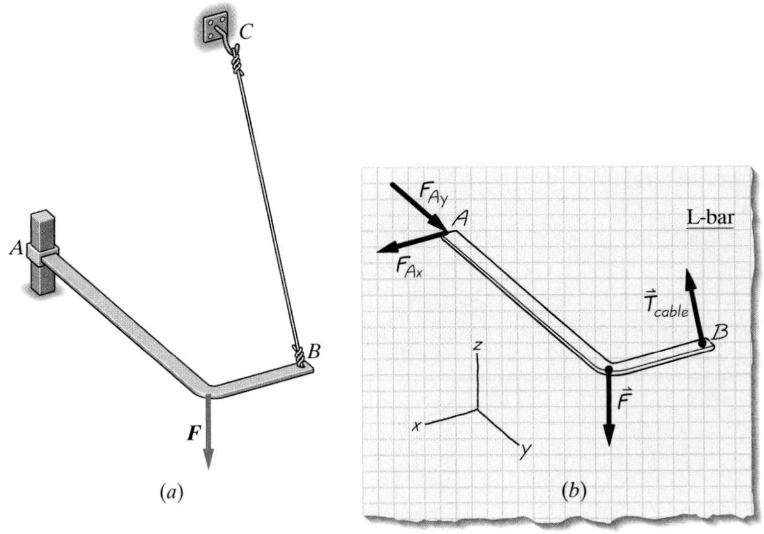

Figure 6.33

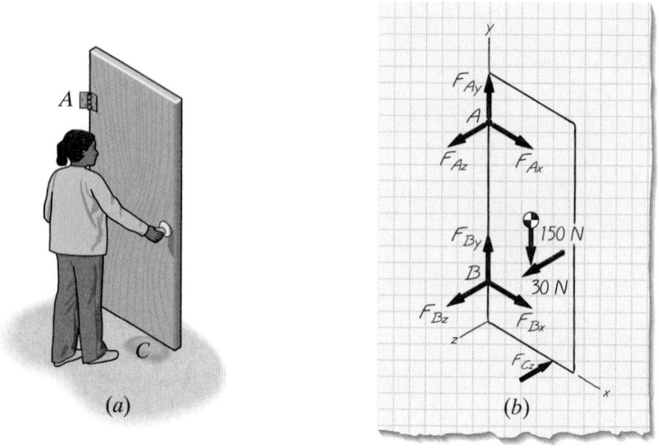

Figure 6.34

Situation F: The 150-N door is supported at A and B by hinges. Someone attempts to open the door by applying a force of 30 N to the handle, but because of a high spot in the floor at C, the door won't open. Both hinges are able to apply forces along their pin axis (**Figure 6.34*a***).

Answer The proposed free-body diagram in **Figure 6.34*b*** is *correct.* Notice that unlike Example 6.3(D), here the door must be treated as a nonplanar system, and therefore the hinge forces in the z direction must be considered.

EXERCISES 6.3

6.3.1. Consider the description of each nonplanar system and determine whether the proposed free-body diagram is correct.

a. A sign of weight W (500 N) with center of gravity as shown is supported by cables and a collar joint (**E6.3.1*a***).

b. A crankshaft is supported by a journal bearing at B and a thrust bearing at D. Ignore the weight of the crank (**E6.3.1*b***).

c. A pulley is used to lift a weight W. The shaft of the pulley is supported by a journal bearing, as shown. Ignore the weights of the pulley and the shaft (**E6.3.1*c***).

d. A pole is fixed at B and tethered by a rope, as shown. Ignore the weight of the pole (**E6.3.1*d***).

e. A triangular plate is supported by a rope at A and a hinge at B. Its weight of 400 N acts at the plate's center of gravity at C, as shown (**E6.3.1*e***).

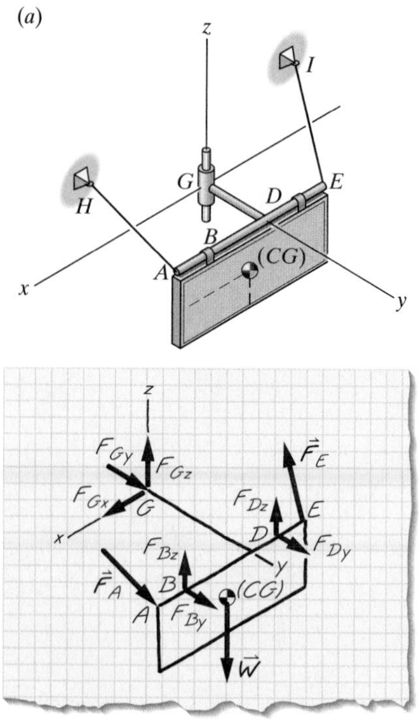

Proposed free-body diagram

E6.3.1

(b)

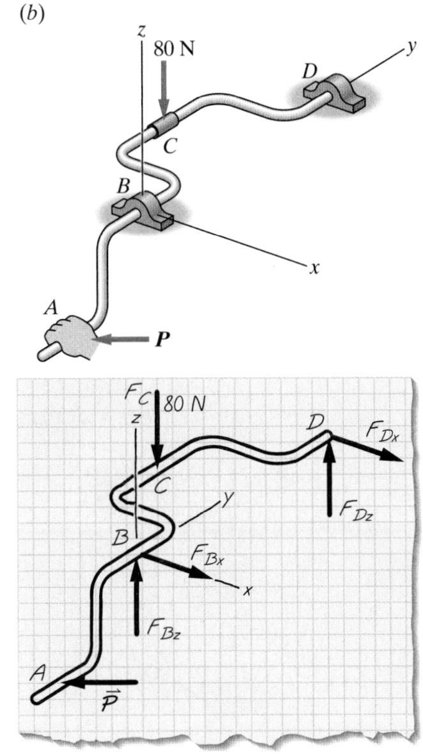

Proposed free-body diagram

(c)

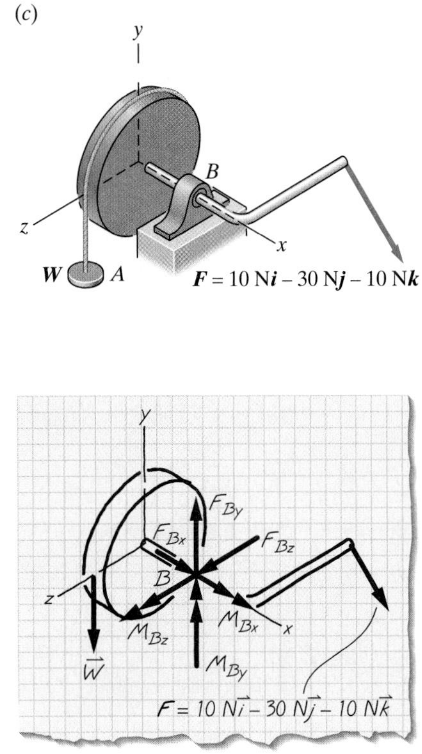

Proposed free-body diagram

(d)

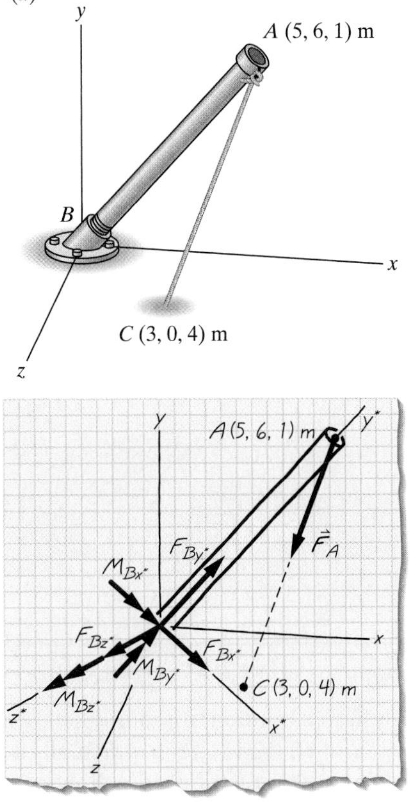

Proposed free-body diagram

(e)

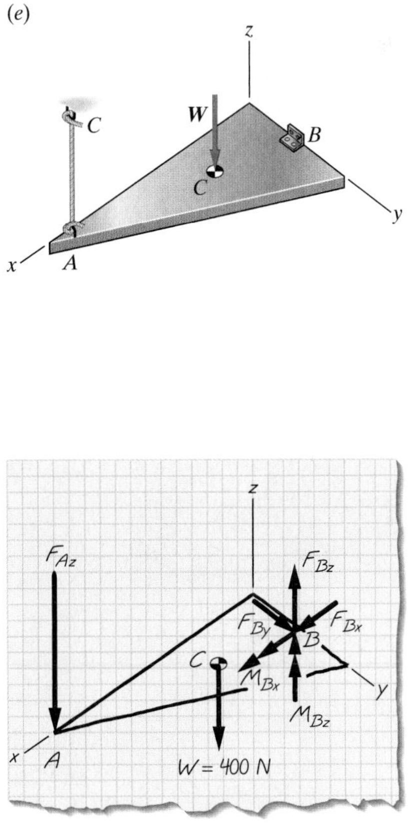

Proposed free-body diagram

E6.3.1 (Cont.)

6.3.2. A tower crane is fixed to the ground at A as shown in E6.3.2. Based on information in **Table 6.2**, what loads do you expect to act on the tower at A? Present your answer in terms of a sketch of the tower that shows the loads acting on it at A. Also comment on whether the sketch you created is or is not a free-body diagram.

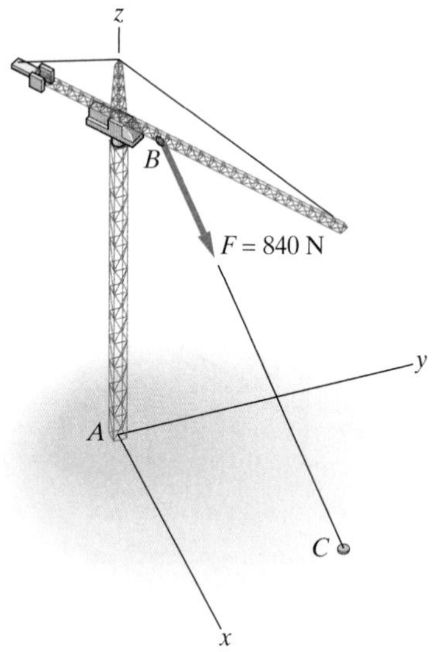

E6.3.2

6.3.3. The uniform 7-m steel shaft in E.6.3.3 is supported by a ball-and-socket connection at A in the horizontal floor. The ball end B rests against the smooth vertical walls, as shown. Based on information in **Table 6.2**, what loads do you expect to act on the shaft at A? What loads do you expect to act on the shaft at B? Present your answer in terms of a sketch of the shaft that shows the loads acting on it at A and B. Also comment on whether the sketch you created is or is not a free-body diagram.

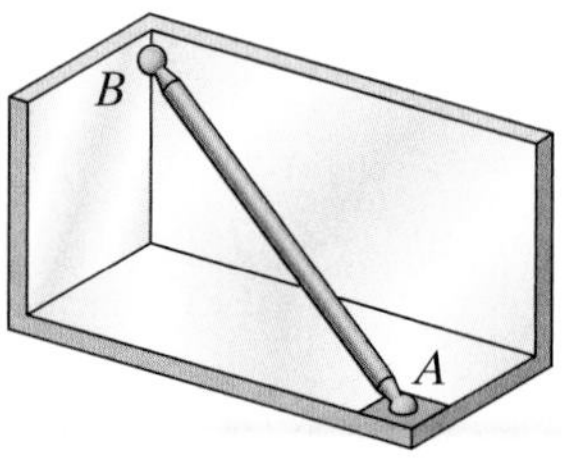

E6.3.3

6.3.4. The welded tubular frame in E6.3.4 is secured to the horizontal xy plane by a ball-and-socket connection at A and receives support from a loose-fitting ring at B. Based on information in **Table 6.2**, what loads do you expect to act on the frame at A? What loads do you expect to act on the frame at B? Present your answer in terms of a sketch of the frame that shows the loads acting on it at A and B. Also comment on whether the sketch you created is or is not a free-body diagram.

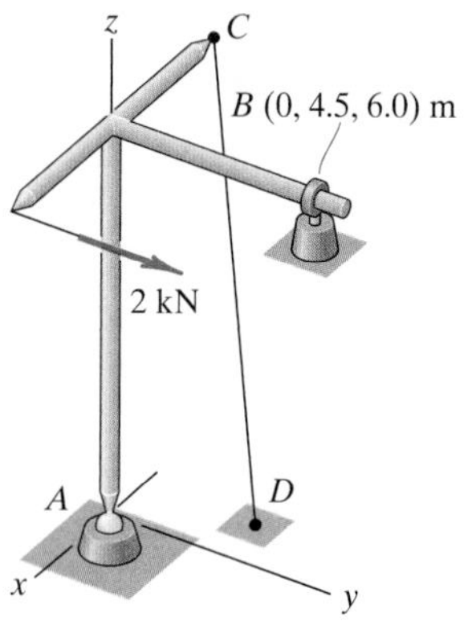

E6.3.4

6.3.5. A bar is supported at A by a hinge, and at B it rests against a rough surface (E.6.3.5). The surface is defined as the x^*z^* plane. Based on information in **Table 6.2**, what loads do you expect to act on the bar at A? What loads do you expect to act on the bar at B? Present your answer in terms of a sketch that shows the loads acting on the bar at A and B. Also comment on whether the sketch you created is or is not a free-body diagram.

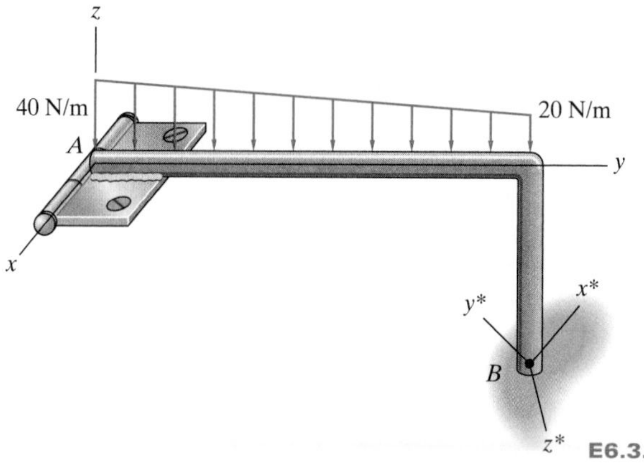

E6.3.5

6.4 PLANAR AND NONPLANAR SYSTEMS

Defining the loads at a system's boundary is simplified if we can classify the system as a planar system, which is one in which all the forces acting on the system lie in the same plane and all moments are about an axis perpendicular to that plane. In this case, cross-boundary loads (e.g., gravity), known loads, fluid boundary loads, and supports are all in a single plane. **Figure 6.35*a*** shows an example of a system that can be classified as a planar system—planar because the gravity force and supports *A* and *B* are all in a single plane. Planar systems are referred to as **two-dimensional systems**. As we saw in Section 6.2, the free-body diagram associated with a planar system typically requires only a single view of the system.

A system in which the loads do not all lie in a single plane can be treated as a planar system for the purpose of static analysis if the system has a plane of symmetry *with regard to its geometry and the loads acting on it*. A **plane of symmetry** is one that divides the system into two sections that are mirror images of each other. None of the forces acting on the system has a component perpendicular to the plane of symmetry, and all moments acting on the system are about an axis perpendicular to the plane. **Figure 6.35*b*** illustrates a system that has a plane of symmetry and therefore can be treated as a planar system. Other examples in which a plane of symmetry was used to classify a system as planar are the Golden Gate Bridge in Chapter 2, and the ladder–person example in Chapter 4 (**Figures 4.19**, **4.20**, and **4.22**, but not **Figure 4.21**). **Figure 6.35*c*** illustrates a system with geometric symmetry, but because the cable forces acting on it have a component perpendicular to the *xy* plane, we are not able to classify the system as planar.

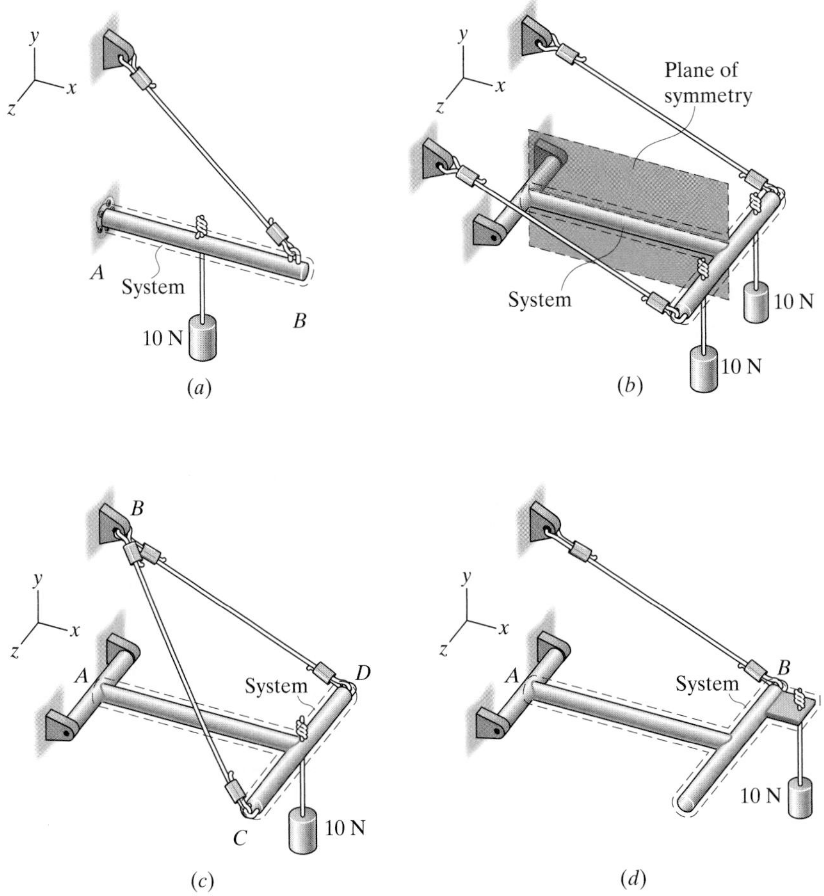

Figure 6.35 (*a*) System that can be modeled as planar; (*b*) system with plane of symmetry can be modeled as planar; (*c*) and (*d*) system that cannot be modeled as planar

If it is not possible to define a single plane in which all forces and moments lie, or there is no plane of symmetry, the system is classified as nonplanar. In the system of **Figure 6.35*d***, for instance, it is not possible to define a single plane that contains the gravity force and supports A and B, and there is no plane of symmetry. Nonplanar systems are referred to as **three-dimensional systems**. The free-body diagram associated with a nonplanar system typically requires an isometric drawing or multiple views.

In Section 6.2 we dealt exclusively with planar systems and in Section 6.3 with nonplanar systems. Drawing the free-body diagram for a planar system is generally more straightforward because only external forces in the plane and external moments about an axis perpendicular to the plane must be considered. In performing analysis in engineering practice you will not be told whether a physical situation can be modeled as a planar system or must be modeled as a nonplanar system—the choice will be up to you. The discussion in this section is intended to give you some guidelines with which to make such a judgment.

EXAMPLE 6.5 IDENTIFYING PLANAR AND NONPLANAR SYSTEMS

Consider the description of each system and determine whether the system can be classified as planar or nonplanar. (No system is really planar, because we live in a three-dimensional world. Even something as thin as a sheet of paper has a third dimension; BUT under certain conditions we can model it as planar for the purpose of static analysis).

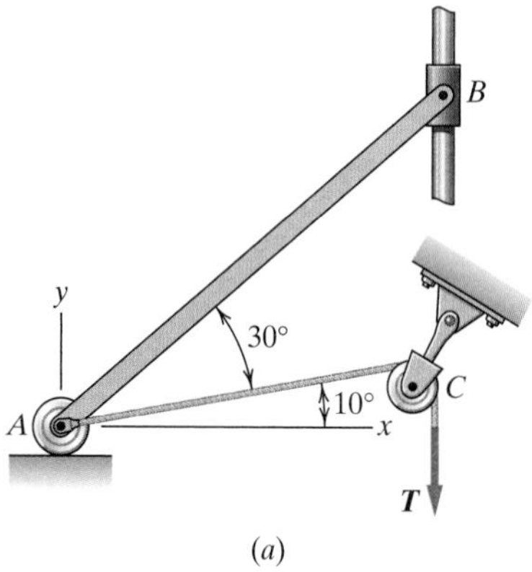

(*a*)

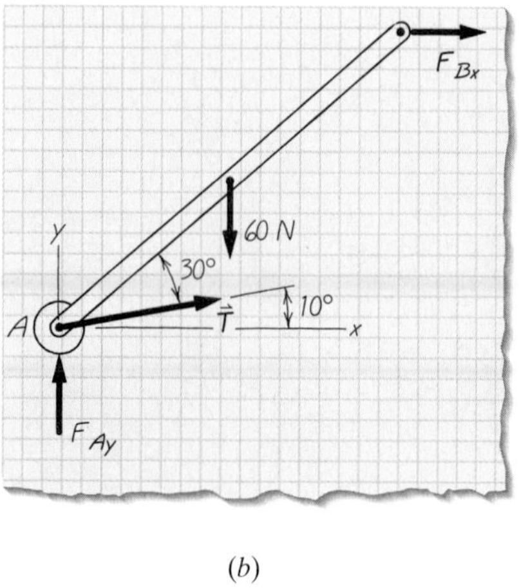

(*b*)

Figure 6.36 Situation A

Goal We are asked to determine, for a number of different cases, whether a system can be classified as planar or nonplanar.

Given We are given a specified system and the loads that act on it.

Assume Unless specified otherwise, assume that gravity is considered and acts in the negative y direction.

Draw An additional drawing is not required to determine the classification of each system; however, you may want to draw a free-body diagram to help you better understand the geometry of the system and the external loads.

Situation A: The uniform bar AB in **Figure 6.36*a*** weighs 60 N and is pulled on by a rope at A. The system is the arm and the wheel at A.

Answer *Planar.* The gravity force of 60 N is in the xy plane. Furthermore, the forces at the supports A (normal force and tension in rope AC) and B (collar guide) are also in the xy plane. Because it is possible to define a single plane that contains all known forces and moments, gravity force, and forces applied at supports, this system can be treated as planar. The free-body diagram of AB is shown in **Figure 6.36*b***.

Question: If gravity acted in the z direction, would we reach the same conclusion? If the pulley at C is not in the xy plane, would we reach the same conclusion?

Situation B: The space truss in **Figure 6.37*a*** is of negligible weight and is supported by rollers at B, C, and D. It supports a vertical 800-N force at A. The system is the space truss.

Answer *Nonplanar.* It is not possible to define a single plane that contains the points of application of supports (B, C, D) and the 800-N force. Therefore, this system must be treated as a nonplanar system. Our answer would be unchanged if we had included gravity forces acting on the space truss. The free-body diagram of the space truss is shown in **Figure 6.37*b***.

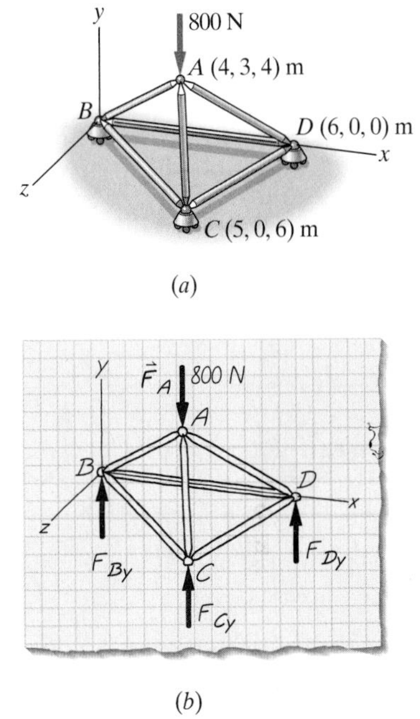

Figure 6.37 Situation B

Situation C: The beam AC in **Figure 6.38*a*** is pinned to its surroundings at A and rests against a rocker at B. Ignore gravity. The system is the beam.

Answer *Planar.* The 800-N · m moment at C and the 500-N force are in the xy plane, as are the points of application of supports at A and B. The free-body diagram of the beam is shown in **Figure 6.38*b***.

Question: If gravity is considered and acts in the negative y direction, would we reach the same conclusion?

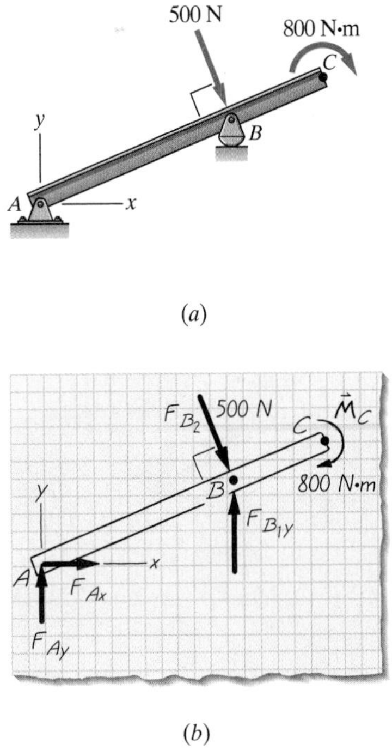

Figure 6.38 Situation C

Situation D: The beam AC in **Figure 6.39*a*** rests on a block at A. In addition, there is a pin connection at A and a rocker at B. The system is the beam.

Answer *Nonplanar.* It is not possible to define a single plane that contains the gravity force and the normal contact force at B and the 500-N force (both in the xz plane). The free-body diagram of beam AC is shown in **Figure 6.39*b***.

Question: If gravity acted in the positive z direction, would we reach the same conclusion? How would our conclusion change if we were to ignore gravity?

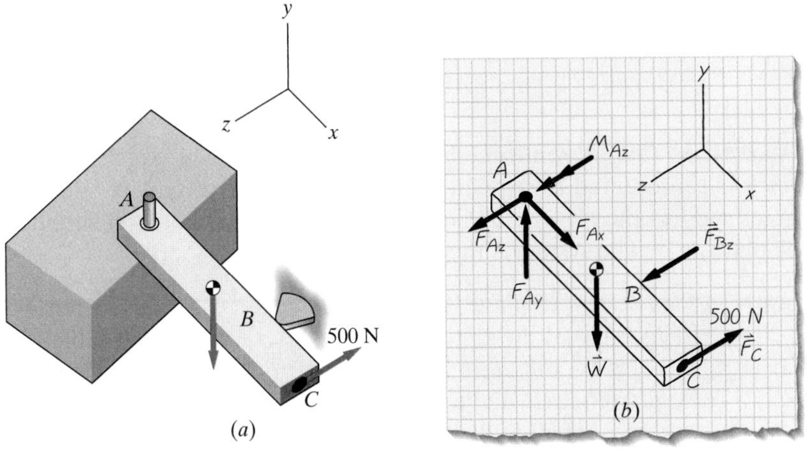

Figure 6.39 Situation D

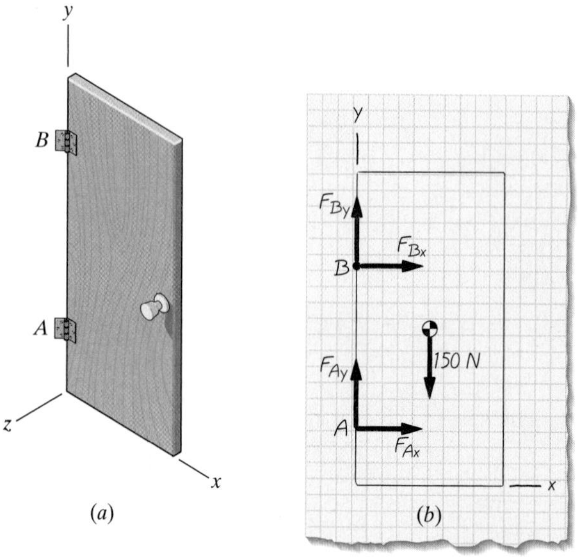

Figure 6.40 Situation E

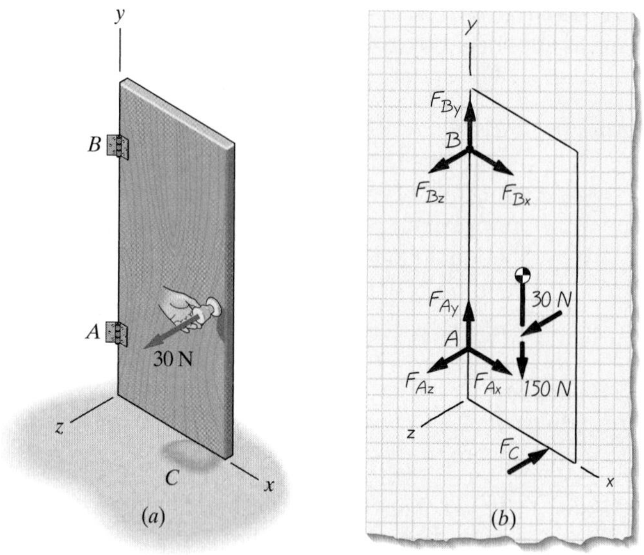

Figure 6.41 Situation F

Situation E: The 150-N door in **Figure 6.40*a*** is supported at *A* and *B* by hinges. The system is the door.

Answer *Planar.* If we assume that the door is of uniform density, we place the 150-N gravity force at the door's center. Furthermore, because the door is thin relative to its other dimensions, this gravity force and the loads due to the supports at *A* and *B* can be assumed to lie in the *xy* plane. The free-body diagram of the door is shown in **Figure 6.40*b***.

Situation F: The 150-N door in **Figure 6.41*a*** is supported at *A* and *B* by hinges. Someone attempts to open the door by applying a force of 30 N to the handle, but because of a high spot in the floor at *C*, the door won't open. The system is the door.

Answer *Nonplanar.* It is not possible to define a single plane that contains all of the forces. The free-body diagram of the door is shown in **Figure 6.41*b***.

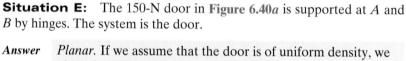

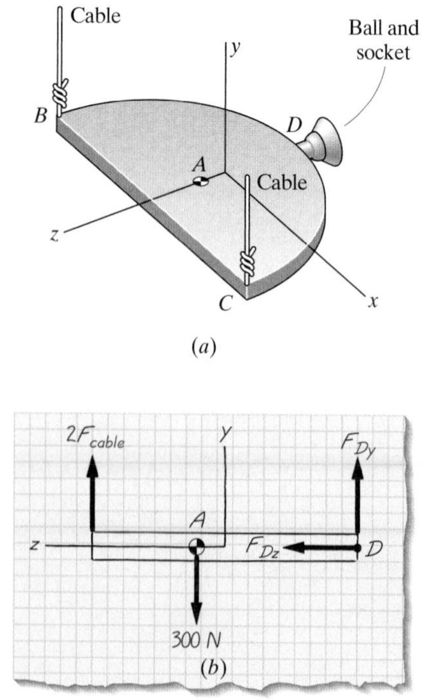

Figure 6.42 Situation G

Situation G: A semicircular plate in **Figure 6.42*a*** weighs 300 N, which is represented by a point force at the center of gravity. Vertical cables support the plate at *B* and *C*, and a ball-and-socket joint supports the plate at *D*. The system is the plate.

Answer *Planar.* The *yz* plane is a plane of symmetry for this system—the portion of the system at $+x$ (a quarter circle and cable force) is the mirror image of the portion of the system at $-x$ (a quarter circle and cable force). Consequently, it is possible to represent all of the forces as projections onto the *yz* plane. The free-body diagram of the plate is shown in **Figure 6.42*b***.

Question: If the 300 N-force acted in the positive x direction, would we reach the same conclusion? If it acted in the positive z direction, what conclusion should be drawn?

Situation H: The same plate as in **G** is supported by diagonal cables at B and C, and a ball-and-socket joint at D, as shown in **Figure 6.43*a***. The system is the plate.

Answer *Nonplanar.* As in **G**, the yz plane is a plane of symmetry for this system. However, since the cable forces have components perpendicular to the plane of symmetry, we cannot represent this as a planar system. The free-body diagram of the plate is shown in **Figure 6.43*b***.

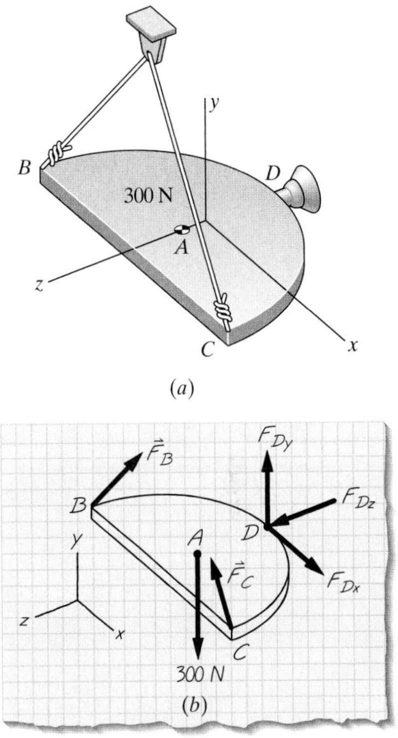

Figure 6.43 Situation H

Situation I: A man weighing 800 N sits at the picnic table halfway between its two ends (**Figure 6.44*a***). His center of gravity (CG_{man}) is noted. The table weighs 200 N, with a center of gravity at CG_{table}. Assume that any friction between the legs and the ground can be neglected. The system is the table.

Answer *Planar.* The xy plane is a plane of symmetry for this system—the portion of the system at $+z$ (half of the picnic table and supports at A and B) is a mirror image of the portion of the system at $-z$ (the other half of the picnic table). The free-body diagram of the table is shown in **Figure 6.44*b***.

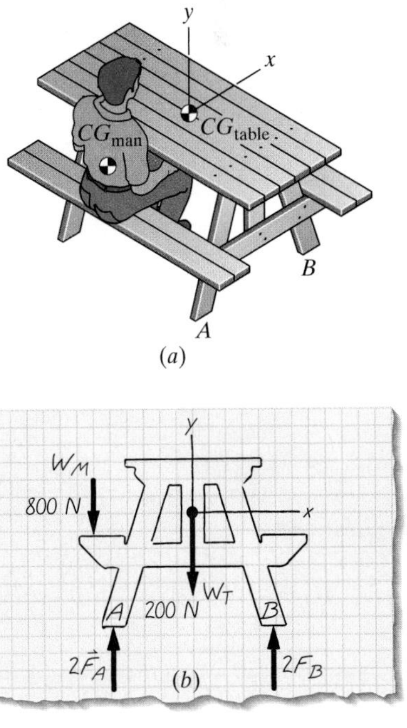

Figure 6.44 Situation I

Situation J: A child sits down next to the man at the picnic table in **I** (**Figure 6.45*a***). The system is again the table.

Answer *Nonplanar.* With the child sitting next to the man, there is no longer any plane of symmetry. The free-body diagram of the table is shown in **Figure 6.45*b***.

Question: If the child sits directly across from the man, could the system be modeled as planar?

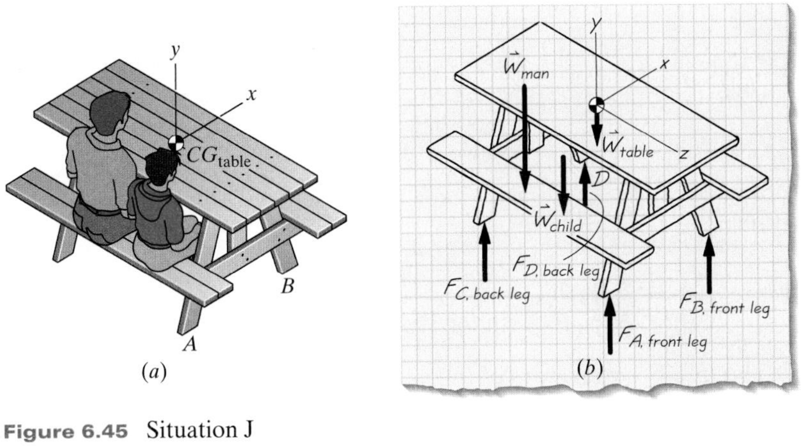

Figure 6.45 Situation J

EXAMPLE 6.6 USING QUESTIONS TO DETERMINE LOADS AT SUPPORTS (A)

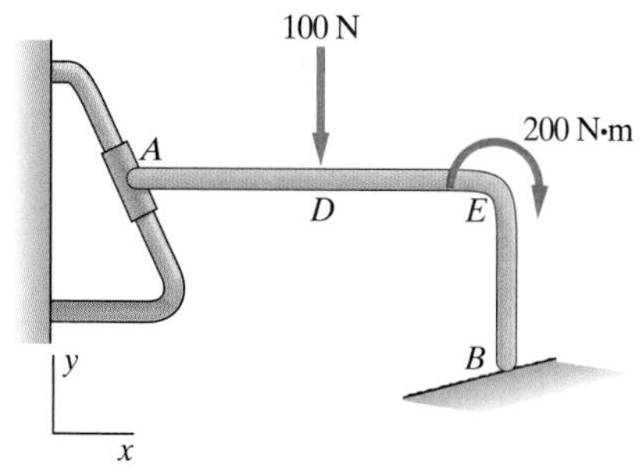

Figure 6.46

A bar is supported at A by a frictionless collar guide. At B it rests against a rough surface. Known forces act at D and E, as shown in Figure 6.46.

(a) What loads act at A and B? Use the general rule about the surroundings preventing translation and/or rotation at each support to answer this question.

(b) Draw a free-body diagram of the bar.

Solution **(a)** The bar (defined as the system) can be classified as planar. This means that in considering motion, we need consider only translations in the xy plane and rotations about the z axis.

At A:

Define the $x'y'$ coordinate system at A (see Figure 6.47*a*).

Possible motion	Answer	Implication
Is x' translation at A possible?	Yes	There is no force acting on the bar in the x' direction, since it is frictionless.
Is y' translation at A possible?	No	There is a force $\boldsymbol{F}_{Ay'}$.
Is z rotation at A possible?	No	There is a moment about the z axis, $\boldsymbol{M}_{Az}$.

At B:

Define the x^*y^* coordinate system at B (see **Figure 6.47*a***).

Possible motion	Answer	Implication
Is x^* translation at B possible?	No, unless the force applied in the x^* direction exceeds the maximum friction force that can be applied by the rough surface	There is a force $\boldsymbol{F}_{Bx^*}$.
Is y^* translation at B possible?	No, it is not possible in the negative y direction. It is possible in the positive y^* direction	There is a force $\boldsymbol{F}_{By^*}$ in the positive y^* direction.
Is z rotation at B possible?	Yes	There is no moment about the z axis.

Answer *At A:* $\boldsymbol{F}_{Ay'}$, $\boldsymbol{M}_{Az}$
At B: $\boldsymbol{F}_{Bx^*}$, $\boldsymbol{F}_{By^*}$ (in the positive y^* direction).

(b) The free-body diagram of the bar is as shown.

Answer See **Figure 6.47*b***.

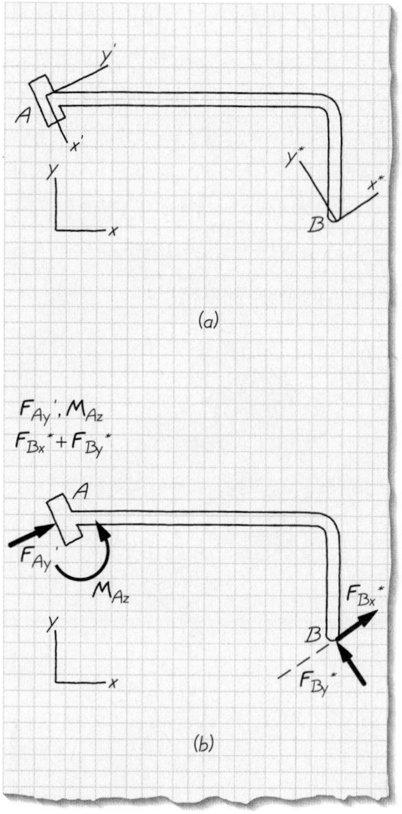

Figure 6.47

EXAMPLE 6.7 USING QUESTIONS TO DETERMINE LOADS AT SOLID SUPPORTS

A crank is supported at B by a pin connection, as shown in Figure 6.48. A cable (in the xy plane) is attached to the bracket at A.

(a) What loads act at A and B? Use the general rule about the surroundings preventing translation and/or rotation at each support to answer this question.

(b) Draw a free-body diagram of the crank.

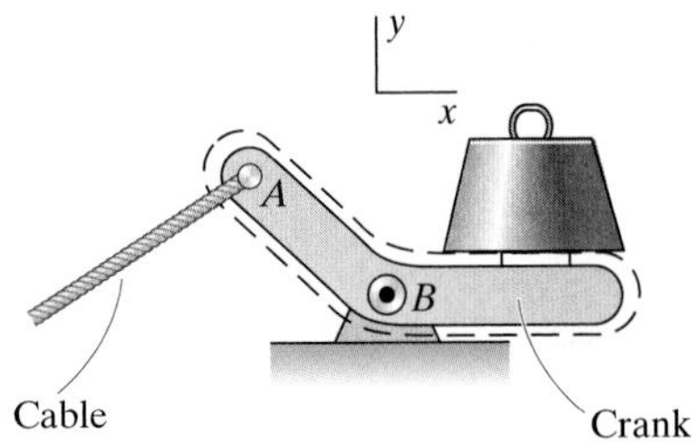

Figure 6.48

Solution **(a)** The crank (defined as the system) can be classified as planar. This means that in considering motion, we need consider only translations in the xy plane and rotations about the z axis.

At B there is a pin connection:

Possible motion	Answer	Implication
Is x translation at B possible?	No	There is a force $\boldsymbol{F}_{Bx}$.
Is y translation at B possible?	No	There is a force $\boldsymbol{F}_{By}$.
Is z rotation at B possible?	Yes	There is no moment about the z axis.

Check The answers can be confirmed with **Table 6.1** for a pin connection.

At A a cable is attached to the system:

Define the x^*y^* coordinate system at A (see Figure 6.49).

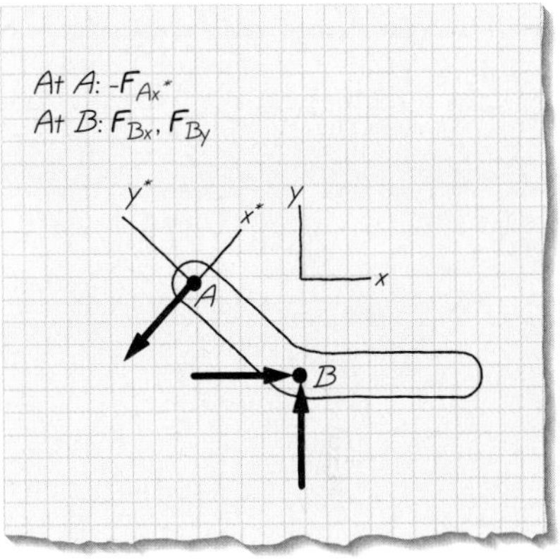

Figure 6.49

Possible motion	Answer	Implication
Is x^* translation at A possible?	No, it is not possible in the positive x^* direction. It is possible in the negative x^* direction.	There is a force $\boldsymbol{F}_{Ax^*}$ in the negative x^* direction.
Is y^* translation at A possible?	Yes	There is no force in the positive y^* direction.
Is z rotation at A possible?	Yes	There is no moment about the z axis.

Answer *At B:* $\boldsymbol{F}_{Bx}, \boldsymbol{F}_{By}$
At A: $\boldsymbol{F}_{Ax^*}$ in the negative x^* direction See **Figure 6.49**.

(b) The free-body diagram of the crank is as shown.

Answer See Figure 6.50.

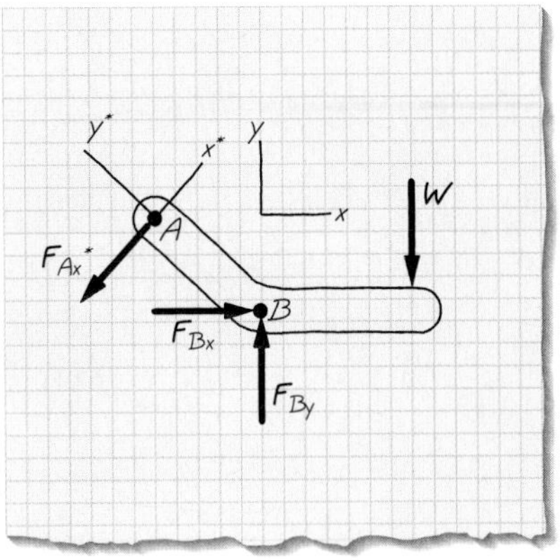

Figure 6.50

EXAMPLE 6.8 USING QUESTIONS TO DETERMINE LOADS AT SUPPORTS (B)

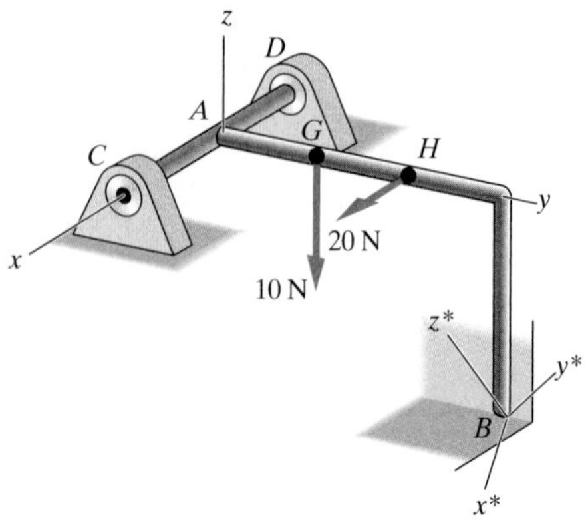

Figure 6.51

An L-shaped bar is supported at A by a hinge and rests against a rough surface at B. Known loads act as shown in **Figure 6.51**. Ignore the weight of the bar.

(a) What loads act at A and B? Use the general rule about the surroundings preventing translation and/or rotation at each support to answer this question.

(b) Draw a free-body diagram of the bar.

Solution (a) The bar (defined as the system) is nonplanar. This means that in considering motion, we must consider translations and rotation in all three directions.

At A:

Define the xyz coordinate system at A (**Figure 6.51**).

Possible motion	Answer	Implication
Is x translation at A possible?	No, if we can assume that the connection at C and/or D prevents motion in the x direction.	There is force in the x direction, F_{Ax}.
Is y translation at A possible?	No	There is force in the y direction, F_{Ay}.
Is z translation at A possible?	No	There is force in the z direction, F_{Az}.
Is rotation about the x axis possible at A?	Yes	There is no moment about x axis.
Is rotation about the y axis possible at A?	No	There is moment about the y axis, M_{Ay}.
Is rotation about the z axis possible at A?	No	There is moment about the z axis, M_{Az}.

At B:

Define the $x^*y^*z^*$ coordinate system at B (**Figure 6.51**).

Possible motion	Answer	Implication
Is x^* translation at B possible?	No, unless the force applied exceeds the maximum friction force that can be applied by the rough surface.	There is a force F_{Bx^*}.
Is y^* translation at B possible?	No, unless the force applied exceeds the maximum friction force that can be applied by the rough surface.	There is a force F_{By^*}.

Possible motion	Answer	Implication
Is z^* translation at B possible?	No, it is not possible in the negative z direction. It is possible in the positive z direction.	There is a force $\boldsymbol{F}_{Bz^*}$ in the positive z^* direction.
Is rotation about the x^* axis possible at B?	Yes	There is no moment about the x^* axis.
Is rotation about the y^* axis possible at B?	Yes	There is no moment about the y^* axis.
Is rotation about the z^* axis possible at B?	Yes	There is no moment about the z^* axis.

Answer *At A:* $\boldsymbol{F}_{Ax}, \boldsymbol{F}_{Ay}, \boldsymbol{F}_{Az}, \boldsymbol{M}_{Ay}, \boldsymbol{M}_{Az}$
At B: $\boldsymbol{F}_{Bx^*}, \boldsymbol{F}_{By^*}, \boldsymbol{F}_{Bz^*}$ (in the positive z^* direction)

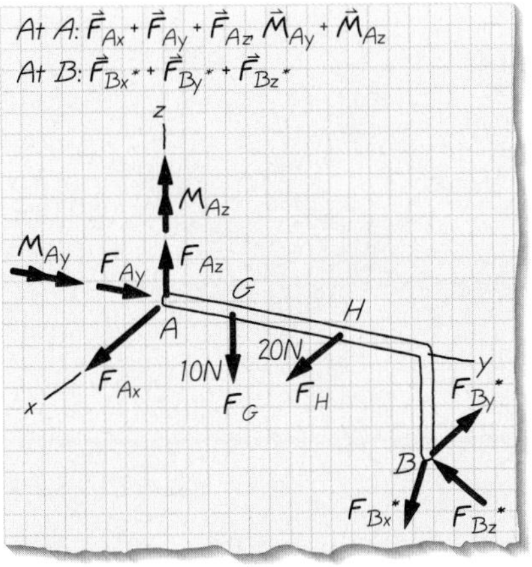

Figure 6.52

(b) The free-body diagram of the bar is shown in **Figure 6.52**.

Answer See **Figure 6.52**.

EXERCISES 6.4

6.4.1. Consider the description of each system in **E6.4.1** and determine whether it can be classified as planar or nonplanar. Describe your reasoning. Unless otherwise stated, ignore the effect of gravity.

a. The uniform L-bar is pinned to its surroundings at A, and slides along a wall at B. A vertical force $\boldsymbol{F}$ acts at C. Gravity acts in the negative y direction. The system is taken as the L-bar (**E6.4.1*a***).

b. The wheelbarrow is loaded, with a center of gravity as shown in **E6.4.1*b***. The system is taken as the wheelbarrow.

c. The wheelbarrow is loaded as shown in **E6.4.1*c***. The system is taken as the wheelbarrow.

d. A tower 70 m tall is tethered by three cables, as shown in **E6.4.1*d***. The system is taken as the tower.

e. A uniform glass rod having a length L is placed in a smooth hemispherical bowl having a radius r. The system is taken as the rod (**E.6.4.1*e***).

f. For the situation shown in **E6.4.1*e***, define the bowl as the system.

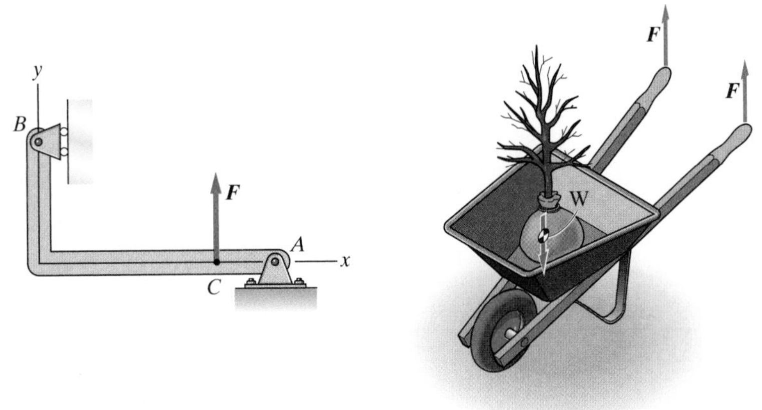

(*a*) (*b*)

E6.4.1

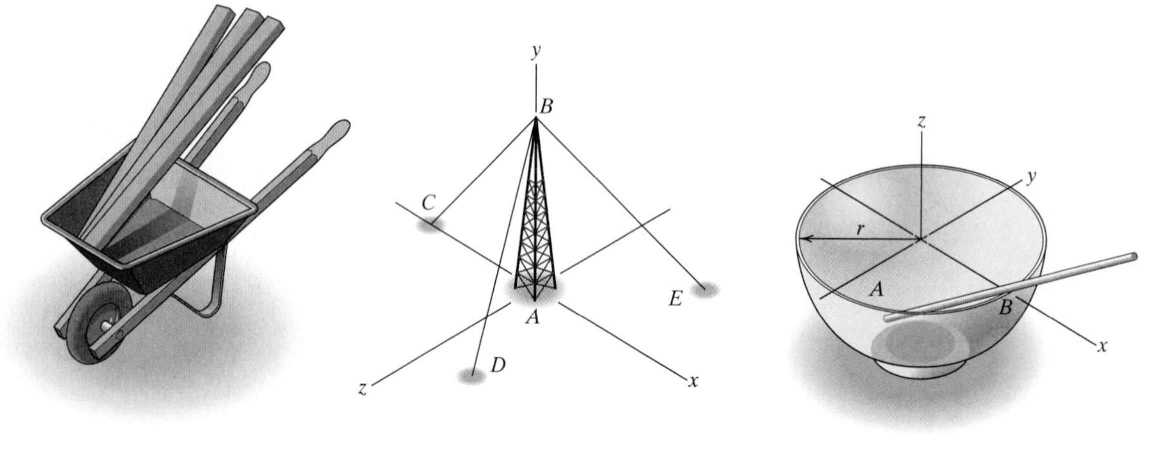

(c) (d) (e)

E6.4.1 (Cont.)

6.4.2. Consider the description of each system in **E6.4.2** and determine whether it can be classified as planar or nonplanar. Describe your reasoning. Unless otherwise stated, ignore the effect of gravity.

a. A bar *AB* is fixed to a wall at end *B*. At end *A*, a force acts, as shown in ***E6.4.2a***. The system is taken as the bar.

b. A bar *AB* is fixed to a wall at end *B*. At end *A* and at *C*, forces act, as shown in ***E6.4.2b***. The system is taken as the bar.

c. A bar *AB* is fixed to a wall at end *B*. At end *A* a force acts, as shown in ***E6.4.2c***. The system is taken as the bar.

d. The frame is supported at *A* and *B*. Loads act at *C*, *D*, *E*, and *F*, as shown in ***E6.4.2d***. The system is taken as the frame.

e. The bracket *ABC* in ***E6.4.2e*** is tethered as shown with cable *CD*. The bracket is taken as the system.

f. The bracket *ABC* in ***E6.4.2f*** is tethered as shown with cable *CD*. The bracket is taken as the system.

g. The airplane in ***E6.4.2g*** weighing 8000 N sits on the tarmac. It has one front wheel and two rear wheels. Its center of gravity is as shown.

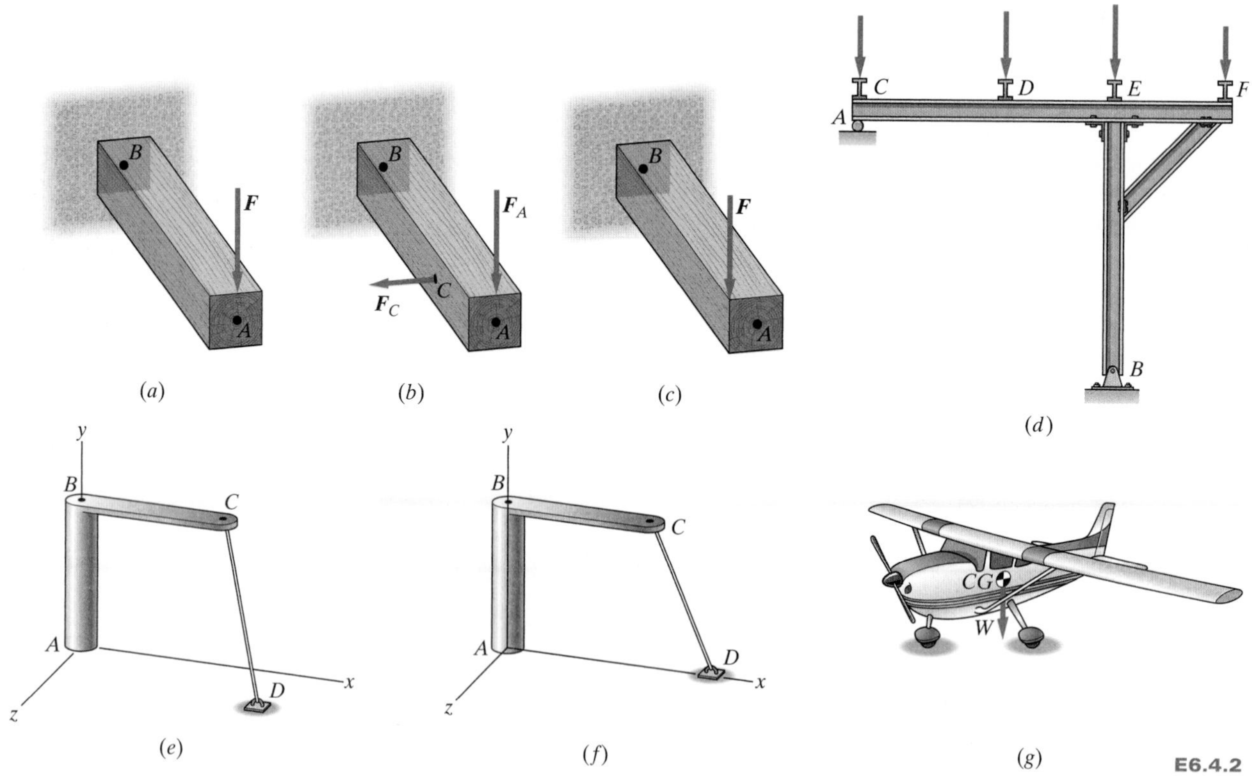

E6.4.2

6.4.3. A cable pulls on the bracket in **E6.4.3** with a force of 1.5 kN. At *A* the bracket is attached to the wall with a pin connection, and at *B* there is a pin-in-slot connection. If the system is defined as the bracket and is considered to be planar, what loads act on the bracket at *A* and *B*? Use the general rule about "prevention of motion" to answer this question (*Strategy:* Review Examples 6.6–6.8.) Confirm that your answers are consistent with the information on loads in **Table 6.1** or **Table 6.2** (whichever is appropriate). Present your answer in words and as a sketch of the bracket that shows these loads acting on it at *A* and *B*. Also comment on whether the sketch you created is or is not a free-body diagram.

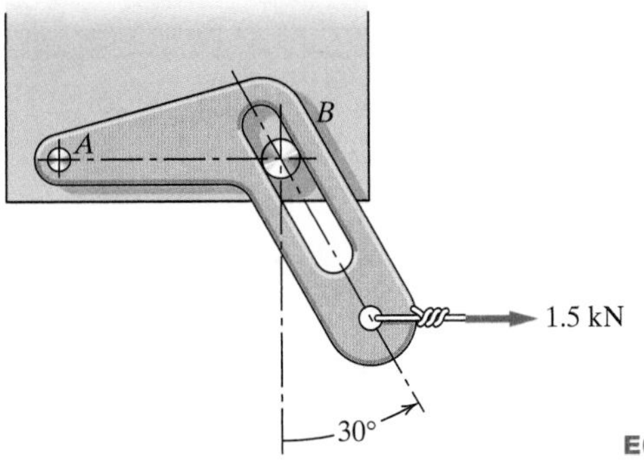

E6.4.3

6.4.4. A steel sphere sits in the groove, as shown in **E6.4.4**. Surfaces *A* and *B* are smooth. If the system is defined as the sphere and is considered to be planar, what loads act on the sphere at *A* and *B*? Use the general rule about "prevention of motion" to answer this question. (*Strategy:* Review Examples 6.6–6.8.) Confirm that your answers are consistent with the information on loads in **Table 6.1** or **Table 6.2** (whichever is appropriate). Present your answer in words and as a sketch of the sphere that shows the loads acting on it at *A* and *B*. Also comment on whether the sketch you created is or is not a free-body diagram.

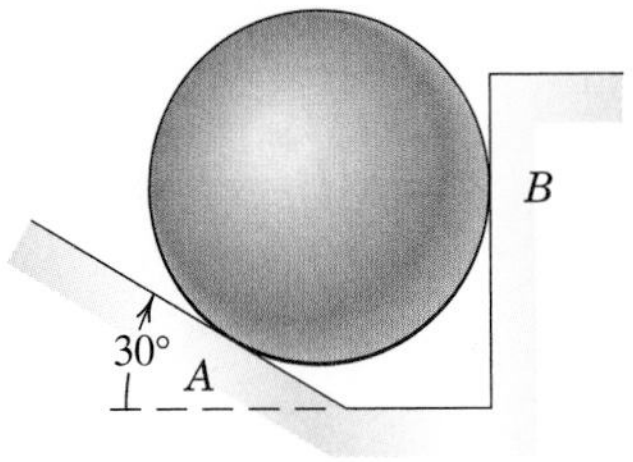

E6.4.4

6.4.5. A cable pulls on the bracket with a force of 2.5 kN (**E6.4.5**). At *A* the bracket rests against a smooth surface, and at *B* it is pinned to the wall. If the system is defined as the bracket and is considered to be planar, what loads act on the bracket at *A* and *B*? Use the general rule about "prevention of motion" to answer this question. (*Strategy:* Review Examples 6.6–6.8.) Confirm that your answers are consistent with the information on loads in **Table 6.1** or **Table 6.2** (whichever is appropriate). Present your answer in words and as a sketch of the bracket that shows the loads acting on it at *A* and *B*. Also comment on whether the sketch you created is or is not a free-body diagram.

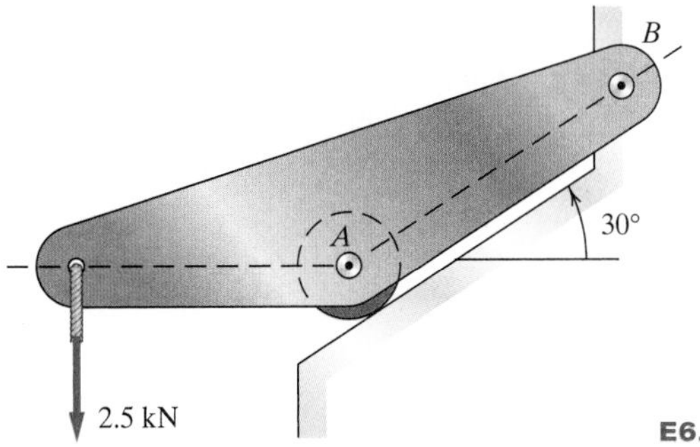

E6.4.5

6.4.6. Because of a combination of soil conditions and the tension in the single power cable, the utility pole in **E6.4.6** has developed the indicated 5° lean. The 9-m uniform pole has a mass per unit length of 25 kg/m, and the tension in the power cable is 900 N. If the system is defined as the power pole, what loads act on the pole at its base where it is fixed into the ground? Use the general rule about "prevention of motion" to answer this question. (*Strategy:* Review Examples 6.6–6.8.) Confirm that your answer is consistent with the information on loads in **Table 6.1** or **Table 6.2** (whichever is appropriate). Present your answer in words and as one or more sketches of the pole that show the loads acting on its base. Also comment on whether the sketch you created is or is not a free-body diagram.

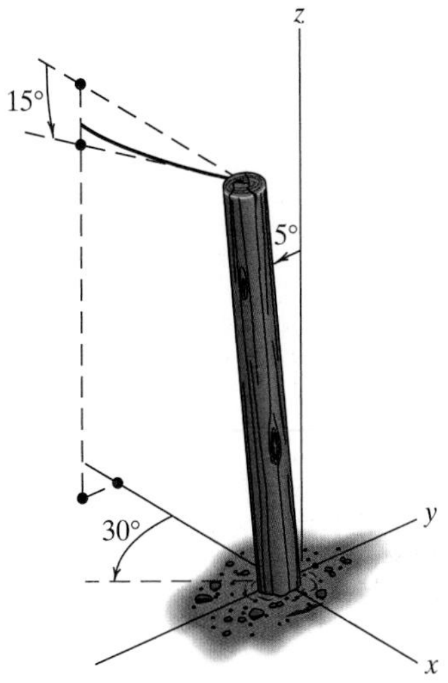

E6.4.6

6.4.7. A steel sphere sits in the grooved trough, as shown in **E6.4.7**. Surfaces A, B, and C are smooth. Gravity acts in the negative y direction. If the system is defined as the sphere, what loads act on the sphere at A, B, and C? Use the general rule about "prevention of motion" to answer this question. (*Strategy:* Review Examples 6.6–6.8.) Confirm that your answer is consistent with the information on loads in **Table 6.1** or **Table 6.2** (whichever is appropriate). Present your answer in words and as one or more sketches of the sphere that show the loads acting on it at A, B, and C.

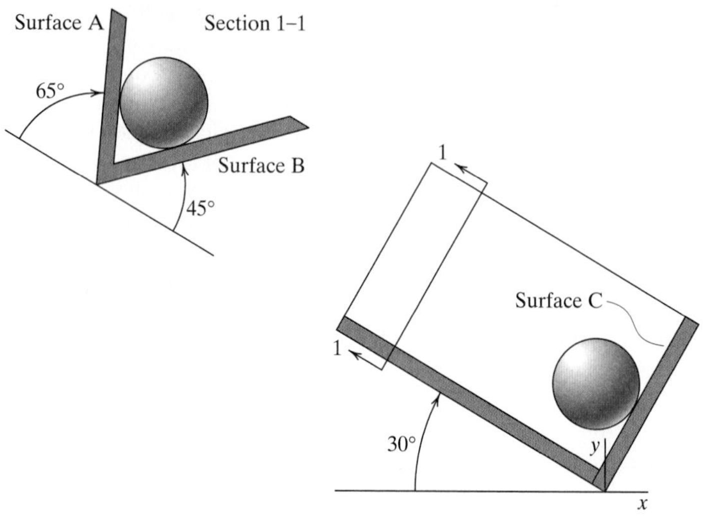

E6.4.7

6.5 DISTRIBUTED FORCES

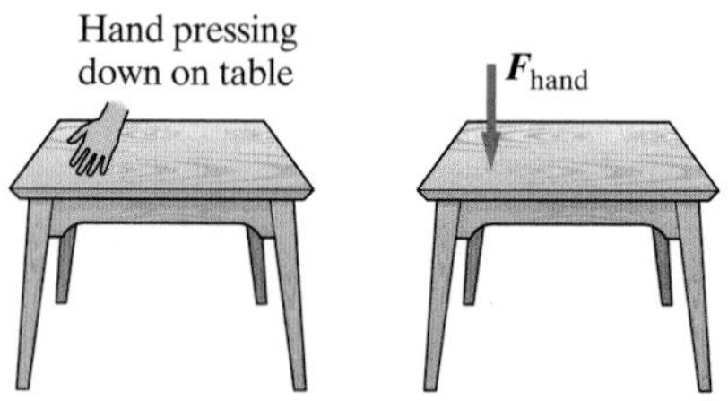

(*a*) (*b*)

Figure 6.53 Hand pressing down on a table modeled as a force

Up to this point we have modeled supports as loads acting at a single location on the system boundary. In actuality, all supports consist of forces distributed over a finite surface area. For example, if you press down on a table with your hand, the force you apply to the table is distributed over a finite area (**Figure 6.53*a***). For many practical applications, we can "condense" this distributed force into a single point force (**Figure 6.53*b***).

There are, however, some situations for which we explicitly consider the loads to be distributed; **Figure 6.54** shows some examples. The key idea we want to get across is that these distributed forces must be included in the system's free-body diagram. They can be represented in the diagram either as distributed forces (**Figure 6.55*b***) or as an equivalent point force (**Figure 6.55*c***). This equivalent point force is the total force represented by the distributed force and is located so as to create the same moment as the distributed force. For the uniformly distributed

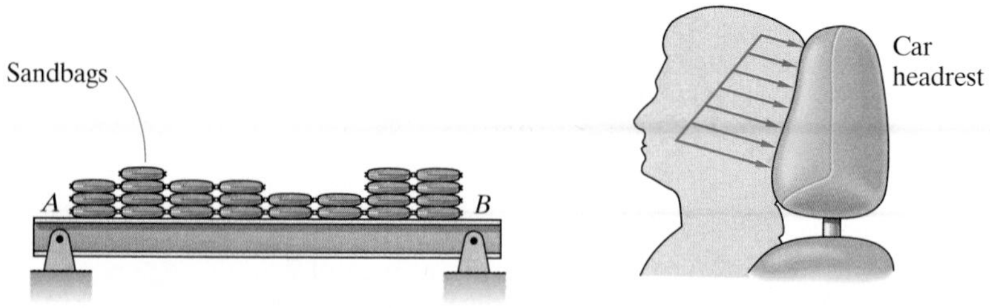

(*a*) (*b*)

Figure 6.54 Distributed loads: (*a*) 60-lb bags of concrete stacked on beam AB; (*b*) a head presses back on a head-rest

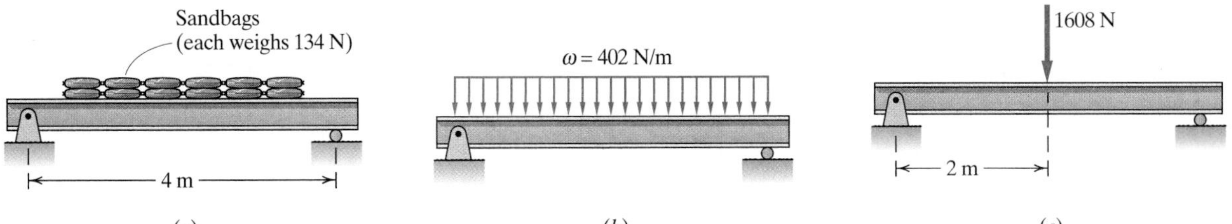

Figure 6.55 (*a*) Sandbags sitting on a beam; (*b*) weight of sandbags represented as distributed load; (*c*) weight of sandbags represented as an equivalent force

force in **Figure 6.55*a*** we are able to find this location by inspection. In Chapter 8 we shall show how to find the location for nonuniformly distributed forces. The important point to remember for your current work is that these distributed forces must be included in the free-body diagram.

By their very nature, fluids acting on a system boundary are distributed. Like distributed loads associated with supports, the loads at fluid boundaries are included in a free-body diagram, either as distributed loads or as an equivalent force. In Chapter 8 we discuss in greater detail distributed loads due to fluids acting on the system.

EXERCISES 6.5

6.5.1 Consider the coat rack in **Exercise 6.1.1**. Redraw the sketch showing the distributed load between the base of the coat rack and the floor.

6.5.2. Consider a building with a nominally flat roof. The actual roof surface is slightly irregular and can be represented by the profile shown in E6.5.2. After a night of heavy rain, an average rainfall total of 2 in. was recorded. Make a sketch of the distributed force that rain water applies to the nominally flat roof. Indicate the magnitude of the forces with the length of the vectors.

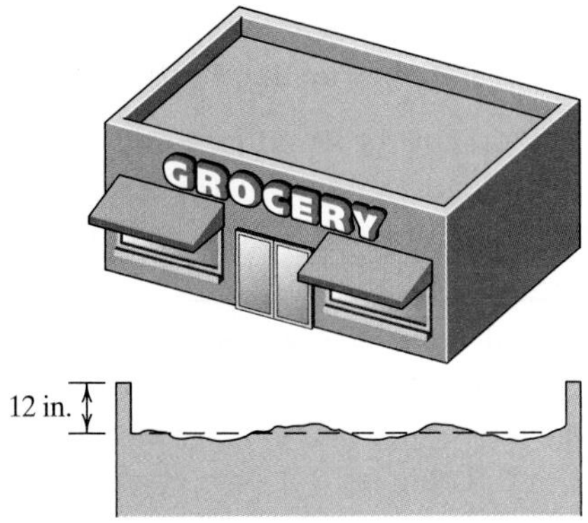

E6.5.2

6.5.3. Consider a cement truck with a tank that is half full in E6.5.3. Draw the distributed load applied to the inside of the tank. Indicate the magnitude of the loads with the length of the vectors.

E6.5.3

6.5.4. Identify three different systems on which distributed forces act. Make a sketch of each system and show what you think the distributed forces look like; indicate the magnitude of the forces with the length of vectors.

6.5.5. Consider a person wanting to cross a frozen pond. She has a choice of going on foot, wearing snow shoes, or using skis.

a. Make three sketches showing the distribution of her body weight on the frozen pond given the three types of footwear.

b. Which type of footwear would you choose? Why?

6.5.6. A hydraulic cylinder works by pumping fluid in and out of a piston assembly as shown in E6.5.6*a*. Draw the loads acting on piston assembly shown in **E6.5.6*b*.**

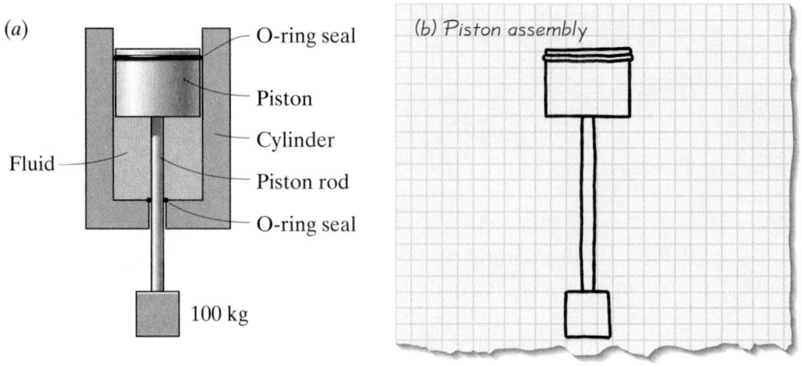

E6.5.6

6.6 FREE-BODY DIAGRAM DETAILS

We now outline a process for drawing a free-body diagram of a system; this is the DRAW step in our engineering analysis procedure.

1. Before diving into drawing, take time to **study the physical situation**. Consider what loads are present at boundaries and ask yourself whether you have ever seen a similar support. Study actual hardware (if available); pick it up or walk around it to really get a sense of how the loads act on the system. This inspection helps in making modeling assumptions. **Classify the system as planar or nonplanar**. If the system can be classified as planar, drawing the free-body diagram and writing and solving the conditions of equilibrium (covered in the next chapter) all become easier. If you are unsure, consider the system to be nonplanar. Also, consider asking for advice and opinions from others.

2. Define (either by imagining or actually drawing) a boundary that isolates the system from the rest of the world, then **draw the system** that is within the boundary. The drawing should contain enough detail so that distances and locations of loads acting on the system can be shown accurately. Sometimes multiple views of the system will be needed, especially if the system is nonplanar. **Establish a coordinate system. State any assumptions** you make.

3. Identify **cross-boundary forces** acting on the system and draw them at appropriate centers of gravity.[1] Include a variable label and the force magnitude (if known). Continue to state any assumptions you make.

4. Identify all **known loads** acting at the boundary and add these to the drawing, placing each known load at its point (or surface area) of application; identify each load on the drawing with a variable label and magnitude.

5. Identify the loads associated with each **support**, including those loads that act at discrete points and those that consist of distributed forces. If possible, classify each support as one of the standard supports (**Table 6.1** for planar systems and **Table 6.2** for nonplanar systems) to

[1]In Chapter 8 we will show how to find the center of gravity of a system.

help in identifying the loads. If this is not possible, consider how the surroundings restrict motion at a particular support (either translation and/or rotation) in order to identify the loads acting on the system that prevent this motion. Add all these loads to the drawing. Identify each load on the drawing with a variable label.

6. Identify **fluid boundaries**. Add loads associated with these boundaries to the drawing, showing them either as distributed or discrete point loads. Add variable labels.

You now have a free-body diagram of a system, as well as a list of the assumptions made in creating it. The diagram consists of a depiction of the system and the external loads acting on the system. The loads are represented in the diagram as vectors and with variable labels, and magnitudes (if known) are indicated.

A free-body diagram is an idealized model of a real system. By making assumptions about the behavior of supports, dimensions, and the material, you are able to simplify the complexity of the real system into a model that you can analyze. You might want the model to describe the real situation exactly, but this is generally not an achievable goal, due to limitations such as information, time, and money. What you do want, however, is a model that you can trust and that gives results that closely approximate the real situation.

In creating a model, an engineer must decide which loads acting on the system are significant. For example, a hinge on a door is often modeled as having no friction about its axis. Yet for most hinges, grease, dust, and dirt have built up, and there is actually some friction—some resistance to rotation. If friction is large enough the engineer should include it in the model. However, if the friction is small enough that the door can still swing freely, the engineer may conclude that it is not significant for the problem at hand, and model the hinge loads as shown in **Figure 6.56**.

Often the significance of loads is judged by their relative magnitude or location. For example, the weight of a sack of groceries is insignificant relative to the weight of an automobile carrying them but very significant if the vehicle is a bicycle. Whenever you are in doubt about the significance of a load, consider it significant. In many of the examples in this book, we will set the stage by making some of the assumptions regarding significance. In others, though, it will be up to you to judge the significance of a load based either on your own experience or on the advice of other engineers. Any loads considered insignificant are not included in the free-body diagram and should be noted in the assumption list.

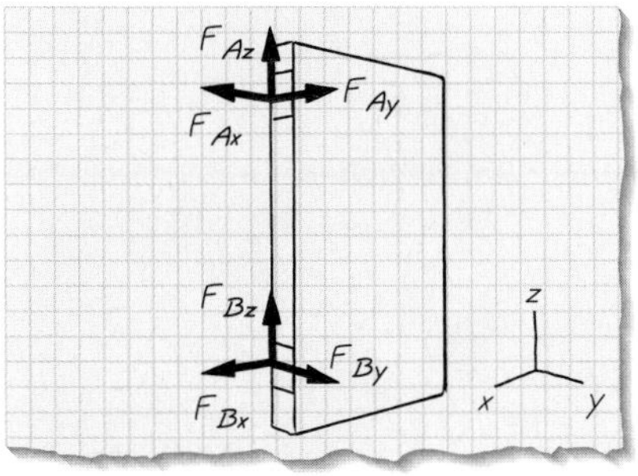

Figure 6.56 Forces acting at hinges A and B when hinges are frictionless

EXAMPLE 6.9 CREATING A FREE-BODY DIAGRAM OF A PLANAR SYSTEM (A)

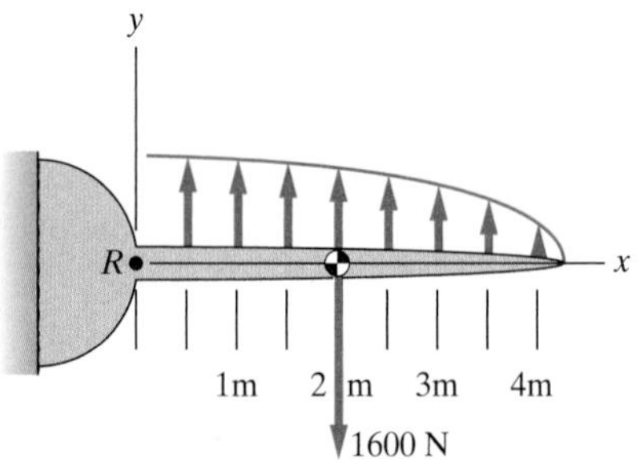

Figure 6.57

The lift force on an airplane wing (which is actually a distributed load) can be modeled by eight forces as shown in Figure 6.57. The magnitude of each force is given in terms of its position x on the wing by

$$F_i = 300\sqrt{1-\left(\frac{x_i}{17}\right)^2}\ \text{N}, \quad \text{with } i = 1,2,\ldots,8 \qquad (1)$$

The location $x = 0$ is at the root of the wing (location R); this is the point where the wing connects with the fuselage.

The weight of the wing $W = 1600$ N can be located at midpoint of the wing's length. Create a free-body diagram of the wing.

Goal Draw a free-body diagram of the wing.

Given We are given a planar view of the wing and some dimensions. In addition, we are told that we can model the lift as eight forces acting upward at locations $x_i = 0.5$ m, 1.0 m, 1.5 m, . . . , 4.0 m. We are also told that the weight of the wing is 1600 N, with a point of application at $x = 2.0$ m.

Assume Ignoring any slight differences in the leading and trailing edges of the wing, we assume there is a plane of symmetry (the xy plane in **Figure 6.57**); therefore we can treat the wing as a planar system. This means we are ignoring any twist on the wings that would occur from asymmetry.

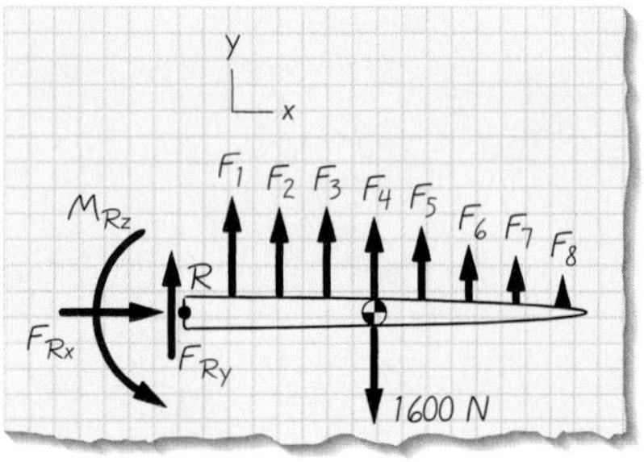

Figure 6.58

Draw Based on the information given in the problem and our assumptions, we isolate the wing at its root. The boundary condition at R can be modeled as a fixed boundary condition; therefore, from **Table 6.1** we find that there will be a force (represented as x and y components) and a moment present. We also determine the values of the lift force at each of the eight locations using function (1), obtaining the values shown in Table 6.3. The free-body diagram is given in Figure 6.58.

Table 6.3 Point Loads Representing Lift on a Wing

x(m)		F(N)
0.5	F_1	300
1	F_2	299
1.5	F_3	299
2	F_4	298
2.5	F_5	297
3	F_6	295
3.5	F_7	294
4	F_8	292

EXAMPLE 6.10 CREATING A FREE-BODY DIAGRAM OF A PLANAR SYSTEM (B)

The ladder in **Figure 6.59** rests against the wall of a building at A and on the roof of an adjacent building at B. If the ladder has a weight of 100 N and length 3 m, and the surfaces at A and B are assumed smooth, create a free-body diagram of the ladder.

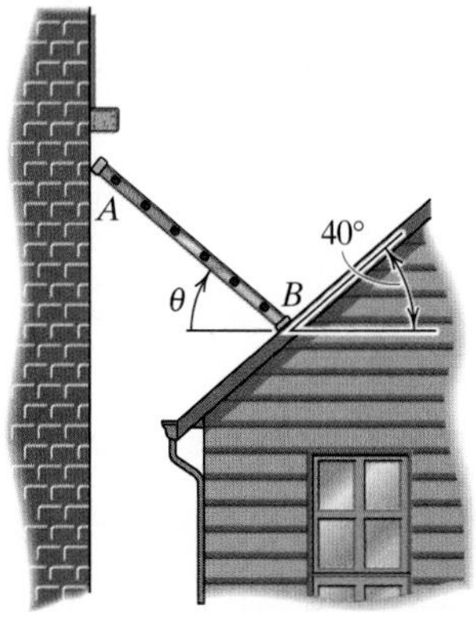

Figure 6.59

Goal Draw a free-body diagram of the ladder.

Given We are given a planar view of the ladder and some spatial information (length of ladder and angle of orientation with respect to the roof and wall). In addition we are told that the roof is at 40° with respect to the horizontal and that the surfaces at A and B are smooth.

Assume First, we assume that the ladder is uniform. By this we mean that the rungs (cross-pieces) are identical to one another, as are the two stringers (sides of ladder to which the rungs are attached). Next we assume that the weight of the ladder is significant and that its center of mass can be located at its midpoint (since it is uniform). We also assume that gravity works downward in the vertical direction. Finally, we assume that there is an xy plane of symmetry; therefore we can treat the ladder as a planar system.

Draw Based on the information given and our assumptions, we isolate the ladder from the wall (at A) and from the roof (at B). At each surface there is a normal force present that pushes on the ladder. There is no frictional force because both surfaces are smooth. The resulting free-body diagram is shown in **Figure 6.60**. Notice that there is a factor of two associated with both normal forces; this factor reflects that there are two stringers.

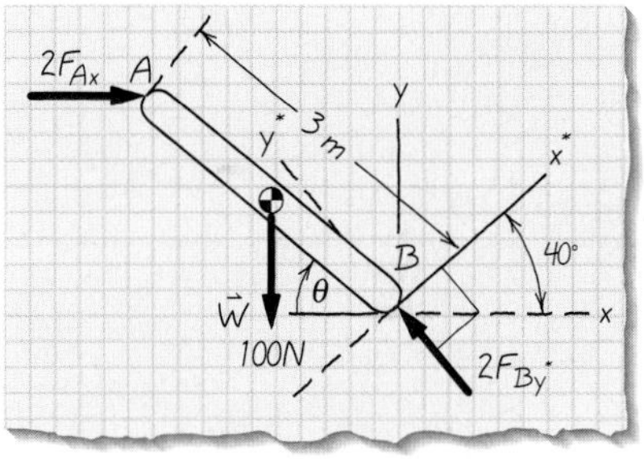

Figure 6.60

EXAMPLE 6.11 CREATING A FREE-BODY DIAGRAM OF A PLANAR SYSTEM (C)

A 500-N crane boom is supported by a pin at A and a hydraulic cylinder at B. At C, the boom supports a stack of lumber weighing W. Create a free-body diagram of the boom (**Figure 6.61**).

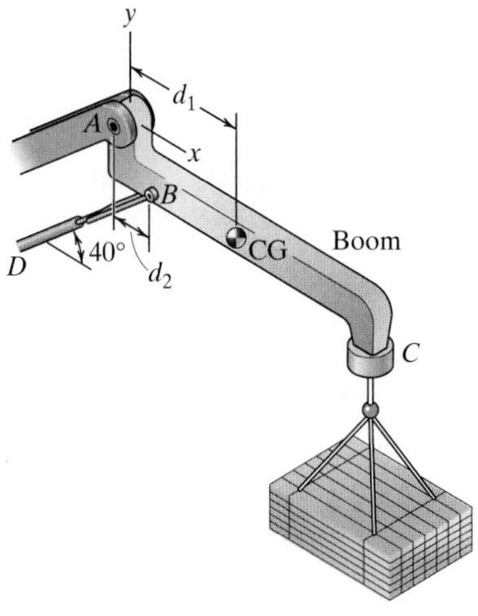

Figure 6.61

Goal Draw a free-body diagram of the boom.

Given We are given an isometric view of the boom. The joint at A is a pin connection, and member DB is a hydraulic cylinder. The hydraulic cylinder acts like a link; therefore there will be a force acting on the boom at B that runs along DB. The boom is being used to lift a load W.

Assume We assume that the weight of the boom can be modeled as a force acting at the point marked CG in **Figure 6.61**. We also assume that gravity works in the negative y direction. Finally, we assume that there is an xy plane of symmetry; therefore we can treat the boom as a planar system.

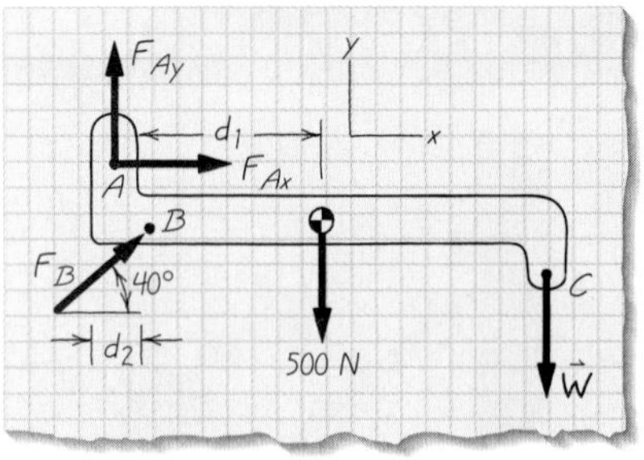

Figure 6.62

Draw Based on the information given and our assumptions, we isolate the boom. At A there is a pin connection, and according to **Table 6.1**, this means that there is a force present (which we represent in terms of its x and y components). The resulting free-body diagram is shown in **Figure 6.62**.

EXAMPLE 6.12 CREATING A FREE-BODY DIAGRAM OF A NONPLANAR SYSTEM (A)

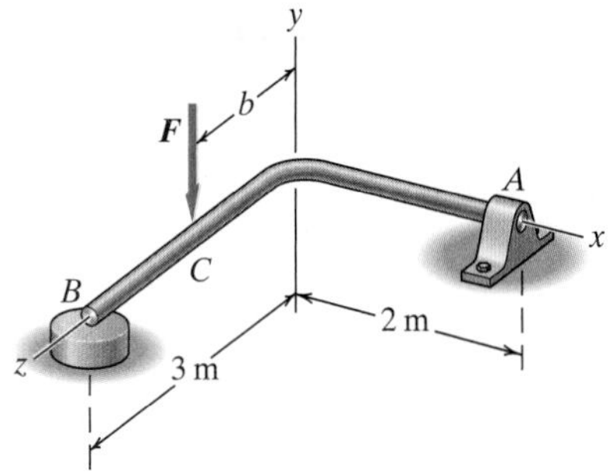

Figure 6.63

The L-shaped bar is supported by a bearing at A and rests on a smooth horizontal surface at B, with $\| \boldsymbol{F} \| = 800$ N and $b = 1.5$ m (**Figure 6.63**). Ignore gravity. Create a free-body diagram of the bar.

Goal Draw a free-body diagram of the bar.

Given We are given the dimensions of the bar and that surface B is smooth. Joint A looks like a journal or thrust bearing. We are told to ignore the weight of the bar.

Assume We will assume that the bearing at A is a journal bearing. Since the loads involved at A, B, and C do not lie in a single plane, we must treat the bar as a nonplanar system.

Draw Based on the information given and our assumptions, we isolate the bar. At A there is a single thrust bearing; according to **Table 6.2**, this means there are forces and moments present. At the smooth surface at B there is a normal force that acts to push up on the bar. The resulting free-body diagram is shown in **Figure 6.64**.

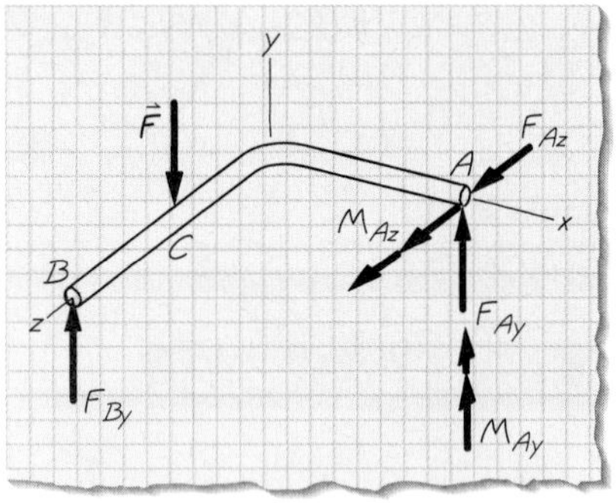

Figure 6.64

EXAMPLE 6.13 CREATING A FREE-BODY DIAGRAM OF A NONPLANAR SYSTEM (B)

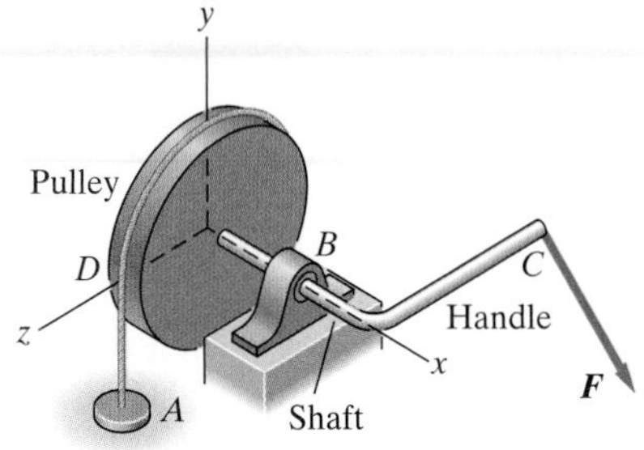

Figure 6.65

The cable in **Figure 6.65** is attached at A to a stationary surface and is wrapped around the pulley. At B is a single thrust bearing, and at C, force $\boldsymbol{F} = (10 \text{ N } \boldsymbol{i} - 30 \text{ N } \boldsymbol{j} - 10 \text{ N } \boldsymbol{k})$ acts. Ignore gravity. Create a free-body diagram of the shaft–handle–pulley system.

Goal We are to draw a free-body diagram of the shaft–handle–pulley system.

Given We are given dimensions of the shaft, pulley, and handle. In addition, we are told the magnitude and direction of the applied force ***F***, and that the bearing at *B* is a thrust bearing.

Assume Since the loads acting on the shaft–handle–pulley system do not lie in a single plane, we must treat the shaft–handle–pulley system as nonplanar. Finally, we assume that the weights of the shaft, handle, and pulley are negligible (because we are not told anything about their weights).

Draw Based on the information given and our assumptions, we isolate the shaft–handle–pulley system. At *D* we show the pull of the cable as $\boldsymbol{F}_{DA}$. At the single bearing at *B* we include a force and moment (**Table 6.2**). Finally, at *C* we show the force ***F***. **Figure 6.66** shows the completed free-body diagram.

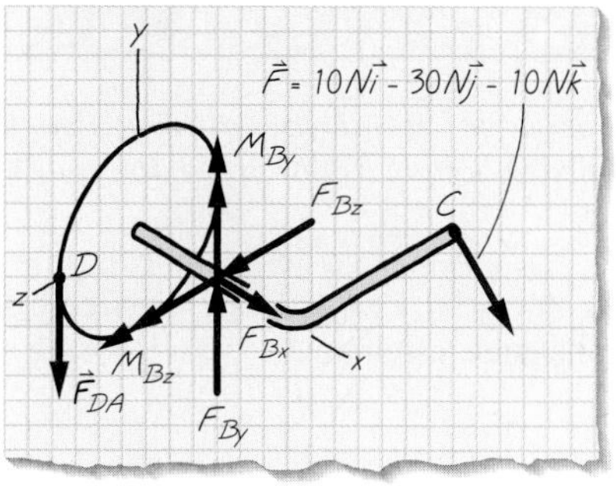

Figure 6.66

EXAMPLE 6.14 CREATING A FREE-BODY DIAGRAM OF A NONPLANAR SYSTEM (C)

Figure 6.67 shows an improved design relative to Example 6.13. The bearing at *B* is a thrust bearing, and the bearing at *E* is a journal bearing. At *C*, force $\boldsymbol{F} = (10\text{ N}\,\boldsymbol{i} - 30\text{ N}\,\boldsymbol{j} - 10\text{ N}\,\boldsymbol{k})$ acts. Ignore gravity. Create a free-body diagram of the shaft–handle–pulley system. Why do you think this design is improved relative to Example 6.13?

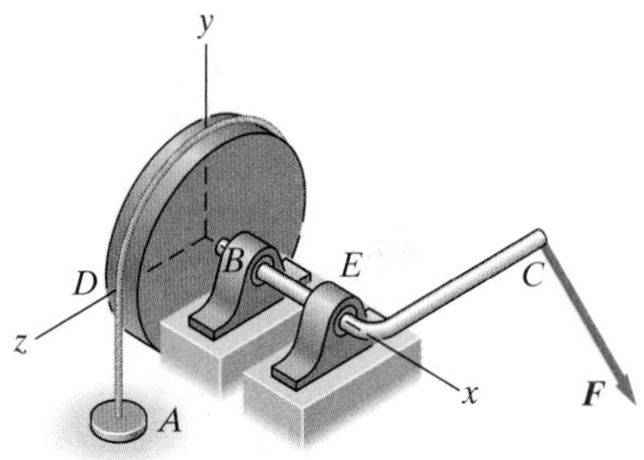

Figure 6.67

Goal Draw a free-body diagram of the shaft–handle–pulley system.

Given We are given the dimensions of the shaft, pulley, and handle. In addition, we are told the magnitude and direction of the applied force ***F***, and to ignore gravity.

Assume We assume that the bearings at *B* and *E* are properly aligned. This means (according to **Table 6.2**) that forces are applied to the bar at the location of each bearing. In addition, we are told that bearing *B* is a thrust bearing; therefore it will apply an axial force to the shaft. Since the loads involved do not lie in a single plane, we must treat the shaft–handle–pulley system as nonplanar.

Draw Based on the information given and our assumptions, we isolate the shaft–handle–pulley system. At *D* we include the pull of the cable. At the thrust bearing at *B* we include radial (F_{By}, F_{Bz}) forces and an axial force (F_{Bx}). At the journal bearing at *E* we include radial (F_{Ey}, F_{Ez}) forces. Finally, at *C* we show the force ***F*** (**Figure 6.68**).

This design is superior to the design in Example 6.13 because the bearings apply only forces to the shaft; this results in better wear of both the shaft and the bearings. In addition, there will be less "radial play" of the system, generally a desirable characteristic of rotating systems. It is generally good practice to design systems with two properly aligned bearings.

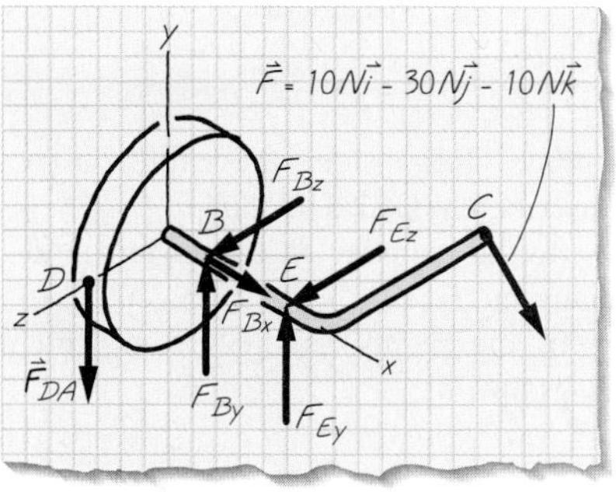

Figure 6.68

EXERCISES 6.6

For each of the exercises in this section, follow the steps outlined in Section 6.6 for presenting your work.

6.6.1. A tape guide assembly is subjected to the loading as shown in E6.6.1. Draw the free-body diagram of the tape guide assembly.

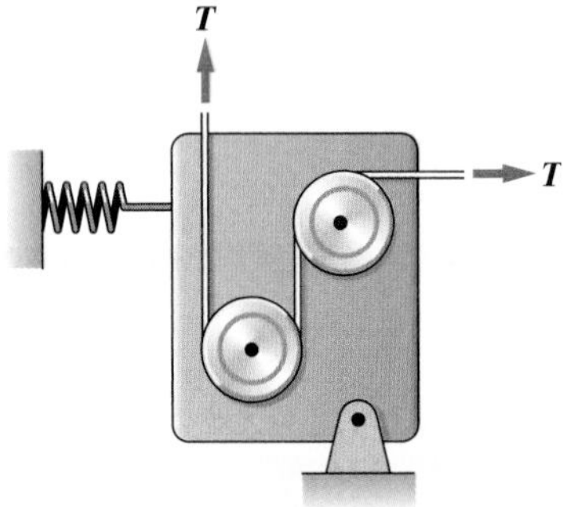

E6.6.1

6.6.2. A wheel and pulley assembly is subjected to the loading as shown in E6.6.2. Draw the free-body diagram of the assembly.

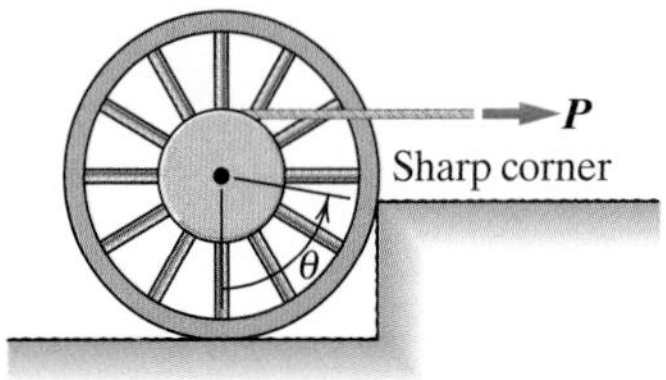

E6.6.2

6.6.3. Reconsider the cable–pulley–cylinder assembly described in **E6.2.5**. Define

a. the cylinder as the system and draw its free-body diagram

b. the pulley as the system and draw its free-body diagram

6.6.4. Reconsider the belt-tensioning device described in **E6.2.4**. Define

a. the pulley *A* as the system and draw its free-body diagram

b. the mass as the system and draw its free-body diagram

c. the pulley, L-arm, and mass as the system and draw its free-body diagram

6.6.5. Reconsider the simple truss described in **E6.2.3**. Define the simple truss as the system and draw its free-body diagram.

6.6.6. Reconsider the beam described in **E6.2.2**. Define

a. the beam as the system and draw its free-body diagram

b. the beam and the 100-N cylinder as the system and draw its free-body diagram

6.6.7. A beam is fixed to the wall at *C* and rests against a block at *D*. Additional loads act on the beam, as shown in E6.6.7. Define the beam as the system. Draw its free-body diagram.

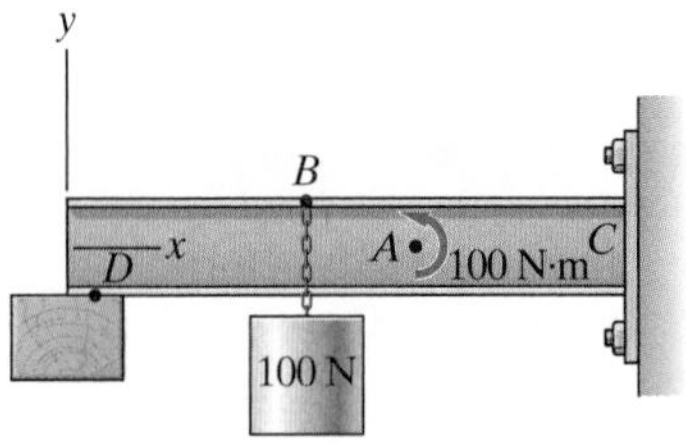

E6.6.7

6.6.8. An engine is lifted with the pulley system shown in E6.6.8. Define

a. the pulley at *B* as the system and draw its free-body diagram

b. the engine and chain as the system and draw its free-body diagram

c. the ring at *C* as the system and draw its free-body diagram

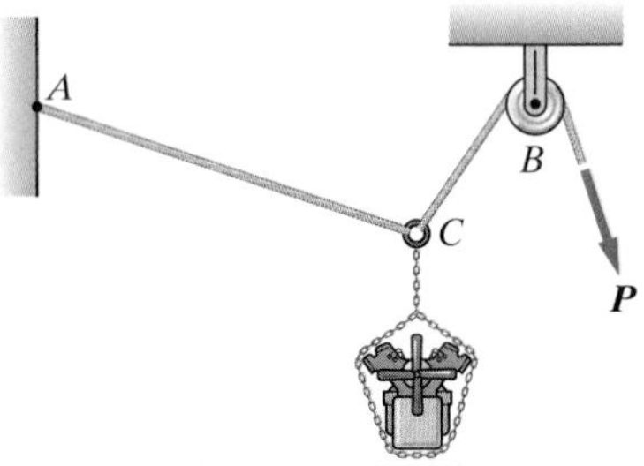

E6.6.8

6.6.9. A concrete hopper and its contents have a combined mass of 400 kg, with mass center at *G* (E6.6.9). The hopper is being elevated at constant velocity along its vertical guide by cable tension ***T***. The design calls for two sets of guide rollers at *A*, one on each side of the hopper, and two sets at *B*. Define

a. the hopper and the triangular guides (including the wheels) at A and B as the system and draw its free-body diagram

b. the triangular guide at A (including the two wheels) as the system and draw its free-body diagram

c. the hopper minus the triangular guides at A and B as the system and draw its free-body diagram

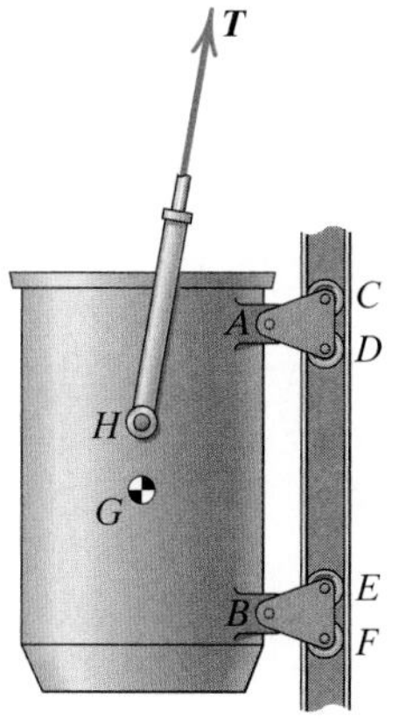

E6.6.9

6.6.10. A portion of a mechanical coin sorter is shown in **E6.6.10*a***. Pennies and dimes roll down the 20° incline, the triangular portion of which pivots freely about a horizontal axis through O. Dimes are light enough (2.28 grams mass each) so that the triangular portion remains stationary, and the dimes roll into the right collection column. Pennies, on the other hand, are heavy enough (3.06 grams mass each) so that the triangular portion pivots clockwise, and the pennies roll into the left collection column. Define

a. the triangular portion in Position 1 (**E6.6.10*a***) as the system and draw its free-body diagram

b. the triangular portion in Position 2 as the system (**E6.6.10*b***) (the triangle has just started to pivot) and draw its free-body diagram

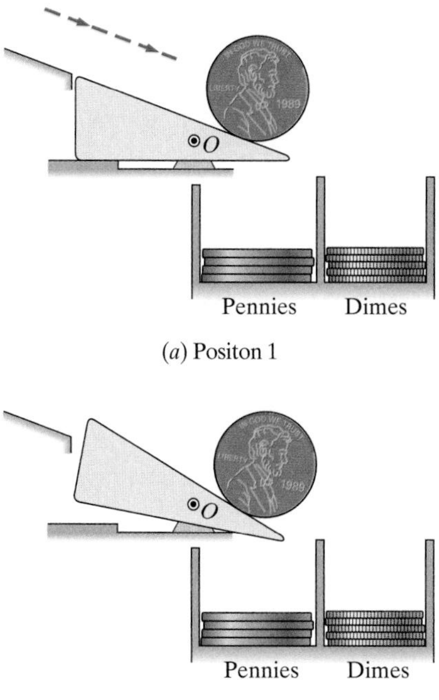

E6.6.10

6.6.11. The throttle-control sector pivots freely at O (**E6.6.11**). An internal torsional spring at O exerts a return moment of magnitude $||\mathbf{M}|| = 2\ \text{N} \cdot \text{m}$ on the sector when in the position shown. Define the sector as the system. Draw its free-body diagram.

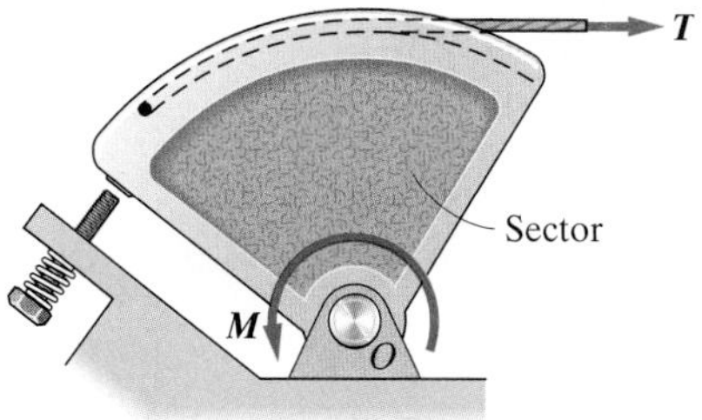

E6.6.11

6.6.12. The following three cases involve a 2 × 4 wooden board and a wrench. In each case a force is applied, and hands are used to react to this force, as shown in **E6.6.12**. Imagine what forces the hands would need to apply to the 2 × 4, then:

a. *Consider Case 1.* Define the system as the 2 × 4, bolt, and wrench. Draw its free-body diagram.

b. *Consider Case 2.* Define the system as the 2 × 4, bolt, and wrench. Draw its free-body diagram.

c. *Consider Case 3.* Define the system as the 2 × 4, bolt, and wrench. Draw its free-body diagram.

d. Repeat **a**, **b**, and **c** if the system is defined as just being the 2 × 4 and the bolt.

6.6.13. Consider the mechanism used to weigh mail in **E6.6.13**. A package placed at A causes the weight pointer to rotate through an angle α. Neglect the weights of the members except for the counterweight at B, which has a mass of 4 kg. For a particular package, $\alpha = 20°$. Define

a. the system as shown in **E6.6.13*a*** and draw its free-body diagram

b. the system as shown in **E6.6.13*b*** and draw its free-body diagram

c. the system as shown in **E6.6.13*c*** and draw its free-body diagram

d. the system as shown in **E6.6.13*d*** and draw its free-body diagram

e. the system as shown in **E6.6.13*e*** and draw its free-body diagram

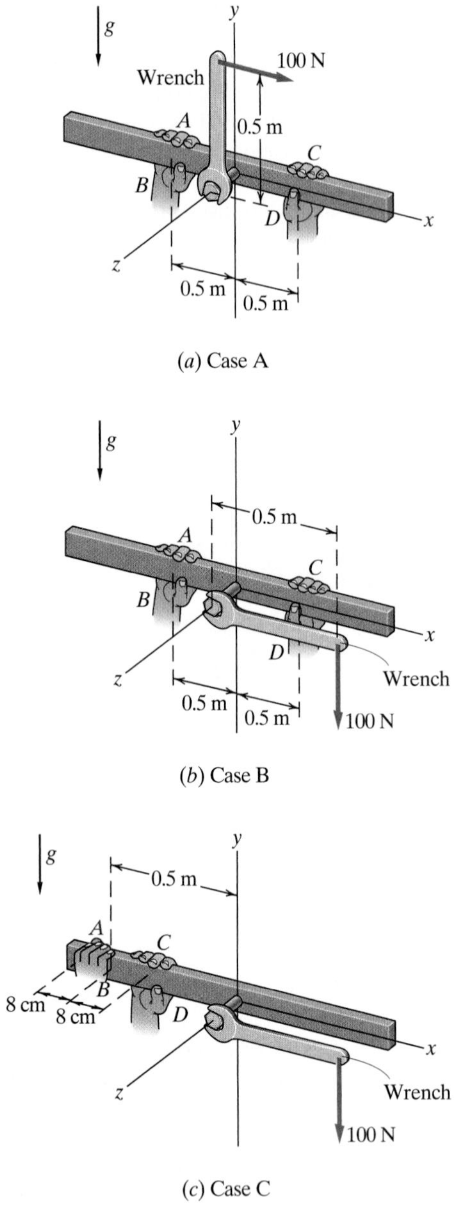
g
y
Wrench
100 N
A
0.5 m
C
B
D
x
z
0.5 m
0.5 m
(a) Case A
g
y
0.5 m
A
C
B
D
x
Wrench
z
0.5 m
0.5 m
100 N
(b) Case B
g
y
0.5 m
A
C
B
D
8 cm
8 cm
x
Wrench
z
100 N
(c) Case C

E6.6.12

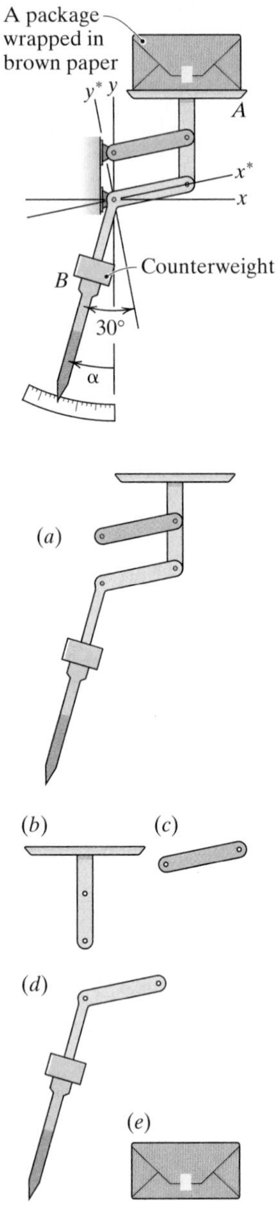
A package
wrapped in
brown paper
y* y
A
x*
x
B
Counterweight
30°
α
(a)
(b)
(c)
(d)
(e)

E6.6.13

6.6.14. Consider the pair of pliers in E6.6.14. Define

a. Member 1 (**E6.6.14*a***) as the system and draw its free-body diagram

b. Member 2 (**E6.6.14*b***) as the system and draw its free-body diagram

c. Member 3 (**E6.6.14*c***) as the system and draw its free-body diagram

d. the entire pair of pliers as the system (minus the shaft it is clamping) and draw its free-body diagram

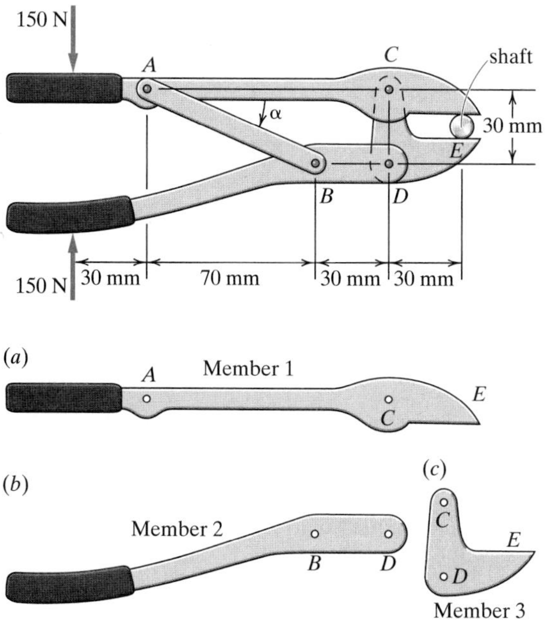

E6.6.14

6.6.15. Consider the frame in E6.6.15. Define

a. the entire frame as the system and draw its free-body diagram

b. Member 1 as the system and draw its free-body diagram

c. Member 2 as the system and draw its free-body diagram

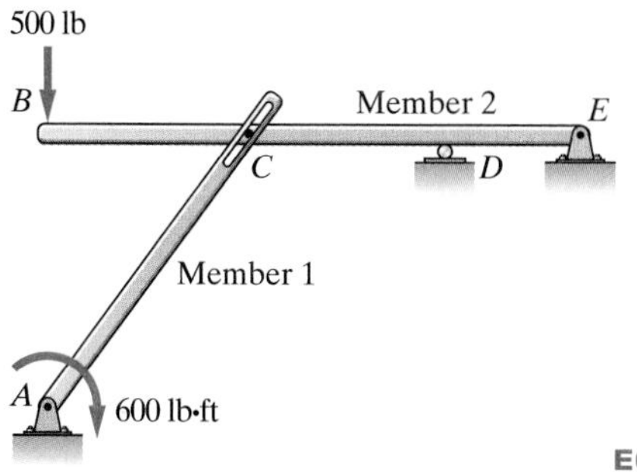

E6.6.15

6.6.16. Consider the frame in E6.6.16. Define

a. the entire frame as the system and draw its free-body diagram

b. Member 1 as the system and draw its free-body diagram

c. Member 2 as the system and draw its free-body diagram

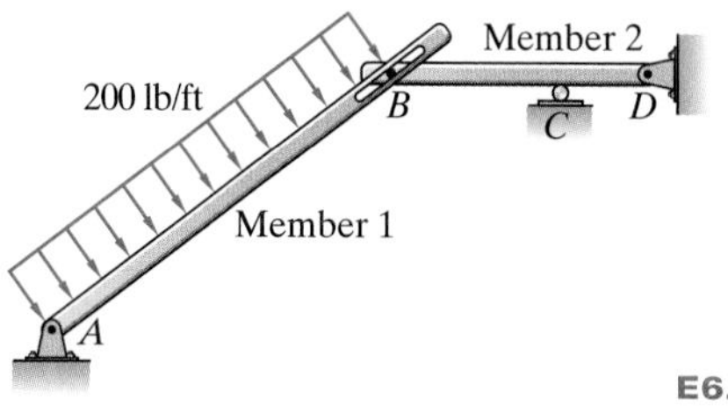

E6.6.16

6.6.17. Consider the frame in E6.6.17. Define

a. the entire frame as the system and draw its free-body diagram

b. Member 1 as the system and draw its free-body diagram

c. Member 2 as the system and draw its free-body diagram

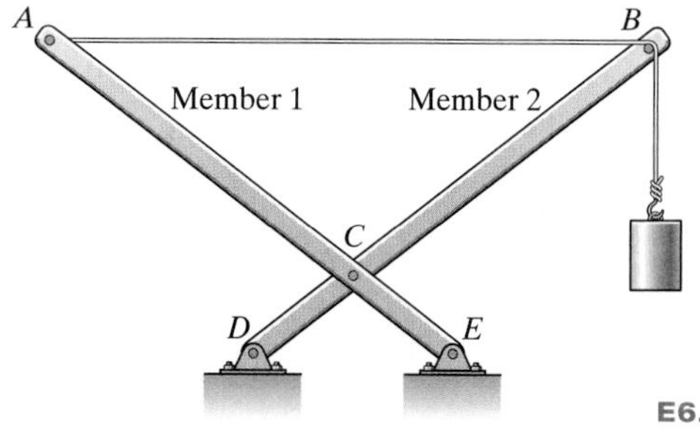

E6.6.17

6.6.18. Consider the frames in E6.6.18. Define

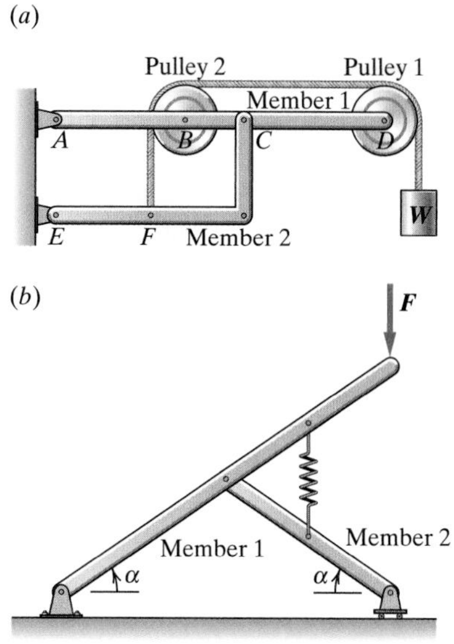

E6.6.18

a. Draw the free-body diagrams for the entire frame (**E6.6.18*a***), Member 1, Member 2, Pulley 1, and Pulley 2.

b. Draw the free-body diagrams for the entire frame (**E6.6.18*b***), Member 1, and Member 2.

6.6.19. Consider the exercise frame in E6.6.19. Define the entire frame as the system and draw its free-body diagram.

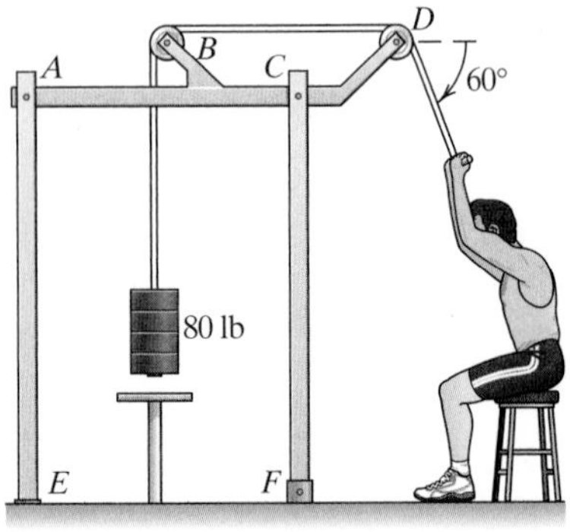

E6.6.19

6.6.20. A roof access cover is buried under a pile of snow as shown in E6.6.20.

a. Draw the free-body diagram of the cover.

b. Describe in words how you showed the snow load and why you chose to show it in this manner.

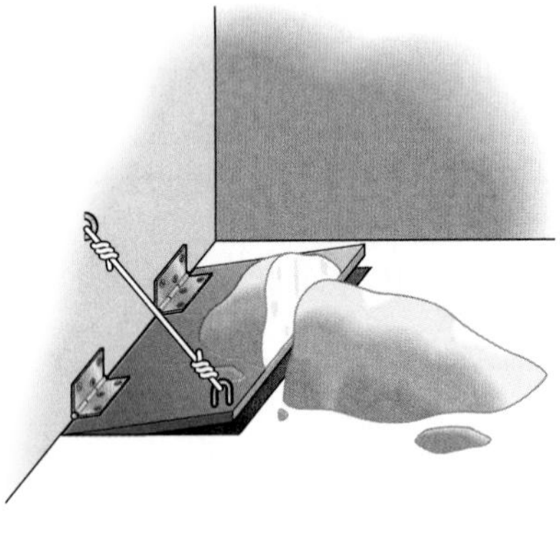

E6.6.20

6.6.21. The bent bar is loaded and attached as shown in E6.6.21. Draw the free-body diagram of the bar.

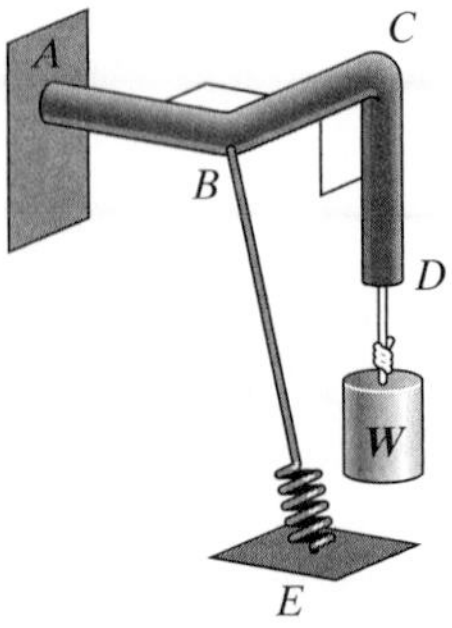

E6.6.21

6.6.22. Reconsider the utility pole described in **E6.4.6** and draw the free-body diagram of the pole.

6.6.23. A 200-N force is applied to the handle of the hoist in the direction shown in E6.6.23. There is a thrust bearing at A and a journal bearing at B. Draw a free-body diagram of the handle–shaft–pulley assembly. Make sure to record any assumptions.

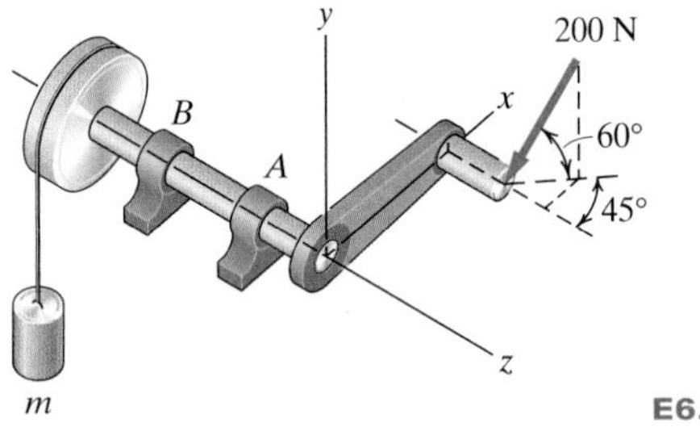

E6.6.23

6.6.24. Reconsider the welded tubular frame described in **E6.3.4** and draw the free-body diagram of the frame.

6.6.25. Reconsider the tower crane described in **E6.3.2** and draw the free-body diagram of the crane.

6.6.26. Reconsider the steel shaft described in **E6.3.3** and draw the free-body diagram of the shaft.

6.6.27. Three workers are carrying a 4-ft by 8-ft panel in the horizontal position shown in E6.6.27. The panel weight is 100 lb. Define the panel as the system and draw a free-body diagram of the panel.

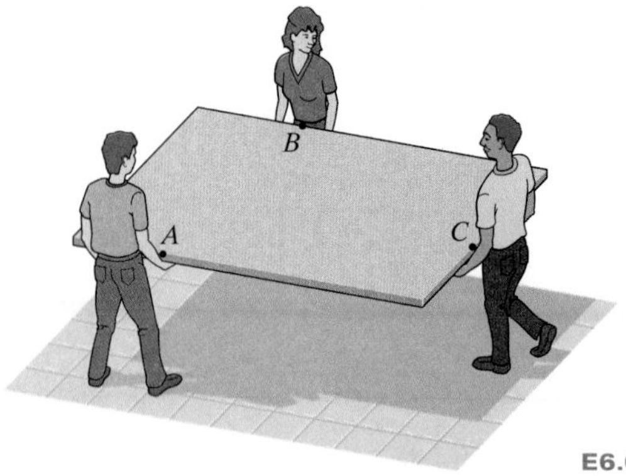

E6.6.27

6.6.28. A bracket is bolted to the shaft at O. Cables load the bracket, as shown in E6.6.28. Define the bracket as the system and draw a free-body diagram.

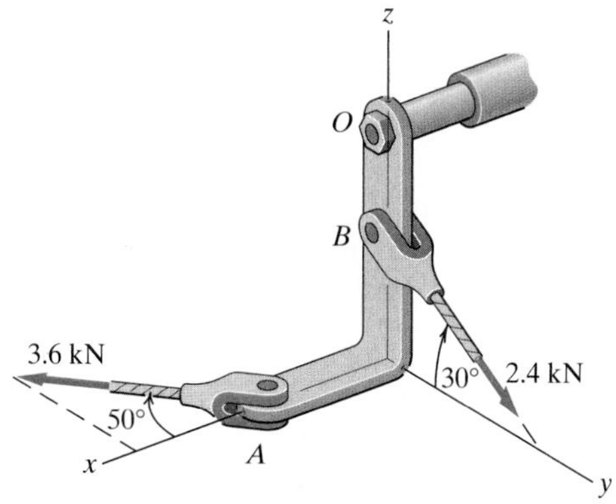

E6.6.28

6.6.29. A carpenter is slowly pushing the 90-kg roof truss into place. In the current position shown in E6.6.29 it is oriented 20° from the vertical. The mass center of the symmetric triangular truss is located up one-third of its 2.25-m dimension from its base. Define the truss of the system and draw its free-body diagram.

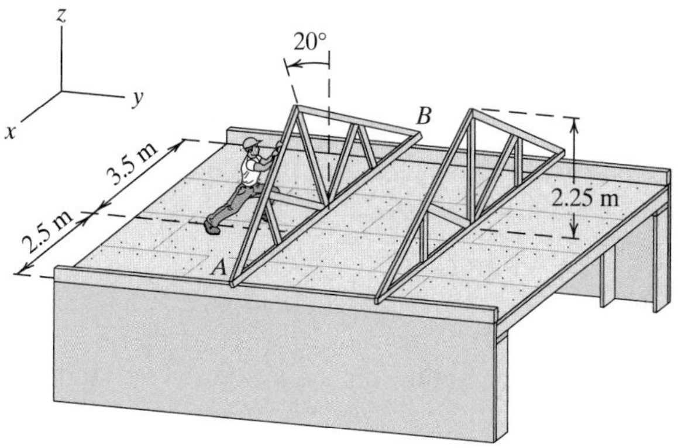

E6.6.29

6.6.30. A hanging chair is suspended, as shown in E6.6.30. A person weighing 800 N is sitting in the chair (but is not shown). Define

a. the ring at D as the system and draw its free-body diagram

b. the chair (including the knot at E) as the system and draw its free-body diagram

c. the eyelet fastener at A as the system and draw its free-body diagram

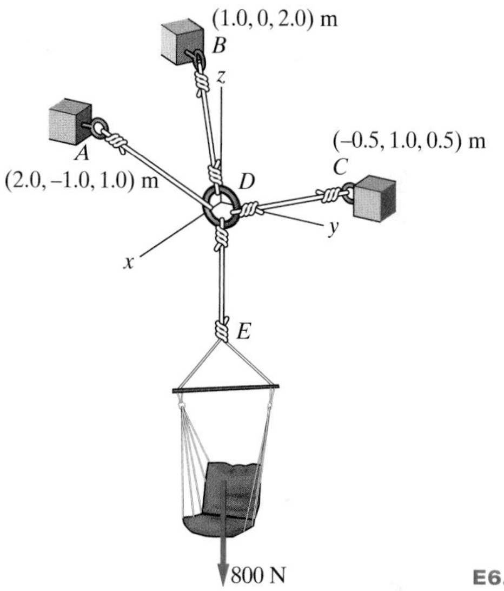

E6.6.30

6.7 JUST THE FACTS

In this chapter we looked at free-body diagrams—what they are and how to create them. To create a free-body diagram:

1. Study the physical situation. Classify the system as planar or nonplanar. A **planar system** is one in which all the forces acting on the system lie in the same plane and all moments are about an axis perpendicular to that plane or there is a **plane of symmetry**. Planar systems are also referred to as **two-dimensional systems**. If a system is not planar, it is **nonplanar** and is also referred to as **three-dimensional**.

2. Define (either by imagining or actually drawing) a boundary that isolates the system from its surroundings, then **draw the system** that is within the boundary. **Establish a coordinate system**. Planar systems typically require a single view drawing, whereas nonplanar systems may require multiple views or an isometric drawing. **State any assumptions** you make.

3. Identify cross-boundary forces (e.g., **gravity**) acting on the system and draw them at appropriate centers of gravity. Include a variable label and the force magnitude (if known). Continue to state any assumptions you make.

4. Identify all **known loads** acting at the boundary and add these to the drawing, placing each known load at its point (or surface area) or application; identify each load on the drawing with a variable label and magnitude.

5. Identify the loads associated with each **support**, both those loads that act at discrete points and those that consist of distributed forces. The loads associated with various planar supports (e.g., normal contact, links, cable) are presented in **Table 6.1**, and those with various nonplanar supports (e.g., hinges, journal bearings) are presented in **Table 6.2**.

Irrespective of whether a system is planar or nonplanar, if a particular support is not described in **Table 6.1** (planar systems) or **Table 6.2** (nonplanar systems), the general rule that describes a boundary's restriction of motion can be used to identify the loads at the support. This rule states:

If a support prevents the translation of the system in a given direction, then a force acts on the system at the location of the support in the opposite direction. Furthermore, if rotation is prevented, a moment opposite the rotation acts on the system at the location of the support.

6. Identify **fluid boundaries**. Add the loads at these boundaries to the drawing, showing them either as distributed or discrete point loads. Add variable labels.

You now have a **free-body diagram** of a system, as well as a list of the assumptions made in creating it. The diagram consists of a depiction of the system and the external loads acting on the system. The loads are represented in the diagram as vectors and with variable labels, and magnitudes (if known) are indicated. In the next chapter we consider how to use a free-body diagram in conjunction with Newton's first law to consider equilibrium of the system.

SYSTEM ANALYSIS (SA) EXERCISES

SA6.1 Checking on the Design of a Chair

To increase the seating capacity during basketball games, collapsible and portable floors on rollers have been installed, as shown in **Figure SA6.1.1**. The chairs that go with this portable flooring system are placed onto the floors but are not connected to the floor. **Figure SA6.1.2** shows the dimensions of one of these chairs.

Situation: The basketball game of Wolfpack against UNC Chapel Hill is underway. As Sierra, an engineering student taking Statics, is cheering her team, she notices a woman sitting on the front edge of one of the chairs described above. Because of the small size of the hinge that holds the chair's seat, Sierra becomes concerned for the safety of the woman.

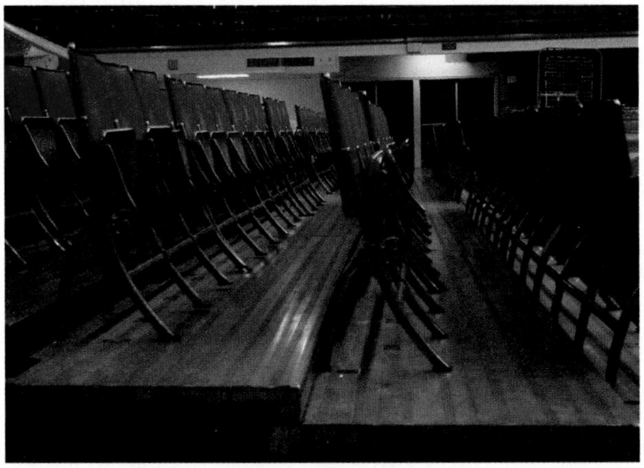

Figure SA6.1.1 Mobile floor and chairs in the Reynolds Coliseum

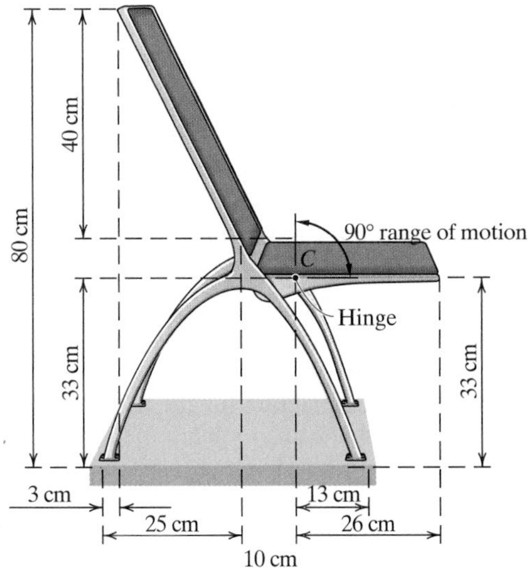

Figure SA6.1.2 Dimensions of the chair

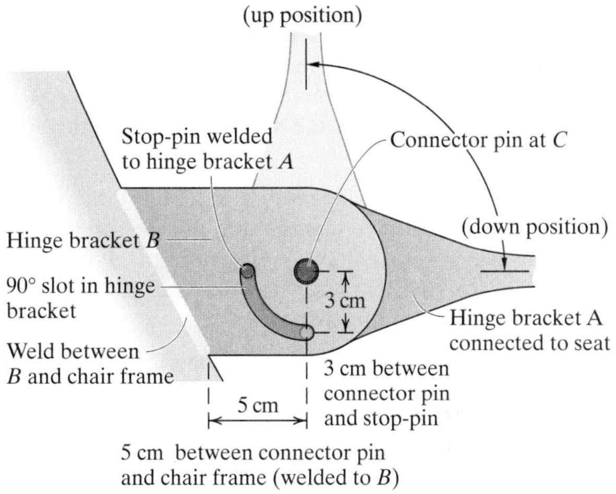

Figure SA6.1.3 Basic design of hinge for seat

Imagine that you are in Sierra's place and re-create what goes through her mind as she has a sudden flashback to moments and the concept of free-body diagrams covered in her Statics class.

(a) Assuming that the mass of the woman is 61 kg, what is the maximum moment that the slotted hinge has to bear? **Figure SA6.1.3** may be helpful.

(b) As indicated in **Figure SA6.1.3**, the entire moment created by the woman has to be held by the connector pins and stop-pins. Consider the chair with the dimensions in **Figures SA6.1.2**. Draw free-body diagrams of the: (1) seat with hinge bracket *A* and (2) hinge bracket *B* with 90° slot. The dimensions in **Figure SA6.1.4** will be useful. Remember that the 90° slot is milled into bracket *B* while the stop-pin is welded to bracket *A* at a distance of 3 cm from *C*, the center of rotation at the connector pin.

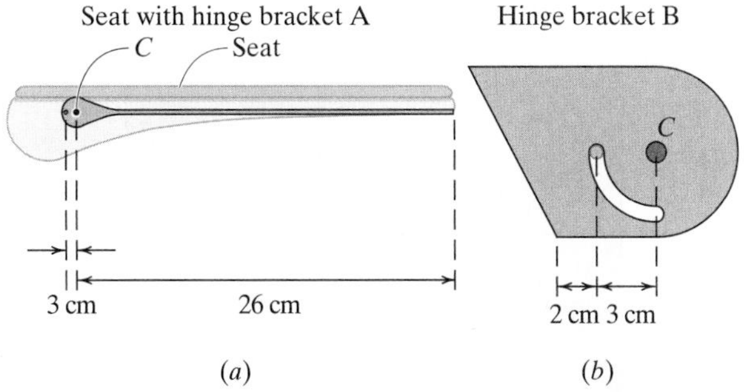

Figure SA6.1.4 Dimensions associated with (*a*) seat with hinge bracket *A*; (*b*) hinge bracket *B* with 90° slot

(c) Based on the free-body diagram of the seat with hinge bracket A created in (b), write expressions for the equivalent load (consisting of an expression for the equivalent force and an expression for the equivalent moment) acting at a moment center at C. If the magnitude of the expression for equivalent moment is zero, what force must the stop-pin apply to the seat plate?

(d) Figure SA6.1.5 provides a plot that shows the maximum allowable force that the stop-pin can safely hold. What size pin is needed to ensure that the woman sitting on the chair is safe? (*Remember:* Each chair has two hinges and stop-pins.)

(e) Assume that the maximum mass of a person for which each chair should be designed is 100 kg. What pin size would you recommend?

(f) Finally, it is very likely that in the process of sitting down, a person might actually "fall" into the chair. Since you did not cover the dynamic effect yet, let's assume you should triple the static force to account for the person's deceleration. What would your final recommendation be to the seat manufacturer regarding the pin size? (*Hint:* You may have to extend the plot by identifying the function that underlies the curve, which depends on the cross-sectional area of the pin and a fixed force per cm^2.)

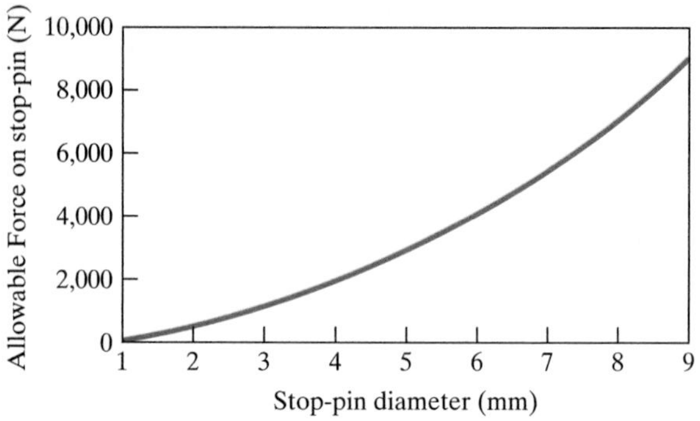

Figure SA6.1.5 Shear strength versus pin diameter

SA6.2 Following the Path of the Gravitational Force

As you are leaving the coliseum after the game, you recognize still another mechanical "beauty" in a corner—a large trash cart ready for action. Here is a picture of it and its dimensions (Figure SA6.2.1).

We can safely assume that the cart would be able to carry a total of 150 kg.

Part I: Lifting a load

(a) *Situation 1:* Draw a free-body diagram for the situation in which the janitor is just starting to dump a full cart by moving the handle on the back of the cart upward. The center of gravity for the full load distributed is shown in Figure SA6.2.2. If the equivalent moment about a moment center at C (the contact point between the front wheels and the ground) is zero, what is the magnitude of the force that the janitor must apply to the bin when the rear wheels just lift off the ground? State any assumptions you make.

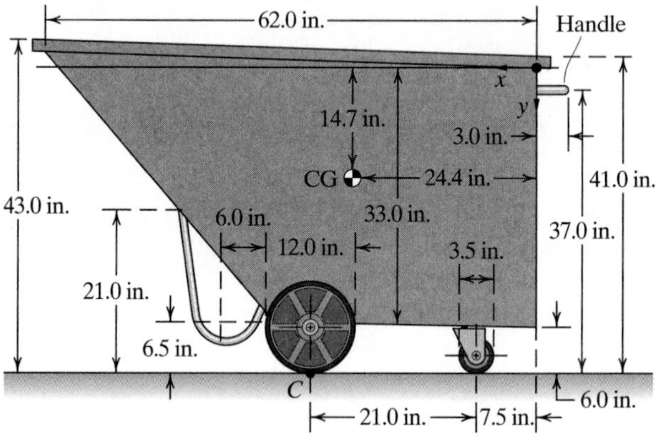

Figure SA6.2.2 Dimensions of trash cart load distributions

Figure SA6.2.1 Trash cart

(b) *Situation 2:* Instead of the load shown in **Figure SA6.2.1**, the cart is loaded with several heavy concrete pieces with a total mass of 150 kg that are placed close to the handle at position A; see Figure SA6.2.3*a*. Draw a free-body diagram for the situation in which the janitor is just starting to dump the cart by moving the handle on the back of the cart upward. If the equivalent moment about a moment center at C is zero, what is the magnitude of the force that the janitor must apply to the bin when the rear wheels just lift off the ground? State any assumptions you make.

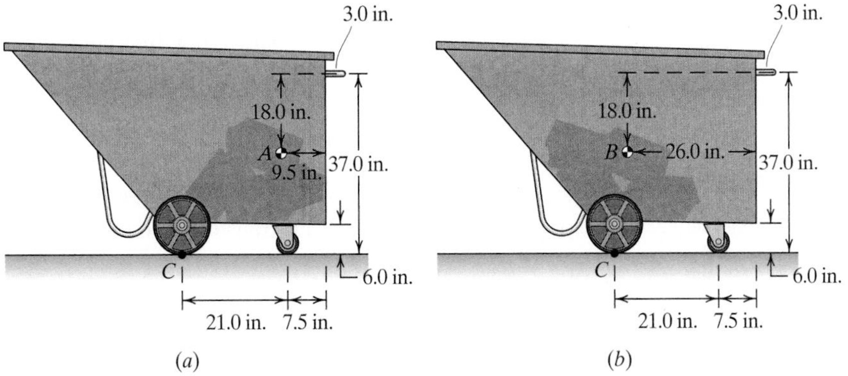

Figure SA6.2.3 The cart is carrying 150-kg mass at (*a*) Position *A*; (*b*) Position *B*

(c) *Situation 3:* If the heavy concrete pieces with a total mass of 150 kg are placed at *B* (**Figure SA6.2.3*b***), what force must the janitor apply to just lift the rear wheels off the ground? Should the janitor worry more about hurting his or her back in Situation 1, 2, or 3, and why?

(d) Will the magnitude of the force that the janitor is required to apply to the cart to move it from Position 1 to Position 2 (see **Figure SA6.2.4**) decrease, increase, or remain the same as the cart goes from Position 1 to Position 2? Include the rationale for your answer (no calculations are required).

Part II: How the load is transferred to the ground

You notice that the cart has large front wheels and a sturdy main axle that connects the wheels to one another. The main axle is attached to a trash bin as shown in **Figure SA6.2.5**. Because the axle is attached to each of the front wheels with a bearing, the wheels rotate and the axle does not. Let's consider how the weight of the trash contained in the bin is transferred to the ground. To do this we will create a series of free-body diagrams.

Consider the following cross-sectional view of the trash bin in **Figure SA6.2.6** with various components labeled.

(a) Draw a free-body diagram of the system defined by Boundary 1 (the cart and the trash it is holding).

(b) Draw a free-body diagram of the system defined by Boundary 2 (the trash).

(c) Draw a free-body diagram of the main axle and wheel assembly.

(d) Now we are ready to separate the main axle from the wheels by pushing the shaft out of the bearing wheel hub. **Figure SA6.2.7** shows details of the axle–wheel

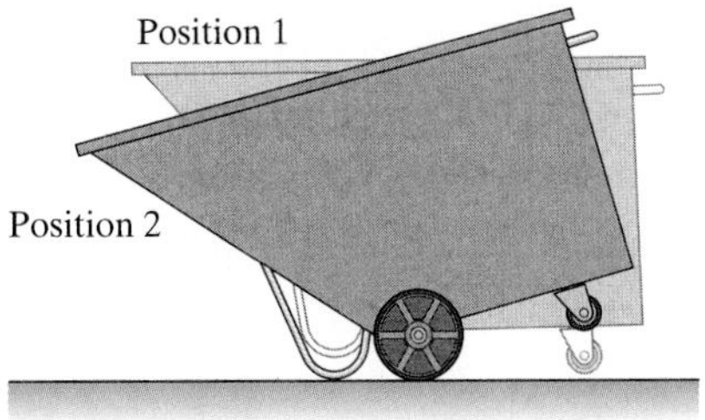

Figure SA6.2.4 From Position 1 to Position 2

Figure SA6.2.5 Cart suspension system

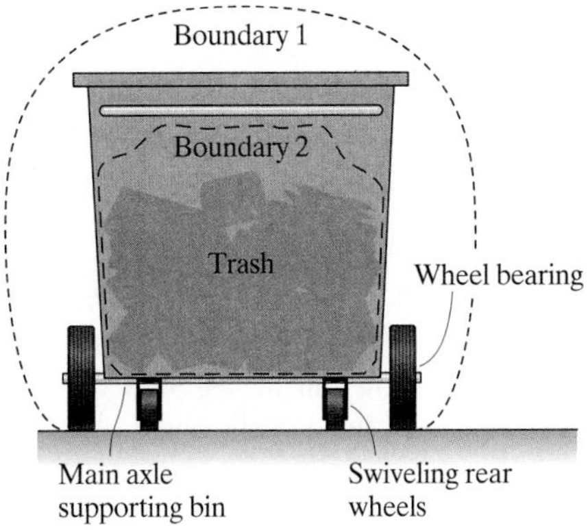

Figure SA6.2.6 Cross section through cart

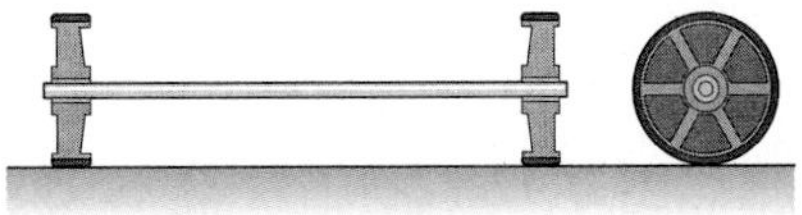

Figure SA6.2.7 Axle and wheels

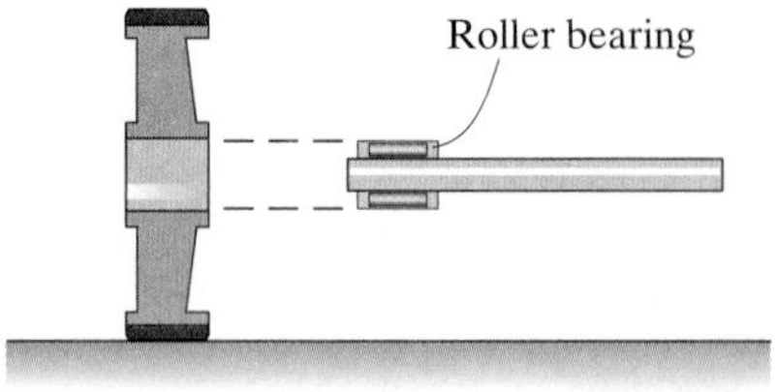

Figure SA6.2.8 Details of axle–wheel connection

connection and the bearings that connect the axle and the wheels. Draw free-body diagrams for the wheels and the axle after they are separated from one another. **Figure SA6.2.8** may be useful in visualizing this.

(e) Based on the free-body diagrams you created in (a)–(d), which of the schematics shown in **Figure SA6.2.9** most accurately depicts how the weight of the trash is transferred to the ground?

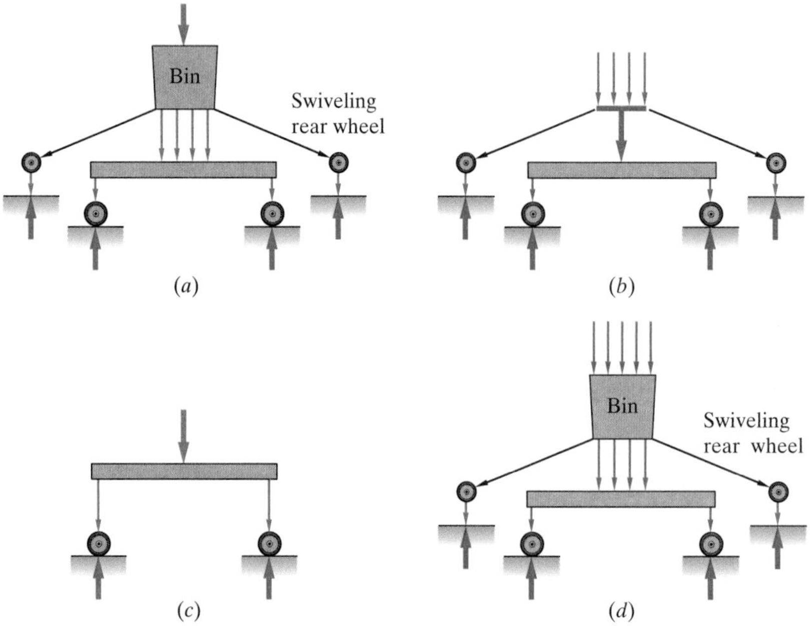

Figure SA6.2.9 Load path schematics

SA6.3 Free-Body Diagram Based on Experimental Evidence

Perform the experiment described below, then follow the steps to create a free-body diagram of the situation.

Materials needed: One wire clothes hanger, rubber band, paper clip, a weight (a candy bar is suggested).

Experiment: Configure the hanger, rubber band, paper clip, and weight as shown in **Figure SA6.3.1**. Using both hands, keep the hanger level. Notice how your hands push or pull on the wire.

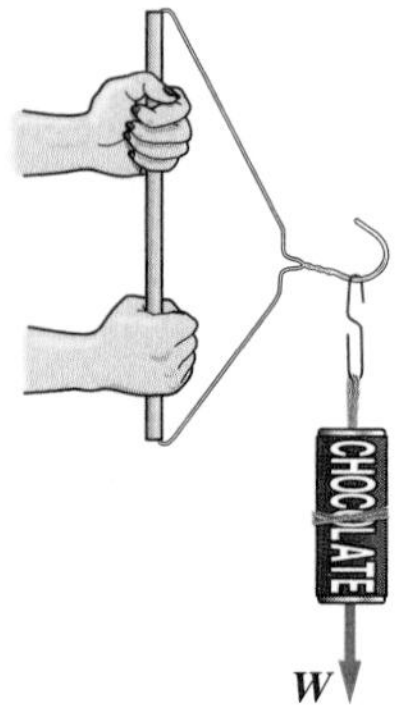

Figure SA6.3.1 Configure hanger and candy bar as shown

(a) Consider the hanger to be your system. **Draw** it. Add a coordinate system such that the y axis is aligned with gravity.

(b) **List** (in words) the forces acting on the hanger when in the configuration shown.

(c) **Draw** each of the forces listed in (b) as a vector on the drawing created in (a). Clearly mark the points of application of each force and add variable labels. If the magnitudes of any of the forces are known, include this information on the drawing. You have now created a free-body diagram of the hanger. Make sure to **list any assumptions** you made and any uncertainties you have.

(d) Study your free-body diagram and **identify any couples** (describe in words).

SA6.4 Free-Body Diagram Based on Experimental Evidence with a Partner

Perform the experiment described below, then follow the steps to create a free-body diagram of the situation.

Materials needed: Two wire clothes hangers, a friend.

Experiment: Hook the two clothes hangers together, as shown in **Figure SA6.4.1**. Have your friend push and pull on her hanger, as shown. At the same time, use your hands (in the positions shown) to keep the hanger level. Notice how your hands and those of your friend push and/or pull on the wire.

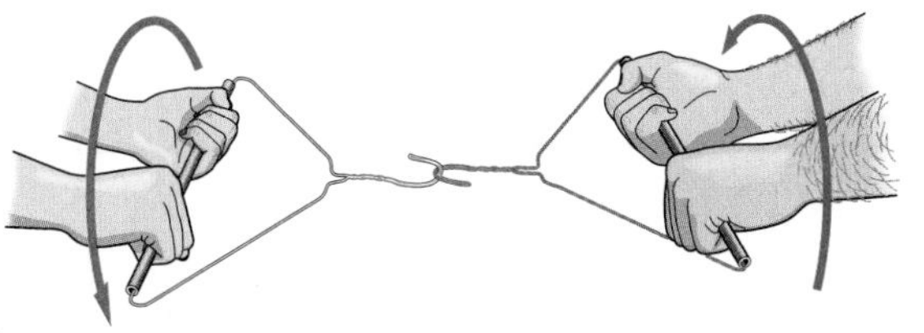

Figure SA6.4.1 Configure two hangers as shown

(a) Consider the two hangers to be your system. **Draw** the structure. Add a coordinate system such that the y axis is aligned with gravity.

(b) **List** (in words) the forces acting on the system.

(c) **Draw** each of the forces listed in (b) as a vector on the drawing created in (a). Clearly mark the points of application of each force and add variable labels. If the magnitudes of any of the forces are known, include this information on the drawing. You have now created a free-body diagram of the system. Make sure to **list any assumptions** you made and any uncertainties you have.

(d) Study your free-body diagram and **identify any couples** (describe in words).

(e) Repeat (a)–(d) if the system is defined as one of the hangers.

SA6.5 The Bicycle Revisited

Consider the bottom bracket assembly of a bicycle, as shown in **Figure SA6.5.1**. The assembly consists of an axle, chain ring, left and right cranks, and left and right pedals. The axle is held in the frame by two sets of ball bearings. For the position shown, draw a free-body diagram of the bottom bracket assembly.

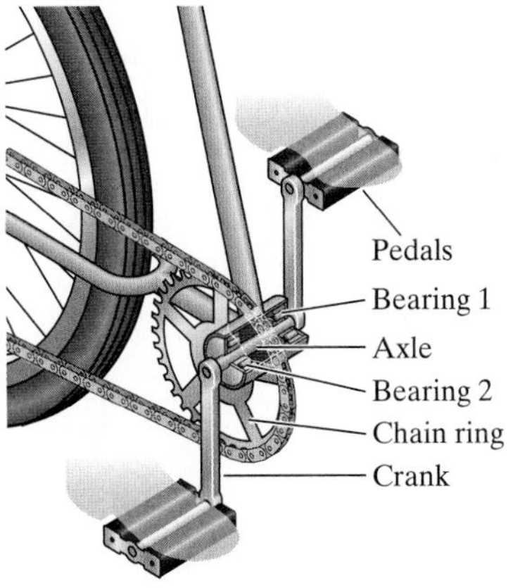

Figure SA6.5.1 Bottom bracket assembly

CHAPTER 7

MECHANICAL EQUILIBRIUM

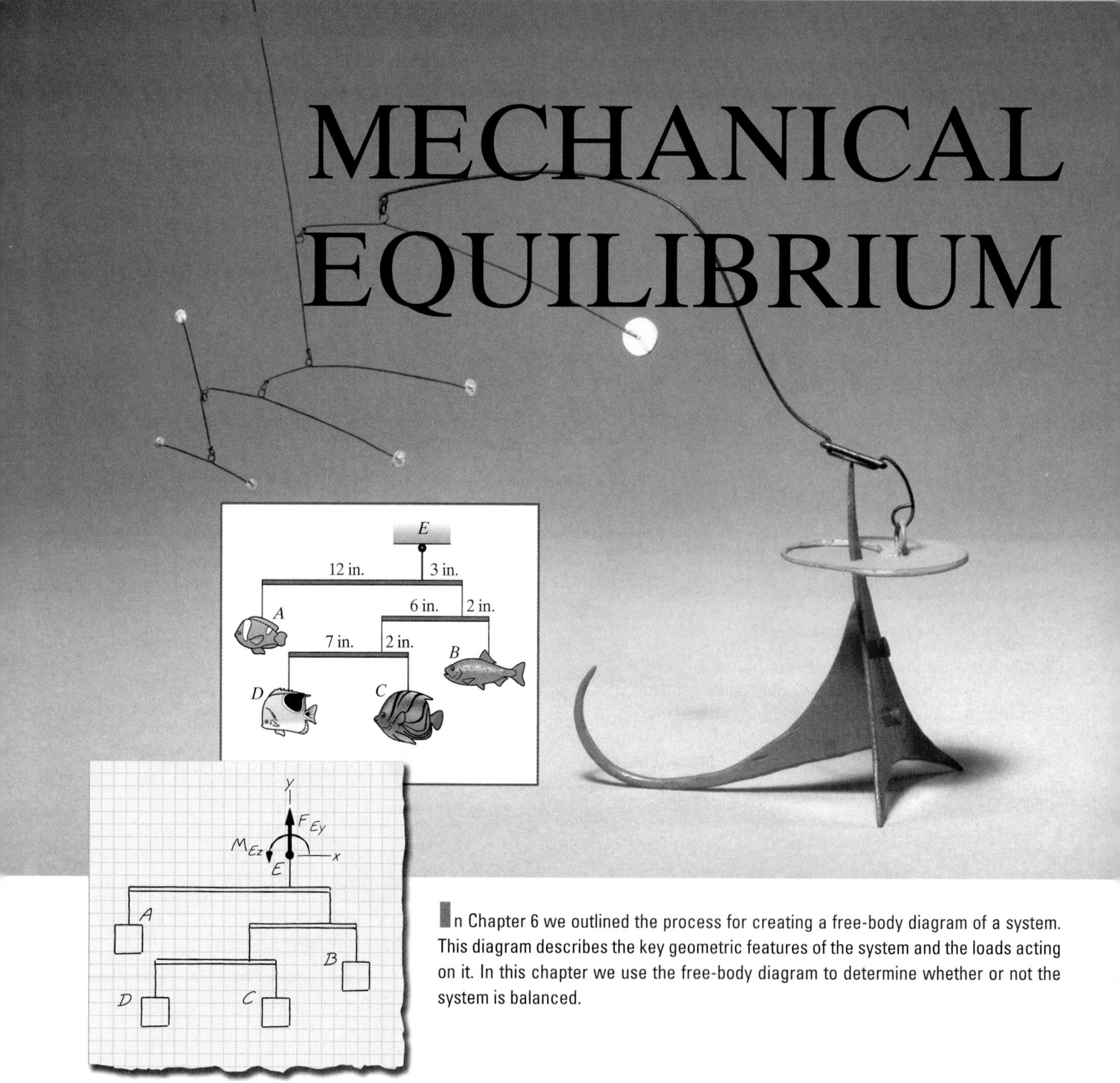

In Chapter 6 we outlined the process for creating a free-body diagram of a system. This diagram describes the key geometric features of the system and the loads acting on it. In this chapter we use the free-body diagram to determine whether or not the system is balanced.

OBJECTIVES

Upon completion of this chapter, you will be able to:

- **Describe the conditions of equilibrium and their associated component equations**
- **Use the conditions of equilibrium to determine the relationships between applied forces and moments acting on a system**
- **Carry out static analysis using the analysis procedure**
- **Define the unique characteristics of *particle, two-force element, three-force element,* and *frictionless pulley,* and be able to identify these elements in analysis**
- **Define *statically determinate, statically indeterminate,* and *underconstrained,* and be able to identify systems having these characteristics**

As an example, consider the system in **Figure 7.1*a*** and its associated free-body diagram (**Figure 7.1*b***). If this system is in **mechanical equilibrium** (where equilibrium is a state of balance), the loads acting on the system have particular relationships to one another. These relationships, called the **equilibrium conditions**, allow us to determine that the normal force from the floor pushing upward on the person in **Figure 7.1*b*** balances her weight and that the friction force balances the rope tension. This chapter develops the equilibrium conditions and then discusses how to apply them consistently in analyzing a system that is in mechanical equilibrium.

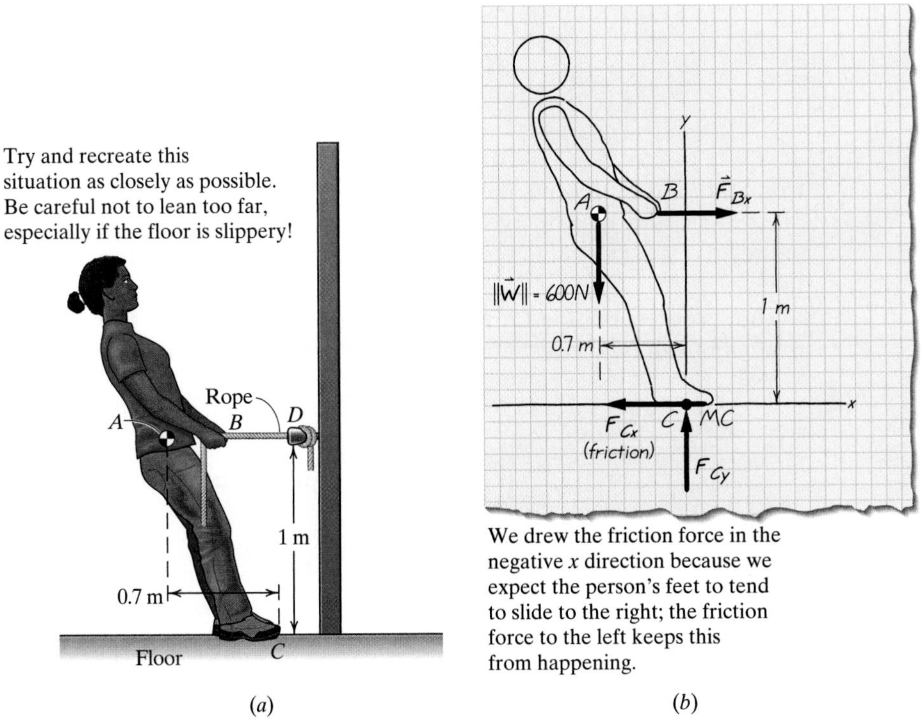

Try and recreate this situation as closely as possible. Be careful not to lean too far, especially if the floor is slippery!
Rope
A
B
D
1 m
0.7 m
Floor
C
y
B
$\vec{F}_{Bx}$
A
$\|\vec{W}\| = 600N$
1 m
0.7 m
C
MC
x
F_{Cx}
(friction)
F_{Cy}
We drew the friction force in the negative x direction because we expect the person's feet to tend to slide to the right; the friction force to the left keeps this from happening.
(a)
(b)

Figure 7.1 (*a*) Situation: person leans back, holding onto a rope. The end of the rope at *D* is tied to a door handle. The person weighs 600 N. *A* is the location of the person's center of gravity. *B* is where the rope if held. *C* is where the person's feet touch the ground. (*b*) Free-body diagram; moment center (MC) has been placed at *C*.

One reason engineers are interested in determining the loads acting on a system in equilibrium is to determine whether any of the loads exceed the capacity of components in the system. For instance, say the manufacturer of the rope in **Figure 7.1** specifies that the rope has a maximum capacity of 500 N (meaning that it breaks when a force exceeding 500 N acts on it). An engineer would use the conditions of equilibrium to determine the force in the rope when a person who weighs 600 N leans backward, then compare this force to the 500-N rope capacity. If the applied force is equal to or greater than the rope capacity, the engineer would warn the person not to attempt to recreate the situation in **Figure 7.1**, lest there be a serious backward tumble.

7.1 CONDITIONS OF MECHANICAL EQUILIBRIUM

A system that is in mechanical equilibrium is one that experiences zero linear acceleration and zero angular acceleration about any axis fixed in an inertial reference frame. With no acceleration, the system either does not move at all or, if it does move, has constant translational and angular velocities. For a system in equilibrium there are restrictions, or conditions, on the forces and moments that act on it. The first of these restrictions is that there is *no net force* acting on the system. This can be described mathematically as

$$\boldsymbol{F}_{\text{net}} = \sum^{\substack{\text{all forces}\\ \text{acting}\\ \text{on system}}} \boldsymbol{F} = 0 \qquad \textbf{force equilibrium condition} \tag{7.1}$$

This condition represents the vector sum of all the forces acting on the system and says that this vector sum is zero. It reflects Newton's first and second laws; a system that is not accelerating has no net force acting on it.

The second condition for a system to be in equilibrium is that there is *no net moment* on the system. This can be described mathematically as

$$\boldsymbol{M}_{\text{net}} = \sum^{\substack{\text{all moments}\\ \text{acting}\\ \text{on system}}} \boldsymbol{M} = 0 \qquad \textbf{moment equilibrium condition} \tag{7.2}$$

This condition represents the vector sum of the moments acting on the system and says that the vector sum of all moments is zero.

Conditions (7.1) and (7.2) are necessary and sufficient conditions for a system to be in mechanical equilibrium.

It is generally more convenient to work with conditions in (7.1) and (7.2) in terms of vector components. When we write each force in the summation in condition (7.1) in component form and then group the *x*, *y*, and *z* components, we can rewrite (7.1) as

$$\boldsymbol{F}_{\text{net}} = \sum^{\substack{\text{all forces}\\ \text{acting}\\ \text{on system}}} \boldsymbol{F} = \left(\sum F_x\right)\boldsymbol{i} + \left(\sum F_y\right)\boldsymbol{j} + \left(\sum F_z\right)\boldsymbol{k} = 0$$

where F_x, F_y, and F_z represent the force scalar components in the x, y, and z directions of an overall coordinate system, respectively. For this component representation of $\boldsymbol{F}_{\text{net}}$ to be zero, each component must be zero. In other words:

$$\sum F_x = 0 \qquad (7.3A)$$
$$\sum F_y = 0 \qquad (7.3B)$$
$$\sum F_z = 0 \qquad (7.3C)$$

These are the **force equilibrium equations**. Each equation states that if we add up all the force components acting on a system in a particular direction, the sum must be zero if there is equilibrium. Equations (7.3A, B, C) comprise the requirements for **force balance**.

Similarly, we can write each moment $\boldsymbol{M}$ in the summation in condition (7.2) in terms of its vector components and rewrite (7.2) as

$$\boldsymbol{M}_{\text{net}} = \sum^{\substack{\text{all moments}\\ \text{acting}\\ \text{on system}}} \boldsymbol{M} = \left(\sum M_x\right)\boldsymbol{i} + \left(\sum M_y\right)\boldsymbol{j} + \left(\sum M_z\right)\boldsymbol{k} = 0$$

where M_x, M_y, and M_z represent the individual moment components about the x, y, and z axes of an overall coordinate system, respectively. For this component representation of $\boldsymbol{M}_{\text{net}}$ to be zero, each of its components must be zero. In other words:

$$\sum M_x = 0 \qquad (7.4A)$$
$$\sum M_y = 0 \qquad (7.4B)$$
$$\sum M_z = 0 \qquad (7.4C)$$

These are the **moment equilibrium equations**. Each equation states that if we add up all of the moments about a particular axis, their sum must be zero if there is equilibrium. Equations (7.4A, B, C) are the three equilibrium conditions for **moment balance**.

The six equations in (7.3) and (7.4) are true for any system in mechanical equilibrium (i.e., a balanced system). We shall refer to them as the **six equilibrium equations**. If a system is in equilibrium, the loads acting on it are such that the six equilibrium equations are satisfied—or, in other words, the conditions are met. We can solve the equations for the unknowns they contain, as illustrated in Example 7.1 for the leaning person in **Figure 7.1**. Alternately, we can work directly with equilibrium conditions in (7.1) and (7.2) to ensure that the vector sums of forces and moments acting on the system are zero.

EXAMPLE 7.1 LEANING PERSON

If the rope that the person in **Figure 7.1*a*** is using to lean back with will break if its tension exceeds 500 N, should the person be concerned about it breaking?

Find We are to find the tension in the rope for the configuration shown in **Figure 7.1*a***, then compare this tension to 500 N. If it exceeds 500 N, the rope exceeds the design capacity and may break.

Given We are given the dimensions associated with the situation, along with the weight of the person and the location of her center of gravity (point A).

Assume We assume that the rope is taut and horizontal, and that there is sufficient friction between the person's feet and the floor for there to be no sliding. We also assume that the person is in equilibrium.

Draw The free-body diagram of the system (taken as the person) is shown in **Figure 7.1*b***.

Formulate Equations **(a)** We set up the force equilibrium equations based on the information in the free-body diagram. Force equilibrium:

y, x (z axis is out of the paper)

$$\left.\begin{aligned}\sum F_x &= 0 = F_{Bx} - F_{Cx} = 0; \text{ based on (7.3A)}\\ \sum F_y &= 0 = F_{Cy} - W = 0; \quad \text{based on (7.3B)}\\ &= F_{Cy} - 600\ N = 0\\ \sum F_z &= 0 = F_{Cz} = 0; \qquad \text{based on (7.3C)}\end{aligned}\right\} \text{see Figure 7.1}b$$

(b) Next we set up the moment equilibrium equations. We start by establishing a moment center (MC). The moment center could be placed anywhere, but placing it at a location where lines of action of forces intersect at C make the algebra easier. Moment equilibrium with moment center at C is:

$$\sum M_{x@C} = 0; \text{ based on (7.4A)}$$
(none of the forces creates a moment about the x axis)

$$\sum M_{y@C} = 0; \text{ based on (7.4B)}$$
(none of the forces creates a moment about the y axis)

$$\sum M_{z@C} = 0 = -(1\text{ m})(F_{Bx}) + (0.70\text{ m})(\underbrace{W}_{600\text{ N}}); \text{ based on (7.4C)}$$

$$= -(1\text{ m})(F_{Bx}) + 420\text{ N}\cdot\text{m} = 0$$

Solve The equations in (a) and (b) for the unknowns:

From (7.3B), $F_{Cy} = 600$ N
From (7.3C), $F_{Cz} = 0$ N
From (7.4C), $F_{Bx} = 420$ N
From (7.3A), $F_{Cx} = 420$ N

The tension in the rope is 420 N when a person weighing 600 N leans backward in the configuration shown. And the further back she leans, the greater the tension. This is less than the 500-N breaking force of the rope.

Answer 420 N (tension in rope) $<$ 500 N (tension that will break the rope). Therefore the person should not be concerned that the rope will break under the given conditions.

Check We add the forces found above to the free-body diagram of the person in mechanical equilibrium, as shown in Figure 7.2. Notice that $\boldsymbol{F}_{Bx}$ and $\boldsymbol{F}_{Cx}$ form a couple that is countered by the couple created by $\boldsymbol{W}$ and $\boldsymbol{F}_{Cy}$. This is a useful visual check of the calculation.

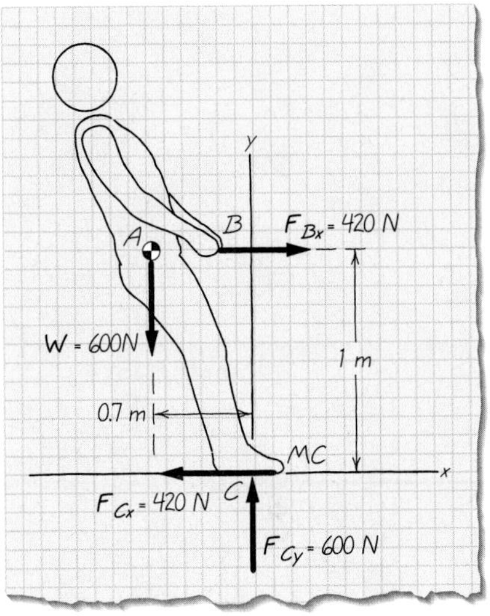

Figure 7.2 Free-body diagram of person, including calculated forces

7.2 APPLICATION OF THE CONDITIONS—PLANAR SYSTEMS

The Basic Equations

Any system that is in equilibrium must satisfy all six equilibrium equations. Three of these equations, however, are automatically satisfied for a planar system. In Chapter 6 a planar system was defined as one where all the forces and moments involved are contained in the same plane. For example, consider the system consisting of triangular frame ABC in Figure 7.3*a*, with its associated free-body diagram in **Figure 7.3*b***. Notice that there are no forces in the z direction in this system—therefore, (7.3C) is automatically satisfied for this planar system. Furthermore, because all the forces are contained in the xy plane, they create no moments about the x or y axis—therefore, (7.4A) and (7.4B) are also automatically satisfied.

We must still confirm that the loads acting on the system satisfy the three remaining equilibrium equations:

$$\sum F_x = 0 \qquad (7.5A)$$
$$\sum F_y = 0 \qquad (7.5B)$$
$$\sum M_z = 0 \qquad (7.5C)$$

We refer to these three equations as the **planar equilibrium equations**.

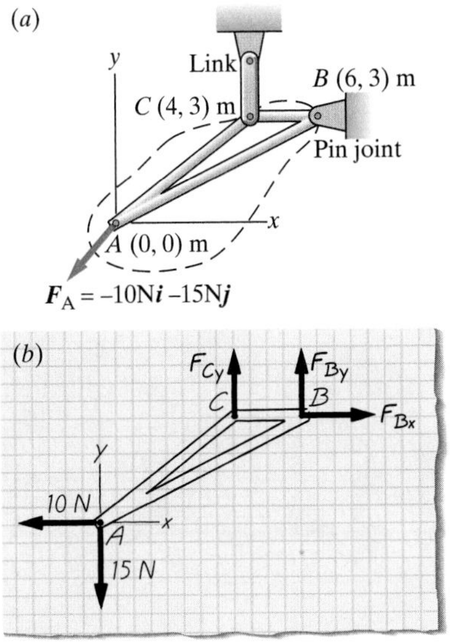

Figure 7.3 (*a*) Frame ABC; (*b*) free-body diagram of frame ABC

The Analysis Procedure Revisited

Now we consider how these planar equilibrium equations fit into the analysis procedure. Recall that a free-body diagram, created as part of the **Draw** step of our analysis procedure, shows the loads acting on a system as well as key dimensions and establishes a Cartesian coordinate system. The free-body diagram serves as the starting point for using the equilibrium equations to determine the magnitudes and/or directions of loads acting on a balanced system. The **Draw** step is followed by the **Formulate Equations**, **Solve**, and **Check** steps, as we discuss next.

Consider the situation and free-body diagram in **Figure 7.4*a***. The weight W of the block is known and we wish to determine the loads acting on the block at A and C.

Formulate Equations. Reading force information from the free-body diagram, write the x component force equilibrium equation (7.5A). Make sure to include all x component forces. Repeat for the y component force equilibrium equation (7.5B). These two equations may contain unknown forces, represented by variable labels:

$$\left.\begin{aligned}\sum F_x &= 0 = F_{Ax} - F_C \sin 60^\circ = 0; && \text{based on (7.5A)}\\ \sum F_y &= 0 = F_{Ay} - W + F_C \cos 60^\circ = 0; && \text{based on (7.5B)}\end{aligned}\right\rangle \quad \textbf{Figure 7.4}\boldsymbol{a}$$

Notice, for example, that in the second term on the right-hand side of (7.5A) we have written $-F_C \sin 60^\circ$. This term is the component of $\boldsymbol{F}_C$ in the x direction. The negative sign indicates that this component is in the negative x direction (based on how $\boldsymbol{F}_C$ is drawn in **Figure 7.4*a***). This point is important when it comes to interpreting answers.

Now write the z component moment equation (7.5C) by first deciding which point to use as the moment center. By choosing a point through which force lines of action pass, you reduce the number of simultaneous equations you deal with in the **Solve** step (because a force does not create a moment at a point that its line of action passes through). Be sure to include all moments about the z axis created by forces and by moments. The equation you write may contain unknowns (moments, forces, and/or dimensions), represented by variable labels.

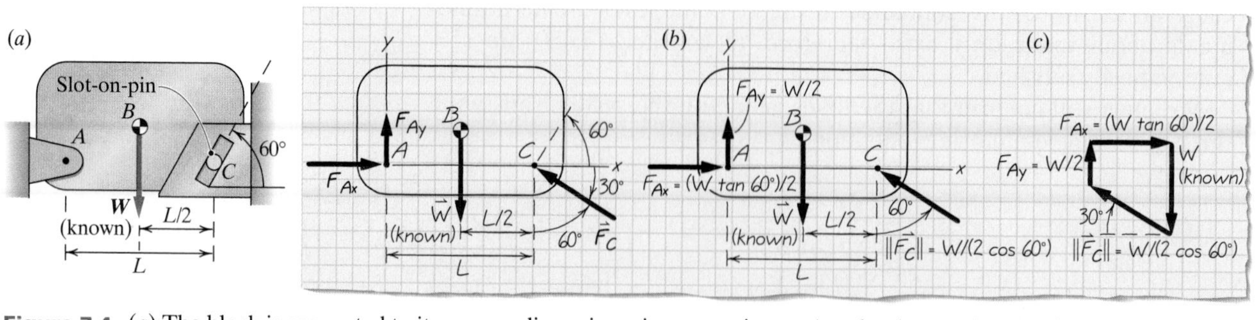

Figure 7.4 (a) The block is connected to its surroundings via a pin connection at A and a slot on pin at C; (b) the associated free-body diagram including calculated forces; (c) checking the solution with graphical force addition

For the situation in **Figure 7.4*a*** we choose the moment center to be at A. Therefore (7.5C) becomes

$$\sum M_{z@A} = 0 = +(L)(F_C \cos 60^\circ) - \left(\frac{L}{2}\right)(W) = 0; \text{ based on (7.5C)}$$

Solve. You now have three equations of equilibrium that can be solved for (at most) three unknowns. Furthermore, these three equations are independent of one another.[1] Most of the systems presented in this book are such that the equilibrium equations can reasonably be solved by hand for the unknowns. (However, if you have the opportunity to learn to use a numerical software package such as MATLAB, we urge you to do so; solving more complex and realistic problems by setting them up in matrix form and using a computer-based solver makes solving a lot easier.)

From the three equations (7.5A), (7.5B), and (7.5C) written based on **Figure 7.4*a***, we find that

Answer

$$F_C = \frac{W}{2 \cos 60^\circ}$$

$$F_{Ax} = \frac{W \tan 60^\circ}{2}$$

$$F_{Ay} = \frac{W}{2}$$

In presenting answers be sure to make them stand out, so that someone reviewing your work can readily identify them.

These values can also be presented as part of the free-body diagram in **Figure 7.4*b***.

Check. Just solving equations and getting numbers is not enough. The results should be checked using technical knowledge, engineering judgment, and common sense. Do the results match your expectations? Verify that the results make sense by looking at the order of magnitude, units, and directions of loads. A negative answer for a load value means that the direction in which the load acts is opposite the direction depicted in the free-body diagram. Also, check your algebra by substituting numerical answers into the equilibrium equations to verify that the equations are satisfied. Still another option for checking your results is to select an alternate moment center and write the moment equilibrium equation, substituting in your numerical answers. Finally, make sure the answers are clearly communicated in the write-up. Compare values with design requirements (if available) and draw conclusions about the adequacy of the design.

[1]Independent equations are such that none of the equations can be derived by scaling and/or combining the other equations. Furthermore, n independent equations written in terms of n variables can generally be solved for values of the n variables. Linearly independent equations are such that all variables are written to the first power. For example, the three linearly independent equations (7.5A), (7.5B), and (7.5C) are written in terms of variables F_{Ax}, F_{Ay}, and F_C.

For the situation in **Figure 7.4*a***, we check that we have correctly calculated the values of F_{Ax}, F_{Ay}, and F_C by selecting a moment center at C, writing out (7.5C), and substituting in the calculated values:

$$\sum M_{z@C} = 0 = -(L)(F_{Ay}) + \left(\frac{L}{2}\right)(W); \text{ based on (7.5C)}$$
$$= -(L)\left(\frac{W}{2}\right) + \left(\frac{L}{2}\right)(W) = 0$$

Indeed, the calculated values give us zero moment at a moment center at C. An alternate check is also presented in **Figure 7.4*c*** that involves application of force equilibrium condition (7.1) using graphical force addition.

EXAMPLE 7.2 CANTILEVER BEAM WITH TWO FORCES AND A MOMENT

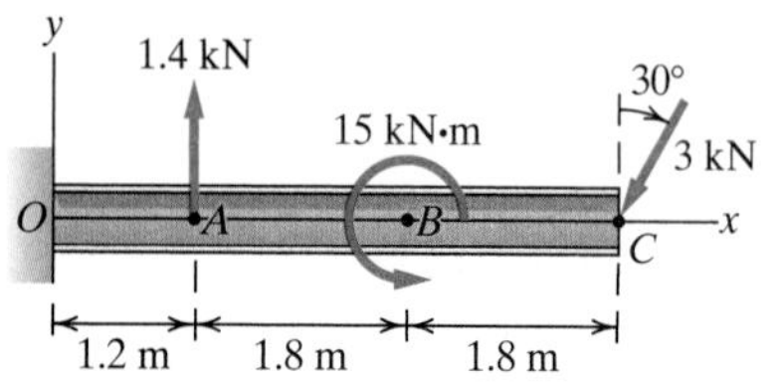

Figure 7.5

A 500-kg uniform beam subjected to the loads shown in **Figure 7.5** is in mechanical equilibrium. Find the loads acting on the beam at support O for the given configuration.

Goal We are to find the loads acting on the beam at O for a beam that is in mechanical equilibrium.

Given We are given information about the geometry and mass of the beam and its mass distribution, as well as the size of the loads acting at A, B, and C.

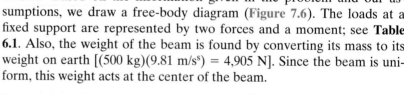

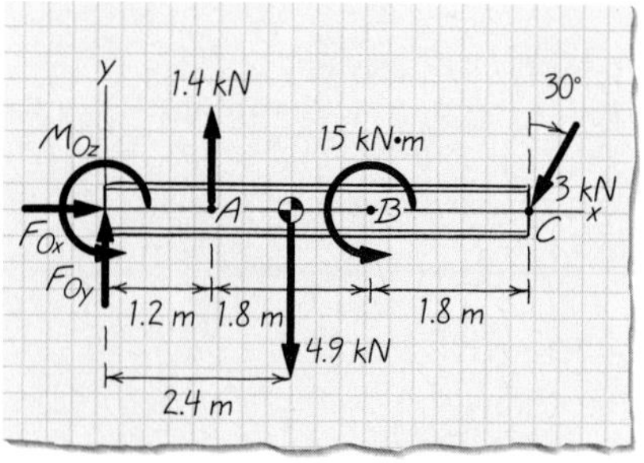

Figure 7.6

Assume We will assume that the support at O is fixed and that we can treat the system (the beam) as planar. Also, the beam is on earth, oriented so that gravity acts in the $-y$ direction.

Draw Based on the information given in the problem and our assumptions, we draw a free-body diagram (**Figure 7.6**). The loads at a fixed support are represented by two forces and a moment; see **Table 6.1**. Also, the weight of the beam is found by converting its mass to its weight on earth [(500 kg)(9.81 m/s^s) = 4,905 N]. Since the beam is uniform, this weight acts at the center of the beam.

Formulate Equations and Solve We set up the equations for planar equilibrium to find the unknown loads at O:

Based on (7.5A):

$$\sum F_x = 0 \; (\rightarrow +)$$
$$F_{Ox} - F_{Cx} = 0$$
$$F_{Ox} - (3 \text{ kN}) \sin 30° = 0$$
$$F_{Ox} = (3 \text{ kN}) \sin 30°$$

$$F_{Ox} = 1.5 \text{ kN}$$

Based on (7.5B):

$$\sum F_y = 0\ (\uparrow +)$$
$$F_{Oy} + 1.4\ \text{kN} - W - F_{Cy} = 0$$
$$F_{Oy} = -1.4\ \text{kN} + 4.91\ \text{kN} + (3\ \text{kN})\cos 30^\circ$$

$$F_{Oy} = 6.1\ \text{kN}$$

Based on (7.5C), with the moment center at O:

$$\sum M_{z@O} = 0\ (\curvearrowleft +)$$
$$M_{Oz} + (1.2\ \text{m})(1.4\ \text{kN}) - (2.4\ \text{m})(4.9\ \text{kN}) + 15\ \text{kN}\cdot\text{m} - (4.8\ \text{m})F_{Cy} = 0$$
$$M_{Oz} = -1.68\ \text{kN}\cdot\text{m} + 11.78\ \text{kN}\cdot\text{m} - 15\ \text{kN}\cdot\text{m} + (4.8\ \text{m})(3\ \text{kN}\cos 30^\circ)$$

$$M_{Oz} = 7.6\ \text{kN}\cdot\text{m}$$

Answer $F_{Ox} = 1.5\ \text{kN} \rightarrow,\ F_{Oy} = 6.1\ \text{kN}\uparrow$
$M_{Oz} = 7.6\ \text{kN}\cdot\text{m} \curvearrowleft$

Check The directions and magnitudes of the loads at the fixed connection are as expected. F_{Ox} balances F_{Cx}. All other forces are acting in the y direction. As an additional check, (7.5C) could be written with the results above for an alternate moment center.

EXAMPLE 7.3 A SIMPLE STRUCTURE

The structure shown in Figure 7.7 is pinned to its surroundings at A and rests against a roller at B. Force $\boldsymbol{P}$ acts at C. The weight of the structure can be ignored because it is much smaller than $\boldsymbol{P}$. Find the loads acting at supports A and B as a function of the geometry and loading.

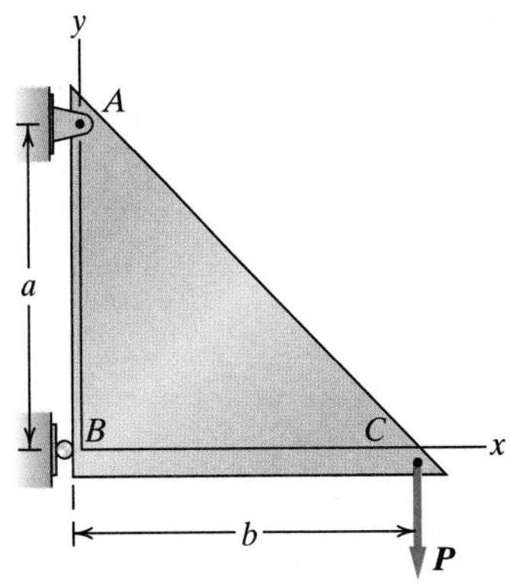

Figure 7.7

Goal We are to find the reactions (loads) at A and B as a function of geometry (dimensions a and b) and force ($\boldsymbol{P}$).

Given We are given information about the geometry of the structure, and are told that A is a pin connection and B is a roller. We are also told that the weight of the structure is negligible compared to $\boldsymbol{P}$.

Assume We will assume that we can treat the system as planar.

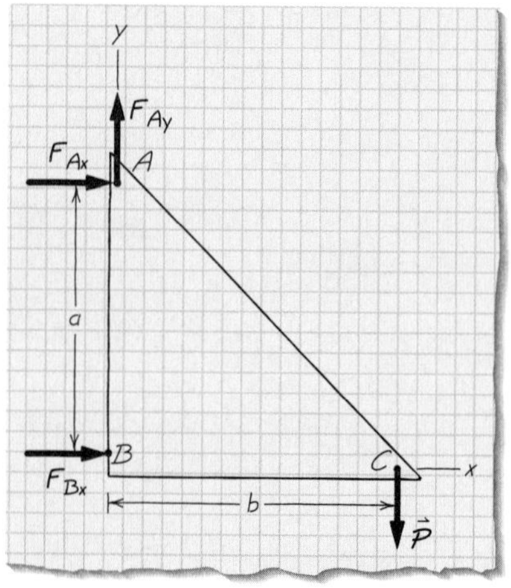

Figure 7.8 Free-body diagram of structure

Draw Based on the information given in the problem and our assumption, we draw a free-body diagram and arbitrarily draw the unknown loads in the directions of the positive axes (**Figure 7.8**). (At the pin connection the loads are two forces, and at the roller a force pushes on the system; see **Table 6.1**.)

Formulate Equations and Solve We set up the equations for planar equilibrium to find the unknown loads at A and B.

Based on (7.5A):

$$\sum F_x = 0\ (\rightarrow +)$$
$$F_{Ax} + F_{Bx} = 0$$
$$F_{Ax} = -F_{Bx} \qquad (1)$$

The minus sign in equation (1) means F_{Ax} will be in the opposite direction of F_{Bx}.

Based on (7.5B):

$$\sum F_y = 0\ (\uparrow +)$$
$$F_{Ay} - P = 0$$
$$F_{Ay} = P \qquad (2)$$

Based on (7.5C), with the moment center at A:

$$\sum M_{z@A} = 0\ (\circlearrowleft +)$$
$$aF_{Bx} - bP = 0$$
$$F_{Bx} = \frac{b}{a}P \qquad (3)$$

Substituting (3) into (1), we find that

$$F_{Ax} = -\frac{b}{a}P.$$

We show three possible ways the answers could be presented.

Answer:

(Approach 1: Scalar components)
$F_{Ax} = -\frac{b}{a}P$ (the minus sign means that it acts in the direction opposite to that defined in **Figure 7.8**)
$F_{Ay} = P$
$F_{Bx} = \frac{b}{a}P$

(Approach 2: Vector notation)
The loads acting on the structure are:
@ A; $\boldsymbol{F}_A = -\frac{b}{a}P\boldsymbol{i} + P\boldsymbol{j}$
@ B; $\boldsymbol{F}_B = \frac{b}{a}P\boldsymbol{i}$

(Approach 3: Narrative description)
The loads acting on the structure:
@ A; a force with a component of $-\frac{b}{a}P$ in the x direction and a component of P in the y direction.
@ B; a force with a component of $\frac{b}{a}P$ in the x direction.

Check Notice that F_{Ay} balances P, and F_{Ax} balances F_{Bx}. Also, F_{Ay} and P form a clockwise couple that balances the counterclockwise couple created by F_{Ax} and F_{Bx}. We can also check that dimension b in the numerators of F_{Ax} and F_{Bx} makes sense; as b increases, the magnitude of the couple created by F_{Ay} and P increases. If dimension a is fixed, the magnitudes of F_{Ax} and F_{Bx} must increase to balance this couple. A similar argument can be made for dimension a being in the denominator.

Four Particularly Important Elements

We conclude this section by highlighting four commonly found elements: the *particle*, the *two-force element*, the *three-force element*, and the *frictionless pulley*. By defining each of these elements as "the system" and applying the equilibrium equations we gain insights that can simplify our static analysis more generally.

A **particle** is an object whose size and shape have negligible effect on the response of the object to forces. Under these circumstances, the mass (if significant) of the object can be assumed to be concentrated at a point. A particle, by definition, can only be subjected to concurrent forces; the point of concurrency is the point that represents the particle, as illustrated in **Figure 7.9**. Recall from Chapter 4 that concurrent forces are any number of forces such that their lines of action intersect at a single point. Therefore, if an object is modeled as a particle, we need only consider the force equilibrium condition since the moment equilibrium condition is automatically satisfied. Example 7.4 illustrates the utility of being able to identify objects as particles when performing static analysis.

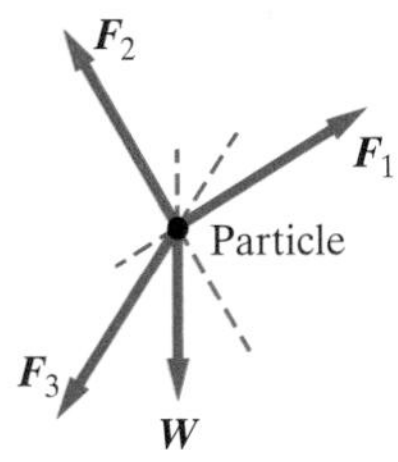

Figure 7.9 A particle is such that all forces acting on it are concurrent

A **two-force element** is an element of negligible weight with only two forces acting on it. An example of a two-force element is illustrated in **Figure 7.10*a***. A force (with two components) acts at each of the pin connections, as shown in the free-body diagram in **Figure 7.10*b***. Now we apply the equations of equilibrium (7.5) to this system:

$$\sum F_x = 0 = F_{Ax} + F_{Bx} = 0; \quad \text{based on (7.5A)}$$
$$\sum F_y = 0 = F_{Ay} + F_{By} = 0; \quad \text{based on (7.5B)}$$
$$\sum M_{z@A} = 0 = (L)(F_{By}) = 0; \quad \text{based on (7.5C)}$$

Solving these three equations, we find:

$$F_{Ax} = -F_{Bx}$$
$$F_{Ay} = F_{By} = 0$$

These findings mean that *a two-force element is in equilibrium when the forces acting on it are equal, opposite, and along the same line of action. The line of action passes through the force points of application at A and B.*

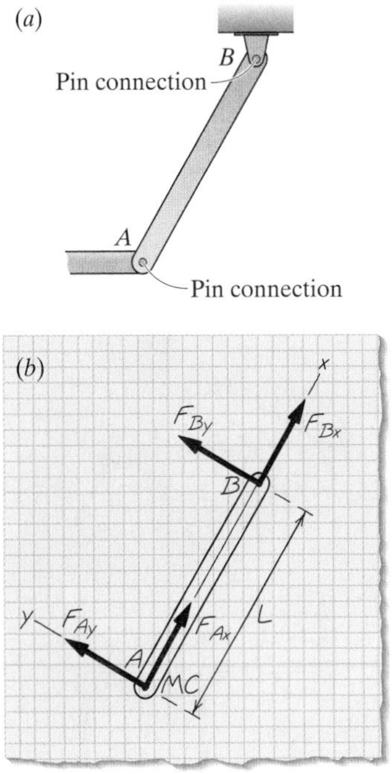

Figure 7.10 (*a*) A two-force element; (*b*) free-body diagram of a two-force element

Two-force elements are not restricted to being straight or to having pin connections at their ends, as illustrated in **Figure 7.11** and Example 7.5. Two-force members with pin connections are commonly referred to as links. Example 7.6 shows how being able to identify two-force elements is incorporated into static analysis.

A **three-force element** is an element with only three forces acting on it. An example of a three-force element is illustrated in **Figure 7.12*a***. *If a three-force element is in equilibrium, the lines of action of the three forces are concurrent.* This means that the three lines of action meet at a single point. If the three forces applied to a three-force element are parallel, as in **Figure 7.12*b***, their point of concurrency is at infinity.

Other examples of three-force elements are shown in **Figure 7.12c**. Convince yourself that the systems in **Figure 7.3** and **Figure 7.4** are three-force elements. Example 7.7 illustrates how information about three-force elements is useful in static analysis.

A **frictionless pulley** is an element that is used to change the direction of a cable or rope (**Figure 7.13*a***). It is connected to its surroundings by a frictionless pin connection. A free-body diagram of a frictionless pulley is shown in **Figure 7.13*b***. Applying the planar moment equilibrium equation with a moment center at *A*,

$$\sum M_{z@A} = 0 = (R)(T_1) - (R)(T_2) = 0; \text{ based on (7.5C)}$$

where R is the pulley radius. From this equation, we find that $\| \boldsymbol{T}_1 \| = \| \boldsymbol{T}_2 \|$. This means *the tension in the cable or rope is the same on both sides of a frictionless pulley*. Example 7.8 illustrates how information about a frictionless pulley is useful in static analysis.

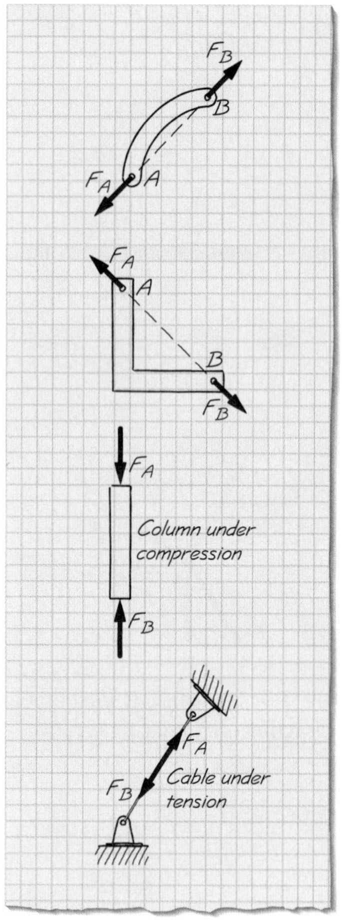

Figure 7.11 Other examples of two-force elements

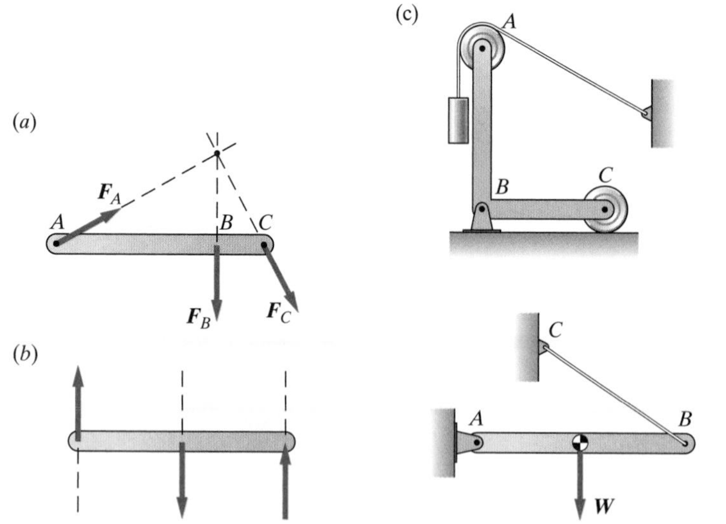

Figure 7.12 (*a*) A three-force element; (*b*) a three-force element in equilibrium requires that the forces be concurrent unless the three forces are parallel; (*c*) L-member *ABC* and beam *AB* are also three-force elements

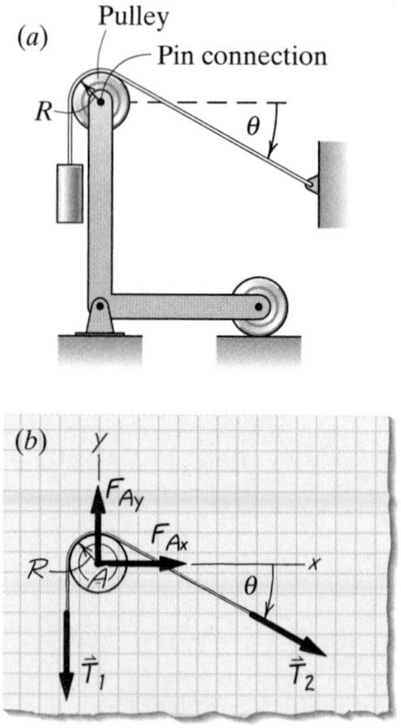

Figure 7.13 (*a*) L-member *ABC*; (*b*) the free-body diagram of frictionless pulley at *A*

EXAMPLE 7.4 PLANAR TRUSS CONNECTION

The gusset plate shown in Figure 7.14 is used to connect four members of a planar truss that is in equilibrium. The loads on members B and D are known. Assume that the weights of the members and the gusset plate are negligible compared to the applied loads. Find the magnitude of the loads $\mathbf{F}_A$ and $\mathbf{F}_C$ acting on members A and C.

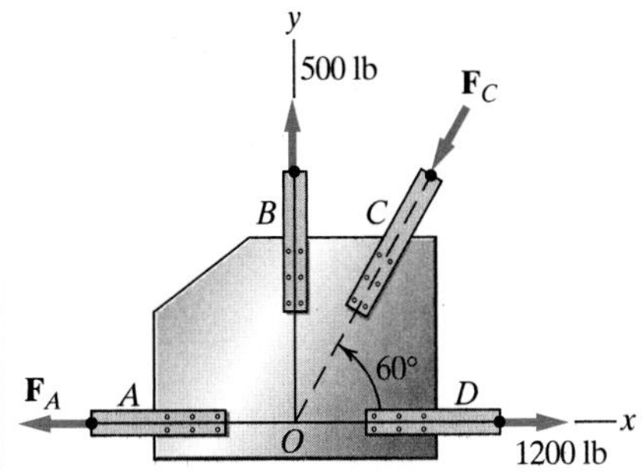

Figure 7.14

Goal We are to find the magnitudes of the forces on members A and C of a truss connection that is in equilibrium.

Given We are given information about the geometry of the connection, and the sizes of the loads on members B and D.

Assume We will assume that we can treat the system as planar and that we can model the point of concurrency of the lines of action of the truss members (O) as a particle.

Draw Based on the information given in the problem and our assumptions, we draw a free-body diagram that consists of a particle at O, acted on by forces (Figure 7.15).

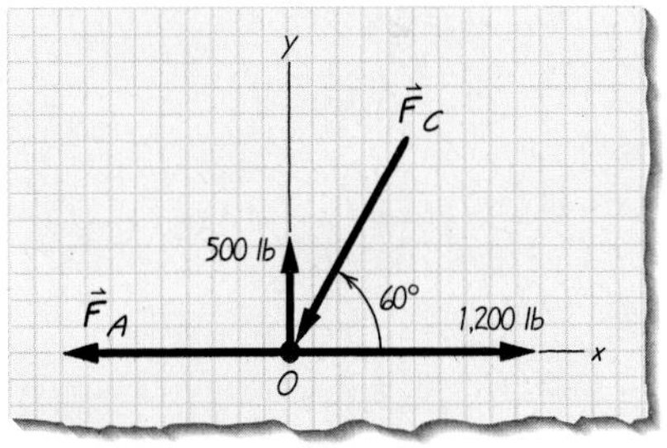

Figure 7.15

We will solve this problem using two approaches to illustrate that there are alternate problem-solving approaches. The first approach uses equations of equilibrium; the second uses the graphical method of vector addition.

Approach 1 (Use Equilibrium Equations):

Formulate Equations and Solve When modeling an object as a particle, we need only to consider force equilibrium equations. The moment equilibrium equations are satisfied because all the applied forces are concurrent at the particle. We set up the force equations for equilibrium of a particle to find the unknown loads:

Based on (7.5A):

$$\sum F_x = 0\ (\rightarrow +)$$
$$-F_A - F_C \cos 60^\circ + 1200 \text{ lb} = 0$$
$$F_A = -F_C \cos 60^\circ + 1200 \text{ lb} \qquad (1)$$

Based on (7.5B):

$$\sum F_y = 0\ (\uparrow +)$$
$$500 \text{ lb} - F_C \sin 60^\circ = 0$$
$$F_C = \frac{500 \text{ lb}}{\sin 60^\circ} = 577 \text{ lb}$$

$$F_C = 577 \text{ lb}$$

Substituting F_C into (1) gives:

$$F_A = -(577 \text{ lb}) \cos 60° + 1200 \text{ lb} = 912 \text{ lb}$$

$$F_A = 912 \text{ lb}$$

Since both F_A and F_B are positive, the forces act in the directions shown on the free-body diagram in **Figure 7.15**.

Answer The magnitudes of the forces are $\| \boldsymbol{F}_A \| = 912$ lb and $\| \boldsymbol{F}_C \| = 577$ lb.

Approach 2 (Use Graphical Vector Addition):

Draw Force Polygon and Solve In Chapter 4 we learned a graphical approach for adding vectors. We placed the vectors head to tail and then closed the force polygon with the resultant force, which was drawn from the tail of the first vector to the head of the last. When a system is in equilibrium, there is no resultant force acting, so when we draw a force polygon using all of the forces acting on the system, the head of the last vector must touch the tail of the first.

Using a piece of graph paper to help us establish a scale (1 square = 100 lb), and referring to the free-body diagram in **Figure 7.15**, we start by drawing the 500-lb vertical force (the choice of this as the first force was arbitrary). Next we add the 1200-lb force by drawing a horizontal arrow 12 squares long from the head of the 500-lb force. Next we draw F_C at a 60° angle to the horizontal. The length of F_C is unknown, but from the geometry of the problem we do know that F_A must be horizontal. This tells us that F_C must be long enough so that its head touches the x axis. Finally, to close the polygon we draw F_A from the head of F_C back to the tail of the 500-lb force (**Figure 7.16**).

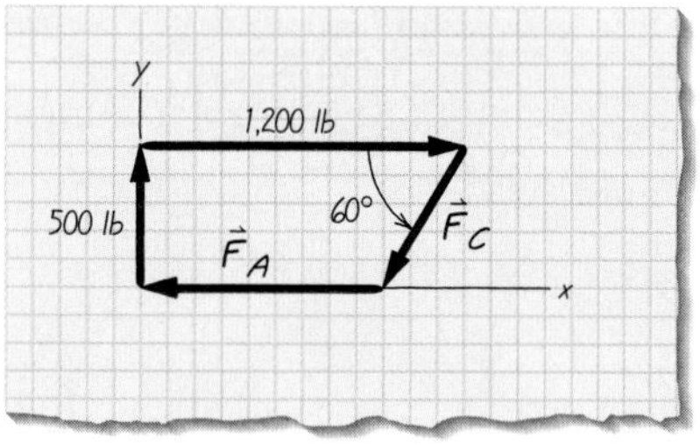

Figure 7.16

Using geometry and trigonometry we find the lengths of F_C and F_A. From the trapezoid in **Figure 7.16** we see that the y component of F_C must be 500 lb:

$$\| \boldsymbol{F}_C \| \sin 60° = 500 \text{ lb}$$
$$\| \boldsymbol{F}_C \| = 577 \text{ lb}$$

The geometry of the trapezoid shows us that the length of F_A can be determined from subtracting the horizontal component of F_C from the 1200-lb force:

$$\| \boldsymbol{F}_A \| = 1200 - \| \boldsymbol{F}_C \| \cos 60°$$
$$\| \boldsymbol{F}_A \| = 1200 - 577 \cos 60° = 912 \text{ lb}$$
$$\| \boldsymbol{F}_A \| = 912 \text{ lb}$$

Answer The magnitudes of the forces are $\| \boldsymbol{F}_A \| = 912$ lb and $\| \boldsymbol{F}_C \| = 577$ lb.

Check Using two different methods to solve a problem is an excellent way to check the accuracy of results.

EXAMPLE 7.5 IDENTIFYING TWO-FORCE AND THREE-FORCE ELEMENTS

Task For each structure shown in Figure 7.17, identify members that can be classified as two-force elements and members that can be classified as three-force elements.

(a) The mass of each box suspended from this pin-connected frame in **Figure 7.17*a*** is 80 kg.

Answer If the weights of members *BD*, *DF*, and *CG* are negligible, we can classify these three members as two-force elements. Notice that *DF* and *CG* are ropes, which would collapse in compression, and so they are also "tension-only" two-force elements.

If member weights are negligible, *AC* and *EC* can be classified as three-force elements.

(b) The lever system shown in **Figure 7.17*b*** is used to support a load at end *A*.

Answer If the weight of member *CD* is negligible, we can classify it as a two-force element. If we considered the weight of member *CD*, why would it not be a two-force element? Could it be considered a three-force element?

If the weights of member *GH* and the T-member are negligible, both are three-force elements.

(c) The toggle clamp in **Figure 7.17*c*** holds the workpiece at *F*.

Answer If the weight of member *CD* is negligible, we can classify it as a two-force element.

If the weights of members *GDB* and *ABE* are negligible, both are three-force elements.

(d) Members *AB* and *BC* in **Figure 7.17*d*** are pinned to one another at *B*.

Answer If the weight of member *AB* is negligible, it can be classified as a two-force element. What do we know about the line of action of the forces on member *AB* at *A* and *B*?

If the weight of member *BC* is negligible, it is a three-force member.

(e) Pipe strut *BC* in **Figure 7.17*e*** is loaded and supported as shown.

Answer If the weight of the cable that runs from *A* to *B* is negligible, we can classify it as a two-force element. In addition, it is a "tension-only" member.

If the weight of member *BC* is negligible, it is a three-force member.

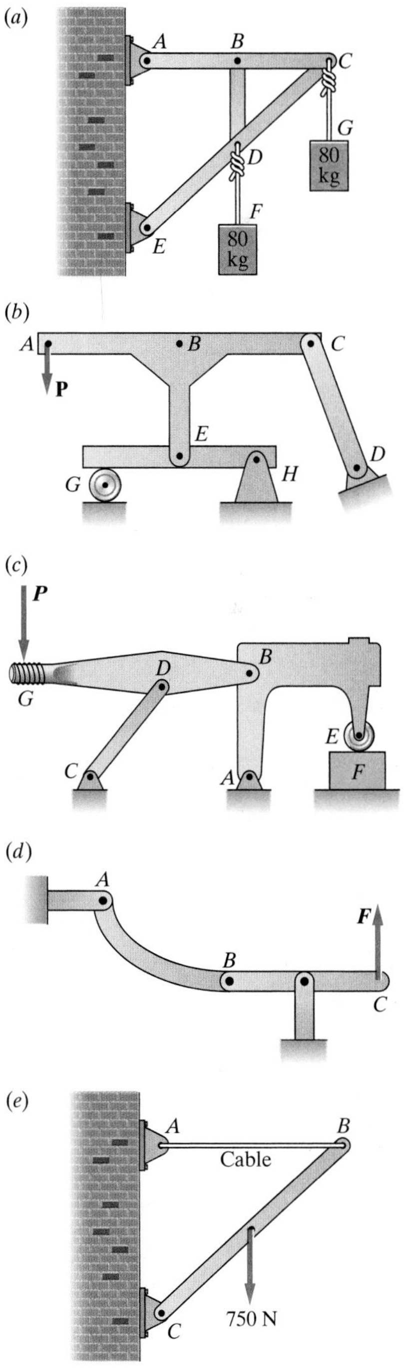

Figure 7.17

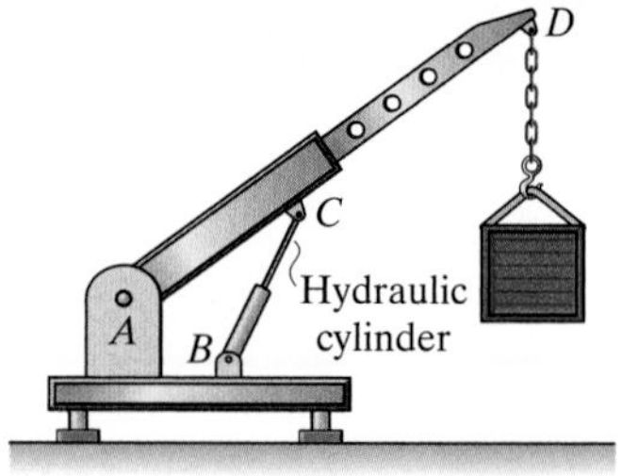

(f) The crane arm AD in **Figure 7.17*f*** is connected to a support base at A with a pin connection. The hydraulic cylinder BC exerts a force on the arm at C in the direction parallel to BC.

Answer The hydraulic cylinder is a two-force element. If the weight of member AD is negligible, it is a three-force member.

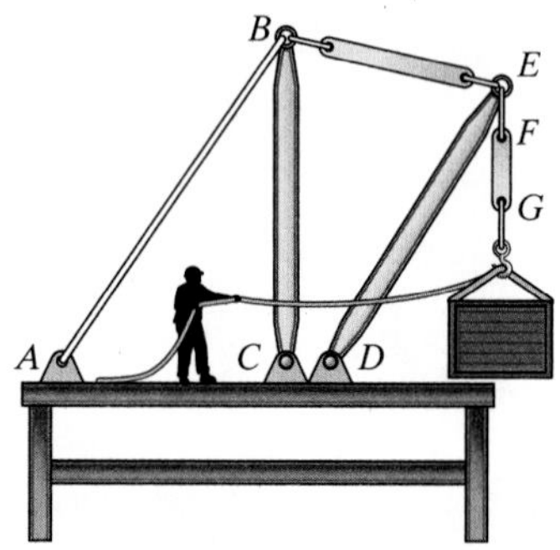

(g) A boom derrick in **Figure 7.17*g*** supports a load suspended at E.

Answer If the weights of the various members are negligible relative to the loads the boom derrick carries, then members BC, BE, DE, and FG are two-force elements. In addition, AB and EF are two-force elements (as well as being "tension-only" members).

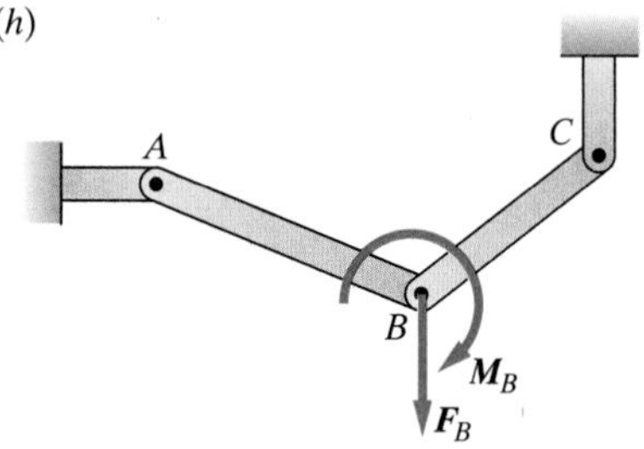

(h) A moment and a force act on the frame at B, as shown in **Figure 7.17*h***.

Answer None of the members act as a two-force element.

If the moment at B was removed, and the weights of the members were negligible, then members AB and BC would be two-force members.

Figure 7.17 (Cont.)

EXAMPLE 7.6 TWO-FORCE ELEMENT ANALYSIS EXAMPLE

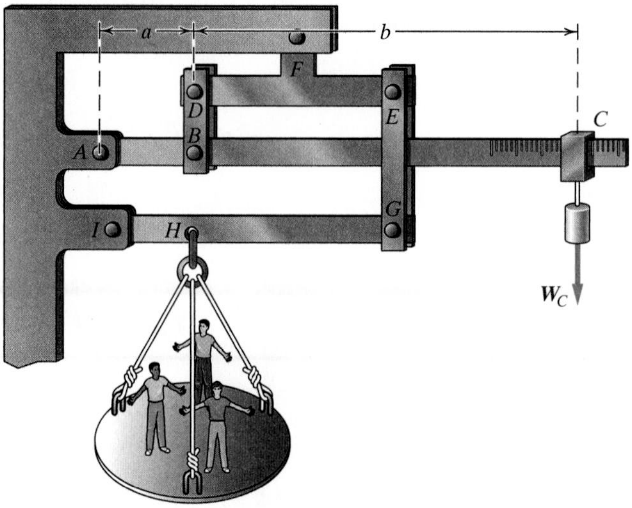

Figure 7.18

Consider the platform scale in **Figure 7.18**, and assume the weights of the various members that make up the lever system are negligible.

(a) Which members act as two-force elements?

(b) If the scale is in equilibrium, determine the force exerted by member BD on member ABC as a function of dimensions a and b, and W_C (the weight of the cylinder that moves along the calibrated bar).

(c) What are the x and y components of the force due to the pin connection at A?

Goal We are to identify which members act as two-force members. In addition, we are to find the force exerted on member *ABC* by member *BD* and the components of the force at *A*.

Given We are given information about the geometry of the lever system and that the scale is in equilibrium.

Assume We will assume that there are pin connections at *A, B, D, E, F, G*, and *I*. Also, we will assume that we can treat the system as planar.

Answer to (a) Each of the two parallel vertical bars joined to the horizontal members by pin connections at *B* and *D* is a two-force element. This is also true for the two vertical bars joined by pin connections at *E* and *G*.

Draw Because we are asked to find the force exerted by member *BD* on *ABC* as a function of dimensions of member *ABC* and weight W_C, we define our system as member *ABC*. The associated free-body diagram is shown in Figure 7.19. For this analysis we are defining member *BD* as the two parallel bars connecting *B* and *D*, and the force $2F_{BD}$ in the drawing represents the force from these two parallel two-force elements. Notice that because member *BD* is a two-force element the force at *B* acts along the long axis of member *BD*.

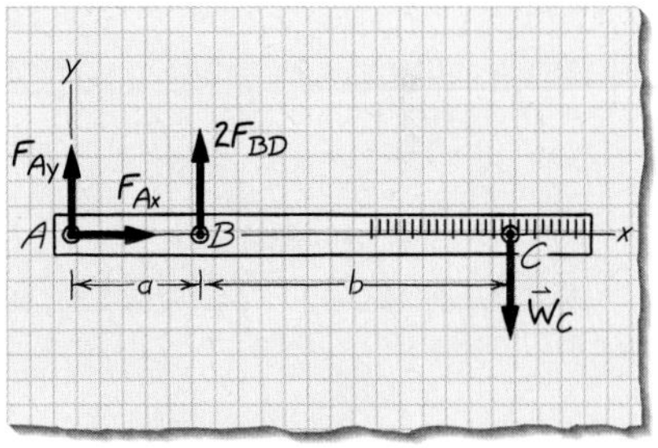

Figure 7.19

Formulate Equations and Solve Based on (7.5A):

$$\sum F_x = 0\ (\rightarrow +)$$
$$F_{Ax} = 0$$

Based on (7.5B):

$$\sum F_y = 0\ (\uparrow +)$$
$$F_{Ay} + 2F_{BD} - W_C = 0$$
$$F_{Ay} = W_C - 2F_{BD} \tag{1}$$

Based on (7.5C), with the moment center at *A*:

$$\sum M_{z@A}\ (\curvearrowleft +)$$
$$+(a)\,2F_{BD} - (a+b)W_C = 0$$
$$2F_{BD} = \frac{(a+b)}{a}W_C \tag{2}$$

Answer to (b) The force that the two members *BD* apply to member *ABC* is an upward force of magnitude $2F_{BD} = \frac{(a+b)}{a}W_C$. We can also say that member *ABC* applies a downward force to member *BD*. This downward force results in member *BD* being in tension.

Now we substitute (2) into (1) to find the y component force at A:

$$F_{Ay} = W_C - \frac{(a + b)}{a} W_C$$

$$F_{Ay} = \frac{-b}{a} W_C$$

(The minus sign tells us that the force of the stand acting on member ABC at A is in the downward direction.)

Answer to (c) $F_{Ax} = 0$ (based on (7.5A))
$F_{Ay} = \frac{-b}{a} W_C$
where F_{Ax} and F_{Ay} are defined in **Figure 7.19**

EXAMPLE 7.7 THREE-FORCE ELEMENT EXAMPLE

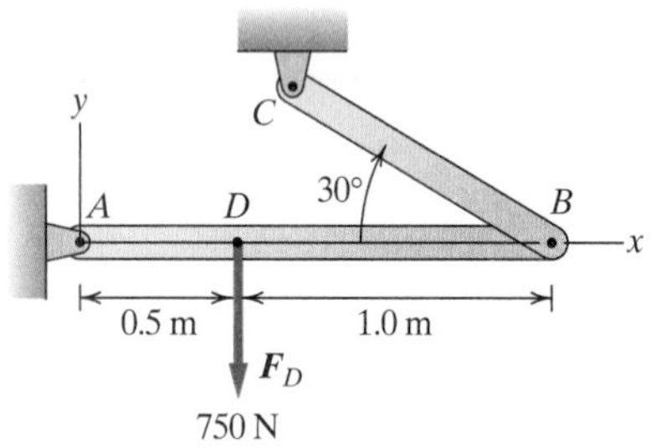

Figure 7.20

Beam AB is loaded and supported as shown in **Figure 7.20**. Find the loads acting on beam AB at A and B. Assume that the weights of the various members are negligible.

Goal We are to find the loads at A and B acting on beam AB.

Given We are given the dimensions of the beam. Furthermore, we are told that the weights of the members are negligible and that a force of 750 N acts at D.

Assume We assume we can treat the system (beam AB) as planar. Furthermore, if B and C are pin connections, member BC will act as a two-force element.

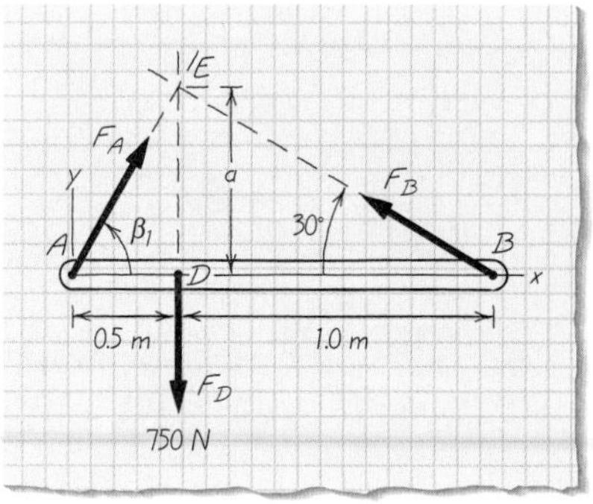

Figure 7.21

Draw We define beam AB as our system. Only three forces act on the beam ($\boldsymbol{F}_A$, $\boldsymbol{F}_B$, and $\boldsymbol{F}_D$); therefore it is a three-force element. This means that the three forces must be concurrent if the beam is in mechanical equilibrium (i.e., the lines of action of three forces intersect at a point). Therefore in constructing the free-body diagram we have the lines of action of the three forces coincide at point E (**Figure 7.21**).

Formulate Equations and Solve Based on **Figure 7.21** (which represents the free-body diagram of a three-force element in equilibrium) we can find:

$$a = (1.0 \text{ m}) \tan 30° = 0.577 \text{ m}$$

$$\tan \beta_1 = \frac{a}{0.5 \text{ m}} = \frac{0.577 \text{ m}}{0.5 \text{ m}}$$

$$\beta_1 = 49.1°$$

The force equilibrium condition (7.1), ($\boldsymbol{F}_{\text{net}} = \Sigma \boldsymbol{F} = 0$), says that if we place the three forces head to tail (as in vector addition; see Chapter 4), they should form a triangle, as depicted in **Figure 7.22**. Based on this figure we can write the law of sines:

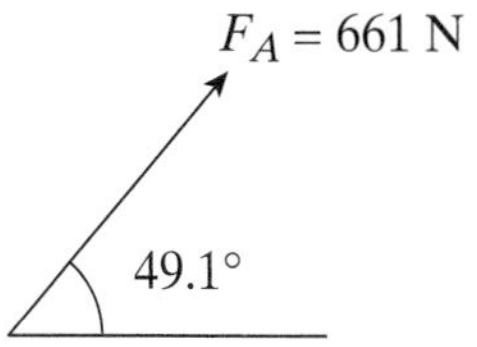

$$\frac{\sin 79.1^\circ}{750 \text{ N}} = \frac{\sin 60^\circ}{F_A}, \text{ from which we find that } F_A = 661 \text{ N}$$

Also based on **Figure 7.22**, we find that $\beta_2 = 40.9^\circ (= 180^\circ - 79.1^\circ - 60^\circ)$ and

$$\frac{\sin 79.1^\circ}{750 \text{ N}} = \frac{\sin 40.9^\circ}{F_B}$$

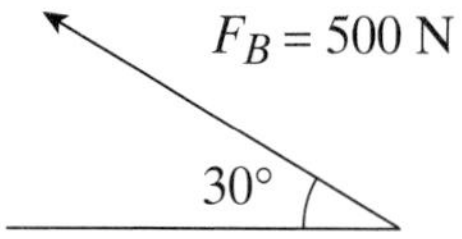

from which we find that $F_B = 500$ N. Therefore, we can write $\boldsymbol{F}_A$ as

$$\boldsymbol{F}_A = \|\boldsymbol{F}_A\| \cos\beta_1 \boldsymbol{i} + \|\boldsymbol{F}_A\| \sin\beta_1 \boldsymbol{j} \underbrace{=}_{\substack{\|\boldsymbol{F}_A\|=661 \text{ N} \\ \beta_1 = 49.1^\circ}} 433 \text{ N}\,\boldsymbol{i} + 500 \text{ N}\,\boldsymbol{j}$$

and $\boldsymbol{F}_B$ as

$$\boldsymbol{F}_B = -\|\boldsymbol{F}_B\| \cos 30^\circ \boldsymbol{i} + \|\boldsymbol{F}_B\| \sin 30^\circ \boldsymbol{j} \underbrace{=}_{\|\boldsymbol{F}_B\|=500 \text{ N}} -433 \text{ N}\,\boldsymbol{i} + 250 \text{ N}\,\boldsymbol{j}$$

Answer $\boldsymbol{F}_A = 433 \text{ N}\,\boldsymbol{i} + 500 \text{ N}\,\boldsymbol{j}$
$\boldsymbol{F}_B = -433 \text{ N}\,\boldsymbol{i} + 250 \text{ N}\,\boldsymbol{j}$

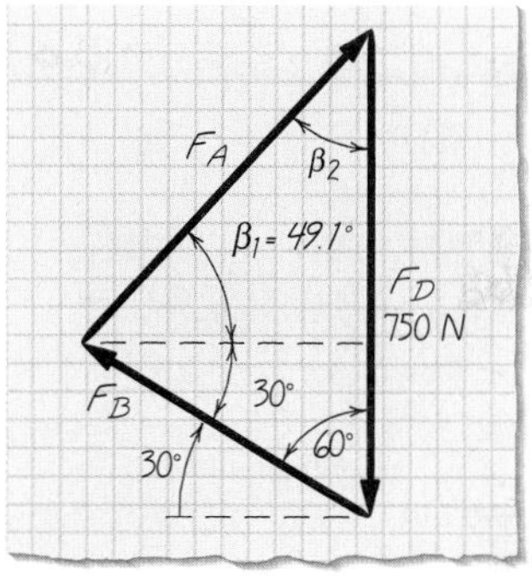

Figure 7.22

Check By recognizing that beam AB is a three-force element, we were able to use vector addition of the three forces (**Figure 7.22**) as the basis for finding the forces at A and B. Alternatively, we could have used (7.5A, B, and C) to find the forces at A and B or to check the values found above. Used as a check, we write (7.5A):

$$\sum F_x = 0\ (\rightarrow +)$$

$$\underbrace{F_A \cos\beta_1}_{\substack{433 \text{ N} \\ \text{(from above)}}} - \underbrace{F_B \cos 30^\circ}_{\substack{433 \text{ N} \\ \text{(from above)}}} = 0$$

$$433 \text{ N} - 433 \text{ N} = 0$$

$$0 = 0 \text{ (Yes!)}$$

Based on (7.5B):

$$\sum F_y = 0\ (\uparrow +)$$

$$F_A \sin\beta_1 + F_B \sin 30^\circ - F_D = 0$$

$$500 \text{ N} + 250 \text{ N} - 750 \text{ N} = 0$$

$$0 = 0 \text{ (Yes!)}$$

Note: Recognition that a member is a three-force element can also be used as a calculation check. For instance, the triangular structure in Example 7.3 is a three-force member. We could confirm that the answers we found for $\boldsymbol{F}_A$ (made up of F_{Ax} and F_{Ay}), F_{Bx}, and $\boldsymbol{P}$ are such that all of these forces are concurrent.

EXAMPLE 7.8 FRICTIONLESS PULLEY

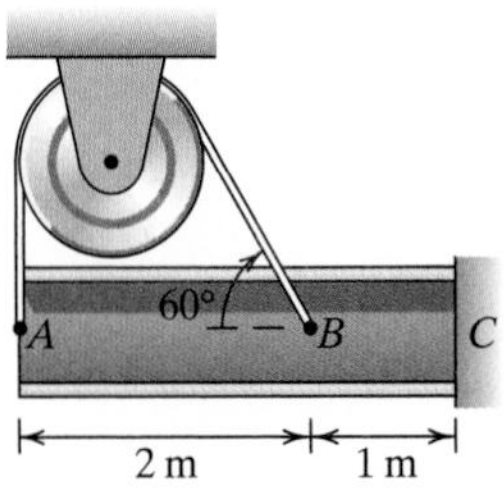

Configuration 1

Figure 7.23

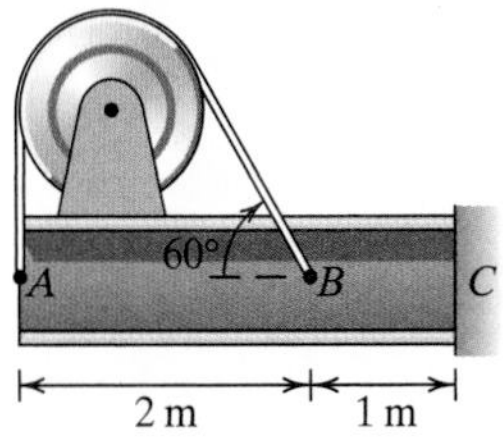

Configuration 2

Figure 7.24

Two beam and pulley systems in **Figure 7.23** and **Figure 7.24** with slightly different configurations are in equilibrium. The tension in the cable is 1.50 kN for both configurations. Assume that you can ignore gravity in the analysis. What loads act on the beam at C in the two configurations?

Goal We are to find the loads at the wall in two different pulley–cable–beam cases.

Given We are given information about the dimensions and geometry of the pulley–cable–beam configurations and the tension in the cable (1.50 kN).

Assume We will assume that we can treat the structure as a planar system and that the support at C is fixed. We will also assume that the pulley is frictionless, which tells us that the force is the same everywhere in the cable.

Draw Based on the information given in the problem and our assumptions, we isolate each configuration with the boundaries shown in **Figure 7.25** and **Figure 7.26**. Then we draw the free-body diagram for each configuration (**Figure 7.27** and **Figure 7.28**). (At the fixed connection at C there is a force $\boldsymbol{F}_C$ $(= F_{Cx}\boldsymbol{i} + F_{Cy}\boldsymbol{j})$ and a moment $\boldsymbol{M}_{Cz}$; see **Table 6.1**.)

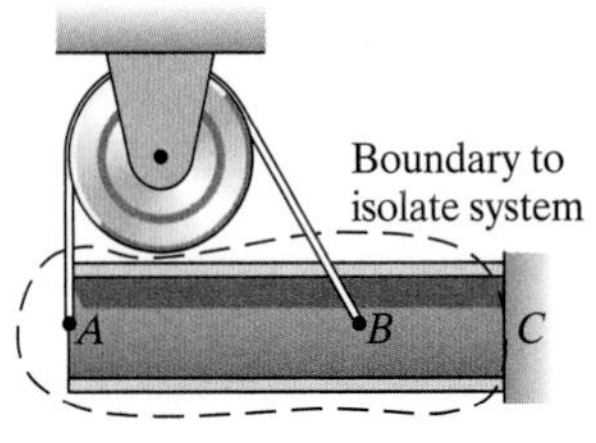

Configuration 1

Figure 7.25

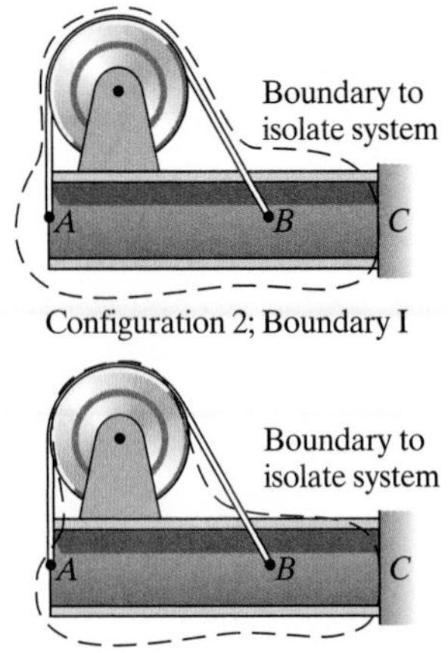

Configuration 2; Boundary II

Figure 7.26

Formulate Equations and Solve

Configuration 1:

Based on (7.5A) and **Figure 7.27**:

$$\sum F_x = 0\ (\rightarrow +)$$
$$-F_B \cos 60^\circ + F_{Cx} = 0$$
$$F_{Cx} = F_B \cos 60^\circ$$
$$F_{Cx} = (1.50\text{ kN}) \cos 60^\circ = 0.750\text{ kN}$$

Based on (7.5B):

$$\sum F_y = 0\ (\uparrow +)$$
$$F_A + F_B \sin 60^\circ + F_{Cy} = 0$$
$$F_{Cy} = -(F_A + F_B \sin 60^\circ)$$
$$F_{Cy} = -(1.50\text{ kN} + 1.50\text{ kN} \sin 60^\circ) = -2.80\text{ kN}$$

Based on (7.5C), with the moment center at C:

$$\sum M_{z@C} = 0\ (\circlearrowleft +)$$
$$-(3\text{ m})F_A - (1\text{ m})F_B \sin 60^\circ + M_{Cz} = 0$$
$$M_{Cz} = (3\text{ m})F_A + (1\text{ m})F_B \sin 60^\circ$$
$$M_{Cz} = (3\text{ m})\ 1.50\text{ kN} + (1\text{ m})\ 1.50\text{ kN} \sin 60^\circ$$
$$M_{Cz} = +5.80\text{ kN}\cdot\text{m}$$

Answer for Configuration 1

$F_{Cx} = 0.750$ kN
$F_{Cy} = 2.80$ kN
$M_{Cz} = 5.80$ kN · m
where F_{Cx}, F_{Cy} and M_{Cz} are defined in Figure 7.27

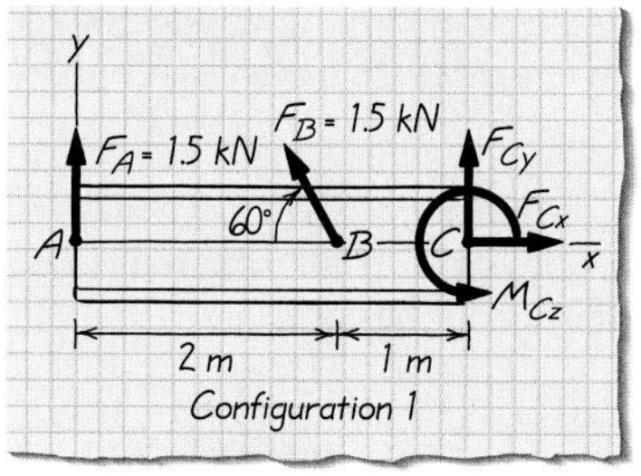

Figure 7.27 Free-body diagram for Configuration 1

Check By summing the moments about a different moment center (e.g., choose A), we can check to see whether our beam is in equilibrium.

$$\sum M_{z@A} = 0 \; (\curvearrowleft+)$$
$$(3\text{ m})F_{Cy} + (2\text{ m})F_B \sin 60^\circ + M_{Cz} = 0$$
$$+(3\text{ m})(-2.80\text{ kN}) + (2\text{ m})\,1.50\text{ kN} \sin 60^\circ + 5.80\text{ kN}\cdot\text{m} = 0$$
$$-8.40\text{ kN}\cdot\text{m} + 2.60\text{ kN}\cdot\text{m} + 5.80\text{ kN}\cdot\text{m} = 0$$
$$0 = 0\text{ (Yes!)}$$

Configuration 2: We will explore two different boundaries to isolate configuration 2, as shown in **Figure 7.26**.

Boundary I cuts through connection C but leaves the pulley and cables attached to the beam. We are ignoring the weight of the beam, pulley, and cable, so inspection of the free-body diagram in **Figure 7.28*a*** reveals that there are no external forces acting on the beam. Therefore equilibrium requires that F_{Cx}, F_{Cy}, and M_{Cz} be zero.

Boundary II cuts through connection C and the cable. We could go through the process of writing (7.5A), (7.5B), and (7.5C) to determine $\mathbf{F}_C$ and M_{Cz}. Alternatively, we can inspect the free-body diagram in **Figure 7.28*b*** to see that boundary II cutting through the cable results in force pairs ($\mathbf{F}_A$ and $\mathbf{F}_A$; $\mathbf{F}_B$ and $\mathbf{F}_B$) that cancel one another in both the force equilibrium and moment equilibrium equations. Because forces and moments acting on the system need to be in equilibrium, this requires that both $\mathbf{F}_C$ and M_{Cz} are equal to zero. Therefore, $\mathbf{F}_C = 0$, and $M_{Cz} = 0$.

Answer for Configuration 2 $\mathbf{F}_C = 0$, $M_{Cz} = 0$

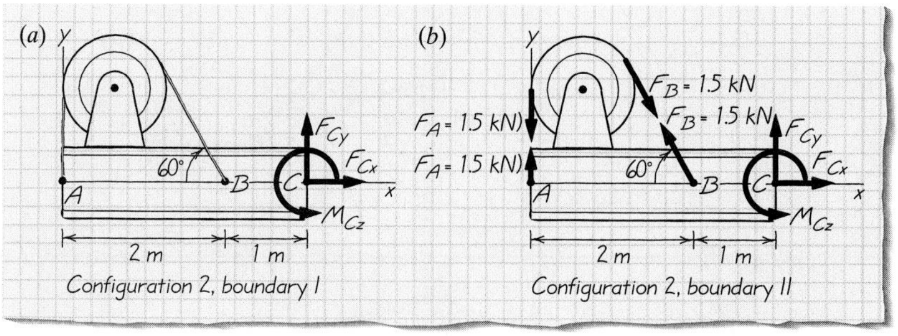

Figure 7.28 Free-body diagrams for Configuration 2: (*a*) Boundary 1; (*b*) Boundary 2

EXERCISES 7.2

7.2.1. The sphere has a mass of 10 kg and is supported as shown in E7.2.1. The cord ABC passes over a frictionless pulley at B. If the sphere is in equilibrium, determine the tension in the cord.

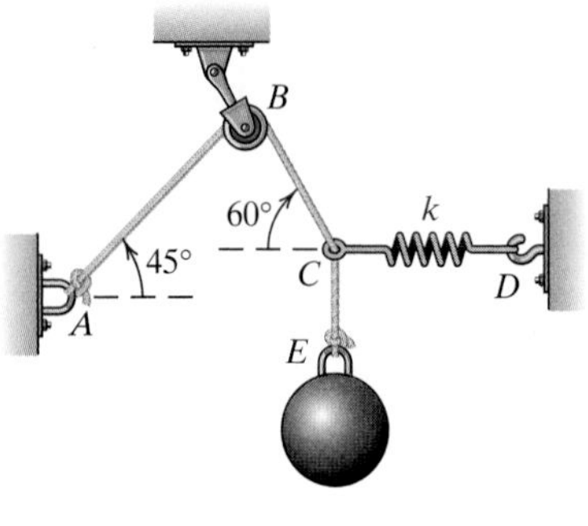

E7.2.1

7.2.2. If the engine in E7.2.2 is in equilibrium, what is the tension in cables AB and AD? The mass of the engine is 225 kg.

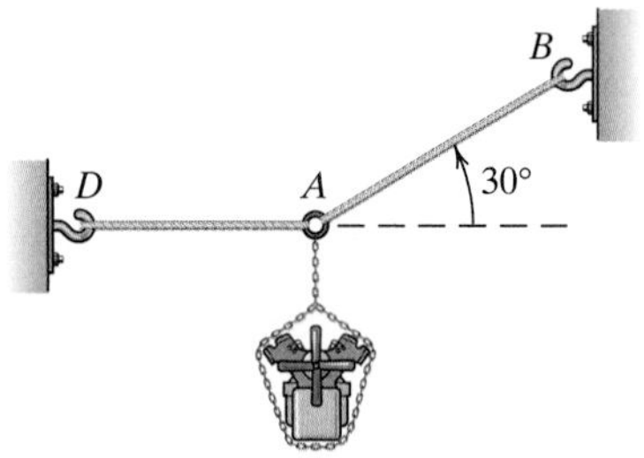

E7.2.2

7.2.3. Members of a truss are connected to the gusset plate, as shown in E7.2.3. If the forces are concurrent at point O, determine the magnitudes of $\mathbf{F}$ and $\mathbf{T}$ for equilibrium.

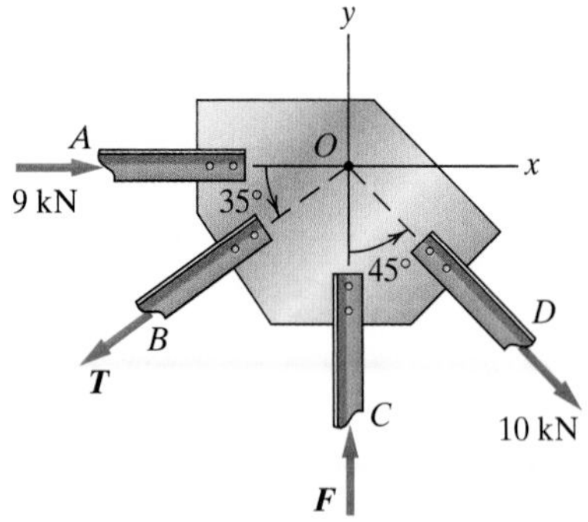

E7.2.3

7.2.4. A 225-kg engine is suspended from a vertical chain at A (E7.2.4). A second chain is wrapped around the engine and held in position by the spreader bar BC. Determine the force acting along the axis of the bar and the tension in chain segments AB and AC. Clearly state whether the bar is in tension or compression.

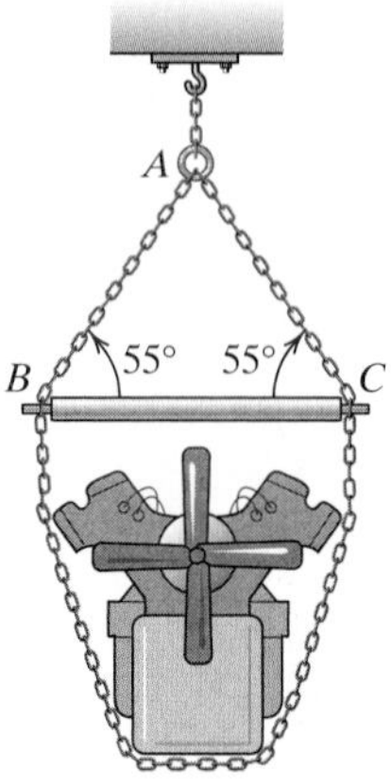

E7.2.4

7.2.5. A ship being towed by a tugboat moves along at constant velocity. Determine the force in each of the cables BD and BC, if the tugboat applies a 40-kN force, as shown in E7.2.5.

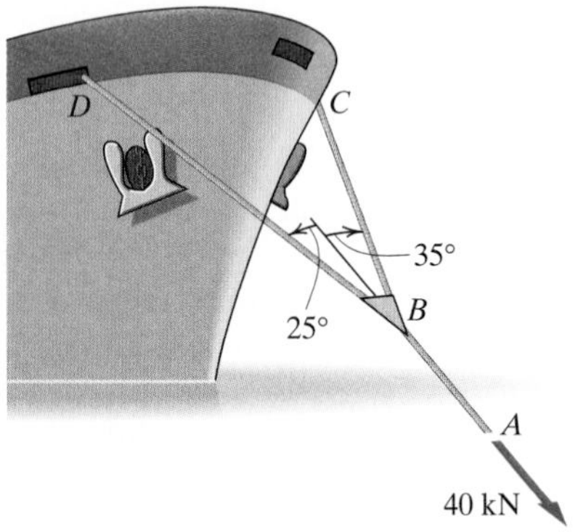

E7.2.5

7.2.6. Determine the forces in cables AC and AB required to hold the 15-kg cylinder D in E7.2.6 in equilibrium.

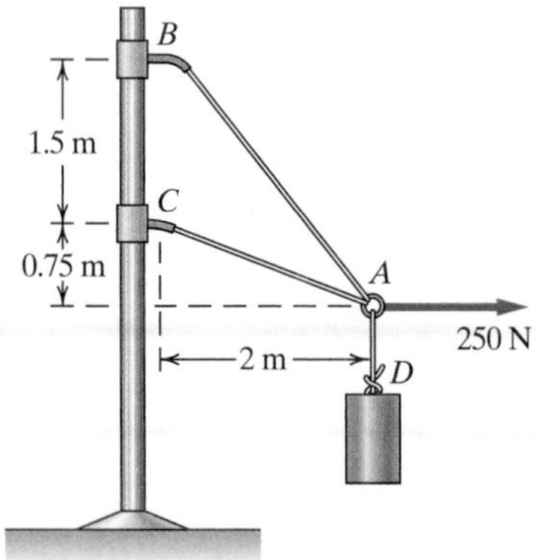

E7.2.6

7.2.7. A steel sphere sits in the smooth groove, as shown in E7.2.7. Its mass is m. If the sphere is in equilibrium, what loads act on the sphere at A? What loads act on the sphere at B?

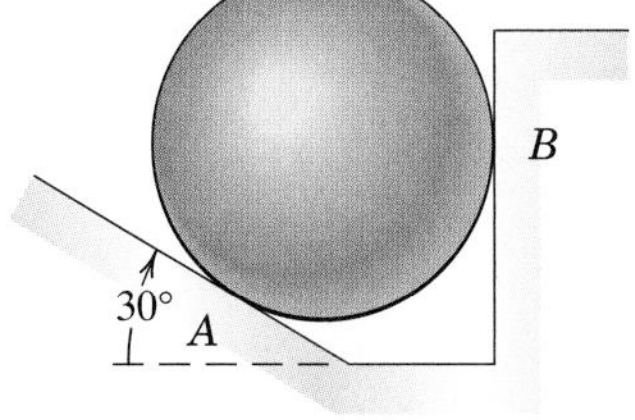

E7.2.7

7.2.8. An engine is lifted with the frictionless pulley shown in E7.2.8. If the engine and pulley are in equilibrium, what is the tension in cable AC? What is the tension in cable CB? What force P is required to hold the engine in equilibrium?

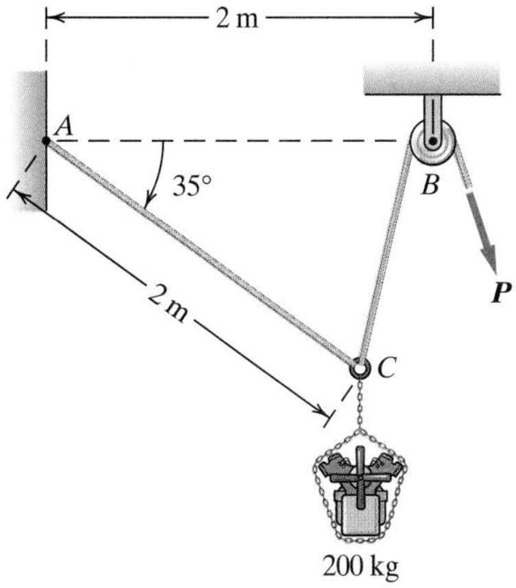

E7.2.8

7.2.9. Cable AB in E7.2.9 passes over the small frictionless pulley C and holds up the metal cylinder that weights 200 N. Determine the tension in cable CD.

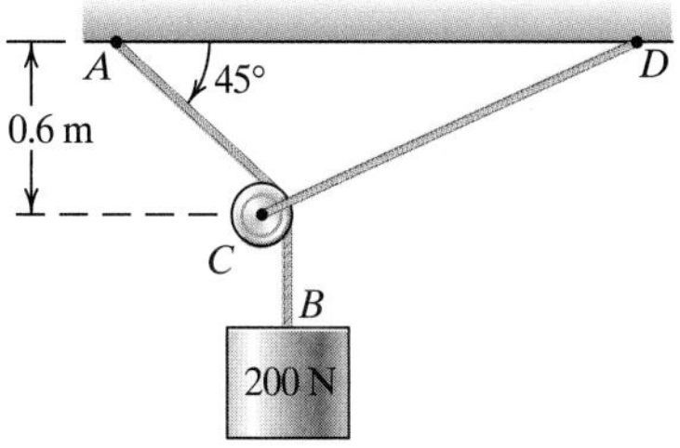

E7.2.9

7.2.10. The L-shaped bar is loaded by a 10-kN force at B as shown in E7.2.10. The weight of the bar is negligible. If the bar is in equilibrium, determine the angle α and the loads acting on the bar at A.

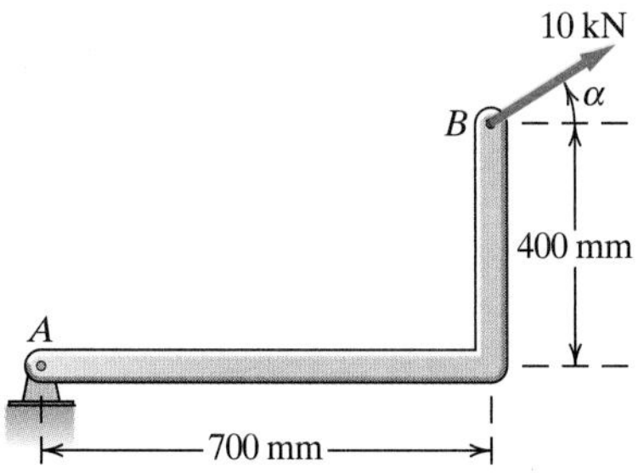

E7.2.10

7.2.11. Consider the 400-N uniform rectangular plate shown in E7.2.11. Determine the loads acting on the plate at B and C. Use the fact that the plate acts as a three-force element to check your answers.

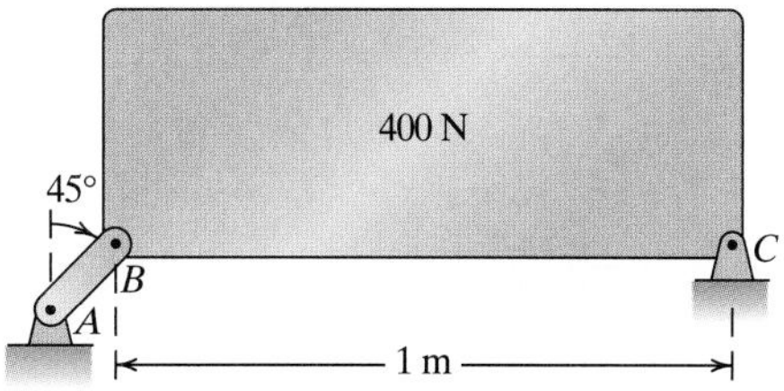

E7.2.11

7.2.12. The uniform beam weighs 10 kN. In addition, a force of 2 kN acts as shown in E7.2.12. Determine the angle α necessary for the beam to be in equilibrium. Is the beam a three-force element, and why or why not?

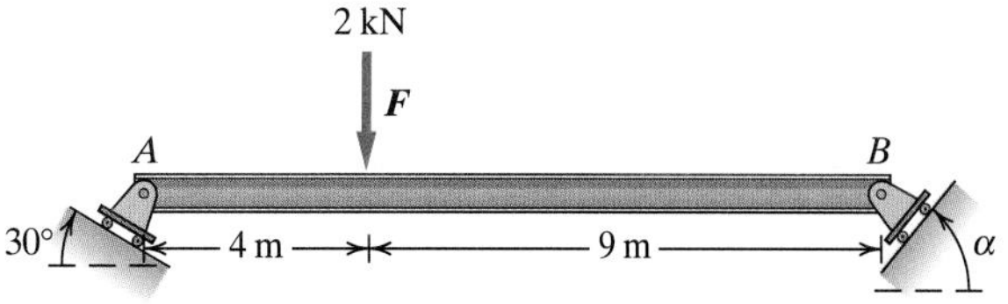

E7.2.12

7.2.13. A force of 2 kN acts on a beam of negligible weight, as shown in E7.2.13. Determine the angle α necessary for the beam to be in equilibrium. Use the fact that the beam acts as a three-force element to check your answer.

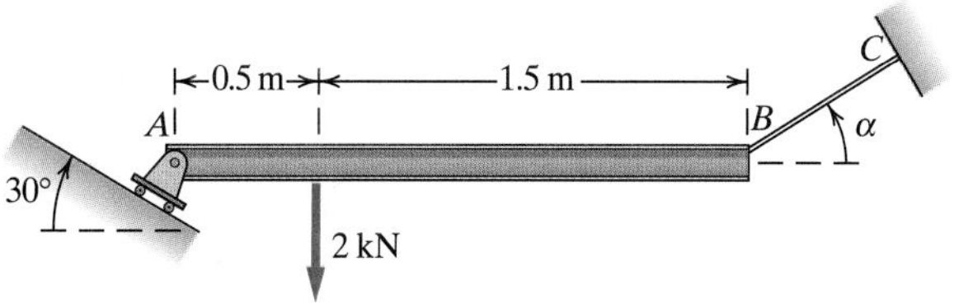

E7.2.13

7.2.14. A weight of 200 lb hangs from a frame of negligible weight, as shown in E7.2.14. If the frame is in equilibrium, determine the loads acting on the frame at A and D. Use the fact that the frame acts as a three-force element to check your answer.

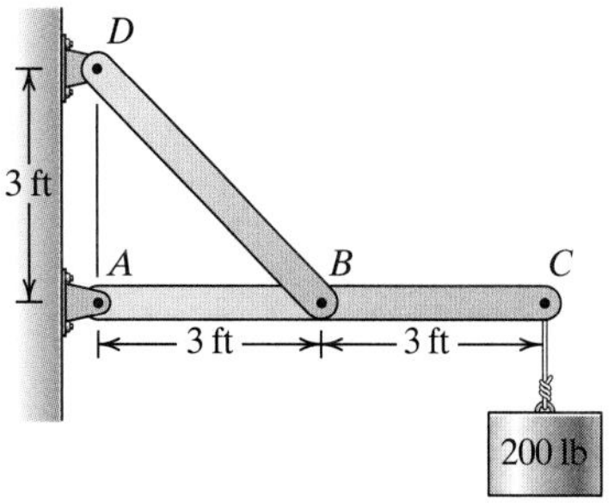

E7.2.14

7.2.15. A force of 4 kN acts on beam *CAB* of negligible weight, as shown in E7.2.15. Determine the angle α necessary for the beam to be in equilibrium. Also determine the loads acting on the beam at *A* and *B*. Finally, use the fact that the beam acts as a three-force element to check your answer.

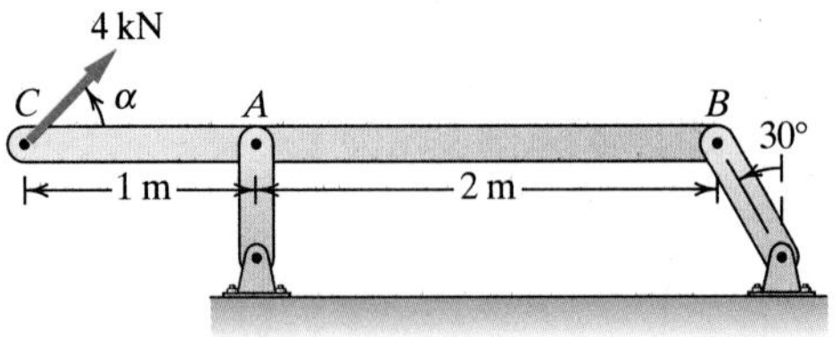

E7.2.15

7.2.16. A force of 4 kN and a moment of 2 kN · m act on frame *ABD* of negligible weight, as shown in E7.2.16. Determine the loads acting on the beam at *A* and *B*. Is frame *ABD* a three-force element, and why or why not?

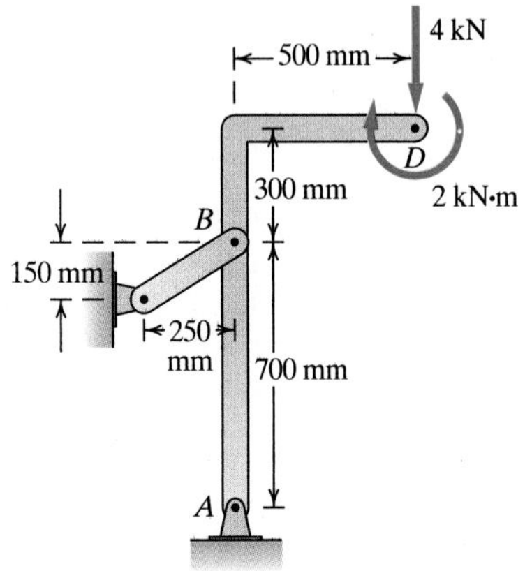

E7.2.16

7.2.17. The frictionless pulley system shown in E7.2.17 is in equilibrium. Find the values of weights W_1 and W_2.

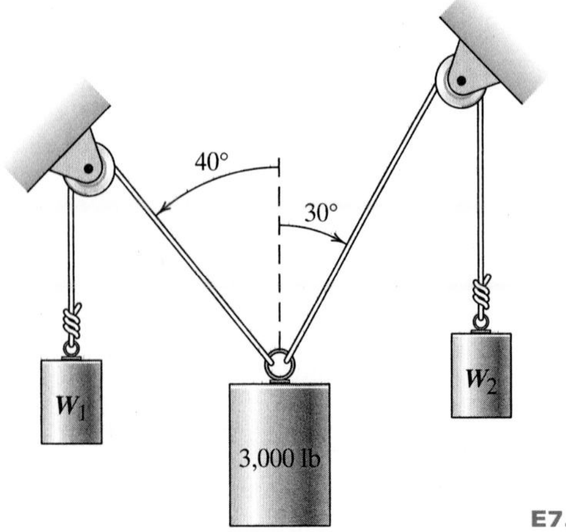

E7.2.17

7.2.18. A block is supported by a system of pulleys, as shown in E7.2.18. Knowing the system is in equilibrium and assuming the pulleys are frictionless and massless, determine the weight of the block.

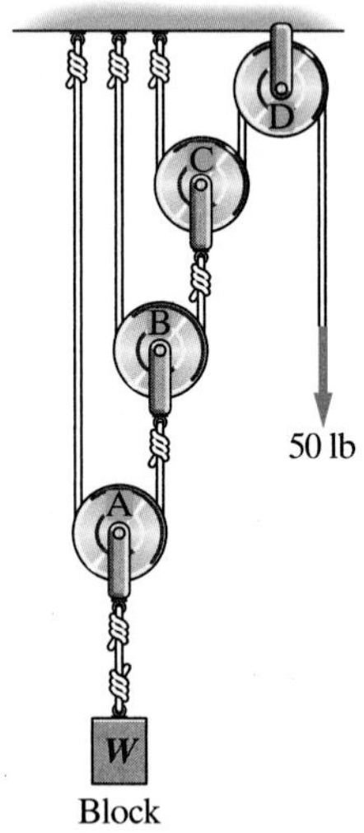

E7.2.18

7.2.19. A pulley system consisting of a rope and pulleys *A*, *B*, *C*, and *D* is being used to lift a 600-N load as shown in E7.2.19. Find the force $\boldsymbol{P}$ in the rope.

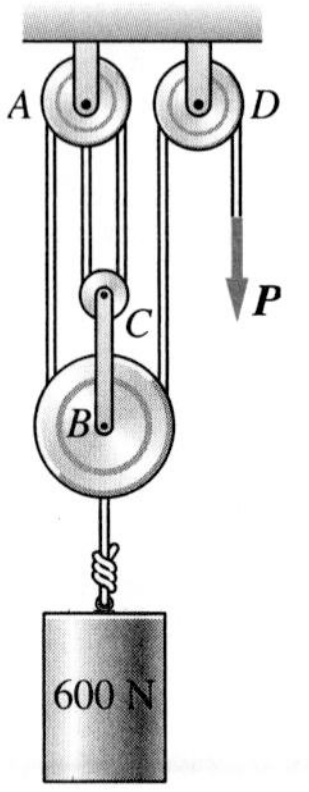

E7.2.19

7.2.20. A cable pulls on the bracket with a force of 2.5 kN as shown in E7.2.20. At *A* the bracket rests against a smooth surface and at *B* it is pinned to the wall. If the bracket is in equilibrium, what loads act on the bracket at *A*? What loads act on the bracket at *B*? Does the bracket act as a three-force member?

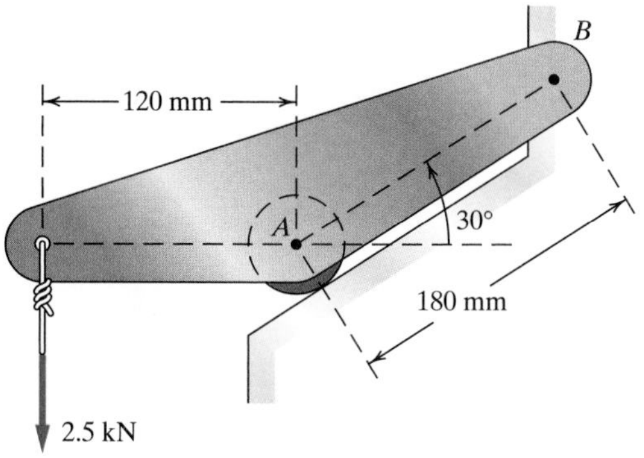

E7.2.20

7.2.21. An airplane rests on the ground (E7.2.21). Its landing gear are at points A, B, and C. Its center of gravity is at (3.0, 0.0, 0.0) m. The airplane weighs 1.56×10^6 N. What are the magnitudes of the normal forces acting on the landing gear? What direction do they act in?

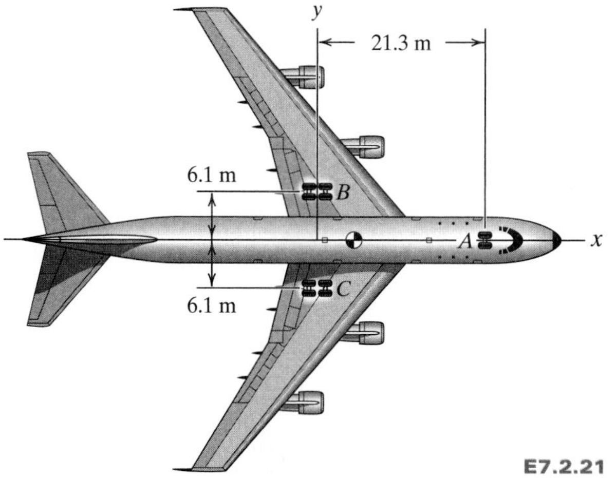

E7.2.21

7.2.22. A gear box is mounted on a beam and is subjected to two moments, as shown in E7.2.22. Determine the loads acting on the beam at A and B if the beam is in equilibrium.

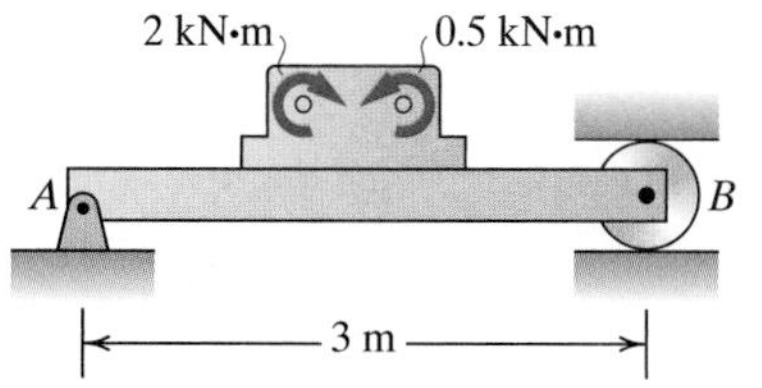

E7.2.22

7.2.23. The forked bar in E7.2.23 is supported by a pin connection at A and a short link (HB) at B. A 200-kg block is suspended from a cable that passes over frictionless pulleys C and D, which are pinned to the bar. Knowing that $\beta = 45°$ and that the bar is in equilibrium, determine the force that the link applies to the bar at B. Also determine the loads acting on the bar at A.

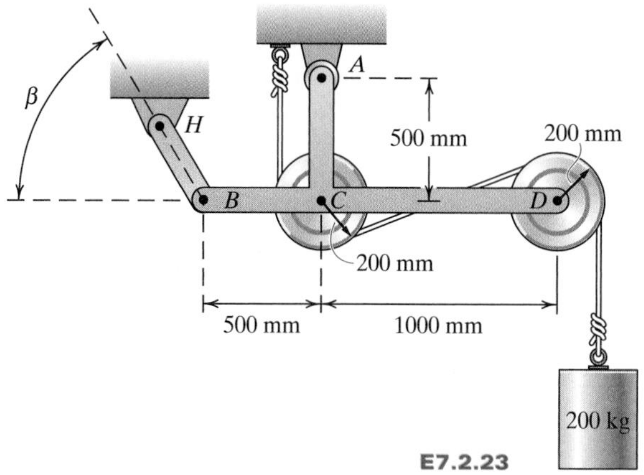

E7.2.23

7.2.24. Consider the semi-circular plate in E7.2.24 that has a radius of 40 cm. The plate weighs 100 N with center of gravity at (d, 0, 0). Determine the tension in the cables at A and B, and the loads acting on the plate at C (a ball-and-socket connection). The cables are at 60° above the horizontal.

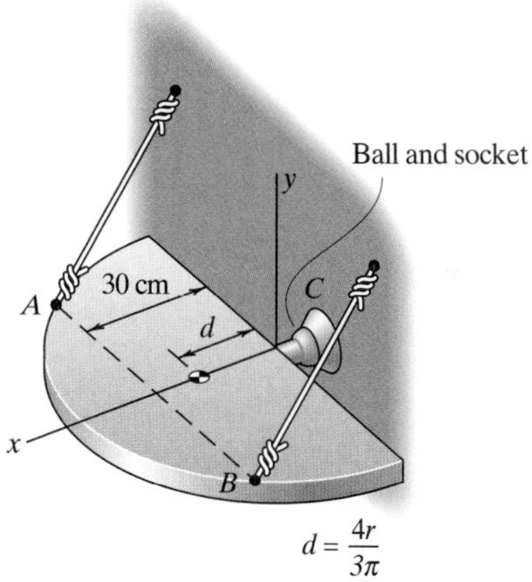

E7.2.24

7.2.25. The uniform L-bar is pinned to its surroundings at A, and slides along a wall at B as shown in E7.2.25. A vertical force $\boldsymbol{F}$ acts at C and the total weight of the L-bar is $\boldsymbol{W}$ with center of gravity at D. Gravity acts in the negative y direction. The system is taken as the L-bar. When the L-bar is in equilibrium, what are the loads acting on the bar at A and at B?

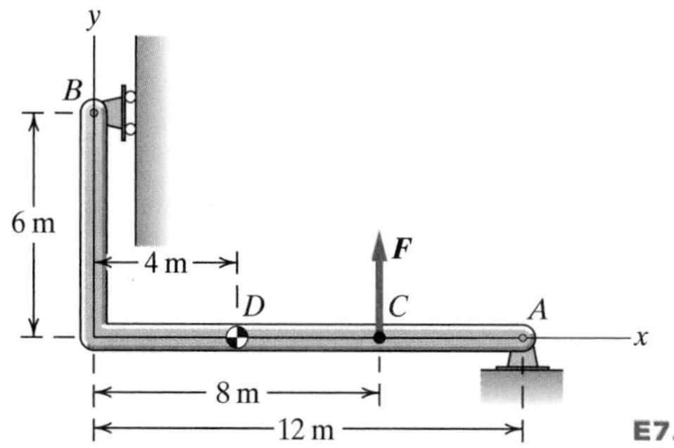

E7.2.25

7.2.26. The frame shown in E7.2.26 is supported at A and B. Forces act at C, D, E, and F, as shown. The system is taken as the frame. If weight of the frame is negligible and the frame is in equilibrium, what loads act on the frame at A and at B?

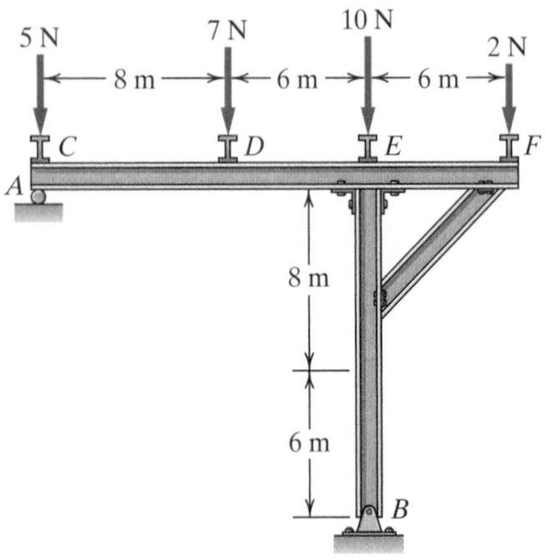

E7.2.26

7.2.27. The bracket ABC in E7.2.27 is tethered as shown with cable CD. The tension in the cable is 750 N. If the weight of the bracket is negligible and the bracket is in equilibrium, what loads act on the bracket at A?

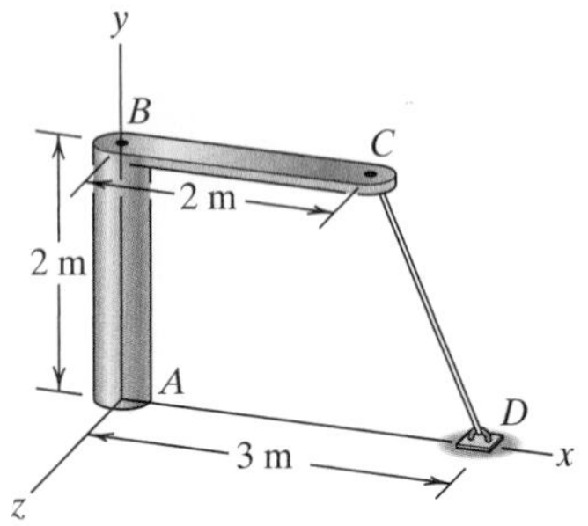

E7.2.27

7.2.28. The airplane weighing 8000 N sits on the tarmac (E7.2.28). It has one front wheel and two rear wheels. If the airplane is in equilibrium, what loads act on the plane at the front wheel and at each of the two rear wheels?

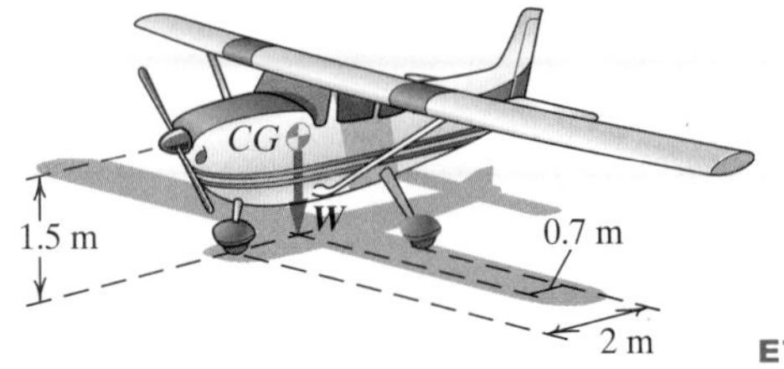

E7.2.28

7.2.29. A uniform curved beam of weight W is supported at A and B, as shown in E7.2.29. If the beam is in equilibrium, what loads act on the beam at A and at B?

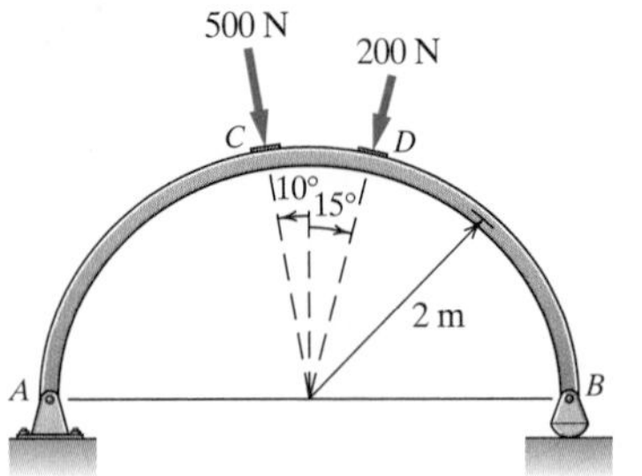

E7.2.29

7.2.30. A uniform beam is pinned at B and rests against a smooth incline at A. Additional loads act as shown in E7.2.30. The total weight of the beam is W. If the beam is in equilibrium, what loads act on the beam at A and at B?

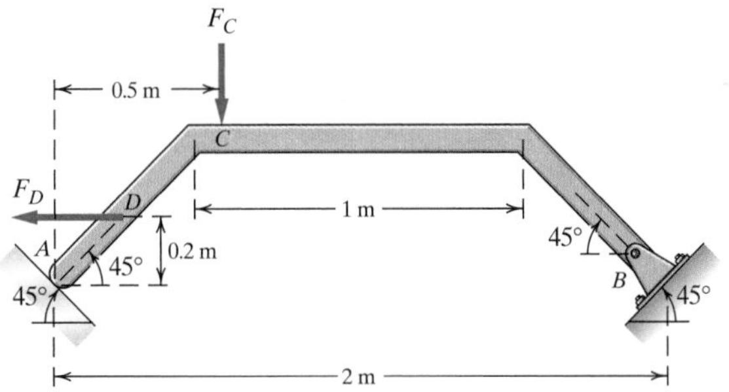

E7.2.30

7.2.31. A forklift is lifting a crate of weight W_1 as shown in E7.2.31. The weight of the forklift is W_2. The front wheels are free to turn, and the rear wheels are locked. If the forklift and crate are in equilibrium, what loads act on the forklift at A and at B?

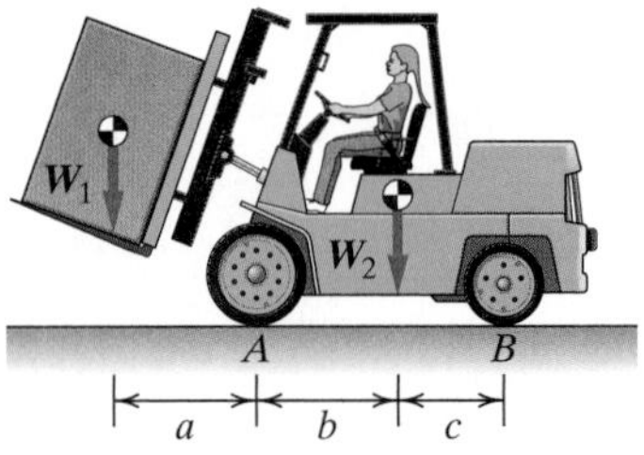

E7.2.31

7.2.32. A mobile hangs from the ceiling at E from a cord (E7.2.32). If the mobile is in equilibrium, what loads act on the mobile at E? Designate the weights of the fish A, B, C, and D as W_A, W_B, W_C, and W_D, respectively.

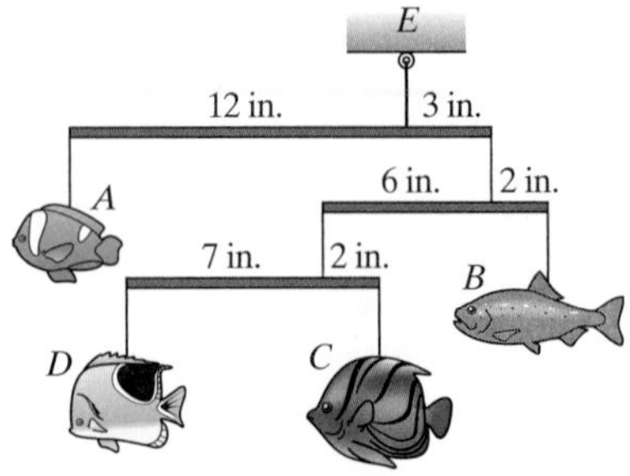

E7.2.32

7.2.33. A 350-N force acts on a brake pedal, as shown in E7.2.33. If the pedal is in equilibrium in the position shown, what is the tension in the spring? What is the magnitude of the force acting at the pin connection at D? If the total deflection at A is to be limited to 5 mm when the 350-N pedal force is applied, what is the minimum required spring stiffness k?

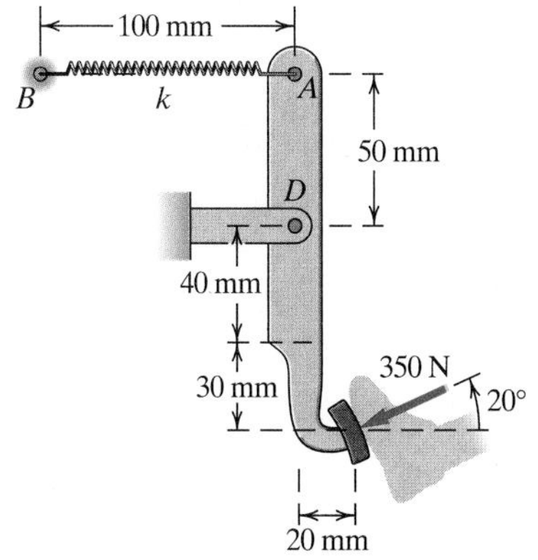

E7.2.33

7.2.34. A child who weighs 230 N balances on the beam in E7.2.34. Planes A and B are smooth. If the beam and child are in equilibrium and the weight of the beam is negligible, what is the tension in the cable at B? What are the loads acting on the beam at A and at B?

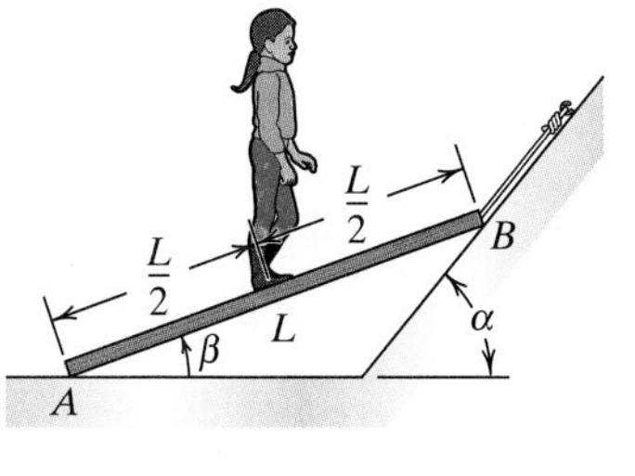

E7.2.34

7.2.35. A beam is fixed at end D. Forces act on the beam at A, B, and C, as shown in E7.2.35. If the weight of the beam is negligible and the beam is in equilibrium, what are the loads acting on the beam at D?

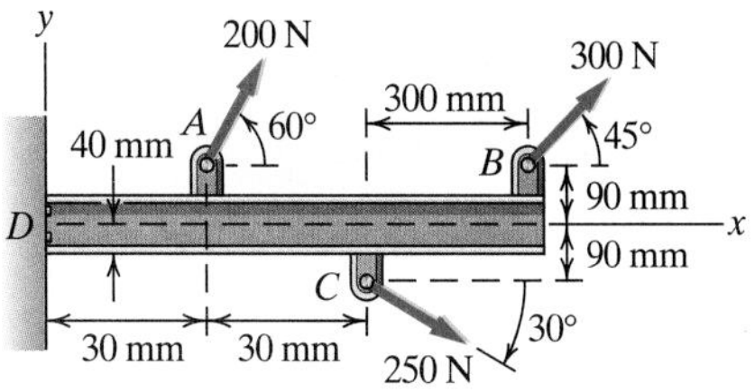

E7.2.35

7.2.36. The truss is attached to the ground at A with a rocker and at D with a pin connection. Additional loads acting on the truss are shown in E7.2.36. If the truss is in equilibrium, what loads act on the truss at A? What loads act on the truss at D?

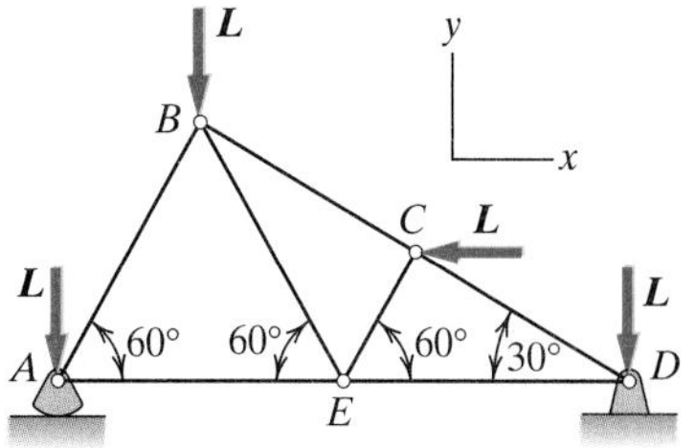

E7.2.36

7.2.37. A belt-tensioning device is shown in E7.2.37. The frictionless pulley A is pinned to the L-arm at B. Assuming that the device is in equilibrium, write an expression for the tension $\boldsymbol{T}$ in the belt as a function of the dimensions l and b, and the mass m of the cylinder attached at C.

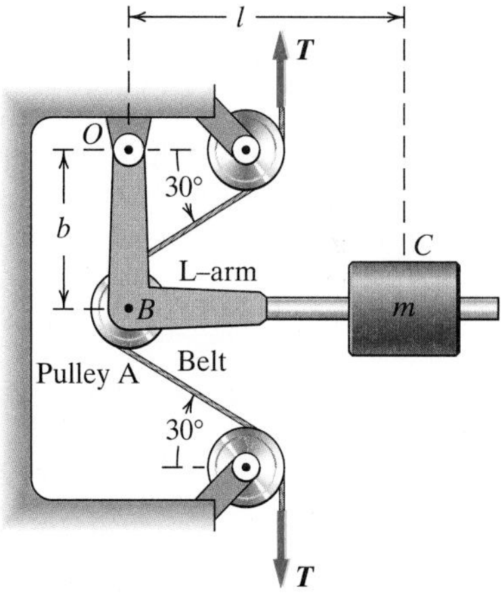

E7.2.37

7.2.38. A cable pulls on the bracket with a force of 1.5 kN (E7.2.38). At A the bracket is attached to the wall with a pin connection, and at B there is a pin-in-slot connection. If the bracket is in equilibrium, what loads act on the bracket at A? What loads act on the bracket at B?

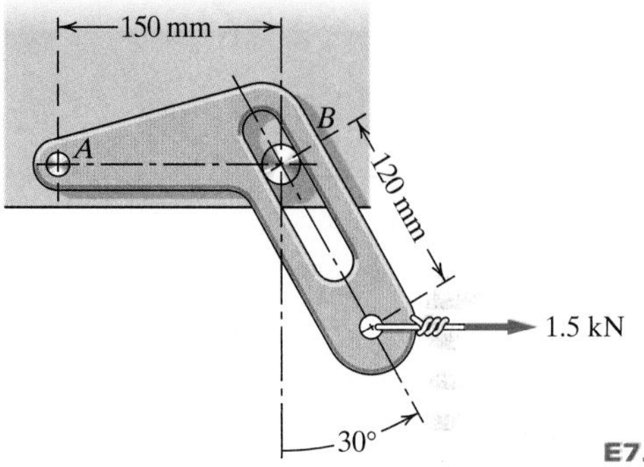

E7.2.38

7.2.39. Forces F_1 and F_2 act on each of the systems shown in E7.2.39. These systems are known as first, second, and third class levers, respectively.

a. For each system in equilibrium determine the ratio of $\| F_2 \| / \| F_1 \|$. This ratio is called the mechanical advantage. Also find the load(s) acting on each lever at A.

b. You are asked to supply F_1 with your arm muscles in order to balance F_2. Which of the lever systems would you choose to work with and why?

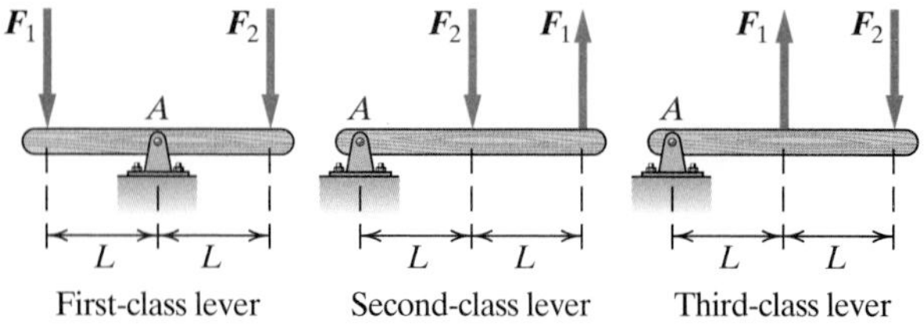

E7.2.39

7.2.40. The truck–trailer assembly in E7.2.40 is in equilibrium on level ground. The hitch at B can be modeled as a pin connection.

a. If the weight of the trailer is 16 kN, determine the normal force exerted on the rear tires at A and the force exerted on the trailer at the pin connection B.

b. If the pin connection at B is designed to support a maximum force of 6.5 kN, what is the weight of the heaviest trailer that should be safely attached to the truck? Clearly state all assumptions in addressing this question.

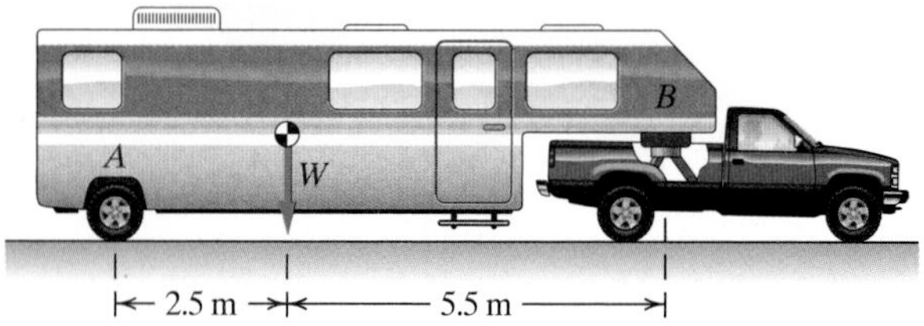

E7.2.40

7.2.41. A wheelbarrow loaded with a tree weighing 400 N sits on level ground as shown in E7.2.41. It is in equilibrium.

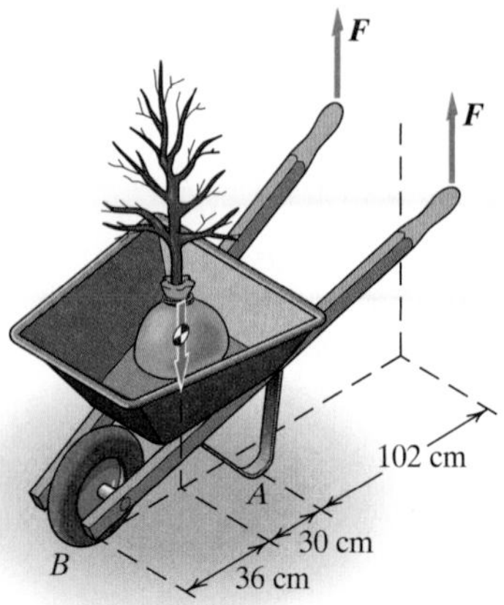

E7.2.41

a. What are the loads at A and B if $\| F \| = 0$?

b. What is the value of $\| F \|$ necessary to just lift the wheelbarrow off the ground at A? Do you think that you could personally supply this magnitude of force?

7.2.42. A beam is bolted to a wall at end A and loaded as shown in E7.2.42.

a. Determine the loads acting on the beam at A if the beam is in equilibrium.

b. Replace the moment you found at end A with a couple, such that the forces act through the centers of the two bolts. What is the magnitude of the force? This couple results in tension in which of the two bolts?

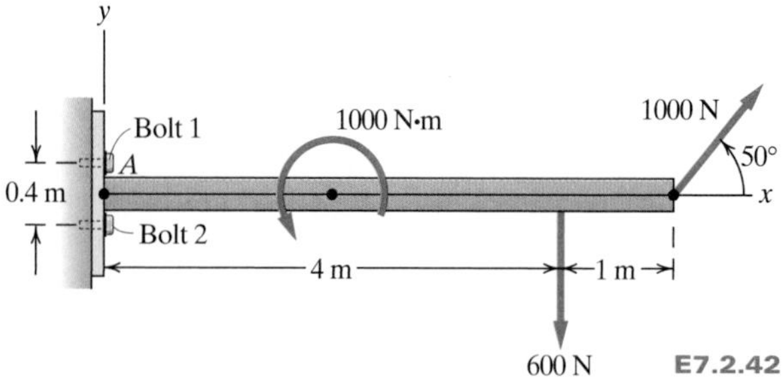

E7.2.42

7.2.43. A child who weighs 400 N is doing push-ups.

a. When the child does standard push-ups and is in the position shown in E7.2.43*a*, determine the loads acting on each hand and on each foot.

b. When the child does modified push-ups and is in the position shown in **E7.2.43*b***, determine the loads acting on each hand and on each knee.

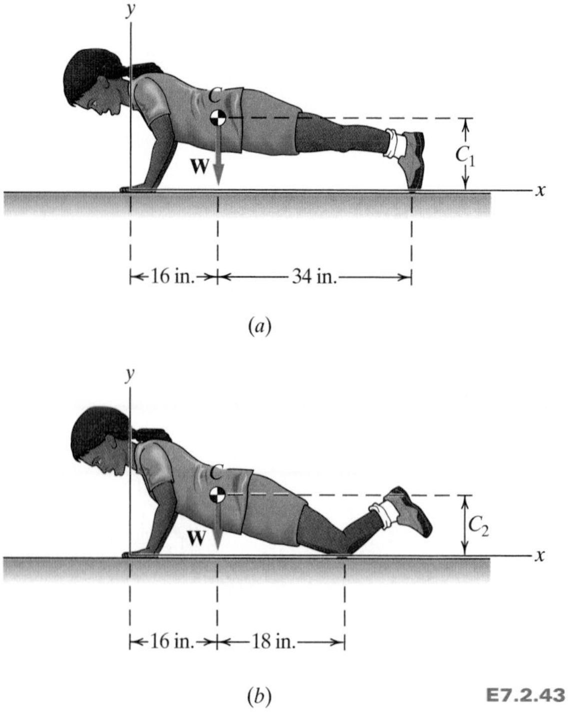

E7.2.43

c. Compare the loads acting on each hand in **a** and **b** by finding the loads in **b** as a percentage of the loads in **a**.

7.2.44. The child in Figure **E7.2.34** is of weight W and is standing at the center of a plank in a playground. The child and plank are in equilibrium. Planes A and B are smooth.

a. Determine the tension in the cable (F_{Bt}) and the ground normal force at B (F_{Bn}), in terms of L, W, α, and β. Present your answers as equations.

b. If $L = 3$ m, $W = 400$ N, and $\beta = 25$ degrees, create plots of F_{Bt} and F_{Bn} as α varies from 10 to 90 degrees. A spreadsheet may be helpful.

c. Use the plots created in **b** to determine at what angle α the magnitudes of F_{Bt} and F_{Bn} are equal.

d. Use the plots created in **b** to determine at what angle α the tension in the cable is 120 N. Call this the limiting value of α_{limit}. If 120 N is the breaking load of the cable, should the angle α be kept to values greater than or less than α_{limit}?

7.2.45. Suppose the cable at B in **E7.2.34** is removed and plane B is rough. Furthermore, plane A is smooth.

a. Determine the friction (F_{Bf}) and the normal force (F_{Bn}) at B in terms of L, W, α, and β. Present your answers as equations.

b. If $L = 3$ m, $W = 400$ N, and $\beta = 25$ degrees, create plots of F_{Bf} and F_{Bn} as β varies from 10 to 90 degrees.

c. If the plank will slide along the plane at B when $F_{Bf} = 0.5\ F_{Bn}$, what is the maximum value of β such that the plank will not slide?

d. If you wanted the plank setup to be in equilibrium at values of β greater than the one determined in **c**, what changes could you recommend to the setup (name at least 2 changes)?

7.2.46. **a.** Consider the free-body diagram of a bicycle in E7.2.46*a* that is traveling at a constant speed. A wind drag force acts at the location shown. Write expressions for the force B (contact between front tire and road; F_{B1}) and force A (contact between rear tire and road; F_{A1}) in terms of the dimensions a_1, b_1, and c_1, and forces W and $F_{\text{wind},1}$.

b. Use the expressions in **a** to answer the questions: If the drag force increases in magnitude, will the normal force at A decrease or increase? How about the normal force at B? Summarize your reasoning.

c. If the person rides in a hunched position, as shown in **E7.2.46*b***, we find that $a_2 > a_1$, $c_2 < c_1$, and $F_{\text{wind},2} < F_{\text{wind},1}$. Based on this, do you expect the normal force at A to decrease, increase, or remain the same relative to the situation in **E7.2.46*a***? How about the normal force at B? (*Hint:* The expressions in **a** may be helpful in thinking through these questions.) Summarize your reasoning.

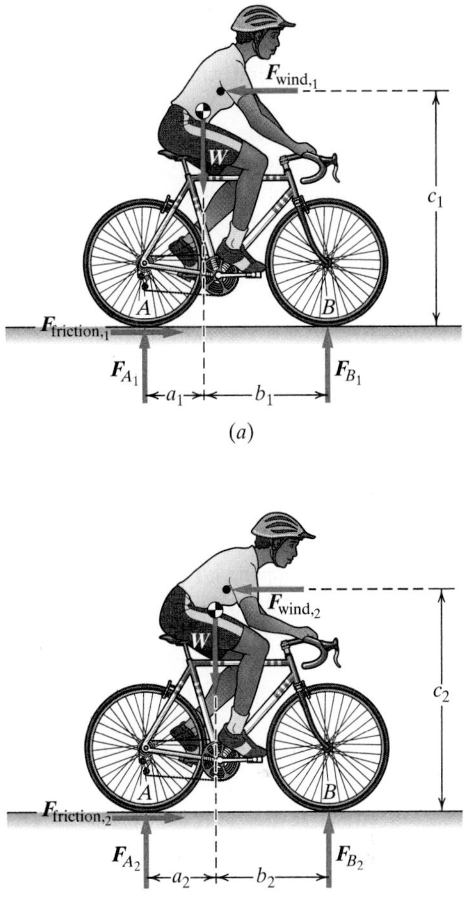

E7.2.46

7.2.47. A student does arm curls with her arm oriented as shown in E7.2.47. Write an expression for F_{muscle} as a function of W and θ, starting with the $\theta = 0°$ orientation shown. Show both the expression and a plot of the expression. Assume that the muscle force remains parallel to the upper arm throughout the motion and that the motion is slow enough for equilibrium to hold.

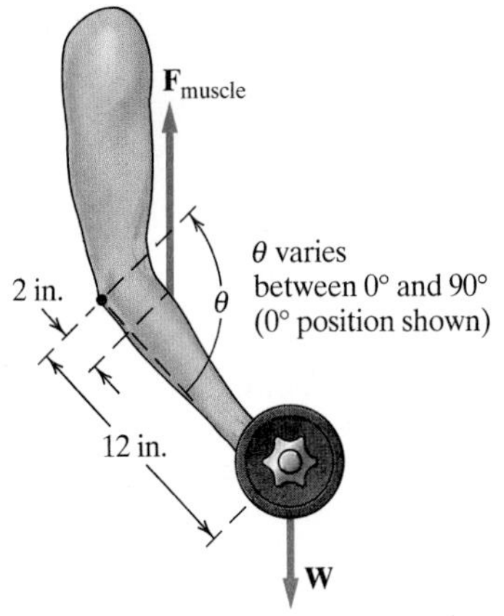

E7.2.47

7.2.48. A student does arm curls with her arm oriented as shown in E7.2.48. Write an expression for F_{muscle} as a function of W and θ, starting with the $\theta = 0°$ orientation shown. Show both the expression and a plot of the expression. Assume that the muscle force remains parallel to the upper arm throughout the motion and that the motion is slow enough for equilibrium to hold.

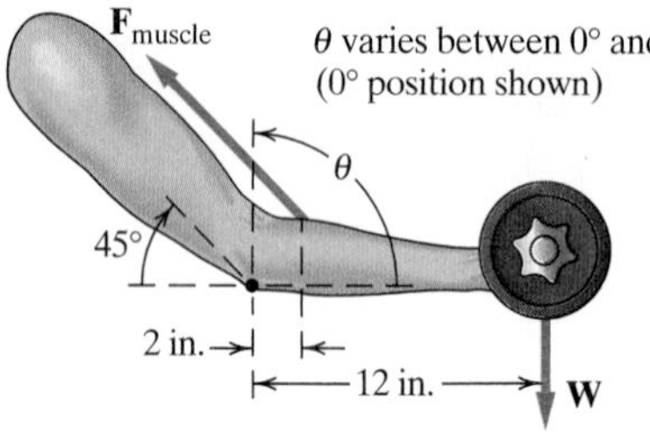

E7.2.48

7.2.49. A concrete hopper and its contents have a combined mass of 400 kg, with mass center at G as shown in E7.2.49. It is being elevated at constant velocity along its vertical guide by cable tension $\boldsymbol{T}$. The design calls for two sets of guide rollers at A, one on each side of the hopper, and two sets at B. What is the tension in the cable? What loads act on the hopper at A? What loads act on the hopper at B?

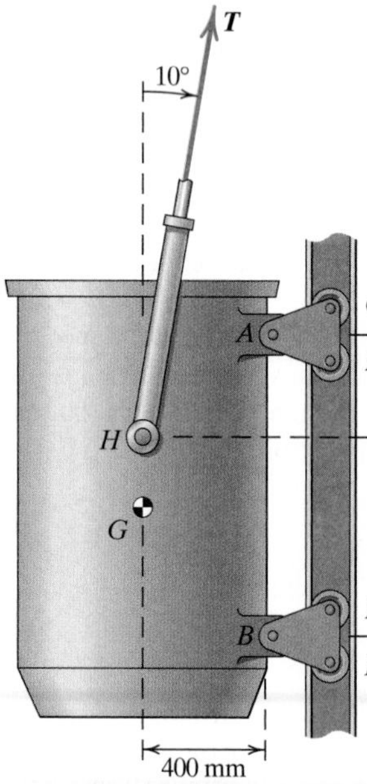

E7.2.49

7.2.50. The throttle-control sector in E7.2.50 pivots freely at O. An internal torsional spring at O exerts a return moment of magnitude $\| \boldsymbol{M} \| = 2$ N · m on the sector when in the position shown. If the sector is in equilibrium, what is the tension in the cable?

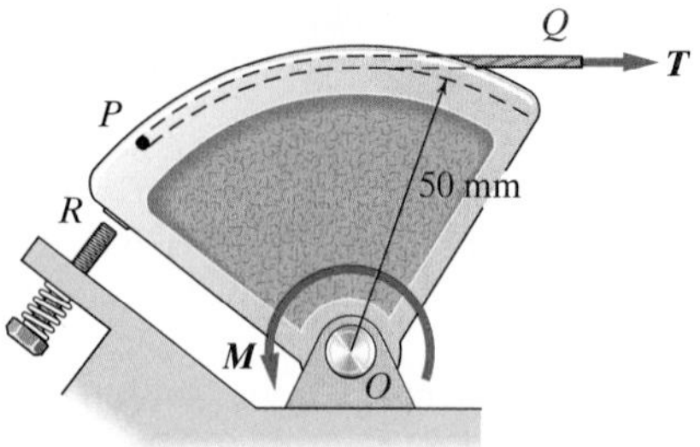

E7.2.50

7.2.51. When the cylinder of weight W in E7.2.51 is 7 m from the pin connection at C, the tension $\boldsymbol{T}$ in the cable has a magnitude of 9 kN. The beam AC is of negligible weight. If the beam is in equilibrium, determine the weight of the cylinder.

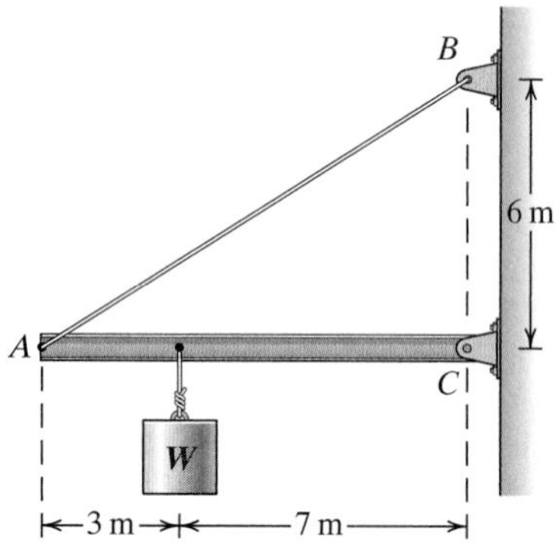

E7.2.51

7.2.52. The device shown in E7.2.52 is designed to apply pressure when bonding laminate to each side of a countertop near the edge. If a 120-N force is applied to the handle, determine the force that each roller exerts on its corresponding surface.

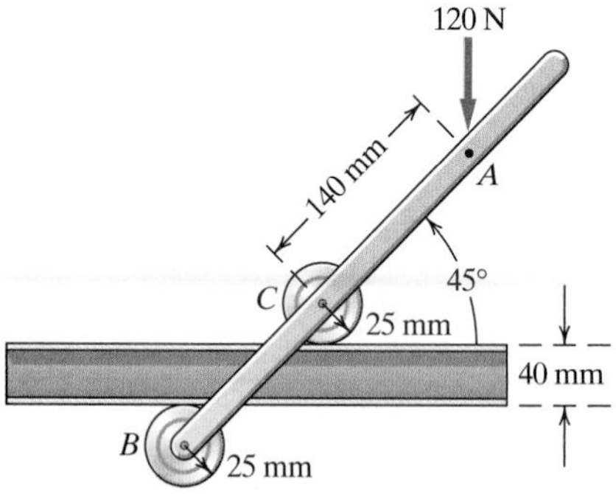

E7.2.52

7.2.53. Determine the forces acting at A and C on frame ABC in E7.2.53 that is in equilibrium.

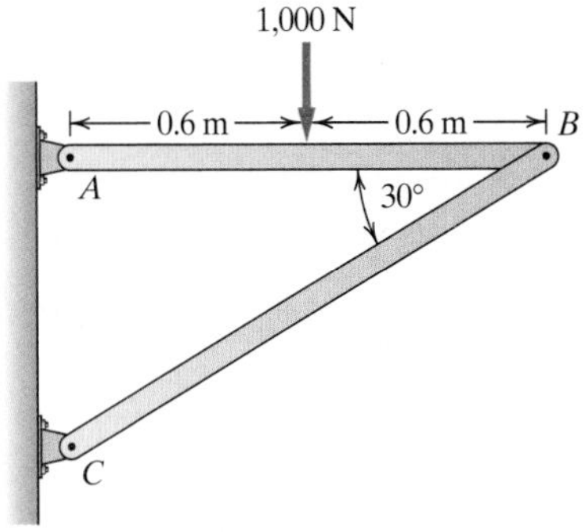

E7.2.53

7.2.54. Determine the forces acting at A and C on frame ABC in E7.2.54 that is in equilibrium.

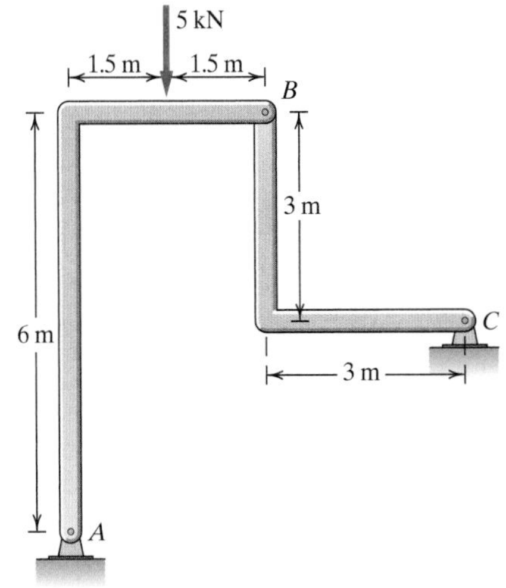

E7.2.54

7.2.55. Consider the frame in E7.2.55. If the weights of Members 1 and 2 are negligible, what loads act on the frame at A and B?

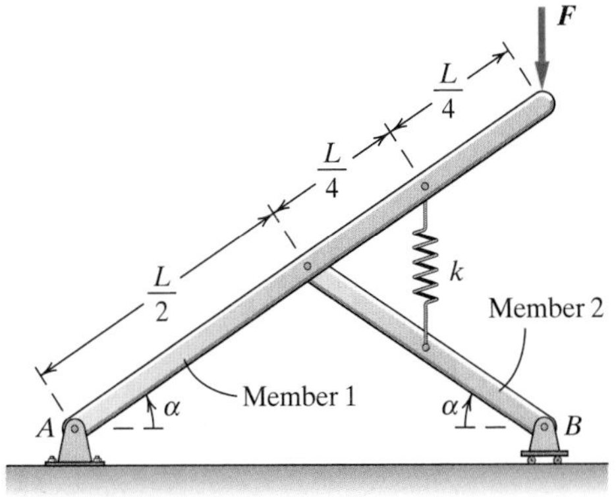

E7.2.55

7.2.56. The exercise frame in E7.2.56 is bolted to the floor at E and pinned at F. Determine the loads acting at E and F if the frame is in equilibrium.

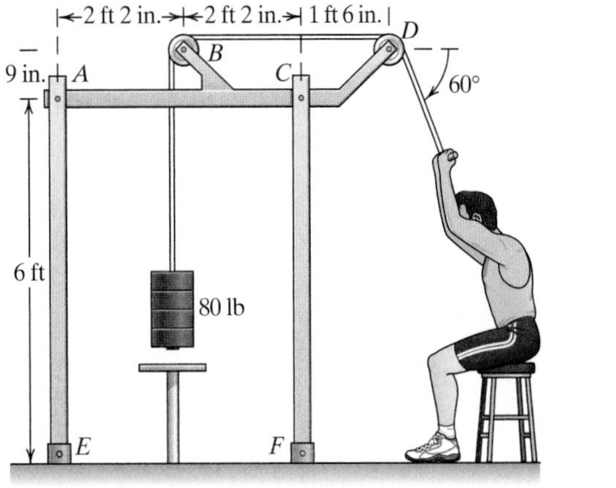

E7.2.56

7.2.57. A portion of a mechanical coin sorter is shown in E7.2.57. Pennies and dimes roll down the 20° incline, the triangular portion of which pivots freely about a horizontal axis through O. Dimes are light enough (2.28 grams mass each) so that the triangular portion remains stationary, and the dimes roll into the right collection column. Pennies, on the other hand, are heavy enough (3.06 grams mass each) so that the triangular portion pivots clockwise, and the pennies roll into the left collection column. The mass of the triangular portion is 5.39 grams, with center of gravity at A. At what position "s" will a penny cause the triangular portion to pivot? Confirm that a dime will not cause the triangular portion to pivot.

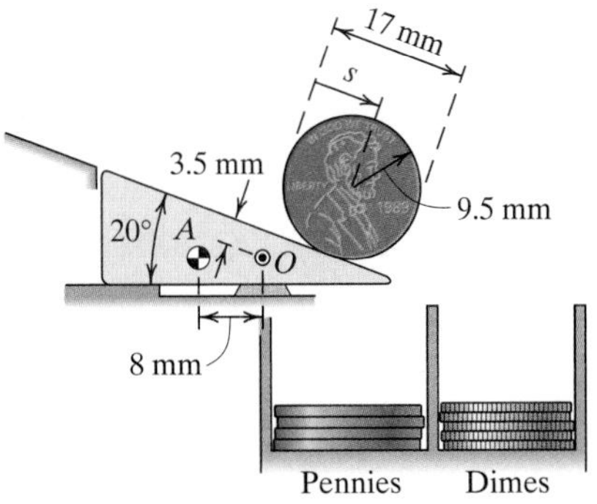

E7.2.57

7.2.58. The following three cases involve a 2 × 4 and a wrench in equilibrium. In each case a force is applied by the wrench and hands are used to react to this force, as shown. Ignore the weight of the 2 × 4.

a. Consider Case A in E7.2.58*a*. Define the system as the 2 × 4, bolt, and wrench and draw its free-body diagram, clearly showing the point of application of each force and stating all assumptions. Then determine the forces the hands apply to the 2 × 4.

b. Consider Case B in **E7.2.58*b***. Define the system as the 2 × 4, bolt, and wrench and draw its free-body diagram, clearly showing the point of application of each force and stating all assumptions. Then determine the forces the hands apply to the 2 × 4.

c. Consider Case C in **E7.2.58*c***. Define the system as the 2 × 4, bolt, and wrench and draw its free-body diagram, clearly showing the point of application of each force and stating all assumptions. Then determine the forces the hands apply to the 2 × 4. (Depending on the assumptions you made, you may not be able to determine all of the loads acting on the 2 × 4.)

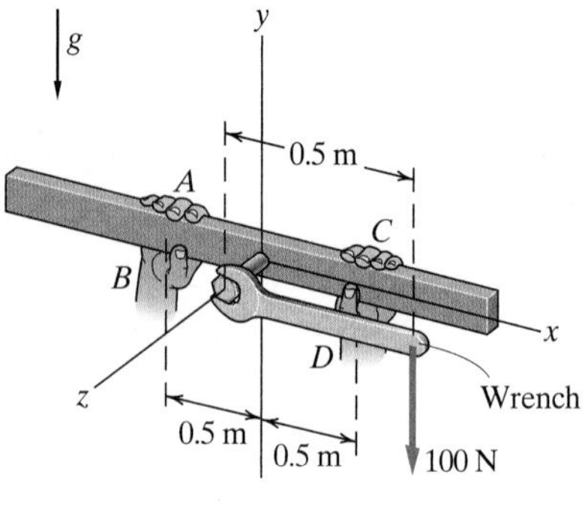

(*b*) Case B

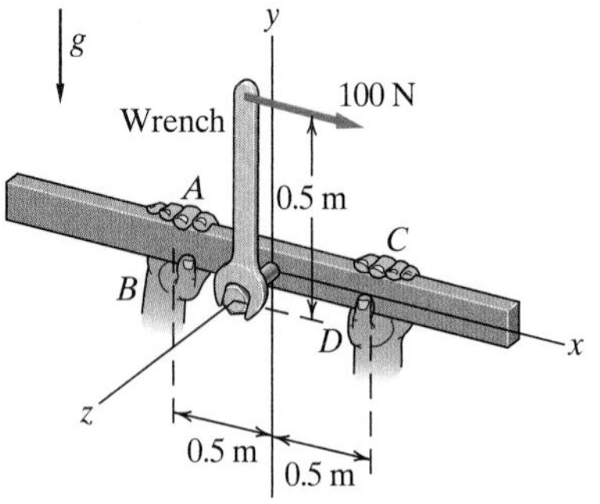

(*a*) Case A

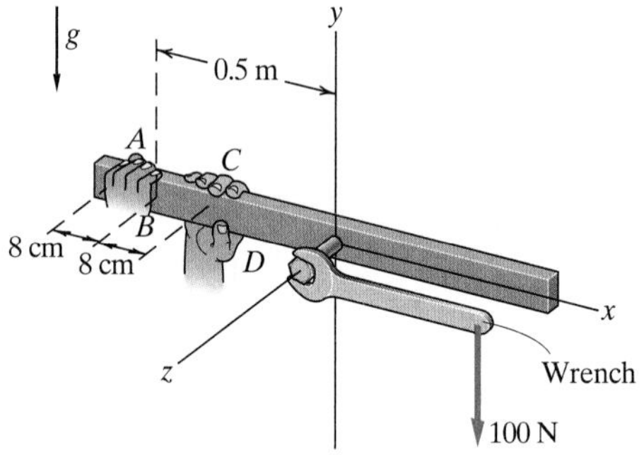

(*c*) Case C

E7.2.58

7.3 APPLICATION OF THE CONDITIONS—NONPLANAR SYSTEMS

The steps **Formulate Equations**, **Solve** and **Check** discussed above for finding moments, forces, and/or dimensions associated with a balanced planar system also apply to a balanced nonplanar system. What differs is that for a nonplanar system we must generally write out all six equilibrium equations. For example, consider the nonplanar system in Figure 7.29*a* and its free-body diagram in **Figure 7.29*b***. If it is

(*a*)

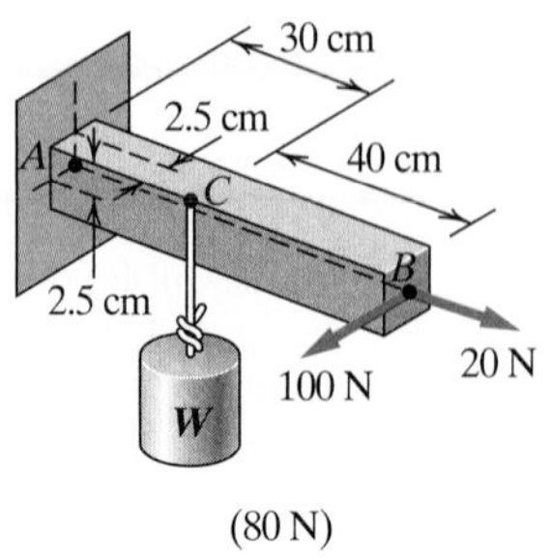

(*b*)

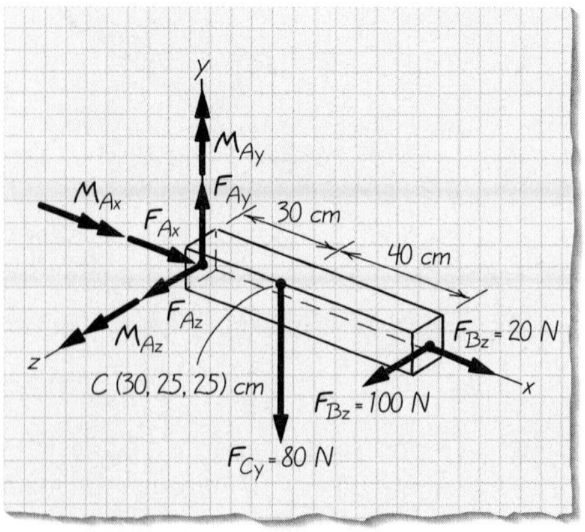

Figure 7.29 (*a*) A beam that is 700 cm long is bolted to a wood pillar at *A*. The cross section of the beam measures 5 × 5 cm. (*b*) Free-body diagram of beam with calculated values of loads shown.

balanced, we must consider forces and moments in the x, y, and z directions, as summarized below.

Formulate Equations

Reading force information from the free-body diagram, write the x component force equilibrium equation (7.3A), making sure to include all x component forces. Repeat for the y component force equilibrium equation (7.3B) and the z component force equilibrium equation (7.3C). You now have written three force equilibrium equations that contain unknown forces. For the system in **Figure 7.29**, these equations are

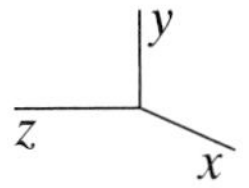

$$\sum \boldsymbol{F} = 0 \qquad (7.1)$$

$$\sum F_x = 0 = F_{Ax} + \underbrace{F_{Bx}}_{20\text{N}} = 0; \text{ based on (7.3A)}$$

$$F_{Ax} + 20\text{ N} = 0$$

$$\sum F_y = 0 = F_{Ay} - \underbrace{F_{Cy}}_{80\text{ N}} = 0; \text{ based on (7.3B)}$$

$$F_{Ay} - 80\text{ N} = 0$$

$$\sum F_z = 0 = F_{Az} + \underbrace{F_{Bz}}_{100\text{ N}} = 0; \text{ based on (7.3C)}$$

$$F_{Az} + 100\text{ N} = 0$$

Now write the x component moment equilibrium equation (7.4A) by first deciding which point to take as the moment center. Be sure to include all moments in the x direction created by forces and by moments. Repeat for the y component equilibrium equation (7.4B) and the z component equilibrium equation (7.4C). These three equations may contain unknown moments, forces, and/or dimensions. For the situation in **Figure 7.29*a*** with MC at A, these equations are

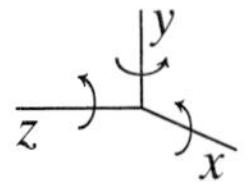

$$\sum \boldsymbol{M} = 0 \qquad (7.2)$$

$$\sum M_{x@A} = 0 = (2.5\text{ cm})\underbrace{F_{Cy}}_{80\text{ N}} + M_{Ax} = 0; \text{ based on (7.4A)}$$

$$200\text{ N}\cdot\text{cm} + M_{Ax} = 0$$

$$\sum M_{y@A} = 0 = -(70\text{ cm})\underbrace{F_{Bz}}_{100\text{ N}} + M_{Ay} = 0; \text{ based on (7.4B)}$$

$$-7000\text{ N}\cdot\text{cm} + M_{Ay} = 0$$

$$\sum M_{z@A} = 0 = -(30\text{ cm})\underbrace{F_{Cy}}_{80\text{ N}} + M_{Az} = 0; \text{ based on (7.4C)}$$

$$-2400\text{ N}\cdot\text{cm} + M_{Az} = 0$$

Solve

There are now six equations of equilibrium that can be solved for (at most) six unknowns. Furthermore, these equations are independent of one another. Most of the systems presented in this book are such that the equilibrium equations can reasonably be solved by hand for the unknowns. Using the six equilibrium equations for the system in **Figure 7.29*a***, we find

$$F_{Ax} = -20 \text{ N (from 7.3A)}$$

(negative sign indicates force in direction opposite to that shown in free-body diagram)

$$F_{Ay} = +80 \text{ N (from 7.3B)}$$
$$F_{Az} = -100 \text{ N (from 7.3C)}$$
$$M_{Ax} = -200 \text{ N} \cdot \text{cm (from 7.4A)}$$

(negative sign indicates moment in direction opposite to that shown in free-body diagram)

$$M_{Ay} = +7000 \text{ N} \cdot \text{cm (from 7.4B)}$$
$$M_{Az} = +2400 \text{ N} \cdot \text{cm (from 7.4C)}$$

Answer The loads acting on the beam at A due to the wall are:
$\boldsymbol{F_A} = -20 \text{ N}\boldsymbol{i} + 80 \text{ N}\boldsymbol{j} - 100 \text{ N}\boldsymbol{k}$
$\boldsymbol{M_A} = -200 \text{ N} \cdot \text{cm } \boldsymbol{i} + 7000 \text{ N} \cdot \text{cm } \boldsymbol{j} + 2400 \text{ N} \cdot \text{cm } \boldsymbol{k}$

Alternately, we can answer $F_{Ax} = -20$ N; $F_{Ay} = 80$ N; $F_{Az} = -100$ N $M_{Ax} = -200$ N · cm; $M_{Ay} = 7000$ N · cm; $M_{Az} = 2400$ N · cm, where the variables $F_{Ax} \ldots M_{Az}$ are defined in **Figure 7.29*b***.

Check

As with planar systems, there is more to do after you have solved the equations and calculated the numbers. The results should be checked using technical knowledge, engineering judgment, and common sense. Compare values with design requirements (if available) and draw conclusions about the adequacy of the design.

An inspection of the equilibrium free-body diagram in **Figure 7.29*b*** in conjunction with calculated force values shows that there is force balance of the system. It is more difficult to check for moment balance simply by inspecting a diagram. One check would be to rewrite the moment equilibrium equations about an alternate moment center. Use the actual values of loads in writing the equations to confirm that there is moment equilibrium.

EXAMPLE 7.9 HIGH-WIRE CIRCUS ACT

Poles AB and EF supporting a circus high wire are each held in their vertical positions by guy wires (Figure 7.30). The tension acting on the high wire (BF) is 9 kN. Assume the weight of the wires and the poles can be ignored. Find the forces acting on guy wires BC and BD and on pole AB at B.

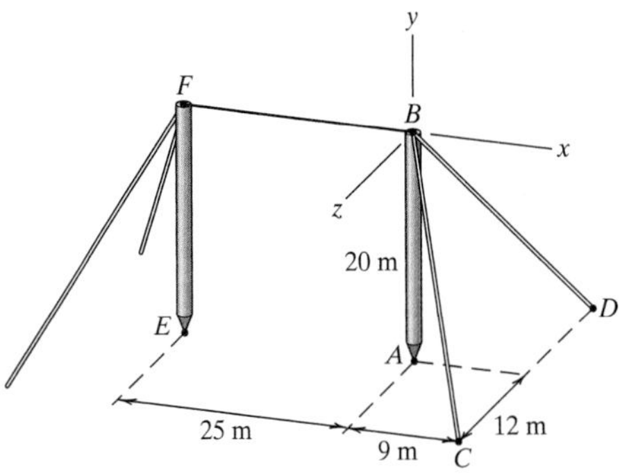

Figure 7.30

Goal We are to find the forces acting on guy wires BC and BD and on pole AB.

Given We are given the dimensions of the system and we are told that the tension force acting on the high wire is 9 kN.

Assume We will assume that the wires are two-force (tension-only) elements, the poles are two-force elements, and joint A is a ball-and-socket connection. We must treat the system as nonplanar. Also, we can model the point of concurrency of the lines of action of the member forces (B) as a particle.

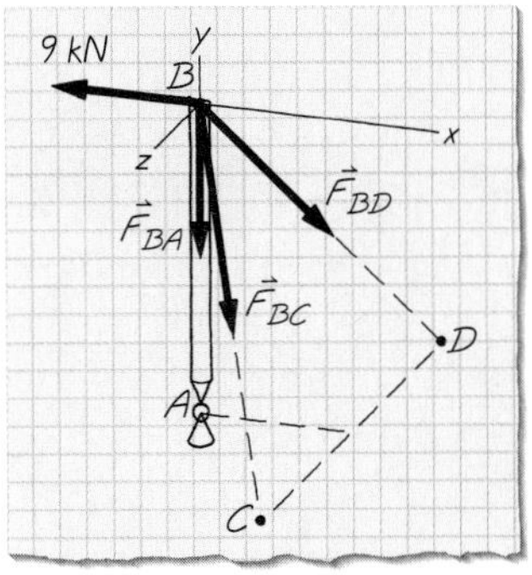

Figure 7.31

Draw Based on the information given in the problem and our assumptions, we draw a free-body diagram of the forces acting on particle B (Figure 7.31). We have arbitrarily assumed that all unknown forces are acting away from B and therefore indicate tension in cables.

Formulate Equations and Solve Before we set up the equilibrium equations, we need to write each force in terms of rectangular components so that we can write one equilibrium equation for each component direction (x, y, and z). Writing the force vector as the product of its magnitude and unit vectors (from (4.11)) we get

$$\begin{aligned}
\boldsymbol{F}_{BA} &= -F_{BA}\boldsymbol{j} \\
\boldsymbol{F}_{BC} &= F_{BC}\boldsymbol{u}_{BC} \\
\boldsymbol{F}_{BD} &= F_{BD}\boldsymbol{u}_{BD} \\
\boldsymbol{F}_{BF} &= -9\text{ kN}\,\boldsymbol{i}
\end{aligned}$$

From (4.10C) we are able to calculate the unit vectors

$$\boldsymbol{u}_{BC} = \frac{9\text{ m}\,\boldsymbol{i} - 20\text{ m}\,\boldsymbol{j} + 12\text{ m}\,\boldsymbol{k}}{\sqrt{(9\text{ m})^2 + (-20\text{ m})^2 + (12\text{ m})^2}} = \frac{9}{25}\boldsymbol{i} - \frac{20}{25}\boldsymbol{j} + \frac{12}{25}\boldsymbol{k}$$

$$\boldsymbol{u}_{BD} = \frac{9\text{ m}\,\boldsymbol{i} - 20\text{ m}\,\boldsymbol{j} - 12\text{ m}\,\boldsymbol{k}}{\sqrt{(9\text{ m})^2 + (-20\text{ m})^2 + (-12\text{ m})^2}} = \frac{9}{25}\boldsymbol{i} - \frac{20}{25}\boldsymbol{j} - \frac{12}{25}\boldsymbol{k}$$

We now set up the equations for nonplanar equilibrium to find the unknown loads at B. Because we are modeling B as a particle, we need only consider force equilibrium equations. The moment equilibrium equations are satisfied because all the applied forces are concurrent at B.

Based on (7.3A):

$$\sum F_x = 0\ (\rightarrow +)$$

$$\frac{9}{25}F_{BC} + \frac{9}{25}F_{BD} - 9\text{ kN} = 0 \qquad (1)$$

Based on (7.3B):

$$\sum F_y = 0\ (\uparrow +)$$

$$-F_{BA} - \frac{20}{25}F_{BC} - \frac{20}{25}F_{BD} = 0 \qquad (2)$$

Based on (7.3C):

$$\sum F_z = 0\ (\text{ out of page } +)$$

$$\frac{12}{25}F_{BC} - \frac{12}{25}F_{BD} = 0$$

$$F_{BC} = F_{BD} \qquad (3)$$

We substitute (3) into (1) to find F_{BC} and F_{BD}:

$$\frac{9}{25}F_{BC} + \frac{9}{25}F_{BC} - 9\text{ kN} = 0$$

$$F_{BC} = 12.5\text{ kN} = F_{BD} \qquad (4)$$

Finally, we substitute (4) into (2) to determine F_{BA}:

$$-F_{BA} - \frac{20}{25}(12.5\text{ kN}) - \frac{20}{25}(12.5\text{ kN}) = 0$$

$$F_{BA} = -20\text{ kN}$$

Answer

$F_{BA} = -20$ kN
$F_{BC} = 12.5$ kN
$F_{BD} = 12.5$ kN
The minus sign for F_{BA} means that the force is acting opposite from the direction we drew in the free-body diagram. Therefore F_{BA} is acting toward point *B*, meaning that member *BA* (which is the pole) is in compression.

Check It makes sense that F_{BC} or F_{BD} are both positive; this indicates tension in the guy wires. We would be suspicious of our answers had either of them been calculated to be negative. Also, because of symmetry of the situation, it makes sense that F_{BC} and F_{BD} are the same size (12.5 kN). Finally, the guy wires pulling on the top of the pole results in the pole being compressed; this is consistent with the negative answer we calculated (−20 kN).

EXAMPLE 7.10 CANTILEVER BEAM WITH OFF-CENTER FORCE AND COUPLE

The uniform I-beam shown in Figure 7.32 is fixed to the wall at 0. Three mutually perpendicular forces are applied at A: 3.0 kN in the x direction, 1.5 kN in the y direction, and 2.0 kN in the negative z direction. At B the beam is subjected to a negative 3.0 kN · m moment about the z axis, and at C a 4.0-kN load in the x direction is applied. Find the loads acting on the beam at the wall (location O).

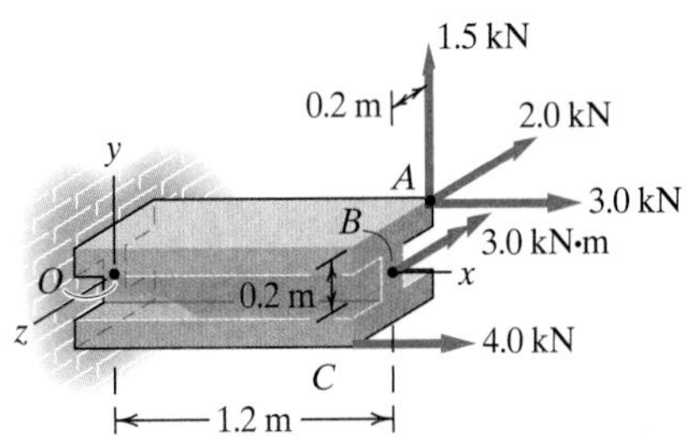

Figure 7.32

Goal We are to find the loads acting on the beam at the wall.

Given We are given information about the geometry of the I-beam and about the known loads acting at A, B, and C.

Assume We will assume that we can ignore the weight of the I-beam (because we are not given any information about it), and treat the structure as a nonplanar system.

Draw Based on the information given in the problem and our assumptions, we draw a free-body diagram. In the free-body diagram in Figure 7.33*a*, the loads acting at the fixed end O are represented in terms of components. In the alternate free-body diagram in **Figure 7.33*b*** the loads at the fixed end are represented in terms of a single force ($\boldsymbol{F}_O$) and a single moment ($\boldsymbol{M}_O$). Both free-body diagrams are correct. We will use **Figure 7.33*a*** with Approach 1, and **Figure 7.33*b*** with Approach 2, as presented below.

Approach 1 (Use Equilibrium Equations):

Formulate Equations and Solve Based on (7.3A) and **Figure 7.33*a***:

$$\sum F_x = 0\ (\rightarrow +)$$
$$F_{Ox} + 3.0\text{ kN} + 4.0\text{ kN} = 0$$
$$F_{Ox} = -\ (3.0 + 4.0)\text{ kN}$$
$$F_{Ox} = -\ 7.0\text{ kN}$$

Based on (7.3B):

$$\sum F_y = 0\ (\uparrow +)$$
$$F_{Oy} + 1.5\text{ kN} = 0$$
$$F_{Oy} = -1.5\text{ kN}$$

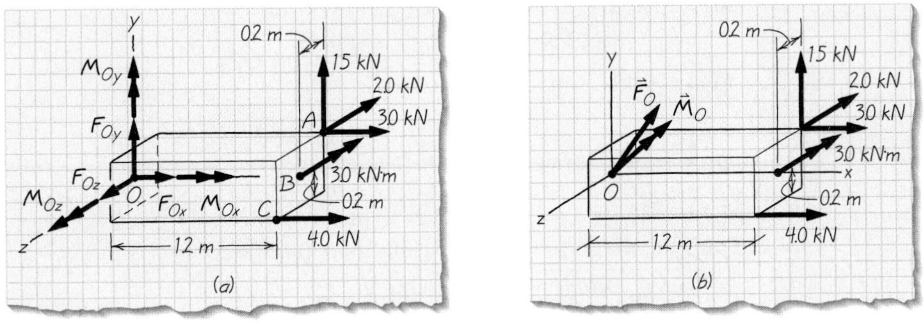

Figure 7.33

Based on (7.3C):

$$\sum F_z = 0 \text{ (out of page +)}$$
$$F_{Oz} - 2.0\text{ kN} = 0$$
$$F_{Oz} = 2.0\text{ kN}$$

Therefore, we can write:

Answer $\boldsymbol{F_O} = F_{Ox}\boldsymbol{i} + F_{Oy}\boldsymbol{j} + F_{Oz}\ \boldsymbol{k} = -7.0\text{ kN}\,\boldsymbol{i} - 1.5\text{ kN}\,\boldsymbol{j} + 2.0\text{ kN}\,\boldsymbol{k}$
We can find the magnitude of $\boldsymbol{F_O}$ using the equation
$\|\boldsymbol{F_O}\| = \sqrt{F_{Ox}^2 + F_{Oy}^2 + F_{Oz}^2}$, giving us
$\|\boldsymbol{F_O}\| = \sqrt{(-7.0)^2 + (-1.5)^2 + (2.0)^2} = 7.43\text{ kN}$

In using the moment equilibrium equations, we define positive moments as those that create a positive moment about an axis. Based on (7.4A) with the moment center at O in **Figure 7.33*a***:

$$\sum M_{x@O} = 0$$
$$M_{Ox} + (0.2\text{ m})(1.5\text{ kN}) - (0.2\text{ m})(2\text{ kN}) = 0$$
$$M_{Ox} = (0.2\text{ m})(2 - 1.5)\text{kN}$$
$$M_{Ox} = 0.1\text{ kN}\cdot\text{m}$$

Based on (7.4B) with moment center at O:

$$\sum M_y = 0$$
$$M_{Oy} + (0.2\text{ m})(4\text{ kN}) - (0.2\text{ m})(3\text{ kN}) + (1.2\text{ m})(2.0\text{ kN}) = 0$$
$$M_{Oy} = -(0.2\text{ m})(4\text{ kN}) + (0.2\text{ m})(3\text{ kN}) - (1.2\text{ m})(2.0\text{ kN})$$
$$M_{Oy} = -2.6\text{ kN}\cdot\text{m}$$

Based on (7.4C) with moment center at O:

$$\sum M_z = 0$$
$$M_{Oz} - (3\text{ kN}\cdot\text{m}) + (1.2\text{ m})(1.5\text{ kN}) - (0.2\text{ m})(3\text{ kN}) + (0.2\text{ m})(4\text{ kN}) = 0$$
$$M_{Oz} = (3\text{ kN}\cdot\text{m}) - (1.2\text{ m})(1.5\text{ kN}) + (0.2\text{ m})(3\text{ kN}) - (0.2\text{ m})(4\text{ kN})$$
$$M_{Oz} = 1\text{ kN}\cdot\text{m}$$

Therefore, we can write:

Answer $\boldsymbol{M_O} = M_{Ox}\boldsymbol{i} + M_{Oy}\boldsymbol{j} + M_{Oz}\boldsymbol{k} = 0.1\text{ kN}\cdot\text{m}\,\boldsymbol{i} - 2.6\text{ kN}\cdot\text{m}\,\boldsymbol{j} + 1.0\text{ kN}\cdot\text{m}\,\boldsymbol{k}$
We can find the magnitude of M_O using the equation
$\|\boldsymbol{M_O}\| = \sqrt{M_{Ox}^2 + M_{Oy}^2 + M_{Oz}^2}$, giving us
$\|\boldsymbol{M_O}\| = \sqrt{(0.1)^2 + (-2.6)^2 + (1.0)^2} = 2.8\text{ kN}\cdot\text{m}$

Approach 2 (Use Equilibrium Conditions):

Formulate Equations and Solve Based on the force equilibrium condition (7.1) and **Figure 7.33*b*** we write:

$$\sum \boldsymbol{F} = 0$$
$$\boldsymbol{F}_O + \boldsymbol{F}_A + \boldsymbol{F}_C = 0$$
$$\boldsymbol{F}_O = -\boldsymbol{F}_A - \boldsymbol{F}_C$$
$$\boldsymbol{F}_O = -(3.0 \text{ kN } \boldsymbol{i} + 1.5 \text{ kN } \boldsymbol{j} - 2.0 \text{ kN } \boldsymbol{k}) - 4.0 \text{ kN } \boldsymbol{i}$$
$$\boldsymbol{F}_O = -7.0 \text{ kN } \boldsymbol{i} - 1.5 \text{ kN } \boldsymbol{j} + 2.0 \text{ kN } \boldsymbol{k}$$

Therefore, we can write

Answer $\boldsymbol{F}_O = -7.0 \text{ kN } \boldsymbol{i} - 1.5 \text{ kN} \boldsymbol{j} + 2.0 \text{ kN } \boldsymbol{k}$

Based on the moment equilibrium condition (7.2), with the moment center at O, we can write:

$$\sum \boldsymbol{M} = 0$$
$$\boldsymbol{M}_O + (\boldsymbol{r}_A \times \boldsymbol{F}_A) + (\boldsymbol{r}_C \times \boldsymbol{F}_C) + \boldsymbol{M}_B = 0$$
$$\boldsymbol{M}_O = -[(\boldsymbol{r}_A \times \boldsymbol{F}_A) + (\boldsymbol{r}_C \times \boldsymbol{F}_C) + \boldsymbol{M}_B]$$
$$\boldsymbol{M}_O = -[(1.2\boldsymbol{i} + 0.2\boldsymbol{j} - 0.2\boldsymbol{k}) \times (3.0\,\boldsymbol{i} + 1.5\,\boldsymbol{j} - 2.0\,\boldsymbol{k})$$
$$+ (1.2\boldsymbol{i} - 0.2\boldsymbol{j} + 0.2\boldsymbol{k}) \times (4.0\,\boldsymbol{i} + 0\,\boldsymbol{j} - 0\,\boldsymbol{k})$$
$$+ (0\boldsymbol{i} + 0\boldsymbol{j} - 3.0\boldsymbol{k})] \text{kN} \cdot \text{m}$$

Alternately, we can use determinants to expand the cross product, as presented in Chapter 5. The determinants for this problem are

$$\boldsymbol{r}_A \times \boldsymbol{F}_A = \begin{vmatrix} \boldsymbol{i} & \boldsymbol{j} & \boldsymbol{k} \\ 1.2 \text{ m} & 0.2 \text{ m} & -0.2 \text{ m} \\ 3.0 \text{ kN} & 1.5 \text{ kN} & -2.0 \text{ kN} \end{vmatrix}$$

$$= [(-0.4 + 0.3)\boldsymbol{i} + (-0.6 + 2.4)\boldsymbol{j} + (1.8 - 0.6)\boldsymbol{k}] \text{ kN} \cdot \text{m}$$
$$= (-0.1\,\boldsymbol{i} + 1.8\,\boldsymbol{j} + 1.2\,\boldsymbol{k}) \text{ kN} \cdot \text{m}$$

$$\boldsymbol{r}_C \times \boldsymbol{F}_C = \begin{vmatrix} \boldsymbol{i} & \boldsymbol{j} & \boldsymbol{k} \\ 1.2 \text{ m} & -0.2 \text{ m} & 0.2 \text{ m} \\ 4.0 \text{ kN} & 0 \text{ kN} & 0 \text{ kN} \end{vmatrix}$$

$$= [(0)\boldsymbol{i} + (0.8 + 0)\boldsymbol{j} + (0 + 0.8)\mathbf{k}] \text{ kN} \cdot \text{m}$$
$$= (0.8\,\boldsymbol{j} + 0.8\,\boldsymbol{k}) \text{ kN} \cdot \text{m}$$

Therefore,

$$\sum \boldsymbol{M} = 0$$
$$\boldsymbol{M}_O + (\boldsymbol{r}_A \times \boldsymbol{F}_A) + (\boldsymbol{r}_C \times \boldsymbol{F}_C) + \boldsymbol{M}_B = 0$$
$$\boldsymbol{M}_O = -[(\boldsymbol{r}_A \times \boldsymbol{F}_A) + (\boldsymbol{r}_C \times \boldsymbol{F}_C) + \boldsymbol{M}_B]$$
$$\boldsymbol{M}_O = -[-0.1\,\boldsymbol{i} + 2.6\,\boldsymbol{j} - 1.0\,\boldsymbol{k}] \text{ kN} \cdot \text{m}$$

Answer $\boldsymbol{M}_O = 0.1 \text{ kN} \cdot \text{m}\,\boldsymbol{i} - 2.6 \text{ kN} \cdot \text{m}\,\boldsymbol{j} + 1.0 \text{ kN} \cdot \text{m}\,\boldsymbol{k}$

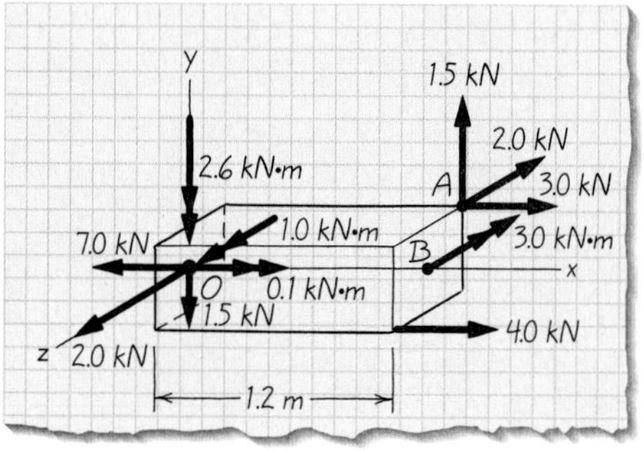

Figure 7.34

Check The two approaches serve as checks for each other. Alternatively, (7.5A), (7.5B), and (7.5C) could be written for another moment center to confirm the values of $\boldsymbol{F}_O$ and $\boldsymbol{M}_O$. Finally, a free-body diagram with all loads drawn in their proper direction with magnitude indicated illustrates that there is force balance of the system (Figure 7.34). Often the approach for checking is up to personal preference.

EXAMPLE 7.11 NONPLANAR PROBLEM WITH UNKNOWNS OTHER THAN LOADS

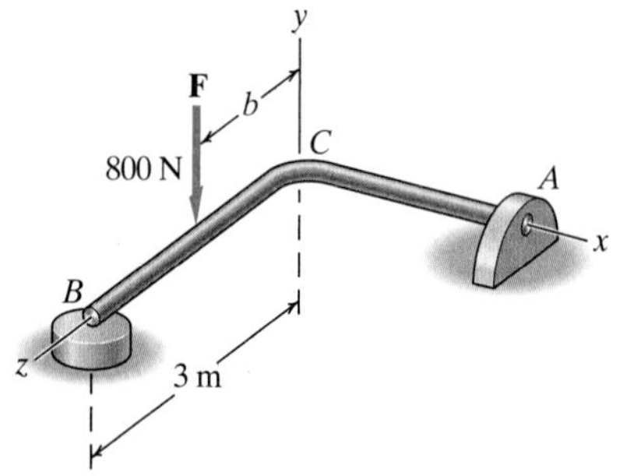

Figure 7.35

The L-shaped bar shown in Figure 7.35 is supported by a thrust bearing at A and rests on a smooth horizontal surface at B. A load $\boldsymbol{F}$ of 800 N is applied at a distance b from point C. If the maximum load that the surface at B can support is 600 N, what is the maximum value of dimension b?

Goal We are to find the largest value of dimension b such that the normal force at B is at most 600 N.

Given We are given information about the geometry of the L-bar and the applied 800-N force. In addition, we are told that the surface at B is smooth and that the support at A consists of a journal bearing.

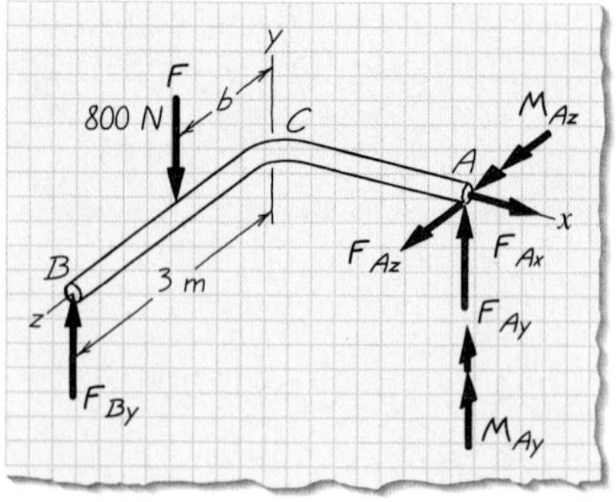

Figure 7.36

Assume We will assume that we can ignore the weight of the L-bar (because we are not given any information about it), and we will treat the L-bar as a nonplanar system. Also, we assume that the surface at B is frictionless, because it is smooth.

Draw Based on the information given in the problem and our assumptions, we draw a free-body diagram (Figure 7.36). Because there is only a single bearing attached to the system, the loads at A consist of forces and moments; see **Table 6.2**.

Formulate Equations and Solve For the L-bar to be in mechanical equilibrium, all six equations in (7.3) and (7.4) must hold. Based on (7.4A) (with the moment center at A):

$$\sum M_{x@A} = 0$$

$$bF - (3.0\text{ m})F_{By} = 0$$

$$F_{By} = \frac{b}{3.0\text{ m}} F$$

with $F = 800$ N, this becomes

$$F_{By} = \frac{b}{3.0\text{ m}}(800\text{ N}) \quad (1)$$

When $F_{By} = 600$ N, we can solve (1) for b:

$$600\text{ N} = \frac{b}{3.0\text{ m}}(800\text{ N})$$

from which we can find that $b = 2.25$ m.

Note:

- (1) tells us that smaller values of b result in smaller values of F_{By}.
- We did not need to write all six equilibrium equations in order to answer the question posed in this example.

Answer At $b = 2.25$ m, $F_{By} = 600$ m. Therefore, with $b \leq 2.25$ m, we have $F_{By} \leq 600$ N.

Check Does it make sense that F_{By} increases as b increases? And that we need to limit b in order to put a 600-N limit on F_{By}? What happens is that as b increases, more of the applied force $\boldsymbol{F}$ is carried by the support at B rather than by the support at A. In fact, from (7.3B) we can write:

$$\sum F_y = 0\ (\uparrow +)$$

$$F_{Ay} + F_{By} - F = 0$$

$$F_{Ay} = F - \frac{b}{3.0\text{ m}}(F)$$

$$F_{Ay} = F\left(1 - \frac{b}{3.0\text{ m}}\right) \qquad (2)$$

If we plot (1) for F_{By} and (2) for F_{Ay}, we can see how the force shifts from support A to support B as dimension b increases (Figure 7.37).

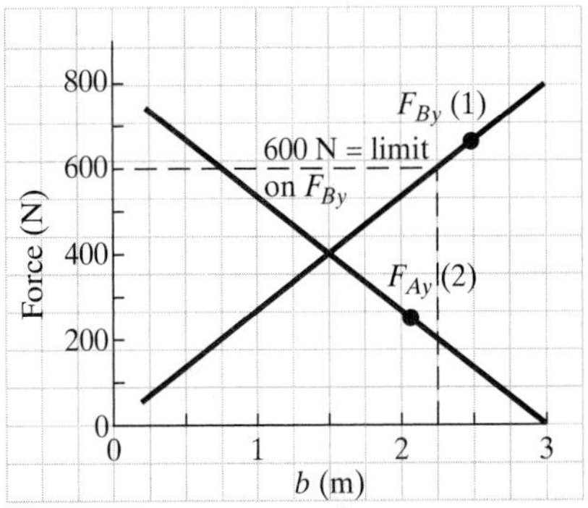

Figure 7.37

EXERCISES 7.3

7.3.1. The rectangular plate of uniform thickness in E7.3.1 has a mass of 500 kg. Determine the tensions in the three cables supporting the plate if the plate is in equilibrium.

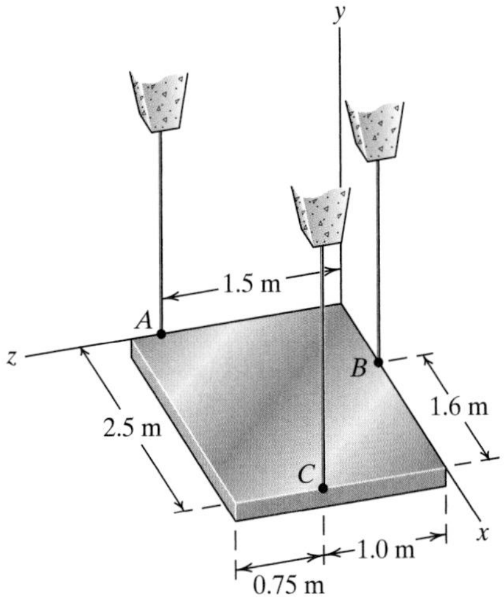

E7.3.1

7.3.2. The bent bar shown in E7.3.2 is supported with three well-aligned journal bearings. If the bar is in equilibrium, determine the loads acting on the bar at the supports

a. at A **b.** at B **c.** at C

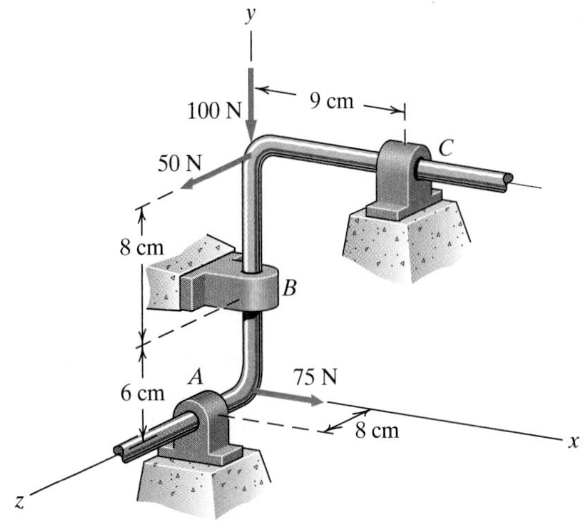

E7.3.2

7.3.3. Bar AB is used to support an 850-N load as shown in E7.3.3. End A of the bar is supported with a ball-and-socket joint. End B of the bar is supported with cables BC and BD. If the bar is in equilibrium, determine

a. the components of the force acting on the bar at support A

b. the tensions in the two cables

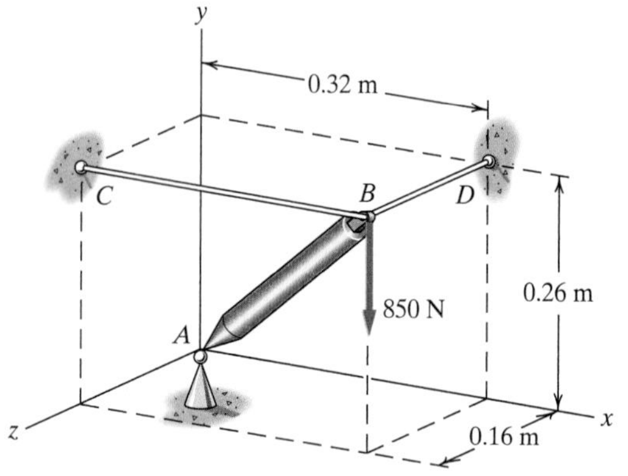

E7.3.3

7.3.4. A shaft is loaded through a pulley and a lever that are fixed to the shaft as shown in E7.3.4. Friction between the belt and pulley prevents the belt from slipping. The support at A is a journal bearing, and the support at B is a thrust bearing. Determine

a. the force $\boldsymbol{P}$ required for equilibrium

b. the loads acting on the shaft at supports at A and at B

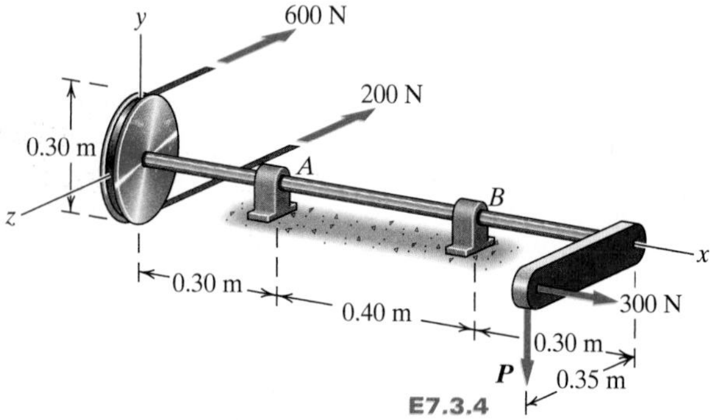

E7.3.4

7.3.5. A 50-N force is applied to the pipe wrench attached to the pipe system shown in E7.3.5. If the pipe is in equilibrium, determine the loads acting on the pipe at support A.

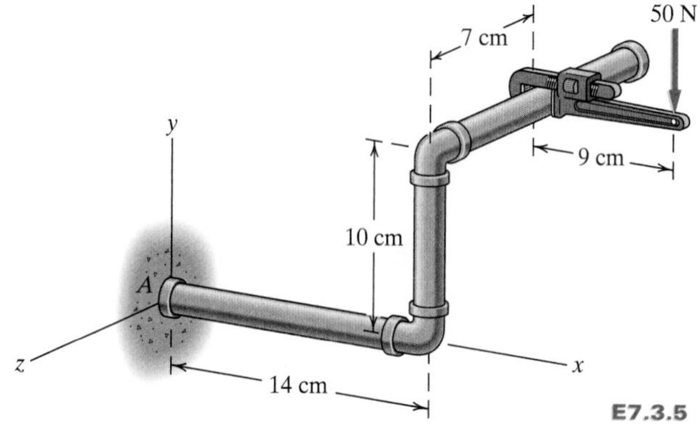

E7.3.5

7.3.6. A package is placed on the three-legged circular table at location (R, θ), as shown in E7.3.6. The magnitudes of the forces the ground applies to the legs are known to be $\|\boldsymbol{F}_A\| = 110$ N, $\|\boldsymbol{F}_B\| = 140$ N, and $\|\boldsymbol{F}_C\| = 130$ N. If the mass of the table is 30 kg and the table is in equilibrium, determine

a. the mass of the package

b. the values of R and θ. In addition, describe the location of the package with a sketch of the tabletop

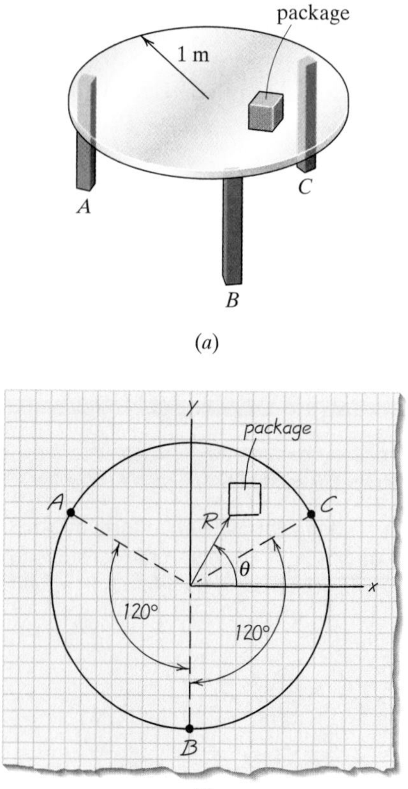

E7.3.6

7.3.7. The winch assembly is holding a 100-kg packet in equilibrium, as shown in E7.3.7. The drum of the winch has a diameter of 400 mm. Determine

a. the force, $\boldsymbol{F}$

b. the loads acting on the winch assembly at the journal bearings, A and B

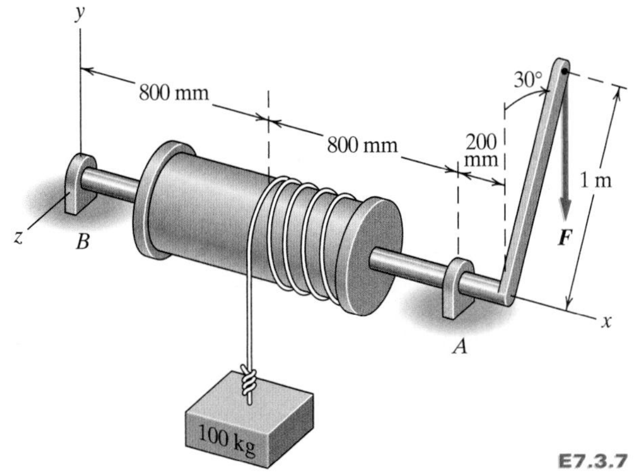

E7.3.7

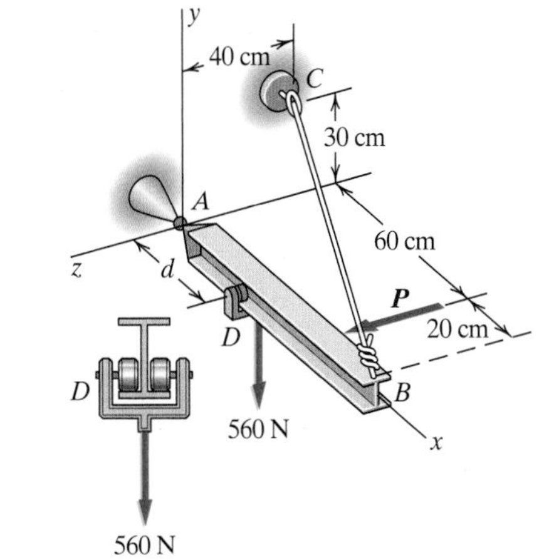

E7.3.9

7.3.8. A bent bar is welded to the wall at end A and carries the loads shown in E7.3.8. Determine the loads acting on the bar at end A, when it is in equlibrium.

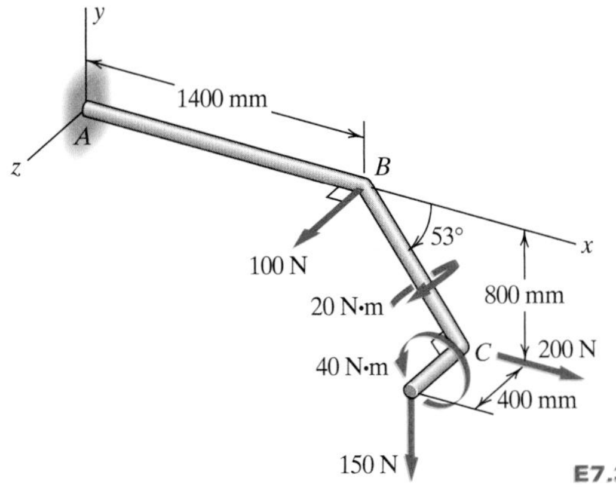

E7.3.8

7.3.9. An I-beam is supported by a ball-and-socket joint at A and a rope BC. A horizontal force, $\boldsymbol{P}$, is applied as shown in E7.3.9. In addition, a 560-N load is suspended from a movable support at D. The uniform beam is 80 cm long and weighs 225 N.

a. Create a graph of T_{BC}, the tension in the rope, as a function of the distance d ($0 \le d \le 80$ cm).

b. On the same graph, plot the magnitude of the force $\boldsymbol{P}$ required to keep the beam horizontal (i.e., aligned with the y axis) as a function of the distance d ($0 \le d \le 80$ cm).

c. If the tension in the rope is not to exceed 1200 N (otherwise it will break), what limit should be specified for d?

d. Come up with two ideas of how the movement of the movable support might be limited so as not to exceed the limit you specified in **c**. Present your idea as sketches.

7.3.10. Ends A and B of the 5-kg rigid bar are attached by lightweight collars that may slide over the smooth fixed rods shown in E7.3.10. The center of mass of the bar is at its midpoint. The length of the bar is 1.5 m. Determine the horizontal force $\boldsymbol{F}$ applied to collar A that will result in static equilibrium as a function of the distance, d ($0.5 \le d \le 1.5$ cm).

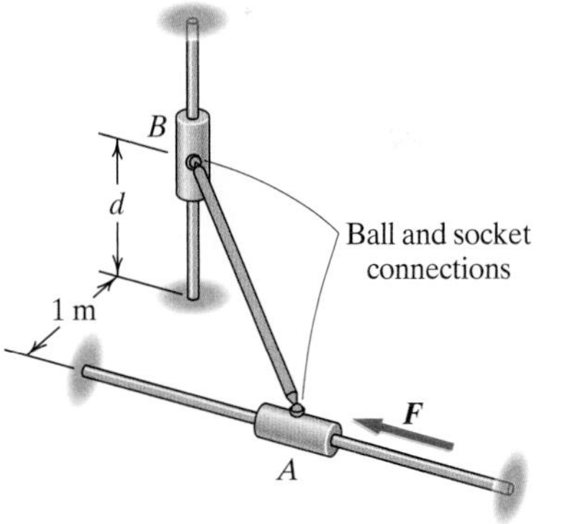

E7.3.10

7.3.11. A 5-kg steel sphere rests between the 45° grooves A and B of the 10° incline, and against a vertical wall at C as shown in E7.3.11. If all three surfaces of contact are smooth, determine the loads acting on the surface on the sphere.

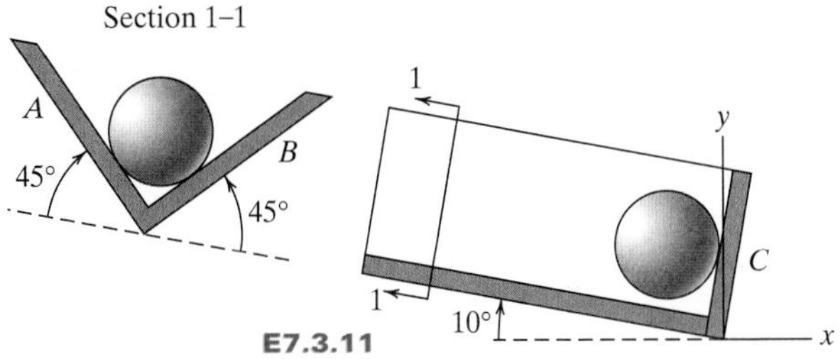

E7.3.11

7.3.12. Two journal bearings A and B and a short link DC support the shaft assembly. If the 250 N · m moment is applied to the shaft as shown in E7.3.12, determine the loads acting on the shaft assembly at the bearings and at C, when it is in equilibrium. The link lies in a plane parallel to the xy plane and the bearings are properly aligned on the shaft.

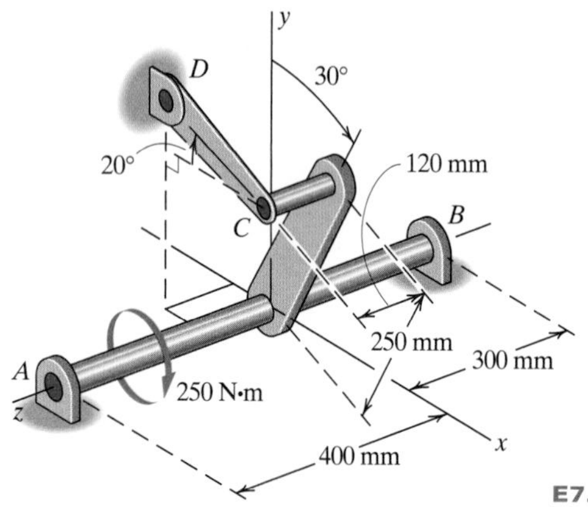

E7.3.12

7.3.13. Point C on the top flange of the I-beam is subjected to the loads shown in E7.3.13. Determine the loads acting on the I-beam at the support at A (which is on the axis of symmetry of the I-beam).

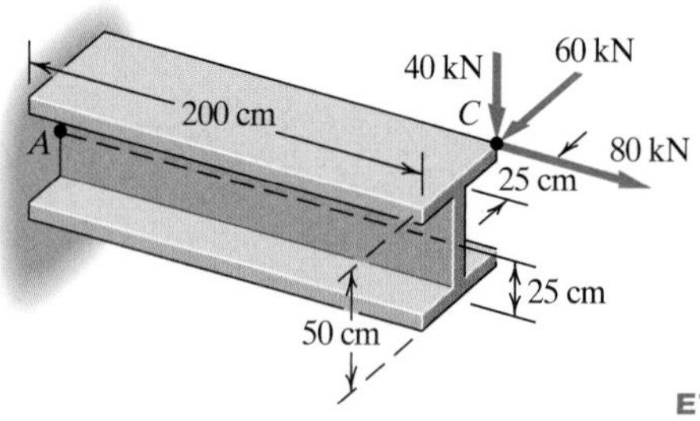

E7.3.13

7.3.14. An airplane rests on the ground, as shown in E7.3.14. Its landing gears are at points A, B, and C. Its center of gravity G is at (3.0, 0.2, −4.6) m, as shown. The airplane weighs 1.56×10^6 N. What are the magnitudes of the normal forces acting on the landing gears?

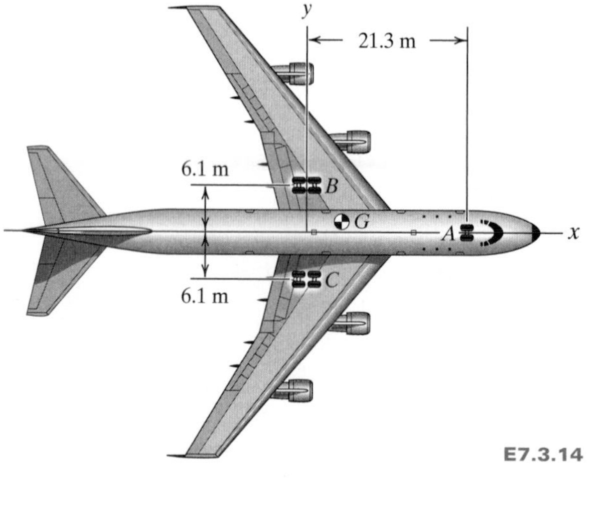

E7.3.14

7.3.15. A bicycle is moving at a constant velocity of 15 mph. The foot force on the right pedal is $5 \text{ N } \boldsymbol{i} - 80 \text{ N } \boldsymbol{j}$ and on the left pedal is $-5 \text{ N } \boldsymbol{j}$ (see E7.3.15). Determine

a. the loads the ball bearings apply to the bottom bracket axle for this pedal configuration

b. the tension force in the chain

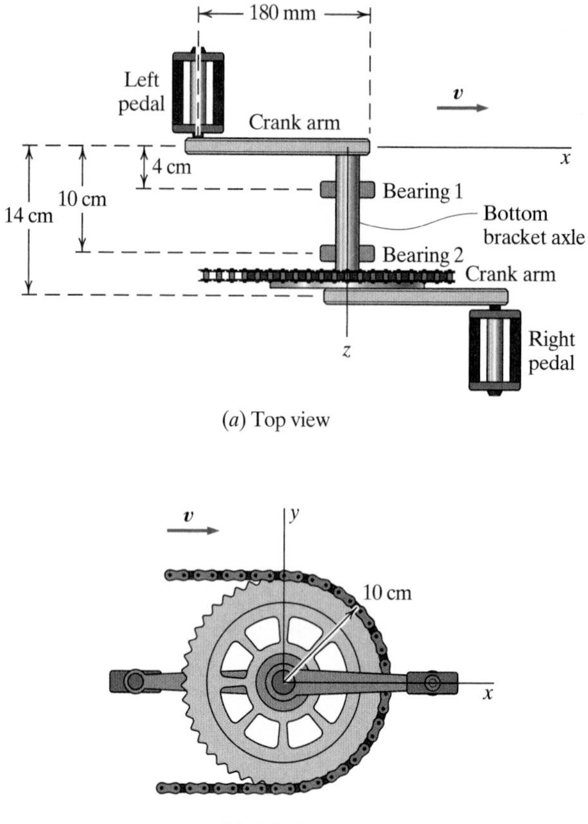

E7.3.15

7.3.16. A bicycle is moving at a constant velocity of 15 mph. The foot force on the right pedal is $5 \text{ N } \boldsymbol{i} - 30 \text{ N } \boldsymbol{j}$ and on the left pedal is $5 \text{ N } \boldsymbol{i} - 5 \text{ N } \boldsymbol{j}$ when in the position shown in E7.3.16. The dimensions of the bottom bracket are as shown in **E7.3.15**. Determine

a. the loads the ball bearings apply to the bottom bracket axle

b. the tension force in the chain

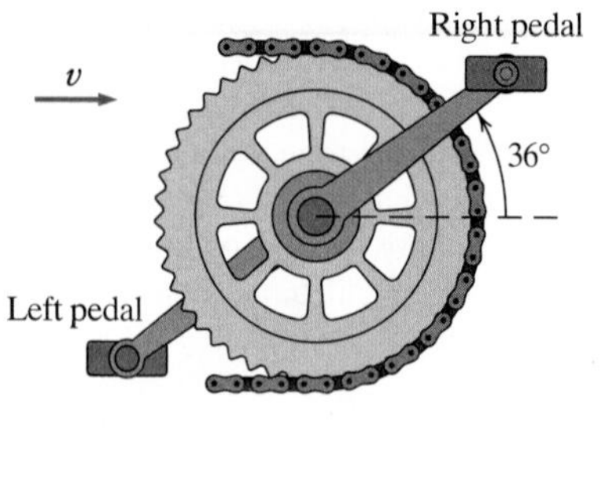

E7.3.16

7.3.17. When Joseph Strauss designed the piers that serve as the foundations for the towers of the Golden Gate Bridge, he determined the forces that the piers would have to support. Based on the approximate analysis we performed on the main cables (summarized in **Figure 3.12**), we can draw the free-body diagram of the north tower (E7.3.17). A summary of the results of the analysis of the approximate model (**Figure 3.12**) indicates that $T_1 = 253.7$ MN and $T_2 = 265.4$ MN. According to Section 3.6, the weight of each tower of the bridge is 196 MN. Using this information and the geometry of the approximate model given to you in **Table 3.1**, find the angles α and β; then find the loads (reactions) acting on the north tower at its base at B.

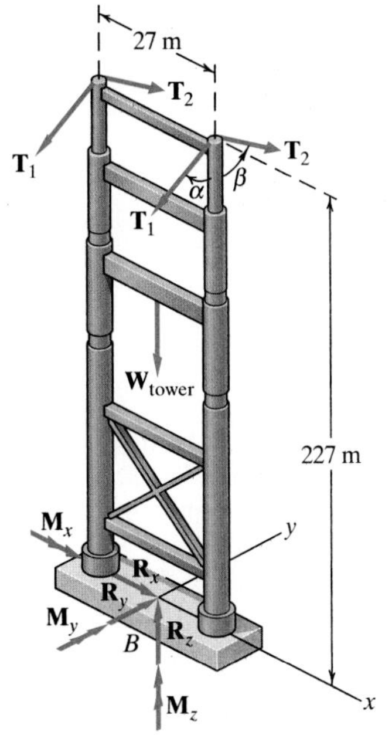

E7.3.17

7.3.18. A tower 70 m tall is tethered by three cables, as shown in E7.3.18. If the tension in each of the cables is 4000 N and the tower is in equilibrium, what loads act on the tower at its base?

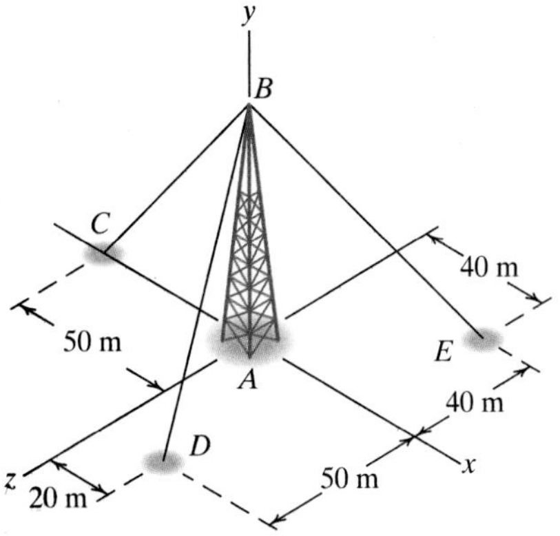

E7.3.18

7.3.19. The bracket ABC is tethered as shown in E7.3.19 with cable CD. The tension in the cable is 750 N. If the weight of the bracket is negligible and the bracket is in equilibrium, what loads act on the bracket at A?

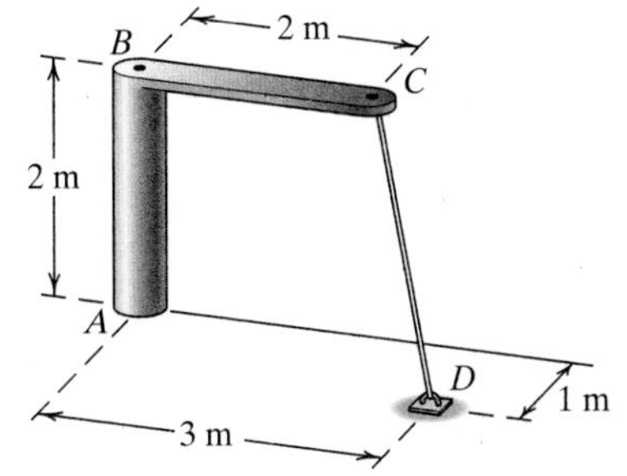

E7.3.19

7.3.20. Cables and a collar joint support a sign of weight W (2000 N) as shown in E7.3.20. If the sign is in equilibrium, determine the tensions in the cables and the loads at the collar joint.

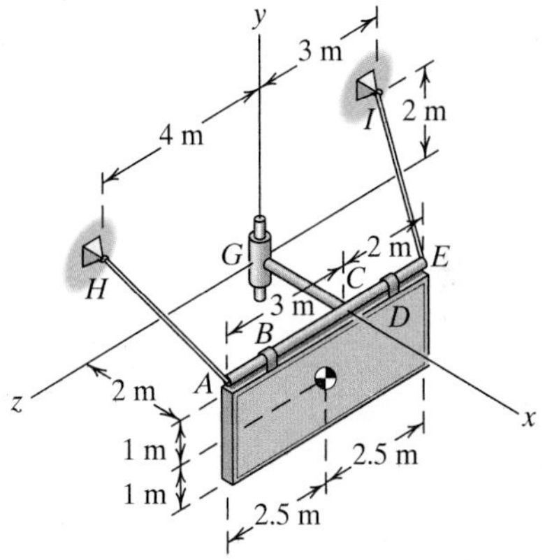

E7.3.20

7.3.21. A crankshaft is supported by a journal bearing at B and a thrust bearing at D as shown in E7.3.21. Ignore the weight of the crank. If the crankshaft is in equilibrium, determine the loads acting on the crankshaft at B and D. The force at A is $\boldsymbol{F}_A = -30\text{ N}\,\boldsymbol{i} + -5\text{ N}\,\boldsymbol{j}$.

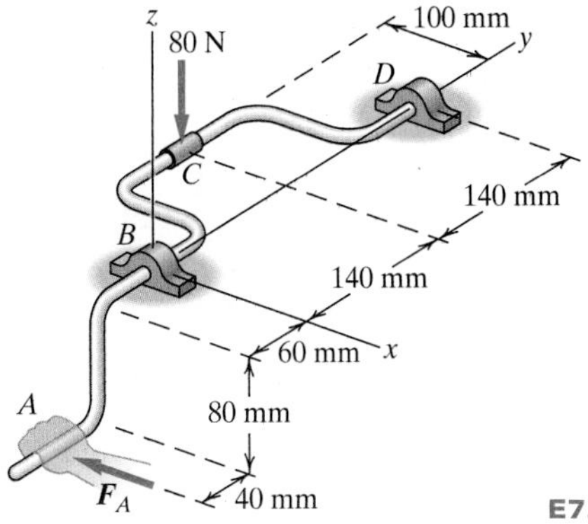

E7.3.21

7.3.22. A pulley is used to lift a weight W. The shaft of the pulley is supported by a thrust bearing, as shown in E7.3.22. Ignore the weights of the pulley and the shaft. If the shaft is in equilibrium, determine the loads acting on the shaft at B. Also determine the value of W.

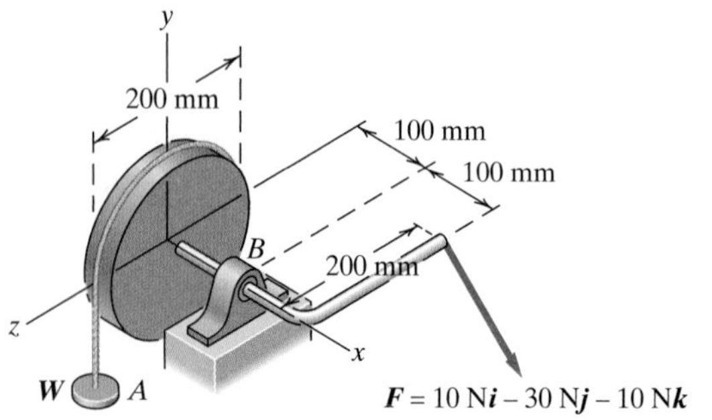

E7.3.22

7.3.23. A pole is fixed at B and tethered by a rope, as shown in E7.3.23. The tension in the rope is 100 N. Ignore the weight of the pole. If the pole is in equilibrium, determine the loads acting at its base.

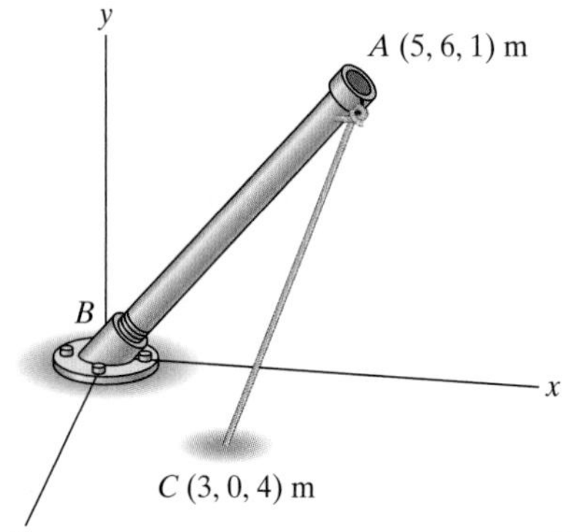

E7.3.23

7.3.24. A triangular plate is supported by a cable at A and a hinge at B. Its weight of 400 N acts at the plate's center of gravity, as shown in E7.3.24. If the plate is in equilibrium, determine the tension in the cable and the loads acting on the plate at the hinge.

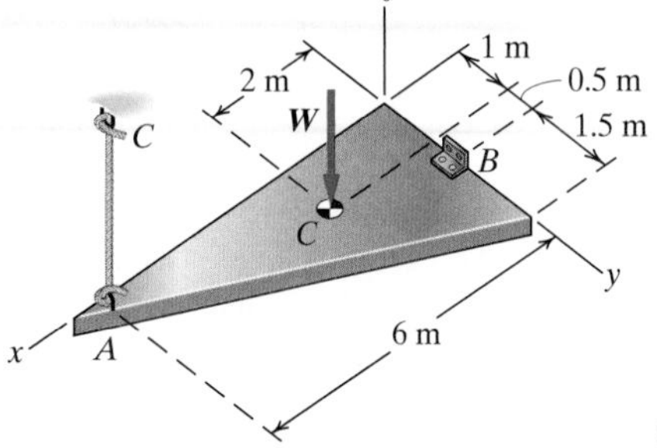

E7.3.24

7.3.25. A tower crane is fixed at the ground at A as shown in E7.3.25. Assume that the weight of the crane is negligible. If the tower crane is in equilibrium, determine the loads acting at its base due to the 4000-N cable tension.

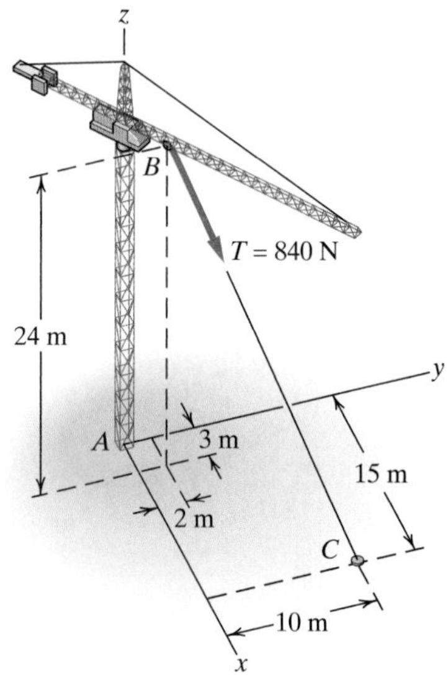

E7.3.25

7.3.26. The uniform 7-m shaft is supported by a ball-and-socket connection at A in the horizontal floor. The ball end B rests against the smooth vertical walls, as shown in E7.3.26. The shaft weighs 2500 N. If the shaft is in equilibrium, what loads act on the shaft at A and at B?

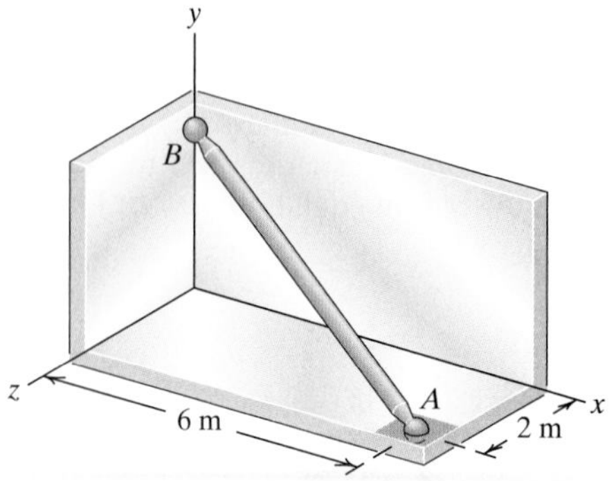

E7.3.26

7.3.27. The welded tubular frame is secured to the horizontal xy plane by a ball-and-socket support at A and receives support from a loose-fitting ring at B as shown in E7.3.27. If the weight of the frame is negligible and it is in equilibrium,

a. what loads act on the frame at A?
b. what loads act on the frame at B?

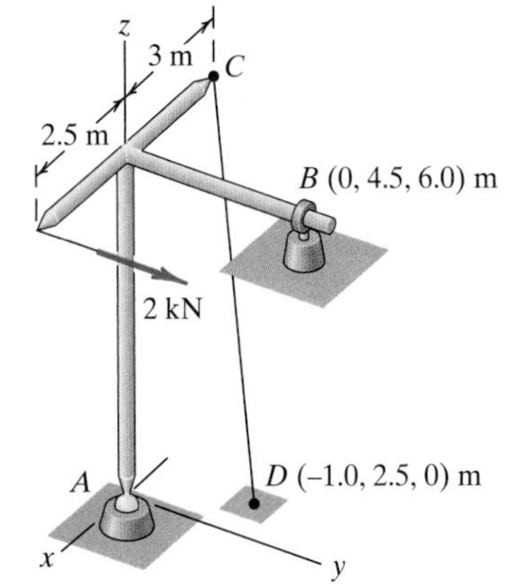

E7.3.27

7.3.28. The bar in E7.3.28 is supported at A by a hinge, and at B it rests against a rough surface. If the bar is in equilibrium, what loads act on the bar at A? What loads act at B?

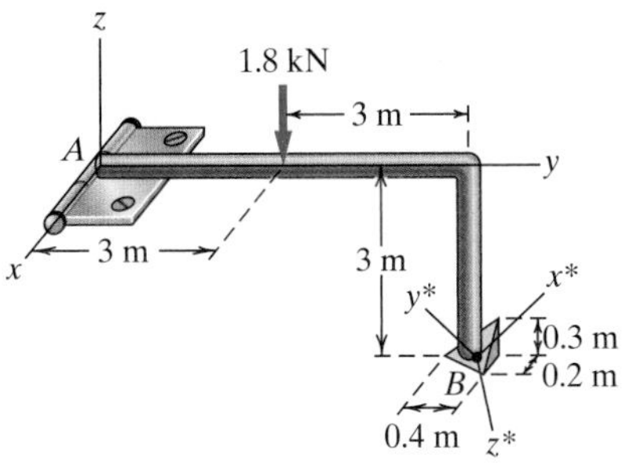

E7.3.28

7.3.29. Because of a combination of soil conditions and the tension in the single power cable, the utility pole has developed the 5° lean indicated in E7.3.29. The 9-m uniform pole has a mass per unit length of 25 kg m and the tension in the power cable is 900 N. If the utility pole is in equilibrium, what loads must act on the base of the pole?

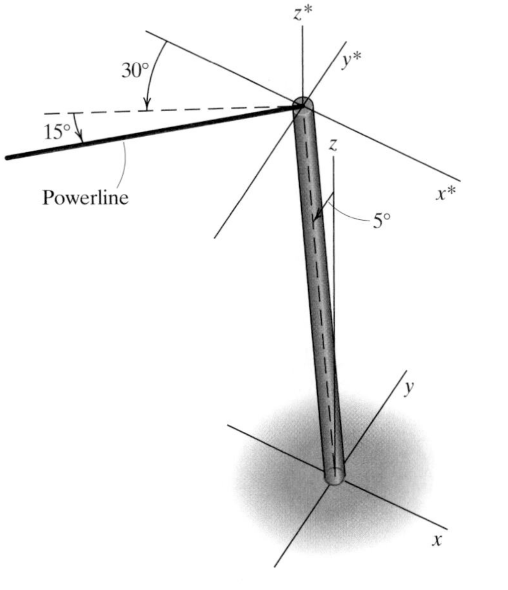

E7.3.29

7.3.30. Three vertical legs, as shown in E7.3.30, support a uniform rectangular table of weight W. Determine

a. the loads the floor applies to the legs

b. the smallest vertical force $\boldsymbol{F}$ that can be applied to the tabletop to cause the table to tip over

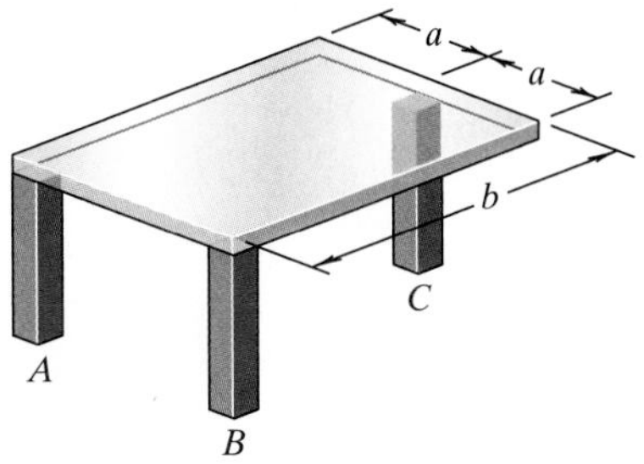

E7.3.30

7.3.31. Consider the semi-circular plate in E7.3.31 that has a radius of 40 cm. The plate weighs 100 N with center of gravity at $(d, 0, 0)$. Determine the tension in the cables at A and B and the loads acting on the plate at C (a ball-and-socket connection).

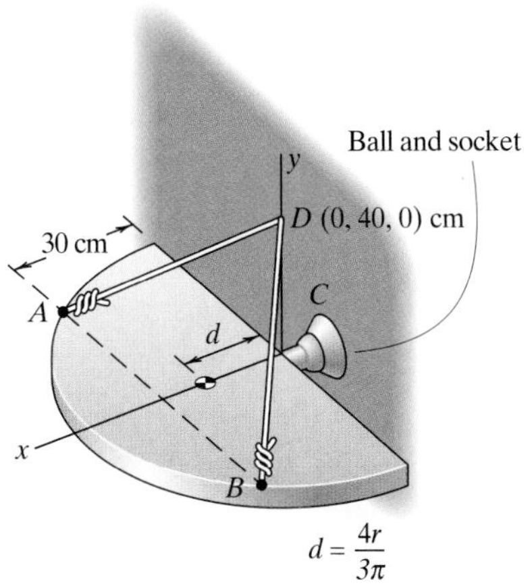

E7.3.31

7.3.32. A 200-N force is applied to the handle of the hoist in the direction shown in E7.3.32. The bearing at A is a thrust bearing, and the bearing at B is a journal bearing. If the hoist is in equilibrium, what loads act on the shaft at A? What loads act on the shaft at B?

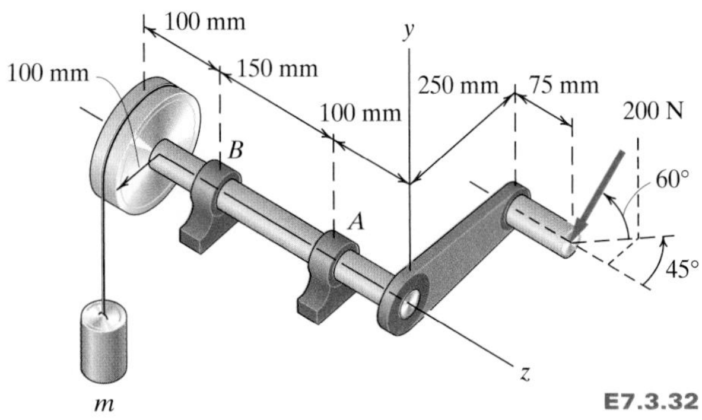

E7.3.32

7.3.33. Three workers are carrying a 4-ft by 8-ft plywood panel by grabbing the panel at the points shown in E7.3.33. The panel weighs 100 lb. If the panel is in equilibrium, what loads must each worker apply to the panel?

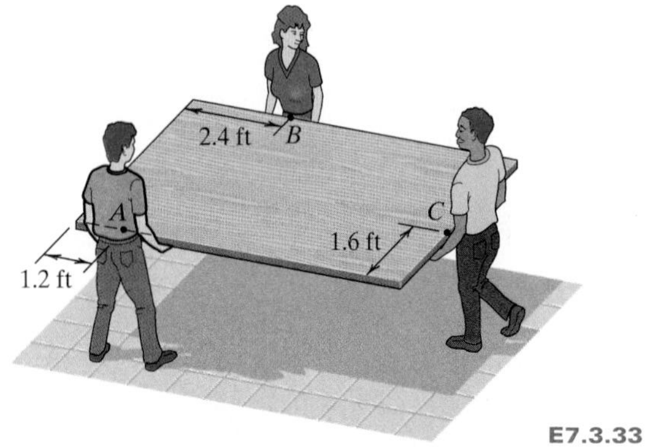

E7.3.33

7.3.34. A bracket is bolted to the shaft at O. Cables load the bracket, as shown in E7.3.34. If the bracket is equilibrium and the weight of the bracket is negligible, what loads act on the bracket at O?

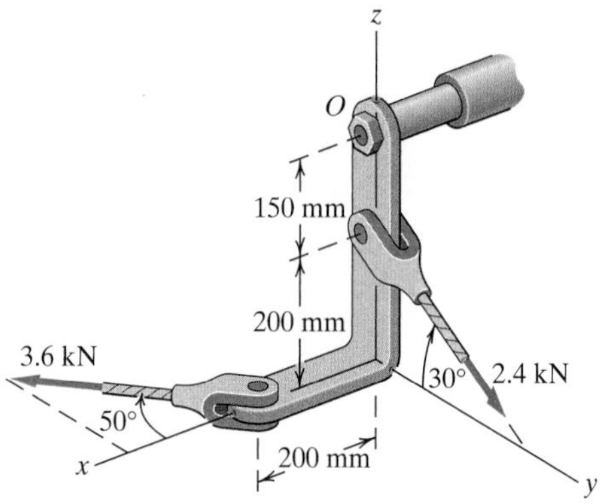

E7.3.34

7.3.35. A carpenter is slowly pushing the 100-kg roof truss into place. In the current position shown in E7.3.35 it is oriented 25° from the vertical. The mass center of the symmetric triangular truss is located one-third of its 2.25-m dimension from its base. To hold the truss in this position, what loads must the carpenter apply? What loads act on the truss at A and at B?

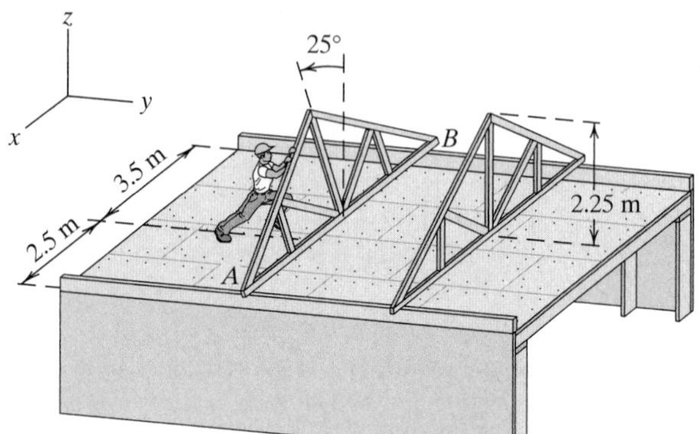

E7.3.35

7.3.36. A hanging chair is suspended, as shown in E7.3.36, and holds a person weighing 800 N. Determine the tension in cables DA, DB, and DC if the chair is in equilibrium.

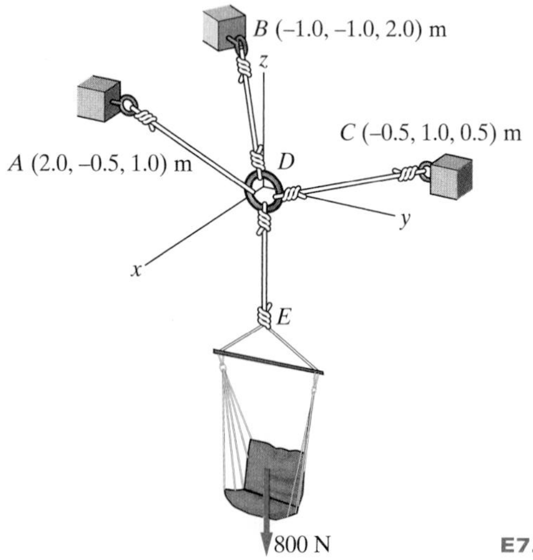

E7.3.36

7.4 ZOOMING IN

A free-body diagram describes both the loads acting on a system and the geometry of the system. We use it as the basis for writing the equilibrium equations for a system that is balanced. For such a system the conditions of equilibrium hold for the system as a whole, and *they must hold for all portions of the system*. In other words, if we zoom in on portions of a system that is in equilibrium, the conditions of equilibrium must be true for each portion. This idea is illustrated in **Figure 7.38**. (This figure should look familiar to you, as it is similar to the ladder example used in Chapter 4 to illustrate external and internal loads.)

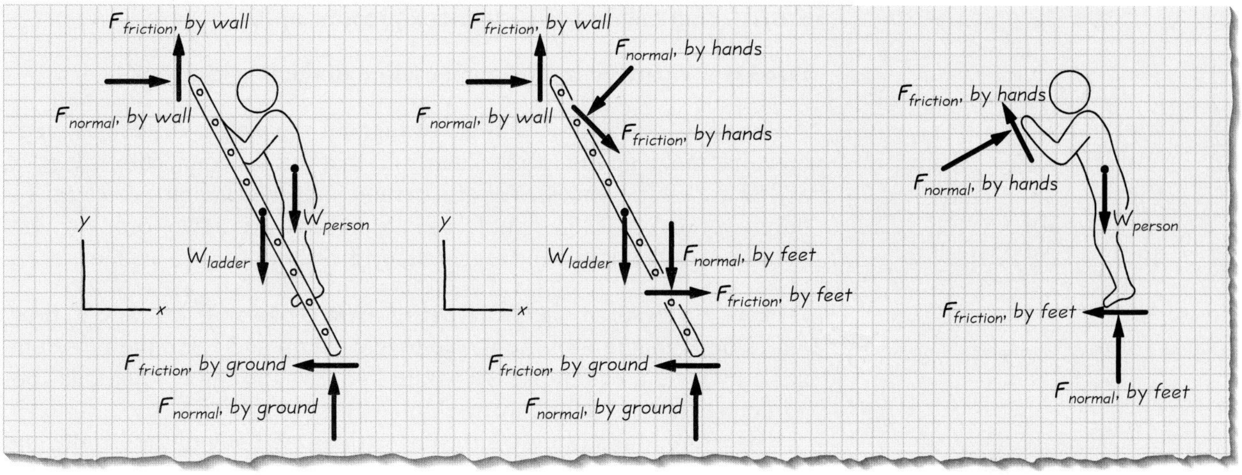

Figure 7.38 (*a*) Person-ladder free-body diagram; (*b*) ladder free-body diagram; (*c*) person free-body diagram

You might well ask, "Why would I be interested in zooming in on portions of a system and checking for equilibrium?" One reason is that sometimes the loads of interest in confirming a design are internal loads in the system you have defined. By zooming in and defining a new, smaller system, you can have these interest loads become external loads acting on your newly defined system, as illustrated in Example 7.12.

Another reason for zooming in is that sometimes when the conditions of equilibrium are applied to a large system there are not enough equations for finding the unknowns, as Example 7.13 shows. If parts of a system are not rigidly attached to one another, you may be able to zoom in on portions of the system in order to generate additional independent equilibrium equations. We will revisit this idea in Chapter 9 when we talk about trusses and frames.

EXAMPLE 7.12 ANALYSIS OF A TOGGLE CLAMP

A toggle clamp holds workpiece H (Figure 7.39). Determine the vertical clamping force at E in terms of the input force $\boldsymbol{P}$ applied to the handle of the toggle clamp.

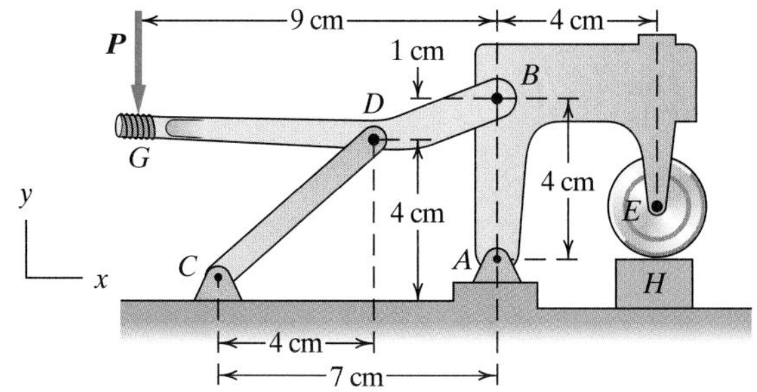

Figure 7.39

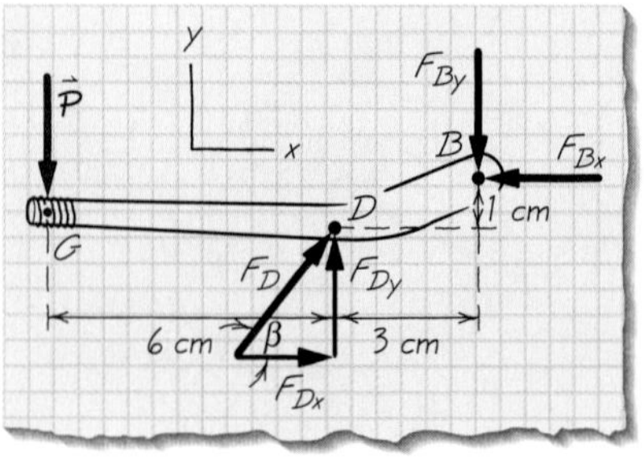

Figure 7.40

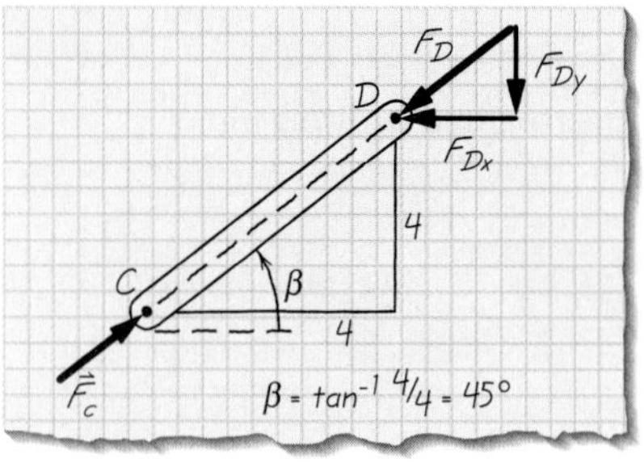

Figure 7.41

Goal Find the force $\boldsymbol{F}_E$ expressed in terms of $\boldsymbol{P}$.

Given We are given the configuration of the toggle clamp.

Assume We assume that we can ignore the weight of various members that make up the toggle clamp and that the pin connections *A, B, C,* and *D* are all frictionless. We also assume that the clamping force at *E* is purely vertical, as is the input force $\boldsymbol{P}$. Member *CD* is a two-force element. Finally, we assume that the toggle clamp is planar.

Draw, Formulate Equations, and Solve If we were to draw a free-body diagram of the whole toggle clamp and write the three planar equilibrium equations, we would find that the three equations contain more than three unknowns.

Therefore, our strategy is to start with a free-body of the handle and relate the input force $\boldsymbol{P}$ to the forces at pin *B*. Then we will work with a free-body diagram of member *ABC* and relate the forces at pin *B* to the clamping force $\boldsymbol{F}_E$. This two-step process will allow us to relate the input force to the clamping force.

The plan is to first zoom in and isolate the handle (**Figure 7.40**); this will allow us to relate the force at *B* ($\boldsymbol{F}_B$) to the input force $\boldsymbol{P}$. Because member *CD* is a two-force element, we can write $\boldsymbol{F}_D = \| \boldsymbol{F}_D \| \cos\beta\, \boldsymbol{i} + \| \boldsymbol{F}_D \| \sin\beta\, \boldsymbol{j}$, where β is found in **Figure 7.41**. Now formulate the planar equilibrium equations for the handle in **Figure 7.40**.

Based on (7.5A):

$$\sum F_x = 0\ (\rightarrow +)$$
$$F_{Dx} - F_{Bx} = 0$$
$$\| \boldsymbol{F}_D \| \cos\beta - F_{Bx} = 0$$
$$F_{Bx} = \| \boldsymbol{F}_D \| \cos\beta \tag{1}$$

Based on (7.5B):

$$\sum F_y = 0\ (\uparrow +)$$
$$F_{Dy} - F_{By} - P = 0$$
$$F_{By} = F_{Dy} - P$$
$$F_{By} = \| \boldsymbol{F}_D \| \sin\beta - P \tag{2}$$

Based on (7.5C) (with the moment center at *D*):

$$\sum M_{z@D} = 0\ (\circlearrowleft +)$$
$$= (6\text{ cm})(P) + (1.0\text{ cm})(F_{Bx}) - (3.0\text{ cm})(F_{By}) = 0 \tag{3}$$

We substitute (1) and (2) into (3) and find that

$$9P - \| \boldsymbol{F}_D \| (3\sin\beta - \cos\beta) = 0$$
$$\| \boldsymbol{F}_D \| = \frac{9}{3\sin\beta - \cos\beta} P \tag{4}$$

We substitute (4) into (1) and (2) to express the force components at B in terms of the input force $\boldsymbol{P}$:

From (1) $$F_{Bx} = \frac{9 \cos \beta}{3 \sin \beta - \cos \beta} P \quad (1')$$

From (2) $$F_{By} = P\left(\frac{9 \sin \beta}{3 \sin \beta - \cos \beta} - 1\right) \quad (2')$$

Next we zoom in and isolate member ABE and draw its free-body diagram (**Figure 7.42**). This allows us to relate $\boldsymbol{F}_B$ to $\boldsymbol{F}_E$. Based on (7.5C) (with the moment center at A):

$$\sum M_{z@A} = 0$$
$$(4.0 \text{ cm})(F_E) - (4.0 \text{ cm})(F_{Bx}) = 0$$
$$F_E = F_{Bx}$$
$$F_E = \frac{9 \cos \beta}{3 \sin \beta - \cos \beta} P \quad (5)$$

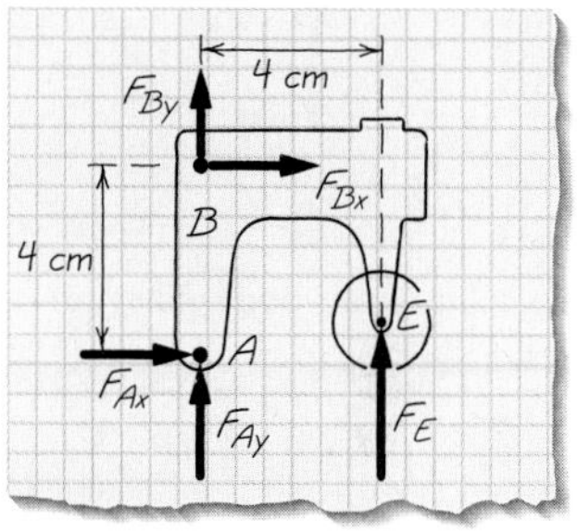

Figure 7.42

From **Figure 7.41** we find that $\beta = 45°$. Therefore, $\cos 45° = \sin 45° = 0.707$. Expression (5) can be written as

$$F_E = \frac{9(0.707)}{3(0.707) - 0.707} P, \; F_E = 4.5 \, P$$

Answer $F_E = 4.5\,P$. The ratio of the input force P to the toggle clamping force F_E is called the mechanical advantage. It says that by applying an input force of magnitude P we can produce a clamping force that is 4.5 times P. The mechanical advantage for this system is 4.5.

EXAMPLE 7.13 MULTIPLE PULLEYS

A cable and pulley system is used to support an object as shown in **Figure 7.43**. The weight of the object is W. Each pulley is free to rotate. One cable is continuous over pulleys A and B, and the other is continuous over pulley C. Determine the ratio W/T, where T is the force you need to apply at D to hold the object in the position shown.

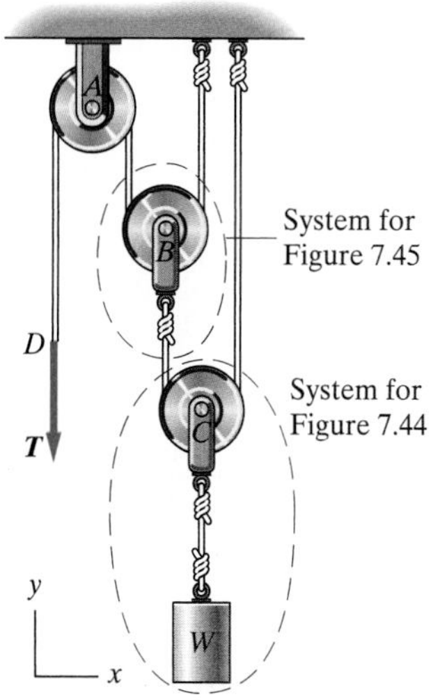

Figure 7.43

Goal Find the ratio W/T, where W is the weight of an object being held by the pulley system and T is the tension in the cable at location D.

Given We are given the configuration of the pulleys. We are told that the cables are free to rotate, allowing us to treat the pulleys as frictionless. This means that the magnitude of the tension T in the cable that goes from D and around pulleys A and B is constant. Similarly, the tension T_C in the cable that goes from B around pulley C is also constant.

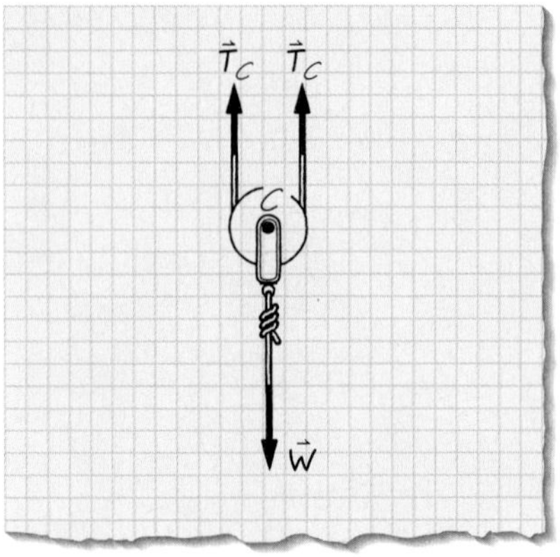

Figure 7.44 Free-body diagram of Pulley C

Assume We will assume that we can ignore the weight of the pulleys and cables. We will also assume that the cable–pulley system is planar.

Draw, Formulate Equations, and Solve Our plan is to first zoom in and isolate the pulley at C with the attached weight W (Figure 7.44); this will allow us to relate the tension T_C in the cable around pulley C to the weight W. Now formulate the equilibrium equations for the planar system in **Figure 7.44**.

Based on (7.5B):

$$\sum F_y = 0\ (\uparrow\ +)$$
$$T_C + T_C - W = 0$$
$$T_C = \frac{W}{2} \qquad (1)$$

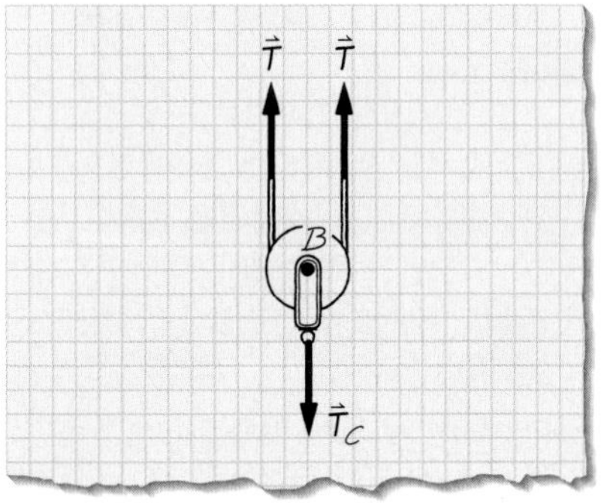

Figure 7.45 Free-body diagram of Pulley B

Because we are able to find T_C in terms of W by just using (7.5B), there is no need to write out (7.5A) and (7.5C).

Next we zoom in and isolate the pulley at B (Figure 7.45); this will allow us to relate T_C to the tension in the cable around pulleys A and B. Now formulate the equilibrium equations for the planar system in **Figure 7.45**. Based on (7.5B):

$$\sum F_y = 0\ (\uparrow\ +)$$
$$-T_C + T + T = 0$$
$$T_C = 2T \qquad (2)$$

Because we are able to find T_C in terms of T by just using (7.5B), there is no need to write out (7.5A) and (7.5C).

By equating (1) and (2), we find that $\frac{W}{T} = 4$. Therefore, we can write:

Answer $\frac{W}{T} = 4$
The mechanical advantage of this system is 4. This indicates that an object weighing W requires a force T that is only $\frac{1}{4}$ of W for the system to be in equilibrium. An alternative interpretation is that with a force of T you can hold up an object that weighs $4T$.

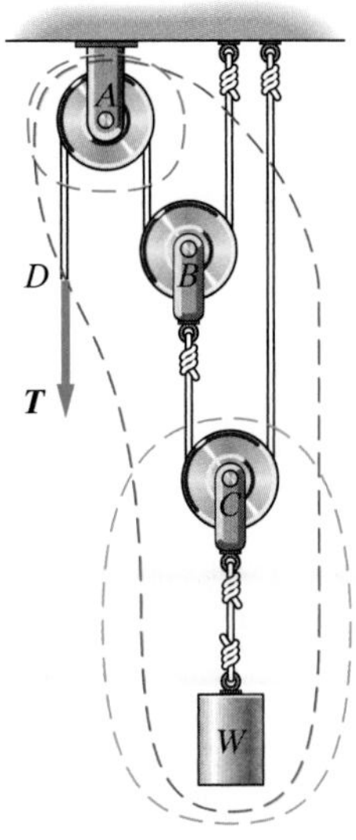

Figure 7.46 Alternate boundaries

Comment: The free-body diagrams shown in **Figures 7.44** and **7.45** are not the only two that could have been used to solve this problem. For example, Figure 7.46 shows other parts of the pulley assembly that could be zoomed in on for the purpose of finding the ratio $\frac{W}{T}$. Having said this, the choice of which parts are zoomed in on affects how complex the calculations are.

EXERCISES 7.4

7.4.1. For the 75-N load supported by the L-bracket, frictionless pulley, and cable shown in E7.4.1, use the technique of "zooming in" as appropriate to determine

a. the loads acting on the L-bracket at supports A and B

b. the force exerted on the bracket by the pin connection at C

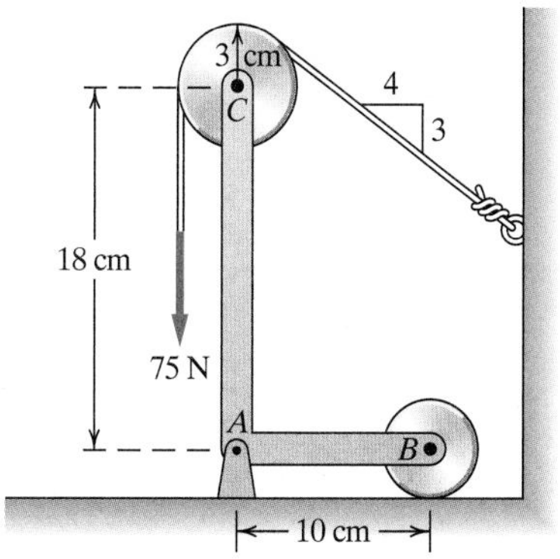

E7.4.1

7.4.2. The pipe strut BC is loaded and supported as shown in E7.4.2. The strut has a uniform cross section and a mass of 50 kg. In addition, the 750-N force acts on the strut, as shown. Determine the loads acting on the structure at C and the tension in the cable AB.

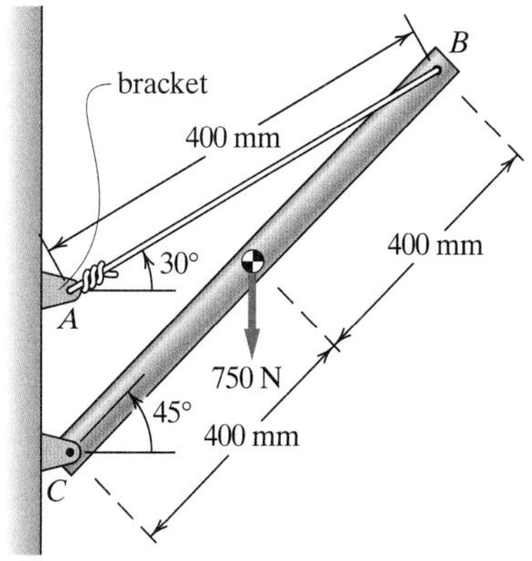

E7.4.2

7.4.3. A cable is attached to a bracket at A as shown in **E7.4.2.** The bracket is bonded to the wall with an adhesive. If the bracket will break away from the wall at this adhesive interface when the tension force at the interface is greater than 1000 N or the shear force at the interface is greater than 950 N, determine whether the bracket will detach from the wall when loaded by the weight of the strut and the 750-N load.

7.4.4. A pin-connected three-bar frame is loaded and supported as shown in E7.4.4. If $\theta = 30°$ and the magnitude of load $\boldsymbol{P}$ is 500 N, determine

a. the loads acting on the frame at A and B

b. the tension or compression in member DE

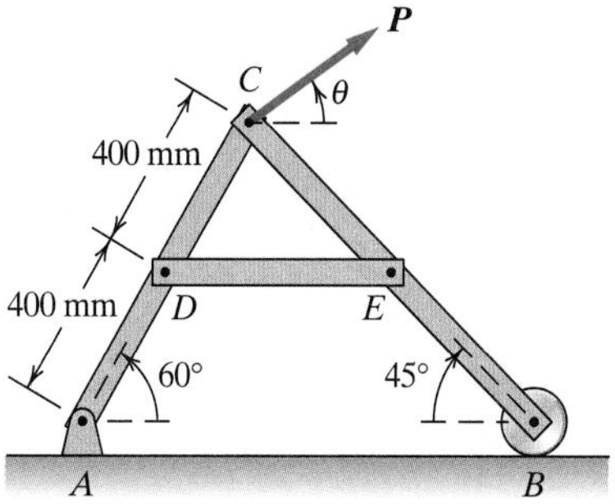

E7.4.4

7.4.5. A pin-connected three-bar frame is loaded and supported as shown in **E7.4.4.** The load $\boldsymbol{P}$ is 500 N. If member DE will fail at a tension force greater than 300 N and a compression force greater than 200 N, what is the range(s) of acceptable values of θ that can be carried by the frame?

7.4.6. A pin-connected three-bar frame is loaded and supported as shown in **E7.4.4.** If $\theta = 40°$ member DE will fail at a tension force greater than 300 N and a compression force greater than 200 N, what is the largest acceptable value of force $\boldsymbol{P}$ that can be safely carried by the frame?

7.4.7. The tape in E7.4.7 is wrapped over frictionless pulleys A and B in the tape guide. The system is oriented in a horizontal plane. Determine the force exerted by the spring at point C. If it is desirable that the system not deflect more than 0.5 mm at point C, what is the minimum spring stiffness you would specify for the spring at C?

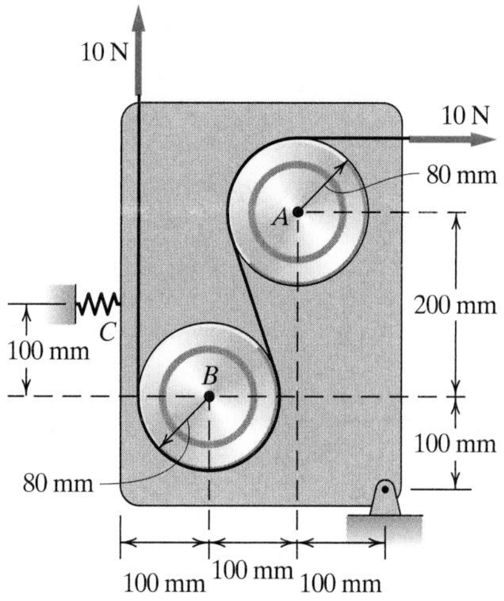

E7.4.7

7.4.8. For the tape unit in **E7.4.7**,

a. determine the loads at A that the tape unit applies to pulley A

b. determine the loads at B that the tape unit applies to pulley B

c. Pulleys A and B are attached to the tape unit by pin connections. Based on your results from **a** and **b**, would you expect the pin at A or B to fail first? Why?

7.4.9. A beam is loaded and supported as shown in E7.4.9. The beam has a uniform cross section and a mass of 20 kg. Determine the loads that act on the beam at A and the tension T in the cable if the beam is in equilibrium.

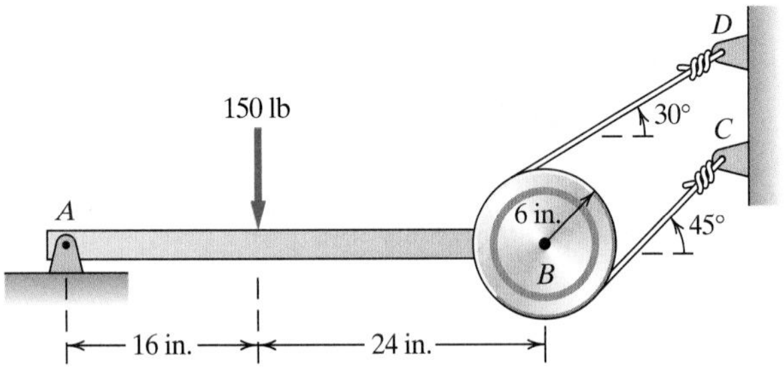

E7.4.9

7.4.10. Determine the loads that act on the beam at B (as applied by the pulley), as shown in **E7.4.9**.

7.4.11. Determine the loads that act on the pulley at B (as applied by the beam), as shown in **E7.4.9**.

7.4.12. A rope and pulley system (**E7.4.12**) is used to support a curtain system in a theater. Each pulley is free to rotate, and the ropes are continuous over the pulleys. If the mass of the block is 40 kg, what force $\boldsymbol{P}$ must be applied for the system to be in equilibrium?

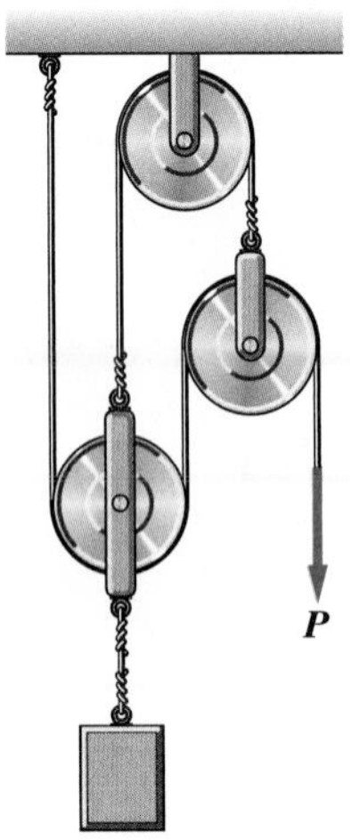

E7.4.12

7.4.13. The person shown in **E7.4.13** has a mass of 80 kg and the uniform beam has a mass of 40 kg. If the beam is in equilibrium, determine the loads acting on the beam at C and the tension in the cable. Comment on whether you think you could pull on the cable with this tension and whether this is a reasonable thing to do.

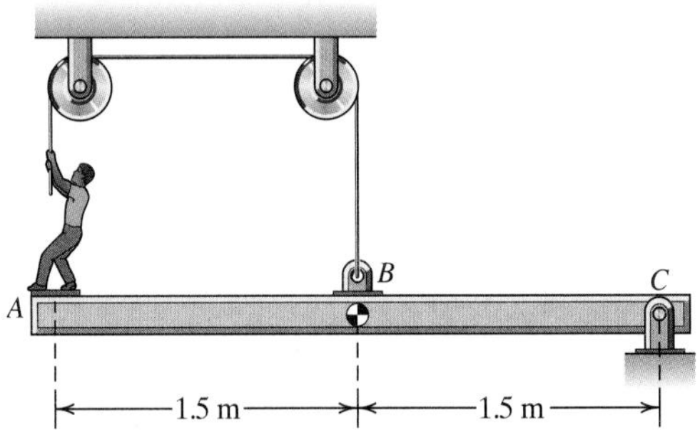

E7.4.13

7.4.14. Pulleys A and B of the chain hoist are connected and rotate as a unit (**E7.4.14**). The chain is continuous, and each of the pulleys contains teeth that prevent the chain from slipping.

a. Determine the magnitude of the force $\boldsymbol{F}$ required to hold an engine block weighing 400 N in equilibrium if the radii of pulleys A and B are 7 and 9 cm, respectively.

b. What is the mechanical advantage of this chain hoist?

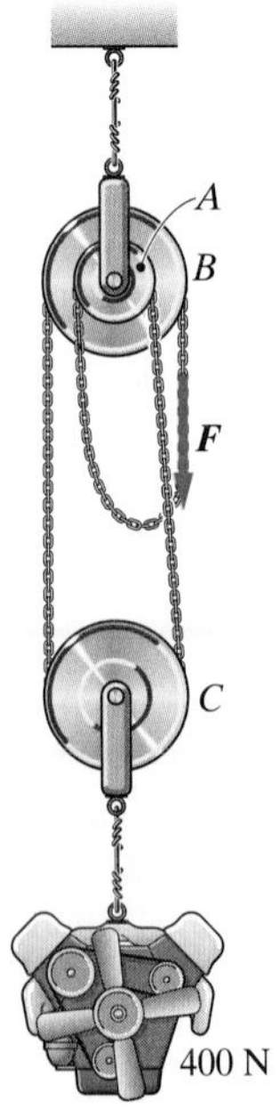

E7.4.14

7.4.15. A bar AB weighing 1000 N is supported by a post (CD) and a cable as shown in E7.4.15. The post weighs 250 N. Assume that all surfaces are smooth. Determine the tension in the cable and the forces acting on the bar at the contacting surfaces if the bar is in equilibrium.

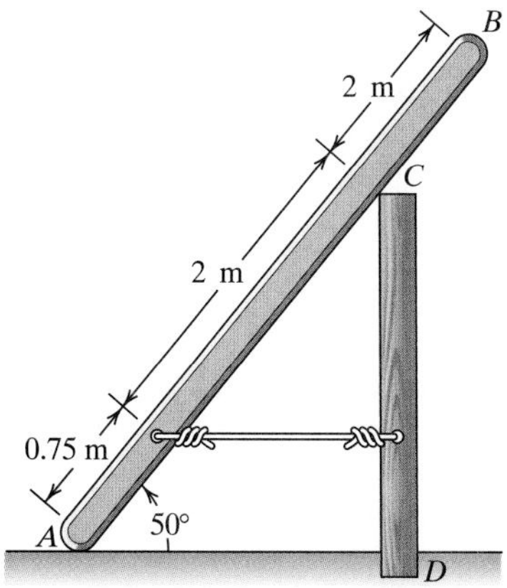

E7.4.15

7.4.16. For the situation in **E7.4.15**, determine the loads acting on the post at D.

7.4.17. A cylinder with a mass of 100 kg is supported on an inclined surface by the pin-connected two-bar frame ABC. Assume that the weight of the frame is negligible and that all surfaces are smooth.

a. Determine the forces exerted on the cylinder by the contacting surfaces.

b. Determine the loads acting on the frame at supports A and C.

c. Comment on whether member AB and/or BC is a two-force element. Comment on whether the cylinder is a three-force element.

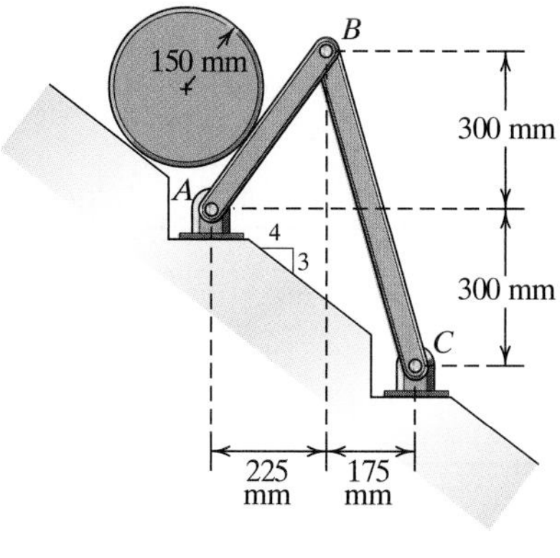

E7.4.17

7.4.18. The mass of the child shown in E7.4.18 is 40 kg. How much force must she exert on the rope of the chair lift in order to support herself?

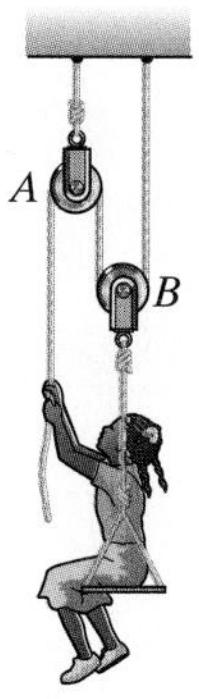

E7.4.18

7.4.19. **a.** Disks 1 and 2 in E7.4.19*a* have weights of 200 N and 100 N, respectively. Determine the maximum horizontal force F that can be applied to Disk 2 without causing Disk 1 to move up the incline. (*Hint:* What is the value of the normal force at B at the instant F gets large enough to cause Disk 1 to move up the incline?) Confirm that the line of action of the net force acting on Disk 2 goes through the center of Disk 1.

b. When the positions of the disks are reversed, as shown in **E7.4.19*b***, is the maximum horizontal force that can be applied to Disk 1 without causing Disk 1 or Disk 2 to move up the incline greater than, the same as, or less than what it was in **E7.4.14*a***? (First use your intuition to reason through an answer, then confirm with calculations, and see **E7.4.19*b***.)

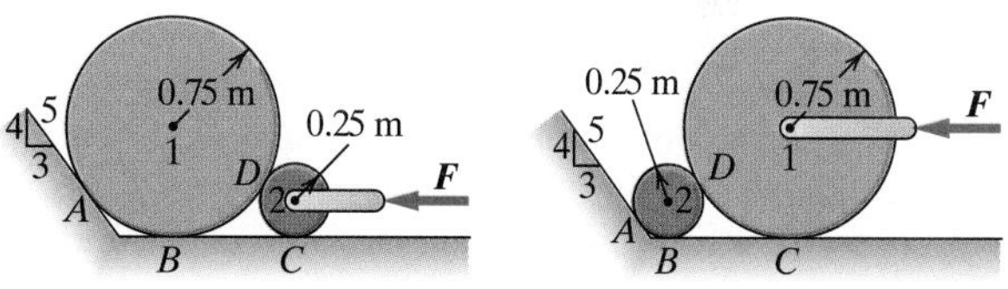

(a) (b) **E7.4.19**

7.4.20. Determine the moment at C (the lower back) when the child in **E7.2.43** is performing

a. a standard push-up and is in the position shown (**E7.2.43*a***)

b. a modified push-up and is in the position shown in (**E7.2.43*b***)

c. Compare the moment at C found in **b** as a percentage of the loads in **a**.

7.4.21. E7.4.21 shows an American Olympic gymnast executing an inverted iron cross. This maneuver is an excellent example of an athlete in mechanical equilibrium who is subjecting his shoulders to extraordinary forces and moments. Calculating these loads will allow us to minimize injuries and develop training regimens that better prepare elite athletes for international competition. The gymnast's arms make an angle of approximately 27° with the horizontal, while the ropes make an angle of 82° with the horizontal.

E7.4.21

a. Draw a free-body diagram of the gymnast's arm and determine the loads acting at his shoulder. (*Hint:* In order to calculate the forces and moments at the gymnast's shoulder, we need to draw a free-body diagram of one of his arms.) If we take a section through the shoulder joint we have to replace the torso with the forces and moments that it exerts on the arm. Similarly, we have to replace the ring with the forces it exerts on the hand (it is reasonable to assume that the rings exert negligible moments on the gymnast's hands). A quick inspection of the free-body diagram should show you that you have more unknowns than you have equations. There are a few ways to calculate the force in the rope, but the most straightforward technique is to draw a free-body diagram of the whole gymnast given that he weighs approximately 154 lb.

b. The magnitude of a moment is a difficult quantity to understand intuitively. Consider the moment exerted on the gymnast's shoulder, as found in **a**. How much force would you have to apply to a standard crescent wrench to generate the same moment? State all assumptions.

7.4.22. In Section 2.2 of Chapter 2 we used the idea of "zooming in" to develop a relationship between foot force ($\boldsymbol{F}_{\text{foot}}$) and friction force ($\boldsymbol{F}_{\text{friction}}$) that pushes a bicycle forward. Develop equation (2.2) by applying the analysis procedure first presented in Chapter 1, as well as the conditions and equations of equilibrium developed in this chapter.

7.4.23. Consider a 10-speed bicycle with two chain rings (one with 40 and one with 52 teeth), and five rear cogs (14, 17, 20, 24, and 28 teeth). The length of the crank is 6 inches. N represents the number of gear teeth.

a. Determine the ratios of $\|\boldsymbol{F}_{\text{friction}}\| / \|\boldsymbol{F}_{\text{foot}}\|$ and $(N_{\text{chainring}}/N_{\text{rear cog}})$ for the various gear combinations. Present your answers in a tabular form, similar to that shown in Table 7.4.23.

b. What gear ratio $(N_{\text{chainring}}/N_{\text{rear cog}})$ results in the greatest force multiplication (i.e., the largest force ratio $(\|\boldsymbol{F}_{\text{friction}}\| / \|\boldsymbol{F}_{\text{foot}}\|)$)? What gear ratio results in the least force multiplication? What gear ratio would you want to use to ascend a hill (include your reasoning)? What gear ratio would you want to use to descend a hill (include your reasoning)?

7.4.24. Find a bicycle with multiple gears in the front and rear. Count the number of teeth on the chainrings and rear cogs; record the numbers using Table 7.4.24 as a template. Also measure the length of the crank arm.

a. Use your counts and measurement to determine the ratios of $\|\boldsymbol{F}_{\text{friction}}\| / \|\boldsymbol{F}_{\text{foot}}\|$ and $N_{\text{chainring}}/N_{\text{rear cog}}$ for the various gear combinations. Present your answers in a tabular form, similar to that shown below. (*Note:* There will be as many columns as there are front gears, and as many rows as there are rear gears.)

b. What gear ratio $(N_{\text{chainring}}/N_{\text{rear cog}})$ results in the largest force ratio $(\|\boldsymbol{F}_{\text{friction}}\| / \|\boldsymbol{F}_{\text{foot}}\|)$ and therefore the greatest force multiplication? What gear ratio results in the least force multiplication? What gear ratio would you want to use to ascend a hill (include your reasoning)? What gear ratio would you want to use to descend a hill (include your reasoning)?

Table 7.4.23 Data on Force Multiplication of Author's Bicycle

	$N_{\text{chainring}} = 40$		$N_{\text{chainring}} = 52$	
$N_{\text{rear cog}} =$	$N_{\text{chainring}}/N_{\text{rear cog}} =$	$\|\boldsymbol{F}_{\text{friction}}\| / \|\boldsymbol{F}_{\text{foot}}\| =$	$N_{\text{chainring}}/N_{\text{rear cog}} =$	$\|\boldsymbol{F}_{\text{friction}}\| / \|\boldsymbol{F}_{\text{foot}}\| =$
14				
17				
20				
24				
28				

Table 7.4.24 Data on Force Multiplication of Student's Bicycle

	$N_{\text{chainring}} =$		$N_{\text{chainring}} =$		$N_{\text{chainring}} =$	
$N_{\text{rear cog}} =$	$N_{\text{chainring}}/N_{\text{rear cog}} =$	$\lVert \boldsymbol{F}_{\text{friction}} \rVert / \lVert \boldsymbol{F}_{\text{foot}} \rVert =$	$N_{\text{chainring}}/N_{\text{rear cog}} =$	$\lVert \boldsymbol{F}_{\text{friction}} \rVert / \lVert \boldsymbol{F}_{\text{foot}} \rVert =$	$N_{\text{chainring}}/N_{\text{rear cog}} =$	$\lVert \boldsymbol{F}_{\text{friction}} \rVert / \lVert \boldsymbol{F}_{\text{foot}} \rVert =$

7.4.25. Consider the bicycle brake hand lever shown in E7.4.25*a*. The givens/assumptions are that all forces are coplanar and 2-dimensional, static analysis is acceptable, and the two forces labeled F_{hand} have the same line of action.

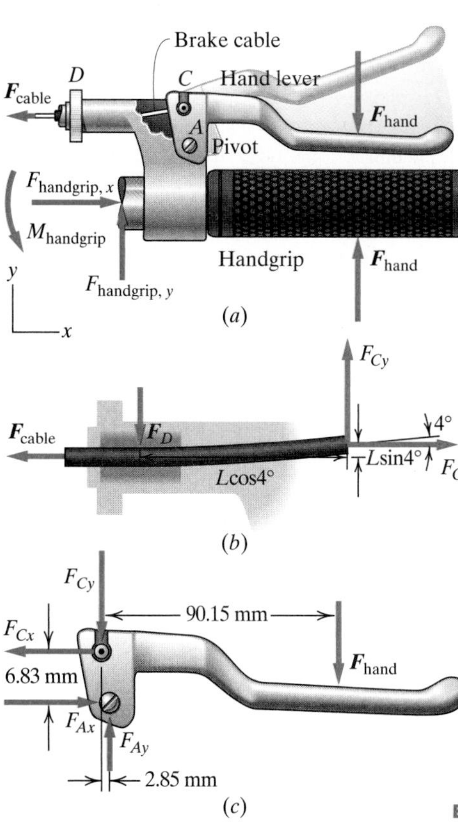

E7.4.25

a. Based on the free-body diagram of the cable housing subassembly shown in **E7.4.25*b*** of the brake cable where it moves through the housing at D, write expressions for F_D, F_{Cy}, and F_{Cx} in terms of F_{cable}. Force F_D is the force that the metal housing of the brake applies to the cable housing.

b. Based on the free-body diagram of the brake lever arm in **E7.4.25*c***, write an expression for F_{hand} in terms of F_{cable}. (Use the results from **a**, as necessary.)

c. If $F_{\text{hand}} = 150$ N, what is the value of F_{cable}? (Use results as appropriate from **b**.)

d. What is the mechanical advantage of the system (i.e., the ratio of $F_{\text{cable}}/F_{\text{hand}}$)?

7.4.26. Consider the mechanism used to weigh mail (E7.4.26*a*). A package placed at A causes the weight pointer to rotate through an angle α. Neglect the weights of the members except for the counterweight at B, which has a mass of 4 kg. For a particular package, $\alpha = 20°$. The mechanism is in equilibrium.

a. Define the system as shown in **E7.4.26*b*** and draw its free-body diagram.

b. Define the system as shown in **E7.4.26*c*** and draw its free-body diagram.

c. Define the system as shown in **E7.4.26*d*** and draw its free-body diagram.

d. Use a combination of the free-body diagrams in **a–c** to find the loads acting on the mechanism at C and F, and the weight of the package.

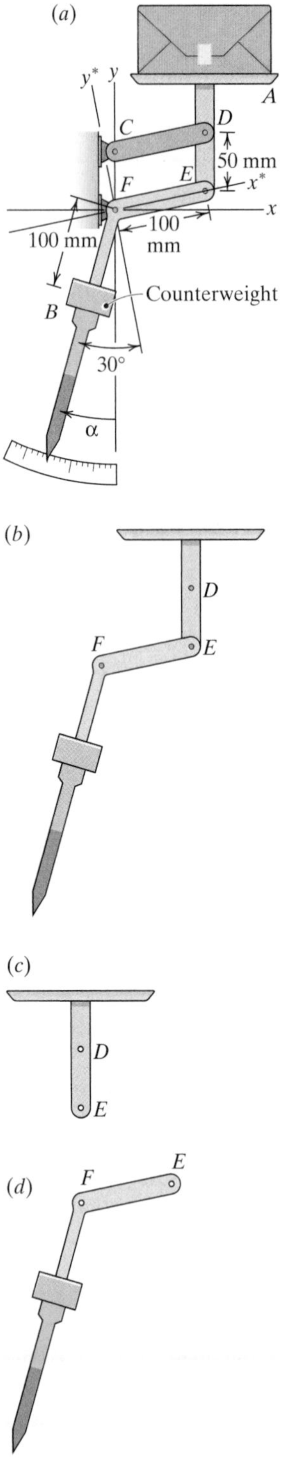

E7.4.26

7.4.27. Consider the pair of pliers in E7.4.27. Determine the ratio of $\| \boldsymbol{F}_E \| / \| \boldsymbol{F}_G \|$ where $\boldsymbol{F}_G$ is 150 N. (*Hint:* First consider equilibrium of Member 1, then of Member 2, then of Member 3.)

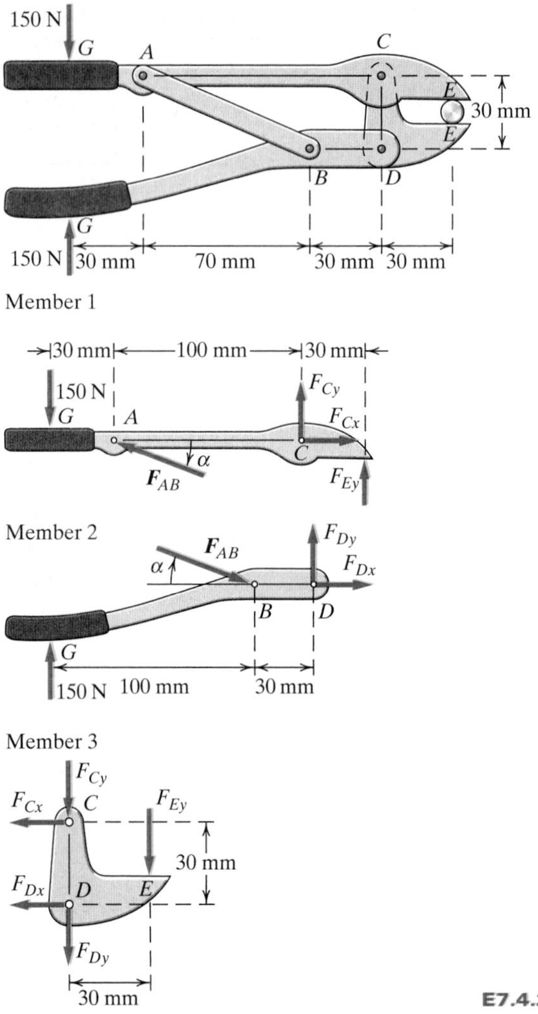

E7.4.27

7.4.28. Consider the two-member frame in E7.4.28. Member AC is connected to member $BCDE$ by a slider connection at C. Determine

a. the loads acting on the frame at A, D, and E
b. the loads of member $BCDE$ acting on member AC
c. the loads of member AC acting on member $BCDE$

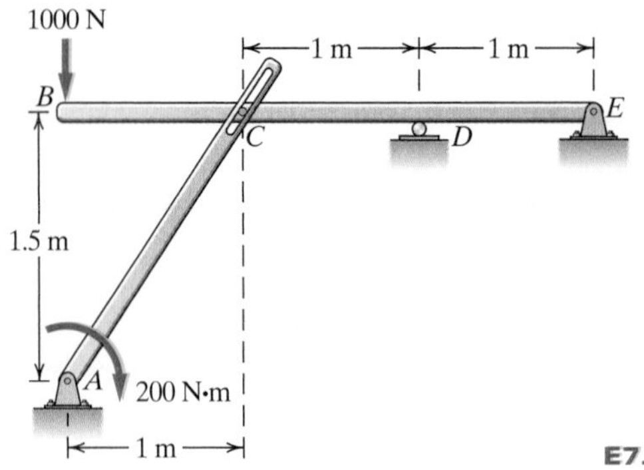

E7.4.28

7.4.29. Consider the two-member frame in E7.4.29. Member AB is connected at B to member BCD by a slider connection. Determine

a. the loads acting on the frame at A, C, and D
b. the loads of member AB acting on member BCD
c. the loads of member BCD acting on member AB

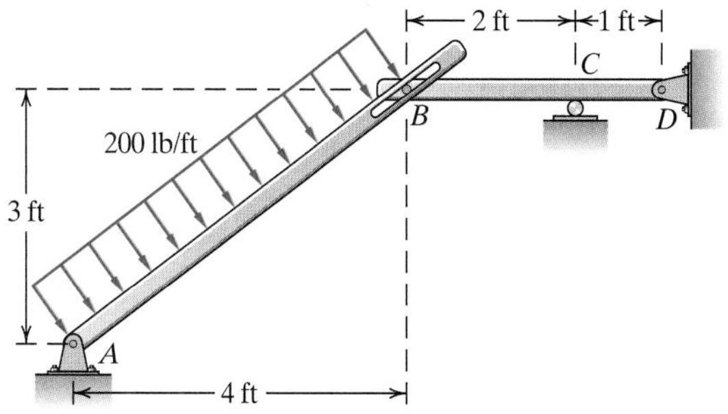

E7.4.29

7.4.30. Consider the two-member frame in E7.4.30. Member AE is connected at C to member BD by a pin connection. The cylinder weighs 200 N. Determine

a. the loads acting on the frame at D and E
b. the loads of member AE acting on member BD at C
c. the loads of member BD acting on member AE at C

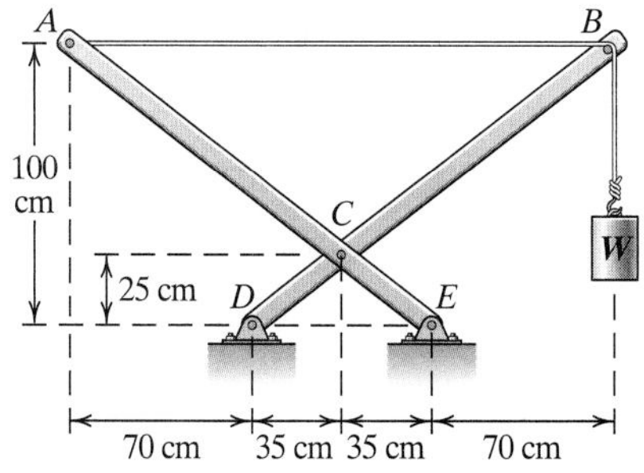

E7.4.30

7.4.31. Consider the frame in E7.4.31. Determine the loads acting on the frame at A and at E. W is 400 N.

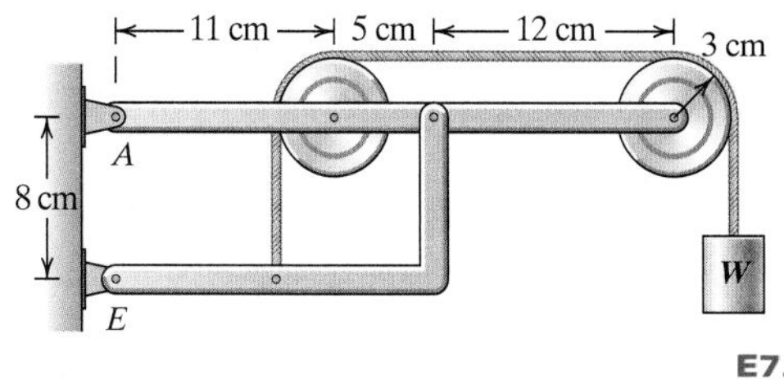

E7.4.31

7.4.32. Consider the frame in E7.4.32. The force at D is vertical and of magnitude F. Determine

a. the loads acting on the frame at A and B
b. the loads of member ACD acting on member CB
c. the loads of member CD acting on member ACD
d. the tension or compression in the spring

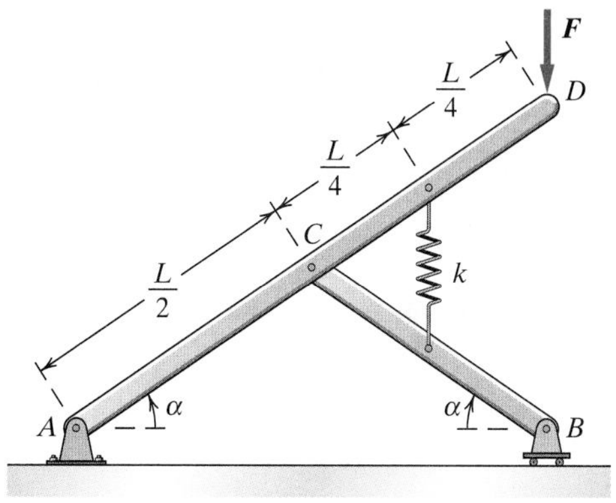

E7.4.32

7.4.33. Consider the exercise frame in equilibrium in E7.4.33. Determine

a. the loads acting on the frame at E and F if the frame is bolted to the floor at E

b. the loads acting on the member ACD at A, B, C, and D

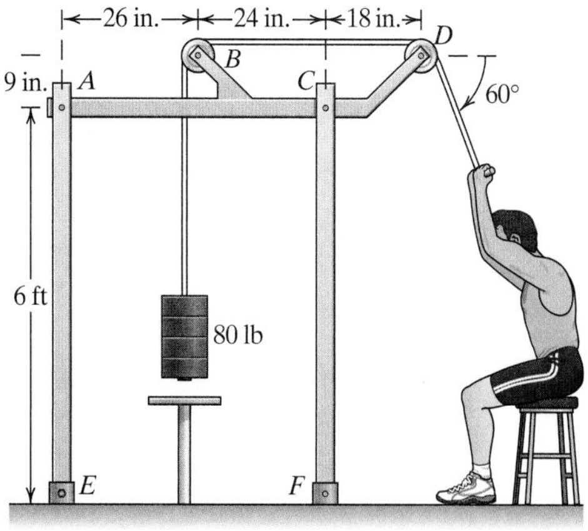

E7.4.33

7.4.34. A mobile hangs from the ceiling at E from a cord, as shown in **E7.2.32**. If the mobile is in equilibrium, write expressions for the weights of the fish B, C, and D (W_B, W_C, and W_D, respectively) in terms of the weight of fish A (W_A).

7.5 DETERMINATE, INDETERMINATE, AND UNDERCONSTRAINED SYSTEMS

In Examples 7.1–7.13 there were as many unknowns as there were independent equilibrium equations. We refer to these problems as **statically determinate** because we were able to determine the unknowns using only the equations of equilibrium. We now look at two types of systems in which the equations of equilibrium are not sufficient for finding the unknowns. These system types are referred to as **statically indeterminate systems** and **underconstrained systems**.

Statically indeterminate systems are systems in which there are more supports than are necessary for the system to be balanced. In other words, there is an excess, or redundancy, of supports. Examples of systems having a redundancy of supports are shown in **Figure 7.47*a***. We could remove a support in each example and create a determinate system that is still balanced under the loads, which is what was done to create **Figure 7.47*b***. The systems in **Figure 7.47*b*** still represent systems in equilibrium. The supports that can be removed and still have a stable system are referred to as **redundant supports**. It is important to note that free-body diagrams of the indeterminate systems in **Figure 7.47*a*** are different from those of the determinate systems in **Figure 7.47*b***.

The additional equation(s) needed for determining redundant support load(s) comes from a description of how the system *deforms* under the loads. This deformation is generally very small for engineered systems, so it does not violate the rigid-body assumption we have been making throughout this book. An understanding of how systems deform is gained

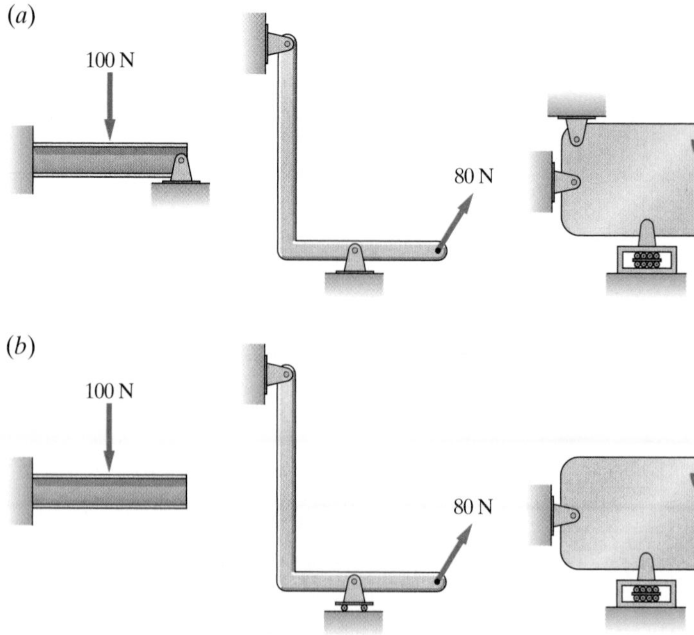

Figure 7.47 (*a*) Examples of systems that are statically indeterminate; (*b*) systems from (*a*) modified so as to be statically determinate

by studying the branch of mechanics called *mechanics of materials* (which is beyond this course, but probably part of your engineering program).

As a general (but not hard-and-fast) rule, a nonplanar system with more than six unknowns is a statically indeterminate system, as is a planar system with more than three unknowns. With any indeterminate system it is possible to reduce the number of unknowns by using the conditions of equilibrium, as illustrated in Example 7.15.

Underconstrained systems are such that the supports are insufficient to keep the system balanced and stable, and are in contrast to statically indeterminate systems (where there are more supports than are necessary to have a balanced system) and in contrast to statically determinate systems (where there are just enough supports to have a balanced system). The conditions of equilibrium, as stated in (7.1) and (7.2), do not hold for an underconstrained system.

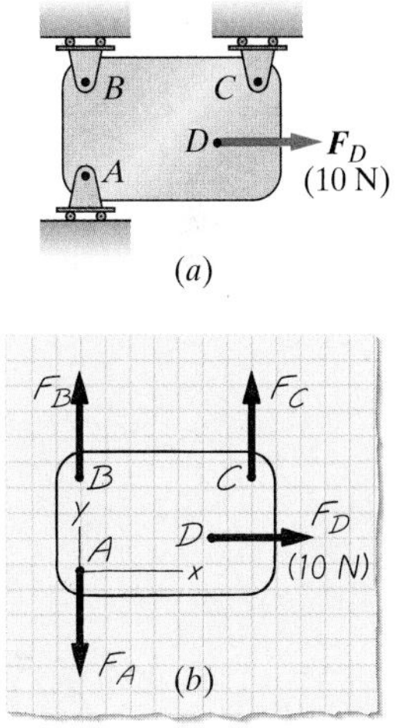

Figure 7.48 (*a*) An example of an underconstrained sytem; (*b*) its associated free-body diagram

Figure 7.48 shows an underconstrained planar system, where there are three unknowns and it would appear that there are three equations with which to find the unknowns. However, one of the equations of equilibrium is an unachievable equilibrium condition because there is no way for the net force to be zero with a nonzero applied load acting at D as we now illustrate. Assuming the system is planar, we write the first planar equilibrium equation as

$$\sum F_x = F_D = 0$$

This equation is "unachievable" if F_D is nonzero. Since F_D is 10 N in **Figure 7.48**, there is a net force on the system in the x direction and it moves and accelerates rightward. Remember, an underconstrained system will move.

Several additional examples of underconstrained systems are presented in **Figure 7.49**. Because $\boldsymbol{F}_C$ in **Figure 7.49*a*** has a horizontal component, the beam is underconstrained and moves to the right. Another way to check this is by noting that the beam is a three-force member but the lines of action of the three forces acting on it ($\boldsymbol{F}_A$, $\boldsymbol{F}_C$, $\boldsymbol{F}_B$) do not intersect at a common point. In **Figure 7.49*b***, it might appear that, because the supports at A and B have been angled relative to their position in (*a*), the system is properly constrained. The lines of action of the three forces still do not intersect at a common point, however. As another example of an understrained system, consider the plate in **Figure 7.49c**. Because the line of action of the link B intersects point A (where the other two links are attached), the force at B cannot counter the moment created by F_D about a moment center at A. The system is again underconstrained and moves into a stable position.

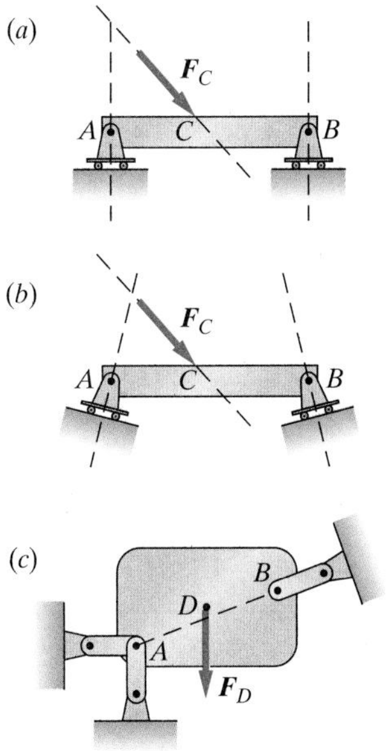

Figure 7.49 Examples of underconstrained systems

Figuring out whether or not a system is statically determinate is done in the early stages of setting up a problem. As you gain experience in solving equilibrium problems, you will find that you can also do some checking for determinacy as you go about setting up the free-body diagram. You may also identify the problem as not being statically determinate as you solve the conditions of equilibrium. Regardless of when in the analysis process you notice that the system is not statically determinate, make sure to document your findings, as illustrated in Example 7.15.

EXAMPLE 7.14 IDENTIFY STRUCTURE

Identify each structure in **Figures 7.50*a*–7.58*a*** as statically determinate, statically indeterminate, or underconstrained. Associated free-body diagrams are shown in **Figures 7.50*b*–7.58*b***.

(a) The beam in **Figure 7.50** is *statically determinate* because there are three unknown loads (at the supports), and by using the equations for planar equilibrium we can find their magnitudes.

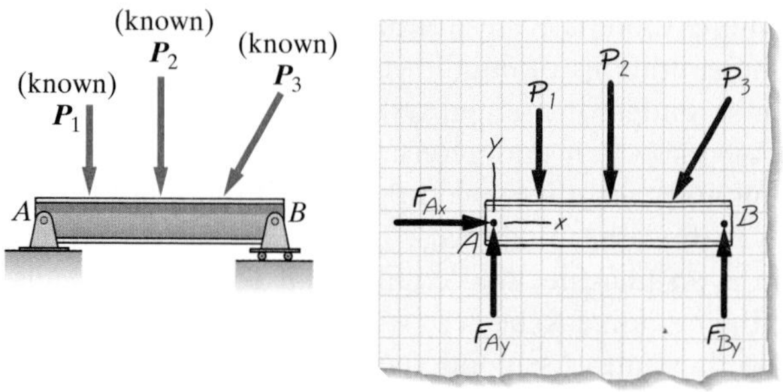

Figure 7.50 (*a*) (*b*)

(b) The beam in **Figure 7.51** is *statically indeterminate* because there are four unknown loads (at the supports), but there are only three equations for planar equilibrium. The number of equilibrium equations is insufficient for finding the unknown loads.

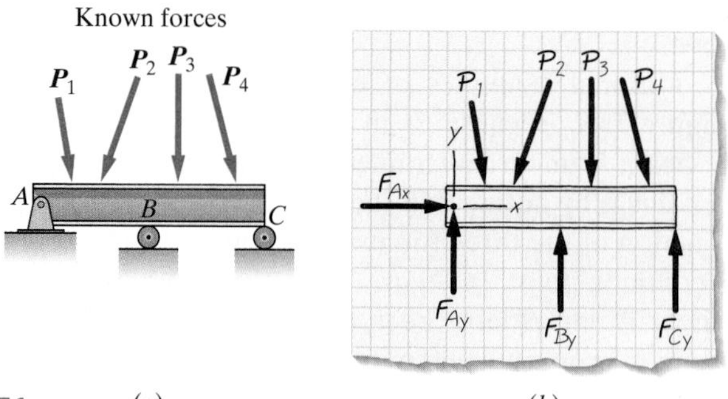

Figure 7.51 (*a*) (*b*)

(c) System *ABCD* in **Figure 7.52** is *statically determinate* because there are three unknown loads (one at *A* and two at *C*), and by using the equations for planar equilibrium we can find their magnitudes. Remember that because member *AB* is a two-force element, there is only one unknown force at *A*.

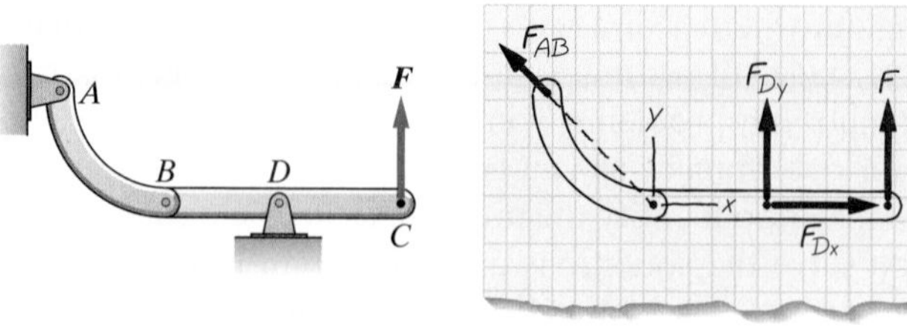

Figure 7.52 (*a*) (*b*)

(d) The system in **Figure 7.53** is *underconstrained* because the applied forces $\mathbf{P}_1$ and $\mathbf{P}_2$ will cause it to shift to a new configuration; therefore equilibrium cannot be achieved in the configuration given.

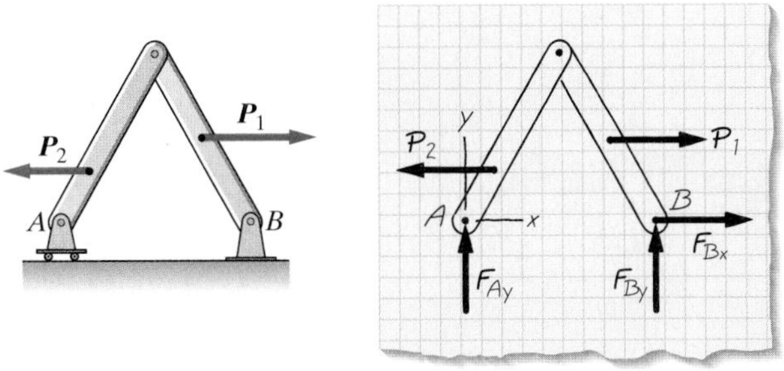

Figure 7.53 (*a*) (*b*)

(e) The system in **Figure 7.54** is *underconstrained* because it will undergo a change in shape once forces are applied. Equilibrium cannot be achieved in this configuration.

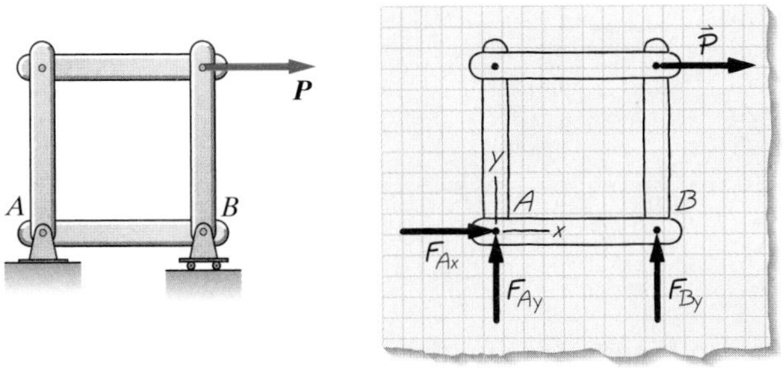

Figure 7.54 (*a*) (*b*)

Note: The systems in **Figures 7.53** and **7.54** both have three unknown loads, and it appears that the three planar equilibrium should be sufficient to find them. However, because these systems are not rigid internally, additional supports are needed. This just goes to show that the general rule of "three unknowns and three equations" is just a guideline for determining if a planar system is statically determinate.

(f) The system in **Figure 7.55** is both *underconstrained* and *statically indeterminate.* A force applied as shown will start the system moving, making the system underconstrained. It is statically indeterminate because there are more rollers than the minimum necessary to keep the slider on the track.

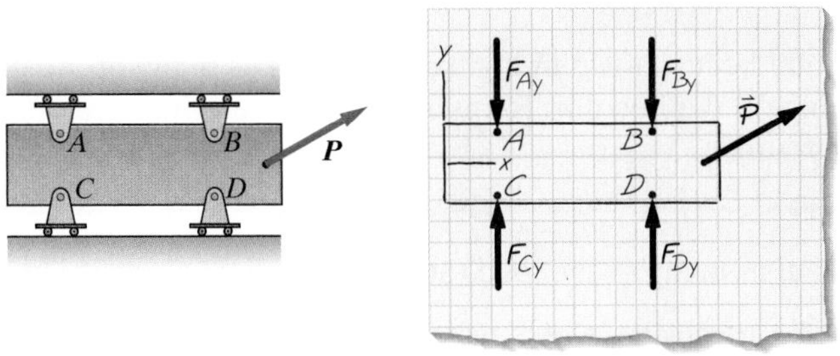

Figure 7.55 (*a*) (*b*)

(g) The L-bar in **Figure 7.56** is a three-force element. Notice that the force at A must be vertical and its line of action cannot pass through point B. Therefore the three forces are not concurrent, resulting in an *underconstrained* system.

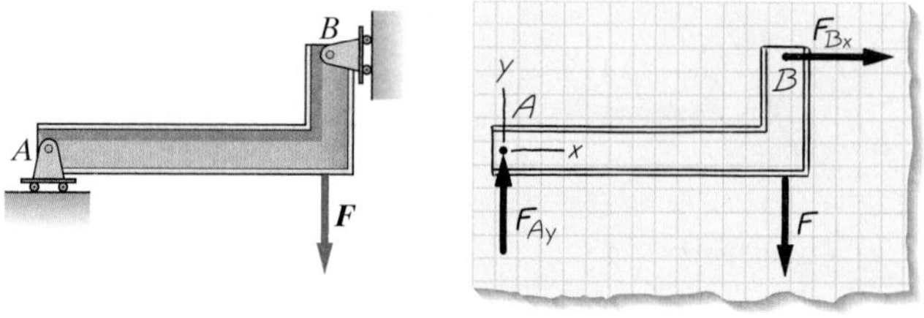

Figure 7.56 (*a*) (*b*)

(h) The L-bar in **Figure 7.57** is a three-force element. The lines of action of the forces at A and B intersect at A, while the line of action of F does not. Instead, $\mathbf{F}$ exerts a moment about A. This system is *underconstrained*.

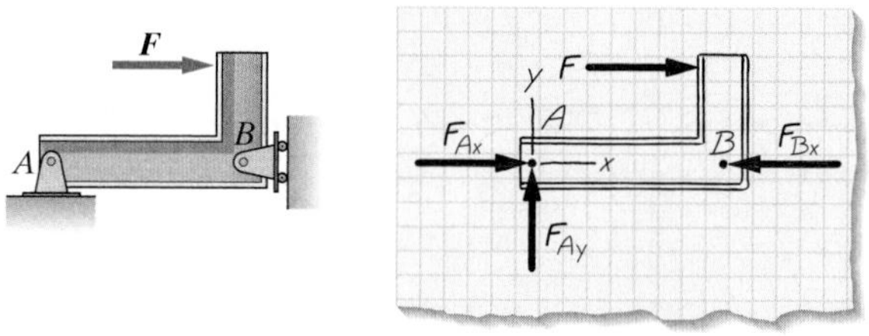

Figure 7.57 (*a*) (*b*)

(i) The L-bar in **Figure 7.58** is a three-force element. The lines of action of the forces at A, B, and C intersect at B. Therefore, this system is *statically determinate*.

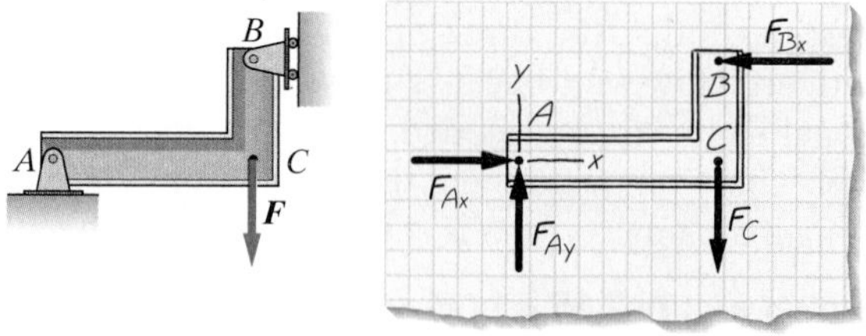

Figure 7.58 (*a*) (*b*)

EXAMPLE 7.15 CONSIDERING A STATICALLY INDETERMINATE SITUATION

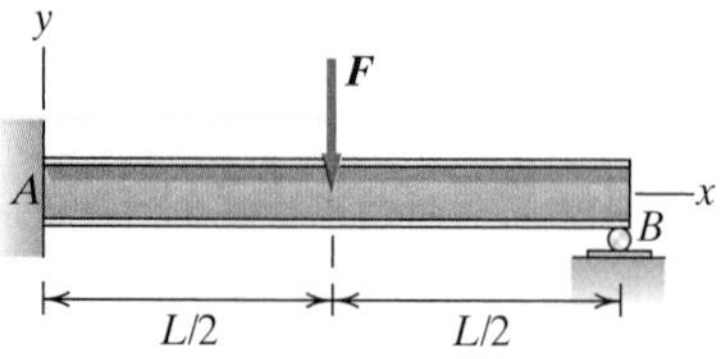

Figure 7.59

Consider the beam shown in **Figure 7.59**. Find as many of the unknown loads acting on the beam as is possible given the principles of mechanical equilibrium.

Goal Find as many of the loads acting on the beam as possible.

Given We are given the dimensions of the beam, and that a known load F acts at midspan. The beam is fixed at end A and sits on a roller at B.

Assume We will assume that we can ignore the weight of the beam and that the beam system is planar.

Draw We isolate the beam from its surroundings and draw a free-body diagram (**Figure 7.60**).

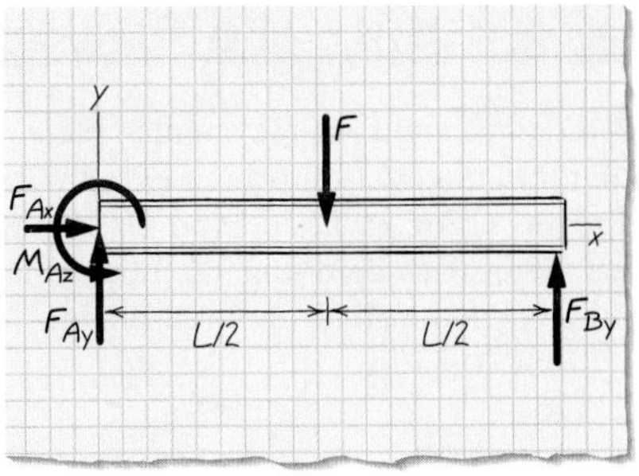

Figure 7.60

Formulate Equations and Solve Now formulate the equilibrium equations for the planar system in **Figure 7.60**. Just by inspecting the free-body diagram we should suspect that this system may be statically indeterminate because there are four unknowns (F_{Ax}, F_{Ay}, F_{By}, and M_{Az}) but only three planar equilibrium equations. This system has one extra support, and we could consider M_{Az}, F_{Ay}, or F_{By} to be the redundant. As we try to analyze this problem we will discover that because the situation depicted in **Figure 7.59** is statically indeterminate, we will be able to use the equilibrium equations to reduce the number of unknowns from four to two, but we will not be able to find these two unknowns in terms of the known force F.

Based on (7.5A):

$$\sum F_x = 0\ (\rightarrow +)$$
$$F_{Ax} = 0$$

Based on (7.5B):

$$\sum F_y = 0\ (\uparrow +)$$
$$F_{Ay} + F_{By} - F = 0$$
$$F_{By} = F - F_{Ay} \quad (1)$$

Based on (7.5C) (with the moment center at A):

$$\sum M_{z@A} = 0(\circlearrowleft +)$$
$$-\left(\frac{L}{2}\right)(F) + (L)(F_{By}) + M_{Az} = 0 \quad (2)$$

Now substitute (1) into (2):

$$\left(\frac{L}{2}\right)(F) - (L)(F_{Ay}) + M_{Az} = 0$$

Equation (3) is an expression containing two unknowns (F_{Ay} and M_{Az}).

You might be tempted to suggest that we generate additional equations by placing the moment center at a location other than A. If we did this, we would find that the equation we could generate would not be linearly independent of (1) and (2). In order to find the unknowns in terms of F, we would need to employ mechanics of materials concepts.

EXERCISES 7.5

7.5.1. A uniform rectangular plate weighs W, with its center of gravity as shown in **E7.5.1**. Gravity acts in the $-y$ direction. For each situation classify the plate (taken as the system) as determinate, indeterminate, or underconstrained, and include your reasoning.

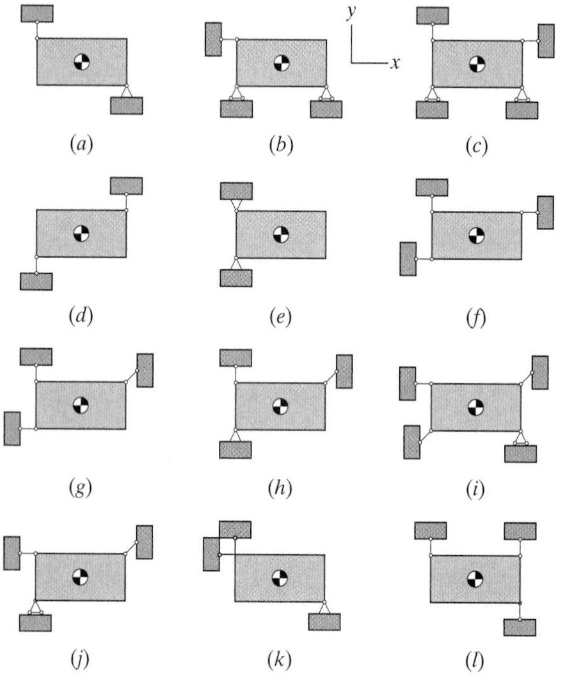

E7.5.1

7.5.2. Consider that gravity acts in the $+x$ direction in the figures shown in **E7.5.1**. For each situation classify the plate (taken as the system) as determinate, indeterminate, or underconstrained, and include your reasoning.

7.5.3. The 200-N box in **E7.5.3** is supported by a knife edge at point A and by a cable that passes over the frictionless pulley, B. The center of mass of the box is at the geometric center, G. For the loading shown, determine the tension in the cable with $d = 25$ cm. Also determine $\mathbf{F}_C$.

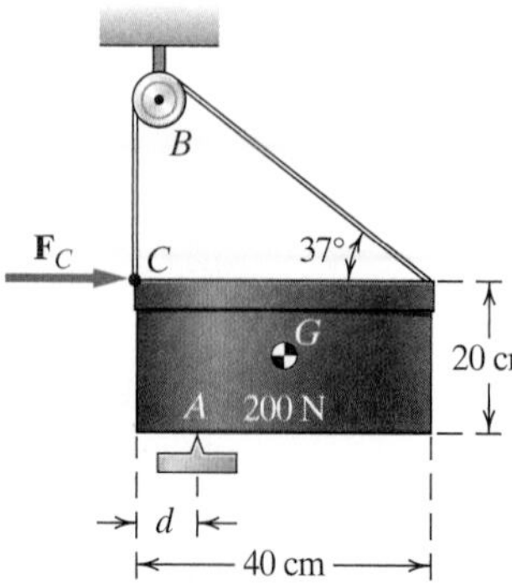

E7.5.3

7.5.4. The 200-N box in **E7.5.3** is supported by a knife edge at point A and by a cable that passes over the frictionless pulley, B. The center of mass of the box is at the geometric center, G. For the loading shown, determine the range of values of d that result in a statically determinate system. For this range, determine

a. the range of F_B (the tension in the cable)

b. the range of F_C (the horizontal force applied at C)

c. the range of F_A (the force the knife edge applies at A)

d. Create a plot that shows F_B, F_C, and F_A as functions of d.

7.5.5. The bent bar is supported by three smooth eyebolts as shown in **E7.5.5**. Is the system statically determinate?

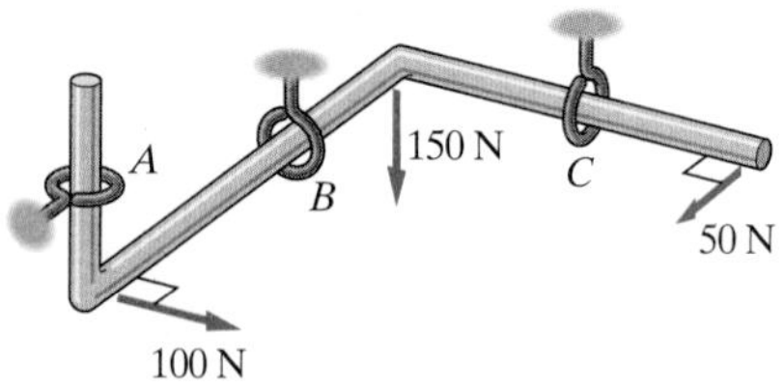

E7.5.5

7.5.6. A rigid bent tube guides a flexible shaft that is transmitting a 20 N · m torque, as shown in **E7.5.6**. The tube is supported by the smooth eyebolts at A, B, and C. Is the system statically determinate?

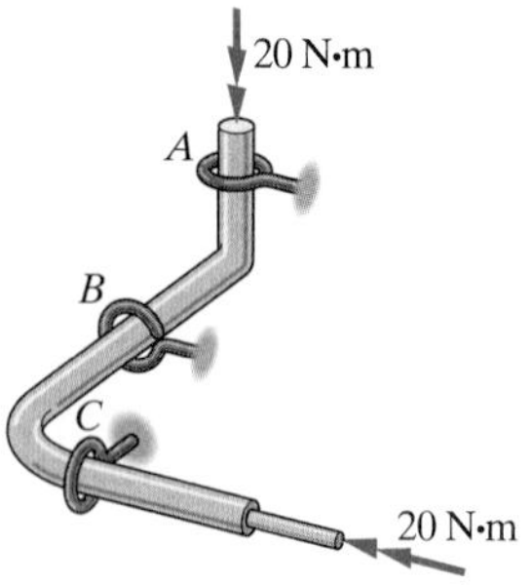

E7.5.6

7.5.7. A 5-kN force and a 4-kN · m moment are applied to the forked bar as shown in **E7.5.7**.

a. Determine the loads acting on the bar at A and B as a function of the angle θ.

b. From the results of **a**, determine whether there are any values of θ for which the bar is underconstrained.

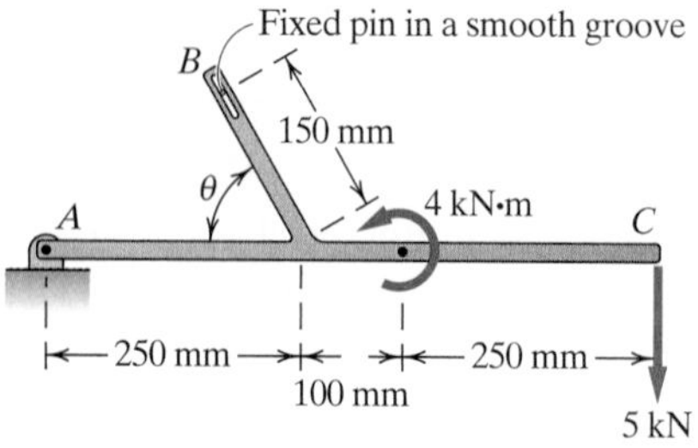

E7.5.7

7.5.8. Determine the value(s) of the angle β for the short link at A that results in the plate in **E7.5.8** being an underconstrained system.

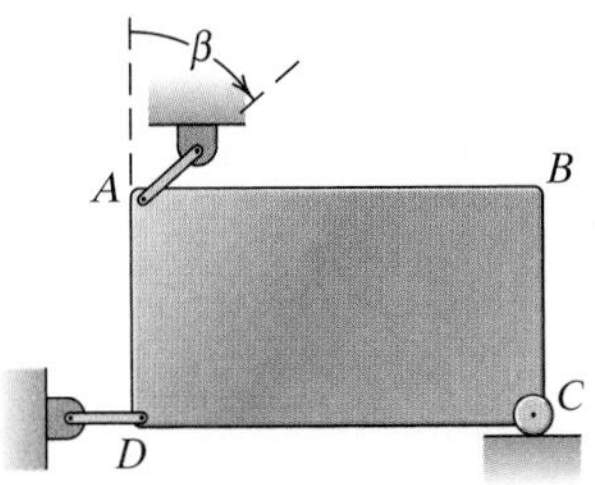

E7.5.8

7.5.9. The beam of uniform weight is fixed at C and rests against a smooth block at A (**E7.5.9**). In addition, a 100-N weight hangs from point B. Classify the beam (taken as the system) as determinate, indeterminate, or underconstrained, and include your reasoning. If you have classified the system as indeterminate or underconstrained, recommend how the boundary connections could be changed to make the system determinate.

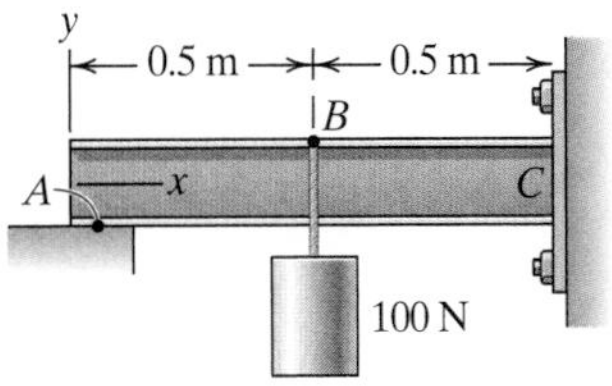

E7.5.9

7.6 JUST THE FACTS

If a system is in **mechanical equilibrium** (where equilibrium is a state of balance), the loads acting on the system have particular relationships to one another. These relationships can be represented mathematically as follows:

$$\boldsymbol{F}_{\text{net}} = \sum^{\substack{\text{all forces}\\\text{acting}\\\text{on system}}} \boldsymbol{F} = 0 \qquad \textbf{force equilibrium condition} \qquad (7.1)$$

$$\boldsymbol{M}_{\text{net}} = \sum^{\substack{\text{all moments}\\\text{acting}\\\text{on system}}} \boldsymbol{M} = 0 \qquad \textbf{moment equilibrium condition} \qquad (7.2)$$

Conditions 7.1 and 7.2 are necessary and sufficient conditions for a system to be in mechanical equilibrium.

It is generally more convenient to work with conditions in 7.1 and 7.2 in terms of vector components. We can rewrite (7.1) as the **force equilibrium equations** that ensure **force balance** and (7.2) as the **moment equilibrium equations** that ensure **moment balance**:

$$\sum F_x = 0 \qquad (7.3\text{A})$$
$$\sum F_y = 0 \qquad (7.3\text{B})$$
$$\sum F_z = 0 \qquad (7.3\text{C})$$

$$\sum M_x = 0 \qquad (7.4\text{A})$$
$$\sum M_y = 0 \qquad (7.4\text{B})$$
$$\sum M_z = 0 \qquad (7.4\text{C})$$

The six equations in (7.3) and (7.4) are true for any system in mechanical equilibrium (i.e., a balanced system). We shall refer to them as the **six equilibrium equations**.

Any system that is in equilibrium must satisfy all six equilibrium equations. Three of these equations, however, are automatically satisfied for a planar system. The three remaining equilibrium equations are

$$\sum F_x = 0 \qquad (7.5A)$$
$$\sum F_y = 0 \qquad (7.5B)$$
$$\sum M_z = 0 \qquad (7.5C)$$

planar equilibrium equations

In applying the equations of equilibrium to a system, it is useful to identify whether the system can be represented as a particle or involves two-force or three-force elements or frictionless pulleys:

A **particle** is an object whose size and shape have negligible effect on the response of the object to forces. Under these circumstances, the mass (if significant) of the object can be assumed to be concentrated at a point. A particle, by definition, can only be subjected to concurrent forces; the point of concurrency is the point that represents the particle.

A **two-force element** is an element of negligible weight with only two forces acting on it. *A two-force element is in equilibrium when the forces acting on it are equal, opposite, and along the same line of action. The line of action passes through the points of application of the two forces.*

A **three-force element** is an element with only three forces acting on it. *If a three-force element is in equilibrium, the lines of action of the three forces are concurrent or parallel to one another.*

A **frictionless pulley** is an element that is used to change the direction of a cable or rope. *The tension in the cable or rope is the same on both sides of a frictionless pulley.*

We should note that the equations of equilibrium apply to systems where there are as many unknown moments, forces, and/or dimensions as there were independent equilibrium equations. We refer to these systems as **statically determinate** because we were able to determine the unknowns using only the equations of equilibrium.

In contrast, **statically indeterminate systems** are systems in which there are more supports than are necessary for the system to be balanced. The supports that can be removed and still have a balanced system are referred to as **redundant supports**. To find the loads acting at the indeterminate system we need information about the deformation of the system, in addition to the conditions of equilibrium.

Underconstrained systems are such that the supports are insufficient to keep the system balanced and the conditions of equilibrium do not hold.

SYSTEM ANALYSIS (SA) EXERCISES

SA7.1 Bracing Against Moving Loads

During the construction of the Reynolds Coliseum in 1948 the contractor realized that he could speed up the placing of concrete for the floor slab by creating temporary trestles consisting of 40-cm-wide wooden planks able to support the 20-ton trucks carrying the "flowable" concrete to the place where it is needed. At that time this was an ingenious approach since concrete pump trucks that can be seen today deploying a concrete pipe with a foldable boom were not invented until the 1960s, as illustrated in Figure SA7.1.1.

As shown in Figure SA7.1.2 the concrete floor of the Reynolds Coliseum was supported by joists, I-beams, and columns. Assume that you were working for the contractor at that time and were asked to figure out what the maximum *additional* load would be on the columns due to the concrete trucks. You are also to recommend to the contractor how to position the temporary trestles. The steps that follow will enable you to make a recommendation.

(a) Draw the free-body diagram for each of the three planks under the front and rear tires (Figure SA7.1.3*a*). Assume that the distance between the front axle and the middle of the rear axles of a concrete truck is 6.0 m while the tandem axle spread is 1.5 m (distance between the two rear axles). The length of the planks laid on top of the joists is 4 m, as shown in **Figure SA7.1.3*a***.

If the rear tires combined support 66% of the total weight of the truck (20 ton), determine all of the loads acting on each of the three planks. You recognize that plank 3 is a statically indeterminate system, so you ask Mat, the design engineer responsible for the Reynolds Coliseum, how you should deal with finding the loads acting on it since you have not yet had a course in mechanics of materials. After consulting a handbook[2] and writing down some calculations, Mat says that the vertical force between the joist and plank at G is 11.3% of the vertical force between the joist and plank at F.

(b) Draw a free-body diagram of each of the 8.2-m-long joists A–G (as shown in **Figure SA7.1.3*b***), assuming that the pair of planks is positioned in the center of the joist span and that each joist can be modeled as if supported by a pin connection at one end (y = 8.2 m)

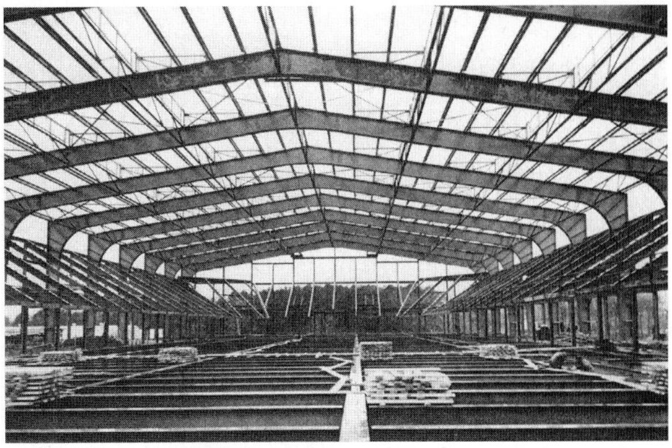

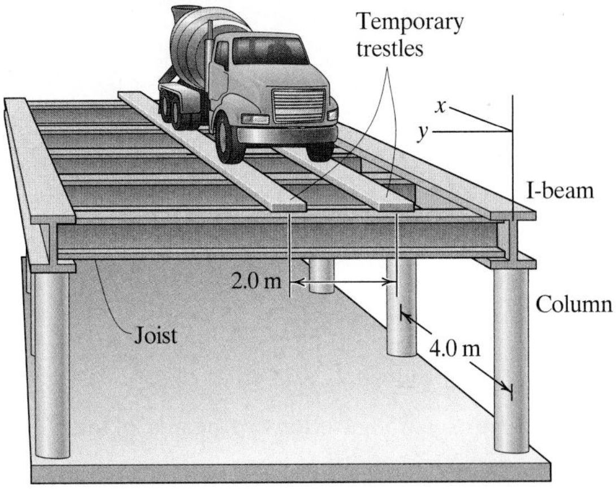

Figure SA7.1.2 Floor structure of Reynolds Coliseum (CCD = Center-to-Center Distance)

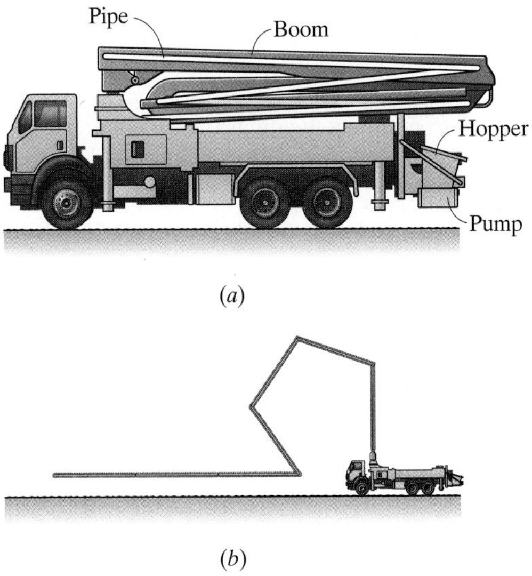

Figure SA7.1.1 Concrete pump trucks have not changed much since 1960s

[2]Mat used the book *Formulas for Stress and Strain* by Roark and Young (McGraw-Hill, 1975). He modeled plank 3 between joists F and G as a beam that is fixed at one end (F) and on a roller at the other end (G) to come up with the 11.3% number. See page 97, Example 1c in Roark and Young.

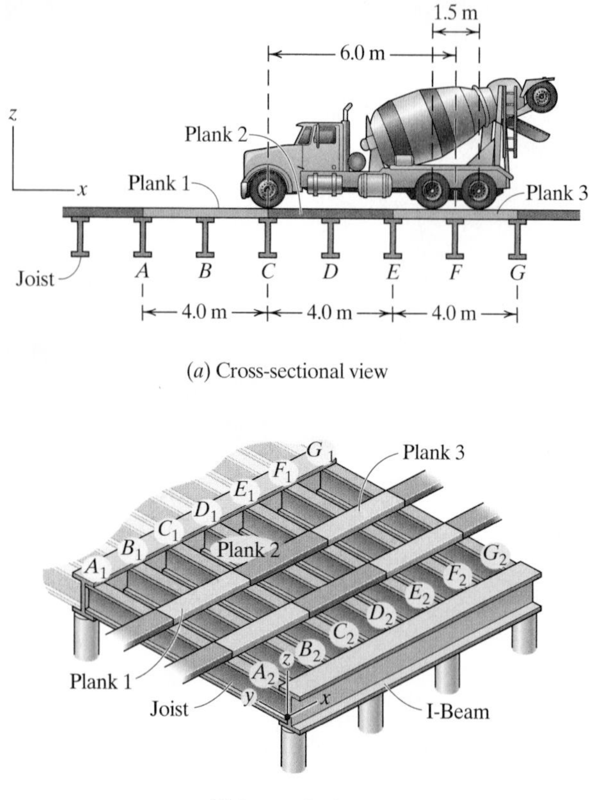

(a) Cross-sectional view

(b) Isometric view

Figure SA7.1.3 Configuration of I-beams, planks, and joists

and a rocker at the other end ($y = 0.0$ m). This is truck Position 2, as shown in **Figure SA7.1.4**. Calculate all the loads acting on each of the joists.

(c) For the truck in Position 2 with its front tire at $x = +4.0$ m (see **Figure SA7.1.5**), what is the additional load on Columns I, II, and III along $y = 8.2$ m due to the truck?

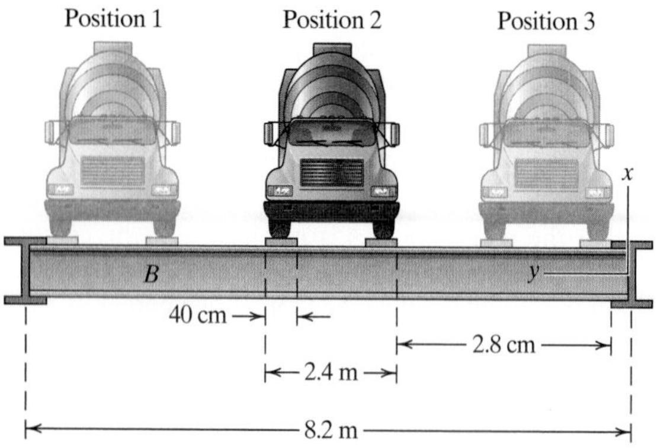

Figure SA7.1.4 Dimensions for center truck joists and beams

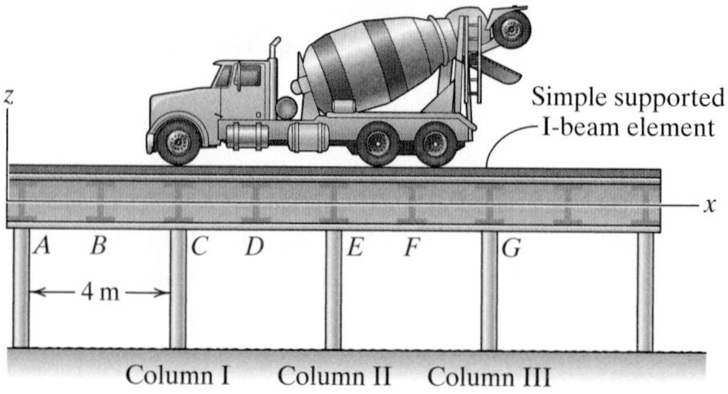

Figure SA7.1.5 Row of Columns in a cross-sectional view

(d) Before being able to answer the question "Will the columns be able to carry the extra load from the 20-ton truck when in Position 2?" you find out from Mat that:

- In addition to the truck load acting on the columns, there is the weight of the formwork for the concrete and the joints. This can be modeled as a distributed load of approximately 18 tons/m onto the I-beams, as shown in **Figure SA7.1.6**.
- He designed the columns for a maximum load of 289 tons each. However, because the concrete is not hardened enough to accept his design load he will allow only 50% of 289 tons being used during construction.

Now answer the question, "Will the columns be able to carry the extra load from the 20-ton truck when in Position 2?"

(e) **Figure SA7.1.4** indicates that the truck can't always stay in the center of the joists. Since the truck is not allowed to drive over an I-beam, the worst loading case arises when the 40-cm-wide temporary trestles are positioned immediately next to its upper flange (Positions 1 and 3 in **Figure SA7.1.4**). Repeat steps (c) and (d) for Positions 1 and 3. Summarize your findings as a recommendation on how to position the temporary trestles to minimize loadings on the columns. Support your recommendation with your calculations.

(f) Repeat steps (c), (d), and (e) for alternate column spans of 6 m and 8 m.

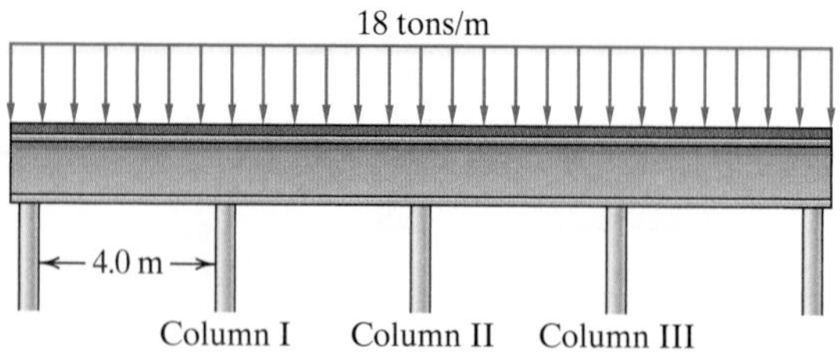

Figure SA7.1.6 Simplified loading diagram for a center I-beam

SA7.2 Keeping the Scoreboard in the Air

George's questions related to the score board inside the coliseum in **SA4.3** required the use of some basic principles of force-vectors. What you have learned since Chapter 4 has prepared you for some forensic engineering that you need to analyze an unexpected accident. As you watch the game, the tie cable between C and D suddenly breaks. The rigging arrangement for the score board is shown in Figure SA7.2.1.

(a) What will you observe after the cable breaks (assuming that everything else stays intact)? Describe in words and diagrams.

(b) What were the forces in the tie cable DC and suspension cables DG and DH before the accident? Recall that the score board weighs 4500 N.

(c) How large are F_{DG} and F_{DH} after the accident? Before the accident, points C and D were located 1.8 m between their respective winches.

(d) What do you recommend should be the sequence of actions to repair the rigging? In particular, would you lower the board to the floor for repair, or should it be fixed up in the air without moving it? Are there special risks before and during the repair?

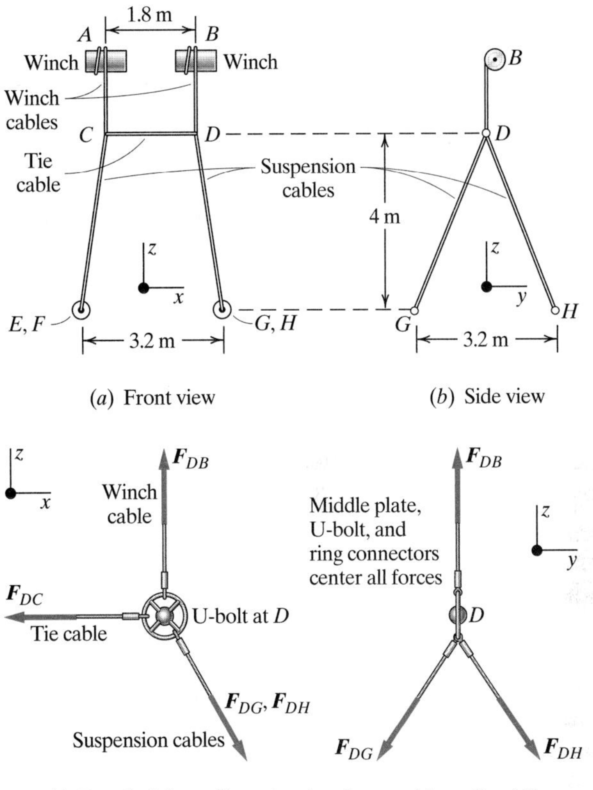

Figure SA7.2.1 Rigging arrangement for board

SA7.3 Will the Chair Flip?

The basketball game of Wolfpack against UNC Chapel Hill is underway (as it was in **SA6.1**). In **SA6.1** Sierra was worried about the size of the hinge pin. Now she is concerned that the chair the woman is sitting in might flip over. The chair is shown in Figure SA7.3.1.

(a) Assuming that the mass of the chair is 5 kg with its center of mass 10 cm to the left of hinge, how close to the front of the edge of the seat can the woman sit and not have the chair flip over when 30% of her weight is supported by her legs? Recall that the mass of the woman is 61 kg.

(b) How will the answer to (a) change when Bill, who sits in the chair behind the woman, grabs the back of the chair and puts his total weight onto it? Bill's mass is 80 kg.

(c) The Wolfpack made another basket, and the woman jumps from her seat while Bill still hangs onto the chair. What happens to the chair?

(d) Will your answer in (a) change when Casey, Bill's girlfriend, grabs the top of the back support and pulls herself out of her chair using a force of approximately 150 N? (Bill is no longer leaning on the chair with his 80-kg mass when Casey does this.)

(e) Suggest at least two changes to the chair design that would further reduce the risk of its flipping over.

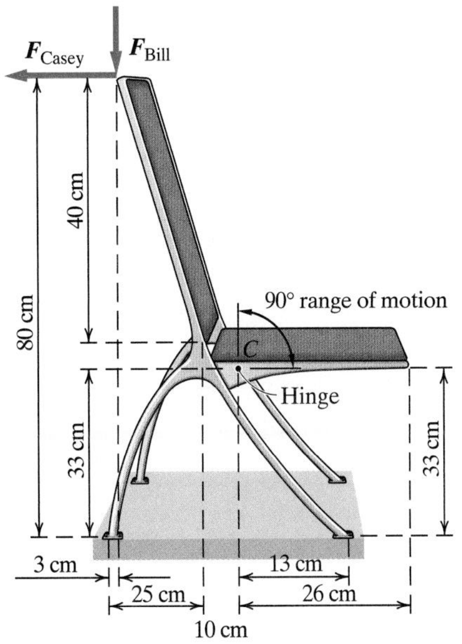

Figure SA7.3.1 Dimensions of movable chair

SA7.4 Analysis of a System in Various Configurations

Materials needed: one wire clothes hanger, rubber band, paper clip, a weight (a candy bar is suggested), a scale (ruler).

Configure the hanger, rubber band, paper clip, and weight in each of the positions shown in **Figure SA7.4.1**. (*Note:* Some of the positions may not be achievable if the hanger is underconstrained.) In Positions 1–3 the hanger is oriented vertically, and in Positions 4–6 it is oriented horizontally. Draw a free-body diagram for the hanger in each position, noting whether it represents a statically determinate, indeterminate, or underconstrained system. Take any measurements necessary to make the free-body diagrams complete. Clearly state your assumptions.

For those positions that are statically determinate or indeterminate, find the loads acting on the hanger. For any positions that are statically indeterminate, you will not be able to find all of the loads, but will be able to reduce the number of unknown loads.

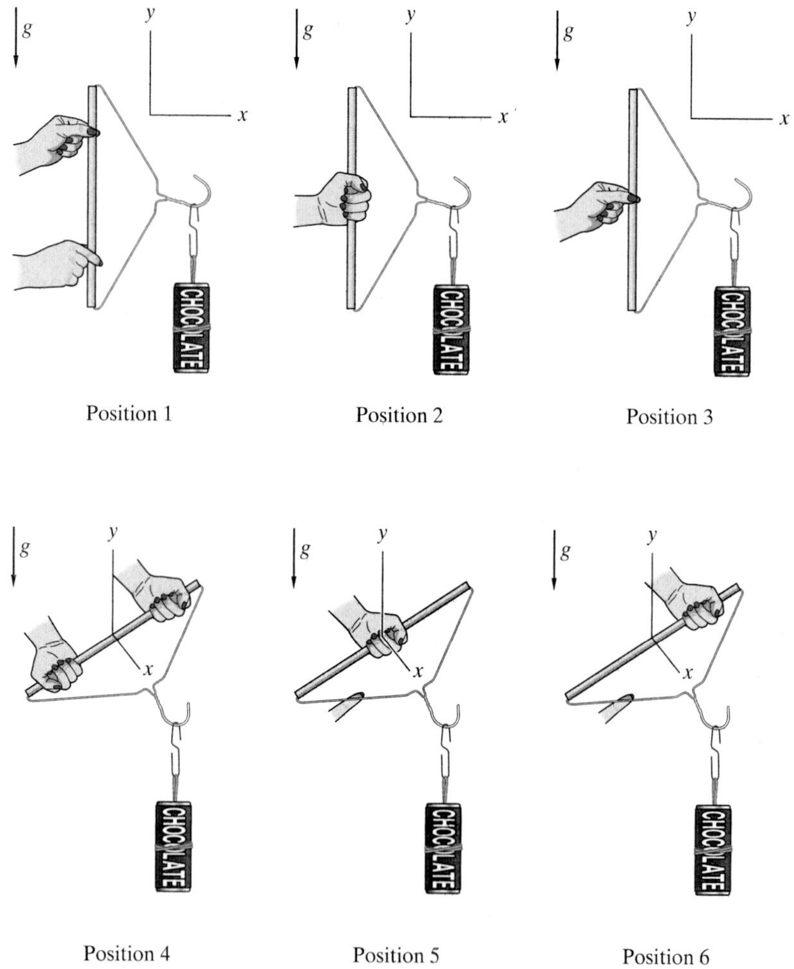

Figure SA7.4.1 Various hanger positions

SA7.5 Arm Strength

Using only a bathroom scale and a fixed surface (such as the edge of a kitchen or laboratory counter, or a push-bar door handle) design a procedure for finding the maximum force a person can exert in the upward direction with a single arm (**Figure SA7.5.1**). Present your procedure in a manner that a person not taking Statics could easily follow. The procedure could be a combination of text and figures.

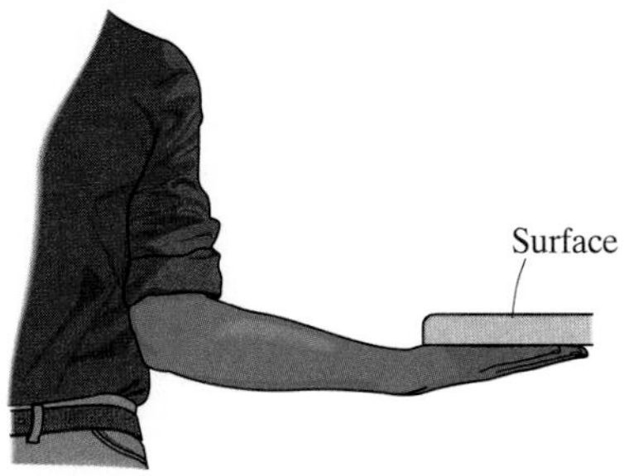

Figure SA7.5.1 Configuration of arm pushing on surface

Follow the procedure you developed to determine the maximum upward force that you can deliver. Compare this number with the data presented in **SA4.1**.

Find 3 or 4 other students and have them follow your procedure to determine their maximum arm forces. Present their maximum arm force data in a table that also includes the average maximum force.

SA7.6 Friction on Golden Gate Bridge Anchorage

When we performed our analysis of the anchorage of the Golden Gate Bridge, we drew the free-body diagram shown in **Figure 3.12**. Based on having determined the force in member OH to be 251.2 MN and assuming a coefficient of static friction of $\mu_{static} = 0.6$, we found that the anchorage had to weigh 479 MN to prevent both uplift and sliding.

(a) If the anchorage were on an incline of 10° (**Figure SA7.6.1**), considering prevention of both uplift and sliding, what would the required weight of the anchorage be?

(b) If the anchorage were tilted as shown in **Figure SA7.6.2**, would it have to be heavier or lighter than if it were on level ground? Explain your reasoning.

(c) Write a general equation for the friction force needed to prevent sliding. Use Excel, MATLAB, or another program to make a plot of friction force versus angle of incline.

(d) Write a general equation for the weight of the anchorage that controls the design. Use Excel, MATLAB or another program to make a plot of anchorage weight versus angle of incline.

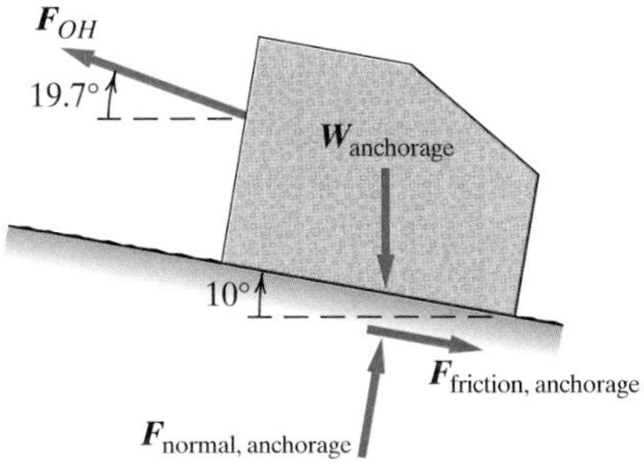

Figure SA7.6.1 10° clockwise tilt of ground

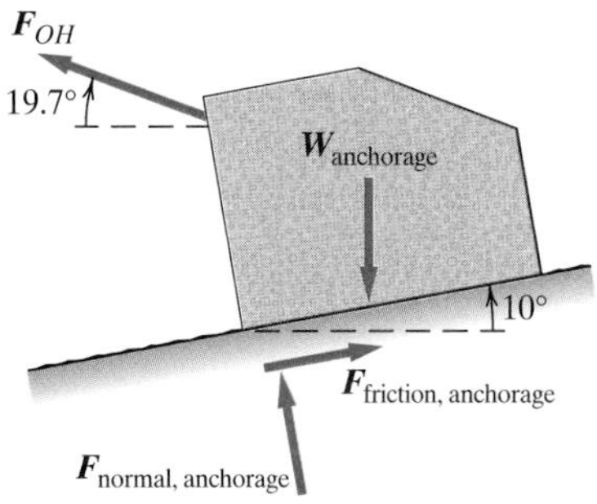

Figure SA7.6.2 10° counterclockwise tilt of ground

SA7.7 Mechanical Equilibrium of a Vehicle

Remote-controlled uncrewed reconnaissance vehicles are commonly used in civilian and military applications. As with any other technologically advanced system many factors are considered in their design. **Figure SA7.7.1** shows two hypothetical vehicles. They are nearly identical, with the primary difference being that the vehicle in **Figure SA7.7.1*a*** has tracks and the one in **Figure SA7.7.1*b*** has wheels. Both axles are drive axles for the two systems.

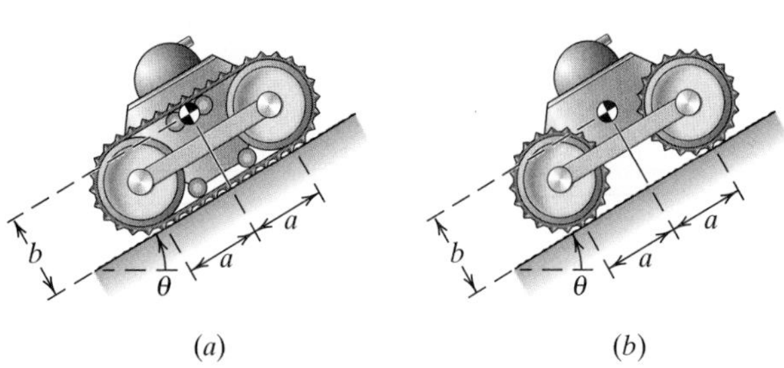

Figure SA7.7.1 Stationary slope comparison: (*a*) track vehicle; (*b*) wheeled vehicle

(a) Draw a complete free-body diagram for each vehicle resting on the slope shown.

(b) Assume dimension a to be 0.5 m and dimension b to be 0.7 m. If the coefficient of static friction is approximately 0.8 for both systems, calculate the maximum angle that each vehicle can rest on without sliding or tipping down the slope.

(c) Discuss your answers to (b). Are the answers what you expected? Comment on the assumptions in (b). Are these assumptions realistic?

(d) How does the location of the center of gravity affect the vehicle's ability to remain stable on the slope without sliding or tipping? As each dimension is changed, how does it affect the maximum angle of the slope?

(e) Discuss your personal thoughts on whether wheels or tracks would be better for this system. Which would you recommend and why?

SA7.8 Ancient Siege Engines

Ancient siege engines provided military commanders with the ability to engage an enemy from a distance; essentially it was the artillery of the armies past. Unfortunately, what is known of these medieval siege engines is limited to crude artist renditions and manuscript references. Hence, the hypothetical analysis of siege engines is still intriguing and challenging. Let us analyze one of the simplest forms of a siege engine, the catapult, as shown in **Figure SA7.8.1**. The siege engine fires a missile using the energy gained from a dropping counterweight and the advantage of a lever arm, the throwing arm. Once the throwing arm is cocked it is held in place by a vertical cable and supported by pin A (**Figure SA7.8.1**). Pin A rests on two identical frames, as shown in the overhead view in **Figure SA7.8.2**.

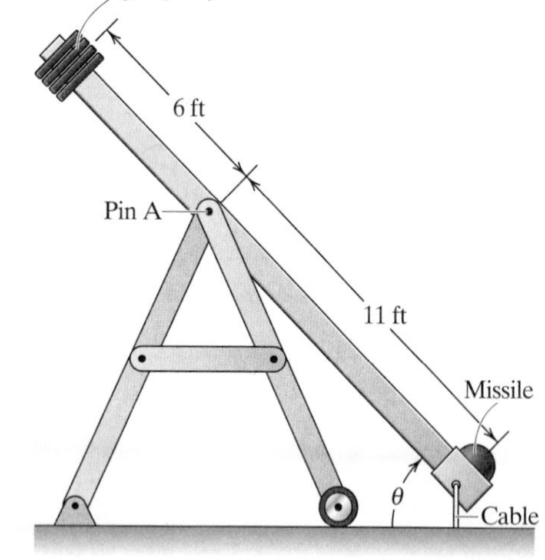

Figure SA7.8.1 Catapult

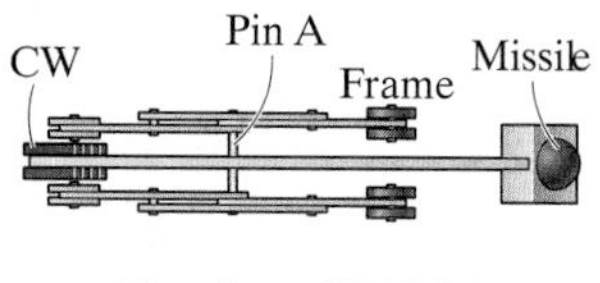

Figure SA7.8.2 Overhead view

Assume that the counterweight (CW) is 400 lb and the missile weight 50 lb. The throwing arm is generally uniform in density and shape with a weight of 100 lb, centered along its 17-foot length.

(a) Calculate the cable tension (T).

(b) Assume that the cable is no longer vertical but is attached to the throwing arm at angle $(90° - \theta)$, as shown in **Figure SA7.8.3**. Calculate the loads (reactions) at pin A and the tension in the cable when $\theta = 50°$.

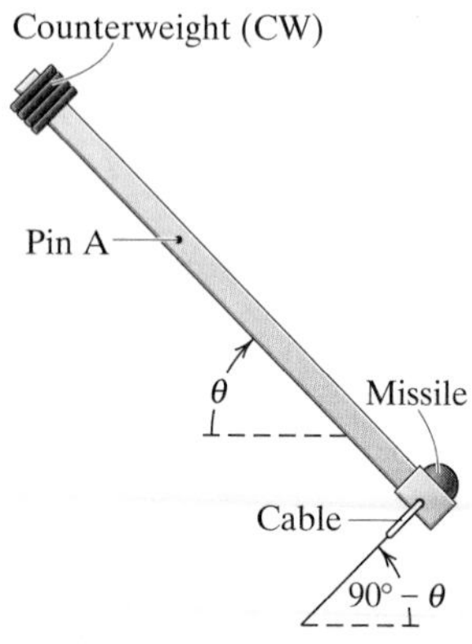

Figure SA7.8.3 Modified throwing arm

SA7.9 Ancient Siege Engines—Other Design Ideas

(a) In loading the catapult of **Figure SA7.9.1*a***, a pulley system is used to lower the throwing arm and raise the counterweight (CW). Calculate the force P required to begin lowering the throwing arm when θ starts at 50 degrees. Assume the counterweight is 400 lb.

(b) How does force P vary as the angle θ varies? Show the relationship with calculations.

(c) Consider the alternate two-pulley configuration (both frictionless and weightless), shown in **Figure SA7.9.1*b***. Calculate which system two-pulley configuration is the most efficient in lowering the throwing arm and raising the counterweight.

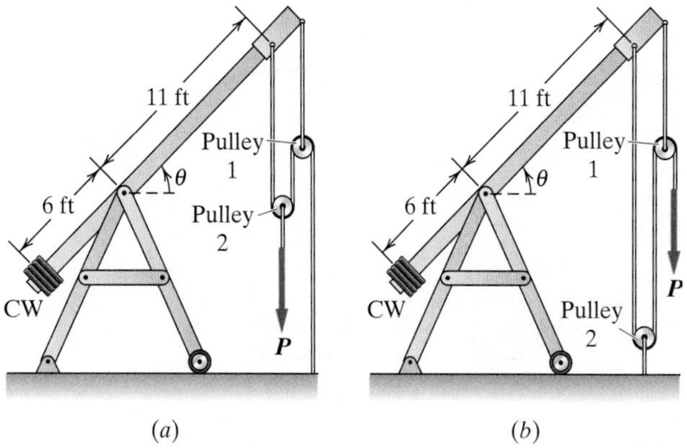

Figure SA7.9.1 (*a*) Pulley System 1; (*b*) Pulley System 2

SA7.10 Evaluation of a Lattice Boom Crane

Link-Belt is a company that makes lattice boom crawler cranes such as the 50-ton, LS-108H model, shown in **Figure SA7.10.1**. The design and analysis of these mechanical systems require all areas of engineering mechanics.

(a) Draw a complete free-body diagram of the crane for a hypothetical static load (LOAD) as shown. Consider the crane weight to be a composite system made up of the tractor weight (TRC) and the counterweight (CW). Write three equilibrium equations for the crane at the moment of impending tip about the track front (point O).

(b) If the TRC is 3000 lb and the CW is 20,000 lb, what is the maximum load the crane can lift and maintain in static equilibrium when θ is 40 degrees?

(c) How does the value of θ affect the results in (b)?

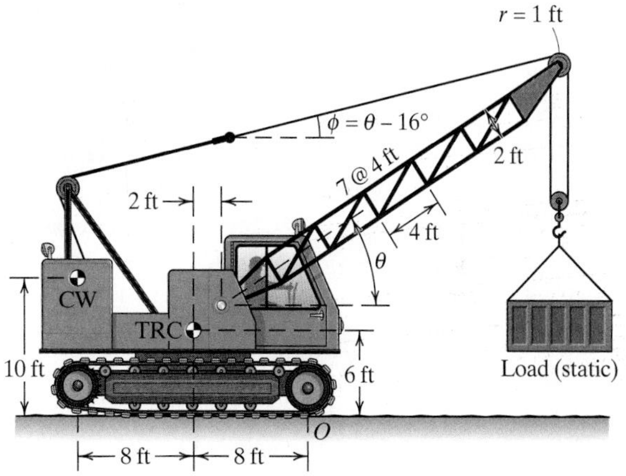

Figure SA7.10.1 Simplified crane system

DISTRIBUTED FORCE

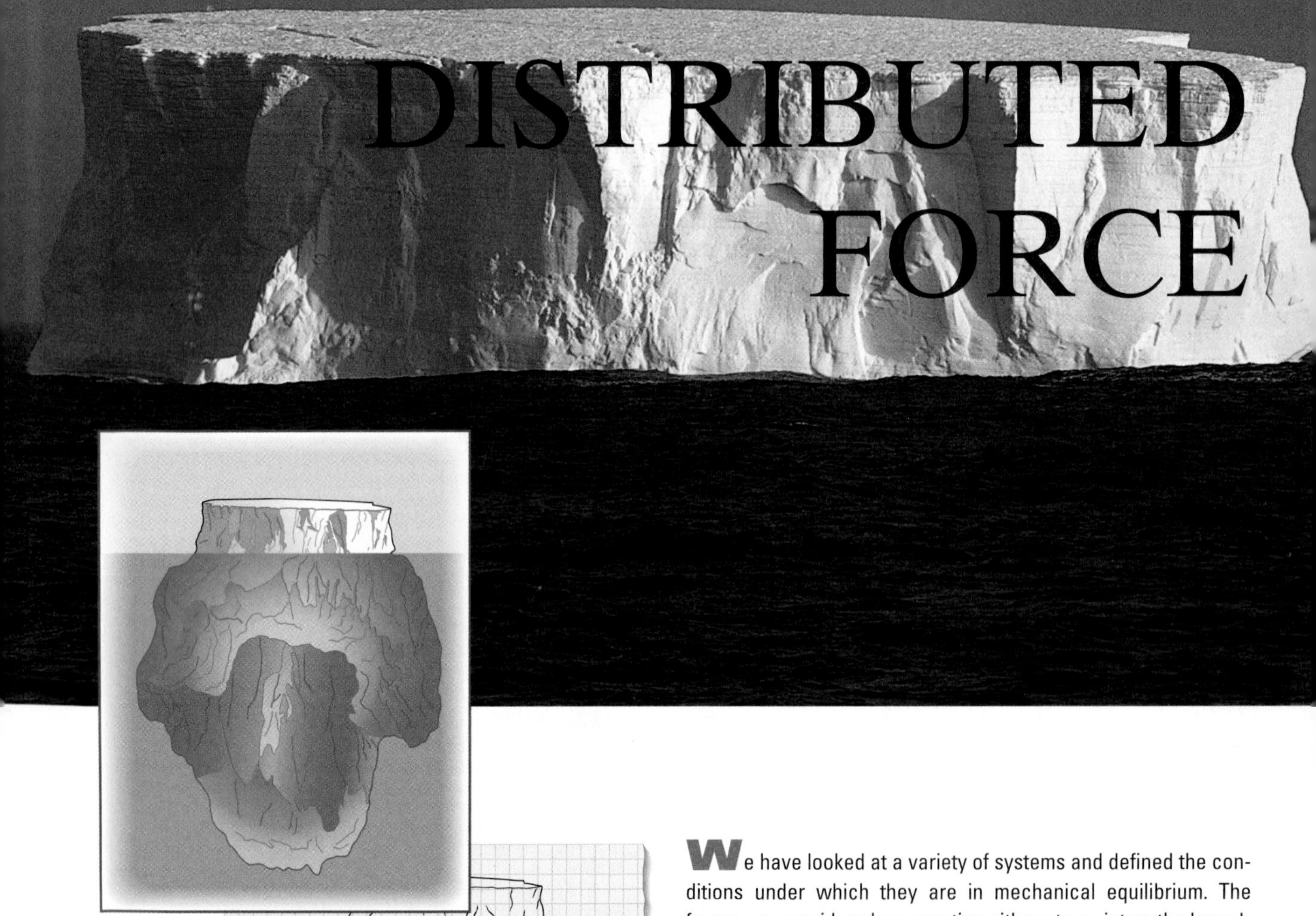

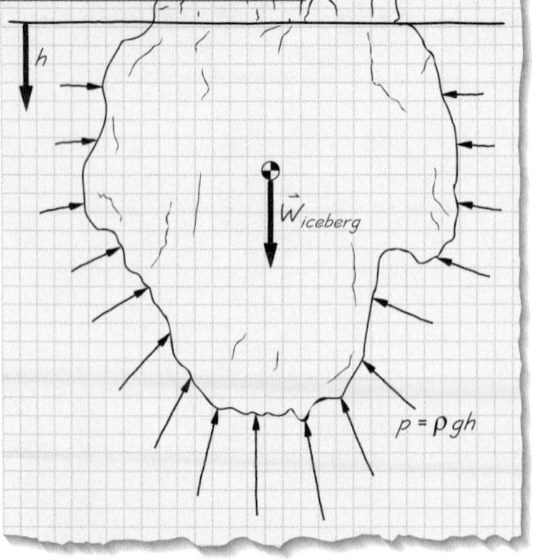

We have looked at a variety of systems and defined the conditions under which they are in mechanical equilibrium. The forces we considered were acting either at a point on the boundary of a system or at the system's center of mass. In contrast, the forces depicted above and in **Figure 8.1** are distributed and act not at a single point but rather over a line, area, or volume.

This chapter illustrates how to represent a **distributed force** by an equivalent force that acts at a single point. This equivalent point force is then included in static analysis. We present procedures for including

- gravitational forces acting across the boundary of a system by determining the total weight of a system and locating this weight at the system's center of gravity, and
- distributed forces acting on the boundary of a system along a line or over an area.

All of the procedures presented in this chapter involve first summing all the individual forces exerted at various points on the line, area, or volume to find what is called the **total force** acting and then locating the point of application of this total force such that the moment it creates is equivalent to the net moment created by all the individual forces. You may recognize this as the concept of equivalent loads, introduced in Chapter 5 and illustrated in **Figure 8.2**.

OBJECTIVES

On completion of this chapter, you will be able to:

- **Calculate the center of gravity for simple and composite volumes**
- **Calculate the center of mass for simple and composite volumes**
- **Calculate the centroid for simple and composite areas and volumes**
- **Represent a distributed line or area load by an equivalent point force, and use the equivalent point force in static analysis**
- **Write an expression for the distribution of hydrostatic pressure and include this distribution in performing static analysis**
- **Calculate the buoyancy forces acting on an object and include them in the static analysis**

8.1 CENTER OF MASS, CENTER OF GRAVITY, AND THE CENTROID

In examples that we have considered up to this point where weight of the system was important, the location of the center of mass of the system was always given. In this section we present how to locate the center of mass (as well as the center of gravity, the centroid of a volume, and the centroid of an area). Knowing how to locate the center of mass as part of carrying out static analysis is important since you will generally not be told in advance its location (it is up to you to find it!).

Volumes

The particles that make up an object are distributed throughout the object's volume. The individual particle masses summed together are the **total mass** of the object. It is generally unnecessary in equilibrium analysis to work with individual particle masses—instead we generally work with the total mass of the object. In addition, we locate this total mass at a point so that the way the object behaves is equivalent to the way the distributed particles, acting in concert, behave. This location in space is referred to as **center of mass**. We now outline how to find the total mass. Then, using the concept of equivalent loads, we find the location of the center of mass.

1. *Total mass M of an object.* The total mass is

$$M = \int_{\text{volume}} \rho \, dV \text{ (general case)} \tag{8.1}$$

where ρ is the object's density in mass/volume, $\rho\, dV$ is the mass of a volume element dV of the object, and integration involves integration throughout the object's volume.

If this total mass M is acted on by gravity, the associated weight W (the gravitational force) of the object is equal to Mg, where g is the gravitational constant, as discussed in Chapter 4. We rewrite (8.1) in terms of weight W as

$$Mg = W = \int_{\text{volume}} \rho g \, dV \tag{8.2A}$$

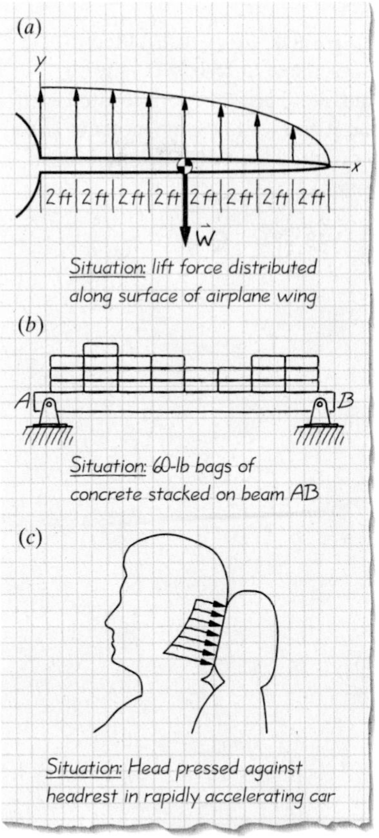

Figure 8.1 (*a*) Lift force is distributed along the surface of an airplane wing; (*b*) sand bags are stacked along beam *AB*; (*c*) pressure distribution between head and headrest in a rapidly accelerating automobile

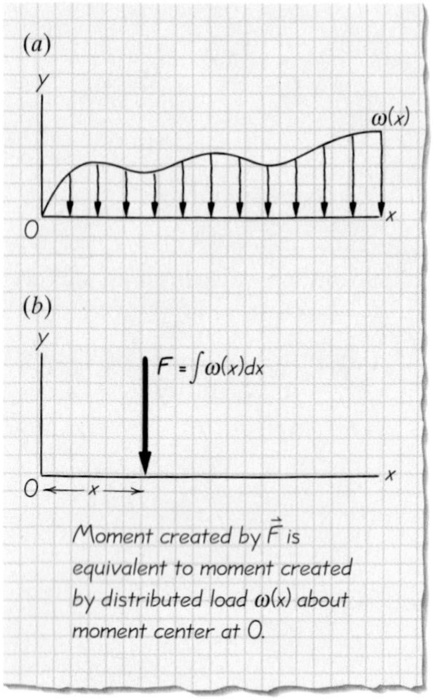

Figure 8.2 (*a*) A distributed load acting along an *x* axis; (*b*) the equivalent point force is located such that it creates the same moment as the distributed load ω(x) about moment center O

The integrand represents the weight of an infinitesimally small volume of the object; we denote it as $dw = \rho g\, dV$. Alternatively, (8.2A) can be rewritten in terms of specific weight γ (where γ is weight per unit volume, and is ρg) as

$$Mg = W = \int_{\text{volume}} \gamma\, dV \tag{8.2B}$$

Values of density and specific weight of commonly used engineering materials are presented in Appendix A2.1.

2. *Location of center of mass.* To find the location of the center of mass of any object we apply the concept of equivalent loads. More specifically, we find the location of the object's total mass M such that the mass placed at that location creates a moment equivalent to the net moment created by all of the particles' masses.

First, consider that the object of mass M located at X_M, Y_M, Z_M in **Figure 8.3*a*** is acted on by a uniform gravity field in the negative z direction. We can find X_M by requiring that the moment about the y axis created by Mg must be equal to the sum of the moments created by the distributed weights dW ($= \rho g\, dV$) (**Figure 8.3*b***):

$$\underbrace{(X_M\, Mg)}_{\substack{\text{moment}\\ \text{created}\\ \text{by } Mg}} = \underbrace{\int_{\text{volume}} x\rho g\, dV}_{\substack{\text{moment created}\\ \text{by distributed}\\ \text{weights}}}$$

$$X_M = \frac{\displaystyle\int_{\text{volume}} x\rho g\, dV}{Mg} \underbrace{=}_{\substack{\text{if gravity constant,}\\ \text{expression can}\\ \text{be simplified}}} \frac{\displaystyle\int_{\text{volume}} x\rho\, dV}{M} \tag{8.3A}$$

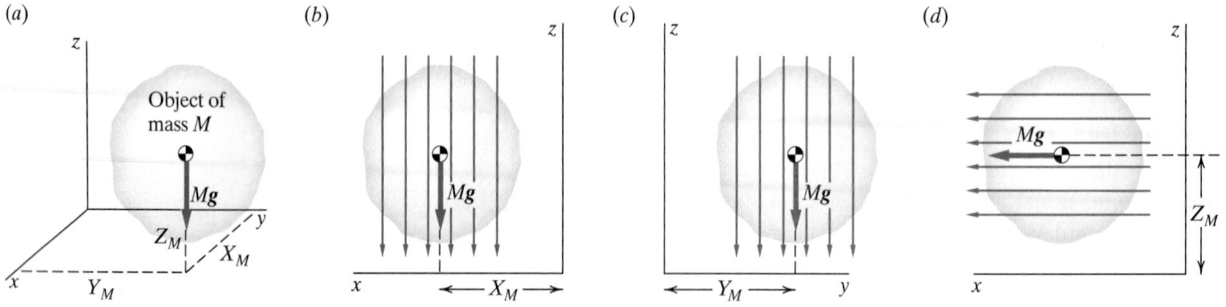

Figure 8.3 (*a*) Object of mass M located at X_m, Y_m, and Z_m; (*b*) finding X_m; (*c*) finding Y_m; (*d*) finding Z_m

Similarly, by equivalent moments about the x axis (**Figure 8.3*c***) we determine

$$Y_M = \frac{\displaystyle\int_{\text{volume}} y\rho g\, dV}{Mg} \underbrace{=}_{\substack{\text{if gravity constant,}\\ \text{expression can}\\ \text{be simplified}}} \frac{\displaystyle\int_{\text{volume}} y\rho\, dV}{M} \tag{8.3B}$$

Finally, if we consider a uniform gravity field in the x direction (**Figure 8.3*d***) and require equivalent moments about the y axis, we determine

$$Z_M = \frac{\displaystyle\int_{\text{volume}} z\rho g\, dV}{Mg} \underbrace{=}_{\substack{\text{if gravity constant,}\\ \text{expression can}\\ \text{be simplified}}} \frac{\displaystyle\int_{\text{volume}} z\rho\, dV}{M} \tag{8.3C}$$

In summary, the center of mass of an object in a uniform gravitational field is at location:

$$X_M = \frac{\displaystyle\int_{\text{volume}} x\rho\, dV}{M}; \quad Y_M = \frac{\displaystyle\int_{\text{volume}} y\rho\, dV}{M}; \quad Z_M = \frac{\displaystyle\int_{\text{volume}} z\rho\, dV}{M} \tag{8.4}$$

In many problems the integration required in (8.4) for finding the center of mass may be simplified by a prudent choice of reference axes. An important clue in locating reference axes may be taken from considerations of symmetry. Whenever there exists a line or plane of symmetry in an object of uniform density, the center of mass will be along the line or in the plane. The coordinate axes should be aligned with the line or plane. **Figure 8.4** shows examples of objects with lines or planes of symmetry.

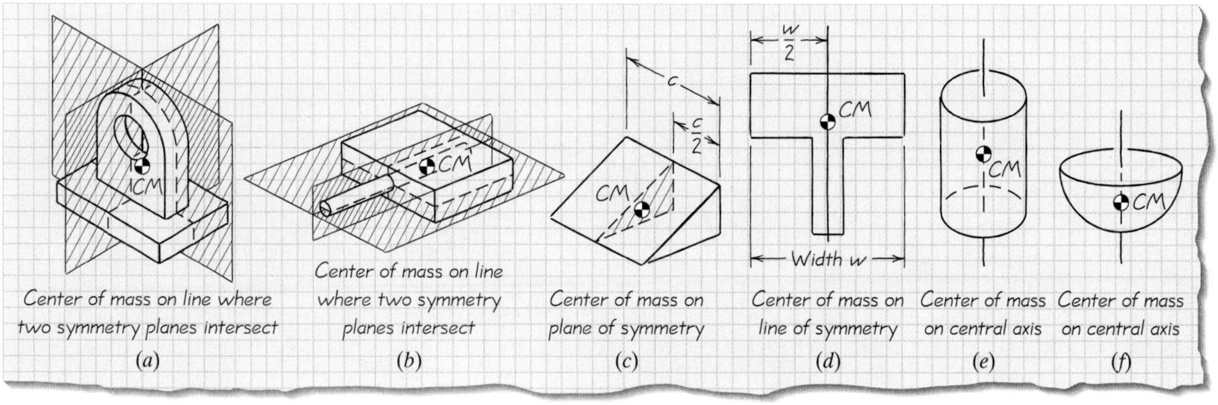

Figure 8.4 Examples of volumes with planes and/or lines of symmetry

If we can treat the gravity field as uniform (which is approximately true for small objects on the earth), the center of mass (as defined in (8.4)) is also the location of the **center of gravity**, defined as the point in an object where we represent the total weight of the object in order to treat the object's weight as a single point force (location X_G, Y_G, Z_G).

The coordinates described by (8.4) locate the **centroid of a volume** if the volume is **homogeneous** (meaning that it is composed of a material of uniform density). For a general shape the centroid is located at

$$X_C = \frac{\int_{\text{volume}} x\,dV}{\int_{\text{volume}} dV}; \quad Y_C = \frac{\int_{\text{volume}} y\,dV}{\int_{\text{volume}} dV}; \quad Z_C = \frac{\int_{\text{volume}} z\,dV}{\int_{\text{volume}} dV} \tag{8.5}$$

The locations of the centroids of several **standard volumes** are presented in Appendix A.3.2. This is also the location of the center of mass and center of gravity if the volume is homogeneous. For these standard volumes, you are urged to use Appendix A.3.2 to locate the center of gravity as a labor-saving alternative to carrying out the integration called for in (8.4) and (8.5).

If we can decompose a composite volume into one made up of N standard volumes, we can use knowledge of the location of the centers of gravity of the N standard volumes to find the location of the center of gravity (X_G, Y_G, Z_G) of the composite volume (**Figure 8.5*a***). Call W_i the weight of an individual volume, and call X_{iG}, Y_{iG}, Z_{iG} the location of its center of gravity (**Figure 8.5*b***). Requiring that the moment of the composite volume be equal to the sum of the individual moments, we write

$$X_G \sum_{i=1}^{N} W_i = \sum_{i=1}^{N} W_i X_{iG}; \quad Y_G \sum_{i=1}^{N} W_i = \sum_{i=1}^{N} W_i Y_{iG}; \quad Z_G \sum_{i=1}^{N} W_i = \sum_{i=1}^{N} W_i Z_{iG} \tag{8.6}$$

Each term to the left of the equal sign is the moment created by the composite volume, and each term to the right is the summation of the moments created by the individual volumes. The total weight of the composite volume is $W_{\text{tot}} = \Sigma_{i=1}^{N} W_i$.

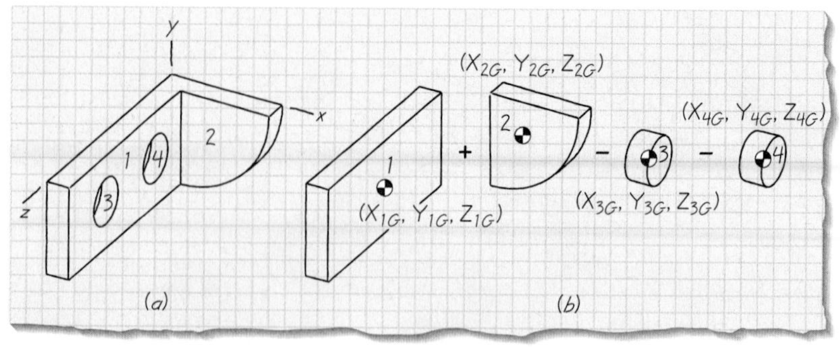

Figure 8.5 A volume decomposed into four separate volumes. Because volume 3 and 4 are holes, they are negative volume.

Substituting W_{tot} for each $\Sigma_{i=1}^{N} W_i$ term in (8.6) and solving for X_G, Y_G, and Z_G we have

$$X_G = \frac{\sum_{i=1}^{N} W_i X_{iG}}{W_{tot}}; \quad Y_G = \frac{\sum_{i=1}^{N} W_i Y_{iG}}{W_{tot}}; \quad Z_G = \frac{\sum_{i=1}^{N} W_i Z_{iG}}{W_{tot}} \tag{8.7A}$$

Equation (8.7A) gives the location of the **center of gravity of the composite volume**.

In a similar manner, we can write the coordinates of the **center of mass of a composite volume** as

$$X_M = \frac{\sum_{i=1}^{N} M_i X_{iM}}{M_{tot}}; \quad Y_M = \frac{\sum_{i=1}^{N} M_i Y_{iM}}{M_{tot}}; \quad Z_M = \frac{\sum_{i=1}^{N} M_i Z_{iM}}{M_{tot}} \tag{8.7B}$$

where M_i is the mass of an individual volume and M_{tot} is the total mass of the composite volume. Furthermore, the coordinates of the **centroid of the composite volume** are

$$X_C = \frac{\sum_{i=1}^{N} V_i X_{iC}}{V_{tot}}; \quad Y_C = \frac{\sum_{i=1}^{N} V_i X_{iC}}{V_{tot}}; \quad Z_C = \frac{\sum_{i=1}^{N} V_i Z_{iC}}{V_{tot}} \tag{8.7C}$$

where V_i is the volume of an element of the composite volume and V_{tot} is the total volume. In a constant gravitational field, (8.7A) and (8.7B) yield the same coordinates. Furthermore, if the composite volume is homogeneous, (8.7A), (8.7B), and (8.7C) yield identical coordinates.

EXAMPLE 8.1 CENTROID OF A VOLUME

Figure 8.6 shows a right circular cone of height h and base radius r. Show that the location of the centroid is $X_C = 0$, $Y_C = 3h/4$, $Z_C = 0$ as indicated in Appendix A3.2.

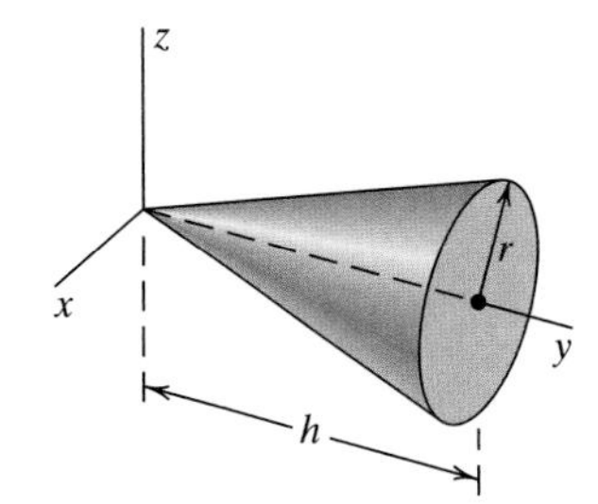

Figure 8.6

Goal We are to find the x, y, and z coordinates of the centroid of a right circular cone.

Given The cone is of height h and base radius r.

Draw We draw an infinitesimal element dV at a distance y from the origin to use in integration of (8.5) (**Figure 8.7**).

Formulate Equations and Solve From symmetry we know that the centroid must lie on the y axis, as defined by $X_C = 0$, $Z_C = 0$. From (8.5) we find Y_C:

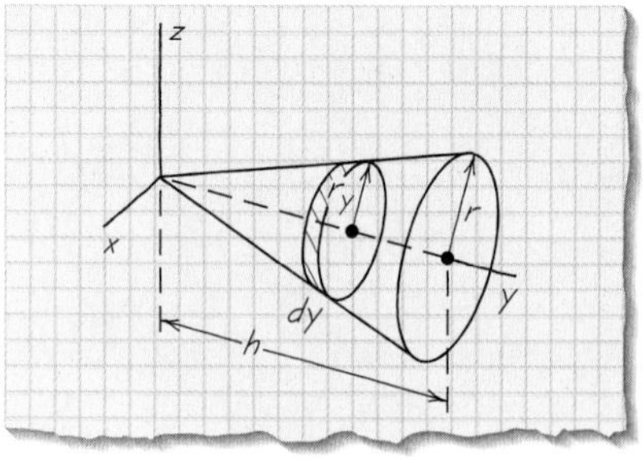

Figure 8.7

$$Y_C = \frac{\int_{\text{volume}} y\, dV}{\int_{\text{volume}} dV}$$

Our infinitesimal slice of cone at a distance y from the origin has a volume

$$dV = \pi (r_y)^2 dy \tag{1}$$

We use similar triangles to solve for r_y as a function of y, r, and h

$$\frac{r_y}{y} = \frac{r}{h}$$

$$r_y = \frac{r}{h} y \tag{2}$$

Substituting (2) into (1) and integrating to find the volume of the cone we have

$$V = \int dV = \int_0^h \pi\, r_y^2\, dy = \int_0^h \pi \frac{r^2}{h^2} y^2\, dy = \frac{\pi r^2}{h^2} \frac{y^3}{3}\Bigg]_0^h = \frac{\pi r^2 h}{3}$$

$$V = \frac{\pi r^2 h}{3}$$

We now solve for Y_C:

$$Y_C = \frac{\int_{\text{volume}} y\, dV}{V} = \frac{\int_0^h y\left(\pi \frac{r^2}{h^2} y^2\right) dy}{V} = \frac{\frac{\pi r^2}{h^2} \frac{y^4}{4}\Big]_0^h}{V}$$

$$Y_C = \frac{\frac{\pi r^2 h^2}{4}}{\frac{\pi r^2 h}{3}} = \frac{3}{4} h$$

We could use (8.5) to find X_C and Z_C. Alternately, by inspection of **Figure 8.6**, we see that symmetry requires that $X_C = 0$ and $Z_C = 0$.

Answer $X_C = 0$
$Y_C = \frac{3h}{4}$
$Z_C = 0$

This agrees with the location of the centroid of a cone as shown in Appendix A3.2.

EXAMPLE 8.2 CENTER OF MASS WITH DISTRIBUTED MASS

The right circular cone of Example 8.1 is made from a material of variable density. The density varies linearly from $3\rho_o$ at the point to ρ_o at the base. Find the location of the center of mass of the cone.

Goal We are to find the x, y, and z coordinates of the center of mass of a right circular cone with linearly varying density.

Given The cone is of height h and base radius r, and the density varies from $3\rho_o$ at the point to ρ_o at the base.

Draw We use the same approach as in Example 8.1, in which we looked at a slice of width dy at a distance y from the origin to find the volume dV (**Figure 8.7**). We develop an equation for the density as a function of y and then use (8.4) to find the center of mass.

Formulate Equations and Solve From symmetry we know that the center of mass must lie on the y axis as defined by $X_M = 0$, $Z_M = 0$. We find Y_M from (8.4).

$$Y_M = \frac{\int_{\text{volume}} y\,\rho(y)\,dV}{M}$$

We need to develop a linear relationship between density and location in the cone. The form of the relationship will be $\rho = my + b$. The constants m and b are found by imposing the boundary conditions

$$\text{At } y = 0, \rho = 3\rho_o$$
$$\text{At } y = h, \rho = \rho_o$$

resulting in the relationship

$$\rho(y) = 3\rho_o - \frac{2\rho_o}{h} y \tag{1}$$

Furthermore, recall from Example 8.1 that

$$dV = \pi \frac{r^2}{h^2} y^2\,dy \tag{2}$$

Substituting (1) and (2) into (8.1) and integrating to find the mass of the cone,

$$M = \int \rho(y)\,dV = \int_0^h \left(3\rho_o - \frac{2\rho_o}{h} y\right)\pi \frac{r^2}{h^2} y^2\,dy = \frac{\rho_o \pi r^2}{h^2}\int_0^h \left(3y^2 - \frac{2}{h} y^3\right) dy$$

$$M = \frac{\rho_o \pi r^2}{h^2}\left[\frac{3y^3}{3} - \frac{2y^4}{4h}\right]_0^h = \frac{\rho_o \pi r^2 h}{2}$$

$$M = \frac{\rho_o \pi r^2 h}{2}$$

Note: If the cone were of a constant density ρ_o, the mass of the cone would be smaller. It would be $\rho_o V_{\text{cone}} = \frac{\rho_o \pi r^2 h}{3}$, as found from Appendix A3.2. We present this as an intermediate check that our calculations are moving along in the right direction.

We now solve (8.4) for Y_M:

$$Y_M = \frac{\int_{\text{volume}} y\rho(y)\,dV}{M} = \frac{\int_0^h y\left(3\rho_o - \frac{2\rho_o}{h}y\right)\left(\pi\frac{r^2}{h^2}y^2\right)dy}{M}$$

$$= \frac{\frac{\rho_o \pi r^2}{h^2}\int_0^h \left(3y^3 - \frac{2}{h}y^4\right)dy}{M} = \frac{\frac{\rho_o \pi r^2}{h^2}\left[\frac{3y^4}{4} - \frac{2y^5}{5h}\right]_0^h}{M}$$

$$Y_M = \frac{\frac{7\rho_o \pi r^2 h^2}{20}}{\frac{\rho_o \pi r^2 h}{2}} = \frac{7}{10}h$$

Answer $X_C = 0$
$Y_C = \frac{7h}{10}$
$Z_C = 0$

Check If the density were constant, the center of mass would coincide with the centroid of the cone ($3h/4$). Because the density varies linearly, with large densities toward the tip of the cone, the center of mass is closer to the tip ($7h/10$) than for the constant density case, $7h/10 < 3h/4$.

EXAMPLE 8.3 LOCATING THE CENTROID OF A COMPOSITE VOLUME

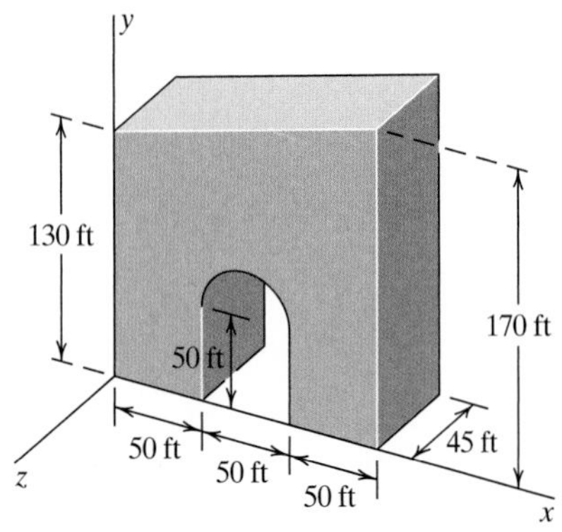

Figure 8.8

Figure 8.8 shows a concrete anchorage from a suspension bridge with a pedestrian archway formed by a semicircle of 25 ft radius at a height of 50 ft from the base. The specific gravity of concrete is $\gamma_C = 150\,\frac{\text{lb}}{\text{ft}^3}$.

Find the x, y, and z coordinates of the centroid of the anchorage.

Goal We are to find the x, y, and z coordinates of the centroid of the anchorage.

Given We are given the dimensions of the anchorage and the specific weight of concrete.

Assume We assume that the anchorage is homogeneous.

Draw We decompose the anchorage into standard volumes and subtract the archway from the anchorage, as in **Figure 8.9**.

Formulate Equations and Solve The x, y, and z coordinates of a composite volume centroid are given by (8.7C) as

$$X_C = \frac{\sum_{i=1}^{N} V_i X_{iC}}{V_{\text{tot}}};\qquad Y_C = \frac{\sum_{i=1}^{N} V_i Y_{iC}}{V_{\text{tot}}};\qquad Z_C = \frac{\sum_{i=1}^{N} V_i Z_{iC}}{V_{\text{tot}}} \tag{1}$$

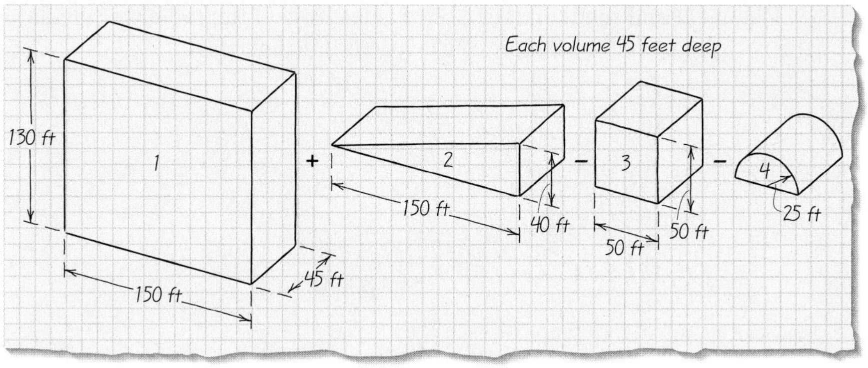

Figure 8.9 Anchorage composed of volumes

To apply (8.7C) we begin by calculating the volume of the four volumes shown in **Figure 8.9**:

$$V_1 = (l_1)(h_1)(w_1) = (150\text{ ft})(130\text{ ft})(45\text{ ft}) = 877{,}500\text{ ft}^3$$

$$V_2 = \frac{bh}{2}w = \frac{(150\text{ ft})(40\text{ ft})}{2}45\text{ ft} = 135{,}000\text{ ft}^3$$

$$V_3 = -(l_3)(h_3)(w_3) = -(50\text{ ft})(50\text{ ft})(45\text{ ft}) = -112{,}500\text{ ft}^3$$

$$V_4 = -\frac{\pi r^2}{2}w = -\frac{\pi(25\text{ ft})^2}{2}45\text{ ft} = -44{,}180\text{ ft}^3$$

Notice that volumes 3 and 4 are negative since they represent removal of volume, whereas volumes 1 and 2 represent addition of volume to the overall object.

From symmetry we know that the centroid of the anchorage is on its midplane defined by $Z_G = -22.5$ ft. We need only calculate the x and y coordinates of the centroids, referring to Appendix A3.2 as needed.

$$X_{C1} = \frac{150\text{ ft}}{2} = 75\text{ ft}; \qquad Y_{C1} = \frac{130\text{ ft}}{2} = 65\text{ ft}$$

$$X_{C2} = \frac{2}{3}150\text{ ft} = 100\text{ ft}; \qquad Y_{C2} = 130\text{ ft} + \frac{1}{3}40\text{ ft} = 143.3\text{ ft}$$

$$X_{C3} = 50\text{ ft} + \frac{50\text{ ft}}{2} = 75\text{ ft}; \qquad Y_{C3} = \frac{50\text{ ft}}{2} = 25\text{ ft}$$

$$X_{C4} = 50\text{ ft} + \frac{50\text{ ft}}{2} = 75\text{ ft}; \qquad Y_{C4} = 50 + \frac{4(25\text{ ft})}{3\pi} = 60.6\text{ ft}$$

It is convenient to summarize the data on volumes and centroid locations in a table:

Component	Vol (ft^3)	X_C (ft)	Y_C (ft)	X_CV (ft^4)	Y_CV (ft^4)
1	877,500	75	65	65.81×10^6	57.04×10^6
2	135,000	100	143.3	13.50×10^6	19.35×10^6
3	−112,500	75	25	-8.44×10^6	-2.81×10^6
4	−44,180	75	60.6	-3.31×10^6	-2.68×10^6
Σ	**855,820**			**67.56×10^6**	**70.90×10^6**

Using (1) and reading data from the last row of the table we calculate the coordinates of the center of gravity of the anchorage:

$$X_C = \frac{\sum_{i=1}^{N} V_i X_{iC}}{V_{\text{tot}}} = \frac{67.56 \times 10^6 \text{ ft}^4}{855{,}820 \text{ ft}^3} = 78.9 \text{ ft}$$

$$Y_C = \frac{\sum_{i=1}^{N} V_i Y_{iC}}{V_{\text{tot}}} = \frac{70.90 \times 10^6 \text{ ft}^4}{855{,}820 \text{ ft}^3} = 82.8 \text{ ft}$$

Answer $X_C = 78.9$ ft
$Y_C = 82.8$ ft
$Z_C = -22.5$ ft

Areas

An **extruded homogeneous volume** is a volume with a constant cross section that lies along an axis; **Figure 8.10*a*** shows an example of an extruded volume. Two of the coordinates of the center of gravity of a homogeneous extruded volume lie in the cross section of the volume at the centroid of a plane area (**Figure 8.10*b***). Using the idea of equivalent loads we find the centroid of this plane area.

1. *Total area* A_{tot}. The total area is

$$A_{\text{tot}} = \int_{\text{area}} dx\, dy \tag{8.8}$$

2. *Location of centroid.* To find the location of the centroid of a plane area we apply the concept of equivalent loads by imagining that the area represents a thin sheet of constant thickness t, made up of a uniform material of density ρ (**Figure 8.11*a***). If gravity acts in the negative y direction, we can write

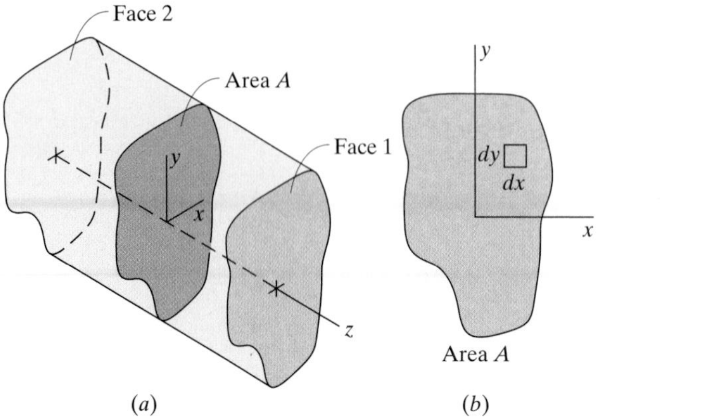

Figure 8.10 (*a*) A homogeneous extruded volume with constant cross section of area A; (*b*) the cross section A

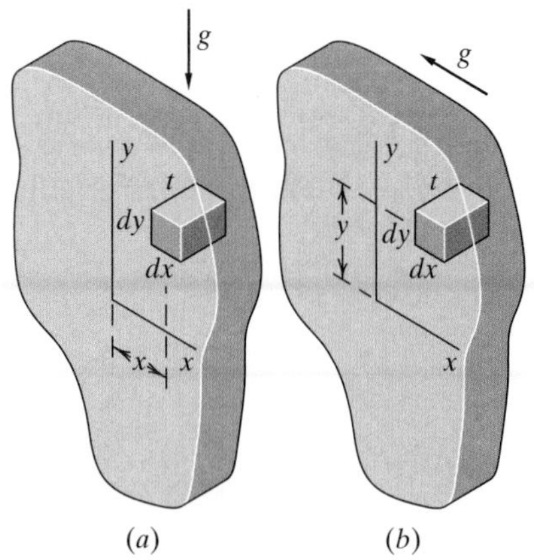

Figure 8.11 (*a*) Finding X_C of the centroid; (*b*) finding Y_C of the centroid

$$X_C\,(t\rho g A_{tot}) = \iint_{area} x(t\rho g dx\, dy) \tag{8.9}$$

moment about z axis produced by total weight $t\rho g A_{tot}$ — total moment about z axis produced by distributed weight

where A_{tot} is defined in (8.8). Since t and ρ are constants, they cancel from both sides of the equation. With rearranging we are left with

$$X_C = \frac{\int_{area} x\, dx\, dy}{A_{tot}} \tag{8.10A}$$

In a similar manner, if we consider gravity to act in the negative x direction (**Figure 8.11*b***) we determine

$$Y_C = \frac{\int_{area} y\, dx\, dy}{A_{tot}} \tag{8.10B}$$

Equations (8.10A) and (8.10B) define two of the coordinates (X_C, Y_C) of the centroid of a plane area.

The integrals in the numerator of (8.10A) and (8.10B) are called the **first area integrals** and are one of a family of integrals used to describe the properties of areas.[1]

[1]The family of integrals related to area are:
Area integral (area)

$$A = \int_{area} dA \tag{8.8}$$

First area integrals

$$\int_{area} x\, dA; \qquad \int_{area} y\, dA \qquad \text{(as used in (8.10A) and (8.10B))}$$

Second area integrals

$$I_x = \int_{area} y^2\, dA$$

$$I_y = \int_{area} x^2\, dA$$

Second area integrals are often referred to as **moments of inertia of the area** or **area moments of inertia**. The terminology "moment of inertia" is a misnomer (since no inertial concepts are involved). The second area integrals reflect the distribution of the area relative to coordinate axes in the plane of the area. We will use them in Chapter 10 when we analyze beams. Details on their calculation are presented in Appendix C.

The locations of centroids of several standard areas are presented in Appendix A3.1. For these standard areas, you are urged to use Appendix A3.1 to locate the centroid of an area as a labor-saving alternative to carrying out the integration called for in (8.8) and (8.10).

Equations (8.10A) and (8.10B) (or alternately, Appendix A3.1) define two of the coordinates of the center of gravity of a homogeneous extruded volume. The coordinate Z_C is at the midplane between the two faces of the extruded volume (i.e., midway between Face 1 and Face 2 in **Figure 8.10**).

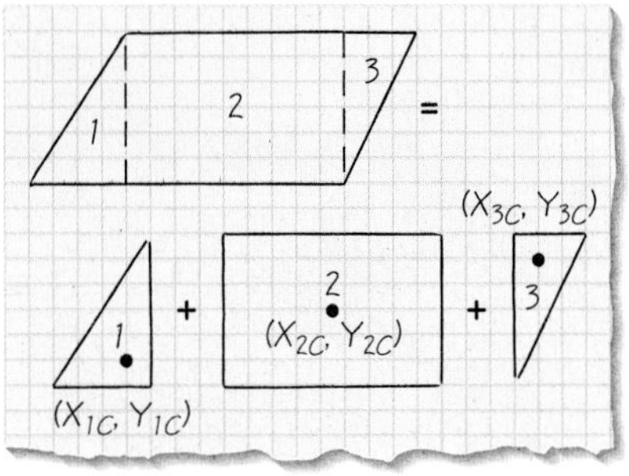

Figure 8.12 An area decomposed into three separate areas

If we can decompose a composite area into one made up of N standard areas, we can use knowledge of the location of the centroids of the N standard areas to find the location of the centroid (X_C, Y_C) of the composite area (**Figure 8.12**). Call A_i the area of an individual area, and call X_{iC}, Y_{iC} the location of its centroid. By requiring that the moments be equivalent we find

$$X_C \sum_{i=1}^{N} A_i = \sum_{i=1}^{N} A_i X_{iC}; \qquad Y_C \sum_{i=1}^{N} A_i = \sum_{i=1}^{N} A_i Y_{iC} \tag{8.11}$$

Each term to the left of the equal sign reflects the moment created by the composite area, and each term to the right reflects the summation of the moments created by the individual areas. The total area of the composite area is $A_{tot} = \Sigma_{i=1}^{N} A_i$.

Substituting A_{tot} for each $\Sigma_{i=1}^{N} A_i$ term in (8.11) and solving for X_C, Y_C we have

$$X_C = \frac{\sum_{i=1}^{N} A_i X_{iC}}{A_{tot}}; \qquad Y_C = \frac{\sum_{i=1}^{N} A_i Y_{iC}}{A_{tot}} \tag{8.12}$$

Equation (8.12) gives the location of the centroid of the composite area.

EXAMPLE 8.4 FINDING THE CENTROID OF AN AREA

Figure 8.13*a* shows a homogeneous extruded volume, with cross-sectional area shown in detail in **Figure 8.13*b***. Determine the centroid (X_C, Y_C) and the area of the shaded cross-section.

Goal We are to find the centroid and area of the shaded cross-section.

Given We are given information about the geometry and boundaries of the shaded cross-section.

Assumptions None needed.

Formulate Equations and Solve One boundary of the cross-section is described by the curve $x = ky^3$. The value of k can be determined from evaluating $x = ky^3$ at a known point on the curve. At $x = a$, $y = b$, giving $a = kb^3$ and $k = a/b^3$. This results in

$$y = \frac{bx^{1/3}}{a^{1/3}}$$

We use (8.10A) and (8.10B) to find the centroid.

$$X_C = \frac{\int_{\text{area}} x\,dx\,dy}{A_{\text{tot}}} \quad \text{and} \quad Y_C = \frac{\int_{\text{area}} y\,dx\,dy}{A_{\text{tot}}}$$

First let's calculate the total area, which is the integral $A_{\text{tot}} = \int_{\text{area}} dA$. **Figure 8.14*a*** shows the element dA.

$$A_{\text{tot}} = \int_{\text{area}} dA = \int y(x)\,dx$$

$$A_{\text{tot}} = \int_0^a \frac{b}{a^{1/3}} x^{1/3}\,dx = \frac{b}{a^{1/3}}\left[\frac{3}{4}x^{4/3}\right]_0^a = \frac{3}{4}ab$$

$$A_{\text{tot}} = \frac{3}{4}ab$$

Based on (8.10A) and using the differential area dA shown in **Figure 8.14*a***, we compute X_C:

$$X_C = \frac{\int_0^a x\,y(x)\,dx}{A_{\text{tot}}} = \frac{\int_0^a x\left(\frac{b}{a^{1/3}}x^{1/3}\right)dx}{A_{\text{tot}}}$$

$$X_C = \frac{\frac{b}{a^{1/3}}\left(\frac{3}{7}x^{7/3}\right)_0^a}{A_{\text{tot}}} = \frac{\frac{3b}{7a^{1/3}}a^{7/3}}{\frac{3}{4}ab} = \frac{4}{7}a$$

$$X_C = \frac{4}{7}a$$

Based on (8.10B) and using the differential area $dA = [a - x(y)]dy$ shown in **Figure 8.14*b***, we compute Y_C:

$$Y_C = \frac{\int_{\text{area}} y(a - x(y))dy}{A_{\text{tot}}} = \frac{\int_0^b y\left(a - \frac{a}{b^3}y^3\right)dy}{A_{\text{tot}}}$$

$$Y_C = \frac{\left[\frac{ay^2}{2} - \frac{a}{b^3}\frac{1}{5}y^5\right]_0^b}{A_{\text{tot}}} = \frac{\frac{ab^2}{2} - \frac{a}{5b^3}b^5}{\frac{3}{4}ab} = \frac{2}{5}b$$

$$Y_C = \frac{2}{5}b$$

Answer $A_{\text{tot}} = \frac{3}{4}ab$
$X_C = (4/7)a$
$Y_C = (2/5)b$

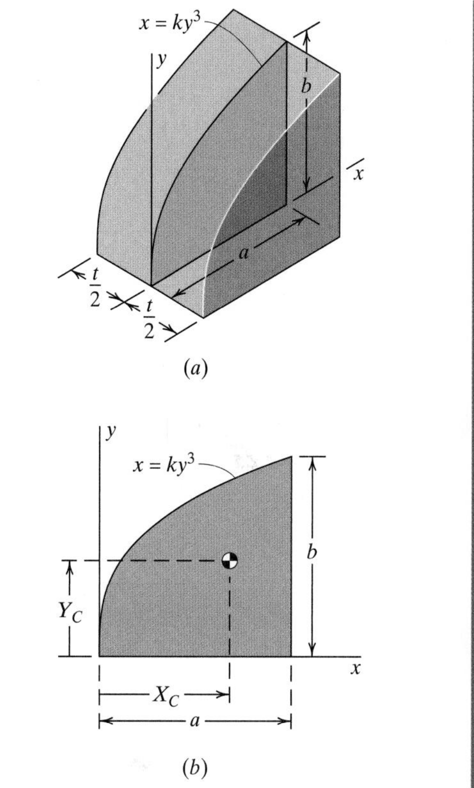

Figure 8.13

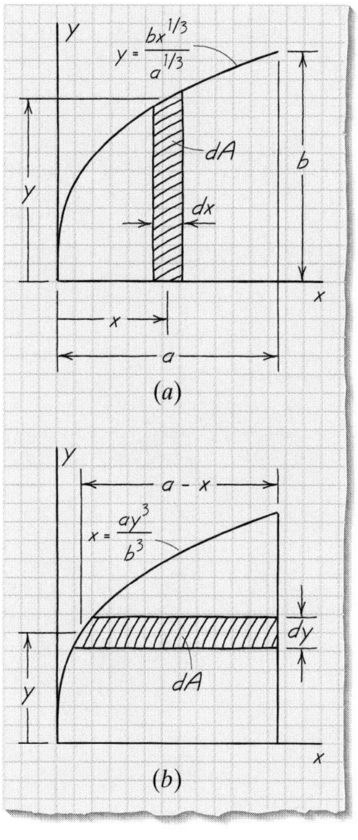

Figure 8.14

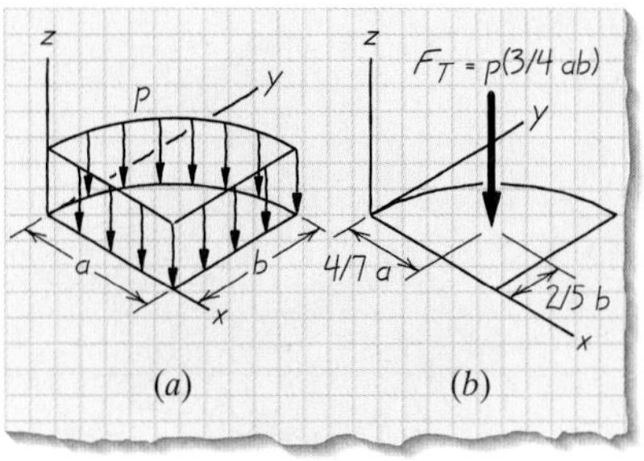

Figure 8.15

By finding the centroid of an area in this example, we were able to find two of the coordinates of the centroid of an extruded volume. As we will see in detail in the next section, being able to find the centroid of an area also enables us to model a distributed force acting on the boundary of a system as a single point force. For example, if a uniform pressure p acts on a horizontal surface as shown in **Figure 8.15*a***, we could determine the total equivalent load and point of application. The total equivalent load F_T is

$$F_T = pA_T = p(3/4ab)$$

The total force would act at the centroid of the area $X_C = (4/7)a$ and $Y_C = (2/5)b$ (**Figure 8.15*b***).

EXAMPLE 8.5 CENTER OF MASS

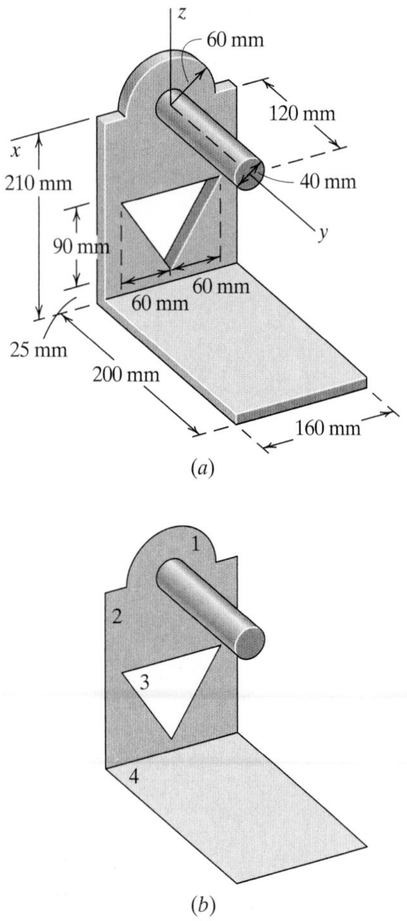

Figure 8.16

Consider the assembly in **Figure 8.16*a***. Both the vertical face and the horizontal base are constructed of thin sheet metal. The vertical face is made from sheet metal with a mass per unit area of 22 kg/m^2, the material of the horizontal base has a mass per unit area of 45 kg/m^2, and the aluminum shaft has a density of 2.71 Mg/m^3. Determine the x, y, and z coordinates of the center of mass of the assembly.

Goal We are to find the x, y, and z coordinates of the center of mass of the assembly.

Given We are given the dimensions of the assembly and the properties of the materials it is built of. The assembly is composed of a vertical face (mass per area of 22 kg/m^2), a horizontal base (mass per area of 45 kg/m^2), and an aluminum shaft (density of 2.71 Mg/m^3).

Assume We assume that the material properties are uniform (homogeneous) within the vertical face, horizontal base, and aluminum shaft. We also assume that we can treat the vertical face and horizontal base as extruded volumes. In addition, because they are made of very thin sheet metal we can present them as areas along their midplane, as shown in **Figure 8.16*b***.

Draw We first decompose the assembly into simpler parts and analyze them separately (**Figure 8.17**).

Formulate Equations and Solve We will use (8.7B) to find the center of mass of the assembly. We begin by using Appendix A3 to find volumes and areas of standard shapes so that we can calculate the mass of each part. Based on areas in Appendix A3.1:

$$m_1 = \left(22\,\frac{\text{kg}}{\text{m}^2}\right)\frac{\pi r^2}{2} = \left(22\,\frac{\text{kg}}{\text{m}^2}\right)\frac{\pi(0.06\text{ m})^2}{2} = 0.124\text{ kg}$$

$$m_2 = \left(22\,\frac{\text{kg}}{\text{m}^2}\right)bh = \left(22\,\frac{\text{kg}}{\text{m}^2}\right)(0.160\text{ m})(0.210\text{ m}) = 0.739\text{ kg}$$

$$m_3 = \left(-22\,\frac{\text{kg}}{\text{m}^2}\right)\frac{bh}{2} = \left(-22\,\frac{\text{kg}}{\text{m}^2}\right)\frac{(0.120\text{ m})(0.090\text{ m})}{2} = -0.119\text{ kg}$$

(This mass is negative because this is a cutout area on the assembly.)

$$m_4 = \left(45\,\frac{\text{kg}}{\text{m}^2}\right)bh = \left(45\,\frac{\text{kg}}{\text{m}^2}\right)(0.160\text{ m})(0.200\text{ m}) = 1.44\text{ kg}$$

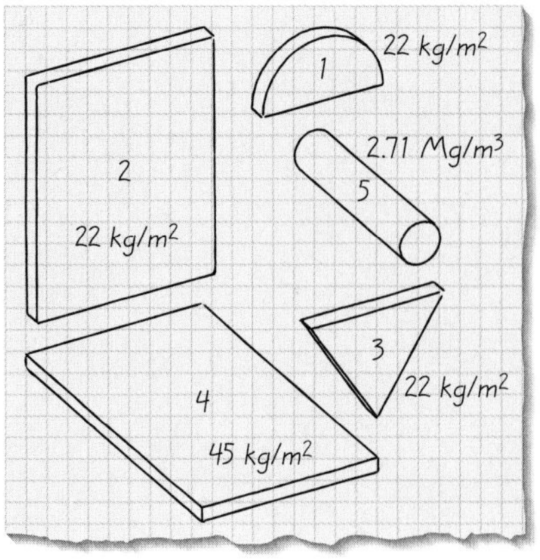

Figure 8.17

Based on volumes in Appendix A3.2:

$$m_5 = \left(2.71\,\frac{\text{Mg}}{\text{m}^3}\right)\pi r^2 L = \left(2710\,\frac{\text{kg}}{\text{m}^3}\right)\pi(0.020\text{ m})^2(0.120\text{ m}) = 0.409\text{ kg}$$

When an object has a plane of symmetry, the center of mass of that object occurs on that plane. Therefore we know that our assembly has its center of mass in the plane defined by $x = 0$ (see **Figure 8.16**). We need only calculate the y and z coordinates of the center of mass. We calculate the coordinates of the center of mass of each individual part based on coordinate axes as located in **Figure 8.16**, using Appendix A3.1 and A3.2 as references:

Component #1:

$$Y_{M1} = 0\text{ (by inspection)};\qquad Z_{M1} = \frac{4r}{3\pi} = \frac{4(60\text{ mm})}{3\pi} = 25.5\text{ mm}$$

Component #2:

$$Y_{M2} = 0\text{ (by inspection)};\qquad Z_{M2} = -\frac{210.0\text{ mm}}{2} = -105.0\text{ mm}$$

Component #3:

$$Y_{M3} = 0\text{ (by inspection)};\qquad Z_{M3} = -95.0\text{ mm} - \frac{1}{3}(90.0\text{ mm}) = -125.0\text{ mm}$$

Component #4:

$$Y_{M4} = \frac{200.0\text{ mm}}{2} = 100.0\text{ mm};\qquad Z_{M4} = -210.0\text{ mm (by inspection)}$$

Component #5:

$$Y_{M5} = \frac{120.0\text{ mm}}{2} = 60.0\text{ mm};\qquad Z_{M5} = 0\text{ mm (by inspection)}$$

We then organize this information into a table:

Component	M_i Mass [kg]	Y_{M_i} [mm]	Z_{M_i} [mm]	M_iY_M [kg mm]	$M_iZ_{M_i}$ [kg mm]
1	0.124	0	25.5	0	3.16
2	0.739	0	−105.0	0	−77.6
3	−0.119	0	−125.0	0	14.9
4	1.44	100.0	−210.0	144.0	−302.4
5	0.409	60.0	0	24.5	0
Σ	**2.593**			**168.5**	**−361.9**

Finally, we compute the location of the center of mass of the assembly using (8.7B) and the data in this table.

$$X_M = 0 \text{ (by symmetry)}$$

$$Y_M = \frac{\sum_{i=1}^{N} M_i Y_{iM_i}}{M} = \frac{168.5 \text{ kg mm}}{2.593 \text{ kg}} = 65.0 \text{ mm}$$

$$Z_M = \frac{\sum_{i=1}^{N} M_i Z_{iM_i}}{M} = \frac{-361.9 \text{ kg mm}}{2.593 \text{ kg}} = -139.6 \text{ mm}$$

Answer $X_M = 0$
$Y_M = 65.0$ mm
$Z_M = -139.6$ mm

Comment: We note that the location of the center of mass is a point in space that is not on the assembly. This is not unusual. For example, a symmetric object with a hole in the middle such as a washer or a pipe also has a mass center that is not on the object. We also note that the center of gravity of this assembly has the same location as its center of mass (as long as the gravity field is uniform).

EXAMPLE 8.6 CENTROID OF A BUILT-UP SECTION

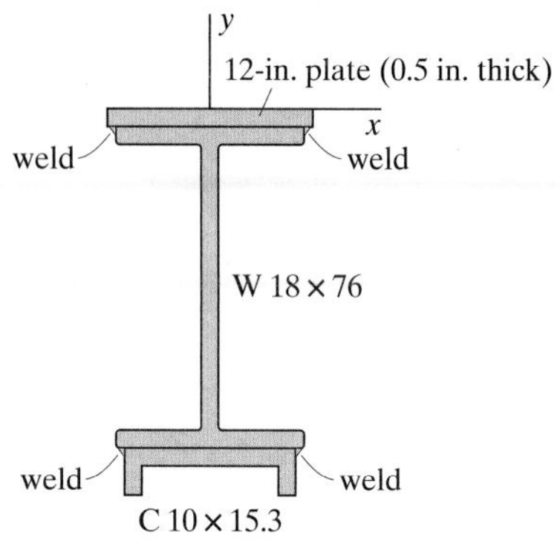

Figure 8.18

A beam used in a three-story building is made from three standard steel sections welded together (**Figure 8.18**). A wide flange section (W18 × 76) is welded to a channel section (C10 × 15.3) at the bottom and a 1/2-in.-thick, 12-in.-wide plate at the top. Determine the coordinates of the centroid of the built-up section (area).

Goal We are to find the location of the centroid of the built-up structural section.

Given We are given the dimensions of the steel plate and the specifications of the standard steel sections.

Assume We assume that the welds are so small they can be ignored in the centroid calculation. We also assume that we can treat the beam as a homogeneous extruded volume; therefore the centroid of the built-up structural section (centroid of the area) defines two of the coordinates of the centroid of the beam.

Draw We look up the dimensions of the standard steel sections in the steel manual[2] and draw the individual pieces with their dimensions (**Figure 8.19**).

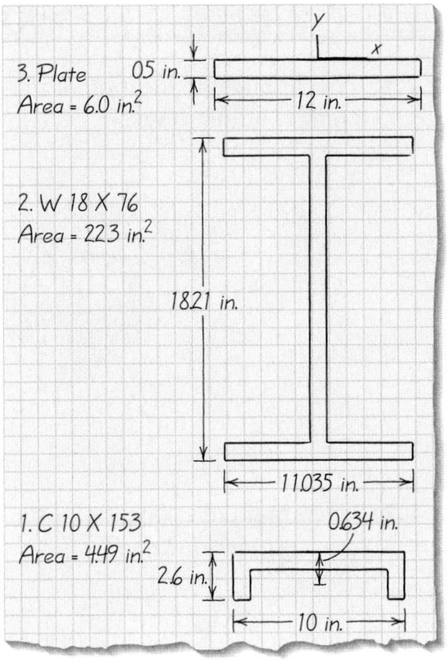

Figure 8.19

Formulate Equations and Solve Because the section is symmetric about the y axis, we know that the centroid lies on the y axis and that $X_C = 0$. We only need to find the y coordinate of the centroid using (8.12).

We calculate the total area of the built-up section,

$$A_{\text{tot}} = A_{C1} + A_{C2} + A_{C3}$$
$$A_{\text{tot}} = 4.49 \text{ in.}^2 + 22.3 \text{ in.}^2 + 6.0 \text{ in.}^2 = 32.79 \text{ in.}^2$$

and the y coordinate of the centroid of each piece (based on coordinate axes, as located in **Figure 8.18**)

$$Y_{C1} = -0.5 \text{ in.} - 18.21 \text{ in.} - 0.634 \text{ in.} = -19.334 \text{ in.}$$
$$Y_{C2} = -0.5 \text{ in.} - \frac{18.21 \text{ in.}}{2} = -9.605 \text{ in.}$$
$$Y_{C3} = -0.25 \text{ in.}$$

We substitute into (8.12) to get

$$Y_C = \frac{\sum_{i=1}^{3} A_i Y_{iC}}{A_{\text{tot}}} = \frac{\begin{array}{c}(4.49 \text{ in.}^2)(-19.334 \text{ in.}) \\ + (22.3 \text{ in.}^2)(-9.605 \text{ in.}) \\ + (6.0 \text{ in.}^2)(-0.25 \text{ in.})\end{array}}{32.79 \text{ in.}^2} = \frac{-302.5 \text{ in.}^3}{32.79 \text{ in.}^2}$$
$$Y_C = -9.23 \text{ in.}$$

Answer $X_C = 0.0$ in.
$Y_C = -9.23$ in.

Check This answer is reasonable. If the built-up section were doubly symmetric, the centroid would be at the mid-height of the wide flange section. Since the plate on the top has a larger area than the channel on the bottom, the centroid is above the mid-height of the wide flange, which is what we found with $Y_C = -9.23$ in.

[2]American Institute of Steel Construction (2001), *Manual of Steel Construction, Load and Resistance Factor Design*, 3rd edition. Chicago, Illinois.

EXERCISES 8.1

8.1.1. Determine the location of the centroid of the paraboloid of revolution shown in **E8.1.1**.

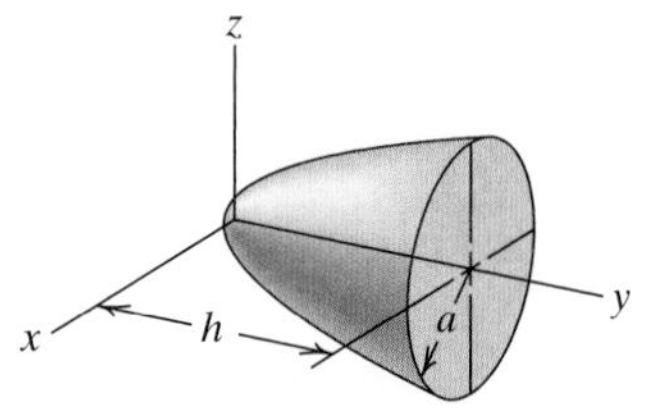

$$x^2 + z^2 = \frac{a^2 y}{h}$$

E8.1.1

8.1.2. Determine the location of the centroid of the ellipsoid of revolution shown in **E8.1.2**.

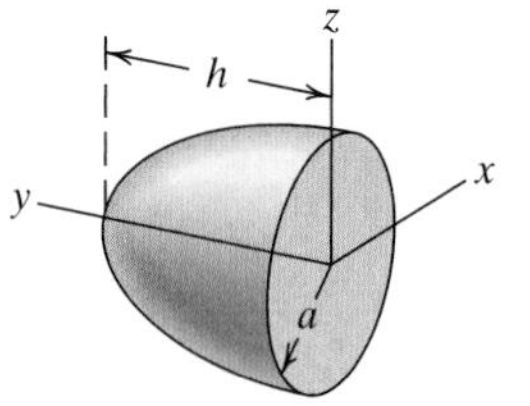

$$x^2 + z^2 + \left(\frac{ay}{h}\right)^2 = a^2$$

E8.1.2

8.1.3. The quarter cone in **E8.1.3** is homogeneous with material of density ρ. Determine its mass and the location of the center of mass.

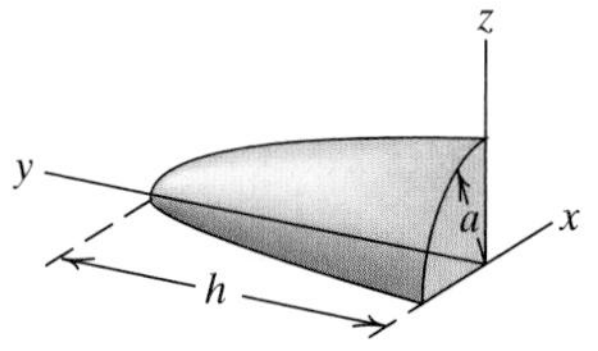

$$x^2 + z^2 = a^2\left(1 - \frac{y}{h}\right)$$

E8.1.3

8.1.4. The conical shell in **E8.1.4** is homogeneous with material of density ρ and is of uniform thickness t. Determine its mass and the location of the center of mass.

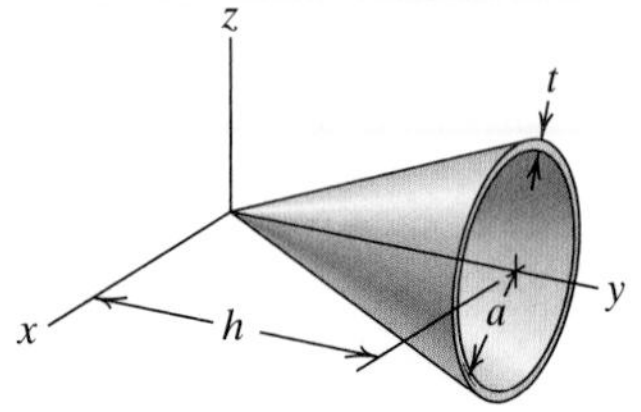

$$x^2 + z^2 = \left(\frac{ay}{h}\right)^2$$

E8.1.4

8.1.5. The hemispherical shell in **E8.1.5** is homogeneous with material of density ρ and is of uniform thickness t. Determine its mass and the location of the center of mass.

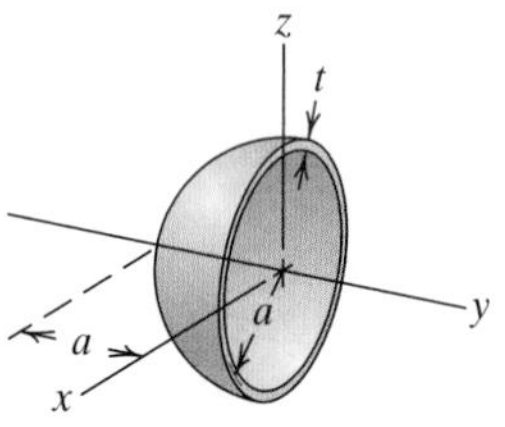

$$x^2 + y^2 + z^2 = a^2$$

E8.1.5

8.1.6. The density of the rectangular volume in **E8.1.6** varies according to

$$\rho = \rho_0\left(1 + \frac{x}{a}\frac{y}{b}\right)$$

Determine

- **a.** the mass of the volume
- **b.** the z coordinate of the center of mass of the volume
- **c.** the x coordinate of the center of mass of the volume
- **d.** the y coordinate of the center of mass of the volume
- **e.** the location of the centroid of the volume

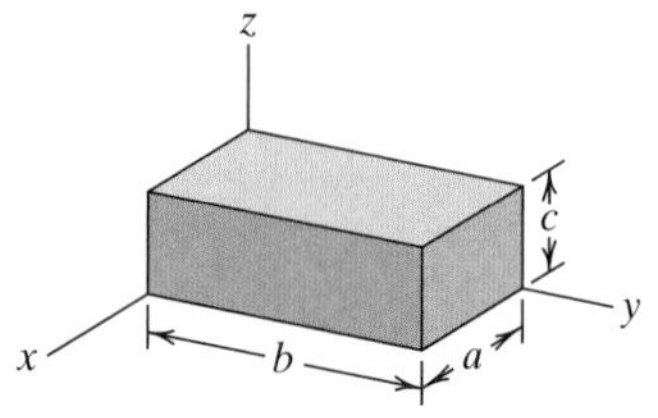

E8.1.6

8.1.7. The volume of the orthogonal tetrahedron in **E8.1.7** is homogeneous with material density ρ. Determine its mass and the location of the center of mass.

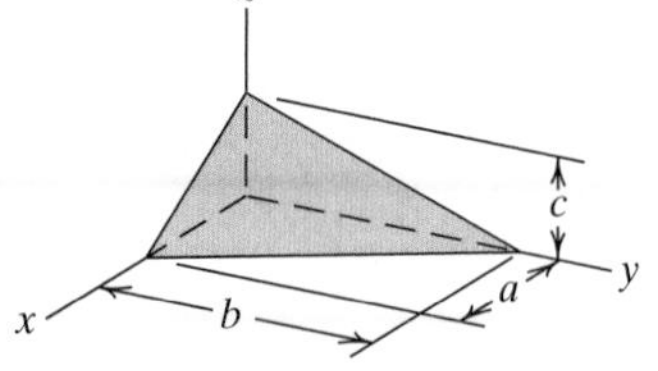

$$z = c\left(1 - \frac{x}{a} - \frac{y}{b}\right)$$

E8.1.7

8.1.8. The volume of the hyperbolic paraboloid in **E8.1.8** is homogeneous with material density ρ. Determine its mass and the location of the center of mass.

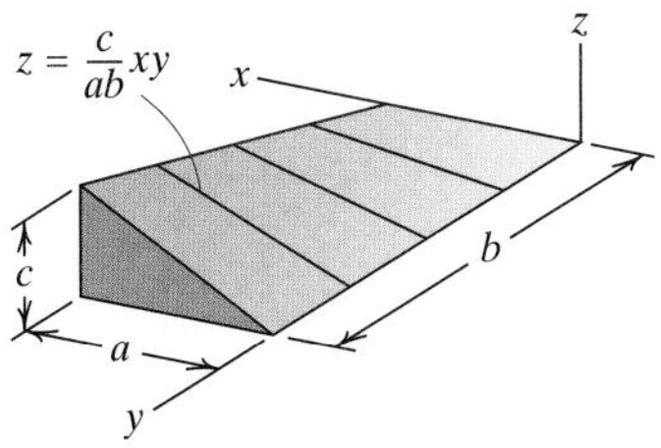

E8.1.8

8.1.9. The volume in **E8.1.9** is homogeneous with material density ρ. Determine its mass and the location of the center of mass.

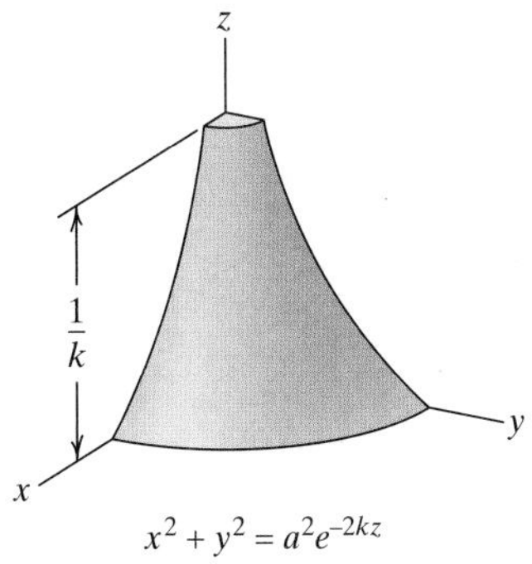

E8.1.9

8.1.10. The density in the cone in **E8.1.10** varies from $3\rho_o$ at the point to $6\rho_o$ at the base. Determine the mass and the location of the center of mass of the cone.

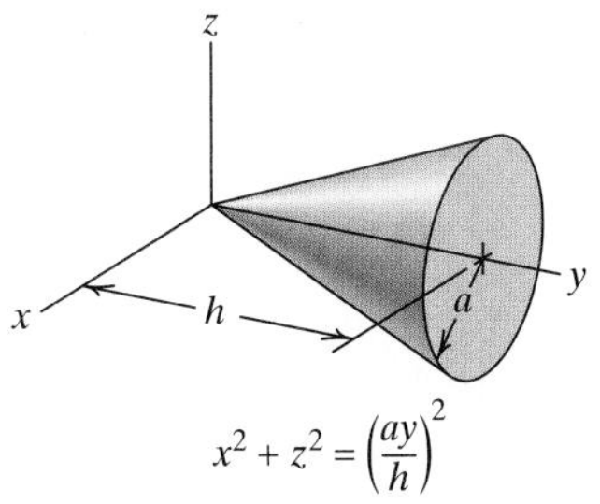

E8.1.10

8.1.11. A solid (quadrant of a cylinder) is shown in **E8.1.11**. It is homogeneous. Find the x, y, and z coordinates of the center of mass by

a. using integration (*Hint:* Rewrite the mass center equations in terms of cylindrical coordinates.)

b. using Appendix A3.2

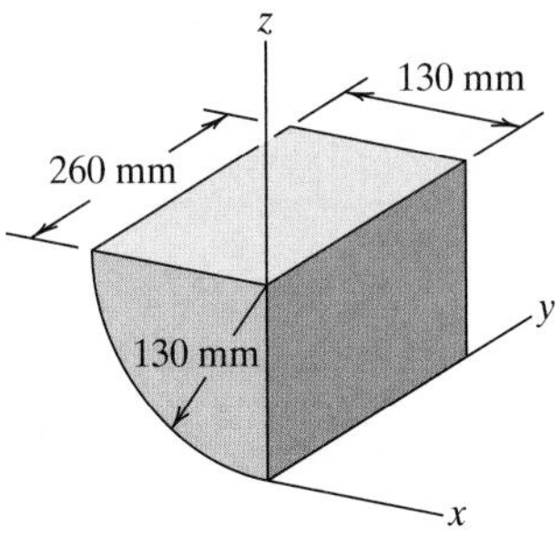

E8.1.11

8.1.12. The object's volume (**E8.1.12**) is determined by revolving the shaded area through 360° about the z axis. It is homogeneous with material of density ρ. Find the location of the center of mass.

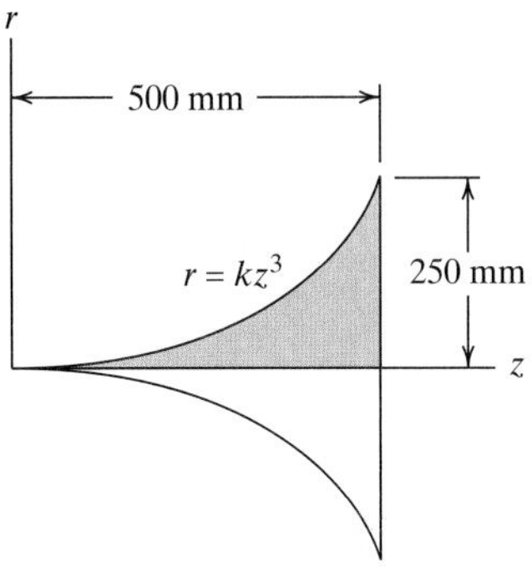

E8.1.12

8.1.13. The mass per unit length of the slender rod in **E8.1.13** varies with position according to $m = m_0(1 - x/2)$, where x is in meters. Determine the location of the center of mass of the rod. Compare your answer to the case of a homogeneous rod of mass per unit length of m_0.

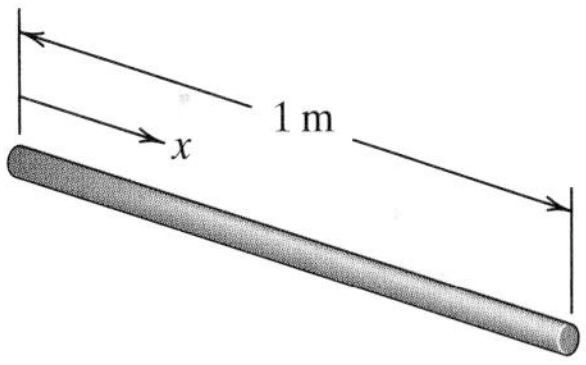

E8.1.13

8.1.14. Consider a quarter-spherical shell of thickness t and mean radius a (**E8.1.14**). The shell is homogeneous, with material density ρ. Determine the z coordinate of the center of mass by

a. using integration

b. using information in Appendix A3.2

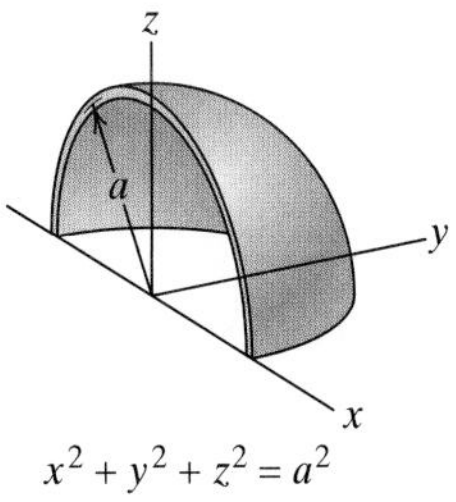

E8.1.14

8.1.15. Consider a semiconical shell of thickness t and mean radius a (**E8.1.15**). The shell is homogeneous, with material density ρ. Determine the mass and location of the center of mass by

a. using integration

b. using information in Appendix A3.2

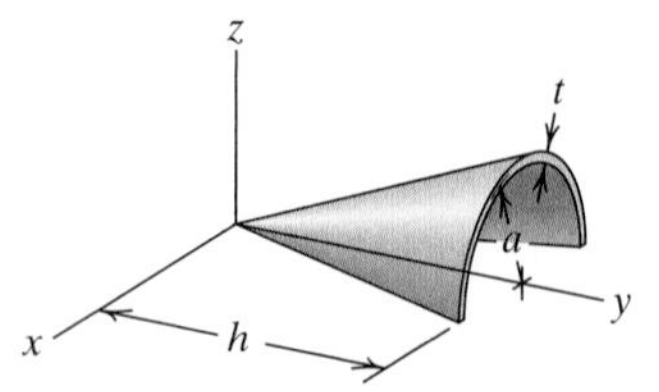

E8.1.15

8.1.16. Determine the x coordinate of the center of mass of the tapered steel rod of length L where the larger diameter end is twice the smaller diameter end (E8.1.16).

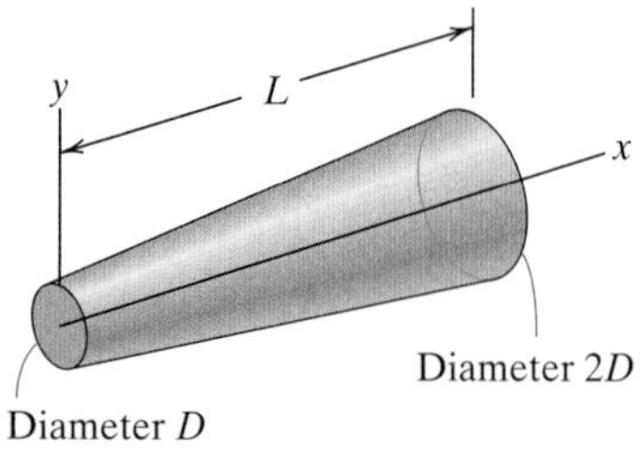

E8.1.16

8.1.17. Copper wire 1/16 in. in diameter is bent into the semicircular configuration shown in E8.1.17. Determine the mass and the location of the center of mass of configuration.

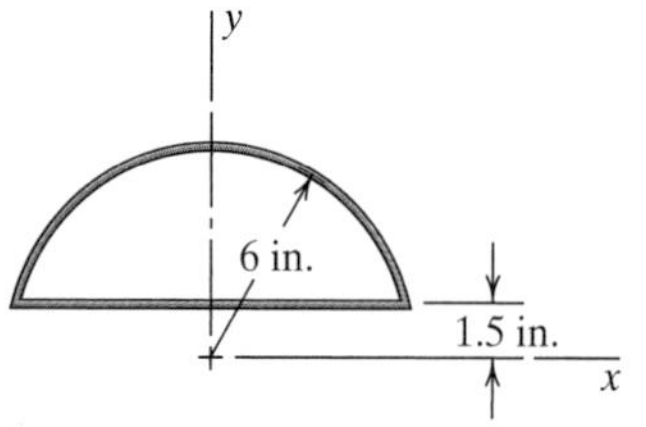

E8.1.17

8.1.18. Aluminum wire 6 mm in diameter is bent into the triangular confirmation shown in E8.1.18. Determine the mass and the location of the center of mass of configuration.

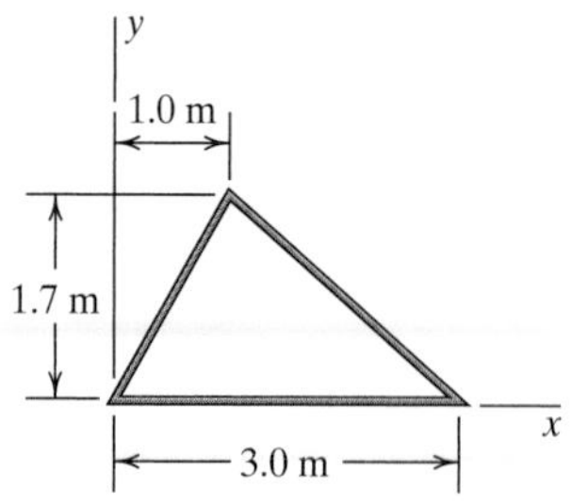

E8.1.18

8.1.19. An aluminum wire 4 mm in diameter is bent into the confirmation shown in E8.1.19. Determine the mass and the location of the center of mass of configuration.

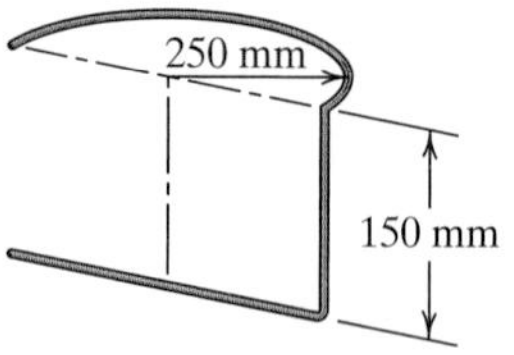

E8.1.19

8.1.20. Consider the object in E8.1.20 that is made of concrete. Determine its mass and the location of its center of mass.

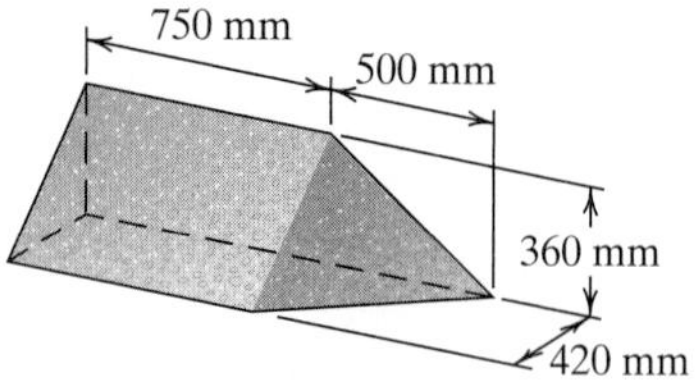

E8.1.20

8.1.21. An object consists of a steel cylinder with a hemispherical cavity at the top, and the cavity is filled with aluminum (E8.1.21).

a. Determine the mass of the object and the location of its center of mass.

b. If the cavity is filled with steel instead of aluminum, determine the mass of the object and the location of its center of mass.

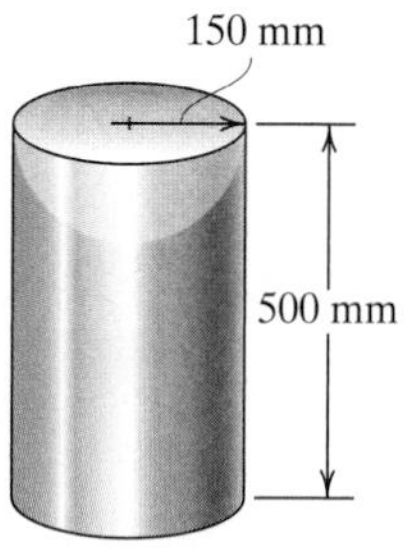

E8.1.21

8.1.22. Calculate the volume and the location of the centroid of the volume in E8.1.22. If the volume is made of aluminum, what is its weight?

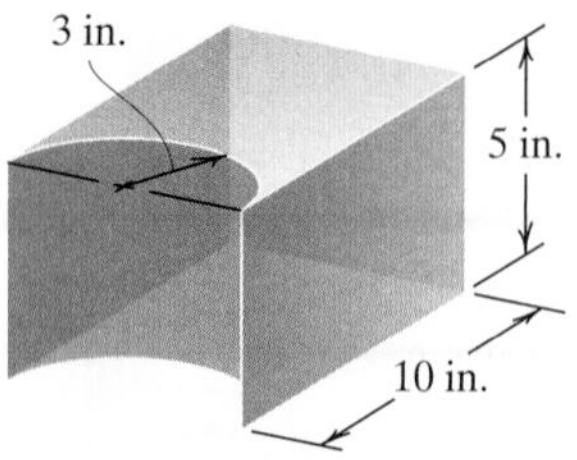

E8.1.22

8.1.23. Calculate the volume and the location of the centroid of the volume in E8.1.23. If the volume is made of glass, what is its weight?

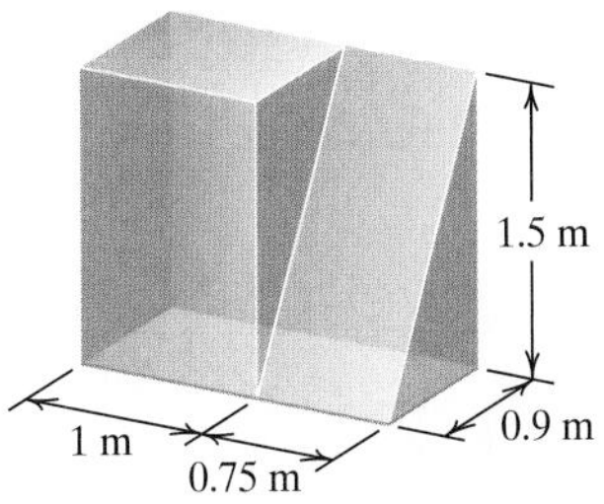

E8.1.23

8.1.24. Calculate the volume and the location of the centroid of the volume in E8.1.24. If the volume is made of steel, what is its weight?

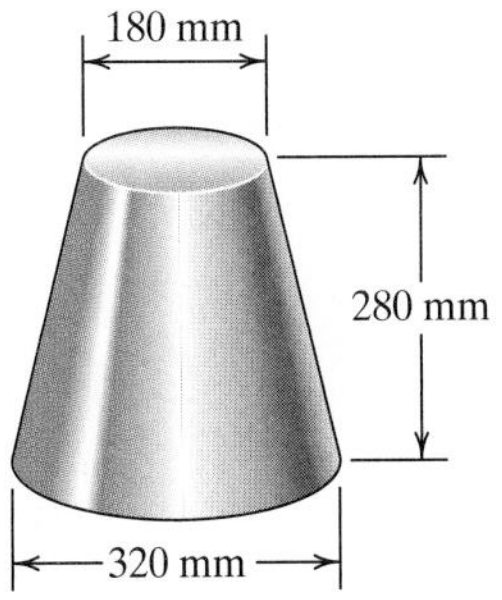

E8.1.24

8.1.25. Calculate the volume and the location of the centroid of the volume in E8.1.25. If the volume is made of concrete, what is its weight?

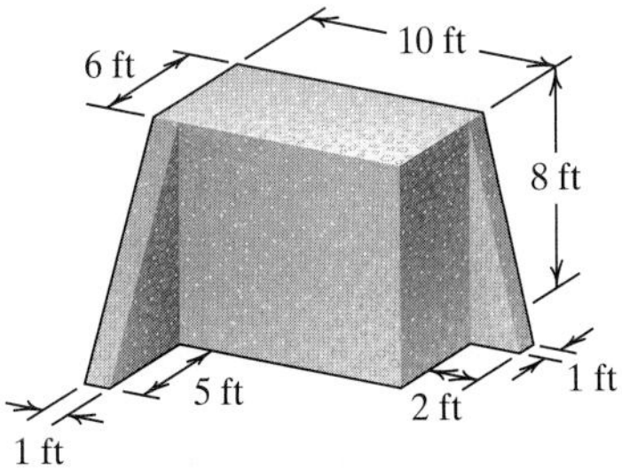

E8.1.25

8.1.26. If the materials in the cylinder with a hemispherical cavity in **E8.1.21** are reversed so that the cylinder is made of aluminum and the cavity is filled with steel, determine

a. the mass of the object

b. the location of its center of mass

c. the weight of the object

8.1.27. The steel disk in E8.1.27 has an aluminum insert whose faces are flush with the faces of the disk. The thickness of the disk is 30 mm. Determine

a. the mass of the disk with insert and the location of its center of mass

b. the centroid of the disk with insert

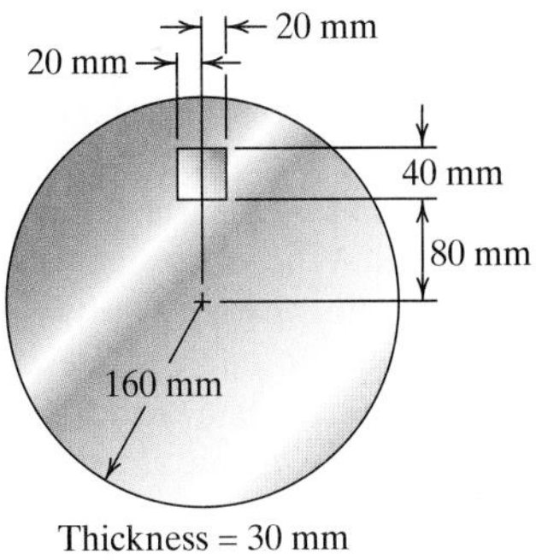

E8.1.27

8.1.28. The copper disk in E8.1.28 has a glass insert whose faces are flush with the faces of the disk. The thickness of the disk is 0.50 in. Determine

a. the mass of the disk with insert and the location of its center of mass

b. the centroid of the disk with insert

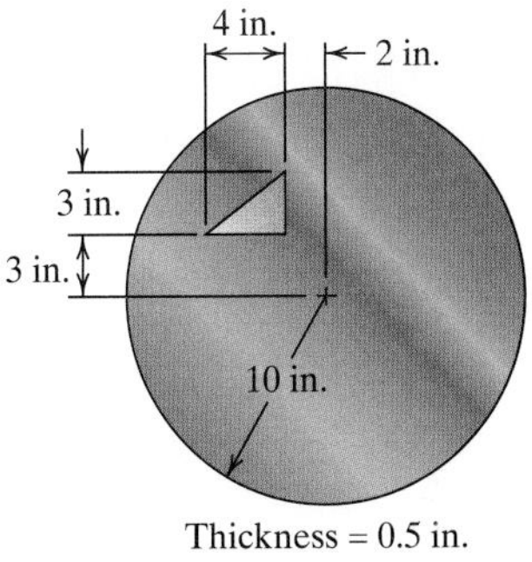

E8.1.28

8.1.29. The L-shaped steel plate in E8.1.29 is 10 mm thick. Determine its mass and the location of its center of mass and show the results on a scale drawing of the plate.

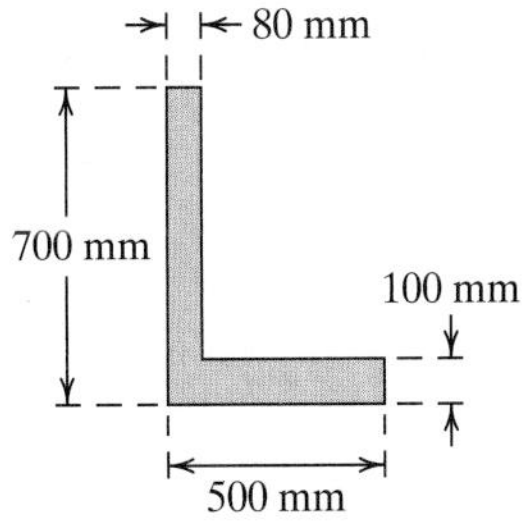

E8.1.29

8.1.30. The aluminum plate in E8.1.30 is 0.25 in. thick. Determine its mass and the location of its center of mass and show the results on a scale drawing of the plate.

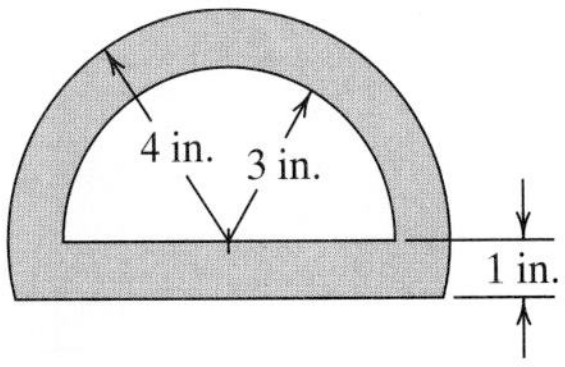

E8.1.30

8.1.31. **E8.1.31** shows the dimensions of the rear stabilizer of a commercial aircraft.

a. Assuming that the stabilizer is 1.5 cm thick (dimension into the paper), find its centroid.

b. If the rear stabilizer is constructed of steel ($\rho = 7830$ kg/m^3), what is its mass? Where is its center of mass?

c. A mechanical engineer wants to divide the stabilizer into three sections as shown and replace one or more of them with a carbon fiber-resin composite ($\rho = 1400$ kg/m^3). What is the weight savings if section 3 is replaced with a composite material of the same dimensions? Where is the new position of the center of mass?

d. What is the weight savings if both sections 2 and 3 are replaced with the carbon fiber-resin composite material? Where is the new position of the center of mass?

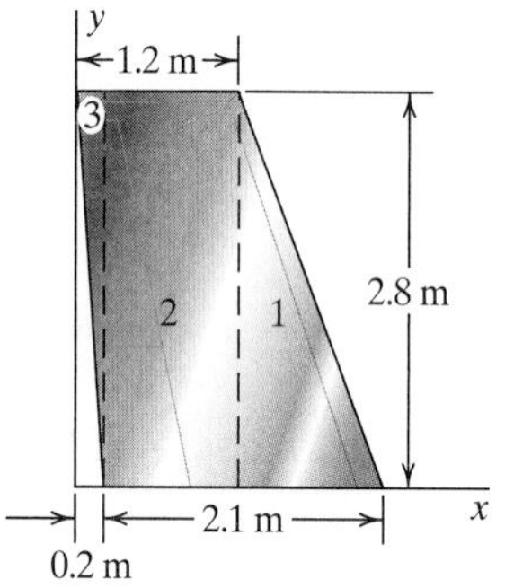

E8.1.31

8.1.32. The Defense Department has commissioned a computer manufacturer to produce for its navigational computers a mother board that can survive high-impact forces. Because standard processing techniques were too unreliable for these purposes, the engineers tried explosively welding a highly conductive gold film (0.250 cm thick) to a silicon substrate. Before they test the strength of the interface they want to find the centroid of the component. Assuming that gold has a density of 19.302 g/cm^3, silicon has a density of 2.33 g/cm^3, and the structure has a uniform thickness of 1.00 cm (into the page), calculate

a. the centroid of the component relative to the coordinate system shown in **E8.1.32**

b. the center of mass of the component relative to the coordinate system shown

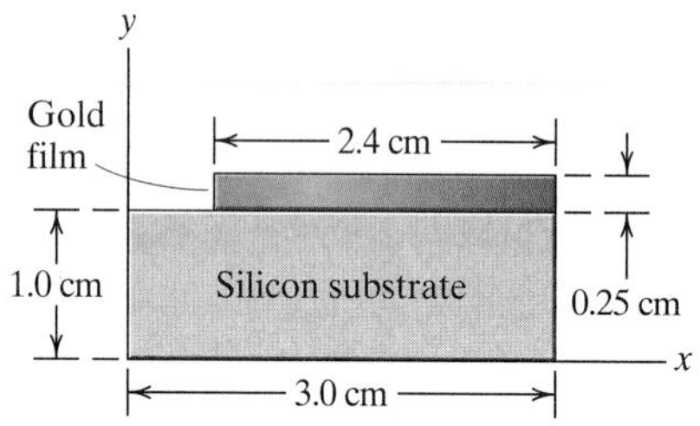

E8.1.32

8.1.33. The container in **E8.1.33** was created from bent sheet steel 2 mm thick. Determine the container's mass and the location of its mass center.

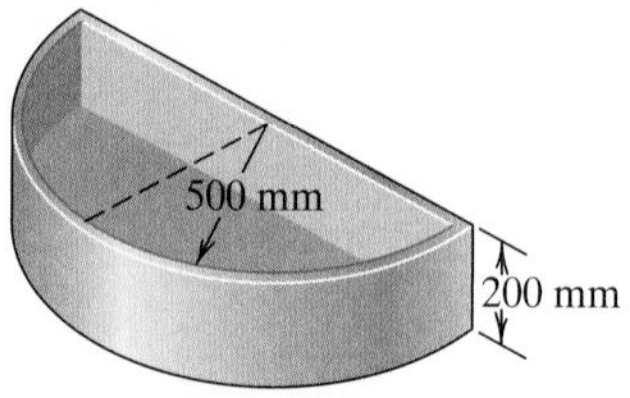

E8.1.33

8.1.34. The container in **E8.1.34** was created from bent sheet aluminum, 0.25 in. thick. Determine the container's mass and the location of its mass center.

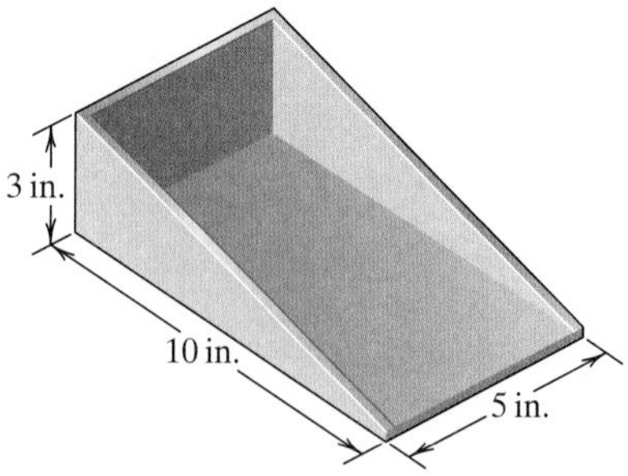

E8.1.34

8.1.35. A copper wire is bent into the configuration shown in **E8.1.35**. The diameter of the wire is 4 mm. Assuming that its density is 8900 kg/m^3 determine its mass and the location of its center of mass.

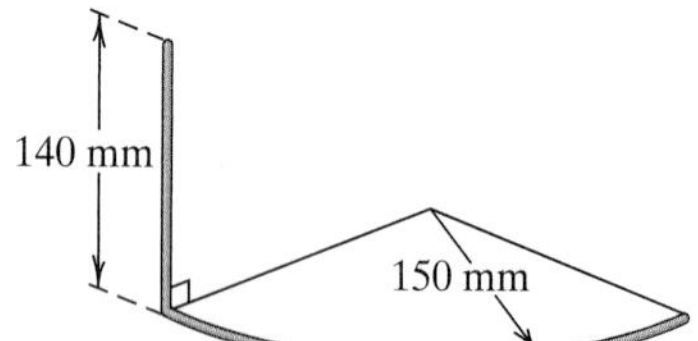

E8.1.35

8.1.36. A sheet of aluminum has been bent into the shape shown in **E8.1.36**.

a. If the thickness of the sheet is 2 mm, determine its mass and the location of its center of mass. The density of aluminum is 2690 kg/m^3.

b. It is desired to make the shape out of sheet steel with the same mass as was found in **a**. What thickness of sheet steel should be specified? The density of steel is 7830 kg/m^3.

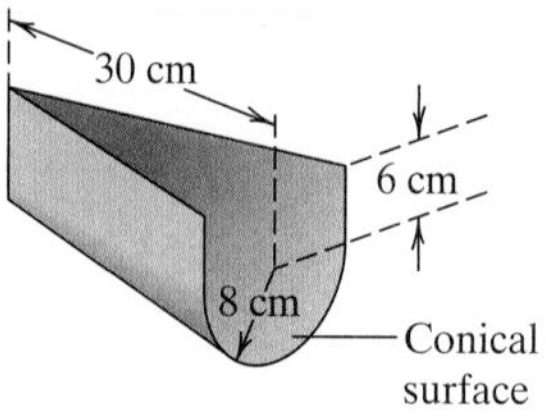

E8.1.36

8.1.37. The three legs of a small glass-topped patio table are equally spaced and are made of aluminum tubing, which has an outside diameter of 24 mm and a cross-sectional area of 150 mm^2 (**E8.1.37**). The diameter and the thickness of the tabletop are 600 mm and 10 mm, respectively.

a. Knowing that the density of aluminum is 2690 kg/m^3 and of glass is 2190 kg/m^3, find the center of mass of the table.

b. Now imagine that a book with mass of 2 kg is placed at point A. Determine the forces of the floor acting on the legs at B, C, and D.

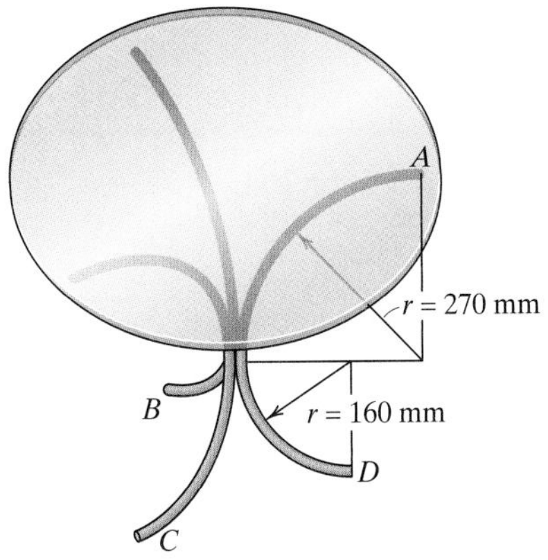

E8.1.37

8.1.38. The ends of the park bench in **E8.1.38** are made of concrete, while the seat and the back are wooden boards. Each piece of wood is 1.5 × 5 × 48 in.

a. Knowing that the specific weight of concrete is 0.084 lb/in.^{3} and of wood is 0.017 lb/in.^{3}, determine the weight of the park bench and the coordinates of its center of gravity.

b. What forces act on the concrete ends of the bench at A, B, C, and D?

c. If a child weighing 60 lb sits at either end of the bench, what forces act on the concrete ends of the bench at A, B, C, and D?

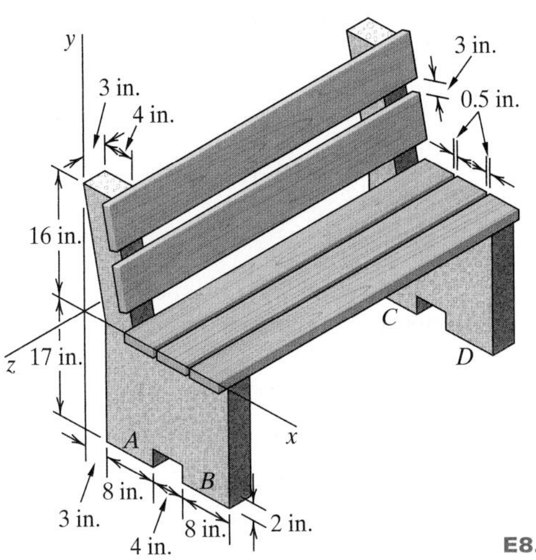

E8.1.38

8.1.39. In **E8.1.39** a uniform semicircular rod of weight W and radius r is attached to a pin at A and bears against a frictionless surface at B. Determine

a. the center of mass of the semicircular rod

b. the loads acting on the rod at A and at B

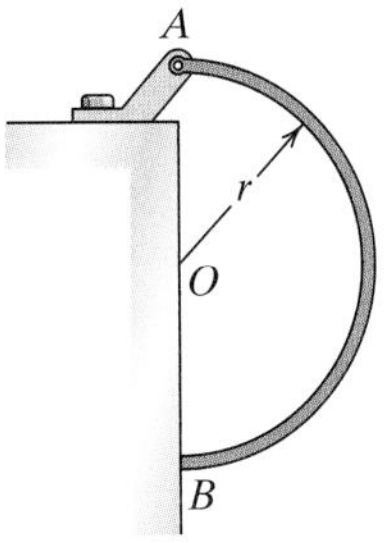

E8.1.39

8.1.40. In **E8.1.40** a uniform semicircular rod of weight W and radius r is supported by a bearing at its upper end and is free to swing about O in a vertical plane. If the rod is in equilibrium, write an expression for the angle θ as a function of W and r.

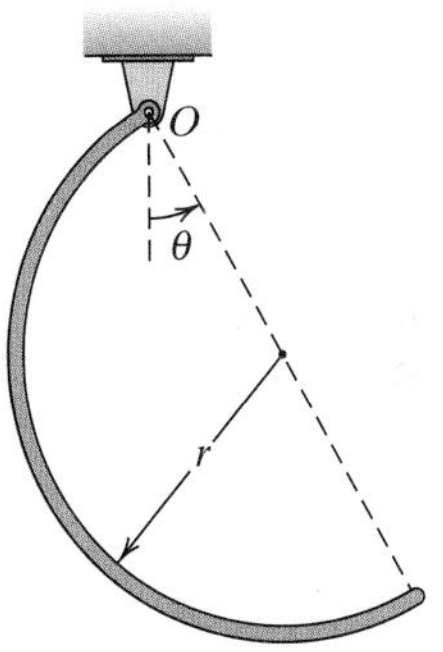

E8.1.40

8.1.41. Calculate the area of the shaded region between the two curves in **E8.1.41**. Also locate the centroid of the shaded region. Present your answer in terms of a scale drawing of the shaded region.

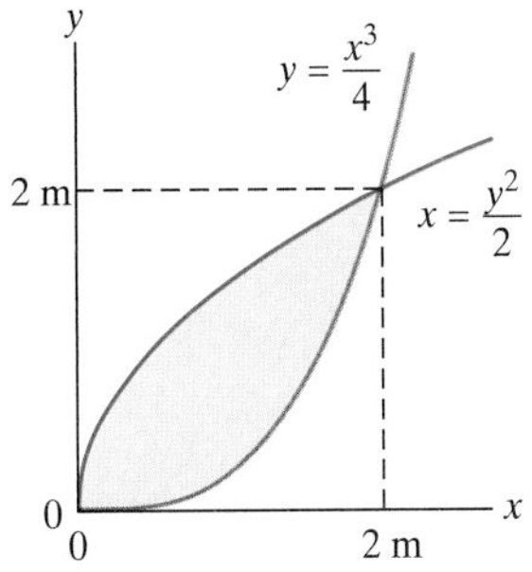

E8.1.41

8.1.42. Calculate the area of the shaded region in **E8.1.42**. Also locate the centroid of the shaded region. Present your answer in terms of a scale drawing of the shaded region.

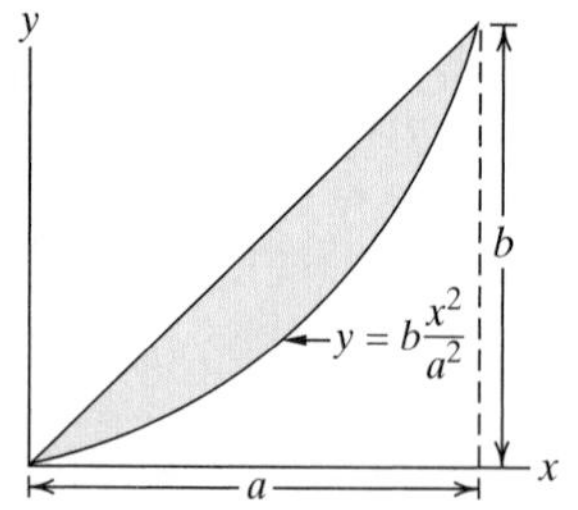

E8.1.42

8.1.43. Calculate the area of the shaded region in E8.1.43. Also locate the centroid of the shaded region. Present your answer in terms of a scale drawing of the shaded region.

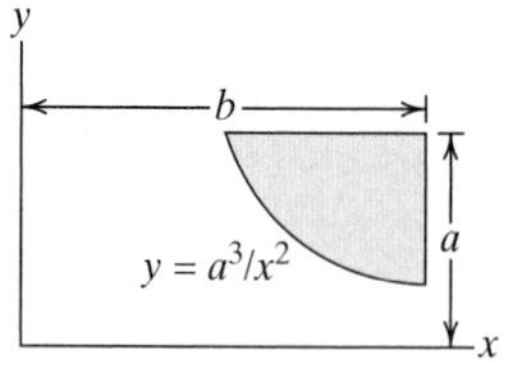

E8.1.43

8.1.44. Calculate the area of the shaded region in E8.1.44. Also locate the centroid of the shaded region. Present your answer in terms of a scale drawing of the shaded region.

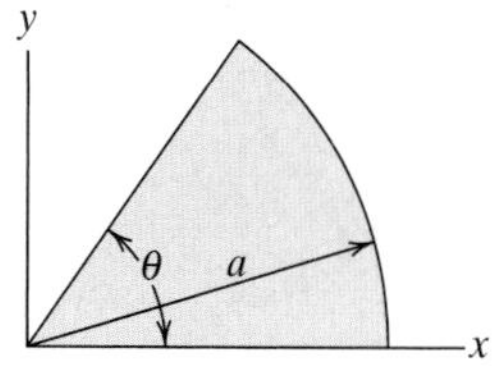

E8.1.44

8.1.45. Calculate the area of the shaded region in E8.1.45. Also locate the centroid of the shaded region. Present your answer in terms of a scale drawing of the shaded region.

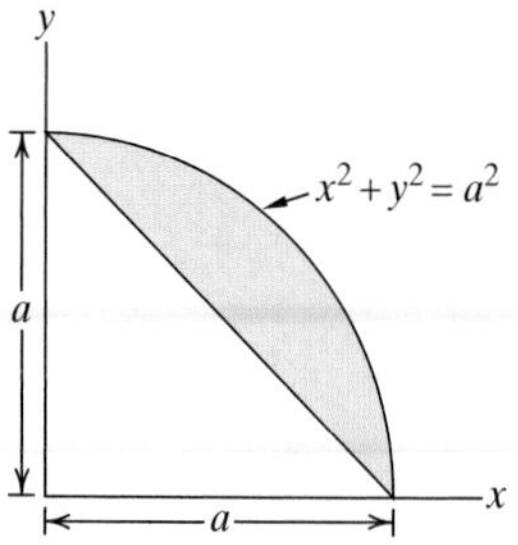

E8.1.45

8.1.46. Calculate the area of the shaded region in E8.1.46. Also locate the centroid of the shaded region. Present your answer in terms of a scale drawing of the shaded region.

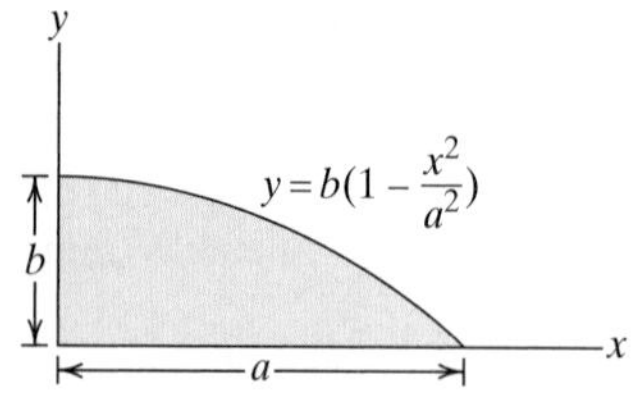

E8.1.46

8.1.47. Locate the centroid of the circular arc in E8.1.47.

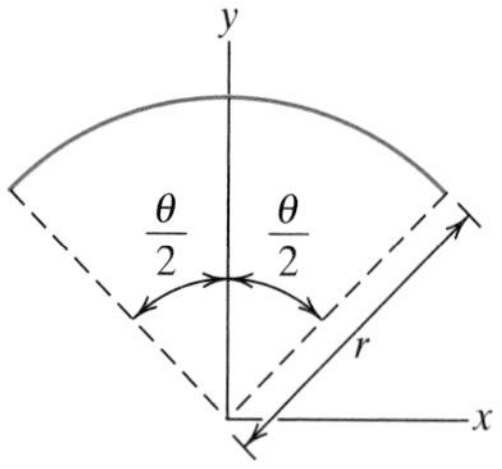

E8.1.47

8.1.48. Locate the centroid of the parabolic arc in E8.1.48.

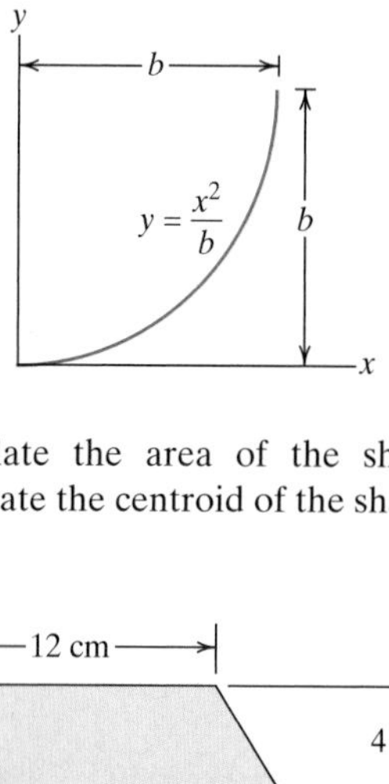

E8.1.48

8.1.49. Calculate the area of the shaded region in E8.1.49. Also locate the centroid of the shaded region.

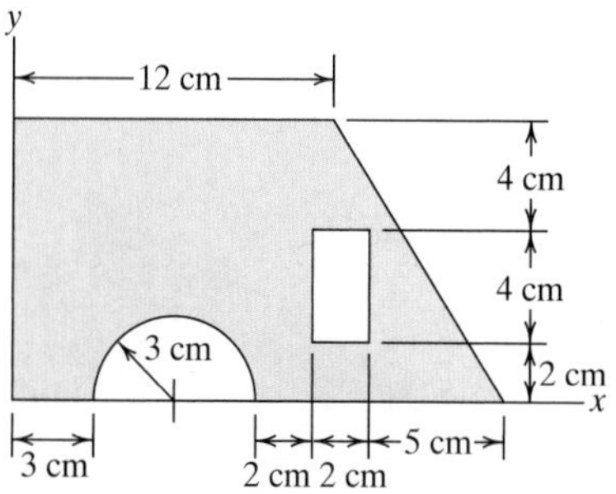

E8.1.49

8.1.50. Calculate the area of the shaded region in E8.1.50. Also locate the centroid of the shaded region.

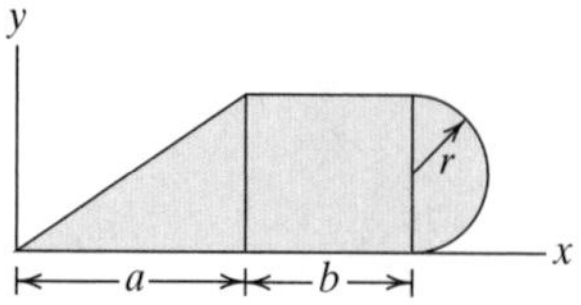

E8.1.50

8.1.51. A scale model of a B2 bomber is being constructed for testing purposes. E8.1.51 shows a simplified diagram of one of the wings, which is made out of a 1/8-in.-thick sheet of plastic. Calculate the centroid of the model's wing relative to the given coordinate system.

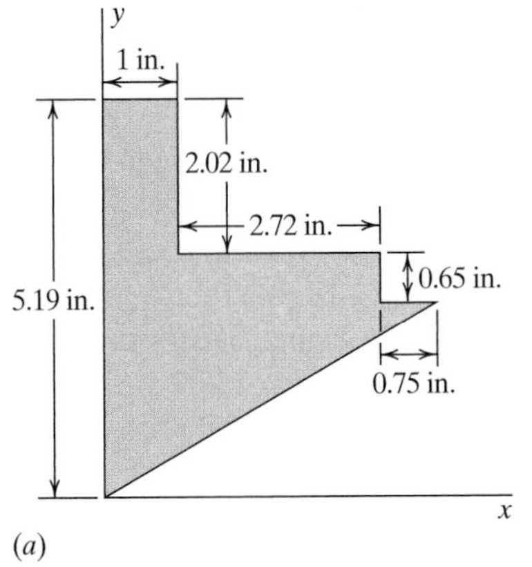

E8.1.51

8.1.52. Calculate the area of the steel surface in **E8.1.27**, excluding the area of the aluminum insert. Also locate the centroid of the steel portion.

8.1.53. Calculate the area of the copper surface in **E8.1.28**, excluding the area of the glass insert. Also locate the centroid of the copper portion.

8.1.54. Calculate the area of the shaded region in E8.1.54. Also locate the centroid of the shaded region.

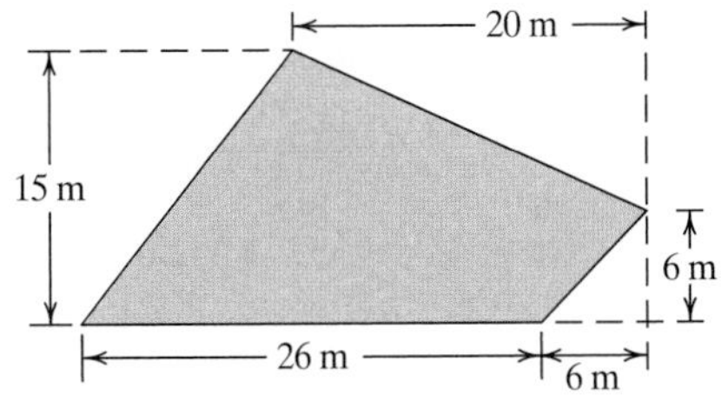

E8.1.54

8.1.55. Calculate the area of the shaded region in E8.1.55. Also locate the centroid of the shaded region.

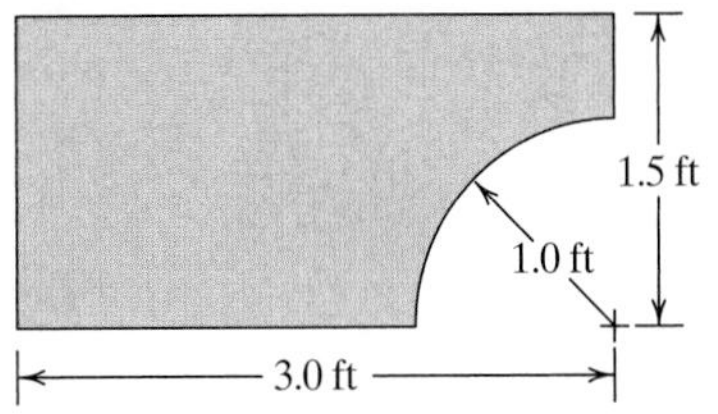

E8.1.55

8.1.56. Calculate the area of the shaded region in E8.1.56. Also locate the centroid of the shaded region. It is reasonable to ignore the fillets (the rounding of the interior and exterior corners).

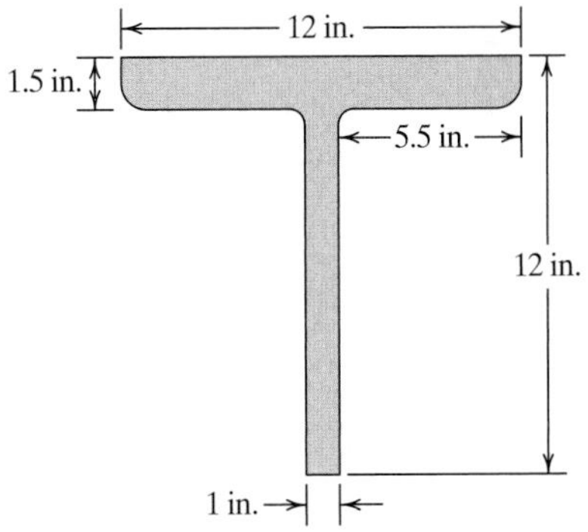

E8.1.56

8.1.57. Calculate the area of the shaded region in E8.1.57. Also locate the centroid of the shaded region. It is reasonable to ignore the fillets (the rounding of the interior and exterior corners).

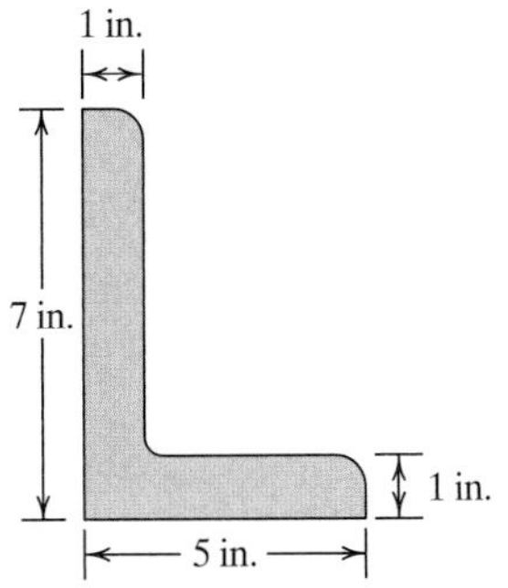

E8.1.57

8.2 DISTRIBUTED FORCE ACTING ON A BOUNDARY

We now consider how to include distributed forces acting on the boundary in static analysis. This involves representing the distributed force by an equivalent single point force. First, by looking at distributed loads with widths so small compared to the object they are acting on that they can be considered one-dimensional and acting along a line, we'll learn

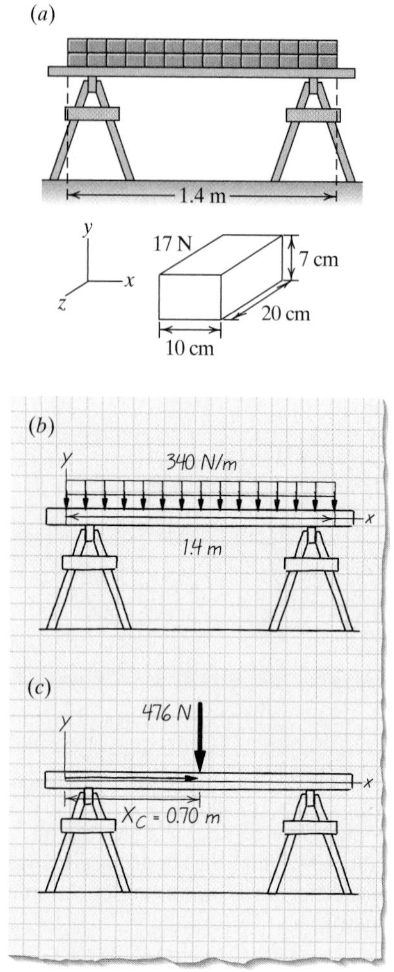

Figure 8.20 (*a*) Twenty-eight bricks stacked as shown, along with detail of brick; (*b*) the distributed force of the bricks acting on the plank; (*c*) distributed load represented as a single point force

how to find the equivalent force acting along that line and the point at which it acts. This point can be defined with only one coordinate. Then we consider a pressure acting over an area, which is two-dimensional, and therefore we'll need two coordinates to locate the point at which the total equivalent force acts.

Distributed Force Along a Line and Its Centroid

Consider a plank with two rows of bricks stacked over a span of 1.4 m, as shown in **Figure 8.20*a***. If each brick weighs 17 N, we can represent the load created by the weight of the bricks on the plank as a distributed line load of 340 N/m (= 17 N/brick × 28 bricks/1.4 m) over the 1.4-m span, as shown in **Figure 8.20*b***. This is referred to as a uniformly distributed **line load**. Notice that the units of a line load are force/length to reflect that the force is distributed over a line. This is in contrast to the forces acting on boundaries that we have considered prior to this chapter which were all point forces (and had units of force).

To replace this distributed line load with a single point force, we first find the total force. We then place this total force as a point force at the location where it creates the same equivalent moment as the distributed line load. For the uniformly distributed line load under consideration, the total force is 17 N/brick × 28 bricks = 476 N, and it seems reasonable that this total force should be placed as a single point force at the center of the distribution (see **Figure 8.20*c***). We could figure this out mathematically by finding first the total force represented by the distributed force and then the location at which the total force acts:

1. *Total force.* This force is found by integrating the distributed force over the span:

$$\text{Total force in } y \text{ direction} = F_y = \int_{\text{span}} \omega \, dx \tag{8.13}$$

where ω is the distributed line load oriented in the y direction and has dimensions force/length, the integrand ($\omega\, dx$) is the force acting on segment dx, and "span" refers to the length over which ω is distributed (**Figure 8.21*a***).

2. *Location.* The point at which the equivalent total force acts is called the **centroid of a line load** and is found by requiring that the moment produced by the total force be the same as the total moment produced by the distributed load (in doing this we are creating a moment that is equivalent to the one created by the distributed force). We can write this requirement as

$$\underbrace{X_C F_y}_{\substack{\text{moment}\\ \text{produced by}\\ \text{the total } y\\ \text{force}}} = \underbrace{\int_{\text{span}} x(\omega\, dx)}_{\substack{\text{total moment}\\ \text{produced by}\\ \text{distributed}\\ \text{load}}} \tag{8.14}$$

where X_C is the location of the centroid measured from a moment center. The integrand represents the moment created by the total force, as illustrated in **Figure 8.21*b***. By integrating over the span, we find the total moment created by the distributed load.

Equation (8.14) solved for X_C is

$$X_C = \frac{\int_{\text{span}} x(\omega\, dx)}{F_y} \underbrace{=}_{\substack{\text{substituting in} \\ \text{from (8.13) for } F_y}} \frac{\int_{\text{span}} x(\omega\, dx)}{\int_{\text{span}} (\omega\, dx)} \tag{8.15}$$

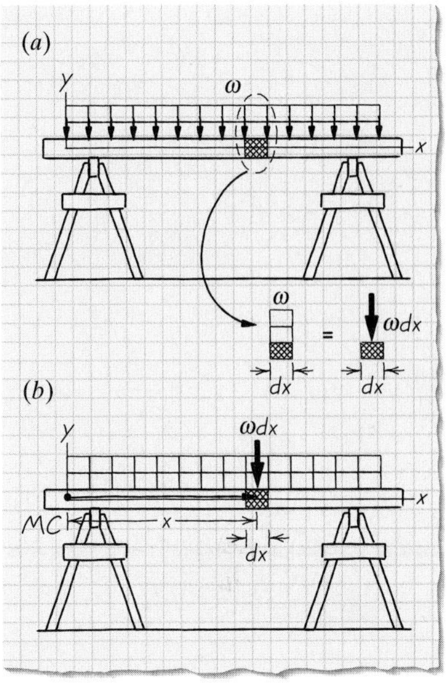

Figure 8.21 (*a*) The force represented by ($\omega\, dx$); (*b*) the momnet created by ($\omega\, dx$) at moment center (MC)

Using (8.13) and (8.15) for the uniform line load in **Figure 8.20*a***, we find

$$\text{Total force in } y \text{ direction} = F_y = \int_0^{1.4\text{ m}} -\,340\text{ N/m}\, dx = -476\text{ N}$$

(the minus sign indicates that the force is in the negative y direction) and

$$X_C = \frac{\int_0^{1.4\text{ m}} x(-340\text{ N/m}\, dx)}{\int_0^{1.4\text{ m}} (-340\text{ N/m}\, dx)} = \frac{-\frac{x^2}{2}\,340\text{ N/m}\Big]_0^{1.4\text{ m}}}{-x\,340\text{ N/m}\Big]_0^{1.4\text{ m}}} = 0.70\text{ m}$$

Equations (8.13) and (8.15) are also valid for a line load that is not distributed uniformly. Consider the brick stacking shown in **Figure 8.22*a***. This stacking can be approximated by a linear distribution of the form $\omega = \beta_1 x$, where β_1 equals 1821 N/m^2, and $x = 0$ is as indicated. The distribution $\omega = \beta_1 x$ spans $0 < x < 1.4$ m. If this distribution is inserted into (8.13) and (8.15), we find that the magnitude of the total force is 1.78×10^3 N, with the force located at $X_C = 0.93$ m (**Figure 8.22*b***).

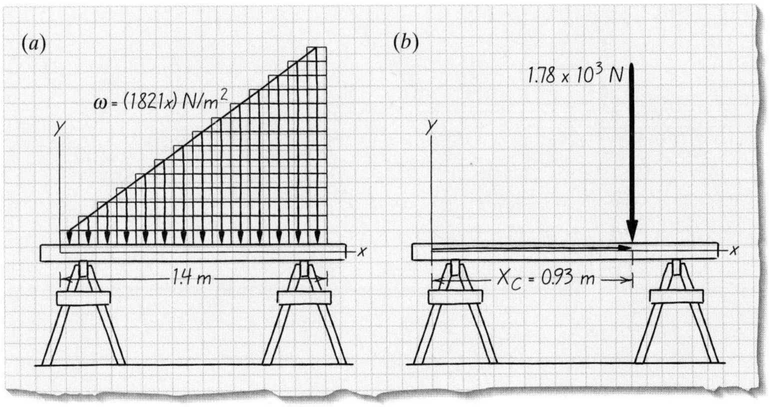

Figure 8.22 (*a*) Bricks stacked in a triangular pattern on top of the plank; (*b*) distributed load represented as a single point force

If a line load distribution is a **standard line load distribution** (which is one that can be described by a simple geometric shape), the data in Appendix A3.1 can be used to locate the centroid. For example, consider the triangularly distributed force shown in **Figure 8.23**. From the triangular configuration in Appendix A3.1, we are able to determine that

$$F_y = \frac{bh}{2} = \frac{40\text{ cm}(-70\text{ N/cm})}{2} = -1400\,N \quad \text{and}$$

$$X_C = \frac{2b}{3} = \frac{2(40\text{ cm})}{3} = 26.7\text{ cm}$$

Alternately we could carry out the integration in (8.13) and (8.15) to find the same answers (but this would involve more work!).

If a distributed force can be decomposed into standard distributions, we can use the centroid locations of the standard distributions as the basis for finding the centroid of the composite distribution. For example, the distribution in **Figure 8.24a** can be approximated by the four distributions shown in **Figure 8.24b**. The total force of the distribution is the sum of the total forces of the standard distributions. If the compos-

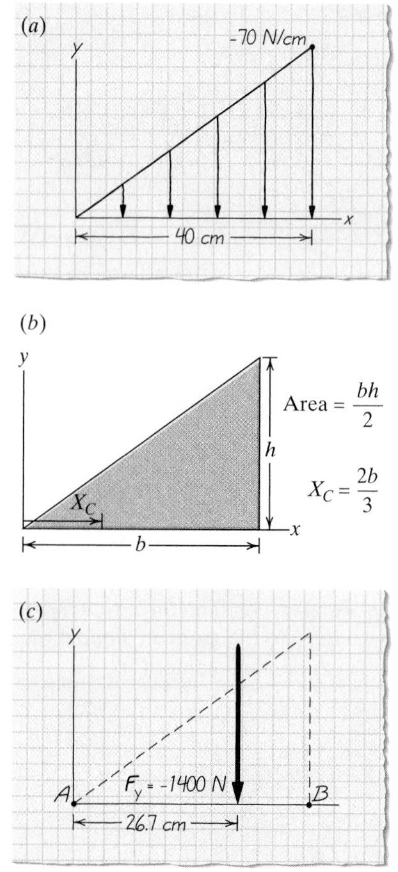

Figure 8.23 (*a*) Distributed triangular line load; (*b*) Table A3.1 data on right triangles; (*c*) distributed line load represented as a single point force

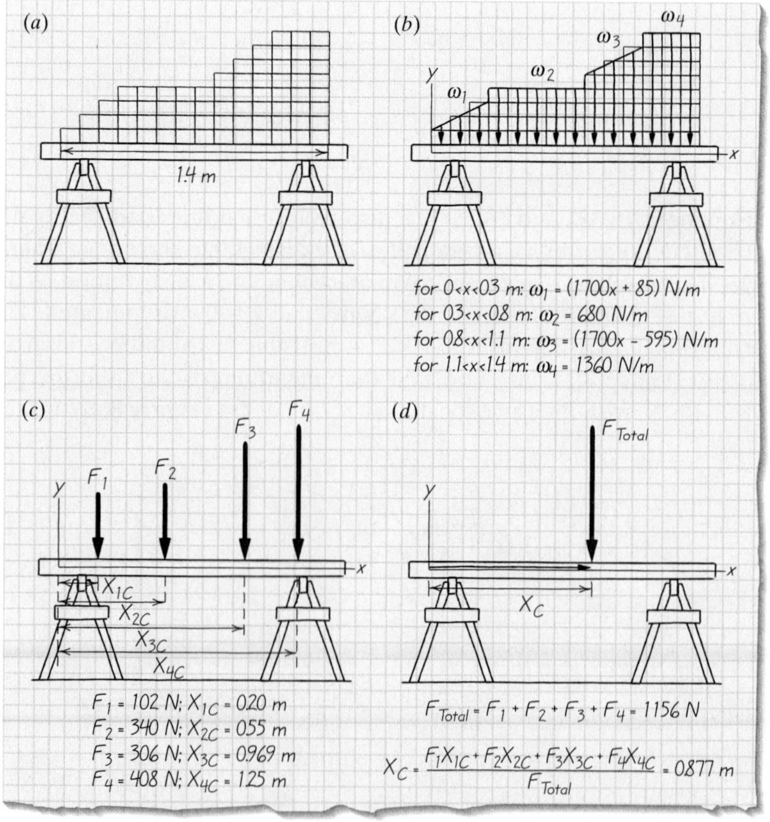

Figure 8.24 (*a*) Bricks stacked on a plank; (*b*) bricks represented as multiple line loads; (*c*) line loads represented as point forces; (*d*) single point force that represents the multiple line loads

ite distribution consists of N standard distributions, we can write the magnitude of the total force F_{total} as

$$F_{\text{total}} = \sum_{i=1}^{N} \underbrace{F_i}_{\substack{\text{force of each}\\ \text{standard}\\ \text{distribution}}} \tag{8.16A}$$

where F_i is the total force associated with distribution i. The location (X_C) of the centroid of the composite distribution is then found by finding the equivalent moment as

$$\underbrace{\underbrace{X_C}_{\substack{\text{centroid of}\\ \text{composite line}\\ \text{distribution}}} \underbrace{\sum_{i=1}^{N} F_i}_{\text{total force}}}_{\substack{\text{moment created}\\ \text{by total force}}} = \underbrace{\sum_{i=1}^{N} \underbrace{F_i X_{iC}}_{\substack{\text{force and centroid}\\ \text{of each standard}\\ \text{distribution}}}}_{\substack{\text{sum of moments}\\ \text{created by forces of}\\ \text{standard distributions}}}$$

or by substituting from (8.16A) and solving for X_C, we have:

$$X_C = \frac{\sum_{i=1}^{N} F_i X_{iC}}{F_{\text{total}}} \tag{8.16B}$$

where N is the number of standard distributions and X_{iC} and F_i are the centroid and total load associated with the ith standard distribution (**Figure 8.24c**). The application of (8.16B) is illustrated in **Figure 8.24*d*** and in Example 8.7.

EXAMPLE 8.7 BEAM WITH COMPLEX DISTRIBUTION OF LINE LOADS

A beam is subjected to a distributed load that can be divided into three standard line loads as shown in Figure 8.25. For each line load segment, determine the total force and the location of its centroid. Also determine the loads acting on the beam at A and B, assuming the beam is in equilibrium.

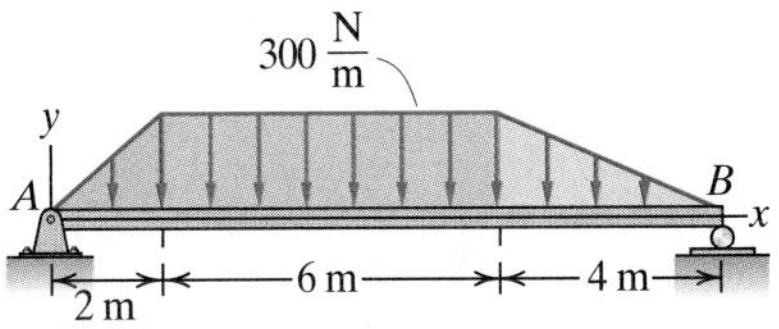

Figure 8.25

Goal For each of the three standard line load distributions we are to find the total force and the location of the centroid. In addition, we are to determine the loads acting at A and B using static analysis.

Given A 12-m beam supported by a pin joint at A and a roller at B is acted upon by a load with a complex distribution. We are given information about the shape of the distributed load.

Assume We assume that an upward force is positive and that the weight of the beam is negligible.

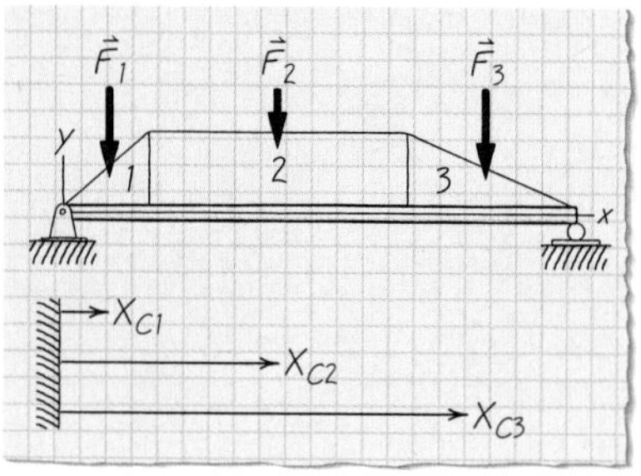

Figure 8.26

Draw We decompose the complex distributed load into three standard distributed loads to facilitate the analysis (**Figure 8.26**).

Formulate Equations and Solve For each line load we use (8.13) and (8.15) to find the equivalent total force and its location. Alternatively, we can recognize that each segment is a "standard distribution" and use the information in Appendix A3.1 for determining the total force and its location. We show both of these approaches. Finally, we use the conditions of equilibrium to determine the loads acting at A and B.

Approach 1: For each segment we use (8.13) to find the total force and (8.15) to find its location.

Segment 1 ($0 < x < 2$ m) (**Figure 8.27**): The distributed load ω_1 (in N/m) is described by

$$\omega_1 = \frac{-300 \text{ N/m}}{2 \text{ m}} x = -150 \frac{\text{N}}{\text{m}^2} x$$

(with x given in meters). Substituting ω_1 into (8.13) results in

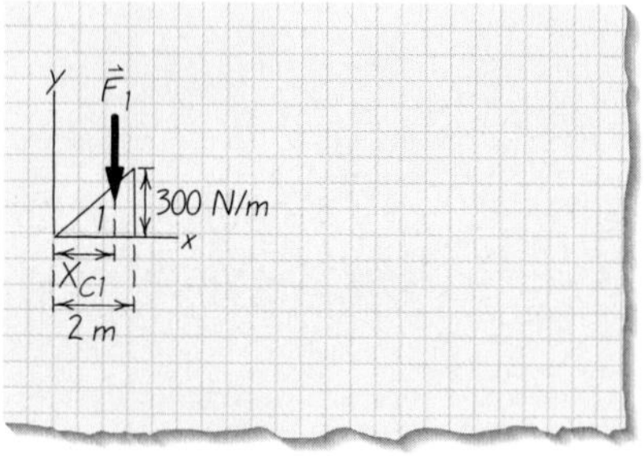

Figure 8.27 Segment 1 ($0 < x < 2$ m)

$$F_1 = \int_0^{2\text{ m}} \left(-150 \frac{\text{N}}{\text{m}^2} x\right) dx = \left. \frac{-150 \frac{\text{N}}{\text{m}^2} x^2}{2} \right]_0^{2\text{ m}} = -300 \text{ N}$$

$$F_1 = -300 \text{ N}$$

Substituting ω_1 into (8.15) gives:

$$X_{C1} = \frac{\int_0^{2\text{ m}} x\left(-150 \frac{\text{N}}{\text{m}^2} x\right) dx}{F_1} = \left. \frac{-150 \frac{\text{N}}{\text{m}^2} x^3}{3F_1} \right]_0^{2\text{ m}} = \left. \frac{-150 \frac{\text{N}}{\text{m}^2} x^3}{3(-300 \text{ N})} \right]_0^{2\text{ m}} = \frac{4}{3} \text{ m}$$

$$X_{C1} = \frac{4}{3} \text{ m}$$

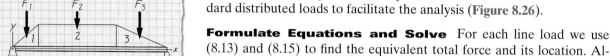

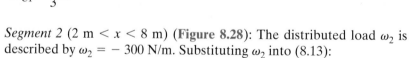

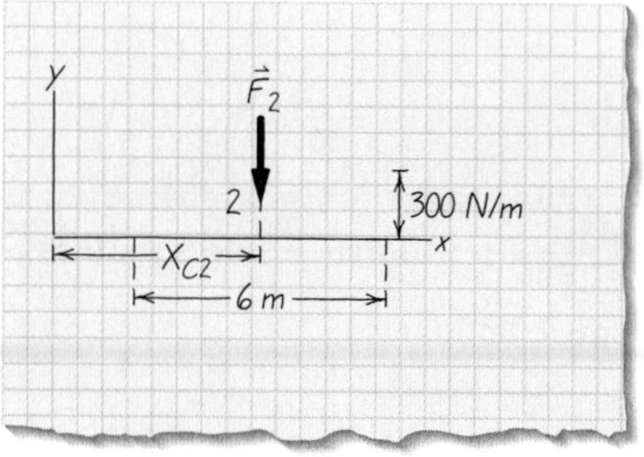

Figure 8.28 Segment 2 (2 m $< x <$ 8 m)

Segment 2 (2 m $< x <$ 8 m) (**Figure 8.28**): The distributed load ω_2 is described by $\omega_2 = -300$ N/m. Substituting ω_2 into (8.13):

$$F_2 = \int_{2\text{ m}}^{8\text{ m}} \left(-300 \frac{\text{N}}{\text{m}}\right) dx = -300 \frac{\text{N}}{\text{m}} x \Big]_{2\text{ m}}^{8\text{ m}} = -1800 \text{ N}$$

$$F_2 = -1800 \text{ N}$$

Substituting ω_2 into (8.15):

$$X_{C2} = \frac{\int_{2\text{ m}}^{8\text{ m}} x\left(-300 \frac{\text{N}}{\text{m}}\right) dx}{F_2} = \left. \frac{-300 \frac{\text{N}}{\text{m}} x^2}{2F_2} \right]_{2\text{m}}^{8\text{ m}} = \left. \frac{-300 \frac{\text{N}}{\text{m}} x^2}{2(-1800 \text{ N})} \right]_{2\text{ m}}^{8\text{ m}} = 5 \text{ m}$$

$$X_{C2} = 5 \text{ m}$$

Segment 3 (8 m < x < 12 m) (**Figure 8.29**): The distributed load ω_3 is described by

$$\omega_3 = -300\,\frac{\text{N}}{\text{m}} + \frac{300\,\frac{\text{N}}{\text{m}}}{4\text{ m}}(x-8) = -900\,\frac{\text{N}}{\text{m}} + 75\,\frac{\text{N}}{\text{m}^2}x$$

(with x given in meters). Substituting ω_3 into (8.13):

$$F_3 = \int_{8\text{ m}}^{12\text{ m}} \left(-900\,\frac{\text{N}}{\text{m}} + 75\,\frac{\text{N}}{\text{m}^2}x\right)dx = -900\,\frac{\text{N}}{\text{m}}x + \frac{75\,\frac{\text{N}}{\text{m}^2}x^2}{2}\Bigg]_{8\text{ m}}^{12\text{ m}}$$

$$F_3 = -600\text{ N}$$

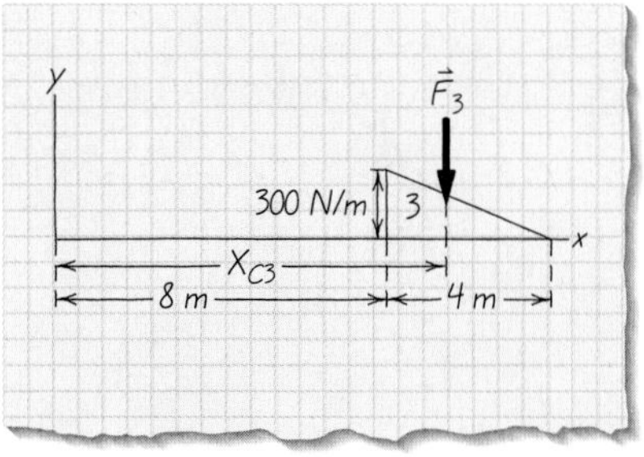

Figure 8.29 Segment 3 (8 m < x < 12 m)

Substituting ω_3 into (8.15):

$$X_{C3} = \frac{\int_{8\text{ m}}^{12\text{ m}} x\left(-900\,\frac{\text{N}}{\text{m}} + 75\frac{\text{N}}{\text{m}^2}x\right)dx}{F_3} = \frac{-900\frac{\text{N}}{\text{m}}\frac{x^2}{2} + 75\frac{\text{N}}{\text{m}^2}\frac{x^3}{3}}{-600\text{ N}}\Bigg]_{8\text{ m}}^{12\text{ m}} = \frac{28}{3}\text{ m}$$

$$X_{C3} = \frac{28}{3}\text{ m}$$

Answer $F_1 = -300\text{N}, X_{C1} = 4/3\text{m}$
$F_2 = -1800\text{N}, X_{C2} = 5\text{m}$
$F_3 = -600\text{N}, X_{C3} = 28/3\text{m}$

Approach 2: We find the forces and their locations for the three standard line loads using Appendix A3.1.

Segment 1: Triangular line load. The total force is the area of the triangle ($A = bh/2$), with $b = 2$ m and $h = -300$ N/m:

$$F_1 = (2\text{ m})(-300\text{ N/m})/2$$
$$F_1 = -300\text{ N}$$

The location of the total force is 2/3 of the distance from the triangle vertex, as shown in **Figure 8.30**.

$$X_{C1} = \frac{2}{3}(b) = \frac{2}{3}(2) = \frac{4}{3}\text{ m}$$

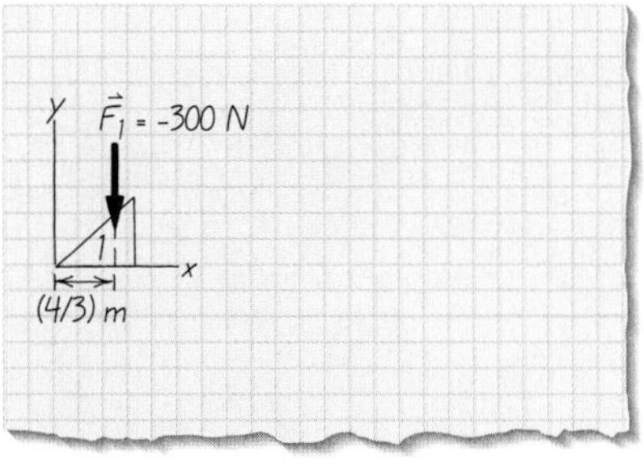

Figure 8.30 Segment 1: Triangular line load

Segment 2: Rectangular line load. The total force is the area of the rectangle ($A = bh$), with $b = 6$ m and $h = -300$ N/m:

$$F_2 = (6\text{ m})(-300\text{ N/m})$$
$$F_2 = -1800\text{ N}$$

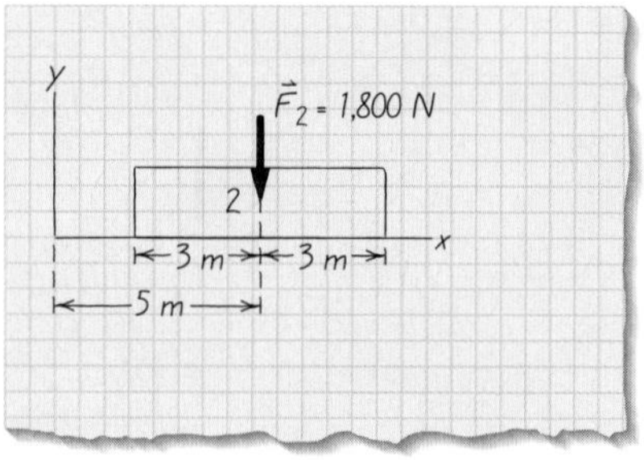

Figure 8.31 Segment 2: Rectangular line load

The location of the total force is 1/2 of the distance along edge b, as shown in **Figure 8.31**. We must include the 2 m from the left end of the beam to the edge of the rectangle when calculating the location of X_{C2}.

$$X_{C2} = 2\text{ m} + \frac{6}{2}\text{ m} = 5\text{ m}$$

Segment 3: Triangular line load. The total force is the area of the triangle:

$$F_3 = (4\text{ m})(-300\text{ N/m})/2$$
$$F_3 = -600\text{ N}$$

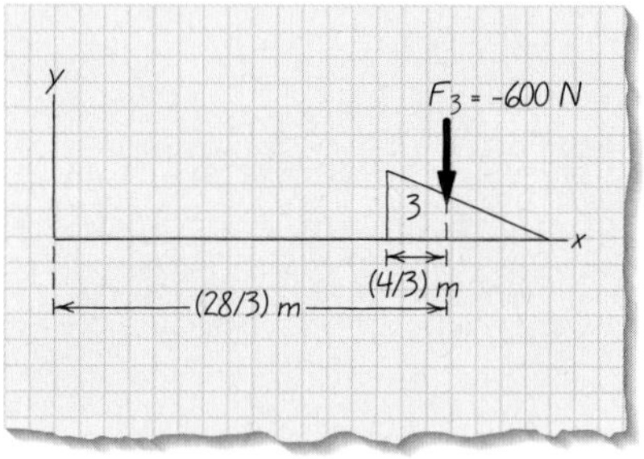

Figure 8.32 Segment 3: Triangular line load

The location of this force is the 8 m from the left end of the beam to the triangle plus 1/3 of the distance form the large side of the triangle, as shown in **Figure 8.32**.

$$X_{C3} = 8\text{ m} + \frac{1}{3}(4\text{ m}) = \frac{28}{3}\text{ m}$$

This alternate approach to finding the forces and their locations results in the same answer as we found using (8.13) and (8.15), and requires a much simpler set of calculations. Therefore, if it is reasonable to model a distributed load as one of the standard distributions in Appendix A3.1, by all means do so.

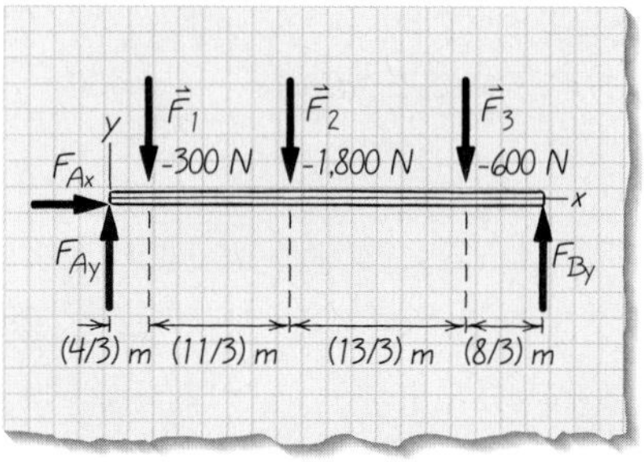

Figure 8.33 Free-body diagram of Beam AB

Draw To find the forces at A and B we start by drawing a free-body diagram (**Figure 8.33**) with the distributed load modeled by the equivalent total forces we calculated previously.

Formulate Equations and Solve When we set up the equations for planar equilibrium to find the unknown loads at A and B, we arbitrarily choose A as the moment center. We could choose B as the moment center and obtain the same result. We solve the equation for moment equilibrium before we solve the equation for equilibrium in the y direction because the moment equation has only one unknown while the y-force equilibrium equation has two unknowns.

Based on (7.5A):

$$\sum F_x = 0\ (\rightarrow +)$$

$$F_{Ax} = 0$$

$$\boxed{F_{Ax} = 0}$$

Based on (7.5C) with moment center at A:

$$\sum M_{z@A} = 0(\circlearrowleft +)$$
$$-(300\text{ N})(\tfrac{4}{3}\text{ m}) - (1800\text{ N})(5\text{ m}) - (600\text{ N})(\tfrac{28}{3}\text{ m}) + F_{By}(12\text{ m}) = 0$$
$$12F_{By} = 15{,}000$$

$$\boxed{F_{By} = 1250\text{ N}}$$

Based on (7.5B):

$$\sum F_y = 0\,(\uparrow\ +)$$

$$F_{Ay} - 300\text{ N} - 1800\text{ N} - 600\text{ N} + F_{By} = 0$$
$$F_{Ay} - 300\text{ N} - 1800\text{ N} - 600\text{ N} + 1250\text{ N} = 0$$

$$F_{Ay} = 1450\text{ N}$$

Answer
$F_{Ax} = 0$
$F_{Ay} = 1450$ N
$F_{By} = 1250$ N

Check To check our solution we use (7.5C) to sum moments about a different moment center. For example

$$\sum M_{z@B} = 0\,(\curvearrowleft +)$$
$$-F_{Ay}(12\text{ m}) + (300\text{ N})(\tfrac{32}{3}\text{ m}) + (1800\text{ N})(7\text{ m}) + (600\text{ N})(\tfrac{8}{3}\text{ m}) = 0$$
$$-(1450\text{ N})(12\text{ m}) + (300\text{ N})(\tfrac{32}{3}\text{ m}) + (1800\text{ N})(7\text{ m})$$
$$+ (600\text{ N})(\tfrac{8}{3}\text{ m}) = 0$$
$$0 = 0$$

Yes, the beam is in equilibrium.

EXAMPLE 8.8 SLANTED SURFACE WITH NONUNIFORM DISTRIBUTION

The inclined beam in Figure 8.34 is subjected to the vertical force distribution shown. The value of the load distribution at the right end of the beam is ω_0 (force units per horizontal length unit). Determine the loads acting on the beam at supports A and B.

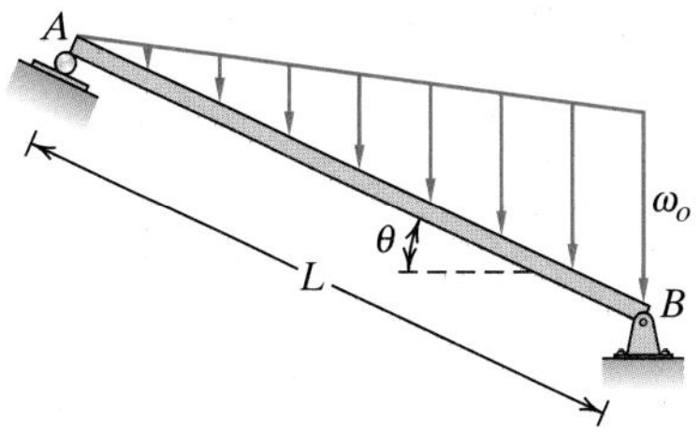

Figure 8.34

Goal Calculate the loads acting on the beam at A and B.

Given A beam of length L supported by a roller at A and a pin joint at B makes an angle θ with the horizontal. The distributed load oriented vertically increases linearly, from a value of zero at A to ω_0 at B.

Assume We assume that the system is planar and that the weight of the beam is negligible.

Draw A free-body diagram of the beam is shown in Figure 8.35. Notice that we have established two coordinate systems, one aligned with the horizontal and vertical (xy), the other along the slope of the inclined beam (x^*y^*).

Formulate Equations and Solve First we will find the total equivalent force and its location. Next we apply equilibrium conditions to find the loads acting at A and B. We use two approaches. In the first we use (8.13) and (8.15) to find the equivalent total force and its location. In the second approach we use the information in Appendix A3.1 for a standard line load distribution to determine the total equivalent force and its location.

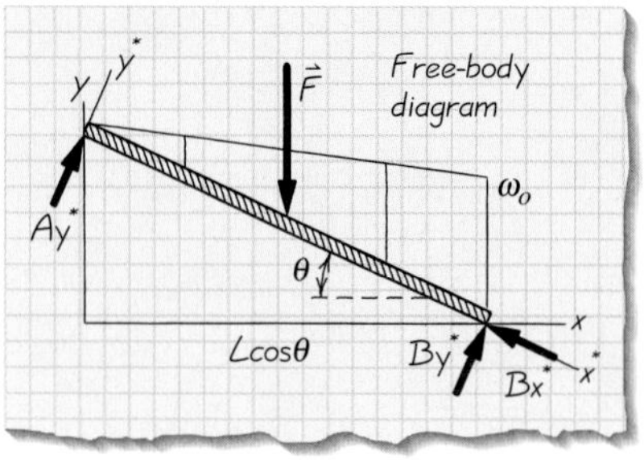

Figure 8.35

Approach 1: We begin by determining $\omega(x)$, the equation for the distributed load as a function of horizontal distance x.

Using a linear distribution with unknown constants k_o and k_1 to represent the load distribution:

$$\omega(x) = k_o + k_1 x$$

and applying the boundary conditions to determine the constants

$$\omega(0) = 0; \qquad \omega(L\cos\theta) = -\omega_o$$

results in

$$\omega(0) = k_o + k_1(0) = 0 \qquad k_o = 0$$
$$\omega(L\cos\theta) = k_o + k_1(L\cos\theta) = -\omega_o \qquad k_1 = -\omega_o/(L\cos\theta)$$

Therefore, $\omega(x) = \dfrac{\omega_o}{L\cos\theta}x$

Using (8.13) to find the total force,

$$F = \int_0^{L\cos\theta} \omega(x)\,dx = \int_0^{L\cos\theta} \frac{-\omega_o}{L\cos\theta}x\,dx = \frac{-\omega_o}{L\cos\theta}\frac{x^2}{2}\bigg]_0^{L\cos\theta}$$
$$= \frac{-\omega_o}{2}\frac{L^2\cos^2\theta}{L\cos\theta}$$
$$F = \frac{-\omega_o L\cos\theta}{2} \qquad (1)$$

Using (8.15) to find centroid,

$$X_C = \frac{\int_0^{L\cos\theta} \omega(x)x\,dx}{F} = \frac{\int_0^{L\cos\theta} \frac{-\omega_o}{L\cos\theta}x^2dx}{F}$$
$$= \frac{\frac{-\omega_o}{L\cos\theta}\frac{x^3}{3}\Big]_0^{L\cos\theta}}{F} = \frac{\frac{-\omega_o}{3}\frac{L^3\cos^3\theta}{L\cos\theta}}{F}$$
$$X_C = \frac{\frac{-\omega_o L^2\cos^2\theta}{3}}{\frac{-\omega_o L\cos\theta}{2}} = \frac{2}{3}L\cos\theta$$

Approach 2: Recognizing that the load distribution is triangular, we use Appendix A3.1 to find the total equivalent force and its location. Because the load distribution ω_0 is given in force units per horizontal length, we must use the horizontal length ($L\cos\theta$) as the base of our triangle.

$$F = \frac{bh}{2} = \frac{(L\cos\theta)(-\omega_o)}{2} = \frac{-\omega_o L\cos\theta}{2}$$

The centroid is located at 2/3 the distance from the vertex:

$$X_C = \frac{2}{3}L\cos\theta$$

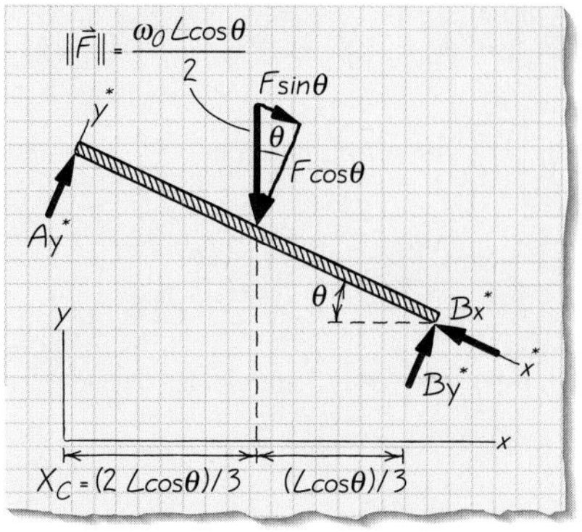

Figure 8.36

These values, the same as we calculated using the integral approach, are shown in **Figure 8.36**, which is also a free-body diagram of the beam.

We apply static equilibrium equations perpendicular and parallel to the beam to find the loads at supports A and B. To simplify the use of the planar equilibrium equations, we use the coordinate system that has its x* axis along the beam (see **Figure 8.36**).

Based on (7.5A):

$$\sum F_{x*} = 0$$

$$F\sin\theta - B_{x*} = 0$$

Substituting for F from (1):

$$B_{x*} = F\sin\theta = \left(\frac{\omega_o L\cos\theta}{2}\right)\sin\theta = \frac{\omega_o L\cos\theta\sin\theta}{2} \qquad (2)$$

Using B as the moment center, (7.5C) gives

$$\sum M_{z@B} = 0\ (\circlearrowleft)$$

$$-A_{y*}L + F\cos\theta\,(L/3) = 0$$

$$A_{y*} = \frac{F\cos\theta}{3} = \frac{\omega_o\,L\cos\theta}{2}\,\frac{\cos\theta}{3} = \frac{\omega_o\,L\cos^2\theta}{6} \qquad (3)$$

Based on (7.5B):

$$\sum F_{y*} = 0$$

$$A_{y*} + B_{y*} - F\cos\theta = 0$$

$$B_{y*} = F\cos\theta - A_{y*}$$

Substituting for F from (1) and for A_{y*} from (3) gives

$$B_{y*} = \frac{\omega_o L\cos\theta}{2}\cos\theta - \frac{\omega_o L\cos\theta}{6} = \frac{\omega_o L\cos^2\theta}{3} \qquad (4)$$

Answer
$A_{y*} = (\omega_o/6)L\cos^2\theta$
$B_{y*} = (\omega_o/3)L\cos^2\theta$
$B_{x*} = (\omega_o/2)L\cos\theta\sin\theta$

Check The solution can be checked by summing the moments about A. In addition, note that when the beam is horizontal, $\theta = 0°$ and B_{y*} is twice as big as A_{y*}; this is as expected since $\boldsymbol{F}$ is located two times closer to B than to A.

EXAMPLE 8.9 COMPLEX LINE LOAD DISTRIBUTION

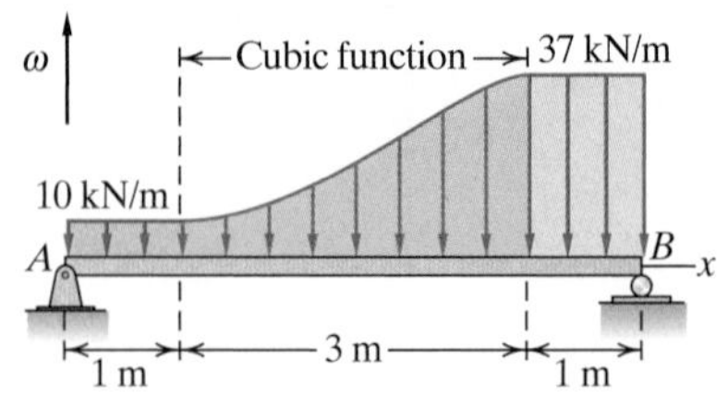

Cubic:

$\omega(x) = k_0 + k_1 + k_2x^2 + k_2x^3$

Slope = 0 at x = 1 m and x = 4 m

Figure 8.37

A beam is subjected to the complex load shown in Figure 8.37. The given cubic polynomial transitions between the 10-kN/m load and the 37-kN/m load. Determine the loads acting on the beam at supports A and B.

Goal We are to find the loads at supports A and B.

Given A beam of length L is supported by a pin joint at A and a roller at B. We are given the magnitude and shape of the load distribution on the beam.

Assume We assume that the system is planar, the weight of the beam is negligible, and a positive load acts upward.

Draw We decompose the distributed load into three segments representing the two uniform loads and the cubic load (Figure 8.38).

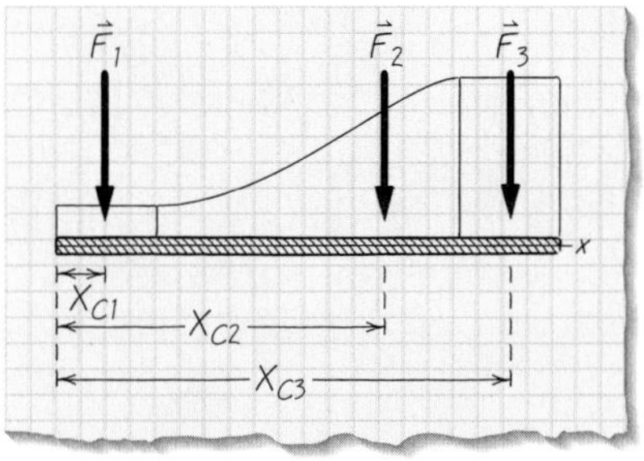

Figure 8.38

Formulate Equations and Solve

Segment 1 (rectangle) (Figure 8.39):

$$F_1 = (1\ \text{m})(-10\ \text{kN/m})$$
$$F_1 = -10.0\ \text{kN}$$

The point at which the total force F_1 is applied is

$$X_{C1} = \frac{1\ \text{m}}{2} = 0.50\ \text{m}$$

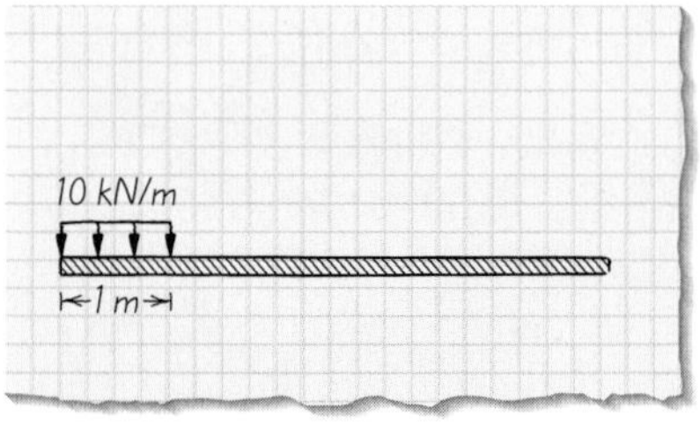

Figure 8.39 Segment 1 (rectangular)

Segment 2 (cubic line load) (Figure 8.40): We first solve for the coefficients of the cubic polynomial using the boundary conditions given for the distributed load:

$$\text{At } x = 1\ \text{m} \rightarrow \omega = -10\ \text{kN/m} \quad \text{and} \quad \frac{d\omega}{dx} = 0$$

$$\text{At } x = 4\ \text{m} \rightarrow \omega = -37\ \text{kN/m} \quad \text{and} \quad \frac{d\omega}{dx} = 0$$

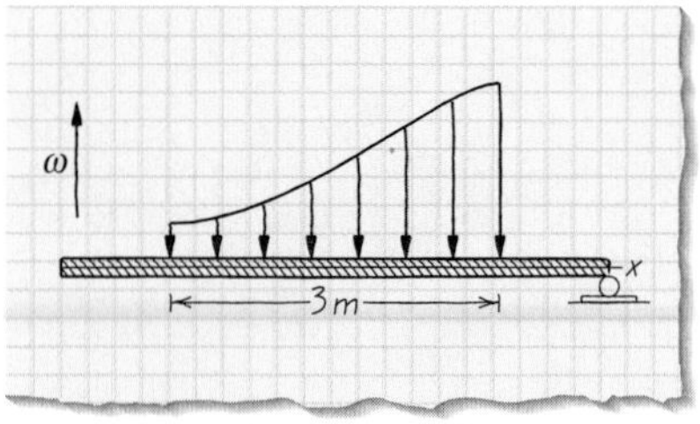

Figure 8.40 Segment 2 (cubic)

Plug the conditions into the polynomial and its first derivative:

$$\omega(x) = k_0 + k_1x + k_2x^2 + k_3x^3$$
$$\frac{d\omega}{dx}(x) = k_1 + 2k_2x + 3k_3x^2$$
$$\text{At } x = 1\ \text{m} \rightarrow \omega(1) = -10\ \text{kN/m}$$
$$\frac{d\omega}{dx}(1) = 0$$
$$k_o + k_1 + k_2 + k_3 = -10 \qquad (1)$$
$$k_1 + 2k_2 + 3k_3 = 0 \qquad (2)$$

At $x = 4\text{ m} \rightarrow \omega(4) = -37\text{ kN/m}$

$$\frac{d\omega}{dx}(4) = 0$$

$$k_0 + 4k_1 + 16k_2 + 64k_3 = -37 \qquad (3)$$

$$k_1 + 8k_2 + 48k_3 = 0 \qquad (4)$$

Write equations (1), (2), (3), and (4) in matrix form as a system of four equations and four unknowns, then solve for the four unknowns:

$$\boldsymbol{Ak} = \boldsymbol{b}$$

where $\boldsymbol{k}$ is a vector of our four unknown constants

$$\begin{bmatrix} 1 & 1 & 1 & 1 \\ 0 & 1 & 2 & 3 \\ 1 & 4 & 16 & 64 \\ 0 & 1 & 8 & 48 \end{bmatrix} \begin{bmatrix} k_0 \\ k_1 \\ k_2 \\ k_3 \end{bmatrix} = \begin{bmatrix} -10 \\ 0 \\ -37 \\ 0 \end{bmatrix}$$

Solving the system by inverting the A matrix, we get

$$\boldsymbol{k} = \boldsymbol{A}^{-1}\boldsymbol{b}$$

$$\boldsymbol{k} = \begin{bmatrix} k_0 \\ k_1 \\ k_2 \\ k_3 \end{bmatrix} = \begin{bmatrix} -21 \\ 24 \\ -15 \\ 2 \end{bmatrix}$$

Therefore, for segment 2:

$$\omega(x) = -21 + 24x - 15x^2 + 2x^3$$

(x in meters and $\omega(x)$ in kN/m). We now calculate the total equivalent force for segment 2 and its point of action along the beam using (8.13) and (8.15):

$$F_2 = \int_{1\text{m}}^{4\text{m}} \omega(x)dx = \int_{1\text{m}}^{4\text{m}} (-21 + 24x - 15x^2 + 2x^3)\, dx$$

$$F_2 = \left[-21x + 12x^2 - 5x^3 + \frac{x^4}{2}\right]_{1\text{m}}^{4\text{m}} = -70.5\text{ kN}$$

$$F_2 = -70.5\text{ kN}$$

$$X_{C2} = \frac{\int_{1\text{m}}^{4\text{m}} x\omega(x)\, dx}{\int_{1\text{m}}^{4\text{m}} \omega(x)\, dx} = \frac{\int_{1\text{m}}^{4\text{m}} x(-21 + 24x - 15x^2 + 2x^3)\, dx}{-70.5}$$

$$X_{C2} = \frac{\left.\frac{-21}{2}x^2 + 8x^3 - \frac{15}{4}x^4 + \frac{2}{5}x^5\right]_{1\text{m}}^{4\text{m}}}{-70.5 \text{ kN}}$$

$$= \frac{-200.55 \text{ kN}\cdot\text{m}}{-70.5 \text{ kN}}$$

$$X_{C2} = 2.85 \text{ m}$$

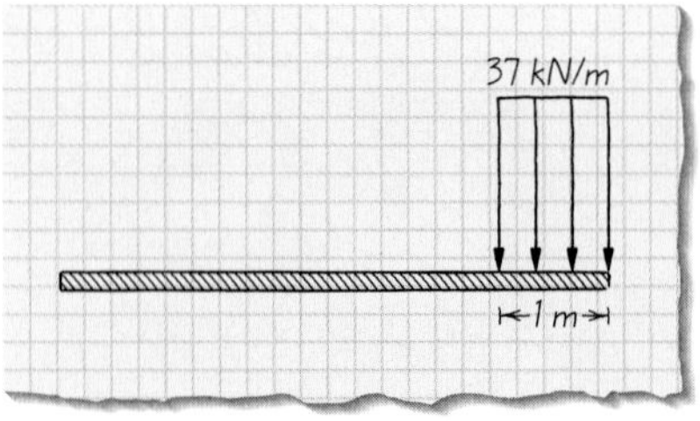

Figure 8.41 Segment 3 (rectangular)

Segment 3 (rectangle) (**Figure 8.41**):

$$F_3 = (1 \text{ m})(-37 \text{ kN/m})$$
$$F_3 = -37.0 \text{ kN}$$

The point of application is the distance from A to the start of the uniform load, plus the distance to the centroid of the rectangle.

$$X_{C3} = 4 \text{ m} + 0.5 \text{ m} = 4.5 \text{ m}$$

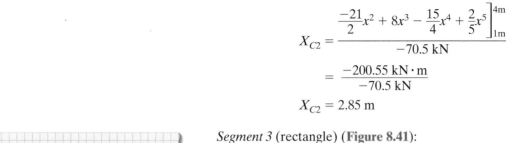

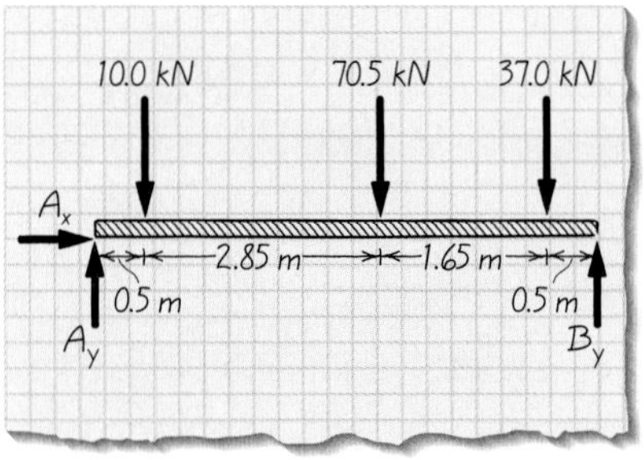

Figure 8.42 Free-body diagram of Beam AB

We now draw a free-body diagram of the beam with the three total equivalent loads so that we have a clear picture of the dimensions to use in the equilibrium equations (7.5) (**Figure 8.42**). From (7.5A)

$$\Sigma F_x = 0\ (\rightarrow +)$$
$$A_x = 0$$

From (7.5C) with MC at A

$$\sum M_{z@A} = 0\ (\curvearrowleft +)$$
$$B_y(5) - 10 \text{ kN}(0.5 \text{ m}) - 70.5 \text{ kN}(2.85 \text{ m}) - 37 \text{ kN}(4.5 \text{ m}) = 0$$
$$B_y = \frac{[(10)(0.5) + (70.5)(2.85) + (37)(4.5)]}{5} = \frac{372.4}{5}$$
$$B_y = 74.5 \text{ kN}$$

And from (7.5B)

$$\sum F_y = 0\ (\uparrow +)$$
$$A_y + B_y - 10 \text{ kN} - 70.5 \text{ kN} - 37 \text{ kN} = 0$$
$$A_y = 117.5 - 74.5$$
$$A_y = 117.5 - 74.41 = 43.0 \text{ kN}$$

Answer $A_y = 43.0$ kN
$B_y = 74.5$ kN

Check We check the results by summing the moments about B.

$$\sum M_{z@B} = 0\ (\curvearrowleft +)$$
$$-43.0 \text{ kN } (5 \text{ m}) + 10.0 \text{ kN } (4.5 \text{ m})$$
$$+ 70.5 \text{ kN } (2.15 \text{ m}) + 37.0 \text{ kN } (0.5 \text{ m}) = 0.075 \text{ kN/m}$$

We note that the moments don't sum to exactly zero; instead, there is a small residual of 0.075 kN/m. This is due to round-off error because we carried only three significant digits in our solution. This residual is quite small, about 0.1% of the moments we are calculating, so we can accept our solution as correct.

While we could carry more significant digits and reduce the residual, it is probably not justified. In most engineering applications there are many sources of uncertainty and error such as the magnitude of the loads, the properties of the materials, and the exact dimensions of the system; thus we tolerate some inaccuracy in our solutions. Acceptable tolerance levels are often specified by the system designer. Tolerances for the space shuttle would be much smaller than those for a desk chair.

EXAMPLE 8.10 BEAM WITH MULTIPLE LINE LOADS

A beam is subjected to the loads represented by the load diagram in **Figure 8.43**. Determine the total force acting on the beam and locate its line of action with respect to support A.

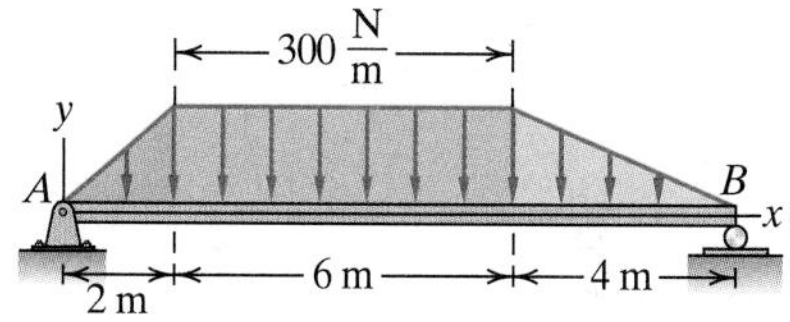

Figure 8.43

Goal Calculate the total equivalent force acting on the beam and determine the point at which it acts.

Given The 12-m beam is supported by a pin joint at A and a roller at B. We are given information about the geometry of the distributed load.

Assume We assume that the system is planar, the weight of the beam is negligible, and the upward force is positive.

Draw From Example 8.7 we have the magnitudes and locations of the total equivalent loads for the three different segments (**Figure 8.44*a***).

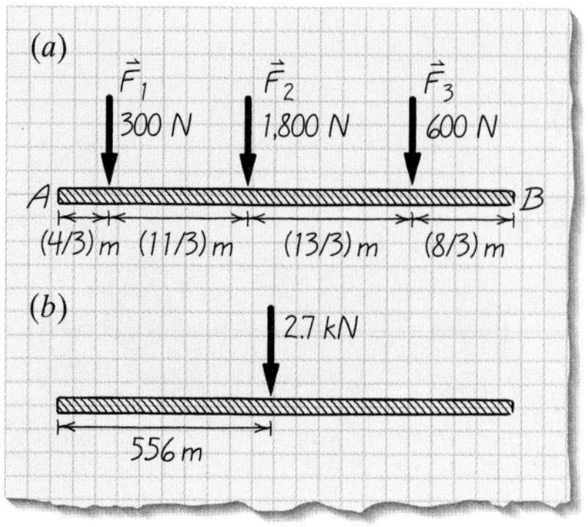

Figure 8.44

Formulate Equations and Solve For each standard line load distribution, we could use (8.13) and (8.15) to find the total equivalent force and its location. Alternately we could use the area information in Appendix A3.1 to find the equivalent force and its location. This step was completed in Example 8.7. We use the results from Example 8.7 in (8.16A) and (8.16B) to find the total equivalent force and its location for this complex load distribution. From **Figure 8.44*a***,

$$F_1 = -300\text{ N},\quad X_{C1} = 4/3\text{ m}$$
$$F_2 = -1800\text{ N},\quad X_{C2} = 5\text{ m}$$
$$F_3 = -600\text{ N},\quad X_{C3} = 28/3\text{ m}$$

We use these values in (8.16A) and (8.16B) to find F_{tot} and X_C:

$$F_{\text{total}} = \sum_{i=1}^{3} F_i = -300\text{ N} - 1800\text{ N} - 600\text{ N} = -2700\text{ N}$$

$$X_C = \frac{\sum_{i=1}^{3} F_i X_{Ci}}{F_{\text{total}}} = \frac{-300\text{ N}\left(\frac{4}{3}\text{ m}\right) - 1800\text{ N}(5\text{ m}) - 600\text{ N}\left(\frac{28}{3}\text{ m}\right)}{-2700\text{ N}}$$

$$= 5.56\text{ m}$$

$$X_C = 5.56\text{ m}$$

Answer $F_{\text{total}} = 2.7\text{ kN}$
$X_C = 5.56\text{ m}$

Check The equivalent total force is shown in **Figure 8.44*b***. It is reasonable that the location of the equivalent load should be in the left half of the beam. The load is not symmetric with respect to the centerline of the beam, and a larger portion of the load sits on the left half.

We could check our answer by calculating the centroid of the distribution measured with respect to *B*.

Distributed Force Over an Area and Its Centroid

Consider a piece of plywood stacked with 32 bricks over an area measuring 0.80 m by 0.40 m, as shown in **Figure 8.45*a***. If each brick weighs 17 N, we can represent the weight of the bricks acting on the plywood as a 544-N force distributed over the 0.32-m^2 (= 0.8 m × 0.4 m) area (**Figure 8.45*b***), so that the distributed force is 544 N/0.32 m^2 = 1700 N/m^2. This is referred to as a **uniform pressure load**, where uniform means uniformly distributed.

To represent this pressure load by a single point force, we first need to find the *total force*. We place this total force as a point force at the *location* where the total moment it creates on the system is the same as the moment created by the pressure load. For the uniform pressure load under consideration, the total load is 544 N, and it would be placed, by inspection, as a point load at the center of the distribution (**Figure 8.45*c***). We could figure this out mathematically by finding first the total force represented by the pressure load and then the location at which the total force acts.

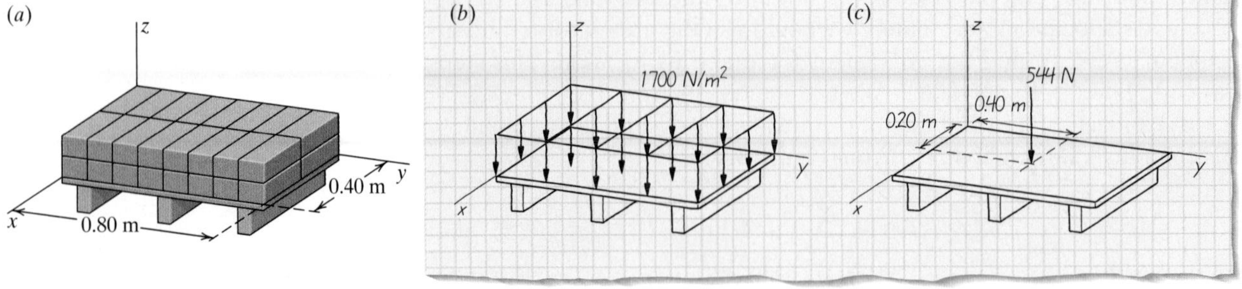

Figure 8.45 (*a*) Thirty-two bricks stacks over an area; (*b*) bricks represented as pressure applied to the area; (*c*) pressure represented as a single point force

1. *Total force.* This force is found by integrating the load over the surface area:

$$\text{Total force in } z \text{ direction} = F_z = \int_{\substack{\text{surface}\\ \text{area}}} p\,dx\,dy \qquad (8.17)$$

where p is the pressure load in dimensions of force/area, the integrand $p\,dx\,dy$ is the force acting on the small area $dx\,dy$, and "surface area" in the limit refers to the total area over which the pressure acts (**Figure 8.46*a***). Notice that a double integral is needed because we are integrating over an area.

2. *Location.* This location at (X_C, Y_C) is called the **center of pressure** (or pressure center) and is found by requiring that the moment the total force produces is the same as the moment produced by the pressure load. The moment will have a component about the x axis and one about the y axis:

Moment about the x axis:

$$\underbrace{Y_C F_z}_{\substack{\text{moment}\\ \text{about } x \text{ axis}\\ \text{produced by}\\ \text{total } z \text{ force}}} = \underbrace{\int_{\substack{\text{surface}\\ \text{area}}} y(p\,dx\,dy)}_{\substack{\text{total moment about}\\ x \text{ axis produced}\\ \text{by pressure load}}} \qquad (8.18\text{A})$$

where Y_C is the distance in the y direction to the moment center and the integrand represents the sum of the infinitesimal moments about the x axis created by the infinitesimal forces $p\,dx\,dy$, as illustrated in **Figure 8.46*b***.

Moment about the y axis:

$$\underbrace{X_C F_z}_{\substack{\text{moment}\\ \text{about } y \text{ axis}\\ \text{produced by}\\ \text{total } z \text{ force}}} = \underbrace{\int_{\substack{\text{surface}\\ \text{area}}} x(p\,dx\,dy)}_{\substack{\text{total moment about}\\ y \text{ axis produced}\\ \text{by pressure load}}} \qquad (8.18\text{B})$$

where X_C is the distance in the x direction to the moment center and the integrand represents the sum of the infinitesimal moments about the y axis created by the infinitesimal forces $p\,dx\,dy$, as illustrated in **Figure 8.46*b***.

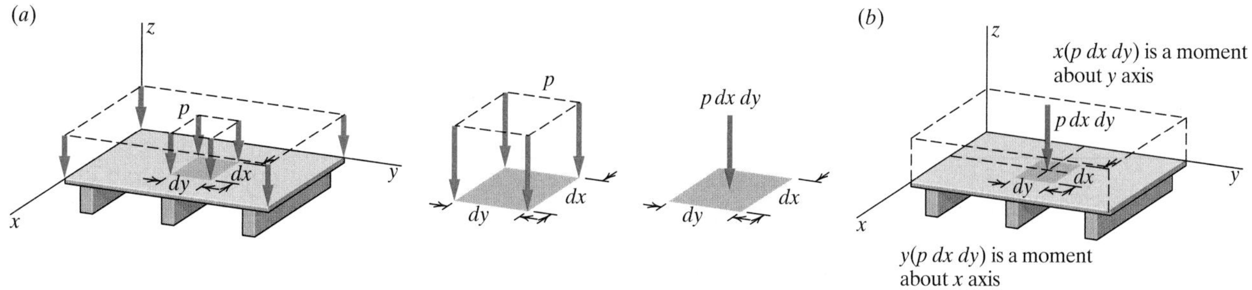

Figure 8.46 (*a*) Force over area $dx\,dy$ represented as pressure (p) or force $p\,dx\,dy$; (*b*) force over area $dx\,dy$ creates moments about x and y axes.

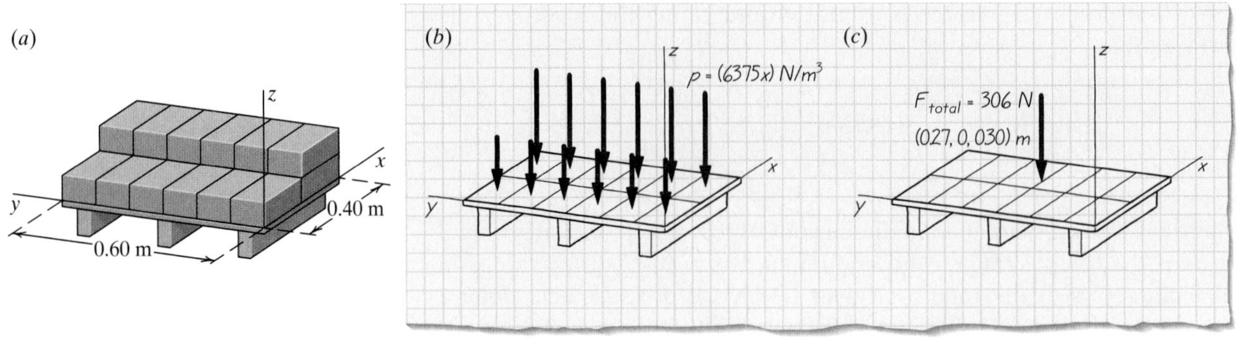

Figure 8.47 (*a*) Triangular brick stacking over area; (*b*) pressure distribution representing triangular brick stacking; (*c*) single point force that represents triangular stacking

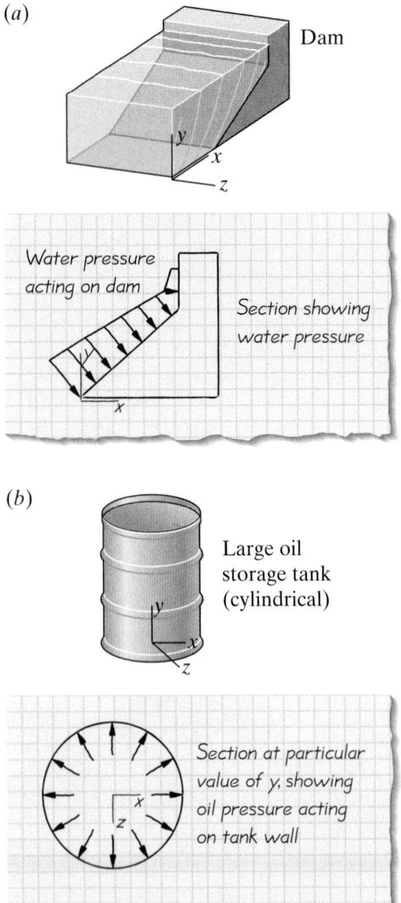

Figure 8.48 Examples of pressure acting on areas not aligned with *xyz* coordinate axes

Equations (8.18A) and (8.18B) can be solved for Y_C and X_C, respectively.

$$Y_C = \frac{\displaystyle\int_{\substack{\text{surface}\\\text{area}}} y\,p\,dx\,dy}{F_z} \underbrace{=}_{\substack{\text{substituting in}\\\text{from (8.17)}}} \frac{\displaystyle\int_{\substack{\text{surface}\\\text{area}}} y\,p\,dx\,dy}{\displaystyle\int_{\substack{\text{surface}\\\text{area}}} p\,dx\,dy} \qquad (8.19A)$$

$$X_C = \frac{\displaystyle\int_{\substack{\text{surface}\\\text{area}}} x\,p\,dx\,dy}{F_z} \underbrace{=}_{\substack{\text{substituting in}\\\text{from (8.17)}}} \frac{\displaystyle\int_{\substack{\text{surface}\\\text{area}}} x\,p\,dx\,dy}{\displaystyle\int_{\substack{\text{surface}\\\text{area}}} p\,dx\,dy} \qquad (8.19B)$$

The coordinates (X_C, Y_C) found with (8.19A) and (8.19B) are the location of the center of pressure.

Equations (8.17) and (8.19) are also valid for a nonuniform pressure load. Consider the brick stacking shown in Figure 8.47*a*. This stacking can be approximated by a linear distribution of the form $p = \beta_2 x$, where β_2 equals 6375 N/m^3 with the $x = 0$, $y = 0$ location as indicated in **Figure 8.47*b***. If this pressure distribution is inserted into (8.17) and (8.18), we find that the total load of the 18 bricks is 306 N and the point at which this total force acts is $X_C = 0.27$ m, $Y_C = 0.30$ m (see **Figure 8.47*c***).

The examples in **Figures 8.45** and **8.47** have the pressure aligned with one of the coordinate axes. Situations in which this is not the case are illustrated in Figure 8.48. To find the equivalent point force and its location, we must carefully select the orientation of the coordinate system.

Finding the location of the center of pressure can be simplified if basic geometric shapes are involved. If, for example, a uniform pressure acts over a standard rectangular or triangular area, the information in

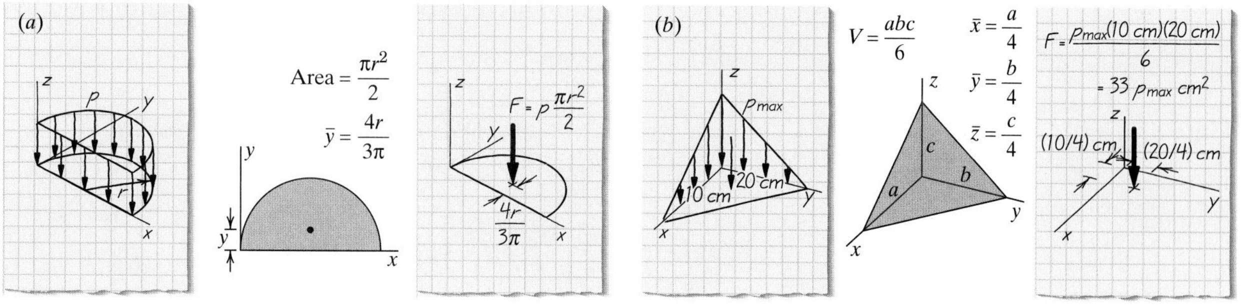

(Appendix A.2) (Appendix A.3)

Figure 8.49 (*a*) A uniform pressure *p* acting over a semicircular area; (*b*) nonuniform pressure represented as single point force

Appendix A3.1 on area and centroid is useful (**Figure 8.49*a***). The total force is simply the product of the magnitude of the uniform pressure and the area, and is located at the centroid of the area. If, on the other hand, a nonuniform pressure forms a standard volume, the information in Appendix A3.2 on volume and centroid is useful (**Figure 8.49*b***).

If we can decompose a composite pressure distribution into one made up of *N* standard distributions, we can use knowledge of the location of pressure centers of the *N* standard pressure distributions to find the pressure center of the composite pressure distribution. This works for both uniform pressure distributions and for nonuniform pressure distributions.

EXAMPLE 8.11 CALCULATING CENTER OF PRESSURE OF A COMPLEX PRESSURE DISTRIBUTION

The area shown in **Figure 8.50** is bounded by the *y* and *x* axes and the line $x = 4 - 2y$. The area is acted upon by a pressure distribution that varies linearly from the origin of the axes to a maximum value of p_o at $x = 4 - 2y$. Find the total equivalent force F_T and the location at which it acts.

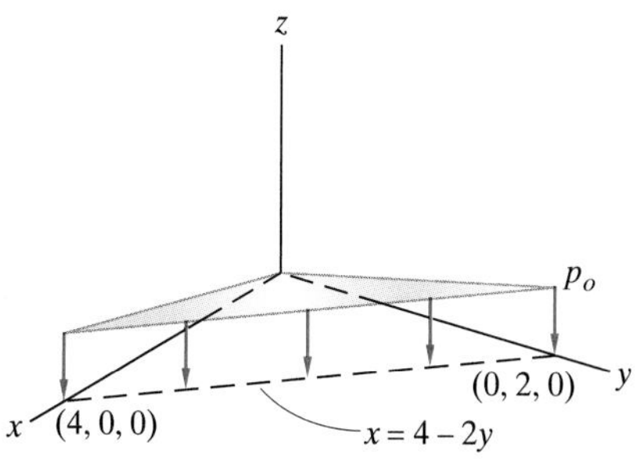

Figure 8.50

Goal We are to find the total equivalent force acting on the area and the center of pressure.

Given We are told that a linearly varying pressure distribution acts on an area bounded by the *y* axis, the *x* axis, and the line $x = 4 - 2y$. The maximum value of the pressure is p_o, along the line $x = 4 - 2y$. The length dimension in **Figure 8.50** is meter, and pressure is given in N/m^2.

Formulate Equations and Solve We use (8.17) to find the total force. We describe the pressure distribution by the linear equation

$$p(x, y) = p_o\left(\frac{y}{2} + \frac{x}{4}\right)$$

First integrating with respect to x and then with respect to y we find the total force.

$$F_T = \int_{\text{surface area}} p(x, y)\, dx\, dy = \int_0^2 \int_0^{4-2y} p_o\left(\frac{y}{2} + \frac{x}{4}\right) dx\, dy$$

$$F_T = p_o \int_0^2 \left[\frac{xy}{2} + \frac{x^2}{8}\right]_0^{4-2y} dy = p_o \int_0^2 \left(2 - \frac{y^2}{2}\right) dy = p_o\left[2y - \frac{y^3}{6}\right]_0^2 = \frac{8p_o}{3}$$

We use (8.19A) and (8.19B) to find the centroid of the pressure distribution. Based on (8.19A) we write

$$Y_C = \frac{\int_{\text{surface area}} y\, p(x, y)\, dx\, dy}{F_T} = \frac{\int_0^2 \int_0^{4-2y} p_o y\left(\frac{y}{2} + \frac{x}{4}\right) dx\, dy}{F_T}$$

$$Y_C = \frac{p_o}{F_T} \int_0^2 \left[\frac{y^2 x}{2} + \frac{yx^2}{8}\right]_0^{4-2y} dy = \frac{p_o}{F_T} \int_0^2 \left(\frac{-y^3}{2} + 2y\right) dy$$

$$= \frac{p_o}{F_T}\left[\frac{-y^4}{8} + y^2\right]_0^2 = \frac{2p_o}{F_T}$$

$$Y_C = \frac{2p_o}{8p_o/3} = \frac{3}{4}$$

Based on (8.19 B) we write

$$X_C = \frac{\iint_{\text{surface area}} xp\,(y, x)\, dx\, dy}{F_T} = \frac{\int_0^2 \int_0^{4-2y} p_o x\left(\frac{y}{2} + \frac{x}{4}\right) dx\, dy}{F_T}$$

$$X_C = \frac{p_o}{F_T} \int_0^2 \left[\frac{yx^2}{4} + \frac{x^3}{12}\right]_0^{4-2y} dy$$

$$X_C = \frac{p_o}{F_T} \int_0^2 \left(\frac{y^3}{3} - 4y + \frac{16}{3}\right) dy = \frac{p_o}{F_T}\left[\frac{y^4}{12} - 2y^2 + \frac{16y}{3}\right]_0^2 = \frac{4p_o}{F_T}$$

$$X_C = \frac{4p_o}{8p_o/3} = \frac{3}{2}$$

Answer $F_T = 8p_o/3$; this is the magnitude of the force that is equivalent to the pressure load.

To represent this as a vector, we write:

$$\boldsymbol{F_T} = \frac{-8p_o}{3}\ \text{N/m}^2\ \boldsymbol{k}$$

$Y_C = (3/4)$ m

$X_C = (3/2)$ m

Check We check to see whether our results seem reasonable by comparing them to those for a constant pressure p_o acting over the triangular area, which we can quickly calculate from Appendix A3.1 or A3.2. If the pressure were constant over the triangular area, the total force would be $4p_o$ and we know that the total force for the linearly varying pressure must be smaller. The total load of $8p_o/3$ seems reasonable for our linearly varying pressure, since it is a little more than half of $4p_o$.

For the constant pressure p_o the center of pressure is located at (x = 4/3, y = 2/3). It seems reasonable that the center of pressure for the linearly varying pressure (x = 3/2, y = 3/4) would be farther from the origin of the coordinate system, since the load is distributed so that a larger portion of the load is away from the origin.

EXAMPLE 8.12 RECTANGULAR WATER GATE

Figure 8.51 shows a cross section through a rectangular gate that is h = 30 ft high and w = 8 ft wide (where the width dimension is perpendicular to the plane of the page). The gate is subjected to a load generated by fresh water currently stored to a depth of d = 25 ft in a reservoir behind the gate. The pressure that the water applies to the gate varies in a linear manner, as indicated in **Figure 8.52**. Determine the magnitude of the total force $\boldsymbol{R}$ exerted on the gate by the water and its location with respect to the hinge.

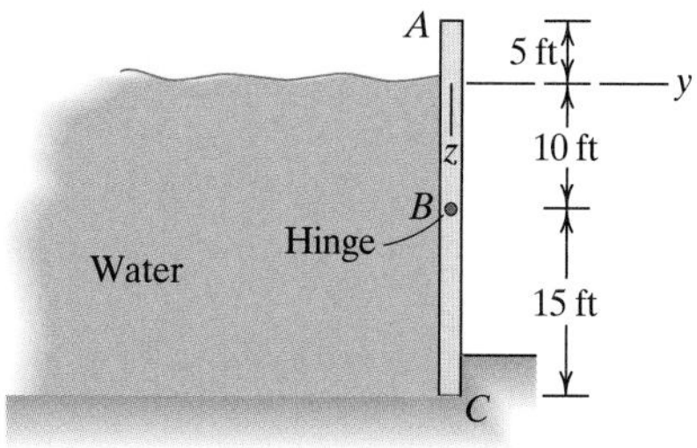

Figure 8.51

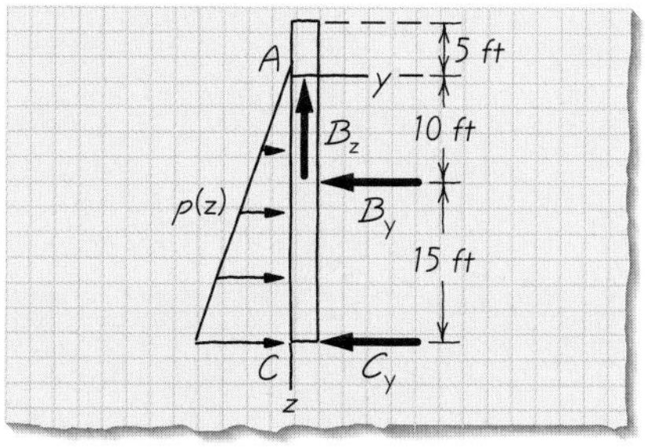

Figure 8.52

Goal We are to find the magnitude of the total force of the water on the gate and the location of the centroid of the pressure distribution with respect to the hinge.

Given We are given information about the geometry of the rectangular water gate and the depth of the water.

Assume We assume that the weight of the gate is negligible, the hinge is frictionless, the fluid is static, and the system is in equilibrium.

Draw Based on the information given in the problem and our assumptions, we draw a free-body diagram (**Figure 8.52**) of the gate. In drawing our free-body diagram we have selected positive z acting downward and locate the origin of the coordinate system at the top of the water. The x axis is pointing into the page. We have also added labels to the top and bottom of the pressure distribution and to the hinge. B_y and B_z are the forces of the hinge acting on the gate, and C_y is the force due to the sill pushing on the bottom of the gate.

Formulate Equations and Solve We determine the pressure distribution to vary according

$$p(z) = \gamma_{\text{wat}} z$$

The specific weight of water, ρg = 62.4 lb/ft^3. Therefore, $p(z)$ = 62.4 z (lb/ft^3).

From (8.17), the total force of the water acting on the gate is

$$R = \int_0^{25\text{ ft}}\int_0^{8\text{ ft}} p(z)dx\,dz = \int_0^{25\text{ ft}}\int_0^{8\text{ ft}} \left(62.4\,\frac{\text{lb}}{\text{ft}^3}\right) z\,dx\,dz = \int_0^{25\text{ ft}} \left(62.4\,\frac{\text{lb}}{\text{ft}^3}\right)[xz]_0^{8\text{ft}}\,dz$$

$$R = \left(62.4\,\frac{\text{lb}}{\text{ft}^3}\right)(8\text{ ft}) \int_0^{25\text{ ft}} z\,dz = \left(499.2\,\frac{\text{lb}}{\text{ft}^2}\right)\left[\frac{z^2}{2}\right]_0^{25\text{ ft}} = 156{,}000\text{ lb}$$

$R = 156$ kip ("kip" stands for kilo-pounds)

We now calculate Z_C, the distance to center, by rewriting (8.19B) to be in terms of z, instead of y.

$$Z_C = \frac{\int_0^{25\text{ ft}}\int_0^{8\text{ ft}} zp(z)\,dx\,dz}{R} = \frac{\int_0^{25\text{ ft}}\int_0^{8\text{ ft}} \left(62.4\,\frac{\text{lb}}{\text{ft}^3}\right) z^2\,dx\,dz}{R}$$

$$Z_C = \frac{\left(62.4\,\frac{\text{lb}}{\text{ft}^3}\right)(8\text{ ft}) \int_0^{25\text{ ft}} z^2\,dz}{R} = \frac{\left(499.2\,\frac{\text{lb}}{\text{ft}^2}\right)\left[\frac{z^3}{3}\right]_0^{25\text{ ft}}}{\text{R}}$$

$$= \frac{2{,}600{,}000\text{ lb}\cdot\text{ft}}{156{,}000\text{ lb}} = 16.67\text{ ft}$$

Z_C is measured with respect to the top of the water.

We find Z_{hinge} (the distance between the hinge and the centroid) using

$$Z_{\text{hinge}} = Z_C - \text{distance from top of water to hinge}$$
$$= 16.67\text{ ft} - 10\text{ ft} = 6.67\text{ ft}$$

Therefore, $Z_{\text{hinge}} = 6.67$ ft.

Answer $R = 156$ kip (acting in the y-direction, as shown in **Figure 8.52**), $Z_{\text{hinge}} = 6.67$ ft

Check We can use Appendix A3.1 to check our results because the pressure distribution can be modeled as a standard line load distribution. We multiply the pressure distribution by the width of the gate to calculate the force per unit length along the height of the gate.

$$\omega_{\text{max}} = (62.4\text{ lb/ft}^3)(25\text{ ft})(8\text{ ft}) = 12{,}480\text{ lb/ft}$$

Based on the triangular distribution in Appendix A3.1, we find

$$R = \frac{(25\text{ ft})(12{,}480\text{ lb/ft})}{2} = 156{,}000\text{ lb}$$

$$Z_C = \frac{2(25\text{ ft})}{3} = 16.67\text{ ft}$$

$$Z_{\text{hinge}} = 16.67\text{ ft} - 10\text{ ft} = 6.67\text{ ft}$$

Yes, our answer checks!

EXERCISES 8.2

8.2.1. A line load acts on the top of beam AB, as shown in E8.2.1. Use integration to determine the point force and its location (centroid) that are equivalent to the line load.

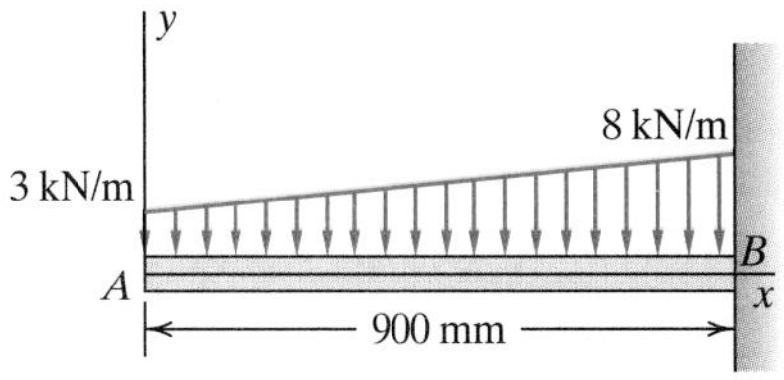

E8.2.1

8.2.2. A line load acts on the top of beam AB, as shown in E8.2.2. Use integration to determine the point load and its location (centroid) that are equivalent to the line load.

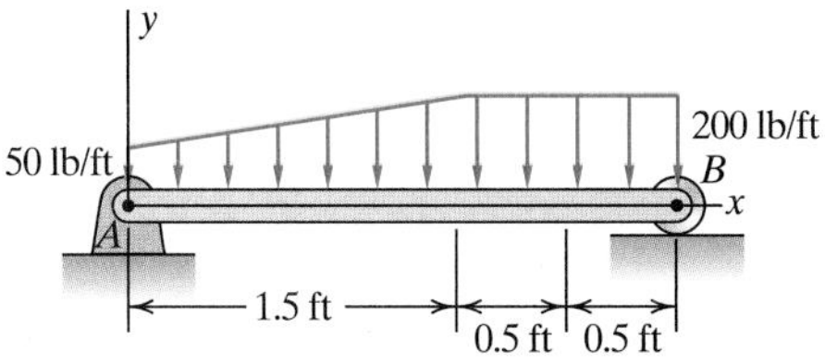

E8.2.2

8.2.3. Use integration to determine the point load and its location (centroid) that are equivalent to the line load in E8.2.3.

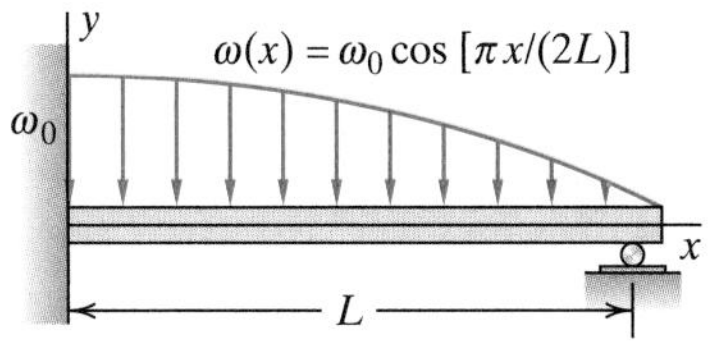

E8.2.3

8.2.4. Use integration to determine the point load and its location (centroid) that are equivalent to the line load in E8.2.4.

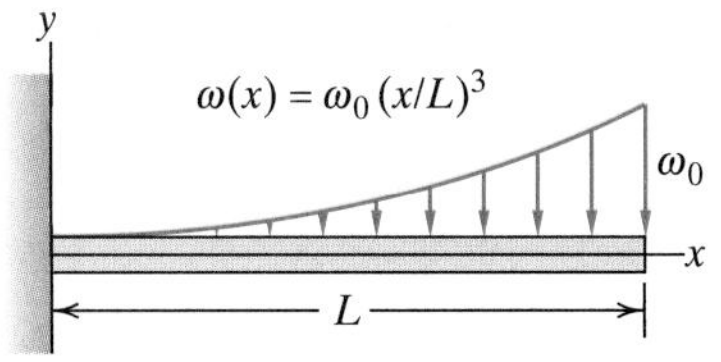

E8.2.4

8.2.5. Use integration to determine the point load and its location (centroid) that are equivalent to the line load in E8.2.5.

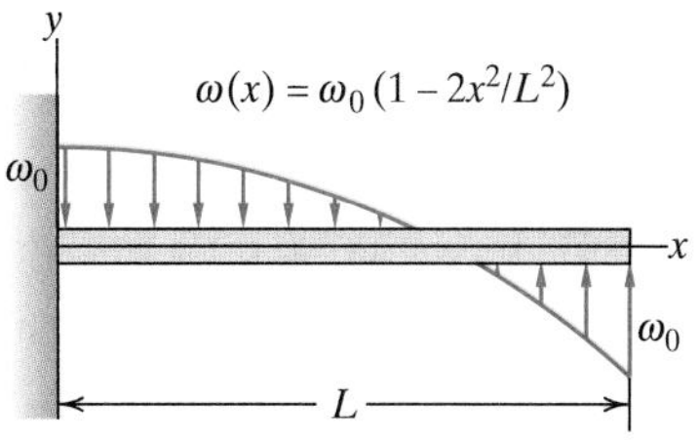

E8.2.5

8.2.6. Use integration to determine the point load and its location (centroid) that are equivalent to the line load in E8.2.6.

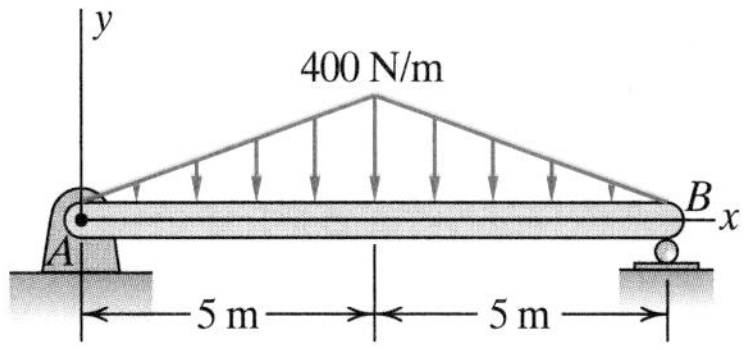

E8.2.6

8.2.7. Use integration to determine the point load and its location (centroid) that are equivalent to the line load in E8.2.7.

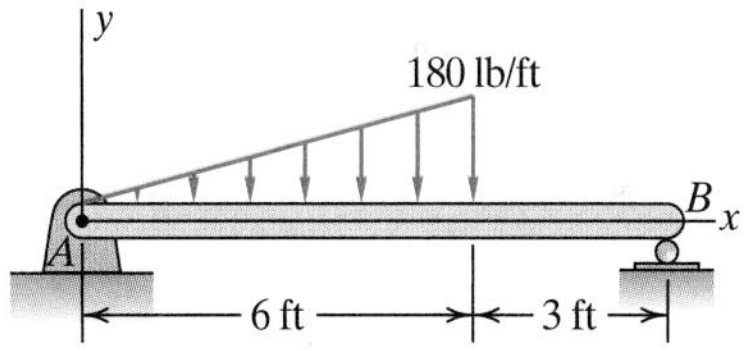

E8.2.7

8.2.8. Use integration to determine the point load and its location (centroid) that are equivalent to the line load in E8.2.8.

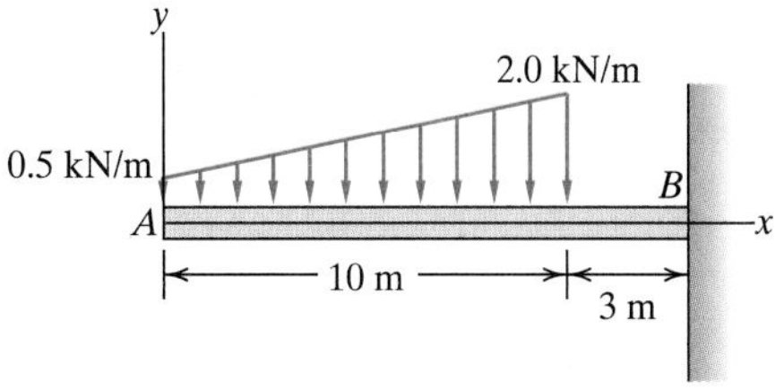

E8.2.8

8.2.9. Use integration to determine the point load and its location (centroid) that are equivalent to the line load in E8.2.9.

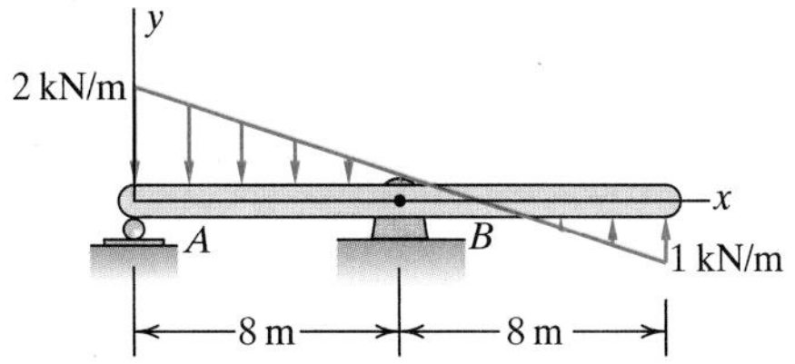

E8.2.9

8.2.10. Consider the beam in E8.2.10. Acting along the top of the beam between 3 ft $< x <$ 8 ft is a uniform line load. Use integration to determine the point force and its location (centroid) that are equivalent to the line load.

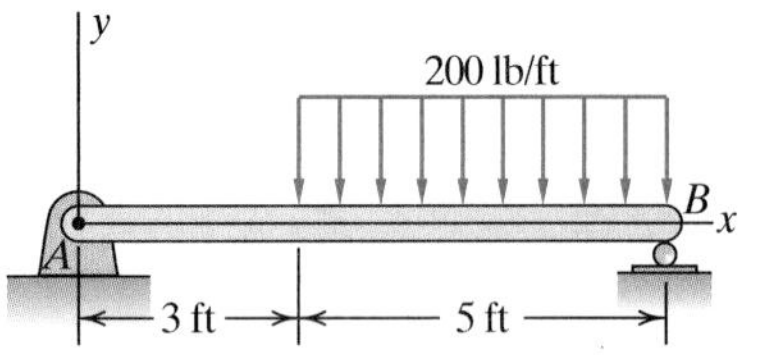

E8.2.10

8.2.11. Use the information in Appendix A3.1 to determine the point load and its location (centroid) that are equivalent to the line load in **E8.2.1**.

8.2.12. Use the information in Appendix A3.1 to determine the point load and its location (centroid) that are equivalent to the line load in **E8.2.2**.

8.2.13. **a.** Use the information in Appendix A3.1 to determine the point load and its location (centroid) that are equivalent to the line load in E8.2.13.

b. Determine the loads acting on the beam at A and B if the beam is in equilibrium. Assume that the weight of the beam is negligible.

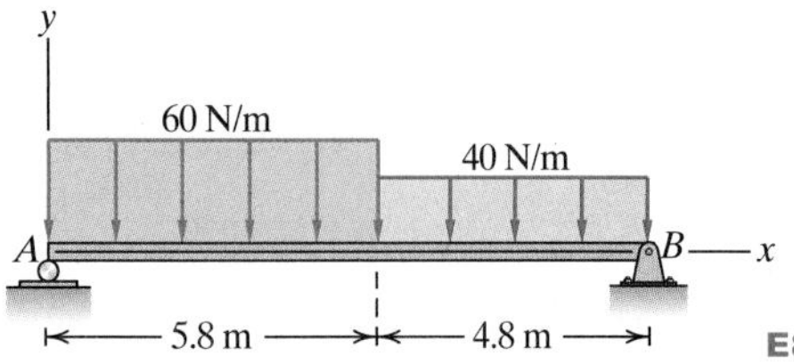

E8.2.13

8.2.14. **a.** Use the information in Appendix A3.1 to determine the point load and its location (centroid) that are equivalent to the line load in E8.2.14.

b. Determine the loads acting on the beam at the wall if the beam is in equilibrium. Assume that the weight of the beam is negligible.

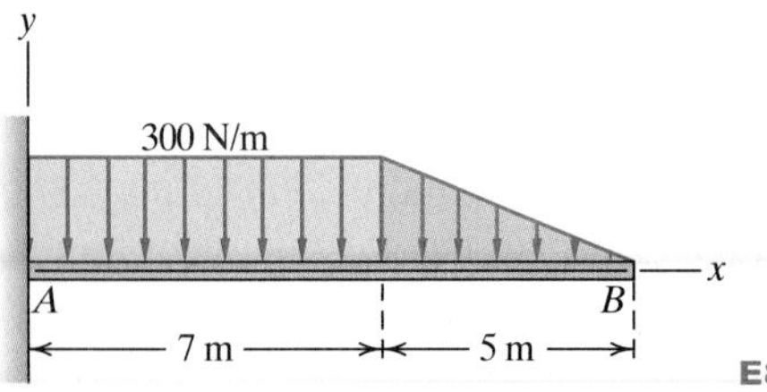

E8.2.14

8.2.15. **a.** Use the information in Appendix A3.1 to determine the point load and its location (centroid) that are equivalent to the line load in E8.2.15.

b. Determine the loads acting on the beam at the wall if the beam is in equilibrium. Assume that the weight of the beam is negligible.

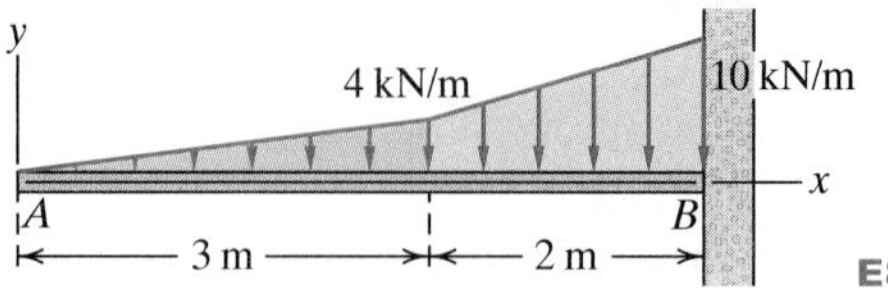

E8.2.15

8.2.16. **a.** Use the information in Appendix A3.1 to determine the point load and its location (centroid) that are equivalent to the line load in **E8.2.7**.

b. Determine the loads acting on the beam at the wall if the beam is in equilibrium. Assume that the weight of the beam is negligible.

8.2.17. Consider the beam in E8.2.17. Acting along the top of the beam between $0 < x < 3$ m is a line load. A 2-kN point force acts at $C(x = 4.5$ m), as shown. Determine

a. the point load and its location (centroid) that are equivalent to the line load

b. the point force and its location that are equivalent to the distributed force AND the point force at C

c. the loads acting on the beam at the wall if the beam is in equilibrium (Assume that the weight of the beam is negligible.)

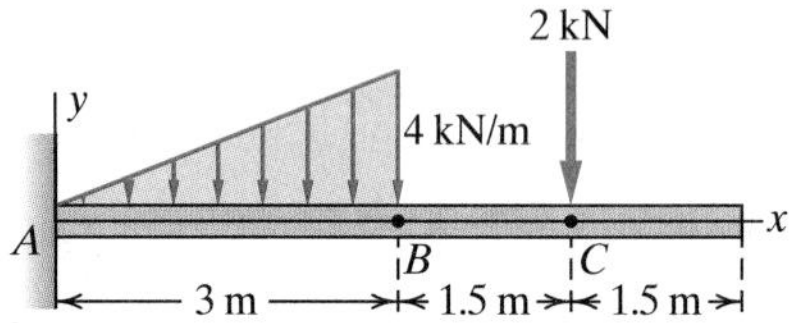

E8.2.17

8.2.18. A line load acting on the top of a horizontal surface between $0 < x < 10$ m is as shown in E8.2.18. Determine the point force and its location that are equivalent to the line load. Use information provided in Appendix A3.2.

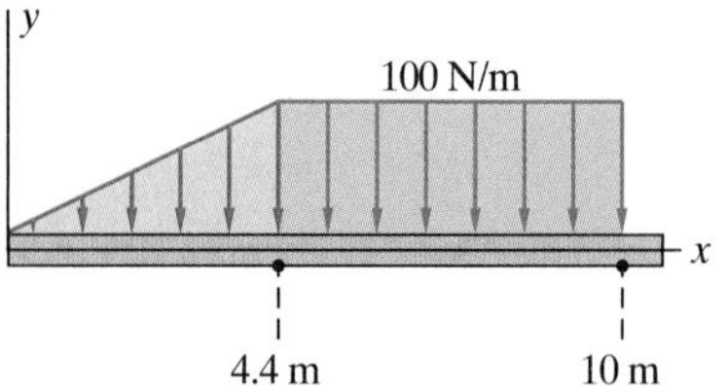

E8.2.18

8.2.19. A line load acting on the top of a horizontal surface is described by $\omega = kx^3$, where $k = 100$ N/m^4 (E8.2.19). Determine the point force and its location (centroid) that are equivalent to the line load.

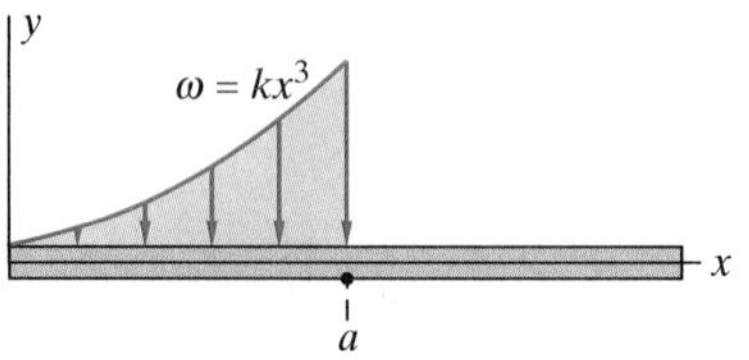

E8.2.19

8.2.20. A line load acting on the top of a horizontal surface between $4 < x < 10$ m is described by $\omega = \beta(x - 4)$, where $\beta = 100$ N/m^2 (E8.2.20). Determine the point force and its location (centroid) that are equivalent to the line load.

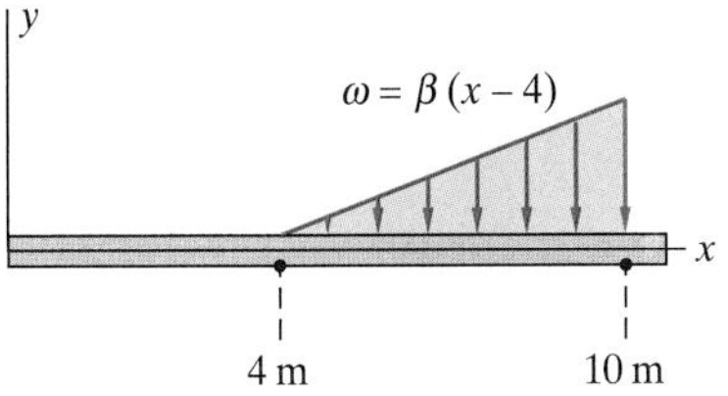

E8.2.20

8.2.21. A carpenter holds a 2 × 12 board. The board has a weight of 6 pounds/linear foot. In addition, bricks piled on the 2 × 12 result in the triangular distributed load.

a. Determine the vertical force that the carpenter must apply to the 2 × 12 for there to be equilibrium for the position in E8.2.21*a*. Is this a reasonable force for the carpenter to apply?

b. Determine the vertical force that the carpenter must apply to the 2 × 12 for there to be equilibrium for the position in **E8.2.21*b***. In addition, find the loads acting on the board at *A*.

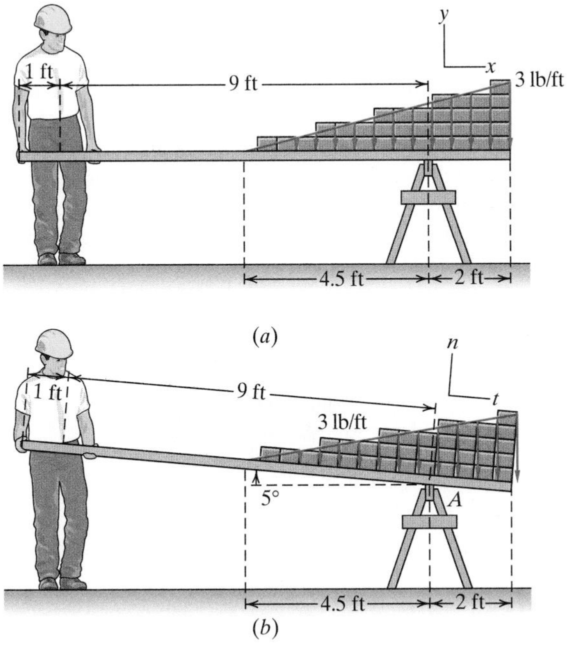

E8.2.21

8.2.22. The line loads shown in E8.2.22 act on the top of a beam.

a. Find the point force and its location that are equivalent to the line loads.

b. Determine the loads acting on the beam at *A* and *B* if the beam is in equilibrium.

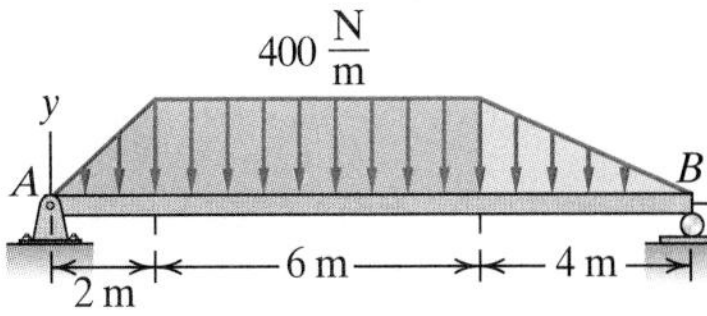

E8.2.22

8.2.23. The line loads shown in E8.2.23 act on the top of a beam.

a. Find the point force and its location that are equivalent to the line loads.

b. Determine the loads acting on the beam at *A* and *B* if the beam is in equilibrium.

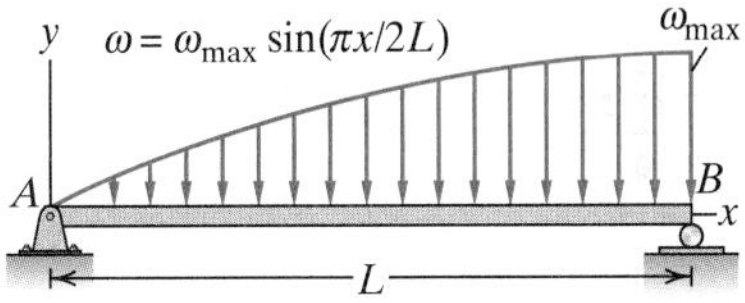

E8.2.23

8.2.24. The lift forces on the airplane's wing are represented by eight forces (E8.2.24). The magnitude of each force is given in terms of its x position on the wing by $200\sqrt{[1 - (x/17)^2]}N$. The weight of the wing W is 1600 N, and the wing has a width of 1 m.

a. Find the point force and its location that are equivalent to the eight forces shown in **E8.2.24**.

b. Determine the loads acting on the wing at the root R if the wing is in equilibrium.

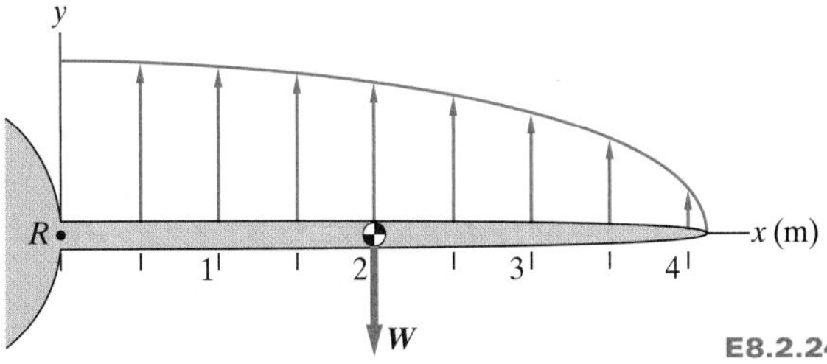

E8.2.24

8.2.25. An ice house is sitting on top of the ice covering Lake Sisabagema in northern Minnesota and the wind is gusting out of the north at speeds up to 40 mph.

a. The wind load is approximated by the distributed load shown in E8.2.25, with $\omega_0 = 40$ lb/ft. Assume that $L = 7$ ft, $H = 8$ ft, and the force due to gravity, W, is 800 lb (a crate, a portable television, a small stove, and three fishermen). Calculate the total force that the wind exerts on the ice house and the loads acting on the runners.

b. How do the forces in **a** change if the wind load is uniform over the entire side of the ice house at $\omega_0 = 40$ lb/ft?

c. If the wind load is uniformly distributed over the entire side of the ice house, how strong does it have to be to cause the ice house to tip?

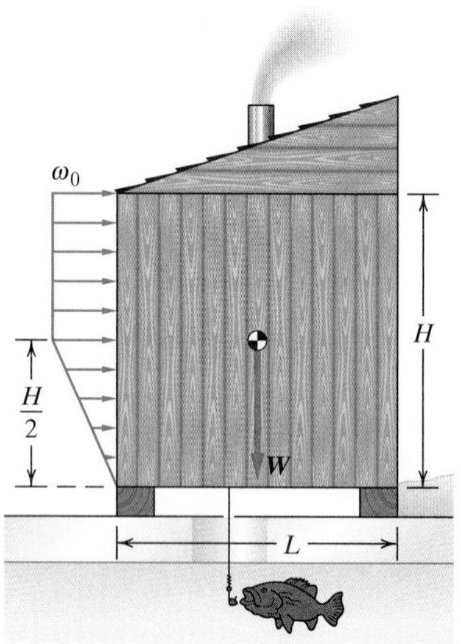

E8.2.25

8.2.26. Determine by integration the point force and its location (center of pressure) for the pressure distribution shown in E8.2.26.

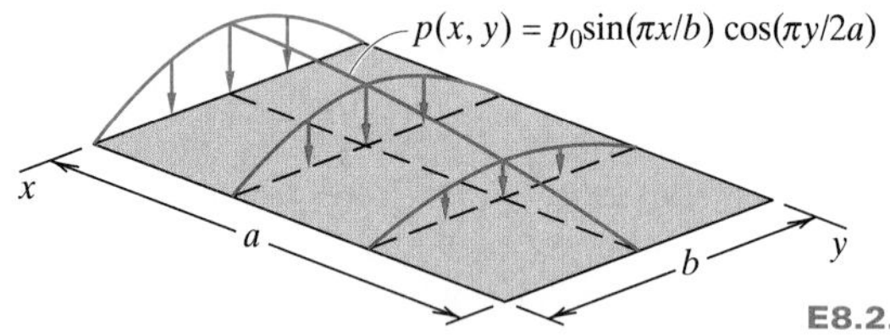

E8.2.26

8.2.27. Determine by integration the point force and its location (center of pressure) for the pressure distribution in E8.2.27.

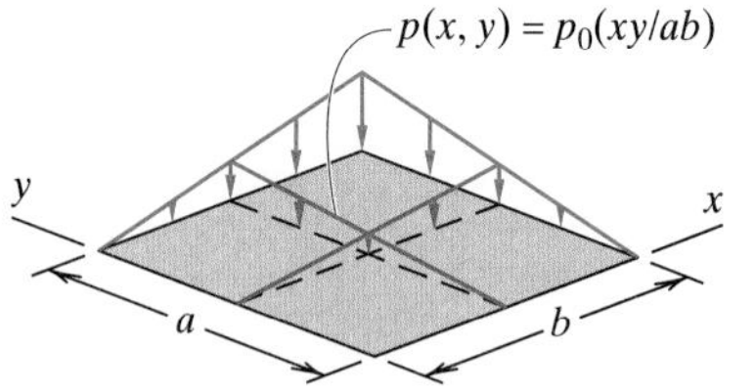

E8.2.27

8.2.28. As part of a design safety study, the effects of wind loads on an 800-ft-tall building are being investigated. The wind pressure has a parabolic distribution as shown in E8.2.28, and the depth of the building is 300 ft.

a. Determine the loads acting on the building at its base due to the wind load.

b. If the moment acting on the building at A may not be greater than 1.0×10^{12} ft·lb, suggest two changes that might be made to the design to meet this requirement.

8.2.29. A uniform pressure (p) acting on the top of a horizontal surface is shown in E8.2.29.

a. Use integration to find the total (equivalent) force and the center of pressure.

b. Use the data in Appendix A3.1 or A3.2 to find the total (equivalent) force and the center of pressure.

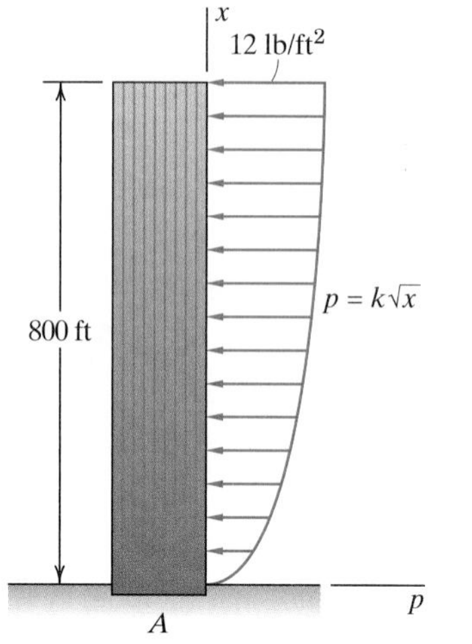

E8.2.28

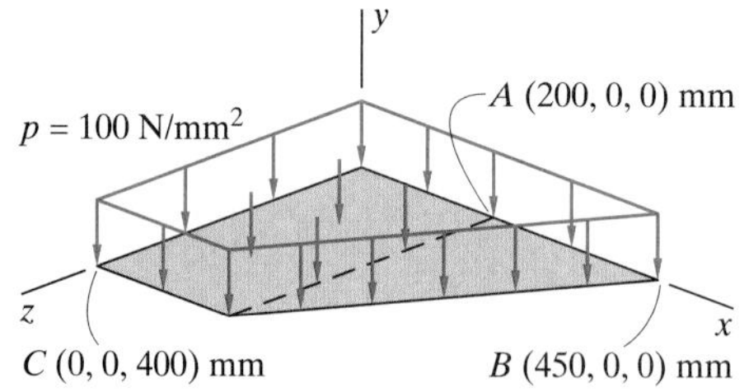

E8.2.29

8.2.30. A uniform wind pressure distribution acts on the shaded area of the sign in E8.2.30.

a. Determine the point force and its location (center of pressure) that are equivalent to this distribution.

b. If the sign is in equilibrium and its pole is fixed into the ground at A, what loads act on the pole at A due to the wind load?

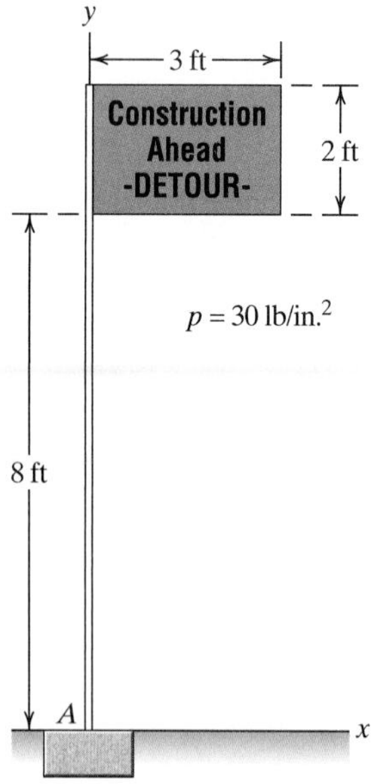

E8.2.30

8.2.31. A uniform wind pressure distribution acts on the shaded area of the sign in E8.2.31.

a. Determine the point force and its location (center of pressure) that are equivalent to this distribution.

b. If the sign is in equilibrium and its pole is fixed into the ground at A, what loads act on the pole at A due to the wind load?

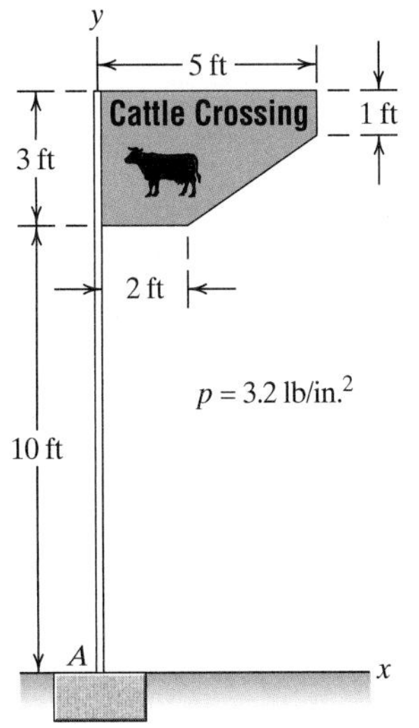

E8.2.31

8.2.32. A uniform wind pressure distribution acts on the shaded area of the sign in E8.2.32.

a. Determine the point force and its location (center of pressure) that are equivalent to this distribution.

b. If the sign is in equilibrium and its pole is fixed into the ground at A, what loads act on the pole at A due to the wind load?

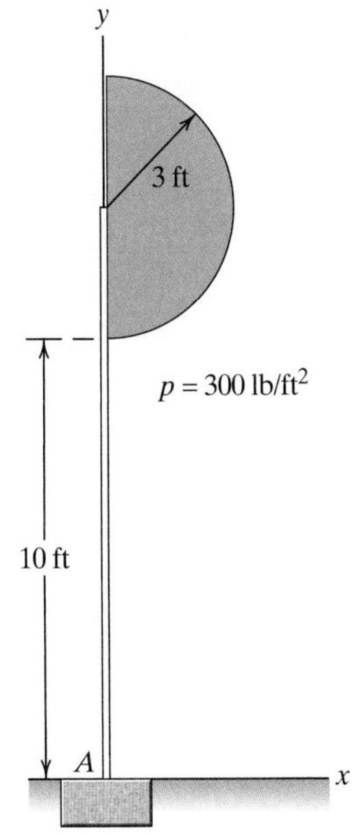

E8.2.32

8.2.33. A uniform wind pressure distribution acts on the shaded area of the sign in E8.2.33.

a. Determine the point force and its location (center of pressure) that are equivalent to this distribution.

b. If the sign is in equilibrium and its pole is fixed into the ground at A, what loads act on the pole at A due to the wind load?

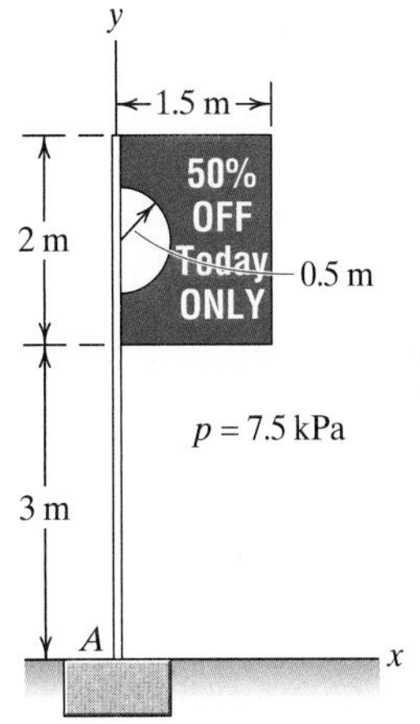

E8.2.33

8.2.34. A semicircular plate is supported in a wind tunnel by a hinge along CD and by a cable running from A to B (E8.2.34). The plate is made of steel, and the lateral wind pressure on the plate is 40 psi. Determine

a. the point force and its location (center of pressure) that are equivalent to the wind pressure

b. the tension in the cable and the loads acting at the hinge if the plate is in equilibrium

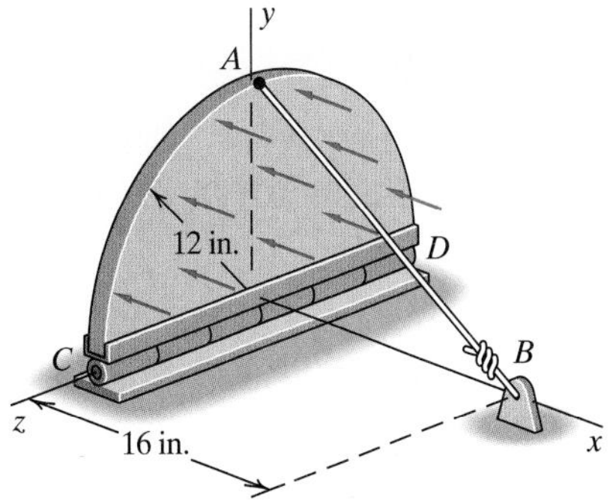

E8.2.34

8.2.35. The semicircular plate of E8.2.35 is 0.25 in. thick and is made of steel. The wind pressure on the plate is 40 psi. The angle θ by which the plate rotates from the vertical is such that $0 < \theta < 45°$. Write

a. expressions in terms of θ for the point force and its location (center of pressure) that are equivalent to the wind pressure

b. expressions in terms of θ for the tension in the cable if the plate is in equilibrium (Present your answer as equations and as a plot of tension versus angle θ.)

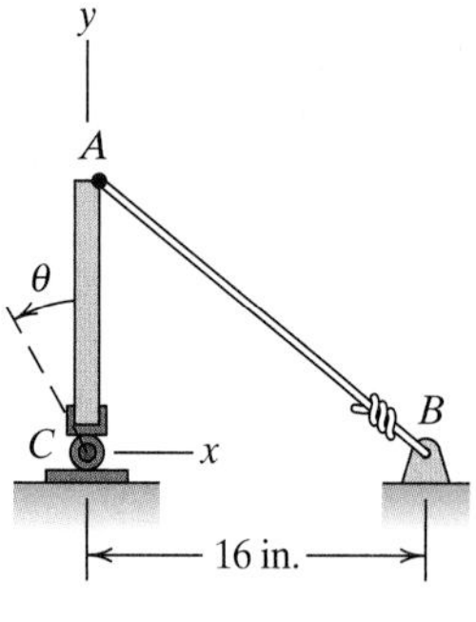

E8.2.35

8.2.36. The pressure distribution on a rectangular plate is as shown in E8.2.36. Determine the point force and its location (center of pressure) that are equivalent to this distribution.

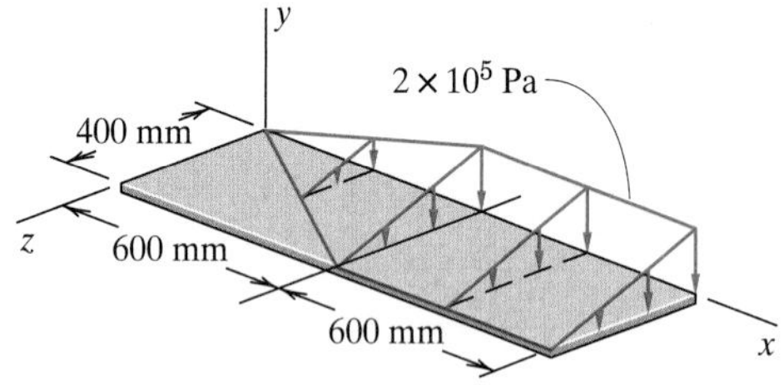

E8.2.36

8.2.37. The snow load on a roof is as shown in E8.2.37. Determine the point force and its location (center of pressure) that are equivalent to the distributed snow load.

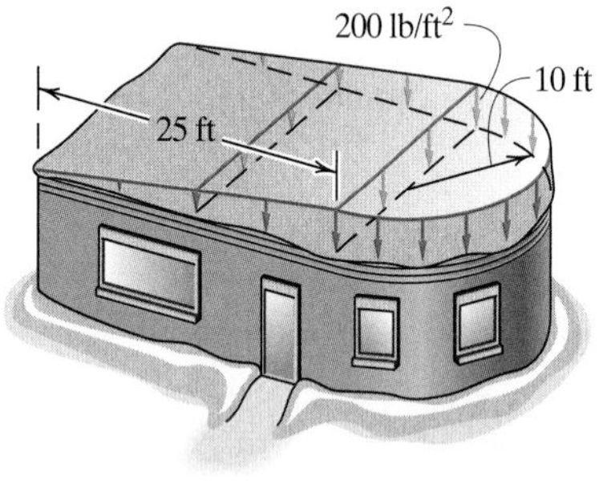

E8.2.37

8.2.38. The snow load on a roof is as shown in E8.2.38. Determine the point force and its location (center of pressure) that are equivalent to the distributed snow load.

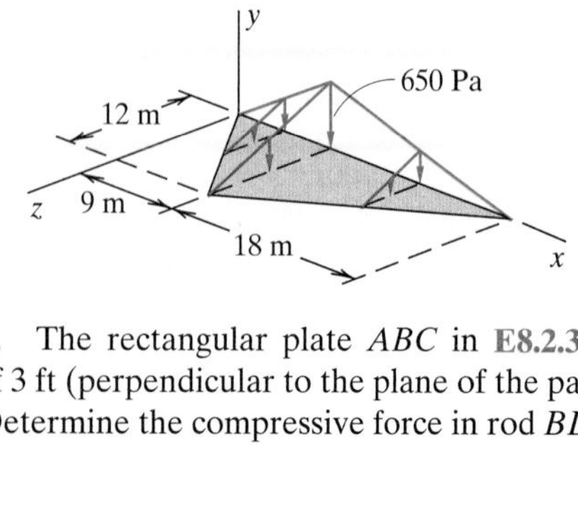

E8.2.38

8.2.39. The rectangular plate *ABC* in E8.2.39*a* has a width of 3 ft (perpendicular to the plane of the page).

a. Determine the compressive force in rod *BD*.

b. Figure **E8.2.39*b*** shows the compressive force that causes 8.57-ft steel rods of various diameters to buckle. Buckling is a form of failure of long slender members loaded in compression. Based on the force found in **a**, what is the minimum diameter that should be specified for rod *BD* to ensure that it will not buckle? Explain your reasoning.

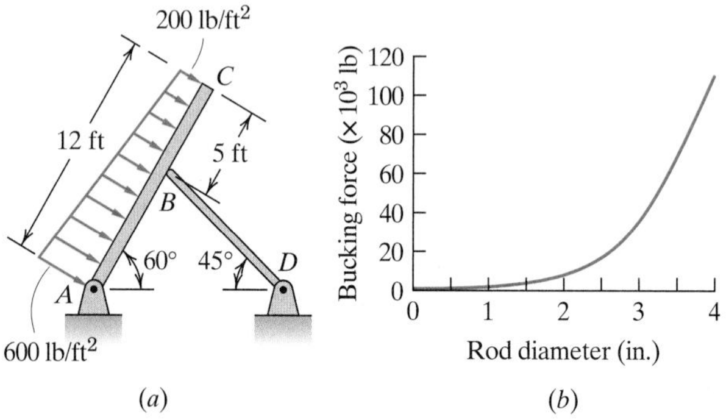

E8.2.39

8.2.40. Water pressure acts on the vertical freshwater aquarium window shown in E8.2.40. If water pressure varies as a linear function of the depth (measured by *y* in the figure), determine the point force and its location (center of pressure) that are equivalent to the water pressure acting on the window. Your answer should include a scale drawing of the window, showing the point force and its location.

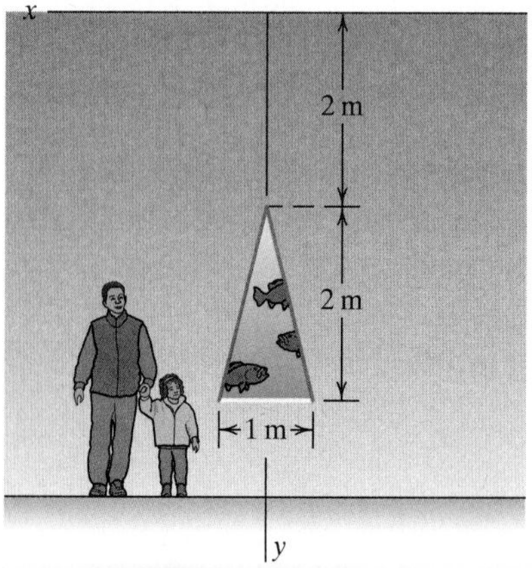

E8.2.40

8.2.41. Water pressure acts on the vertical freshwater aquarium window shown in E8.2.41. If water pressure varies as a linear function of the depth (measured by *y* in the figure), determine the point force and its location (center of pressure) that are equivalent to the water pressure acting on the window. Your answer should include a scale drawing of the window, showing the point force and its location.

E8.2.41

8.2.42. The flat plate in E8.2.42 is used as an access port in an oil tank. It is hinged along AB and is held in place by force $\boldsymbol{P}$ acting at C. Oil pressure p varies as a linear function of depth (measured by y in the figure); the relationship is of the form $p = \rho_{\text{oil}}\, gy$, where ρ_{oil} is the density of oil and is 800 kg/m^3. Determine the force $\boldsymbol{P}$ applied at C required to keep the plate in place.

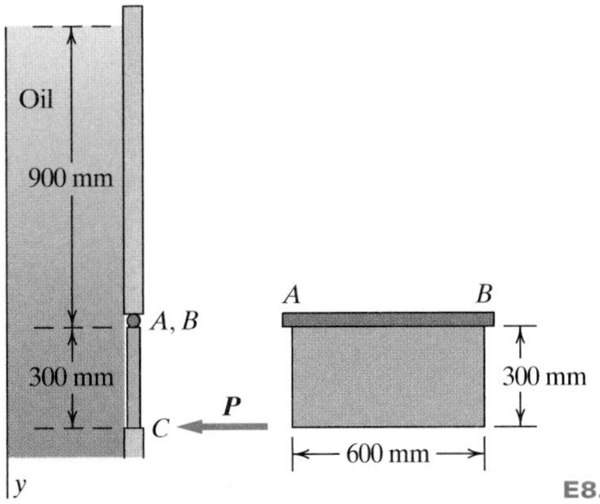

E8.2.42

8.3 HYDROSTATIC PRESSURE

If you have ever been at the bottom of a swimming pool and had to clear your ears or if you have ever used a pressure regulator attached to a scuba tank to swim the ocean depths, you have had personal experience with hydrostatic pressure. And if you've ever seen how an unopened potato chip bag expands when you travel from sea level to altitude, you have had personal experience with aerostatic pressure. The pressure within a static fluid is called the **hydrostatic pressure** when the fluid is a liquid and **aerostatic pressure** when the fluid is a gas. The pressure is in units of force/area. Hydrostatic pressure and aerostatic pressure are two examples of distributed pressure loads, as discussed in Section 8.2. These loads, if deemed significant by the analyst, must be included in static analysis of a system. Now we consider how hydrostatic pressure changes with depth.

If a liquid at rest acts at the boundary of a system, the pressure exerted by the hydrostatic pressure is normal to the boundary and oriented so as to push on the boundary. Furthermore, at any particular location within the fluid, the magnitude of the pressure acting at that location is the same in all directions, as we illustrate in Example 8.13. Even though the magnitude of the pressure acting at a particular location in a liquid is the same in all directions, the magnitude of the pressure is not the same everywhere in the liquid. In a liquid at rest we find that the hydrostatic pressure at depth h (Figure 8.53) is given by the expression:

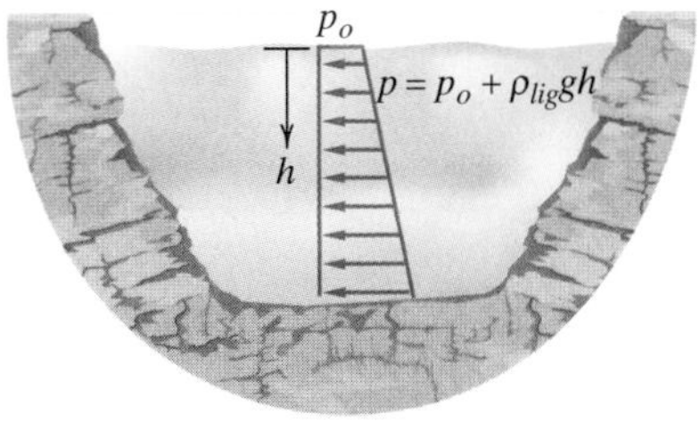

Figure 8.53 Hydrostatic pressure increases with distance below the surface

$$p = p_o + \rho_{\text{liq}} gh \tag{8.20A}$$

where p_o is the atmospheric pressure acting on the top surface of the liquid, ρ_{liq} is the density of the liquid, and g is the gravitational constant. The quantity ρg is called the specific weight of the liquid. Equation (8.20A) says that *hydrostatic pressure increases with depth in a linear manner* and is valid when density (or specific weight) is constant everywhere in the liquid. The derivation of (8.20A) is shown in Example 8.14. Equation (8.20A) can be rewritten in terms of specific weight γ_{liq} (where γ_{liq} is weight per unit volume as is $\rho_{liq}g$) as

$$p = p_o + \gamma_{liq}h \tag{8.20B}$$

Values of density and specific weight for water are given in Appendix A2.1.

The pressure p in (8.20A) and (8.20B) is referred to as the **absolute pressure**. Sometimes fluid pressure is measured with instruments that read pressure at depth relative to atmospheric pressure. This relative pressure is called **gage pressure** and is expressed as

$$p_{gage} = \rho_{liq}gh \tag{8.21A}$$

or in terms of specific weight as

$$p = \gamma_{liq}h \tag{8.21B}$$

Note that the term on the right of (8.21A) and (8.21B) is also the rightmost term in (8.20A) and (8.20B). Thus (8.20A) tells us that the absolute pressure at depth h in any fluid is atmospheric pressure p_o plus the gage pressure ρgh at that depth.

For fluid pressure acting on a flat surface of any shape, the equivalent point force is the pressure acting on the surface at the centroid of the surface multiplied by the surface area. We write this as

$$F_{liq} = \rho_{liq}gh_{cent}A \tag{8.22}$$

where h_{cent} is the distance from the surface to the centroid of the surface area A (**Figure 8.54**). The center of pressure is at the centroid of the hydrostatic pressure distribution acting on the surface, as shown in **Figure 8.55**. Note that the center of pressure (CP) and the centroid (C) of the surface area are not at the same location.

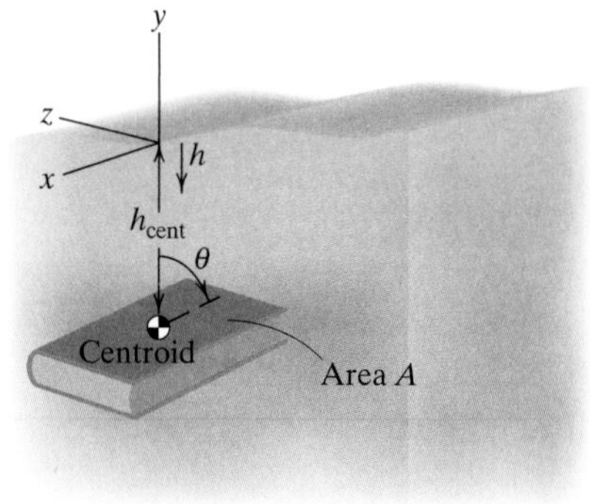

Figure 8.54 Water pressure acting on a flat submerged surface

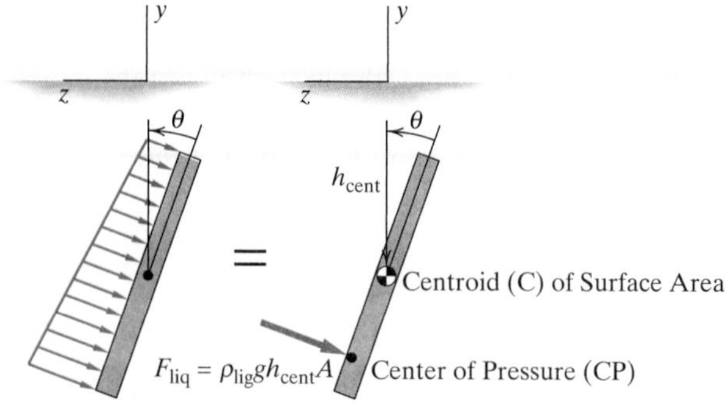

Figure 8.55 Equivalent point force acts at center of pressure

EXAMPLE 8.13 PROOF OF NONDIRECTIONALITY OF FLUID PRESSURE

Figure 8.56 shows an arbitrary infinitesimal triangular prism of liquid taken from an arbitrary point in a liquid at rest. The faces of the prism are numbered as follows: face 1 is parallel to the yz plane, face 2 is on the bottom, face 3 is on the top, and face 4 is the triangle at the front of the prism running parallel to the xy plane. Prove that the pressure exerted on the prism by the surrounding liquid is the same in all directions.

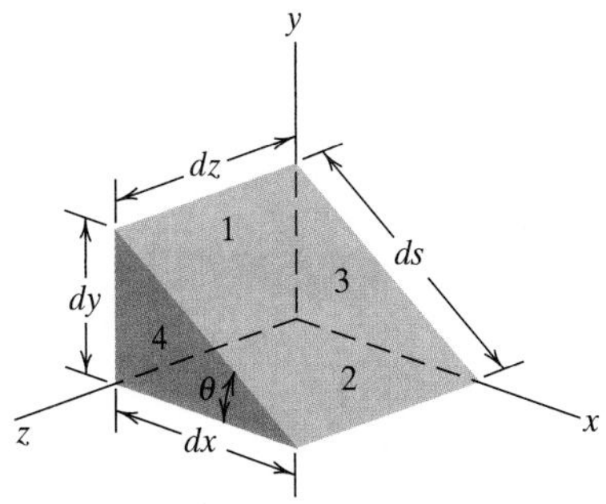

θ = Arbitrary direction

Figure 8.56

Goal We are to prove for a liquid at rest that the pressure at an arbitrary point is the same in all directions.

Given We are told the fluid is at rest, the liquid prism is taken from an arbitrary point, and the angle θ defining the top plane of the prism is defined as shown in **Figure 8.56**.

Assume We assume that the prism represents a particle since it is infinitesimal. Also, we assume that the particle is in equilibrium because it is at rest. The weight of the particle can be ignored. We also assume that the pressure on each of the faces is different (p_1, p_2, p_3, p_4) and that the pressure on any face acts perpendicular to that face.

Formulate Equations and Solve Because the particle is in equilibrium we know the forces must sum to zero. Based on equilibrium equation (7.3A) we write:

$$\sum F_x = 0\ (\rightarrow +)$$

Since the pressure must be perpendicular to any surface, the only two surfaces with pressure in the x direction are 1 and 3.

$$\sum F_x = p_1\, dy\, dz - p_3\, ds\, dz \sin\theta = 0$$

$$p_1\, dy = p_3 ds \sin\theta \qquad (1)$$

Based on equilibrium equation (7.3B) we consider equilibrium in the y direction. The only two surfaces with pressure in the y direction are 2 and 3.

$$\sum F_y = 0\ (\uparrow +)$$

$$\sum F_y = p_2\, dx\, dz - p_3\, ds\, dz \cos\theta = 0$$

$$p_2\, dx = p_3 ds \cos\theta \qquad (2)$$

From **Figure 8.56** we see that

$$ds \cos\theta = dx \qquad (3)$$

$$ds \sin\theta = dy \qquad (4)$$

Substituting (4) in (1) and (3) into (3) we get

$$p_1\,dy = p_3\,dy$$
$$p_2\,dx = p_3\,dx$$
$$p_1 = p_3 = p_2$$

If we were to rotate the prism 90° about the y axis, we could perform the same equilibrium analysis to show that

$$p_4 = p_2 = p_1$$

Combining our results we can conclude that

Answer $p_1 = p_2 = p_3 = p_4$

EXAMPLE 8.14 PROOF THAT HYDROSTATIC PRESSURE INCREASES LINEARLY WITH DEPTH

Figure 8.57 shows a vertical column of motionless liquid that is part of a larger body of liquid of constant density ρ_{liq}. The column, of height h, extends from depth h_1 to depth h_2, and has a cross-sectional area dA. The pressure is p_1 at depth h_1 and p_2 at depth h_2. Use this column to prove that hydrostatic pressure increases linearly in a homogeneous motionless liquid as a function of depth h.

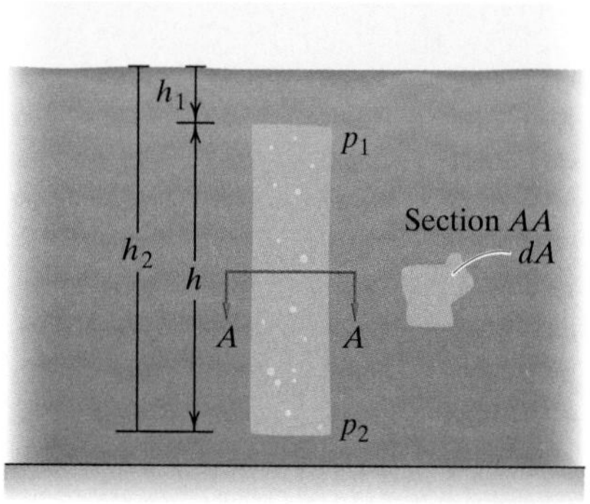

Figure 8.57

Goal We are to prove that hydrostatic pressure increases linearly as a function of depth.

Given We are told that the liquid is motionless and of constant density and are given the dimensions of an arbitrary column of liquid.

Assume No assumptions necessary.

Draw We draw a free-body diagram of the column of liquid, keeping in mind that in a liquid at rest, the pressures must act perpendicular to any surface (Figure 8.58*a*).

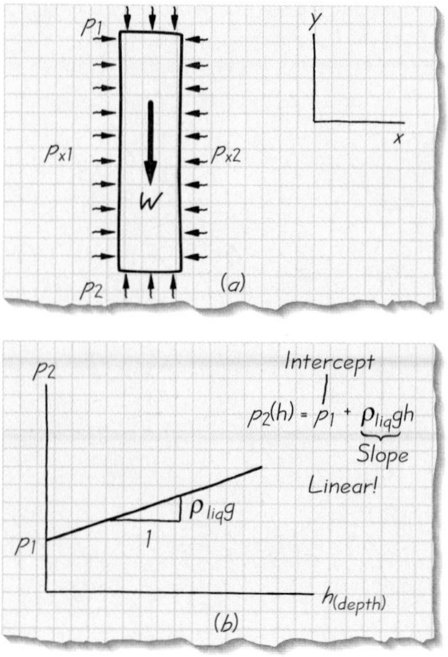

Figure 8.58

Formulate Equations and Solve We apply equilibrium equations to the column of fluid in the vertical direction.

$$\sum F_y = 0\;(\uparrow +)$$

$$p_2\,dA - p_1\,dA - W = 0 \qquad (1)$$

where W is the weight of the fluid in the column, which is given by

$$W = \rho_{liq} g h\,dA \qquad (2)$$

Substituting (2) into (1) gives

$$p_2\,dA - p_1\,dA - \rho_{liq} g h\,dA = 0$$

Solving for p_2 gives

$$p_2 = p_1 + \rho_{\text{liq}} g h \qquad (3)$$

This is the equation of a line representing p_2 as a function of depth, with intercept p_1 and slope $\rho_{\text{liq}} g$ (**Figure 8.58*b***). Therefore we can conclude that the pressure varies linearly with depth.

Note: Equation (8.20A) is a special case of (3), when p_1 is located at the surface of the liquid and is therefore p_o, the atmospheric pressure.

EXAMPLE 8.15 WATER RESERVOIR

Figure 8.59*a* shows a cross section through a rectangular gate that is 8 m high and 3 m wide (where the width dimension w is perpendicular to the plane of the page). The gate blocks the end of a freshwater channel and opens automatically when the water reaches a certain depth. It opens by rotating, as shown in **Figure 8.59*b***. Determine the depth d at which the gate just begins to open.

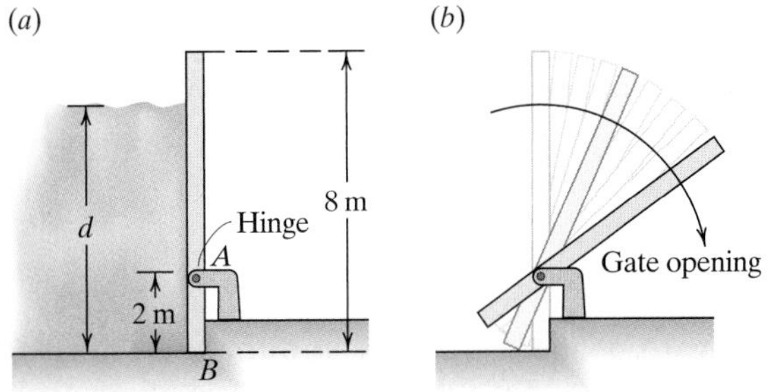

Figure 8.59

Goal We are to find the depth of water that causes the gate to open.

Given We are given information about the dimensions of the rectangular water gate and the location of the hinge. The gate holds back fresh water ($\rho = 1000$ kg/m^3).

Assume We assume that the weight of the gate is negligible and the hinge is frictionless. We also assume that the fluid is static and the system is in equilibrium. We can treat the system as planar.

Draw Included on the free-body diagram of the gate are the hydrostatic gage pressure (based on (8.21)), the forces at the frictionless hinge, and the force of the sill at B pushing on the bottom of the gate (**Figure 8.60*a***). We then represent the hydrostatic pressure by an equivalent total force at the center of pressure (**Figure 8.60*b***). Because we have chosen to work with gage pressure (as opposed to absolute pressure from (8.20)), we do not need to include the atmospheric pressure on the back side of the gate. If we had chosen to work with absolute pressure, the free-body diagram of the gate would look as in **Figure 8.61**.

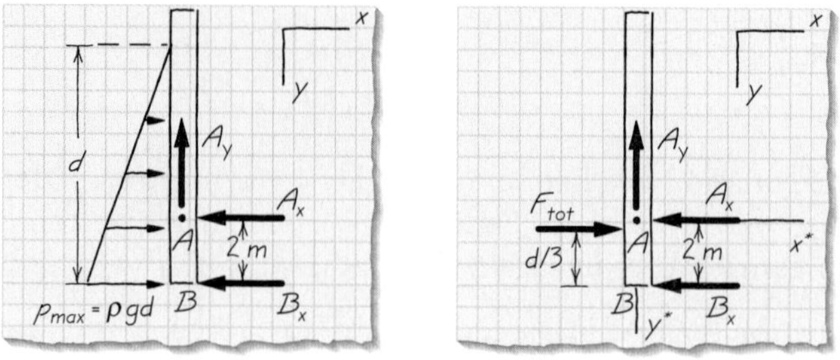

(*a*) (*b*)

Figure 8.60

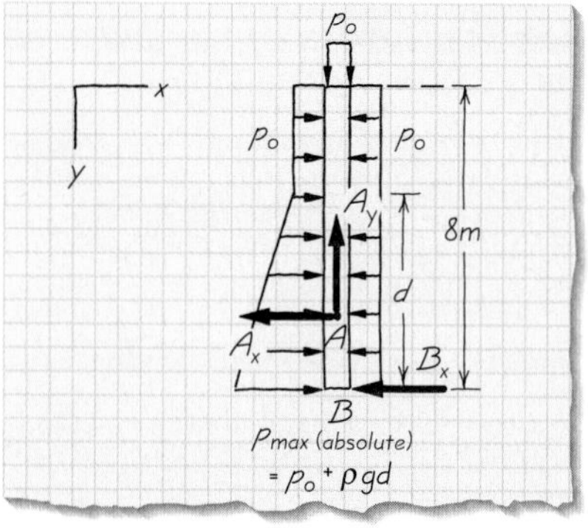

Figure 8.61 Free-body diagram drawn in terms of absolute pressure

Formulate Equations and Solve We use Appendix A3.1 to find the magnitude and location of the total hydrostatic gage force

$$F_{\text{tot}} = \frac{p_{\text{max}} dw}{2} = \frac{(\rho g d)dw}{2} = \frac{\rho g d^2 w}{2}$$

at a distance $\frac{d}{3}$ from the bottom of the gate. To find the force B_x, we sum the moment about the z axis, using the hinge at A as the moment center. We define an x^* y^* coordinate system with its origin at the hinge (**Figure 8.60*b***).

$$\sum M_{z@A} = 0 \; (\circlearrowleft +)$$

$$F_{\text{tot}}\, y^* - B_x(2\text{ m}) = 0$$

$$y^* = \frac{B_x(2\text{ m})}{F_{\text{tot}}}$$

Just as the gate is about to open, the force B_x must be zero, giving $y^* = 0$. This tells us that just as the gate is about to open, the total force F_{tot} must act through the hinge at 2 m from the base of the gate. Therefore when the gate is about to open, the location of the center of pressure is

$$\frac{d}{3} = 2\text{ m}$$

The gate opens when $d = 6$ m.

Note: An alternative approach to finding the depth at which the gate opens is to recognize that the gate opens when the moment created by the portion of the hydrostatic gage pressure above the hinge (which wants to open the gate) just equals the moment created by the hydrostatic gage pressure below the hinge (which wants to close the gate) (**Figure 8.61**). At larger depths, the opening moment will be greater than the closing moment, so the gate will open. This approach is a lot more work, because it means finding the total force and centroid for two separate distributed loads (as summarized in **Figure 8.62**).

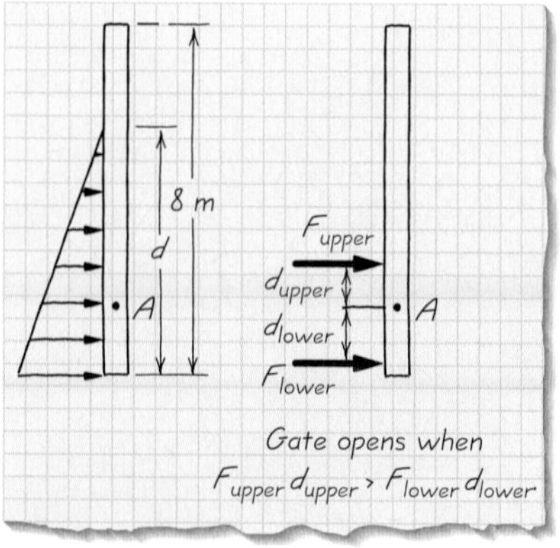

Figure 8.62

EXAMPLE 8.16 SLOPED GATE WITH LINEAR DISTRIBUTION

Figure 8.63 shows a cross section through a tank with a rectangular sloping gate that is 10 m long and 4 m wide (width is measured into the page). The gate is hinged along its top edge and held closed by a force at its bottom edge at A. Friction in the hinge and the weight of the gate can be neglected. Find the total load on the gate and the magnitude of the force acting on the bottom edge of the gate.

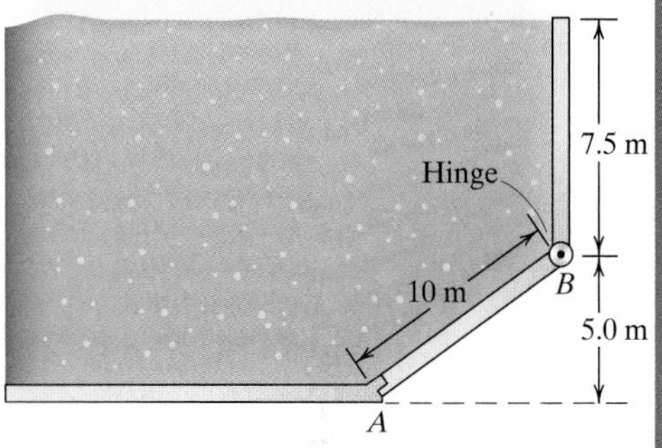

Figure 8.63

Goal We are to find the total load on the gate due to the liquid and the value of $\boldsymbol{P}$ needed to keep the gate closed.

Given We are given the dimensions of the rectangular water gate and the depth of water. We are told that we can ignore the weight of the gate and any friction at the hinge.

Assume We assume that the system can be treated as planar. The surface the gate rests against at A is frictionless, requiring $\boldsymbol{F}_A$ to be perpendicular to the gate. Also, we assume that the liquid is static and system is in equilibrium.

Draw We draw a free-body diagram of the sloped gate with the linear force distribution and include the loads at the supports (**Figure 8.64*a***). We have called the force acting on the gate at its bottom edge $\boldsymbol{P}$. Because we will work with gage pressure (8.21), we do not need to include the atmospheric pressure that acts on the back side of the gate. We then represent the distributed load by an equivalent point force $\boldsymbol{R}$ at a distance d_R from A (**Figure 8.64*b***).

Formulate Equations and Solve We could integrate this load distribution; however, since we recognize it as a complex line load made up of standard line load distributions, we will use Appendix A3.1 to calculate the total load due to the liquid pressure and its location. Since the pressure distribution is a trapezoid, we will break it into a rectangle and a triangle.

The magnitude of the load/unit length at point B is the hydrostatic gage pressure at B multiplied by the width of the gate:

$$\omega_{\min} = \rho_{\text{liq}} g h_B w = (1000 \text{ kg/m}^3)(9.81 \text{ m/s}^2)(7.5 \text{ m})(4 \text{ m})$$

$$\omega_{\min} = 294.3 \text{ kN/m}$$

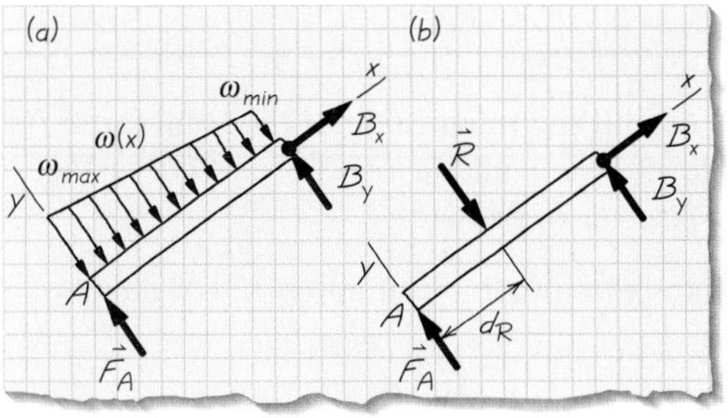

Figure 8.64

The magnitude of the load/unit length at point A is the hydrostatic gage pressure at A multiplied by the width of the gate:

$$\omega_{max} = \rho_{liq} g h_A w = (1000\ \text{kg/m}^3)(9.81\ \text{m/s}^2)(12.5\ \text{m})(4\ \text{m})$$
$$\omega_{max} = 490.5\ \text{kN/m}$$

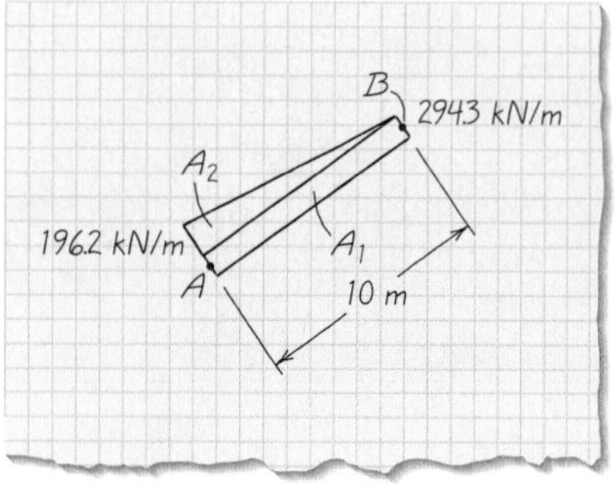

Figure 8.65

Figure 8.65 shows how we divide the load into two standard line load distributions: a rectangle (A_1) of height 294.3 kN/m and a triangle (A_2) of height 490.5 − 294.3 = 196.2 kN/m. We then calculate the total force for each standard line load and locate them at their centers of pressure as shown in **Figure 8.66**.

Using Appendix A3.1 we find that

$$F_1 = (10\ \text{m})(294.3\ \text{kN/m}) = 2943\ \text{kN, and } d_1 = 10/2 = 5.0\ \text{m}$$
$$F_2 = (10\ \text{m})(196.2\ \text{kN/m})/2 = 981\ \text{kN, and } d_2 = 10/3 = 3.33\ \text{m}$$

Then values of F_1, F_2, d_1, and d_2 are shown in **Figure 8.66**.

Using (8.16A), we find the magnitude of the total force on the gate due to the hydrostatic gage pressure is

$$R = F_1 + F_2$$
$$R = 2943\ \text{kN} + 981\ \text{kN} = 3924\ \text{kN}$$

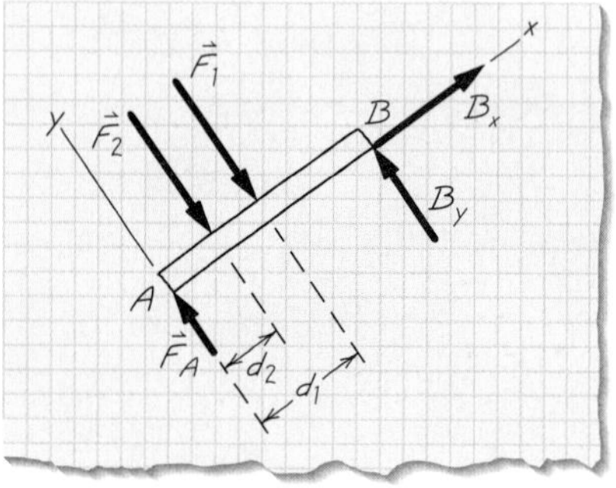

Figure 8.66

To find the magnitude of force $\boldsymbol{P}$, we use the equilibrium equation (7.5C), selecting the moment center at B:

$$\sum M_{z@B} = 0\ (\curvearrowleft +)$$
$$2943\ \text{kN}\ (10.0\ \text{m} - 5.0\ \text{m}) + 981\ \text{kN}\ (10.0\ \text{m} - 3.33\ \text{m}) - P(10\ \text{m}) = 0$$
$$P = 2126\ \text{kN}$$

Answer $R = 3924\ \text{kN}, F_A = 2126\ \text{kN}$

EXAMPLE 8.17 PRESSURE DISTRIBUTION OVER A CURVED SURFACE

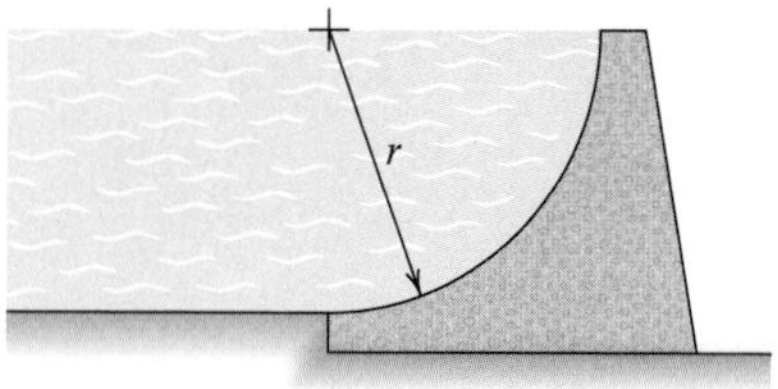

Figure 8.67

Figure 8.67 shows a cross section through a 100-ft-long concrete seawall. The face of the seawall is a quarter circle with a radius of 25 ft and is subjected to a load generated by the pressure of the seawater (specific weight = 64.0 lb/ft³). Determine the magnitude of the total force $\boldsymbol{F}$ that is exerted by the seawater on the seawall and the center of pressure.

Goal We are to find the magnitude of the total force $\boldsymbol{F}$ represented by hydrostatic gage pressure acting on the curved surface of the seawall and its location (center of pressure).

Given We are given information about the geometry of the seawall.

Assume We will assume that the water is at rest, the system is in equilibrium, and we can treat the system as planar.

Draw Based on the information given in the problem and our assumptions, we draw the water acting on the seawall (**Figure 8.68**). We define y positive downward. We chose to work with gage pressure; therefore we don't draw atmospheric pressure on the back side of the seawall.

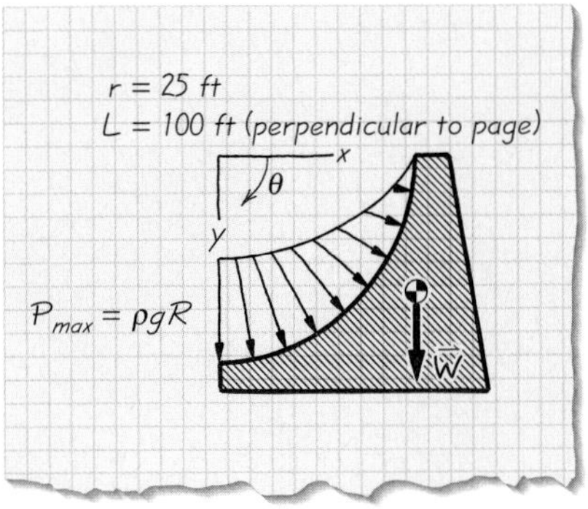

Figure 8.68

Formulate Equations and Solve Because hydrostatic pressure must always act perpendicular to a surface, the direction of the pressure varies along the height of the wall. As shown in **Figure 8.68**, at the top of the wall the pressure acts horizontally and at the bottom the pressure acts vertically. To account for this we need to integrate the horizontal components and the vertical components separately and combine them to calculate the total force $\| \boldsymbol{F} \|$:

$$\| \boldsymbol{F} \| = \sqrt{F_x^2 + F_y^2} \tag{1}$$

where $F_x = \| \boldsymbol{F} \| \cos \theta$ and $F_x = \| \boldsymbol{F} \| \sin \theta$.

At any point on the seawall the differential force dF is the pressure at that point multiplied by the differential area dA:

$$dF = p(x, y)\, dA$$

Breaking dF into its rectangular components gives

$$dF_x = p(x, y)\, dA \cos \theta \tag{2}$$
$$dF_y = p(x, y)\, dA \sin \theta \tag{3}$$

We represent the hydrostatic gage pressure in terms of cylindrical coordinate system so that ultimately we can integrate (2) and (3) with respect to θ. Based on (8.21A) (written in terms of y)

$$p(x, y) = \rho g y = \rho g r \sin \theta \tag{4}$$

We define a differential element $r\, d\theta$ along the length of the seawall (L) as

$$dA = L(r\, d\theta) \tag{5}$$

We substitute (4) and (5) into (2) and (3) to get

$$dF_x = \rho g r \sin \theta (L r\, d\theta) \cos \theta \tag{6}$$
$$dF_y = \rho g r \sin \theta (L r\, d\theta) \sin \theta \tag{7}$$

and integrate (6) and (7) from 0 to $\pi/2$ to find the total force in the horizontal and vertical directions:

$$F_x = \int_0^{\pi/2} \rho g L r^2 \cos \theta \sin \theta \, d\theta$$

$$F_x = -\rho g L r^2 \left[\frac{\cos^2 \theta}{2} \right]_0^{\pi/2}$$

$$F_x = \frac{\rho g L r^2}{2} \tag{8}$$

$$F_y = \int_0^{\pi/2} \rho g L r^2 \sin^2\theta \, d\theta$$

$$F_y = \rho g L r^2 \left[\frac{\theta}{2} - \frac{\sin 2\theta}{4}\right]_0^{\pi/2}$$

$$F_y = \frac{\rho g L r^2 \pi}{4} \tag{9}$$

The total force is found from substituting into (1):

$$\| \boldsymbol{F} \| = \sqrt{F_x^2 + F_y^2} = \sqrt{\left(\frac{\rho g L r^2}{2}\right)^2 + \left(\frac{\rho g L r^2 \pi}{4}\right)^2}$$

$$\| \boldsymbol{F} \| = \frac{\rho g L r^2}{2}\sqrt{1 + \frac{\pi^2}{4}}$$

Substituting numerical values gives

$$\| \boldsymbol{F} \| = \frac{\left(64.0 \frac{\text{lb}}{\text{ft}^3}\right)(100 \text{ ft})(25 \text{ ft})^2}{2}\sqrt{1 + \frac{\pi^2}{4}} = 3720 \text{ kip}$$

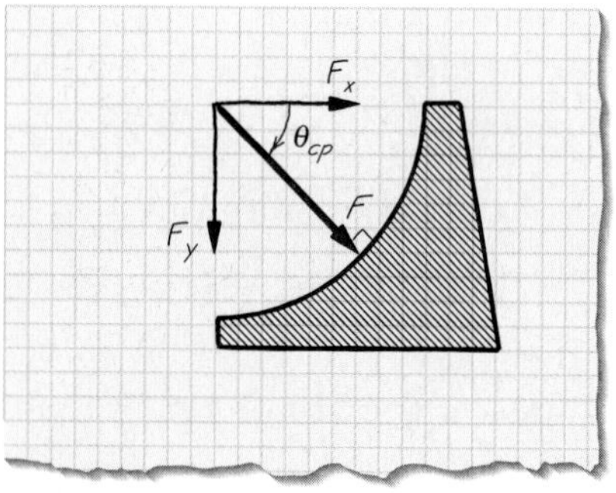

Figure 8.69

The total force $\boldsymbol{F}$ acts perpendicular to the quarter-circle surface at angle θ_{cp}, as shown in Figure 8.69. The center of pressure (cp) will then be at

$$X_{cp} = r \cos \theta_c$$
$$Y_{cp} = r \sin \theta_c$$

To find θ_{cp} we recognize that its tangent can be defined in terms of F_y and F_x as

$$\theta_{cp} = \tan^{-1}\frac{F_y}{F_x} = \tan^{-1}\frac{\frac{\rho g L r^2 \pi}{4}}{\frac{\rho g L r^2}{2}} = \tan^{-1}\frac{\pi}{2} = 57.5° \tag{10}$$

Therefore

$$X_{cp} = r \cos \theta_{cp} = 25 \text{ ft} \cos(57.5°) = 13.4 \text{ ft}$$
$$Y_{cp} = r \sin \theta_{cp} = 25 \text{ ft} \sin(57.5°) = 21.1 \text{ ft}$$

Answer $\| \boldsymbol{F} \| = 3720$ kip, $X_{cp} = 13.4$ ft, $Y_{cp} = 21.1$ ft

Note: We were able to avoid needing to integrate to find the center of pressure, because we were able to identify the relationship in (10) between the force components and their resultant being perpendicular to the quarter-circle surface.

Buoyancy

Having developed an expression for how hydrostatic pressure changes with depth (8.20) and the concept of the center of gravity (Section 8.1), we are in a position to discuss the principle of buoyancy. Buoyancy governs whether an object sinks or floats in a fluid. The discovery of this principle is credited to Archimedes. To illustrate the concept, we create a free-body diagram of an object partially submerged in a liquid (**Figure 8.70**). Acting on the object are atmospheric pressure, hydrostatic pressure, and gravity, as shown. If the object is in mechanical equilibrium, we can write the force equilibrium equation in the *y* direction as

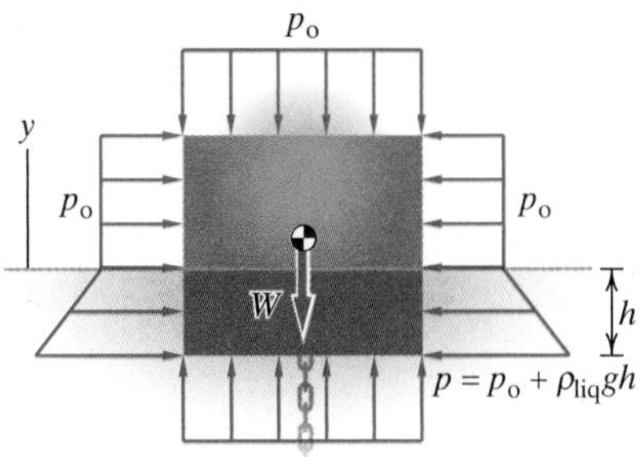

Figure 8.70 A partially submerged buoy (block)

$$\sum F_y = -W + (p_o + \rho_{liq}gh)A - p_o A = 0$$
$$-W + \rho_{liq}ghA = 0$$

where ρ_{liq} is the density of the fluid. The product hA is the volume of the submerged portion of the object, V_{subm}. Therefore

$$\sum F_y = -W + \rho_{liq} g V_{subm} = 0 \tag{8.23}$$

The term $\rho_{liq} g\, V_{subm}$ is an upward force acting on the object and is commonly referred to as the **buoyancy force**. The buoyancy force is always directed upward, and we shall use the notation F_{buoy} to represent it. The centroid of V_{subm} (the volume of the submerged portion of the object) is known as the **center of buoyancy**. The buoyancy force passes through the center of buoyancy. The center of buoyancy, along with the center of gravity of the object, play a key role in determining stability of partially submerged objects, as illustrated in **Figure 8.71**.

Equation (8.23) says that if the weight of an object is balanced by the buoyancy force, the object is in mechanical equilibrium. Being in mechanical equilibrium, the object will move neither up nor down in the fluid. Alternately, if the weight and buoyancy force do not balance each other, the object either sinks or rises in the fluid:

$$F_{buoy} < W; \quad \text{object sinks} \tag{8.24A}$$
$$F_{buoy} > W; \quad \text{object rises} \tag{8.24B}$$

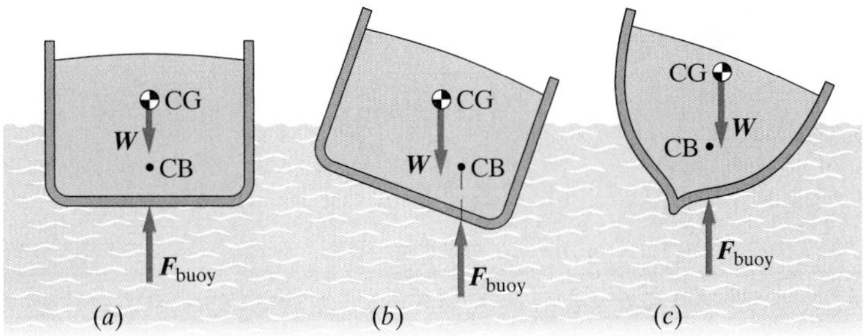

Figure 8.71 (*a*) *W* aligned with $\boldsymbol{F}_{buoy}$, floating volume is stable; (*b*) *W* and $\boldsymbol{F}_{buoy}$ form a couple that works to return volume to upright position (stable); (*c*) *W* and $\boldsymbol{F}_{buoy}$ form a couple that does not work to return volume to upright position (unstable)

EXAMPLE 8.18 CENTER OF BUOYANCY AND STABILITY

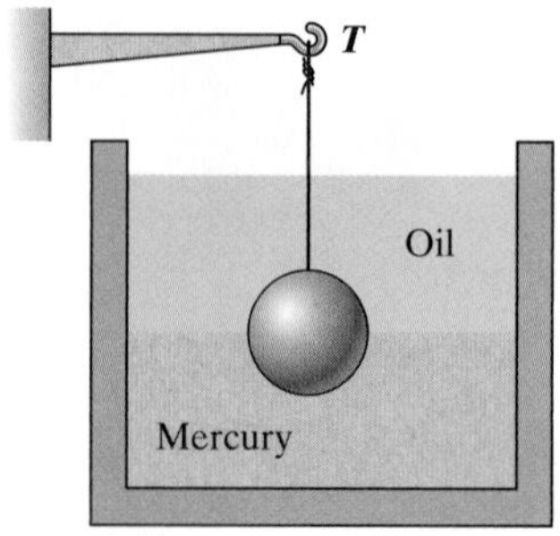

Figure 8.72

A solid steel ball of radius $r = 50.0$ mm and specific weight $\gamma_s = 76.8$ kN/m^3 is suspended by a thin wire and lowered into a tank containing oil and mercury ($\gamma_{\text{oil}} = 7.80$ kN/m^3, $\gamma_{\text{Hg}} = 133$ kN/m^3) until it is submerged half in the oil and half in the mercury (**Figure 8.72**). Determine the tension T in the wire.

Goal We are to find the tension in the wire that is holding up the ball.

Given We are given properties of the steel ball and the liquid it is suspended in.

Assume We assume that the specific weights are uniform and the ball is at rest.

Draw We draw a free-body diagram of the steel ball. The hydrostatic forces acting in the horizontal direction sum to zero, so we are not including them on the free-body diagram (**Figure 8.73**).

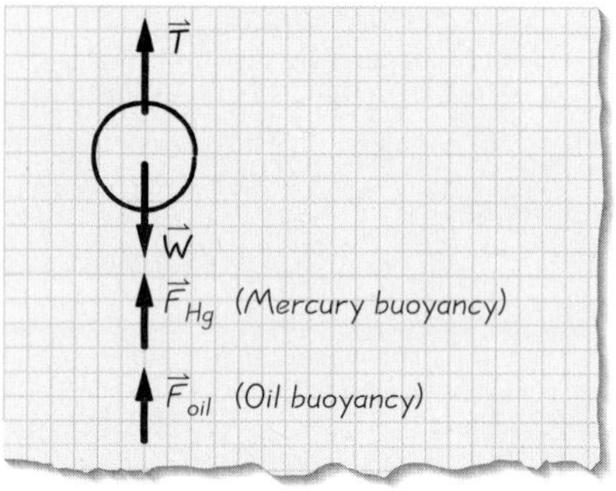

Figure 8.73

Formulate Equations and Solve We apply equilibrium in the vertical direction:

$$\sum F_y = 0(\uparrow +)$$

$$T + F_{\text{oil}} + F_{\text{Hg}} - W = 0$$

$$T = W - F_{\text{oil}} - F_{\text{Hg}} \qquad (1)$$

where

$$W = \gamma_{\text{ball}} V_{\text{ball}} = \gamma_s \frac{4}{3}\pi r^3$$

$$= \frac{4}{3}(76.8 \text{ kN/m}^3)\pi(0.0500 \text{ m})^3 = 40.21 \text{ N}$$

$$F_{\text{oil}} = V_{\text{ball submerged}}(\gamma_{\text{oil}}) = \frac{1}{2}V_{\text{ball}}(\gamma_{\text{oil}})$$

$$= \frac{1}{2}\left(\frac{4}{3}\pi r^3\right)\gamma_{\text{oil}} = \frac{1}{2}\frac{4}{3}\pi(0.0500 \text{ m})^3(7.80 \text{ kN/m}^3) = 2.042 \text{ N}$$

$$F_{\text{Hg}} = V_{\text{ball submerged}}(\gamma_{\text{Hg}}) = \frac{1}{2}V_{\text{ball}}(\gamma_{\text{Hg}})$$

$$= \frac{1}{2}\left(\frac{4}{3}\pi r^3\right)\gamma_{\text{Hg}} = \frac{1}{2}\frac{4}{3}\pi(0.0500 \text{ m})^3(133 \text{ kN/m}^3) = 34.82 \text{ N}$$

Find T using (1) and the calculated values of F_{oil} and F_{Hg}:

$$T = W - F_{\text{oil}} - F_{\text{Hg}}$$

$$T = 40.21 \text{ N} - 2.042 \text{ N} - 34.82 \text{ N}$$

$$T = 3.35 \text{ N}$$

Answer $T = 3.35$ N

If T were negative, it would mean that the wire is in compression. A thin wire would not support a compression force, so it would buckle and the steel ball would rise. A downward force would therefore be needed to sink the steel ball halfway into each liquid.

EXERCISES 8.3

8.3.1. Determine the point force and its location (center of pressure) that are equivalent to the hydrostatic pressure acting on the seawall in E8.3.1. The seawater is 2.5% denser than fresh water. The width of the seawall (dimension into the paper) is 20 ft. Your answer should include a scale drawing of the seawall, showing the point force and its location.

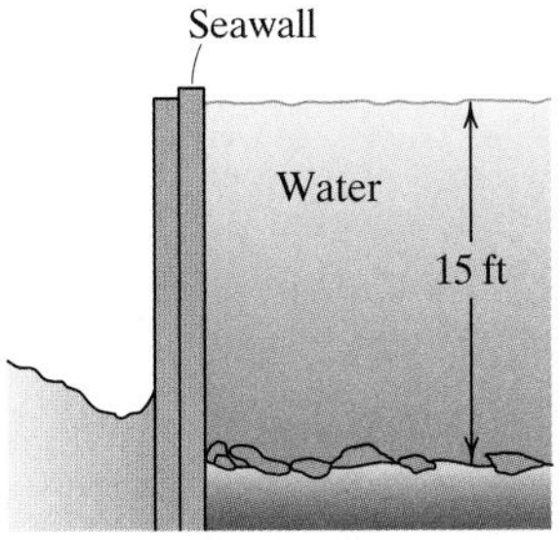

E8.3.1

8.3.2. Determine the point force and its location (center of pressure) that are equivalent to the hydrostatic pressure acting on the seawall in E8.3.2. The seawater is 2.5% denser than fresh water. The width of the seawall (dimension into the paper) is 6 m. Your answer should include a scale drawing of the seawall, showing the point force and its location.

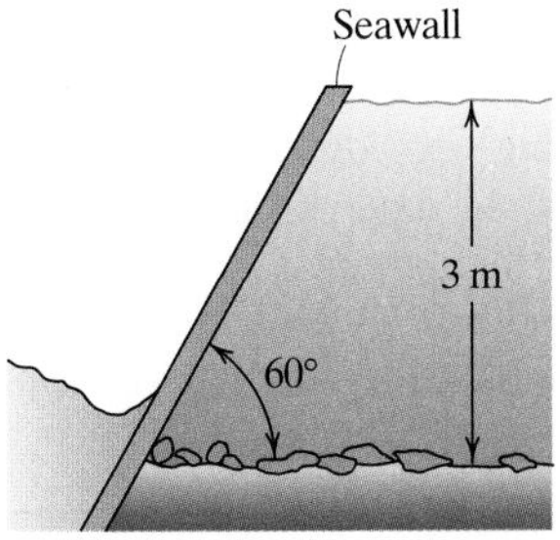

E8.3.2

8.3.3. Determine the point force and its location (center of pressure) that are equivalent to the hydrostatic pressure acting on the gate in E8.3.3. The width of the gate (dimension into the paper) is 2 ft, and the water is fresh water. Your answer should include a scale drawing of the gate, showing the point force and its location.

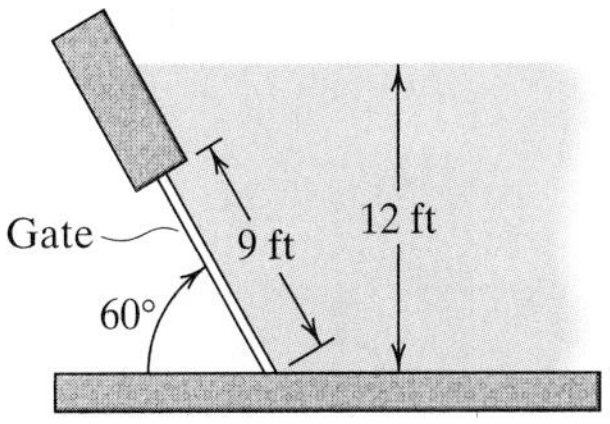

E8.3.3

8.3.4. The freshwater fish tank in E8.3.4 holds water to a depth of 1.25 m. For each face, including the bottom, determine the point force and its location (center of pressure) that are equivalent to hydrostatic pressure. Your answer should include a scale drawing of the fish tank, showing the point forces and their locations.

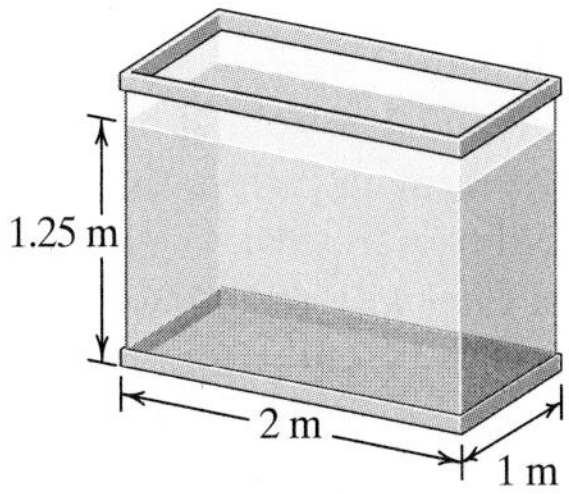

E8.3.4

8.3.5. One section of the upstream side of an arched dam has the form of a vertical cylindrical surface of 400-ft radius and subtends at an angle of 50° (E8.3.5). If the fresh water is 80 feet deep, determine the total force $\boldsymbol{F}$ exerted by the water on the cylindrical surface.

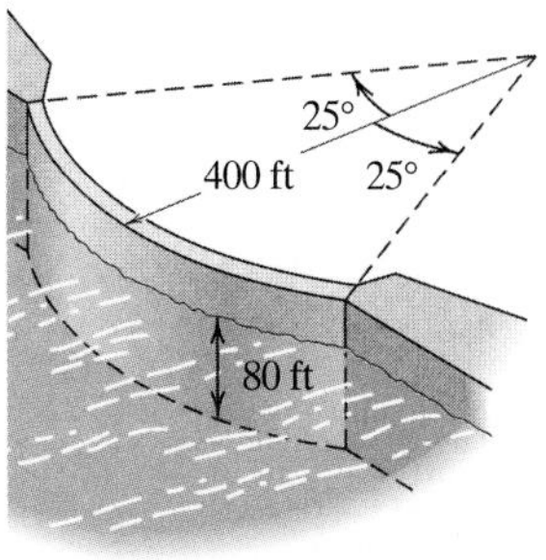

E8.3.5

8.3.6. The center of the rectangular window in a freshwater aquarium is at a distance d below the surface of the water (**E8.3.6**). The horizontal width of the window is 2 ft.

a. Write expressions (as functions of θ and d) for the point force and its location (center of pressure) that are equivalent to the water pressure acting on the window.

b. The seal around the perimeter of the window will leak at pressures greater than 60 psi, and the window will break if a force perpendicular to its surface greater than 5000 lb is applied. What is the maximum depth d that the window should be installed below the surface of the water to ensure that it will work adequately?

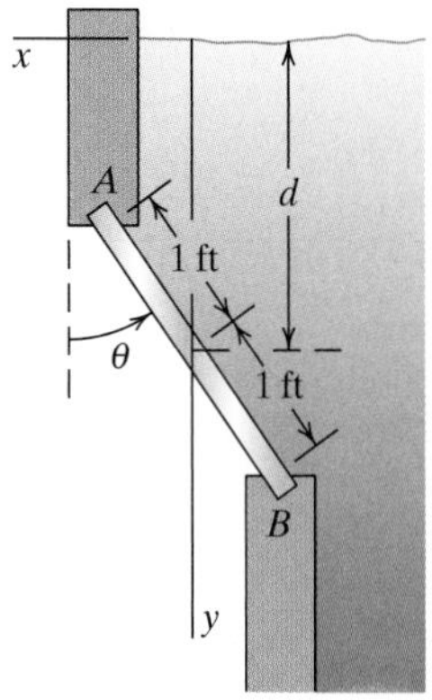

E8.3.6

8.3.7. The automatic gate valve AB in **E8.3.7** consists of a 4.5 × 3.0 ft rectangular plate that pivots about a horizontal hinge that runs through C (into the paper). The valve is part of a dam holding a freshwater reservoir. To open, the valve rotates in a clockwise direction about the hinge at C.

a. Neglecting the weight of the gate valve, determine the depth d in the reservoir at which the gate valve will begin to open.

b. If the weight of the plate is included in the analysis, will the depth of water necessary to open the valve increase, decrease, or remain the same relative to the answer in **a**? (Don't do any calculations, just reason through how various moments act on the valve.)

c. The rectangular plate is made of steel and is 0.5 in. thick. Determine the depth d in the reservoir at which the gate valve will begin to open.

d. Compare your answers from **a**, **b**, and **c**.

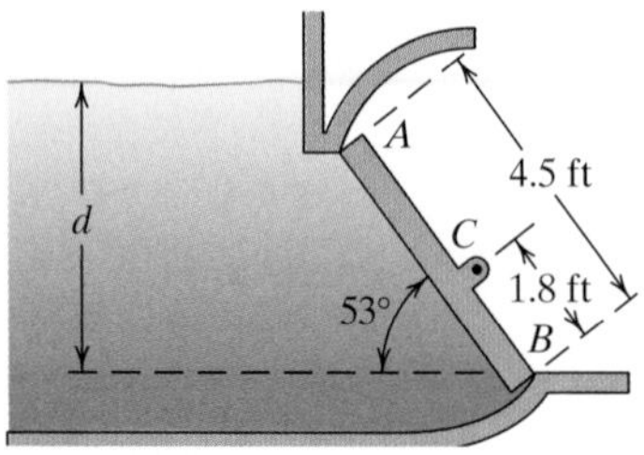

E8.3.7

8.3.8. The 750-mm-wide gate valve pivots about a horizontal shaft at A in **E8.3.8**. It is held in the closed position by a preloaded spring. The density of the oil in the tank behind the gate is 800 kg/m^3. Determine the preload in the spring for which the valve will open when d = 2.0 m. Does your answer indicate that the spring is preloaded in tension or compression?

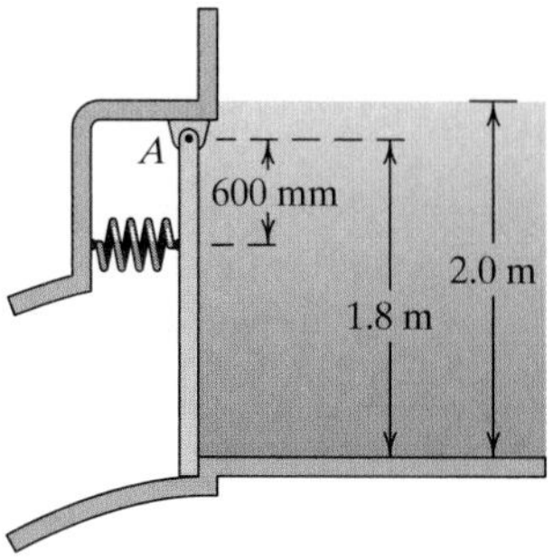

E8.3.8

8.3.9. The cross section of a 30-ft-long section of concrete formwork is shown in **E8.3.9**. The upright panels are connected to one another by 30 equally spaced tie rods. Determine the tension in each tie rod when the concrete is in a liquid state. The specific weight of liquid concrete is approximately 150 lb/ft^3. Assume that the lower edges of the upright panels (A and B) can be modeled as hinges.

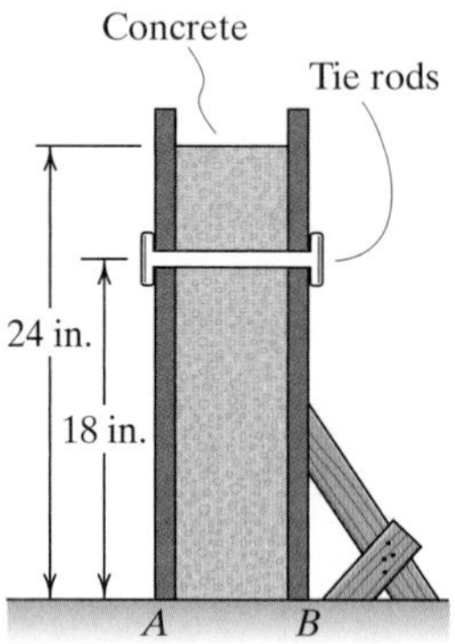

E8.3.9

8.3.10. The triangular and rectangular sections in **E8.3.10** are being considered for the design of a small freshwater concrete dam. From the perspective of resistance to overturning about C, which section will require less concrete, and how much less per foot of dam length? (The specific weight of concrete is approximately 150 pounds/ft^3.)

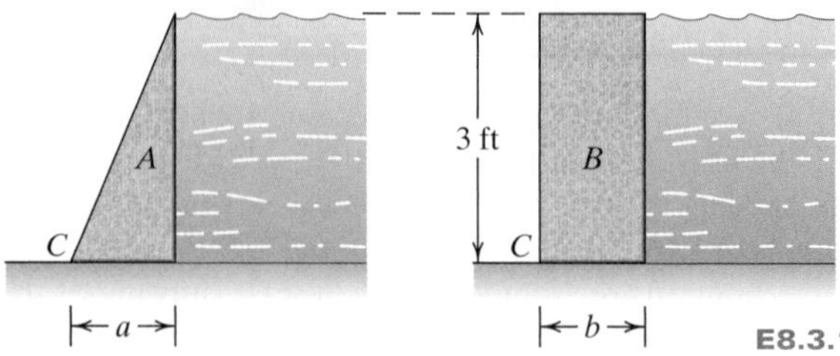

E8.3.10

8.3.11. The rectangular gate in **E8.3.11** has a width of 8 ft (dimension into the page). Determine the point force and its location (center of pressure) that are equivalent to the water pressure acting on the gate. Express the location with respect to the hinge at the top of the gate.

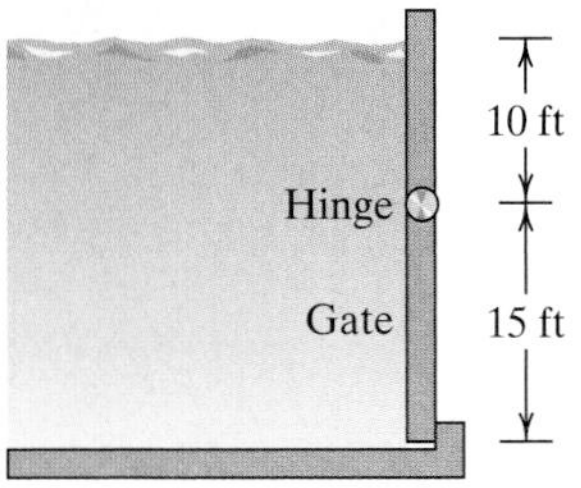

E8.3.11

8.3.12. The water gate whose cross section is shown in **E8.3.12** has a uniform density of 4000 kg/m^3. The gate is 5 m wide. If the gate is in equilibrium, determine the loads acting on the gate at A and C. Assume that the contact at C is frictionless.

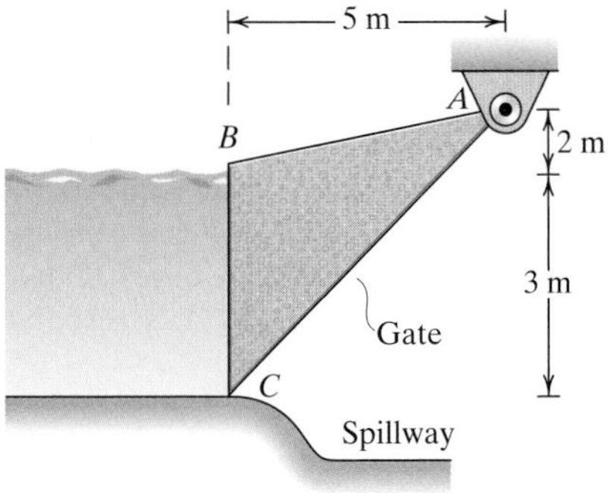

E8.3.12

8.3.13. A flat plate seals a triangular opening in the vertical wall of a tank of liquid with density ρ (**E8.3.13**). The plate is hinged to the upper edge O of the triangle. Determine the force $\boldsymbol{P}$ required to hold the gate in a closed position against the pressure of the liquid.

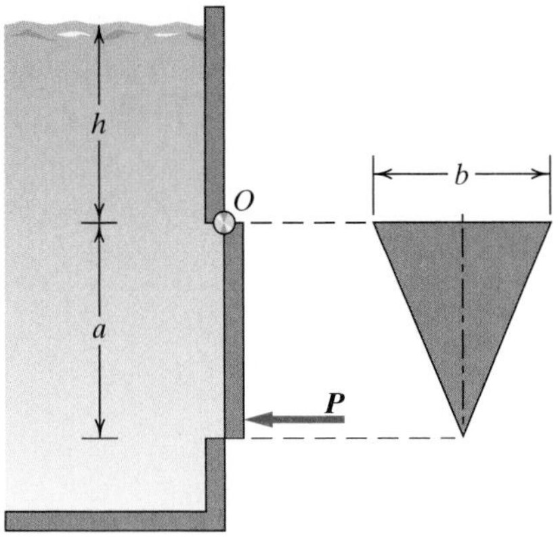

E8.3.13

8.3.14. A rectangular vertical plate in the hull of an aircraft carrier is submerged with its top edge 7 ft below the water surface and its bottom edge 36 ft below the top (**E8.3.14**). The plate is 9 ft wide.

a. Determine the magnitude of the normal pressure force that the sea water ($\gamma = 10$ kN/m^3) exerts on the plate.

b. Locate the center of pressure on the plate.

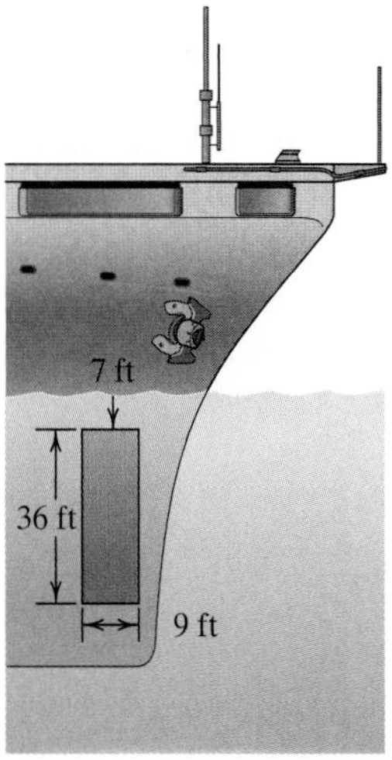

E8.3.14

8.3.15. The triangular water trough in **E8.3.15** consists of panel $ABEF$, which is supported by a horizontal hinge along AE. In addition, 10 horizontal cables, each 1.5 ft long, support the panel along CD and are equally spaced along the 20 ft length of the trough. If the tension in any one cable is to be limited to 60 lb, what is the maximum depth d that the water trough should be filled?

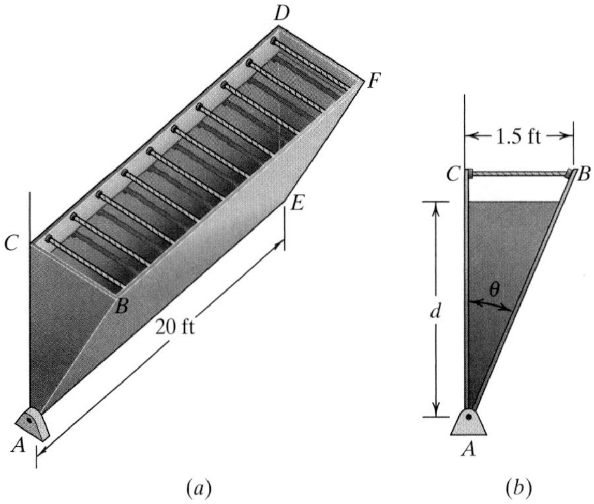

E8.3.15

8.3.16. Suppose that cable length of the trough in **E8.3.15** is adjustable, as shown in **E8.3.16**.

a. How does the volume in the trough usable for storing water vary with the angle θ? Present your answer as an equation and as a plot of volume versus angle θ. What value of θ results in the maximum water storage (call this θ_{max})?

b. How does the tension in any one of the cables vary with the angle θ? (Assume that in any given position the trough is filled just to the point of overflowing.) Present you answer as an equation and as a plot of tension versus angle θ.

c. If you wish to limit the tension in any one cable to 100 lb at θ_{max} (as found in **a**), how many cables should be used to support the trough?

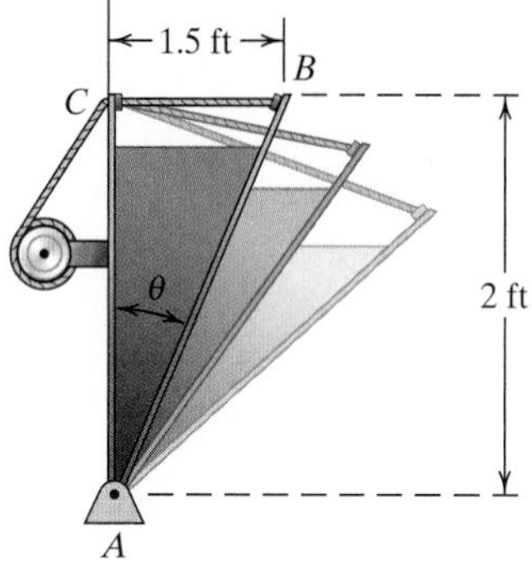

E8.3.16

8.3.17. Gate AB in **E8.3.17** is 3 m tall and 1 m wide (dimension perpendicular to page) and holds back water of depth 1.5 m. The density of water is 1000 kg/m^3.

a. What is the minimum force $\boldsymbol{F}$ that must be applied to just get the gate to swing open (i.e., for the bottom edge of the gate to just move away from the stop at B)? The pulleys at C and D are each 0.4 m in diameter and their centers are spaced 0.5 m apart. Assume that the pin joint at A is frictionless and that when the gate rests against the stop, there is normal contact.

b. If the average student can pull with a force of 600 N, how many students will be needed to just get the gate to swing open?

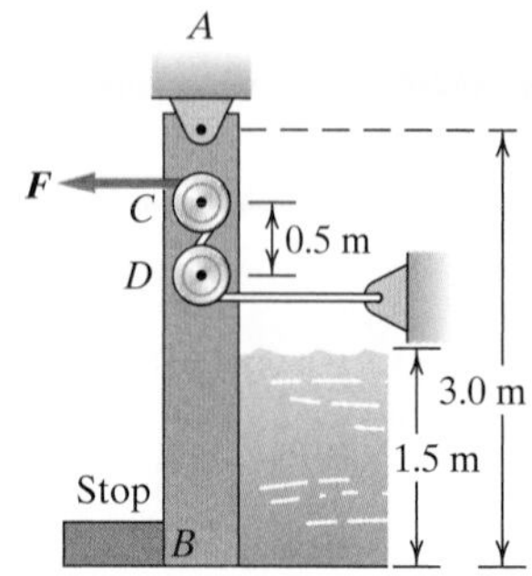

E8.3.17

8.3.18. The gate AB in **E8.3.18** is located at the end of a 2-m-wide water channel (dimension perpendicular to page) and is supported by hinges along its top edge A.

a. If the floor of the channel is frictionless, determine the reactions at A and B when $d = 2.4$ m.

b. What is the minimum force that must be applied to just get the gate to open? Where would this force be applied?

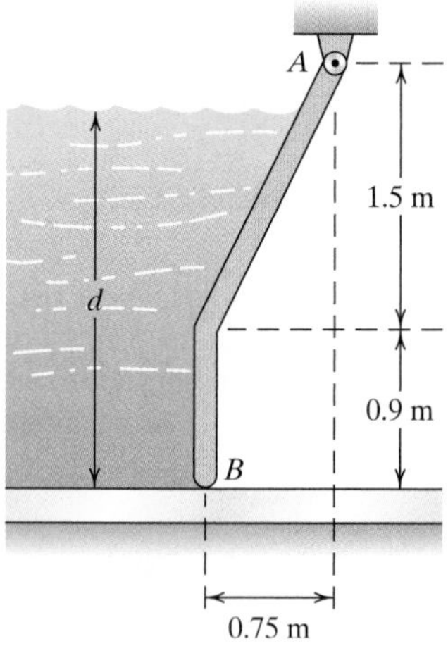

E8.3.18

8.3.19. The end of the freshwater channel in **E8.3.19** consists of a plate $ABCD$ that is hinged at B and is 0.4 m wide (dimension perpendicular to page). Determine the length a for which the reaction at A is zero. Neglect the weight of the plate.

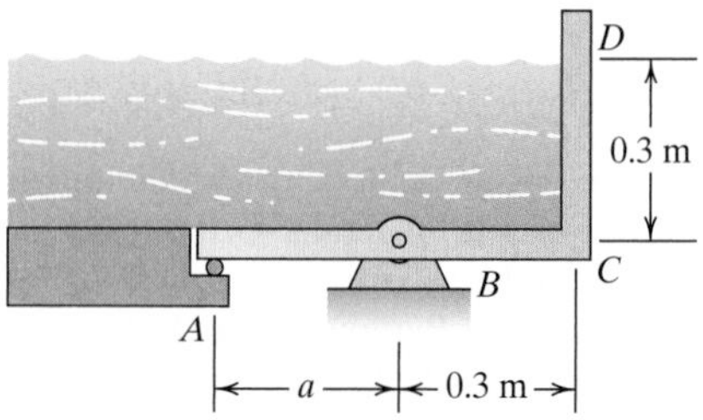

E8.3.19

8.3.20. Determine the point force and its location (center of pressure) that are equivalent to the water pressure acting on the curved surface AB (**E8.3.20**).

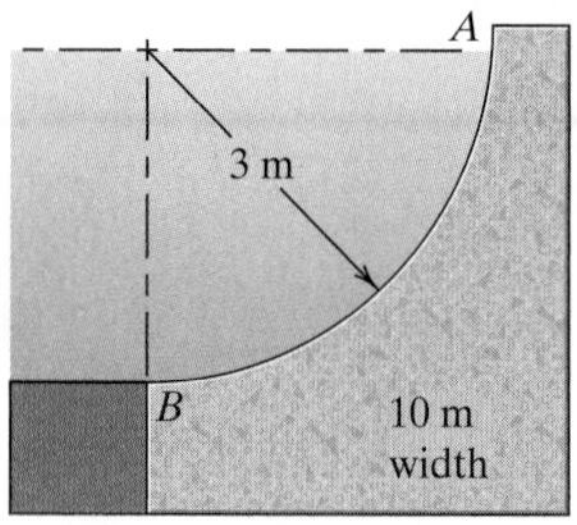

E8.3.20

8.3.21. Determine the point force and its location (center of pressure) that are equivalent to the water pressure acting on the curved surface AB (E8.3.21).

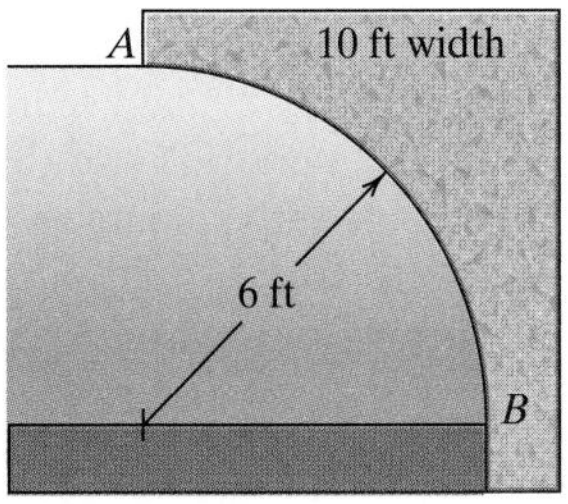

E8.3.21

8.3.22. Determine the point force and its location (center of pressure) that are equivalent to the water pressure acting on the curved surface AB (E8.3.22).

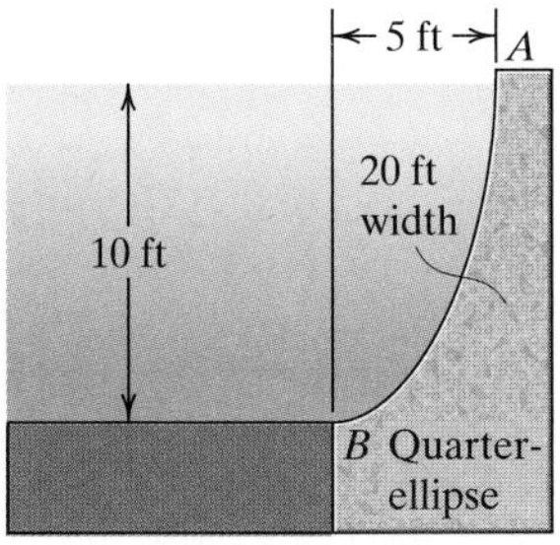

E8.3.22

8.3.23. Determine the point force and its location (center of pressure) that are equivalent to the water pressure acting on the curved surface AB (E8.3.23).

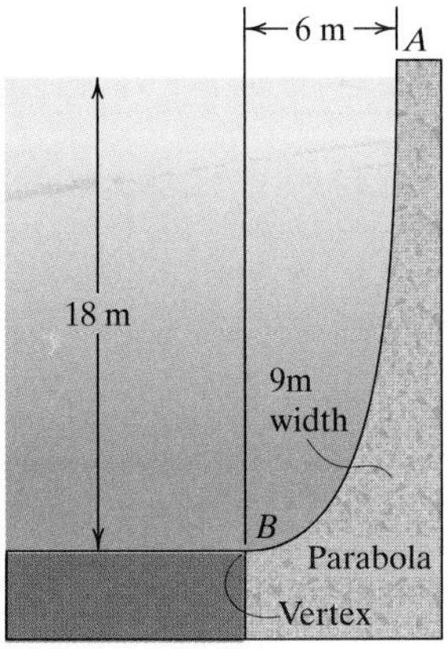

E8.3.23

8.3.24. A buoy consists of a 10-ft hollow steel cylinder that is 15 in. in diameter, weighs 200 lb, and is anchored to the bottom with a cable as shown in E8.3.24. The specific weight of sea water is 65 lb/ft^3. Assume the buoy is weighted at its base so that it remains vertical.

a If $h = 3$ ft at high tide, determine the tension $\boldsymbol{T}$ in the cable.

b. Find the value of h when the cable goes slack as the tide drops.

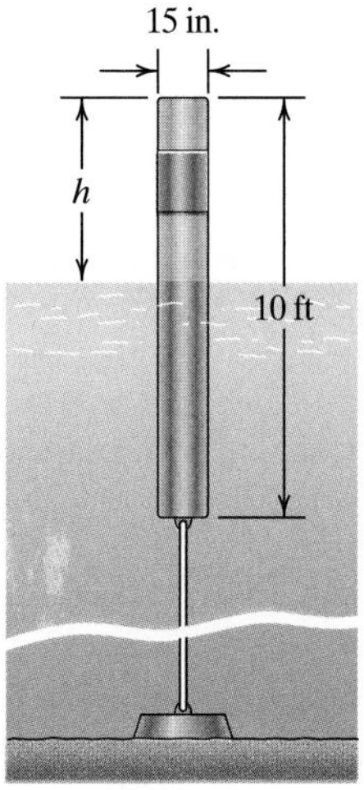

E8.3.24

8.3.25. The uniform 31-kg pole of 100-mm diameter is hinged at A, and its lower end is immersed in fresh water, as shown in E8.3.25. Determine the tension $\boldsymbol{T}$ in the vertical cable required to maintain C at a depth of 0.5 m.

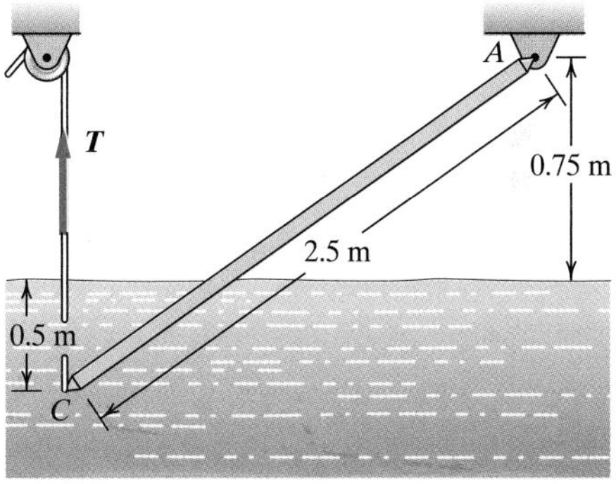

E8.3.25

8.3.26. A buoy in the form of a uniform 8-m-long pole that is 0.2 m in diameter has a mass of 200 kg and is secured at its lower end to the bottom of a freshwater lake with 10 m of cable, as shown in E8.3.26. If the depth of the water is 15 m, determine the angle θ made by the pole with the horizontal.

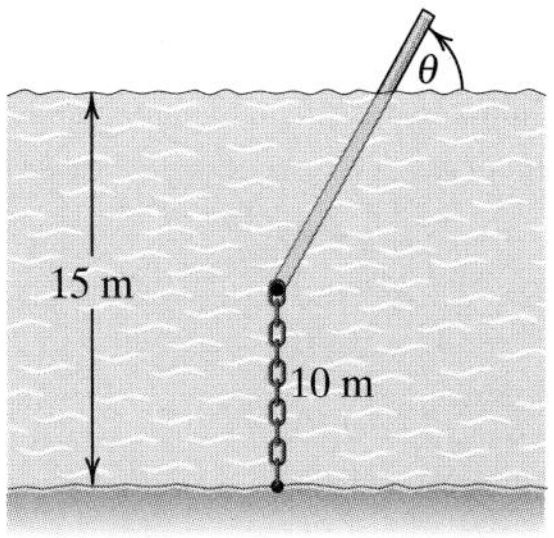

E8.3.26

8.3.27. The hinged gate ABC closes an opening of width w (perpendicular to the paper) in a water channel (E8.3.27). The water has access to the underside as well as the right side of the gate. When the water level rises above a certain value of depth d, the gate will open (by rotating counterclockwise about the hinge at B).

a. Determine the critical value of d that will cause the gate to rotate. Neglect the weight of the gate.

b. If the weight of the gate is not ignored, would the depth d at which it would open increase, decrease, or remain the same? (Don't carry out any calculations; instead, base your answer on reasoning about the moments about B.)

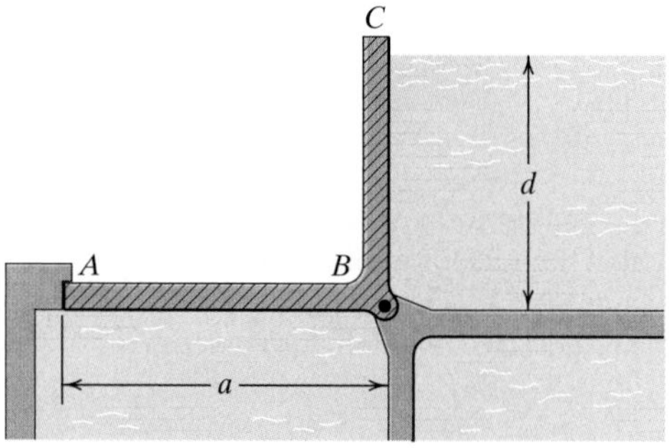

E8.3.27

8.3.28. The sides of a V-shaped freshwater trough, shown in cross section in E8.3.28, are hinged along their common intersection through O and are supported on both sides by vertical struts hinged at each end and spaced every 2 m. Determine the compressive force $\boldsymbol{F}$ in each strut. Neglect the weight of the members.

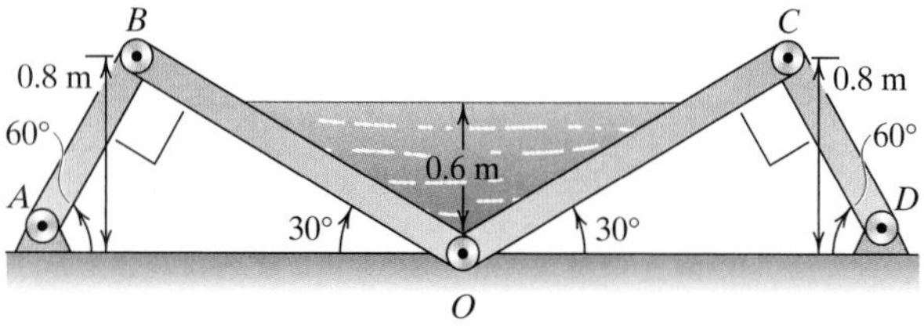

E8.3.28

8.3.29. A steel sphere of radius $r = 30$ mm and of specific weight $\gamma_s = 76.8$ kN/m^3 is suspended by a thin wire.

a. When the sphere hangs from the thin wire in air, what is the tension $\boldsymbol{T}$ in the wire?

b. The sphere is lowered into a tank containing oil of specific weight $\gamma_{oil} = 7.8$ kN/m^3 and mercury of specific weight $\gamma_{Hg} = 133$ kN/m^3, until it is two-thirds submerged in oil and one-third submerged in mercury. Determine the tension $\boldsymbol{T}$ in the wire.

c. Express your answer in **b** as a percentage of your answer in **a**.

8.3.30. E8.3.30 shows the underwater cross-sectional area A (m^2) of the bow at the waterline of a sailboat hull. The variation of A with x is shown in the graph for a particular hull. Determine the distance x to the center of buoyancy of the hull (centroid of the displaced volume of water). The location of the center of buoyancy is a critical parameter in the design of the hull.

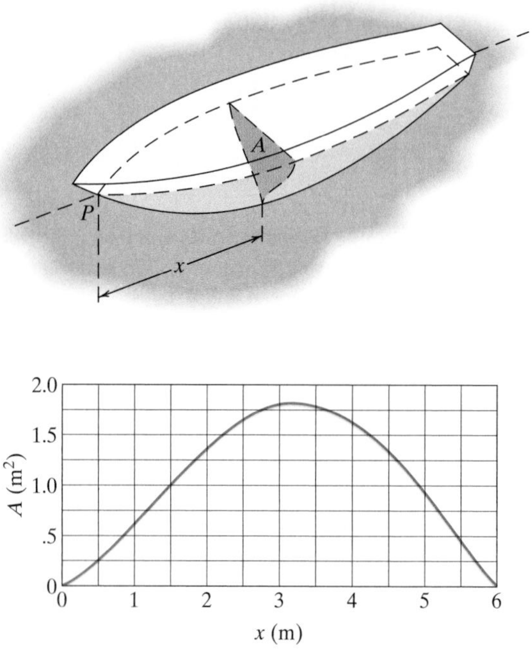

E8.3.30

8.4 JUST THE FACTS

Volumes

The individual particle masses summed together are the **total mass** of the object. The total mass is

$$M = \int_{\text{volume}} \rho\, dV \text{ (general case)} \tag{8.1}$$

where ρ is the object's density in mass/volume, $\rho\, dV$ is the mass of a volume element dV of the object, and integration is throughout the object's volume.

The weight of the object is given by

$$Mg = W = \int_{\text{volume}} \rho g \, dV \tag{8.2A}$$

Alternately, (8.2A) can be rewritten in terms of **specific weight** γ, (where γ is weight per unit volume as is ρg) as

$$Mg = W = \int_{\text{volume}} \gamma \, dV \tag{8.2B}$$

The **center of mass** of an object in a uniform gravitational field is at location

$$X_M = \frac{\int_{\text{volume}} x\rho \, dV}{M}; \quad Y_M = \frac{\int_{\text{volume}} y\rho \, dV}{M}; \quad Z_M = \frac{\int_{\text{volume}} z\rho \, dV}{M} \tag{8.4}$$

If we can treat the gravity field as uniform and parallel (which is approximately true for small objects on the earth), the center of mass (as defined in (8.4)) is also the location of the **center of gravity**.

The **centroid of a volume** is the location in a volume composed of a **homogeneous** material (meaning that it is uniform throughout the volume) and is at

$$X_C = \frac{\int_{\text{volume}} x \, dV}{\int_{\text{volume}} dV}; \quad Y_C = \frac{\int_{\text{volume}} y \, dV}{\int_{\text{volume}} dV}; \quad Z_C = \frac{\int_{\text{volume}} z \, dV}{\int_{\text{volume}} dV} \tag{8.5}$$

The locations of the centroid of several standard volumes are presented in Appendix A3.2. The centroid is also the location of the center of mass and center of gravity if the volume is of uniform density.

If we can decompose a composite volume into one made up of N standard volumes, we can use knowledge of the location of the centers of gravity of the N standard volumes to find the location of the center of gravity (X_G, Y_G, Z_G) of the composite volume. Call W_i the weight of an individual volume, and call X_{iG}, Y_{iG}, Z_{iG} the location of its center of gravity. Therefore the location of the **center of gravity of the composite volume** is

$$X_G = \frac{\sum_{i=1}^{N} W_i X_{iG}}{W_{\text{tot}}}; \quad Y_G = \frac{\sum_{i=1}^{N} W_i Y_{iG}}{W_{\text{tot}}}; \quad Z_G = \frac{\sum_{i=1}^{N} W_i Z_{iG}}{W_{\text{tot}}} \tag{8.7A}$$

Similarly, the **center of mass of a composite volume** is

$$X_M = \frac{\sum_{i=1}^{N} M_i X_{iM}}{M_{tot}}; \qquad Y_M = \frac{\sum_{i=1}^{N} M_i Y_{iM}}{M_{tot}}; \qquad Z_M = \frac{\sum_{i=1}^{N} M_i Z_{iM}}{M_{tot}} \tag{8.7B}$$

and the **centroid of the composite volume** is at

$$X_C = \frac{\sum_{i=1}^{N} V_i X_{iC}}{V_{tot}}; \qquad Y_C = \frac{\sum_{i=1}^{N} V_i Y_{iC}}{V_{tot}}; \qquad Z_C = \frac{\sum_{i=1}^{N} V_i Z_{iC}}{V_{tot}} \tag{8.7C}$$

Areas

The **centroid of an area** is at

$$X_C = \frac{\int_{\text{area}} x\, dx\, dy}{A_{tot}} \tag{8.10A}$$

$$Y_C = \frac{\int_{\text{area}} y\, dx\, dy}{A_{tot}} \tag{8.10B}$$

where

$$A_{tot} = \int_{\text{area}} dx\, dy \tag{8.8}$$

The locations of centroids of several standard areas are presented in Appendix A3.1.

If we can decompose a composite area into one made up of N standard areas, we can use knowledge of the locations of the centroids of the N standard areas to find the location of the centroid (X_C, Y_C) of the composite area by

$$X_C = \frac{\sum_{i=1}^{N} A_i X_{iC}}{A_{tot}}; \qquad Y_C = \frac{\sum_{i=1}^{N} A_i Y_{iC}}{A_{tot}} \tag{8.12}$$

Distributed Force Along a Line and Its Centroid

The single point force equivalent to a distributed line load is

$$\text{Total force in } y \text{ direction} = F_y = \int_{\text{span}} \omega\, dx \tag{8.13}$$

where ω is the distributed line load oriented in the y direction and has dimension force/length, and "span" refers to the length over which ω is distributed. The point at which the single point force acts is called the **centroid of a line load** and is at

SA8.4 The Freedom Ship

As of this printing, the Freedom Ship has not been completed, but it has been featured on the Discovery Channel's "Engineering the Impossible." It's projected cost is over \$9 billion and, if it is ever built, it will be approximately 1 mile long, 750 feet wide, and nearly 25 stories (~340 ft.) tall. Originally, it was planned to be a floating city, but it is currently being designed to continuously circle the globe, visiting most of the earth's inhabited coastal regions every two years. Figure SA8.4.1 is a concept drawing of the ship.

Perhaps the most impressive statistic is the proposed weight of the ship, approximately 2.7–3.0 million tons. It will be able to house nearly 50,000 passengers with a crew of 15,000. Your goal is to figure out how deep this boat will sit in the water. A very simple analysis you might say. But it is a good starting point. The most difficult aspect is the actual construction. It is quite likely that completely new manufacturing techniques will have to be developed before the ship ever leaves port.

First, recall Archimedes' principle: the upward acting buoyancy force acting on a submerged (or partially submerged) object is equal to the weight of the water that is displaced (density of water = ~1000 kg/m^3). A relatively simple free-body diagram then yields an expression for the depth that the ship sits in the water in terms of its weight and the overall dimensions of the ship.

When faced with such large numbers, it is useful to do a few simple calculations to develop some intuition for the problem. How much higher or lower will the boat sit if you decrease or increase its overall weight by 10%? What would the change be if you lengthened or shortened the overall dimensions by 10%? This is a simple example of a sensitivity analysis and, in addition to helping an engineer better understand the problem, it may alert him or her to a previously unconsidered design parameter or problem constraint.

www.freedomship.com

Figure SA8.4.1 A computer-generated overhead image of the proposed Freedom Ship

SA8.5 Center of Mass Calculations

As we all know, Spider-Man is able to stick to virtually any surface. In Figure SA8.5.1, he is contacting the wall with only his feet. Given that Peter Parker is 5 ft, 10 in. tall and weighs approximately 170 lb, calculate his center of gravity and estimate the reaction forces exerted on his feet.

At first glance, this is a very involved problem and a number of assumptions need to be made in order to obtain a good approximation. As always, start with a free-body diagram of Spider-Man, assuming that the center of gravity of each body part occurs at its midpoint. **Figure SA8.5.1*b*** may be helpful. Then think very carefully about the kinds of reactions that may be supported by our superhero's feet. If it helps, you can assume that the loads are evenly distributed between each foot.

The data in Tables SA8.5.1 and SA8.5.2 will be useful.

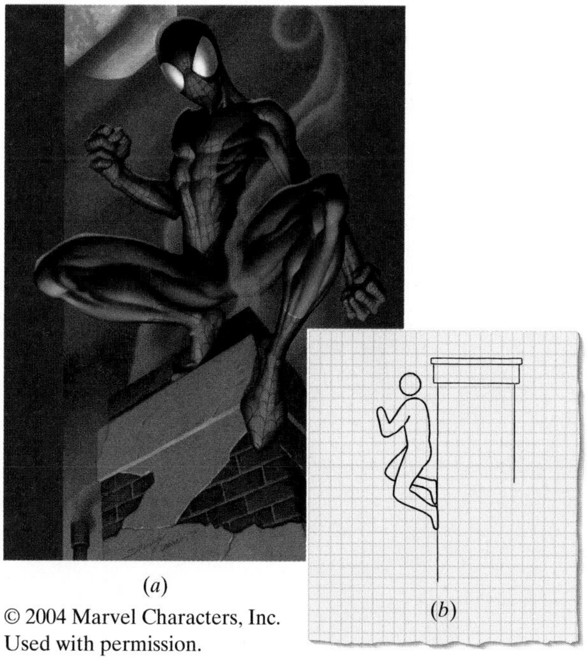

(*a*)

Figure SA8.5.1 (*a*) Spider-Man sticking to a wall; (*b*) a simplified two-dimensional view

Table SA8.5.1 Mean segment weights expressed as percentages of total body weight

Segment	Weight for Males	Weight for Females
Head and trunk	55.10	53.20
Total arm	5.77	4.97
Thigh	10.50	11.75
Shank and foot	6.18	6.68

Adapted from Williams and Lissner, *Biomechanics of Human Motion* (Philadelphia: W. B. Saunders, 1962).

Table SA8.5.2 Mean segment lengths expressed as percentages of total body height

Segment	Length for Males	Length for Females
Head and trunk	40.75	39.75
Total arm	32.9	33.3
Thigh	23.2	24.9
Shank and foot	28.95	29.95

Adapted from Williams and Lissner, *Biomechanics of Human Motion* (Philadelphia: W. B. Saunders, 1962).

SA8.6 Fighter Jet Design

The design of a high-performance jet, especially one that will see combat, is an extremely long process that requires input from teams of mechanical, electrical, chemical, and biomedical engineers. This problem will help you get started and give you an idea of some of the many design considerations involved. Choose one of the planes from **Table SA8.6.1** and use the information provided in the table and in **Figure SA8.6.1** to estimate the forces and moments that keep the wing attached to the fuselage. Suggested steps are as follows:

Step 1: Choose an operating mass and calculate the weight of the aircraft.

Step 2: Assume that all the lift force that keeps the plane in the air is generated by the wings. Let's further assume that the pressure is distributed evenly over the entire underside of the wing. In that case, the lift force acts through the centroid of the wing. Estimate the location of the wing's centroid. If the plane is moving at constant velocity, what is the pressure exerted on the underside of the wing?

Step 3: If we assume that the wing is constructed from a uniform piece of steel (7830 kg/m^3) 2 cm thick, what is the gravitational force exerted on the wing?

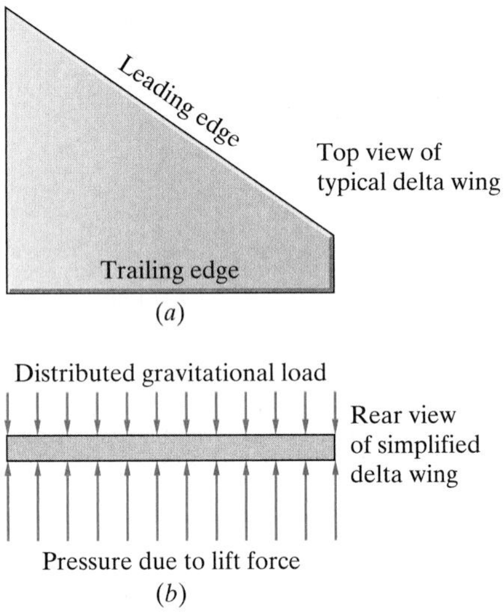

Figure SA8.6.1 A delta wing design commonly used on military fighter jets (e.g., the F-15E Eagle)

Step 4: Assume that the wing is cantilevered to the fuselage. Calculate the forces and moments that act on the wing at the fuselage. Keep in mind that there are several ways to do this. Does your answer reflect a best-case scenario, a worst-case scenario, or a relatively accurate estimate?

Table SA8.6.1 Aircraft Specifications

Aircraft	Wingspan (m)	Length (m)	Empty	Mass (kg) Normal	Maximum	Powerplant (kN)
Mirage 2000 C	9.13	14.36	7,500	10,680	17,000	1 × 64
F15-E Eagle	13.05	19.43	14,379		36,741	2 × 65
F/A 18 C	11.43	17.07	10,455	16,652	25,401	2 × 71
F 14A Tomcat	19.54	19.10	18,191	26,632	74,349	2 × 93
F 104 Starfighter	6.68	16.69	6,387	9,840	13,054	1 × 44
YF 22 Lightning II	13.11	19.56	13,608		26,308	2 × 156
F-16 Fighting Falcon	9.45	15.03	8,663	9,791	19,187	1 × 123
MIG 29 Fulcrum	11.36	17.32		15,300	19,700	2 × 49
A-10A Thunderbolt II	17.53	16.26	9,771	14,865	22,680	2 × 40
Sukhoi Su-37	15.16	22.18		26,000	34,000	2 × 130
Tornado ADV	13.91	18.68	14,502		27,986	2 × 40
B-1B Lancer Bomber	41.67	44.81	87,091		216,365	4 × 65
AV-SB Harrier	9.25	14.12	6,336	10,410	14,061 (STO) 8,596 (VTO)	1 × 106
SR-71 Blackbird	16.94	31.17			53,049	2 × 133

Adapted from *The Encyclopedia of World Military Aircraft*, edited by David Donald and Jon Lake (New York: Barnes & Noble, 2000).

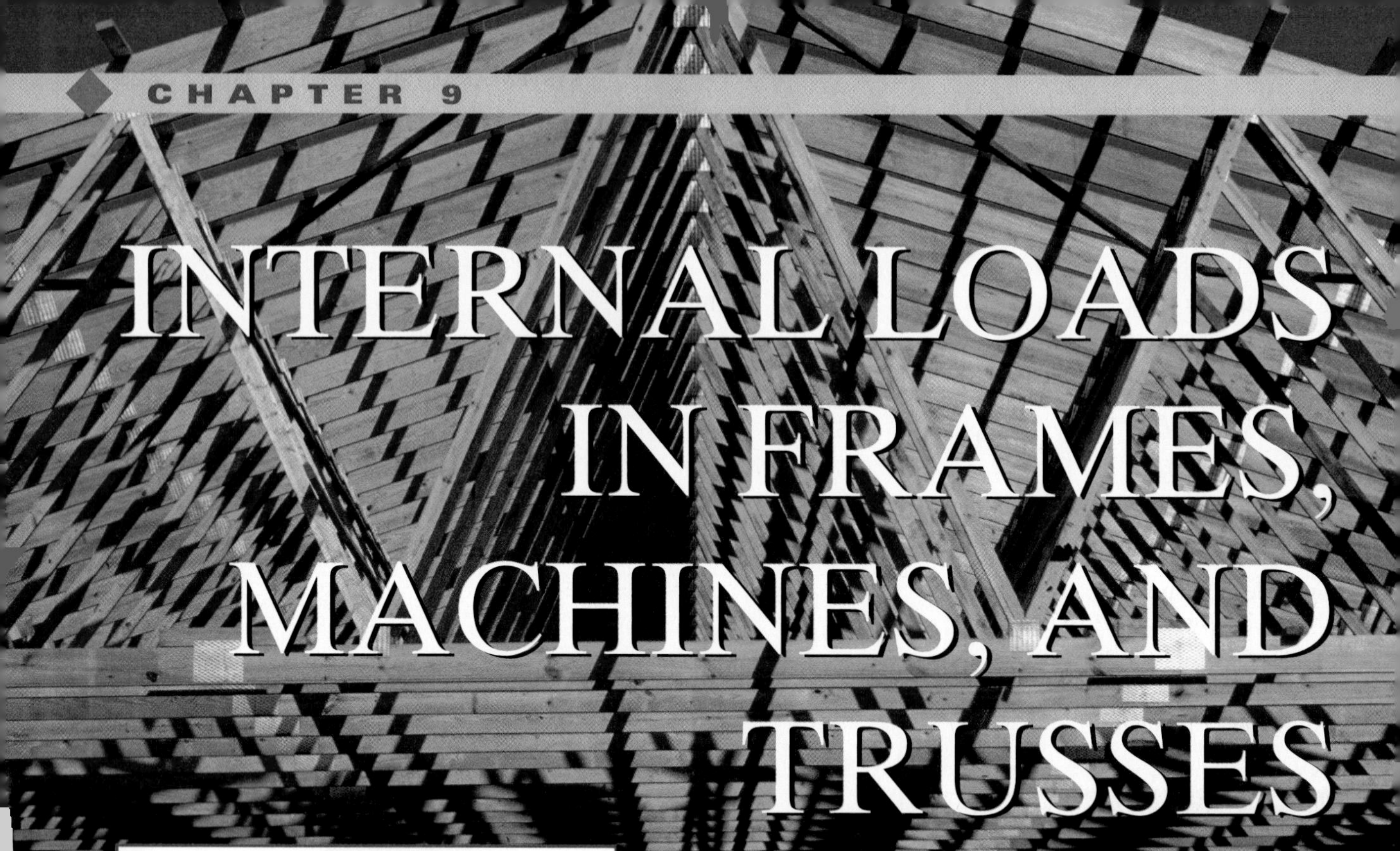

INTERNAL LOADS IN FRAMES, MACHINES, AND TRUSSES

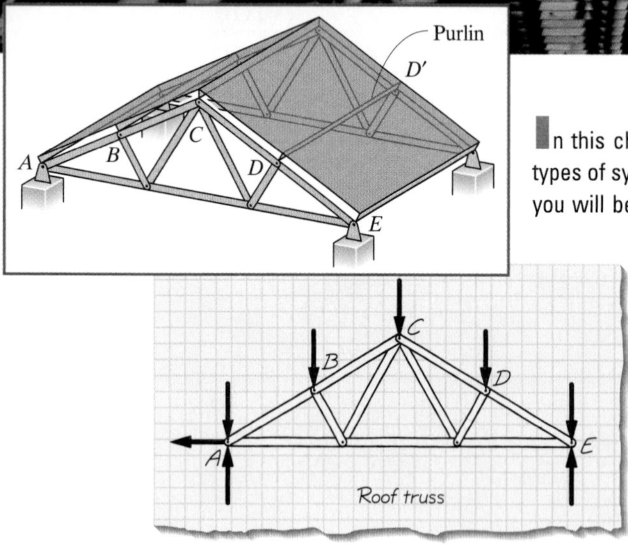

In this chapter we use these two ideas to look at loads internal to three types of systems: frames, machines, and trusses. By the end of this chapter, you will be able to systematically find the loads acting on members within such systems. You will also be able to identify whether the circus troupe in **Figure 9.1** pose constitutes a frame, truss, or machine! (The answer to this is somewhere in the chapter.)

OBJECTIVES

On completion of this chapter, you will be able to:

- **Define and identify frames, machines, and trusses**
- **Carry out equilibrium analysis of a frame and its members**
- **Carry out equilibrium analysis of a truss or its members, using the method of joints, the method of sections, or a combination of these two methods**
- **Determine whether a frame is internally stable with or without redundancy**
- **Determine whether a truss is internally stable with or without redundancy**
- **Include the effects of bearing friction and rolling resistance in equilibrium analysis of machines**

Consider the pose of six members of the circus performance troupe shown in **Figure 9.1*a***.[1] Defining these three performers as the system, we can draw a free-body diagram for the system and setup and solve the equations of equilibrium in order to define the loads acting on the system (**Figure 9.1*b***). Now suppose that we want to find the loads acting on

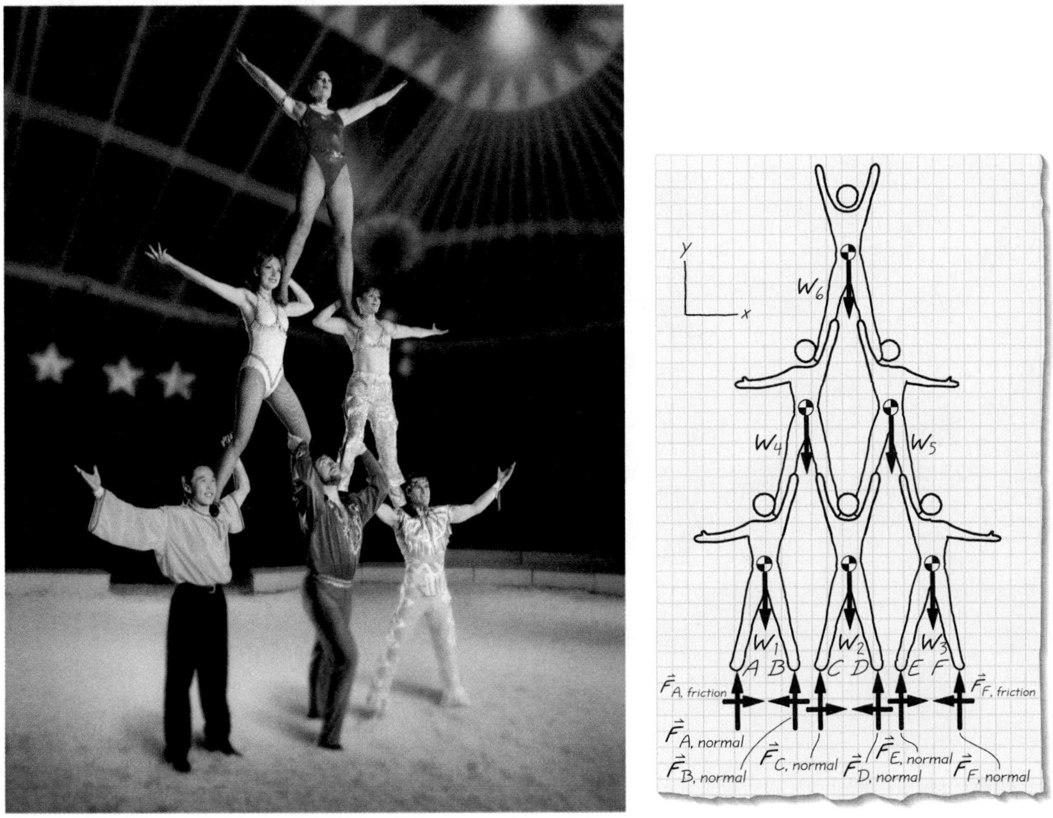

(*a*) (*b*)

Figure 9.1 (*a*) Six performers in the circus; (*b*) free-body diagram of the performers

[1]Unlike the physical poses presented at the beginning of prior chapters, it is not recommended that the reader try reproducing the pose in this figure!

the shoulders of member A or member B. To find these loads, we apply two key ideas:

- Loads internal to a system exist as equal and opposite pairs and are therefore self-canceling (this is Newton's third law, and we refer to the pairs as third-law force pairs). When we draw a boundary around a member (or group of members) in a system, one component of each force pair at the boundary becomes an external force acting on the member.
- If the system as a whole is in equilibrium, then each member (or group of members) is in equilibrium. This means that equilibrium equations can be written for each member in terms of the external loads acting on that member.

9.1 FRAME ANALYSIS

A **frame** is a system designed to support loads, both forces and moments. It consists of members that are connected together by a variety of methods: welds, pins, rivets, glue, nails, wedging, and/or bolts. Frames may be made up of a few members or many, but at least one member must be a **multiforce member**, which is any member that is not a two-force member (was defined in Section 7.2). **Figure 9.2** shows

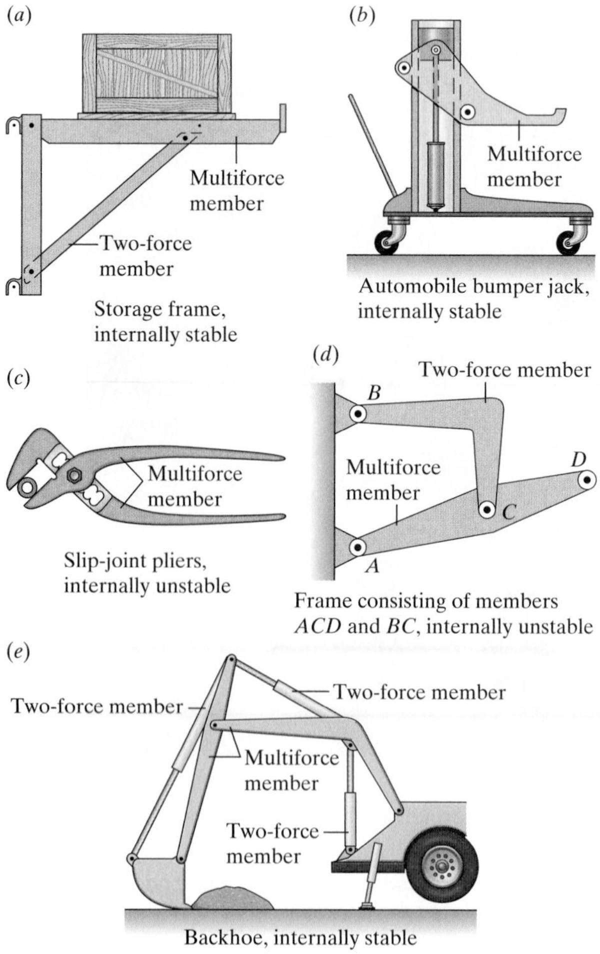

Figure 9.2 (*a*) Storage frame is an internally stable frame. (*b*) Automobile jack is an internally stable frame. (*c*) Slip-joint pliers are an internally unstable frame. (*d*) Frame consisting of members ACD and BC is an internally unstable frame. (*e*) Back hoe is an internally stable frame.

examples of frames you may have passed on your way to class, in a workshop, at an airport, or at a construction site.

Frames are classified as being either **internally stable** or **internally unstable**. When you disconnect an internally stable frame from the rest of the world, its members remain in the same configuration relative to one another. This is in contrast to an internally unstable frame—when it is disconnected from the rest of the world, its members are movable relative to one another. Internally stable and unstable frames are noted in the caption of **Figure 9.2**.

Frames can also be classified as **planar** or **nonplanar**. Planar frames are ones in which all forces and members lie in a single plane (or it is reasonable to treat them as such) and all moments act about an axis perpendicular to this plane. A frame that is not planar is nonplanar. All of the frames in **Figure 9.2** are planar, whereas those shown in Figure 9.3 are nonplanar.

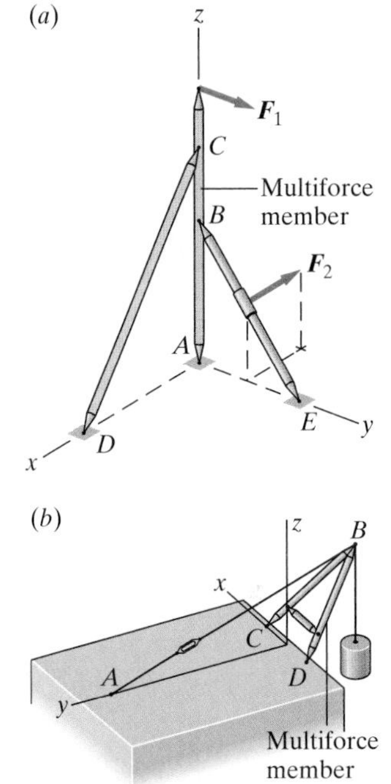

Figure 9.3 (*a*) Tower held by members *CD* and *BE* is an internally unstable frame. (*b*) Frame *CBD* is an internally stable frame.

Determining Internal Loads in Frames

Suppose we want to know about the loads acting at the boundary of and within a frame. For example, let's say we want to find the force carried by member *BC*'s end pins in Figure 9.4*a* or the moment required at *A* to hold the door *OD* in the position shown in **Figure 9.4*b***.

Our analysis will consist of making assumptions, drawing free-body diagrams, setting up and solving equilibrium equations, and summarizing results. This is exactly the analysis procedure we have developed in the preceding chapters. Routinely in the analysis of frames we are concerned with finding the loads acting on the boundary of the frame and the loads internal to the frame. For example, suppose that the frame in **Figure 9.4*a*** is a design idea for a set of lifting tongs. As part of evaluating the design idea we might be interested in finding the force acting on the pins at *B* and *C* in **Figure 9.4*a*** when the frame is lifting an 800-lb crate. The calculated forces acting on the pins would then be compared with their force capacity (i.e., the force level that causes the pin to fail) to see whether they would be up to their proposed task. As another example, suppose that **Figure 9.4*b*** shows the design idea for a ventilation door. As part of specifying the diameter of the pins at *A*, *B*, and *C* so that they are large enough to prevent failure, we need to find the shear forces acting on the pins.

As part of carrying out the analysis in **Figure 9.4*a*** (to find the force in the pins at *B* and *C*) we will draw free-body diagrams of the system and parts of the system, and write and solve multiple sets of linearly independent equilibrium equations. This analysis procedure works for frames that are both statically determinate and internally stable without redundancy (Figure 9.5*a*, Category A). Recall from Chapter 7 that a statically determinate system is one in which there are just enough boundary conditions (supports) for the system to be in equilibrium, and therefore it is possible to determine all external loads acting at the boundary by using just equilibrium conditions. A frame that is **internally stable without redundancy** has just enough members for it to be rigid; with any fewer the frame would not be rigid, and with any more it would have more members than necessary to be rigid. Examples 9.1 and 9.4 illustrate analysis of frames that are internally stable without redundancy.

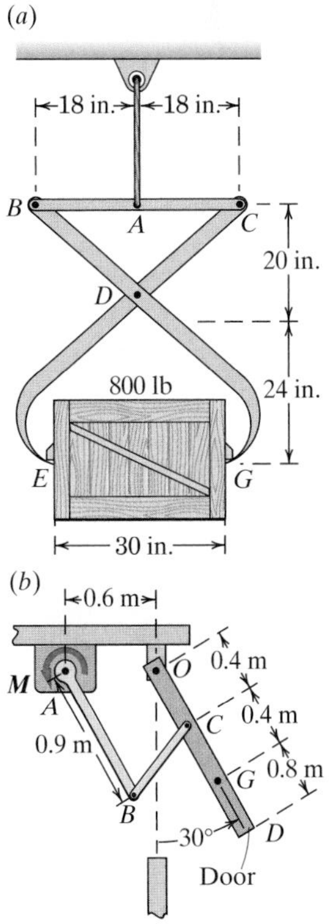

Figure 9.4 Examples of two frames: (*a*) lifting tongs, and (*b*) powered door.

The analysis procedure outlined above also works for **internally unstable frames** when the combination of boundary conditions and frame

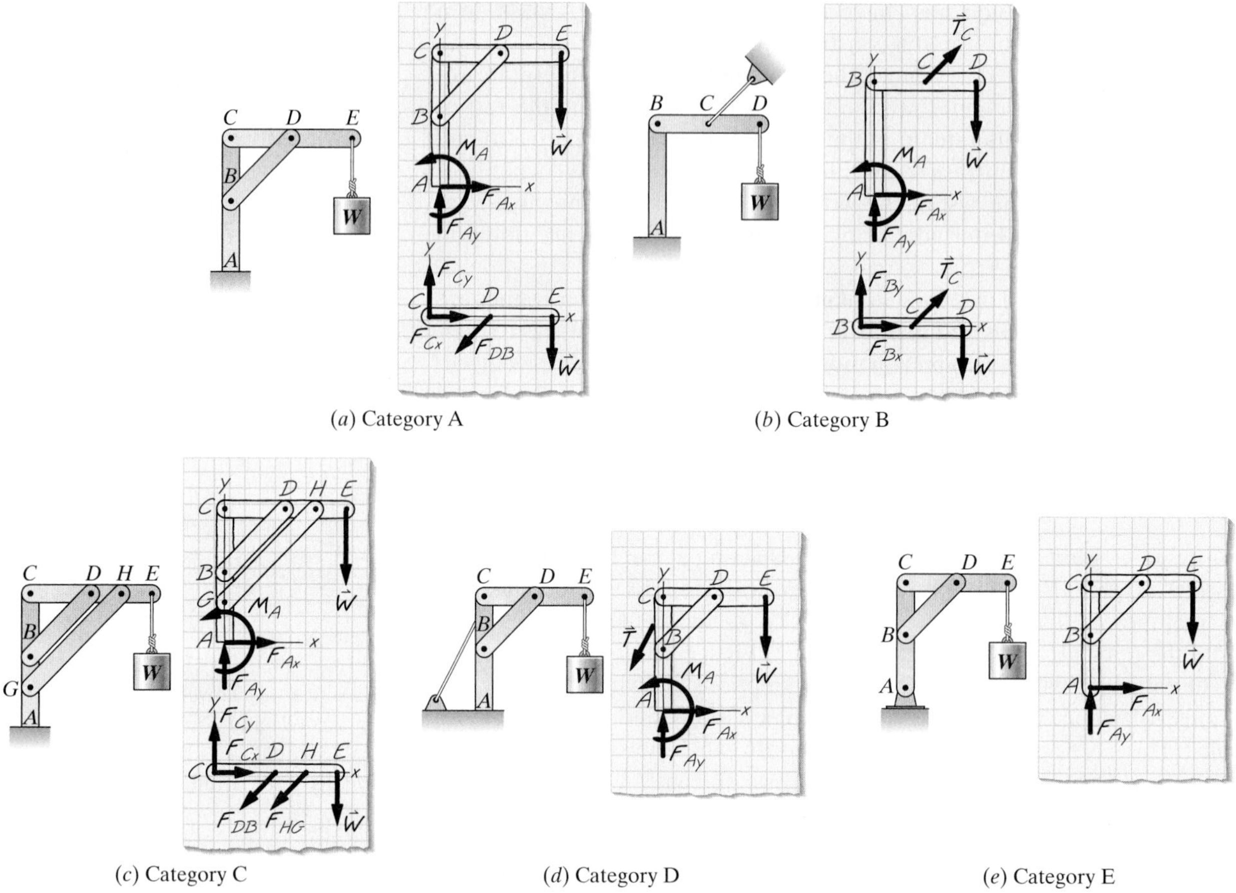

Figure 9.5 Frame that is (*a*) statically determinate and internally stable without redundancy (Category A); (*b*) internally unstable with sufficient boundary conditions to be statically determinate (Category B); (*c*) statically determinate and internally stable with redundancy (Category C); (*d*) statically indeterminate and internally stable with or without redundancy (Category D); (*e*) underconstrained and internally stable (Category E).

members are just enough for the frame to be statically determinate as illustrated by the frame in **Figure 9.4*b*** and in **Figure 9.5*b***, Category B. We are able to generate enough linearly independent equilibrium equations to find boundary and internal loads. Examples 9.2 and 9.3 illustrate analysis of frames that are internally unstable but that have just enough boundary conditions to prevent them from moving and/or collapsing.

EXAMPLE 9.1 PLANAR FRAME ANALYSIS

Three hydraulic cylinders control the motion of the backhoe arm and bucket in **Figure 9.6**. In the position shown, a 14.0-kN horizontal force is applied to the bucket at point *L*.

(a) Assuming the weight of the backhoe arm is negligible compared to forces acting on it, find the force acting on pin *A*.

(b) Determine whether the force acting on hydraulic cylinder *HG* exceeds its capacity of 40 kN.

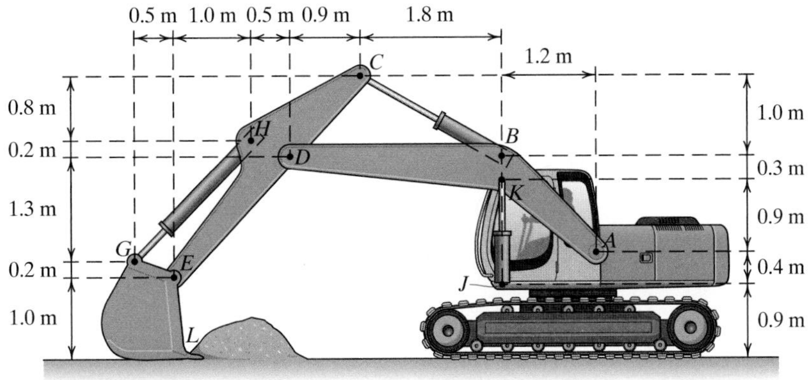

Figure 9.6

Goal We are to find the force acting on the pin at A (a) and determine whether the load acting on HG exceeds its capacity (b).

Given We are given information about the geometry of the backhoe, the load at L, and the capacity of HG.

Assume We assume that we can treat the system as planar because the forces are all acting in the plane of the page. We also assume that all of the hydraulic cylinders act as two-force members because they are pinned at both ends and have no other external forces acting on them.

Draw For part (a) we draw an imaginary boundary around the backhoe arm, which isolates it from the cab at A and J. We draw a free-body diagram (Figure 9.7) of the arm realizing that since hydraulic cylinder JK is a two-force member, the unknown force at J must act along the axis of the member. We represent the unknown force at the pin connection A by two mutually perpendicular forces (see **Table 6.1**). For part (b) we isolate the bucket and draw a free-body diagram (Figure 9.8).

Formulate Equations and Solve **(a)** We set up the equations for planar equilibrium to find the unknown load at A.

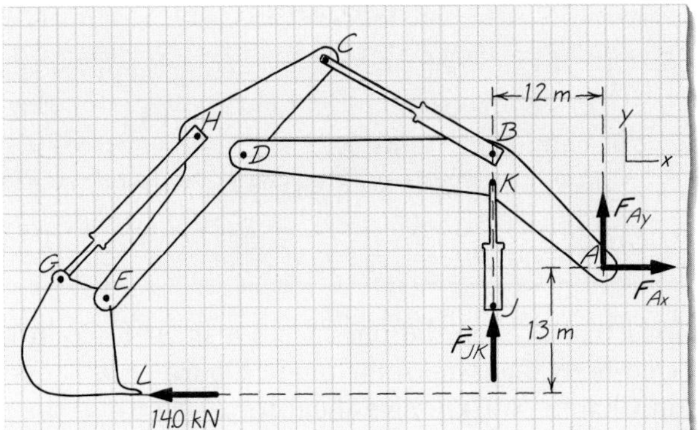

Figure 9.7

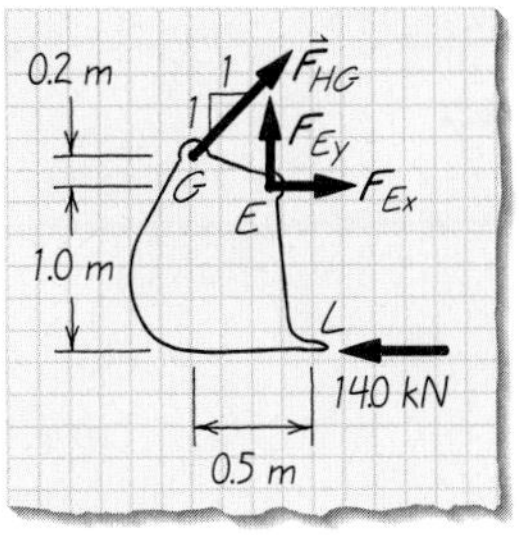

Figure 9.8

We start with the moment equation (7.5C) because (when we place the moment center at A), there is only one unknown in the equation. Based on (7.5C) and the free-body diagram in **Figure 9.7**, with the moment center at A:

$$\sum M_{z@A} = 0(\curvearrowleft +)$$
$$-F_{JK}(1.2\text{ m}) - 14.0\text{ kN}(1.3\text{ m}) = 0$$
$$F_{JK} = -14.0\text{ kN}\left(\frac{1.3\text{ m}}{1.2\text{ m}}\right) = -15.2\text{ kN}$$
$$F_{JK} = -15.2\text{ kN}$$

The negative sign indicates that the force at J is acting in the opposite direction from our free-body diagram, meaning that the hydraulic cylinder is actually in tension.

Based on (7.5A):

$$\sum F_x = 0(\rightarrow +)$$
$$-14.0\text{ kN} + F_{Ax} = 0$$
$$F_{Ax} = 14.0\text{ kN}$$

Based on (7.5B):

$$\sum F_y = 0(\uparrow +)$$
$$F_{JK} + F_{Ay} = 0$$
$$F_{Ay} = -F_{JK} = 15.2\text{ kN}$$

Now we have all of the information to find the magnitude of $\boldsymbol{F}_A$ (the force acting on the pin at A):

$$\|\boldsymbol{F}_A\| = (F_{Ax}^2 + F_{Ay}^2)^{1/2} = [(14.0\text{ kN})^2 + (15.2\text{ kN})^2]^{1/2}$$
$$\|\boldsymbol{F}_A\| = 20.7\text{ kN}$$

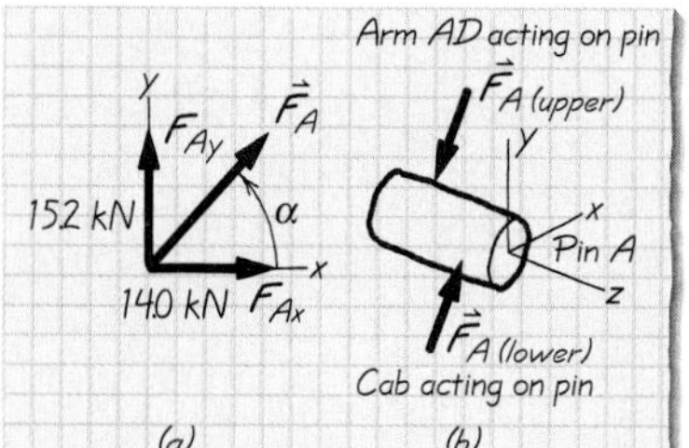

Figure 9.9

The orientation angle α of $\boldsymbol{F}_A$ (**Figure 9.9*a***) is given by

$$\alpha = \tan^{-1}\left(\frac{15.2\text{ kN}}{14.0\text{ kN}}\right) = 47.4°$$

Answer to (a) The magnitude of the force acting on pin A is $\|\boldsymbol{F}_A\| = 20.7$ kN. Its orientation is given by $\alpha = 47.4°$ where α is defined in **Figure 9.9**. Alternately, we could give the answer in vector notation as $\boldsymbol{F}_A = 14.0\text{ kN}\boldsymbol{i} + 15.2\text{ kN}\boldsymbol{j}$, or in terms of space angles as $\|\boldsymbol{F}_A\| = 20.7$ kN, $\theta_x = 47.4°$, $\theta_y = 42.6°$, $\theta_z = 90°$.

Note: The load that acts on the pin causes a shear force in the pin, as shown in a free-body diagram of the pin in **Figure 9.9*b***. Shear force was introduced in Chapter 4. Excessive shear force will cause the pin to fail. Can you think of an alternate design of the pin connection at A that would reduce the likelihood of failure?

(b) Using the free-body diagram drawn in **Figure 9.8**, we set up the equations for planar equilibrium to find the load acting on *HG*.

Based on (7.5C), with the moment center at *E*:

$$\sum M_{z@E} = 0(\curvearrowleft+)$$

$$-F_{HG} \cos 45°(0.2 \text{ m}) - F_{HG} \sin 45°(0.5 \text{ m}) - 14.0 \text{ kN}(1.0 \text{ m}) = 0$$

$$F_{HG} = -14.0 \text{ kN}\left(\frac{1.0 \text{ m}}{0.495 \text{ m}}\right) = -28.3 \text{ kN}$$

Answer to (b) $\|\boldsymbol{F}_{HG}\| = 28.3$ kN (compression) < 40 kN, indicating that the hydraulic cylinder has not exceeded capacity.

Check **(a)** Using an alternate moment center, the results could be substituted into (7.5C) as a check. It is also useful to redraw the free-body diagram with the values of the loads at the supports (**Figure 9.10**); this shows us that the couple created by the vertical forces at *J* and *A* is balanced by the couple created by the horizontal forces at *L* and *A*.

(b) We could solve the planar equilibrium equations with *G* as the moment center, to find F_{Ex}, F_{Ey}, and F_{HG} and compare these results with our previous answer.

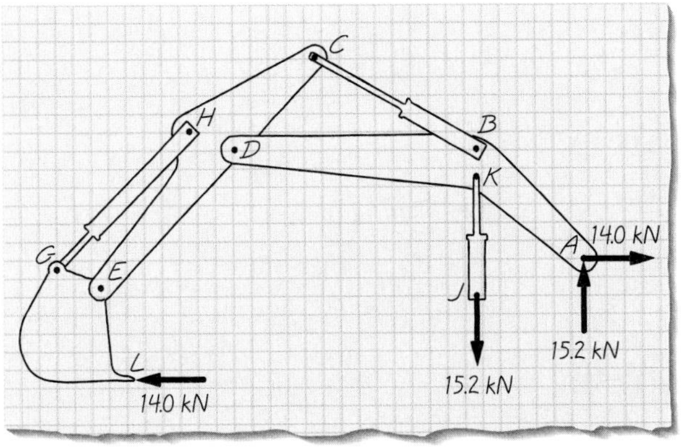

Figure 9.10

EXAMPLE 9.2 INTERNALLY UNSTABLE FRAME

The internally unstable frame in **Figure 9.11** is subjected to the force and moment shown. Assuming the weights of the frame members are negligible, find the loads acting on the frame at the supports at *A* and *C*. Also express these loads in terms of shear forces acting on pins *A* and *C*.

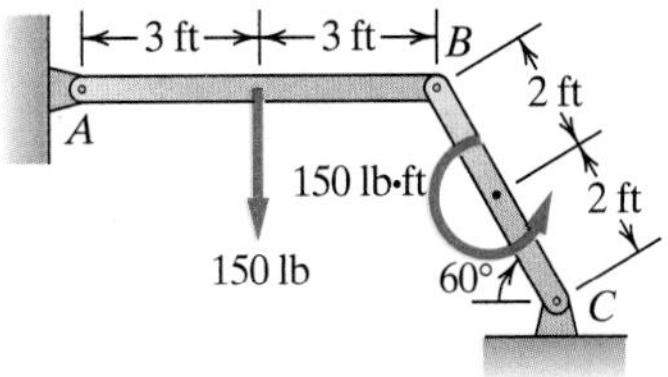

Figure 9.11

Goal We are to find the loads acting on the frame at the supports *A* and *C*.

Given We are given the dimensions of the frame and told that a 150-lb force acts at the midspan of member *AB* and a 150-lb · ft moment acts at the midspan of member *BC*.

Assume We assume that we can treat the system as planar because the 150-lb force acts in the plane of the frame and the moment acts about an axis perpendicular to the frame.

Draw We draw free-body diagrams for the two members, making sure that forces at B on member AB are equal and opposite to the forces at B on member BC (**Figure 9.12** and **Figure 9.13**).[2]

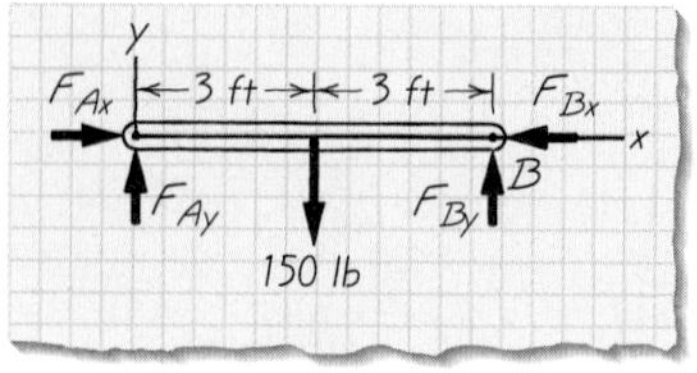

Figure 9.12 Free-body diagram of member AB

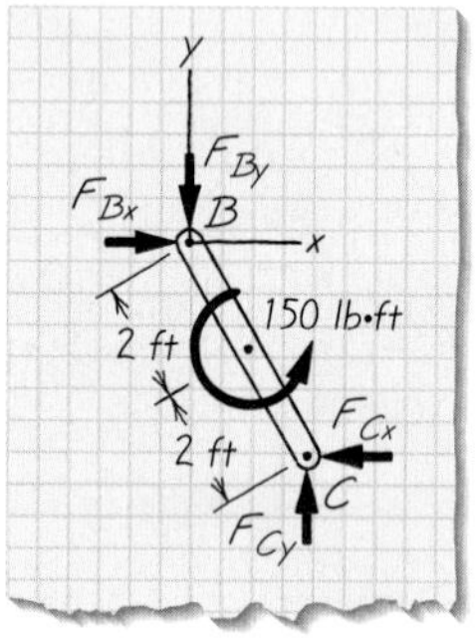

Figure 9.13 Free-body diagram of member BC

Formulate Equations and Solve If, instead of the free-body diagrams in **Figures 9.12** and **9.13**, we drew a single free-body diagram of the whole frame ABC, we would find that our three equlibrium equations would contain four unknowns (F_{Ax}, F_{Ay}, F_{Cx}, F_{Cy}). By taking the frame apart and analyzing each member we have six equations with which to find six unknowns (and some of these unknowns are the loads at A and C).

For member *AB* (Figure 9.12): Based on (7.5C), with the moment center at B:

$$\sum M_{z@B} = 0(\curvearrowleft +)$$
$$-(6\text{ ft})F_{Ay} + (150\text{ lb})(3\text{ ft}) = 0$$
$$F_{Ay} = 150\text{ lb}\left(\frac{3\text{ ft}}{6\text{ ft}}\right)$$

$$F_{Ay} = 75.0\text{ lb} \tag{1}$$

Based on (7.5A):

$$\sum F_x = 0(\rightarrow +)$$
$$F_{Ax} - F_{Bx} = 0$$
$$F_{Ax} = F_{Bx} \tag{2}$$

Based on (7.5B):

$$\sum F_y = 0(\uparrow +)$$
$$F_{Ay} + F_{By} - 150\text{ lb} = 0$$

Substituting for F_{Ay} from (1) gives

$$F_{By} = 75.0\text{ lb} \tag{3}$$

For member *BC* (Figure 9.13): Based on (7.5C), with the moment center at C (see **Figure 9.14*a*** for geometry):

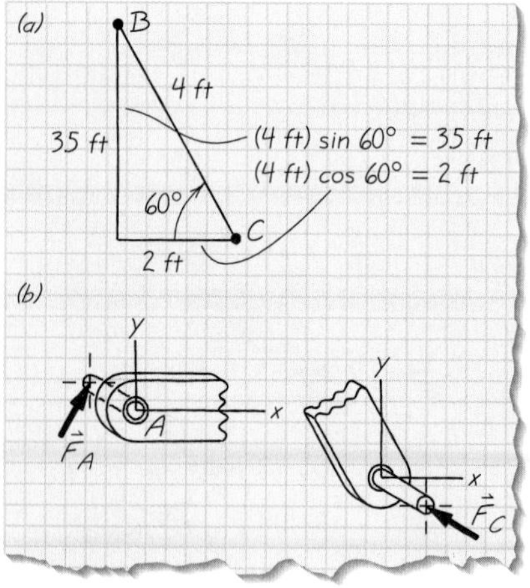

Figure 9.14

$$\sum M_{z@C} = 0(\curvearrowleft +)$$
$$150\text{ lb}\cdot\text{ft} + F_{By}(2\text{ ft}) - F_{Bx}(3.5\text{ ft}) = 0 \tag{4}$$

[2]In **Figure 9.13** the two forces F_{Bx} and F_{By} are drawn in the opposite direction from **Figure 9.12**. These are the halves of the internal force pairs at B.

We substitute for F_{By} from (3) to get

$$F_{Bx}(3.46\text{ ft}) = 150\text{ lb}\cdot\text{ft} + (75.0\text{ lb})(2\text{ ft})$$
$$F_{Bx} = 86.7\text{ lb} \qquad (5)$$

Based on (7.5A):

$$\sum F_x = 0(\rightarrow +)$$
$$F_{Bx} - F_{Cx} = 0$$

$$F_{Cx} = F_{Bx} = 86.7\text{ lb}$$

Based on (7.5B):

$$\sum F_y = 0(\uparrow +)$$
$$F_{Cy} - F_{By} = 0$$

$$F_{Cy} = F_{By} = 75.0\text{ lb} \qquad (6)$$

Finally, we substitute (5) into (2) and solve for F_{Ax}:

$$F_{Ax} = F_{Bx} = 86.7\text{ lb}$$

The force components F_{Ax}, F_{Ay} and F_{Cx}, F_{Cy} are forces from the rest of the world acting on pins A and C (and therefore on the frame). We combine (F_{Ax}, F_{Ay}) and (F_{Cx}, F_{Cy}) into the shear forces acting on the pins:

$$\text{Pin } A \text{ shear force} = \|\boldsymbol{F}_A\| = \sqrt{F_{Ax}^2 + F_{Ay}^2} = 115\text{ lb}$$
$$\text{Pin } C \text{ shear force} = \|\boldsymbol{F}_C\| = \sqrt{F_{Cx}^2 + F_{Cy}^2} = 115\text{ lb}$$

These shear forces are illustrated in **Figure 9.14*b***.

Answer $F_{Ax} = F_{Cx} = 86.7$ lb, $F_{Ay} = F_{Cy} = 75.0$ lb
Pins A and C shear forces: $\|\boldsymbol{F}_A\| = \|\boldsymbol{F}_C\| = 115$ lb

Check As a check, the moment equlibrium equation (7.5C) could be written with the above results for a free-body diagram of the entire frame using either A, C, or even B as a moment center. Another check would be to draw the results on each of the member free-body diagrams and apply the equilibrium equations, being sure to use a different moment center than was used in the **Formulate Equations and Solve** step. Finally, note that the couple formed by the vertical forces at A and C balance the couple formed by the horizontal forces at A and C.

Note: In this solution we generated six linearly independent equations that contained the six unknowns (F_{Ax}, F_{Ay}, F_{Bx}, F_{By}, F_{Cx}, F_{Cy}). The basis for these six equations were the free-body diagrams of members AB (**Figure 9.12**) and member BC (**Figure 9.13**). Alternately, we could have generated six equations based on a free-body diagram of the whole frame (not shown) and either the free-body diagram in **Figure 9.12** or **9.13**.

EXAMPLE 9.3 INTERNALLY UNSTABLE FRAME WITH FRICTION

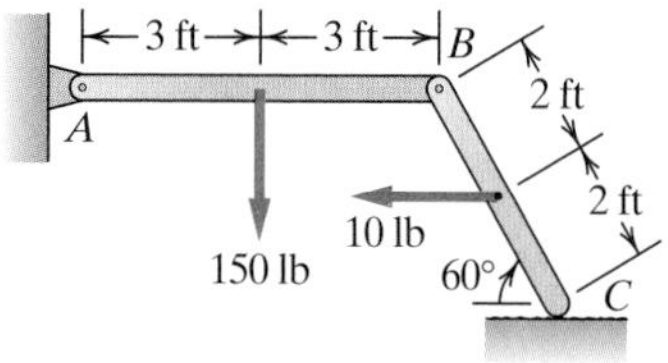

Figure 9.15

The internally unstable frame in Figure 9.15 is subjected to the forces shown. End C of the frame rests against a rough surface.

(a) Assuming that the weight of the members can be neglected, find the loads acting on the frame at supports A and C.

(b) If the component of force acting on the frame parallel to the surface at C is less than $\mu F_{C\text{ normal contact}}$, where $\mu = 0.6$ is the coefficient of friction, the frame will not slide. Based on your findings in (a), will the frame slide?

Goal We are to find the loads at the supports A and C (a) and determine whether or not the frame will slide (b).

Given We are given information about the geometry of the structure and the loading placed on it (150 lb at the midspan of member AB and 10 lb at the midspan of member BC). In addition, we are told that the coefficient of friction is 0.6.

Assume We assume that we can treat the system as planar because all the loads act in a plane. We also assume that the frame is stationary where it rests on the ground at C (therefore, it is not sliding). This means we are assuming that the component of force acting on the frame parallel to the surface at C is less than or equal to the product of the coefficient of friction and the normal force at C. We will check this assumption in (b).

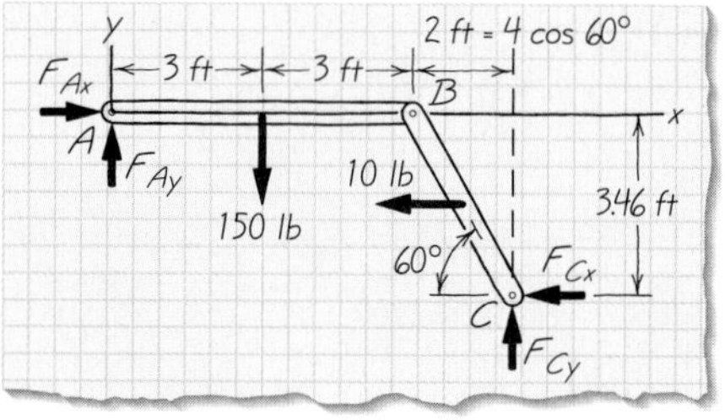

Figure 9.16 Free-body diagram of entire frame

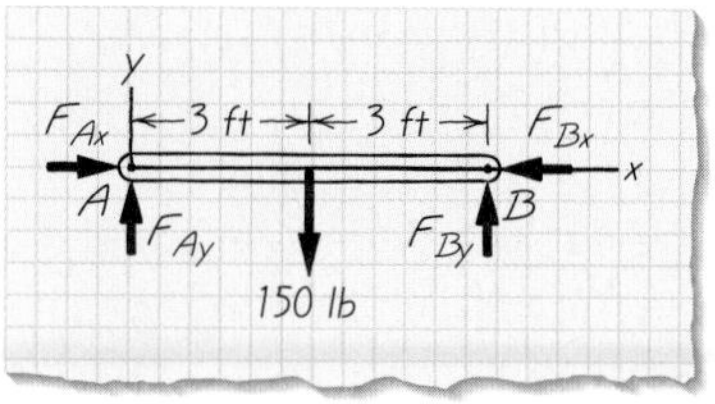

Figure 9.17 Free-body diagram of members AB

Draw We draw the free-body diagram for the entire frame (Figure 9.16) and for member AB (Figure 9.17).

Formulate Equations and Solve As with Example 9.2, if we try to solve for the forces at supports A and C without taking the frame apart we find that we have four unknowns (F_{Ax}, F_{Ay}, F_{Cx}, F_{Cy}) and only three equations of equilibrium. For (a) we use the free-body diagram for member AB to write the moment equilibrium equation with a moment center at B. This allows us to solve for F_{Ay}. We then put the frame back together and write the three equations of planar equilibrium for the entire frame to solve for the three remaining unknowns. Alternately, for (a) we could work with a free-body diagram of member AB and a free-body diagram of BC, as we did in Example 9.2.

For (b) we compare the friction force at C found in (a) with the product of the coefficient of friction and the normal force at C to confirm that the frame will not slide.

(a) For member AB (Figure 9.17): Based on (7.5C), with the moment center at B:

$$\sum M_{z@B} = 0(\curvearrowleft+)$$

$$-(6\text{ ft})F_{Ay} + (150\text{ lb})(3\text{ ft}) = 0$$

$$F_{Ay} = 150\text{ lb}\left(\frac{3\text{ ft}}{6\text{ ft}}\right)$$

$$F_{Ay} = 75.0\text{ lb} \quad (1)$$

For the entire frame (Figure 9.16): Based on (7.5C), with the moment center at C and referring to **Figure 9.14*a*** as an aid in calculating the moment arms:

$$\sum M_{z@C} = 0(\curvearrowleft +)$$
$$-F_{Ay}(6\text{ ft} + 2\text{ ft}) - F_{Ax}(3.46\text{ ft})$$
$$+ 150\text{ lb}(3\text{ ft} + 2\text{ ft}) + 10\text{ lb}(2\text{ ft})\sin 60° = 0$$

We substitute for F_{Ay} from (1) to get

$$-75.0\text{ lb}(8\text{ ft}) - F_{Ax}(3.46\text{ ft}) + 150\text{ lb}(5\text{ ft}) + 10\text{ lb}(1.73\text{ ft}) = 0$$

$$F_{Ax} = \frac{167.3\text{ lb}\cdot\text{ft}}{3.46\text{ ft}} = 48.4\text{ lb} \qquad (2)$$

Based on (7.5A):

$$\sum F_x = 0(\rightarrow +)$$
$$F_{Ax} - F_{Cx} - 10\text{ lb} = 0$$
$$F_{Cx} = F_{Ax} - 10\text{ lb}$$

We substitute for F_{Ax} from (2) to get

$$F_{Cx} = 38.4\text{ lb}$$

Based on (7.5B):

$$\sum F_y = 0(\uparrow +)$$
$$F_{Ay} + F_{Cy} - 150\text{ lb} = 0$$
$$F_{Cy} = -F_{Ay} + 150\text{ lb}$$

We substitute for F_{Ay} from (1) to get

$$F_{Cy} = 75.0\text{ lb}$$

Answer to (a) $F_{Ax} = 48.4$ lb, $F_{Cx} = 38.4$ lb,
$F_{Ay} = F_{Cy} = 75.0$ lb

Check Equation (7.5C) could be written with the results above for an alternate moment center as a check. For example, if we choose the moment center to be at point D (**Figure 9.18**) we find

$$\sum M_{z@D} = 0(\curvearrowleft +)$$
$$-75.0\text{ lb}(3\text{ ft}) - 10\text{ lb}\left(\frac{3.46\text{ ft}}{2}\right)$$
$$+ 75.0\text{ lb}(3\text{ ft} + 2\text{ ft}) - 38.4\text{ lb}(3.46\text{ ft}) \overset{??}{=} 0$$

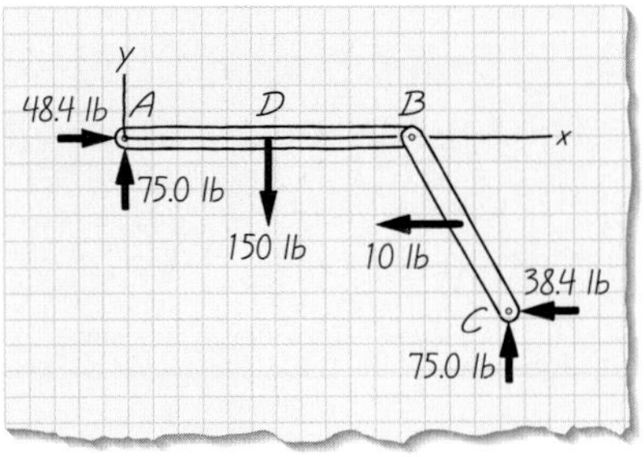

Figure 9.18

The left-hand side of the equation sums to -0.16 lb·ft. This is very close to zero, and we can say our answer is correct. (If we were to carry one more significant digit, using $F_{Ay} = 38.35$ lb, then the left-hand side would decrease to 0.009.)

(b) We check that sliding is not predicted at C. To do this we need to confirm that F_{Cx} (the friction force) is less than μF_{Cy} (the maximum allowable friction force before sliding occurs). In other words, if $|F_{Cx}| \leq \mu\,|F_{Cy}|$, then the support will not slide at C. Otherwise, the support will slide. This is basically a statement of the Coulomb Friction Law, as described in detail in Appendix B.

If we carry out this check we find that F_{Cx}(38.4 lb) is less than μF_{Cy} (0.6(75.0 lb) = 45.0 lb). Therefore, we conclude that the frame is stationary at C and we have properly modeled the presence of friction. If we had found that $F_{Cx} = \mu F_{Cy}$, we would also have concluded that the frame is stationary. If we had found that $F_{Cx} > \mu F_{Cy}$ we would have concluded that the frame slides along the ground and the frame is therefore not in equilibrium (and the analysis we carried out in (a) is not valid).

Answer to (b) The frame will not slide along the surface at C.

EXAMPLE 9.4 NONPLANAR FRAME ANALYSIS

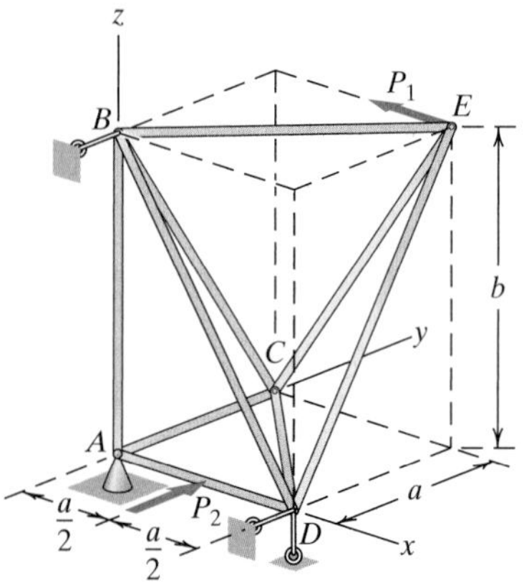

Figure 9.19

The nonplanar frame shown in **Figure 9.19** consists of a rigid tetrahedron $ABCD$ anchored by a ball-and-socket joint at A. It is prevented from rotating by the links attached to the ground at joints B and D. Assuming the weight of the frame members can be ignored, find an expression for

(a) the load at each support, and

(b) the forces acting on members BE, CE, DE, and AD (written in terms of a and b).

(c) For $a = 0.8$ m, $b = 1.0$ m, $P_1 = 15.0$ kN, $P_2 = 20.0$ kN, find the forces in members BE, CE, DE, and AD.

(d) Define the **safety factor** as the ratio

$$\frac{\text{force capacity of a member}}{\text{force in a member}}$$

where the force capacity of a member is the force at which the member fails and is a function of the member's cross-sectional area and the material from which it is constructed. If the force capacity of members BE, CE, DE, and AD is an axial force of 30 kN, calculate safety factors for these members for the conditions defined in (c). Confirm that this ratio is greater than unity for each member (this means that failure is not predicted in the member).

Goal We are to find expressions for the load at each support (a) and the forces acting on members CE, BE, DE, and the axial force in AD (b) and the forces in those members for specific dimensions and forces (c). Also calculate the safety factors for these members (d).

Given We are given information about the supports and geometry of the nonplanar frame. Also, we are told that the force that causes failure in members CE, BE, DE, and AD is an axial force of 30 kN, and we are given a definition of safety factor.

Assume We assume that members AB, AC, BC, BD, BE, CD, CE, and DE act as two-force members because they are pinned at both ends and no forces act on them between the ends.

Draw We draw a free-body diagram of the entire frame (Figure 9.20).

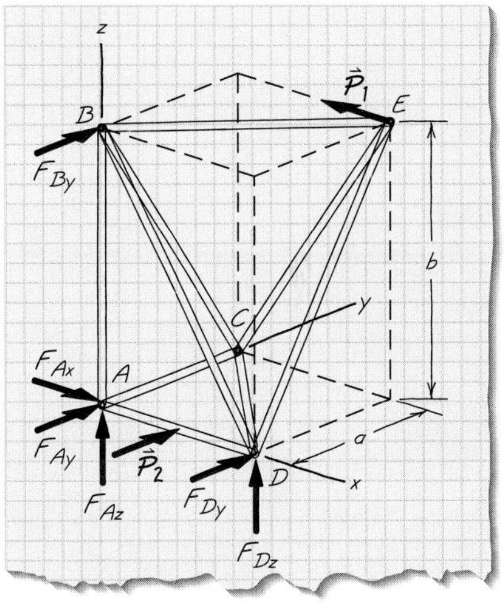

Figure 9.20 Free-body diagram of entire frame

Formulate Equations and Solve For (a) we draw a free-body diagram of the whole frame, from which we write six equilibrium equations to find six unknown loads at the supports. For (b) we draw a free-body diagram of joint E (treating it as a particle), from which we write three equilibrium equations to find expressions for the forces in members CE, BE, and DE. We substitute the given dimensions and applied loads into these expressions to find the forces in members CE, BE, and DE for (c). We also draw free-body diagrams of the joint at A to find the axial force in AD. For (d) we find the ratio of the force capacity of each of these members (30 kN) to the axial force in the member (found in part (c)) to confirm that the ratio is greater than unity.

(a) To solve for the loads at the supports, we first analyze the entire structure.

Based on (7.1),

$$\sum \boldsymbol{F} = 0 \text{ (vector sum of forces)}$$
$$\boldsymbol{F}_A + \boldsymbol{F}_B + \boldsymbol{F}_D + \boldsymbol{P}_1 + \boldsymbol{P}_2 = 0$$

where:

$$\boldsymbol{F}_A = F_{Ax}\boldsymbol{i} + F_{Ay}\boldsymbol{j} + F_{Az}\boldsymbol{k}$$
$$\boldsymbol{F}_B = F_{By}\boldsymbol{j}$$
$$\boldsymbol{F}_D = F_{Dy}\boldsymbol{j} + F_{Dz}\boldsymbol{k}$$
$$\boldsymbol{P}_1 = -P_1\boldsymbol{i}$$
$$\boldsymbol{P}_2 = P_2\boldsymbol{j}$$

Summing the force components in the $\boldsymbol{i}$, $\boldsymbol{j}$, and $\boldsymbol{k}$ directions we get three equations:

$$\text{In the } \boldsymbol{i} \text{ direction: } F_{Ax} - P_1 = 0 \quad (1)$$
$$\text{In the } \boldsymbol{j} \text{ direction: } F_{Ay} + F_{By} + F_{Dy} + P_2 = 0 \quad (2)$$
$$\text{In the } \boldsymbol{k} \text{ direction: } F_{Az} + F_{Dz} = 0 \quad (3)$$

Based on (7.2),

$$\sum \boldsymbol{M} = 0 \qquad \text{(vector sum of moments)}$$

With a moment center at A^3 this can be written in terms of cross-products as

$$\sum M_{@A} = (\boldsymbol{r}_{AB} \times \boldsymbol{F}_B) + (\boldsymbol{r}_{AD} \times \boldsymbol{F}_D) + (\boldsymbol{r}_{P1} \times \boldsymbol{P}_1) + (\boldsymbol{r}_{P2} \times \boldsymbol{P}_2) = 0 \quad (4)$$

where, the forces in vector notation were given above, and the position vectors are given as:

$$\boldsymbol{r}_{AB} = b\boldsymbol{k}$$
$$\boldsymbol{r}_{AD} = a\boldsymbol{i}$$
$$\boldsymbol{r}_{P_1} = a\boldsymbol{i} + a\boldsymbol{j} + b\boldsymbol{k}$$
$$\boldsymbol{r}_{P_2} = \frac{a}{2}\boldsymbol{i}$$

Substituting these position vectors and forces into (4), we have:

$$(b\boldsymbol{k} \times F_{By}\boldsymbol{j}) + (a\boldsymbol{i} \times (F_{Dy}\boldsymbol{j} + F_{Dz}\boldsymbol{k})) + ((a\boldsymbol{i} + a\boldsymbol{j} + b\boldsymbol{k}) \times -P_1\boldsymbol{i}) + \left(\frac{a}{2}\boldsymbol{i} \times P_2\boldsymbol{j}\right) = 0 \quad (5)$$

Calculating the cross products:

$$-bF_{By}\boldsymbol{i} + (-aF_{Dz}\boldsymbol{j} + aF_{Dy}\boldsymbol{k}) + (-bP_1\boldsymbol{j} + aP_1\boldsymbol{k}) + \frac{a}{2}P_2\boldsymbol{k} = 0 \quad (6)$$

Summing the moment components in the $\boldsymbol{i}$, $\boldsymbol{j}$, and $\boldsymbol{k}$ directions gives three more equations:

In the $\boldsymbol{i}$ direction: $-bF_{By} = 0$ (7)

In the $\boldsymbol{j}$ direction: $-aF_{Dz} - bP_1 = 0$ (8)

In the $\boldsymbol{k}$ direction: $aF_{Dy} + aP_1 + \frac{a}{2}P_2 = 0$ (9)

Equations (1)–(3) and (7)–(9) are six equilibrium equations. Solve them for F_{Ax}, F_{Ay}, F_{Az}, F_{Dy}, F_{Dz}, and F_{By}:

Answer to (a) $F_{Ax} = P_1,\ F_{Ay} = P_1 - \frac{P_2}{2},\ F_{Az} = P_1\frac{b}{a},$

$F_{By} = 0,\ F_{Dy} = -P_1 - \frac{P_2}{2},\ F_{Dz} = -P_1\frac{b}{a}$

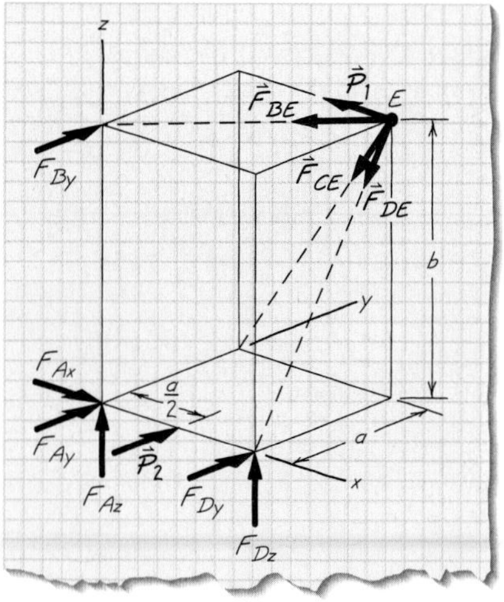

Figure 9.21 Free-body diagram of particle E

(b) We analyze joint E to find the forces acting on members BE, CE, and DE (**Figure 9.21**). We have assumed that all of the members are in tension and thus are pulling away from the joint.

Based on (7.1) applied at particle E,

$$\sum \boldsymbol{F}_x = 0$$
$$\boldsymbol{P}_1 + \boldsymbol{F}_{BE} + \boldsymbol{F}_{CE} + \boldsymbol{F}_{DE} = 0$$

[3]The choice of a moment center is a matter of convenience. By selecting A as the moment center we will generate equations that are easier to solve for unknown loads because they do not contain F_{Ax}, F_{Ay}, and F_{Az}.

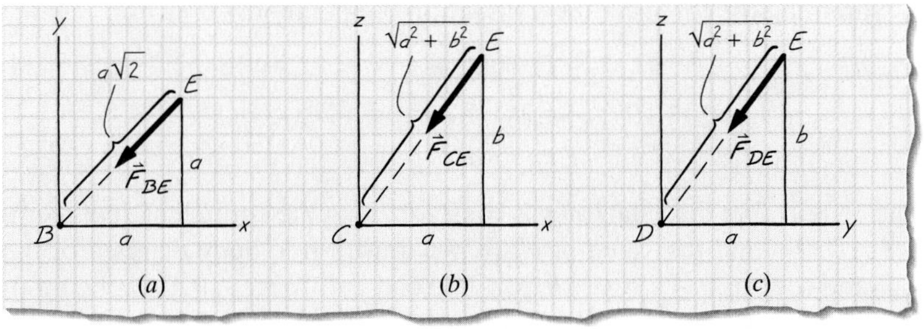

Figure 9.22

where

$$P_1 = -P_1 \boldsymbol{i}$$

$$\boldsymbol{F}_{BE} = -\frac{F_{BE}}{\sqrt{2}}\boldsymbol{i} - \frac{F_{BE}}{\sqrt{2}}\boldsymbol{j} \qquad \text{(Figure 9.22a)}$$

$$\boldsymbol{F}_{CE} = -\frac{aF_{CE}}{\sqrt{a^2+b^2}}\boldsymbol{i} - \frac{bF_{CE}}{\sqrt{a^2+b^2}}\boldsymbol{k} \qquad \text{(Figure 9.22b)}$$

$$\boldsymbol{F}_{DE} = -\frac{aF_{DE}}{\sqrt{a^2+b^2}}\boldsymbol{j} - \frac{bF_{DE}}{\sqrt{a^2+b^2}}\boldsymbol{k} \qquad \text{(Figure 9.22c)}$$

Summing the forces in the ***i***, ***j***, and ***k*** directions, we get three equations:

$$\text{In the } \boldsymbol{i} \text{ direction: } -P_1 - \frac{F_{BE}}{\sqrt{2}} - \frac{aF_{CE}}{\sqrt{a^2+b^2}} = 0 \qquad (10)$$

$$\text{In the } \boldsymbol{j} \text{ direction: } -\frac{F_{BE}}{\sqrt{2}} - \frac{aF_{DE}}{\sqrt{a^2+b^2}} = 0 \qquad (11)$$

$$\text{In the } \boldsymbol{k} \text{ direction: } -\frac{bF_{CE}}{\sqrt{a^2+b^2}} - \frac{bF_{DE}}{\sqrt{a^2+b^2}} = 0 \qquad (12)$$

Using (10), (11), and (12), we solve for F_{BE}, F_{CE}, and F_{DE}:

From (12): $$F_{CE} = -F_{DE} \qquad (13)$$

Using (10) and (11), we solve for F_{BE}:

$$F_{BE} = \frac{-P_1}{\sqrt{2}}$$

Substituting this into (11) we find

$$F_{CE} = F_{DE} = \frac{-P_1\sqrt{a^2+b^2}}{2a}$$

To find the axial force in member AD we start by drawing a free-body diagram of member AD (Figure 9.23). The axial force in member AD is aligned with the x axis. Notice that member AD has forces acting on it that are perpendicular to its axis; therefore it is a multiforce member.

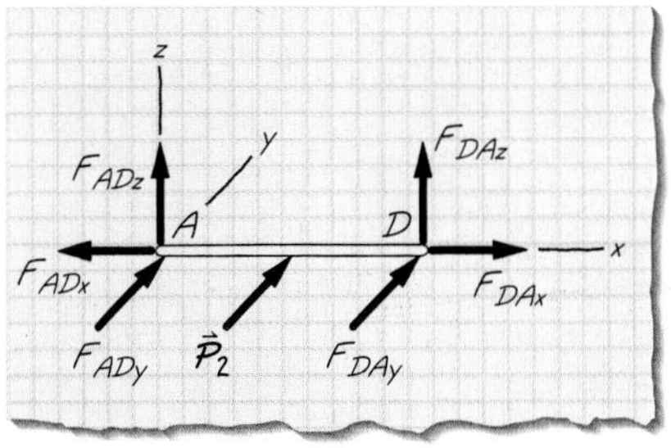

Figure 9.23

Summing the forces in the x direction on member AD gives

$$\sum F_x = 0(\rightarrow +)$$
$$-F_{ADx} + F_{DAx} = 0$$
$$F_{ADx} = F_{DAx}$$

This tells us that the axial force is constant along the entire length of member AD. To find the value of F_{ADX}, we draw a free-body diagram of the ball-and-socket at A (Figure 9.24). The ball-and-socket at A can be modeled as a particle because all of the forces are concurrent. Based on (7.3A):

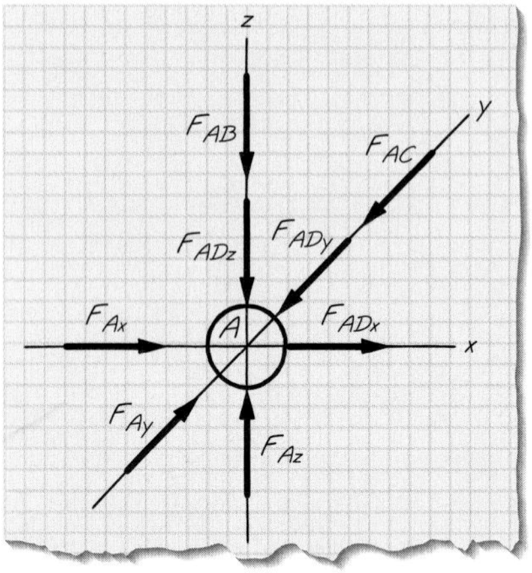

Figure 9.24 Free-body diagram of particle A

$$\sum F_x = 0(\rightarrow +)$$
$$F_{Ax} + F_{ADx} = 0$$

Based on finding that $F_{Ax} = P_1$ in (a), then

$$F_{ADx} = -F_{Ax} = -P_1$$

Answer to (b) Force in members: $F_{BE} = \dfrac{-P_1}{\sqrt{2}}$, $F_{CE} = \dfrac{-P_1\sqrt{a^2+b^2}}{2a}$,

$F_{DE} = \dfrac{P_1\sqrt{a^2+b^2}}{2a}$, $F_{ADx} = -P_1$

(c) We now consider a specific geometry and loading of the frame; namely, $a = 0.8$ m, $b = 1.0$ m, and $P_1 = 15.0$ kN.

The forces in members BE, CE, DE, and AD are found from substituting into the results of (b).

Answer to (c)

$$F_{BE} = \frac{-P_1}{\sqrt{2}} = \frac{-15\text{ kN}}{\sqrt{2}} = -10.6\text{ kN},$$

$$F_{CE} = \frac{-P_1\sqrt{a^2+b^2}}{2a} = -12.0\text{ kN},$$

$$F_{DE} = \frac{P_1\sqrt{a^2+b^2}}{2a} = +12.0\text{ kN},$$

$$F_{ADx} = -15.0\text{ kN}$$

Note: This result indicates that members AD, BE, and CE are in compression (the negative sign), and member DE is in tension (the positive sign).

(d) For members AD, BE, CE, and DE, we have been told that the force capacity of each member is 30 kN (we will assume that this is its capacity in both tension and compression). Therefore,

$$(\text{Safety factor})_{BE} = \text{SF}_{BE} = \frac{-30\text{ kN}}{-10.6\text{ kN}} = 2.8$$

$$\text{SF}_{CE} = \frac{-30\text{ kN}}{-12.0\text{ kN}} = 2.5$$

$$\text{SF}_{DE} = \frac{+30\text{ kN}}{+12.0\text{ kN}} = 2.5$$

$$\text{SF}_{AD} = \frac{-30\text{ kN}}{-15.0\text{ kN}} = 2.0$$

Answer to (d) $SF_{BE} = 2.8$, $SF_{CE} = 2.5$, $SF_{DE} = 2.5$, $SF_{AD} = 2.0$.
This ratio is greater than unity for each member, which means failure is not predicted in any of these members.

Examples 9.1–9.4 illustrate the use of static analysis to determine the loads acting on a frame as a whole, and on portions of the frame. In Examples 9.1 and 9.4, the frames are internally stable without redundancy and are statically determinate. In Examples 9.2 and 9.3, the frames are internally unstable with just enough boundary conditions to prevent the frame from moving and/or collapsing.

For a frame that is statically determinate and **internally stable with redundancy** (meaning there are more members than are necessary for the frame to be rigid), application of the equilibrium conditions enable us to find the loads acting on the frame at its boundaries. In order to find the internal loads, we must use concepts from mechanics of materials in addition to equilibrium conditions (**Figure 9.5*c***, Category C). Internal redundancy is sometimes created in frames when you want the frame to "fail safe"; this means that when one member fails the whole structure will not immediately collapse, thereby allowing people to get out of harm's way.

For a statically indeterminate, internally stable frame (with or without redundancy), concepts from mechanics of materials and the equilibrium conditions are required to find boundary and internal loads (**Figure 9.5*d***, Category D). In contrast, for an underconstrained, internally stable frame (with or without redundancy), the concepts from dynamics are required to describe boundary and internal loads and the motion of the frame (**Figure 9.5*e***, Category E).

EXAMPLE 9.5 DETERMINING STATUS OF A FRAME

For the frame configurations referenced below, determine which of these labels apply to each frame: internally stable with or without redundancy, internally unstable, statically determinate, statically indeterminate, underconstrained.

Goal We are to determine the status of frame configurations.

Frame 1 (Figure 9.25): The frame is *internally stable without redundancy* because the frame members form a rigid shape that will maintain its basic geometry when loads are applied; no additional members are needed to make it rigid. If we remove a member from the frame, it cannot maintain its triangular shape and will be internally unstable. In addition, the frame is *statically determinate* because we can determine the three unknown forces at the supports using three planar equilibrium equations. Because the frame is statically determinate and internally stable without redundancy, we could use the planar equilibrium equations to analyze the forces acting at pin connections and in the members.

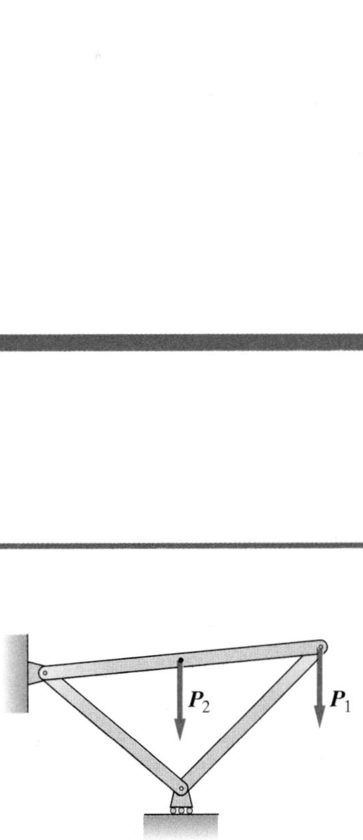

Figure 9.25 Frame 1

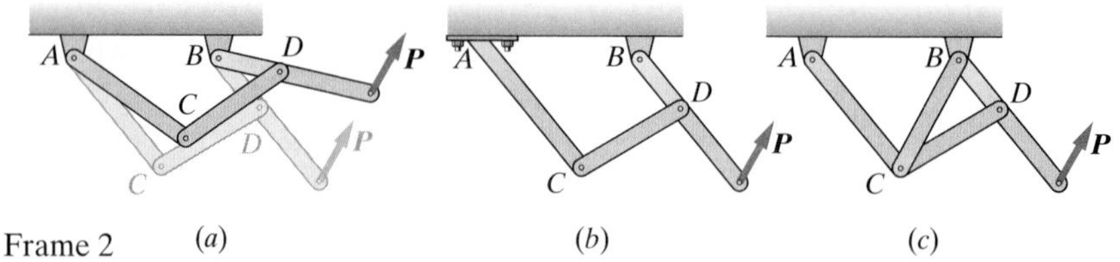

Figure 9.26 Frame 2

Frame 2 (Figure 9.26*a*): The frame is *internally unstable* because the force ***P*** causes members to rotate, as shown by the ghosted image. A frame in which members intentionally move is called a **mechanism**. An internally unstable frame with insufficient boundary conditions to prevent movement is also *underconstrained*. More members and/or boundary conditions are needed for this frame to be in equilibrium. For example, by changing the pin connection at A to a fixed boundary condition, as in **Figure 9.26*b***, we now have an internally unstable frame that is statically determinate. Alternatively, by adding a diagonal member, as in **Figure 9.26*c***, we have created an internally stable frame without redundancy that is statically determinate.

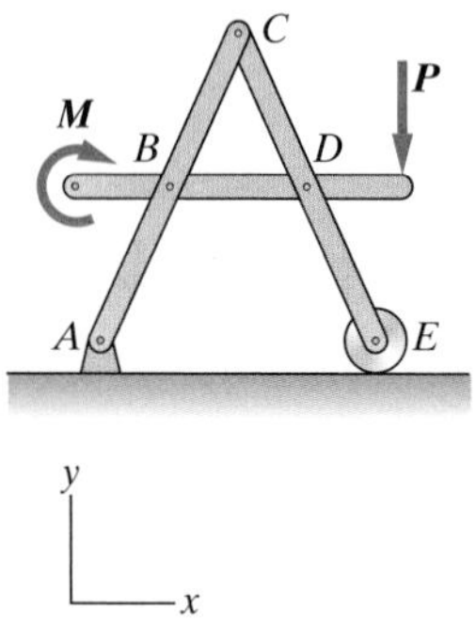

Figure 9.27 Frame 3

Frame 3 (Figure 9.27): The three members are sufficient to form a rigid frame, and removal of a member would make the frame unstable; thus the frame is *internally stable without redundancy*. The system has nine unknown loads (F_{Ax}, F_{Ay}, F_{Bx}, F_{By}, F_{Cx}, F_{Cy}, F_{Dx}, F_{Dy}, F_{Ey}) and nine linearly independent equilibrium equations can be written, making it *statically determinate*.

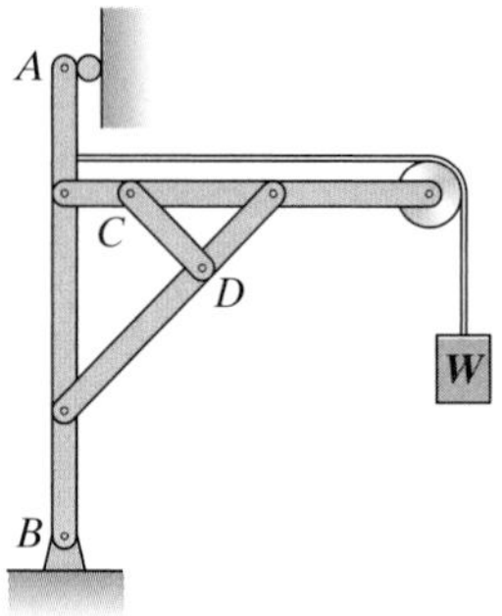

Figure 9.28 Frame 4

Frame 4 (Figure 9.28): There are more than enough members to form a rigid frame. If member CD were removed, the frame would remain stable, thus the frame is *internally stable with redundancy*. This means we cannot find all of the loads at the pin joints that connect the various members using equilibrium equations alone (we also need concepts from strength of materials). The frame is externally *statically determinate*; this means that by applying the conditions of equilibrium to the frame as a whole we would be able to find the loads acting on the frame at A and B.

Frame 5 (Figure 9.29*a*): The three members AB, BC, CD are not sufficient to form a rigid frame; thus the frame is *internally unstable*. In addition, there are as many equilibrium equations as there are unknown loads, making it *statically determinate*. The free-body diagrams of the various members show that there are six unknown loads, and we are able to write six equilibrium equations (**Figure 9.29*b***).

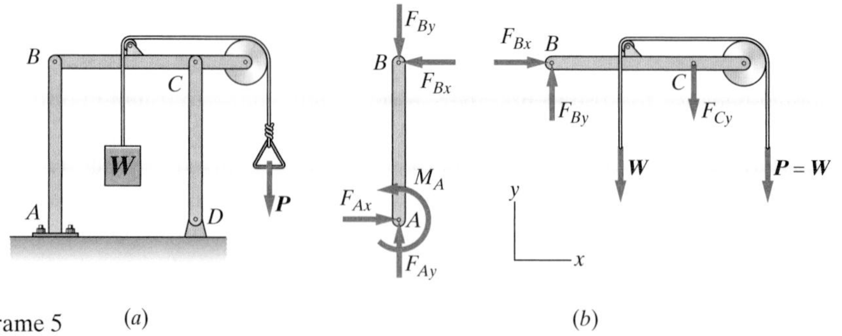

Figure 9.29 Frame 5

EXAMPLE 9.6 SPECIAL CONDITIONS IN FRAMES

Presented here are examples of frames with special conditions. Recognizing these conditions and understanding how they affect the behavior of the structure are key in both the equation formulation and check steps of an analysis.

These examples consist of:

- Case 1a (Zero-Force Members)
- Case 1b (More on Zero-Force Members)
- Case 2 (Unstable Configurations)
- Case 3 (Using Cables as Diagonals)
- Case 4 (Structures Not Fixed in Space)

Case 1a (Zero-Force Members) Identify the zero-force members in the frame in Figure 9.30.

The frame in **Figure 9.30** supports a Wearhouse roof and could be subjected to the weight of the roof and loads due to snow, wind, and earthquakes, as well as the load due to a crane attached to member *CE*. Under certain of these loading conditions, a member may have no load acting on it. That member is called a **zero-force member**. Identifying zero-force members early in the analysis is important, as it simplifies the analysis. For the frame in **Figure 9.30**, identify the zero-force members.

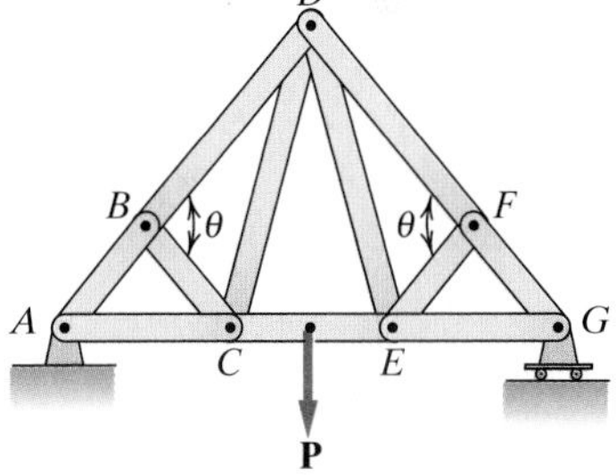

Figure 9.30 Case 1a

A zero-force member occurs when three two-force members are pinned together at a pin connection as at *J* in Figure 9.31.

1. No external loads are applied to the pin connection (*J*).
2. Two of the members have the same line of action (Members 1 and 2), and the third member (Member 3) is at an angle to that line of action.

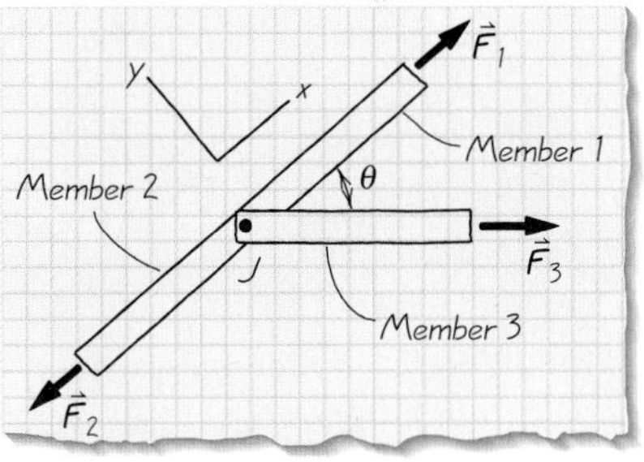

Figure 9.31 Member 3 is a zero-force member

Then the member at an angle to the line of action of the other two is a zero-force member ($F_3 = 0$). The magnitude of the forces acting on the other two members are equal ($\| \boldsymbol{F}_1 \| = \| \boldsymbol{F}_2 \|$). To prove this we analyze pin connection *J* in **Figure 9.31**.

$$\sum F_x = 0$$

$$F_1 - F_2 + F_3 \cos\theta = 0 \tag{1}$$

$$\sum F_y = 0$$

$$-F_3 \sin\theta = 0 \tag{2}$$

Because (2) must be valid for all values of θ, F_3 must be a zero-force member. Substituting $F_3 = 0$ into (1) gives

$$F_1 = F_2$$

indicating that Members 1 and 2 are subjected to equal forces.

We now evaluate the frame in **Figure 9.30** and identify joints that have the same characteristics as pin connection *J* in **Figure 9.31**. Joints *B* and *F* both meet the requirements.

Answer to Case 1a Members *BC* and *EF* are zero-force members. In addition, $F_{AB} = F_{BD}$ and $F_{FD} = F_{FG}$.

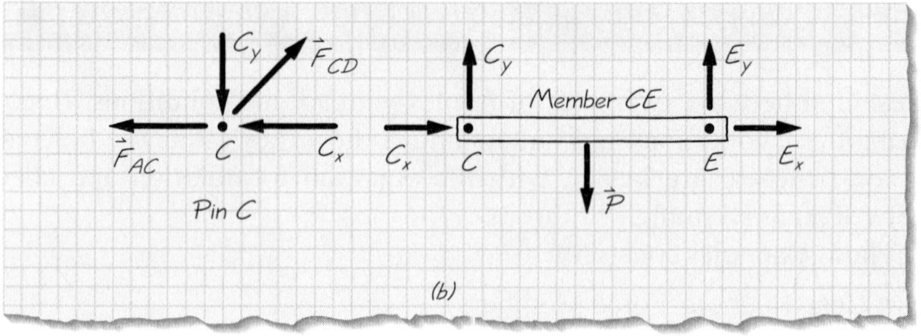

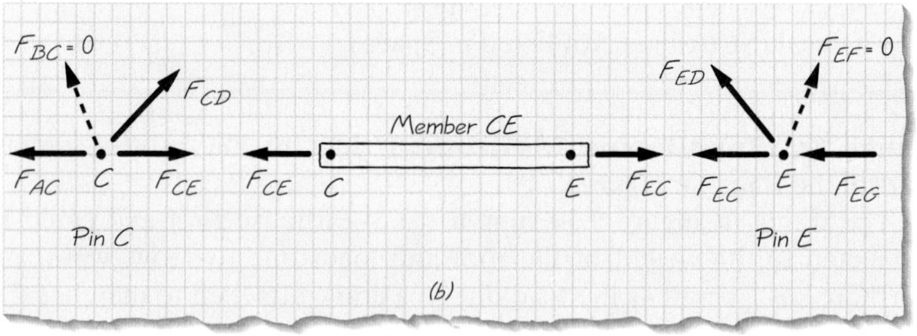

Figure 9.32 (*a*) Free-body diagrams of member *CE* and pin *C* when force *P* is applied midspan on member *CE*; (*b*) free-body diagrams of member *CE* and pins *C* and *E* when force is applied at *D*

There are no other zero-force members in the frame. The loads at pin *C* on member *CE* (Figure 9.32*a*) are transferred to other locations on the frame by members *AC* and *CD*, which are both two-force members. Summing forces in the *x* and *y* directions at pin *C* in **Figure 9.32*a*** shows that C_y is transferred by *CD* and C_x is transferred by both *AC* and *CD*. A similar situation occurs at pin *E*, indicating that *DE* and *EG* are both carrying loads.

An interesting situation occurs if we move the load *P* from *CE* to joint *D* so that horizontal member *CE* acts as a two-force member. Again, members *BC* and *EF* are zero-force members, which for analysis purposes we can then treat as if they are not there. Pin connections *C* and *E* now have the characteristics of pin connection *J* in **Figure 9.31**, indicating that members *CD* and *DE* must be zero-force members (**Figure 9.32*b***).

Keep in mind that zero-force members exist because of two factors: the geometry of the system and the loading. For different loading cases, *BC* and *EF* may not be zero-force members; for this reason, one would be ill-advised to remove these members!

Case 1b (More on Zero-Force Members) For the frame in Figure 9.33, identify the zero-force members.

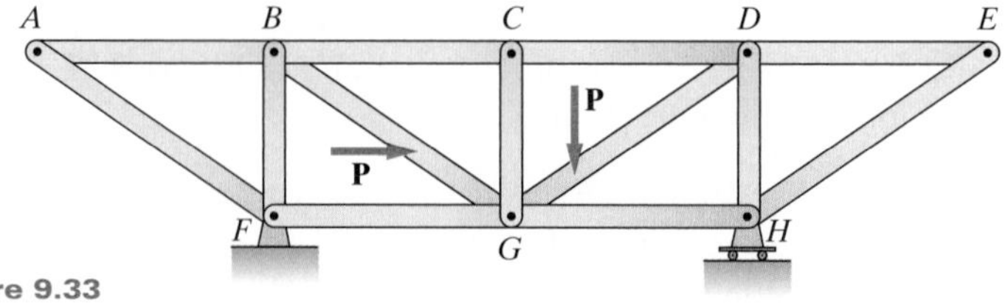

Figure 9.33

In looking at Case 1*a*, we established one set of characteristics of zero-force members. Here we consider another set. A zero-force member occurs when two two-force members are pinned together at a joint as in pin connection *K* in **Figure 9.34**.

1. No external loads are applied to the pin connection.
2. The members do not have the same line of action.

Then both members are zero-force members. To prove this we analyze pin connection *K* in **Figure 9.34**.

$$\sum F_x = 0$$

$$F_1 + F_2 \cos\theta = 0 \qquad (3)$$

$$\sum F_y = 0$$

$$F_2 \sin\theta = 0 \qquad (4)$$

Because (4) must be valid for all values of θ, F_2 must be a zero-force member. Substituting $F_2 = 0$ into (3) gives

$$F_1 = 0$$

indicating that Member 1 is also a zero-force member.

We now evaluate the frame in **Figure 9.33** and identify joints that have the same characteristics as pin connection *J* in **Figure 9.31** and pin connection *K* in **Figure 9.34**. Pin connection *C* has three two-force members and no external loads; therefore member *CG* must be a zero-force member. Pin connection *A* and *E* both meet the requirements of pin connection *K*, indicating that *AB*, *AF*, *DE*, and *EH* are zero-force members.

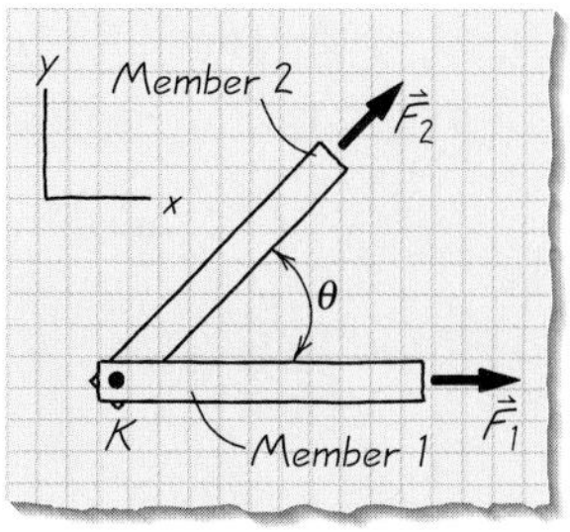

Figure 9.34 Members 1 and 2 are zero-force members

Answer to Case 1b Members *AB*, *AF*, *CG*, *DE*, and *EH* are zero-force members.

Case 2 (Unstable Configurations) Explain why the frame shown in **Figure 9.35** is unstable.

The upper part of the frame constructed from members *AB*, *AC*, *BD*, and *CD* pinned together is internally unstable (it is in fact what is called a four-bar mechanism).

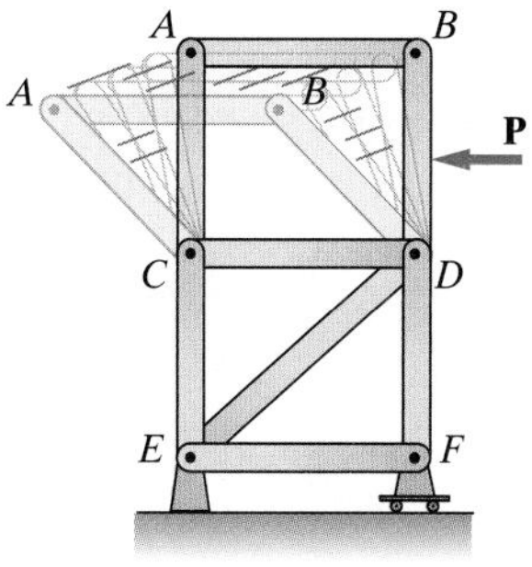

Figure 9.35

Answer to Case 2 Under the load shown it will move as shown by the ghosted lines. Mathematically speaking, the number of constraints is fewer than required to achieve static equilibrium. By adding an additional two-force member between joints *A* and *D* or between joints *B* and *C*, you could make the frame stable. You could also stabilize the frame by replacing members *AC* and *CE* with one continuous member *ACE*. What other ideas do you have for stabilizing the frame?

Case 3 (Using Cables as Diagonals) What kind of loadings can cables support?

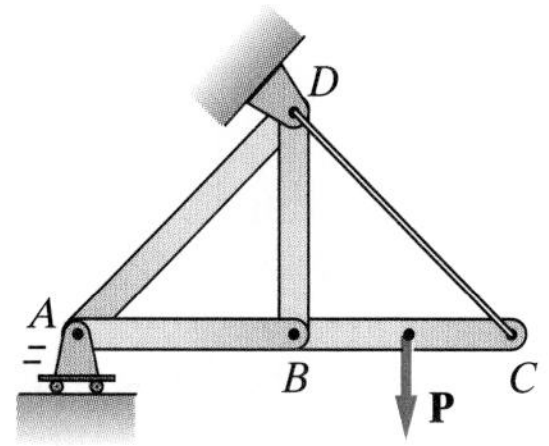

Figure 9.36

Answer to Case 3 Whenever cables are part of a structure, it is important to remember that cables can support only tension and not compression; that is, $T_{\text{cable}} > 0$. Therefore, the frame shown in **Figure 9.36** is in equilibrium for the loading ***P*** shown. (If ***P*** were directed upward, what would happen to the cable?)

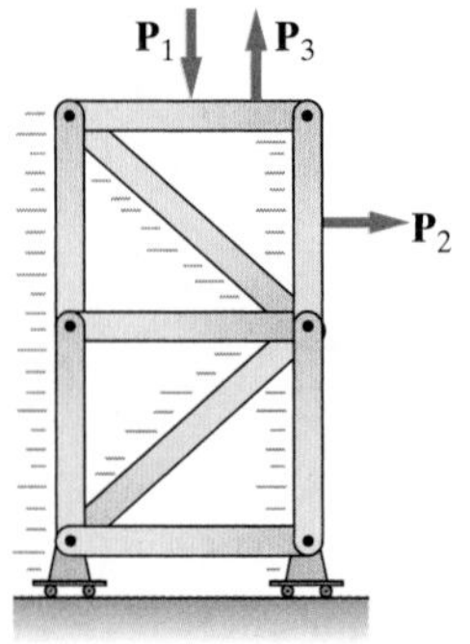

Figure 9.37

Case 4 (Structures Not Fixed in Space) What happens to the structure shown in Figure 9.37*a* when only load P_1 is applied; when only P_2 is applied; and when only P_3 is applied?

Answer to Case 4 This structure can support a downward force (such as P_1) that creates normal contact between the rollers and the ground. When a force is applied horizontally (P_2), the structure slides along the surface. Therefore, this system is classified as **underconstrained.** When an upward force is applied (P_3), the response of the structure depends on the actual condition the rollers represent. A roller is an idealized boundary connection that represents the ability to resist movement normal to, but not parallel to, the contact surface. A roller can be a representation of a real connection that prevents movement both away from and toward the contact surface. If this is the real condition, the frame is stable when P_3 is applied. On the other hand, the roller can represent a real connection that would allow the frame to lift from the contact surface when an upward force is applied (similar to a skateboard). In this case, the frame is unstable when P_3 is applied, and the system is **underconstrained.** When you are modeling a real structure, you must know the behavior of the connection you are modeling with a roller.

EXERCISES 9.1

Unless otherwise stated, assume that the frames below are in equilibrium and are of negligible weight.

9.1.1. Consider the frame in E9.1.1 from which a block weighing 600 lb hangs.

a. Draw the free-body diagram of pin *A*.

b. Use the free-body from **a** to determine the axial forces in members *AB* and *AC*. Indicate whether they are in tension (T) or compression (C).

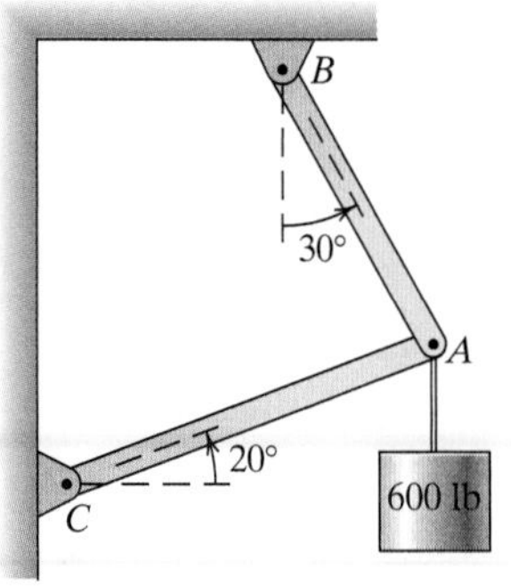

E9.1.1

9.1.2. Consider the frame in E9.1.2. Determine

a. the loads acting on the frame at *A* and *C*

b. the loads acting on member *AB*

9.1.3. Consider the frame in **E9.1.2**. Determine

a. the loads acting on member *BC*

b. the shear force acting on the pin at *B*

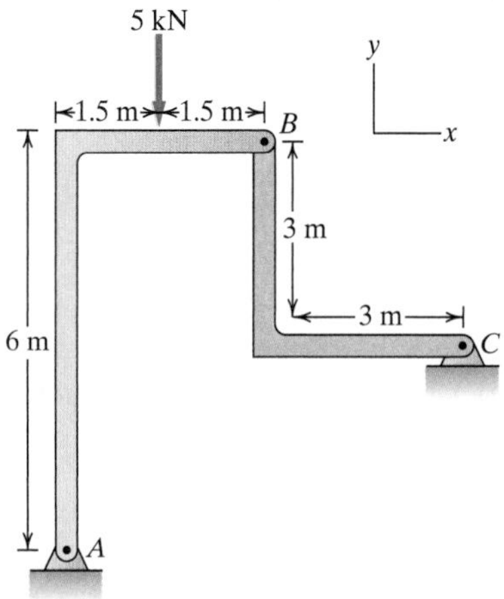

E9.1.2

9.1.4. Consider the frame shown in E9.1.4. Determine the shear force acting on the pin at *B* by

a. ignoring that *BD* is a two-force member in your calculations

b. incorporating *BD* acting as a two-force member in your calculations

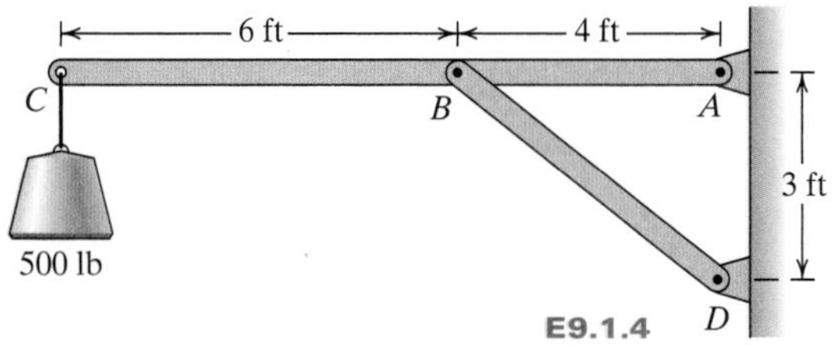

E9.1.4

9.1.5. Consider the frame shown in E9.1.5. Determine

a. the loads acting on the frame at A and D

b. the loads acting on member AC

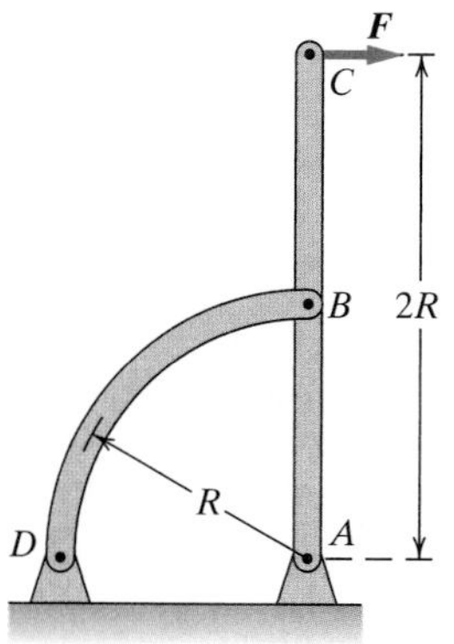

E9.1.5

9.1.6. Consider the frame shown in **E9.1.5**.

a. Determine the shear force acting on the pin at B.

b. If $R = 200$ mm and the value of the shear force should not exceed 500 N, what is the maximum allowable magnitude of the force $\boldsymbol{F}$?

For **Exercises 9.1.7 and 9.1.8** consider the frame in E9.1.7. Member AB is pinned at end B to a collar that may slide over the smooth bar CD. The frame is in equilibrium.

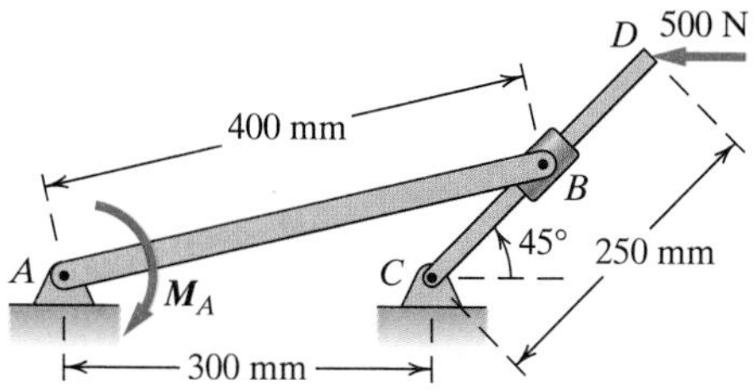

E9.1.7

9.1.7. Determine

a. the magnitude of the moment at A

b. the other loads that act on the frame at A

9.1.8. Determine the loads that act on the frame at C.

9.1.9. Consider the frame in E9.1.9. Determine the loads acting on members ABC and EC.

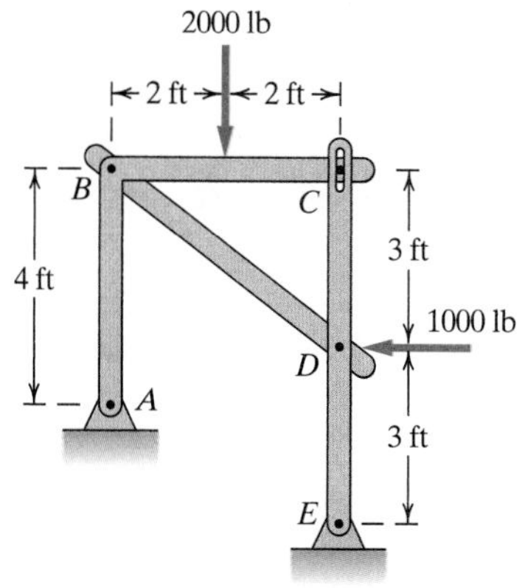

E9.1.9

9.1.10. Consider the frame in **E9.1.9**. Determine the loads acting on the frame at A and E.

9.1.11. Consider the frame in **E9.1.9**. Determine the shear force acting on the pin at B.

9.1.12. Consider the frame in E9.1.12. Determine the loads acting on members AB and BC.

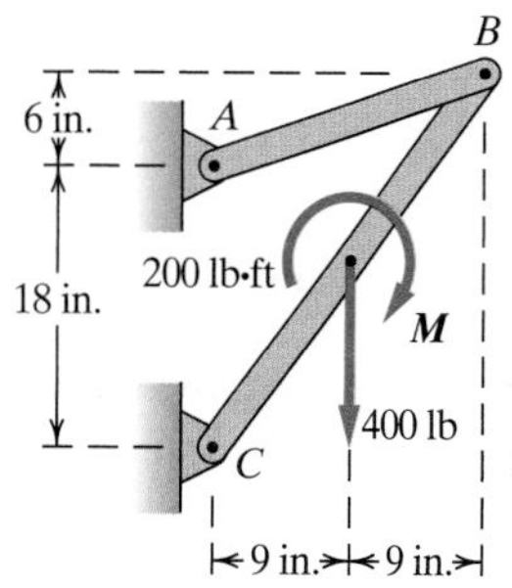

E9.1.12

9.1.13. Consider the frame in **E9.1.12**. Determine the loads acting on the frame at A and C.

9.1.14. Consider the frame in **E9.1.12**. Determine the shear force acting on the pin at B.

9.1.15. Consider the frame in E9.1.15. Determine the loads acting on members AB and BC.

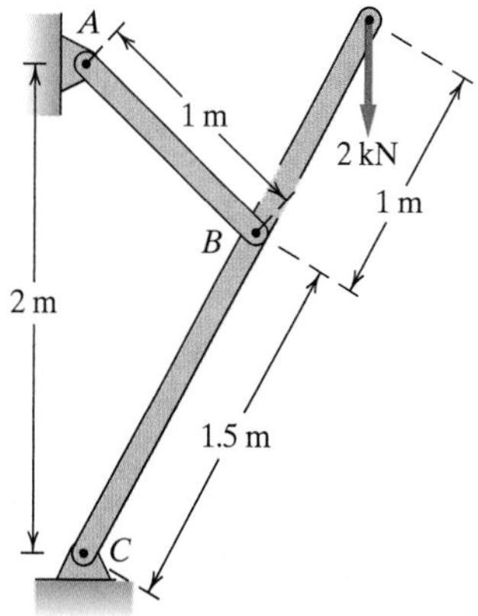

E9.1.15

9.1.16. Consider the frame in **E9.1.15**. Determine the loads acting on the frame at A and C.

9.1.17. Consider the frame in **E9.1.15**. Determine the shear force acting on the pin at B.

9.1.18. Consider the frame in E9.1.18.

a. Determine the force in members DE, BE, and CE. Note that the curved members act as two-force members.

b. If the magnitude of the force (compression or tension) in members DE, BE, and CE, is not to exceed 500 lb, what is the maximum allowable magnitude of the force $\boldsymbol{F}$?

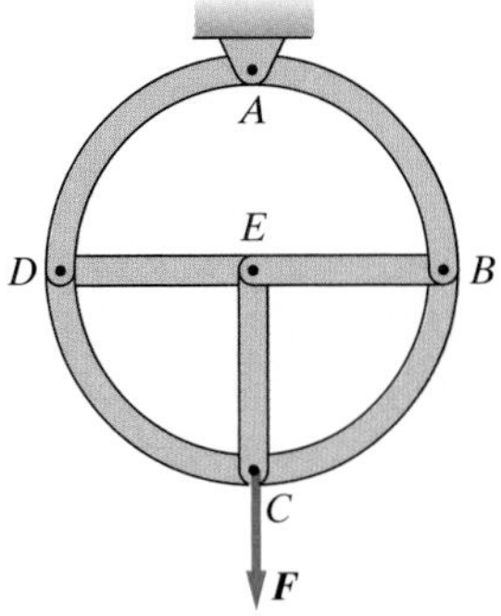

E9.1.18

9.1.19. Consider the frame in E9.1.19.

a. Determine the forces in all members of the frame. All interior angles are 60° or 120°.

b. If the magnitude of the force (compression or tension) in any member is not to exceed 500 lb, what is the maximum allowable magnitude of the force ***F***?

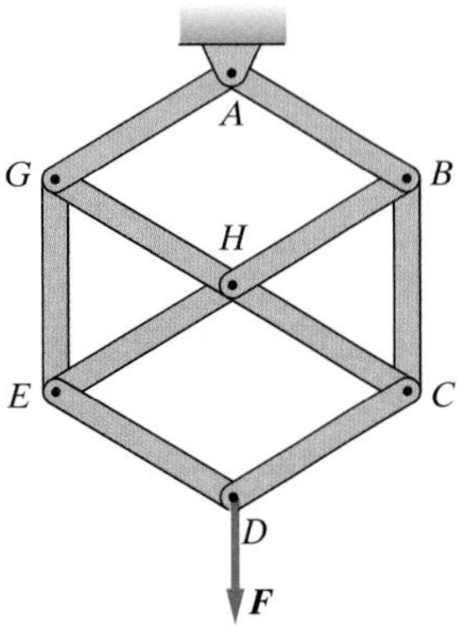

E9.1.19

9.1.20. Consider the frame in E9.1.20. Determine the loads acting on members *AB* and *BC*.

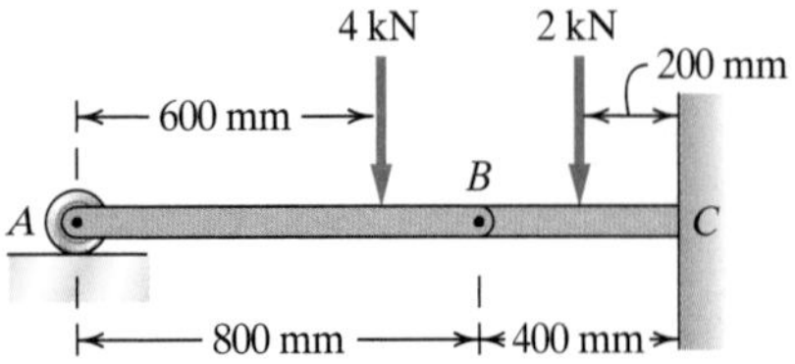

E9.1.20

9.1.21. Consider the frame in **E9.1.20**. Determine the loads acting on the frame at *A* and *C*.

9.1.22. Consider the frame in **E9.1.20**. Determine the shear force acting on the pin at *B*.

9.1.23. Consider the frame in E9.1.23. Determine the loads acting on members *AB*, *BC*, and *CD*.

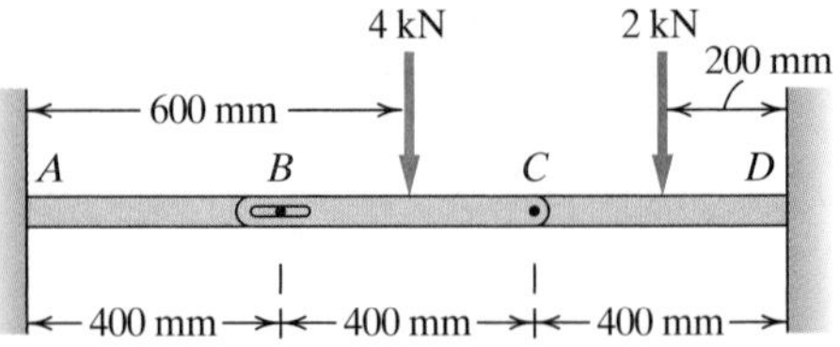

E9.1.23

9.1.24. Consider the frame in **E9.1.23**. Determine the loads acting on the frame at *A* and *D*.

9.1.25. Consider the frame in **E9.1.23**. Determine the shear forces acting on the pins at *B* and *C*.

9.1.26. A bag of laundry resting on a chair is shown in E9.1.26*a*. The forces on the chair by the laundry are shown in **E9.1.26*b***. Determine

a. the loads acting on the chair at *A* and *B*

b. the shear forces acting on pins *F*, *C*, and *D*

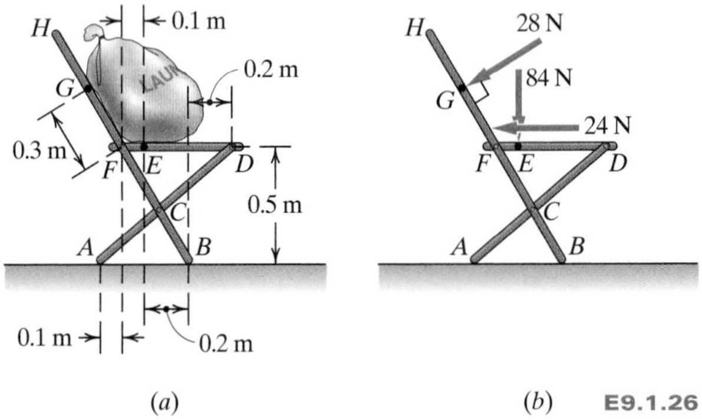

E9.1.26

9.1.27. Consider the frame in E9.1.27. Determine the loads acting on member *DH* and on the frame at *A* and *C*.

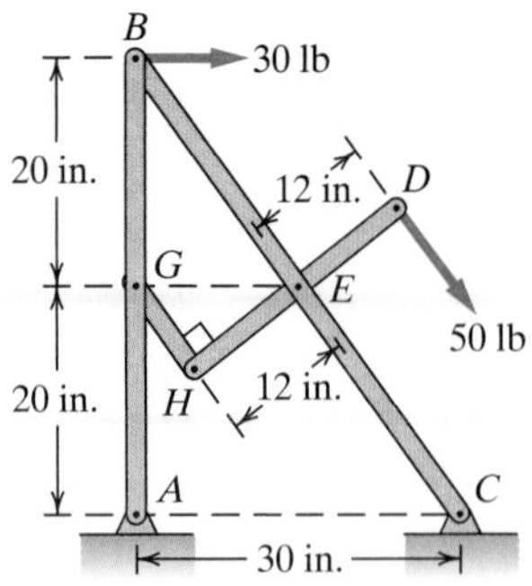

E9.1.27

9.1.28. Consider the frame in **E9.1.27**. Determine the shear forces acting on the pins at *G* and *E*.

For **Exercises 9.1.29 and 9.1.30**, the frame supports the 600-kg mass in the manner shown in E9.1.29.

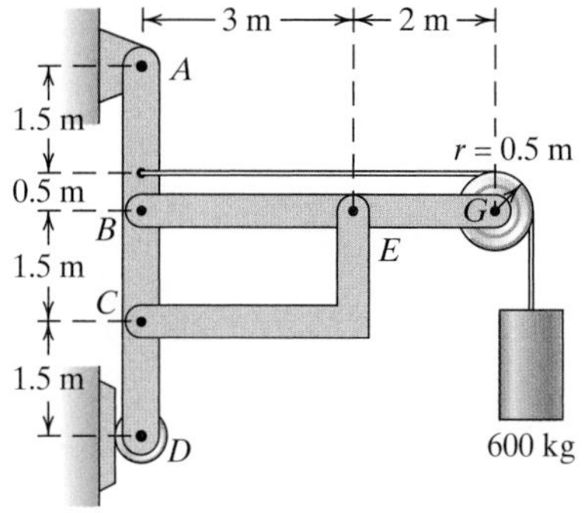

E9.1.29

9.1.29. Determine the forces acting on each of the members, and on the frame at A and D.

9.1.30. Determine the shear force acting on the pin at E.

9.1.31. Consider the frame in E9.1.31. Determine the loads acting on the frame at A and C.

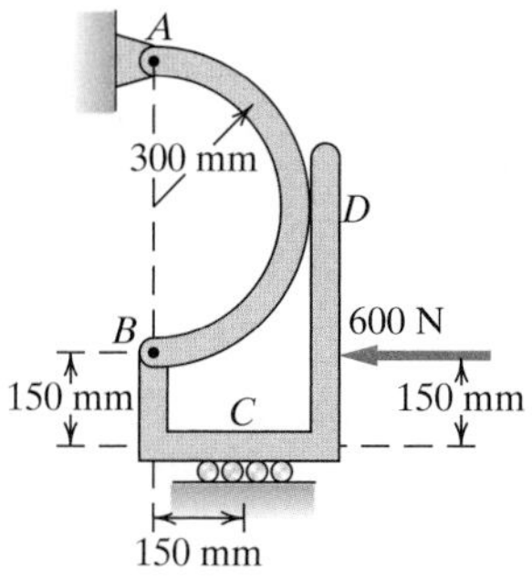

E9.1.31

9.1.32. Consider the frame in **E9.1.31**. Determine the loads acting on members AB and BCD.

9.1.33. Consider the frame in **E9.1.31**. Determine the shear force acting on the pin at B.

9.1.34. Link BC in E9.1.34 prevents rotation of link ACD under the action of the 100 N · m moment applied at D. Determine the shear forces acting on the pins at A, B, and C.

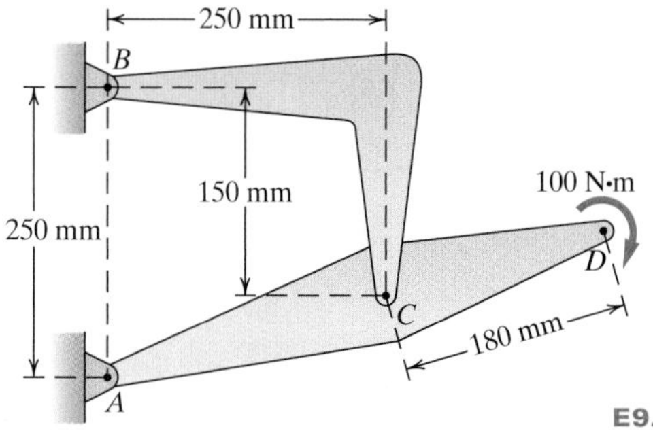

E9.1.34

9.1.35. Consider the frame in E9.1.35. Determine the loads acting on the frame at A and C.

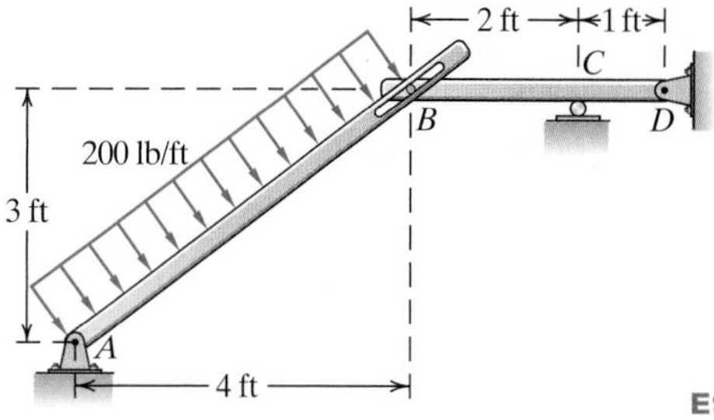

E9.1.35

9.1.36. Consider the frame in **E9.1.35**. Determine the shear force acting on the pin at B.

9.1.37. Consider the frame in E9.1.37. Determine the loads acting on the frame at A, D, and E.

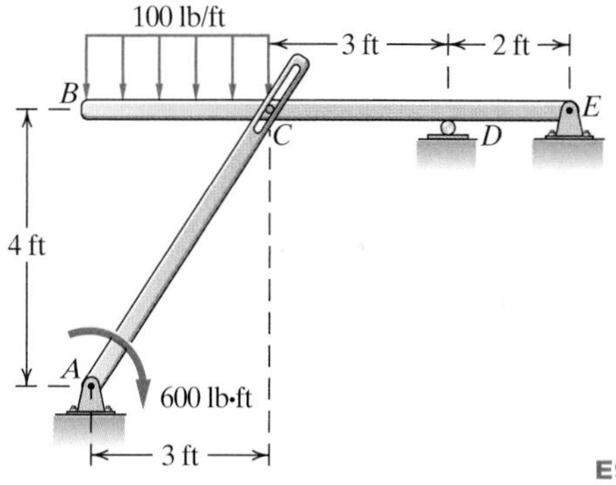

E9.1.37

9.1.38. Consider the frame in **E9.1.37**. Determine the shear force acting on the pin at C.

9.1.39. Consider the frame in E9.1.39. The magnitudes of $\boldsymbol{F}_1$ and $\boldsymbol{F}_2$ are 200 N and 0 N, respectively. Determine the tension in the cable AC.

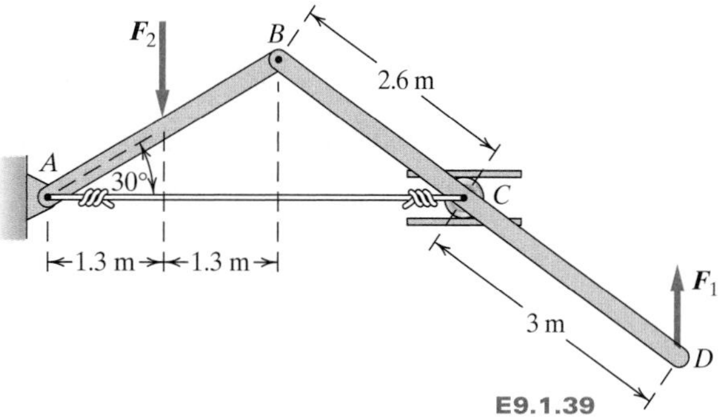

E9.1.39

9.1.40. Consider the frame in **E9.1.39**. The magnitudes of $\boldsymbol{F}_1$ and $\boldsymbol{F}_2$ are 200 N and 0 N, respectively. Determine the shear forces acting on the pin at B.

9.1.41. Consider the frame in **E9.1.39**. The magnitudes of $\boldsymbol{F}_1$ and $\boldsymbol{F}_2$ are 0 N and 3000 N, respectively. Determine the tension in the cable *AC*.

9.1.42. Consider the frame in **E9.1.39**. The magnitudes of $\boldsymbol{F}_1$ and $\boldsymbol{F}_2$ are 0 N and 3000 N, respectively. Determine the shear force acting on the pin at *B*.

9.1.43. Consider the frame in **E9.1.43**. Determine the loads acting on members *AE* and *BD*.

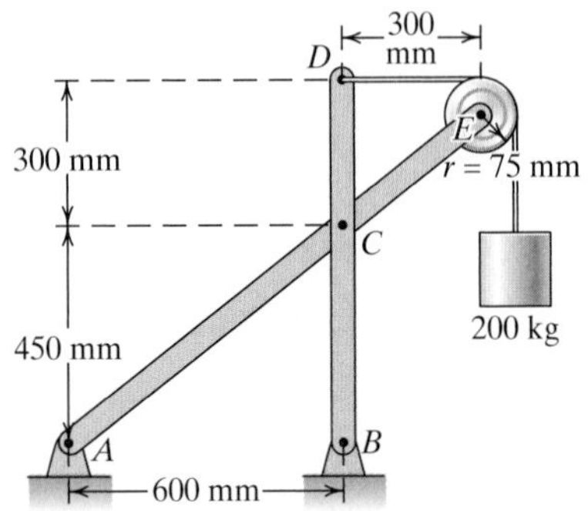

E9.1.43

9.1.44. Consider the frame in **E9.1.43**. Determine the loads acting on the frame at *A* and *B*.

9.1.45. Consider the frame in **E9.1.43**. Determine the shear forces on the pins at *C* and *E*.

9.1.46. Consider the frame in **E9.1.46**. Determine the loads acting on the frame at *A* and *G*, and the shear forces acting on the pins at *B* and *C*.

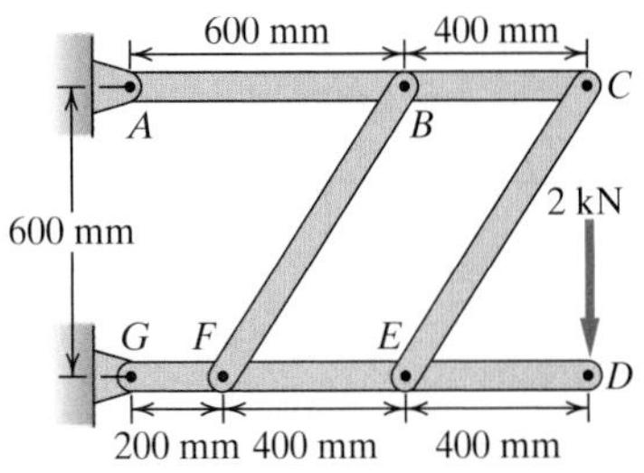

E9.1.46

9.1.47. Consider the frame in **E9.1.47**. Determine the loads acting on the frame at *A* and *F*, and the shear forces acting on the pins at *B* and *C*.

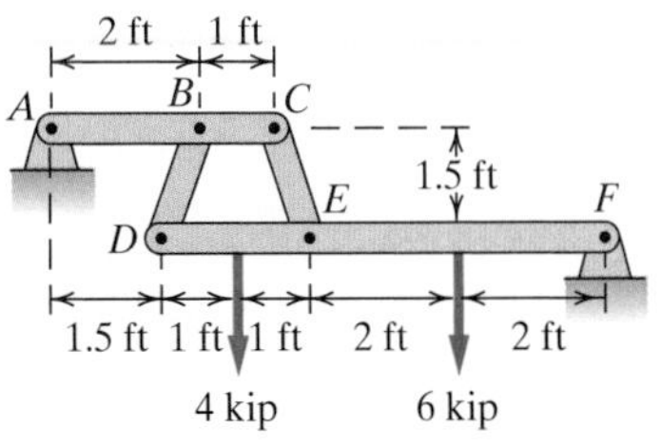

E9.1.47

9.1.48. Consider the frame in **E9.1.48**. The spring has a stiffness of 8 kN/m and an unstretched length of 220 mm. If $\alpha = 36.87°$ in the position shown (which is in equilibrium), determine the magnitude of the force $\boldsymbol{F}$.

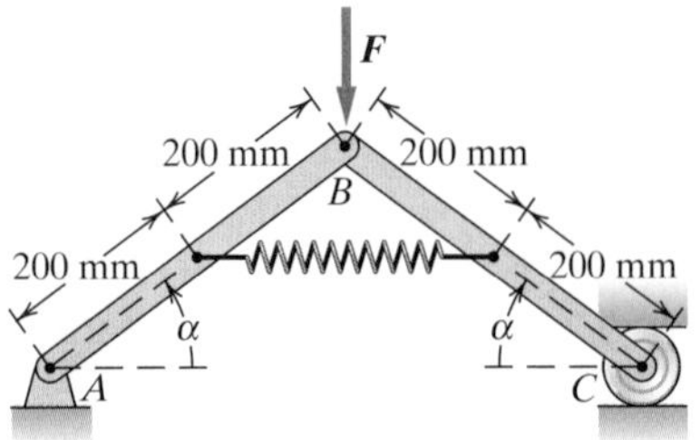

E9.1.48

9.1.49. Consider the frame in **E9.1.49**. The spring has a stiffness of 60 lb/in. and an unstretched length of 15 in. Determine the loads acting on the frame at *A* and *D*.

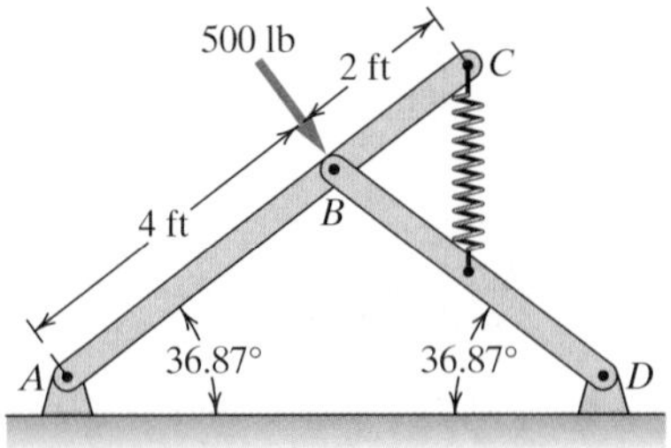

E9.1.49

9.1.50. The top of a folding workbench (along with associated materials on top of the workbench) has a mass of *M*. The center of mass is located at point *G*, a distance *d* from the pivot point at *D*, as shown in **E9.1.50**.

a. Write a relationship between the force in cable *FH* and the distance *d*.

b. What values of *d* would cause the top of the workbench to tip over?

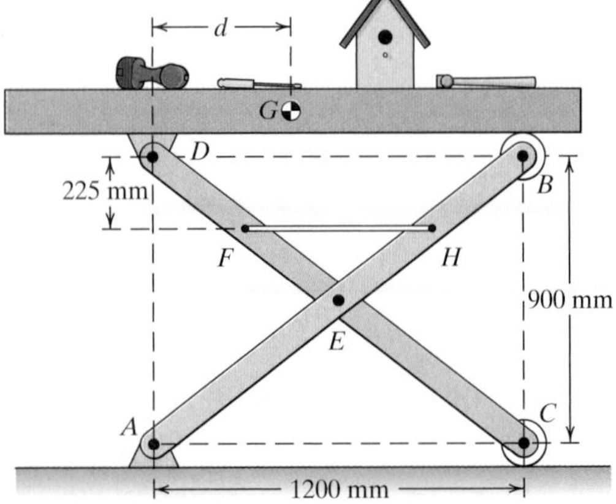

E9.1.50

9.1.51. Figure E9.1.51 depicts a lifting device for transporting steel drums of weight W.

a. Determine the magnitude of the tension or compression in member DC as a function of W.

b. Determine the shear forces acting in the pins at A and C, as functions of W.

c. If the force in member DC exceeds 2000 N, member DC will fail, and if the shear force in any pin exceeds 1000 N, that pin will fail. Given this, what is the maximum drum weight you would recommend lifting? Support your answer with calculations.

d. At the weight you determine in **c**, what force is exerted on the drum at E and F?

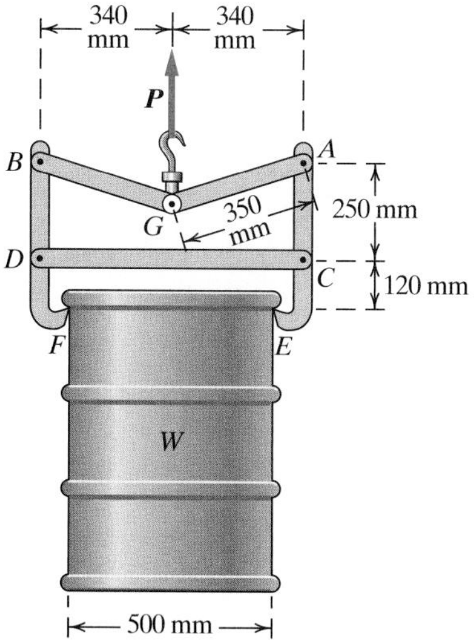

E9.1.51

9.1.52. Consider the frame in E9.1.52. Determine the loads acting on the frame at A and D.

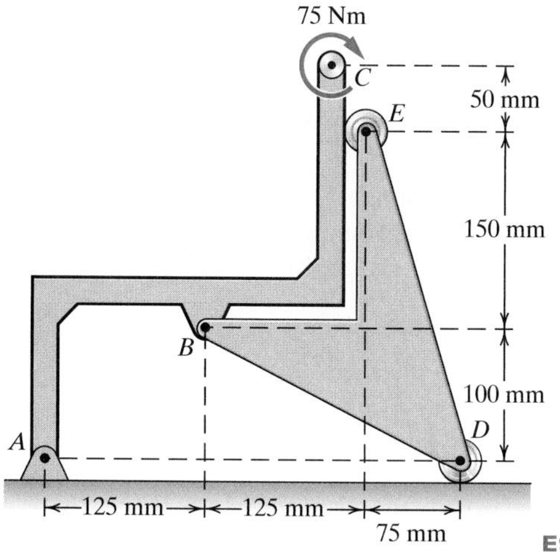

E9.1.52

9.1.53. Consider the frame in **E9.1.52**. Determine the loads acting on members AC and BED.

9.1.54. Consider the frame in **E9.1.52**. Determine the shear force acting on the pin at B.

9.1.55. A typical bicycle rack is shown in E9.1.55.

a. Determine the loads acting on the rack at A and E if the weight being carried on the rack is 160 N. (The total weight is 320 N, but you need only consider 1/2 of the design because of symmetry.) The connections at A, C, D, and E are pin connections.

b. Buckling will occur in member CE or DE if the compressive force in either member exceeds 100 N. Are you concerned about buckling as a failure mode?

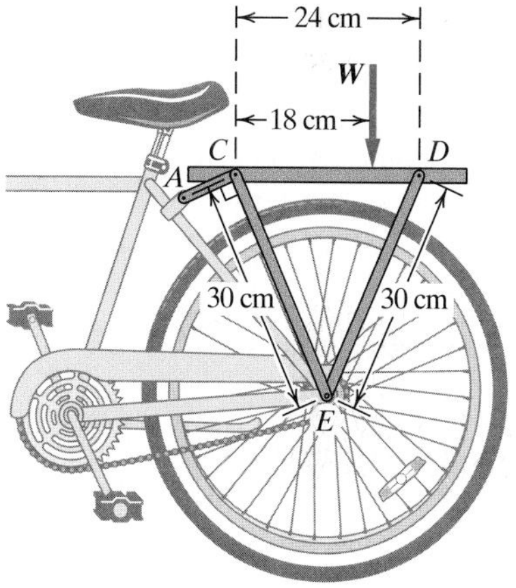

E9.1.55

9.1.56. Determine the loads acting on the frame in E9.1.56 at A, D, and E. Treat each connection as a ball-and-socket joint.

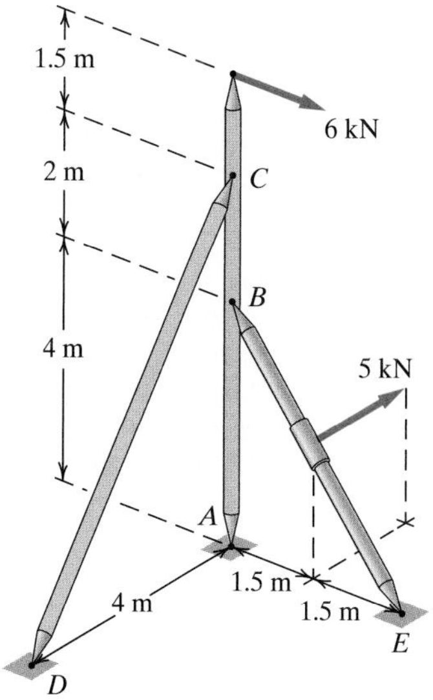

E9.1.56

9.1.57. In **E9.1.57**, T represents a turnbuckle, and C and D are hinges whose axes are along the line CD. Pin connections attach member EF to members CB and DB. Determine the tension T in the turnbuckle and the force in member EF.

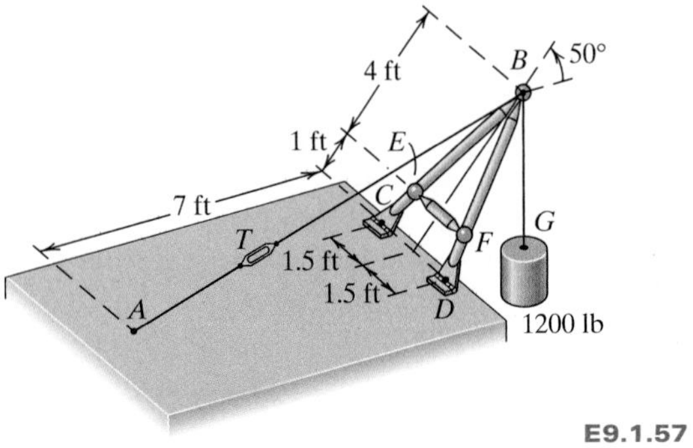

E9.1.57

9.1.58. The cabling system in **E9.1.58** holds up the 840-lb hopper. Determine the tensions in all segments of the cables.

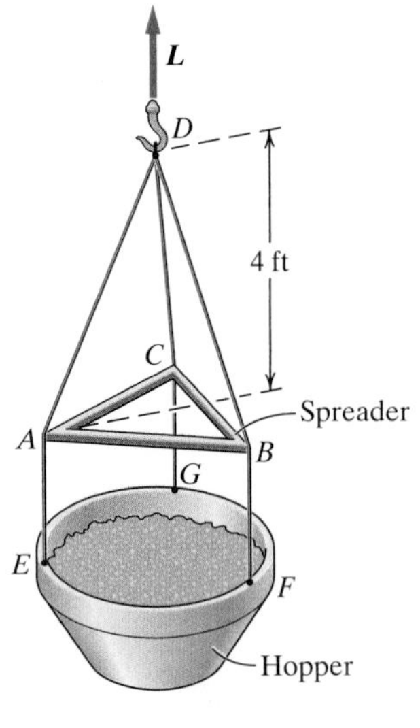

E9.1.58

For **Exercises 9.1.59–9.1.62**: Ancient siege engines provided military commanders with the ability to engage an enemy from a distance; essentially it was the artillery of the armies past. What is known of these medieval siege engines is limited to crude artist renditions and manuscript references. Hence, the hypothetical analysis of siege engines is still intriguing and challenging. Let us analyze one of the simplest forms of siege engine, the catapult, as shown in **E9.1.59*a***. This siege engine fires a missile using the energy gained from dropping a weight and the advantage of a lever arm.

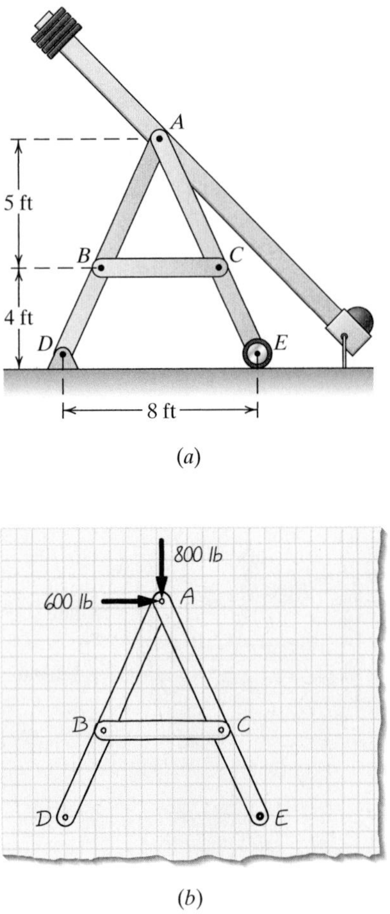

E9.1.59

9.1.59. If the lever arm applies the 600 lb and 800 lb loads to the pin at A, as shown in **E9.1.59*b***, calculate the loads acting on the frame at D and E.

9.1.60. Calculate the shear forces acting on the pins at A, B, and C in **E9.1.59**.

9.1.61. Determine the loads acting on members AD, AE, and BC. Based on these results, which member would you be most concerned about failing, and why?

9.1.62. If the boundary connection at E was changed from a roller to a pin connection, how would your calculations in (a) and (b) change? (Describe in words and with sketches.)

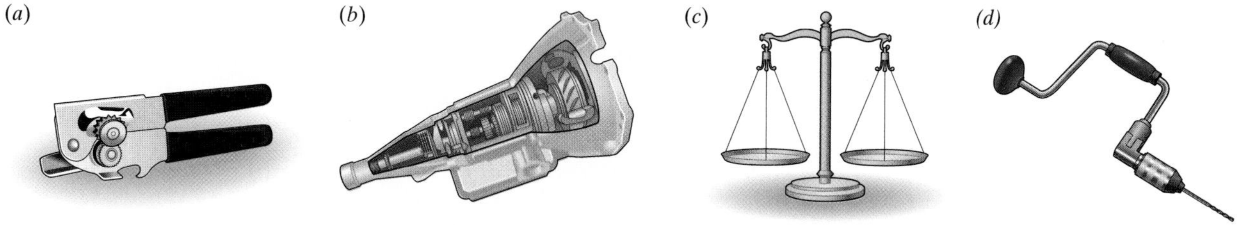

Figure 9.38 Some common machines

9.2 MACHINES

A **machine** is a system designed to change the direction and/or magnitude of loads or motion. In considering machines, we often think in terms of the load *into* the system, the load *out of* the system, and the ratio of output to input. Hand tools such as pliers and tweezers are examples of machines for which we want the ratio of output force (clamping force) to input force (hand force) to be greater than unity. Other examples of machines are can opener (multiples force); car transmission (multiples force or torque); balance scale (equalizes moments); hand mixer/hand drill (multiplies speed) (Figure 9.38).

Like frames, machines support loads. In fact, many machines can be classified as frames because they are assemblies of various members (including at least one multiforce member). However, because the primary purpose of a machine is to modify loads or motion, we have created a separate classification.

Finding Internal Loads in a Machine

The basic analysis of a machine in mechanical equilibrium is identical to that of a frame in terms of analyzing separate members by drawing free-body-diagrams, setting up the equilibrium equations, and then using these equations to find the unknown loads. Often the analysis of a machine also involves the calculation of the ratio of output load to input load (referred to as the **mechanical advantage**) or output motion to input motion. Sometimes the biggest challenge in analyzing a machine is understanding how it works. For this reason, you should always take the time to gain this understanding *before* diving into calculations (this same advice holds for analysis of any system!).

EXAMPLE 9.7 FORCE MULTIPLICATION

For the grippers shown in Figure 9.39, find the magnitude of the gripping force $\boldsymbol{Q}$ as a function of the magnitude of the input force $\boldsymbol{P}$ and the geometric parameters given. Assume that the pins at N and N' slide freely in the slots.

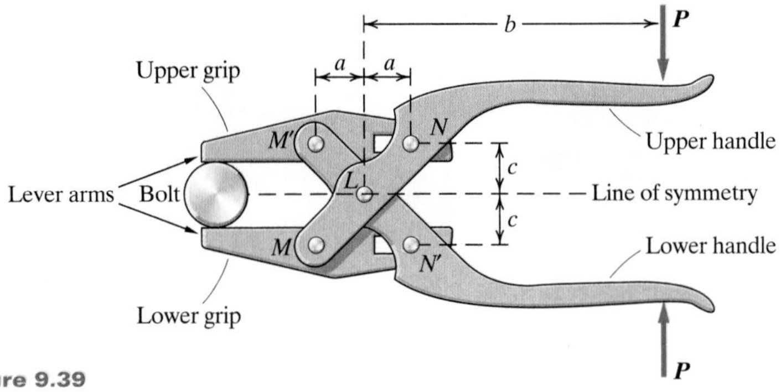

Figure 9.39

Goal We are to find the magnitude of the gripping force $\boldsymbol{Q}$ as a function of the magnitude of the input force $\boldsymbol{P}$ and the various lengths given.

Given We are given information about the geometry of the gripper and the loading placed on it.

Assume We assume that the weights of the various members that make up the gripper are negligible compared to the loads acting on it and that we can treat the system as planar. Also, we are told to assume that the slots at N and N' are frictionless so there is no force acting parallel to the slots.

Draw The free-body diagram for the upper handle is shown in Figure 9.40, and that for the lower grip is shown in Figure 9.41. We draw the forces at pin M on the lower grip equal and opposite to those at M on the upper handle. The symmetry of the gripper tells us that the forces perpendicular to the line of symmetry on the upper handle are equal and opposite to those on the lower handle. Because of this symmetry, we know that the upward force $\boldsymbol{R}_5$ acting at N on the upper handle means there is a downward force acting at N' on the lower handle. The lower grip is connected to the lower handle at N', which means $\boldsymbol{R}_5$ acts upward at N' on the lower grip.

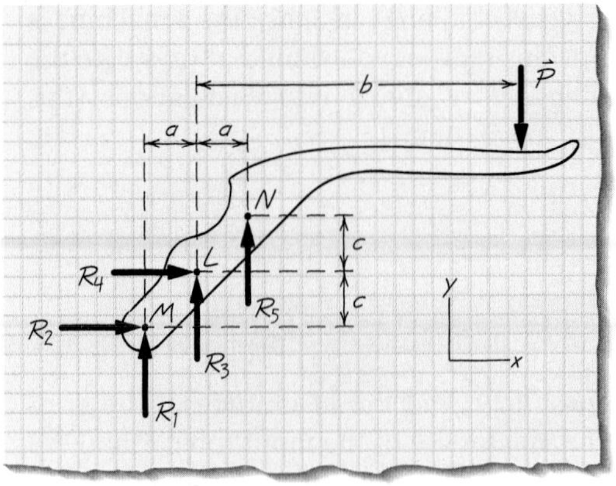

Figure 9.40 Upper handle free-body diagram

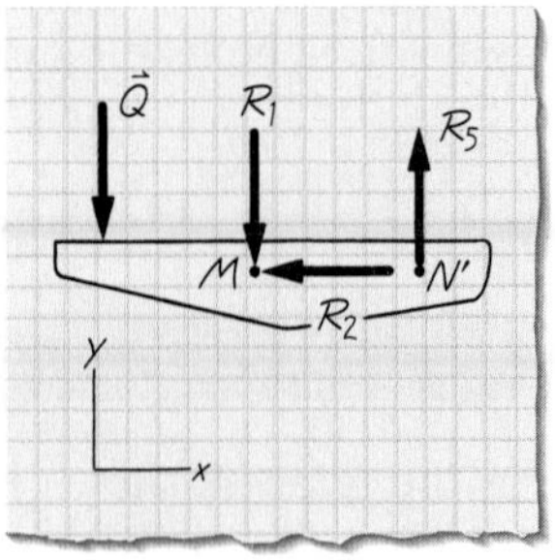

Figure 9.41 Lower grip free-body diagram

Formulate Equations and Solve We first define the upper handle as our system, and from its free-body diagram generate three equilibrium equations that contain five unknowns. We then examine a second system, the lower grip. From this we generate two equilibrium equations and no additional unknowns. Based on these five equations, we are able to relate $\| \boldsymbol{Q} \|$ (gripping force) to $\| \boldsymbol{P} \|$ (input force).

For the upper handle (**Figure 9.40**), based on (7.5A):

$$\sum F_x = 0\,(\rightarrow +)$$
$$R_2 + R_4 = 0$$
$$R_2 = -R_4 \qquad (1)$$

Based on (7.5B):

$$\sum F_y = 0(\uparrow +)$$
$$R_1 + R_3 + R_5 - P = 0 \qquad (2)$$

Based on (7.5C), with the moment center at L:

$$\sum M_{z@L} = 0(\curvearrowleft +)$$
$$-R_1(a) + R_2(c) + R_5(a) - P(b) = 0 \qquad (3)$$

For the lower grip (**Figure 9.41**), based on (7.5A):

$$\sum F_x = 0(\rightarrow +)$$
$$-R_2 = 0$$
$$R_2 = 0 \qquad (4)$$

Based on (7.5B):

$$\sum F_y = 0(\uparrow +)$$
$$-R_1 + R_5 - Q = 0$$
$$Q = R_5 - R_1 \qquad (5)$$

We rearrange (3), substitute (4) and (5) into (3), and solve for Q:

$$(R_5 - R_1)a + R_2(c) - P(b) = 0$$
$$Q(a) - P(b) = 0 \qquad (6)$$

Answer From (6), $Q = P(b/a)$. Notice that the gripping force depends on the ratio of the dimensions b/a, and on no other geometric parameters. If $b/a > 1$, the tool is a force multiplier. The ratio (b/a) is the mechanical advantage provided by the gripper.

EXAMPLE 9.8 ANALYSIS OF A TOGGLE CLAMP

Table 9.1 Dimensions of toggle

$d = 4$ in.
$e = 4$ in.
β is such that $\cos\beta = 4/5$, $\sin\beta = 3/5$
Due to size restrictions:
1 in. $\leq a \leq$ 2 in.
4 in. $\leq b \leq$ 7 in.

A toggle clamp holds workpiece G (Figure 9.42) and has dimensions shown in Table 9.1.

(a) Assuming the weight of the clamp members can be ignored, determine the vertical clamping force at E in terms of the input force $\boldsymbol{P}$ applied to the clamp handle.

(b) Given the dimensions in **Table 1**, determine four possible toggle clamp geometries that result in a vertical clamping force at E that is 10 times greater than input force $\boldsymbol{P}$.

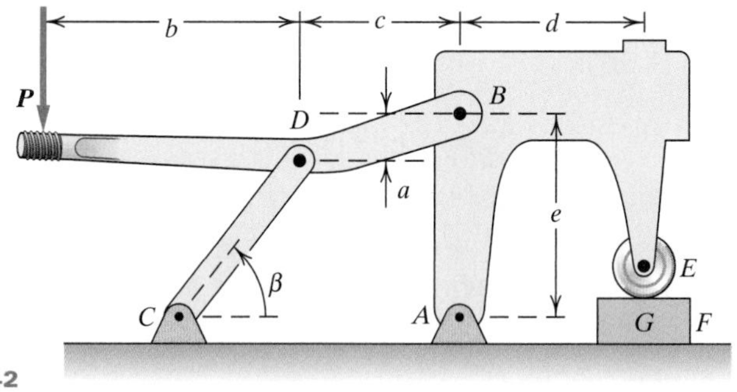

Figure 9.42

Goal Find the force at E ($\boldsymbol{F}_E$) in terms of $\boldsymbol{P}$ (a), then determine four possible toggle geometries that give a mechanical advantage of 10 (b).

Given We are given the configuration of the toggle clamp and tolerances for dimensions a and b.

Assume We assume that the pin connections A, B, C, D are all frictionless. We also assume that the clamping force at E is purely vertical, as is the input force $\boldsymbol{P}$. Member CD is a two-force member (because it has pin connections at its ends and is loaded by force through its ends). Finally, we assume that the system is planar.

Draw We draw free-body diagrams of individual parts of the toggle clamp, recognizing that $\boldsymbol{F}_D$ is at angle β with the horizontal because member CD is a two-force member (Figure 9.43 and Figure 9.44).

Formulate Equations and Solve **(a)** We first isolate the handle (**Figure 9.43**); doing this allows us to relate $\boldsymbol{F}_B$ to $\boldsymbol{P}$. Next we isolate member ABE (**Figure 9.44**); doing this allows us to relate $\boldsymbol{F}_B$ to $\boldsymbol{F}_E$. Finally, we relate $\boldsymbol{F}_E$ to $\boldsymbol{P}$.

We can write $\boldsymbol{F}_D = F_{Dx}\boldsymbol{i} + F_{Dy}\boldsymbol{j}$. Furthermore, because member CD is a two-force member, we can write $F_{Dx} = \| \boldsymbol{F}_D \| \cos\beta$ and $F_{Dy} = \| \boldsymbol{F}_D \| \sin\beta$.

Now we formulate the equilibrium equations for the planar system in **Figure 9.43**. Based on (7.5C) (with the movement center at B):

$$\sum M_{z@B} = 0(\curvearrowleft+)$$

$$P(b + c) - \| \boldsymbol{F}_D \| \sin\beta(c) + \| \boldsymbol{F}_D \| \cos\beta(a) = 0$$

$$\| \boldsymbol{F}_D \| = \frac{b + c}{c\sin\beta - a\cos\beta} P \tag{1}$$

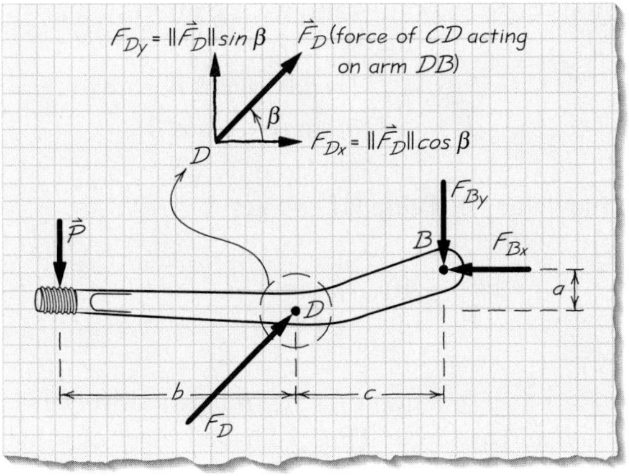

Figure 9.43 Free-body diagram of handle

Based on (7.5A):

$$\begin{aligned} &\sum F_x = 0(\rightarrow +) \\ &-F_{Bx} + \|\boldsymbol{F}_D\| \cos\beta = 0 \\ &F_{Bx} = \|\boldsymbol{F}_D\| \cos\beta \end{aligned} \tag{2}$$

Based on (7.5B):

$$\begin{aligned} &\sum F_y = 0(\uparrow +) \\ &-F_{By} + \|\boldsymbol{F}_D\| \sin\beta - P = 0 \\ &F_{By} = \|\boldsymbol{F}_D\| \sin\beta - P \end{aligned} \tag{3}$$

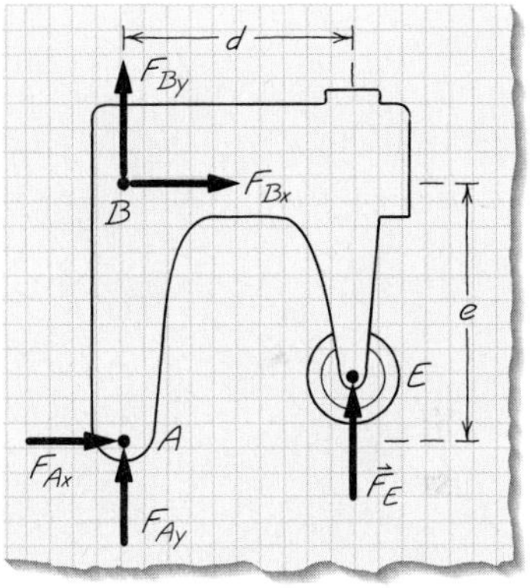

Figure 9.44 Free-body diagram of member ABE

We substitute (1) into (2) and (3) to express the force components at B in terms of the input force $\boldsymbol{P}$:

From (2) $$F_{Bx} = \frac{(b + c)\cos\beta}{c\sin\beta - a\cos\beta} P \tag{2*}$$

From (3) $$F_{By} = \left(\frac{(b + c)\sin\beta}{c\sin\beta - a\cos\beta} - 1\right)P \tag{3*}$$

For member ABE (**Figure 9.44**) we have, based on (7.5C) (with the moment center at A):

$$\begin{aligned} &\sum M_{z@A} = 0(\circlearrowleft) \\ &-F_{Bx}e + F_E d = 0 \\ &F_E = \frac{e}{d} \underbrace{F_{Bx} =}_{\text{substituting from (2*)}} \left(\frac{e}{d}\right) \frac{(b + c)\cos\beta}{c\sin\beta - a\cos\beta} P \end{aligned} \tag{4}$$

Answer to (a) $$F_E = \left(\frac{e}{d}\right) \frac{(b + c)\cos\beta}{c\sin\beta - a\cos\beta} P$$

(b) Rearranging (4) gives us the ratio we are interested in:

$$\frac{F_E}{P} = \left(\frac{e}{d}\right) \frac{(b + c)\cos\beta}{c \sin\beta - a\cos\beta}$$

Substituting the values in **Table 9.1**, we get

$$\frac{F_E}{P} = \frac{10}{1} = \left(\frac{4 \text{ in.}}{4 \text{ in.}}\right) \frac{(b + c)\frac{4}{5}}{c\frac{3}{5} - a\frac{4}{5}} \tag{5}$$

We solve (5) for c as a function of a and b because we were given tolerances for a and b and not for c. Thus we have ranges of values of a and b we can substitute into this equation. With simplifying and rearranging (5) becomes

$$c = \frac{2}{13}(10a + b) \tag{6}$$

This equation tells us that the dimensions a, b, and c are not independent. Instead, only certain combinations of a, b, and c will result in a 10 to 1 ratio between the clamping force and the input force.

We now need to find four combinations of (a, b, c) that work in (6) and that meet the restrictions on a and b given in **Table 9.1**. One straightforward approach is to solve for c when a and b are at their limits. In other words, use (6) to find c when

$$a = 1 \text{ in.}, \qquad b = 4 \text{ in.}$$
$$a = 1 \text{ in.}, \qquad b = 7 \text{ in.}$$
$$a = 2 \text{ in.}, \qquad b = 4 \text{ in.}$$
$$a = 2 \text{ in.}, \qquad b = 7 \text{ in.}$$

When we make these substitutions into (6) we find that the combinations of (a, b, c) listed in Table 9.2 result in $F_E/P = 10$:

Answer to (b)

Table 9.2

	a (in.)	b (in.)	c (in.)
Solution 1	1	4	2.15
Solution 2	1	7	2.62
Solution 3	2	4	3.69
Solution 4	2	7	4.15

Note: The combinations in **Table 9.2** are just four of an infinite number of combinations that result in $F_E/P = 10$. Additional solutions are shown in Figure 9.45, which also shows the four solutions in **Table 9.2**. Do you see any advantages or disadvantages to the combinations shown in **Table 9.2**?

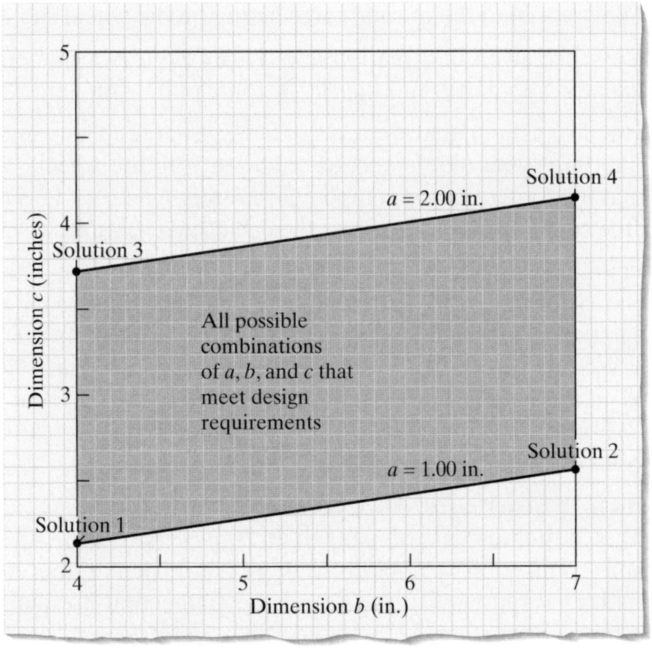

Figure 9.45 Combinations of *a*, *b*, *c* that give a mechanical advantage of 10

EXAMPLE 9.9 BICYCLE BRAKE

When a bicycle rider squeezes the brake lever on the handlebars, he pulls on a cable that is indirectly attached to the brake calipers. The cable pulls up at *G* on a wire attached to the calipers at *B* and *C*, causing the two calipers to rotate about pin *A* and engage with the wheel (Figure 9.46). A normal force and friction force are generated at both *D* and *E* between the brake pad and the wheel rim (the friction force is acting perpendicular to the plane of the page). When the cable is pulling upward with a force of 200 N on the center-pull bicycle brake shown in **Figure 9.46**, find the normal force due to the brake pad acting on the wheel rim at *E*. Ignore the force generated at *A* by the spring (not shown in **Figure 9.46**) that restores the brake to its open position when the brake lever is released.

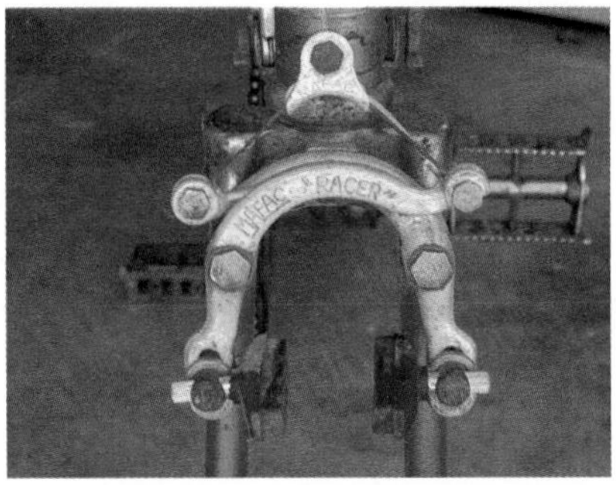

(*a*)

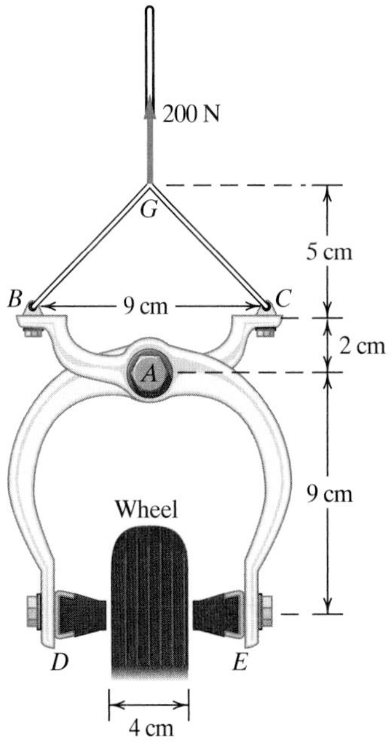

(*b*)

Figure 9.46

Goal We are to find the normal force acting on the wheel rim at E.

Given We are given information about how the brake calipers work, their associated geometry, and the 200-N force pulling on wire BC at G.

Assume We assume that we can treat the system as planar and that it is in equilibrium (with the brake pads engaged with the wheel). We also assume that the weights of the components are negligible, pin A is frictionless, and the connection at G between the cable and the wire is frictionless.

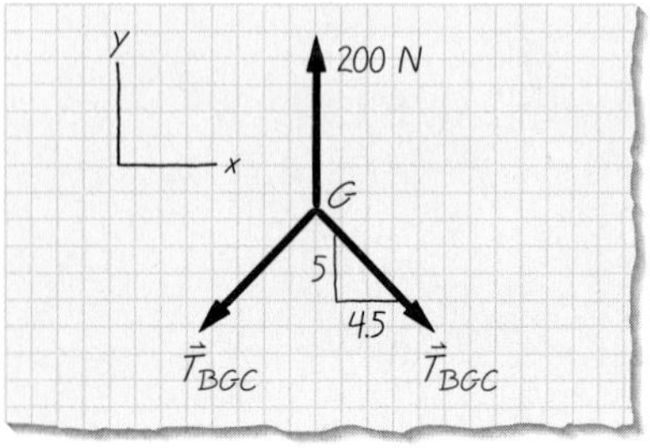

Figure 9.47 Free-body diagram of particle G

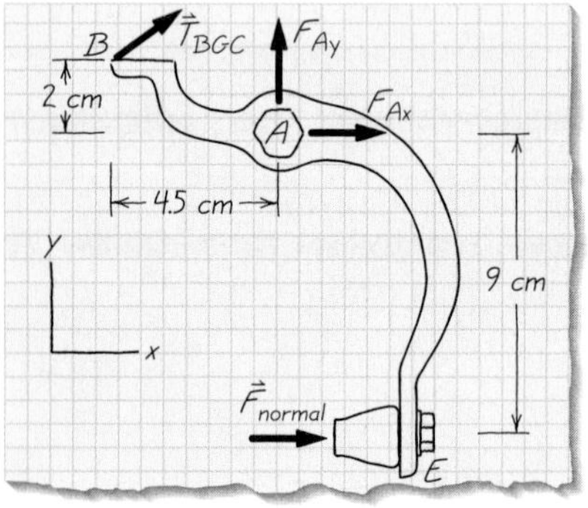

Figure 9.48 Free-body diagram of caliper BAE

Draw First we draw a free-body diagram of the connection of the cable and wire at G (**Figure 9.47**) to determine the tensile force acting on wire BC. Due to the symmetry of the brake, just one of the calipers needs to be analyzed. We draw a free-body diagram of caliper BAE (**Figure 9.48**) and apply the planar equilibrium equations to determine F_{normal}, the normal force acting on the brake pad at E. F_{normal} acting on the brake pad is equal and opposite to the normal force acting on the wheel rim.

Formulate Equations and Solve First we find T_{BGC}, the wire tension that is pulling on the caliper at B. The wire can slide freely through the connection at G because there is no friction, therefore the tension acting on the wire to the left of the connection is equal to that acting on the right portion.

Based on (7.5B) and **Figure 9.47** we can write:

$$\sum F_y = 0(\uparrow +)$$
$$-T_{BGCy} - T_{BGCy} + 200\text{ N} = 0$$

Using the geometry of the wire to calculate the y component, we get:

$$-T_{BGC}\left(\frac{5\text{ cm}}{6.73\text{ cm}}\right) - T_{BGC}\left(\frac{5\text{ cm}}{6.73\text{ cm}}\right) + 200\text{ N} = 0$$
$$2T_{BGC} = 269.2\text{ N}$$
$$T_{BGC} = 135\text{ N} \qquad (1)$$

We now analyze caliper BAE (**Figure 9.48**) to find F_{normal}. Based on (7.5C), with the moment center at A:

$$\sum M_{z@A} = 0(\circlearrowleft +)$$
$$-T_{BGCy}(4.5\text{ cm}) - T_{BGCx}(2\text{ cm}) + F_{\text{normal}}(9\text{ cm}) = 0$$

Recognizing that $T_{BGCy} = \frac{5}{6.73}T_{BGC}$, $T_{BGCx} = \frac{4}{6.73}T_{BGC}$, and that $T_{BGC} =$ 135N (from (1)) we get:

$$-135\text{ N}\left(\frac{5\text{ cm}}{6.73\text{ cm}}\right)(4.5\text{ cm}) - 135\text{ N}\left(\frac{4.5\text{ cm}}{6.73\text{ cm}}\right)(2\text{ cm}) + F_{\text{normal}}(9\text{ cm}) = 0$$
$$F_{\text{normal}}(9\text{ cm}) = 451.3\text{ N}\cdot\text{cm} + 180.5\text{ N}\cdot\text{cm}$$
$$F_{\text{normal}} = 70.2\text{ N}$$

Answer $F_{\text{normal}} = 70.2\text{ N}$

Check To check our result, we reanalyze the problem by first finding F_{Ay} and F_{Ax} and then F_{normal}.

Based on (7.5B) and **Figure 9.48**:

$$\sum F_y = 0(\uparrow +)$$
$$T_{BGCy} + F_{Ay} = 0$$
$$F_{Ay} = -T_{BGC}\left(\frac{5 \text{ cm}}{6.73 \text{ cm}}\right) = -100 \text{ N}$$

Based on (7.5C), with the moment center at E:

$$\sum M_{z@E} = 0(\circlearrowleft +)$$
$$-T_{BGCy}(6.5 \text{ cm}) - T_{BGCx}(11 \text{ cm}) - F_{Ay}(2 \text{ cm}) - F_{Ax}(9 \text{ cm}) = 0$$
$$-135 \text{ N}\left(\frac{5 \text{ cm}}{6.73 \text{ cm}}\right)(6.5 \text{ cm}) - 135 \text{ N}$$
$$\left(\frac{4.5 \text{ cm}}{6.73 \text{ cm}}\right)(11 \text{ cm}) - (-100 \text{ N})(2 \text{ cm}) - F_{Ax}(9 \text{ cm}) = 0$$
$$F_{Ax} = -160.5 \text{ N}$$

Based on (7.5A):

$$\sum F_x = 0(\rightarrow +)$$
$$T_{BGCx} + F_{Ax} + F_{\text{normal}} = 0$$
$$135 \text{ N}\left(\frac{4.5 \text{ cm}}{6.73 \text{ cm}}\right) + (-160.5 \text{ N}) + F_{\text{normal}} = 0$$
$$F_{\text{normal}} = 70.2 \text{ N}$$

We get the same result!

Rotating Machines

A machine may involve moving members, which may be rotating, translating, or both. This is in contrast to frames and trusses, which are generally stationary. As long as each member in a machine is rotating about an axis of fixed orientation at a constant angular velocity, the conditions of mechanical equilibrium developed in Chapter 7 apply. Examples 9.10 and 9.11 illustrate how the conditions of equilibrium can be used to study the performance of a gear train in an automobile and the power train in a bicycle.

EXAMPLE 9.10 ANALYSIS OF A GEAR TRAIN

A single-stage gear train is shown in **Figure 9.49** and **Figure 9.50**. The input force $\boldsymbol{F}_1$ causes Shaft 1 to rotate counterclockwise. Gear 2 meshes with Gear 3 and causes Shaft 2 to rotate clockwise and exert the output force $\boldsymbol{F}_4$ on an output device (not shown).

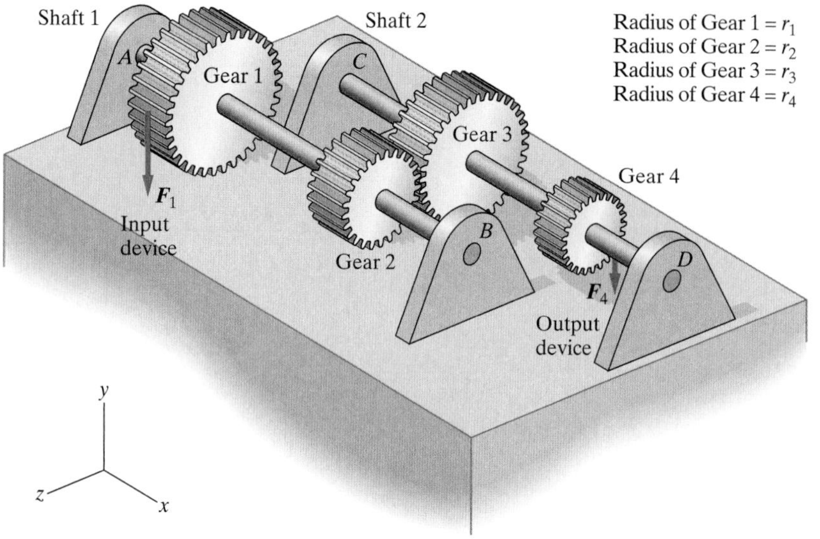

Figure 9.49

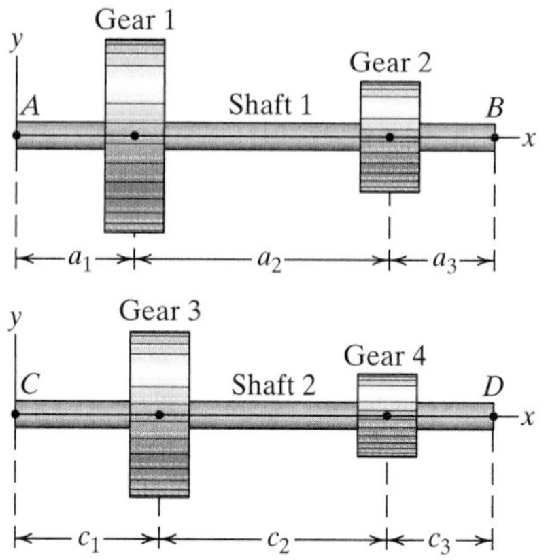

Figure 9.50

(a) Determine the forces applied to the shafts at journal bearing A, B, C, and D (see **Table 6.2**) in terms of the input force $\boldsymbol{F}_1$.

(b) For the shafts rotating at constant speed, determine the ratio of input moment to output moment in terms of the radii of various gears.

Note: Gear trains are used in rotating systems to either increase or decrease moment or to decrease or increase rotational speed. In an automotive transmission, which is an example of a gear train, the moment from the engine is increased and the rotational speed is decreased.

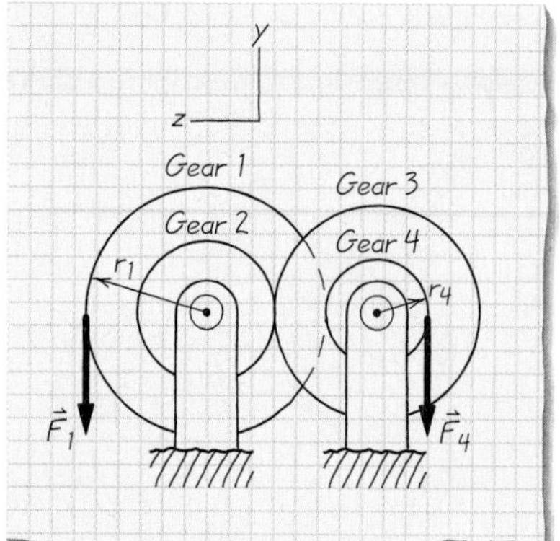

Figure 9.51

Goal We are asked to find the loads at the journal bearings at A, B, C, and D in terms of $\boldsymbol{F}_1$. In addition, we are asked to find a relationship between the input and output moments when the gear train is in equilibrium. The input is via Gear 1, and the output is via Gear 4. Therefore based on **Figure 9.51**, the requested ratio is r_1F_1/r_4F_4 It is common when dealing with gear trains to refer to this as the ratio of input to output *torque*, where torque is the particular component of moment that is aligned with the axes of the gear shafts.

Given We are given the geometry of the gear train.

Assume We assume that we can ignore the weight of various members that make up the gear train and that the journal bearings are frictionless (we will check this assumption in Example 9.12). Because there must be a boundary connection to prevent the shafts from sliding in the x direction, and we are not told whether any bearing acts as a thrust bearing (**Table 6.2**), we arbitrarily assume that bearings A and C act as thrust bearings (and will check this assumption below).

Draw First we draw free-body diagrams for Shaft 1 (**Figures 9.52**). Based on these free-body diagrams we write equilibrium equations that enable us to find a relationship between the input force and the force exerted by Gear 3 on Gear 2, as well as the loads acting at bearings at

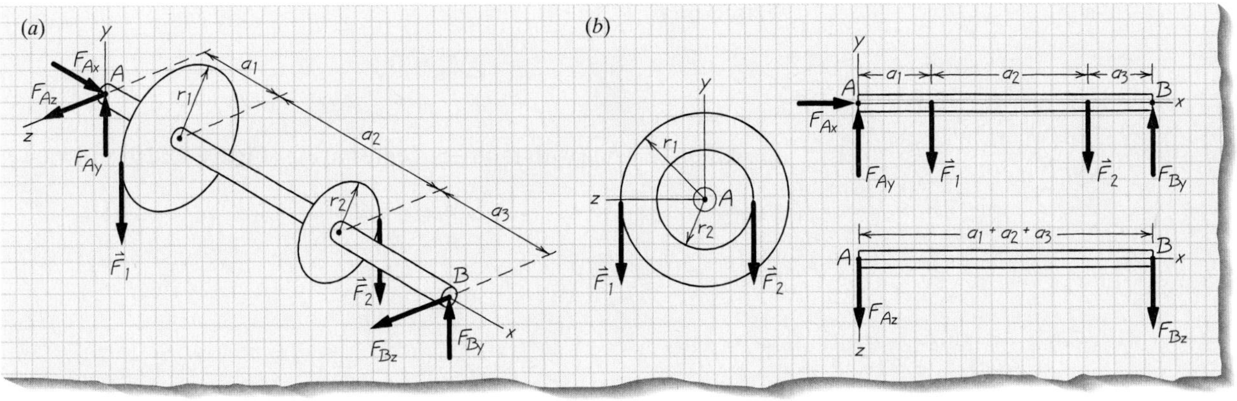

Figure 9.52 Free-body diagram of shaft AB

A and B. Next we draw free-body diagrams for Shaft 2 (**Figures 9.53**), allowing us to write associated equilibrium equations, and find the relationship between the output force and the force exerted on Gear 2 by 3, and bearing loads at C and D. Finally we use the expression for the various forces to find the input–output moment ratio in terms of gear radii.

Formulate Equations and Solve **Figure 9.52*a*** is a free-body diagram of Shaft 1 in isometric view, and **Figure 9.52*b*** shows the free-body diagram for this shaft based on end, side, and top views.

Based on (7.3A):

$$\sum F_x = 0(\rightarrow +)$$

$$F_{Ax} = 0 \tag{1}$$

(This says that for Shaft 1, no force acts in the x direction; therefore there is no thrust force acting on Shaft 1.)

Based on (7.3B):

$$\sum F_y = 0(\uparrow\ +)$$

$$-F_1 - F_2 + F_{Ay} + F_{By} = 0 \tag{2}$$

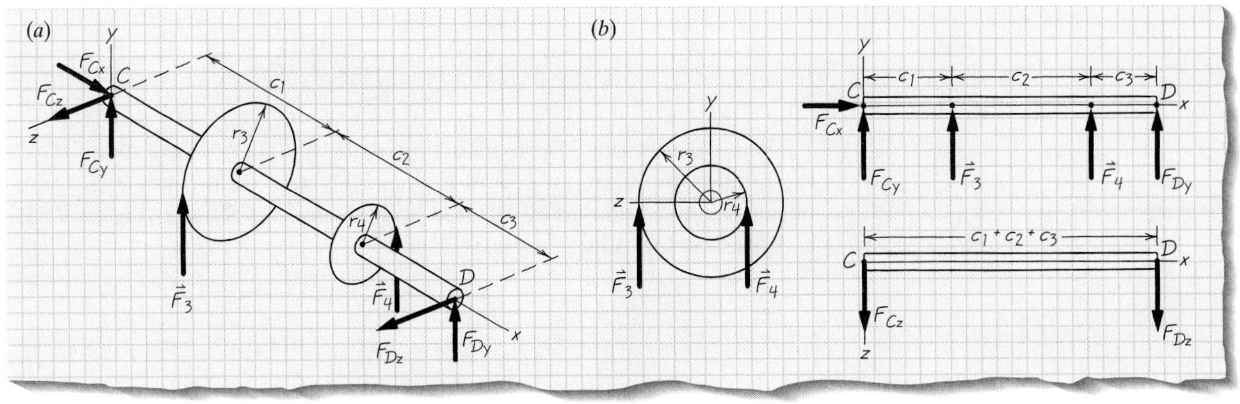

Figure 9.53 Free-body diagram of shaft CD

Based on (7.3C):

$$\sum F_z = 0$$

$$F_{Az} + F_{Bz} = 0 \tag{3}$$

Based on (7.4A) (with moment center at A) (**Figure 9.52*b***, end view):

$$\sum M_{x@A} = 0(\curvearrowleft +)$$

$$F_1 r_1 - F_2 r_2 = 0$$

$$\boxed{F_2 = \frac{r_1}{r_2} F_1} \tag{4}$$

Based on (7.4B) (with moment center at A) (**Figure 9.52*b***, zx view):

$$\sum M_{y@A} = 0(\curvearrowleft +)$$

$$F_{Bz}(a_1 + a_2 + a_3) = 0 \Rightarrow \boxed{F_{Bz} = 0}$$

Substituting this into (3), we find that $\boxed{F_{Az} = 0}$.

Based on (7.4C) (with moment center at A) (**Figure 9.52*b***, xy view):

$$\sum M_{z@A} = 0(\curvearrowleft +)$$

$$-F_1 a_1 - F_2(a_1 + a_2) + F_{By}(a_1 + a_2 + a_3) = 0$$

$$F_{By} = \frac{F_1 a_1 + F_2(a_1 + a_2)}{a_1 + a_2 + a_3}$$

Substituting from (4), we obtain:

$$F_{By} = \frac{F_1 a_1 + F_1 \frac{r_1}{r_2}(a_1 + a_2)}{a_1 + a_2 + a_3}$$

$$\boxed{F_{By} = \frac{r_2 a_1 + r_1 a_1 + r_1 a_2}{r_2(a_1 + a_2 + a_3)} F_1} \tag{5}$$

We substitute (5) and (4) into (2) to find F_{Ay}:

$$F_{Ay} = F_1 + \frac{r_1}{r_2} F_1 - \frac{r_2 a_1 + r_1 a_1 + r_1 a_2}{r_2(a_1 + a_2 + a_3)} F_1$$

$$\boxed{F_{Ay} = \frac{r_1 a_3 + r_2 a_2 + r_2 a_3}{r_2(a_1 + a_2 + a_3)} F_1}$$

We have now found the loads acting on the shaft at bearings A and B and the relationship between F_1 and F_2 (see (4)). Next we formulate the equilibrium equations based on the free-body diagram of Shaft 2 (**Figure 9.53**).

Based on (7.3A):

$$\sum F_x = 0(\rightarrow +)$$

$$\boxed{F_{Cx} = 0} \tag{6}$$

(This says that there is no thrust force acting on Shaft 2.)

Based on (7.3B):

$$\sum F_y = 0(\uparrow +)$$
$$F_3 + F_4 + F_{Cy} + F_{Dy} = 0 \tag{7}$$

Based on (7.3C):

$$\sum F_z = 0$$
$$F_{Cz} + F_{Dz} = 0 \tag{8}$$

Based on (7.4A) (with moment center at C) (**Figure 9.53*b***, end view):

$$\sum M_{x@C} = 0(\curvearrowleft +)$$
$$-F_3 r_3 + F_4 r_4 = 0$$
$$F_4 = \frac{r_3}{r_4} F_3 \tag{9}$$

Based on (7.4B) (with moment center at C) (**Figure 9.53*b***, *xz* view):

$$\sum M_{y@C} = 0(\curvearrowleft +)$$
$$-F_{Dz}(c_1 + c_2 + c_3) = 0 \Rightarrow \boxed{F_{Dz} = 0}$$

Substituting this into (8), we find that $\boxed{F_{Cz} = 0}$.

Based on (7.4C) (with moment center at C) (**Figure 9.53*b***, *xy* view):

$$\sum M_{z@C} = 0(\curvearrowleft +)$$
$$F_3 c_1 + F_4(c_1 + c_2) + F_{Dy}(c_1 + c_2 + c_3) = 0$$
$$F_{Dy} = \frac{-F_3 c_1 - F_4(c_1 + c_2)}{c_1 + c_2 + c_3}$$

Substituting from (9), we obtain:

$$F_{Dy} = \frac{-F_3 c_1 - F_3 \frac{r_3}{r_4}(c_1 + c_2)}{c_1 + c_2 + c_3}$$
$$F_{Dy} = \frac{-(r_4 c_1 + r_3 c_1 + r_3 c_2)}{r_4(c_1 + c_2 + c_3)} F_3 \tag{10}$$

According to Newton's third law,

$$|F_2| = |F_3| \tag{11}$$

(i.e., the force of Gear 3 pushing on Gear 2 is equal to and opposite the force of Gear 2 pushing on Gear 3).

Combining (4) and (11), we find:

$$F_3 = \frac{r_1}{r_2} F_1 \tag{12A}$$

Combining (9) and (11), we find:

$$F_4 = \frac{r_3}{r_4}\frac{r_1}{r_2} F_1 \tag{12B}$$

We substitute (12A) into (10) to get an expression for F_{Dy} in terms of F_1:

$$F_{Dy} = -\frac{r_1(r_4c_1 + r_3c_1 + r_3c_2)}{r_2r_4(c_1 + c_2 + c_3)} F_1 \tag{13}$$

Substituting (9) and (10) into (7) to find F_{Cy}:

$$F_{Cy} = -\frac{r_3c_3 + r_4c_2 + r_4c_3}{r_4(c_1 + c_2 + c_3)} F_3$$

Substituting from (12A) gives

$$F_{Cy} = -\frac{r_1(r_3c_3 + r_4c_2 + r_4c_3)}{r_2r_4(c_1 + c_2 + c_3)} F_1$$

Answer to (a)

Summary of Bearing Loads on Shaft 1:	Summary of Bearing Loads on Shaft 2:
$F_{Ax} = 0$	$F_{Cx} = 0$
$F_{Ay} = \frac{r_1a_3 + r_2a_2 + r_2a_3}{r_2(a_1 + a_2 + a_3)} F_1$	$F_{Cy} = \frac{r_1(r_3c_3 + r_4c_2 + r_4c_3)}{r_2r_4(c_1 + c_2 + c_3)} F_1$
$F_{Az} = 0$	$F_{Cz} = 0$
$F_{By} = \frac{r_2a_1 + r_1a_1 + r_1a_2}{r_2(a_1 + a_2 + a_3)} F_1$	$F_{Dy} = -\frac{r_1(r_4c_1 + r_3c_1 + r_3c_2)}{r_2r_4(c_1 + c_2 + c_3)} F_1$
$F_{Bz} = 0$	$F_{Dz} = 0$

By definition, the magnitude of the output moment is r_4F_4, and the magnitude of the input moment is r_1F_1. Based on (12B) we can write the input–output ratio as

Answer to (b)

$$\frac{\text{input moment}}{\text{output moment}} = \frac{r_1F_1}{r_4F_4} = \frac{r_1F_1}{r_4\frac{r_3}{r_4}\frac{r_1}{r_2}F_1} = \frac{r_2}{r_3}$$

This means that if the radius of Gear 2 is greater than that of Gear 3, the output moment is less than the input moment; therefore the gear train functions to reduce the moment. If the radius of Gear 2 is less than that of Gear 3, the output moment is greater than the input moment and the gear train functions to increase the moment. Notice that the radii of Gears 1 and 4 are not included in this ratio.

Question: How is the $\frac{\text{input moment}}{\text{output moment}}$ related to the mechanical advantage of the gear train?

EXAMPLE 9.11 ANALYSIS OF A BICYCLE POWER TRAIN

For the bicycle shown in Figure 9.54, derive an expression that relates the magnitude of $\boldsymbol{F}_{\text{foot}}$ to the magnitude of $\boldsymbol{F}_{\text{friction}}$, where $\boldsymbol{F}_{\text{foot}}$ is the force of a foot pressing down on the pedal and $\boldsymbol{F}_{\text{friction}}$ is the friction force pushing forward on the rear wheel. Your final solution should be expressed in terms of $N_{\text{chain ring}}$, $N_{\text{rear cog}}$, L_{crank}, and $R_{\text{rear wheel}}$ (where these variables are defined in **Figure 9.54**).

Bicycle specifications:

- L_{crank} (length of crank) = 17.8 cm
- $R_{\text{rear wheel}}$ (radius of rear wheel) = 34.3 cm
- $R_{\text{chain ring}}$ (radius of chain ring)
- $R_{\text{rear cog}}$ (radius of rear cog)
- $N_{\text{chain ring}}$ (number of teeth on chain ring) = 50 teeth
- $N_{\text{rear cog}}$ (number of teeth on rear cog) = 14 teeth
- C_D (coefficient of drag) = 0.9
- A_{frontal} (frontal area of cyclist + bicycle) = 0.50 m^2
- W_{total} (weight of cyclist + bicycle) = 860 N

Figure 9.54

Goal We are asked to find a relationship between the input force $\boldsymbol{F}_{\text{foot}}$ and the output force $\boldsymbol{F}_{\text{friction}}$. The "connection" between these two forces is the bicycle's power train.

Given We are given the geometry of the bicycle.

Assume We assume that we can ignore the weights of various members that make up the bicycle, that the bearings at the bottom bracket and rear sprocket are frictionless, and that the chain is well lubricated. We can ignore the effect of rolling resistance (we will check this assumption in Example 9.13). We also assume that $\boldsymbol{F}_{\text{foot}}$ is perpendicular to the crank arm, $\boldsymbol{F}_{\text{foot}}$ and $\boldsymbol{F}_{\text{friction}}$ lie in the same plane, and this plane is along the centerline of the bicycle. Therefore we will assume that we can treat the situation as planar. Furthermore, we assume the bicycle is moving at constant velocity, the chain ring is rotating at constant rotational velocity ω_1, and the rear wheel is rotating at constant rotational velocity ω_2. Each is rotating about an axis whose orientation is fixed. Therefore we can assume that the equilibrium conditions hold.

Draw We draw separate free-body diagrams of the chain ring (Figure 9.55), a portion of the chain (Figure 9.56), and the rear wheel assembly (Figure 9.57). Applying the conditions of equilibrium to each of these free-body diagrams enables us to relate the foot force to the friction force.

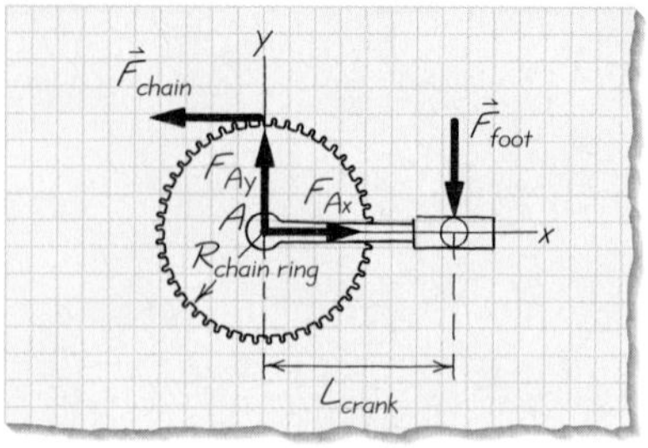

Figure 9.55 Free-body diagram of chain ring

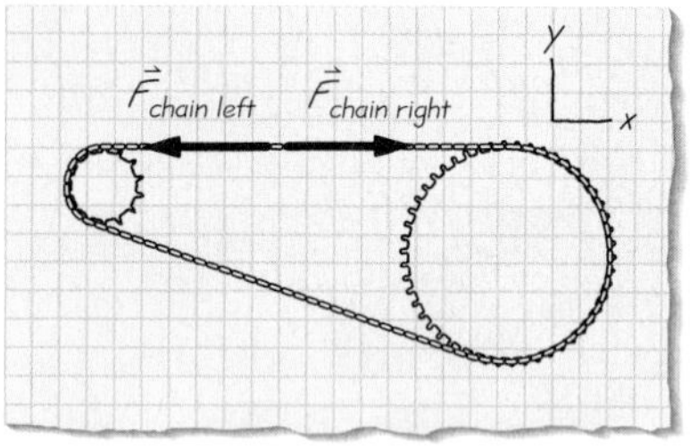

Figure 9.56 Free-body diagram of chain

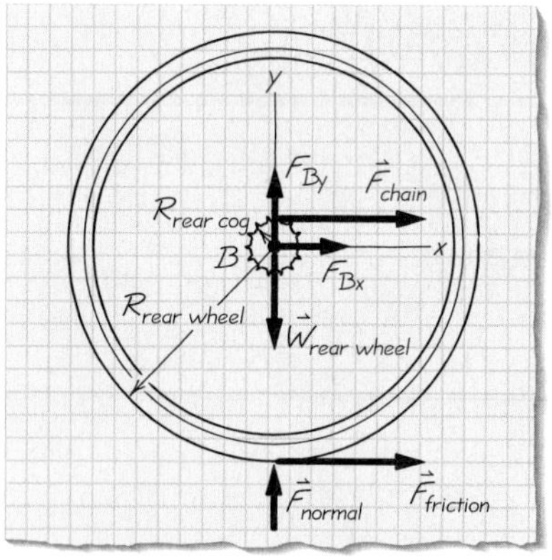

Figure 9.57 Free-body diagram of rear-wheel assembly

Formulate Equations and Solve

Step 1: We formulate the equilibrium equations based on the free-body diagram of the chain ring (**Figure 9.55**) and (7.5C) (with moment center at A):

$$\sum M_{z@A} = 0(\curvearrowleft +)$$
$$F_{\text{chain}} R_{\text{chain ring}} - F_{\text{foot}} L_{\text{crank}} = 0$$
$$F_{\text{chain}} = \frac{L_{\text{crank}}}{R_{\text{chain ring}}} F_{\text{foot}} \tag{1}$$

Because we are not interested in the bearing load at A, we don't need to write the two force equilibrium equations.

Step 2: We formulate the equilibrium equations based on the free-body diagram of a portion of the chain (**Figure 9.56**) and (7.5A):

$$\sum F_x = 0(\rightarrow +)$$
$$F_{\text{chain right}} - F_{\text{chain left}} = 0$$
$$F_{\text{chain right}} = F_{\text{chain left}}$$

We could have concluded this result by inspecting **Figure 9.56** because the chain acts like a cable in tension.

Step 3: We now formulate the equilibrium equations based on the free-body diagram of the rear wheel assembly (**Figure 9.57**) and (7.5C) (with moment center at B):

$$\sum M_{z@B} = 0(\curvearrowleft +)$$
$$-F_{\text{chain}} R_{\text{rear cog}} + F_{\text{friction}} R_{\text{rear wheel}} = 0$$
$$F_{\text{friction}} = \frac{R_{\text{rear cog}}}{R_{\text{rear wheel}}} F_{\text{chain}}$$

We substitute (1) into this expression to get

$$F_{\text{friction}} = \left(\frac{R_{\text{rear cog}}}{R_{\text{rear wheel}}}\right)\left(\frac{L_{\text{crank}}}{R_{\text{chain ring}}}\right) F_{\text{foot}} \tag{2}$$

This expression for F_{friction} as a function of F_{foot} can be expressed in terms of the number of teeth of the chain ring ($N_{\text{chain ring}}$) and number of teeth of the rear cog ($N_{\text{rear cog}}$), if we recognize that for both ring and cog, the number of teeth is proportional to the circumference. Therefore we can write

$$\frac{R_{\text{rear cog}}(\cancel{2\pi})}{R_{\text{rear wheel}}(\cancel{2\pi})} = \frac{N_{\text{rear cog}}}{N_{\text{chain ring}}}$$

Now we substitute (3) into (2) to obtain:

Answer $$F_{\text{friction}} = \left(\frac{N_{\text{rear cog}}}{N_{\text{chain ring}}}\right)\left(\frac{L_{\text{crank}}}{R_{\text{rear wheel}}}\right)F_{\text{foot}}$$

Notice that the factor $(\frac{N_{\text{rear cog}}}{N_{\text{chain ring}}})(\frac{L_{\text{crank}}}{R_{\text{rear wheel}}})$ is a product of the geometry of the bicycle (crank length, number of teeth of chain wheel, radius of rear wheel, and number of teeth of rear cog). For the bicycle described in **Figure 9.54**, this factor has a value of

$$\left[\left(\frac{14 \text{ teeth}}{50 \text{ teeth}}\right)\left(\frac{17.8 \text{ cm}}{34.3 \text{ cm}}\right)\right] = 0.145$$

This means that the magnitude of the frictional force pushing the bicycle forward is only 14.5% of the magnitude of the foot force pushing down on the pedal.

When we model a system as an idealized machine, we assume that all associated bearings are frictionless. In addition, any wheels in the machine are perfectly rigid and therefore cause no rolling resistance. In real machines, bearing friction and rolling resistance do exist, and their presence may sometimes be significant. We now consider how they are included in the analysis.

When machines contain shafts that are held by dry or only partially lubricated journal bearings, the analysis may need to account for the presence of **journal bearing friction**. To see how to account for bearing friction, consider the journal bearing shown in Figure 9.58 with bearing load L. The force $\boldsymbol{L}$ is the radial force to be supported by the bearing. If the bearing is well lubricated, there is negligible friction and the bearing support force $\boldsymbol{F}$ is aligned with $\boldsymbol{L}$ (Figure 9.59*a*). In contrast, if the bearing is dry or only partially lubricated, friction between the shaft (which is rotating clockwise) and the bearing sleeve result in the shaft "riding up" the inner surface of sleeve just to the point of slipping (**Figures 9.59*b*** and ***c***). At the contact point A we consider the components of $\boldsymbol{F}$ in the normal and tangential directions, with F_{tangent} providing the friction (**Figure 9.59*c***). For a shaft of radius r, the friction force F_{tangent} creates a moment about a moment center at O of

$$M_O = (r \cdot F_{\text{tangent}}) \tag{9.1}$$

in the direction *opposite* the clockwise rotation of the shaft. According to our discussion of friction in Chapter 7 and in Appendix B, we can write the friction force when there is impending slip in terms of the normal force as $F_{\text{friction}} = F_{\text{tangent}} = \mu F_{\text{normal}}$, where μ is the coefficient of friction. If the angle θ in **Figure 9.59** is small, then F_{normal} is approximately equal to the magnitude of the bearing support force $\boldsymbol{F}$. Therefore, we write (9.1) as

$$M_O = r \cdot \mu F \tag{9.2}$$

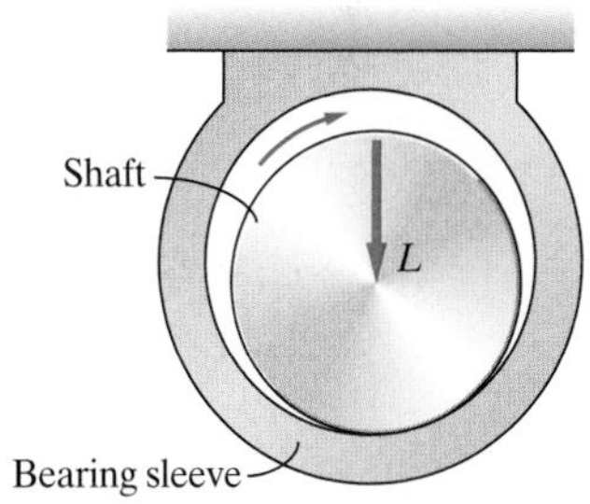

Figure 9.58 A journal bearing consists of a shaft within a bearing sleeve. The sleeve is fixed and the shaft rotates within the sleeve.

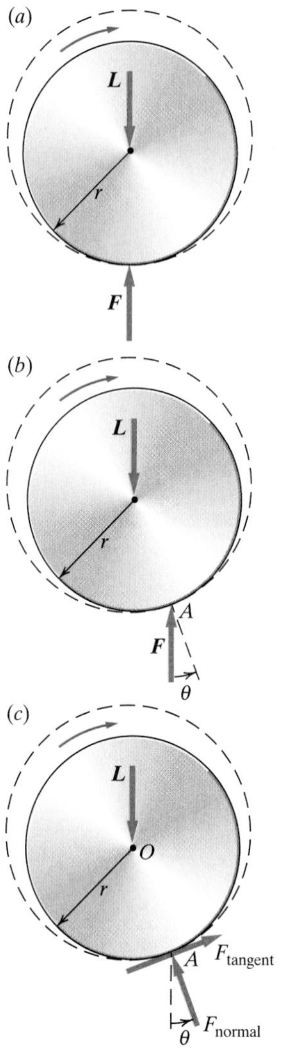

Figure 9.59 (*a*) A stationary shaft; (*b*) shaft rotates in clockwise direction (force ***F*** is sleeve acting on shaft); (*c*) force ***F*** shown in terms of its components

In most engineering design situations, one would work to minimize M_O by making μ as small as is practical. For a shaft that is stationary relative to a journal bearing, the coefficient of static friction (μ_s) should be used in (9.2). If the shaft is rotating relative to the journal bearing, the coefficient of kinetic friction (μ_k) should be used in (9.2). Sample values of μ_s and μ_k are presented in Appendix B. Generally, $\mu_s > \mu_k$. Additional information on incorporating the presence of friction in static analysis is presented in Appendix B.

The analysis of machines that contain wheels may need to account for the presence of **rolling resistance**. You have experienced firsthand the presence of rolling resistance if you have ever attempted to bicycle with underinflated tires. To see how we account for the presence of rolling resistance in static analysis, first consider a perfectly rigid wheel and the surface on which it is rolling (**Figure 9.60**). Acting on the wheel is ***L***, the weight of the wheel plus any vertical load exerted by whatever the wheel is attached to. The contact point between wheel and surface is directly below the center of the wheel, and equilibrium of moments shows us that ***P***, the force required to keep the wheel rolling, is zero when there is no rolling resistance. In contrast, when a real wheel rolls on a surface, the wheel and surface both deform slightly, thereby creating resistance to rolling. For example, a rubber wheel rolling at constant speed on a paved road is shown in exaggerated form in **Figure 9.61*a*** and **9.61*b***. If the wheel speed is constant, equilibrium of moments about the wheel center requires that the normal contact force $\boldsymbol{F}_{normal}$, shown in **Figures 9.61*b*** and ***c*** acting at point A, must act through the wheel center as shown. By summing moments about A, we find

$$M_A = La - Pb = 0 \tag{9.3}$$

Generally the wheel deformation involved is so small that b can be replaced by r (the radius of the wheel), and (9.3) becomes

$$P = (a/r)L \tag{9.4}$$

Remember that P is the magnitude of the force required to keep the wheel rolling because of the presence of rolling resistance.

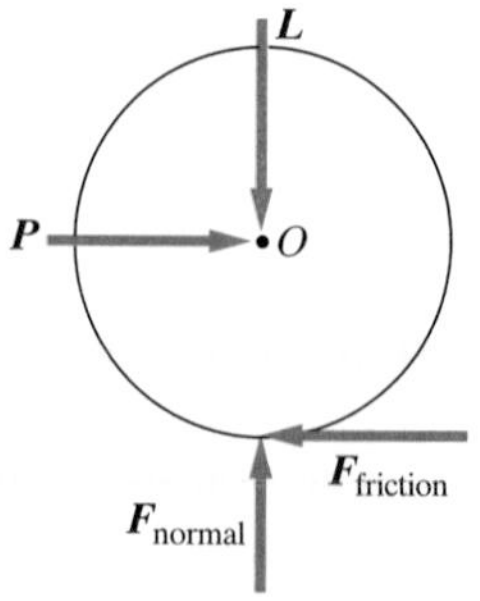

Figure 9.60 Perfectly rigid wheel on perfectly rigid surface, rolling to the right. According to the equilibrium conditions, if the wheel is moving to the right at a constant speed then $P = F_{friction} = 0$.

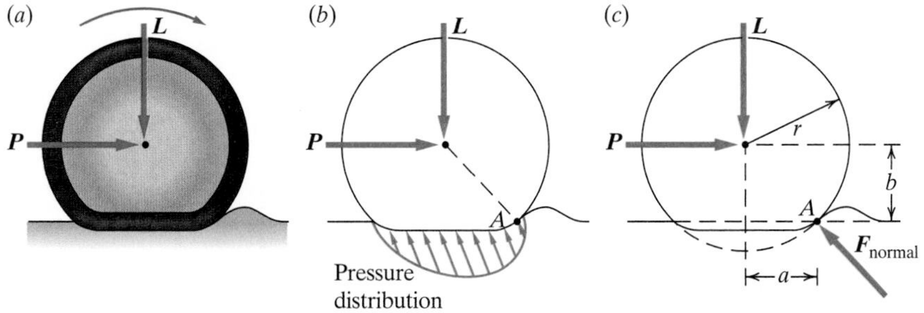

Figure 9.61 (*a*) Real wheel on real surface, rolling to the right; (*b*) depiction of pressure distribution between surface and wheel; (*c*) if the wheel is moving to the right at constant speed, $P = (a/r)L$

The ratio of (a/r) is referred to as the **coefficient of rolling friction** (f_r).[4] Values of f_r reported in the literature for loaded pneumatic rubber tires that are properly inflated range from 0.01 (hard road) to 0.05 (well-packed gravel) to 0.40 (loose sand). The coefficient of rolling friction is dependent on the size of the wheel radius; for two identically constructed wheels, the one with a larger radius would have a smaller coefficient of rolling friction.

[4]Some texts refer to the distance *a* as the coefficient of rolling resistance. We go with the definition presented in Marks' *Standard Handbook for Mechanical Engineers*, 8th Edition.

EXAMPLE 9.12 ANALYSIS OF A PULLEY SYSTEM WITH BEARING FRICTION

The pulley in **Figure 9.62** consists of a pulley that fits loosely over a shaft. The shaft is rigidly supported at its ends and does not turn. The coefficient of static friction is 0.40, and the coefficient of kinetic friction is 0.35. The diameter of the wheel is 150 mm, and that of the shaft is 20 mm.

(a) Determine the minimum tension in the rope to hold the 300-N block in a stationary position.
(b) Determine the minimum tension in the rope to just start raising the block. Also express as a percentage of the tension found in (a).
(c) Determine the tension in the rope to raise the block at a constant rate. Also express as a percentage of the tension found in (a).
(d) Determine the tension in the rope to lower the block at a constant rate. Also express as a percentage of the tension found in (a).

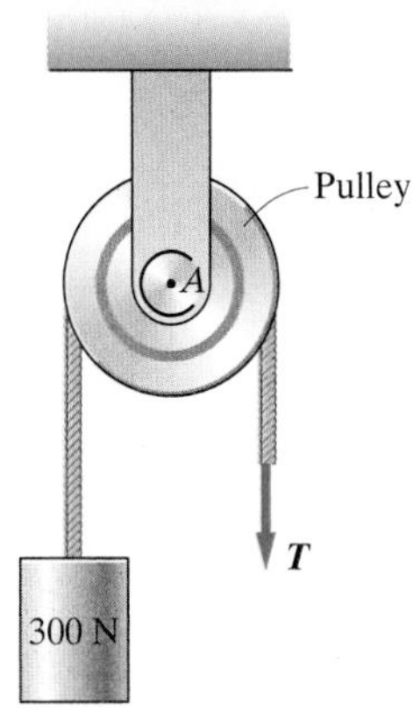

Figure 9.62

Goal We are asked to find the tension T in the rope to keep the weight stationary (a), to just begin raising it (b), to raise it at a constant rate (c), and to lower it at a constant rate (d).

Given We are given the diameters of the pulley (150 mm) and the shaft (20 mm) on which it is loosely fit. In addition, we are given the coefficients of static ($\mu_s = 0.40$) and kinetic ($\mu_k = 0.35$) friction and the weight of block (300 N) being held by the pulley system.

Assume We assume that we can ignore the weight of the pulley and that the rope does not slide in the pulley groove.

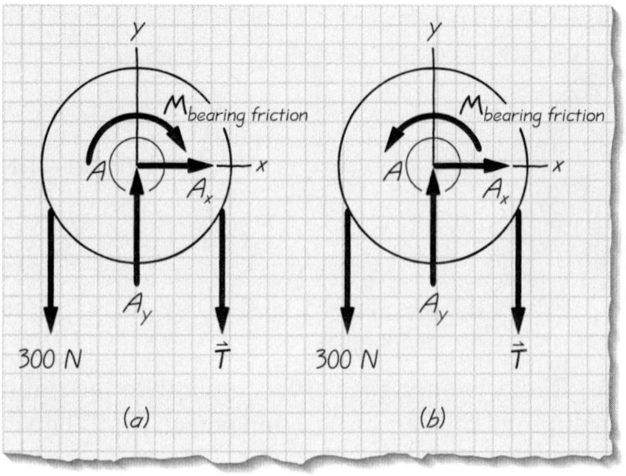

Figure 9.63 Free-body diagrams: (*a*) stationary pulley; (*b*) just-about-to-move-upward pulley

Draw For **(a)** First we draw the free-body diagram for the pulley in a stationary position (**Figure 9.63*a***). The clockwise moments created about a moment center at A by the rope tension and the bearing friction are balanced by the counterclockwise moment created by the block's weight about A. Bearing friction in the clockwise direction means that it is helping to keep the block stationary (and therefore, the required tension $\boldsymbol{T}$ is at its minimum).

Formulate Equations and Solve For **(a)** From (9.2), the moment created by the bearing friction is $M = r \cdot \mu F$, where F is the magnitude of the force supported by the bearing and r_{shaft} is the radius of the shaft. For the situation in **Figure 9.63*a***, the force supported by the bearing is A_y, and it is appropriate for us to use the static coefficient of friction μ_s (because the pulley is not turning relative to the shaft for part (a)). Therefore,

$$M_{\text{bearing friction}} = r_{\text{shaft}} \cdot \mu_s A_y \tag{1}$$

Now we find the value of A_y by using equilibrium equation (7.5B):

$$\begin{aligned} &\sum F_y = 0(\uparrow +) \\ &A_y - 300\text{ N} - T = 0 \\ &A_y = 300\text{ N} + T \end{aligned} \tag{2}$$

Substituting (2) into (1), we find that the magnitude of the moment created by the bearing friction is

$$M_{\text{bearing friction}} = r_{\text{shaft}} \cdot \mu_s(300\text{ N} + T) \tag{3}$$

Based on equilibrium equation (7.4A) and **Figure 9.63*a*** (with moment center at A) we write:

$$\sum M_{z@A} = 0(\curvearrowleft +)$$

$$300\text{ N}(150/2\text{ mm}) - T(150/2\text{ mm}) - \underbrace{M_{\text{bearing friction}}}_{\text{from (3)}} = 0 \tag{4}$$

$$300\text{ N}(150/2\text{ mm}) - T(150/2\text{ mm}) - r_{\text{shaft}} \cdot \mu_s(300\text{ N} + T) = 0$$

Rearranging and substituting in for $r = 10$ mm and $\mu_s = 0.40$, we solve for the minimum tension T in the rope to hold the 300-N block in a stationary position:

Answer to (a) $T = 270$ N

Question: If the tension in the rope is less than 270 N, what will happen?

Draw For **(b)** Now we draw the free-body diagram for the block just as it is about to start moving upward (**Figure 9.63*b***). The clockwise moment created about a moment center at A by the rope tension is balanced by counterclockwise moments created by the block's weight and the bearing friction. The moment created by the bearing friction being in the counterclockwise direction means that it is working to prevent the impending upward movement of the block.

Formulate Equations and Solve For **(b)** As in the calculations for (a), the magnitude of the moment created by the bearing friction is

$$M_{\text{bearing friction}} = r_{\text{shaft}} \cdot \mu_s(300 \text{ N} + T)$$

but its direction is in the counterclockwise direction. It is appropriate to use the static coefficient of friction because the block is just about to start moving upward.

Based on equilibrium equation (7.4A) and **Figure 9.63*b*** (with moment center at A) we write:

$$\sum M_{z@A} = 0(\curvearrowleft\!+)$$

$$300 \text{ N}(150/2 \text{ mm}) - T(150/2 \text{ mm}) + M_{\text{bearing friction}} = 0 \quad (5)$$

$$300 \text{ N}(150/2 \text{ mm}) - T(150/2 \text{ mm}) + r_{\text{shaft}} \cdot \mu_s(300 \text{ N} + T) = 0 \quad (6)$$

Rearranging and substituting in for r_{shaft} and μ_s, we solve for the minimum tension T in the rope when the block is just about to start moving upward.

Answer for (b) $T = 334$ N. This is 23.8% greater than the tension required to hold the block stationary.

Draw, Formulate, and Solve For **(c)** Now we consider the block when it is moving upward at a constant rate. The free-body diagram that describes this situation is identical to the one in **Figure 9.63*b***. What is different is that now the moment created by the bearing friction uses the coefficient of kinetic friction (because the pulley and shaft are moving relative to each another). Therefore the magnitude is

$$M_{\text{bearing friction}} = r_{\text{shaft}} \cdot \mu_k(300 \text{ N} + T) \quad (7)$$

Substituting (7) into (5) (the moment equilibrium equation) and solving for T, we find that $T = 330$ N.

Answer to (c) $T = 330$ N. This is 22.2% greater than the tension required to hold the block stationary. Notice that it is slightly less than the tension before the upward movement started; this reflects the fact that the kinetic coefficient of friction is less than the static coefficient of friction. So after getting the block moving upward, there is a slight reduction in the tension needed to keep it moving upward at a constant speed.

Draw, Formulate, and Solve For **(d)** Now we consider the block as it is moving downward at a constant rate. The free-body diagram that describes this situation is identical to the one in **Figure 9.63*a***. What is different is that now the moment created by the bearing friction uses the coefficient of kinetic friction (because the pulley and shaft are moving relative to one another), as described by (7).

Substituting (7) into (4) (the moment equilibrium equation) and solving for T, we find that $T = 273$ N.

Answer to (d) $T = 273$ N. This is 1.1% greater than the tension required to hold the block stationary.

Check The results are summarized in Table 9.3.

Table 9.3

Condition	Tension
Block stationary	270 N
Block just starting to move upward	334 N
Block moving upward at a constant rate	330 N
Block moving downward at a constant rate	273 N

These results say that

- the most tension is required to just start raising the block (which makes sense, since the moment created by the tension must balance the moments created by the block's weight and the bearing friction);
- the least tension is required to hold the block stationary (which makes sense, since the moment created by tension and bearing friction balance the moment created by the block's weight);
- to hold the block stationary requires a tension of 270 N, whereas to move the block downward at a constant rate requires tension of 273 N. Does this make physical sense?

EXAMPLE 9.13 ANALYSIS OF A GEAR TRAIN WITH FRICTION

Consider the single-stage gear train in Figure 9.64, which is in static equilibrium.

(a) Determine the ratio of input-to-output moment in terms of the radii of various gears, and include the effect of friction in the journal bearings. The coefficient of kinetic friction is μ_k.

(b) Compare the result with that found in Example 9.10, where the bearings were assumed to be frictionless.

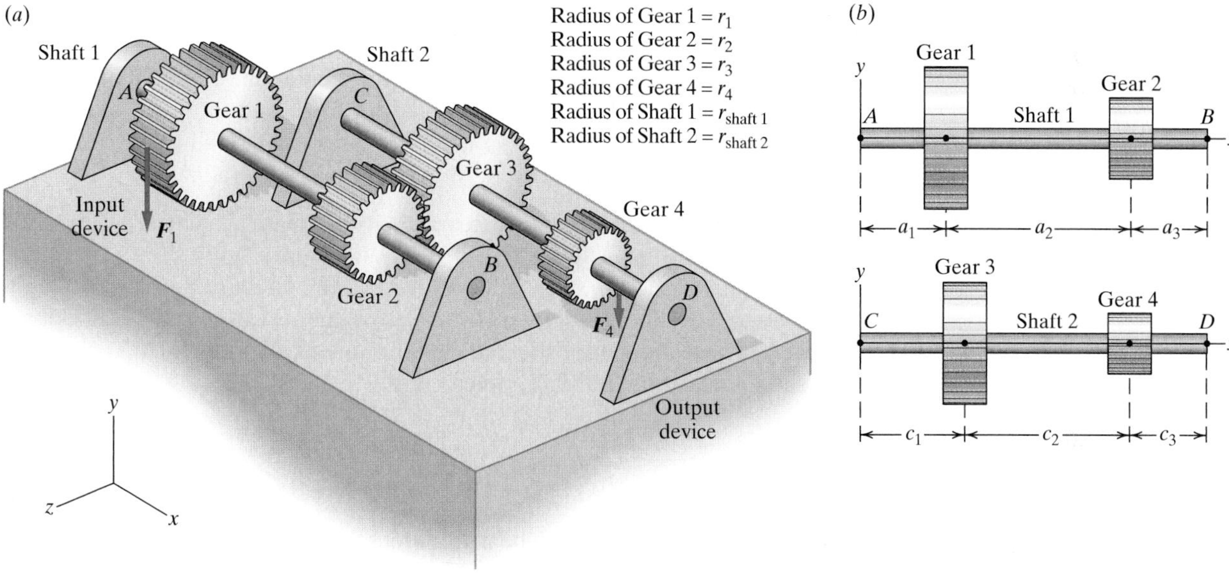

Figure 9.64

Goal We are asked to find a relationship between the input moment and output moment for this gear train in equilibrium (turning at a constant speed). The input is via Gear 1, and the output is via Gear 4. Therefore, based on Figure 9.65, the requested ratio is r_1F_1/r_4F_4. We are also asked to compare this result with the friction free analysis in Example 9.10.

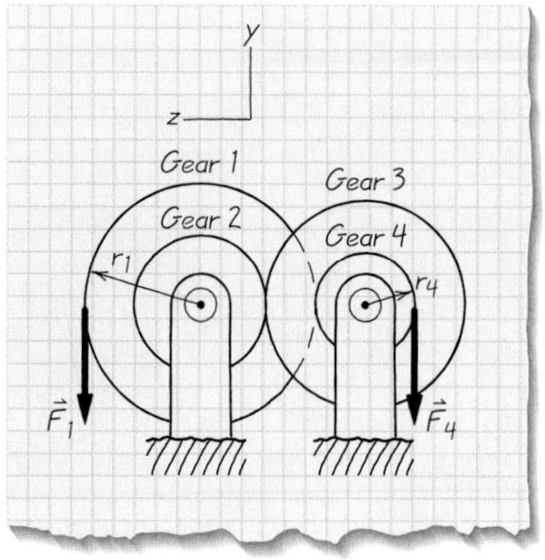

Figure 9.65

Given We are given the geometry of the gear train and the coefficient of kinetic friction.

Assume We assume that we can ignore the weight of various members that make up the gear train. Also, because we are not told that any bearing acts as a thrust bearing, we arbitrarily assume that bearings A and C act as thrust bearings.

Draw First we draw free-body diagrams for Shaft 1 (Figures 9.66). Based on these free-body diagrams we write equilibrium equations that enable us to find a relationship between the input force and the force exerted by Gear 2 on Gear 3. Next we draw free-body diagrams for Shaft 2

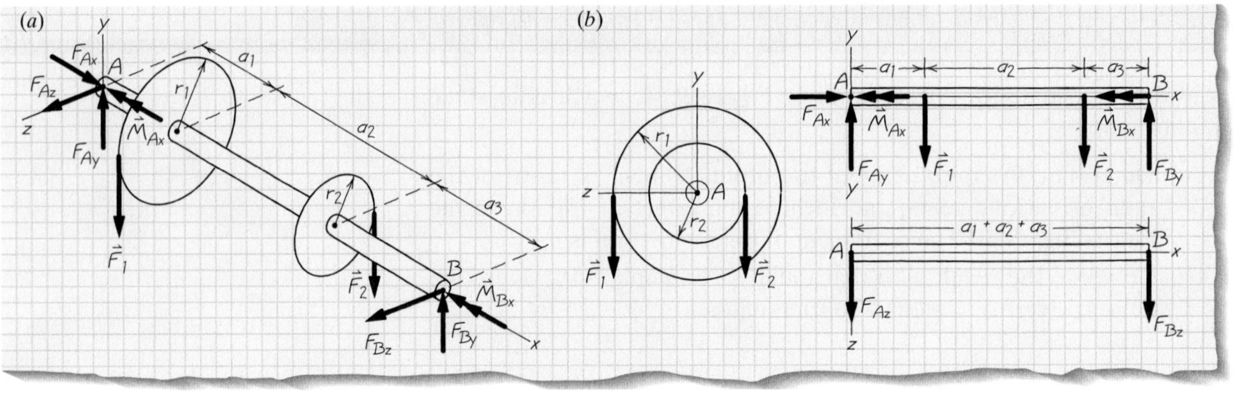

Figure 9.66 Free-body diagram of shaft AB

(**Figures 9.67**), allowing us to write associated equilibrium equations, and find the relationship between the output force and the force exerted on Gear 2 by Gear 3. Finally we use the expression for the various forces to find the input–output moment ratio in terms of gear radii.

Formulate Equations and Solve **Figure 9.66*a*** is a free-body diagram of Shaft 1 in isometric view, and **Figure 9.66*b*** shows free-body diagrams based on three views. Note that the effect of bearing friction is included in the free-body diagrams as moments (M_{Ax}, and M_{Bx}) that act in the direction opposite the input moment created by $\boldsymbol{F}_1$ about the shaft axis. Based on (9.2), we can write

$$\boldsymbol{M}_{Ax} = -F_{A\,\text{friction}} r_{\text{shaft }1}\boldsymbol{i} = -\mu_k F_A r_{\text{shaft }1}\boldsymbol{i}$$
$$\boldsymbol{M}_{Bx} = -F_{B\,\text{friction}} r_{\text{shaft }1}\boldsymbol{i} = -\mu_k F_B r_{\text{shaft }1}\boldsymbol{i}$$

where F_A and F_B are the bearing forces at A and B.

Now we formulate the equilibrium equations based on the free-body diagram of Shaft 1 (**Figure 9.66**). Based on (7.4A) (with moment center at A):

$$\sum M_{x@A} = 0(\circlearrowleft +)$$
$$F_1 r_1 - F_2 r_2 - \mu_k F_A r_{\text{shaft }1} - \mu_k F_B r_{\text{shaft }1} = 0 \quad (1)$$

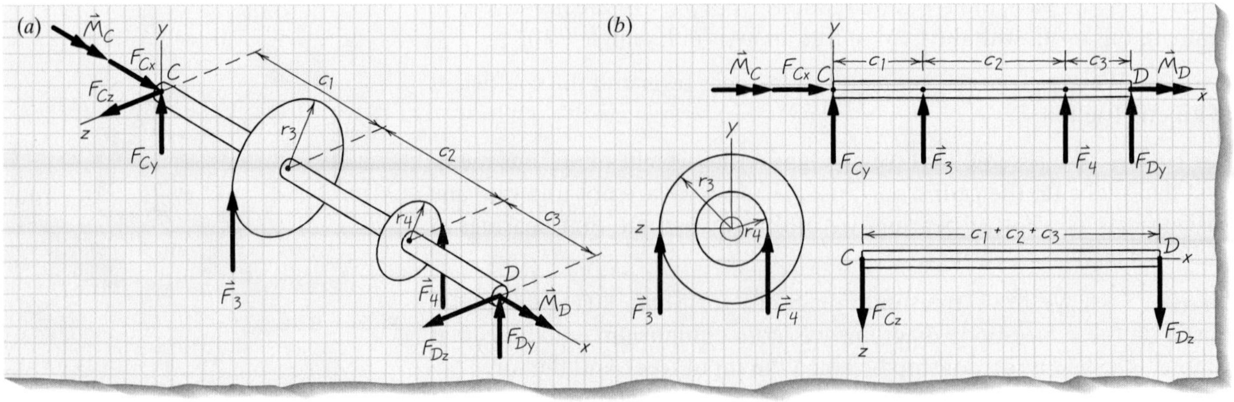

Figure 9.67 Free-body diagram of shaft CD

To find the bearing forces F_A and F_B, either we can write the other five equilibrium equations for Shaft 1 or we can recognize that we can use the values for the y, and z components of F_A and F_B found in Example 9.10, because the bearing loads are not affected by the presence of friction. Thus, after remembering that

$$F_A = \sqrt{F_{Ay}^2 + F_{Az}^2};\ F_B = \sqrt{F_{By}^2 + F_{Bz}^2},$$

and substituting in values for F_{Ay}, F_{Az}, F_{By} and F_{Bz} from Example 9.10, we can rewrite (1) as

$$F_2 = \frac{r_1}{r_2}F_1 - \mu r_{\text{shaft 1}}\frac{(r_1a_1 + r_1a_2 + r_1a_3 + r_2a_1 + r_2a_2 + r_2a_3)}{r_2^2(a_1 + a_2 + a_3)}F_1$$

which, with some simplification, becomes

$$F_2 = \left(\frac{r_1}{r_2} - \mu r_{\text{shaft 1}}\frac{(r_1 + r_2)}{r_2^2}\right)F_1 \tag{2}$$

Next we formulate the equilibrium equations based on the free-body diagram of Shaft 2 (**Figure 9.67**). Note that the effect of bearing friction is included in the free-body diagram as moments (M_{Cx}, and M_{Dx}) that act in the direction opposite the input moment created by $\boldsymbol{F}_3$ about the shaft axis. Based on (9.2), we can write

$$\boldsymbol{M}_{Cx} = F_{C\,\text{friction}}r_{\text{shaft 2}}\boldsymbol{i} = \mu_k F_C r_{\text{shaft 2}}\boldsymbol{i}$$
$$\boldsymbol{M}_{Dx} = F_{D\,\text{friction}}r_{\text{shaft 2}}\boldsymbol{i} = \mu_k F_D r_{\text{shaft 2}}\boldsymbol{i}$$

Based on (7.4A) and the free-body diagram in **Figure 9.67** (with moment center at C):

$$\sum M_{x@C} = 0(\curvearrowleft)$$
$$-F_3r_3 + F_4r_4 + \mu_k F_C r_{\text{shaft 2}} + \mu_k F_D r_{\text{shaft 2}} = 0 \tag{3}$$

As with Shaft 1, we can use our y and z bearing force components from Example 9.10 to find F_C and F_D. Using these components and substituting into (3), we find

$$F_4 = \frac{r_3}{r_4}F_3 - \mu_k r_{\text{shaft 2}}\frac{r_1(r_3c_1 + r_3c_2 + r_3c_3 + r_4c_1 + r_4c_2 + r_4c_3)}{r_2r_4^2(c_1 + c_2 + c_3)}F_1$$

$$F_4 = \frac{r_3}{r_4}F_3 - \mu_k r_{\text{shaft 2}}\frac{r_1(r_3 + r_4)}{r_2r_4^2}F_3 \tag{4}$$

Because $F_2 = F_3$ (Newton's third law), we combine (2) and (4) (ignoring terms involving μ_k^2, since they will be very small) to find

$$F_4 = \frac{F_1}{r_2r_4}\left[r_1r_3 - \mu_k\left(r_{\text{shaft 1}}\frac{r_3(r_1 + r_2)}{r_2} + r_{\text{shaft 2}}\frac{r_1^2(r_3 + r_4)}{r_2r_4}\right)\right] \tag{5}$$

By definition, the magnitude of the output moment is r_4F_4 and the magnitude of the input moment is r_1F_1. Based on (5), we write the input–output ratio as

Answer to (a)

$$\frac{\text{input torque}}{\text{output torque}} = \frac{r_1 F_1}{r_4 F_4} = \frac{r_2}{r_3 - \mu_k\left(r_{\text{shaft 1}}\dfrac{r_3(r_1 + r_2)}{r_1 r_2} + r_{\text{shaft 2}}\dfrac{r_1(r_3 + r_4)}{r_2 r_4}\right)} \tag{6}$$

This means that if the radius of Gear 2 is greater than that of Gear 3, the output moment is less than the input moment; therefore the gear train functions to reduce the moment. If, on the other hand, the radius of Gear 2 is less than that of Gear 3, the output moment is greater than the input moment and the gear train functions to increase the moment.

The effect of the bearing friction is to reduce the output moment.

Answer to (b) Comparison with the results from Example 9.10: Equation (6) is identical to the result in Example 9.10 when $\mu = 0$. This would be the case when you have very well-lubricated journal bearings or are using some form of ball or roller bearing. For nonzero values of friction, (6) illustrates that the presence of friction results in less output moment.

EXAMPLE 9.14 ROLLING RESISTANCE

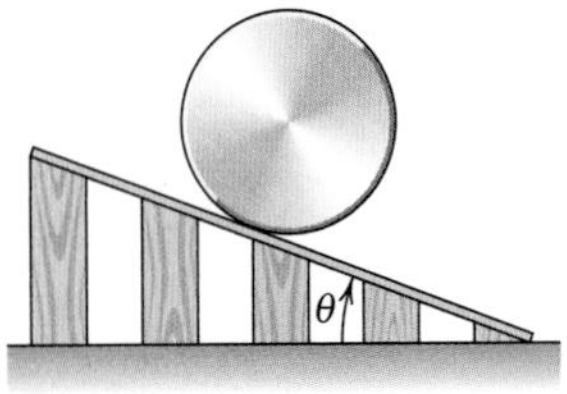

Figure 9.68

The 100-kg steel wheel in **Figure 9.68** has a radius of 50 mm and rests on a ramp made of wood. At what angle θ will the wheel begin to roll down the ramp with constant velocity if the coefficient of rolling resistance is 0.02?

Goal We are asked to find the ramp angle θ at which the steel wheel will begin to roll down the wooden ramp.

Given We are given the mass and radius of the steel wheel (100 kg and 50 mm, respectively). We are also told that the coefficient of rolling resistance between the steel and the wood is 0.02.

Assume We assume that the wheel will roll and not slide down the ramp.

Draw We draw the free-body diagram of the wheel (**Figure 9.69*a***). **Figure 9.69*b*** shows an alternate free-body diagram, where the weight W of the wheel is drawn in terms of its components ($W \sin\theta$ and $W \cos\theta$).

Formulate Equations and Solve Now we find the value of R (the force of the ramp pushing on the wheel) by using equilibrium equation (7.5B):

$$\sum F_y = 0(\uparrow +)$$
$$-W + R = 0$$
$$R = W \tag{1}$$

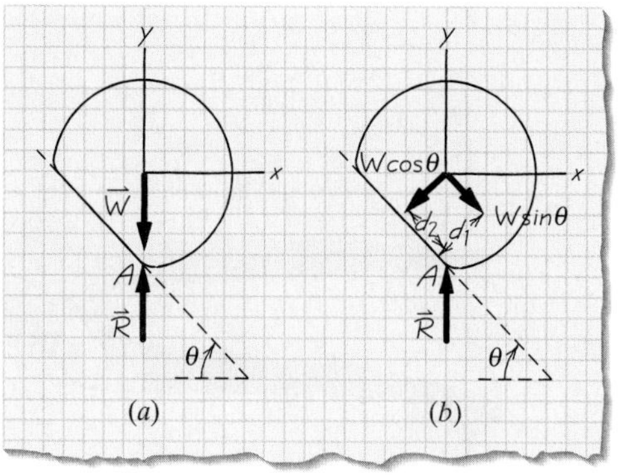

Figure 9.69 Free-body diagrams of wheel with wheel deformation highly exaggerated

Based on **Figure 9.69*b*** and equilibrium equation (7.5C) (with moment center at A):

$$\sum M_{z@A} = 0(\curvearrowleft +)$$

$$-W \sin \theta(d_1) + W \cos \theta(d_2) = 0, \qquad (2)$$

where d_1 and d_2 are shown in **Figure 9.69*b***. For small wheel deformations $d_1 \approx r$ and $d_2 = a$ (as shown in **Figure 9.61c**). Therefore (2) can be rewritten as

$$-W \sin \theta\ (r) + W \cos \theta\ (a) = 0 \qquad (3)$$

Rearrange (3), to

$$\frac{\sin \theta}{\cos \theta} = \tan \theta = \frac{a}{r} = f_r \qquad (4)$$

By definition, the coefficient of rolling resistance is $f_r = a/r$. With $f_r = 0.02$ we solve (4) for θ to find $\theta = 1.14°$

Answer Therefore $\theta = 1.14°$.

At $\theta = 1.14°$, the steel wheel will roll down the ramp at a constant speed. (*Note:* At angles greater than 1.14°, the wheel will accelerate down the ramp. Also notice that the answer is independent of the weight of the wheel).

Check If the wheel–ramp interface had larger rolling resistance, the value of a would increase. Substituting a larger value of a into (2) would result in a larger angle θ. This makes intuitive sense—the larger the rolling resistance, the greater the force required to roll the wheel. In this case this force is the component of the wheel's weight parallel to the ramp, and this component gets bigger with ramp angle.

EXERCISES 9.2

9.2.1. A pair of pliers grips the bolt as shown in E9.2.1. For the input force ***P***, determine the force exerted on the bolt.

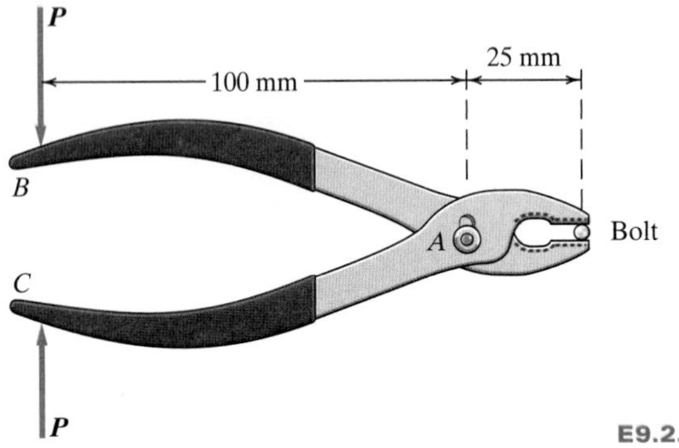

E9.2.1

9.2.2. A pair of pliers grips the bolt as shown in **E9.2.1**. Determine the mechanical advantage of this machine.

9.2.3. A pair of pliers grips the bolt as shown in **E9.2.1**. Determine the shear force acting on the pin at *A*.

9.2.4. A pair of pliers grips the bolt as shown in E9.2.4. For the input force ***P***, determine the force exerted on the bolt.

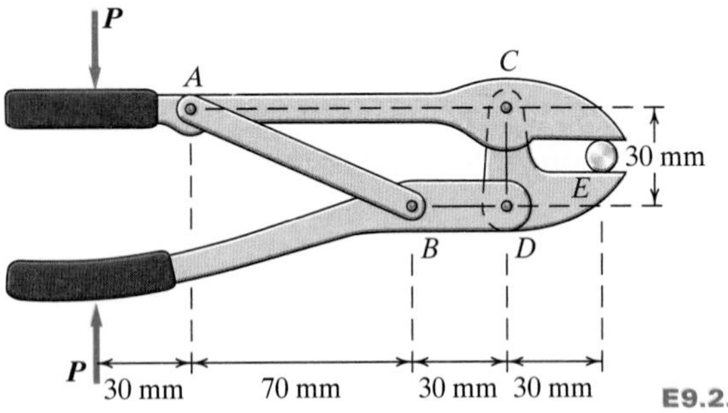

E9.2.4

9.2.5. A pair of pliers grips the bolt as shown in **E9.2.4**. Determine the mechanical advantage of this machine.

9.2.6. A pair of pliers grips the bolt as shown in E9.2.6. For the input force ***P***, determine the force exerted on the bolt.

9.2.7. A pair of pliers grips the bolt as shown in **E9.2.6**. Determine the mechanical advantage of this machine.

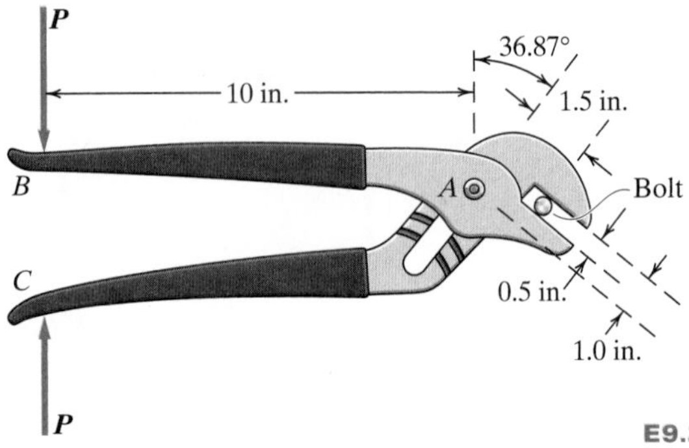

E9.2.6

9.2.8. A pair of pliers grips the bolt as shown in E9.2.8. For the input force ***P***, determine the force exerted on the bolt.

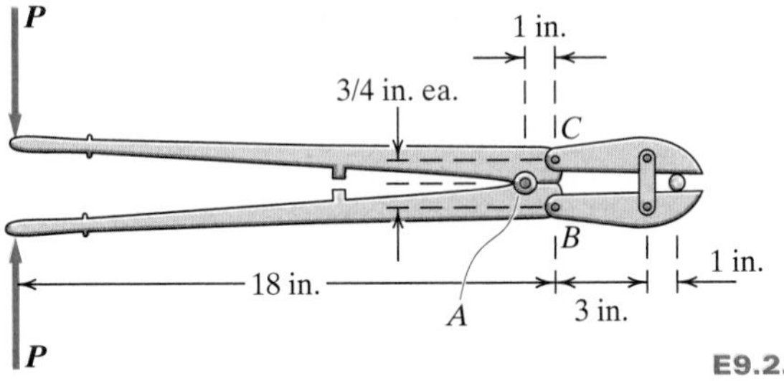

E9.2.8

9.2.9. A pair of pliers grips the bolt as shown in **E9.2.8**. Determine the mechanical advantage of this machine.

9.2.10. A pair of pliers grips the bolt as shown in E9.2.10. For the input force ***P***, determine the force exerted on the bolt.

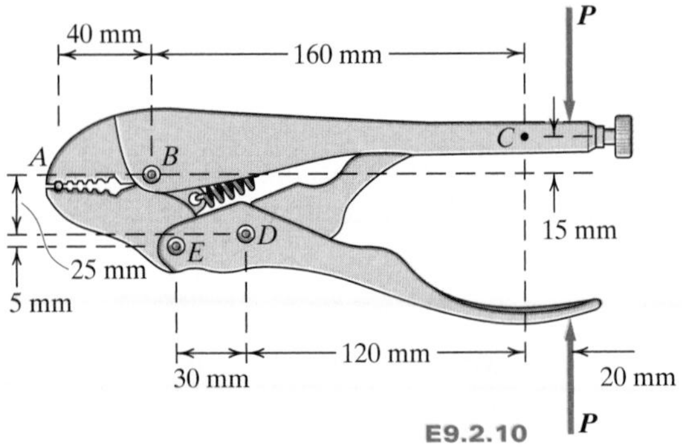

E9.2.10

9.2.11. A pair of pliers grips the bolt as shown in **E9.2.10**. Determine the mechanical advantage of this machine.

9.2.12. The high-pressure hand pump in E9.2.12 is used for boosting oil pressure in a hydraulic line. Consider the position shown, with the handle at $\theta = 15°$.

a. Determine the mechanical advantage of this machine.

b. Determine the oil pressure p that acts on the 50-mm-diameter piston (pressure on top of the piston is atmospheric).

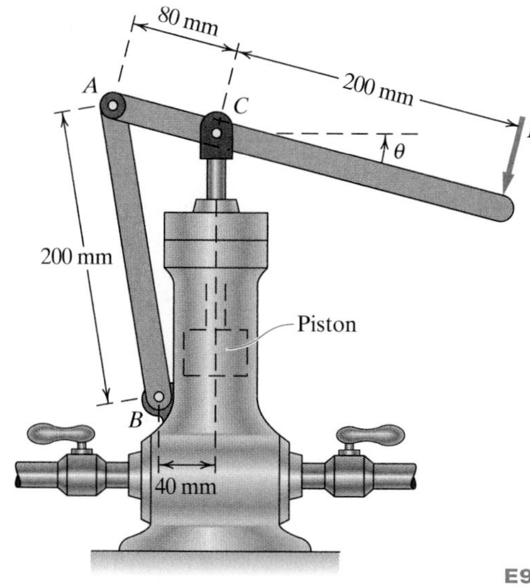

E9.2.12

9.2.13. A dual-grip clamp is shown in E9.2.13. First the pieces are clamped with force P_{clamp} by tightening the screw at A. Then the screw at B is tightened against arm ABC, thereby providing additional clamping force. For this additional clamping force of ΔP_{clamp}, determine

a. the force that acts on the arm ABC at B

b. the shear force acting on the pin at C due to ΔP_{clamp}

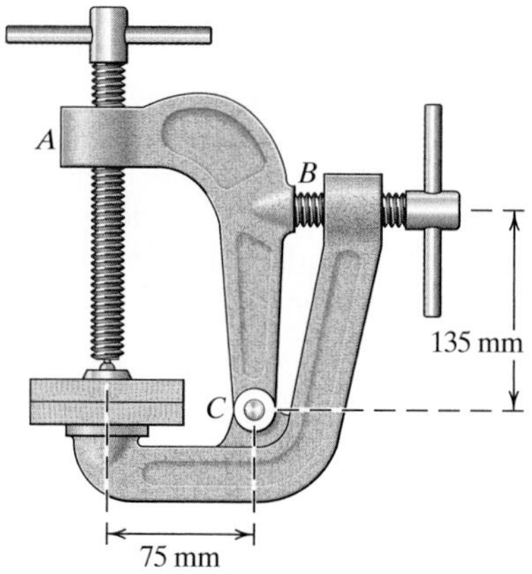

E9.2.13

9.2.14. The foot-operated lift in E9.2.14 is used to raise a platform of mass $m = 100$ kg a small amount. For the position shown, determine

a. the magnitude of the necessary force $\boldsymbol{P}$

b. the mechanical advantage provided by the lift

c. the force acting in member AB and the shear force acting on the pin at C

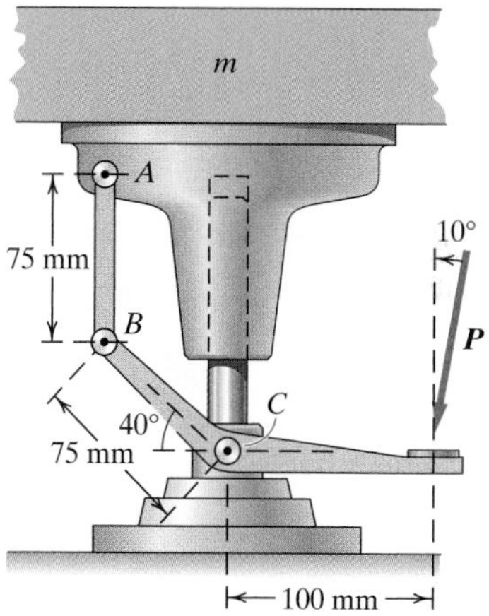

E9.2.14

9.2.15. A pneumatic cylinder in E9.2.15 pivoted at A operates the lever BC of the toggle clamp, which holds the work piece in position while it is machined. For an air pressure of 250 kPa above atmospheric pressure against the 45-mm-diameter piston, determine

a. the clamping force at G

b. the mechanical advantage provided by the lever system

c. the force acting in member DE and the shear force acting on the pin at C

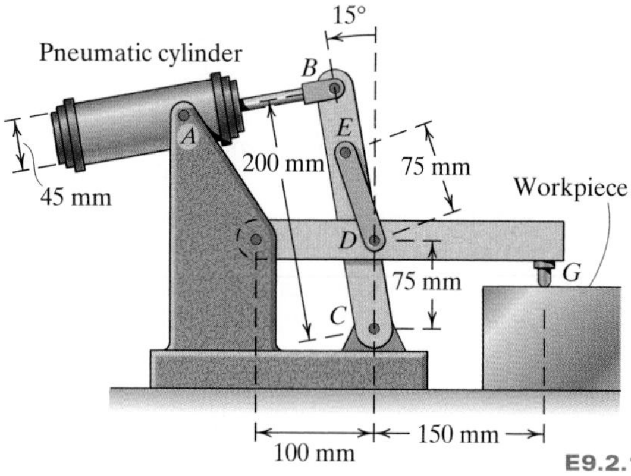

E9.2.15

9.2.16. The diagram of the bones and biceps muscle of a person's arm supporting a mass m is shown in E9.2.16*a*. Tension in the muscle holds the forearm in the horizontal position, as illustrated in the model in **E9.2.16*b***. The weight of the forearm is 9 N. Determine

a. the magnitude of the tension in the biceps muscle, represented by member AB

b. the loads acting on the elbow joint at C

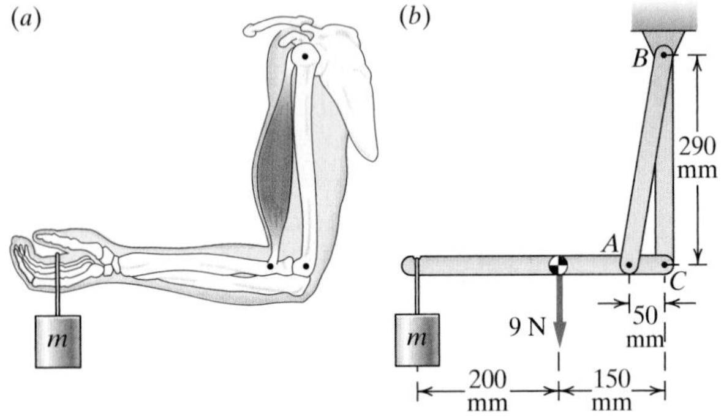

E9.2.16

9.2.17. A grinding wheel weighing 3 lb is supported by a journal bearing at each end of the axle, as shown in E9.2.17. The coefficients of static (μ_s) and kinetic (μ_k) friction are 0.15 and 0.10, respectively, and the diameter of the axle is 0.75 in. Determine

a. the moment required to just get the wheel–axle assembly to rotate

b. the moment required to rotate the wheel at a constant speed

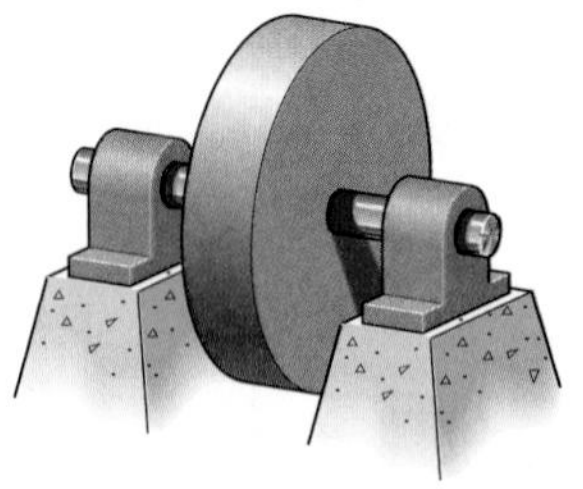

E9.2.17

9.2.18. The crank used to control a ventilation door is shown in E9.2.18. The crank fits loosely over the shaft, which is fixed and cannot rotate. A force of 75 N is required to operate the door.

a. Determine the maximum and minimum values (magnitude) T may have without causing the crank to rotate in either direction. The coefficient of static friction is 0.15.

b. At values of T greater than found in **a**, what happens?

c. At values of T less than found in **a**, what happens?

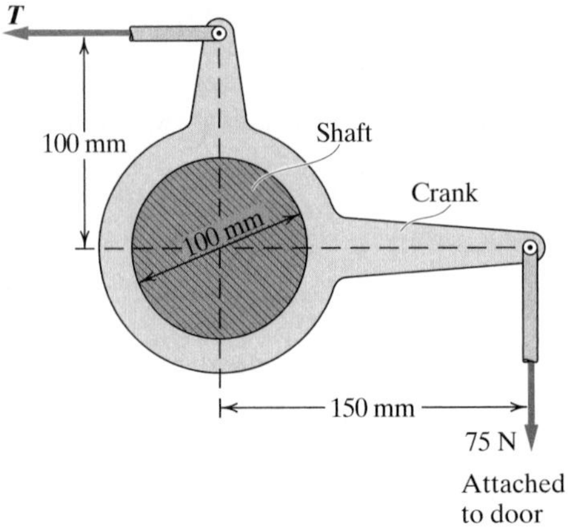

E9.2.18

For **Exercises 9.2.19 and 9.2.20** each pulley in E9.2.19 has a radius of 110 mm and a mass of 5 kg. They are mounted on shafts of 15-mm radius supported by journal bearings and are used to raise block A of mass 15 kg.

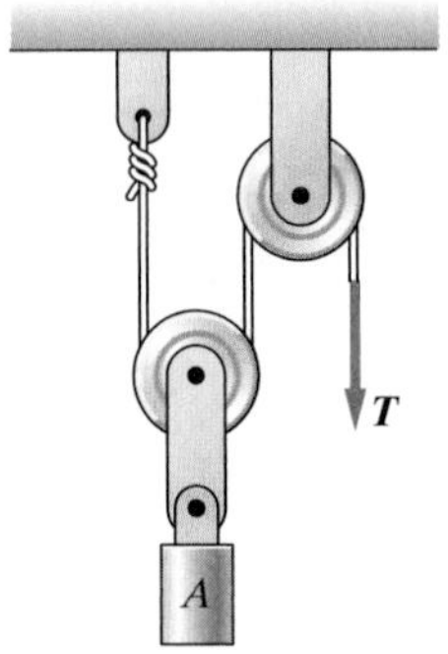

E9.2.19

9.2.19. Determine the force T that is required to raise the block at a constant rate if the bearings are frictionless. Also express in terms of mechanical advantage.

9.2.20. Determine the force T that is required to raise the block at a constant rate if the coefficient of kinetic friction between the shafts and bearings is 0.18. Also express your answer in terms of mechanical advantage.

For **Exercises 9.2.21 to 9.2.24**, the pulley in E9.2.21 consists of a 100-mm-diameter wheel that fits loosely over a 15-mm-diameter shaft. A rope passes over the pulley and is attached to a 200-N weight. The static and kinetic coefficients of friction between the pulley and the shaft are 0.35 and 0.25, respectively. The weight of the pulley is 10 N. The angle $\theta = 90°$.

9.2.21. Determine the minimum tension T in the rope needed to hold the 200-N weight in a stationary position.

9.2.22. Determine the minimum tension in the rope needed to just start raising the 200-N weight. Also express as a percentage of the tension found in **Exercise 9.2.21**.

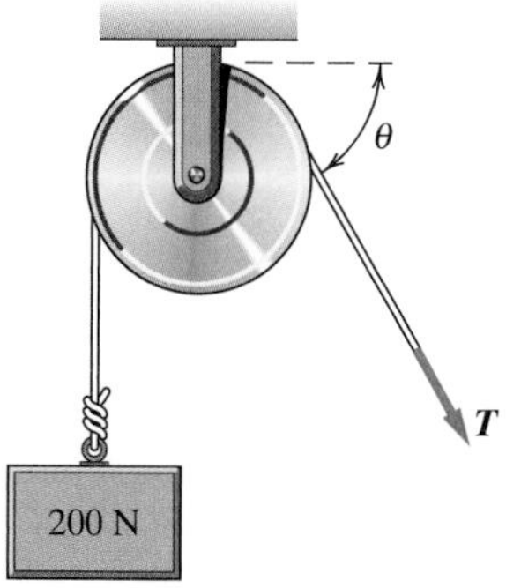

E9.2.21

9.2.23. Determine the tension in the rope needed to raise the 200-N weight at a constant rate. Also express as a percentage of the tension found in **Exercise 9.2.21**.

9.2.24. Determine the tension in the rope needed to lower the 200-N weight at a constant rate. Also express as a percentage of the tension found in **Exercise 9.2.21**.

For **Exercises 9.2.25 to 9.2.28** use **E9.2.21** and assume angle $\theta = 60°$.

9.2.25. Determine the minimum tension T in the rope needed to hold the 200-N weight in a stationary position.

9.2.26. Determine the minimum tension in the rope needed to just start raising the 200-N weight. Also express as a percentage of the tension found in **Exercise 9.2.25**.

9.2.27. Determine the tension in the rope needed to raise the 200-N weight at a constant rate. Also express as a percentage of the tension found in **Exercise 9.2.25**.

9.2.28 Determine the tension in the rope needed to lower the 200-N weight at a constant rate. Also express as a percentage of the tension found in **Exercise 9.2.25**.

9.2.29. The two-wheel cart in E9.2.29 is used to haul rocks along a dirt road. It is equipped with 600-mm-diameter high-pressure bicycle tires that fit loosely over a 15-mm-diameter fixed axle. The coefficient of kinetic friction between the wheel and the axle is 0.30. The cart weighs 200 N and is carrying 1000 N of rocks; the center of mass of the cart and the rocks is 75 mm in front of the axle. Determine the force F (both horizontal and vertical components) that must be applied to the handle of the cart to pull it at a constant speed. Assume that there is negligible rolling resistance.

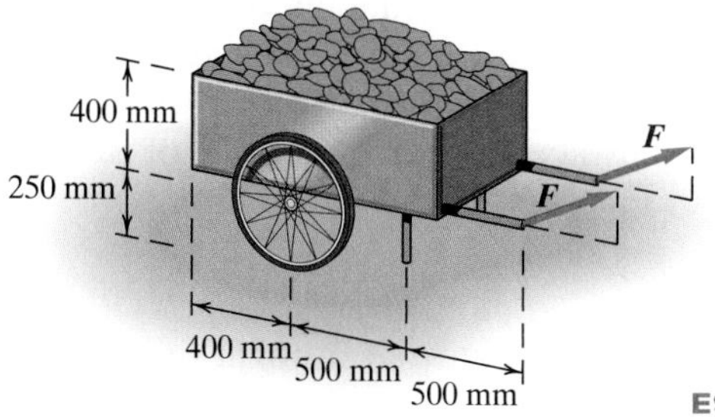

E9.2.29

For **Exercises 9.2.30 and 9.2.31**, consider the single-stage gear train in E9.2.30. Both gears are mounted on shafts of 5-mm radii that are supported by journal bearings, as shown. The gear train is rotating at a constant rate. The gear diameters are shown in the figure.

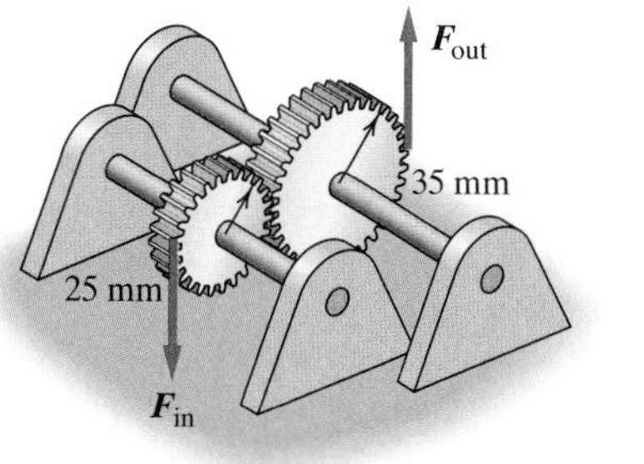

E9.2.30

9.2.30 Determine the ratio of input to output moments in terms of the diameters of various gears and shafts when the journal bearings are frictionless.

9.2.31. Determine the ratio of input to output moments in terms of the diameters of various gears and shafts when the coefficient of kinetic friction associated with the journal bearings is 0.25.

9.2.32. A 2000-lb automobile has four 23-in.-diameter tires. Neglect bearing friction. Determine the horizontal force required to push the automobile

a. on level pavement, where the coefficient of rolling friction is $f_r = 0.03$

b. on loose sand, where the coefficient of rolling friction is $f_r = 0.35$

9.2.33. A 1100-kg automobile is observed to roll at a constant speed down a 1° incline. The auto has 550-mm-diameter tires. Determine the coefficient of rolling friction. Neglect bearing friction.

9.2.34. Reconsider the two-wheel cart in **E9.2.29**. Determine the force (both horizontal and vertical components) that must be applied to the handle of the cart to pull it at a constant speed due to rolling resistance. The coefficient of rolling resistance is 0.0002. Ignore bearing friction.

9.2.35. Reconsider the two-wheel cart in **E9.2.29**. Determine the force (both horizontal and vertical components) that must be applied to the handle of the cart to pull it at a constant speed. Include the effects of rolling resistance ($f_r = 0.0002$) and bearing friction ($\mu_k = 0.30$).

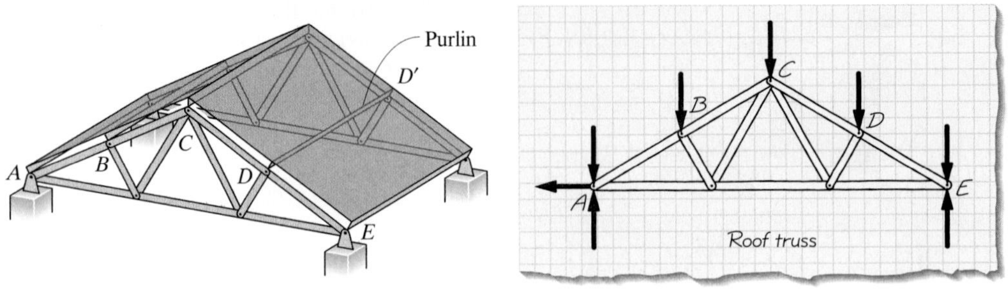

(*a*)

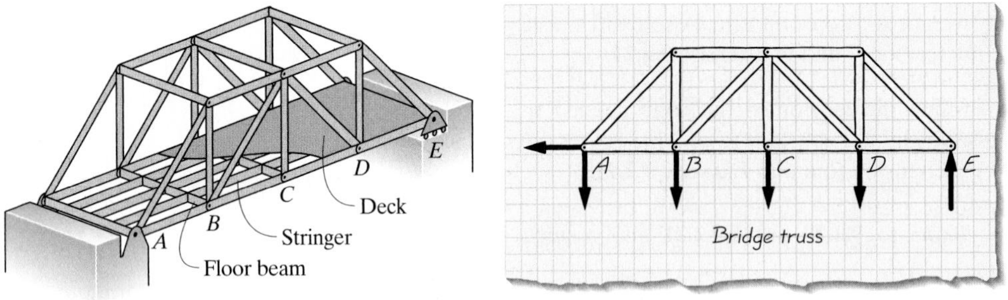

(*b*)

Figure 9.70 (*a*) A truss used as part of a roof structure; (*b*) a truss used as part of a bridge structure

9.3 TRUSS ANALYSIS

A **truss** is a structural system, made up of two-force members, that is generally lightweight compared to the loads it can support. Trusses are commonly seen in bridges, roof supports, cranes, derricks, towers, and amusement park structures. For example, the truss shown in **Figure 9.70*a*** carries the weight of the roof to the walls, and the truss in **Figure 9.70*b*** carries the weight of a bridge roadway to the vertical supports.

For a system to be classified as a truss, it must:

1. Consist exclusively of *straight members joined at their ends* to form a rigid frame.
2. Have joints that can be *represented as pin connections*, even though the actual joints may consist of welds, rivets, large bolts, or pins (often in conjunction with a gusset plate). We idealize these joints into the representation shown in **Figure 9.71*a***—that of a simple pin connection. In this idealized joint, the pin fits smoothly into holes at the ends of the straight members and is therefore capable of transferring forces between members but not capable of transferring moments. Furthermore, the centerline of each straight member is assumed to intersect the center axis of the pin (**Figure 9.71*b***).
3. Carry external *forces* (and no moment), *exclusively at the joints*. This requirement is not as restrictive as it might first appear. For

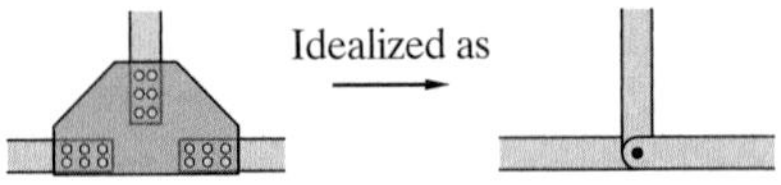

(*a*)

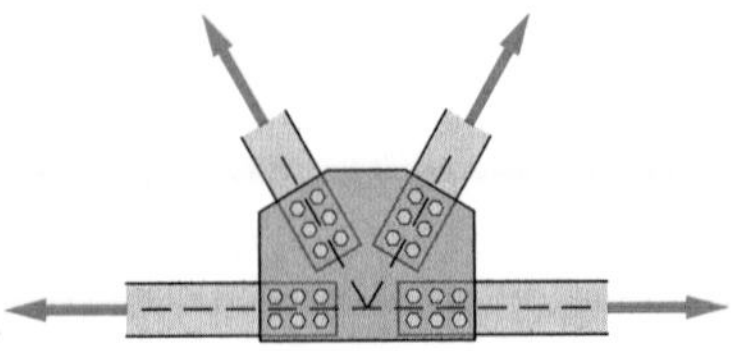

(*b*)

Figure 9.71 (*a*) A connection in a truss, idealized as a pin connection; (*b*) forces of two-force members connected at a pin connection are concurrent

example, in **Figure 9.70*a***, the roof force is transferred to purlins, then to the truss, and finally to the walls. In **Figure 9.70*b***, the weight of the roadway is transferred to the stringers and floor beams, then to the trusses, and finally to the vertical supports.

As a consequence of these three requirements, the members that make up a truss behave as two-force members. This means that if we isolate any member from the rest of the truss and draw a free-body diagram of that member, the only way for the member to be in equilibrium is for the two forces acting on it to be collinear (as we discussed in Chapter 7 on two-force members).

A truss is made up *entirely* of two-force members. This is in contrast to a frame, where at least one of the members is a multiforce member. The system in Figure 9.72 is a truss, whereas the system in Figure 9.73 is a frame. Convince yourself of the correctness of these labels.

Trusses are classified as planar or space (i.e., nonplanar). A **planar truss** is one in which all forces and members lie in a single plane (or it is reasonable to assume that they all lie in a single plane). The basic building block of a planar truss is a triangle, composed of three two-force members connected by pins (Figure 9.74*a*). The triangle is the simplest **rigid structure** that can be created with two-force members; by rigid we mean that the structure is internally stable. Systems built up from a number of basic triangles are simple planar trusses. Various planar truss structures are illustrated in **Figures 9.74*b*–*e***. Notice that two of these

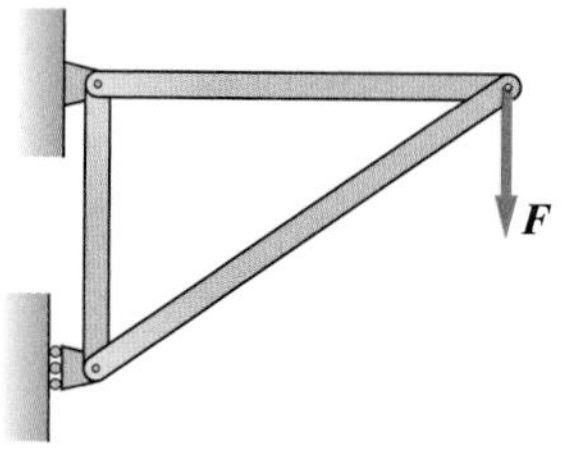

Figure 9.72 Force ***F*** acts at a pin connection, so structure is a truss.

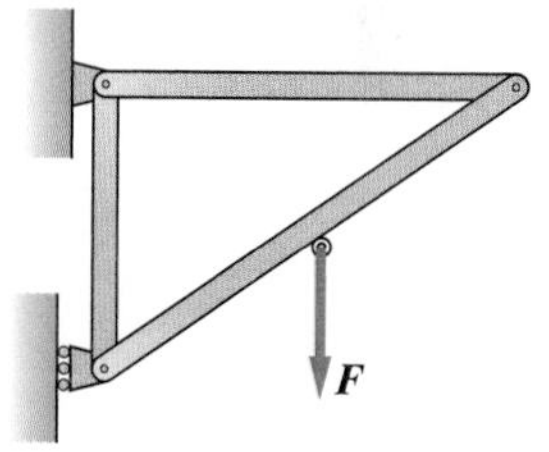

Figure 9.73 Force ***F*** acts along a member, so structure is a frame.

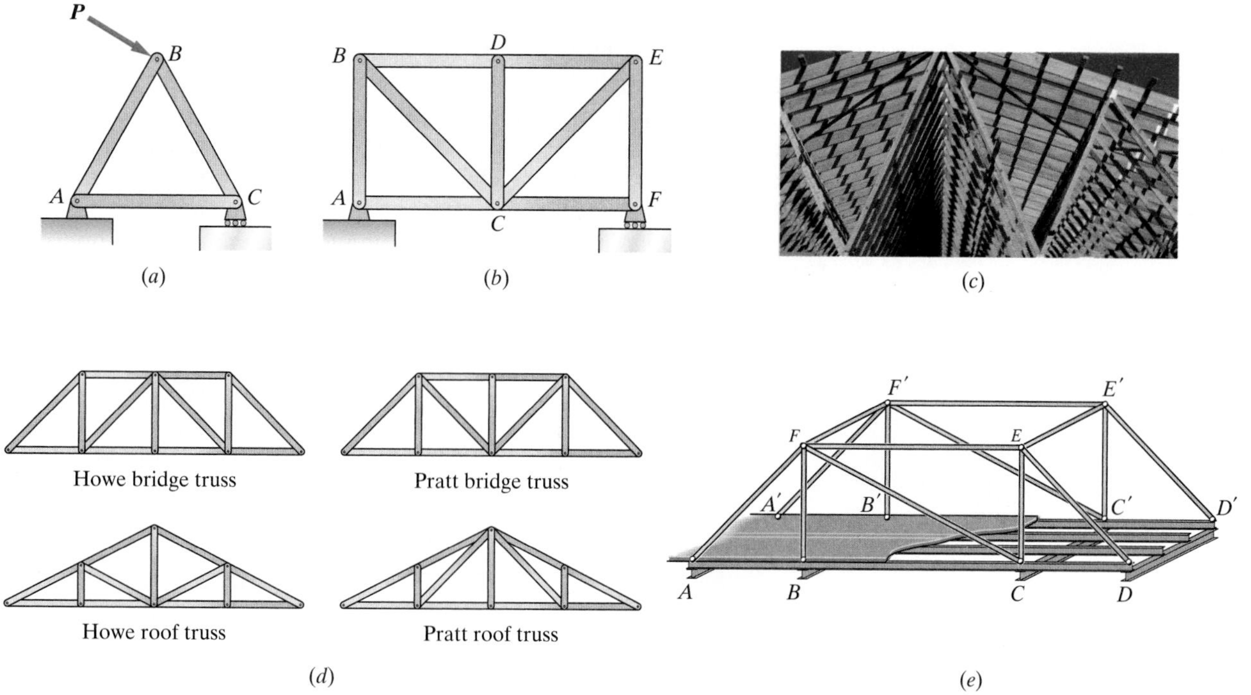

Figure 9.74 Examples of various planar trusses

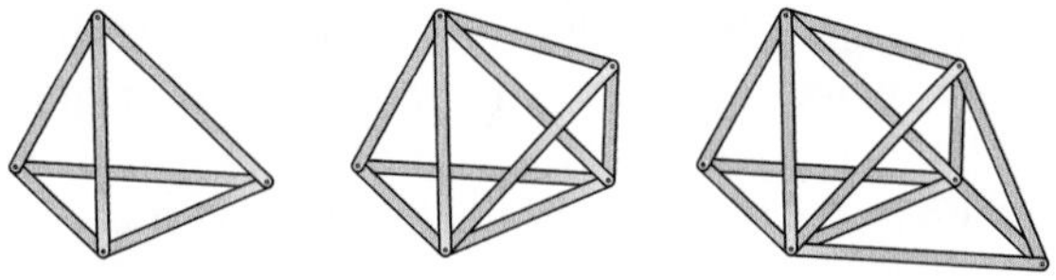

(a) (b) (c)

Figure 9.75 Examples of space trusses of increasing complexity

structures are made up of several planar trusses and use cross-pieces to distribute the force to the trusses; in this manner, planar trusses can be incorporated into three-dimensional structures.

If a truss is not planar, it is a **space truss**. The basic building block of a space truss is a tetrahedron, which is an assembly of six two-force members joined by ball-and-socket joints (**Figure 9.75**).

Trusses are typically designed so that they are connected either to the ground or to a supporting structure with pin joints and/or rocker connections. A rocker or roller connection allows a truss to expand or contract as the temperature changes and as applied forces deform the members. This configuration is illustrated in **Figure 9.76**.

Truss analysis involves determining the force that each two-force member and pin connection must carry for given external forces. The calculated forces acting on the two-force members are then used to check for the adequacy of the cross section and material of each member. For a two-force member in tension, the check is to ensure that calculated tension is less than a critical level for tensile failure. For a two-force member in compression, the check is to ensure that the calculated compression is less than a critical level for compressive failure (which includes buckling). Buckling is a failure mode that only occurs when a member is in compression; when the compressive load becomes too large the member becomes unstable, causing it to bow and ultimately collapse. The calculated forces acting on the pins (which are shear forces) are checked to ensure adequacy of the cross section and material of each pin; an inadequate pin would undergo a shear failure.

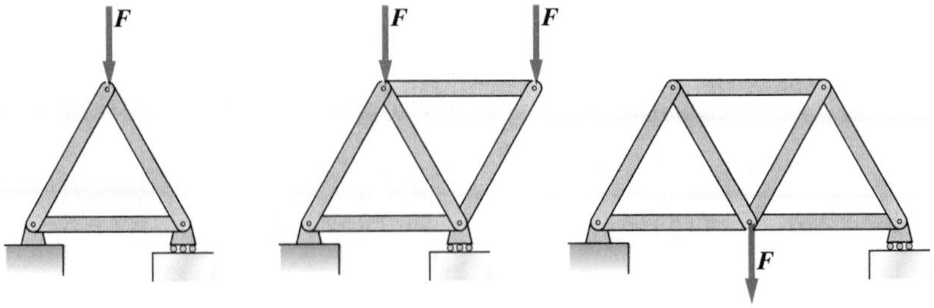

(a) (b) (c)

Figure 9.76 Examples of rollers and pins as truss boundary connections

EXAMPLE 9.15 IDENTIFY SYSTEMS AS TRUSSES OR FRAMES

Identify each system as either a truss or frame. Justify your answer.

System 1 (Figure 9.77*a*) The system is a *truss*. Assuming we ignore their weights, all of the members are two-force members because they are all connected by pins and the forces are applied at the pins.

System 2 (Figure 9.77*b*) The system is a *frame* because it has at least one multiforce member (the member from which the weight is hanging).

Question: Based on inspection, will the vertical member carry any load?

System 3 (Figure 9.77*c*) The system is a *truss*. If we ignore the weights of the members, they are all two-force members.

System 4 (Figure 9.77*d*) The system is a *frame* because DE is a multiforce member. All other members are two-force members if their weights are negligible. Notice that portions of the system are trusses (ACD and BEG form trusses).

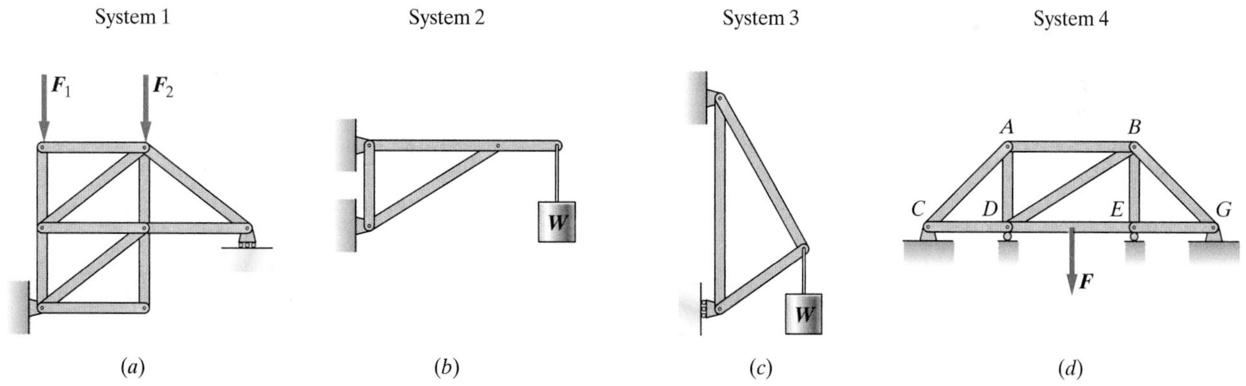

Figure 9.77

Determining Internal Forces in Trusses

We outline two procedures for finding the forces carried by two-force members in a truss—the **method of joints** and the **method of sections**. In general, the method of joints is preferable for determining the force acting on every two-force member in a truss, and the method of sections is preferable for determining the forces for only a few members. We shall return to this point later in this section.

Both methods assume that the reason for analyzing the system has been defined and that information about the problem has been recorded, including whether it is reasonable to categorize the system as a planar or space truss. The first step in both methods is creating a free-body diagram of the entire truss, as in Figure 9.78*a*. Next the equilibrium equations are set up and solved (if possible) to find the external forces acting on the entire truss.

The Method of Joints

The next step in this method is to isolate each pin and draw its free-body diagram (**Figure 9.78*b***). For each two-force member connected to a pin, there is a third-law force pair. One component of this pair is the force exerted by the member on the pin. Draw this force on the pin free-body diagram as an arrow with its line of action along the centerline of the member. Label the force with a subscript that describes the two-force member by the two pins connecting it to the rest of the truss and the pin (e.g., $\boldsymbol{F}_{AB,B}$ is the force of member AB, which runs between pins A and B acting on pin B. In **Figure 9.78** we have shortened this to $\boldsymbol{F}_{AB}$). Be sure to include any external forces acting on each pin in the pin's free-body diagram.

The other component of the third-law force pair is the force exerted by the pin on the member ($\boldsymbol{F}_{B,AB}$). Because the force on the pin points toward the center of the member, the other component ($\boldsymbol{F}_{B,AB}$) (not shown in the free-body diagram of the pin, of course) points away from the member center, meaning the member is under tension. The assumption that the member is under tension will be verified by the equilibrium analysis of the pin. Furthermore, because $\boldsymbol{F}_{B,AB}$ and $\boldsymbol{F}_{AB,B}$ form a third-law force pair,

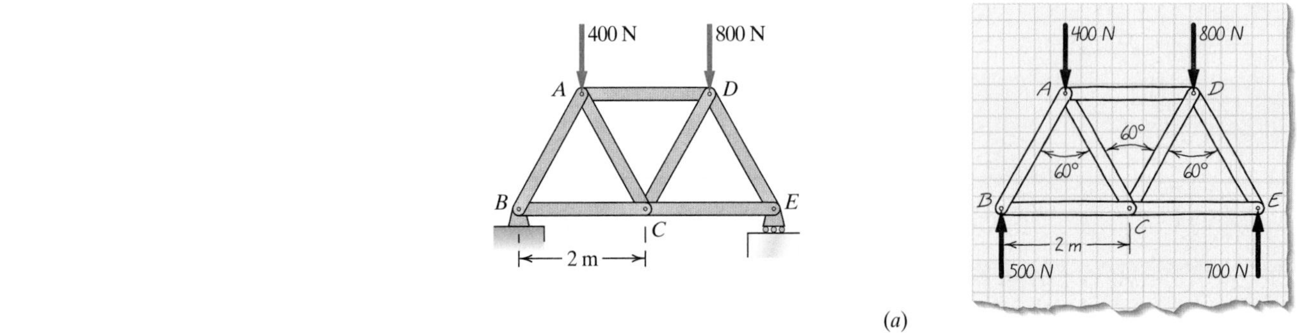

(*a*)

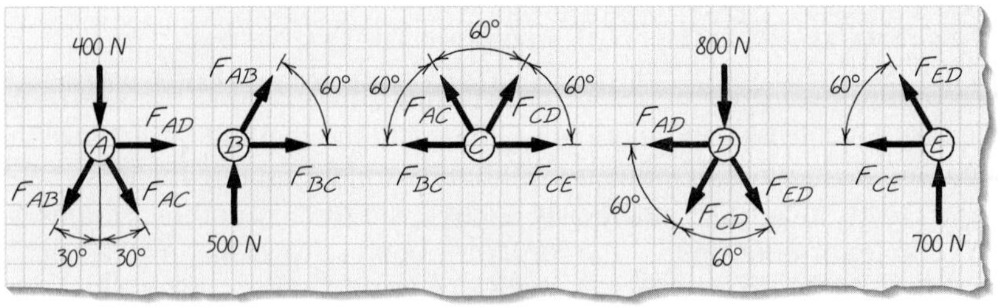

(*b*)

Figure 9.78 (*a*) A truss and its free-body diagram; (*b*) free-body diagrams of pins *A*, *B*, *C*, *D*, and *E*

they are equal and opposite. This means that $\| \boldsymbol{F}_{B,AB} \| = \| \boldsymbol{F}_{AB,B} \|$, and as we mentioned above, we choose to simplify the subscript label to AB.

The steps of zooming–isolating–drawing are repeated for each pin. Be consistent in defining force labels and force directions. When done, you should have as many free-body diagrams as you have pins, as shown in **Figures 9.78*b***.

Now set up the force equilibrium equations for each pin. For a planar truss there are two force equations per pin, and for a space truss there are three. Because the lines of action of the forces acting on a particular pin intersect at a common point, they produce no moment—therefore we do not need to consider the moment equilibrium conditions. The pin is, in fact, treated as a particle. If there are N pins, there will be $2N$ force equilibrium equations for a planar truss and $3N$ for a space truss. If equilibrium conditions have been applied to the truss as a whole (as was done in **Figure 9.78*a***), only $2N - 3$ of the equations generated in considering the N pins for a planar truss will be linearly independent of one another ($3N - 6$ will be linearly independent for a nonplanar truss).

We solve these equations for unknown forces. It is easiest to start at a pin that has at least one known force and at most two unknowns for a planar truss (three for a space truss)—this means that we can find the two unknowns (three for a space truss) using the associated force equilibrium equations. We proceed to the next pin that has at least one known force and at most two (three) unknowns and solve its equilibrium equations, until all pins (and therefore equilibrium equations) have been considered. A positive force value indicates that the associated two-force member is in tension, and a negative force value indicates that it is in compression.

After solving the equilibrium equations, check your results. Consider which two-force members are in tension and which are in compression—are these answers consistent with your intuition? Finally, present your answers in a manner that clearly indicates the forces acting on the truss and on each two-force member.

EXAMPLE 9.16 TRUSS ANALYSIS USING METHOD OF JOINTS

The truss shown in **Figure 9.79** is supported by a pin connection at C and a roller on a frictionless inclined plane at G. Use the method of joints to find the force acting on each member.

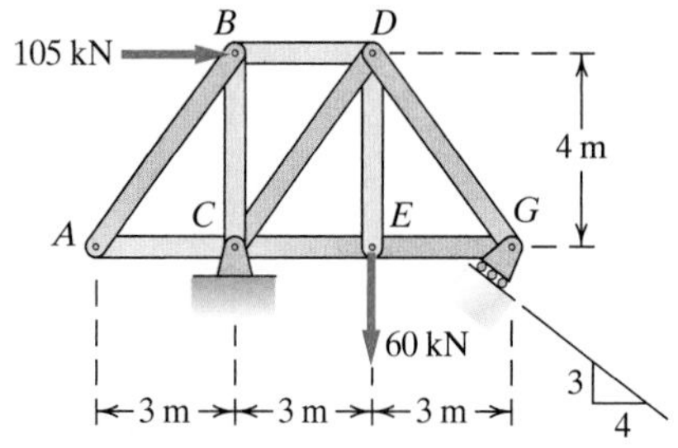

Figure 9.79

Goal We are to find the force supported by each member using the method of joints.

Given We are given information about the geometry of the truss, the supports, and the loads acting at B and E.

Assume We assume that we can treat the system as planar and that the weight of the members is negligible.

Draw We first draw a free-body diagram of the entire truss (**Figure 9.80**); this will allow us to find the loads at supports C and G. Then we

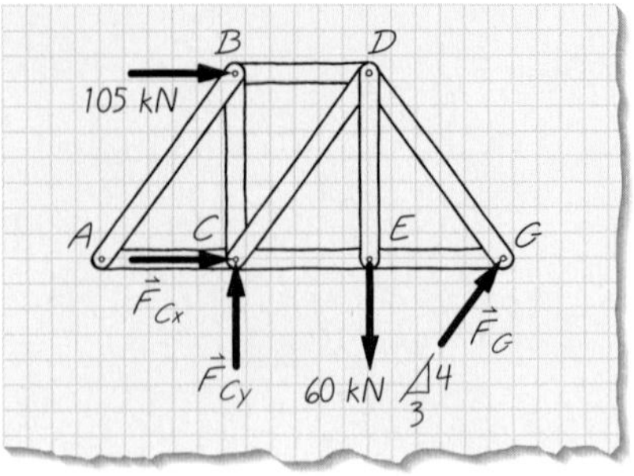

Figure 9.80 Free-body diagram of entire truss

draw separate free-body diagrams for pins A (two equations, two unknowns), B (two equations, two additional unknowns), C (two equations, two additional unknowns), D (two equations, two additional unknowns), and E (two equations, two additional unknowns).

Formulate Equations and Solve Before diving in and writing equations, consider how many unknowns there are—12: (F_{Cx}, F_{Cy}, F_G, F_{AB}, F_{AC}, F_{BC}, F_{BD}, F_{CD}, F_{CE}, F_{DE}, F_{DG}, F_{EG}). We will therefore be writing 12 equilibrium equations to find these 12 unknowns. For the truss as a whole, based on (7.5C) and **Figure 9.80** with the moment center at C, we have

$$\sum M_{z@C} = 0\ (\circlearrowleft +)$$

$$\frac{4}{5}F_G(6\text{ m}) - 60\text{ kN}(3\text{ m}) - 105\text{ kN}(4\text{ m}) = 0$$

$$\frac{4}{5}F_G(6\text{ m}) = 600\text{ kN}\cdot\text{m}$$

$$F_G = 125\text{ kN} \tag{1}$$

Based on (7.5A):

$$\sum F_x = 0\ (\rightarrow +)$$

$$F_{Cx} + \frac{3}{5}F_G + 105\text{ kN} = 0$$

$$F_{Cx} + \frac{3}{5}(125\text{ kN}) + 105\text{ kN} = 0$$

$$F_{Cx} = -180\text{ kN} \tag{2}$$

Based on (7.5B):

$$\sum F_y = 0\ (\uparrow +)$$

$$F_{Cy} + \frac{4}{5}F_G - 60\text{ kN} = 0$$

$$F_{Cy} + \frac{4}{5}(125\text{ kN}) - 60\text{ kN} = 0$$

$$F_{Cy} = -40.0\text{ kN} \tag{3}$$

To summarize, so far we have

$$F_{Cx} = -180\text{ kN}, \qquad F_{Cy} = -40.0\text{ kN}, \qquad F_G = 125\text{ kN}$$

Next we analyze each pin and solve for forces. Good candidate pins to start with are joints A and G; both have only two unknown forces. We arbitrarily choose joint A.

In drawing the free-body diagram of pin A, we choose to draw the unknowns F_{AB} and F_{AC} so that they are pulling on the joint. In response, the joint is pulling on the members so that they are in tension. When analyzing the joint, if we get a positive value for either force, the member is in tension, and a negative value means the member is in compression.

We proceed to apply planar equilibrium to pin A. Based on (7.5B) and the free-body diagram in **Figure 9.81**, we write:

$$\sum F_y = 0\,(\uparrow\ +)$$
$$F_{AB}\sin\theta = 0$$
$$F_{AB} = 0 \qquad (4)$$

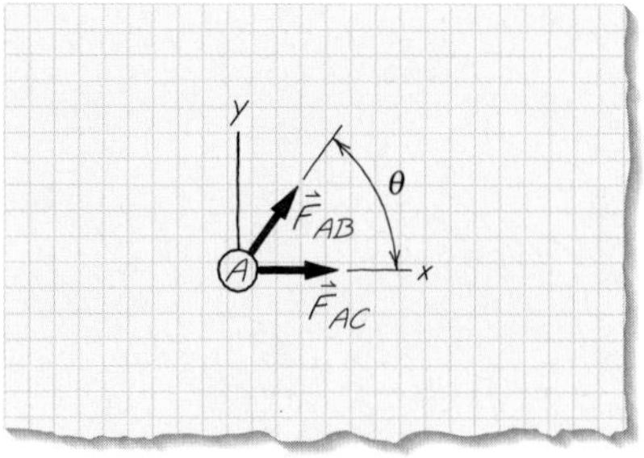

Figure 9.81 Free-body diagram of pin A

Based on (7.5A):

$$\sum F_x = 0(\rightarrow +)$$
$$F_{AB}\cos\theta + F_{AC} = 0$$
$$F_{AC} = -F_{AB}\cos\theta = 0 \qquad (5)$$

In summary, we have

$$F_{AB} = F_{AC} = 0\text{ kN}$$

Notice that pin A (**Figure 9.81**) is one of the special conditions we examined in Example 9.6 (Case 1b) in which only two members connect at A; they do not have the same line of action, and no external force is applied at A. In this case, they both must be zero-force members.

We now look at pin B as it has only two unknowns (**Figure 9.82**). Based on (7.5A):

$$\sum F_x = 0\,(\rightarrow +)$$
$$105\text{ kN} - F_{AB}\sin\alpha + F_{BD} = 0$$

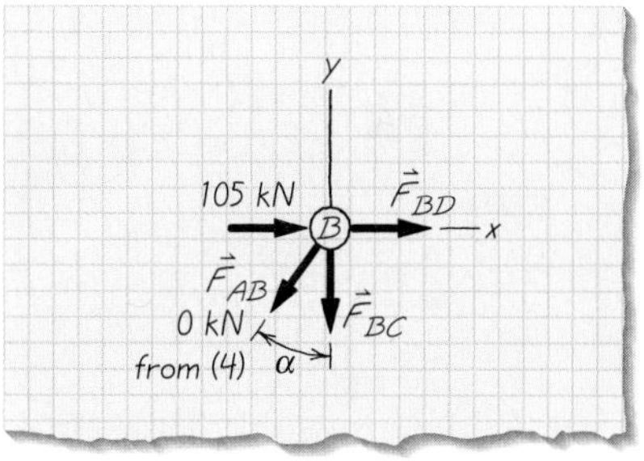

Figure 9.82 Free-body diagram of pin B

Substituting from (4) for F_{AB} (0 kN):

$$F_{BD} = -105\text{ kN (compression)} \qquad (6)$$

Based on (7.5B):

$$\sum F_y = 0\,(\uparrow\ +)$$
$$-F_{BC} - F_{AB}\cos\alpha = 0$$

Substituting from (4) for F_{AB}(0 kN):

$$F_{BC} = -F_{AB}\cos\alpha = 0\text{ kN} \qquad (7)$$

In summary, we now have

$$F_{BC} = 0\text{ kN}, \qquad F_{BD} = -105\text{ kN} \quad \text{(compression)}$$

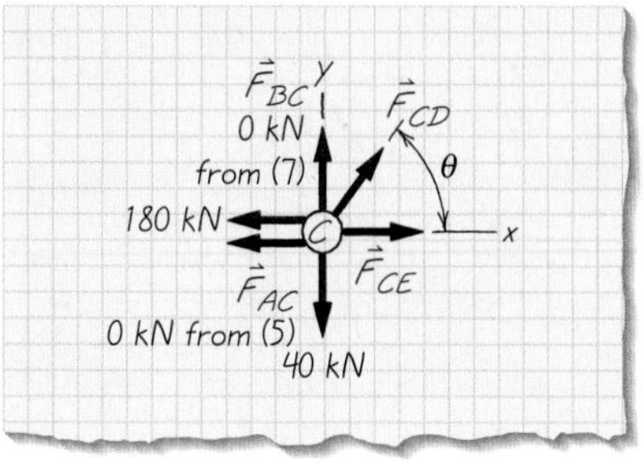

Figure 9.83 Free-body diagram of pin C

(Note that our calculation says that member BC is a zero-force member. Based on Case 1a in Example 9.6 does this make sense?)

We next analyze pin C (**Figure 9.83**). In **Figure 9.83**, we include the support forces acting on pin C. Based on (7.5B):

$$\sum F_y = 0(\uparrow +)$$
$$-40.0\text{ kN} + F_{CD}\sin\theta = 0$$
$$F_{CD}\left(\frac{4}{5}\right) = 40.0\text{ kN}$$
$$F_{CD} = 50.0\text{ kN} \quad \text{(tension)} \tag{8}$$

Based on (7.5A):

$$\sum F_x = 0(\rightarrow +)$$
$$-180\text{ kN} + F_{CD}\cos\theta + F_{CE} = 0$$
$$F_{CE} = 180\text{ kN} - F_{CD}\left(\frac{3}{5}\right) \tag{9}$$

Substitute (8) into (9) and solve for F_{CE}:

$$F_{CE} = 180\text{ kN} - 50.0\text{ kN}\left(\frac{3}{5}\right) = 150\text{ kN} \quad \text{(tension)} \tag{10}$$

In summary:

$$F_{CD} = 50.0\text{ kN} \quad \text{(tension)}, \qquad F_{CE} = 150\text{ kN} \quad \text{(tension)}$$

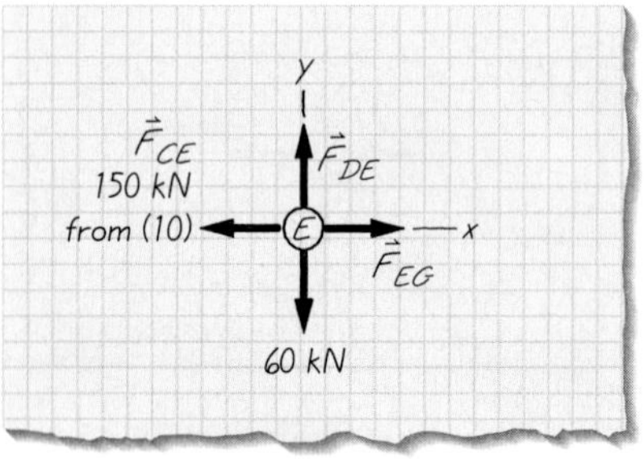

Figure 9.84 Free-body diagram of pin E

We now look at either pin E or pin D. We arbitrarily choose E (**Figure 9.84**). Based on (7.5A):

$$\sum F_x = 0(\rightarrow +)$$
$$-F_{CE} + F_{EG} = 0$$

Substituting from (10)

$$F_{EG} = 150\text{ kN} \quad \text{(tension)} \tag{11}$$

Based on (7.5B):

$$\sum F_y = 0(\uparrow +)$$
$$F_{DE} - 60.0\text{ kN} = 0$$
$$F_{DE} = 60.0\text{ kN} \quad \text{(tension)} \tag{12}$$

In summary:

$$F_{EG} = 150\text{ kN} \quad \text{(tension)}, \qquad F_{DE} = 60.0\text{ kN} \quad \text{(tension)}$$

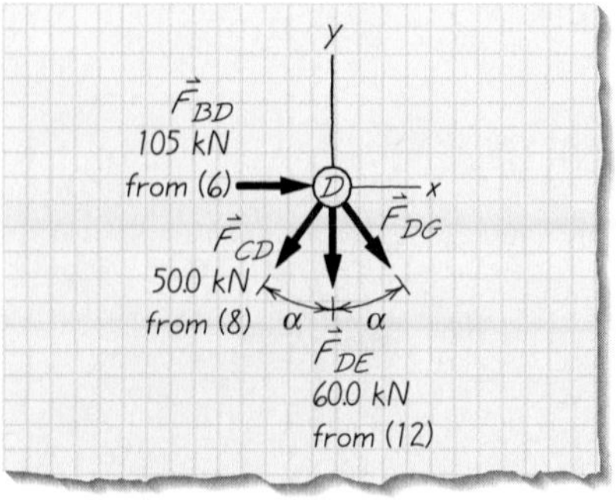

Figure 9.85 Free-body diagram of pin D

Finally we consider pin D (**Figure 9.85**). Based on (7.5A):

$$\sum F_x = 0(\rightarrow +)$$
$$F_{BD} - F_{CD}\sin\alpha + F_{DG}\sin\alpha = 0$$

Substituting from (6) and (8),

$$105 \text{ kN} - 50.0 \text{ kN}\left(\frac{3}{5}\right) + F_{DG}\left(\frac{3}{5}\right) = 0$$

$$F_{DG}\left(\frac{3}{5}\right) = -105 \text{ kN} + 30.0 \text{ kN} = -75.0 \text{ kN}$$

$$F_{DG} = -125 \text{ kN} \quad \text{(compression)}$$

In summary:

$$F_{DG} = -125 \text{ kN} \quad \text{(compression)}$$

Answer Forces acting on the truss structure at the supports:

$F_{Cx} = -180$ kN, $F_{Ey} = -40.0$ kN, $F_G = 125$ kN

Forces acting on the various two-force members:

$F_{AB} = 0$ kN	$F_{AC} = 0$ kN
$F_{BC} = 0$ kN	$F_{BD} = -105$ kN (compression)
$F_{CD} = 50.0$ kN (tension)	$F_{CE} = 150$ kN (tension)
$F_{DE} = 60.0$ kN (tension)	$F_{DG} = -125$ kN (compression)
$F_{EG} = 150$ kN (tension)	

Check One check on our results is to analyze pin G (**Figure 9.86**), for which we now know all of the forces, and confirm that G is in equilibrium. If pin G is not in equilibrium, then we have made a mistake in our analysis at one or more of the joints.

Based on (7.5A):

$$\sum F_x = 0(\rightarrow +)$$

$$-150 \text{ kN} + 125 \text{ kN}\left(\frac{3}{5}\right) + 125 \text{ kN}\left(\frac{3}{5}\right) = 0$$

$$-150 \text{ kN} + 75.0 \text{ kN} + 75.0 \text{ kN} = 0$$

$$0 = 0$$

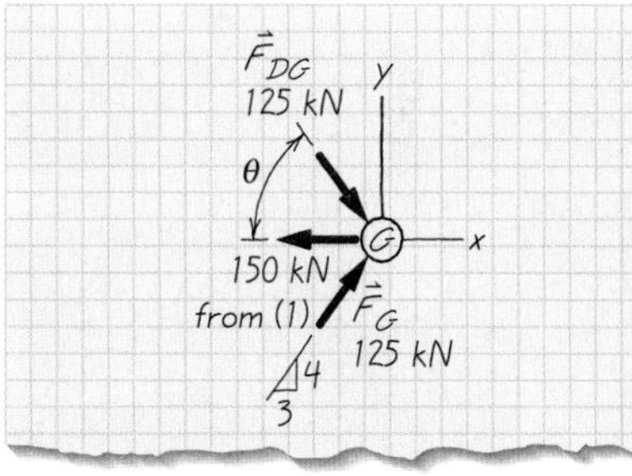

Figure 9.86 Free-body diagram of pin G

Based on (7.5B):

$$\sum F_y = 0(\uparrow +)$$

$$125 \text{ kN}\left(\frac{4}{5}\right) - 125 \text{ kN}\left(\frac{4}{5}\right) = 0$$

$$0 = 0$$

Yes, pin G is in equilibrium!

Questions: How might we have proceeded to solve Example 9.16 if the weight of the members was not negligible? Well, one way would be to determine the weight of each member, then place one-half of this weight at the pins at the end of the member. Can you think of any limitations to this approach to modeling member weight?

Summary In the solution to Example 9.16, we started by enforcing equilibrium on the truss as a whole, then proceeded to consider equilibrium of joints (pins). Alternately, we could have started by considering equilibrium of joints. What is key is that we ended up generating enough linearly independent equilibrium equations to find the unknowns.

Our analysis was such that:

1. All truss members are two-force members.
2. Forces at the ends of a two-force member act along a line of action coincident with the member axis and are equal in magnitude but opposite in sense.
3. Because we consistently drew the force exerted by a member on a pin so as to indicate that the member is in tension, we could interpret any negative answers as indicating compression in the member.

The Method of Sections

The strength of this method is that the force in almost any desired two-force member can be found directly from an analysis of a free-body diagram that includes a boundary cut through the member. The method provides a straightforward means of finding the force acting on a truss member if it is possible to pass a boundary cut through at most three two-force members for a planar truss (six for a space truss). Thus it is not necessary to proceed from pin to pin as in the method of joints.

The method of sections involves isolating a portion of the truss such that the boundary cuts through at most three two-force members for a planar truss (six for a space truss) (**Figure 9.87*a***) and drawing a free-body diagram for the isolated portion (**Figure 9.87*b***). Wherever the boundary cuts a two-force member, there is a force pair. One component of this pair is the force exerted by the part of the two-force member

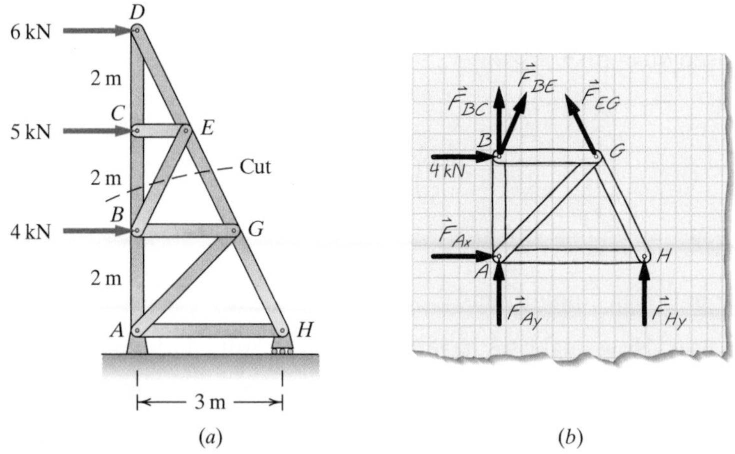

Figure 9.87 (*a*) A truss under consideration; (*b*) free-body diagram of a portion of the truss, based on cut boundary defined in (*a*)

outside of the boundary on the part inside the boundary (e.g., F_{BC}, F_{BE}, and F_{EG} in **Figure 9.87*b***).

Draw each component on the free-body diagram as an arrow with its line of action along the centerline of the cut two-force member, as shown in **Figure 9.87**. Drawn in this manner, this force indicates tension in the cut member. If we get a negative value for this force when we solve the equilibrium equations, the member is actually in compression. Label the force using a subscript that names the pins the two-force member connects (e.g., F_{AB} represents the force in member AB that connects pin A and pin B). Now set up the equilibrium equations (three if planar, six if space) and solve for the unknowns. Finally, as with the method of joints, check and present your answers.

EXAMPLE 9.17 METHOD OF SECTIONS (A)

Given the truss and loading shown in **Figure 9.88**, (a) use the method of sections to find the forces supported by members DE and DG. (b) The allowable load for any of the truss members is 20 kip. If the magnitude of the force acting on any truss member is greater than this allowable load, damage will occur. Therefore, if the force found in (a) is greater than 20 kip for either of the members, how much should the forces at B and E be reduced to ensure that the allowable load is not exceeded? Assume that you will reduce all forces acting on the truss by the same amount.

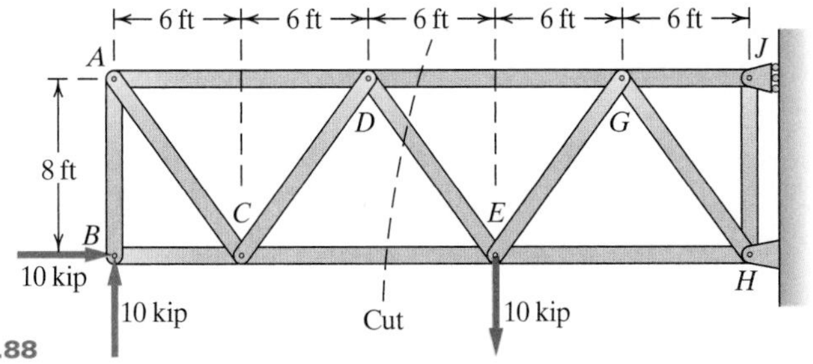

Figure 9.88

Goal We are to find the forces supported by members DE and DG using the method of sections (a). We are to determine whether the force acting on either of these members exceeds 20 kip (b). If it does, we are to recommend a reduction in the applied forces.

Given We are given information about the geometry of the truss and the loading on it.

Assume We assume that we can treat the system as planar and that the weight of the members can be neglected.

Draw We run a boundary cut through members CE, DE, and DG to isolate a portion of the truss (**Figure 9.89**). Because we have three planar equilibrium equations, the maximum number of members (with unknown forces) we should cut is three. We can choose to analyze the

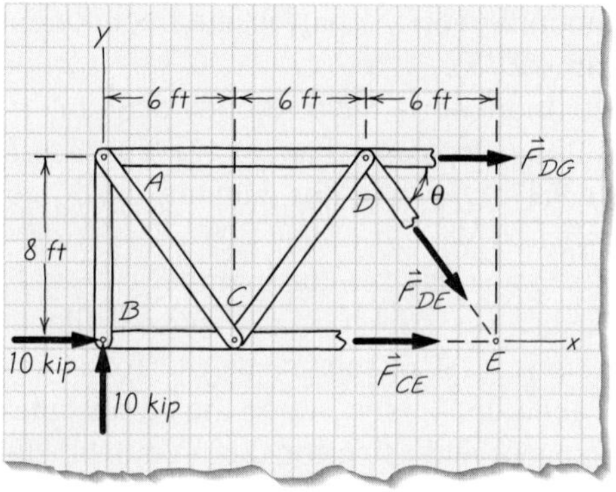

Figure 9.89 Free-body diagram of portion of truss to the left of the cut

portion of the truss to the left or right of the cut. We choose the left portion, because then we do not need to find the forces at supports *J* and *H* in order to determine the forces acting on the members *DE* and *DG*. We draw the unknown forces acting on cut members *CE*, *DE*, and *DG*, with the assumption that the members are in tension and therefore the forces are pulling on the members.

Formulate Equations and Solve **(a)** We use the planar equilibrium equations for the left portion of the truss (**Figure 9.89**) to find the forces acting on *DE* and *DG*.

We first use (7.5C) with the moment center at *E* to find the force acting on *DG*. We choose *E* as the moment center because F_{CE} and F_{DE} act through *E* and therefore have no moment arm with respect to *E*. Thus the moment equilibrium equation will have only one unknown, F_{DG}. Notice that the moment center does not have to be on the portion of the truss we are analyzing.

$$\sum M_{z@E} = 0(\curvearrowleft +)$$
$$-10\text{ kip}(18\text{ ft}) - F_{DG}(8\text{ ft}) = 0$$
$$F_{DG} = -10\text{ kip}\left(\frac{18\text{ ft}}{8\text{ ft}}\right)$$
$$F_{DG} = -22.5\text{ kip (compression)} \quad (1)$$

Based on (7.5B):

$$\sum F_y = 0(\uparrow +)$$
$$10\text{ kip} - F_{DE}\sin\theta = 0$$

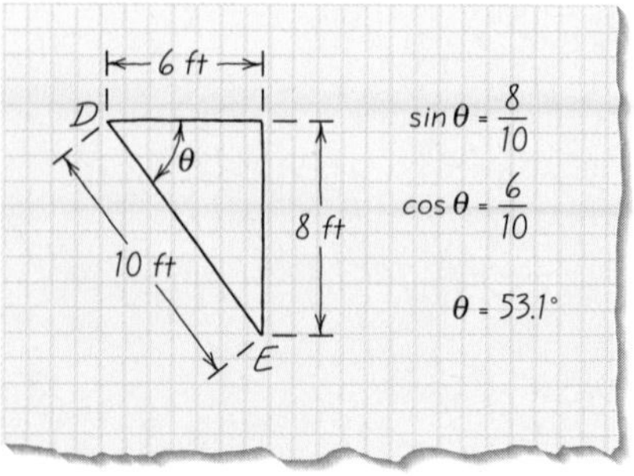

Figure 9.90

We determine the angle θ from the geometry in **Figure 9.90** to be 53.1°.

$$10\text{ kip} - F_{DE}\sin 53.1° = 0$$
$$F_{DE} = 12.5\text{ kip (tension)} \quad (2)$$

Answer to (a) The magnitude of the forces supported by members *DE* and *DG*:
$\| \mathbf{F}_{DE} \| = 12.5$ kip. Member *DE* is in tension.
$\| \mathbf{F}_{DG} \| = 22.5$ kip. Member *DG* is in compression.

Using the method of sections, we were able to determine the forces acting on DE and DG by solving two equilibrium equations. We would have analyzed eight equations to obtain the same result using the method of joints.

Check **(a)** To check our result, we can draw a free-body diagram of the portion of the truss to the right of the cut and apply our equilibrium equations. To do this we would first need to find the loads at supports H and J.

(b) Because the magnitude of the force acting on member DG is greater than the allowable load of 20 kip, the magnitude of the force at B must be reduced. We base the reduction on the ratio

$$\frac{F_{DG}}{F_B} = \frac{22.5 \text{ kip}}{\sqrt{(10 \text{ kip})^2 + (10 \text{ kip})^2}} = \frac{F_{DG,\text{allowable}}}{F_{B,\text{max}}} = \frac{20 \text{ kip}}{F_{B,\text{max}}} \tag{3}$$

Solving (3) for $F_{B,\text{max}} = 12.57$ kip.

Answer to (b) To eliminate the possibility of damage to member DG, the magnitude of the force acting at B should be reduced to 12.57 kip.

Our recommendation was to reduce all applied force at B to 12.57 kip. This would mean that a horizontal force of 8.89 kip and a vertical force of 8.89 kip would act at B. Why did we not consider the 10 kip force acting at E?

EXAMPLE 9.18 METHOD OF SECTIONS (B)

A stiffening truss for a bridge deck is shown in **Figure 9.91**. Use the method of sections to find the forces supported by members DE, EM, MV, and UV.

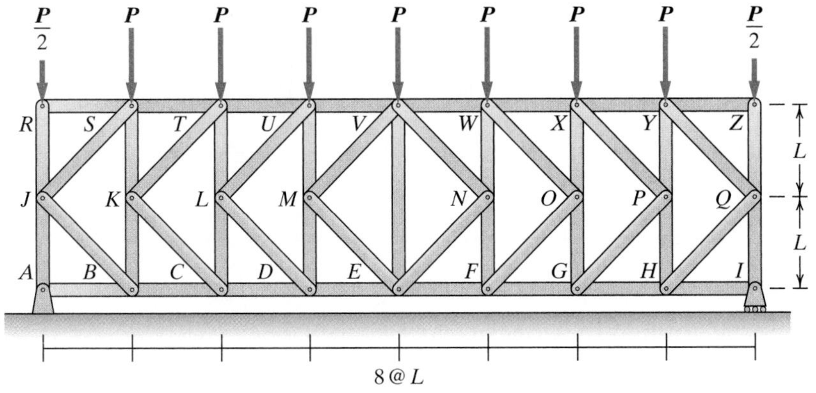

Figure 9.91

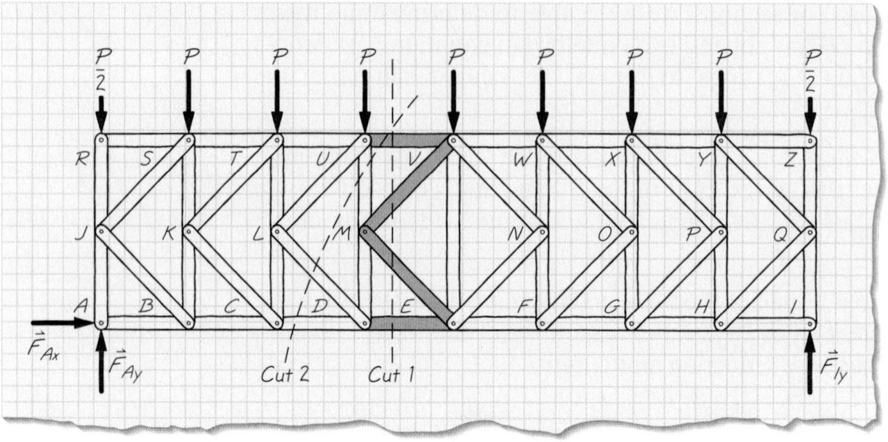

Figure 9.92 Free-body diagram of entire truss with members of interest shaded

Goal We are to find the forces supported by members *DE*, *EM*, *MV*, and *UV* using the method of sections.

Given We are given information about the geometry and loading of the truss.

Assume We assume that weight of the members can be neglected and that we can treat the system as planar.

Draw We draw a free-body diagram of the entire truss (**Figure 9.92**), which we use to determine the forces at the supports.

When we cut a section (Cut 1) through the members of interest (*DE*, *EM*, *MV*, and *UV*), we have cut four members with unknown forces. We can draw a free-body diagram of a section to the left or right of this cut. We choose to analyze the left portion (**Figure 9.93**). Unfortunately, no matter where we place our moment center, we have at least two unknowns in the moment equilibrium equation. Similarly, if we try to sum forces in the x or y direction, we have two or more unknowns in the equilibrium equation. Therefore, cut 1 by itself does not allow us to find the magnitude of any forces. Instead we make a cut through *CD*, *DL*, *MU*, and *UV* (Cut 2) and draw a free-body diagram of the left portion of

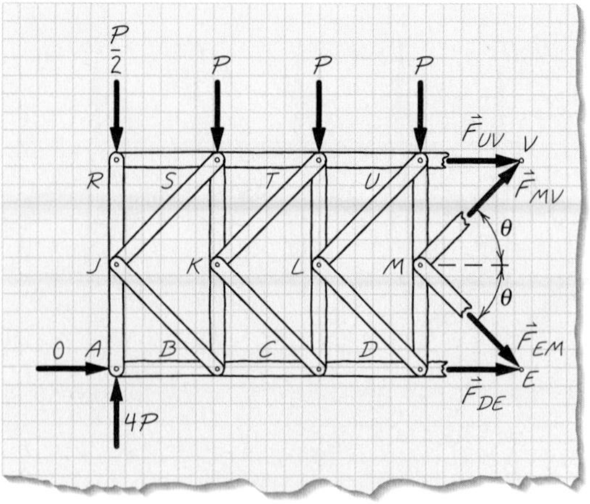

Figure 9.93 Free-body diagram of portion to the left of Cut 1

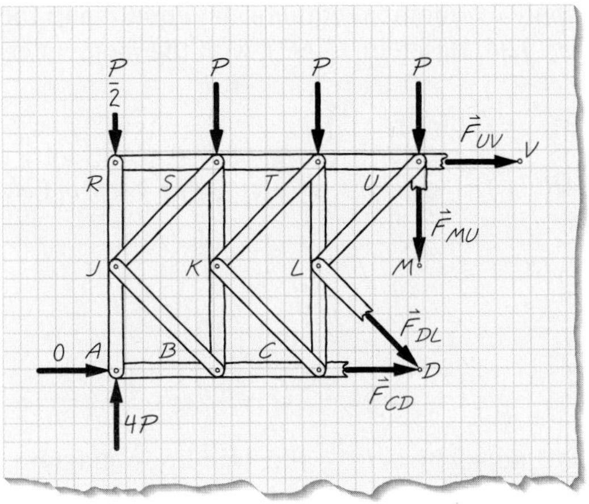

Figure 9.94 Free-body diagram of portion to the left of Cut 2

the truss (**Figure 9.94**). We see on this free-body diagram that although we have cut four members with unknown forces, three of the member forces are concurrent at D (F_{CD}, F_{DL}, and F_{MU}). Therefore we can use the moment equilibrium equation with a moment center at D to find F_{UV}. We then use the portion of the truss we isolated with Cut 1 to find the remaining three member forces.

Formulate Equations and Solve We start by finding the forces acting at the supports. None of the applied loads are acting horizontally; therefore F_{Ax} must be zero. The truss and the loading are symmetric, so we know that the vertical force at each support must be equal to one-half the applied load ($F_{Ay} = F_{Iy} = 4P$). To prove the two statements we made here, we apply our planar equilibrium equations to the free-body diagram in **Figure 9.92**.

Based on (7.5A):

$$\sum F_x = 0(\rightarrow +)$$

$$F_{Ax} = 0$$

Based on (7.5C), with the moment center at A:

$$\sum M_{z@A} = 0(\circlearrowleft +)$$

$$-P(L) - P(2L) - P(3L) - P(4L) - P(5L)$$

$$- P(6L) - P(7L) - \frac{P}{2}P(8L) + F_{Iy}(8L) = 0$$

$$F_{Iy}(8L) = 32PL$$

Therefore, $F_{1y} = 4P$.

Based on (7.5B):

$$\sum F_y = 0(\uparrow +)$$

$$F_{Ay} - 8P + F_{Iy} = 0$$

$$F_{Ay} - 8P + 4P = 0$$

$$F_{Ay} = 4P$$

Next we apply the moment equilibrium equation to the free-body diagram in **Figure 9.94**. Based on (7.5C), with the moment center at D:

$$\sum M_{z@D} = 0(\curvearrowleft +)$$

$$-4P(3L) + \frac{P}{2}(3L) + P(2L) + P(L) - F_{UV}(2L) = 0$$

$$F_{UV}(2L) = -\frac{15}{2}PL$$

$$F_{UV} = -\frac{15}{4}P \qquad (1)$$

Because we now know F_{UV}, the free-body diagram in **Figure 9.93** only has three unknown forces, which we can determine using the planar equilibrium equations. We start with a moment center at E because then we have only one unknown in the equation. Based on (7.5C):

$$\sum M_{z@E} = 0(\curvearrowleft +)$$

$$-4P(4L) + \frac{P}{2}(4L) + P(3L) + P(2L) + P(L)$$
$$- F_{MV}\cos\theta(L) - F_{MV}\sin\theta(L) - F_{UV}(2L) = 0$$

Substituting from (1) for F_{UV} and noting that θ is 45°:

$$-4P(4L) + \frac{P}{2}(4L) + P(3L) + P(2L) + P(L) - \frac{\sqrt{2}}{2}F_{MV}(L)$$
$$-\frac{\sqrt{2}}{2}F_{MV}(L) - \left(-\frac{15}{4}P\right)(2L) = 0$$

$$\sqrt{2}\,F_{MV}L = -\frac{P}{2}L$$

$$F_{MV} = -\frac{P}{2\sqrt{2}} \qquad (2)$$

Based on (7.5B):

$$\sum F_y = 0(\uparrow +)$$

$$4P - \frac{P}{2} - P - P - P + \frac{\sqrt{2}}{2}F_{MV} - F_{EM}\sin\theta = 0$$

Substituting from (2) for F_{MV} and noting that θ is 45°:

$$4P - \frac{P}{2} - P - P - P + \frac{\sqrt{2}}{2}\left(-\frac{P}{2\sqrt{2}}\right) - \frac{\sqrt{2}}{2}F_{EM} = 0$$

$$\frac{\sqrt{2}}{2}F_{EM} = \frac{P}{4}$$

$$F_{EM} = \frac{P}{2\sqrt{2}} \qquad (3)$$

Based on (7.5A):

$$\sum F_x = 0(\rightarrow +)$$
$$F_{DE} + F_{UV} + F_{MV} \cos\theta + F_{EM} \cos\theta = 0$$

Substituting from (1), (2) and (3):

$$F_{DE} + \left(-\frac{15}{4}P\right) + \frac{\sqrt{2}}{2}\left(-\frac{P}{2\sqrt{2}}\right) + \frac{\sqrt{2}}{2}\left(\frac{P}{2\sqrt{2}}\right) = 0$$

$$F_{DE} = \frac{15}{4}P$$

Answer $F_{DE} = \frac{15}{4}P$ (tension), $F_{EM} = \frac{P}{2\sqrt{2}}$ (tension),

$F_{MV} = -\frac{P}{2\sqrt{2}}$ (compression), $F_{UV} = -\frac{15}{4}P$ (compression)

Comment: **Figure 9.95** shows an exaggerated sketch of the deflected shape of the loaded truss. We expect the members that form the top chord (*RS, ST, TU, UV, VW, WX, XY, YZ*) to be in compression and the members that form the lower chord (*AB, BC, CD, DE, EF, FG, GH, HI*) to be in tension.

Check We can check our results by substituting the values we have found for the unknown member forces on the free-body diagram of the portion of the truss shown in **Figure 9.93**. We then solve the moment equilibrium equation with a different moment center such as point *T*. If the moments sum to zero, the member forces are correct.

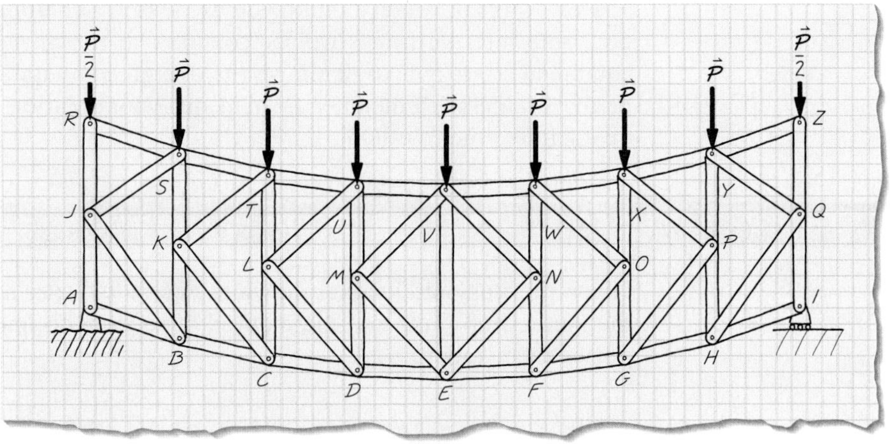

Figure 9.95

Combining Methods

It is possible to combine the method of joints and the method of sections. Some of the internal forces in a truss can be found with one method and other internal forces with the other method. For example, suppose we wish to find the force in a central member of a large planar

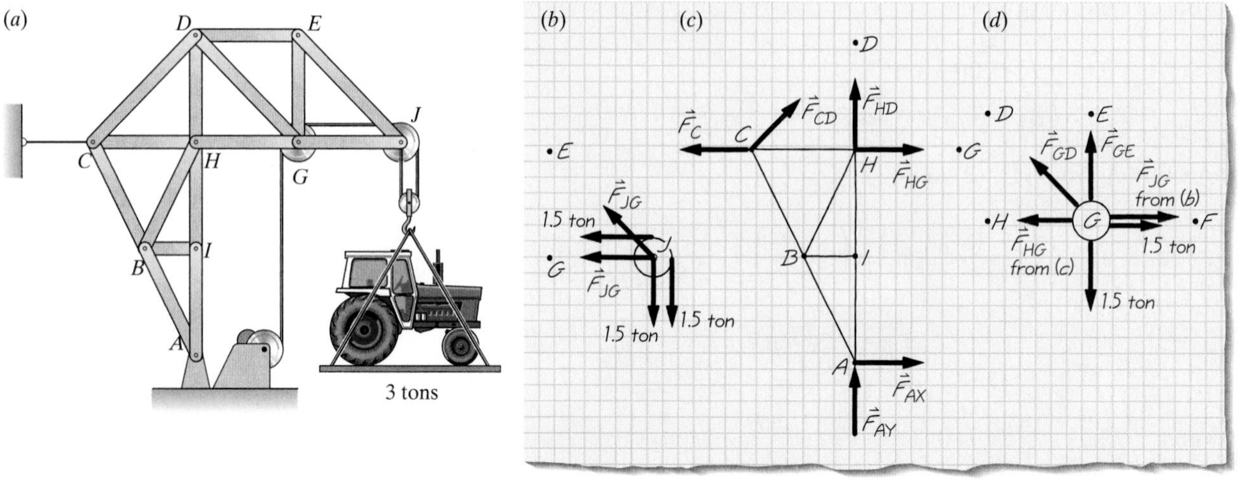

Figure 9.96 (*a*) If it is of interest to find the forces in members *CD*, *HD*, *GD*, *GE*, and *EJ*, begin by using equilibrium applied to truss as a whole to find loads acting at *C* (F_C) and *A* (F_{Ax}, F_{Ay}). (*b*) Use method of joints at pin *J* to define forces in members *JG* and *JE*, then (*c*) use the method of sections to find the force in members *CD*, *HD*, and *HG*. Finally, (*d*) use the method of joint at pin *G* to find the force in member *GD*.

truss, and it is not possible to pass the boundary through the member without passing through more than three two-force members that have unknown internal forces. Just such a situation is shown in **Figure 9.96**. It may be possible to determine the forces in nearby members by the method of joints and then proceed to the unknown member by the method of sections (or vice versa). Such a combination of the two methods may be more expedient (and less calculation intensive) than exclusive use of either method. This is illustrated in Example 9.19.

Regardless of which method you use first in a combination approach, the same internal forces will be found. The reason for considering a combined strategy is to minimize the overall computational complexity of the problem. Remember to map out your solution strategy before diving into either method.

EXAMPLE 9.19 COMBINATION OF METHOD OF JOINTS AND METHOD OF SECTIONS

Given the truss and loading shown in **Figure 9.97**, use a combination of the method of joints and the method of sections to find the force supported by member *DJ*.

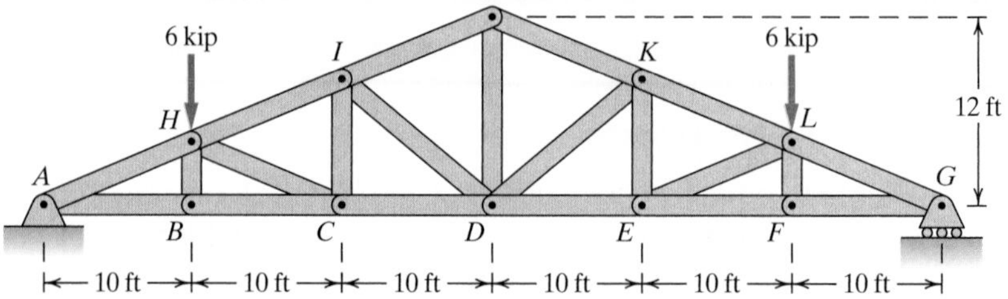

Figure 9.97

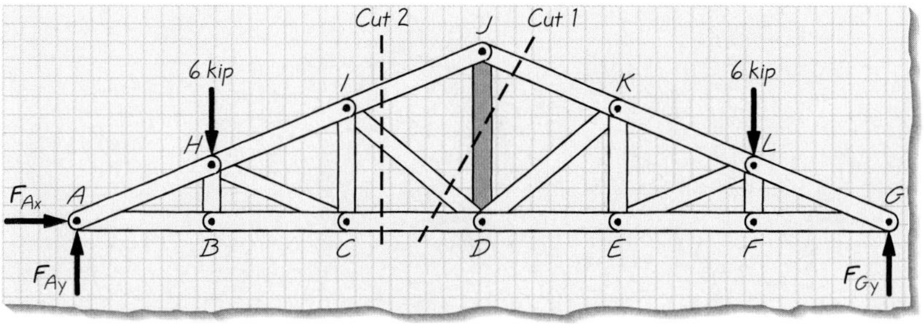

Figure 9.98 Free-body diagram of entire truss

Goal We are to find the force supported by *DJ* using a combination of the method of joints and the method of sections.

Given We are given information about the geometry and loading of the truss.

Assume We assume that the weight of the members can be neglected and that we can treat the system as planar.

Draw We first draw a free-body diagram of the entire truss in order to find the loads at support *A* (**Figure 9.98**). If we use the method of sections to solve for *DJ*, we find that we must cut at least four members (Cut 1) when we cut through *DJ* (**Figure 9.99**). Because there are four unknowns, the three equations of planar equilibrium are not sufficient to determine the forces acting on all four members. However, because three of the members are concurrent at *D*, we can sum the moments about *D* to find F_{JK}. This leaves only two unknowns at joint *J*. Therefore, we draw a free-body diagram of pin *J* (**Figure 9.100**) and use the method of joints to find the force acting on member *DJ*.

Alternatively, we could cut through *CD*, *ID*, and *IJ* (Cut 2 in **Figure 9.98**) and sum the moments about *D* to find the force acting on member *IJ* (**Figure 9.101**). We would then sum forces at joint *J* to find F_{DJ}.

Formulate Equations and Solve We begin with the free-body diagram in **Figure 9.98** and (7.5C) with moment center at *G*:

$$\sum M_{z@G} = 0(\curvearrowleft +)$$
$$6 \text{ kip}(10 \text{ ft}) + 6 \text{ kip}(50 \text{ ft}) - F_{Ay}(60 \text{ ft}) = 0$$
$$F_{Ay}(60 \text{ ft}) = 360 \text{ kip} \cdot \text{ft}$$
$$F_{Ay} = 6 \text{ kip}$$

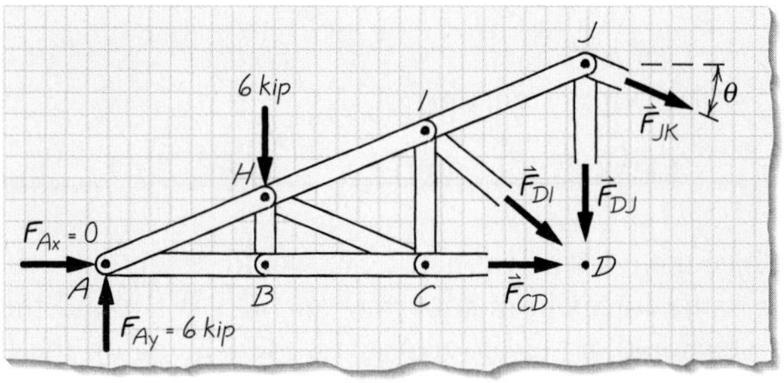

Figure 9.99 Free-body diagram of portion to the left of Cut 1

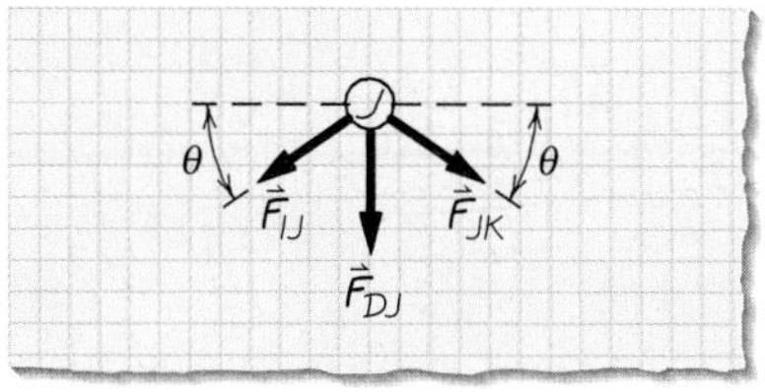

Figure 9.100 Free-body diagram of pin *J*

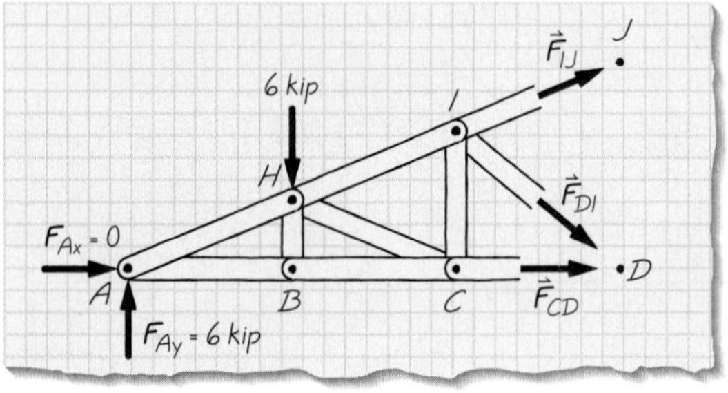

Figure 9.101 Free-body diagram of portion to the left of Cut 2

Based on (7.5A):

$$\sum F_x = 0(\rightarrow +)$$
$$F_{Ax} = 0$$

Now we use the method of sections for the portion of the truss to the left of Cut 1 to calculate F_{JK} (**Figure 9.99**). Based on (7.5C), with the moment center at D:

$$\sum M_{z@D} = 0(\curvearrowleft +)$$
$$-6\text{ kip}(30\text{ ft}) + 6\text{ kip}(20\text{ ft}) - F_{JK}\cos\theta(12\text{ ft}) = 0$$

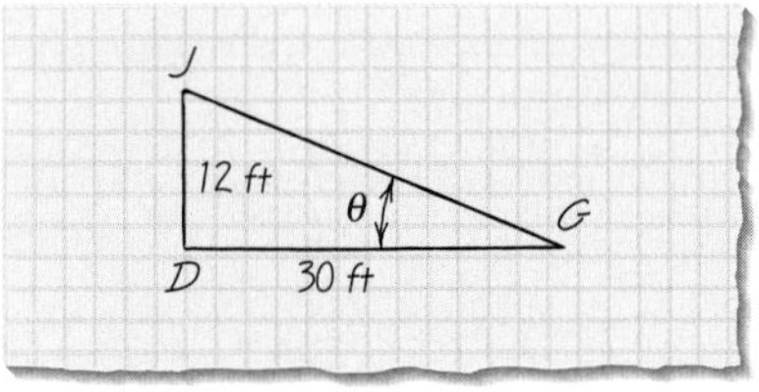

Figure 9.102

Based on **Figure 9.102** we find

$$\theta = \tan^{-1}\left(\frac{12\text{ ft}}{30\text{ ft}}\right) = 21.8°$$

then

$$F_{JK} = \frac{-6\text{ kip}(10\text{ ft})}{\cos(21.8°)(12\text{ ft})} = -5.39\text{ kip} \quad \text{(compression)} \tag{1}$$

Using the method of joints at joint D, we solve for F_{DJ} (**Figure 9.100**): Based on (7.5A) and having found that $F_{JK} = -5.39$ kip in (1):

$$\sum F_x = 0(\rightarrow +)$$
$$-F_{IJ}\cos\theta - (5.39\text{ kip})\cos\theta = 0$$
$$F_{IJ} = -5.39\text{ kip} \quad \text{(compression)} \tag{2}$$

Based on (7.5B):

$$\sum F_y = 0(\uparrow +)$$
$$-F_{IJ}\sin\theta - F_{JK}\sin\theta - F_{DJ} = 0$$

With F_{JK} given in (1), F_{IJ} given in (2), and $\theta = 21.8°$, we find

$$-(-5.39\text{ kip})\sin 21.8° - (-5.39\text{ kip})\sin 21.8° - F_{DJ} = 0$$
$$F_{DJ} = 4.00\text{ kip} \quad \text{(tension)}$$

Answer $F_{DJ} = 4.00$ kip (tension)

Check Since the truss is symmetric, we would expect our calculations to result in the same forces in members *IJ* and *JK*. We found this to be the case. Also, we should anticipate that given the loading on the truss, members *IJ* and *JK* should be in compression. This was also found to be the case.

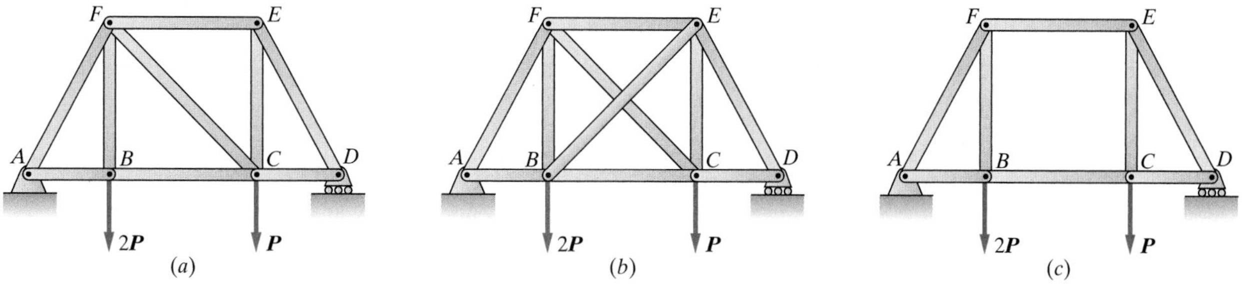

Figure 9.103 (*a*) Internally stable truss without redundancy that is also statically determinate; (*b*) internally stable truss with redundancy that is also statically determinate; (*c*) internally unstable truss

The solution to Example 9.19 involved writing six equilibrium equations; we were able to get the solution down to so few equations by combining the method of joints and the method of sections, and by judicious selection of the moment center. Compare this solution with one using the method of joints requiring 12 equations. The conclusion? Plan your strategy before undertaking the calculations.

The method of joints and the method of sections work only for trusses that are both externally statically determinate and internally stable without redundancy. A truss that is internally stable without redundancy has just enough two-force members to be stable (**Figure 9.103*a***). A truss with more members than are necessary to be stable is labeled internally stable with redundancy—concepts from mechanics of materials in addition to mechanical equilibrium conditions are required to determine the internal loads for the system (**Figure 9.103*b***). With any fewer members, the truss would collapse and is labeled internally unstable—concepts from dynamics are required to describe the internal loads and motion of the system (**Figure 9.103*c***).

Checks to determine whether a truss is internally stable without redundancy, internally stable with redundancy, or internally unstable are outlined in **Box 9.1** for planar trusses and **Box 9.2** for space trusses.

Box 9.1: Check for Internal Stability in a Planar Truss with Three Support Forces

Count the number of two-force members (m) and the number of joints (j). The quantity $(2j - 3)$ represents the number of linearly independent equations remaining after three equations have been used to enforce equilibrium on the truss as a whole. Check to see which inequality or equality listed below holds.

Number of two-force members: $m = 2j - 3$ (9.5A)

The truss is **internally stable without redundancy** if all two-force members contribute to configuration stability. This is illustrated in **Figure 9.103*a***. If the truss is internally stable without redundancy, the equilibrium equations are sufficient to solve for internal forces. All trusses that are internally stable without redundancy obey the equality $m = 2j - 3$; however, not all trusses that obey this equality are internally stable without redundancy (therefore the equilibrium condition $m = 2j - 3$ is referred to as a *necessary but not sufficient* equilibrium condition).

Number of two-force members: $m > 2j - 3$ (9.5B)

The truss is **internally stable with redundancy**. There is an excess of two-force members, and you cannot solve for internal forces with equilibrium conditions alone (**Figure 9.103*b***).

Number of two-force member: $m < 2j - 3$ (9.5C)

There is a deficiency of two-force members, and the truss is **internally unstable**. The truss may be part of an engineering device called a mechanism, and concepts from dynamics are required to solve for the internal forces (**Figure 9.103*c***). A structure that is internally unstable is, by definition, also underconstrained.

Box 9.2: Check for Internal Stability in a Space Truss with Six Support Forces

Count the number of two-force members (m) and the number of joints (j). The quantity ($3j - 6$) represents the number of linearly independent equations remaining after six equations have been used to enforce equilibrium on the truss as a whole. Therefore, check to see which inequality or equality listed below holds.

Number of two-force members: $m = 3j - 6$ (9.6A)

The truss is **internally stable without redundancy** if all two-force members contribute to configuration stability. If the truss is internally stable without redundancy, the equilibrium equations are sufficient to solve for internal forces. All trusses that are internally stable without redundancy obey the equality $m + 6 = 3j$; however, not all trusses that obey this equality are internally stable without redundancy (therefore the equilibrium condition $m + 6 = 3j$ is referred to as a *necessary but not sufficient* equilibrium condition).

Number of two-force members: $m > 3j - 6$ (9.6B)

The truss is **internally stable with redundancy**. There is an excess of two-force members, and you cannot solve for internal forces with equilibrium conditions alone.

Number of two-force members: $m < 3j - 6$ (9.6C)

There is a deficiency of two-force members and the truss is **internally unstable**. The truss may be part of an engineering device called a mechanism, and concepts from dynamics are required to solve for the internal forces. A structure that is internally unstable is, by definition, also underconstrained.

EXAMPLE 9.20 CHECKING THE STATUS OF PLANAR TRUSSES

By using **Box 9.1** and counting the number of joints (pins), members, and equations, determine whether each system is internally stable without redundancy, internally stable with redundancy, or internally unstable. Also, describe why the truss is externally statically determinate, statically indeterminate, or underconstrained.

Truss 1 (Figure 9.104*a*)

Number of joints $= j = 4$
Number of two-force members $= m = 5$
Number of equations $= 2j - 3 = 5$

Because the number of two-force members (five) matches exactly the number of linearly independent equations (five), the truss is *internally stable without redundancy* (9.5A). We could have also reached this conclusion by noting that it is constructed of a series of rigid triangles and has no more members than are needed to make it stable. Furthermore, the truss is externally *statically determinate* because the number of planar equilibrium equations (three) matches exactly the number of unknown boundary support (three).

Truss 2 (Figure 9.104*b*) It might appear that the truss as a whole is statically determinate (three support loads and three equilibrium equations). Before reaching that conclusion, we use the checks in **Box 9.1** to find:

Number of joints $= j = 4$
Number of two-force members $= m = 4$
Number of equations $= 2j - 3 = 5$

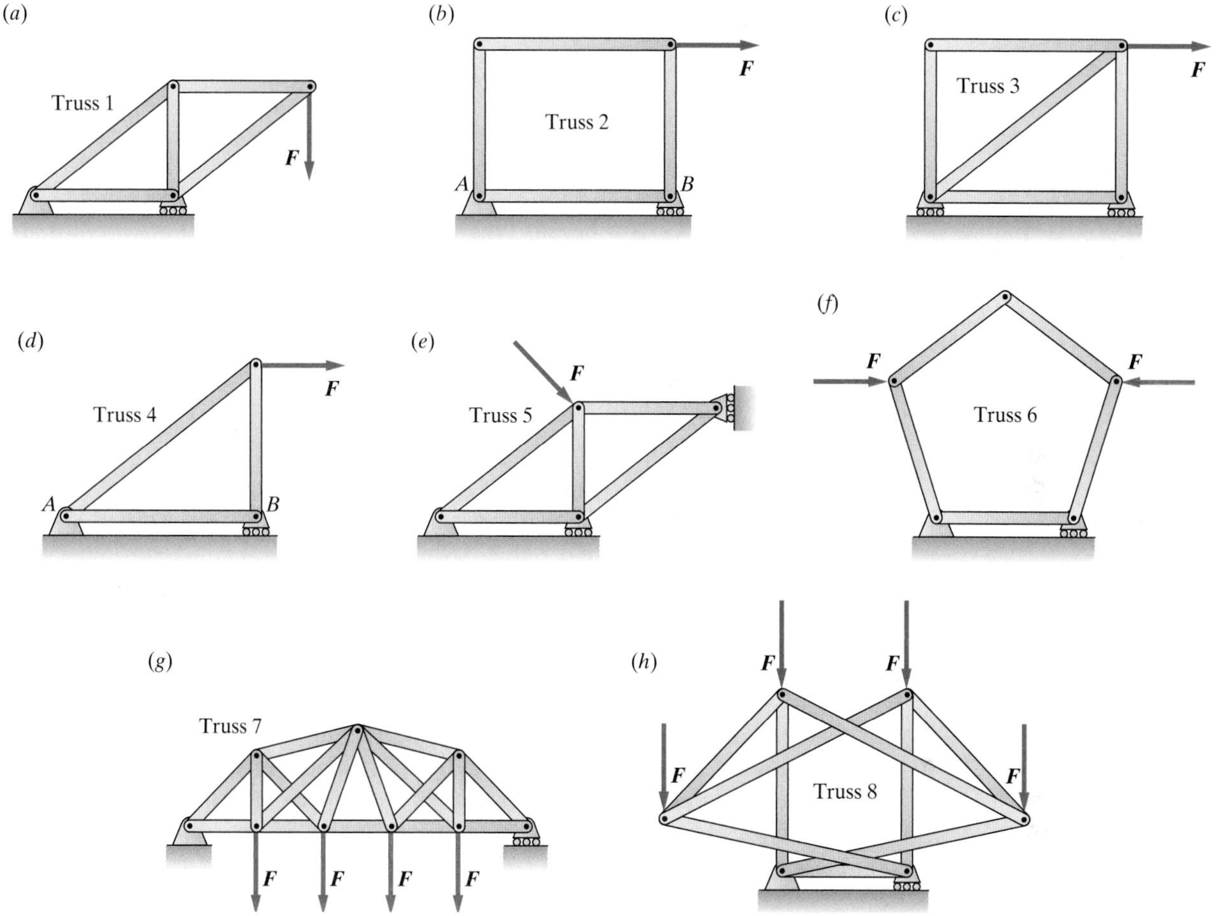

Figure 9.104

According to **Box 9.1**, since the number of equations (five) is greater than the number of members (four), this truss is *internally unstable* (9.5C). There are fewer members than there are equations of equilibrium. The three boundary connection forces at A and B are insufficient to form a rigid structure; therefore, we also say that the truss is *underconstrained*. The addition of a diagonal member would make the truss internally stable without redundancy and statically determinate. Alternatively, the truss would become statically determinate if an additional boundary condition was added; what ideas do you have for this?

Truss 3 (Figure 9.104*c*) We first note that for the truss as a whole, there are three equilibrium equations and only two boundary loads; per the discussion in Chapter 7, this means that this truss is *underconstrained*. An additional boundary condition would be needed for the truss to be statically determinate. Because it is *underconstrained*, we cannot use the conditions of equilibrium to find the loads acting at the supports.

Now to check to see if the truss is internally stable or unstable we use the checks in **Box 9.1**:

Number of joints = $j = 4$

Number of two-force members = $m = 5$

Number of equations = $2j - 3 = 5$

Since the number of two-force members (five) equals the number of remaining linearly independent equations, the truss is *internally stable without redundancy*; it has just enough members to form a rigid structure (9.5A). But because the truss is underconstrained, we are not able to use the conditions of equilibrium to find the forces in each member.

Truss 4 (Figure 9.104*d*)

Number of joints = $j = 3$

Number of two-force members = $m = 3$

Number of equations = $2j - 3 = 3$

Per **Box 9.1**, the number of members (three) equals the number of linearly independent equations (three); therefore, the three members form a rigid structure that is *internally stable without redundancy* (9.5A). For the truss as a whole, there are three boundary loads and three equilibrium equations; therefore the truss is statically determinate *if* it is also internally stable. Because the truss is internally stable, we can therefore conclude that it is also externally *statically determinate.*

Truss 5 (Figure 9.104*e*) On inspecting the truss as a whole, note that there are four boundary support loads; this might mean that the truss is statically indeterminate *if* the truss is internally stable. To check to see if the truss is internally stable or unstable, use the checks in **Box 9.1**:

Number of joints = $j = 4$

Number of two-force members = $m = 5$

Number of equations = $2j - 3 = 5$

We find that the number of members (five) equals the number of linearly independent equations; therefore the truss is *internally stable without redundancy* (9.5A). Because the truss is internally stable, the presence of four boundary loads acting on the truss as a whole means that it is also *statically indeterminate.*

Truss 6 (Figure 9.104*f*) For the truss as a whole, there are three boundary loads and three equilibrium equations; therefore the truss would be statically determinate *if* it was also internally stable. Per **Box 9.1**:

Number of joints = $j = 5$

Number of two-force members = $m = 5$

Number of equations = $2j - 3 = 7$

Because the number of two-force members (five) is less than the number of equations (seven), the truss is *internally unstable* (9.5C). Because the truss is internally unstable, it is also classified as *underconstrained*, meaning that there are not enough boundary supports to prevent the truss from moving.

Truss 7 (Figure 9.104*g*)

Number of joints = $j = 9$
Number of two-force members = $m = 17$
Number of equations = $2j - 3 = 15$

The truss is *internally stable with redundancy* and *statically determinate* because the number of unknowns is greater than the number of equations (9.5B), and the truss is constructed of rigid triangles.

Truss 8 (Figure 9.104*h*)

Number of joints = $j = 6$
Number of two-force members = $m = 9$
Number of equations = $2j - 3 = 9$

The truss is *internally stable without redundancy* and *statically determinate* because the number of unknowns equals the number of equations (9.5A). Imagine the challenge of constructing this truss with its overlapping members.

EXAMPLE 9.21 STATUS OF SPACE TRUSSES

Use **Box 9.2** to identify each spatial truss as being internally stable without redundancy, internally stable with redundancy, or internally unstable. Also, describe why the truss is statically determinate, statically indeterminate, or underconstrained.

Truss 1 (Figure 9.105*a*)

Number of joints = $j = 4$
Number of two-force members = $m = 6$
Number of equations = $3j - 6 = 6$

The truss is *internally stable without redundancy* as the number of members (six) equals the number of equations (six) (9.6A). Also, it is *statically determinate* because the number of equilibrium equations matches exactly the number of unknown supports.

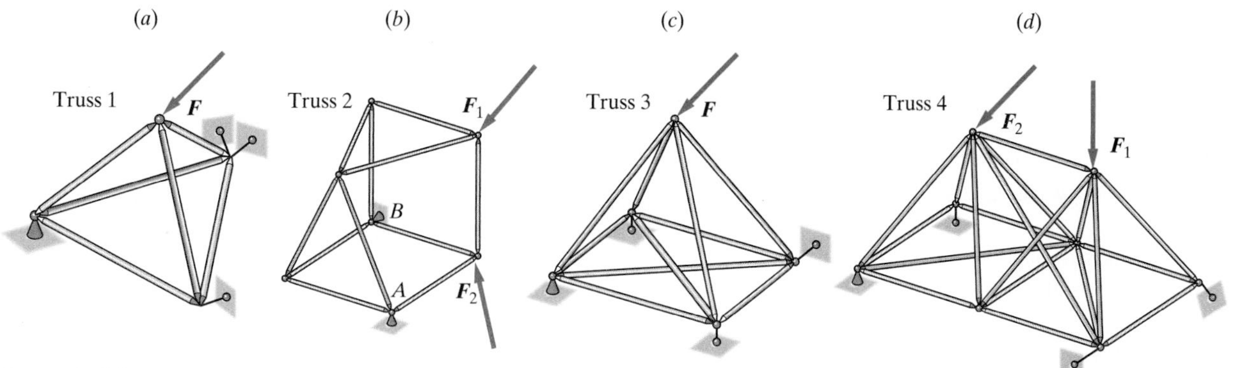

Figure 9.105

Truss 2 (Figure 9.105*b*)

Number of joints = $j = 7$
Number of two-force members = $m = 11$
Number of equations = $3j - 6 = 15$

The truss is *internally unstable* because the forces $\boldsymbol{F}_1$ and $\boldsymbol{F}_2$ would cause the truss to deform and collapse; we see this in that the number of members (eleven) is less than the number of equations (fifteen) (9.6C). The addition of four appropriately placed bars would make the truss internally stable. *If* the truss were internally stable (which it is not), the six boundary supports would result in a structure that is externally statically determinate. But because it is internally unstable, we also conclude that it is *underconstrained*; the six boundary connection forces at A and B are insufficient to prevent the truss from moving.

Truss 3 (Figure 9.105*c*)

Number of joints = $j = 5$
Number of two-force members = $m = 10$
Number of equations = $3j - 6 = 9$

The truss is *internally stable with redundancy* because we have ten members and only nine equations (9.6B). Using the six equations of nonplanar equilibrium, we can, however, solve for the six loads at the supports; therefore we conclude that this truss is externally *statically determinate.*

Truss 4 (Figure 9.105*d*)

Number of joints = $j = 8$
Number of two-force members = $m = 19$
Number of equations = $3j - 6 = 18$

The truss is *internally stable with redundancy* because we have one more member than equilibrium equations (19 vs. 18, 9.6B). Furthermore, using the six equations of nonplanar equilibrium, we can solve for the six loads at the supports; therefore we also conclude that the structure is externally *statically determinate.*

EXERCISES 9.3

Unless otherwise stated, assume that the trusses below are in equilibrium and are of negligible weight.

9.3.1. Classify each of the structures in **E9.3.1** as a frame or a truss. Include your reasoning.

9.3.2. Classify each of the structures in **E9.3.2** as a frame or a truss. Include your reasoning.

9.3.3. Inspect the truss in **E9.3.3.**

a. List those members you think are in tension, those you think are in compression, and those for which you are unsure whether they are in tension or compression.

b. Determine the force in each member and state whether each member is in tension or compression.

c. Note how your list on tension/compression members in **a** differs from your findings in **b**.

d. If the magnitude of the tension or compression in any member is not to exceed 400 N, what is the maximum allowable magnitude of the force $\boldsymbol{F}$?

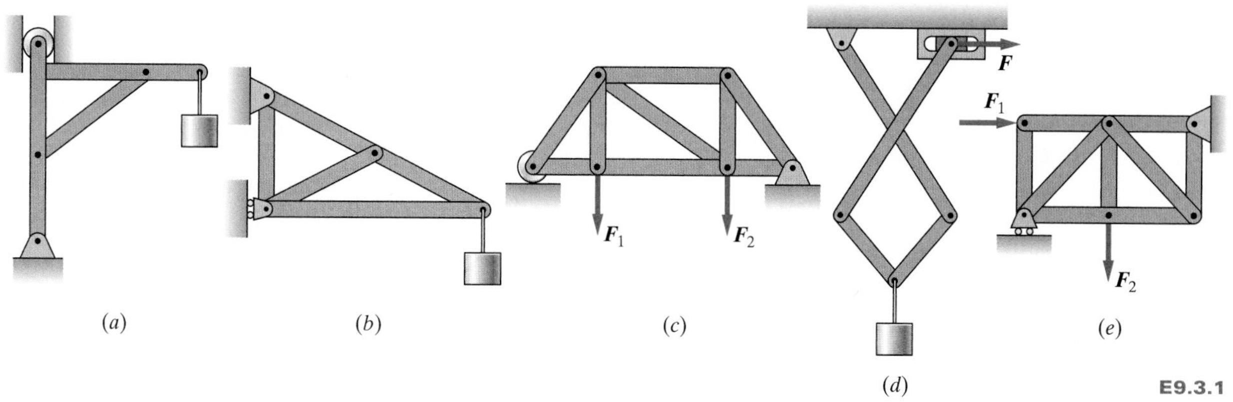

E9.3.1

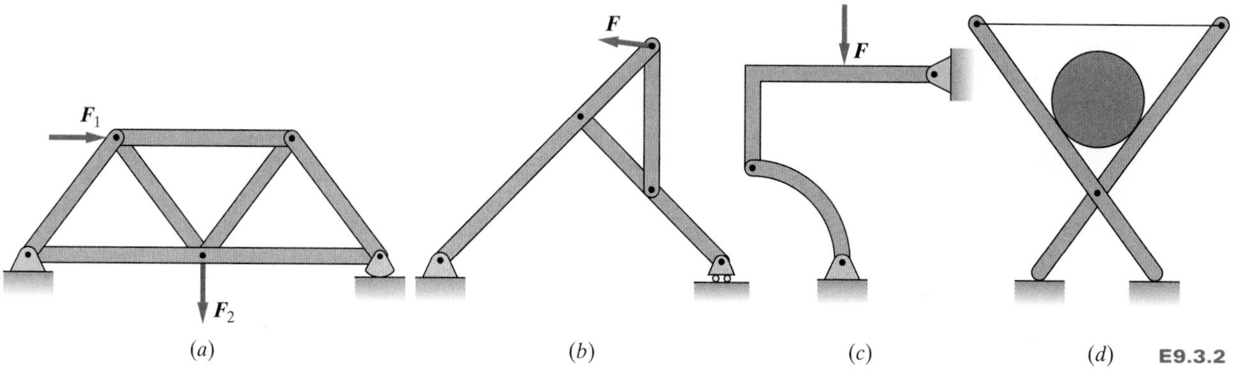

E9.3.2

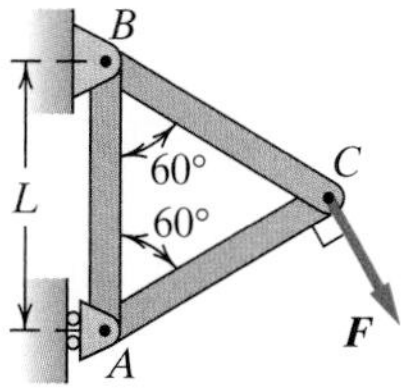

E9.3.3

9.3.4. Determine the force in each member of the truss shown in E9.3.4. State whether each member is in tension or compression.

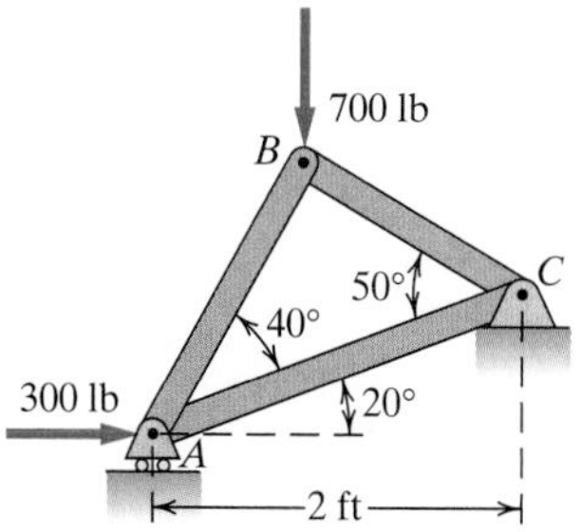

E9.3.4

9.3.5. Inspect the truss in E9.3.5.

a. List those members you think are in tension, those you think are in compression, and those for which you are unsure whether they are in tension or compression.

b. Determine the force in each member and state whether each member is in tension or compression.

c. Note where your list on tension/compression members in **a** differs from your findings in **b**.

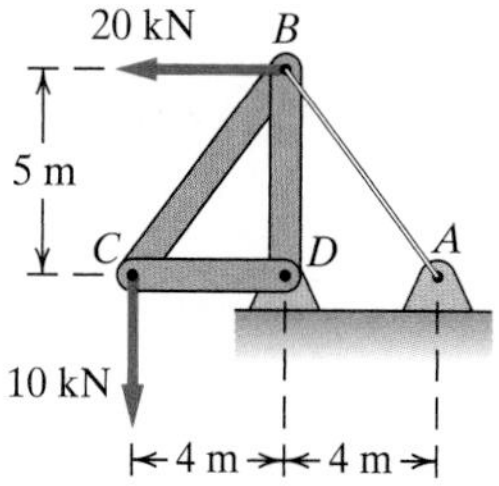

E9.3.5

9.3.6. Inspect the truss in E9.3.6.

a. List those members you think are in tension, those you think are in compression, and those for which you are unsure whether they are in tension or compression.

b. Determine the force in each member and state whether each member is in tension or compression.

c. Note where your list on tension/compression members in **a** differs from your findings in **b**.

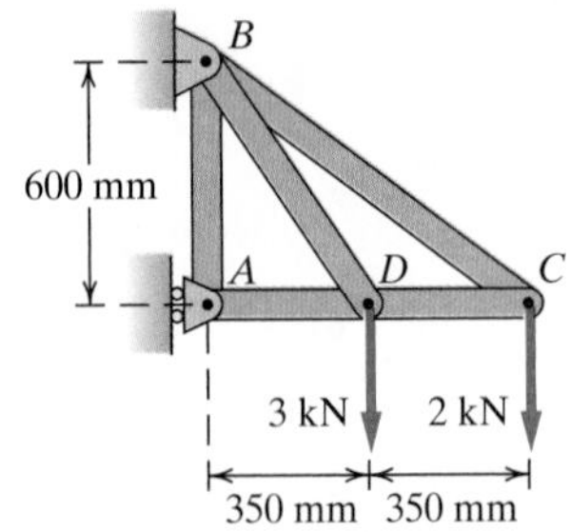

E9.3.6

9.3.7. Inspect the truss in E9.3.7.

a. List those members you think are in tension, those you think are in compression, and those for which you are unsure whether they are in tension or compression.

b. Determine the force in each member and state whether each member is in tension or compression.

c. Note where your list on tension/compression members in **a** differs from your findings in **b**.

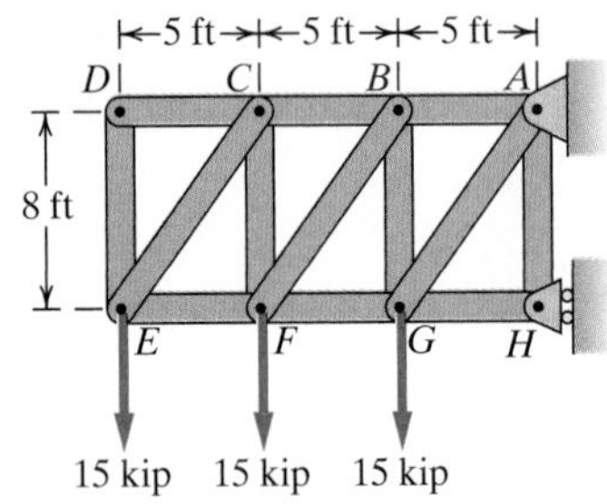

E9.3.7

9.3.8. Inspect the truss in E9.3.8.

a. List those members you think are in tension, those you think are in compression, and those for which you are unsure whether they are in tension or compression.

b. Determine the force in each member and state whether each member is in tension or compression.

c. Note where your list on tension/compression members in **a** differs from your findings in **b**.

d. If the magnitude of the tension or compression in any member is not to exceed 2 kN, what is the largest mass m that can hang at D?

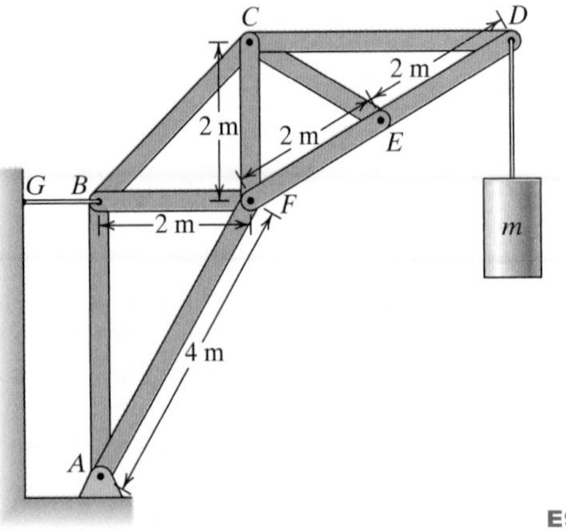

E9.3.8

9.3.9 Inspect the truss in E9.3.9.

a. List those members you think are in tension, those you think are in compression, and those for which you are unsure whether they are in tension or compression.

b. Determine the force in each member and state whether each member is in tension or compression.

c. Note where your list on tension/compression members in **a** differs from your findings in **b**.

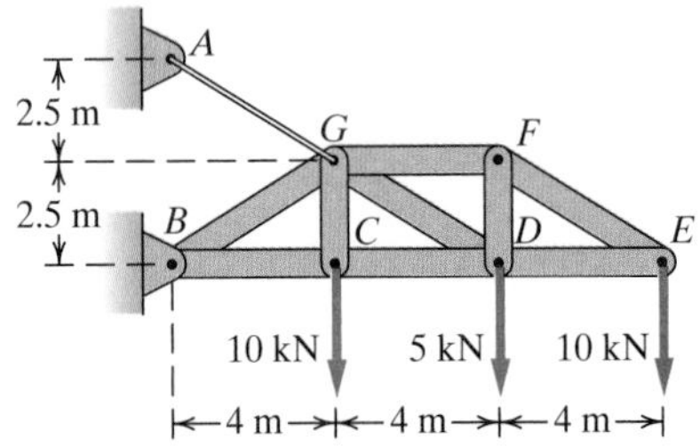

E9.3.9

9.3.10. Determine the force in each member of the truss shown in E9.3.10. State whether each member is in tension or compression.

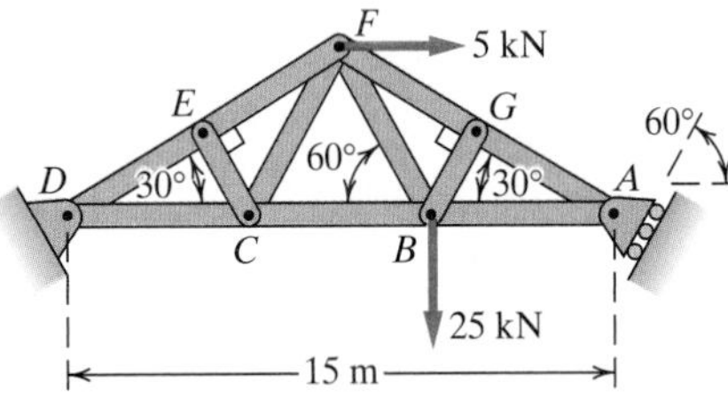

E9.3.10

9.3.11 Inspect the truss in E9.3.11.

a. List those members you think are in tension, those you think are in compression, and those for which you are unsure of whether they are in tension or compression.

b. Determine the force in each member and state whether each member is in tension or compression.

c. Note where your list on tension/compression members in **a** differs from your findings in **b**.

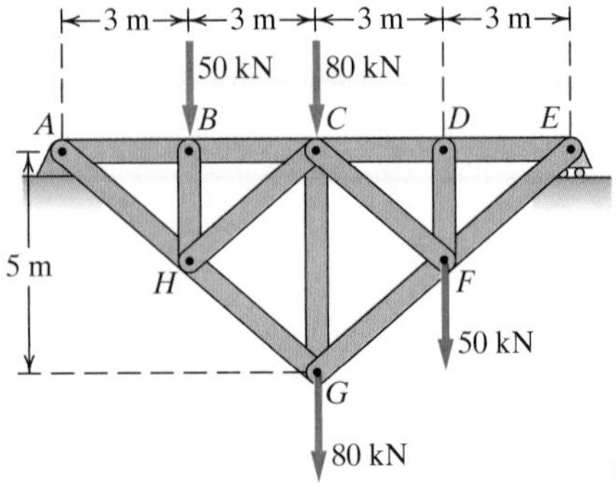

E9.3.11

9.3.12. Determine the force in each member of the truss shown in **E9.3.12**. State whether each member is in tension or compression.

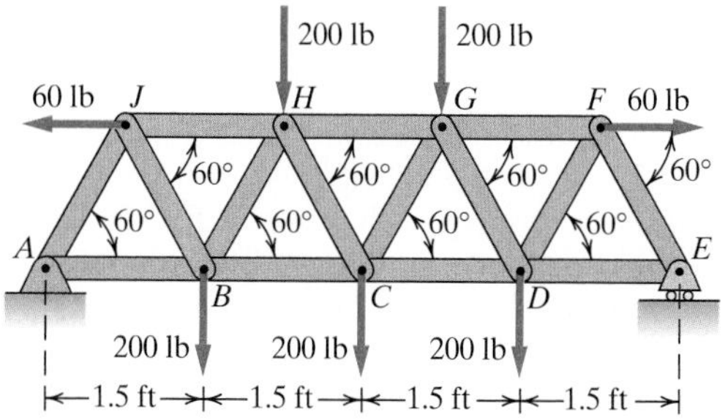

E9.3.12

9.3.13. Determine the force in each member of the truss shown in **E9.3.13**. State whether each member is in tension or compression.

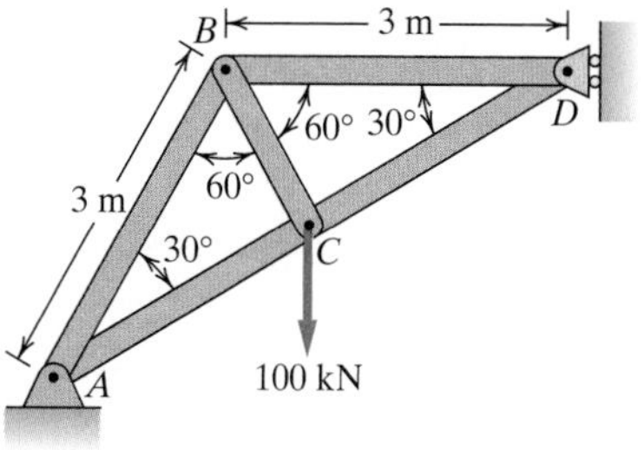

E9.3.13

9.3.14. Consider the truss shown in **E9.3.14**.

a. Determine the force in each member. State whether each member is in tension or compression.

b. If the maximum tension in any member is to be limited to 1000 N and the maximum compression to 500 N, what is the largest allowable magnitude of the force ***F***?

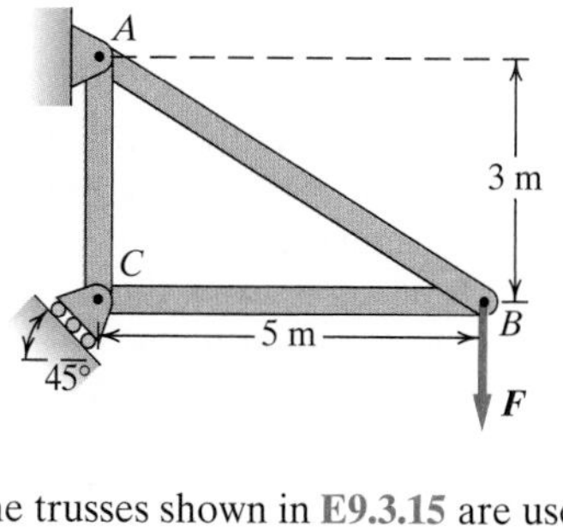

E9.3.14

9.3.15. The trusses shown in **E9.3.15** are used as part of a footbridge structure (the Forth Bridge in Scotland, built in 1890, uses a similar configuration of trusses).

a. Determine the force in each of the five members *AB*, *AC*, *BC*, *BD*, and *CD* of the truss *ABCD* shown in **E9.3.15**. State whether each member is in tension or compression.

b. If the magnitude of ***F*** is 10 kN, what are the values of the forces in members *AB*, *AC*, *BC*, *BD*, and *CD*?

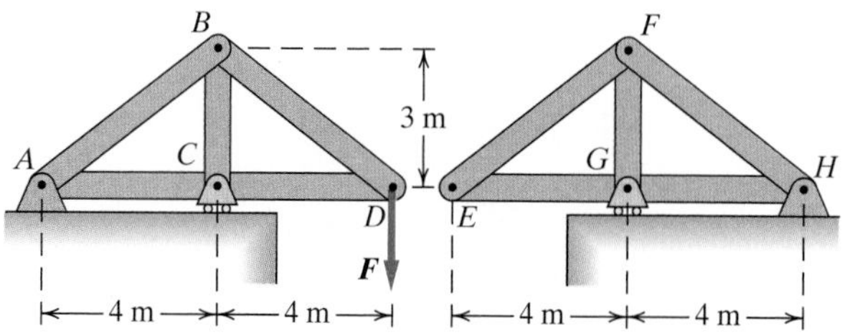

E9.3.15

9.3.16. The asymmetric simple truss is loaded as shown in **E9.3.16**.

a. Determine the loads acting on the truss at *A* and *D*.

b. Determine the forces in all members, clearly stating whether they are in tension or compression.

c. If the maximum tension in any member is to be limited to 2000 lb and the maximum compression to 1500 N, what is the largest allowable magnitude of the force ***L***?

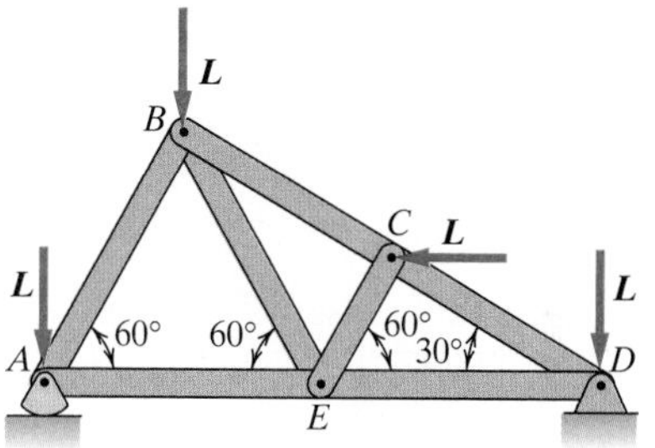

E9.3.16

9.3.17. A mass of 200 kg hangs from the truss, as shown in **E9.3.17**.

a. Find the loads acting on the truss at *A* and *C*, and find the force in each member.

b. Which member(s) might buckle? Which members could be replaced with cables?

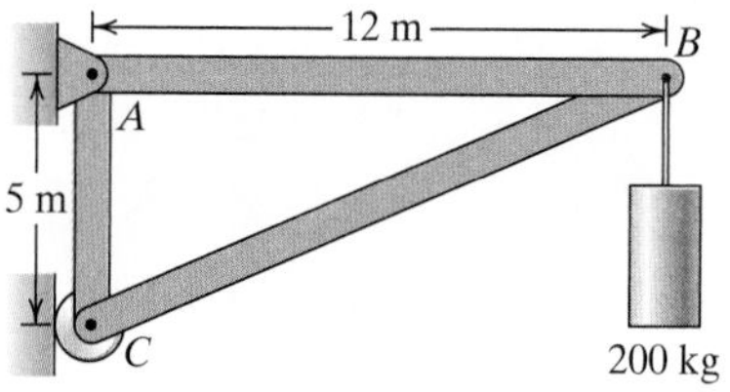

E9.3.17

9.3.18. A 3-kN force is applied to a truss, as shown in **E9.3.18**.

a. Find the loads acting on the truss at *A* and *D*, and find the force in each member.

b. Which member(s) might buckle? Which member(s) could be replaced with cables?

c. Are there any zero-force members?

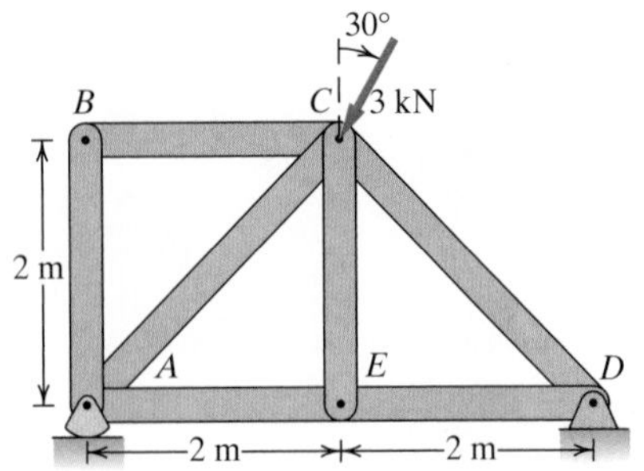

E9.3.18

9.3.19 Forces are applied to a truss, as shown in **E9.3.19**.

a. Find the loads acting on the truss at *D* and *E*, and find the force in each member.

b. Which member(s) might buckle?

c. Which member(s) could be replaced with cables?

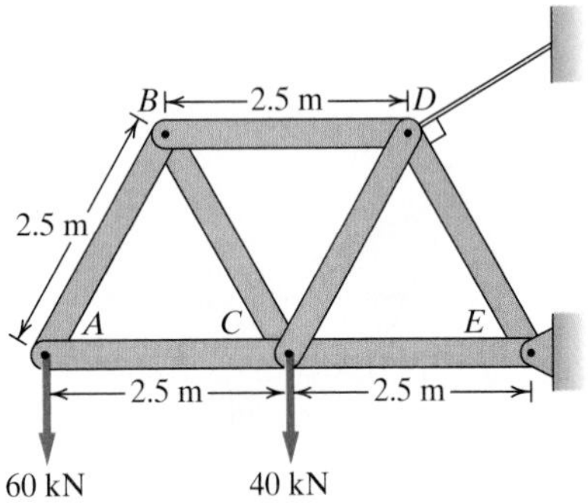

E9.3.19

9.3.20. Forces are applied to a truss, as shown in **E9.3.20**.

a. Find the loads acting on the truss at *A* and *D*, and find the force in each member.

b. Which member(s) might buckle? Which member(s) could be replaced with cables?

c. Are there any zero-force members?

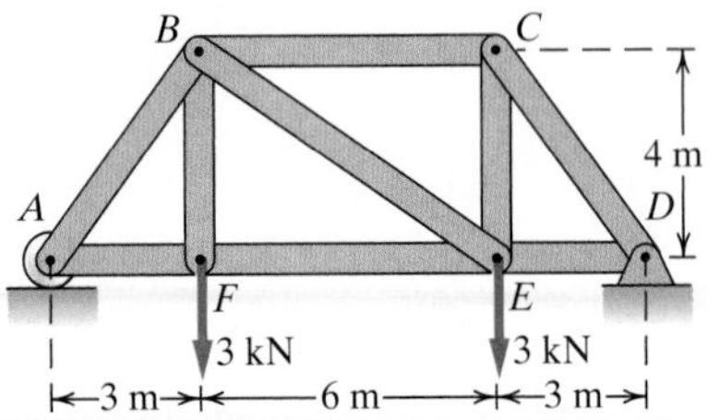

E9.3.20

9.3.21. Repeat **Exercise 9.3.17**, except include the weight of the members, which is 38 N/m.

9.3.22. Repeat **Exercise 9.3.18**, except include the weight of the members, which is 1000 N/m.

9.3.23. Repeat **Exercise 9.3.6**, except include the mass of the members, which is 400 kg/m.

9.3.24. Repeat **Exercise 9.3.13**, except include the mass of the members, which is 700 kg/m.

9.3.25. Consider the truss in **E9.3.25**. The magnitudes of $\boldsymbol{F}_1$ and $\boldsymbol{F}_2$ are 40 kN and 0 kN, respectively.

a. Based on inspecting the truss in **E9.3.25**, list those members you think are in tension, those you think are in compression, and those for which you are unsure whether they are in tension or compression.

b. Determine the force in each member. State whether each member is in tension or compression.

c. Note where your list on tension/compression members in **a** differs from your findings in **b**.

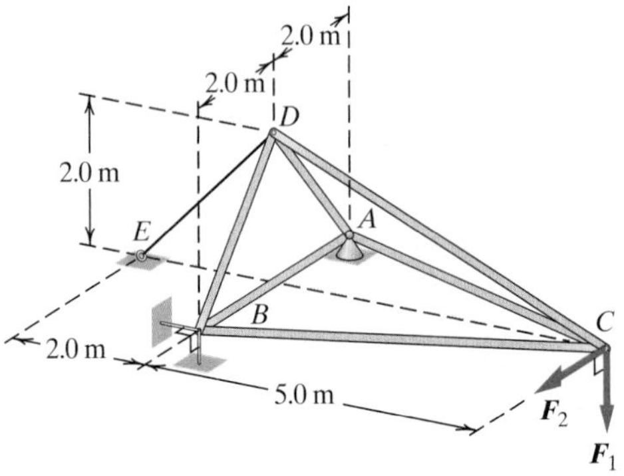

E9.3.25

9.3.26. Determine the force in each member of the space truss in **E9.3.25** if the magnitudes of $\boldsymbol{F}_1$ and $\boldsymbol{F}_2$ are 40 kN and 30 kN, respectively. State whether each member is in tension or compression.

9.3.27. Consider the truss in **E9.3.27**. The magnitudes of $\boldsymbol{F}_1$ and $\boldsymbol{F}_2$ are 8 kip and 0 kip, respectively.

a. Based on inspecting the truss in **E9.3.27**, list those members you think are in tension, those you think are in compression, and those for which you are unsure whether they are in tension or compression.

b. Determine the force in each member. State whether each member is in tension or compression.

c. Note where your list on tension/compression members in **a** differs from your findings in **b**.

9.3.28. Determine the force in each member of the space truss in **E9.3.27** if the magnitudes of $\boldsymbol{F}_1$ and $\boldsymbol{F}_2$ are 8 kip and 4 kip, respectively. State whether each member is in tension or compression.

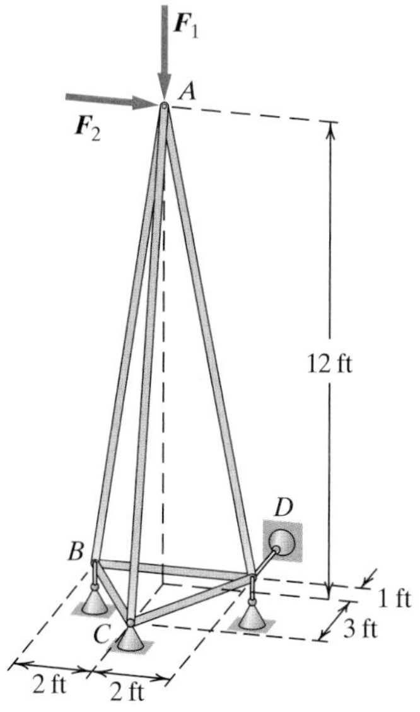

E9.3.27

9.3.29. Determine the forces in members *BC*, *BE*, and *EF* of the truss in E9.3.29 using the method of sections.

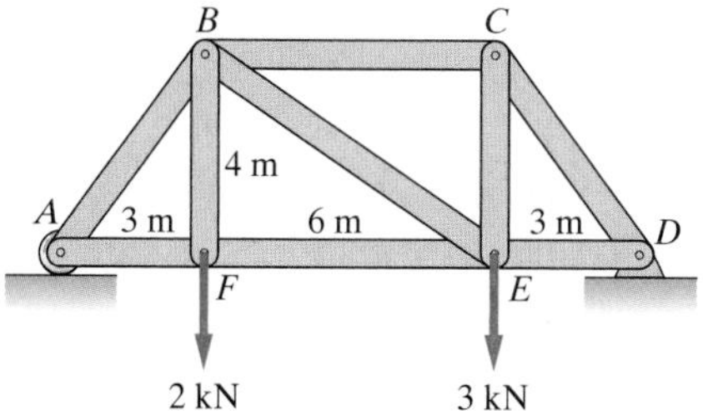

E9.3.29

9.3.30. Determine the force in member *BE* of the loaded truss in E9.3.30 using the method of sections.

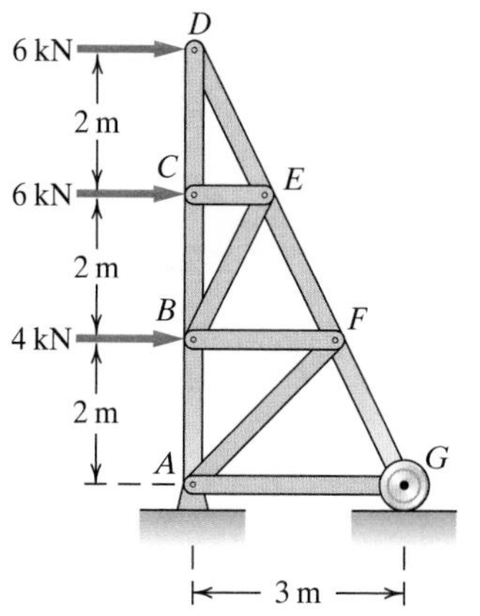

E9.3.30

9.3.31. Determine the force in member *BF* of the loaded truss in **E9.3.30** using the method of sections.

9.3.32. Consider the truss in E9.3.32.

a. Determine the force in member *DG* in terms of the load *L* using the method of sections. All internal angles are 60°.

b. If the magnitude of the force in member *DG* must be equal to or less than 1 kN, what is the largest allowable value (magnitude) of ***L***?

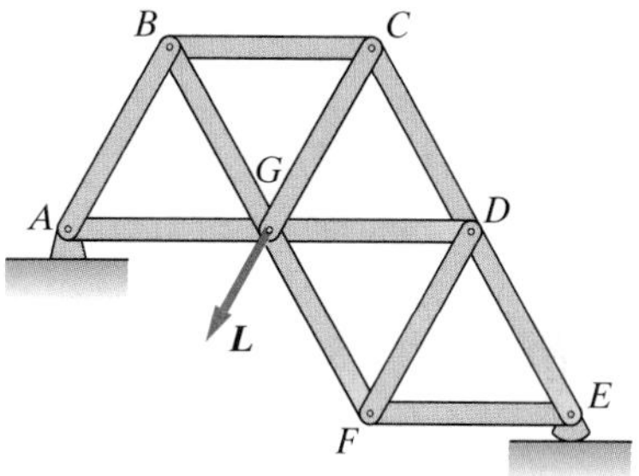

E9.3.32

9.3.33. Determine the forces in members *AB*, *BG*, and *FG* of the truss shown in E9.3.33 using the method of sections.

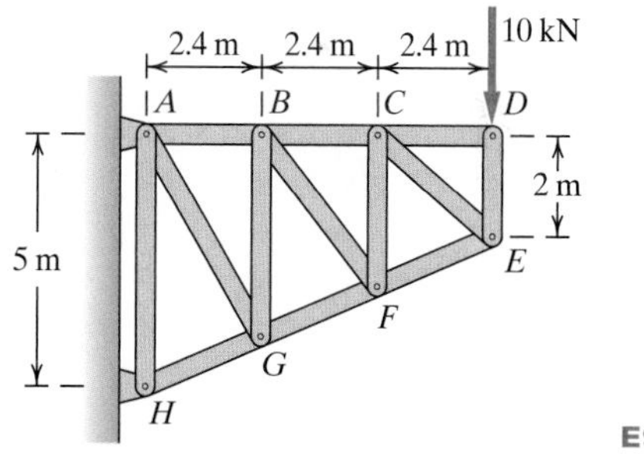

E9.3.33

9.3.34. The Warren truss supports a walkway between two buildings, as shown in E9.3.34. The walkway exerts vertical 12,000-lb loads at *B*, *D*, *F*, and *H*. Model the support at *A* as a roller connection and the support at *J* as a pin connection. Determine the forces in members *BC*, *CD*, and *CE*.

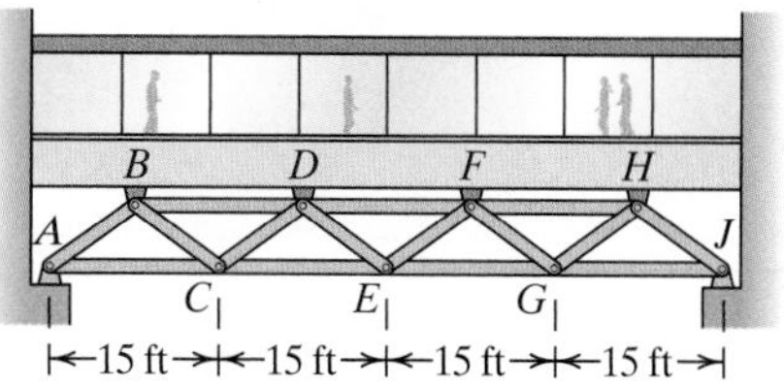

E9.3.34

9.3.35. For the walkway shown in **E9.3.34**, determine the forces in members *BD*, *DF*, *AC*, and *CE*.

9.3.36. Determine the forces in member CE of the loaded truss in **E9.3.36**.

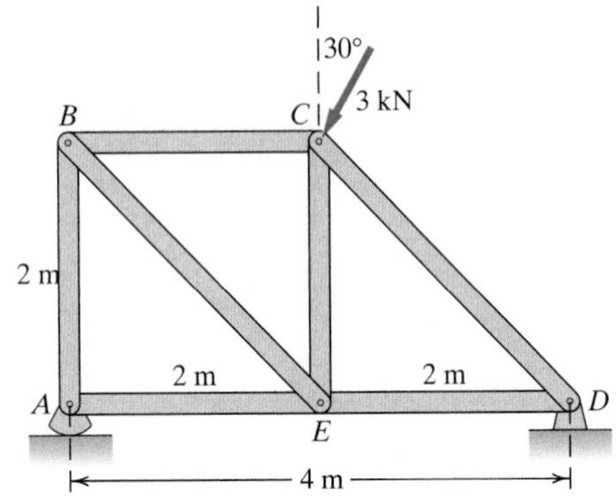

E9.3.36

9.3.37. Determine the forces in member BE of the loaded truss in **E9.3.37**.

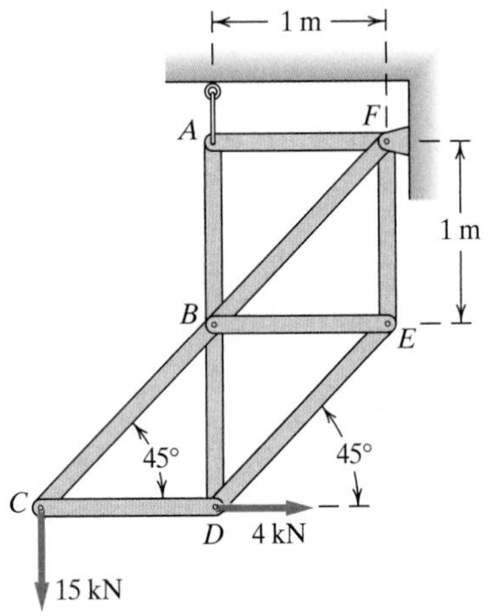

E9.3.37

9.3.38. Determine the forces in member CG of the truss shown in **E9.3.38**.

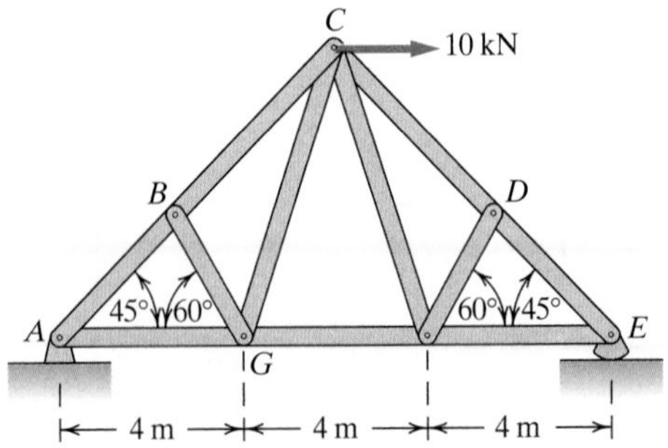

E9.3.38

9.3.39. The truss in **E9.3.39** is composed of equilateral triangles of sides b and is loaded and supported as shown. Determine the forces in members EF, DE, and DF.

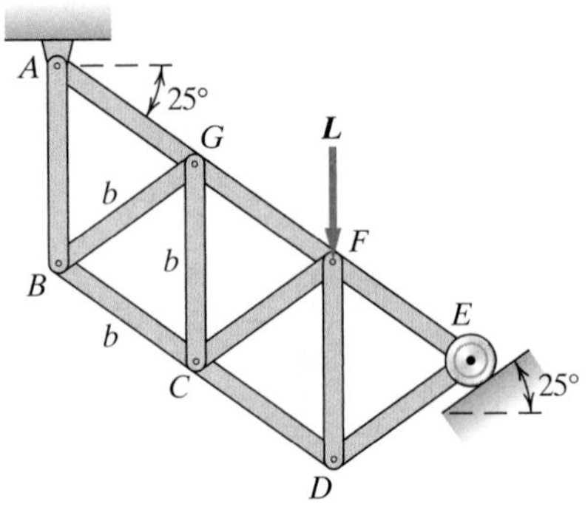

E9.3.39

9.3.40. Determine the forces in members DK and DL of the truss shown in **E9.3.40**.

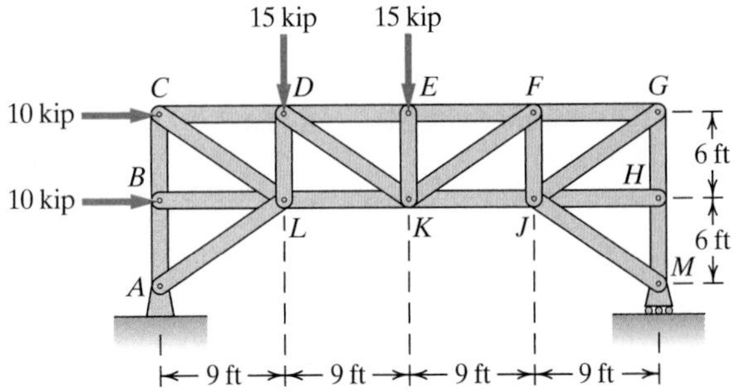

E9.3.40

9.3.41. Determine the forces in members EF and QR of the truss shown in **E9.3.41**.

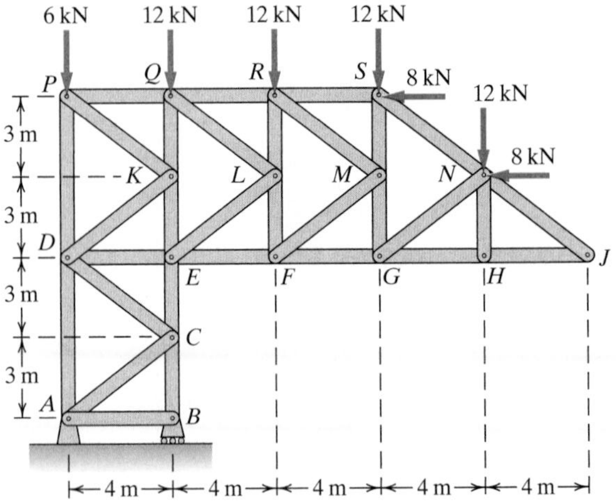

E9.3.41

9.3.42. Determine the forces in members EL and LM of the truss shown in **E9.3.42**.

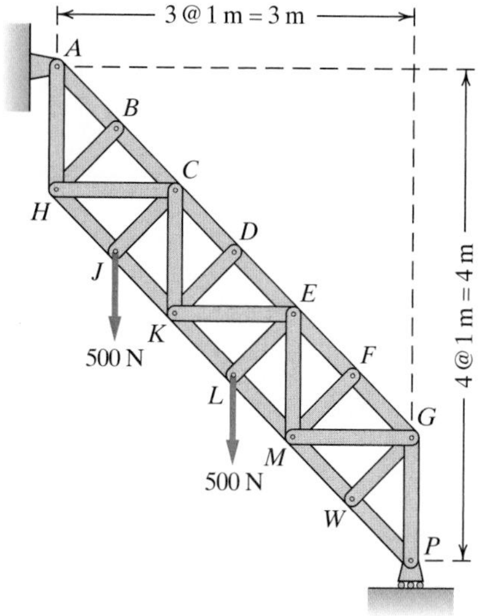

E9.3.42

9.3.43. Determine the force in member *JK* of the truss shown in E9.3.43.

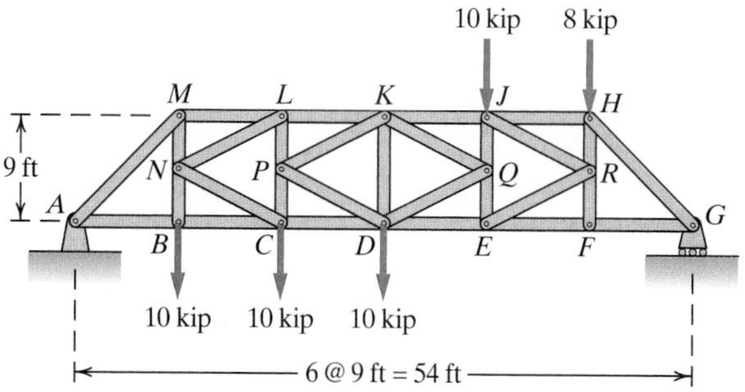

E9.3.43

9.3.44. Determine the forces in members *BC* and *JK* of the truss shown in E9.3.44.

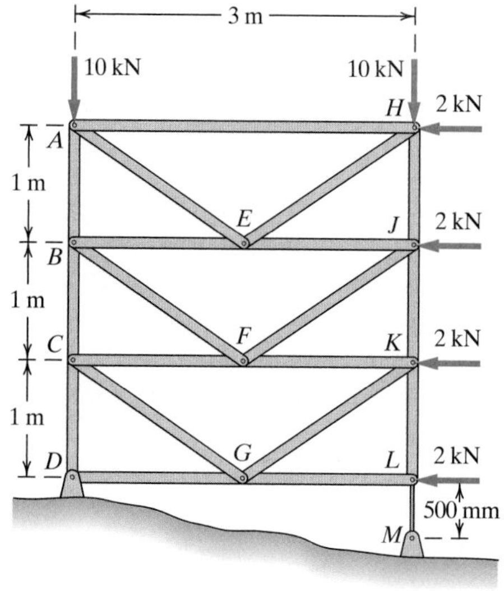

E9.3.44

9.3.45. Determine the forces in members *DC* and *FG* of the truss shown in E9.3.45.

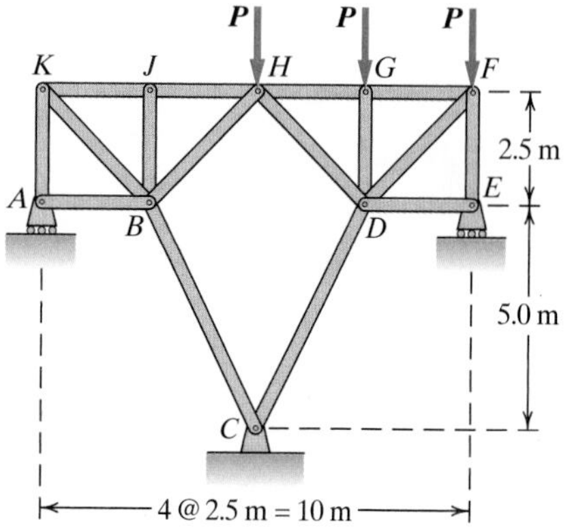

E9.3.45

9.3.46. Using the method of sections, determine the force in each member of the loaded truss in **E9.3.36**.

9.3.47. Using the method of sections, determine the force in each member of the loaded truss in **E9.3.37**.

9.3.48. The base of an automobile jack stand in E9.3.48*a* forms an equilateral triangle of side length 250 mm and is centered under the collar *A*. Model the structure with ball and sockets at all joints.

a. Determine the forces in members *BC*, *BD*, and *CD*. Neglect any horizontal reaction components under the feet *B*, *C*, and *D*.

b. In addition to the 2-kN vertical force, a horizontal force (F_{horiz}) acts at a height of 500 mm, as shown in **E9.3.48*b***. At what magnitude will $\mathbf{F}_{horiz}$ cause the jack stand to tip over?

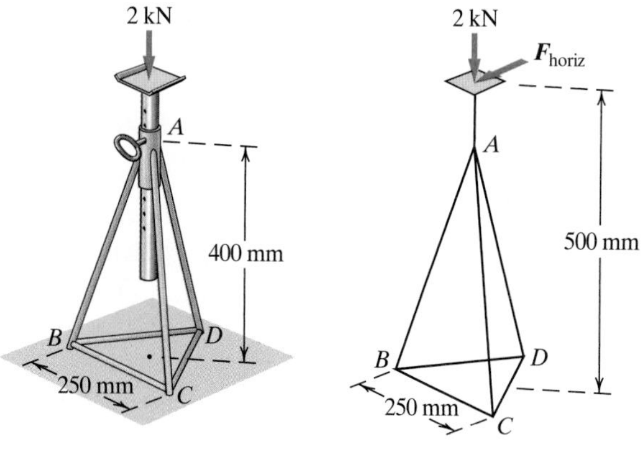

(*a*) (*b*)

E9.3.48

9.3.49. The long boom of an overhead construction crane, a portion of which is shown in E9.3.49, is an example of a periodic structure, where identical structural units are repeated. Determine the forces in members *FK* and *GK*.

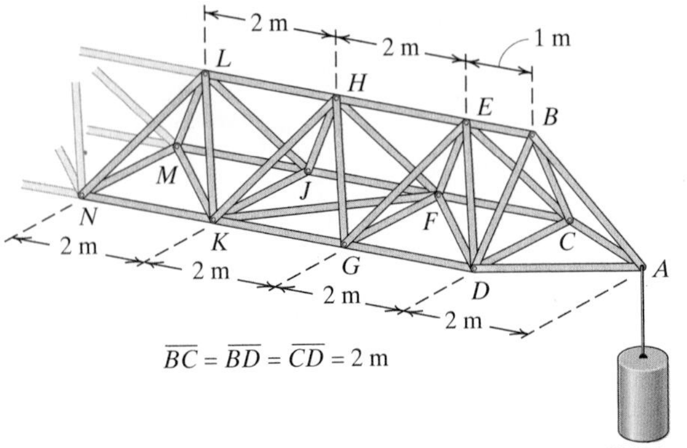

E9.3.49

9.3.50 Consider the boom shown in **E9.3.49.**

a. Determine the forces in members *HL*, *EH*, *EG*, *KN*, *GK*, *DG*, *JM*, *FJ*, and *CF*. Note which members are in tension and which are in compression.

b. Is there a pattern of compression and tension in your answers in **a**? If so, explain why the pattern does or does not make intuitive sense.

9.3.51. Select the terms that describe each of the structures shown in E9.3.51: statically determinate, statically indeterminate, underconstrained, internally stable without redundancy, internally stable with redundancy, internally unstable. Also note any zero-force members in a sketch.

9.3.52. Why does it make sense to classify a bicycle frame as a frame and not as a truss? How would the design need to change to be classified as a truss? What would be the major disadvantage of a truss structure in this situation? Are there any advantages?

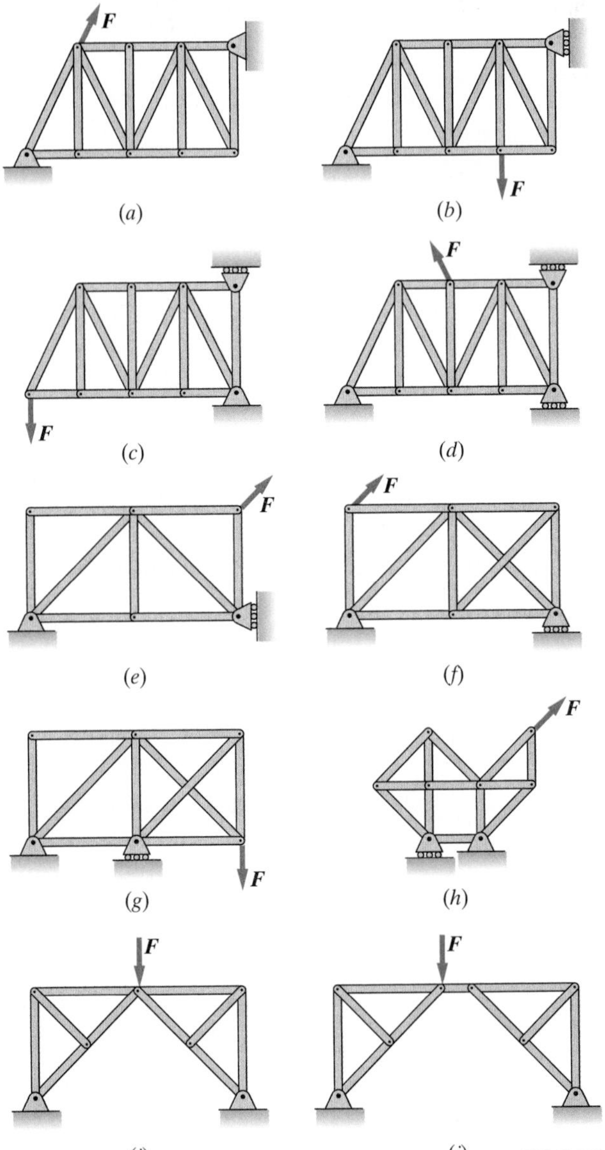

E9.3.51

9.4 JUST THE FACTS

In this chapter we laid out how to find the loads acting on and internal to three types of systems that are in mechanical equilibrium: frames, machines, and trusses.

A **frame** is a system designed to support loads, both forces and moments. It may be made up of a few members or many, but at least one member must be a **multiforce member**, which is any member that is not a two-force member. In addition, frames are classified as being either **internally stable** or **internally unstable**, and as **planar** or **nonplanar.**

The analysis of frames to find the loads acting on their boundaries and internal to the frame (between members) consists of making assumptions, drawing free-body diagrams, setting-up and solving equilibrium equations, and summarizing results. This is exactly the analysis procedure we have developed in the preceding chapters, and it works for frames that are statically determinate and **internally stable without redundancy**. For a frame that is statically determinate and **internally stable with redundancy** (meaning there are more members than are necessary for the frame to be rigid), we must use concepts from mechanics of materials in addition to equilibrium conditions.

A **machine** is a system designed to change the direction and/or magnitude of loads or motion. In considering machines, we often think in terms of the load *into* the system, the load *out of* the system, and the ratio of output to input. The basic analysis of a machine in equilibrium is identical to that of a frame in terms of analyzing separate members by drawing free-body-diagrams, setting up the equilibrium equations, and then using these equations to find the unknown loads. Often the analysis of a machine also involves the calculation of the ratio of output load to input load (referred to as the **mechanical advantage**) or output motion to input motion.

When machines contain shafts that are held by dry or only partially lubricated journal bearings, the analysis may need to account for the presence of **journal bearing friction**. For a shaft of radius r, the friction creates a moment about a moment center at the center of the shaft of

$$M_O = r \cdot \mu F \tag{9.2}$$

in the direction *opposite* the rotation of the shaft. The force F in (9.2) is the radial force being carried by the bearing. For a shaft that is stationary relative to a journal bearing, the coefficient of static friction (μ_s) should be used in (9.2). If the shaft is rotating relative to the journal bearing, the coefficient of kinetic friction (μ_k) should be used in (9.2).

The analysis of machines that contain wheels may need to account for the presence of **rolling resistance**. Rolling resistance is caused by deformation of the rolling wheel and/or the surface on which it rolls and results in an additional force P being required to keep the wheel rolling at a constant speed. The additional force is

$$P = (a/r)L \tag{9.4}$$

where L is the gravity force acting on the rolling wheel. The ratio of (a/r) is referred to as the **coefficient of rolling friction** (f_r).

A **truss** is a structural system, made up exclusively of two-force members, that is generally lightweight compared to the loads it can support.

Trusses

1. consist exclusively of straight members,
2. have joints that can be represented as pin connections, and
3. carry external forces exclusively at joints.

A **planar truss** is one in which all forces and members lie in a single plane (or it is reasonable to assume that they all lie in a single plane). The basic building block of a planar truss is a triangle, composed of three two-force members connected by pins. The triangle is the simplest **rigid structure** that can be created with two-force members; by rigid we mean that the structure is internally stable. If a truss is not planar, it is a **space truss**. The basic building block of a space truss is a tetrahedron composed of two-force members.

The two procedures for finding the forces carried by two-force members in a truss are the **method of joints** and the **method of sections**, or some combination of the two procedures. Both methods assume that the reason for analyzing the system has been defined and that information about the problem has been recorded, including whether it is reasonable to categorize the system as a planar or space truss. The method of joints consists of considering equilibrium of the truss as a whole (in order to find the forces acting on the boundary of the truss), and of each of the pin connections that make up the truss. The method of sections consists of considering equilibrium of the truss as a whole, and of portions of the truss with a boundary cut through the two-force members themselves.

The method of joints and the method of sections work only for trusses that are both statically determinate and internally stable without redundancy. A truss with more members than are necessary to be stable is labeled internally stable with redundancy—concepts from mechanics of materials in addition to mechanical equilibrium conditions are required to determine the internal loads for the system. With any fewer members, the truss would collapse and is labeled internally unstable—concepts from dynamics are required to describe the internal loads and motion of the system. Checks to determine whether a truss is internally stable without redundancy, internally stable with redundancy, or internally unstable are outlined in **Box 9.1** for planar trusses and in **Box 9.2** for space trusses.

SYSTEM ANALYSIS (SA) EXERCISES

SA9.1 The Marvelous Truss

During construction, the large field-spanning girders for the Reynolds Coliseum had to be kept vertical before the roof could be put on. This was mainly accomplished with the help of trusses, beams, and horizontal cross bracings connecting the girders. **Figure SA9.1.1** presents a picture taken in 1944 during World War II (WWII). The construction started in 1942 but was interrupted in 1943 because of a lack of skilled workers. When completed, at 48 m × 98 m, it represented the largest such building in the southeast. On December 2, 1949, the opening game in the brand new William Neal Reynolds Coliseum (named after William Neal Reynolds of Winston-Salem), NC State's Wolfpack basketball team won 67-47 over Washington.

At the same time, the trusses fulfilled other functions such as supporting electrical conduits and light fixtures. It was common at that time to create the connections between truss and beam with rivets and steel plates, as shown in **Figure SA9.1.2**. Today, the same connections would be welded or bolted. **Figure SA9.1.3** shows a model that represents a truss and its loading conditions.

Figure SA9.1.1 Trusses keep girders in place during WWII

Figure SA9.1.2 View of truss-beam connections

(a) Develop a free-body diagram of the truss for the case where the side load (SL) is 0 kN. Start by replacing the distributed loads on the top truss members into point loads (in the joints) and add the light fixtures. Assume that the weight of the truss elements is negligible.

(b) Use hand calculations or develop a quick spreadsheet to compute the forces in members *BC*, *AC*, *CD*, *CE*, *AD*, *DE*, *CF*, and *EF*.

(c) Consider that during construction (see **Figure SA9.1.1**) a strong wind is blowing, resulting in a side load (SL = 2.16 kN/m, which represents a wind pressure of 0.24 kN/m^2 resulting from a wind of approximately 90 mph). Speculate which truss members will be impacted. Verify your speculations by revising your calculation from (b).

(d) Assume that the wind suddenly changes its direction during construction. Use the method of sections to assess the force in member *CE* if SL changes to −2.16 kN/m.

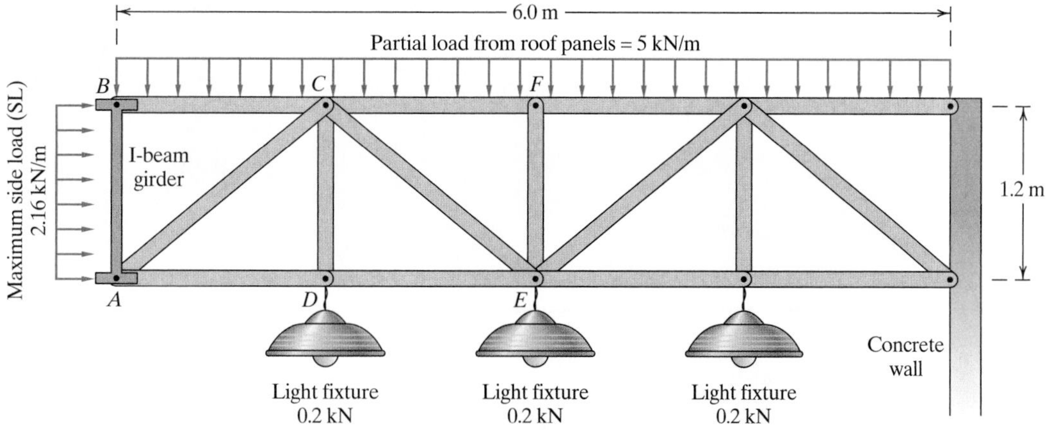

Figure SA9.1.3 Model of one truss supporting I-beam steel girder

SA9.2 A Self-Erecting Basketball Goal for the Reynolds Coliseum

There are two sets of portable basketball goals used in Reynolds Coliseum. One type has to be erected by hand and the other self-erects using a hydraulic cylinder (Figure SA9.2.1).

The basketball team found out that your class is studying the design of mechanical systems and asks for your help. They are tired of having to erect the large goal by hand and wonder whether you could develop a better solution. Figure SA9.2.2 shows the goal that they would like to have modified so it would erect itself with the assistance of a single hydraulic cylinder.

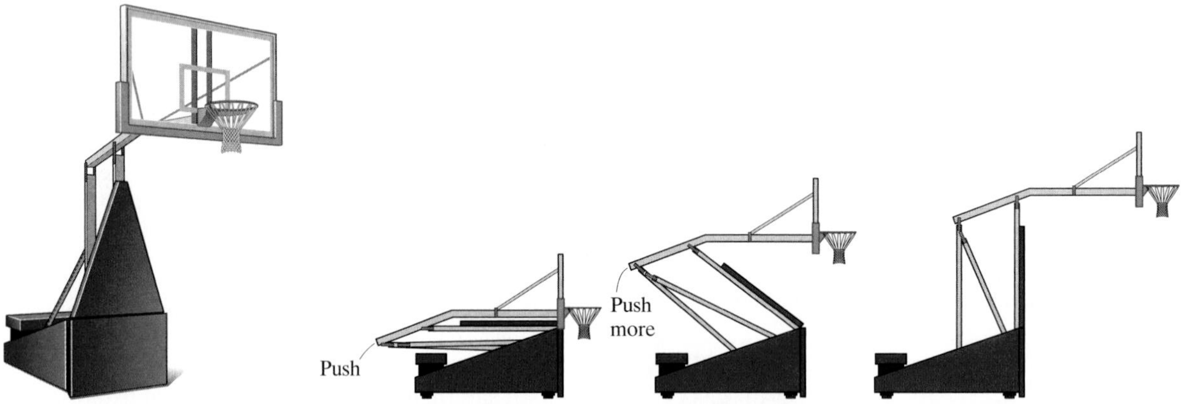

Figure SA9.2.1 Raising a basketball goal from a portable platform by hand

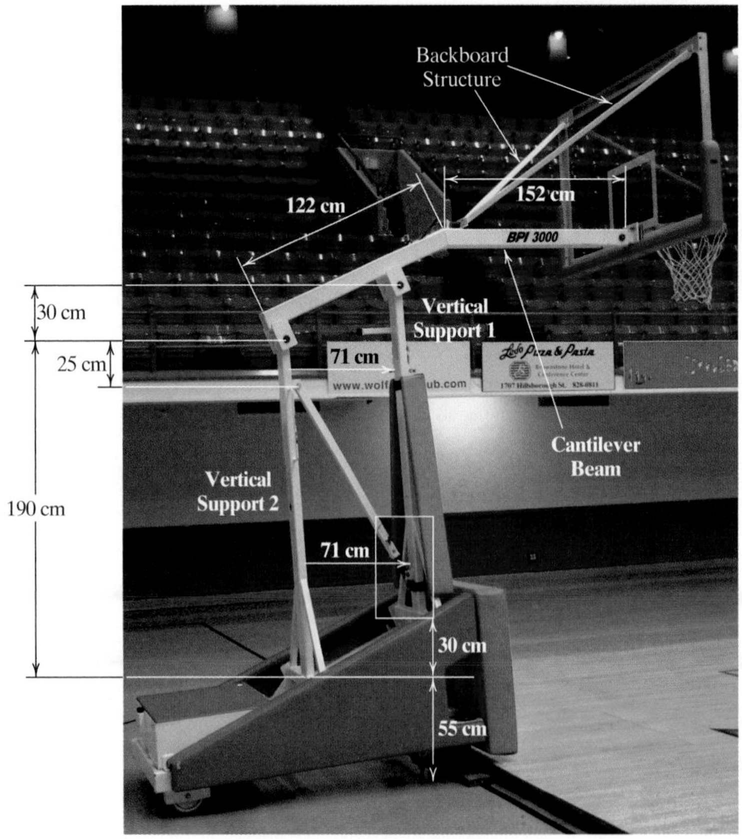

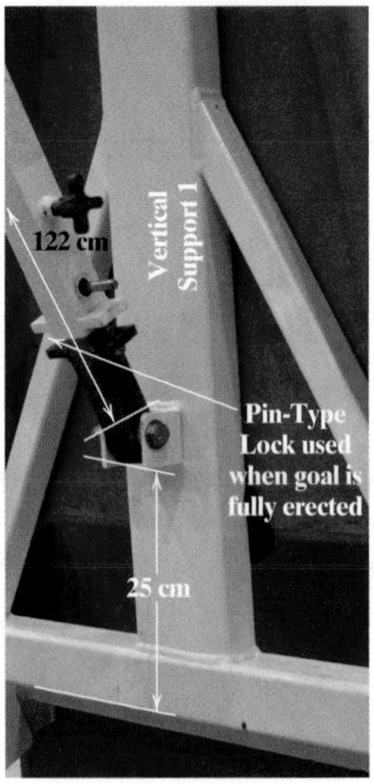

Figure SA9.2.2 A "cumbersome" basketball goal

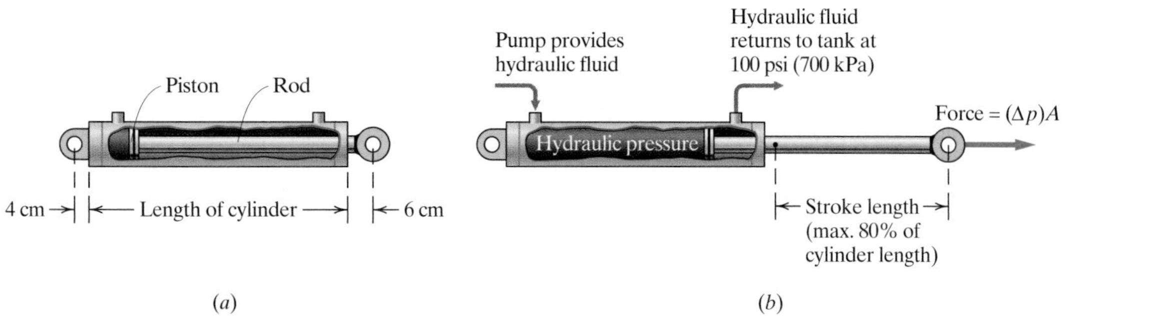

Figure SA9.2.3 Basics of force from a hydraulic cylinder: (*a*) totally retracted cylinder; (*b*) pressure difference creating a force at the end of rod

(a) Draw to scale the geometrical envelope for the entire deployment process of the goal shown in **Figure SA9.2.2**. Even better, use popsicle sticks and brads (serving as pins) to build a scale model.

(b) Now comes the critical decision—which structural element should be replaced with a hydraulic cylinder?

A note about how hydraulic cylinders work: the motion distance for erecting the goal is at most 80% of the length of the cylinder (see Figure SA9.2.3*b*). Thus, the more stroke length you require, the longer the retracted cylinder has to be. (*Be aware that you need to have enough space to store the cylinder in the base of the goal.*)

(c) Finally, you need to make sure that the cylinder generates enough force. One way to do this is to find the position of your mechanism that requires the most force. You also need to assume the loads that impact your "machine." Assume that the backboard structure weighs 0.40 kN, the metal cantilever beam 0.55 kN, and the two vertical supports 0.15 kN each. At this time it should be possible to calculate the required *diameter of the hydraulic cylinder/piston* assuming that the hydraulic pump will provide a maximum of 2000 psi or 14 mPa. Don't forget to include the 100 psi back-pressure that is needed to keep the piston from a sudden jerk. Is the diameter you calculate the maximum or minimum that is required? (Include the reason behind your answer.)

SA9.3 Analysis of Bicycle Performance[5]

In this problem you will be considering shifting on a 3-speed bicycle. More specifically, you will consider what shifting speeds would be good and why there are multiple gears on a bicycle. To this end, consider the bicycle depicted in Figure SA9.3.1, with supporting date in Table SA9.3.1

Table SA9.3.1

Bicycle specifications
L_{crank} (length of crank) = 17.8 cm
$R_{rear\ wheel}$ (radius of rear wheel) = 34.3 cm
$N_{rear\ cog}$ (number of teeth on rear cog) = 14 teeth, 20 teeth, 28 teeth
$N_{chain\ ring}$ (number of teeth on chain ring) = 50 teeth
C_D (coefficient of drag) = 0.9
$A_{Frontal}$ (frontal area of cyclist + bicycle) = 0.05 m^2
$\boldsymbol{W}_{Total}$ (weight of bicycle)

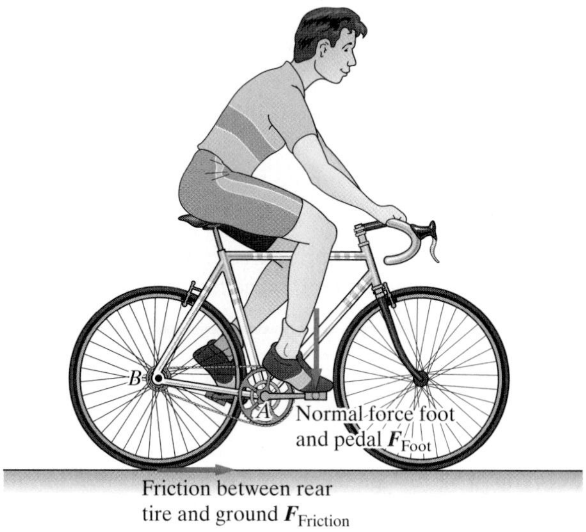

Figure SA9.3.1 Basics of bicycle under consideration

[5]Suggested resource material: Chapter 2 of *Bicycling Science*, 3rd edition, by D. G. Wilson (MIT Press, April 1, 2004).

(a) Derive an expression that relates the magnitude of $\boldsymbol{F}_{\text{friction}}$ to the velocity of the bicycle for the case where there is rolling resistance. Clearly state your source of data on the coefficient of rolling friction, along with all of your assumptions. Plot this expression. (This plot should look similar to the plot of $\boldsymbol{F}_{\text{friction}}$ versus velocity in **Figure 2.13**, except that your plot includes rolling resistance.)

(b) For the bicycle in first gear, derive an expression that relates the magnitude of $\boldsymbol{F}_{\text{foot}}$ to the velocity of the bicycle. Add to the plot from (a).

(c) For the bicycle in second gear, derive an expression that relates the magnitude of $\boldsymbol{F}_{\text{foot}}$ to the velocity of the bicycle. Add to the plot from (a).

(d) For the bicycle in third gear, derive an expression that relates the magnitude of $\boldsymbol{F}_{\text{foot}}$ to the velocity of the bicycle. Add to the plot from (a). After completing part (d), you should have a plot that looks something like **Figure SA9.3.2**.

(e) Assuming that a cyclist in reasonable shape pedals at a rate of 80–110 rpm (revolutions per minute), at what bicycle velocities would you recommend that the cyclist shift from first to second, and from second to third gear? Mark these shifting velocities on the plot that you created above; this is your recommended shifting pattern. In addition, check that the foot forces required by the cyclist for your recommended shifting pattern are reasonable.

(f) Using the plot you created as a "teaching prompt" (along with any figures that may be useful), write a description of why there are multiple gears on a bicycle. Your targeted audience for this description is a middle-school student.

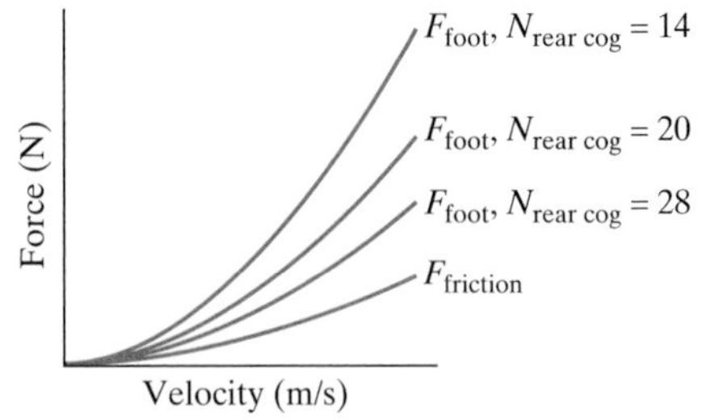

Figure SA9.3.2 Form of velocity versus force curves in defining bicycle performance

SA9.4 Review of Chapter 2 Analysis

In Chapter 2 a static analysis of bicycle performance was presented. More specifically, the question, "How fast can Merrill sprint toward the finish line?" was asked and was addressed by answering three interrelated sub-questions:

- What is the maximum force that Merrill can apply to the pedal?
- How is this force related to the friction force between the rear tire and the ground?
- How does the friction force relate to the drag force on the bicycle?

(a) Review the analysis presented in Chapter 2. For the bicycle considered (with specifications given in **Figure 2.7**), calculate the rate at which the cyclist would need to be pedaling when traveling at the maximum speed. Express your answer in revolutions per minute (rpm).

(b) If your answer in (a) is greater than 110 rpm (which it should be), it is not likely that Merrill will be able to achieve the maximum speed. This is because the ratio of $N_{\text{chain ring}}/N_{\text{rear cog}} = 50/14$ results in Merrill having to pedal too fast. (In general, bicyclists prefer to pedal at 80–110 rpm.) Specify whether the ratio of $N_{\text{chain ring}}/N_{\text{rear cog}}$ would need to increase or decrease in order to reduce Merrill's pedaling speed when he is traveling at the maximum speed. Provide an explanation of your answer.

(c) Based on calculations, recommend a ratio of $N_{\text{chain ring}}/N_{\text{rear cog}}$ that is more appropriate than 50/14.

SA9.5 Designing a Bridge[6]

Scenario: Your design company has been solicited to submit a bridge design for the rural community of Hector, Arkansas. Hector is a small Ozark town in northwest Arkansas in the Arkansas River Valley. Located near Hector is the Illinois Bayou (**Figure SA9.5.1**), one of the few bayous in the world with Class II/III rapids. Prone to springtime flooding, the Bayou recently took out the bridge near Scottsville, five miles from Hector. The bridge is critical in serving the Hector area's principal agricultural interest, chicken farming. Without the bridge, the current chicken truck traffic has slowed to a standstill, leaving 500,000 chickens waiting on the other side of the bridge. The noise of clucking, coupled with the smell, is causing the citizens to demand a quick solution.

[6]This design exercise was originally proposed by John Feland, at the time a Ph.D. candidate in Mechanical Engineering at Stanford University. John is from Hector, Arkansas, and grew up on a chicken farm.

Figure SA9.5.1

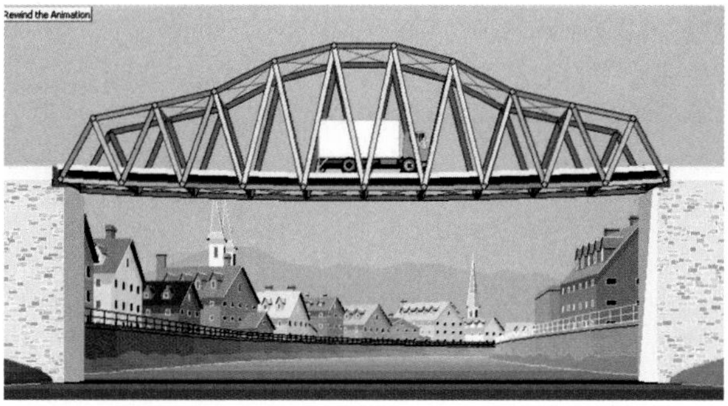

Figure SA9.5.2

Hector has limited means to support the construction of this new bridge. Their budget is limited to $20,000, including the cost of labor. Should a team deliver a bridge under cost, a bonus may be awarded by the town of Hector. The bridge should be a through/overhead truss. This means that the roadbed will be laid through the truss as shown in Figure SA9.5.2.

Your design company submits a proposal containing initial site analysis and bridge design for consideration by the town of Hector. To assist you in the development of your proposal to Hector, a state-of-the-art bridge design simulation software is being made available to you free of charge. *WestPoint Bridge Designer* (WPBD) is a realistic tool that allows users to rapidly design and test a variety of bridge structures. Unfortunately it only runs on the PC. It can be downloaded at *http://bridgecontest.usma.edu.* When you use WPDB please limit your materials to carbon steel and ensure that every member measures 140 mm × 140 mm. Additional criteria that your design must meet are listed in Table SA9.5.1.

Cost Analysis Guidance: Teams should utilize the cost analysis functions in WPBD. The software automatically calculates the cost of your bridge. All you need to do is include the cost report from WPBD in your final report.

Table SA9.5.1 Bridge Design Criteria

Design Criteria	Real Bridge
Span	40 meters
Width	5 meters
Maximum height	14.5 meters
Maximum load	~1.5 MN
Minimum factor of safety	1.75
Maximum cost	$20,000
Material	Carbon steel
Dimensions of cross section	140 mm × 140 mm
Max internal compressive load	Varies with length (buckling)
Max internal tensile load	4655 kN

Deliverable: The cover sheet for your proposal is shown in Figure SA9.5.3. Your proposal should contain an executive summary and bridge design printouts. The executive summary should be about a page in length and should briefly discuss the requirements of the bridge and how your particular design met those requirements. The bridge design printouts come from WPBD. WPBD allows you to print out labeled versions of your truss. In addition, you can obtain a listing of the materials used for each member and the maximum loads in each member during the load test. You should also printout a screen shot of the bridge under no truck loading and under truck loading. The printouts should consist of about 4 pages.

Presented to the citizens of Hector by:

Team: ____________________

Member 1: ____________________

Member 2: ____________________

Final estimated cost: ____________________

Number of joints: ____________________

Number of members: ____________________

Calculation to determine whether the final design is internally stable without or with redundancy (calculations go here):

Contents	**Page Number**
Summary of results (1 page maximum executive summary)	___
Bridge design printouts	___
Bridge schematic (labeled max tension and compression members)	___
Load report (with highlighted minimum factor of safety)	___
Screenshot of unloaded bridge	___
Screenshot of loaded bridge	___
Cost analysis (printed from WPBD)	___

Figure SA9.5.3 Illinois Bayou Bridge proposal cover sheet

SA9.6 Internal Loads in a Crane

Link-Belt is a company that makes lattice boom crawler cranes such as the 50-ton, LS-108H model (**Figure SA9.6.1**). The analysis and design of these mechanical systems require knowledge of all areas of engineering mechanics. A simplified lattice boom crawler crane is shown in **Figure SA9.6.2**. The analysis outlined below looks at various configurations of the crane and considers which configurations result in the most severe loadings of members of the crane.

(a) Assume that the crane is holding a 10,000-lb load in static equilibrium, as shown in **Figure SA9.6.2**. Write an expression as a function of θ for the tension in the cable that runs from the center of pulley B through pulley A, and back around pulley B to the winch at C.

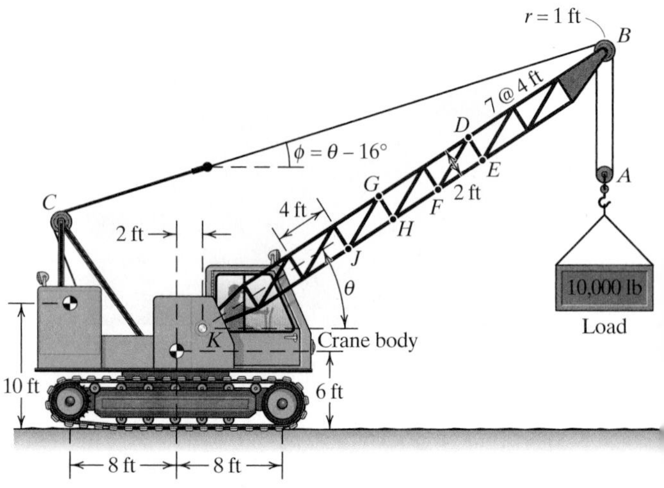

Figure SA9.6.2 Simplified lattice boom crawler crane

Figure SA9.6.1 A sample Link-Belt 50-ton crane

Assume that all pulleys are frictionless and weightless. For a reasonable range of the angle θ, at what angle is the tension maximum? What is the maximum tension?

(b) Write expressions for the forces in members EF and HJ as functions of the angle θ. For a reasonable range of the angle θ, at what angle is the tension or compression in member EF largest? For a reasonable range of the angle θ, at what angle is the tension or compression in member HJ largest?

(c) Organize your findings in (a) and (b) in a table that would allow a fellow engineer to readily understand the maximum loads in the cable, and in members EF and HJ.

SA9.7 A Heavy Load

The weightlifter shown in **Figure SA9.7.1** was in the process of starting the second phase of a clean-and-jerk maneuver when his patellar tendon ruptured. The goal of this problem is to determine the forces exerted on the patellar tendon and develop some intuition for the forces that we subject our bodies to. In the interest of completeness, it should be noted that the patellar tendon is not actually a tendon. It connects bone (patella) to bone (tibia) and is, more properly speaking, a ligament. The strength of the patellar tendon may depend on a variety of factors including its size, training history, rate of loading (tendons, ligaments, and most other soft tissues stiffen and strengthen when they are loaded quickly), orientation (which is why an athlete's form is crucial), and the presence or absence of injury. Strength estimates for patellar tendons from non-athletes, loaded relatively slowly, top out at approximately 10,000 N (per leg).

(a) Let's begin by assuming that the weightlifter has a mass $m_w = 100$ kg and is lifting a mass of 175 kg. Calculate the upward force, G_Y, that the ground exerts on the athlete's feet.

(b) Calculate the average force in each patellar tendon as a function of the angle θ (as shown in **Figure SA9.7.2a**). At what position is the force the greatest? What happens if we change the angle θ (as shown in **Figure SA9.7.2b**) at which the patellar tendon connects to the tibia by $\pm 10\%$?

From *Sports Biomechanics* by Roger Barlett (New York: E&FN, 1999).

Figure SA9.7.1 Initiation and progression of a patellar tendon rupture by weightlifter in Olympic competition

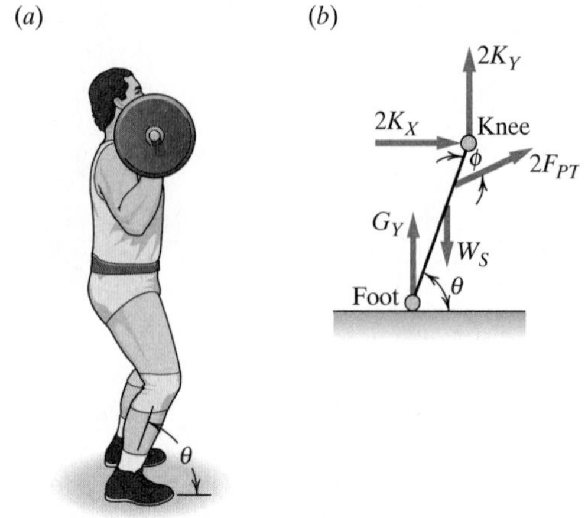

Figure SA9.7.2 (*a*) A schematic of a weightlifter just before he attempts to press the weight above his head. (*b*) A schematic of a weightlifter's lower legs just before he attempts to press the weight above his head. Note that we have included the reactions at both knees and both patellar tendons. Realistically, this problem requires a dynamic analysis, but we can learn a lot by starting with a static representation.

A few things you need to know and assume in order to get these calculations underway:

When combined, the lower legs and feet account for 13% of the weightlifter's total weight. We'll assume that this weight, W_s, acts in the middle of the lower leg (sometimes called the shank) as shown in the simplified model in **Figure SA9.7.2*b***. In addition, we'll assume that the shank has a total length, L_s, = 40 cm and that the patellar tendon attaches approximately 10 cm below the knee joint's center. To start the calculation, fix the angle between the patellar tendon's insertion point and the long axis of the tibia at $\theta = 15°$ and assume that it remains constant throughout the motion. In addition, allow the tibia angle θ, as shown in **Figure SA9.7.2**, to go from 45° to 90° at the completion of the lift.

(c) Using the 10,000 N patellar tendon number presented above, determine at what angle θ the weightlifter's patellar tendon fails? If we assume that, as a result of his training, his patellar tendon is much stronger than the average person's—perhaps on the order of 15,000 N—what is the factor of safety for this particular activity?

CHAPTER 10

"OUT ON A LIMB" AND "HUNG OUT TO DRY": A LOOK AT INTERNAL LOADS IN BEAMS AND CABLES

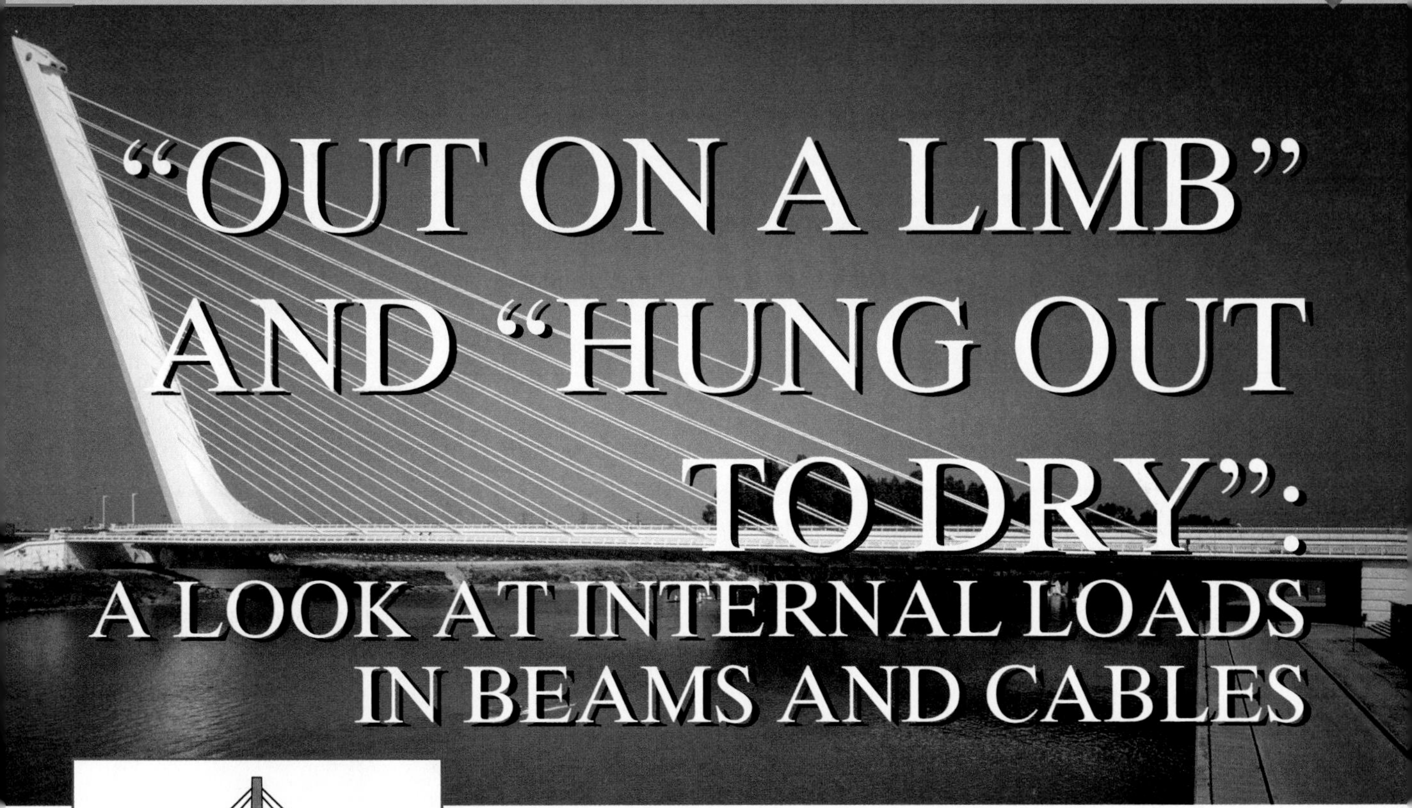

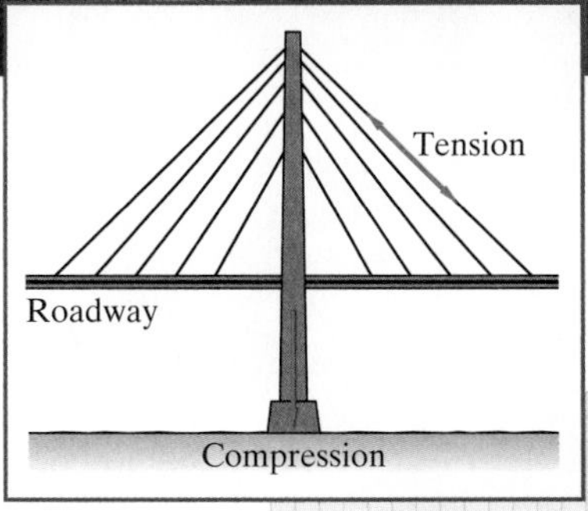

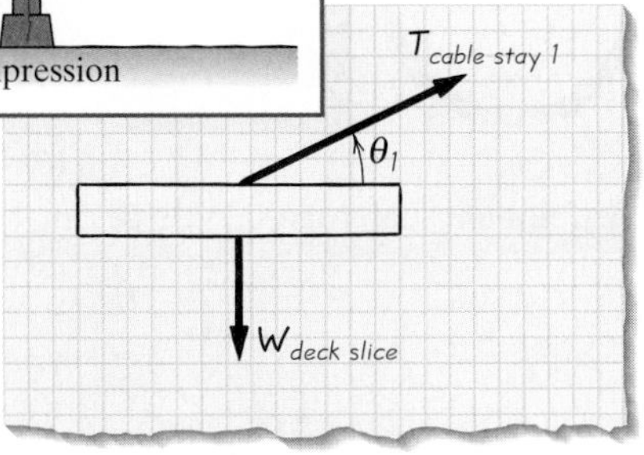

"Out on a limb" and "hung out to dry" may seem like unlikely phrases to include in the final chapter title of a book on equilibrium, but in fact they are apropos phrases. More specifically:

If you look at the bicycle in **Figure 10.1*a*** (certainly not your typical touring bicycle), you see that the cyclist's weight is applied at the end of the horizontal member *AB*—the cyclist is literally "out on a limb" (if we call the horizontal member the limb). This member is more commonly referred to as a beam.

If you look at the bridge in **Figure 10.1*b***, you see that the weight of the roadway deck hangs from the main cables—the roadway literally is "hung out" on the main cables.

Engineers are concerned with selecting beam and cable sizes and materials that will ensure adequate structural performance. Part of this selection process involves calculating the loads internal to a beam or a cable. In this chapter we lay out how to systematically calculate the loads acting externally and internally on beams and cables.

OBJECTIVES

Upon completion of this chapter, you will be able to:

- **Identify a beam**
- **Carry out equilibrium analysis of a beam, finding both external and internal loads**
- **Present the results from beam equilibrium analysis as a series of shear, bending moment, and axial force diagrams**
- **Relate the loads internal to a beam to one another**
- **Identify the loading on a cable**
- **Carry out equilibrium analysis of a cable to determine internal loads and cable shape**

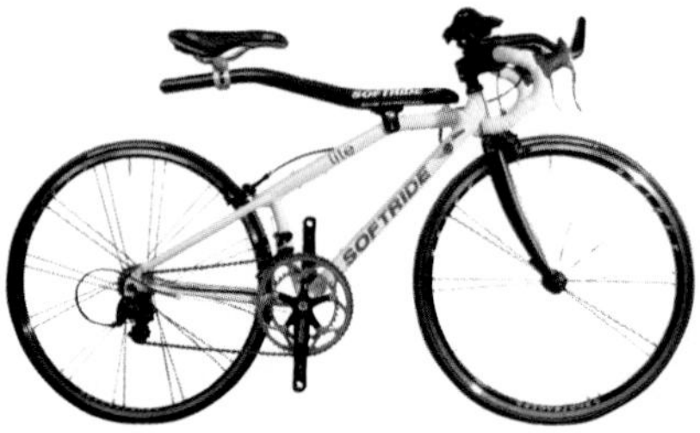

(*a*)

(*b*)

Figure 10.1 (*a*) Modern, if not unusual, road-bicycle design; (*b*) a bridge held up by cables

10.1 BEAMS

A member is called a **beam** if loads are applied perpendicular to its long axis. These loads, which may consist of forces and/or moments, are referred to as **lateral loads**. Because the loads are perpendicular to the long axis of the beam, they cause the beam to bend, as illustrated in Figure 10.2.

A beam is a particular type of multiforce member. The frame in Figure 10.3 contains two beams (CG and AE). The two-force member, BG, is not a beam because there are no loads acting perpendicular to its long axis.

Beams are a common structural member—examples of systems that incorporate beams are trees (trunk and branches), the human skeleton

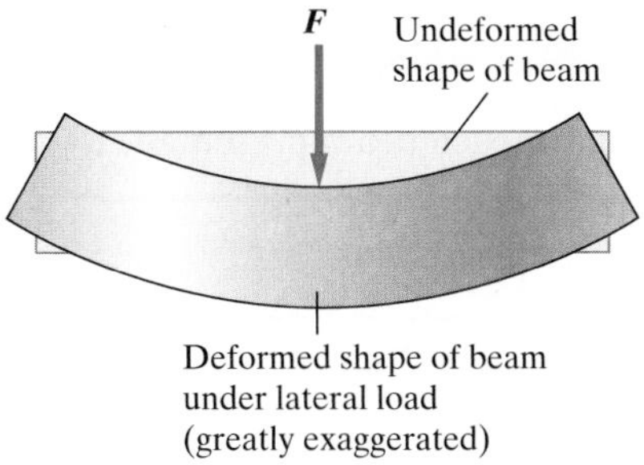

Figure 10.2 Loads applied to a beam cause it to bend

(tibia, femur), swing sets (the metal tubes or wooden posts), building and bridge frames, automotive suspensions, crutches, and diving boards. In designing and evaluating beams, engineers must consider the beam's length and cross-sectional shape and the loads it must carry, as well as its material, weight, connection to the rest of the structure, cost, and availability of prefabricated beams. Beams are so ubiquitous in our world that specialized analysis procedures have been developed to calculate the loads internal to a beam in order to assess the beam's capacity. In this section we present these procedures. This presentation will involve some new vocabulary, a slight adaptation of several analysis steps, and some insights into beam behavior gained by revisiting the free-body diagram.

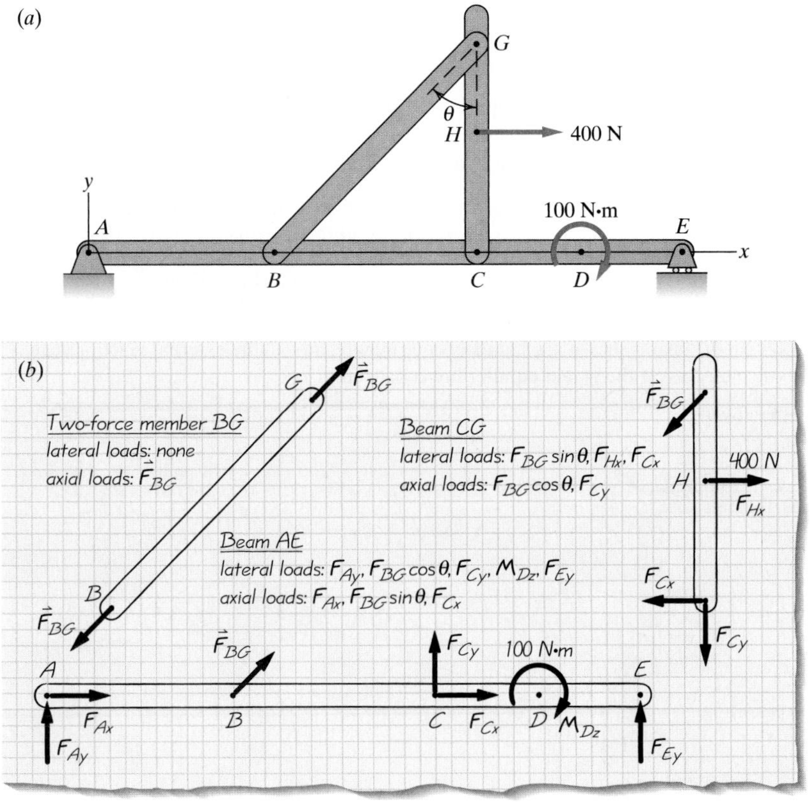

Figure 10.3 (*a*) A frame consisting of members *AE*, *BG*, and *CB*; (*b*) free-body diagrams of the members

EXAMPLE 10.1 BEAM IDENTIFICATION

Describe the beam type and the loading for each beam presented.

Beam and Loading	Description
Beam 1 (**Figure 10.4*a***)	Simply supported beam with a single concentrated force.
Beam 2 (**Figure 10.4*b***)	Combination beam with two concentrated loads, one at each end. This beam can also be described as simply supported beam with overhangs.
Beam 3 (**Figure 10.4*c***)	Cantilever beam supporting a uniformly distributed load along its entire length.
Beam 4 (**Figure 10.4*d***)	Force-loaded curved beam. This beam is actually a two-force member. The internal loads for the straight two-force members that we studied in our analysis of trusses are axial loads. A bent or curved two-force member also has bending moments and shear internal loads. Examples of curved beams include clamps, hooks, and bows.
Beam 5 (**Figure 10.4*e***)	Tapered airplane wing with a varying distributed load.
Beam 6 (**Figure 10.4*f***)	Cantilever beam of varying cross section with a concentrated load at the narrow end.
Beam 7 (**Figure 10.4*g***)	Round solid beam that acts as a shaft for the pulley system shown. The loading consists of concentrated loads at the locations of the pulleys.
Beam 8 (**Figure 10.4*h***)	L-beam with a concentrated load, a triangular distributed load, and a uniformly distributed load.
Beam 9 (**Figure 10.4*i***)	Round, hollow, thin-walled cantilever beam with a concentrated load at the end.

Note: All the beams shown are statically determinate, and all the external reactions can be calculated using the planar equilibrium equations.

(*a*) Beam 1

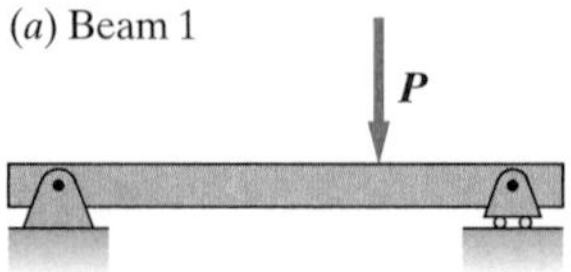

(*b*) Beam 2

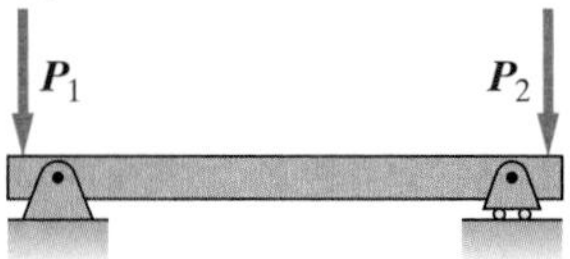

(*c*) Beam 3

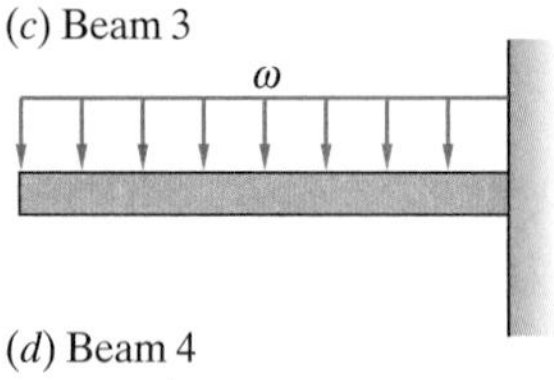

(*d*) Beam 4

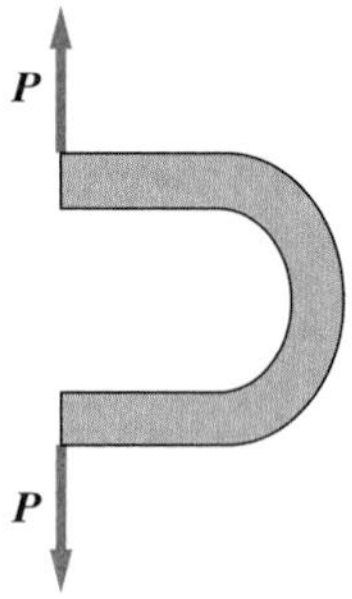

(*e*) Beam 5

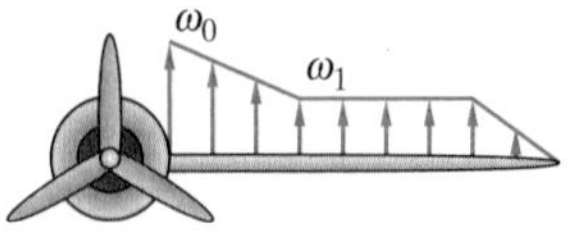

(*f*) Beam 6

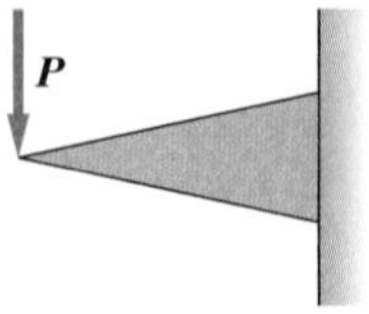

(*g*) Beam 7

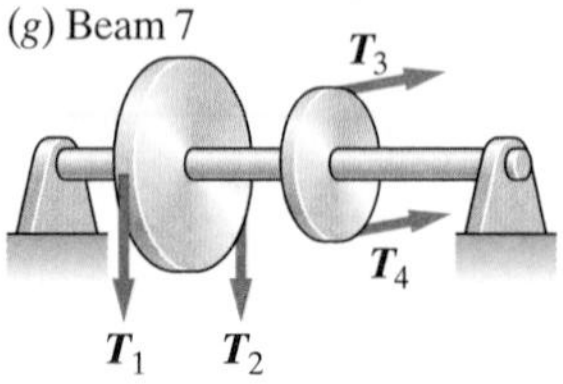

(*h*) Beam 8

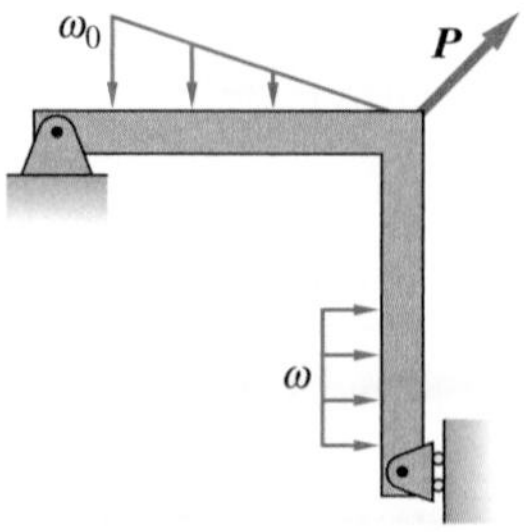

(*i*) Beam 9

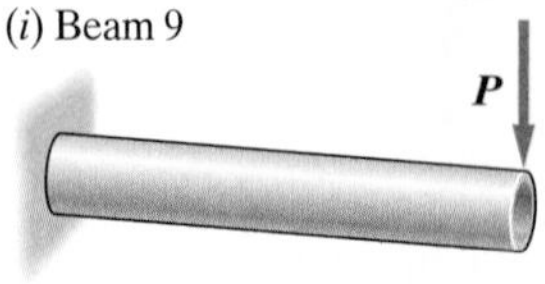

Figure 10.4

Beam Vocabulary

Beam Configurations. Geometric features that are important in describing a beam are its length and its cross-sectional area and shape. We set up a **beam coordinate system** with an x_b axis along the length (**Figure 10.5**)—we refer to this axis as the longitudinal or long axis. The other two axes (y_b and z_b) lie in the cross section of the beam, as shown. If the long axis of the beam is aligned with the x axis of the overall coordinate system, we can omit the b subscript in describing bending moments, shear forces, and axial force; otherwise the b subscript helps in differentiating the beam's orientation from the global coordinate system.

Some beam configurations are given names based on how they are connected to the rest of the world. For example:

- A **cantilever beam** is fixed to the rest of the world at one end while free at the other end. A cantilever beam is generally represented as in **Figure 10.6*a***, where end (B) is fixed and end (A) is free.
- A **simply supported beam** is pinned to the rest of the world at one end and attached via a roller or rocker at the other end. A simply supported beam is generally represented as in **Figure 10.6*b***, where end (C) is pinned and end (D) is supported by a roller.
- A **fixed-fixed beam** is fixed to the rest of the world at each end and is generally represented as in **Figure 10.6*c***.

Figure 10.7 illustrates commonly found beam configurations. Some of the configurations are a combination of simply supported and cantilever beams. The configurations in (a) are statically determinate, meaning that the conditions of equilibrium are sufficient for determining the loads acting on the beam. The configurations in (b) are statically indeterminate, and the conditions of equilibrium are not sufficient for determining the loads acting on the beam; additional relationships from mechanics of materials are needed.

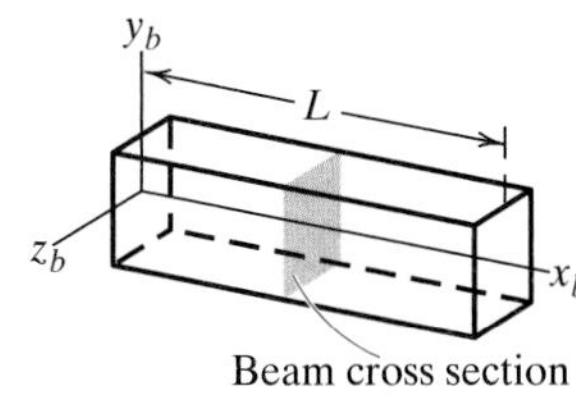

Figure 10.5 Beam coordinate system

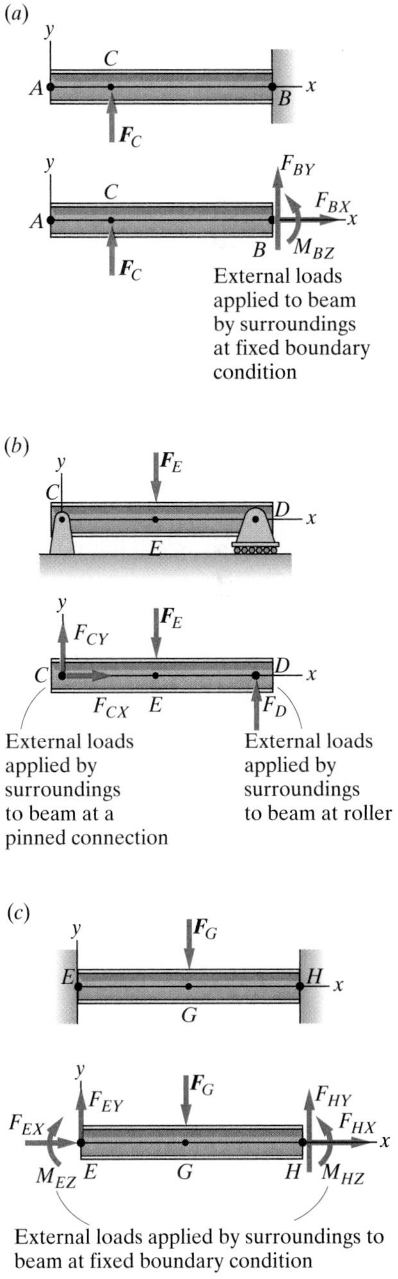

Figure 10.6 (a) Cantilever beam; (b) simply supported beam; (c) fixed-fixed beam

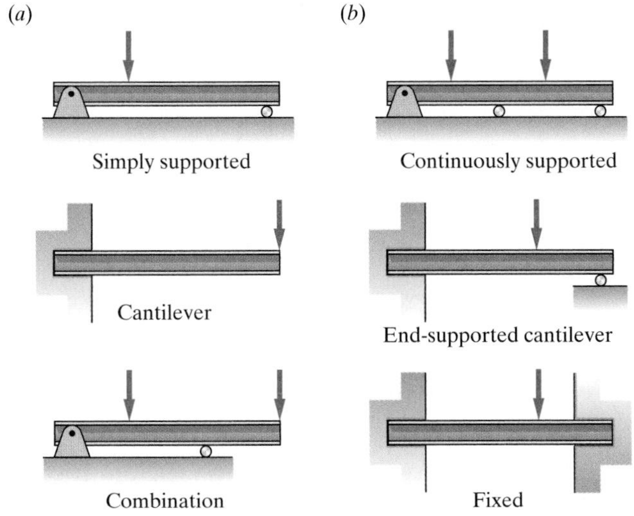

Figure 10.7 Various beam configurations: (a) statically determinate beams; (b) statically indeterminate beams

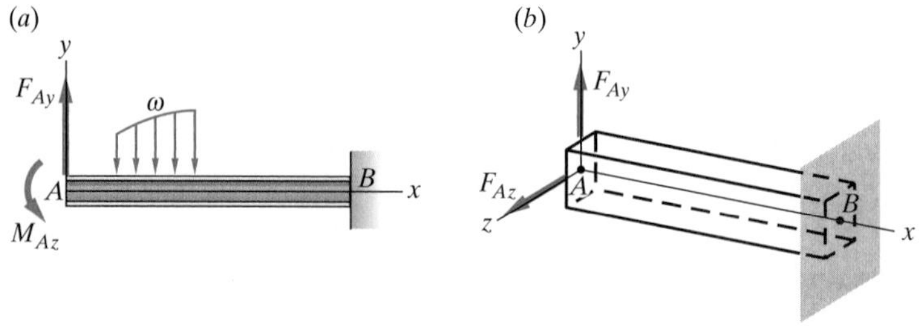

Figure 10.8 Fixed boundary condition: (*a*) planar beam; (*b*) nonplanar beam

If all of the external forces acting on a beam (single point forces and distributed forces) lie in a single plane and all external moments are about an axis perpendicular to this plane, the beam is a **planar beam** (also called a **two-dimensional beam**); otherwise it is a **nonplanar beam** (also called a **three-dimensional beam**). **Figure 10.8*a*** illustrates a planar cantilever beam, and **Figure 10.8*b*** illustrates a nonplanar cantilever beam.

Generally the first step in finding loads within a beam is to find the loads at its supports. As it is common for a beam to be part of a larger system (for example, **Figure 10.3**), this may also involve finding the loads acting on the larger system as illustrated in Example 10.2.

EXAMPLE 10.2 A BEAM WITHIN A FRAME

Beam *BDG* is part of the frame shown in **Figure 10.9** and is supported by pins at *B* and *D*. Determine the loads acting on *BDG* at pins *B* and *D*.

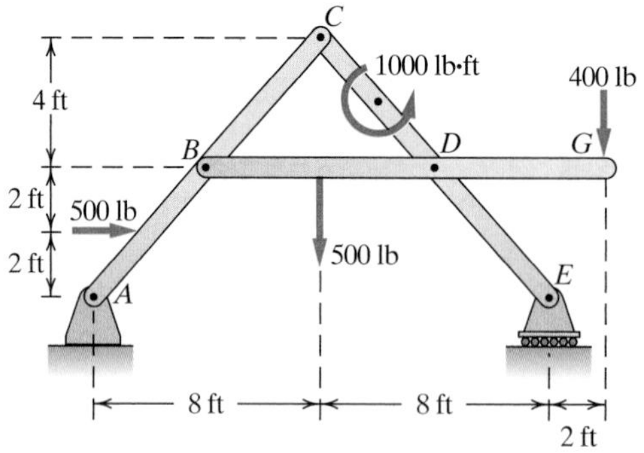

Figure 10.9

Goal We are to find the loads acting on beam *BDG* at pins *B* and *D*.

Given We are given information about the geometry and loading of the frame.

Assume We assume that the pins at *B* and *D* are frictionless and that we can treat the system as planar. We also assume that the weights of the members are negligible.

Draw If we first isolate *BDG* and analyze it for the loads at *B* and *D* we find that we have four unknown forces (F_{Bx}, F_{By}, F_{Dx}, F_{Dy}) and only three equations of planar equilibrium. Thus we need to analyze members that are connected to *BD* so we can reduce the number of unknowns for member *BDG*. The first step is to find the loads at supports *A* and *E*. To accomplish this we draw a free-body diagram (**Figure 10.10**) of the entire structure. We then disassemble the frame and analyze individual members to find the loads at *B* and *D* (**Figure 10.11** and **Figure 10.12**).

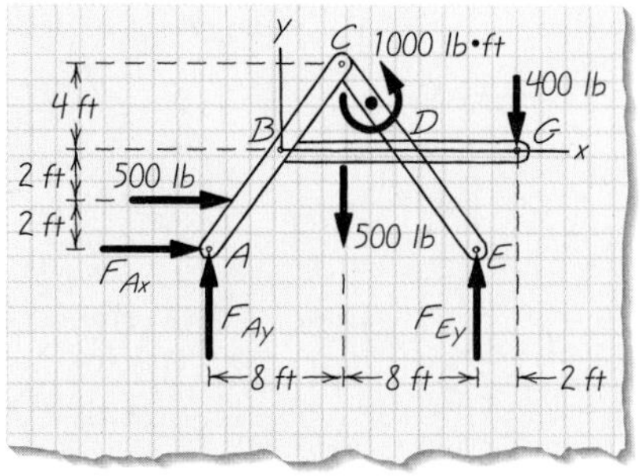

Figure 10.10 Free-body diagram of entire frame

Formulate Equations and Solve Based on the free-body diagram in **Figure 10.10**, we set up the equations for planar equilibrium to find the loads at supports *A* and *E*. Using (7.5C) with the moment center at *E* we write:

$$\sum M_{z@E} = 0\ (\circlearrowleft +)$$
$$-F_{Ay}\,(16\text{ ft}) - 500\text{ lb}(2\text{ ft}) + 500\text{ lb}(8\text{ ft}) - 400\text{ lb}(18\text{ ft}) + 1000\text{ lb}\cdot\text{ft} = 0$$
$$F_{Ay}\,(16\text{ ft}) - 11200\text{ lb}\cdot\text{ft} = 0$$
$$F_{Ay} = 200\text{ lb} \qquad (1)$$

Based on (7.5A);

$$\sum F_x = 0\ (\rightarrow +)$$
$$F_{Ax} + 500\text{ lb} = 0$$
$$F_{Ax} = -500\text{ lb} \qquad (2)$$

We next analyze beam *BDG* to solve for the loads at *B* and *D* (**Figure 10.11**). Based on (7.5C), with the moment center at *B*:

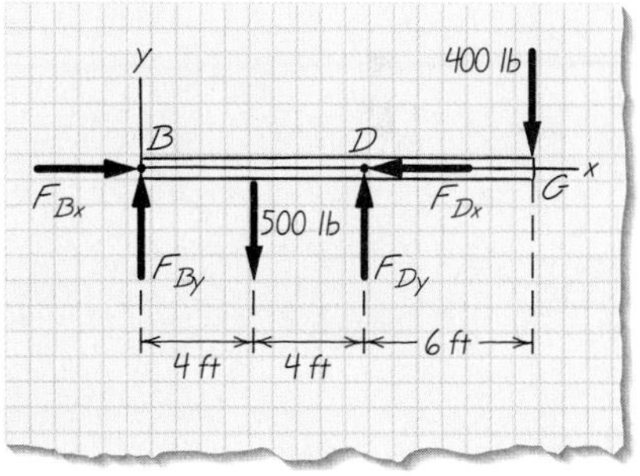

Figure 10.11 Free-body diagram of beam *BDG*

$$\sum M_{z@B} = 0\ (\circlearrowleft +)$$
$$-500\text{ lb}(4\text{ ft}) + F_{Dy}(8\text{ ft}) - 400\text{ lb}(14\text{ ft}) = 0$$
$$F_{Dy}(8\text{ ft}) = 7600\text{ lb}\cdot\text{ft}$$
$$F_{Dy} = 950\text{ lb} \qquad (3)$$

Based on (7.5B):

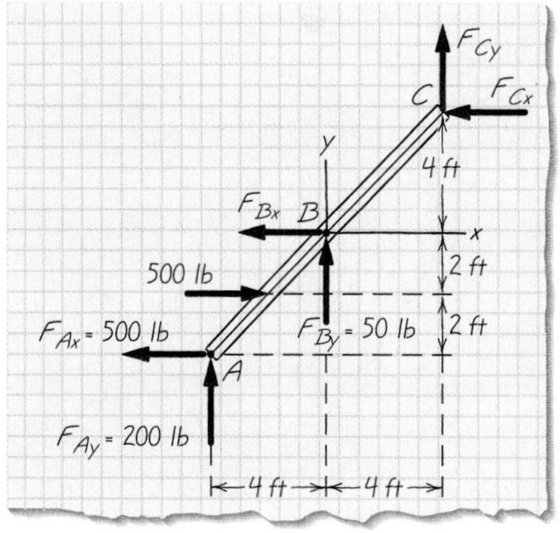

Figure 10.12 Free-body diagram of beam *ABC*

$$\sum F_y = 0\ (\uparrow +)$$
$$F_{By} + F_{Dy} - 500\text{ lb} - 400\text{ lb} = 0$$

Substituting from (3) for F_{Dy},

$$F_{By} + 950\text{ lb} - 900\text{ lb} = 0$$
$$F_{By} = -50\text{ lb} \qquad (4)$$

The minus sign indicates that F_{By} is acting downward on *BDG* and therefore upward on *ABC*.

Based on (7.5A):

$$\sum F_x = 0\ (\rightarrow +)$$
$$F_{Bx} - F_{Dx} = 0 \quad (5)$$

We do not have another equation of planar equilibrium that we can apply to BDG to solve for F_{Bx} and F_{Dx}. However, since F_{Bx} also acts on ABC and F_{Dx} also acts on CDE, we can analyze either of these members to complete the solution. We choose to analyze member ABC to solve for F_{Bx} (**Figure 10.12**).

We choose C as the moment center because then there will only be one unknown in the moment equilibrium equation. Based on (7.5C):

$$\sum M_{z@C} = (\circlearrowleft +)$$
$$-200\text{ lb}(8\text{ ft}) - 500\text{ lb}(8\text{ ft}) + 500\text{ lb}(6\text{ ft}) - 50\text{ lb}(4\text{ ft}) - F_{Bx}(4\text{ ft}) = 0$$
$$-2800 - 4F_{Bx} = 0$$
$$F_{Bx} = -700\text{ lb} \quad (6)$$

Finally, substitute (6) into (5) and solve for F_{Dx}:

$$-700\text{ lb} - F_{Dx} = 0$$
$$F_{Dx} = -700\text{ lb}$$

Answer $F_{Bx} = -700$ lb, $F_{By} = -50$ lb, $F_{Dx} = -700$ lb, $F_{Dy} = 950$ lb

Alternately, we could give the answers as:

at B on beam BDG

$$\boldsymbol{F}_B = -700\text{ lb}\,\boldsymbol{i} - 50\text{ lb}\,\boldsymbol{j}$$

at D on beam BDG

$$\boldsymbol{F}_D = 700\text{ lb}\,\boldsymbol{i} + 950\text{ lb}\,\boldsymbol{j}$$

Comment: F_{By} is pulling down on the beam to counteract rotation about pin D that is caused by the 400-lb force at G. Also note that beam BDG is in tension between the pins due to the axial forces pulling on the beam at B and D.

Check To check our solution we can reconsider the free-body diagram of BDG in **Figure 10.11** and sum the moments about any point other than B. We do not want to sum the moments around B because we used B as the moment center in calculating the results. We want our check to consist of equations that are independent of those used in the analysis.

Beam Internal Loads. Consider a planar cantilever beam in **Figure 10.13*a***; a distributed force ω acts along the top surface of the beam. At point P along the long axis (x) indicated in the figure, we "cut" the beam to investigate the internal loads. At this location there are third-law internal load pairs consisting of forces and moments, as illustrated in **Figure 10.13*b***.

If we now isolate the portion of the beam between A and P, the loads at the cut at P consist of a moment and forces that are half of internal load pairs (**Figure 10.13*c***). The moment, which is internal to a beam, is called a **bending moment**, the force in the plane of the cross section of the beam is called a **shear force**, and the force perpendicular to the cross section is called an **axial force**. Remember that these loads are internal to the beam when it is considered as a whole.

Bending moment is an internal moment about the z axis of the beam, and we denote it as M_{bz} (the b in the subscript stands for bending). If M_{bz} is positive, the beam is bent such that its top surface tends to concavity and its bottom surface tends to convexity (**Figure 10.14*a***). Because the x axis runs along the long central axis of the beam, the beam's top surface is a plane defined by positive y values and its bottom surface is a plane defined by negative y values. This means that a positive M_{bz} leads to compression in the $+y$ surface and tension in the $-y$ surface. Conversely, a negative M_{bz} leads to tension in the $+y$ surface and compression in the bottom $-y$ surface (**Figure 10.14*b***).

Shear force is an internal force in the y direction, and we denote it as V_y. It lies in the plane of the beam's cross section. For the left-hand side of the portion of the beam between A and P, the shear force is positive if it acts in the negative y direction (**Figure 10.13*c***).

Axial force is an internal force in the x direction, and we denote it as N_x. It is perpendicular to the beam's cross section and is positive if it acts to pull (thereby creating tension) on the cross section, as shown in **Figure 10.13*c***.

This discussion of bending moment, shear force, and axial force has been in terms of the portion of the beam from A to P, with positive as defined in **Figure 10.13*c***. We also could have discussed the portion of the beam from P to B, with positive as defined in **Figure 10.13*d***. Notice that in both **Figures 10.13*c*** and **10.13*d*** the bending moment and shear force act so as to create compression in the $+y$ surface and tension in the $-y$ surface.

With nonplanar beams we are also concerned with finding the internal loads M_{bz}, V_y, and N_x, but in addition there may be another bending moment (M_{by}), another shear force (V_z), and a moment about the x axis (M_{bx}, which is commonly called torque), as illustrated in **Figure 10.15**.

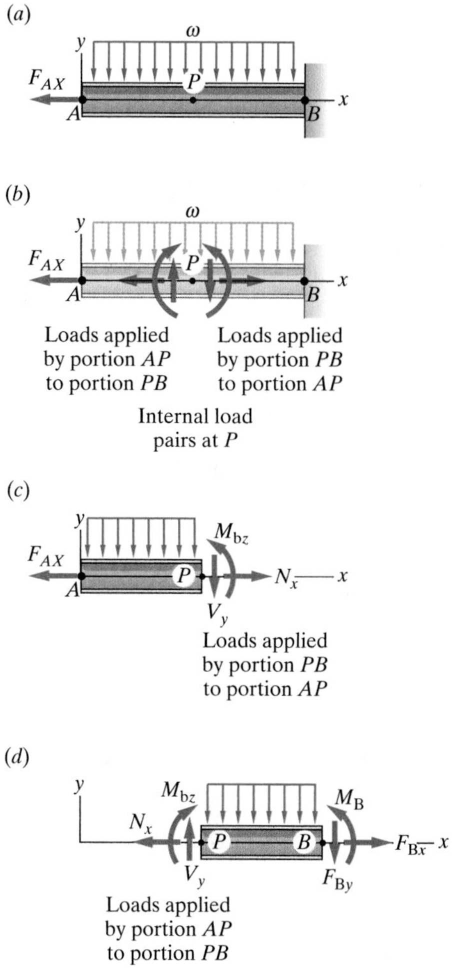

Figure 10.13 Loads at point P within a beam: (a) cantilever beam; (b) internal load pairs at point P; (c) free-body diagram of left-hand portion of beam; (d) free-body diagram of right-hand portion of beam

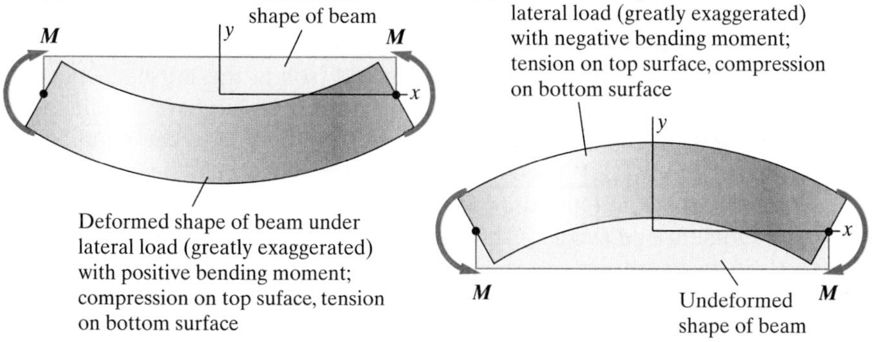

Figure 10.14 (a) Positive bending moment M_b; (b) negative bending moment M_b

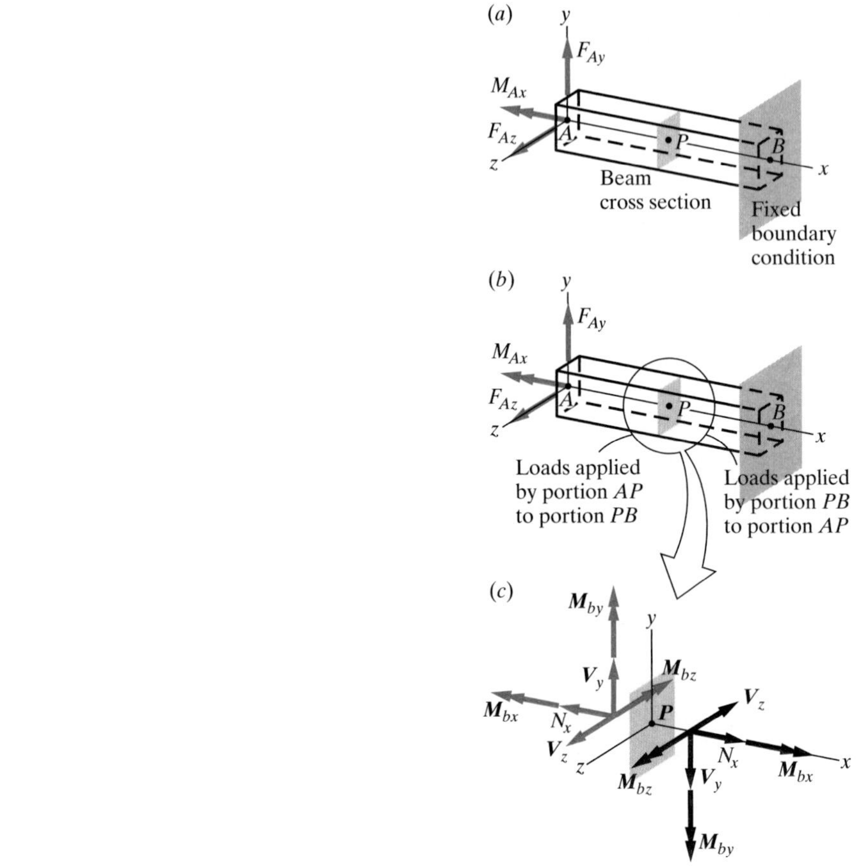

Figure 10.15 Loads at point P within a beam: (a) cantilever nonplanar beam; (b) internal load pairs at point P; (c) detail of internal load pairs at P

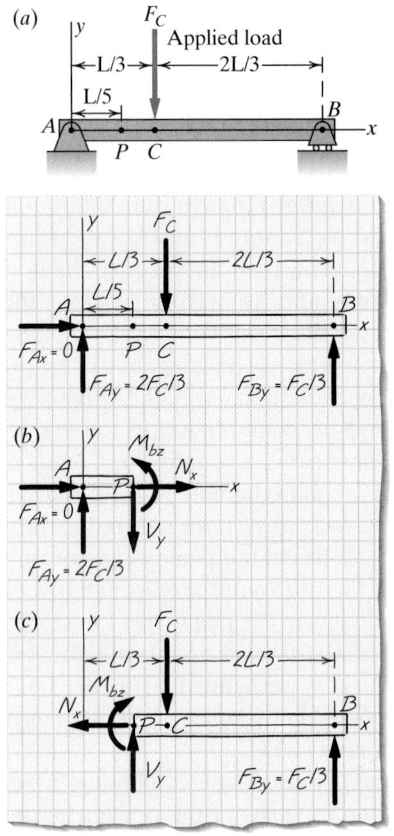

Figure 10.16 (a) A simply supported beam and its free-body diagram; (b) free-body diagram of left-hand portion; (c) free-body diagram of right-hand portion

Procedure for Finding Internal Loads in Beams

Let's say that we are interested in finding the internal loads (bending moment, shear force, and axial force) at a specified location P in a beam in equilibrium (Figure 10.16a). The analysis procedure consists of isolating a portion of the beam that includes a boundary cut through the beam at the specified position P, drawing a free-body diagram of the portion, then writing and solving equilibrium equations for the portion. For the beam in **Figure 10.16*a*** we could isolate the left-hand portion AP, draw the associated free-body diagram (with positive bending moment and shear as shown in **Figure 10.16*b***), and solve the equilibrium equations for M_{bz}, V_y, and N_x. As an alternative, we could work with the right-hand portion and draw its free-body diagram (with positive bending moment and shear as shown in **Figure 10.16*c***). Both approaches result in the same values of internal loads. You are urged to choose to analyze the portion that requires the fewest calculations to find the internal loads; just be sure to correctly draw positive bending moments and shear forces.

EXAMPLE 10.3 INTERNAL LOADS IN A PLANAR BEAM (A)

A simply supported beam is loaded as shown in **Figure 10.17**. Determine the axial force, shear force, and bending moment at cross sections at B and D.

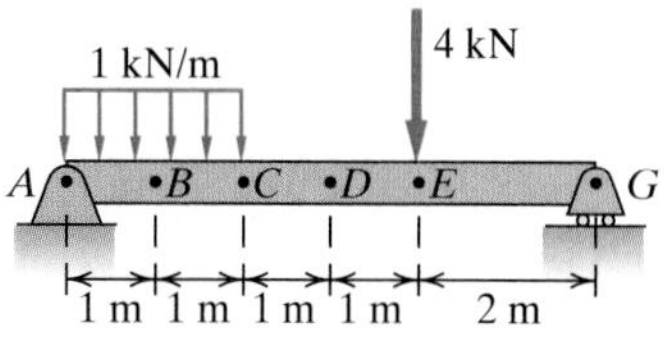

Figure 10.17

Goal We are to find the axial force, shear force, and bending moment at two locations (B and D).

Given We are given information about the dimensions and loading of the beam.

Assume We assume that we can treat the system as planar and that the weight of the beam is negligible.

Draw First we need to find the loads at supports A and G. To accomplish this we draw a free-body diagram (**Figure 10.18**) of the entire beam. We then cut the beam at B and D and draw free-body diagrams of portions of the beam to find the internal loads at B and D.

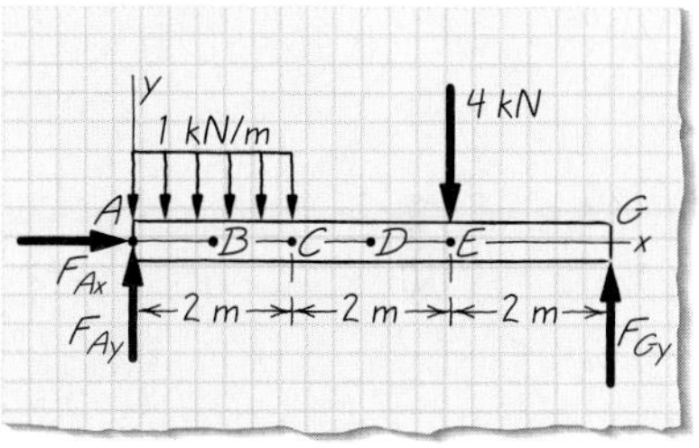

Figure 10.18 Free-body diagram of entire beam

Formulate Equations and Solve First we use the equations for planar equilibrium to find the loads at supports A and G ($F_{Ax} = 0$, $F_{Ay} = 3$ kN, and $F_{Gy} = 3$ kN). (Calculation not shown.)

We next make a cut at B and draw a free-body diagram of the left portion of the beam (**Figure 10.19**). When drawing the diagram, we assume that the unknown internal loads are positive. First we sum moments about the cut at B. Using B as the moment center eliminates V_y from the equilibrium equation so that we have only one unknown.

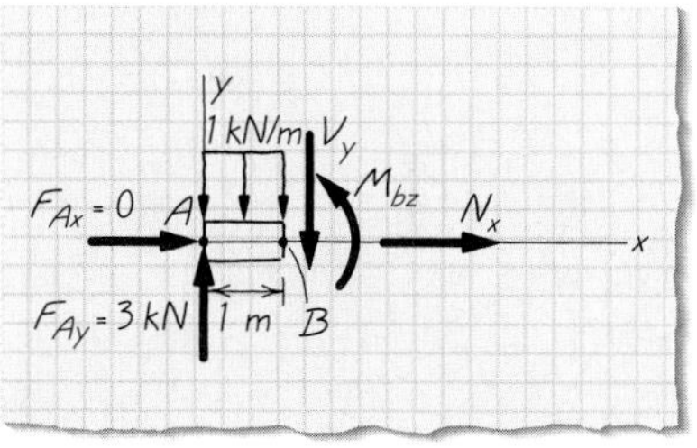

Figure 10.19 Free-body diagram of portion AB

As we apply (7.5C), with the moment center at B, each position vector is measured from the cut to the load. For the distributed load, we measure the position vector from the cut to the centroid of the load.

$$\sum M_{z@B} = 0 \ (\curvearrowleft +)$$

$$-3\text{ kN}(1\text{ m}) + 1\frac{\text{kN}}{\text{m}}(1\text{ m})(0.5\text{ m}) + M_{bz} = 0$$

$$M_{bz} = 2.5\text{ kN}\cdot\text{m}$$

Based on (7.5A):

$$\sum F_x = 0 \ (\rightarrow +)$$

$$N_x = 0$$

Based on (7.5B):

$$\sum F_y = 0 \ (\uparrow +)$$

$$-V_y + 3\text{ kN} - 1\frac{\text{kN}}{\text{m}}(1\text{ m}) = 0$$

$$V_y = 2\text{ kN}$$

An alternative method of solving for the internal loads, and also a check on our solution, is to analyze the right portion of the beam

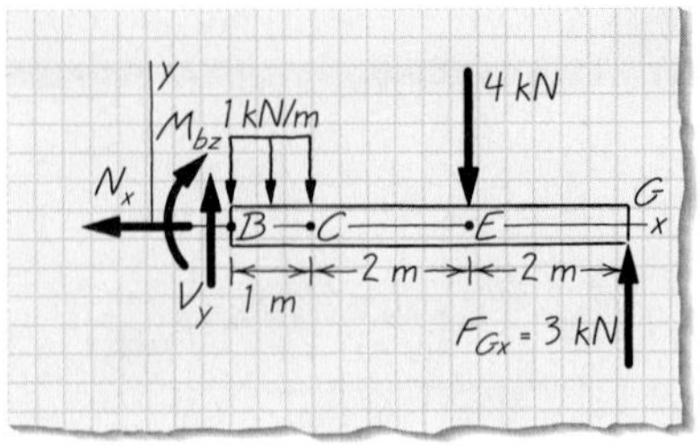

Figure 10.20 Free-body diagram of portion *BG*

(**Figure 10.20**). Based on (7.5C) and **Figure 10.20** with the moment center at *B*:

$$\sum M_{z@B} = 0 \,(\circlearrowleft +)$$

$$-1\frac{\text{kN}}{\text{m}}(1\text{ m})(0.5\text{ m}) - 4\text{ kN}(3\text{ m}) + 3\text{ kN}(5\text{ m}) - M_{bz} = 0$$

$$M_{bz} = 2.5\text{ kN}\cdot\text{m}$$

Based on (7.5A):

$$\sum F_x = 0\,(\rightarrow +)$$

$$-N_x = 0$$

Based on (7.5B):

$$\sum F_y = 0\,(\uparrow +)$$

$$V_y - 1\frac{\text{kN}}{\text{m}}(1\text{ m}) - 4\text{ kN} + 3\text{ kN} = 0$$

$$V_y = 2\text{ kN}$$

These are the same results we obtained earlier.

We proceed, using the free-body diagram in **Figure 10.21** as a reference, to calculate the internal forces at *D* using the same approach we used for cross section *B*. Based on (7.5C), with the moment center at *D*:

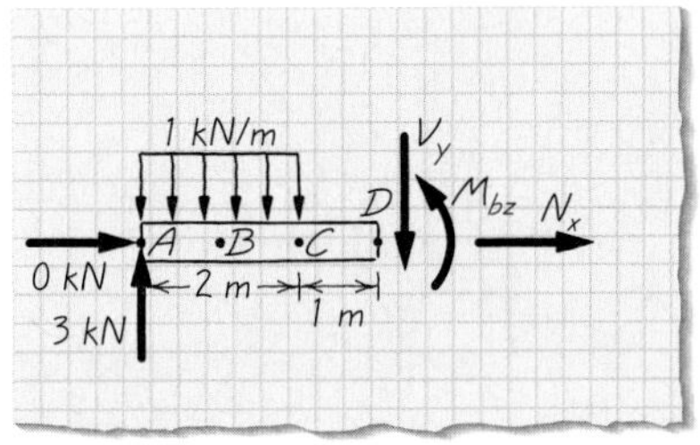

Figure 10.21 Free-body diagram of portion *AD*

$$\sum M_{z@D} = 0\,(\circlearrowleft +)$$

$$-3\text{ kN}(3\text{ m}) + 1\frac{\text{kN}}{\text{m}}(2\text{ m})(2\text{ m}) + M_{bz} = 0$$

$$M_{bz} = 5\text{ kN}\cdot\text{m}$$

Based on (7.5A):

$$\sum F_x = 0\,(\rightarrow +)$$

$$N_x = 0$$

Based on (7.5B):

$$\sum F_y = 0\,(\uparrow +)$$

$$-V_y + 3\text{ kN} - 1\frac{\text{kN}}{\text{m}}(2\text{ m}) = 0$$

$$V_y = 1\text{ kN}$$

Answer At *B*: $N_x = 0$, $V_y = 2$ kN, $M_{bz} = 2.5$ kN·m
At *D*: $N_x = 0$, $V_y = 1$ kN, $M_{bz} = 5$ kN·m

Check To check the answer, we can analyze the portion of the beam to the right of the cut at *D*.

EXAMPLE 10.4 INTERNAL LOADS IN A PLANAR BEAM (B)

The cantilever beam in Figure 10.22 is loaded with a point force at the free end and a concentrated moment at a distance of $L/2$ from the free end. Determine the axial force, shear force, and bending moment at cross sections at B and D.

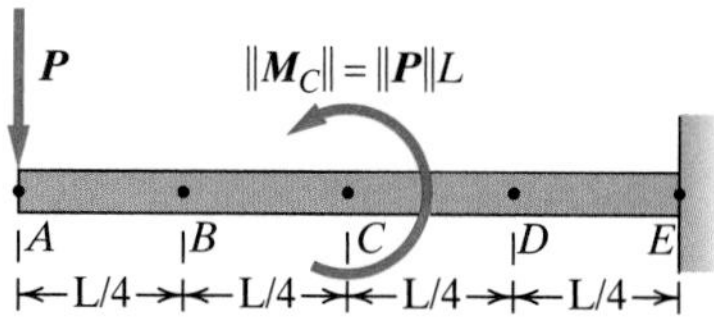

Figure 10.22

Goal We are to find the axial force, shear force, and bending moment at cross sections at B and D.

Given We are given information about the dimensions and loading of the beam.

Assume We assume that we can treat the system as planar and that the weight of the beam is negligible.

Draw We can determine the internal loads without calculating the loads at support E; therefore we do not need to start with a free-body diagram of the entire beam. Instead we cut the beam at B and choose to analyze the portion of the beam to the left of the cut. We do the same for the cut at D. The free-body diagrams of the left portions of the beam for each cut are shown in Figure 10.23 and Figure 10.24.

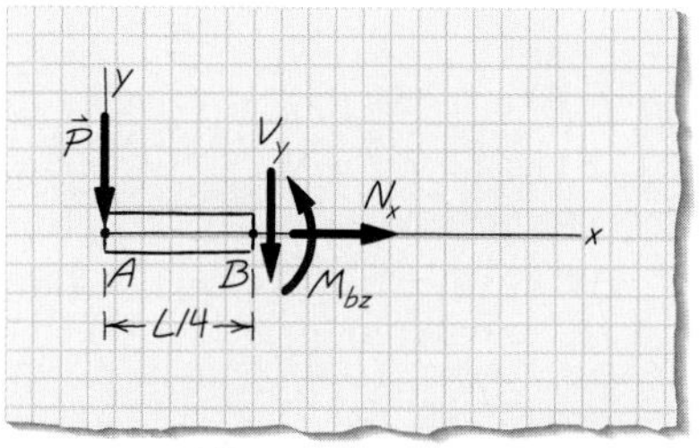

Figure 10.23 Free-body diagram of portion AB

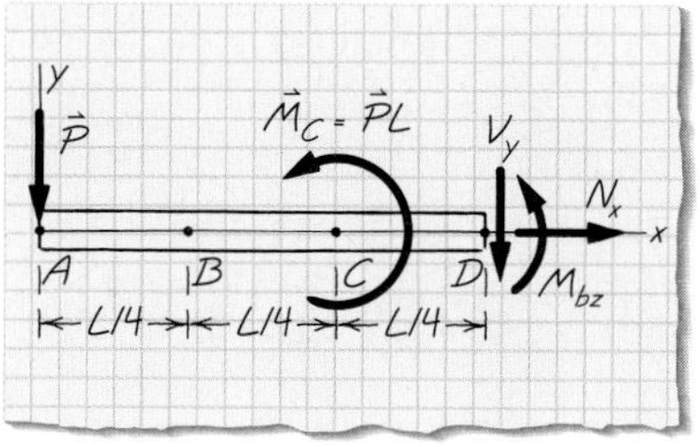

Figure 10.24 Free-body diagram of portion AD

Formulate Equations and Solve Using the free-body diagram of the left portion of the beam (**Figure 10.23**) as a reference we first sum moments about the cut at B. Based on (7.5C) with the moment center at B:

$$\sum M_{z@B} = 0 \;(\curvearrowleft +)$$

$$P\left(\frac{L}{4}\right) + M_{bz} = 0$$

$$M_{bz} = -\frac{PL}{4}$$

The minus sign indicates that the bending moment is in the opposite direction from what we assumed. Since we assumed a positive bending moment, the bending moment is actually negative. This means that the beam is curving downward with the bottom surface of the beam in compression and the top in tension.

Based on (7.5A):

$$\sum F_x = 0 \;(\rightarrow +)$$

$$N_x = 0$$

Based on (7.5B):

$$\sum F_y = 0 \;(\uparrow +)$$

$$-P - V_y = 0$$

$$V_y = -P$$

The minus sign means that the direction of the shear force is opposite from that which was assumed and therefore is acting upward.

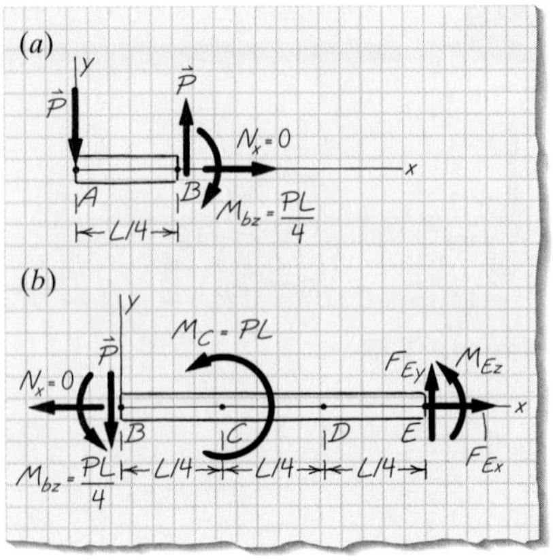

Figure 10.25

Before going on to calculate the internal loads at D, we reflect on the answers we obtained for the loads at B. **Figure 10.25** is a free-body diagram of the left and right portions of the beam when it is cut at B and shows our calculated values from above. At the cut we see the equal and opposite force pairs and moment pairs that we expect because of Newton's third law. Looking at the portion of the beam to the left of B, we notice that the applied load at A and the shear force at B form a counterclockwise couple with a magnitude of $\frac{PL}{4}$. The clockwise internal bending moment of magnitude $\frac{PL}{4}$ at B maintains equilibrium for this portion of the beam.

Now we find the internal loads at the cross section at D. Using **Figure 10.24** as a reference, we calculate the internal forces at D using the same approach we used for the cross section at B. Based on (7.5C) with the moment center at D:

$$\sum M_{z@D} = 0 \; (\curvearrowleft +)$$

$$P\left(\frac{3L}{4}\right) + M_c + M_{bz} = 0$$

Substituting for M_C (we are told in **Figure 10.22** that $M_C = PL$), we get

$$P\left(\frac{3L}{4}\right) + PL + M_{bz} = 0$$

$$M_{bz} = -\frac{7PL}{4}$$

Based on (7.5A):

$$\sum F_x = 0 \; (\rightarrow +)$$

$$N_x = 0$$

Based on (7.5B):

$$\sum F_y = 0 \; (\uparrow +)$$

$$-P - V_y = 0$$

$$V_y = -P$$

Answer At B: $N_x = 0$, $V_y = -P$, $M_{bz} = -\frac{PL}{4}$

At D: $N_x = 0$, $V_y = -P$, $M_{bz} = -\frac{7PL}{4}$

There are no axial loads applied to the beam, and therefore the internal axial force is zero throughout the beam.

Check To check the results, we can analyze the portions of the beam to the right of the cuts at B and D. This requires us to first calculate the loads at support E.

EXAMPLE 10.5 LOADS IN A NONPLANAR BEAM

A rider is pushing on each arm of the bicycle handlebars in **Figure 10.26** with a force of 50.0 lb. Each force is applied at 30.0° to the horizontal and at 2.00 in. from the end of the arm. Find the internal loads at the intersection of the handlebars and the stem.

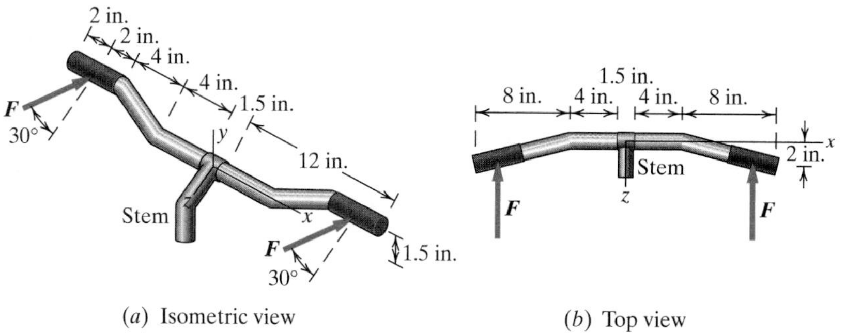

Figure 10.26

Goal We are to find the internal forces and bending moments in the x, y, and z directions at the intersection of the handlebars and the stem.

Given We are given information about the dimensions and loading of the handlebars.

Assume We assume that the weight of the handlebars is negligible.

Draw Each arm of the handlebars is a cantilever beam supported by the stem, loaded by forces in the y and z directions. We cut the handlebars at the stem, isolate the left arm, and draw the internal loads (**Figure 10.27**). To clarify the loading for the analysis we break the 50-lb force into components in the y and z directions ($F_y = -50 \sin 30° = 25.0$ lb and $F_z = -50 \cos 30° = -43.3$ lb). Because the system and its loading are symmetric, we know that both arms of the handlebars have the same internal loads.

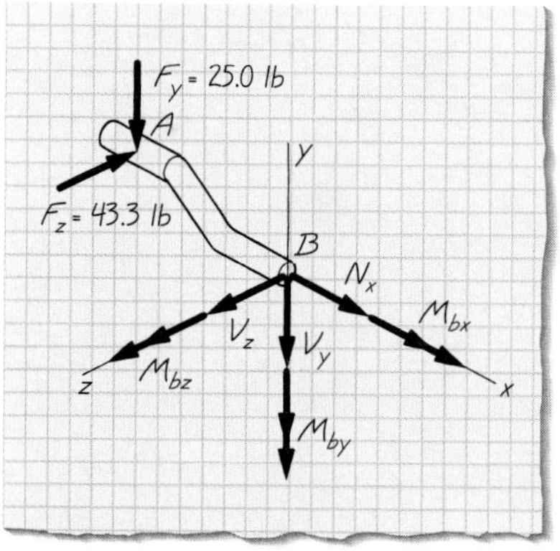

Figure 10.27 Free-body diagram of left handlebar

Formulate Equations and Solve We apply the six nonplanar equilibrium equations to the free-body diagram in **Figure 10.27** to solve for the internal loads. Analyzing force equilibrium gives the following three results.

Based on (7.3A):

$$\sum F_x = 0$$
$$N_x = 0$$

Based on (7.3B):

$$\sum F_y = 0$$
$$-25.0 \text{ lb} - V_y = 0$$
$$V_y = -25.0 \text{ lb}$$

Based on (7.3C):

$$\sum F_z = 0$$
$$-43.3 \text{ lb} + V_z = 0$$
$$V_z = 43.3 \text{ lb}$$

Because the handlebars are bent out of the x–y plane the force F_y is offset from that plane by a distance d (**Figure 10.28**), creating a moment about the x axis. Similarly F_z is offset from the x–z plane by 1.50 in., creating a moment about the x axis. To determine the distance d, we use similar triangles. From **Figure 10.26*b*** we know that the end of the arm is offset 2 in. from the x axis. In addition, $\boldsymbol{F}$ is applied 2 in. from the end of the arm; therefore analysis of similar triangles gives

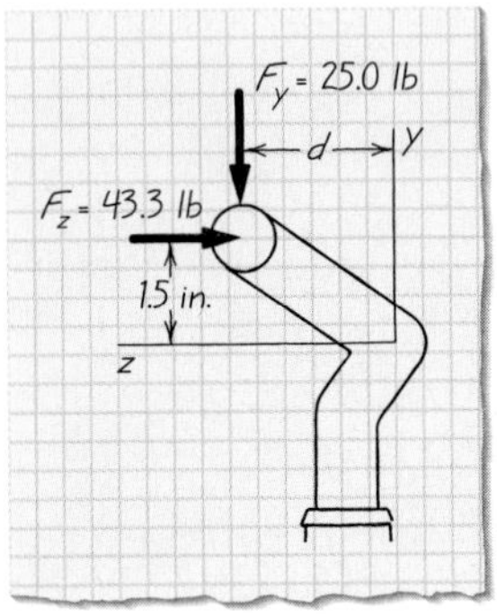

Figure 10.28

$$\frac{d}{6 \text{ in.}} = \frac{2 \text{ in.}}{8 \text{ in.}} \Rightarrow d = \frac{(6 \text{ in.})2 \text{ in.}}{8 \text{ in.}} = 1.5 \text{ in.}$$

We now analyze moment equilibrium using (7.4) with the moment center at B in **Figure 10.27**. When writing our equations, we define positive moments as those that create a positive moment about an axis.

$$\sum M_{x@B} = 0$$
$$F_y d - F_z(1.5 \text{ in.}) + M_{bx} = 0$$
$$-(-25.0 \text{ lb})(1.5 \text{ in.}) + (-43.3 \text{ lb})(1.5 \text{ in.}) + M_{bx} = 0$$
$$M_{bx} = 27.5 \text{ lb} \cdot \text{in.}$$

$$\sum M_{y@B} = 0$$
$$-F_z(10 \text{ in.}) - M_{by} = 0$$
$$(-43.3 \text{ lb})(10 \text{ in.}) - M_{by} = 0$$
$$M_{by} = -433 \text{ lb-in.}$$

$$\sum M_{z@B} = 0$$
$$F_y(10 \text{ in.}) + M_{bz} = 0$$
$$(25.0 \text{ lb})(10 \text{ in.}) + M_{bz} = 0$$
$$M_{bz} = -250 \text{ lb} \cdot \text{in.}$$

Answer $N_x = 0$ lb $V_y = -25.0$ lb, $V_z = 43.3$ lb, $M_{bx} = 27.5$ lb · in.
$M_{by} = -433$ lb · in., $M_{bz} = -250$ lb · in.

M_{bx} (often called a torque) causes twisting about the x axis. The moment M_{by} causes the handlebars to bend toward the front of the bicycle and M_{bz} causes them to bend downward. The negative signs in the results indicate that the internal loads are in the opposite direction from those shown in **Figure 10.27**.

Check To check our result, we can apply the internal loads to the free-body diagram in **Figure 10.27** and perform an equilibrium analysis.

Question: Can you think of how using the cross product might have made the solution easier?

Procedure for Creating Axial Force, Shear Force, and Bending Moment Diagrams

Examples 10.3–10.5 illustrate finding internal loads at specified locations along the long axis of beams. It is just as likely that in analyzing a beam you will be called upon to find the maximum bending moment and its location along the length of a beam because this is where the beam is likely to fail. We now illustrate how to use the by now very familiar static analysis procedure to find both the maximum bending moment *and* its location for a cantilever beam loaded by a uniformly distributed load along its top surface and an axial force at A (**Figure 10.29*a***).

Figure 10.29*a* shows a cantilever beam of length L extending from $x = 0$ (at A) to $x = L$ (at B). Consider a portion AP of the beam that extends from end A to some point $P < L$. Isolate this portion and draw its free-body diagram (**Figure 10.29*b***). The loads acting on the portion are the distributed force on the top and the point force at $x = 0$, at the left end of the portion. In addition, there are a bending moment, axial force, and shear force on the right—these loads are one-half of the internal third-law load pairs between the rest of the beam and portion AP. We will consider M_{bz}, V_y, and N_x positive as drawn in **Figure 10.29*b***. (Notice that this same free-body diagram describes any portion of the beam that extends from the origin to some length $x < L$.)

We write the three equilibrium equations for a planar system based on this free-body diagram in order to create expressions for the bending moment, the shear force, and the axial force as functions of x:

$$\sum F_x \Rightarrow -F_{Ax} + N_x = 0 \qquad (10.1A)$$

$$\sum F_y \Rightarrow \underbrace{-\omega x}_{\text{force created by distributed load}} - V_y = 0 \qquad (10.1B)$$

In order to write the moment equilibrium equation, we must select a moment center. Although the origin of the coordinate system shown in **Figure 10.29*a*** might seem like the obvious choice, we can simplify the math involved by choosing P, because then neither N_x nor V_y are included in the equation. We write the moment equilibrium equations about P as

$$\sum M_{z@P} \Rightarrow + \underbrace{\frac{x}{2}\,\omega x}_{\text{moment created by distributed load}} + M_{bz} = 0 \qquad (10.1C)$$

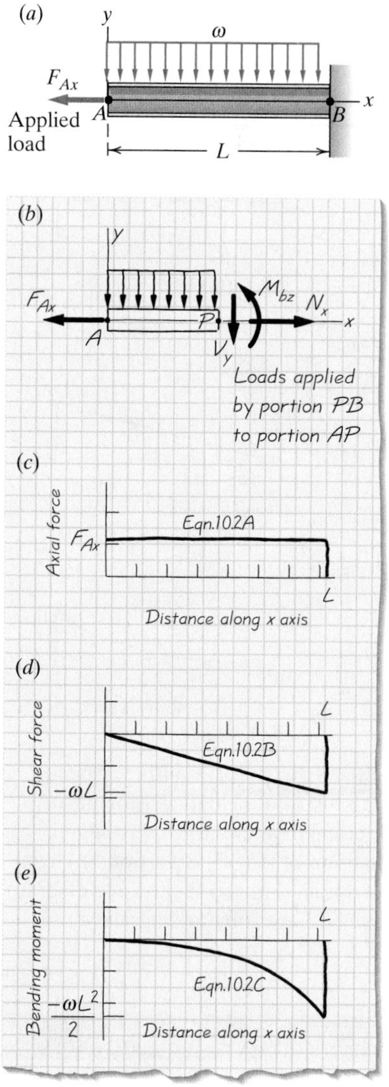

Figure 10.29 (*a*) Cantilever beam with distributed load causing negative bending moment; (*b*) free-body diagram of left-hand portion; (*c*) axial force diagram; (*d*) shear force diagram; (*e*) bending moment diagram

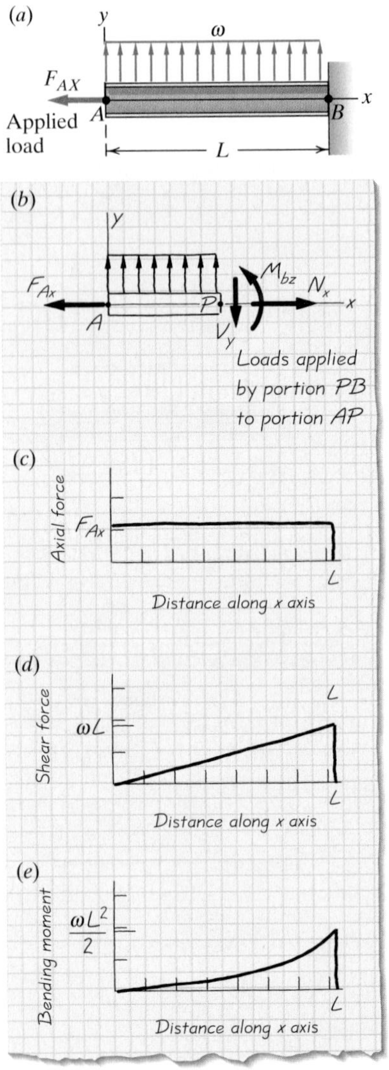

Figure 10.30 (*a*) Cantilever beam with distributed load causing positive bending moment; (*b*) free-body diagram of left-hand portion; (*c*) axial force diagram; (*d*) shear force diagram; (*e*) bending moment diagram

Both terms in this equation are positive because both create a positive (counterclockwise) moment about the z axis.

Some rearranging of (10.1A), (10.1B), and (10.1C) allows us to write expressions for the shear and axial forces and bending moment as functions of x:

$$N_x = F_{Ax} \quad \text{axial force} \qquad (10.2A)$$

$$V_y = -\omega x \quad \text{shear force} \qquad (10.2B)$$

$$M_{bz} = -\frac{1}{2}\omega x^2 \quad \text{bending moment} \qquad (10.2C)$$

These expressions are valid for any x such that $0 < x < L$. For this loading scheme, the shear force and the bending moment are functions of x, but the axial force is independent of x, being the same everywhere along the length of the beam.

We plot these expressions in **Figures 10.29*c*, *d*, and *e*** to gain insight into how they change with x. These plots are referred to as the **axial force diagram**, **shear force diagram**, and **bending moment diagram**. Engineers are often interested in finding the maximum bending moment a beam must carry. The bending moment diagram is a convenient tool for identifying this maximum value and at what point along the length of the beam it occurs. From **Figure 10.29*e*** the maximum for the cantilever in our example occurs at $x = L$; is this consistent with where you thought it would be? Is this where you would expect the beam to fail?

This cantilever beam example is repeated in **Figure 10.30** for the case in which the distributed load is acting upward. The solution is identical to that in **Figure 10.29**, except that the signs for the shear and bending moment are reversed.

The procedure we have illustrated in **Figure 10.29** for creating axial force, shear force, and bending moment diagrams works for any beam. In applying the procedure more generally, remember that it is necessary

(a) to create an additional free-body diagram wherever along the length a new load is introduced, and

(b) to be consistent in defining positive and negative bending moments and shear forces.

EXAMPLE 10.6 SHEAR, MOMENT, AND AXIAL FORCE DIAGRAM FOR A SIMPLY SUPPORTED BEAM

The simply supported beam in **Figure 10.31*a*** is loaded with a point load at a distance $\frac{L}{3}$ to the right of support A. Determine the axial force, shear force, and bending moment diagrams for the beam.

Goal We are to find the axial force, shear force, and bending moment diagrams for beam AB.

Given We are given the dimensions, support conditions, and loading of the beam.

Assume We assume that we can treat the system as planar and that the weight of the beam is negligible.

Draw By creating a free-body diagram of the entire beam and applying the conditions of equilibrium, we find the forces at A and B in terms of the applied force to be $F_{Ax} = 0$, $F_{Ay} = \frac{2Q}{3}$, $F_{By} = \frac{Q}{3}$ (**Figure 10.31*b***). We then follow the same procedure we did with the cantilever beam and create a free-body diagram of a portion of the beam. However, because a point load is applied to the simply supported beam, we need to create two free-body diagrams, one with the right-hand boundary at $x < \frac{L}{3}$ (**Figure 10.31*c***) and the other with the boundary at $x > \frac{L}{3}$ (**Figure 10.31*d***).

Formulate Equations and Solve Using the free-body diagram of the portion of the beam with the boundary at $x < \frac{L}{3}$ (**Figure 10.31*c***) as a reference, we write the planar equilibrium equations with the moment center at P (which is at $0 \le x \le \frac{L}{3}$):

$$\sum F_x = 0\ (\rightarrow +) \Rightarrow F_{Ax} + N_x = 0 \qquad \text{(1A)}$$

$$\sum F_y = 0\ (\uparrow +) \Rightarrow \frac{2Q}{3} - V_y = 0 \qquad \text{(1B)}$$

$$\sum M_{z@P} = 0\ (\circlearrowleft +) \Rightarrow -\frac{2Q}{3}x + M_{bz} = 0 \qquad \text{(1C)}$$

Equations (1A)–(1C) are valid for any x such that $0 \le x \le \frac{L}{3}$ and can be rearranged as

$$N_x = -F_{Ax} = 0 \quad \text{axial force} \qquad \text{(2A)}$$

$$V_y = \frac{2Q}{3} \quad \text{shear force} \qquad \text{(2B)}$$

$$M_{bz} = \frac{2Q}{3}x \quad \text{bending moment} \qquad \text{(2C)}$$

Referring to the free-body diagram of the portion of the beam with the boundary at $x > \frac{L}{3}$ (**Figure 10.31*d***), we write a second set of planar equilibrium equations with moment center at P (where P is now at $\frac{L}{3} \le x \le L$):

$$\sum F_x = 0\ (\rightarrow +) \Rightarrow F_{Ax} + N_x = 0 \qquad \text{(3A)}$$

$$\sum F_y = 0\ (\uparrow +) \Rightarrow \frac{2Q}{3} - Q - V_y = 0 \qquad \text{(3B)}$$

$$\sum M_{z@P} = 0\ (\circlearrowleft +) \Rightarrow -\frac{2Q}{3}x + Q\left(x - \frac{L}{3}\right) + M_{bz} = 0$$

which can be rearranged as

$$\sum M_{z@P} = 0 \Rightarrow \frac{Q}{3}(x - L) + M_{bz} = 0 \qquad \text{(3C)}$$

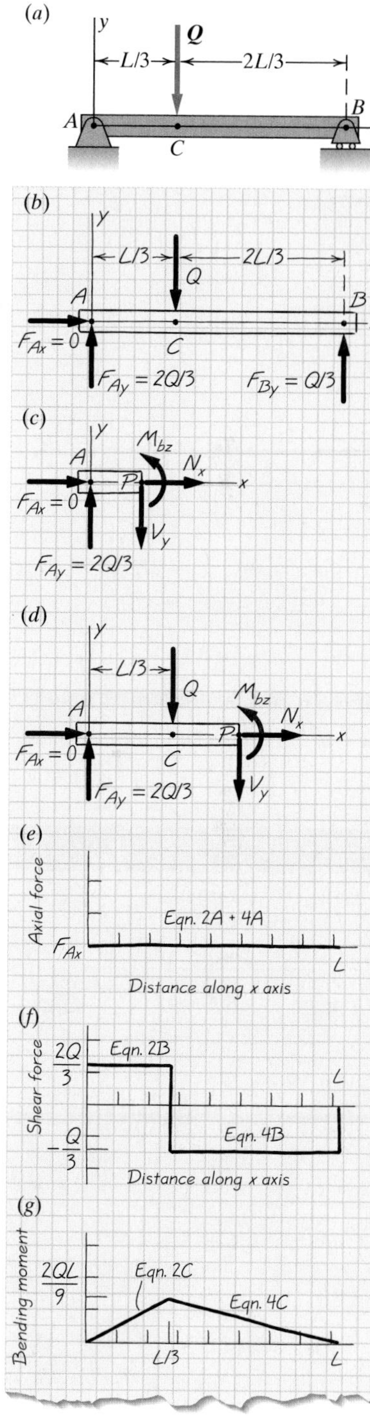

Figure 10.31

Equations (3A)–(3C) are valid for any x such that $\frac{L}{3} \leq x \leq L$ and can be rearranged as

$$N_x = -F_{Ax} = 0 \qquad \text{axial force} \qquad (4A)$$

$$V_y = -\frac{Q}{3} \qquad \text{shear force} \qquad (4B)$$

$$M_{bz} = \frac{Q}{3}(L - x) \qquad \text{bending moment} \qquad (4C)$$

We now create axial force, shear force, and bending moment diagrams for the simply supported beam. We use (2A)–(2C) to draw the diagrams for $0 \leq x \leq \frac{L}{3}$, and (4A)–(4C) for $\frac{L}{3} \leq x \leq L$, as shown in **Figure 10.31*e, f,*** and ***g***.

We notice that at $x = \frac{L}{3}$, the bending moment curves intersect at $M_{bz} = \frac{2QL}{9}$ (this is the value of bending moment given by both (2C) and (4C) at $x = \frac{L}{3}$). At the same point, there is a step change from positive to negative in the shear force.

From **Figure 10.31*g*** the largest magnitude of bending moment occurs at $x = \frac{L}{3}$; is this consistent with where you thought it would be? Is this where you would expect the beam to fail?

EXAMPLE 10.7 A SIMPLE BEAM

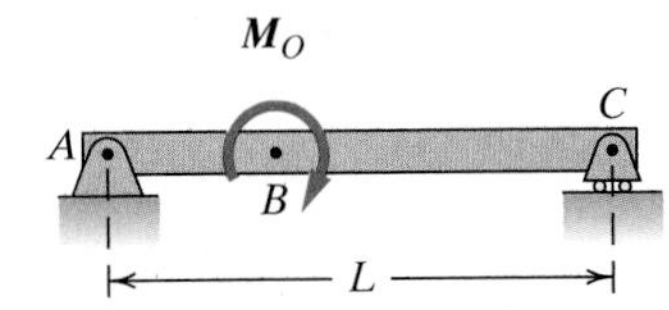

Figure 10.32

A moment is applied at point B on the simply supported beam of length L in **Figure 10.32**. Determine the shear force and bending moment diagrams.

Goal We are to find the shear and bending moment diagrams for beam AC.

Given We are given information about the geometry and loading of the beam.

Assume We assume that we can treat the system as planar and that the weight of the beam is negligible. We also assume that there are no forces in the x direction.

Draw We draw a free-body diagram of the entire beam and solve for loads at the supports (**Figure 10.33*a***). Next, we create a coordinate system with its origin at A and make two different cuts in the beam. We first place point P to the left of the applied moment, isolate portion AP, and draw a free-body diagram (**Figure 10.33*b***). Then we relocate P, cut the beam to the right of the applied load, and draw a free-body diagram of an isolated portion AP (**Figure 10.33*c***).

Formulate Equations and Solve Analyzing the equilibrium of the beam portions, we create equations that describe the shear and moment as a function of x.

For $0 \le x \le B$ (**Figure 10.33*b***):

$$\sum F_y = 0\ (\uparrow +)$$

$$-\frac{M_o}{L} - V_y = 0$$

$$V_y = -\frac{M_o}{L} \tag{1}$$

$$\sum M_{z@P} = 0\ (\curvearrowleft +)$$

$$\frac{M_o}{L}x + M_{bz} = 0$$

$$M_{bz} = -\frac{M_o}{L}x\ (0 \le x \le B) \tag{2}$$

For $B \le x \le L$ (**Figure 10.33*c***)

$$\sum F_y = 0\ (\uparrow +)$$

$$-\frac{M_o}{L} - V_y = 0$$

$$V_y = -\frac{M_o}{L} \tag{3}$$

$$\sum M_{z@P} = 0\ (\curvearrowleft +)$$

$$\frac{M_o}{L}x + -M_o + M_{bz} = 0$$

$$M_{bz} = M_o - \frac{M_o}{L}x$$

With some rearranging this becomes

$$M_{bz} = M_o\left(1 - \frac{x}{L}\right) \qquad (B \le x \le L) \tag{4}$$

Using (1)–(4) we create the shear force and bending moment diagrams for the beam (**Figures 10.33*d*** and **10.33*e***).

The shear force is constant between A and C, because no forces are applied between the supports. The bending moment diagram exhibits a discontinuity at the location of the applied moment. The magnitude of the discontinuity is equal to the applied moment (M_o).

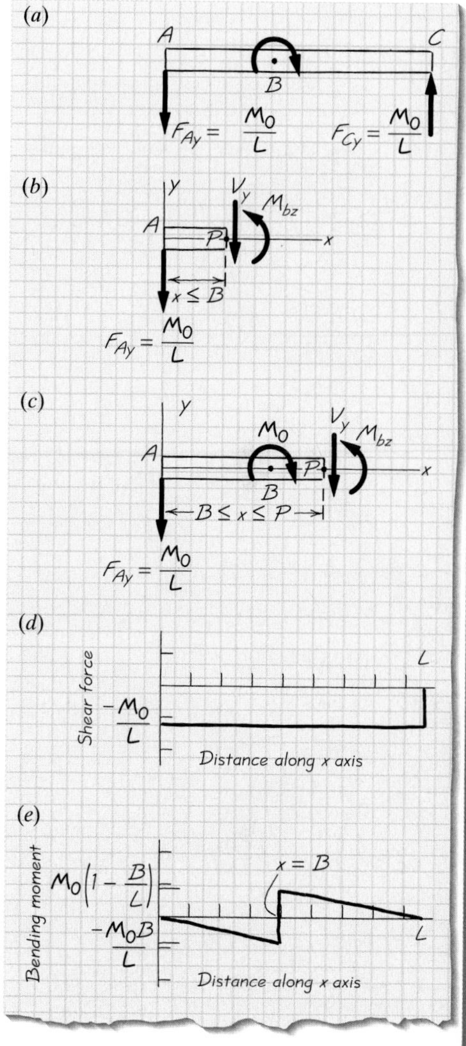

Figure 10.33

EXAMPLE 10.8 BEAM ANALYSIS

For the beam and loading shown in **Figure 10.34*a*** draw the shear force and bending moment diagrams.

Goal We are to draw the shear force and bending moment diagrams for beam ABC.

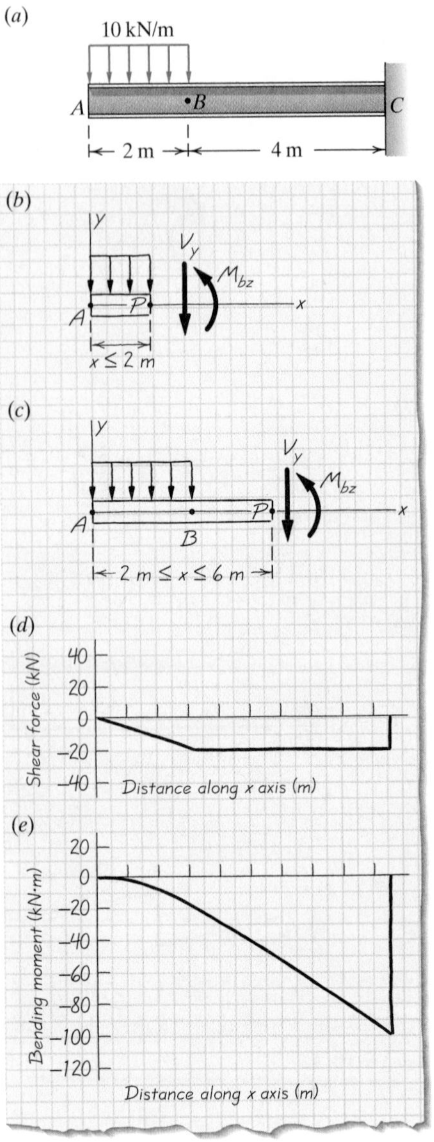

Figure 10.34

Given We are given information about the dimensions and loading of the beam.

Assume We assume that we can treat the system as planar and that the weight of the beam is negligible.

Draw We require two free-body diagrams to develop the equations that describe the shear and moment diagrams for the beam. First we cut the beam between A and B to isolate a portion of the beam (**Figure 10.34*b***). This free-body diagram is valid for a cut anywhere between A and B—that is, for $0 \leq x \leq 2$ m. Then we cut between B and C (**Figure 10.34*c***) and draw a free-body diagram that is valid for a cut line such that $2 \text{ m} \leq x \leq 6$ m.

Formulate Equations and Solve Analyzing the equilibrium of the beam portions, we create equations that describe the shear and moment as a function of x.

For $0 \leq x \leq 2$m (**Figure 10.34*b***): In our equilibrium analysis, we must include only the portion of the distributed load that is acting on the segment of the beam we are considering. The length of the beam segment we have isolated to the left of the cut at P is x. Therefore the load is $10\frac{\text{kN}}{\text{m}}x$, and we write (7.5A) as

$$\sum F_y = 0 \;(\uparrow +)$$

$$-10\frac{\text{kN}}{\text{m}}x - V_y = 0$$

$$V_y = -10\frac{\text{kN}}{\text{m}}x \tag{1}$$

This equation describes a line from 0 kN at $x = 0$ m to -20 kN at $x = 2$ m.

We choose P as our moment center in writing (7.5C). This means that we measure the moment arm from P to the centroid of the distributed load. For free-body diagram in **Figure 10.34*b*** that distance is $\frac{x}{2}$.

$$\sum M_{z@P} = 0 \;(\curvearrowleft +)$$

$$\left(10\frac{\text{kN}}{\text{m}}x\right)\frac{x}{2} + M_{bz} = 0$$

$$M_{bz} = -\left(5\frac{\text{kN}}{\text{m}}\right)x^2 \qquad (0 \leq x \leq 2 \text{ m}) \tag{2}$$

This equation describes a curve that starts from 0 kN · m at $x = 0$ m and bends parabolically down to -20 kN · m at $x = 2$ m.

For $2 \text{ m} \leq x \leq 6$ m (**Figure 10.34*c***): The entire distributed load is acting on the isolated portion of the beam shown in **Figure 10.34*c***. Therefore the load is $10\frac{\text{kN}}{\text{m}}$ (2 m) = 20 kN, and we write (7.5A) as

$$\sum F_y = 0 \;(\uparrow +)$$

$$-20 \text{ kN} - V_y = 0$$

$$V_y = -20 \text{ kN} \tag{3}$$

We measure the moment arm from P to the centroid of the distributed load, which is 1 m to the right of A in writing (7.5C). Therefore the length of the moment arm is $(x - 1\text{ m})$.

$$\sum M_{z@P} = 0\ (\curvearrowleft +)$$
$$20\text{ kN}(x - 1\text{ m}) + M_{bz} = 0$$
$$M_{bz} = 20\text{ kN}\cdot\text{m} - (20\text{ kN})x \qquad (2\text{ m} \le x \le 6\text{ m}) \qquad (4)$$

For this portion of the beam, the bending moment diagram is a straight line extending from $-20\text{ kN}\cdot\text{m}$ to $-100\text{ kN}\cdot\text{m}$.

Using (1)–(4) we create the shear force and bending moment diagrams for the beam (**Figures 10.34*d*** and **10.34*e***).

Check We illustrate a method for checking the shear and bending moment relationships following the next example.

EXAMPLE 10.9 A SIMPLY SUPPORTED BEAM WITH AN OVERHANG

Given the beam and loading shown in **Figure 10.35*a***, draw the shear force and the bending moment diagrams. Locate the maximum bending moment for the beam.

Goal We are to draw the shear force and bending moment diagrams and locate the maximum bending moment.

Given We are given information about the dimensions and loading of the beam.

Assume We assume that we can treat the system as planar and that the weight of the beam is negligible. We also assume that the pin at B is frictionless and imparts no moment to the beam, and that the roller at D is frictionless and imparts no moment or axial force to the beam.

Draw We draw a free-body diagram of the beam and solve for the loads at B and D (**Figure 10.35*b***). We then create three free-body diagrams: one for a cut between A and B, one for a cut between B and C, and one for a cut between C and D (**Figures 10.35*c***, **10.35*d***, and **10.35*e***).

Formulate Equations and Solve We perform an equilibrium analysis of each of the three free-body diagrams to create equations that describe the shear and bending moment as a function of x.

For $0 \le x \le 2$ ft (**Figure 10.35*c***), equlibrium equation (7.5A) is:

$$\sum F_y = 0\ (\uparrow +)$$
$$-2\frac{\text{kip}}{\text{ft}}x - V_y = 0$$
$$V_y = -2\frac{\text{kip}}{\text{ft}}x \qquad (1)$$

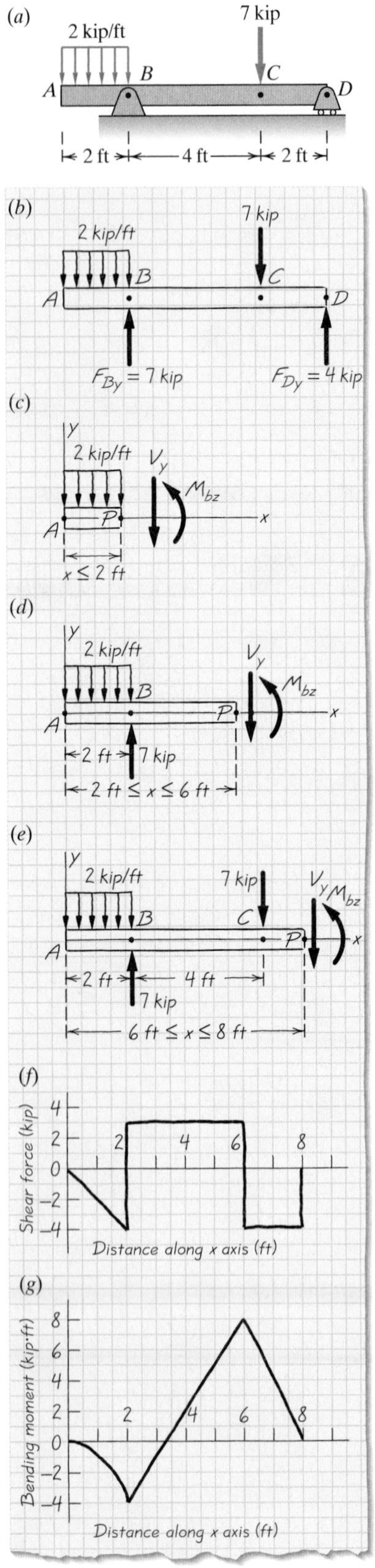

Figure 10.35

This equation describes a line from 0 kip at $x = 0$ ft to -4 kip at $x = 2$ ft. Equilibrium equation (7.5C) is:

$$\sum M_{z@P} = 0 \ (\curvearrowleft +)$$
$$\left(2\frac{\text{kip}}{\text{ft}}x\right)\frac{x}{2} + M_{bz} = 0$$
$$M_{bz} = -\left(1\frac{\text{kip}}{\text{ft}}\right)x^2 \qquad (0 \le x \le 2 \text{ ft}) \tag{2}$$

This equation describes a parabolic curve from 0 kip · ft at $x = 0$ ft to $-$ 4 kip · ft at $x = 2$ ft.

For 2 ft $\le x \le$ 6 ft (**Figure 10.35*d***), equilibrium equation (7.5A) is:

$$\sum F_y = 0 \ (\uparrow +)$$
$$-\left(2\frac{\text{kip}}{\text{ft}}\right)(2 \text{ ft}) + 7 \text{ kip} - V_y = 0$$
$$V_y = 3 \text{ kip} \tag{3}$$

Equilibrium equation (7.5C) is:

$$\sum M_{z@P} = 0 \ (\curvearrowleft +)$$
$$2\frac{\text{kip}}{\text{ft}}(2 \text{ ft})(x - 1 \text{ ft}) - 7 \text{ kip } (x - 2 \text{ ft}) + M_{bz} = 0$$

With rearranging and simplifying this becomes

$$M_{bz} = -10 \text{ kip} \cdot \text{ft} + (3 \text{ kip})x \quad (2 \text{ ft} \le x \le 6 \text{ ft}) \tag{4}$$

For this portion of the beam, the bending moment diagram is a straight line extending from -4 kip · ft at $x = 2$ ft to 8 kip · ft at $x = 6$ ft.

For 6 ft $\le x \le$ 8 ft (**Figure 10.35*e***), equilibrium equation (7.5A) is:

$$\sum F_y = 0 \ (\uparrow +)$$
$$-\left(2\frac{\text{kip}}{\text{ft}}\right)(2 \text{ ft}) + 7 \text{ kip} - 7 \text{ kip} - V_y = 0$$
$$V_y = -4 \text{ kip} \tag{5}$$

Equilibrium equation (7.5C) is:

$$\sum M_{z@P} = 0 \ (\curvearrowleft +)$$
$$2\frac{\text{kip}}{\text{ft}}(2 \text{ ft})(x - 1 \text{ ft}) - 7 \text{ kip}(x - 2 \text{ ft}) + 7 \text{ kip}(x - 6 \text{ ft}) + M_{bz} = 0$$

With rearranging and simplifying this becomes

$$M_{bz} = 32 \text{ kip} \cdot \text{ft} - (4 \text{ kip})x \qquad (6 \text{ ft} \le x \le 8 \text{ ft}) \tag{6}$$

The bending moment diagram is a straight line extending from 8 kip · ft at $x = 6$ ft to 0 kip · ft at $x = 8$ ft.

Using (1)–(6) we create the shear force and bending moment diagrams for the beam (**Figures 10.35*f*** and **10.35*g***).

Answer The shear and bending moment diagrams for the beam in **Figure 10.35*a*** are presented in **Figures 10.35*f*** and **10.35*g***, respectively. We see from the bending moment diagram that the maximum moment is 8 kip · ft and it occurs at point C.

Check We will check the results in Example 10.10 using the soon-to-be-developed relationships between ω, V_y, and M_{bz}.

The beams considered in **Figures 10.29** and **10.30** and Examples 10.5–10.9 are planar beams because the applied forces are all in a single plane and the moments are about an axis perpendicular to that plane. In contrast, the beam illustrated in **Figure 10.36** is a nonplanar beam. At a point P along the length of the beam there is bending moment M_{bz} about the z axis and shear force V_y in the y direction, but also bending moment M_{by} about the y axis, moment M_{bx} (sometimes referred to as torque) about the x axis, and shear force V_z in the z direction. Bending moment diagrams for M_{bx}, M_{by}, and M_{bz}, shear diagrams for V_y and V_z, and an axial force diagram for N_x are required to get a complete picture of the loads internal to the beam.

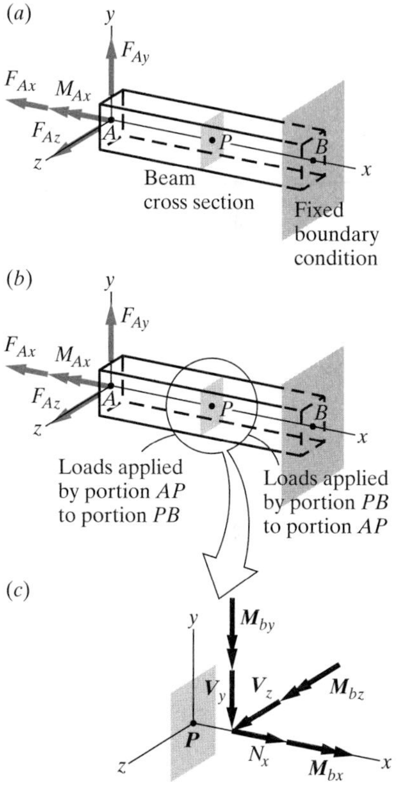

Figure 10.36 Loads internal to a nonplanar beam

Bending Moment Related to Shear Force

Figure 10.37*a* shows a beam that extends from A to B and is loaded by a distributed load ω. If we zoom in and consider equilibrium for a portion Δx of the beam (**Figure 10.37*b***) we can determine that the relationship between shear force and bending moment is

$$V_y = \frac{dM_{bz}}{dx} \tag{10.3A}$$

This expression says that the shear force is equal to the slope of the bending moment curve. Alternately, it can be interpreted as saying that the bending moment is the integral of the shear force. Equation (10.3A) is a useful check on bending moment and shear force equations, as illustrated in Example 10.10.

Equilibrium applied to the free-body diagram in **Figure 10.37*b*** also results in a relationship between the distributed force ω and the shear force:

$$\omega = \frac{-dV_y}{dx} \tag{10.3B}$$

This expression says that the distributed force ω is the negative slope of the shear force curve. If we combine (10.7A) and (10.7B) we get

$$-\omega = \frac{dV_y}{dx} = \frac{d^2M_{bz}}{dx^2} \tag{10.3C}$$

This expression can be used to find the bending moment if ω is a known and continuous function of x; two integrations are required.

More generally, expressions (10.3A)–(10.3C) show that equilibrium imposes specific relationships between the shear force, bending moment, and distributed force acting along the length of a beam.

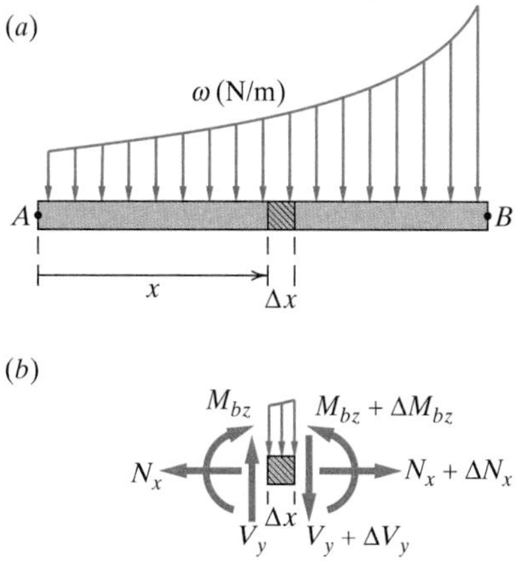

Figure 10.37 (*a*) Distributed load ω applied to beam AB; (*b*) free-body diagram of portion Δx of AB

EXAMPLE 10.10 USING EQUATIONS 10.3(A, B, AND C)

Check the solution for Example 10.9 using (10.7A, B, and C).

$$V_y = \frac{dM_{bz}}{dx} \quad (10.3\text{A}) \qquad \omega = -\frac{dV_y}{dx} \quad (10.3\text{B}) \qquad -\omega = \frac{d^2M_{bz}}{dx^2} \quad (10.3\text{C})$$

Goal We are to examine the equations that describe the shear force and bending moment diagrams for Example 10.9 and verify that they satisfy equations (10.7A, B, and C).

Given We are given information about the dimensions and loading of the beam.

Assume We assume that we can treat the system as planar and that the weight of the beam is negligible. We also assume that the pin at B is frictionless and therefore imparts no moment to the beam and that the roller at D is frictionless and therefore imparts no moment or axial force to the beam.

Section A–B **($0 \leq x \leq 2$ ft)** and free-body diagram in **Figure 10.35*c***:

$$M_{bz} = -\left(1\,\frac{\text{kip}}{\text{ft}}\right)x^2$$

Based on (10.3A) we write:

$$V_y = \frac{dM_{bz}}{dx} = \frac{d}{dx}\left(-1\,\frac{\text{kip}}{\text{ft}}\,x^2\right) = -\left(2\,\frac{\text{kip}}{\text{ft}}\right)x$$

This agrees with (1) in Example 10.9.

Based on (10.3B) we write

$$\omega = -\frac{dV_y}{dx} = -\frac{d}{dx}\left(-2\,\frac{\text{kip}}{\text{ft}}\,x\right) = 2\,\frac{\text{kip}}{\text{ft}}$$

Yes, this agrees with the $2\frac{\text{kip}}{\text{ft}}$ distributed load that is applied to the beam between A and B.

Another way to examine these relationships is to look at the integrals. Integrating (10.3B),

$$\int dV_y = -\int \omega(x)dx$$

$$V_{y@B} - V_{y@A} = -\int_0^{2\text{ ft}} 2\,\frac{\text{kip}}{\text{ft}}\,dx$$

This equation says the change in the shear diagram between A and B is the area under the distributed load curve between A and B.

$$V_{y@B} = -2\,\frac{\text{kip}}{\text{ft}}\,x\Big|_0^{2\text{ ft}} + \cancelto{0}{V_{y@A}} = -4\text{ kip}$$

The shear is represented by a linear equation with a value of -4 kip at $x = 2$ ft. Integrating (10.7A),

$$M_{bz@B} - M_{bz@A} = \int V_y \, dx$$

This equation says the change in the bending moment diagram between A and B is the area under the shear curve between A and B.

$$M_{bz@B} = -\int_0^{2\text{ ft}} 2\frac{\text{kip}}{\text{ft}} x \, dx + \cancelto{0}{M_{bz@A}} = -1\frac{\text{kip}}{\text{ft}} x^2 \Big|_0^{2\text{ ft}} = -4 \text{ kip} \cdot \text{ft}$$

The bending moment is represented by a quadratic equation with a value of -4 kip · ft at $x = 2$ ft.

Section B–C **(2 ft ≤ *x* ≤ 6 ft)** and free-body diagram in **Figure 10.35*d***: Along this section,

$$\omega = 0$$

Integrating (10.3B),

$$V_{y@C} - V_{y@B} = -\int 0 \, dx = 0$$

indicating that the shear is a constant between B and C. In fact the shear is a constant 3 kip between B and C.

Now consider integration of (10.3A):

$$M_{bz@C} - M_{bz@B} = \text{area under the shear} = \int_{2\text{ ft}}^{6\text{ ft}} 3\frac{\text{kip}}{\text{ft}} dx$$

$$= 3\frac{\text{kip}}{\text{ft}} x \Big|_{2\text{ ft}}^{6\text{ ft}} = 12 \text{ kip} \cdot \text{ft}$$

This shows that M_{bz} is changing linearly between B and C, and the change is 12 kip · ft. It increases from -4 kip · ft at B to 8 kip · ft at C.

Section C–D **(6 ft ≤ *x* ≤ 8 ft)** and free-body diagram in **Figure 10.35*e***: Along this section,

$$\omega = 0$$

Integrating (10.3B),

$$V_{y@D} - V_{y@C} = -\int 0 \, dx = 0$$

indicating that the shear is a constant between C and D. The shear diagram indicates a constant -4 kip between C and D.

Now consider integration of (10.3A):

$$M_{bz@D} - M_{bz@C} = \text{areas under the shear} = \int_{6\text{ ft}}^{8\text{ ft}} -4\frac{\text{kip}}{\text{ft}} dx$$

$$= -4\frac{\text{kip}}{\text{ft}} x \Big|_{6\text{ ft}}^{8\text{ ft}} = -8 \text{ kip} \cdot \text{ft}$$

M_{bz} is changing linearly between C and D, from 8 kip · ft at C to 0 kip · ft at D.

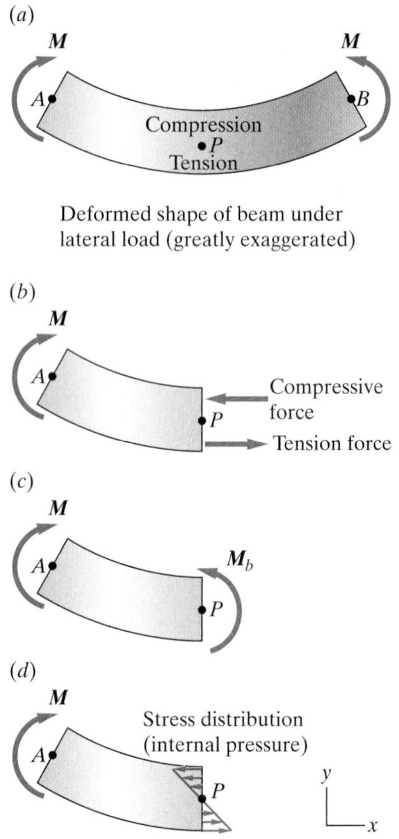

Figure 10.38 (*a*) Moment M applied to beam AB; (B) portion of beam AP showing compression force in top and tension force in bottom at P (these forces form a couple); (*c*) couple in (*b*) represented as the bending moment M_b; (*d*) bending moment created by distributed internal pressure (stress)

Bending Moment Related to Stress

Apply two moments to a beam as in **Figure 10.38*a*** and the bottom of the beam stretches (because it is in tension) while the top contracts (because it is in compression). The tension force in the bottom of the beam and the compression force in the top form a couple (**Figure 10.38*b***). This couple is, in fact, what we have been calling "bending moment" (see **Figure 10.38*c***). Therefore we can legitimately replace the two anti-parallel arrows representing the bending moment M_b in **Figure 10.38*b*** with the curved single arrow for M_b in **Figure 10.38*c***.

The tension and compression that make up the bending moment are not actually internal point forces as depicted in **Figure 10.38*b***, but rather are an internal "pressure" commonly referred to as **stress**. The tension force is created within the beam as a distribution of stresses that tend to pull the material, and the compression force is created within the beam as a distribution of stresses that then push the material. The complete distribution of stresses in the beam's cross section is shown in **Figure 10.38*d*** and is described by the relationship

$$\text{stress} = \frac{-M_{bz}y}{\displaystyle\iint_{\substack{\text{beam}\\\text{cross}\\\text{section}}} y^2 dy\, dz} \tag{10.4}$$

where M_{bz} is the bending moment at a particular location x along the length of the beam and y in the numerator is the y coordinate within the cross section where the stress is being calculated. The denominator is a scalar quantity that reflects the distribution of material in the beam cross section and is called the **area moment of inertia**—its calculation is detailed in Appendix C. Equation (10.8) is derived by considering equilibrium in a portion of a beam, and you can find its complete development in a mechanics of materials text. Stress is commonly expressed in N/mm^2 or MPa in the SI system, and in $pounds/in^2$ (psi) or kip/in^2 (ksi) in the English system. Notice that it has the same units as pressure.

Equation (10.8) shows that the stress is largest

- at the cross section of the beam where the bending moment M_{bz} is largest;
- at the location on that cross section with the largest value of y; (Because y is measured as shown in **Figure 10.38*d***, this means that stress is highest at the beam surface.)
- for equal bending moments, at the cross section with the smallest area moment of inertia.

Determining stress values in beams, and in systems more generally, is the focus of courses on mechanics of materials (James M. Gere and Stephen P. Timoshenko, *Mechanics of Materials*, PWS Publishing, 2002). Values of stress are compared with material capacity to determine whether the system is adequate (i.e., will not fail).

EXAMPLE 10.11 EXPLORING EQUATION (10.4)

The cross section of the beam of Example 10.8 is a rectangle 0.4 m high and 0.1 m wide. Determine the largest stress in the beam.

Goal We are to determine the maximum stress in beam ABC of **Figure 10.34*a***.

Given The beam cross section and the shear and bending moment diagrams we developed in Example 10.8.

Assume We assume that the cross section of the beam is constant throughout its length.

Formulate Equations and Solve Using (10.4),

$$\text{stress} = \frac{-M_{bz}y}{\displaystyle\iint_{\substack{\text{beam}\\ \text{cross}\\ \text{section}}} y^2 dy\, dz} = \frac{-M_{bz}y}{I}$$

We can examine the variation of stress throughout the beam. In (10.4) I is the area moment of inertia, and y is the distance from the centroid of the cross section to the beam surface. If the cross section remains constant throughout the beam, then I and y are the same everywhere along the length of the beam. Thus the maximum stress will occur where $\| \boldsymbol{M}_{bz}(x) \|$ is maximum. Examining the moment diagram in the results of Example 10.8 (**Figure 10.34*e***), we see that the maximum moment of -100 kN · m occurs at C.

According to Appendix C, the area moment of inertia for a rectangular beam is $\frac{bh^3}{12}$, where b is the beam width and h is the height.

$$I = \frac{0.1 \text{ m } (0.4 \text{ m})^3}{12} = 0.533 \times 10^{-3} \text{ m}^4$$

$$\text{Maximum stress at } C = -\frac{(-100 \text{ kN} \cdot \text{m})(0.2 \text{ m})}{0.533 \times 10^{-3} \text{ m}^4} =$$

$$37.5 \times 10^3 \frac{\text{N}}{\text{m}^2} = 37.5 \text{ MPa}$$

Answer $(\text{Stress})_{max} = 37.5$ MPa and occurs at $x = 6$ m. Because the beam is curving downward, the top of the beam experiences tensile stresses and the bottom of the beam experiences compressive stresses, as illustrated in **Figure 10.39**.

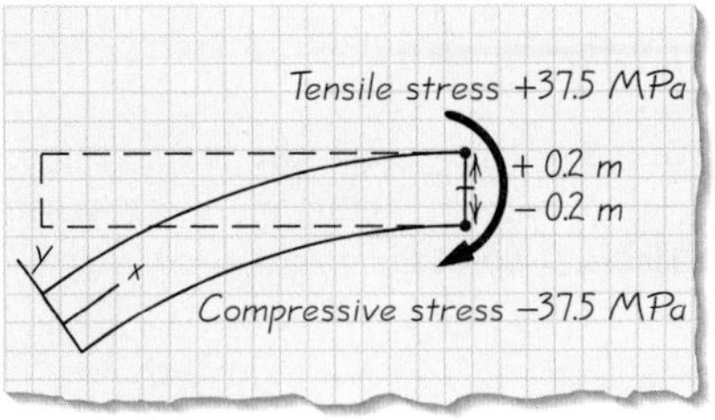

Figure 10.39

Beam Summary

A member is called a **beam** if the external loads that act on it are perpendicular to its long axis. These loads, which may consist of forces and/or moments, are referred to as **lateral loads**. The loads *internal* to a beam are called axial force ($\boldsymbol{N}$), shear force ($\boldsymbol{V}$), and bending moment ($\boldsymbol{M}_b$).

These internal loads are created by tension, compression, and shear forces (as described in Chapter 4). By isolating portions of a beam and then applying equilibrium conditions, we are able to describe how these internal loads vary along the length of the beam. These loads are not independent of one another—for example, the shear force is related to the bending moment by $V_y = dM_{bz}/dx$ (10.7A).

The tension and compression that make up the bending moment are not actually internal point forces, but rather are an internal "pressure" commonly referred to as **stress**. The relationship between bending moment and stress is given in (10.8).

EXERCISES 10.1

10.1.1. Determine the axial force, shear force, and bending moment acting on the cross section of the beam in **E10.1.1** at point *C*.

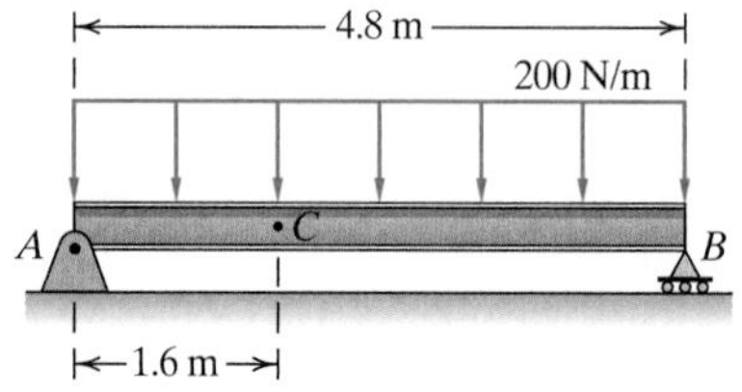

E10.1.1

10.1.2. Determine the axial force, shear force, and bending moment acting on the cross section of the beam in **E10.1.2** at point *C*.

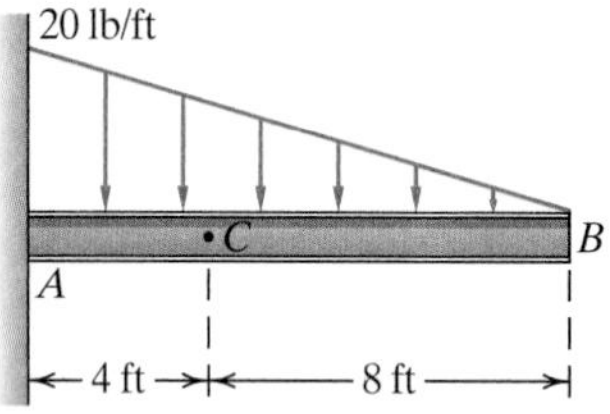

E10.1.2

10.1.3. Determine the axial force, shear force, and bending moment acting on the cross section of the beam in **E10.1.3** at point *C*.

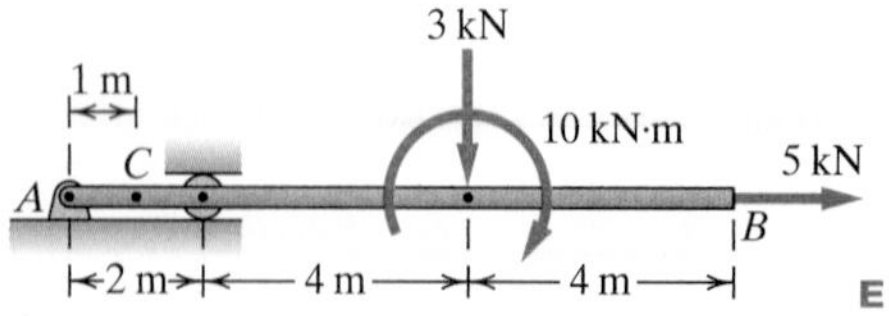

E10.1.3

10.1.4. Determine the axial force, shear force, and bending moment acting on the cross section of the beam in **E10.1.4** at

a. point *C*
b. point *D*

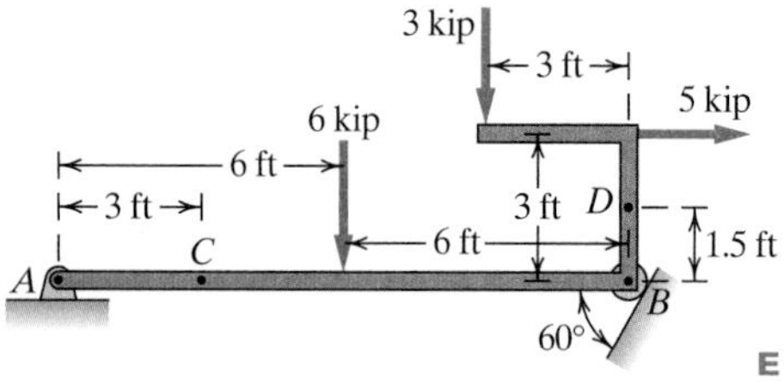

E10.1.4

10.1.5. Determine the axial force, shear force, and bending moment acting on the cross section of the beam in **E10.1.5** at point *C*.

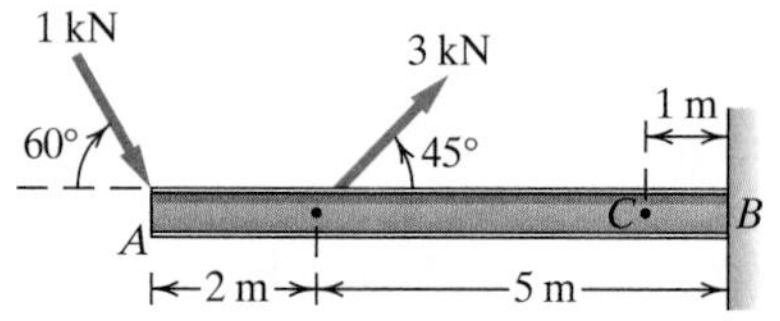

E10.1.5

10.1.6. Determine the axial force, shear force, and bending moment acting on the cross section of the beam in **E10.1.6** at point *D*. Note that member *BC* acts as a two-force member.

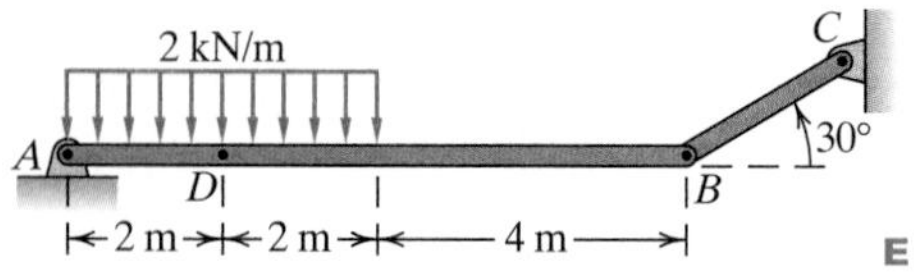

E10.1.6

10.1.7. Determine the axial force, shear force, and bending moment acting on the cross section of the beam *BC* at *D* (**E10.1.7**).

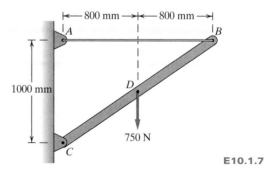

E10.1.7

10.1.8. Determine the axial force, shear force, and bending moment acting on the cross section of the beam *AB* at *D* (**E10.1.8**).

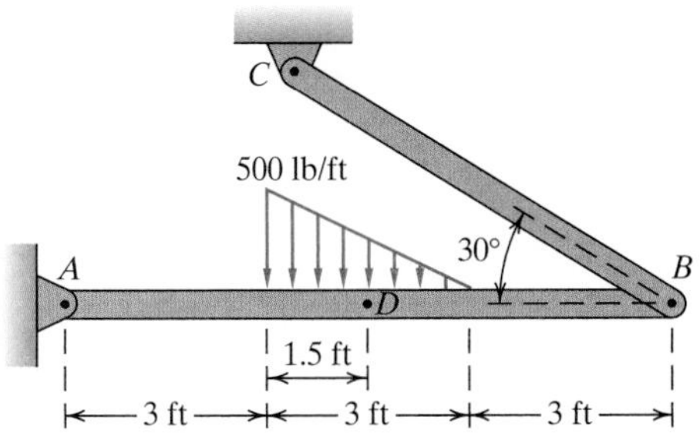

E10.1.8

10.1.9. The bulldozer shown in **E10.1.9** pushes a dirt load with a resultant horizontal force of 500 lb. The frame consists of push-arm *ABC*, hydraulic arms *BD* and *EG*, and blade *CDE*. All connections (*A*, *B*, *C*, *E*, *D*, and *G*) are frictionless pins. Ignore the weight of the members. Determine the axial force, shear force, and bending moment acting on the cross section of the beam *ABC* just to the right of point *B*.

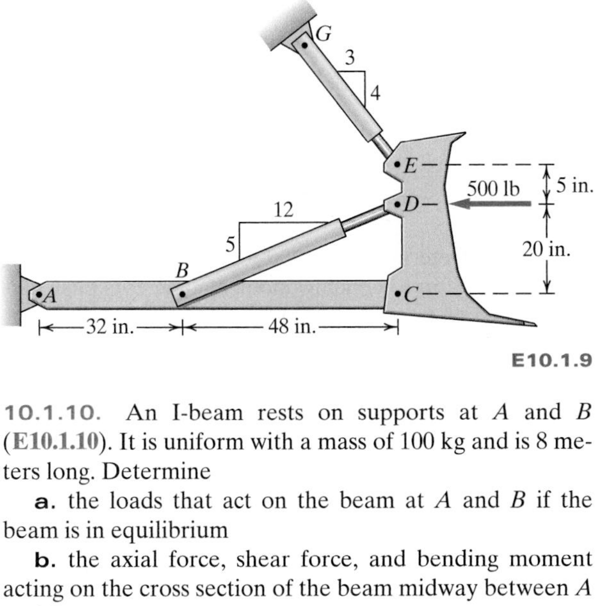

E10.1.9

10.1.10. An I-beam rests on supports at *A* and *B* (**E10.1.10**). It is uniform with a mass of 100 kg and is 8 meters long. Determine

a. the loads that act on the beam at *A* and *B* if the beam is in equilibrium

b. the axial force, shear force, and bending moment acting on the cross section of the beam midway between *A* and *B*

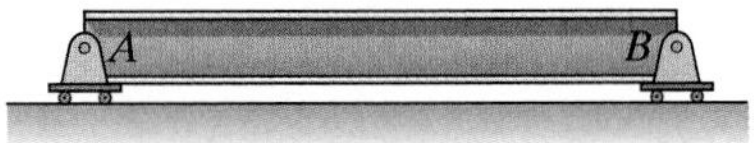

E10.1.10

10.1.11. An overhead crane consisting on the three pulleys shown in **E10.1.11** has been attached to the I-beam at *C* to move the beam. The beam is uniform with a mass of 100 kg. In the position shown, end *B* is 3 m off the ground.

a. Determine the tension ***P*** acting at *C* if the I-beam is in equilibrium.

b. Determine the axial force, shear force, and bending moment acting on the cross section of the beam midway between *A* and *B*.

c. Estimate how many statics students would be required to apply the force ***T*** to hold the I-beam in the position shown.

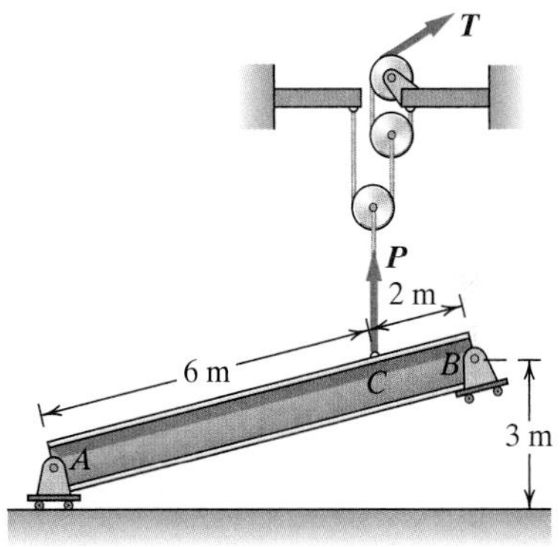

E10.1.11

10.1.12. Determine the axial force, shear force, and bending moment acting on the cross section of the beam *AB* in **E10.1.12** at point *C*.

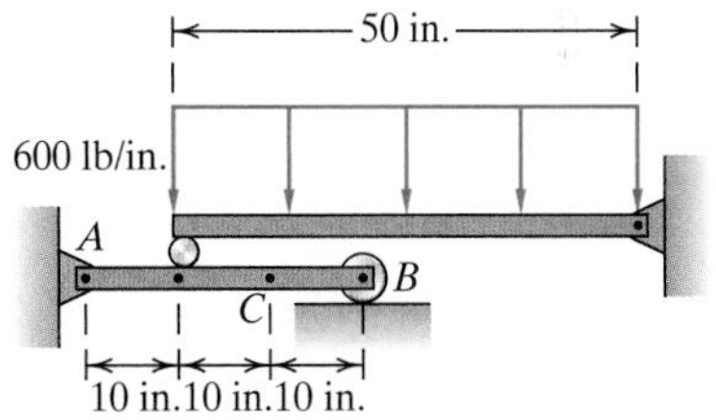

E10.1.12

10.1.13. Determine the axial force, shear force, and bending moment acting on the cross section of the beam *ABC* in **E10.1.13** just below point *B*.

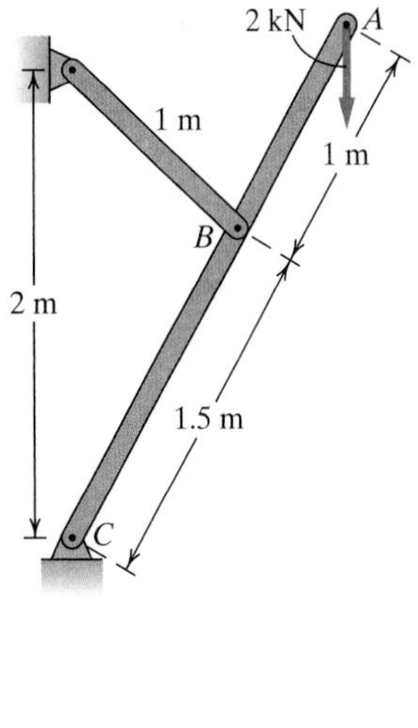

E10.1.13

10.1.14. Determine the axial force, shear force, and bending moment acting on the cross section of the beam *AB* in **E10.1.14** at point *C*.

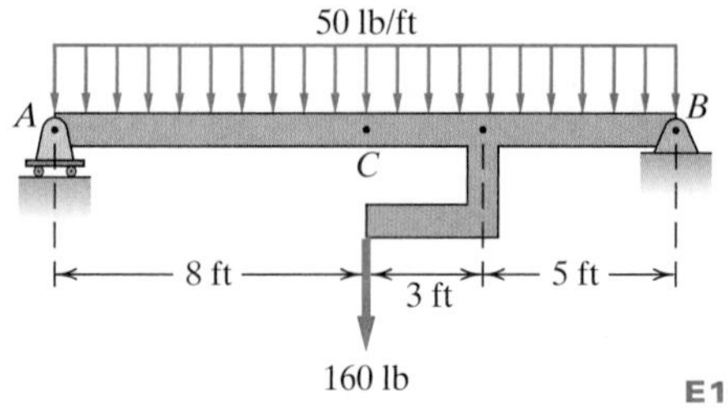

E10.1.14

10.1.15. Determine the axial force, shear force, and bending moment acting on the cross section of the beam *AB* in **E10.1.15** at point *D*, which is located midway between *A* and *B*.

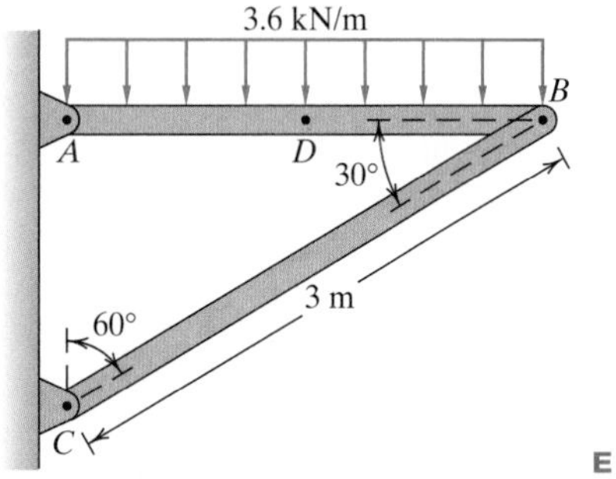

E10.1.15

10.1.16. Determine the axial force and the magnitudes of the shear force, bending moment, and torsional moment acting on the cross section of the bent beam in **E10.1.16** at

a. location just to the right of *C*

b. location *D*

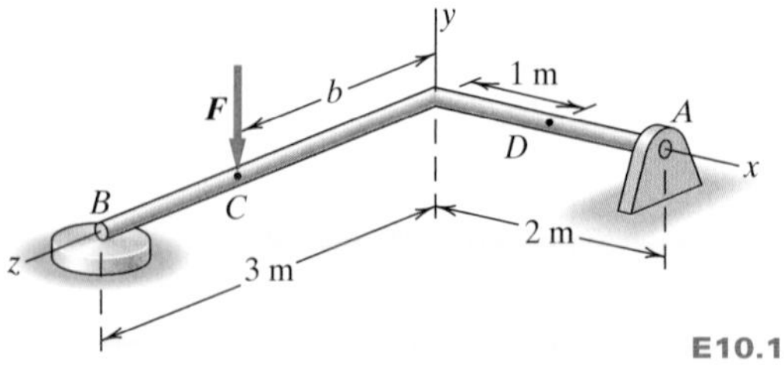

E10.1.16

10.1.17. Determine the axial force and the magnitudes of the shear force, bending moment, and torsional moment acting on the cross section at location *C* of the T-beam in **E10.1.17**.

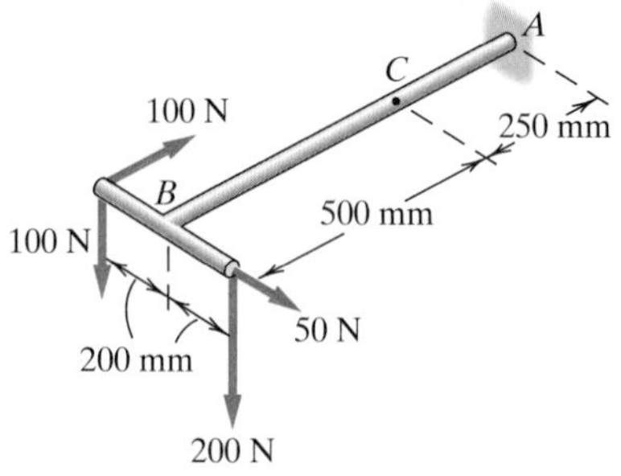

E10.1.17

10.1.18. Determine the axial force and the magnitudes of the shear force, bending moment, and torsional moment acting on the cross section at location *C* of the beam in **E10.1.18**.

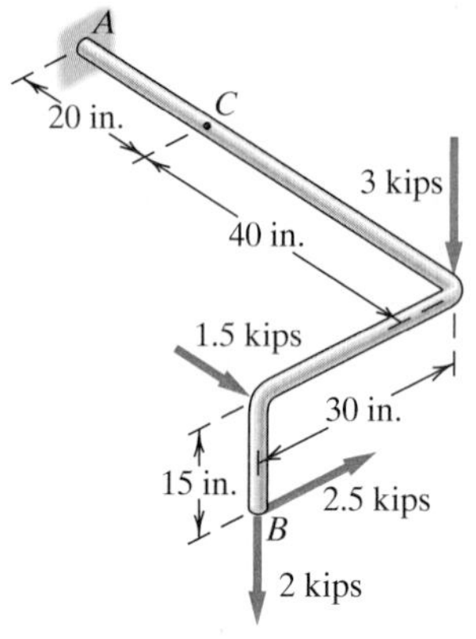

E10.1.18

10.1.19. Determine the axial force and the magnitudes of the shear force, bending moment, and torsional moment acting on the cross section at location *C* of the beam in **E10.1.19**.

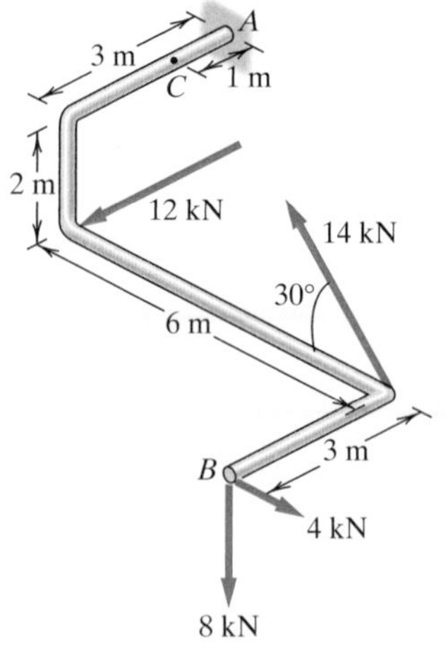

E10.1.19

10.1.20. Consider the freeway exit sign shown in **E10.1.20*a***.

a. Determine the loads that the horizontal boom (as shown in **E10.1.20*b***) applies to the vertical beam *KD* if the system is in equilibrium.

b. Draw the shear, axial, and bending moment diagrams for the beam *KD*.

c. If the beam *KD* is in equilibrium, what loads act on it at its base? (You could use the diagrams in **b** to answer this question, or you could do additional calculations.)

d. Based on the loads you found in **c**, which of the four bolts do you expect to be in greatest tension (and why)?

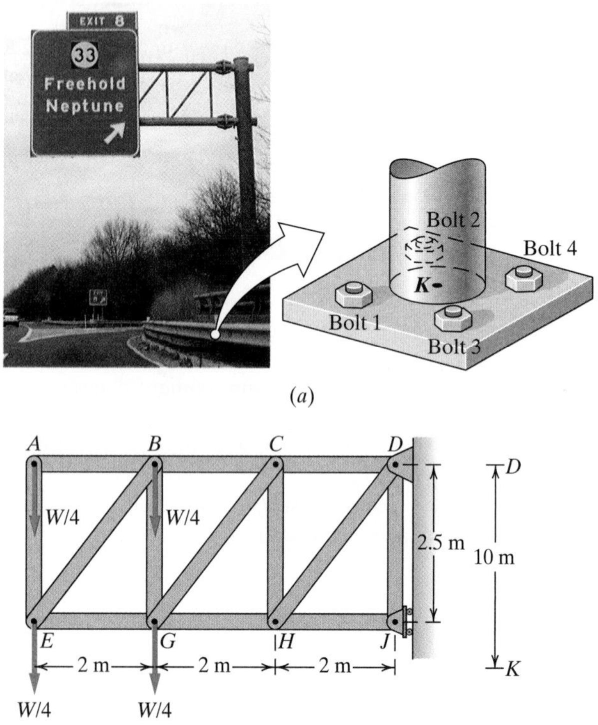

E10.1.20

10.1.21. Consider the beam in **E10.1.21**.

a. Draw the shear and bending moment diagrams for the beam.

b. Confirm that the shear and bending moments follow the relationship in (10.7A).

c. Determine the largest (in terms of magnitude) bending moment and its location.

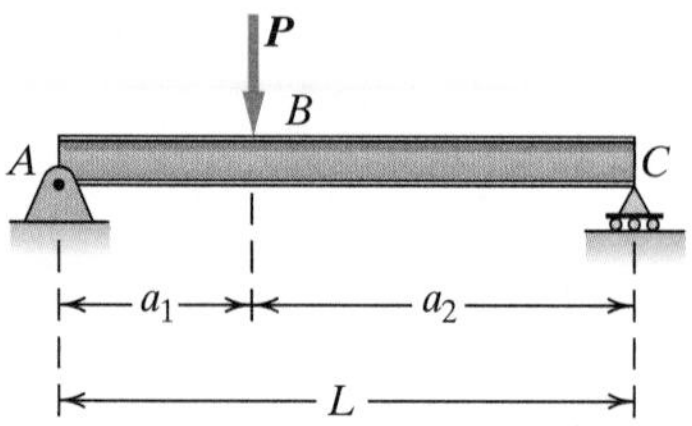

E10.1.21

10.1.22. Consider the beam in **E10.1.22**.

a. Draw the shear and bending moment diagrams for the beam.

b. Confirm that the shear and bending moments follow the relationship in (10.7A).

c. Determine the largest (in terms of magnitude) bending moment and its location.

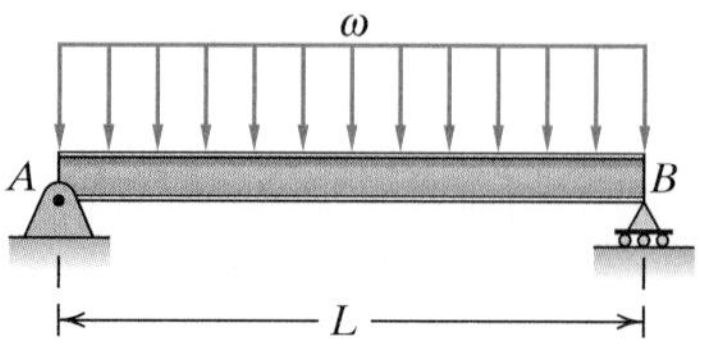

E10.1.22

10.1.23. Consider the beam in **E10.1.23**.

a. Draw the shear and bending moment diagrams for the beam.

b. Confirm that the shear and bending moments follow the relationship in (10.7A).

c. Determine the largest (in terms of magnitude) bending moment and its location.

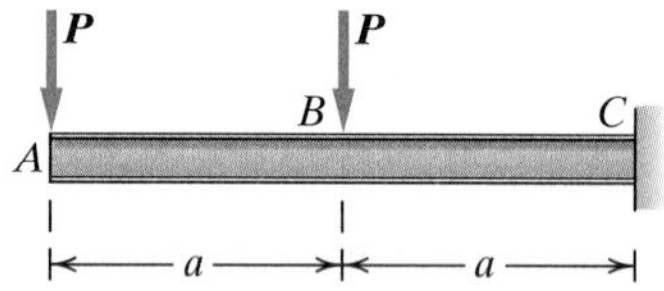

E10.1.23

10.1.24. Consider the beam in **E10.1.24**.

a. Draw the shear and bending moment diagrams for the beam.

b. Confirm that the shear and bending moments follow the relationship in (10.7A).

c. Determine the largest (in terms of magnitude) bending moment and its location.

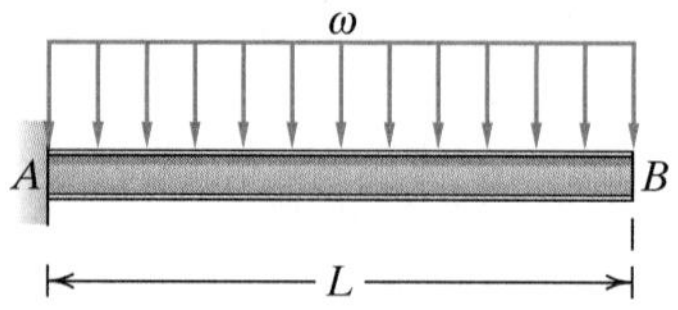

E10.1.24

10.1.25. Consider the beam in **E10.1.25**.

a. Draw the shear and bending moment diagrams for the beam.

b. Confirm that the shear and bending moments follow the relationship in (10.7A).

c. Determine the largest (in terms of magnitude) bending moment and its location.

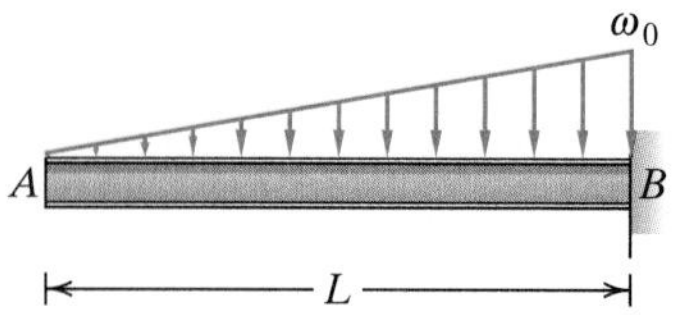

E10.1.25

10.1.26. Consider the beam in **E10.1.26**.

a. Draw the shear and bending moment diagrams for the beam.

b. Confirm that the shear and bending moments follow the relationship in (10.7A).

c. Determine the largest (in terms of magnitude) bending moment and its location.

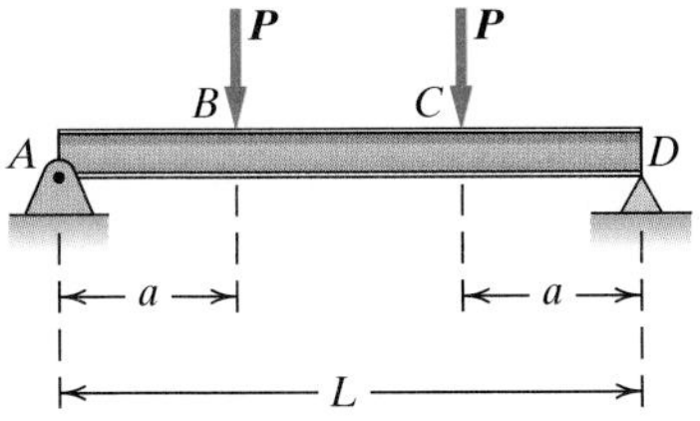

E10.1.26

10.1.27. Consider the beam in **E10.1.27**.

a. Draw the shear and bending moment diagrams for the beam.

b. Confirm that the shear and bending moments follow the relationship in (10.7A).

c. Determine the largest (in terms of magnitude) bending moment and its location.

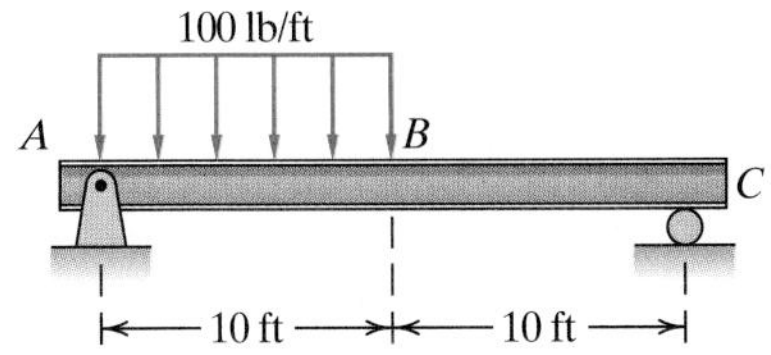

E10.1.27

10.1.28. Consider the beam in **E10.1.28**.

a. Draw the shear and bending moment diagrams for the beam.

b. Confirm that the shear and bending moments follow the relationship in (10.7A).

c. Determine the largest (in terms of magnitude) bending moment and its location.

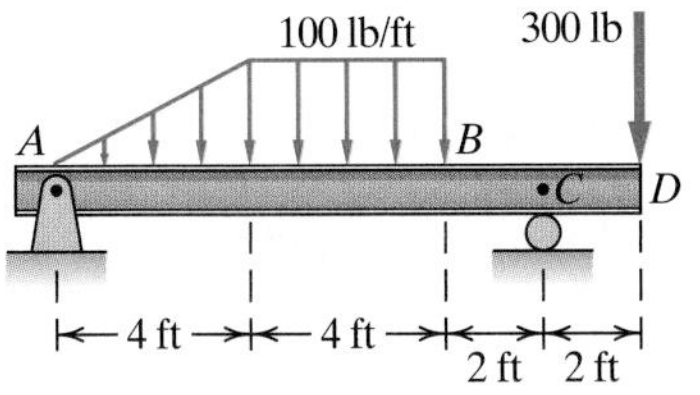

E10.1.28

10.1.29. Consider the beam in **E10.1.29**.

a. Draw the shear and bending moment diagrams for the beam.

b. Confirm that the shear and bending moments follow the relationship in (10.7A).

c. Determine the largest (in terms of magnitude) bending moment and its location.

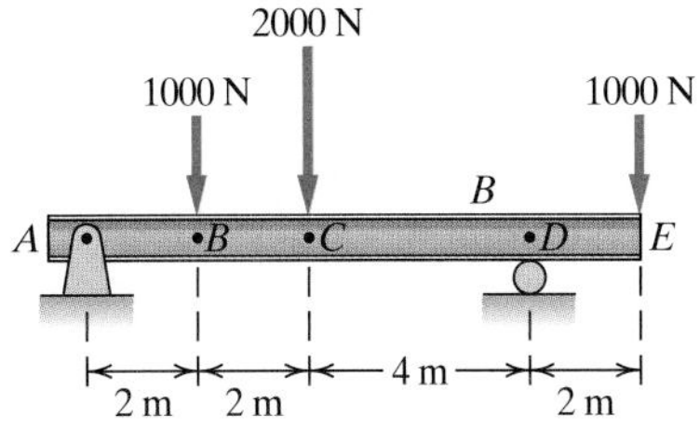

E10.1.29

10.1.30. Consider the beam in **E10.1.30**.

a. Draw the shear and bending moment diagrams for the beam.

b. Confirm that the shear and bending moments follow the relationship in (10.7A).

c. Determine the largest (in terms of magnitude) bending moment and its location.

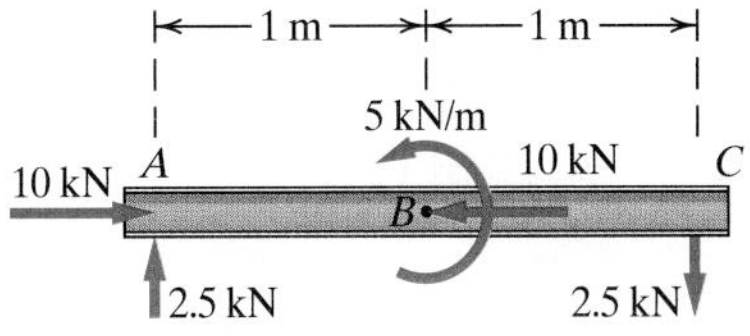

E10.1.30

10.1.31. Consider the beam in **E10.1.31**.

a. Draw the shear and bending moment diagrams for the beam.

b. Confirm that the shear and bending moments follow the relationship in (10.7A).

c. Determine the largest (in terms of magnitude) bending moment and its location.

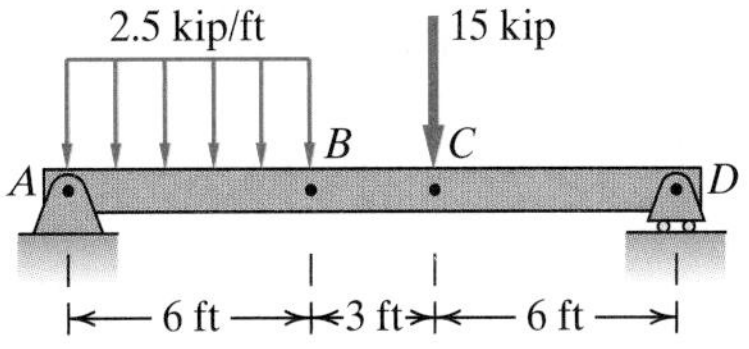

E10.1.31

10.1.32. Consider the beam in **E10.1.32**.

a. Draw the shear and bending moment diagrams for the beam.

b. Confirm that the shear and bending moments follow the relationship in (10.7A).

c. Determine the largest (in terms of magnitude) bending moment and its location.

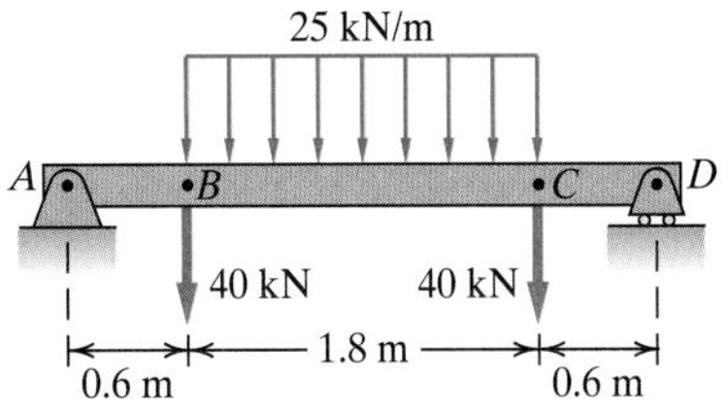

E10.1.32

10.1.33. Consider the airplane wing in E10.1.33.

a. Draw the shear and bending moment diagrams for the airplane wing due to the lift load shown.

b. Confirm that the shear and bending moments follow the relationship in (10.7A).

c. Determine the largest (in terms of magnitude) bending moment and its location.

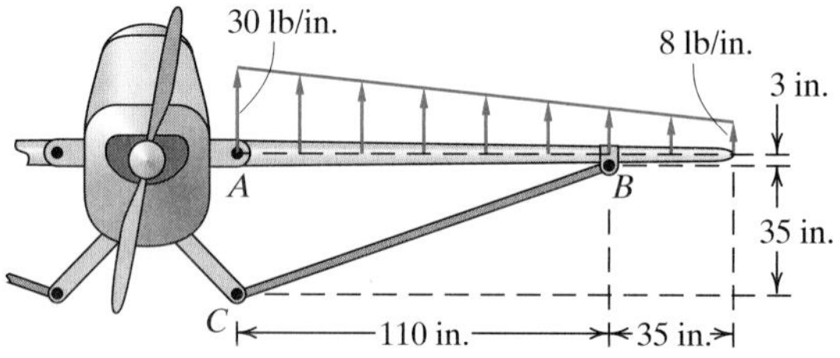

E10.1.33

10.1.34. Consider the L-beam in E10.1.34.

a. Draw the shear and bending moment diagrams for the L-beam. (*Hint:* Consider AB to be one beam and BC to be another beam.)

b. Confirm that the shear and bending moments follow the relationship in (10.7A).

c. Determine the largest (in terms of magnitude) bending moment and its location.

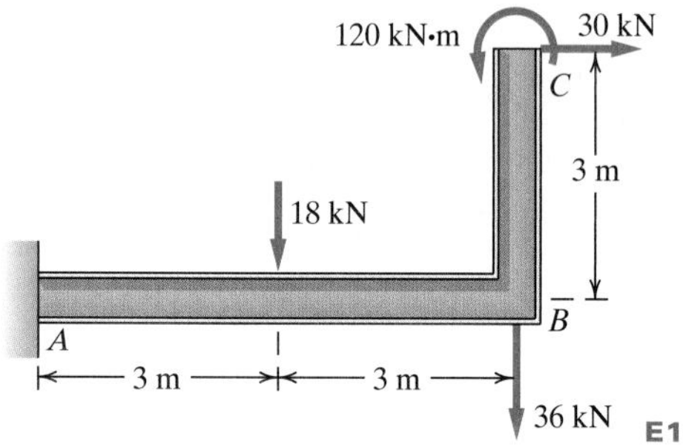

E10.1.34

10.1.35. Consider the overhung diving board in E10.1.35.

a. Draw the shear and bending moment diagrams for the diving board. A 100-kg person is standing on the free end.

b. Confirm that the shear and bending moments follow the relationship in (10.7A).

c. Determine the largest (in terms of magnitude) bending moment and its location.

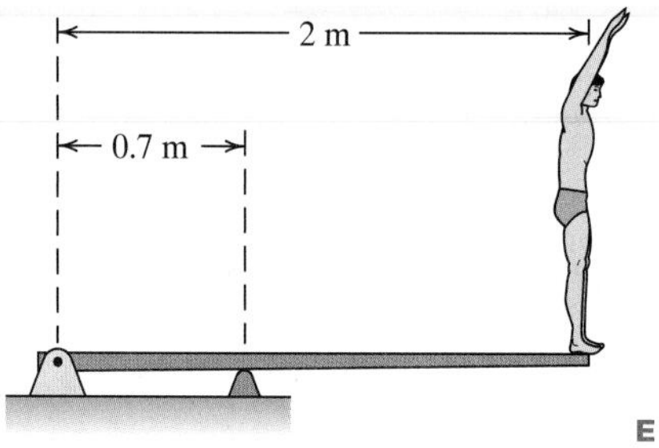

E10.1.35

10.1.36. Consider the cantilevered diving board in E10.1.36.

a. Draw the shear and bending moment diagrams for the diving board. A 100-kg person is standing on the free end.

b. Confirm that the shear and bending moments follow the relationship in (10.7A).

c. Determine the largest (in terms of magnitude) bending moment and its location.

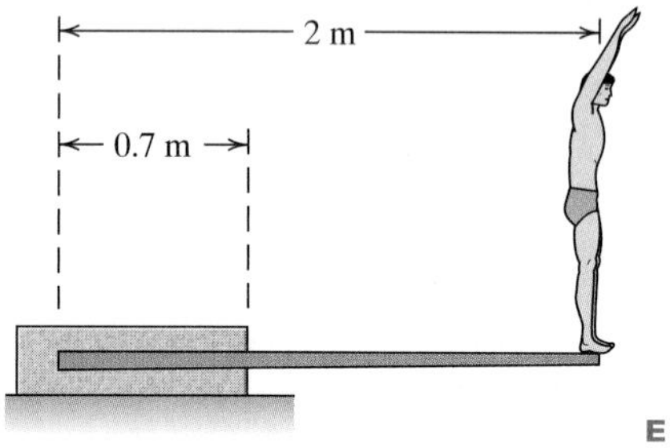

E10.1.36

10.1.37. A clamp holds the block shown in E10.1.37 with a clamping force of 350 N. Determine the shear force, tension, and bending moment at section A of the clamp bar for $x = 150$ mm. Which of these three quantities changes with x?

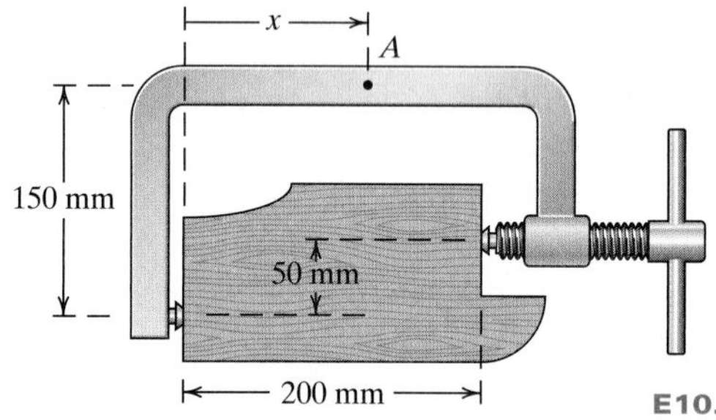

E10.1.37

10.1.38. Consider the beam AB in E10.1.38.

a. Draw the shear and bending moment diagrams for the shaft member AB.

b. Confirm that the shear and bending moments follow the relationship in (10.3A).

c. Determine the largest (in terms of magnitude) bending moment and its location.

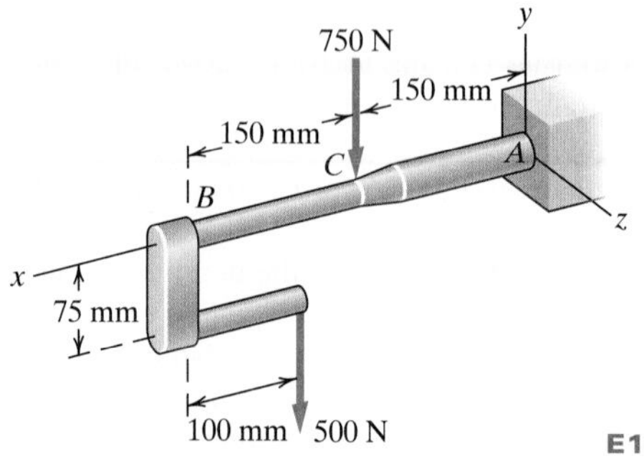

E10.1.38

10.1.39. For the beam shown in **E10.1.6** determine the maximum

a. bending moment (magnitude) and its location
b. shear force (magnitude) and its location
c. axial force and its location

10.1.40. For the beam *BC* shown in **E10.1.7** determine the maximum

a. bending moment (magnitude) and its location
b. shear force (magnitude) and its location
c. axial force and its location

10.1.41. For the beam *AB* shown in **E10.1.8** determine the maximum

a. bending moment (magnitude) and its location
b. shear force (magnitude) and its location
c. axial force and its location

10.1.42. For the beam *ABC* shown in **E10.1.9** determine the maximum

a. bending moment (magnitude) and its location
b. shear force (magnitude) and its location
c. axial force and its location

10.1.43. For the beam *AB* shown in **E10.1.12** determine the maximum

a. bending moment (magnitude) and its location
b. shear force (magnitude) and its location
c. axial force and its location

10.1.44. For the beam *ABC* shown in **E10.1.13** determine the maximum

a. bending moment (magnitude) and its location
b. shear force (magnitude) and its location
c. axial force and its location

10.1.45. For the beam *AB* shown in **E10.1.14** determine the maximum

a. bending moment (magnitude) and its location
b. shear force (magnitude) and its location
c. axial force and its location

10.1.46. Beam *AB* is supported by a thrust bearing at *A* and a journal bearing at *B*, as shown in **E10.1.46**. It is subjected to linearly varying loads in mutually perpendicular planes as shown.

a. Write an expression for the bending moment M_{bz} as a function of x.

b. Write an expression for the bending moment M_{by} as a function of x.

c. Combine the expressions for bending moment into a single expression for bending moment as a function of x. (Remember, moment is a vector quantity.)

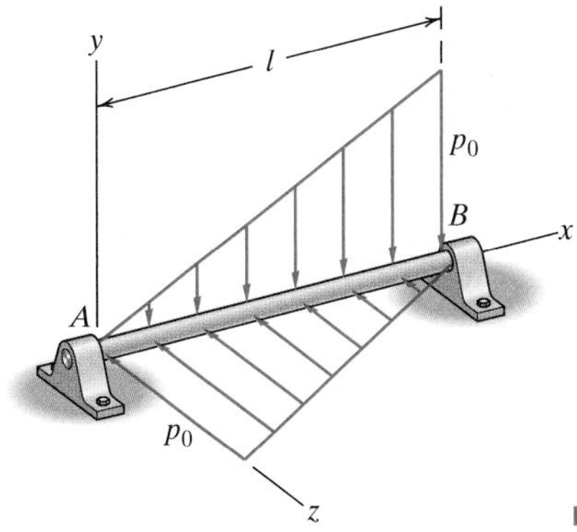

E10.1.46

10.1.47. The shaft shown in **E10.1.47** carries a pulley *A* (diameter of 150 mm) and a gear *B* (diameter of 100 mm). It is supported by frictionless bearings *D* and *E*. The gear *B* is subjected to the force $\boldsymbol{P} = -40\text{ N}\,\boldsymbol{j} + 210\text{ N}\,\boldsymbol{k}$. A moment (torque) of 20 N · m is applied to prevent rotation of the shaft.

a. Determine the magnitude of the tension $\boldsymbol{T}$ in the belt at point *G* and the loads acting on the shaft at the bearing *D* and *E*. Belt force $\boldsymbol{T}$ and the 270 N force are in the z direction.

b. Write an expression for the bending moment M_{bz} as a function of x.

c. Write an expression for the bending moment M_{by} as a function of x.

d. Combine the expressions for bending moment into a single expression for bending moment as a function of x. (Remember, moment is a vector quantity.)

e. Write an expression for the twisting (torque) moment in the shaft as a function of x.

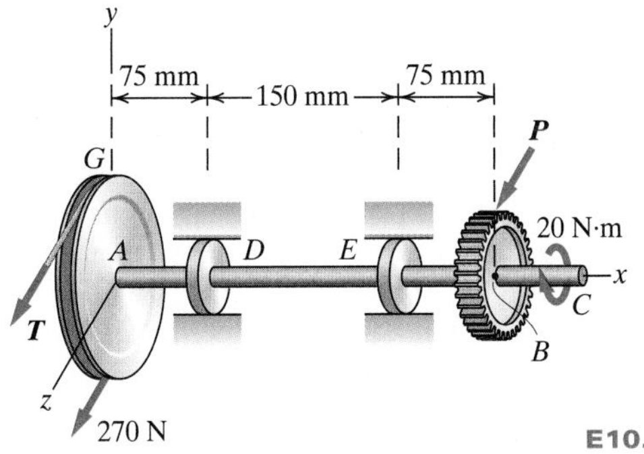

E10.1.47

10.1.48. An NFL goal post consists of the base, crossbar, and uprights (**E10.1.48**). The base rises 10 ft (the bottom 6 ft are usually padded) and extends forward approximately 5 ft. The crossbar is 18.5 ft wide, and the uprights rise 30 ft above the crossbar. The base and crossbar are each approximately 6 in. in diameter, and the uprights are 4 in. in diameter.

a. If the goal post weighs 900 lb and all parts are made of the same material, estimate the goal post's center of gravity relative to the bottom of its base.

b. What are the loads acting at the bottom of the base if the goal post is in equilibrium? What is the maximum bending moment in the crossbar and where is it located?

c. Suppose that two exuberant fans have climbed onto the crossbar. One is positioned immediately adjacent to the left upright, and the other is halfway between the base and the left upright. Assume that each fan weighs W. What are the loads acting at the bottom of the base if the goal post is in equilibrium? What is the maximum bending moment in the crossbar, and where is it located? Compare your answers with those from **b**.

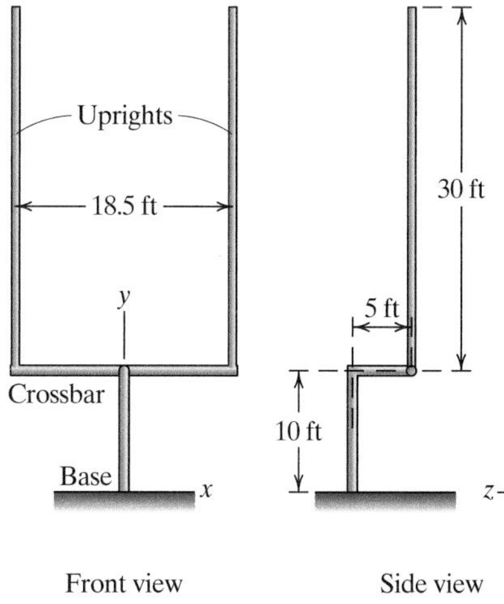

E10.1.48

10.1.49. A beam and loading are shown in **E10.1.49**.

a. Determine the maximum value (magnitude) of the bending moment.

b. If the area moment of inertia I for the cross section shown in the figure is 124 in.^{4} and $y_{max} = \pm 5$ in., calculate the maximum stress due to bending.

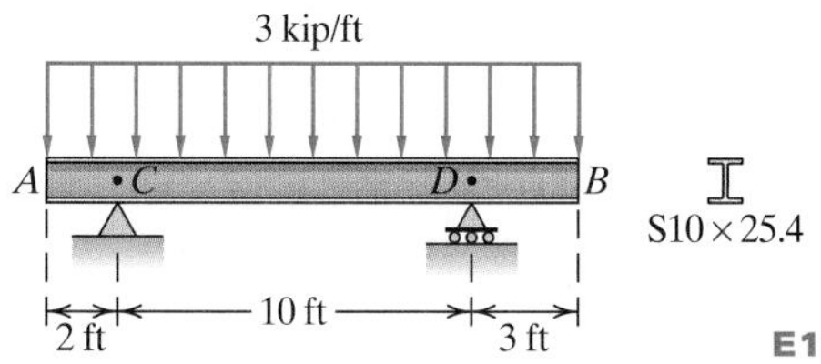

E10.1.49

10.1.50. A beam and loading are shown in **E10.1.50**.

a. Determine the maximum value (magnitude) of the bending moment.

b. If the area moment of inertia I for the cross section shown in the figure is 190 in.^{4} and $y_{max} = \pm 5.75$ in., calculate the maximum stress due to bending.

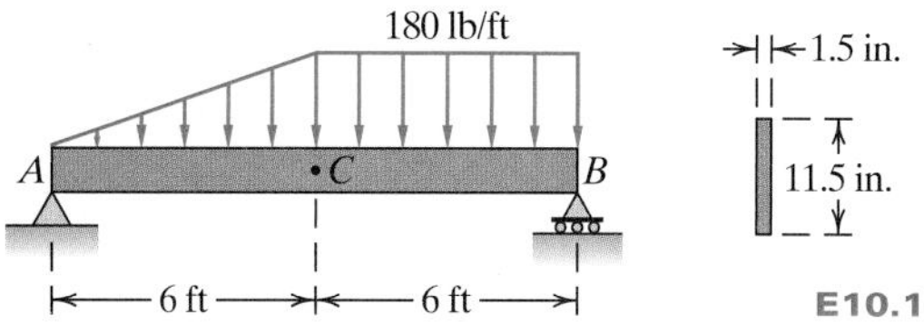

E10.1.50

10.1.51. A storage rack to hold the paper roll is shown in **E10.1.51**.

a. Determine the maximum value (magnitude) of the bending moment for the mandrel that extends 50% into the roll. The paper roll weighs 0.20 N.

b. If the area moment of inertia I for the mandrel's cross section is 3.0E-9 m^4, determine the maximum stress in the mandrel and specify its location.

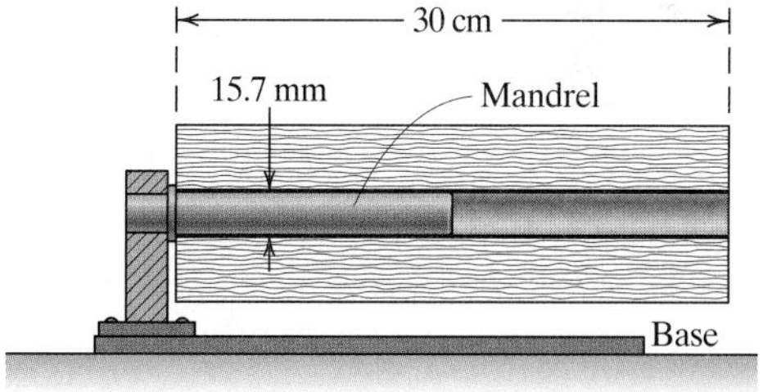

E10.1.51

10.1.52. To get a car weighing 2000 lb over a wall 12 in. high, two boards are laid across the wall in front of the wheels as shown in **E10.1.52**. The driver plans to drive the car onto the boards until they tip, allowing her to drive down the other side. If the bending moment in either board exceeds 35,000 lb · in, the board will break. Determine whether the boards are strong enough to allow the woman to drive the car slowly to the other side of the wall. Neglect the inclination of the board.

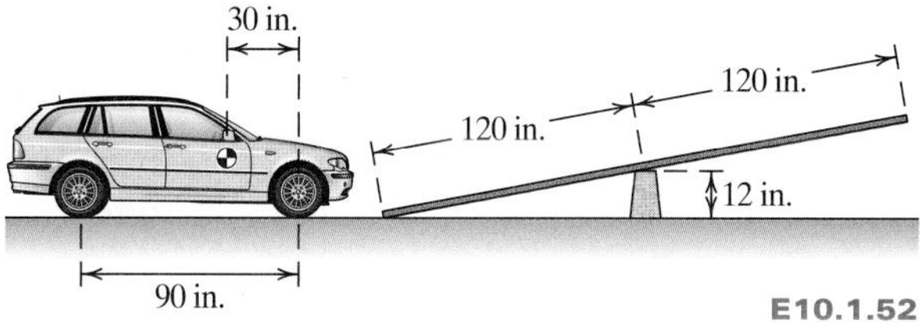

E10.1.52

10.1.53. A 50-ft steel beam that weighs 76 lb per foot is being lifted by cables attached at B and D as shown in **E10.1.53**. Draw the axial force and shear and moment diagrams for the beam.

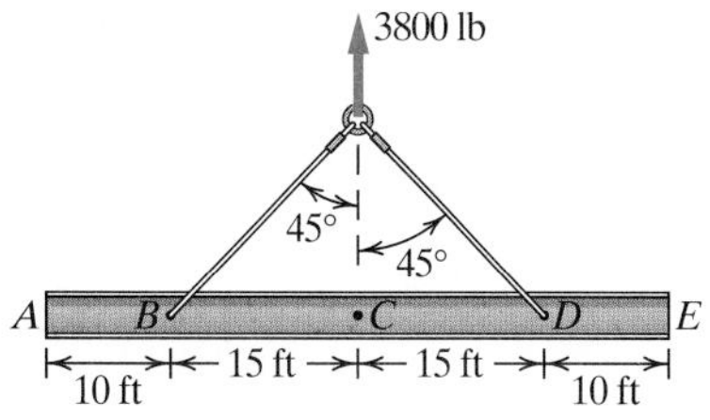

E10.1.53

10.2 FLEXIBLE CABLES

Cables are constructed from a large number of wound wires. Examples of structures that incorporate cables are amusement park rides, circus tents, suspension bridges, railings on boats, and electric power lines.

The forces acting on cables may be distributed along their length or may be point forces. When subjected to compressive forces, cables collapse. Therefore, cables are unlike springs and two-force members in trusses where there can be internal tension and compressive forces. Instead, the internal forces in cables are only tension forces. When a cable is subjected to loads perpendicular to its long axis, it changes shape so that the internal axial forces are aligned to equilibrate the system. Thus cables are also unlike beams, where there can be internal shear force and bending moment. Cables are just one example of tension-only two-force members; other tension-only members are ropes and chains.

Engineers undertake analysis to predict the tension in a cable so that they can select a cable with a cross-sectional area that has enough capacity to transmit the tension without breaking. They also need to be able to predict the shape of the cable so that they will know the geometry of the fully loaded structure.

We will consider four possible cable loadings, as illustrated in **Figure 10.40**. In case I the cable is fully taut and the force is along the cable. In case II, concentrated forces act on the cable (an example of this is traffic lights strung on a cable). In case III, the cable is only loaded by its own weight, and in case IV the cable is carrying a distributed load and the cable's weight is small relative to this distributed load.

Cable Vocabulary

For each of the four cases of a cable of length L suspended between anchors A and B in **Figure 10.40** we are interested in finding the tension force $\boldsymbol{T}$ anywhere along the length. We are also interested in finding the shape of the cable.

- In Case I (taut cable), the shape is simply a straight line between the two anchors (**Figure 10.40*a***).
- In Case II (concentrated forces), the shape is defined by the locations of the concentrated force (**Figure 10.40*b***).

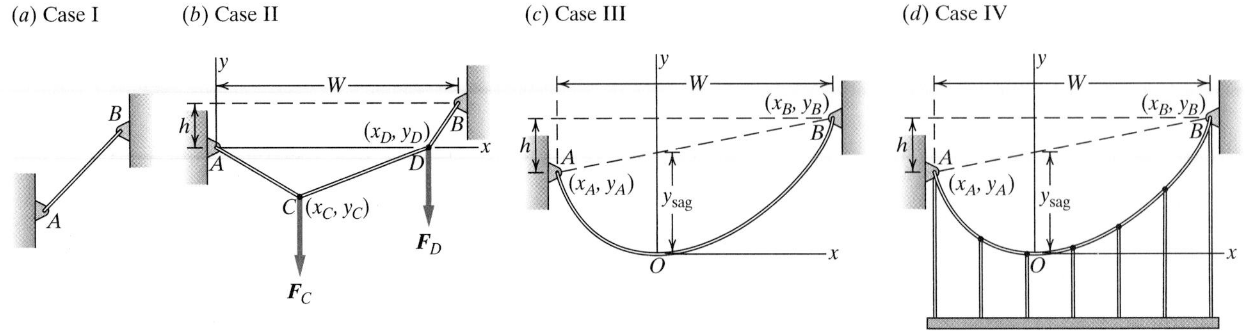

Figure 10.40

- In Cases III and IV, the shape is defined relative to the lowest point in the cable (**Figures 10.40*c*** and **10.40*d***). We define this lowest point as the origin, and supports A and B are at coordinates (x_A, y_A) and (x_B, y_B), respectively. The maximum vertical distance from the line connecting the two anchors to the cable is called the sag, y_{sag}. Also shown in the figures are the span (W) and the difference in height between A and B (h). The ratio of sag to span (y_{sag}/W) is often of interest in designing systems with cables.

Procedures for Finding Shape and Tension in Cables

Case I: Taut Cable. Consider the cable illustrated in **Figure 10.40*a***. With the cable pulled taut between the anchors it behaves like a two-force member. Its shape is simply a line between A and B, and the tension in the cable acts along this line. This case should look familiar as this is an analysis we have done in previous chapters.

Case II: Concentrated Forces. When loaded with concentrated forces, the cable takes the form of several taut line segments (**Figure 10.40*b***). Within each segment there is a constant tensile force aligned with the segment. Considering that we know the distance h and W, coordinates x_C and x_D, the forces F_C and F_D, and the total cable length, L, we can find the tension in each of the segments (T_{AC}, T_{CD}, T_{DB}), and y coordinates y_C and y_D, at C and D, respectively. We need five equations to find these five unknowns (T_{AC}, T_{CD}, T_{DB}, y_C, y_D). Four of these equations come from writing force equilibrium equations (ΣF_x, ΣF_y) at both points C and D (**Figure 10.41**). The fifth equation relates the total cable length L to y_C, y_D, x_C, x_D, h, and W. These five equations can be solved for the five unknowns; unfortunately, this type of problem is hard to solve by hand. If the value of y_C or y_D is specified, the number of unknowns is reduced by one and it becomes easier to solve for the internal forces by hand, as illustrated in Example 10.12.

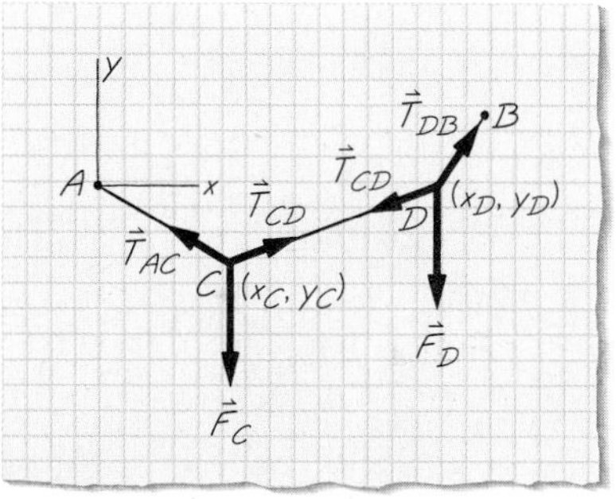

Figure 10.41 Free-body diagram at C and D

EXAMPLE 10.12 FLEXIBLE CABLE WITH CONCENTRATED LOADS

The cable shown in **Figure 10.42** supports a 500-lb load at C and a 250-lb load at D. Support A is 6 ft below B, and C is 3 ft below support A. Determine the vertical location of point D (y_D) and the maximum tension in the cable.

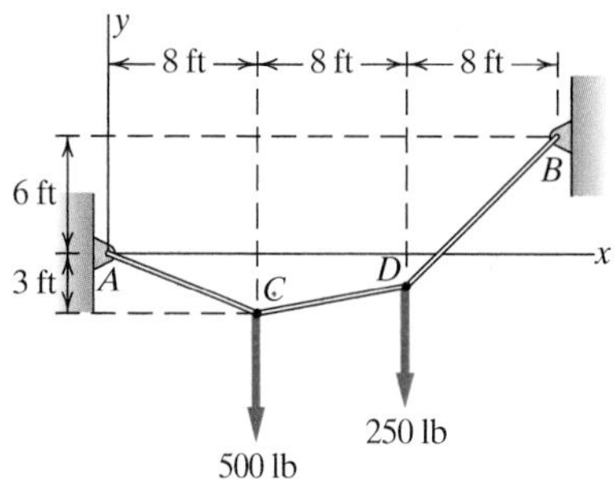

Figure 10.42

Goal We are to find the vertical coordinate of point D (y_D) and the maximum tension in the cable.

Given We are given information about the cable geometry and applied loads, and a coordinate system with origin at A.

Assume We assume that the weight of the cable is negligible compared to the loads and that the cable is uniform.

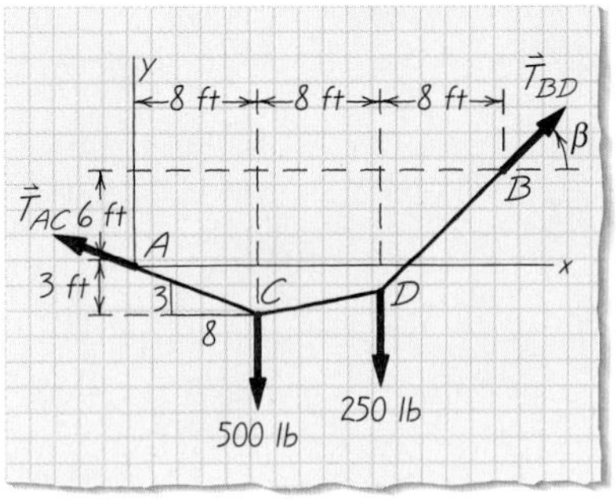

Figure 10.43 Free-body diagram of entire cable

Draw Figure 10.43 is a free-body diagram of the entire system. Because the cable is in tension, it is pulling on the supports at A and B and the supports are pulling back on the cable. This information allows us to draw the tension at A (T_{AC}) in a known direction (we know the direction because we know, based on the information in **Figure 10.42**, that $x_C = 8$ ft and $y_C = -3$ ft). We also draw the tension at B (T_{BD}), but since we do not know y_D based on the information in **Figure 10.42**, we show it at an unknown angle β.

Formulate Equations and Solve Based on the free-body diagram of the entire system (**Figure 10.43**), we write the equilibrium equation (7.5C):

$$\sum M_{z@A} = 0\ (\circlearrowleft+)$$
$$-500\text{ lb}(8\text{ ft}) - 250\text{ lb}(16\text{ ft}) + T_{BD}\sin\beta\,(24\text{ ft}) - T_{BD}\cos\beta\,(6\text{ ft}) = 0$$
$$-4000\text{ lb}-\text{ft} + T_{BD}\sin\beta\,(12\text{ ft}) - T_{BD}\cos\beta\,(3\text{ ft}) = 0 \qquad (1)$$

Also, we can write:

$$\sum M_{z@C} = 0\ (\circlearrowleft+)$$
$$-250\text{ lb}(8\text{ ft}) + T_{BD}\sin\beta\,(16\text{ ft}) - T_{BD}\cos\beta\,(9\text{ ft}) = 0$$
$$-2000\text{ lb}\cdot\text{ft} + T_{BD}\sin\beta\,(16\text{ ft}) - T_{BD}\cos\beta\,(9\text{ ft}) = 0 \qquad (2)$$

Solving (1) and (2) simultaneously gives

$$\beta = 36.9°, \qquad T_{BD} = 834\text{ lb}$$

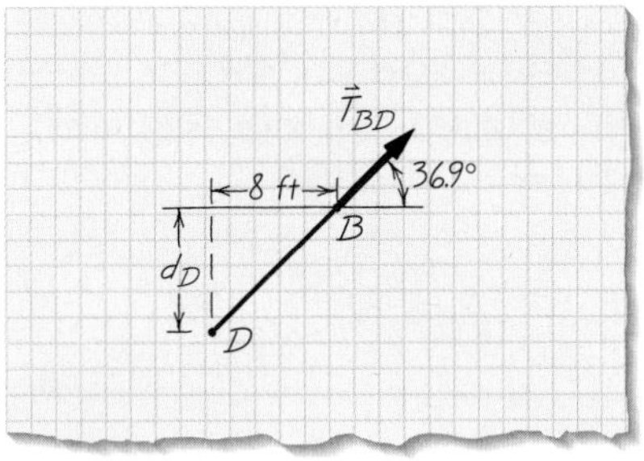

Figure 10.44

We now use this value of β to find the vertical coordinate of D (y_D). Based on the geometry in Figure 10.44, we write

$$\tan\beta = \tan 36.9° = d_D/8\text{ ft}$$

Solving for d_D, we find

$$d_D = 6.00\text{ ft}$$

If point D is 6 feet below anchor B, it is at the same level as anchor A. This means that the y coordinate of point D relative to the xy coordinate system defined in **Figure 10.42** is

$$y_D = 0.0\text{ ft}$$

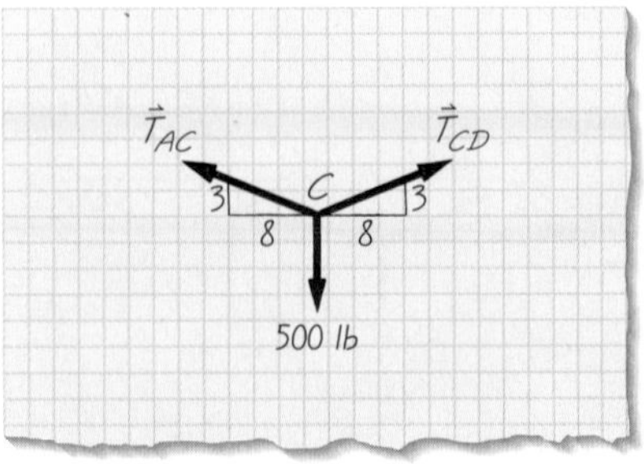

Figure 10.45 Free-body diagram of particle C

Now we find the maximum tension in the cable. We have already found that the tension in cable segment BD is 834 lb. Based on equilibrium conditions applied to the free-body diagram in Figure 10.45 we find that $T_{CD} = T_{AC} = 712$ lb.

Answer $y_D = 0.00$ ft, $T_{max} = T_{DB} = 834$ lb

Note: In general, the tension will be maximum in the portion of the cable with the steepest slope relative to the horizontal. Furthermore, because there are no horizontal forces applied to the cable, the horizontal force component is constant along the cable. In other words, we can show that $T_{CAx} = T_{CDx} = T_{BDx} = 667$ lb. This is also a good check of our results.

Check We can check our results by using the free-body diagram in **Figure 10.43** and the calculated values of T_{BD} and T_{AC} to write the moment equilibrium equation about a moment center at D.

Case III: Cable Weight Only. Consider the cable illustrated in Figure 10.46. We wish to find the cable's shape and tension when it is loaded by its own weight (which we denote as μ in force/length). The shape of the cable is found by considering a free-body diagram of an infinitesimally small portion of the cable (Figure 10.47). Application of equilibrium equations to this free-body diagram results in equations for the shape (curve) and slope of the cable:

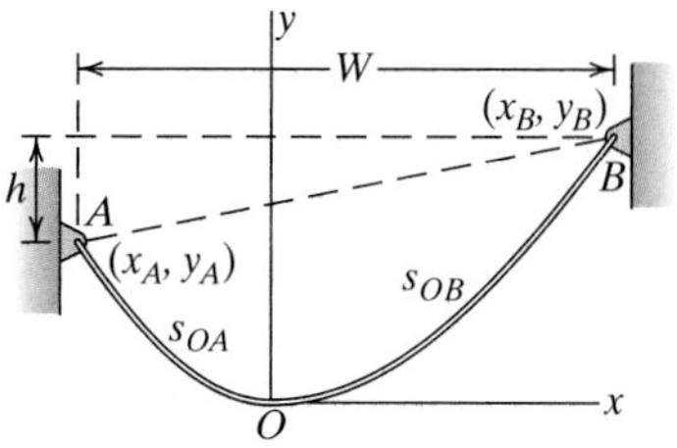

Figure 10.46 Cable of uniform weight μ (force/length)

$$y = \frac{T_o}{\mu}\left(\cosh\frac{\mu x}{T_O} - 1\right) \qquad (10.5A)$$

where cosh is the hyperbolic cosine, and T_O is the constant horizontal force component everywhere in the cable. This curve is defined relative to the origin noted in the figure. The particular shape described by (10.9A) is referred to as a **catenary curve**. The slope of the curve is given as

$$\frac{dy}{dx} = \sinh\frac{\mu x}{T_O} = \frac{e^{\mu x/T_O} - e^{-\mu x/T_O}}{2} \qquad (10.5B)$$

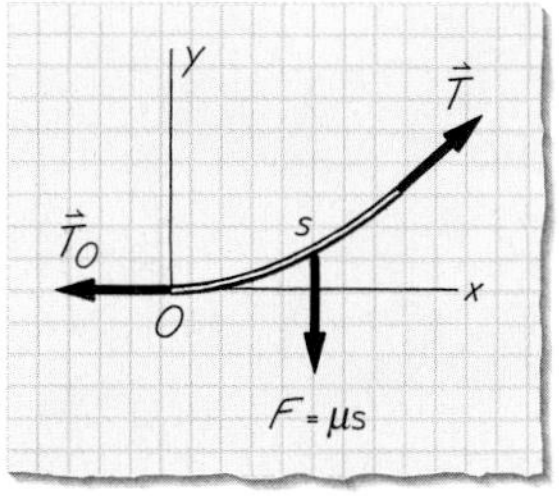

Figure 10.47 Free-body diagram of a portion of the cable

where sinh is the hyperbolic sine. The complete development of (10.5A) and (10.5B) is presented in Note 1 at the end of this chapter.

To use (10.5A) to plot the shape of a particular cable requires that we find the value of T_O and define an origin that is located at the lowest point in the cable (we define it relative to anchors A and B at x_A, y_A and x_B, y_B, respectively). We need five equations to find the five unknowns $(x_A, y_A, x_B, y_B, T_O)$.

Two equations can be written based on the horizontal (W) and vertical (h) spacing of the two anchors:

$$W = -x_A + x_b \quad \text{or} \quad x_A = x_B - W \qquad \text{[Equation III.1]}$$

and

$$h = |y_b - y_A| \qquad \text{[Equation III.2]}$$

Two additional equations result from writing (10.9A) for anchor points A and B follow. At A, $x = x_A$:

$$y_A = \frac{T_o}{\mu}\left(\cosh\frac{\mu x_A}{T_o} - 1\right) \qquad \text{[Equation III.3]}$$

at B, $x = x_B$:

$$y_B = \frac{T_O}{\mu}\left(\cosh\frac{\mu x_B}{T_O} - 1\right) \qquad \text{[Equation III.4]}$$

Finally, based on the geometry depicted in **Figure 10.46**, we write the total cable length (L) as the sum of the cable lengths to the left (s_{OA}) and right (s_{OB}) of the origin. The geometry in **Figure 10.47** shows us that $dy/dx = \tan\theta = \mu s/T_O$, which, when combined with (10.5B), results in an expression for cable length. Therefore,

$$L = -s_{OA} + s_{OB} = \frac{T_O}{\mu}\left(-\sinh\frac{\mu x_A}{T_O} + \sinh\frac{\mu x_B}{T_O}\right) \qquad \text{[Equation III.5]}$$

These five equations (III.1–III.5) can be used to find the five unknowns T_O, x_A, x_B, y_A, and y_B.

When anchors A and B are at different heights (i.e., $h \neq 0$), these five nonlinear equations must be solved to find the five unknowns (x_A, x_B, y_A, y_B, T_O). If the anchor points are at the same height, $-x_A = x_B = W/2$, and $y_A = y_B = y_{sag}$, equation (III.3) or (III.4) in combination with equation (III.5) can then be used to find T_O and y_{sag}, as illustrated in Example 10.13.

Tension in the cable: Equilibrium conditions applied to the free-body diagram in **Figure 10.47** result in an expression for tension anywhere along the length of a cable loaded by its own weight:

$$T = T_O \cosh\frac{\mu x}{T_O} \underbrace{=}_{\substack{\text{using} \\ (10.9A)}} T_O + \mu y \qquad (10.6)$$

This expression describes the tension at any location (x, y) along the cable—it tells us that the tension is maximum at the greatest distance from the origin (i.e., where x and y are maximum, which will be at the anchor farthest from the origin). Details of the development of (10.6) are contained in Note 1 at the end of this chapter.

EXAMPLE 10.13 CATENARY CURVE

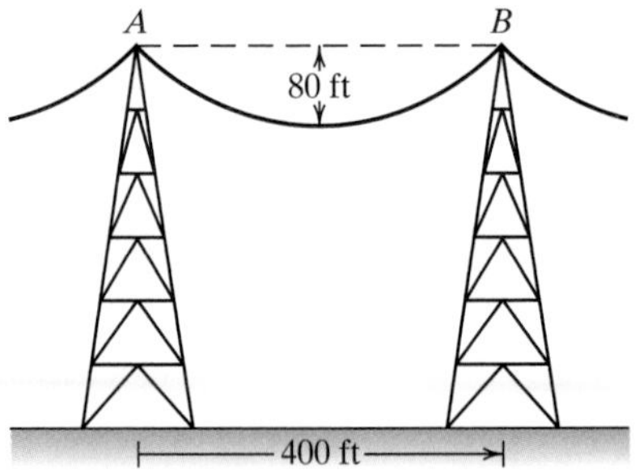

Figure 10.48

An electrical power line weighs 1 lb/ft and is anchored to towers of the same height (**Figure 10.48**). The span between towers is 400 ft, and the maximum design sag is restricted to 80 ft. Determine the length of the cable and the maximum tension.

Goal We are to find the length of the cable and the maximum tension.

Given We are given information about the geometry and weight of the power line as well as a specific design constraint (80 ft maximum sag).

Assume We assume that the cable is inextensible and that the weight of the cable is the only significant loading.

Draw Because of symmetry, the lowest point in the cable occurs halfway between the two towers. This is where we locate the origin of the coordinate system on the free-body diagram (**Figure 10.49**) of a section of the cable.

Formulate Equations and Solve Since the only significant load on the cable is its own weight ($\mu = 1$ lb per foot of cable length), the cable will take on a catenary shape, and the equations developed under case III are relevant ((10.5), (10.6), and equations III).

To determine the maximum tension in the cable we first determine the horizontal force in the cable, T_O, which we calculate using (10.5A):

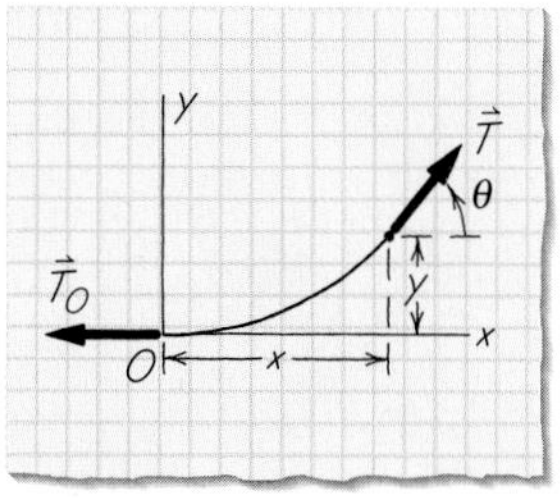

Figure 10.49

$$y = \frac{T_O}{\mu}\left[\cosh\left(\frac{\mu x}{T_O}\right) - 1\right]$$

At the tower at B, $x_B = 200$ ft, $y_B = 80$ ft. Therefore, (10.5A) can be written as

$$80 \text{ ft} = \frac{T_O}{1 \text{ lb/ft}}\left[\cosh\left(\frac{1 \text{ lb/ft}(200 \text{ ft})}{T_O}\right) - 1\right]$$

We solve this equation using a trial-and-error approach with the result $T_O = 262.3$ lb.

Using (10.10), we now solve for T_{max}:

$$T_{max} = T_{\substack{x = 200 \text{ ft} \\ y = 80 \text{ ft}}} = T_O + \mu y = 262.3 \text{ lb} + 1 \text{ lb/ft}(80 \text{ ft}) = 342.3 \text{ lb}$$

$$T_{max} = 342.3 \text{ lb}$$

This maximum tension occurs where the cable attaches to the towers and where its slope is the steepest.

We find the total cable length between any two towers by using (III.5), remembering that tower A is at (-200 ft, 80 ft) and tower B is at (200 ft, 80 ft):

$$L = s_{oA} + s_{oB} = \frac{T_O}{\mu}\left[-\sinh\left(\frac{\mu x_A}{T_O}\right) + \sinh\left(\frac{\mu x_B}{T_O}\right)\right]$$

$$L = \frac{262.3 \text{ lb}}{1 \text{ lb/ft}}\left[-\sinh\left(\frac{1 \text{ lb/ft}(-200 \text{ ft})}{262.3 \text{ lb}}\right) + \sinh\left(\frac{1 \text{ lb/ft}(200 \text{ ft})}{262.3 \text{ lb}}\right)\right]$$

$$\text{L} = \frac{2(262.3 \text{ lb})}{1 \text{ lb/ft}}\sinh\left(\frac{1 \text{ lb/ft}(200 \text{ ft})}{262.3 \text{ lb}}\right) = 439.9 \text{ ft}$$

Therefore, $L = 440$ ft.

Answer Catenary: $T_{max} = 342$ lb and $L = 440$ ft

EXAMPLE 10.14 CATENARY WITH SUPPORTS AT DIFFERENT ELEVATIONS

Two pairs of civil engineering students use surveying tapes that weigh 0.2 N/m to measure a distance W of 29.0 m. Each pair pulls horizontally on the ends of the tape with a force of 50 N. One pair is instructed to have both ends of the tape at the same elevation (Pair 1; **Figure 10.50*a***). The other pair is instructed to have one end 0.1 m above the other end

(Pair 2; **Figure 10.50*b***). What length will Pair 1 measure on their tape? What length will Pair 2 measure on their tape? How much effect does the 0.1 m in elevation difference have on the measurement?

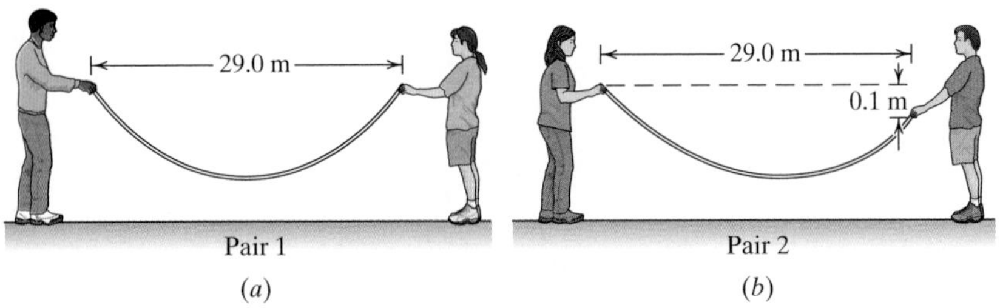

Figure 10.50

Goal We are to find the measurements taken by two pairs of engineering students, then consider how their measurements differ.

Given We are given information about the weight of the tape and the horizontal force on the tape.

Assume We assume that the tape is inextensible (i.e., it does not stretch), the error due to temperature variation can be ignored, and the error due to inexact markings on the tape can be ignored.

Draw Because the surveying tape is supported only at the ends and is loaded along its length only by its own weight, it assumes the shape of a catenary. We compare the measurement error from two cases: the supports at the same elevation (**Figure 10.51**) and the supports at different elevations (**Figure 10.52**).

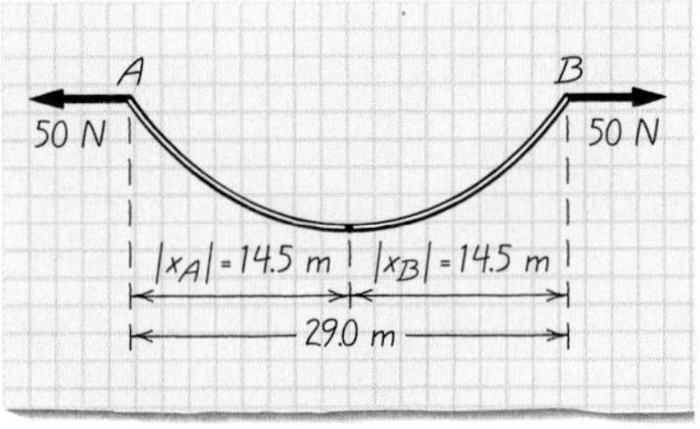

Figure 10.51

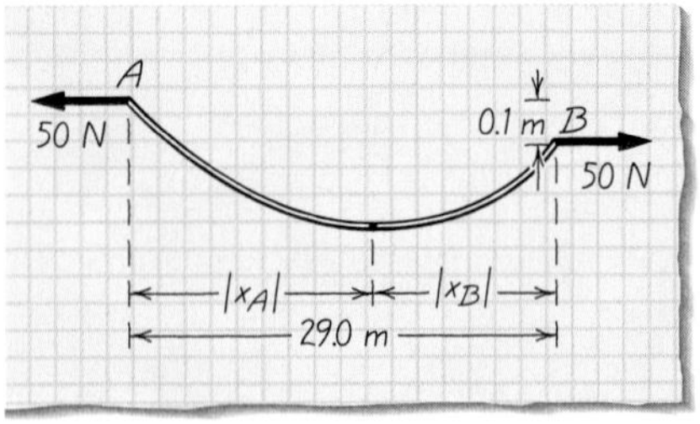

Figure 10.52

Formulate Equations and Solve *Pair 1:* When the supports are at the same elevation, $|x_A| = |x_B| = \frac{W}{2}$, which allows us to write equation (III.5) for the length of the cable in a simplified form

$$L_{\text{pair}} = -s_{OA} + s_{OB} = \frac{2T_O}{\mu}\sinh\left(\frac{\mu\frac{W}{2}}{T_O}\right) \quad (1)$$

We know that $\mu = 0.2$ N/m, $T_O = 50$ N, and $W = 29.0$ m. Therefore, (1) can be solved to find that $L_{\text{pair1}} = 29.0163$ m. This is the measurement that pair 1 will make on their tape.

Pair 2: When the supports are at different elevations, we do not know the location of the lowest point on the catenary curve, meaning that x_A is not known. This introduces an additional unknown, which we find using the difference in the support elevations (equation (III.2)).

$$h = 0.1 \text{ m} = \| y_A - y_B \| \quad (2)$$

Substituting equations (III.3) and (III.4) into (2), and noting that $x_B = W + x_A$, we get

$$0.1 \text{ m} = \frac{T_O}{\mu}\left[\cosh\frac{\mu x_A}{T_O} - \cosh\frac{\mu(W + x_A)}{T_O}\right]$$

$$0.1 \text{ m} = \frac{50 \text{ N}}{0.2 \text{ N/m}}\left[\cosh\frac{0.2 \text{ N/m}\, x_A}{50 \text{ N}} - \cosh\frac{0.2 \text{ N/m}(29.0 + x_A)}{50 \text{ N}}\right]$$

Solving numerically, we find that

$$x_A = -15.36 \text{ m and } x_B = 13.64 \text{ m}$$

Substituting these values into equation (III.5):

$$L_{\text{pair2}} = \frac{50 \text{ N}}{0.2 \text{ N/m}}\left[-\sinh\frac{0.2 \text{ N/m}(-15.36 \text{ m})}{50 \text{ N}} + \sinh\frac{0.2 \text{ N/m}(13.64 \text{ m})}{50 \text{ N}}\right]$$
$$= 29.0164 \, m$$

Therefore, $L_{\text{pair2}} = 29.0164$ m.

If each pair had interpreted its tape measurement as the correct answer for W,

Pair 1 would have had an error of 29.0163 m − 29.0 m = 0.0163 m (= 16.3 mm).

Pair 2 would have had an error of 29.0164 m − 29.0 m = 0.0164 m (= 16.4 mm).

The additional error introduced by the 0.1-m difference in the elevation of the two ends of the tape is 16.4 mm − 16.3 mm = 0.1 mm (this error is more than 100 times smaller than the error due to the sag, and it is reasonable to assume that it is insignificant).

Answer $L_{\text{pair1}} = 29.0163$ m and $L_{\text{pair2}} = 29.0164$ m. Additional error introduced by the 0.1-m difference in elevation = 0.1 mm. The error in measurement due to the sag is 16 mm.

Case IV: Uniform Weight Hanging from Cable. Consider the cable illustrated in **Figure 10.53*a***. We wish to find the cable's shape and tension when a uniform weight hangs from it—this description closely approximates a suspension bridge with a roadway hanging from it. We denote the uniform weight as ω in force/length. The shape of the cable is found by considering a free-body diagram of an infinitesimally small portion of the cable (**Figure 10.53*b***). We can derive an equation for the curve describing the cable configuration:

$$y = \frac{\omega x^2}{2T_O} \qquad (10.7\text{A})$$

where T_O is the constant horizontal force component everywhere in the cable. This curve is defined relative to the origin noted in the figure, which is the lowest point of the cable. The particular shape described by (10.7A) is referred to as a **parabolic curve**. The slope of the curve is given as

$$\frac{dy}{dx} = \frac{\omega x}{T_O} \qquad (10.7\text{B})$$

The complete development of (10.7A) and (10.7B) is contained in Note 2 at the end of this chapter.

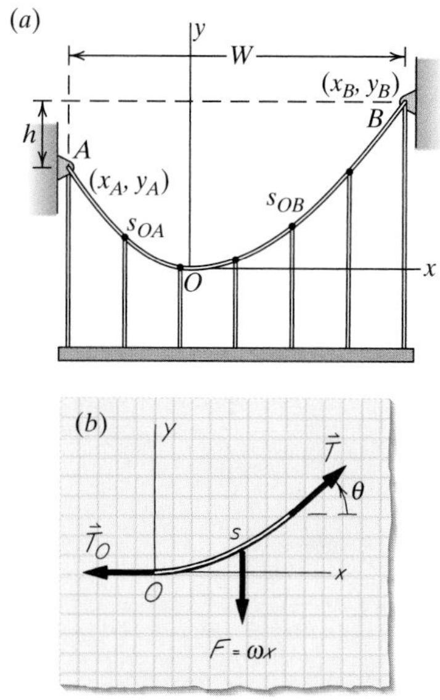

Figure 10.53 (*a*) Uniform load on cable ω; (*b*) free-body diagram of a portion of the cable

To plot the shape of a particular cable using (10.11A) requires that we find a value of T_O and that we define an origin that is located at the lowest point in the cable (we define it relative to anchors A and B at x_A, y_A, x_B, y_B). We need five equations to find the five unknowns (x_A, y_A, x_B, y_B, T_O).

Two equations can be written based on the horizontal (W) and vertical (h) spacing of the two anchors:

$$W = -x_A + x_B \quad \text{or} \quad x_A = x_B - W \quad \text{[Equation IV.1]}$$

and

$$h = |y_B - y_A| \quad \text{[Equation IV.2]}$$

Two additional equations result from writing (10.11A) for anchor points A and B.

At A, $x = x_A$:

$$y_A = \frac{\omega x_A^2}{2T_O} \quad \text{[Equation IV.3]}$$

At B, $x = x_B$:

$$y_B = \frac{\omega x_B^2}{2T_O} \quad \text{[Equation IV.4]}$$

Finally, based on the geometry depicted in **Figure 10.53*b***, we can write the total cable length (L) as the sum of the cable lengths to the left (s_{oA}) and right (s_{oB}) of the origin. The cable lengths are found based on integrating the differential relationship $ds = [(dx)^2 + (dy)^2]^{0.5}$, as detailed in Note 2. Therefore,

$$L = -s_{OA} + s_{OB} = -\left[x_A + \frac{\omega^2 x_A^3}{6T_O^2} - \frac{\omega^4 x_A^5}{40\,T_O^4} + \cdots\right] + \left[x_B + \frac{\omega^2 x_B^3}{6\,T_O^2} - \frac{\omega^4 x_B^5}{40\,T_O^4} + \cdots\right]$$

[Equation IV.5A]

Equation IV.5A can be reformulated if we first rewrite (10.11A) as

$$T_O = \frac{\omega\, x_B^2}{2y_B} = \frac{\omega\, x_A^2}{2y_A}$$

Substituting this into equation IV.5A results in an alternate expression for the length of the cable, written in terms of the x and y coordinates of the supports:

$$L = -s_{OA} + s_{OB} = -x_A\left[1 + \frac{2}{3}\left(\frac{y_A}{x_A}\right)^2 - \frac{2}{3}\left(\frac{y_A}{x_A}\right)^4\right] + x_B\left[1 + \frac{2}{3}\left(\frac{y_B}{x_B}\right)^2 - \frac{2}{3}\left(\frac{y_B}{x_B}\right)^4\right]$$

[Equation IV.5B]

These five equations (IV.1–IV.5) can be solved for the five unknowns T_O, x_A, x_B, y_A, and y_B.

When anchors A and B are at different heights (i.e., $h \neq 0$), these nonlinear five equations must be solved to find the five unknowns (x_A, x_B, y_A, y_B, T_O). If the anchor points are at the same height, $-x_A = x_B = W/2$, and $y_A = y_B = y_{sag}$, equation (IV.3) or (IV.4) and equation (IV.5) can then be used to find T_O and y_{sag}.

Tension in the cable: Equilibrium conditions applied to the free-body diagram in **Figure 10.53*b*** result in an expression for tension anywhere along the length of a cable loaded by uniform load:

$$T^2 = \omega^2 x^2 + T_O^2 \quad \text{or} \quad T = \sqrt{\omega^2 x^2 + T_O^2} \tag{10.8}$$

This expression describes the tension at any location (x, y) along the cable—it tells us that the tension is maximum at the greatest distance from the origin (i.e., where x and y are maximum, which will be at the anchor farthest from the origin).

Comparison of Cases III and IV. Finding the shape and tension of a cable loaded by its own weight (catenary curve) may be approximated by solution for a cable loaded by a constant weight (parabolic) when the sag-to-span ratio (y_{sag}/W) is small. A small sag-to-span ratio means a fairly taut cable, and the uniform distribution of weight along the cable is not much different from the same load intensity distributed uniformly along the horizontal. Example 10.17 compares the solutions for catenary and parabolic cases.

EXAMPLE 10.15 LOADED CABLE (UNIFORMLY)

The cable shown in **Figure 10.54** is a model of one of the main cables on the center span of the Golden Gate Bridge. The cable supports the roadbed, which weighs 165 kN/m of horizontal length. The towers are 1280 m apart, and the sag of the cable is 144 m. Determine the cable tension at midspan, the maximum tension, and the total length of the cable.

Figure 10.54

Goal We are to find the cable tension at midspan, the maximum tension, and the total length of the cable.

Given We are given information about the cable geometry and loading.

Assume We assume that the cable is inextensible and that the weight of the cable is negligible relative to the loading.

Formulate Equations and Solve Given the conditions of the problem, the equations developed for case IV (uniformly distributed load) apply. The curve that the cable takes on will be parabolic in shape. Based on the geometry in **Figure 10.54**, we know that

$$|x_A| = |x_B| = 640 \text{ m}, \quad y_A = y_B = 144 \text{ m}, \quad \text{and} \quad \omega = 165 \text{ kN/m}$$

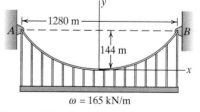

Using (10.11A) we find T_O:

$$y = \frac{\mu x^2}{2T_O} \underbrace{=}_{\substack{\text{at } x = 640 \text{ m} \\ y = 144 \text{ m}}} 144 \text{ m} = \frac{\left(165\,\dfrac{\text{kN}}{\text{m}}\right)(640 \text{ m})^2}{2T_O}$$

which when rearranged gives $T_O = 234{,}667$ kN = 235 MN.

Using (10.8) we find T_{max}:

$$T_{max} = T_{@x=640\text{ m}}$$
$$= \sqrt{\left(165\,\frac{\text{kN}}{\text{m}}\right)^2(640 \text{ m})^2 + (234{,}667 \text{ kN})^2} = 257{,}332 \text{ kN}$$

Therefore $T_{max} = 257{,}332$ kN = 257 MN.

Finally, to find the length of the cable L, substitute $x_A = -640$ m, $x_B = 640$ m, $T_O = 234{,}667$ kN, and $\omega = 165$ kN/m into equation (IV.5A):

$$L = -s_{OA} + s_{OB} = -\left[x_A + \frac{\omega^2 x_A^3}{6\,T_O^2} - \frac{\omega^4 x_A^5}{40\,T_O^4}\right] + \left[x_B + \frac{\omega^2 x_B^3}{6\,T_O^2} - \frac{\omega^4 x_B^5}{40\,T_O^4}\right]$$

We find that $L = 1322$ m. (Alternately, equation (IV.5B) could have been used to find L.)

Answer $T_O = 235$ MN, $T_{max} = 257$ at $x = +-\ 640$ m, $L = 1322$ m

EXAMPLE 10.16 UNIFORMLY LOADED CABLE WITH SUPPORTS AT UNEQUAL HEIGHTS

A 150-ft foot bridge crossing a gorge is suspended from two cables. Due to the natural terrain, the right support is 12 ft higher than the left support. The weight of the bridge deck and handrails supported by the cables is 60 lb/ft (**Figure 10.55**) and the lowest point of the cables is 10 ft below the left support. Determine the maximum and minimum cable tensions, the angle between the cables and the horizontal at the right support, and the total length of each cable.

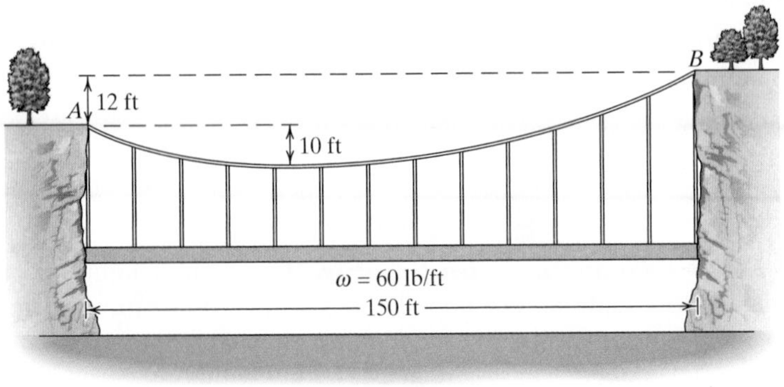

Figure 10.55

Goal For each cable we are to find the maximum and minimum cable tensions, the angle between the cable and the horizontal at the right support, and the total length.

Given We are given information about the cable geometry and loading.

Assume We assume that the cable is inextensible and that the weight of the cable is negligible relative to the loading.

Draw We define a coordinate system with its origin at the lowest point of the cable, the location of which is unknown (**Figure 10.56**). We assume the x coordinate of support B is x_B; then the coordinate x_A is $x_B - 150$ ft.

Formulate Equations and Solve The load is shared equally by the two suspension cables; therefore $\omega = 30$ lb/ft for each cable. We locate the origin by applying (10.7A) at supports A and B:

At A:
$$y_A = 10 \text{ ft} = \frac{\left(30 \frac{\text{lb}}{\text{ft}}\right)(x_B - 150 \text{ ft})^2}{2T_O} \qquad (1)$$

At B:
$$22 \text{ ft} = \frac{\left(30 \frac{\text{lb}}{\text{ft}}\right)(x_B)^2}{2T_O} \qquad (2)$$

We eliminate T_O by dividing (1) by (2), which gives

$$\frac{10 \text{ ft}}{22 \text{ ft}} = \frac{(x_B - 150 \text{ ft})^2}{x_B^2}$$

Rearranging and simplifying results in an equation which can be solved using the quadratic formula:

$$x_B^2 - 550x_B + 41{,}250 = 0$$
$$x_B = \frac{550 \pm \sqrt{(-550)^2 - 4(1)(41{,}250)}}{2(1)} = 275 \text{ ft} \pm 185.4 \text{ ft}$$
$$x_B = 460.4 \text{ ft or } 89.6 \text{ ft}$$

We discard $x_B = 460.4$ ft because it is not a realistic solution when the constraints of this problem are considered. Therefore $x_B = 89.6$ ft.

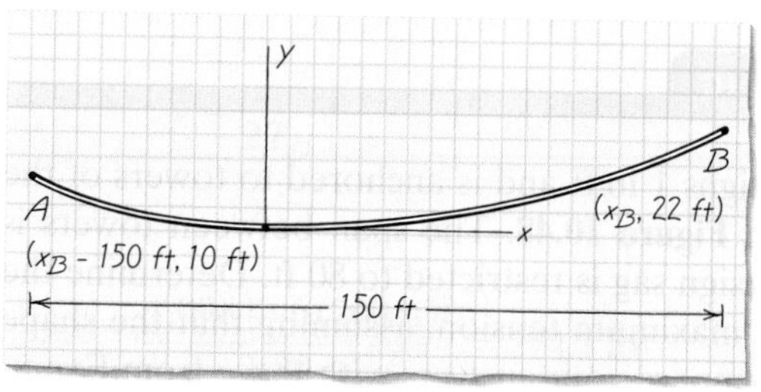

Figure 10.56

Using (10.12), we find T_{max}, which occurs at one of the towers:

$$T_{max} = \sqrt{\omega^2 x_A^2 + T_O^2} = \sqrt{(1.096 \text{ lb/ft})^2(200 \text{ ft})^2 + (274.1 \text{ lb})^2} = 351.0 \text{ lb}$$

Answer Catenary (from Example 10.13): $T_{max} = 342$ lb and $L = 440$ ft.
Parabolic: $T_{max} = 351$ lb and $L = 439$ ft.

In this case (sag-to-span ratio $y_{sag}/W = 80/400 = 0.2$), the difference in the answers was not large. For larger ratios, the catenary approach will generate more accurate solutions than the parabolic approach for this loading case. As a general rule, the parabolic approach is only recommended for sag-to-span ratios of 0.1 or less.

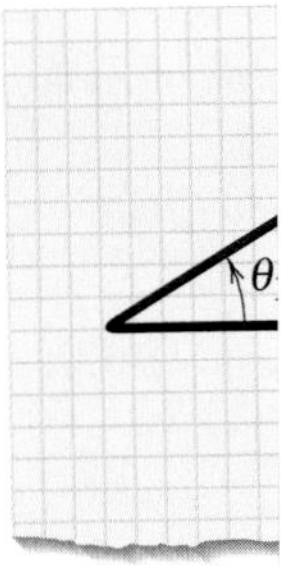

Figure 10.57

EXAM

EXERCISES 10.2

10.2.1. In the cable shown in **E10.2.1**, $d_C = 7$ m. Determine

a. the vector components of the force acting at E
b. the maximum tension in the cable

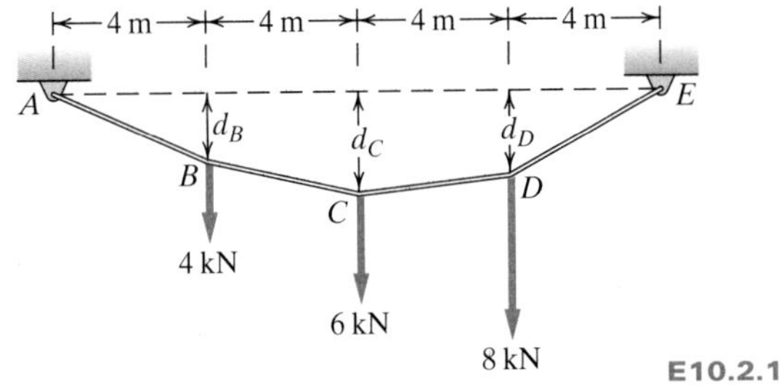

E10.2.1

10.2.2. Three loads are suspended as shown in **E10.2.2** from the cable. Distance $d_C = 7$ m. Determine

a. the vector components of the forces acting at A and E
b. the maximum tension in the cable

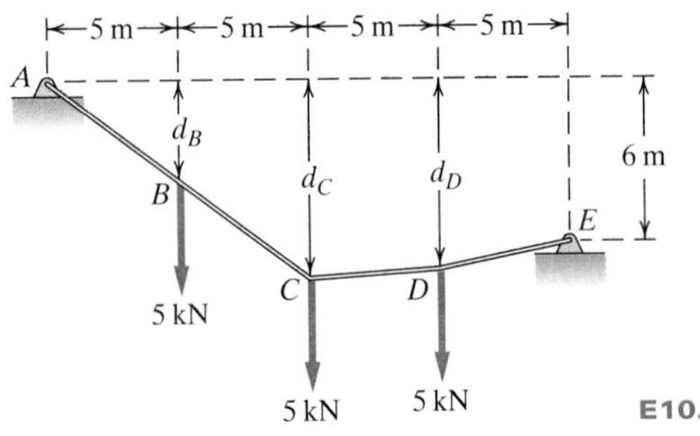

E10.2.2

10.2.3. Cable ABC supports two masses as shown in **E10.2.3**. Distance $b = 7$ m. Determine

a. the required magnitude of the force $\boldsymbol{P}$
b. the corresponding distance a

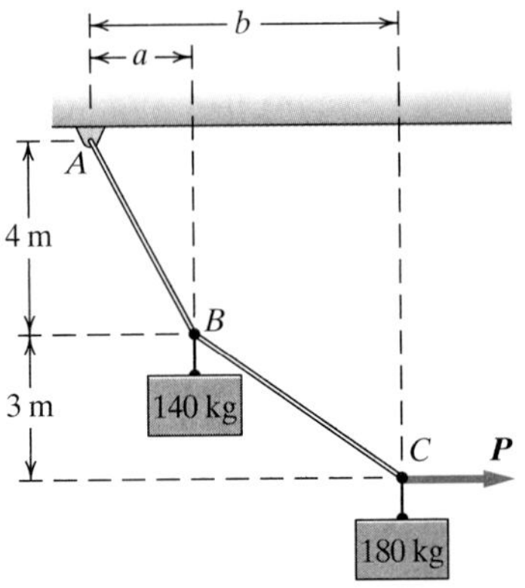

E10.2.3

10.2.4. Cable ABC supports two masses as shown in **E10.2.3**. Determine the distances a and b when a horizontal force $\boldsymbol{P}$ of magnitude 1000 N is applied at C.

10.2.5. The cable $ABCD$ is maintained in the position shown in **E10.2.5** by the 440-lb force applied at C and the block attached at B. Determine

a. the required weight W of the block
b. the tension in each portion of the cable

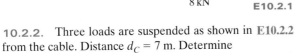

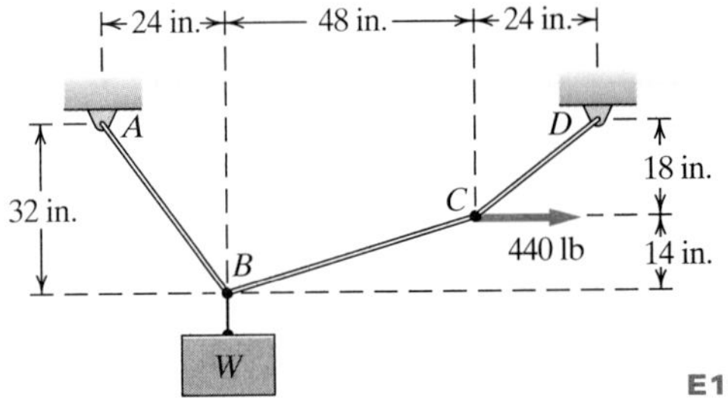

E10.2.5

10.2.6. A cable carries three forces, as shown in E10.2.6. Determine

a. the locations y_B and y_D of points B and D, respectively

b. the maximum cable tension and its location

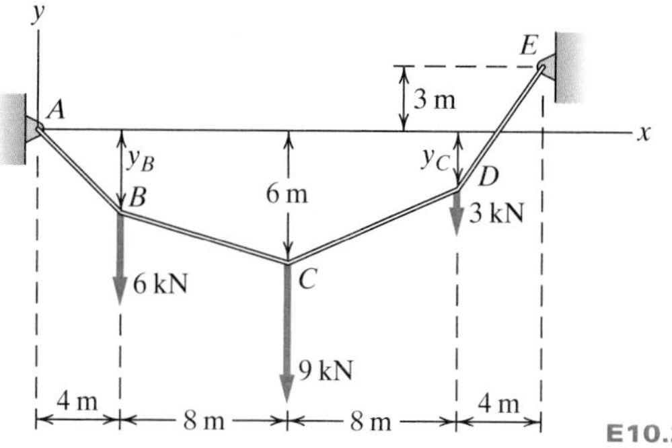

E10.2.6

10.2.7. A uniform cable weighing μ is suspended between two points A and B, as shown in E10.2.7. Values of y_{sag} (the sag) and W (the distance between A and B) are known.

a. Describe, in words, your procedure for finding the minimum and maximum tension in the cable and the length of the cable.

b. Apply your procedure to find minimum and maximum tension and cable length for the case when $\mu = 10$ lb/ft, $y_{sag} = 100$ ft, and $W = 500$ ft.

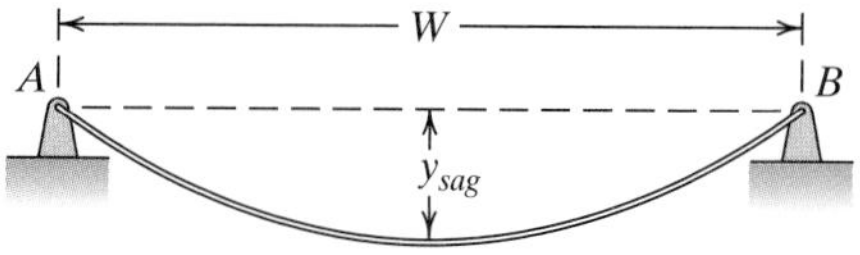

E10.2.7

10.2.8. A uniform 0.5-lb/ft cable depicted in **E10.2.7** is suspended between two supports at equal elevation. The span (W) is 250 ft and the sag is 100 ft. Determine the minimum and maximum tensions in the cable and the length of the cable.

10.2.9. A cable is attached to a support at A, passes over a small ideal pulley at B, and supports a force $\boldsymbol{P}$, as shown in E10.2.9. The sag of the cable (y_{sag}) is 0.5 m, and the mass per unit length of the cable is 1.0 kg/m. Neglect the weight of the cable portion from B to D. Determine

a. the magnitude of the load $\boldsymbol{P}$

b. the slope of the cable just to the left of B

c. the total length of the cable from A to B

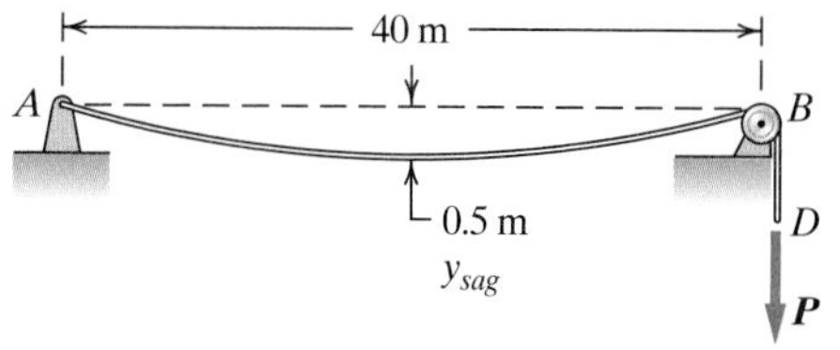

E10.2.9

10.2.10. A cable is attached to a support at A, passes over a small ideal pulley at B, and supports a force $\boldsymbol{P}$, as shown in **E10.2.9**. The sag of the cable (y_{sag}) varies from 0.25 to 5 m. The mass per unit length of the cable is 1.0 kg/m.

a. Create a plot of the required force $\boldsymbol{P}$ as a function of y_{sag}.

b. Create a plot of the cable length (between A and B) as a function of y_{sag}.

c. If the maximum cable tension is to be limited to 1000 N, specify the maximum force $\boldsymbol{P}$ that should be applied to the cable.

10.2.11. A uniform cord 750 mm long passes over a frictionless pulley at B and is attached to a rigid support at A (E10.2.11). Width $W = 250$ mm. Determine the smaller of the two values of y_{sag} for which the cord is in equilibrium.

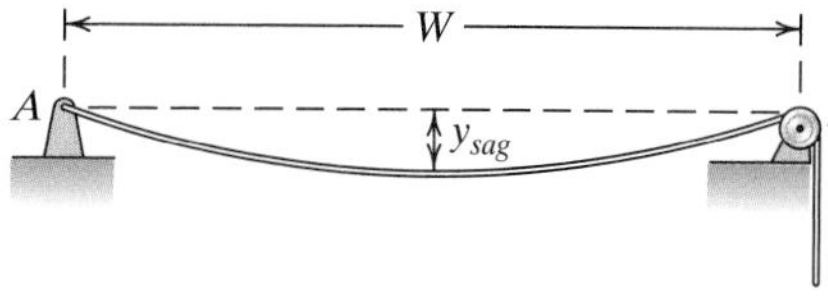

E10.2.11

10.2.12. A uniform 0.4-kg/m cable is attached to collar A, which may slide along the frictionless horizontal bar, as shown in E10.2.12. If a horizontal force of 10 N is required to hold the collar in place when $W/2 = 900$ mm, determine

a. y_{sag}

b. the length of the cable

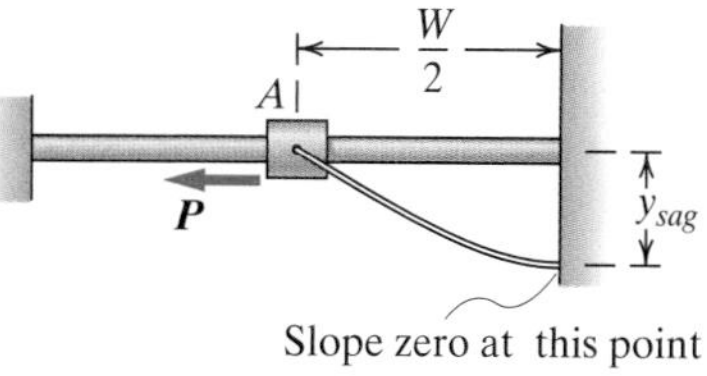

E10.2.12

10.2.13. For the cable in **E10.2.12**, determine the horizontal force $\boldsymbol{P}$ that is required to hold the collar such that $W/2 = 600$ mm, and $y_{sag} = 360$ mm.

10.2.14. A uniform 0.5-lb/ft cable is used to tow the glider A in E10.2.14 in level flight. The glider is 400 ft behind and 100 ft below the tow plane B. Determine the horizontal tension T_O in the cable at the glider. Neglect air resistance.

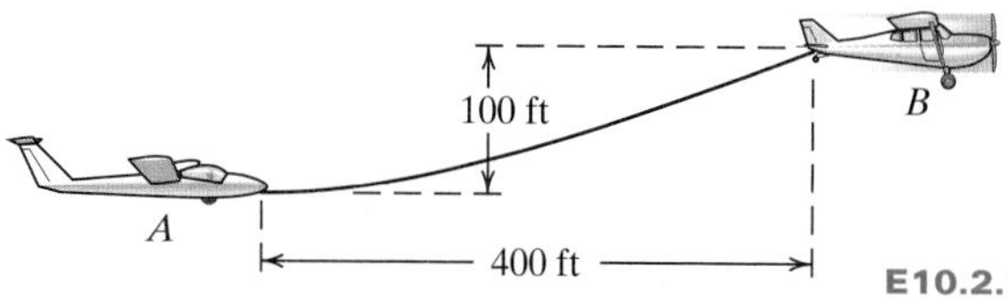

E10.2.14

10.2.15. Determine the total length L of cable that will have the configuration shown in **E10.2.15** when suspended from points A and B. Also determine y_{sag}.

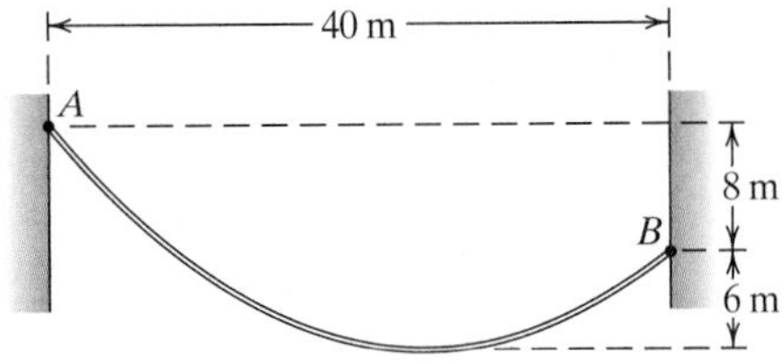

E10.2.15

10.2.16. An electrical transmission cable having a weight per unit length of 2 lb/ft is suspended between and pinned to a series of towers, as shown in **E10.2.16.** If $W =$ 400 ft and $y_{sag} = 5$ ft, determine

a. the horizontal and vertical components of the resultant force exerted by segment AB on tower A

b. the loads acting on the base of tower A

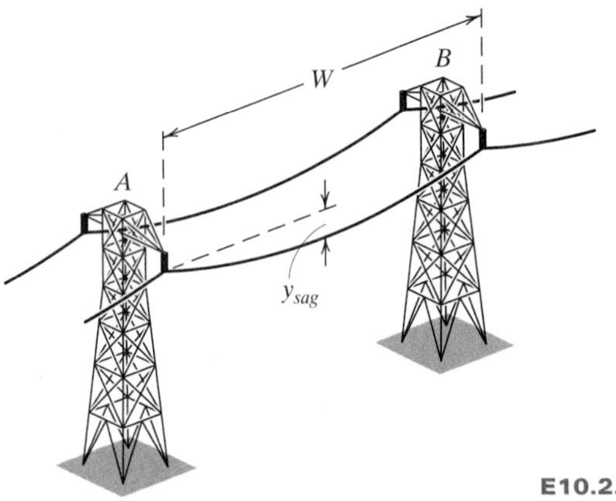

E10.2.16

10.2.17. Determine the sag-to-span ratio y_{sag}/W of a cable whose total weight equals the minimum tension in the cable.

10.2.18. Determine the sag-to-span ratio y_{sag}/W of a cable whose total weight equals the maximum tension in the cable.

10.2.19. The mass per unit length of a cable is 0.4 kg/m. Determine the two values of y_{sag} for which the maximum tension in the cable is 250 N if $W = 70$ m.

10.2.20. A uniformly distributed roadway is carried by the cables of a suspension bridge, as shown in **E10.2.20.** Each cable supports $\omega = 1200$ lb per horizontal foot, with span $W = 880$ ft and $y_{sag} = 48$ ft. Each cable is anchored at its ends at the same elevation. Determine

a. the tension in the cable at a support and the minimum tension in the cable

b. the length of the cable if it takes on a catenary shape

c. the length of the cable if it takes on a parabolic shape (Compare the answer to **b.**)

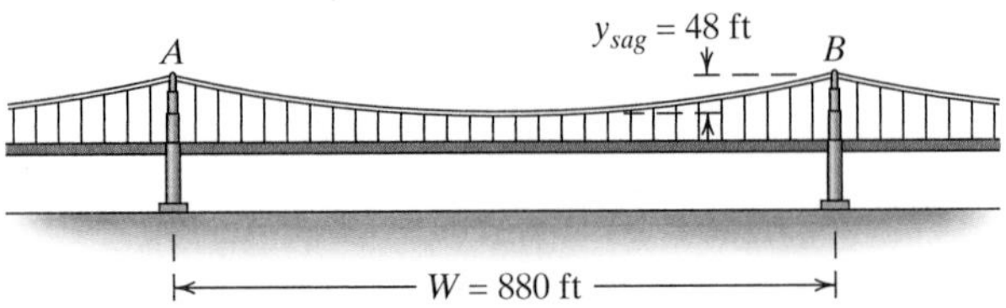

E10.2.20

10.2.21. A suspension bridge roadway is supported by two cables AB and BC, as shown in **E10.2.21.** One end of each cable is pinned to a tower at B. The other ends of the cables are anchored at the same elevation at A and C, respectively, where the cables have horizontal slopes. The roadway load is 200 lb per horizontal foot. Determine

a. the resultant of the forces exerted on the tower by the cables and the forces exerted on the supports at A and C

b. the lengths of cables AB and BC

c. the loads acting on the tower at its base D

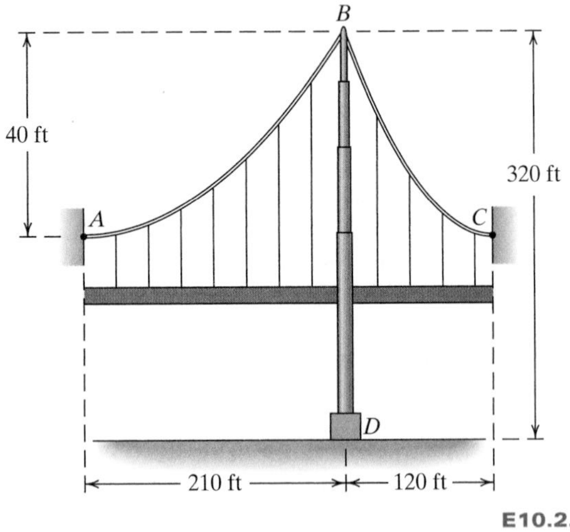

E10.2.21

10.2.22. A suspension bridge roadway is supported by a cable anchored at A and B, as shown in **E10.2.22.** The roadway weighs 3 kN per meter of horizontal length. Determine

a. the minimum and maximum tension in the cable

b. the cable tensions at the supports A and B

c. the length of the cable

d. y_{sag}

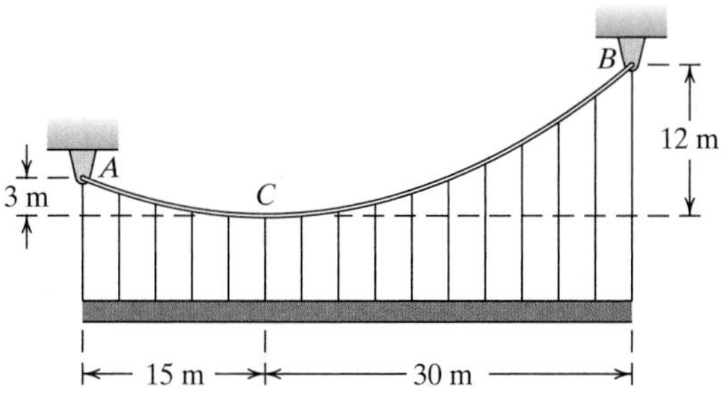

E10.2.22

10.2.23. A suspension bridge supports a 150-kN load that is uniformly distributed along the horizontal (E10.2.23). The slope in the cable is zero at the lower end. Determine

a. the minimum and maximum tension in the cable

b. the cable tensions at the supports A and B

c. the location on the cable where the tension is the average of the values determined in **a**

d. y_{sag}

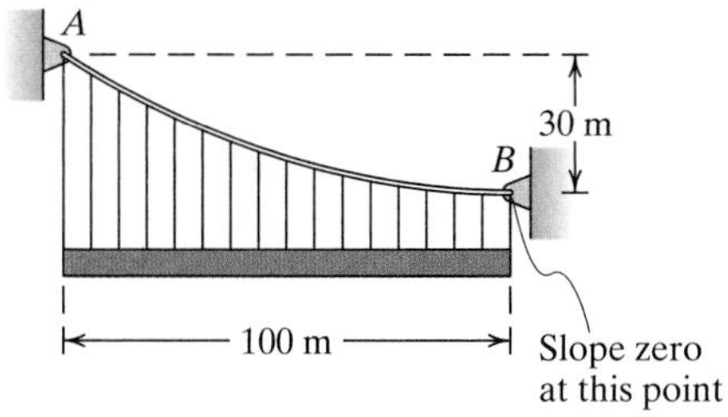

E10.2.23

10.2.24. A cable supports a uniform load of 100 lb/ft and is anchored at A and B, as shown in E10.2.24. At A, the cable is horizontal. Determine

a. the minimum and maximum tension in the cable

b. the length of the cable using both the parabolic and catenary methods (Compare the two solutions.)

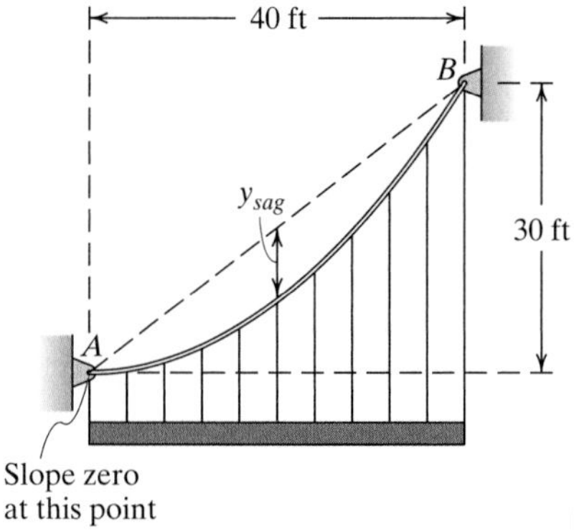

E10.2.24

10.2.25. A water conduit that crosses a canyon is suspended from a cable, as shown in E10.2.25. When full of flowing water, the conduit weighs 30 kN per meter of length.

a. Determine the maximum tension in the cable.

b. Find the length of the cable.

c. When water is not flowing through the conduit, the conduit weighs 15 kN per meter of length. How might the towers A and B be designed in order to maintain the same cable sag as in the fully loaded condition? Why would it be desirable to maintain constant cable sag?

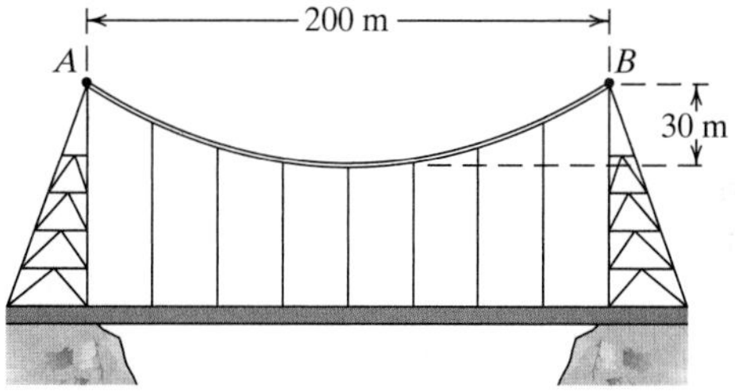

E10.2.25

10.2.26. Cable AB carries a uniform load of 500 N/m and has a span of 300 m, as shown in E10.2.26. If the tangent to the cable at an end is at a 45° angle from the vertical, determine y_{sag} and the maximum tension in the cable.

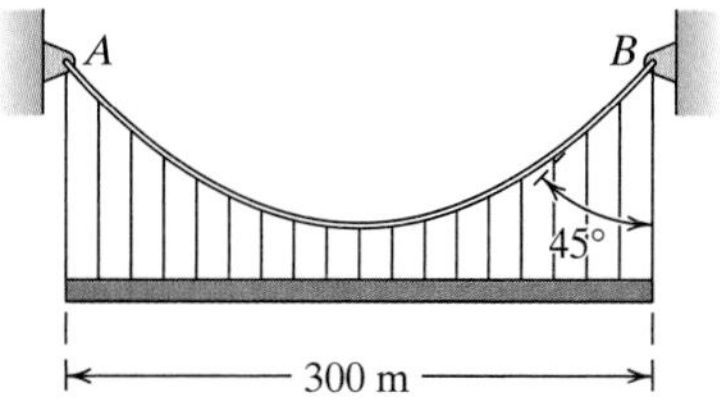

E10.2.26

10.2.27. Cable AB in E10.2.27 carries a pipeline, whose weight per unit length is 220 lb/ft. If $\theta = 20°$ and B is located 10 ft below A, determine

a. the maximum tension in the cable

b. the length of the cable

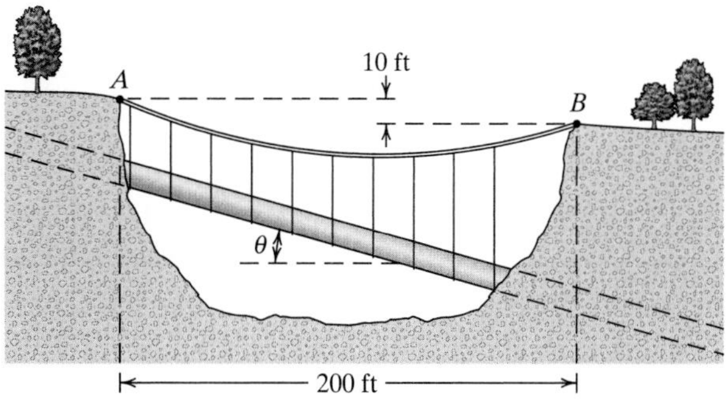

E10.2.27

10.2.28. When Joseph Strauss designed the Golden Gate Bridge, he specified the height of the towers, the length of the center span ($W = 1280$ m), and the sag of the main cable ($y_{sag} = 143$ m). If he had specified shorter towers so that the allowable sag was only 130 m, by what percentage would the horizontal component of the cable tension T_O have changed? (Assume $\omega = 165$ kN/m.)

10.2.29. A cable supports a mass of 12 kg/m of its own length and is suspended between two points at the same level. It supports only its own weight. Determine the tension at midspan (T_O), the maximum tension, and the total cable length for the following conditions:

	Span	Sag
Case 1	300 m	60 m
Case 2	200 m	40 m

10.3 JUST THE FACTS

Engineers are concerned with selecting beam and cable sizes and materials that will ensure adequate structural performance. Part of this selection process involves calculating the loads internal to a beam or a cable. In this chapter we considered how to systematically calculate the loads acting externally and internally on beams and cables.

A member is called a **beam** if loads are applied perpendicular to its long axis. These loads, which may consist of forces and/or moments, are referred to as **lateral loads**. Geometric features that are important in describing a beam are its length and its cross-sectional area and shape. We set up a **beam coordinate system** with an x_b axis along the long axis. The other two axes (y_b and z_b) lie in the cross section of the beam.

Some beam configurations are given names (**cantilever beam, simply supported beam, fixed-fixed beam**). A beam is classified as a **planar beam** if all of the external forces acting on it lie in a single plane and all external moments are about an axis perpendicular to this plane; otherwise it is a **nonplanar beam**.

The loads internal to a beam consist of a **bending moment** (M_b), **shear force** (V), and **axial force** (N). Static analysis of beams consists of finding these internal loads at various locations along the long axis of the beam. These internal loads can be depicted in terms of **bending moment, shear force**, and **axial force diagrams**.

Bending moment and shear are related by

$$V_y = \frac{dM_{bz}}{dx} \tag{10.3A}$$

Furthermore, a distributed force ω acting on the beam is related to the shear force

$$\omega = \frac{-dV_y}{dx} \tag{10.3B}$$

The bending moment is also related to internal "pressure" in the beam, commonly referred to as **stress** by

$$\text{stress} = \frac{-M_{bz}y}{\displaystyle\iint_{\substack{\text{beam}\\ \text{cross}\\ \text{section}}} y^2 dy\, dz} \tag{10.4}$$

where M_{bz} is the bending moment at a particular location x along the length of the beam and y in the numerator is the y coordinate within the cross section where the stress is being calculated. The denominator is a scalar quantity that reflects the distribution of material in the beam cross section and is called the **area moment of inertia**—its calculation is detailed in Appendix C.

Cables are constructed from a large number of wound wires. The forces acting on cables may be distributed along their length or may be point forces. The internal forces in cables are tension forces. Engineers undertake analysis to predict the tension in a cable so that they can select a cable whose cross-sectional area has enough capacity to transmit the tension without breaking. They also need to be able to predict the shape of the cable so that they will know the geometry of the fully loaded structure.

We consider four possible cable loadings. They are

- Case I—the cable is fully taut and the force is along the cable.
- Case II—concentrated forces act perpendicular to the cable and the cable is fully taut between the concentrated forces.
- Case III—the cable is only loaded by its own weight. Its shape is a **catenary curve** and the tension varies along the length.
- Case IV—the cable is carrying a distributed load and the cable's weight is small relative to this distributed load. Its shape is a **parabolic curve** and the tension varies along the length.

Procedures for finding the cable tension and shape (including sag) are outlined for each of the cases in Section 10.2.

NOTES

Note 1: Development of Expressions that Describe Catenary Cables

Development of (10.5A) and (10.5B). Using **Figure 10.47**, we write the equations of equilibrium. Based on (7.5A),

$$\sum F_x = 0$$
$$T\cos\theta - T_O = 0$$
$$T\cos\theta = T_O \tag{1}$$

Based on (7.5B),

$$\sum F_y = 0$$
$$T\sin\theta - \mu s = 0$$
$$T\sin\theta = \mu s \tag{2}$$

Dividing (2) by (1),

$$\tan\theta = \frac{\mu s}{T_O} \tag{3}$$

Recognizing that $\tan\theta = \frac{dy}{dx}$, we can rewrite (3) as

$$\frac{dy}{dx} = \frac{\mu s}{T_O} \tag{4}$$

Differentiating (4) with respect to x,

$$\frac{d^2y}{dx^2} = \frac{\mu}{T_O}\frac{ds}{dx} \tag{5}$$

We now write s as a function of x and y as

$$ds^2 = dx^2 + dy^2$$

$$\frac{ds}{dx} = \sqrt{1 + \left(\frac{dy}{dx}\right)^2} \tag{6}$$

and substitute (6) into (5):

$$\frac{d^2y}{dx^2} = \frac{\mu}{T_O}\sqrt{1 + \left(\frac{dy}{dx}\right)^2} \tag{7}$$

Calling $\frac{dy}{dx} = q$, we can rewrite (7) as

$$\frac{dq/dx}{\sqrt{1 + q^2}} = \frac{\mu}{T_O} \tag{8}$$

Integration of (8) with the boundary condition $\left(x = 0, \frac{dy}{dx} = 0\right)$ results in

$$q = \frac{dy}{dx} = \sinh\frac{\mu x}{T_O}$$

which can be rewritten in terms of exponentials as

$$\frac{dy}{dx} = \sinh\frac{\mu x}{T_O} = \frac{e^{\mu x/T_O} - e^{-\mu x/T_O}}{2} \tag{9}$$

Integration of (9) with the boundary condition $(x = 0, y = 0)$ results in

$$y = \frac{T_O}{\mu}\left(\cosh\frac{\mu x}{T_O} - 1\right) \tag{10}$$

Equation (9) is the same as (10.5B), and (10) is the same as (10.5A).

Now We Develop (10.6) for the Tension in the Cable. With the expression for slope in (9), we can rewrite (4) as

$$s = \frac{T_O}{\mu}\sinh\frac{\mu x}{T_O} \tag{11}$$

Squaring (1) and (2) from above and adding, we have

$$T^2 = T_O^2 + \mu^2 s^2 \tag{12}$$

Substituting (11) into (12) and rearranging we write an expression for the tension T:

$$T = T_O \cosh \frac{\mu x}{T_O} \tag{13A}$$

Equation (13A) can be written in an alternate form by substituting in for the hyperbolic cosine from (10):

$$T = T_O + \mu y \tag{13B}$$

Equations (13A) and (13B) are the same as (10.6).

Note 2: Development of Expressions that Describe Parabolic Cables

Development of (10.7A) and (10.7B). Using **Figure 10.53*b***, we write the equations of equilibrium. Based on (7.5A),

$$\sum F_x = 0$$
$$T\cos\theta - T_O = 0$$
$$T\cos\theta = T_O \tag{1}$$

Based on (7.5B),

$$\sum F_y = 0$$
$$T\sin\theta - \omega x = 0$$
$$T\sin\theta = \omega x \tag{2}$$

Dividing (2) by (1),

$$\tan\theta = \frac{\omega x}{T_O} \tag{3}$$

Recognizing that $\tan\theta = \frac{dy}{dx}$, we can rewrite (3) as

$$\frac{dy}{dx} = \frac{\omega x}{T_O} \tag{4}$$

This is the same as (10.7B).

Integrating (4) with respect to x with the boundary condition ($x = 0$, $y = 0$) results in

$$y = \frac{\omega x^2}{2\,T_O} \tag{5}$$

which is the same as (10.7A).

Developing an Expression for the Cable Length (Equation IV.5A). A segment of the cable ds can be written as

$$ds^2 = dx^2 + dy^2$$

$$\frac{ds}{dx} = \sqrt{1 + \left(\frac{dy}{dx}\right)^2} \tag{6}$$

Now substitute (4) into (6):

$$\frac{ds}{dx} = \sqrt{1 + \left(\frac{\omega x}{T_O}\right)^2} \tag{7}$$

Rearranging and integration for the segment s_{oA} that runs from $x = 0$ to $x = x_A$, we have

$$\int_0^{s_{oA}} ds = \int_0^{x_A} \sqrt{1 + \left(\frac{\omega x}{T_O}\right)^2}\, dx \tag{8}$$

Although this expression can be integrated in closed form, for computational purposes it is more convenient to express the radical as a convergent series and then integrate term by term. For this purpose we use the binomial expression

$$\left(1 + \left(\frac{\omega x}{T_O}\right)^2\right)^n = 1 + n\left(\frac{\omega x}{T_O}\right)^2 + \frac{n(n-1)}{2!}\left(\frac{\omega x}{T_O}\right)^4 + \frac{n(n-1)(n-2)}{3!}\left(\frac{\omega x}{T_O}\right)^6 + \cdots \tag{9}$$

which converges for $\left(\frac{\omega x}{T_O}\right)^2 < 1$. Setting $n = \frac{1}{2}$ in (9) gives the expression

$$\begin{aligned} s_{OA} &= \int_0^{x_A}\left(1 + \frac{\omega^2 x^2}{2\,T_O^2} - \frac{\omega^4 x^4}{8\,T_O^4} + \cdots\right)dx \\ &= \left[x_A + \frac{\omega^2 {x_A}^3}{6\,T_O^2} - \frac{\omega^4 {x_A}^5}{40\,T_O^4} + \cdots\right] \end{aligned} \tag{10}$$

Notice that this is the same as the first term on the right-hand side of equation IV.5A. A similar expression can be created for s_{OB}.

SYSTEM ANALYSIS (SA) EXERCISES

SA10.1 Handle Design for the Money-Maker Plus Water Pump[1]

General Background: ApproTEC (Appropriate Technologies for Enterprise Creation) is a nonprofit engineering firm started by Stanford University graduate Martin Fisher in 1991. It is headquartered in Nairobi, Kenya, and aims to design and mass-produce high-quality devices that focus on the particular needs of Kenyan citizens. Among the products produced by ApproTEC is the Super-MoneyMaker, a one-person, leg-powered irrigation pump. This pump and other mechanical devices produced by ApproTEC have made over 26,000 desperately poor Africans rich entrepreneurs (by the standard of their homeland). As testimony of the impact of the pump on the lives of the people, an article in the *San Francisco Chronicle Magazine* (December 8, 2002) tells the story of Kenyan Janet Ondiak. When her husband died

> *She barely kept her family alive on the food she managed to grow on the 1/8th acre of land. She owned 2 acres, but even with all six children lugging buckets of water, they were able to irrigate no more than a small area. One afternoon, she saw a demonstration of the Super-MoneyMaker irrigation pump in her local village. She worked for six months to scrape together $75 and bought the pump. Today, she has three full-time workers who irrigate her entire 2 acres. Last year, she made $2,500 in profit from selling vegetables grown on her land. She recently opened a small shop from which to sell her food. She can now pay for all six children to attend school.*

Technical Background: ApproTEC designers are currently designing the MoneyMaker Plus water pump, shown in Figure SA10.1.1. This pump is similar to ApproTEC's Super-MoneyMaker pump. In operation, it is capable of pulling water from 6 m (from a water source such as a creek) and pushing it another 12 m (up a hill or to power mechanical sprinklers). However, the MoneyMaker Plus aims to be much cheaper (50 percent less!) so that more farmers will be able to afford to buy it. The cost of the entire pump is extremely constrained—the target cost, excluding manufacturing, is 1600 Kenya Shillings (approx. $20 USD). The ApproTEC pump designers are hoping that they can use a handle consisting of a steel SHS (square hollow section) shaft and sleeve; this handle design has been used successfully for the Super-MoneyMaker pump. The square handle shaft fits snugly in a

[1]This systems analysis problem was created by Dr. Krista Donaldson. At the time she authored this problem she was a Ph.D. student working on her thesis related to increasing manufacturing capacity in developing countries.

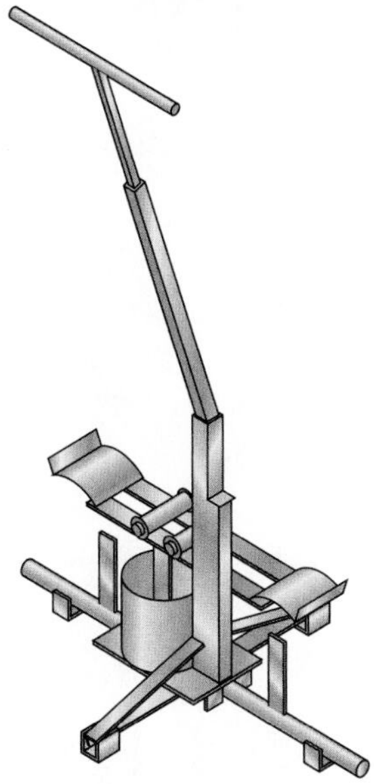

Figure SA10.1.1 MoneyMaker Plus water pump

square sleeve tube with a stopper at the end and is easily removable from the pump for easy transport (by owners).

The proposed geometry of the handle is shown in Figure SA10.1.2 and is based on successfully tested design prototypes. The handle is important because it helps users maintain their balance and shift their weight smoothly from one foot to the other (to operate the pump). This pump is a bit trickier to operate than previous ones as the user tips her weight back and forth as if her feet were on a mini seesaw with the handle coming up from between her

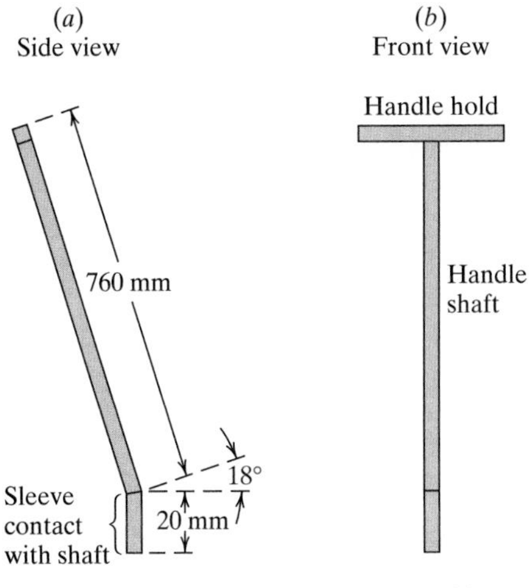

Figure SA10.1.2 Design of pump handle: (*a*) side view; (*b*) front view

Figure SA10.1.3 Operation of pump

feet near the pivot (Figure SA10.1.3). Because the user tips back and forth a bit, the handle is necessary to maintain balance, particularly with new users. Experienced users occasionally lean on the handle (not recommended as that wastes body weight that could be applied to pumping), but otherwise hold the handle lightly during operation.

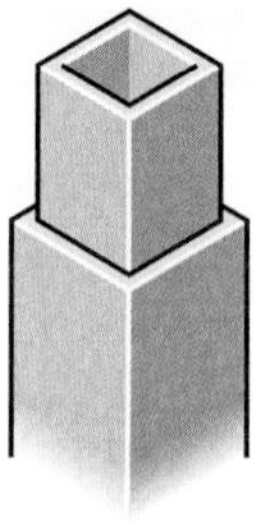

Figure SA10.1.4 Detail of shaft-sleeve connection

Of the available tubing sizes for the handle, only two shaft–sleeve pairs (Figure SA10.1.4) are suitable for ergonomic and pump size constraints, and interface snugly (and somewhat consistently) in practice. They are:

- The $20 \times 20 \times 2$ SHS and the $25 \times 25 \times 2$ SHS pair
- The $25 \times 25 \times 2$ (or $\times$ 3) and the $30 \times 30 \times 2$ pair

The smaller SHS pair is preferable as it is cheaper (cost related to mass of material). The maximum allowable tensile stress for the grade of steel that will be used is 220 MPa; this number takes into account the low grade of steel and a safety factor.

Your Assignment: You are a summer intern working at ApproTEC and have been asked to size the tubing for the handle to ensure that it will not break. More specifically, you must determine whether the preferred $20 \times 20 \times 2/25 \times 25 \times 2$ shaft–sleeve combination is structurally appropriate for this pump. If it is not, you must determine whether any of the available combinations will work (and if not, it is back to the drawing board!).

Based on discussions you have with your fellow ApproTEC engineers about the imagined uses of pumps, you come up with the worst case scenarios for loads on the pump handle listed in Table SA10.1.1.

Pick what you think is the worst of the worst case scenarios and check that the proposed shaft–sleeve pairs will not fail.

Table SA10.1.1 Handle Usage Patterns and Possible Worst Case Scenarios

Case	Side View	Front View	Description
1		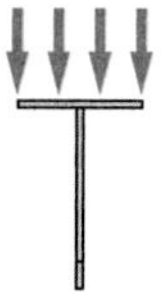	User leans on handle pumping (evident during testing). This would result in a downward force of approximately 1/3 the body weight of the user.*
2		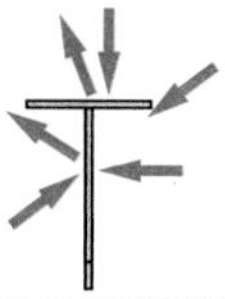	User shakes handle vigorously to determine how sturdy the pump is (observed behavior with customers). The force associated with this is likely to be less than 1/3 of the body weight of the user, since there is no body weight behind it.
3			The pump is cantilevered by its handle when picked up (perhaps dirt is in the sleeve and the handle gets stuck). The mass of the pump with water in the valve box and piston cylinder and mud caked to the bottom is at most 10 kg. In testing, the tester was observed to cantilever it over his shoulder like a hobo carries his stick because he was experimenting with how he might carry it easily while carrying something else with his other hand. In practice, this would likely be the case with a real farmer. The farmer, generally being a woman, would probably also have a baby on her front and a basket holding vegetables on her head.
4		Out of page 	The user loses her/his balance during operation and grasps the handle to keep from toppling. If falling, the user could direct all of her/his mass (75 kg) at the handle. But the pump was not designed to make sure that people don't fall off (this would be too expensive), so assuming the pump would not be fixed into the ground, it is estimated that about 50 kg of the user's mass will be directed horizontally at the top of the handle should it tip over.

*ApproTEC assumes user mass to be around 75 kg. This might seem low, but men in the targeted pump regions tend to be much smaller and thinner than Americans; the average woman, on the other hand, is often bigger than the average man, but is still likely to be less than 75 kg.

SA10.2 Golden Gate Bridge Approximate Analysis 1

Undertaking analysis of structures often requires that engineers make assumptions about the structure's behavior. These approximations are commonly simplifications or idealizations about the behavior made because not everything is known about the structure and/or because of constraints on the resources available for the analysis. In Chapter 3 an analysis of the Golden Gate Bridge was undertaken in which the main cables were approximated as a series of links (**Figure SA10.2.1**). We found the following:

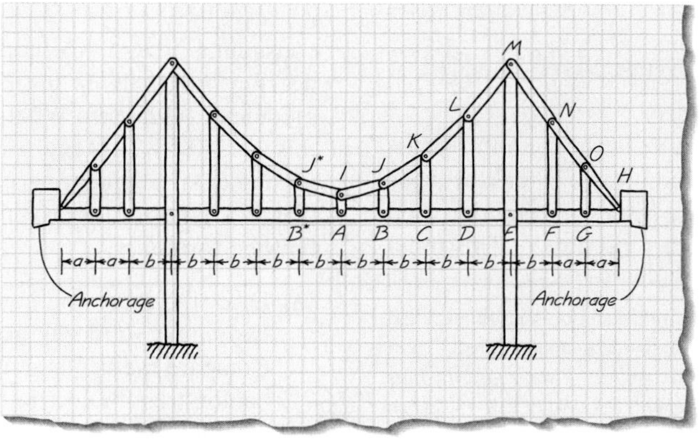

Figure SA10.2.1 Golden Gate cables modeled as two-force members

Member	Force
IJ and *IJ'*	237 MN (tension)
IA	26.4 (tension)
JK	240 MN (tension)
JB	26.4 MN (tension)

The following tasks are designed to further explore the approximation of the main cables as presented in Chapter 3:

(a) Draw a free-body diagram of joint K and use the geometry as defined in **Table 3.1** to find the force in member KL.

(b) Draw a free-body diagram of joint L and use the geometry as defined in **Table 3.1** to find the force in member LM.

(c) By calculating the horizontal components of the forces acting on links IJ, JK, KL, and LM, show that the horizontal force in the main cable remains constant between any two points along its length.

(d) Compare the tension acting on links IJ and IJ' to T_O, and the tension acting on LM to T_{max} presented in **Figure 3.12**. Based on this comparison, do you think that the links serve as a reasonable approximation to the continuous cable? How might the approximation be improved?

SA10.3 Golden Gate Bridge Approximate Analysis 2

Sometimes an engineer wants to approximate a complex structure with a simplified model so that he can perform a "back of the envelope" analysis to get a quick idea of the magnitudes of loads and internal forces in a structure. This could be useful in checking whether the output of a computer analysis looks reasonable or providing a first guess at member sizes for a design.

Undertaking analysis of structures often requires that engineers make assumptions about the structure's behavior. Commonly these approximations are simplifications or idealizations about the behavior made because not everything is known about the structure and/or because of constraints on the resources available for the analysis.

We would like to explore a very simple model of the Golden Gate Bridge. Such a simplified model is shown in **Figure SA10.3.1**, consisting of the bridge deck, a suspender on each side at the middle of the bridge, the two towers, and four truss links to represent each main cable.

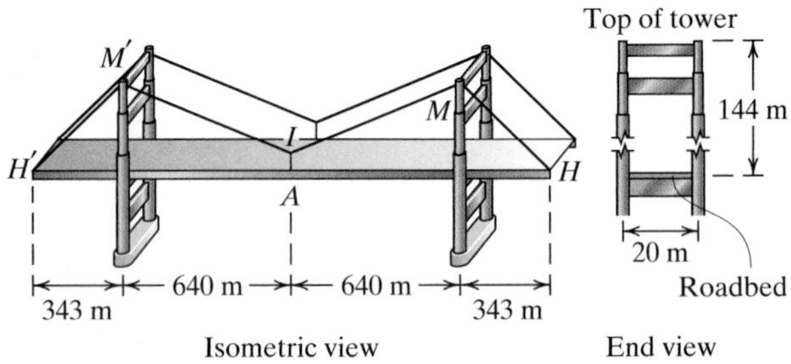

Figure SA10.3.1 Simplified model of Golden Gate cables

Assuming that the bridge deck weighs 330 kN/m and that each main cable carries an equal share of the bridge load, calculate the force in members IA and IM. How do these values compare with

(a) the forces acting on the links in the simplified model in **Figure 3.12**, or

(b) the theoretical force, as shown in **Figure 3.12**?

SA10.4 Form Follows Function[2]

Karl Culmann (1821–1881) was one of the foremost structural engineers of his day. One of his major breakthroughs was the development of graphical statics, a technique used to visualize forces and aid in structural design. This time-consuming analysis method was an integral part of engineering programs up until the introduction of computer modeling. (*Question:* What date/decade would you put to this?. . . . 1960s?) Many truss bridges are built using this fundamental engineering principle in which the moment diagram gives an approximation of the superstructure's finished form.

(a) Draw the shear and bending moment diagrams for the two simply supported beams shown in **Figure SA10.4.1**. Compare the bending moment diagrams with the companion truss bridges in **Figure SA10.4.2**. Does it matter whether the superstructure is on the top or the bottom of the bridge? What does switching the location of the

[2]Phrase coined by the American architect Louis Sullivan in his article, "The Tall Office Building Artistically Considered," first published in 1896. To be found in I. Athey, ed. *Kindergarten Chats (revised 1918) and Other Writings*. New York 1947: 202–13. More on this phrase and its implications for design work can be found at http://www.geocities.com/Athens/2360/jm-eng.fff-hai.html.

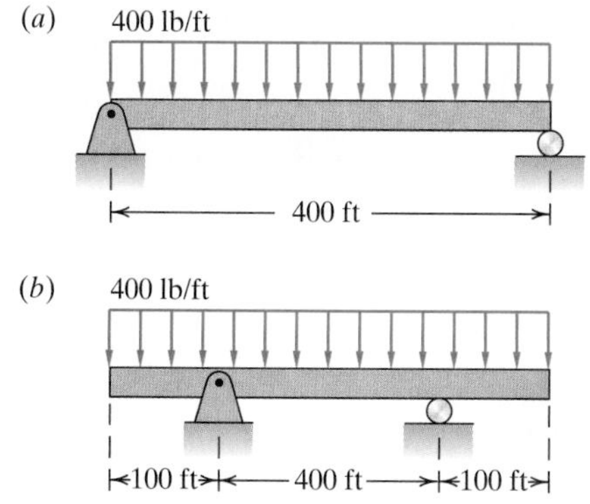

Figure SA10.4.1 Two simply supported beams

superstructure from the top to the bottom do to the individual function of the load-bearing elements?

(b) The spine of quadruped animals is very similar to a cantilever bridge in both form and function. Draw the shear and moment diagrams of the simplified American bison spine in Figure SA10.4.3. Compare your diagrams to the skeletal structure of the animal shown in Figure SA10.4.4. What are the functions of the legs and the processes originating from the spine? Are there any relationships between the size of the processes and the moment diagram? What elements missing in the skeletal structure are needed to complete the "bridge"?

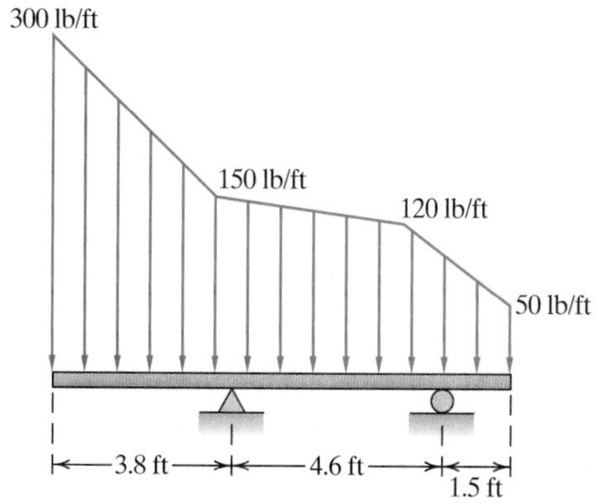

Figure SA10.4.3 A simplified model of an American bison's spine

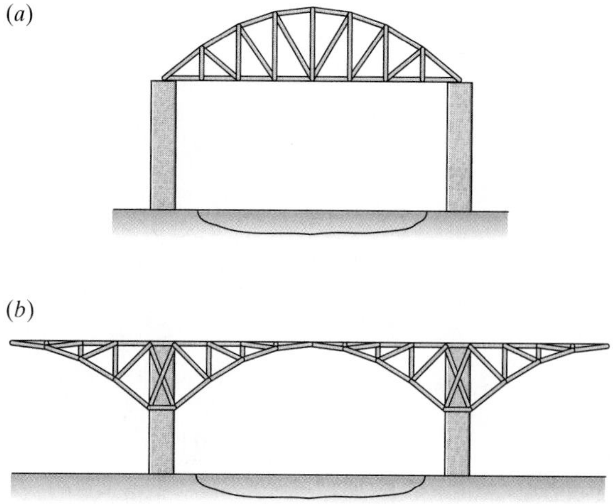

From D'Arcy Tompson's *On Growth and Form* (Dover, 1992).

Figure SA10.4.2 Two common bridge designs that incorporate Culmann's methodology.

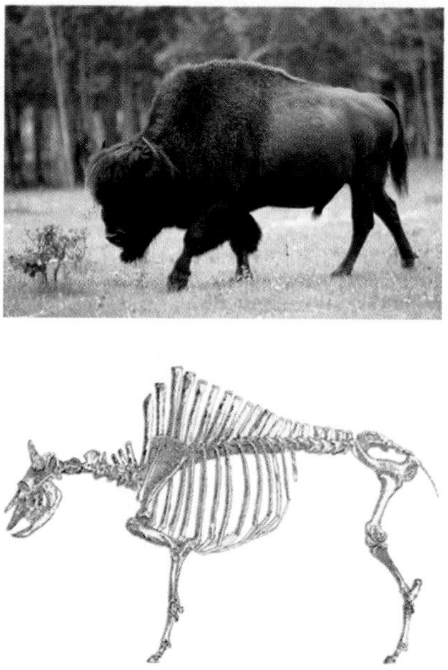

Figure SA10.4.4 The skeletal structure of an American bison

SA10.5 Hoover Dam

Hoover Dam is one of the greatest monuments to industrial strength and human audacity (Figure SA10.5.1). The dam itself rises approximately 726 feet above the floor of the Black Canyon between Nevada and Arizona. It is 1244 feet long, 660 feet thick at its base, 45 feet thick at its crest and weighs 5,500,000 tons. The lake it created, Lake Mead, backs up 110 miles behind the dam and is nearly 500 feet deep. It contains enough water to submerge the state of Connecticut under 10 feet of water or to supply

(*a*)

(*b*)

Figure SA10.5.1 Images of Hoover Dam: (*a*) from above; (*b*) from its base (adapted from government publications)

5000 gallons of water to every person in the world. The concrete dam was cast as a single solid piece. Pouring started in June 1933 and continued 24 hours a day, seven days a week until May 1935.

The engineer responsible for this feat was Frank Crowe. He studied civil engineering at the University of Maine and, after a summer internship working on the drainage basin of the Yellowstone River, joined the U.S. government's Reclamation Service. Crowe was widely considered the best construction engineer in the United States and was eventually hired by the Six Companies to spearhead their efforts to obtain the Hoover Dam contract. He was the first to use cableways to transport people and supplies to the worksite (**Figure SA10.5.2**). In many instances the cables were required to support loads of several tons. In addition, because falling rocks were always a hazard during the construction of the dam, the first hard hats were developed by the workers using baseball caps and tar as raw materials. It is worth noting that after painstaking calculations and recalculations and four years of construction, nobody knew for sure that the dam would work! The only way to be certain was to seal the diversion tunnels and hope for the best.

Figure SA10.5.2 Transporting equipment and materials at Hoover Dam (postcard from Hoover Dam)

(a) Model the dam as a simple cantilever that rises up out of the bottom of Black Canyon. Given the weight of the dam and the depth of the water behind it, calculate the loads acting on the dam at its base.

(b) Consider a simple model of the cableways that Frank Crowe employed during the construction of Hoover Dam (**Figure SA10.5.3**). Model link BE as a particle and assume that $\alpha = 25°$, $\beta = 30°$, the mass of the car and its load is approximately 12,000 kg, and the tension in cable EG is negligible. Find the tension in the support cable ABC and the traction cable DE. What uncertainty in these forces is associated with α?

(c) To support the kind of loads found in (b), locked-coil track strand wire rope (shown in **Figure SA10.5.3**) was used. This particular wrapping style helps to distribute the contact loads exerted by the pulleys on the cable. **Table SA10.5.1** gives cable strengths (adapted from *Marks' Standard Handbook*, page 10-9). Choose a factor of safety and a grade of cable, then select an appropriate diameter for cables ABC and DE.

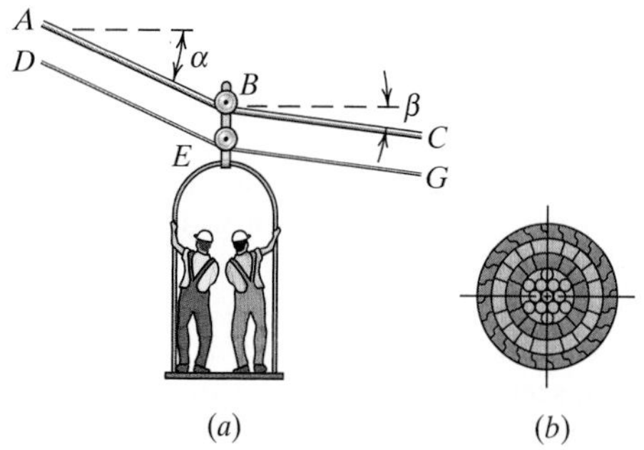

Figure SA10.5.3 (*a*) Cableways similar to those used to build Hoover Dam; (*b*) cross-section of the locked-coil track strand wire rope that was used to support the heavy loads (*Marks' Standard Handbook*)

Table SA10.5.1 Capacity Data for Cables

Diameter		Special Grade		Standard Grade	
in.	mm	Short ton	Tonne	Short Ton	Tonne
0.750	19.1	31.5	28.6	25	22.7
0.875	22.2	41.5	37.6	32	29.0
1.00	25.4	52.5	47.6	42	38.1
1.125	28.6	66.0	59.9	54	49.0
1.250	31.8	81.0	73.5	65	59.0
1.375	34.9	100.0	90.7	78	70.8
1.500	38.1	120.5	109.3	93	84.4
1.625	41.3	140.0	127.0	108	98.0
1.750	44.5	165.0	150	125	113.4
1.875	47.6	187.5	170	138	125.2
2.00	50.8	215	195	158	143
2.250	57.2	280	254		
2.500	63.5	345	313		
2.750	69.9	420	381		
3.00	76.2	500	454		
3.250	82.6	580	526		
3.500	88.9	690	626		
3.750	95.3	785	712		
4.00	101.6	880	798		

SA10.6 How Much Load Does a Main Column Carry?

In Chapter 5 we learned how the 4.0-kN panels were lifted onto the top of the 48.0 m wide and 12.5–15.0 m high Reynolds Coliseum (see **Figure SA5.4.2**). As the panels were placed onto the flanges of the T-beams spanning the 6.0-m distance between the I-beam girders, the column section of each frame experienced increasing loads (**Figures SA10.6.1** and **SA10.6.2**). After the panels were firmly in place, a crew applied tar and several layers of roofing paper to create a water barrier weighing 50 N/m^2.

In order to assess the effects of the different loading conditions you are asked to calculate the effect of the construction sequence on the forces in the columns.

(a) What is the total load in point D, as indicated in **Figure SA10.6.3*a***, right after the field-spanning steel girders have been erected? (You can neglect the lightweight trusses.)

(b) Next, the T-beams spanning the space between the girders are installed. What are the loads at A, B, and C in **Figure SA10.6.2** after they are in place?

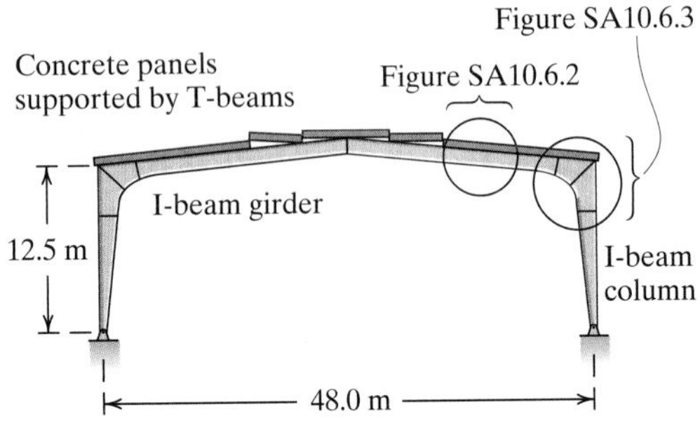

Figure SA10.6.1 Review of Mina Coliseum structure introduced in Chapter 5

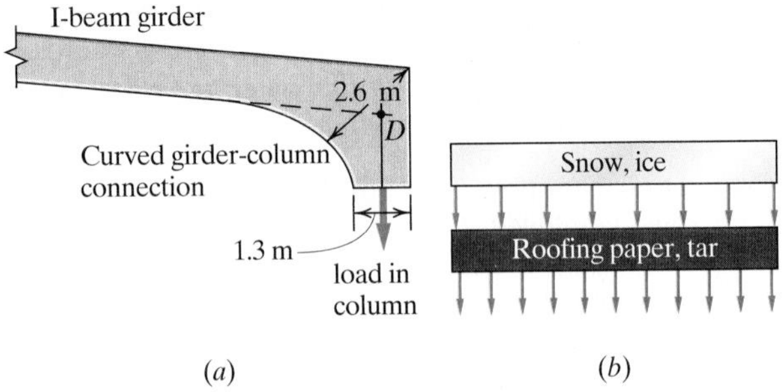

Figure SA10.6.3 (*a*) Main girder connects to column; (*b*) loads on roof

(c) In Chapter 5 we learned how the panels were placed. Develop and sketch a placement sequence from beginning to end for 1949 or today (label rows parallel to the ridge of the roof, spans between adjacent girders, and panels according to their location within a row and a span).

(d) Based on the sequence you proposed in (c), plot the steady increase of the forces at A, B, and C in **Figure SA10.6.2** as panels in three rows and three spans are put in place.

(e) Pick a girder column and plot the load in point D as a function of completed spans and rows per your sequence designed in (c).

(f) If you assume that two crews work in parallel, how would you have to adjust your plots in (e)?

(g) Finally, calculate the loads at A, B, C, and D after the water barrier has been installed.

There are additional loading cases that we need to consider—namely, wind, snow, and ice. Since these loads vary with the location of the building (e.g., coastal region, mountains), one needs to find those values that are valid for the area where the building is to be built.

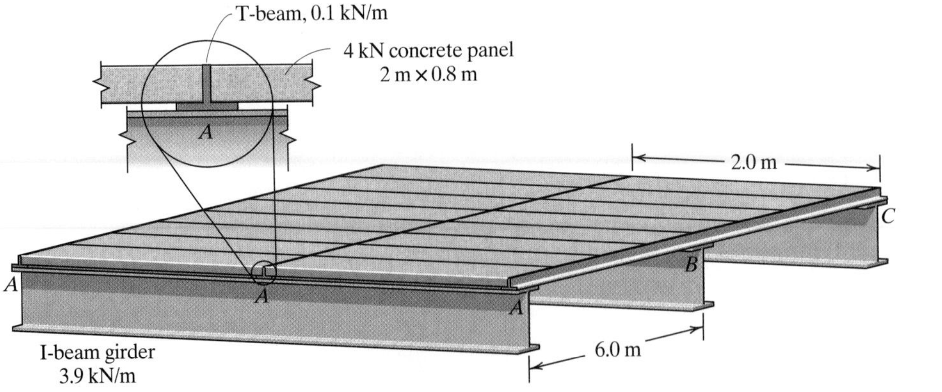

Figure SA10.6.2 Schematic of roof support with girder: T-beams and panels (to locate, see Figure SA10.6.1)

(h) Find the 100-year wind speed and depth of snow and ice accumulation for your area using data from the National Weather Service (i.e., maximum wind speeds, snow, and ice that can be measured once every 100 years). If you live in an area that receives little or no snow and ice, find the data for Houghton, Michigan.

(i) Convert the snow or ice that you found to a snow/ice pressure that would be applied to the roof. Before determining the additional load acting at D in **Figure SA10.6.3*a*** due to the snow and ice, draw the distributed load as shown in **Figure SA10.6.3*b***. (*Hint:* Powder snow = 50–100 kg/m^3, wet snow = 300–350 kg/m^3, glacier ice = 600–700 kg/m^3.)

(j) **Figure SA10.6.4** presents a given wind load of 0.6 kN/m^2 resulting from a wind of approximately 75 mph. (*Comment:* This number is different from the number for a girder because now we are considering a vertical wall that measures 15 meters. As you can find in any building design book or website, wind loads or pressures depend on a variety of different factors that go beyond statics.) Based on what you found in (h), adjust the wind load in **Figure SA10.6.4** up or down.

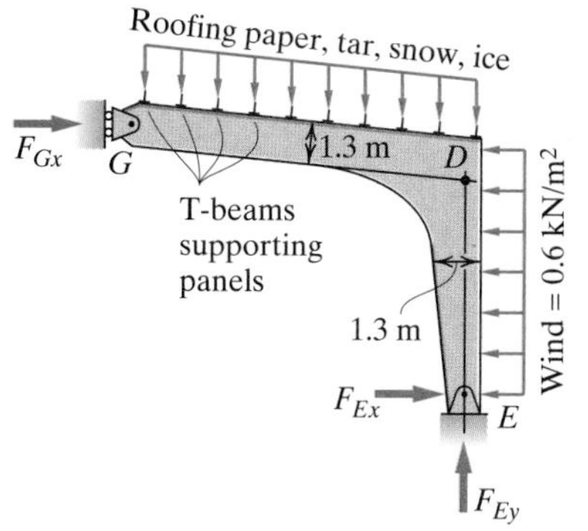

Figure SA10.6.4 Simplified model of symmetrical girder frame with total load calculation model

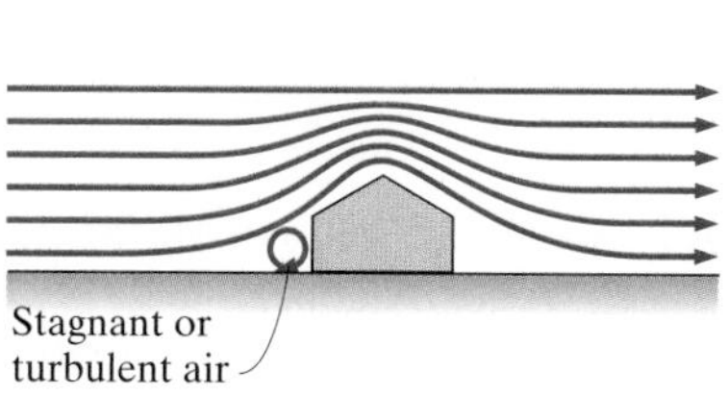

Figure SA10.6.5 Basics of wind speed and wind pressure on a building

(k) How would you introduce the wind forces on the two sides of a sloped roof as shown in **Figure SA10.6.5**? Draw the approximate pressure distribution of a cross section for the totally sealed coliseum due to the wind pressure (assume that the leeward-side pressure is 1/3 of the windward side pressure).

(l) Let's assume that the girder frame shown in **Figure SA10.6.4** behaves as indicated (pin in E and roller in G). Develop the moment, shear, and axial force diagrams for the girder when the building is just completed (without external loads of wind, snow, and ice).

(m) How large and where are the *maximum* moment, shear, and axial forces with and without the external loads for the area that you picked?

(n) Assume that you are the building inspector and the town requires a safety factor of 1.2 (multiplication factor from theoretical to design values of moment, shear, and axial forces). Decide and defend minimum loads, moments, shear, and axial forces that you would expect the designer to use. (Expect the design engineer to argue against you since every additional amount of force that must be considered costs more money. Be sure that you have the proper arguments!)

APPENDIX A

A1 SELECTED TOPICS IN MATHEMATICS

A1.1 Algebra

1. Quadratic equation

$$ax^2 + bx + c = 0$$

$$x = \frac{-b \pm \sqrt{b^2 - 4ac}}{2a}, \; b^2 \geq 4ac \text{ for real roots}$$

2. Logarithms

$b^x = y, x = \log_b y$
Natural logarithms

$$b = e = 2.718\,282$$

$$e^x = y, x = \log_e y = \ln y$$

$$\log(ab) = \log a + \log b$$

$$\log(a/b) = \log a - \log b$$

$$\log(1/n) = -\log n$$

$$\log a^n = n \log a$$

$$\log 1 = 0$$

$$\log_{10} x = 0.4343 \ln x$$

3. Determinants
2nd order

$$\begin{vmatrix} a_1 & b_1 \\ a_2 & b_2 \end{vmatrix} = a_1 b_2 - a_2 b_1$$

3rd order

$$\begin{vmatrix} a_1 & b_1 & c_1 \\ a_2 & b_2 & c_2 \\ a_3 & b_3 & c_3 \end{vmatrix} = \begin{matrix} +a_1 b_2 c_3 + a_2 b_3 c_1 + a_3 b_1 c_2 \\ -a_3 b_2 c_1 - a_2 b_1 c_3 - a_1 b_3 c_2 \end{matrix}$$

A1.2 Analytic Geometry

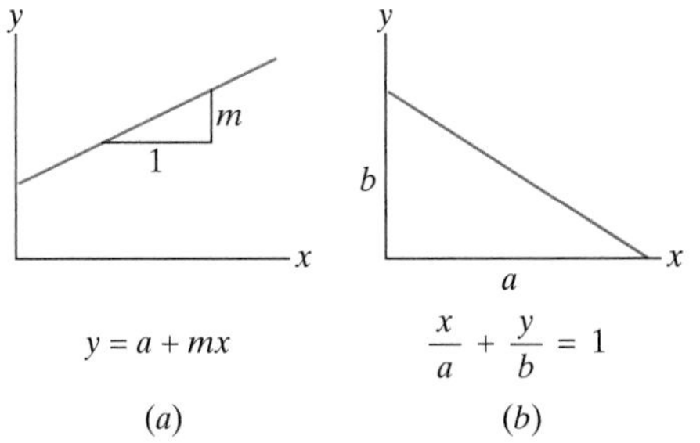

Figure A1.2.1 Straight line

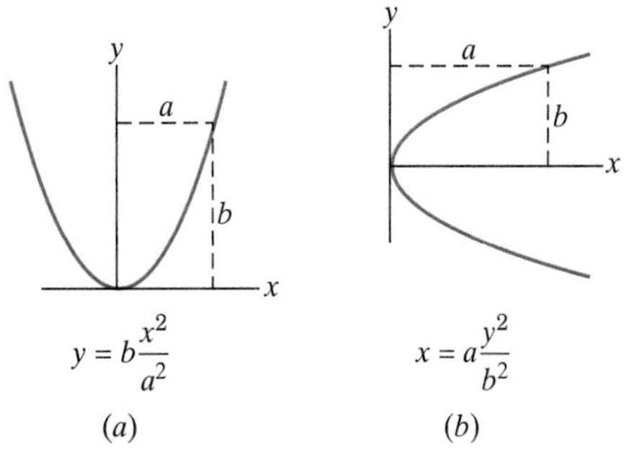

Figure A1.2.3 Parabola

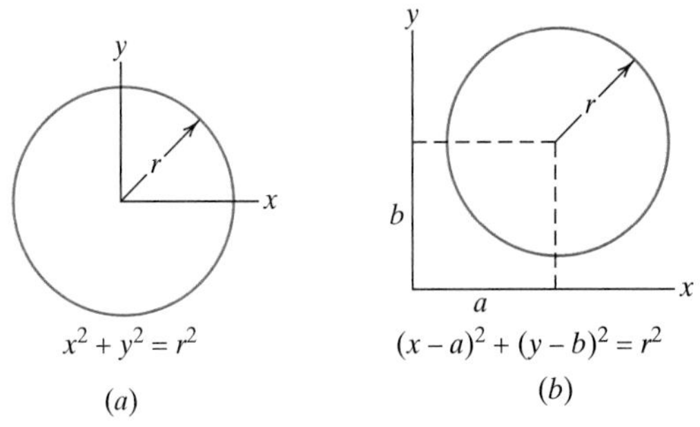

Figure A1.2.2 Circle

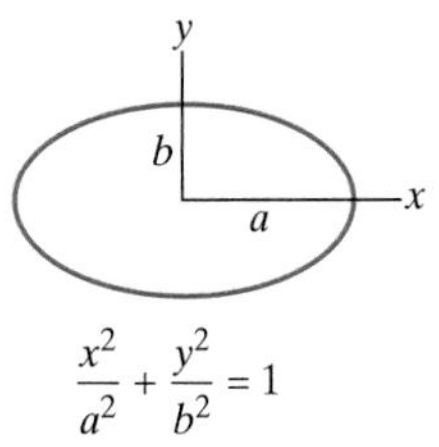

Figure A1.2.4 Ellipse

A1.3 Trigonometry

1. Definitions

$\sin\theta = a/c \quad \csc\theta = c/a$

$\cos\theta = b/c \quad \sec\theta = c/b$

$\tan\theta = a/b \quad \cot\theta = b/a$

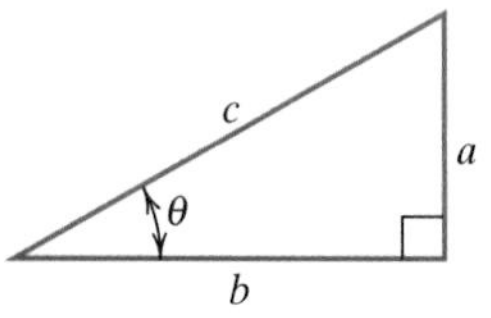

Figure A1.3.1

2. Signs in the four quadrants

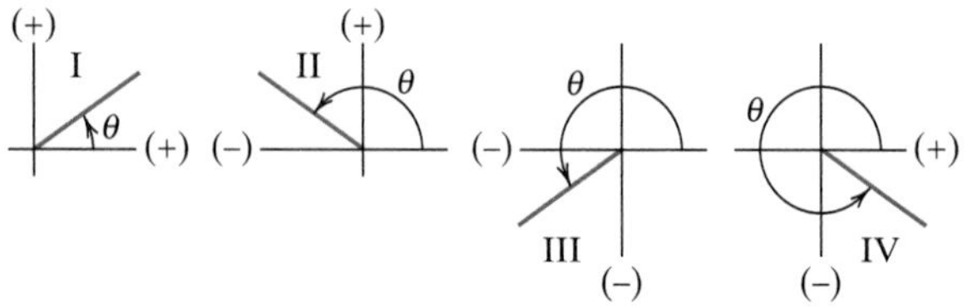

	I	II	III	IV
$\sin\theta$	+	+	−	−
$\cos\theta$	+	−	−	+
$\tan\theta$	+	−	+	−
$\csc\theta$	+	+	−	−
$\sec\theta$	+	−	−	+
$\cot\theta$	+	−	+	−

Figure A1.3.2

3. Miscellaneous relations

$\sin^2\theta + \cos^2\theta = 1$

$1 + \tan^2\theta = \sec^2\theta$

$1 + \cot^2\theta = \csc^2\theta$

$\sin\dfrac{\theta}{2} = \sqrt{\tfrac{1}{2}(1 - \cos\theta)}$

$\cos\dfrac{\theta}{2} = \sqrt{\tfrac{1}{2}(1 + \cos\theta)}$

$\sin 2\theta = 2\sin\theta\cos\theta$

$\cos 2\theta = \cos^2\theta - \sin^2\theta$

$\sin(a \pm b) = \sin a\cos b \pm \cos a\sin b$

$\cos(a \pm b) = \cos a\cos b \mp \sin a\sin b$

4. Law of sines

$$\frac{\sin A}{a} = \frac{\sin B}{b}$$

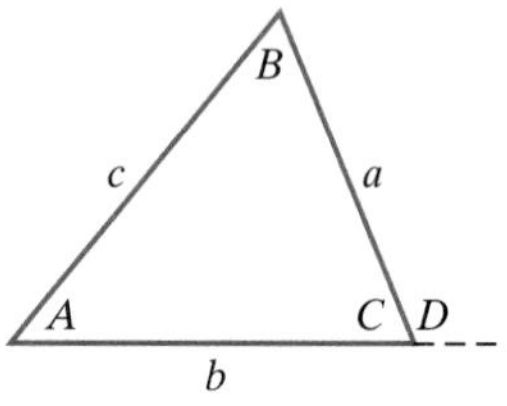

Figure A1.3.3

5. Law of cosines

$c^2 = a^2 + b^2 - 2ab\cos C$

$c^2 = a^2 + b^2 + 2ab\cos D$

A1.4 The Cross Product (Vector Product)

The cross product or vector product of two vectors (in the case at hand, $\boldsymbol{r}$ and $\boldsymbol{F}$, in that order) is, by definition, a third vector that is perpendicular to the plane defined by the two vectors. Call this third vector $\boldsymbol{M}$ (for moment). We can write the cross product in terms of the $\boldsymbol{r}$ (the position vector) and $\boldsymbol{F}$ (the force vector) in terms of their respective components as

$$\boldsymbol{M} = \boldsymbol{r} \times \boldsymbol{F} = (r_x\boldsymbol{i} + r_y\boldsymbol{j} + r_z\boldsymbol{k}) \times (F_x\boldsymbol{i} + F_y\boldsymbol{j} + F_z\boldsymbol{k}) \qquad (5.6\text{B})$$

By applying the distributive and associative laws of vector multiplication to (5.6B) and noting that

$$\begin{array}{lll} \boldsymbol{i} \times \boldsymbol{i} = 0 & \boldsymbol{j} \times \boldsymbol{i} = -\boldsymbol{k} & \boldsymbol{k} \times \boldsymbol{i} = \boldsymbol{j} \\ \boldsymbol{i} \times \boldsymbol{j} = \boldsymbol{k} & \boldsymbol{j} \times \boldsymbol{j} = 0 & \boldsymbol{k} \times \boldsymbol{j} = -\boldsymbol{i} \\ \boldsymbol{i} \times \boldsymbol{k} = -\boldsymbol{j} & \boldsymbol{j} \times \boldsymbol{k} = \boldsymbol{i} & \boldsymbol{k} \times \boldsymbol{k} = 0 \end{array} \qquad (\text{A1.4}a)$$

we can simplify (5.6B) as

$$\boldsymbol{M} = \boldsymbol{r} \times \boldsymbol{F} = \underbrace{(+r_yF_z - r_zF_y)}_{M_x}\boldsymbol{i} + \underbrace{(+r_zF_x - r_xF_z)}_{M_y}\boldsymbol{j} + \underbrace{(+r_xF_y - r_yF_x)}_{M_z}\boldsymbol{k} \qquad (5.7)$$

The cross product of the position vector and the force vector, as represented in (5.7), can be written in matrix form as:

$$\boldsymbol{M} = \boldsymbol{r} \times \boldsymbol{F} = \begin{vmatrix} \boldsymbol{i} & \boldsymbol{j} & \boldsymbol{k} \\ r_x & r_y & r_z \\ F_x & F_y & F_z \end{vmatrix} \qquad (5.12)$$

The form in (5.12) specifies that the **determinate** of the matrix be taken. This determinate consisting of three rows and three columns may be evaluated by repeating the first and second columns and forming products along each diagonal line. The sum of the products obtained along the colored lines is then subtracted from the sum of the products obtained along the black lines; this is identical to the expression on the righthand side of (5.7).

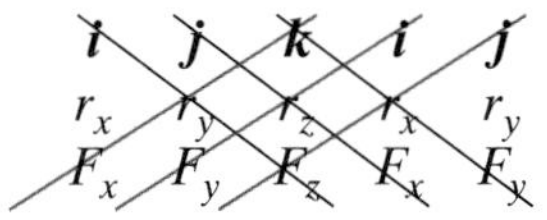

(A1.4*b*)

A2 PHYSICAL QUANTITIES

A2.1 Physical Properties

Density (kg/m^3) *and specific weight* (lb/ft^3)

	kg/m^3	lb/ft^3		kg/m^3	lb/ft^3
Air*	1.2062	0.07530			
Aluminum	2 690	168	Iron (cast)	7 210	450
Concrete (av)	2 400	150	Lead	11 370	710
Copper	8 910	556	Mercury	13 570	847
Earth (wet, av.)	1 760	110	Oil (av.)	900	56
(dry, av.)	1 280	80	Steel	7 830	489
Glass	2 590	162	Titanium	3 080	192.0
Gold	19 300	1205	Water (fresh)	1 000	62.4
Ice	900	56	(salt)	1 030	64
			Wood (soft pine)	480	30
			(hard oak)	800	50

*At 20°C (68°F) and atmospheric pressure

A2.2 Solar System Constants

Universal gravitational constant $G = 6.673(10^{-11})\ m^3/(kg \cdot s^2)$

$= 3.439(10^{-8})\ ft^4/(lbf \cdot s^4)$

Mass of Earth $m_e = 5.976(10^{24})$ kg

$= 4.095(10^{23})\ lbf \cdot s^2/ft$

Period of Earth's rotation (1 sidereal day) = 23 h 56 min 4 s

= 23.9344 h

Angular velocity of Earth $\omega = 0.7292(10^{-4})$ rad/s

Mean angular velocity of Earth–Sun line $\omega' = 0.1991(10^{-6})$ rad/s

Mean velocity of Earth's center about Sun = 107 200 km/h

= 66 610 mi/h

BODY	MEAN DISTANCE TO SUN km (mi)	ECCENTRICITY OF ORBIT e	PERIOD OR ORBIT solar days	MEAN DIAMETER km (mi)	MASS RELATIVE TO EARTH	SURFACE GRAVITATIONAL ACCELERATION m/s^2 (ft/s^2)	ESCAPE VELOCITY km/s (mi/s)
Sun	—	—	—	1 392 000 (865 000)	333 000	274 (898)	616 (383)
Moon	384 398* (238 854)*	0.055	27.32	3 476 (2 160)	0.0123	1.62 (5.32)	2.37 (1.47)
Mercury	57.3×10^6 (35.6×10^6)	0.206	87.97	5 000 (3 100)	0.054	3.47 (11.4)	4.17 (2.59)
Venus	108×10^6 (67.2×10^6)	0.0068	224.70	12 400 (7 700)	0.815	8.44 (27.7)	10.24 (6.36)
Earth	149.6×10^6 (92.96×10^6)	0.0167	365.26	12 742† (7 918)†	1.000	9.821‡ (32.22)‡	11.18 (6.95)
Mars	227.9×10^6 (141.6×10^6)	0.093	686.98	6 788 (4 218)	0.107	3.73 (12.3)	5.03 (3.13)

*Mean distance to Earth (center-to-center)

†Diameter of sphere of equal volume, based on a spheroidal Earth with a polar diameter of 12 714 km (7900 mi) and an equatorial diameter of 12 756 km (7926 mi)

‡For nonrotating spherical Earth, equivalent to absolute value at sea level and latitude 37.5°

A2.3 Conversion Factors from U.S. Customary Units

PHYSICAL QUANTITY	U.S. CUSTOMARY UNIT	= SI EQUIVALENT
	BASIC UNITS	
Length	1 foot (ft)	= $3.048(10^{-1})$ meter (m)*
	1 inch (in.)	= $2.54(10^{-2})$ meter (m)*
	1 mile (U.S. statute)	= $1.6093(10^{3})$ meter (m)
Mass	1 slug ($lb \cdot s^2/ft$)	= 1.4594(10) kilogram (kg)
	1 pound mass (lbm)	= $4.5359(10^{-1})$ kilogram (kg)
	DERIVED UNITS	
Acceleration	1 foot/second^2 (ft/s^2)	= $3.048(10^{-1})$ meter/second^2 (m/s^2)*
	1 inch/second^2 (in./s^2)	= $2.54(10^{-2})$ meter/second^2 (m/s^2)*
Area	1 foot^2 (ft^2)	= $9.2903(10^{-2})$ meter^2 (m^2)
	1 inch^2 (in.^{2})	= $6.4516(10^{-2})$ meter^2 (m^2)*
Density	1 slug/foot^3 ($lb \cdot s^2/ft^4$)	= $5.1537(10^{2})$ kilogram/meter^3 (kg/m^3)
	1 pound mass/foot^3 (lbm/ft^3)	= 1.6018(10) kilogram/meter^3 (kg/m^3)
Energy and Work	(1 joule ≡ 1 meter-newton)	
	1 foot-pound ($ft \cdot lb$)	= 1.3558 joules (J)
	1 kilowatt-hour ($kW \cdot hr$)	= $3.60(10^{6})$ joules (J)*
	1 British thermal unit (Btu)	= $1.0551(10^{3})$ joules (J)
Force	(1 newton ≡ 1 kilogram-meter/second^2)	
	1 pound (lb)	= 4.4482 newtons (N)
	1 kip (1000 lb)	= $4.4482(10^{3})$ newtons (N)
Power	(1 watt ≡ 1 joule/second)	
	1 foot-pound/second ($ft \cdot lb/s$)	= 1.3558 watt (W)
	1 horsepower (hp)	= $7.4570(10^{2})$ watt (W)
Pressure and Stress	(1 pascal ≡ 1 newton/meter^2)	
	1 pound/foot^2 (lb/ft^2)	= 4.7880(10) pascal (Pa)
	1 pound/inch^2 (lb/in.^{2})	= $6.8948(10^{3})$ pascal (Pa)
	1 atmosphere (standard, 14.7 lb/in.^{2})	= $1.0133(10^{5})$ pascal (Pa)
Speed	1 foot/second (ft/s)	= $3.048(10^{-1})$ meter/second (m/s)*
	1 mile/hr	= $4.4704(10^{-1})$ meter/second (m/s)
	1 mile/hr	= 1.6093 kilometer/hour (km/hr)
Volume	1 foot^3 (ft^3)	= $2.8317(10^{-2})$ meter^3 (m^3)
	1 inch^3 (in.^{3})	= $1.6387(10^{-5})$ meter^3 (m^3)
	1 gallon (U.S. liquid)	= $3.7854(10^{-3})$ meter^3 (m^3)

*Denotes an exact factor.

A3 PROPERTIES OF AREAS AND VOLUMES

A3.1 Areas, Centroids, and Area Moments of Inertia

Shape	Area Moment of Inertia

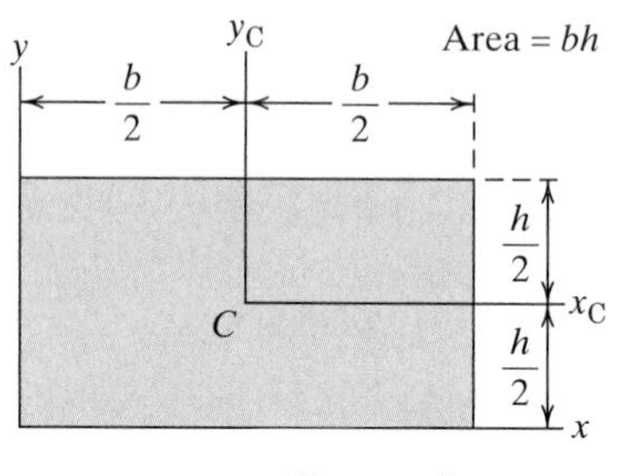

$$I_{x_C} = \frac{bh^3}{12}$$

$$I_{y_C} = \frac{b^3h}{12}$$

$$I_x = \frac{bh^3}{3}$$

$$I_y = \frac{b^3h}{3}$$

Figure A3.1.1 Rectangle

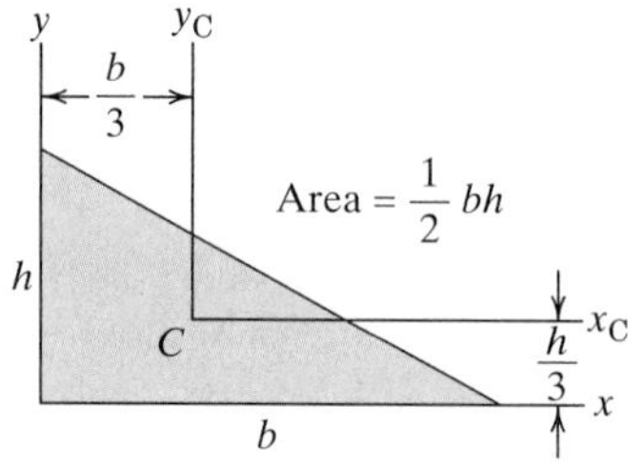

$$I_{x_C} = \frac{bh^3}{36}$$

$$I_{y_C} = \frac{b^3h}{36}$$

$$I_x = \frac{bh^3}{12}$$

$$I_y = \frac{b^3h}{12}$$

Figure A3.1.2 Right triangle

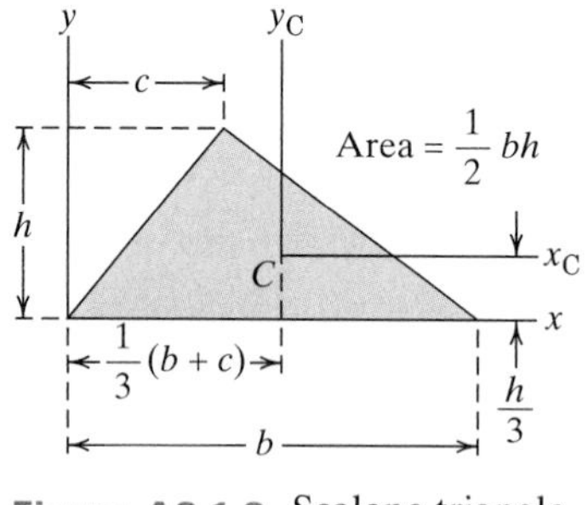

$$I_{x_C} = \frac{bh^3}{36}$$

$$I_{y_C} = \frac{bh}{36}(b^2 + c^2 - bc)$$

$$I_x = \frac{bh^3}{12}$$

Figure A3.1.3 Scalene triangle

Shape	Area Moment of Inertia
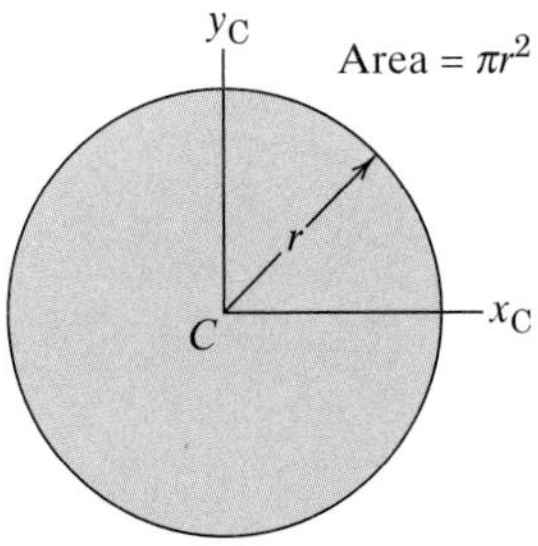	$I_{x_C} = I_{y_C} = \frac{1}{4}\pi r^4$

Figure A3.1.4 Circle

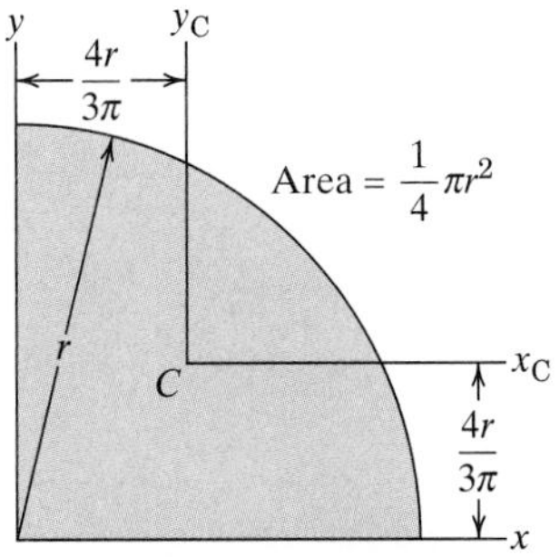

$$I_{x_C} = I_{y_C} = \left(\frac{\pi}{16} - \frac{4}{9\pi}\right)r^4$$

$$I_x = I_y = \frac{\pi r^4}{16}$$

Figure A3.1.5 Quarter circle

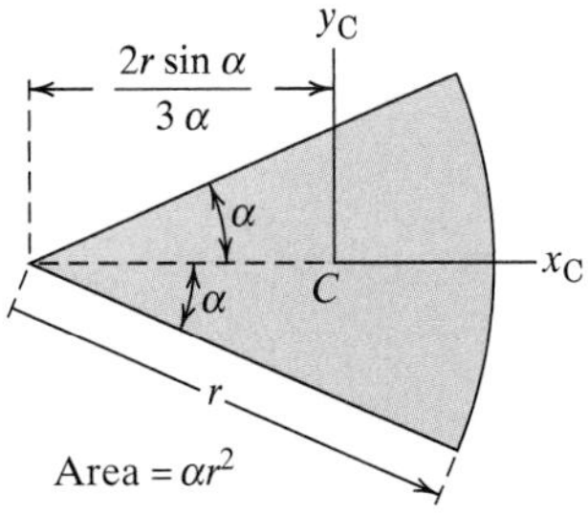

$$I_{x_C} = \frac{r^4}{4}\left(\alpha - \frac{\sin\,2\alpha}{2}\right)$$

$$I_{y_C} = \frac{r^4}{4}\left(\alpha + \frac{\sin\,2\alpha}{2}\right)$$

Figure A3.1.6 Circular sector

Shape	Area Moment of Inertia
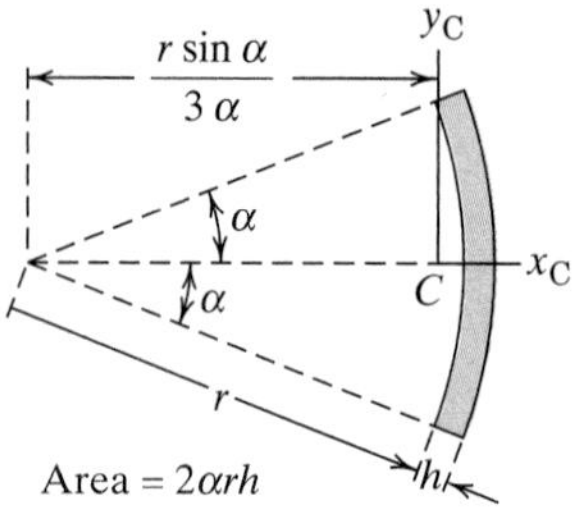 **Figure A3.1.7** Circular arc	$I_{x_C} = \frac{1}{2}(2\alpha - \sin 2\alpha) r^3 h$ $I_{y_C} = \left[\frac{2\alpha + \sin 2\alpha}{2} - \frac{1}{2}(1 - \cos 2\alpha)\right] r^3 h$
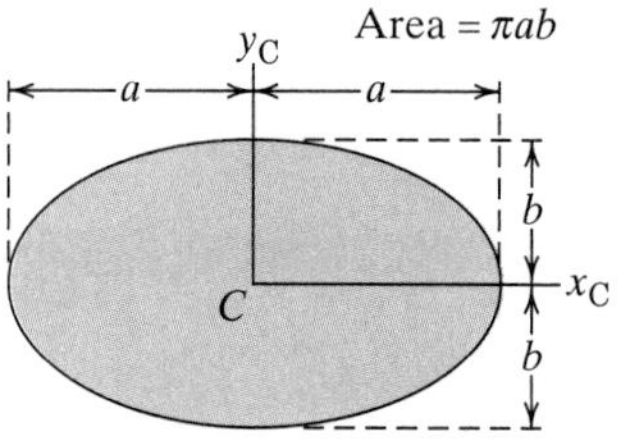 **Figure A3.1.8** Ellipse	$I_{x_C} = \frac{\pi}{4} ab^3$ $I_{y_C} = \frac{\pi}{4} a^3 b$
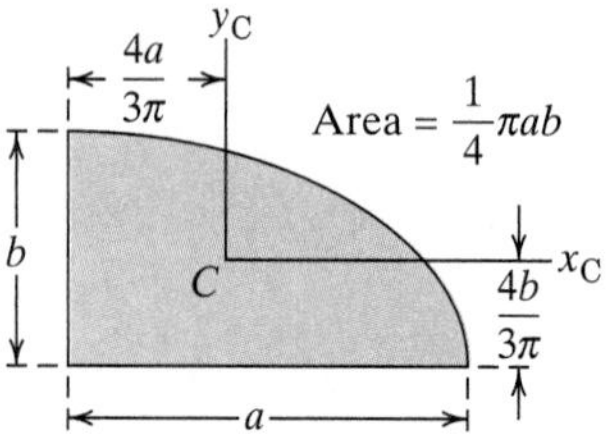 **Figure A3.1.9** Quarter ellipse	$I_{x_C} = \left(\frac{9\pi^2 - 64}{144\pi}\right) ab^3$ $I_{y_C} = \left(\frac{9\pi^2 - 64}{144\pi}\right) a^3 b$
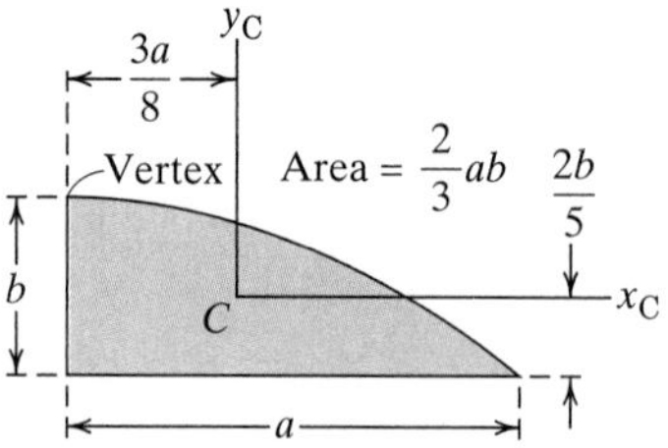 **Figure A3.1.10** Parabolic section	$I_{x_C} = \frac{8}{175} ab^3$ $I_{y_C} = \frac{19}{480} a^3 b$

Shape	Area Moment of Inertia
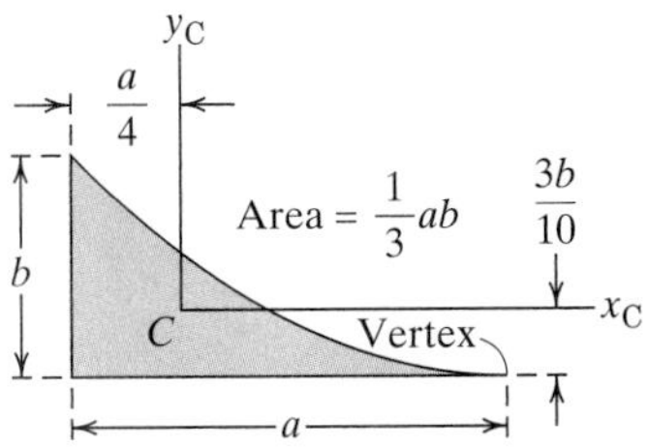	$I_{x_C} = \dfrac{19}{1050} ab^3$ $I_{y_C} = \dfrac{1}{80} a^3 b$

Figure A3.1.11 Parabolic spandrel

A3.2 Volumes and Centroids

Shape

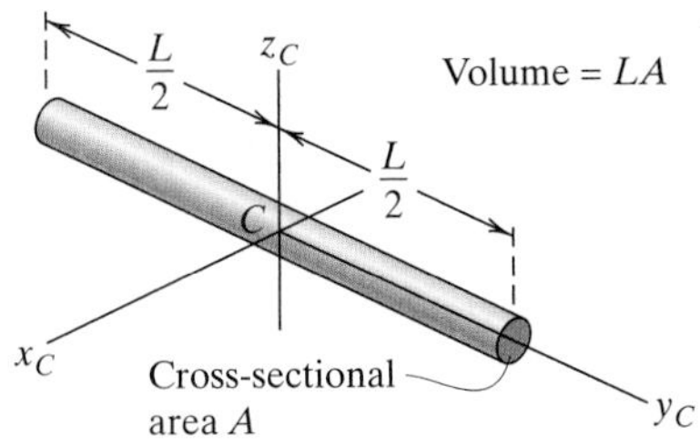

Figure A3.2.1 Uniform slender rod

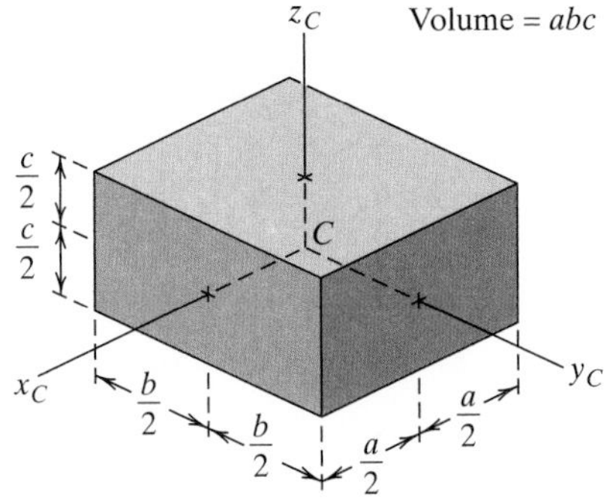

Figure A3.2.2 Rectangular parallelopiped

Shape

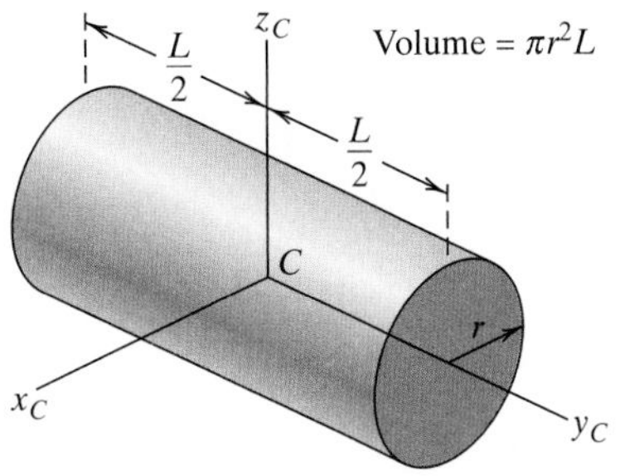

Figure A3.2.3 Cylinder

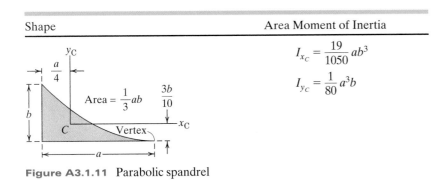

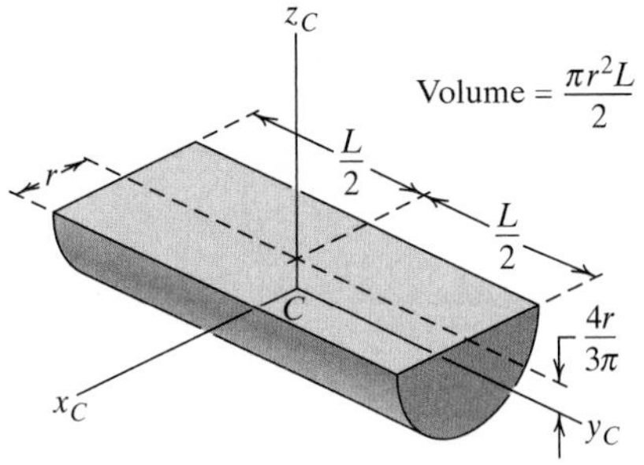

Figure A3.2.4 Semicylinder

Shape	Shape

Volume $= \pi r L h$

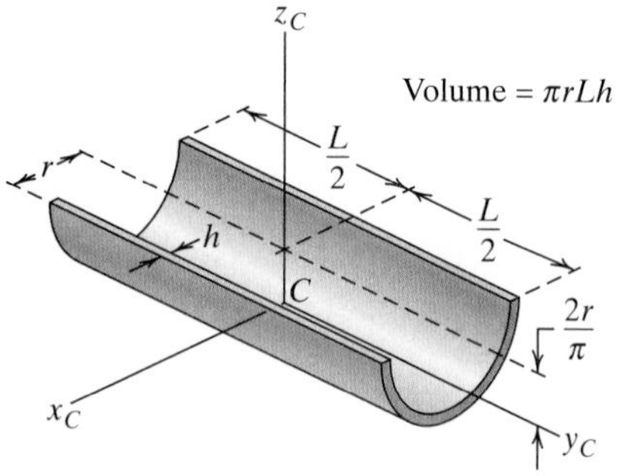

Figure A3.2.5 Semicylindrical shell

Volume $= \dfrac{2\pi abc}{3}$

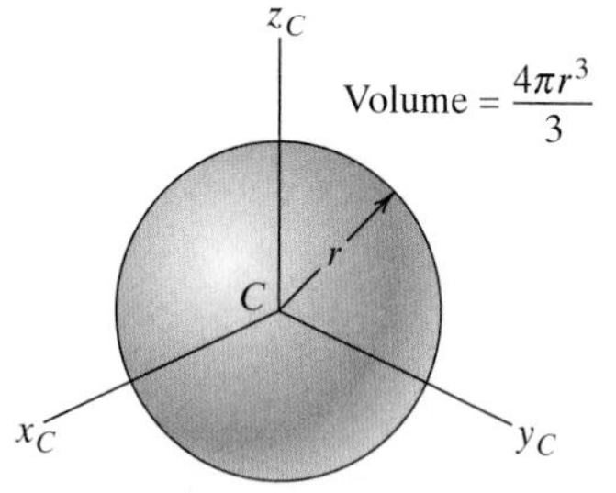

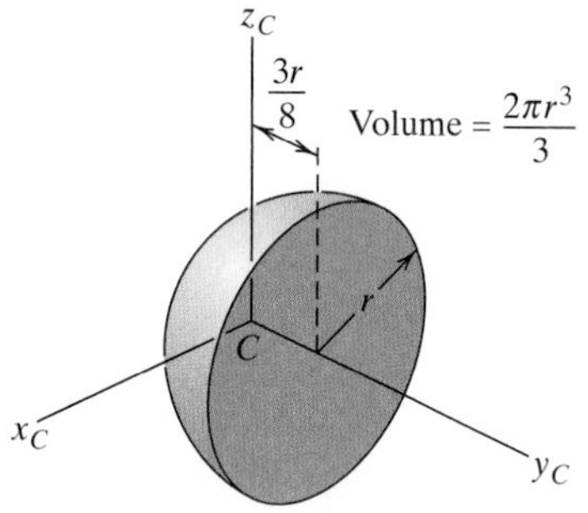

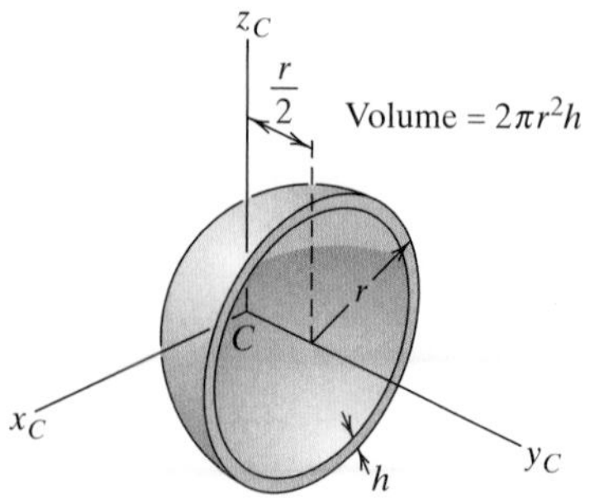

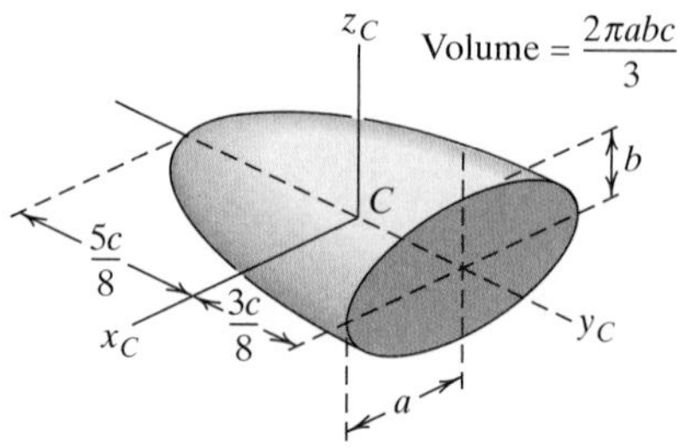

Figure A3.2.9 Semiellipsoid

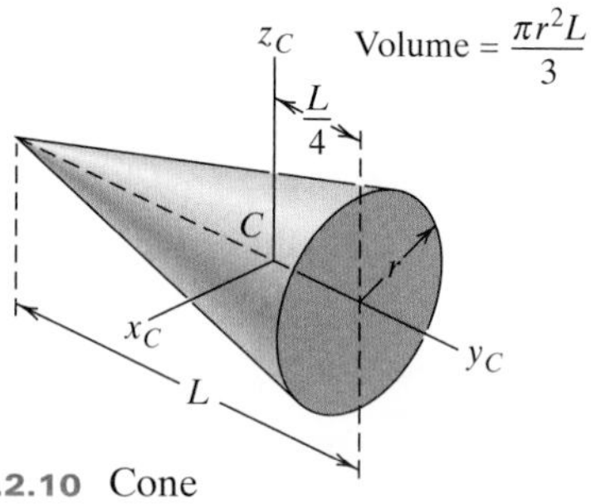

Volume $= \dfrac{4\pi r^3}{3}$

Figure A3.2.6 Sphere

Volume $= \dfrac{\pi r^2 L}{3}$

Figure A3.2.10 Cone

Volume $= \dfrac{2\pi r^3}{3}$

Figure A3.2.7 Hemisphere

Volume $= \dfrac{\pi r^2 L}{6}$

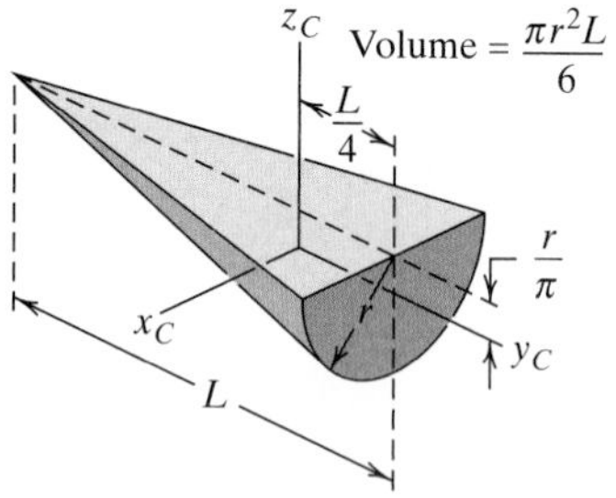

Figure A3.2.11 Semicone

Volume $= 2\pi r^2 h$

Figure A3.2.8 Hemispherical shell

Volume $= \dfrac{\pi h r (r^2 + L^2)^{1/2}}{2}$

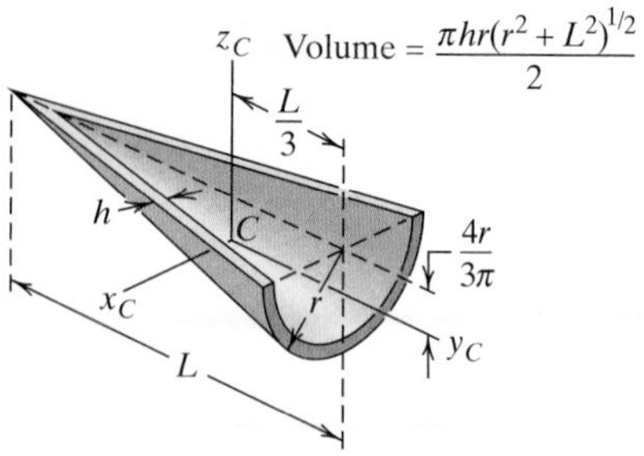

Figure A3.2.12 Semiconical shell

Shape

Shape

$$\text{Volume} = \frac{abc}{6}$$

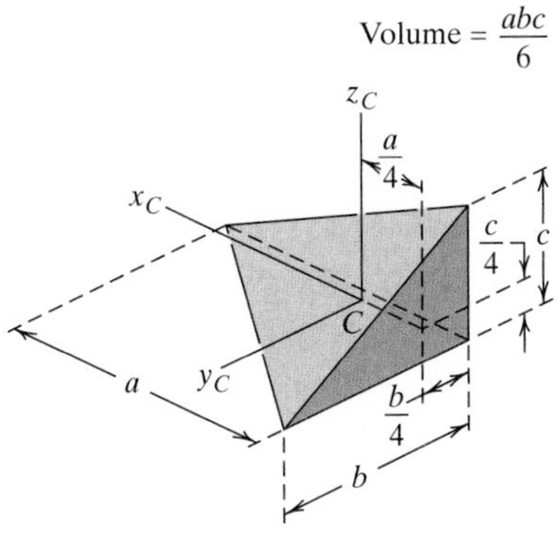

Figure A3.2.13 Orthogonal tetrahedron

$$\text{Volume} = 2\pi rA$$

Cross-sectional area A

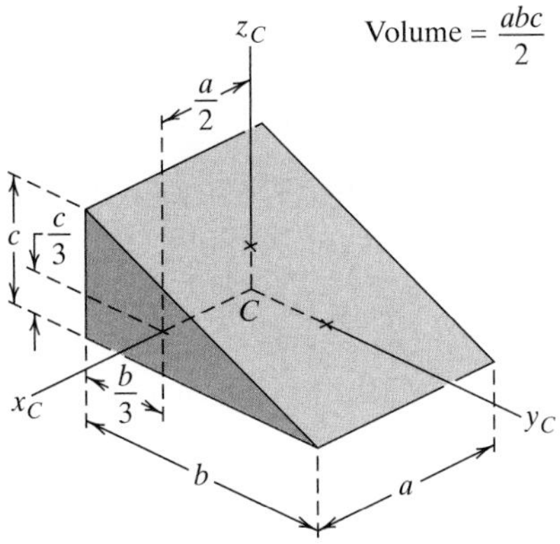

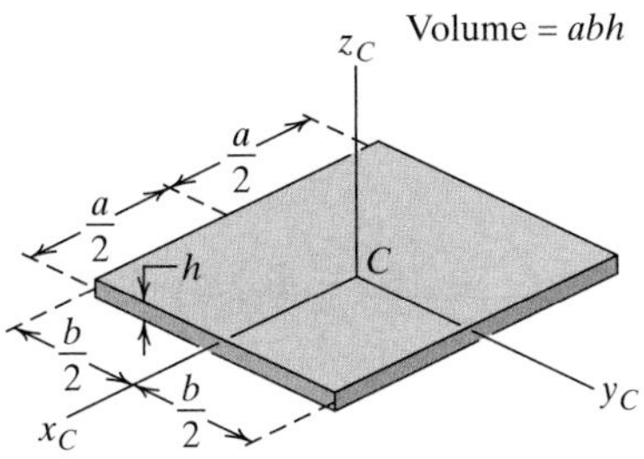

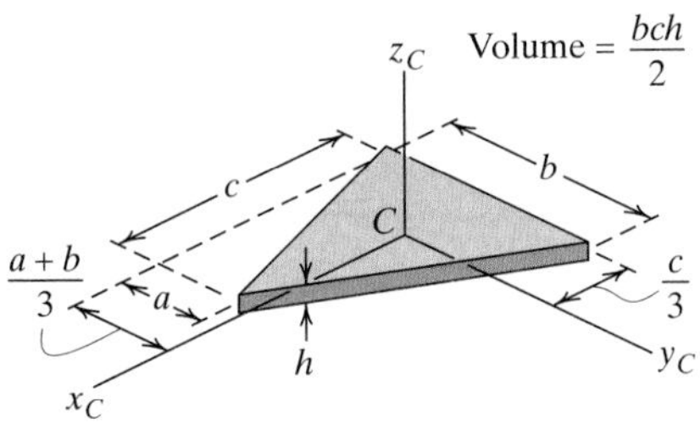

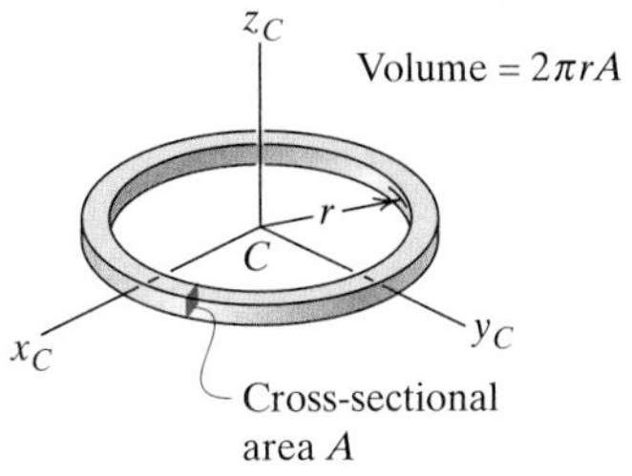

Figure A3.2.17 Thin circular ring

$$\text{Volume} = \frac{abc}{2}$$

Figure A3.2.14 Right triangular prism

$$\text{Volume} = \pi r^2 h$$

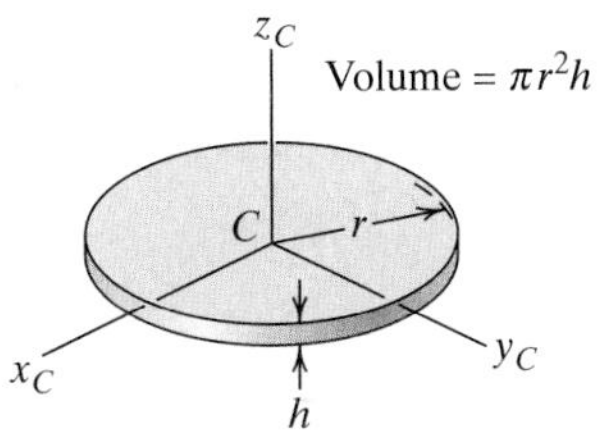

Figure A3.2.18 Circular plate

$$\text{Volume} = abh$$

Figure A3.2.15 Rectangular plate

$$\text{Volume} = \frac{\pi r^2 h}{4}$$

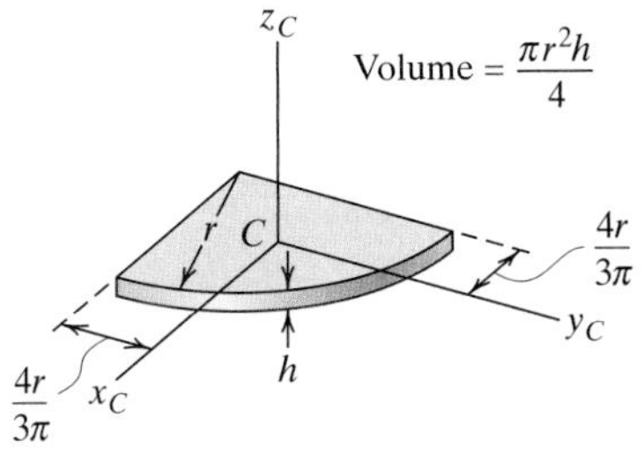

Figure A3.2.19 Quarter circular plate

$$\text{Volume} = \frac{bch}{2}$$

Figure A3.2.16 Triangular plate

$$\text{Volume} = \pi abh$$

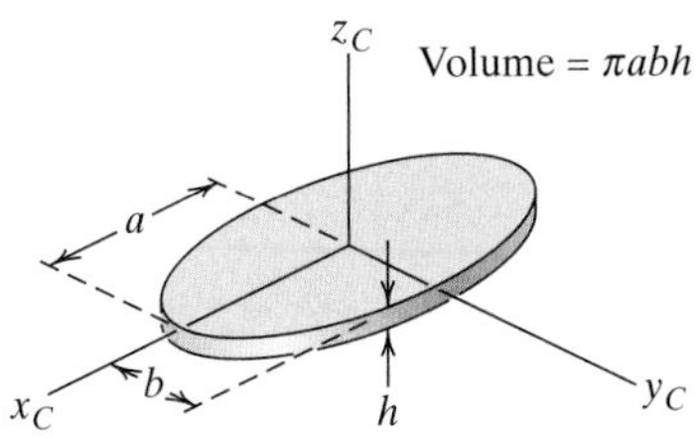

Figure A3.2.20 Elliptical plate

DRY FRICTION

If you attempt to slide one solid object across another, the sliding is resisted by interactions between the surfaces of the two objects. This resistance is referred to as **friction force** and is oriented *parallel* to the two surfaces in a direction opposite the direction of (pending) sliding. For example, if you push on a crate of weight ***W*** as it rests on the ground, as in **Figure B.1*a***, the friction force exerted by the ground on the crate is in the direction opposite to the sliding direction. An equal and opposite friction force (not shown in **Figure B.1*a***) is exerted by the crate on the ground (per Newton's third law).

Coulomb Friction Model

Friction is a complicated phenomenon that is not fully understood and is an area of continuing research.[1] Current thinking is that friction results from the microscopic irregularities (called asperities) present in all

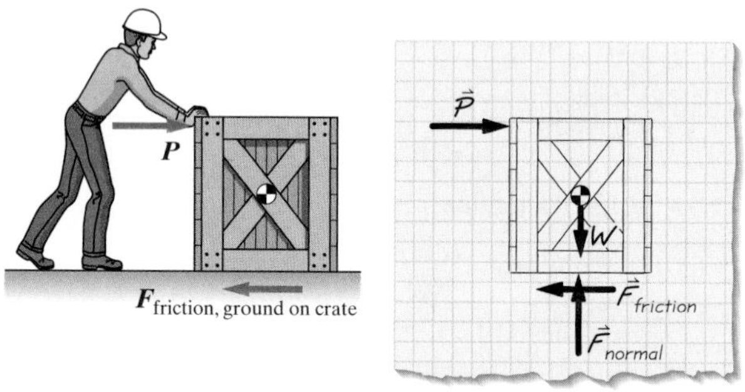

(*a*)

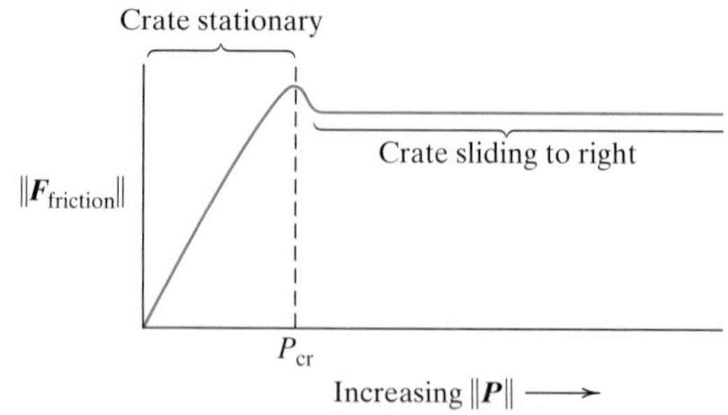

(*b*)

Figure B.1 (*a*) Person pushing a crate and the crate's free-body diagram; (*b*) friction force F_{friction} versus push force P

[1]The area of work involved with studying friction and wear characteristics of material surfaces is called *tribology*.

surfaces and from molecular attraction. On a macroscopic scale, friction between surfaces results in the behavior shown in **Figure B.1*b***; an increase in the magnitude of force $\boldsymbol{P}$ is balanced by an increase in the friction force ($\boldsymbol{F}_{\text{friction}}$) up to a maximum value (P_{cr}) and the crate remains stationary. An attempt to increase the magnitude of force $\boldsymbol{P}$ beyond P_{cr} results in the crate sliding to the right.

We can generalize the behavior in **Figure B.1*b*** to say that the friction force ($\boldsymbol{F}_{\text{friction}}$) is always parallel to the contacting surfaces of the two objects and is directed so as to oppose their relative motion. Furthermore, the friction force is related to and limited by the normal contact force ($\boldsymbol{F}_{\text{normal}}$) and the characteristics (e.g., smoothness) of the objects' surfaces, it is perpendicular to the normal force, and *normal contact force must be present in order for friction force to be present* (but not vice versa). We can model the behavior of contacting surfaces of many systems with the elementary friction law proposed by Coulumb in 1781. The **Coulumb friction law** states:

1. If the two solid objects remain stationary relative to one another (i.e., the two objects do not slide relative to one another), the friction force is such that

$$\| \boldsymbol{F}_{\text{friction}} \| < \mu_s \| \boldsymbol{F}_{\text{normal}} \| \tag{B1}$$

where μ_s is called the **coefficient of static friction**. Sample values of μ_s are given in **Table B.1** and should only be used if values from experiments on the actual system are not available. For specific cases the values in **Table B.1** may be incorrect by more than 100%.

Table B.1 Sample Coefficients of Friction*

Materials	μ_s	μ_k
Mild steel on mild steel	0.74	0.57
Aluminum on mild steel	0.61	0.47
Copper on mild steel	0.53	0.36
Cast iron on cast iron	1.10	0.15
Brake material on cast iron	0.40	0.30
Leather on cast iron	0.60	0.56
Rubber on metal	0.40	0.30
Rubber on wood	0.40	0.30
Rubber on pavement	0.90	0.80
Leather on oak	0.61	0.52
Glass on nickel	0.78	0.56

*Eugene A. Avallone and Theodore Baumeister III, *Marks' Standard Handbook for Mechanical Engineers*, 10th Edition (McGraw-Hill Publishers, 1996).

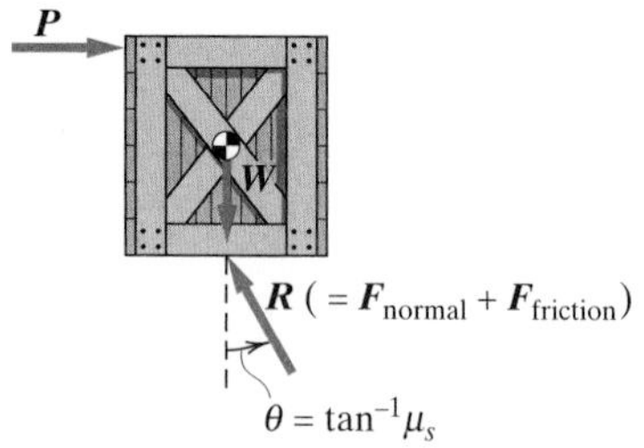

Figure B.2 The resultant force $\boldsymbol{R}$ when friction is at its limit of $\mu_s \parallel \boldsymbol{F}_{\text{normal}} \parallel$

2. If the friction force is such that

$$\parallel \boldsymbol{F}_{\text{friction}} \parallel = \mu_s \parallel \boldsymbol{F}_{\text{normal}} \parallel \tag{B2}$$

there is a state of impending sliding of the two objects relative to one another. Therefore, the product $\mu_s \parallel \boldsymbol{F}_{\text{normal}} \parallel$ is the largest magnitude of friction force that can exist at the contacting surfaces without there being sliding. When this maximum friction force has been developed, the angle θ between $\boldsymbol{F}_{\text{normal}}$ and the resultant $\boldsymbol{R}$ $(= \boldsymbol{F}_{\text{normal}} + \boldsymbol{F}_{\text{friction}})$ is called the **angle of friction**, and $\theta = \tan^{-1} \mu_s$ (**Figure B.2**).

3. If there is sliding between the two contacting surfaces, the friction force is given by

$$\parallel \boldsymbol{F}_{\text{friction}} \parallel = \mu_k \parallel \boldsymbol{F}_{\text{normal}} \parallel \tag{B3}$$

where μ_k is the **coefficient of kinetic friction**. In general, for given contacting surfaces, $\mu_k < \mu_s$ (which is consistent with the behavior depicted in **Figure B.1*b***). Sample values of μ_k are given in **Table B.1**.

Use of the Coulomb Friction Model in Static Analysis

You might be wondering how to use the Coulomb friction law in static analysis. We illustrate its use by considering questions you might ask about the crate sitting on the ground in **Figure B.1**. For example, say that you are interested in

- *Finding* $\boldsymbol{F}_{\text{friction}}$ *for a given value of* $\boldsymbol{P}$ *if the crate is in equilibrium.* You would apply the conditions of equilibrium to calculate the size of $\boldsymbol{F}_{\text{friction}}$, followed by using (B1) as a check; if the calculated value of $\parallel \boldsymbol{F}_{\text{friction}} \parallel$ is less than or equal to $\mu_s \parallel \boldsymbol{F}_{\text{normal}} \parallel$ (i.e., $\parallel \boldsymbol{F}_{\text{friction}} \parallel \leq \mu_s \parallel \boldsymbol{F}_{\text{normal}} \parallel$), the crate is stationary relative to the ground. If, on the other hand, the calculated value of $\parallel \boldsymbol{F}_{\text{friction}} \parallel$ is greater than $\mu_s \parallel \boldsymbol{F}_{\text{normal}} \parallel$ (i.e., $\parallel \boldsymbol{F}_{\text{friction}} \parallel > \mu_s \parallel \boldsymbol{F}_{\text{normal}} \parallel$), the crate is not stationary relative to the ground and will slide (and accelerate) to the right. This means that equilibrium is not possible with the given values of $\boldsymbol{P}$ and $\boldsymbol{W}$ (weight of the crate) and the character of the contacting surfaces (as indicated in the value of μ_s).
- *Finding the maximum allowable magnitude of* $\boldsymbol{P}$ *such that the crate will not slide.* You would apply the conditions of equilibrium to calculate the magnitude of $\boldsymbol{P}$ when $\parallel \boldsymbol{F}_{\text{friction}} \parallel$ is equal to $\mu_s \parallel \boldsymbol{F}_{\text{normal}} \parallel$.
- *Finding the magnitude of* $\boldsymbol{P}$ *such that the crate moves to the right at a constant speed.* You would apply the conditions of equilibrium to calculate the magnitude of $\boldsymbol{P}$ when $\parallel \boldsymbol{F}_{\text{friction}} \parallel$ is equal to $\mu_k \parallel \boldsymbol{F}_{\text{normal}} \parallel$.

The Coulomb friction law was used when we considered the bicycle in Chapter 2. In the free-body diagram of the bicycle in **Figure 2.1** we included $\boldsymbol{F}_{\text{friction}}$ where the ground contacts the rear tire. We were implicitly assuming in the analysis of the bicycle that our calculated value of $\parallel \boldsymbol{F}_{\text{friction}} \parallel$ was less than $\mu_s \parallel \boldsymbol{F}_{\text{normal}} \parallel$ so that the rear tire would not slide relative to ground. This is probably a pretty good assumption (unless we are considering bicycling across a frozen lake!); a check of this assumption should be added to increase the completeness of the analysis. Without dry friction between the tire and the ground, a bicycle would not function. We also factored in the presence of friction when we considered the bridge in Chapter 3. In calculating the required minimum weight of the anchorages that hold the main cables of the Golden Gate Bridge we made an explicit check of weight necessary to prevent sliding by applying (B1). The anchorages are able to do their job because of dry friction. Finally, we considered how friction affects the performance of a gear train in Chapter 9. Dry friction between shafts and bearings increases the moment (often referred to as torque) necessary to turn gear trains. Therefore, in gear trains one would hope to reduce friction between shafts and bearings to a very small value (ideally to $\mu_s = 0$); this can be done with the use of lubricants and/or more sophisticated bearing systems.

The discussion above about contacting surfaces is about dry friction. When a fluid film separates the surfaces, the contact is fully lubricated, and principles from fluid mechanics are required to describe the interaction of one object relative to the other. We don't develop this topic in this book.

Other Examples of Friction in Static Analysis

We now present the analysis of two simple systems that operate because of dry friction.

…es. A wedge is a simple machine used to make adjustments in …on of one object relative to another. Wedges can also be used …large force. For example, the wedge in **Figure B.3** can be used …block. To calculate the maximum force $\boldsymbol{P}$ required to raise …apply equilibrium conditions when the wedge is just about …right (and therefore begins to raise the block). When the …bout to move, we can describe the relationship between …mal force using (B2). More specifically, for the contact…g A and B (**Figure B.4**) we write

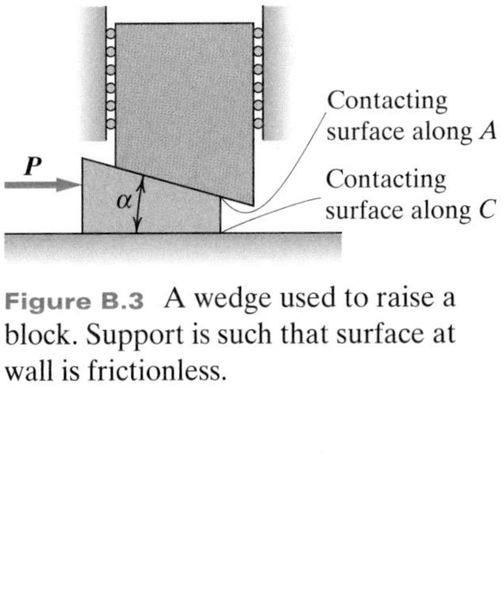

Figure B.3 A wedge used to raise a block. Support is such that surface at wall is frictionless.

$$\| \boldsymbol{F}_{A,\text{friction}} \| = \mu_s \| \boldsymbol{F}_{A,\text{normal}} \| \quad \text{(B4a)}$$

$$\| \boldsymbol{F}_{C,\text{friction}} \| = \mu_s \| \boldsymbol{F}_{C,\text{normal}} \| \quad \text{(B4b)}$$

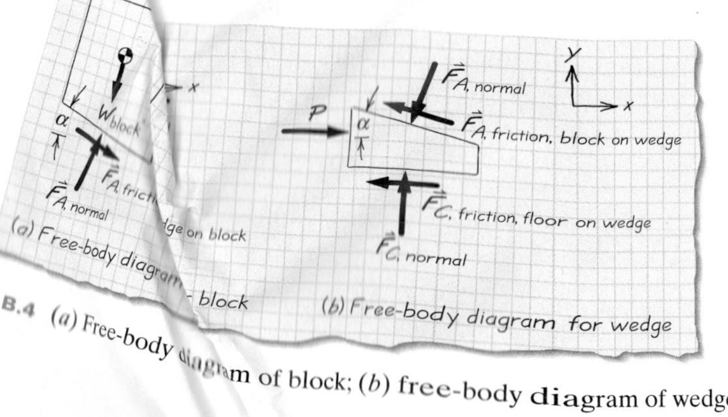

Figure B.4 (*a*) Free-body diagram of block; (*b*) free-body diagram of wedge

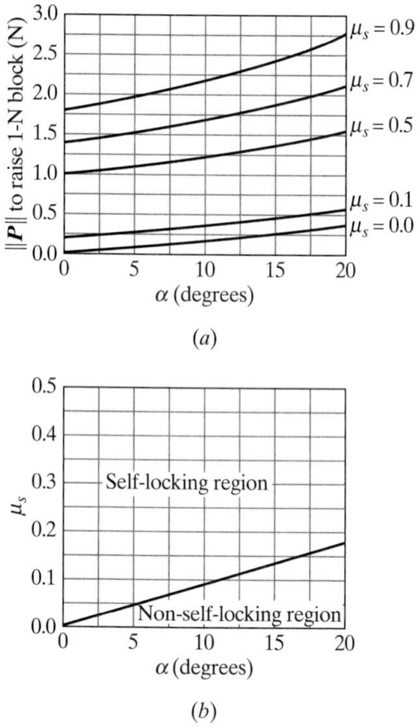

Figure B.5 (*a*) The maximum force $\boldsymbol{P}$ necessary to lift a 1-N block for various coefficients of static friction; (*b*) conditions under which a wedge-block system is self-locking

The application of equilibrium equation (B7.5*b*) applied to the block (**Figure B.4*a***) then allows us to write

$$\sum F_y = -W + F_{A,\text{normal}}\cos\alpha - \mu_s F_{A,\text{normal}}\sin\alpha = 0 \qquad \text{(B5)}$$

which can be solved for $F_{A,\text{normal}}$:

$$F_{A,\text{normal}} = \frac{W}{\cos\alpha - \mu_s \sin\alpha} \qquad \text{(B6)}$$

Application of equilibrium equations (B7.5*a*) and (B7.5*b*) to wedge (assuming that its mass is very small) (**Figure B.4*b***) then allows us to write

$$\sum F_x = P - \mu_s F_{C,\text{normal}} - \mu_s F_{A,\text{normal}}\cos\alpha - F_{A,\text{normal}}\sin\alpha = 0 \qquad \text{(B7a)}$$

$$\sum F_y = F_{C,\text{normal}} + \mu_s F_{A,\text{normal}}\sin\alpha - F_{A,\text{normal}}\cos\alpha = 0 \qquad \text{(B7b)}$$

Equations (B7*a*) and (B7*b*) in combination with (B6) can be solved for *P*:

$$P = \frac{W}{\cos\alpha - \mu_s \sin\alpha}\left[(1-\mu_s^2)\sin\alpha + 2\mu_s\cos\alpha\right] \qquad \text{(B8)}$$

Plots of (B8) are shown in **Figure B.5*a*** for various values of static friction and illustrate that the force $\boldsymbol{P}$ required to lift a 1-N block increases with the static friction coefficient and with the wedge angle. Notice that for a small coefficient of static friction and wedge angle $\|\boldsymbol{P}\|$ is les than one (e.g., $\mu_s = 0.1$, $\alpha = 10°$, $\|\boldsymbol{P}\| = 0.38$), whereas for a larger c efficient of static friction and wedge angle $\|\boldsymbol{P}\|$ is much greater t one (e.g., $\mu_s = 0.7$, $\alpha = 20°$, $\|\boldsymbol{P}\| = 2.10$). This means that the p selection of coefficient of static friction and wedge angle will allow lift a large weight (*W*) with a (relatively) small force $\boldsymbol{P}$.

Under a certain condition, the wedge is **self-locking**; this m the wedge will remain in place (holding up the block) even is no applied force *P*. This might be advantageous if you n an object (like the block) in a raised position for an exten time. We can determine the condition under which a we tem will be self-locking by imposing equilibrium on wedge's free-body diagrams in **Figure B.6** (notice ho diagrams are different from the ones in **Figure B.4**).

For the block we write:

$$\sum F_y = -W + F_{A,\text{normal}}\cos\alpha + \mu_s F_{A,\text{nc}} \ldots \alpha = 0 \qquad \text{(B9a)}$$

For the wedge we write

$$\sum F_x = \mu_s F_{C,\text{normal}} + \mu_s F_{A,\text{normal}}\cos\alpha - F_{A,\text{normal}}\sin\alpha = 0 \qquad \text{(B9b)}$$

$$\sum F_y = F_{C,\text{normal}} - \mu_s F_{A,\text{normal}}\sin\alpha - F_{A,\text{normal}}\cos\alpha = 0 \qquad \text{(B9c)}$$

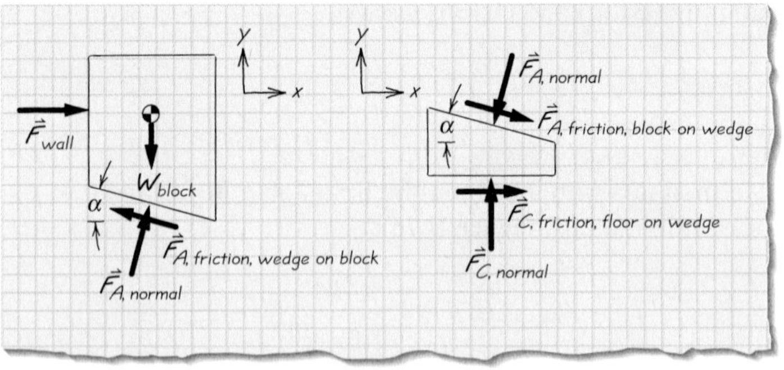

Figure B.6 Free-body diagrams of block and wedge with no force $\boldsymbol{P}$ applied

From (B9*a*)–(B9*c*) we find that for the wedge to be self-locking when

$$\tan \alpha \leq \frac{2\mu_s}{(1-\mu_s^2)} \tag{B10}$$

Figure B.5*b* shows a plot of wedge angle (α) versus static coefficient of friction (as determined from (B10)). Convince yourself that the correct regions of this plot have been marked as the "self-locking region" and the "non-self-locking region."

Belts. Belts are used to connect mechanical components to one another, often with the intention of transferring power from one component to another. For example, a rubber belt connects a pulley on the crankshaft of an automobile engine to a pulley on the water pump (**Figure B.7*a***). A belt operates because of friction between it and the pulleys that it connects. To calculate the condition under which a belt will not slip on the pulleys, we consider equilibrium of a portion of the belt, as shown in **Figure B.7*b***:

$$\sum F_x = -(T+\Delta T)\cos\frac{\Delta\theta}{2} + T\cos\frac{\Delta\theta}{2} + \mu_s\,\Delta F_{\text{normal}} = 0 \tag{B11a}$$

$$\sum F_y = -(T+\Delta T)\sin\frac{\Delta\theta}{2} - T\sin\frac{\Delta\theta}{2} + \Delta F_{\text{normal}} = 0 \tag{B11b}$$

If $\Delta\theta$ is small, we can write

$$\sin\frac{\Delta\theta}{2} \approx \frac{\Delta\theta}{2}; \qquad \cos\frac{\Delta\theta}{2} \approx 1 \tag{B12}$$

Substituting (B12) into (B11*a*) and (B11*b*), and recognizing that in the limit each Δ term can be written as a differential, we have

$$dT = \mu_s\, dF_{\text{normal}} \tag{B13a}$$

$$T\,d\theta = dF_{\text{normal}} \tag{B13b}$$

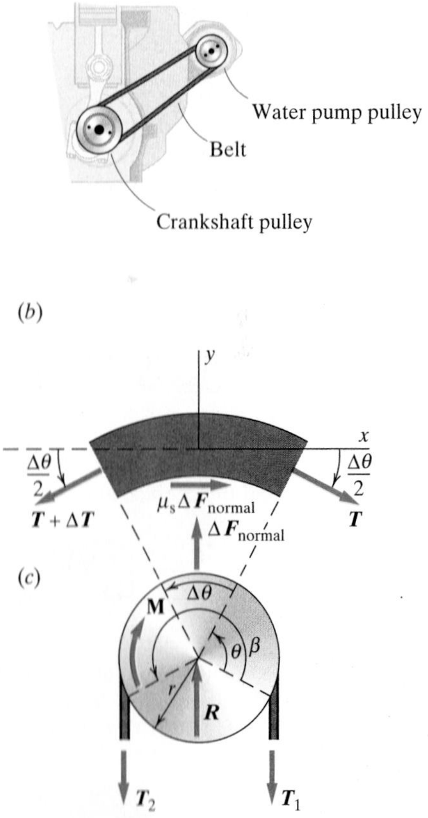

Figure B.7 (*a*) A belt connects the crack shaft and the water pump on an automobile engine; (*b*) free-body diagram of belt segment; (*c*) free-body diagram of pulley and belt

Combining (B13*a*) and (B13*b*) we write

$$\frac{dT}{T} = \mu_s \, d\theta \tag{B14}$$

which we then integrate between the limits of T_1 to T_2 and 0 to β (wrap angle of belt, in radians, **Figure B.7*c***):

$$\int_{T_1}^{T_2} \frac{dT}{T} = \int_{0}^{\beta} \mu_s \, d\theta \tag{B15}$$

$$\ln \frac{T_2}{T_1} = \mu_s \beta \tag{B16a}$$

Equation (B16*a*) can also be presented as

$$e^{\mu_s \beta} = \frac{T_2}{T_1} \tag{B16b}$$

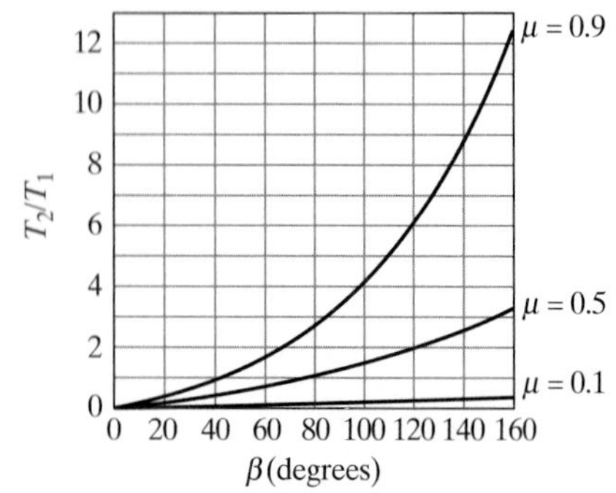

Figure B.8 The ratio T_2/T_1 versus belt wrap angle (β)

Figure B.8 plots this expression and shows that the greater the belt wrap (larger values of β), the greater the ratio of T_2/T_1 before belt slipping occurs. Equations (B16*a*) and (B16*b*) (as well as **Figure B.8**) also work for cables wrapped around drums.

EXERCISES

B1. What is the magnitude of the horizontal force $\boldsymbol{P}$ in **EB1** that must be exerted on the 100-kg block to cause the block to move if the coefficient of static friction is 0.25?

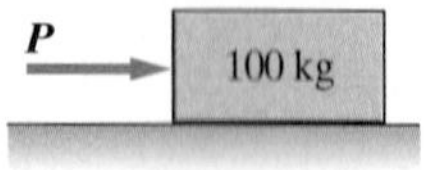

EB1

B2. What is the maximum angle θ for which the block of mass m in **EB2** will not slide down the incline if the coefficient of static friction is 0.30 and $\| \boldsymbol{P} \| = 0$?

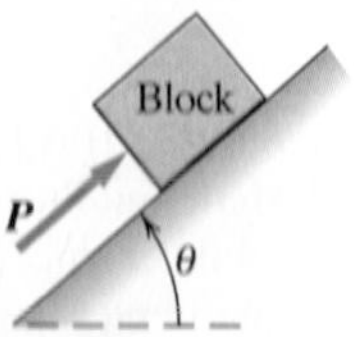

EB2

B3. If the static and kinetic coefficients of friction are 0.35 and 0.20, respectively, and $\| \boldsymbol{P} \| = 0$ determine the friction force acting on the block in **EB2** if

a. $\theta = 10°$
b. $\theta = 25°$

B4. If the static and kinetic coefficients of friction are 0.60 and 0.45, respectively, and the angle of the incline is 15°, determine the friction force acting on the block in **EB2** if

a. $P = 10$ N
b. $P = 60$ N

B5. If the static coefficient of friction between the block and the incline in **EB2** is 0.60 and the angle of the incline is 15°, determine the range of values of the force $\boldsymbol{P}$ for which the block will not slide up or down the incline.

B6. The coefficient of static friction between block A and its incline is 0.25 in **EB6**. What must the minimum coefficient of static friction between block B and its incline be if the blocks are in equilibrium? The mass of block B is twice that of block A. If the coefficient of friction is less than this minimum, in which direction will the blocks slide?

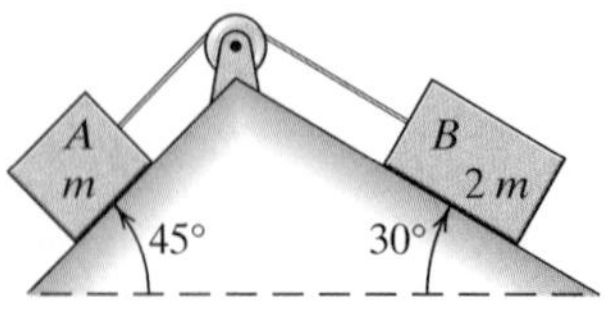

EB6

B7. Determine an expression for the force $\boldsymbol{P}$ as a function of wedge angle α and static coefficient μ_s required to raise the block in **EB7** if there is friction acting along Surfaces 1, 2, and 3. Plot your answer and compare with equation (B8).

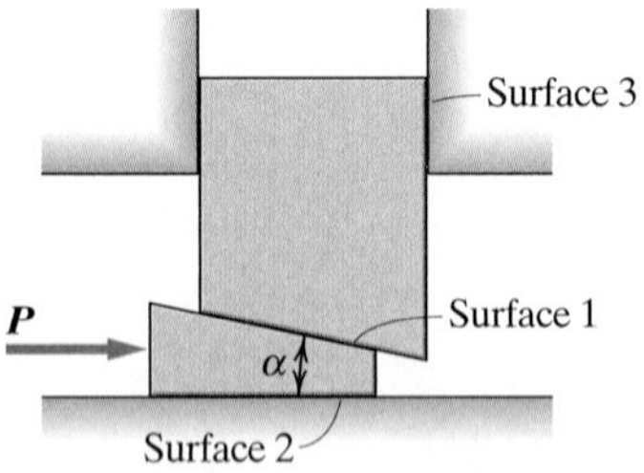

EB7

B8. Determine an expression for the force $\boldsymbol{P}$ as a function of wedge angle α and static coefficient μ_s required to remove the wedge in **EB8** if there is friction acting along surfaces 1 and 2. Plot your answer.

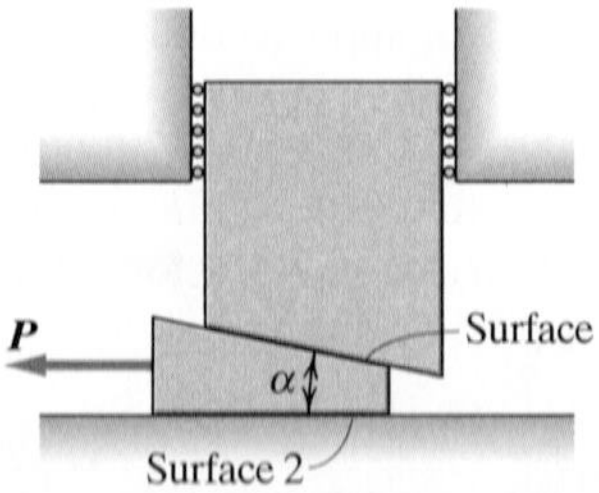

EB8

B9. Determine an expression for the force $\boldsymbol{P}$ as a function of wedge and angle α and static coefficient μ_s required to remove the wedge in **EB9** if there is friction acting along surfaces 1, 2, and 3. Plot your answer and compare with the answer found in **Exercise B8**.

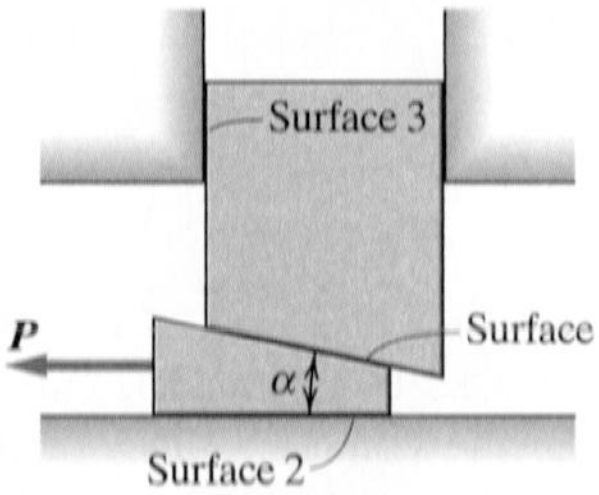

EB9

B10. Determine the magnitude of the minimum horizontal force $\boldsymbol{P}$ that must be applied to the wedge B to raise the block A in **EB10**. The coefficient of friction between all surfaces is 0.15. Is the wedge self-locking?

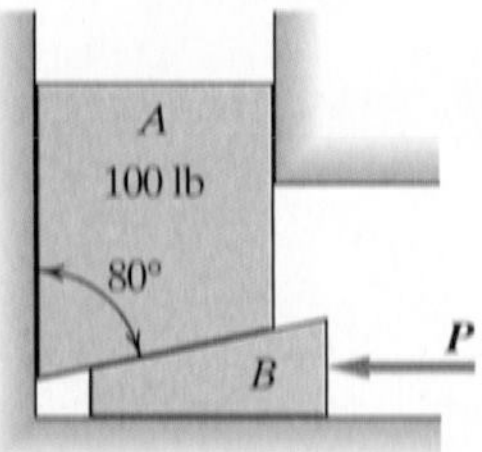

EB10

B11. Determine the magnitude of the minimum downward force $\boldsymbol{P}$ in **EB11** that must be applied to the wedge B to move the block A. The coefficient of friction between all surfaces is 0.15. Is the wedge self-locking?

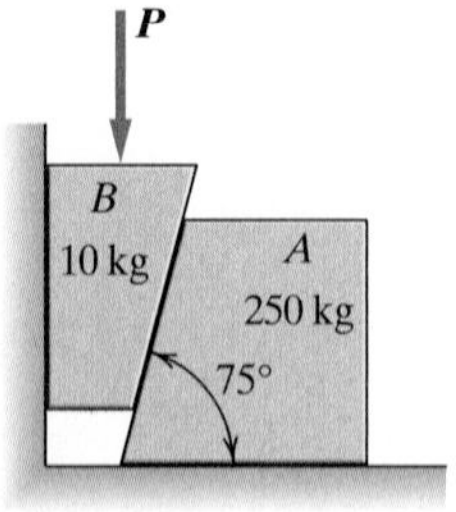

EB11

B12. What is the minimum vertical force $\boldsymbol{P}$ in **EB12**, applied to wedge E, necessary to push end C of the bar CD to the right? The coefficient of static friction at all surfaces is 0.15. Assume that the pins at A and D are frictionless. If the pins at A and D are not frictionless, would the required force increase, decrease, or remain the same? (Do not do any calculations—support your answer with qualitative reasoning.)

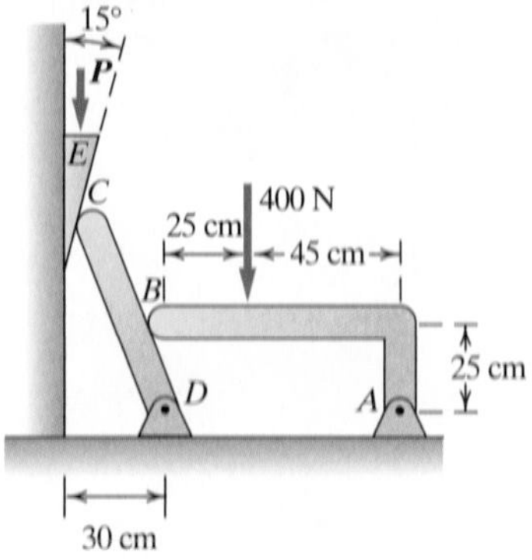

EB12

B13. The rope connecting the 6-kg block A with block B passes over a fixed cylinder in **EB13**. Determine the largest and smallest masses of block B for which static equilibrium is possible if the coefficient of static friction is 0.30.

EB13

B14. A cable is completely wrapped around the horizontal shaft in **EB14**. One end is attached to a 60-kg crate, and a tensile force T pulls the other. The coefficient of static friction between the shaft and the cable is 0.30. Determine

a. the minimum tension T for which the crate will not descend

b. the maximum tension T for which the crate will not rise

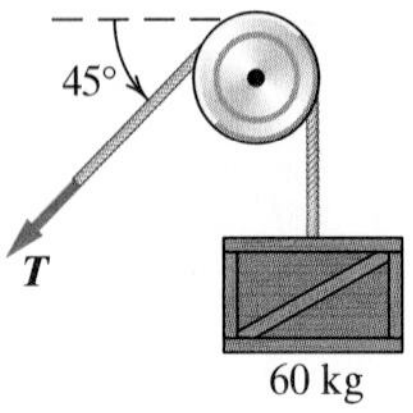

EB14

B15. A ship is secured by wrapping a rope around a capstan in **EB15**. If a dockworker can apply a 170-N force to counteract a 7-kN force by the ship, determine the number of complete turns of the rope about the capstan required to keep the rope from slipping if the coefficient of static friction is 0.30.

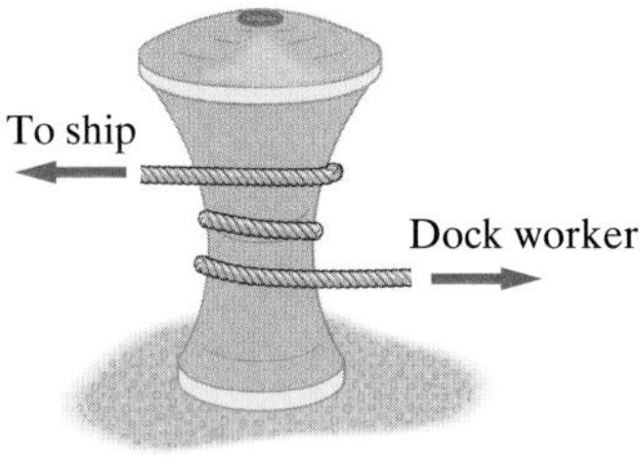

EB15

B16. Determine the maximum moment M_A in **EB16** that a motor may apply to pulley A without exceeding the maximum allowable belt tension of 200 N. Also determine the corresponding moment M_B exerted on pulley B by its drive shaft if the system is in equilibrium. The coefficient of static friction between the belt and the pulleys is 0.30.

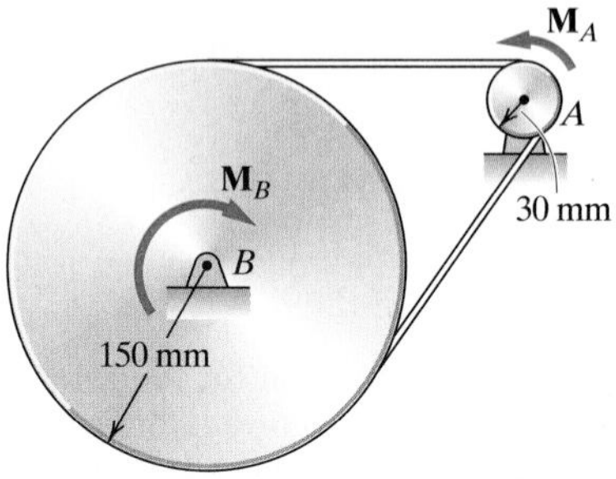

EB16

B17. It is known that the maximum moment M_A in **EB17** that a motor can apply to pulley A without causing the belt to slip over either pulley is 20 N · m. The coefficient of static friction is 0.40. For this moment, determine

a. the tension in the belt on either side of pulley A

b. the moment M_B on pulley B for equilibrium

c. whether slip is impending at pulley A or pulley B

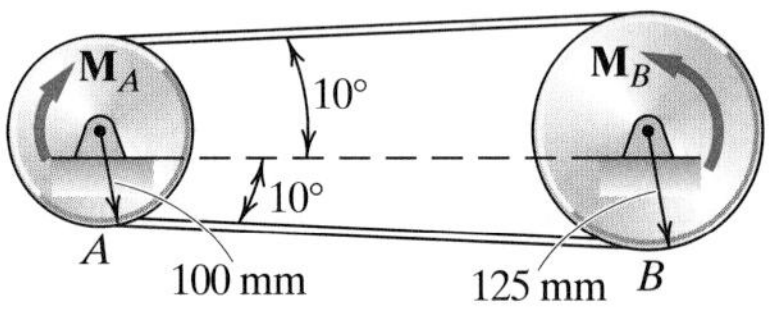

EB17

MOMENT OF INERTIA OF AREA

In this appendix we discuss how to calculate the moment of inertia of an area.[1] The **moment of inertia of an area** was first introduced in a footnote in Chapter 8 as one of the "family of integrals" related to area. It was also mentioned in Chapter 10 as the measure of the cross-sectional area that relates the bending moment in a beam to the stress in the beam (the denominator in (10.8)).

Consider the area shown in **Figure C.1**. The moment of inertia of this area about the x axis is

$$I_x = \int_{\text{area}} y^2\, dA \qquad (\text{C}1a)$$

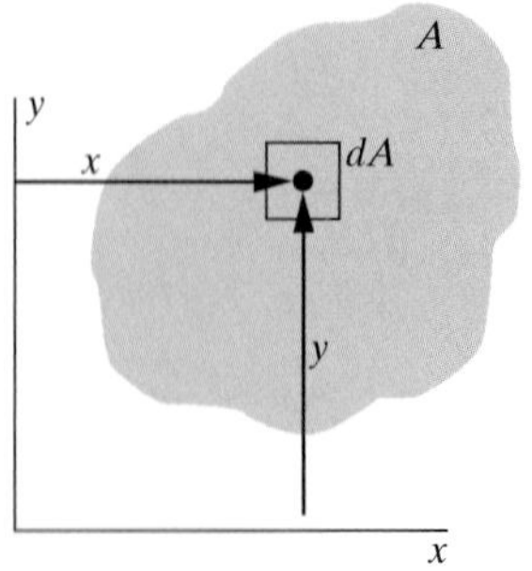

Figure C.1

and about the y axis is

$$I_y = \int_{\text{area}} x^2\, dA \qquad (\text{C}1b)$$

The unit of the moment of inertia of area is the fourth power of length.

The integrals in (C1a) and (C1b) are also called the **second area integrals**. For reference, the **first area integrals** are used to find the geometric center or centroid of an area (x_c, y_c),[2] and the (zero) **area integral** is used to find the area (A).[3]

Now consider that the moments of inertia have been calculated relative to axes located at the area's centroid; call these values I_{xc} and I_{yc}. The **parallel axis theorem** allows us to use these values of I_{xc} and I_{yc} to find the moments of inertia of the area relative to any other parallel axes by

$$I_x = I_{xc} + A(y_1)^2 \qquad (\text{C}2a)$$
$$I_y = I_{yc} + A(x_1)^2 \qquad (\text{C}2b)$$

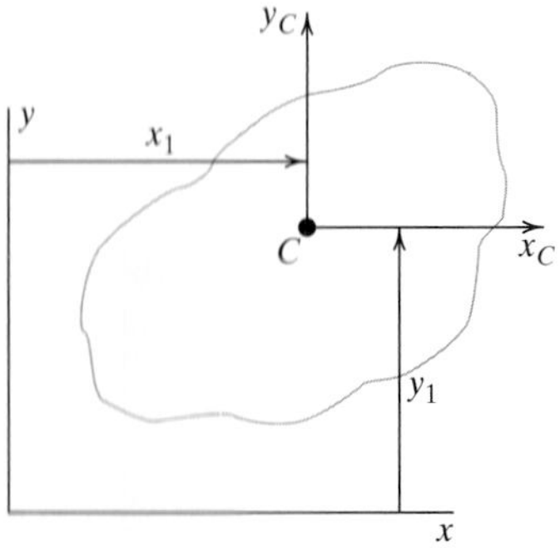

Figure C.2

where x_1 and y_1 are the distances between axes, as depicted in **Figure C.2**.

[1]The terminology "moment of inertia" is a misnomer, since no inertial concepts are involved.

[2]First area integrals are

$$Q_x = \int_{\text{area}} x\, dA;\; Q_y = \int_{\text{area}} y\, dA \qquad \text{(from (8.10A) and (8.10B))}$$

These two integrals are used in finding the geometric center (centroid) of an area.

[3]Area integral (area) $A = \int_{\text{area}} dA \qquad (8.8)$

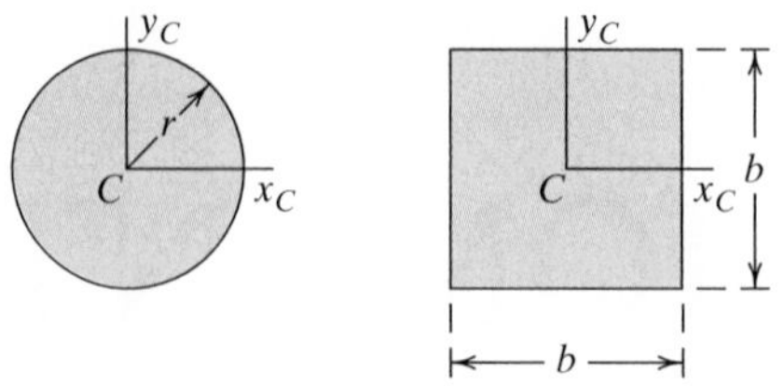

If $A_{\text{circle}} = A_{\text{square}}$, $b = \sqrt{\pi}\, r$, then

- $A = \pi r^2 = A = b^2 = \pi r^2$

- $\underbrace{I_{x_C} = I_{y_C} = \dfrac{\pi r^4}{4}}_{\text{Circle}} < \underbrace{I_{x_C} = I_{y_C} = \dfrac{b^4}{12}}_{\text{Square}} = \dfrac{(\sqrt{\pi}\, r)^4}{12} = \dfrac{\pi^2 r^4}{12}$

Figure C.3 The circle and square have the same cross-sectional area, but different area moments of inertia.

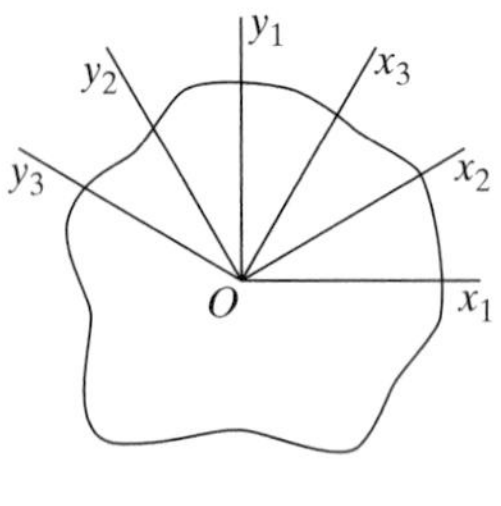

$$I_{x_1} + I_{y_1} = I_{x_2} + I_{y_2} = \bullet\bullet\bullet = \text{constant}$$

Figure C.4 For a given origin, the sum $(I_x + I_y)$ is a constant.

$$I_{x_\square} = I_{x_{\square}b_2} - I_{x_{\square}b_1}$$

$$I_{x_\square} = \dfrac{b_2^4}{12} - \dfrac{b_1^4}{12}$$

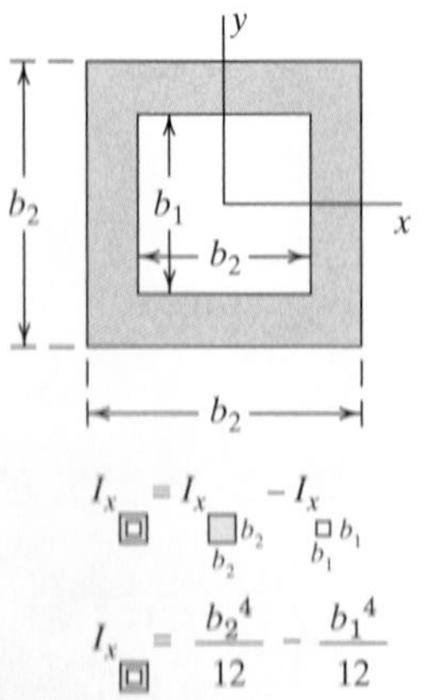

Figure C.5 An example of calculating the moment of inertia of a composite area.

A few notes about the moments of inertia of an area:

- If the area is simple (e.g., circular, rectangular), you can generally find the values of I_{xc} and I_{yc} in a reference table, such as **Table A3.1**. The tabulated data were found based on the application of (C1a) and (C1b).

- Equations (C2a) and (C2b) can be used to find I_x and I_y based on known values of I_{xc} and I_{yc}. Alternately, if I_x and I_y are known, these equations can be rearranged to calculate I_{xc} and I_{yc}.

- The value of I_x and I_y reflect the distribution of the area relative to coordinate axes in the plane of the area. This means that two areas may have the same cross-sectional area, but if their areas are distributed differently, they will have different moments of inertia. For example, the circular and square areas in **Figure C.3** have been sized to have the same areas, but have different moments of inertia of their areas.

- Though not proven here, one finds that the sum $(I_x + I_y)$ is a constant, independent of the orientation of axes (**Figure C.4**). This sum is called the **polar moment of inertia**. Furthermore, there is one particular orientation of the axes in which the value of I_x or I_y is a maximum and the other is a minimum.

- The moment of inertia of an area that is composed of distinct parts of simple areas can be found by using the parallel axis theorem applied to each distinct part relative to the centroid for the composite area. Determining the centroid of a composite area was presented in Section 8.1 (equation (8.12)).

- The ideas presented in this section on calculating the moment of inertia of area apply to both positive and negative areas (i.e., a hole). In the case of a negative area, a negative sign is used in front of the moment of inertia, as illustrated in **Figure C.5**.

- The moments of inertia of an area are related to another area used in calculations, the **radius of gyration of an area**. The radius

of gyration of an area A about the x axis (r_x) and about the y axis (r_y) are defined to be

$$r_x = \sqrt{\frac{I_x}{A}} \tag{C3a}$$

$$r_y = \sqrt{\frac{I_y}{A}} \tag{C3b}$$

EXERCISES

C1–C3. Use integration to evaluate the moments of inertia of area I_x and I_y of the shaded area shown in EC1–EC3.

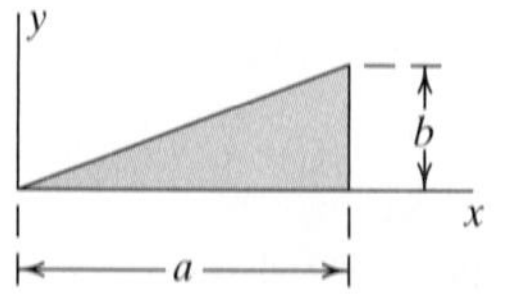

EC1

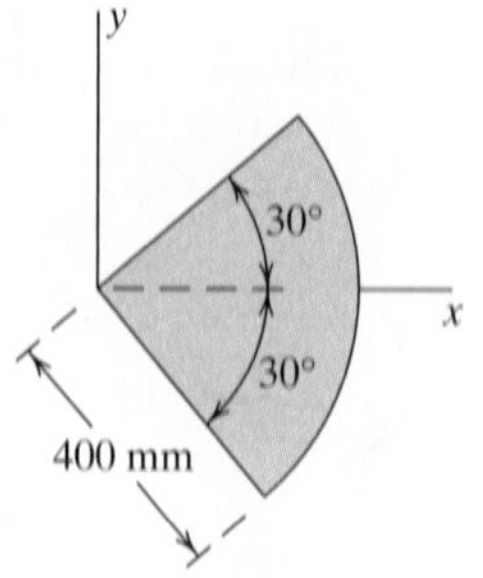

EC2

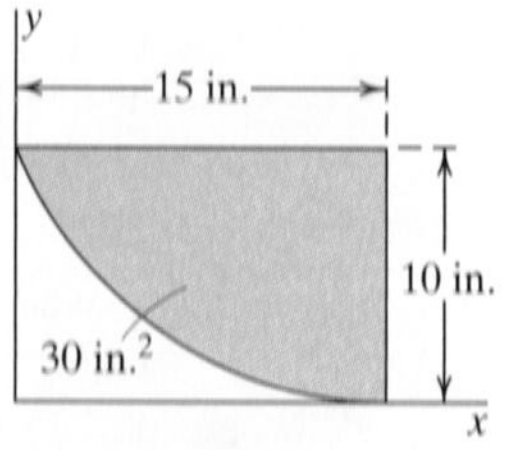

EC3

C4–C6. Use integration to evaluate the moments of inertia I_{xc} and I_{yc} of the shaded area shown in **EC1–EC3**, respectively.

C7–C10. Determine the moments of inertia of area I_x and I_y of the shaded area shown in EC7–EC10 using the properties in **Table A3.1**.

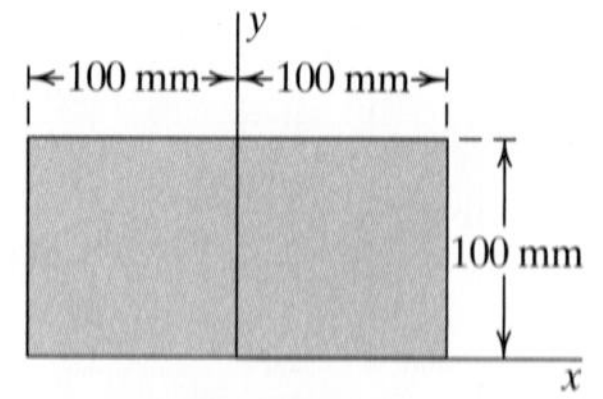

EC7

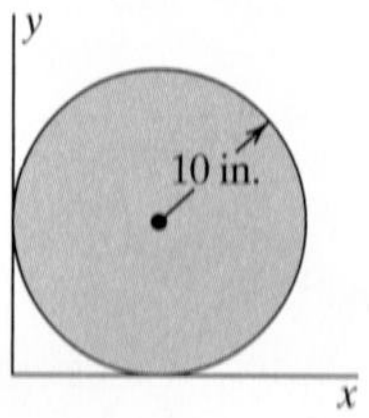

EC8

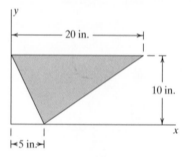

EC9

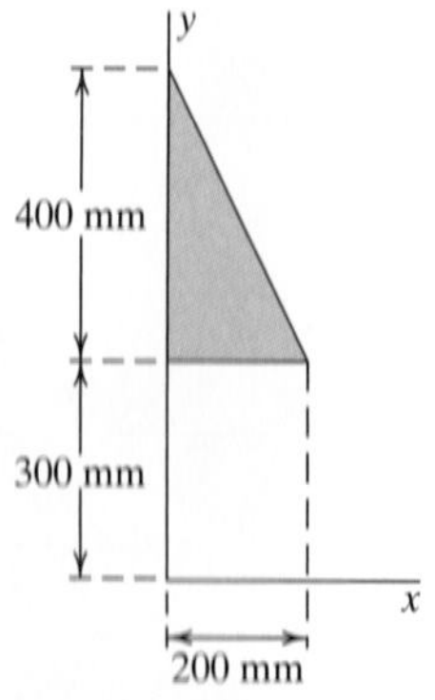

EC10

C11–C15. Determine the moments of inertia of area I_x and I_y of the composite area shown in **EC11–EC15**.

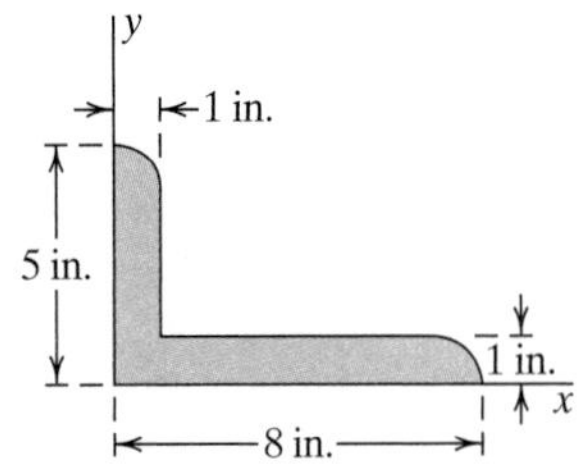

EC11

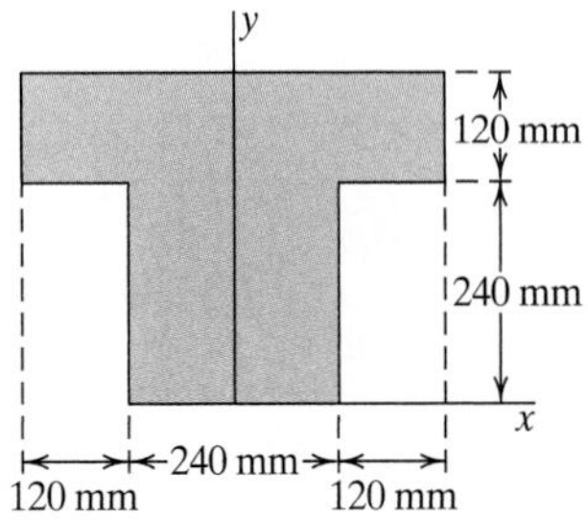

EC12

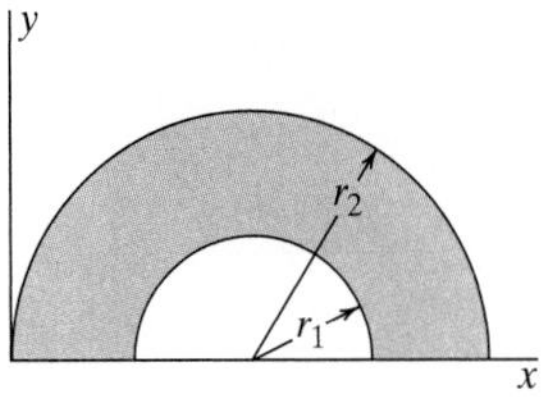

EC13

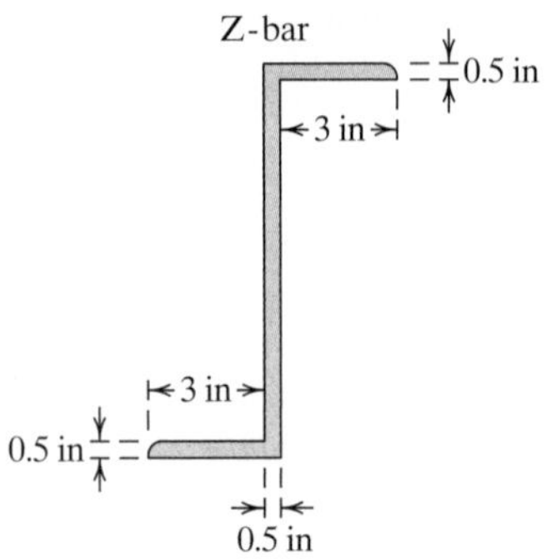

EC14

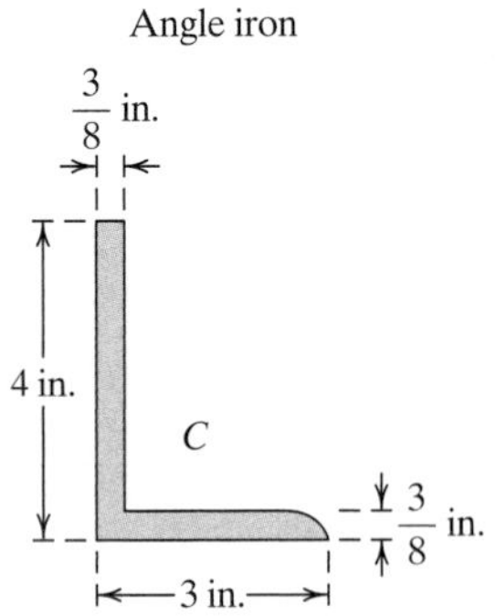

EC15